T0224213

ADVANCES IN PHOTOSYNTHESIS RESEARCH

ADVANCES IN AGRICULTURAL BIOTECHNOLOGY

Related titles previously published

Akazawa T., et al., eds: The New Frontiers in Plant Biochemistry. 1983.
ISBN 90-247-2829-0

Gottschalk W. and Müller H.P., eds: Seed Proteins: Biochemistry, Genetics, Nutritive Value. 1983. ISBN 90-247-2789-8

Marcelle R., Clijsters H. and Van Poucke M., eds: Effects of Stress on Photosynthesis. 1983. ISBN 90-247-2799-5

Veeger C. and Newton W.E., eds: Advances in Nitrogen Fixation Research. 1984.
ISBN 90-247-2906-8

Chinoy N.J., ed: The Role of Ascorbic Acid in Growth, Differentiation and Metabolism of Plants. 1984. ISBN 90-247-2908-4

Witcombe J.R. and Erskine W., eds: Genetic Resources and Their Exploitation – Chickpeas, Faba beans and Lentils. 1984. ISBN 90-247-2939-4

Cover photograph by: Y. Carpels

Advances in Photosynthesis Research

Proceedings of the VIth International Congress on Photosynthesis, Brussels, Belgium, August 1–6, 1983

Volume II

edited by

C. SYBESMA

Biophysics Laboratory
Vrije Universiteit Brussel
B-1050 Brussels
Belgium

1984 Springer-Science+Business Media, B.V.

Distributors

for the United States and Canada: Kluwer Boston, Inc., 190 Old Derby Street, Hingham, MA 02043, USA
for all other countries: Kluwer Academic Publishers Group, Distribution Center, P.O.Box 322, 3300 AH Dordrecht, The Netherlands

Library of Congress Cataloging in Publication Data

```
International Congress on Photosynthesis Research (6th :
    1983 : Brussels, Belgium)
    Advances in photosynthesis research.

    (Advances in agricultural biotechnology)
    Includes index.
    1. Photosynthesis--Congresses.  I. Sybesma, C.
II. Title.  III. Series. [DNLM: 1. Photosynthesis--
Congresses. IN 636 P 6th 1983a /QK 882 I61 1983a]
QK882.I56  1983      561.1'3342      84-1518
```


ISBN 978-90-247-2943-2   ISBN 978-94-017-6368-4 (eBook)
DOI 10.1007/978-94-017-6368-4

Copyright

GENERAL CONTENTS

Volume I

1. Excitation Energy Transfer — 1
2. Primary Reactions — 89
3. Oxygen Evolution — 227
4. Photosynthetic Electron Transport I: Plants and Algae — 399
5. Photosynthetic Electron Transport II: Photosynthetic Bacteria — 621
6. Physical Parameters; Special Methods; Model Systems — 677

Volume II

1. Pigment and Pigment-Protein Complexes — 1
2. Reaction Center and Light-Harvesting Complexes I:
    Plants, Algae and Cyanobacteria — 73
3. Reaction Center and Light-Harvesting Complexes II:
    Photosynthetic Bacteria — 153
4. Membrane Pontentials and Ion Gradients — 233
5. Membrane Bioenergetics; Membrane Transport — 371
6. Coupling Factors and ATPase — 477
7. Cyanobacteria — 625
8. Photosynthesis and Solar Energy Technology — 727

Volume III

1. The Architecture of Photosynthetic Membranes I:
    General; Envelope Membranes; Membrane Fluidity; Stacking — 1
2. The Architecture of Photosynthetic Membranes II:
    Membrane Proteins and Lipids — 91
3. The Architecture of Photosynthetic Membranes III:
    Photosystem Particles; Electron Transport Components — 195
4. The Architecture of Photosynthetic Membranes IV: Dynamic Aspects — 263

5. The Architecture of Photosynthetic Membranes V:
    Prokaryotic Membranes   335
6. Carbon Metabolism I: Metabolic Pathways; Metabolites; Enzymes   381
7. Carbon Metabolism II: Regulation, Induction and Activation; Transport   557
8. Ribulose Bisphosphate Carboxylase/Oxygenase   717
9. Photorespiration; Amino Acid Synthesis; Nitrogen Metabolism   811

Volume IV

1. Herbicides   1
2. The Productivity of Photosynthesis   85
3. Environmental Influences on Photosynthesis   181
4. Photosynthesis under Stress Conditions   345
5. Molecular Genetics of the Photosynthetic Apparatus   483
6. Chloroplast Development   595
7. Light-Controlled Development of the Photosynthetic Apparatus   783

CONTENTS TO VOLUME II

Preface   XXIII

In Memory of Kazuo Shibata   XXV

Sponsors   XXVII

International Photosynthesis Committee 1980–1983   XXIX

Local Organizing Committee   XXXI

1. Pigment and Pigment-Protein Complexes

Long-wavelength Absorbing Forms of Carotenoids   1
S. Wieckowski, E. Wloch, B. Broniowska

Detection and Characteristics of the Pigments of Variegata Leaves   5
M. Bogdanović, A. Rastović

Formulae and Program to Determine Total Carotenoids and Chlorophylls
a and b of Leaf Extracts in Different Solvents   9
A. Wellburn, H. Lichtenthaler

Unusual Pigments in a Primitive Green Alga   13
J. Brown

Artificial Chlorophyll-protein Complexes   17
E. Tombácz, Z. Várkonyi, L. Szalay

Model Systems for Studying Chlorophyll-protein Interactions   21
P. Pančoška, K. Vacek, L. Skála

Comparative Biochemistry of Chlorophyll-protein Complexes   25
J. Thornber, R. Cogdell, R. Seftor, B. Pierson, E. Tobin

Fluorescence Properties of Chlorophyll-protein Complexes from Spinach
Chloroplasts – on the Specificity of the Emission Bands   33
N. Fuad, K. Kinosita, M. Mimuro, Y. Inoue, Y. Fujita

Supramolecular Structure of Pigment-protein Complexes in Relation to
the Chlorophyll a Fluorescence of Chloroplasts at 25 C or at -196 C:
Cation Effect, pH Effect and Trypsinolysis   37
J. Argyroudi-Akoyunoglou, G. Akoyunoglou

The Chlorophyll-protein Complexes of Prochloron   41
A. Larkum, R. Hiller

Pigment-Protein Complexes of Codium, a Marine Green Alga   45
J. Anderson

Pigment Composition and Stability in the Intertidal Alga Codium fragile   49
A. Cobb, E. Benson

Changes in Chlorophyll Protein Complexes during the Life Cycle of
Scenedesmus   53
H. Senger, K. Krupinska, G. Akoyunoglou

Characterization of Chlorophyll-protein Complexes from a Cryptophycea:
Cryptomonas rufescens   57
C. Lichtlé, J. Duval

Chlorophyll- and Heme-binding Proteins in Porphyridium cruentum
Thylakoids   61
T. Redlinger, E. Gantt

Effect of Hydrolytic Enzymes on Cyanobacterial Pigment-protein
Complexes of Photosystem 1   65
J. Hladík, R. Bísková, D. Sofrová

Luminescence Study of PS I Pigment-protein Complexes Isolated from
the Cyanobacterium Plectonema boryanum   69
P. Pančoška, M. Ambrož, J. Hladík, D. Sofrová

2. Reaction Center and Light-Harvesting Complexes I:
   Plants, Algae and Cyanobacteria

A 160-kilodalton Photosystem-I Reaction-center Complex: Low-
temperature Difference Absorption, EPR and Fluorescence Spectroscopy
of the Primary Electron Acceptors   73
I. Ikegami, B. Ke

The Novel Chlorophyll RC I Associated with PS I   77
D. Dörnemann, H. Senger

Structure of a Chlorophyll-RC1   81
H. Scheer, H. Wieschhoff, W. Schaefer, E. Cmiel, B. Nitsche, H. Schiebel,
H. Schulten

Photosystem I Reaction Center from Cyanobacterium, Green Algae and
Higher Plants   85
R. Nechushtai, N. Nelson

Evidence that CP47 (CPa-1) is the Reaction Centre of Photosystem II   95
B. Green, E. Camm

Characterization of the Photosystem II Reaction Center Polypeptide
(CP47)   99
H. Nakatani, C. Arntzen, Y. Inoue

Lifetime Studies by Phase Fluorimetry in Isolated Light-harvesting
Complexes (LHC): Dependence on the Agregation State   103
I. Moya, P. Tapie

Chlorophyll A/b-proteins in their Relation to the Light-harvesting
Complex   107
O. Machold

Magnetic Resonance and Picosecond Laser Spectroscopy of Light-
harvesting Chlorophyll A/b-protein   115
G. Searle, J. Fraaije, T. Schaafsma

Structure of Two-dimensional Crystals of the Light-harvesting
Chlorophyll A/b-protein Complex   119
W. Kühlbrandt, T. Thaler, E. Wehrli

Biochemical Analysis of Polypeptide Composition of the Light-harvesting
Chlorophyll A/b-protein Complex from Pea Chloroplasts   121
T. Thaler, W. Kühlbrandt, K. Mühlethaler

Detached Light-harvesting Complex in Chloroplasts of Cold-grown
Triticale Seedlings   125
B. Elfman, W. Hopkins

Polypeptides of the Light-harvesting Complex in a Chlorophyll b-Less
Mutant of Chlamydomonas Reinhardtii   129
H. Michel, M. Tellenbach, A. Boschetti

Lipid Induced Transformation of the Monomeric to Oligomeric Forms
of the LHCP   133
R. Remy, F. Ambard-Bretteville, J. Dubacq, A. Tremolieres

Function of Chlorophylls and Carotenoids in Thylakoid Membranes.
Pigment Bleaching in Relation to PS I and PS II Activity of Sub-
chloroplast Particles Prepared with Digitonin   137
T. Braumann, T. Gröpper, I. Damm, L. Grimme

Function of Chlorophylls and Carotenoids in Thylakoid Membranes:
Chlorophylls between Pigment-protein Complexes Might Function by   141
Stabilizing the Membrane Structure
L. Grimme, L. Horst, J. Brown

Function of Chlorophylls and Carotenoids in Thylakoid Membranes:
Pigment Composition of LHCP-complexes of Spinach and Chlorella fusca   145
W. Menger, J. Knötzel, T. Braumann, L. Grimme

The Beta-carotene-protein and Other Carotenoid-protein of Brown Algae
and of Cyanophytes   149
J. Barrett

3. Reaction Center and Light-Harvesting Complexes II:
   Photosynthetic Bacteria

Proposal for the Nomenclature of Reaction Centres and Light-harvesting
Complexes from the Purple Photosynthetic Bacteria   153

Structure and Function of the Reaction Center from Rhodopseudomonas
sphaeroides   155
G. Feher, M. Okamura

The Bacterial Reaction Center: Where is the Center?   165
V. Wiemken, R. Bachofen

Studies on the Iron-binding Site of Bacterial Reaction Centers   169
S. Bodmer, R. Bachofen

Low Temperature Magnetic Resonance and Optical Spectroscopy of
Chromatophores and (Crystalline) Reaction Centers of Rhodopseudo-
monas viridis   173
F. van Wijk, T. Schaafsma, H. Michel

Polarized Infrared Spectroscopy of Bacterial Reaction Centers: the
LMH and LM Complexes in Reconstituted Membrane   177
E. Nabedruk, D. Tiede, P. Dutton, J. Breton

A Light-harvesting Pigment-protein Complex Associated with the
Reaction Center of Green Bacteria   181
H. Vasmel, T. Swarthoff, H. Kramer, J. Amesz

Reaction Center Components of the Green Photosynthetic Bacterium
Chloroflexus Aurantiacus   185
H. Vasmel, R. Meiburg, H. Kramer, H. den Blanken, L. de Vos, J. Amez

Regulation of Photosynthetic Unit Structure in Rhodospirillum rubrum
Whole Cells   189
P. Loach, P. Parkes, P. Bustamante

More than Two Structurally Distinct Types of Antenna in Rhodospirillales   199
B. Robert, A. Vermeglio, M. Lutz

Pigment Organization in the Light-harvesting B800-B850 Antenna
Complex of Rhodopseudomonas sphaeroides   203
H. Kramer, R. van Grondelle, C. Hunter, J. Amesz

Linear Dichroism and Fluorescence Emission of Antenna Complexes
during Photosynthetic Unit Assembly in Rhodopseudomonas sphaeroides   207
H. Kramer, C. Hunter, R. van Grondelle, W. Bakker

Resonance Raman Spectroscopy of Light-harvesting Bacteriochlorophyll
in Purple Photosynthetic Bacteria   211
H. Hayashi, M. Tasumi

Associations of Pigment-proteins and Phospholipids into Specific
Domains in Rhodopseudomonas sphaeroides Photosynthetic Membranes
as Determined by Lithium Dodecyl Sulfate/Polyacrylamine Gel
Electrophoresis   215
C. Radcliffe, J. Pennoyer, R. Broglie, R. Niederman

Reversible pH Induced Absorption Change in the Bchl B-LH-protein
of the Alkalophile Ectothiorhodospira halochloris   221
R. Steiner, H. Scheer

On the Effect of Proteinases, Phospholipases and Group-specific
Reactants on Carotenoid Spectra and Electrochromic Response of
Membrane Preparations of Photosynthetic Purple Bacteria   225
C. Swysen, M. Symons, C. Sybesma

Radical Formation by Dark Oxidation of Antenna Bacteriochlorophyll
in Rhodopseudomonas   229
I. Gomez, F. del Campo

4. Membrane Potentials and Ion Gradients

Is the Transmembrane Electrochemical Potential a Competent Intermediate
in Membrane Associated ATP Synthesis?   233
B. Melandri, G. Venturoli, R. Casadio, G. Azzone, D. Kell,
H. Westerhoff

Electrical Events and P515 Response in Thylakoid Membranes   241
W. Vredenberg, O. van Kooten, R. Peters

Protolytic Reactions in the Thylakoid Interior   247
W. Junge, Y. Hong, S. Theg, L. Qian, A. Viale

The Inner- and Inter-thylakoidal Electric Potential of Isolated
Chloroplasts in the Dark   257
D. Walz

Diffusion Barriers for Protons at the External Surface of Thylakoids   261
A. Polle, W. Junge

Indications for the Chloroplast as a Tri-compartment System: Micro-
electrode and P515 Measurements Imply Semi-localized Chemiosmosis   265
D. van Kooten, F. Leermakers, R. Peters, W. Vredenberg

The Kinetics of the Flash-induced P515 Electrochromic Bandshift in
Dependence of the Energetic State of the Thylakoid Membrane and the
Ion Concentration in the Outer Aqueous Phase   269
R. Peters, O. van Kooten, W. Vredenberg

Redox Titrations of the Fast and Slow Phase of the 515 nm Electrochromic Absorption Change in Chloroplasts   273
L. Giorgi, N. Packham, J. Barber

Redox Titration of the Slow Electrochromic Phase in Chloroplasts   277
M. Girvin, W. Cramer

On the Origin of the Slow Electric Potential Component Induced by
Ferredoxin-mediated Cyclic Electron Transfer in Photosystem I-enriched
Subchloroplast Vesicles   281
F. Peters, G. Smit, R. Kraayenhof

Cation Binding to Thylakoid Membranes: Association Constant and Effect
on Transmembrane Potentials   285
S. Mauro, P. Botte, R. Lannoye, H. Hurwitz

Temperature Dependence of Electric Events in Thylakoid Membranes:
Comparison of Membrane Potential, Surface Potential and Aminoacridine
Binding   289
R. Kraayenhof, J. Torres-Pereira, F. Peters, H. Wong Fong Sang

Microchemiosmotic Coupling in Thylakoids: I. Dependence of Local
Delta mu H on the Topography of the H^+ Entry and Exit Points   293
F. Haraux, C. Sigalat, A. Moreau, Y. de Kouchkovsky

Microchemiosmotic Coupling in Thylakoids: II. Dependence of Local
Delta mu H on the Membrane Conductance   297
C. Sigalat, Y. de Kouchkovsky, A. Moreau, F. Haraux

Regulation of the Energization of Thylakoids in Leaves and in Isolated
Chloroplasts   301
G. Garab, C. Barabás, L. Zimányi, J. Farineau

Protolytic Reactions at the Donor Side of PSII: Proton Release in Tris-
washed Chloroplasts with t 1/2 $\cong$ 100 Microsec. Implications for the
Interpretation of the Proton Release Pattern in Untreated, Oxygen-
evolving Chloroplasts   305
V. Forster, W. Junge

Influence of a Proton Gradient on the Activity of the Reconstituted
Chloroplast Phosphate Translocator   309
U. Flügge, H. Heldt

Complex Kinetics of Proton Deposition Inside Chloroplast Thylakoids   313
A. Hope, D. Matthews

Permeability Properties of the Chloroplast Envelope for Monovalent
Inorganic Cations and their Distribution between the Intact Chloroplast
and the External Medium in the Light and Dark   317
B. Demmig, H. Gimmler

Flash-induced pH Changes in Photosystem I-enriched Vesicles Monitored
with Neutral Red and Cresol Red   321
F. de Wolf, L. van Houte, F. Peters, R. Kraayenhof

The Effect of Trypsin and Chymotrypsin on the 519 nm Field-indicating
Absorption Change in Isolated Chloroplasts   325
C. Raines, M. Hipkins

The Effects of an Electrical Field on the Primary Reactions of System II   329
H. van Gorkom, R. Meiburg, R. van Dorssen

Electric Measurements of the Kinetics of Photosynthetic Events   333
P. Gräber, H. Bauermeister

Membrane Potential Measurements in C3 Protoplasts   337
K. Valles, M. Proudlove, R. Beechey, A. Moore

The Ionic Conductance but not the Electric Capacitance of Membranes
from Photosynthetic Bacteria Depends on the Membrane Potential   341
J. Jackson, A. Clark

The Electrochemical Proton Gradient and Solute Transport in Rhodopseudomonas sphaeroides   347
W. Konings, M. Elferink, K. Hellingwerf

Simultaneous Measurements of Electrical Field Changes in the Outer
Surface and Central Regions of Chromatophore Membranes of Rhodo-
pseudomonas Sphaeroides by Responses of Merocyanine Dyes and
Carotenoids   355
S. Itoh

Measurements of Surface Potentials in Chromatophore Preparations of
Photosynthetic Bacteria in Agreement with the Gouy-Chapman Theory;
Estimation of the Binding Constants   359
P. Botte, M. Symons, C. Swysen, S. Mauro, H. Hurwitz, R. Lannoye,
C. Sybesma

Proton Pumping in Vesicle-reconstituted Bacterial Reaction Centre/bc1
and Photosystem I/bf Systems   363
P. Rich, P. Heathcote

Reaction Centers from Rhodopseudomonas sphaeroides in Reconstituted
Phospholipid Vesicles. Structural Properties and Light-dependent Proton
Translocation   367
K. Hellingwerf

5. Membrane Bioenergetics; Membrane Transport

Protons, Thiols and Coupling Factor 1 in Relation to Photophosphorylation   371
R. McCarty, J. Davenport, S. Ketcham, J. Moroney, C. Nalin, W. Patrie,
K. Warncke

Flash-induced ATP Synthesis in Pea Chloroplasts   379
J. Galmiche, G. Girault, C. Lemaire

The Dynamics of Millisecond Delayed Light Emission and its Relation
with Photophosphorylation   381
Y. Shen, C. Xu, J. Wei, D. Li, Y. Feng, Z. Huang

Inhibition of PS I and PS II-dependent Flash-induced ATP Synthesis by
Triphenyltin Chloride and DCCD   387
S. Flores, D. Ort

Chemical Models of Pyrophosphate Bond Synthesis and Ion-radical
Mechanism of Photophosphorylation   391
A. Yasnikov, N. Volkova, N. Kanivets, L. Vasilenok

Characterization of ATP Formation from Acetyl Phosphate and ADP by
Photosynthetic Membranes   395
H. Sakurai, K. Shinohara, R. Yamagishi

Enzyme Kinetics of ATP-ase and Energy Balance in Thylakoids   399
L. Jahn, R. Strasser

Enhancement of the Rate of Photophosphorylation in Spinach Chloroplasts
by low Concentrations of Carboxylic Ionophores and Amines   403
C. Giersch

Differential Inhibition of Pi-ATP Exchange in Relation to ATP Synthesis
and Hydrolysis by Glutaraldehyde Modification of Chloroplast
Membranes   407
D. Bar-zvi, N. Shavit

Differential Effects of Short Time Glutaraldehyde Treatment on Light
Induced Thylakoid Membrane Conformational Changes, Proton Pumping
and Electron Transport Properties   411
S. Coughlan, U. Schreiber

Two Phosphorylating Cycles Associated with Photosystem I   415
J. Hosler, C. Yocum

Coupling between Redox and Acid-base Anergies by Chloroplast
Cytochrome b-559   419
M. Hervas, F. de la Rosa, M. de la Rosa, M. Losada

The Effect of Flash-induced ATP-hydrolysis on the 515 Absorbance
Change   423
U. Siggel

ATP Synthesis Catalized by Reconstituted CFOF1 Liposomes Driven by
an Artificially Generated Delta PS I and Delta pH   427
P. Gräber, M. Rögner, D. Samoray, G. Hauska

The Rate of ATP-synthesis as a Function of Delta pH and Delta PS I
in Preactivated and Non-preactivated Chloroplasts   431
U. Junesch, P. Gräber

Role of Hydrogen Atoms in the First Chemical Steps of the Photosynthesis   437
E. Tyszkiewicz, E. Roux

Electric Potential and pH Gradient Formation by Reconstituted ATPase
Proteoliposomes from the Thermophilic Cyanobacterium synechococcus
6716   441
H. van Walraven, H. Lubberding, H. Marvin, R. Kraayenhof

The Effects of Light-dark Transition and of Specific Inhibitors on the
ATP Level in Some Cyanobacteria   445
H. Lubberding, W. Schroten

The Inhibition of Nitrate Reduction by Light in Rhodopseudomonas
capsulata is Mediated by the Membrane Potential, but Inhibition by
Oxygen is not   449
A. McEwan, N. Cotton, S. Ferguson, J. Jackson

Glycerate Transport Across the Chloroplast Envelope   453
S. Robinson

A Glycolate Transporter in the Chloroplast Envelope   457
K. Howitz, R. McCarty

Regulation of the Export of the Reducing Power from Chloroplast to
Cytosol in Relation with Molecular Properties of Chloroplastic
NADP-MDH   461
N. Ferte, J. Meunier, J. Buc

Pi and G6P Translocation in Chloroplasts of Codium fragile   465
J. Rutter, A. Cobb

The Lactose Carrier of Escherichia coli Functionally Incorporated in
Rhodopseudomonas sphaeroides Obeys the Regulatory Conditions of the
Phototrophic Bacterium   469
M. Elferink, K. Hellingwerf, F. Nano, S. Kaplan, W. Konings

Characterization of Adenylate Kinases from Photosynthetic Purple
Bacteria   473
K. Knobloch, H. Neufang, H. Müller

6. Coupling Factors and ATPase

Functional Properties of Chloroplast ATPase: the Fate of Tightly Bound
ADP and ATP in Illuminated Chloroplasts   477
H. Strotmann

Correlation between Membrane-Localized Protons and Flash-driven ATP
Formation in Chloroplasts   485
R. Dilley, J. Laszlo, U. Schreiber

Modulation of the Chloroplast ATP Synthetase: Conformational States,
Nucleotide Binding and Limited Accessibility to the Active Site   493
N. Shavit, C. Aflalo, D. Bar-zvi, M. Tiefert

Conservation and Organization of Subunits of the Chloroplast Proton
ATPase Complex   501
R. Rott, N. Nelson

Mode of Action and Regulation of Chloroplast H^+-ATPase   511
R. Hillel, A. Jagendorf, C. Carmeli

Light Activation in vivo of the Chloroplast Proton ATPase: Effect on
the Pi-ATP Exchange Reaction   519
R. Vallejos, R. Ravizzini

A Dual pH Optimum Model for Activation of the Chloroplast ATPase,
CF0-CF1   523
J. Mills, P. Mitchell

Regulation of the H^+-ATPase in Intact and Osmotically Shocked
Chloroplasts   527
Y. Shahak

ATP-induced Delta pH in CF0-CF1 Proteoliposomes   531
A. Admon, U. Pick, M. Avron

The Role of ADP in Regulation of Chloroplast Coupling Factor   535
A. Loehr, B. Huchzermeyer

Interaction of Membrane-bound and Soluble CF1 with the Photoreactive
Nucleotide 3'-0-(4-benzoyl)benzoyl ADP   539
D. Bar-zvi, M. Tiefert, N. Shavit

Nucleotide Binding to the Membrane-bound Chloroplast Coupling Factor
(CF1) in the Light   543
J. Schumann

Binding and Exchange of Nucleotides and Magnesium on the Chloroplast
Coupling Factor CF1   547
G. Girault, J. Galmiche, C. Lemaire

Loose and Tight Binding of Adenine Nucleotides by Membrane-associated
Chloroplast ATPase   551
S. Bickel-Sandkötter

Activation of the Light Triggered Chloroplast ATPase-role of Nucleotides
and an Inhibitory Peptide   555
P. Andralojc, K. Simpson, D. Harris

Limited Access of Nucleotides to the Active Site of ATP Synthetase
during Photophosphorylation   559
C. Aflalo, N. Shavit

Accessibility and Function of CF0-subunits in Chloroplast Thylakoids   563
L. Klein-Hitpass, R. Berzborn

A Partial Characterization of the Halotolerant, Green Algal Dunaliella
Bardawil CF1 ATPase and its Modification by Fluorescein Isothiocyanate   567
S. Selman-Reimer, M. Finel, U. Pick, B. Selman

Immunological Studies on the Cross-binding and Cross-reconstitution of
Maize and Spinach Coupling Factors CF1   571
N. Pucheu, R. Berzborn

Detection of Conformational Changes in Chloroplast Coupling Factor 1
by ANS Fluorescence Changes   575
U. Pick, M. Finel

Spectroscopic Studies on the Conformation of Isolated and Membrane
Bound CF1   579
R. Wagner, C. Andreo, W. Junge

Identification and Purification of the Alpha and Beta Subunits of the
Chloroplast Coupling Factor One from Chlamydomonas Reinhardtii   583
S. Merchant, B. Selman

Stoichiometry and Function of the Delta-subunit of CF1   587
R. Berzborn, P. Roos, G. Bonnekamp

Analysis of Amino Acids at the Active Site of Chloroplast Coupling
Factor One: a Spin-echo nmr Study   591
W. Frasch, R. Sharp

Chemical Modification of Essential Amino-acid Residues in the
Chromatophore F1-ATPase and its Isolated Beta-subunit   595
Z. Gromet-Elhanan, D. Khananshvili

The Coupling Factor (Ca-ATPase) of the Cyanobacterium Spirulina
platensis   599
D. Hicks, C. Yocum

Properties and Intracellular Localization of a Cyanobacterial (Ca_2^+,
Mg_2^+)-ATPase   603
W. Lockau

A (Mg_2^+-Ca_2^+)-stimulated ATPase in Spinach Chloroplast Envelopes:
Isolation by Calmodulin Affinity Chromatography 607
T. Nguyen, P. Siegenthaler

Mode of Action of the ATP-ase Inhibitor Tri-n-butyl-tin in Rhodopseudomonas sphaeroides Cells 611
A. Vermeglio, J. Galmiche, G. Girault

Arsenylation and Phosphorylation of ADP and GDP in Rhodospirillum
rubrum Chromatophores 615
L. Slooten, A. Nuyten

Characterization and Localization of the Pea Chloroplast Envelope
Mg_2^+-ATPase 619
D. McCarty, B. Selman, K. Keegstra

Effect of Chemical Modification of Primary Amino Group of Salt Washed
Thylakoid Membrane by Fluorescamine on Different Parameters of Energy
Transduction in Chloroplast 623
J. Parkash, G. Singhal

7. Cyanobacteria

Salt Adaptation Mechanisms in the Cyanobacterium Synerchococcus 6311 627
E. Blumwald, R. Mehlhorn, L. Packer

Interaction of Photosynthetic and Respiratory Electron Transport in
Blue-green Algae: Proton-efflux Studies 631
S. Scherer, E. Stürzl, P. Böger

Respiratory and Photosynthetic Electron Transport in Anabaena
variabilis: Light-dark Activities of Pyridine-nucleotide Dehydrogenases 635
E. Stürzl, S. Scherer, P. Böger

Redox Reactions of Cytochrome b-564 and f-556 in Isolated Heterocysts
of Anabaena variabilis (ATCC 29413) 639
H. Böhme, A. Ernst

Presence of Cytochromes in the Thylakoid and Cell Membranes of the
Cyanobacterium Plectonema boryanum Cultured in Normal and Copper
Depleted Growth Medium 643
H. Matthijs, A. van Hoek, H. Löffler, R. Kraayenhof

Membrane Organisation of Anacystis nidulans Following Iron Deprivation
and Heme Deficiency 647
J. Guikema

Properties of Purified Oxygen-evolving Photosystem II Particles from
the Blue-green Alga, Phormidium laminosum 651
A. Stewart, J. Bowes, D. Bendall

Characterization of an Oxygen-evolving Photosystem II Complex from a Thermophilic Cyanobacterium Synechococcus sp.   655
G. Schatz, H. Witt

A Calcium/Sodium Requirement near Reaction Center II in Anacystics nidulans   · 659
D. Becker, J. Brand

Effects of Ca_2^+ Ions on the Light-induced Electron Transport Activities of Anacystis nidulans Permeaplasts and Spheroplasts   663
G. Sotiropoulou, T. Lagoyanni, G. Papageorgiou

The Single Subunit P-700 Reaction Center of a Thermophilic Cyanobacterium   667
A. Binder, P. Muster, R. Bachofen

Electron Transfer around Photosystem I in Cyanobacterial Heterocyst Membranes   671
M. Hawkesford, J. Houchins, G. Hind

Transient and Steady-state Kinetics of the Reaction between Cytochrome c and the Photosystem I Reaction Centre in Cyanobacteria   675
A. Hadberg, L. Olsen, R. Cox

Kinetics of a Chlorophyll a Long Wavelength Fluorescing Form: Quenching of F750 in Relation to PSI Oxido-reduction State in a Cyanobacterium (Pseudoanabaena sp.)   679
J. Duval, J. Thomas, H. Jupin

Measurement of Conformational Changes in Pigments of Aphanocapsa 6714 by Fluorescence Lifetime Following Transition from Dark to Light Growth   683
J. Manwaring, B. May, R. Brown, E. Evans

Phycoerythrin: Spectroscopic Analysis of its Subunits and Aggregates from Monomer to Dodecamer   687
B. Zilinskas, J. Grabowski, S. Campbell

Photocontrol of Phycoerythrocyanin Synthesis in an Oscillatoria Strain (Cyanobacteria)   691
J. Thomas, A. Mousseau, N. Hauswirth

Phycobilisome Composition and Relationship to Reaction Centers in Anacystis nidulans   695
R. Khanna, J. Graham, J. Myers, E. Gantt

Bioenergetics of Nitrogenase in Blue-green Algae. I. Physiological Conditions for Nitrogenase Activity in Phormidium foveolarum   699
H. Weisshaar, H. Almon, P. Böger

Bioenergetics of Nitrogenase in Blue-green Algae. II. Electron Transport
to Nitrogenase in Heterocysts of Anabaena variabilis   703
B. Schrautemeier, H. Böhme, P. Böger

Surface Electric Properties of Thylakoid Fragments Isolated from Vege
Vegetative and Heterocystous Cyanobacteria   707
K. Kalosaka, G. Papageorgiou

Studies on Glutathione Reductase Activity in the Cyanobacteria
Anabaena sp. Strain 7119   711
A. Serrano, J. Rivas, M. Losada

Regulation of Nitrate Utilization by CO_2 Fixation Products in the
Cyanobacterium Anacystis nidulans   715
C. Lara, J. Romero, E. Flores, M. Guerrero, M. Losada

Possible Role of an Amino Acid Oxidase in Photosystem II of Anacystis
nidulans   719
E. Pistorius

A Post-illumination CO_2 Burst in Anabaena variabilis as a Measure of
Bicarbonate Transport Driven by Cyclic Photophosphorylation   723
T. Ogawa, Y. Inoue, R. McLilley, W. Ogren

8. Photosynthesis and Solar Energy Technology

Photosynthesis for Energy   727
D. Hall

Specially Developed Photosynthetic Organisms as Future Possibility for
Large-scale Low-cost Conversion of Solar Radiation into Useful High-
grade Energy   741
L. Duysens

The Outdoor Cultivation of the Halotolerant Alga Dunaliella – a Model for
Biosolar Energy Utilization for Useful Chemical Products   745
M Avron

Industrial Mass Algae Cultures in Luke-warm Water   755
E. Dujardin

Environmental Effects of Ethanol and Methanol Production from Biomass   761
H. Egneús

H_2-photoproduction of Green Algae: Water Serves as the Main Source
of Electrons   769
B. Mahro, L. Grimme

Photoproduction of Hydrogen by Immobilized "Adapted"Algae   773
M. Brouers, J. Jeanfils, F. Collard

Photoproduction of Hydrogen from Water Using Immobilized Biological
and Synthetic Catalysts
K. Rao, D. Hall, P. Cuendet, M. Grätzel
777

H_2-photoproduction of Green Algae: Changes in the Energy State of
Chlorella fusca under H_2-photoproductive Conditions
A. Küsel, B. Mahro, L. Grimme
781

Simultaneous Production of Hydrogen and Oxygen as Affected by Light
Intensity in Unicellular Aerobic Nitrogen Fixing Blue-green Alga
Synechococcus sp. Miami BGO43511
K. Reddy, A. Mitsui
785

Continuous Hydrogen Photoproduction from Sulfide by an Immobilized
Marine Photosynthetic Bacterium, Chromatium sp. Miami PBS 1071
H. Ikemoto, A. Mitsui
789

Biophotolysis: Generation of Low-potential Reducing Equivalents by
Photosystem I-enriched Subchloroplast Vesicles
K. Krab, R. Boog, F. Peters
793

Photoproduction of H_2 and $NADPH_2$ by Polyurethane-immobilized
Cyanobacteria
A. Muallem, D. Bruce, D. Hall
797

Development of Hydrogen Production Activity in the Marine Blue-green
Alga Oscillatoria sp. Miami BG7 under Natural Sunlight Conditions
E. Phlips, A. Mitsui
801

Effect of Seawater Quality on Biomass and Hydrogen Photoproduction by
a Marine Blue-green Alga Oscillatoria sp. (Miami BG7)
S. Ramachandran, A. Mitsui
805

Mixotrophic Growth of Nostoc 268
T. Vaara, K. Sivonen, S. Kurkela
809

Immobilisation of PSI and PSII on Semiconducting Particles
P. Cuendet, M. Grätzel
813

Cyanobacterial Electrode, its Properties and Functions
H. Ochiai, H. Shibata, Y. Sawa, K. Takata
817

Characterization of a 680-nm Absorbing Hydrated Chlorophyll Dimer as
Photocatalyst for the Water Splitting Reaction in vitro
M. Showell, J. You, F. Fong
821

Index of Names
824

PREFACE

The Sixth International Congress on Photosynthesis took place from 1 to 6 August 1983, on the Campus of the "Vrije Universiteit Brussel", in Brussels, Belgium. These Proceedings contain most of the scientific contributions offered during the Congress.

The Brussels Congress was the largest thus far held in the series of International Congresses on Photosynthesis. It counted over 1100 active participants. The organizers tried to minimize the disadvantages of such a large size by making maximum use of the facilities available on a university campus. Most contributions were offered in the form of posters which were displayed in a substantial number of classrooms. The discussion sessions, twice a day, four or five in parallel, took place in lecture rooms in the very vicinity of these classrooms. In this way it was attempted to generate the atmosphere of a small meeting. The unity of the subject Photosynthesis was preserved in the ten plenary lectures, organised in such a way that a general overview of two diverse topics was given every day. In addition, there were the five times four parallel symposia dealing with some sixteen general topics.

Every editor of proceedings of a congress is faced with the problem of editing and arranging the contributions, a problem compounded by the wide diversity and the large number of the 753 manuscripts. This editor did very little in the way of editing the papers: all papers were prepared, camera-ready, by the authors themselves and there was no proof-reading. The main reason for this was the need to ensure speedy publication. The contributions are arranged in four volumes but the Proceedings form one set. Although some attempts were made to bring related topics together in one volume, the volumes I to IV should be seen as a succession of chapters, rather than as volumes in their own right. Thus, artificial and arbitrary subdivisions were avoided. A page limit was imposed in order to prevent oversized volumes.

The contributions are arranged in chapters which have no direct relation to the sessions or symposia in which they were presented. The sole criterium for putting a contribution into a certain chapter was its contents. The contributions offered during the Round Table Discussion on Light-Controlled Development of the Photosynthetic Apparatus, July 29 to 30, 1983 in Antwerp, are also included in these Proceedings. They comprise most of the contents of Chapter 7 of Volume IV.

The early publication date of these Proceedings could not have been realised without the efforts of, and the pleasant cooperation with, Mr. Ad Plaizier of Martinus Nijhoff Publishing House. Thanks are due to all Congress members, whose active participation made the Congress a success and these volumes an important document on the state of photosynthesis research. The very much needed assistance of the Local Organizing Committee is gratefully acknowledged. The Photosynthetic Community is indebted to the "Vrije Universiteit Brussel" for making available its premises, facilities and staff. Thanks are also due to the administrative staff of the Congress: secretaries, hostesses, technicians and the two diligent computer programmers, Mr. W. Dierickx and Mr. B. Philips. Special appreciation goes to Ms Blanche van den Haute for her dedicated work in the preparation and the management of the Congress and her help in editing these volumes.

Brussels, March 1984

C. Sybesma, Editor

In Memory of

KAZUO SHIBATA

Only a few days before the start of this Sixth International Photosynthesis Congress our research community was saddened to learn of the
loss of one of our most prominent members, Dr. Kazuo Shibata. His un-
timely death on July 27, 1983 followed a long fight against cancer.
Dr. Shibata contributed signficantly to several different research areas
in the study of photosynthesis. In addition, we will long remember him
for his warmth and charm, infectious enthusiasm, and his eminently hono-
rable and friendly approach to friends and scientific colleagues.

Kazuo Shibata was born in Kyoto, Japan on November 15, 1918. He was
the 5th son of the famous artist and painter Seiho Takeuchi. He graduated
from Waseda University in 1942 and subsequently was drafted into military
service. While at the 7th military research institute he met Hiroshi
Tamiya whose influence and interests in biological research were to shape
Kazuo's future.

From January of 1946 to early 1970 Kazuo Shibata was associated with
the Tokugawa Institute for Biological Research. It was early during this
period that he developed an interest in photosynthesis and cell physiology.
His strong interest in the spectral characterization of living algal cells
and chloroplasts led to the discovery of the "opal glass" technique, which
permitted accurate absorption measurements of scattering samples; this was
rapidly accepted as a standard technique in all aspects of biological research. He also began another major activity : the training of students
in physical an biophysical chemistry. He served as a lecturer and, subsequently, an associate professor at Waseda University between 1946 and
1957.

From 1953 to 1955 Kazuo Shibata joined the research group of Melvin
Calvin and Andrew Benson at Radiation Laboratory at Berkeley, California.
He contributed to discoveries of the photosynthetic carbon reduction cycle.
In addition, the opal glass method for the spectral characterization of
chloroplasts was firmly established during this stay at Berkeley. Invited
by Stacy French, he spent 1955-56 at the Carnegie Foundation Research
Laboratories, Stanford, California studying the greening of etiolated

seedlings. It was at this time that he discovered the discrete shifts in the absorption maxima as protochlorophyll(ide) is converted to chlorophyll(ide). These changes are now universally called the "Shibata Shift"; their characterization has stimulated research in many laboratories and has broadened our understanding of chloroplast membrane assembly. In 1957 Kazuo Shibata received his doctorate of science from the University of Tokyo. His thesis was entitled "Spectroscopic Studies on Chlorophyll Formation in Intact Leaves".

From 1957 to 1969 Kazuo served as a professor at the Tokyo Institute of Technology. His research continued in photosynthesis and protein chemistry and led to several incisive discoveries. These included : the development of site-specific amino acid modifying reagents to characterize tertiary structures of proteins; the discovery of chloroplast structural (shrinkage) changes; and, the very important application of polyacrylamide gel electrophoresis to separate chloroplast pigment proteins. In 1966 he was invited to join a voyage of the Alpha Helix into the Great Barrier Reef to characterize photosynthetic productivity of symbiotic algae of reef organisms. He generalized the principle of his opal glass method into the theory of "Integral Attenuance Measurement" and developed various spectrophotometric equipment for biological use based on this theory. During the 1970's he applied various biophysical techniques to investigate photosynthesis in intact leaves and established the relationship of thermoluminescence to the oxygen evolving system of plants.

Kazuo's contributions to science have been recognized in many ways. He was chosen to be the director of the RIKEN Research Laboratory of Plant Physiology from 1970 to 1979, and served as the honorary invited director from 1979 until his death. He was the vice president of International Association of Photobiologists from 1976 to 1980 and was the president of the Japan Optical Society from 1981 to 1983. In 1980 the Medal with Purple Ribbon was awarded by the Japanese Government to him for his contribution to spectrophotometric analysis of biological materials.

Kazuo worked tirelessly to promote basic research on photosynthesis, both at home in the creation of "Green Energy" and "Biomass Conversion" projects, as well as internationally to promote binational or multinational research programs including the "U.S.-Japan Program of Cooperation in Photoconversion and Photosynthesis".

Besides his scientific publications, Kazuo left one other major legacy : a collection of paintings which reflect his artistic skills as well as his sensitivity to his friends and surroundings. These works remain to remind us of a man who believed in the unity of man's creative activities in art and science. To these concepts we dedicate the Proceedings of the Sixth International Photosynthesis Congress to the memory of Kazuo Shibata.

Friends of K. Shibata

The Congress has been sponsored by the "Koninklijke Academie voor
Wetenschappen, Letteren en Schone Kunsten van België", the "Académie
Royale des Sciences, des Lettres et des Beaux-Arts de Belgique",
and the "International Union for Pure and Applied Biophysics".

The Congress has been supported by contributions from:

- The Commission of the European Communities
- The "Ministerie van Onderwijs" (Ministery of Education)
- The Commissariat-General for International Cultural Cooperation,
  Flemish Community
- The "Nationaal Fonds voor Wetenschappelijk Onderzoek" (National
  Fund for Scientific Research)
- The "Nationale Loterij van België"
- The SABENA, Belgian World Airlines
- The "Algemene Spaar-en Lijfrentekas/Caisse Générale d'Epargne et
  de Retraite"
- The "Generale Bankmaatschappij N.V./Société Générale de Banque, S.A."
- The "Kredietbank, N.V."
- The Belgian Biophysical Society
- The OCE Belgium S.A.

INTERNATIONAL PHOTOSYNTHESIS COMMITTEE 1980 - 1983.

C. Sybesma,   Biophysics Laboratory, Vrije Universiteit Brussel,
                     Pleinlaan 2, 1050 Brussels, Belgium, Chairman.

J. Amesz,   Department of Biophysics, Huygens Laboratory,
                     University of Leiden, P.O. Box 9504,
                     2300 RA Leiden, The Netherlands.

J.M. Anderson,   CSIRO, Division of Plant Industry,
                     P.O. Box 1600, Canberra, ACT, 2601, Australia.

J. Barber,   Department of Botany, Imperial College,
                     Prince Consort Road, London SW7 2BB, United Kingdom.

N.I. Bishop,   Department of Botany, Oregon State University,
                     Corvallis, Or. 97331, U.S.A.

R. Douce,   DRF/Biologie Végétale, 38041 Grenoble, France.

U.W. Heber,   Institute of Botany, University of Würzburg,
                     Mittlerer Dallenbergweg 64, 8700 Würzburg, F.R.G.

B.A. Melandri,   Department of Botany, University of Bologna,
                     Via Irnerio 42, 40126 Bologna, Italy.

N. Murata,   Department of Biology, College of General Education,
                     University of Tokyo, Komaba Neguro-Ku,
                     Tokyo 153, Japan.

N. Nelson,   Department of Biology, Technion, Haifa, Israel.

A. Shlyk,   Institute of Photobiology, Academy of Sciences of
                     the USSR, 27 Academicheskaya, 220733 Minsk, USSR.

D. Von Wettstein,   Department of Physiology, Carlsberg Laboratory,
                     Gl. Carlsbergvej 10, 2500 Copenhagen, Denmark.

I. Zelitch,   The Connecticut Experimental Station, P.O.Box 1106,
                     New Haven, Con. 06504, U.S.A.

LOCAL ORGANIZING COMMITTEE.

C. Sybesma,   Biophysics Laboratory, Vrije Universiteit Brussel,
Pleinlaan 2, 1050 Brussels, Belgium. Chairman.

R. Lannoye,   Laboratoire de Physiologie Végétale, Université
Libre de Bruxelles, 28 Av. Paul Héger, 1050 Brussels,
Belgium. Vice-Chairman.

J. De Greef,   Department of Biology, Universitaire Instelling
Antwerpen, Universiteitsplein 1, 2610 Wilrijk, Belgium.

C. Sironval,   Institut de Botanique, Université de Liège,
Sart-Tilman, 4000 Liège, Belgium.

M. Van Poucke,   Department SBM, Limburgs Universitair Centrum,
Universitaire Campus, 3610 Diepenbeek, Belgium.

J. Aghion,   Institut de Botanique, Université de Liège,
Sart-Tilman, 4000 Liège, Belgium;

H. Clijsters,   Department SBM, Limburgs Universitair Centrum,
Universitaire Campus, 3610 Diepenbeek, Belgium.

R. Marcelle,   Laboratory of Plant Physiology, Opzoekingscentrum
van Gorsum, 3800 St.-Trond, Belgium.

LONG-WAVELENGTH ABSORBING FORMS OF CAROTENOIDS

STANISŁAW WIECKOWSKI, EWA WŁOCH, BARBARA BRONIOWSKA

INTRODUCTION

It has been recognized that under certain conditions carotenoids with red-
-shifted absorption spectra for about 50 nm as compared with those in lipid
solvents may occur. This shifting is believed to be due to interaction of
pigment (mainly β-carotene) molecules with protein (e.g. Nishimura, Takamatsu,
1957, 1960, Ji et al., 1968, Vernon et al., 1969) or to aggregation that may
take place in artificial lipid membranes (Litvin et al., 1965) as well as in
aqueous organic solvents (Hager, 1970). A suspension of crystalline carotene
also exhibits an absorption band within the range 520-530 nm (Shibata, 1956).
It is still by no means clear, however, whether any particular pigments may
occur in one or more LWAFs[+] or whether they represent true natural forms or
merely appear in vitro under appropriate conditions. We have undertaken
studies in order to answer some of these questions. This communication deals
with the comparison of the absorption spectrum of LWAF of β-carotene
obtained by two different methods with that of the crystalline form. We have
also attempted to find out whether other carotenoids may form LWAFs.

MATERIALS AND METHODS

β-carotene and xanthophylls were isolated from spinach purchased at the
local market or collected from a field. The absorption spectrum of commercial
β-carotene (Fluka AG) suspension was also analysed.
The form of β-carotene investigated was isolated either by the method
described previously (Więckowski, Droba, 1978), or by the procedure of
Ji et al. (1968).
The methods of pigment extraction and saponification of lipids were described
previously (Więckowski, 1961). The pigments solubilized in a mixture of light
petroleum (b.t. 40-55°C) and acetone (10:0.6, v/v) were subjected to column
chromatography on aluminium oxide (Mecherey Nage, W.Germany). The chromato-
gram was developed with light petroleum containing acetone (to 17%). The
isolated xanthophylls were rechromatographed on silica gel (60F254, Merck,
W.Germany) columns whereas carotenes were purified chromatographically on
magnesium oxide (BDH, England) column. A mixture of light petroleum and
acetone (10:2.5, v/v) or light petroleum and benzene (9:1, v/v) was used to
purify xanthophylls investigated or β-carotene, respectively.
β-carotene was crystallized out from a mixture of benzene-methanol, or from
80% acetone, while xanthophylls were crystallized out from methanol.
The carotene crystals were pulverized with sucrose in a Potter type
homogenizer prior to suspending in phosphate buffer (0.06 M, pH 7.2)
containing 0.5% Triton X-100. The suspension was subjected to DEAEcc.
The absorption spectra were recorded on a spectrophotometer Specord UV VIS
(K.Zeiss, Jena).

[+] Abbreviations: LWAF - long-wavelength absorbing form, DEAEcc - DEAE
cellulose column chromatography.

Sybesma, C. (ed.), Advances in Photosynthesis Research, Vol. II. ISBN 90-247-2943-2.
© *1984 Martinus Nijhoff/Dr W. Junk Publishers, The Hague/Boston/Lancaster.*

RESULTS

Comparison of the absorption spectrum of the aqueous suspension of red
pigment isolated by DEAEcc (Więckowski, Droba, 1978) with that by the
procedure of Ji et al. (1968) indicates that either method leads to

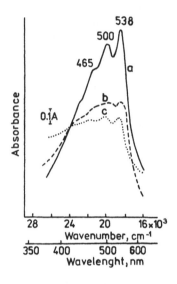 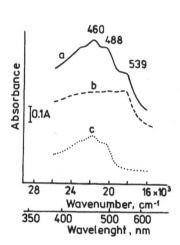

Fig.1 (left). Absorption spectra of suspensions of β-carotene form isolated
chromatographically on a DEAE cellulose column (a), or by sequential
extraction (b). The absorption spectrum of a mixture of chromatographically
isolated pigment and residue after extraction of lipids with 100% acetone
is also shown (c).
Fig.2 (right). Absorption spectra of suspensions of crystalline β-carotene:
a - before purification on a DEAE cellulose column, b - red fraction
collected from the top of the column, c - red fraction passed through the
column. Pigments were suspended in phosphate buffer (0.06 M, pH 7.4)
containing 1% Triton X-100.

the isolation of the same LWAF of the pigment which exhibited four bands
within the visible part of the spectrum: at about 538, 500, 465 and 430 nm
(Fig.1a, b). Flattening of the absorption spectrum of the pigment isolated
by the procedure of Ji et al. (1968) is due to impurities, since adding
the residue of the lipid extraction with 100% acetone to the suspension
of pigment collected from the top of the column also caused a flattening
effect on the absorption spectrum (Fig.1c). Thus either method leads to
the separation of the same form of β -carotene although of various degrees
of purity. This form also shows weak luminescence with maximum at 568 nm
(unpublished).
We have found that the crystalline β-carotene suspension with the absorption
maximum at 460 nm and shoulders at 539, 488 and 433 nm (Fig.2a) is composed
of two components separable from each other on DEAEcc. One red fraction
passes through the column and the other remains on the top. The first shows
absorption bands at 486, 460 and 433 nm, and the second at 539, 496 and

463 nm. Thus the pulverized commercial β-carotene, like that obtained by crystallization from acetone or benzene-methanol, consists of two fractions of which presumably only one (that remaining on top of the column) represents the crystalline form and resembles the LWAF described above.

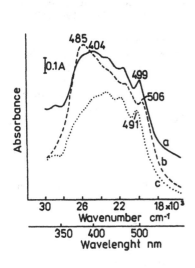

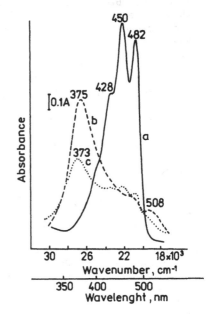

Fig.3 (left). Absorption spectra of aqueous suspensions of violaxanthin (a), lutein (b), or neoxanthin (c) crystals.
Fig.4 (right). Absorption spectra of lutein in 100% acetone (a), 50% acetone (c), or 40% acetone (b).

The crystalline α-carotene suspension behaves similarly to that of β-carotene (unpublished). The fraction remaining on top of the column shows an absorption spectrum with main bands at 521, 481 and 448 nm while the fraction passing through shows maxima at 479, 451 and a shoulder at 429 nm.
We have established that the absorption spectra of the suspension of xanthophyll crystals are also complex (Fig.3). This may indicate that under our conditions each xanthophyll occurred in more than one form. Their separation cannot be achieved by DEAEcc. Short-wavelength absorbing forms may, however, be partially removed by extensive washing with the solvents used for crystallization. It may then be assumed that the LWAFs represent crystals of appropriate xanthophylls.
The gradual addition of water to an acetone solution of lutein causes marked changes in the structure of the absorption spectrum (Fig.4). In the first step only a short-wavelength absorbing form with one maximum is formed. The LWAF appears in the second step. It may then be concluded that the crystalline lutein and presumably some other xanthophylls are accompanied by non-crystalline forms of the pigments.

DISCUSSION

Analysis of the absorption spectra indicates that β-carotene isolated either chromatographically on DEAE cellulose or by sequential extraction presents the same form. Its absorption spectrum also coincides well with that of purified crystalline β-carotene. We may therefore admit that the LWAF of β-carotene may represent a fine-crystalline form of the pigment. The crystals can easily be adsorbed on DEAE cellulose chains or on protein aggregates (unpublished). Presumably such artificialy formed β-carotene-protein complexe have been described by some authors (Nishimura, Takamatsu, 1956, 1960, Vernon et al.,1969).

Not only β-carotene but also ℒ-carotene, lutein, violaxanthin, neoxanthin and presumably other carotenoids may form LWAFs. It seems likely that the LWAFs of pigments are formed mainly under appropriate conditions in artificial systems. However, the possibility cannot be excluded that under suitable conditions they may also be encountered in chloroplasts. For example, when using DEAEcc we were able to isolate the β-carotene LWAF only from winter spinach (Więckowski et al., 1981). So it seems likely that it occurred in the chloroplasts prior to treatment with Triton X-100.

REFERENCES

Hager A (1970) Ausbildung von Maxima im Absorptionspectrum von Carotenoiden im Bereich um 370 nm, Planta (Berl.) 90, 38-53.
Ji TH, Hess JL and Benson AA (1968) Studies on chloroplast membrane structure. I.Association of pigments with chloroplast lamellar protein, Biochim.Biophys.Acta 150, 676-685.
Litvin FF, Gulyaev BA, Sineshchekov VA (1965) Agriegirowanne formy chloro-filla a, chlorofilla b, β-karotina w monoslojach i plenkach; migracja ener-gii mieshdu nimi i w kompleksach (chlorofill a + β-karotin), Dokl.Akad.Nauk SSSR 162, 1184-1187.
Nishimura M and Takamatsu K (1957) A carotene-protein complex isolated from green leaves, Nature (London) 180, 699-700.
Nishimura M and Takamatsu K (1960) Studies on a carotene-protein complex isolated from green leaves, Plant Cell Physiol. 1, 305-309.
Shibata K (1956) Absorption spectra of suspension of carotene crystals, Biochim.Biophys.Acta 22, 398-399.
Vernon LP, Ke B, Mollenhauer HH and Shaw ER (1969) Composition and structure of spinach subchloroplast fragments obtained by the action of Triton X-100. In Metzner H, ed. Progress in Photosynthesis Research, vol.1,pp 137-148. Tübingen.
Więckowski S (1961) Changes in the content of carotenoids in growing bean leaves, Bull.Acad.Polon.Sc. ser.biol. 9, 325-332.
Więckowski S, Droba M (1978) Heterogeneity of the thylakoid membrane fractions derived from spinach chloroplasts by the action of Triton X-100 in low salt medium. II.Isolation of β-carotene by the DEAE cellulose column chromatography. Plant Sc.Lett. 13, 397-404.
Więckowski S, Droba M, Włoch E, Majewska G and Pająk S (1981) On the release of carotene from spinach thylakoid membranes treated with Triton X-100. In Akoyunoglou G, ed. Photosynthesis, vol.3, pp 655-663. Philadelphia: Balaban Inter. Sc.Ser.

Authors address: Institute of Molecular Biology, Jagellonian University, Al.Mickiewicza 3, 31-120 Kraków, Poland.

DETECTION AND CHARACTERISTICS OF THE PIGMENTS OF VARIEGATA
LEAVES

M. BOGDANOVIĆ AND A. RASTOVIĆ

1. INTRODUCTION

Varigata leaves have several zones of coloration in the same
leaf. This different coloration is geneticaly determined. Our
previous results (Bogdanović, Rastović, 1982) have shown that
crude pigments from diferent parts of box elder leaves exibited
different absorbtion maxima in Soret region (aceton solutions)
and in red region (petrol ether solutions. So it was of interest
to study the pigment distribution and pigment characteristics in
those different tissues.

2. MATERIAL AND METHODS

Leaves of Acer negundo L. forma variegata and leaves of
Chlorophytum elatum Variegatum were used as an experimental
material. Extraction of the pigments was performed in 90%
Aceton water solution. To prevent oxydation $MgCO_3$ was added.
Crude pigments were transfered in to the petrol ether and
aceton was washed with water. Ascendent paper chromatography
was used for pigment separation in system petrol ethar : metanol
(30 : 1) (Marcus, 1952). Absorbtion spectra were obtained by the
spectrophotometar Aminco DW-2A operated at split beam mode.
Chlorophyll concentration was determined according to Vernon
(1960) and carotenoid content according to Holme (1954).

3. RESULTS AND DISCUSSION

Leaf tissues of box elder and Chlorophytum contain high
concentration of oxydative substances. So it was necessery to
performe the extraction in 90% aceton with the addition of
Mg CO_3 to prevent oxidation. Variegata leaves of box elder
exhibit ver- high diversity of pigment distribution in the same
leaf (Tab. 1.). Green part of the leaf has very high concentratio
of chlorophylls and carotenoids. High ratio of Chl. $\underline{a}$ / Chl. $\underline{b}$
indicates rather low concentration of Chlorophyll $\underline{b}$. Yellow
part of the leaf has ten times lower concentration of chlorophyll
than green part, however, very high concentration of carotenoids
in this part is registered. High ratio of Chl. $\underline{a}$ / Chl. $\underline{b}$
indicates a very low concentration of chlorophyll $\underline{b}$. At the same
time a ratio of chls. / carot. is lower than one indicating very
high concentration of carotenoids in this part of the leaf. In
white part of the leaf only traces of chlorophylls were detected,
and carotenoids are dominant. Chlorophytum leaves consist of
three distinctive tissues: green, pale green and white. Content
of chlorophyll in green part was 18 times higher than in pale
green part, and 490 times higher than in white part. Low ration
of Chlorophyll a/b indicates very high concentration of
chlorophyll b.

Sybesma, C. (ed.), Advances in Photosynthesis Research, Vol. II. ISBN 90-247-2943-2.
© *1984 Martinus Nijhoff/Dr W. Junk Publishers, The Hague/Boston/Lancaster.*

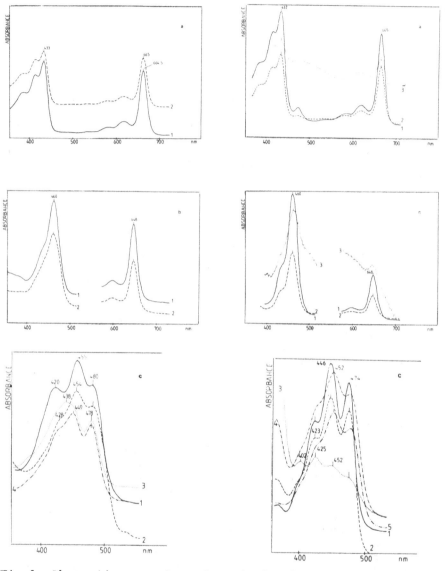

Fig.1. Absorption spectra of box elder pigments acetone solution(a) Chl.a̲, (b) Chl.b̲, (c) Carot. and X̅a̅nt. Curves 1,3: green part. pigments

Curves 2,4: White part pigments.

Fig.2. Absorption spectra of Chlorophytum pigments acetone solution.(a) Chl.a̲, (b) Chl.b̲ (c) Carot. and X̅a̅nt. Curves 1,4: green part. pigments

Curves 2,5: Pale green part pigments,
Curve 3: White part pigments.

Table 1. Distribution of the pigments in different parts of
the leaves of box elder and Chlorophytum

Part of the leaf	Chl. (a+b) ug/g ($\overline{f.m.}$)	Chl $\underline{a}$ $\overline{Chl\ b}$	Total Carot. ug/g(f.m.)	Total Chl. $\overline{Total\ Carot}$
Box elder				
green	1273,00	8,22	507,00	2,51
yellow	137,00	26,40	153,00	0,89
white	4,24	3,82	79,52	0,05
Chlorophytum				
Green	1395,85	2,12	65,64	21,26
Pale green	76,52	2,38	14,30	5,35
White	2,82	2,14	2,36	1,20

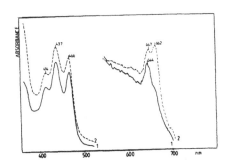

Fig. 3. Absorbtion spectra of Chlorophytum unidentified
pigment dissolved in acetone. Curve 1 : Pigment of
green part.
Curve 2 : Pigment of pale green part.

Absorption spectra of pigments of box eledr separated by the
paper chromatography (Fig 1) indicate that chlorophyls in
diferent tissues are the same, when carotenoids seam to be
different. The shifts of absorbtion maxima of carotenoids
and xanthophylls could be either caused by different chemical
forms or by different concentrations of pigments. Similar
results were obtained with pigments of Chlorophytum (Fig. 2).
Further more in leaf of this plant (Fig. 3) the distinct
pigment with a new absorption maxima is detected.

REFERENCES

Bogdanović M and Rastović A (1982) Absorption spectra
characteristics of Acer negundo L. leaf pigments. Periodicum
biologorum 84, 2, 117-118.
Holm g. (1954) Chlorophyll mutations in barley. Acta Agric.
Scand. 4, 457-471.
Marcus L. (1952) Agrokemia Talajtan, 1. 291.
Vernon P. L (1960) Spectrophotometric determination of
Chlorophylls and Pheophytins in Plant Extracts. Anal. Chem.
32, 1144-1150.

Autor's adress: Institute for the aplication of Nuclear Energy
                in Agriculture, Veterinary Medicine and
                Forestry, Banatska 31, b, 11080 Zemun, Yugosl-
                avia.

FORMULAE AND PROGRAM TO DETERMINE TOTAL CAROTENOIDS AND CHLOROPHYLLS A
AND B OF LEAF EXTRACTS IN DIFFERENT SOLVENTS

A.R. WELLBURN and H. LICHTENTHALER

1. INTRODUCTION

Various equations for the determination of total chlorophyll and individual amounts of chlorophylls a and b in extracts from plant tissues
exist (see Holden 1976) and some of them (e.g. Arnon 1949) have been
widely used. Additional modifications to the equations have also been
developed so as to permit an estimate of total carotenoids to be made
from the spectrum of the same mixture in di-ethyl ether (Ziegler, Egle
1965; Gaudillère 1974).

2. PROCEDURE

During the course of studies which involved the use of various solvents
we noted large discrepancies (> 40%) between estimations made using the
different published equations for different solvents, all of which were
known to contain similar amounts of pigments. Taking advantage of a
thin-layer chromatography method (Lichtenthaler and Pfister 1978) which
permits a distinct separation between the two chlorophylls and also the
major carotenoids using light petroleum (b.p. 40-60°C)/dioxane/iso-propanol
(7:3:1, by vol.) as developing solvent, fresh samples of chlorophyll a
and b uncontaminated with each other were readily available for re-
evaluation of the specific absorption coefficients.

3. RESULTS AND DISCUSSION

Those spectroscopic values published by Smith and Benitez (1955) using
di-ethyl ether were found to be still the most acceptable and relative
specific absorption coefficients to these values were established (Table 1)
for various other solvents. The red peak maxima of the chlorophylls
were shifted to longer wavelengths with increasing polarity of the
solvents. The red absorption peaks of the chlorophylls were also broadened in the same sequence and the values for the specific absorption
coefficients decreased. At the same time suitable $E_{1cm}^{\%}$ values for total
carotenoids at 470nm were also determined. On the basis of these coefficients the following equations were derived to determine the individual
levels of both chlorophyll a (C_a) and chlorophyll b (C_b) and the total
amounts of carotenoids (C_{x+c}) and chlorophylls ($C_a + C_b$) in $\mu g.ml^{-1}$ of
plant extract using the measured extinction values (E) at the different
wavelengths. A micro-computer program written in BASIC which allows
selection of solvent and the use of these different formulae is given
alongside the formulae which also includes formulae for light petroleum
(b.p. 40-60). This option should be used with caution and only when the
solvent is freshly dried and redistilled.

Each formula was tested against each other and also with the original
published equation using the same amounts of total leaf pigments in
different solvents. The variation between determinations using the
formulae above was reduced below 5% across the 5 different solvents and
within the experimental error of reading the values from the spectrophotometer. Some published equations notably those for 96% (v/v) ethanol

Sybesma, C. (ed.), Advances in Photosynthesis Research, Vol. II. ISBN 90-247-2943-2.
© *1984 Martinus Nijhoff/Dr W. Junk Publishers, The Hague/Boston/Lancaster.*

di-ethyl-ether:

$$C_a = 10.05E_{662} - 0.766E_{644}$$

$$C_b = 16.37E_{644} - 3.14E_{662}$$

$$C_{x+c} = \frac{1000E_{470} - 1.28C_a - 56.7C_b}{230}$$

Methanol:

$$C_a = 15.65E_{666} - 7.34E_{653}$$

$$C_b = 27.05E_{653} - 11.21E_{666}$$

$$C_{x+c} = \frac{1000E_{470} - 2.86C_a - 129.2C_b}{245}$$

96% (v/v) ethanol:

$$C_a = 13.95E_{665} - 6.88E_{649}$$

$$C_b = 24.96E_{649} - 7.32E_{665}$$

$$C_{x+c} = \frac{1000E_{470} - 2.05C_a - 114.8C_b}{245}$$

80% (v/v) acetone:

$$C_a = 12.21E_{663} - 2.81E_{646}$$

$$C_b = 20.13E_{646} - 5.03E_{663}$$

$$C_{x+c} = \frac{1000E_{470} - 3.27C_a - 104C_b}{229}$$

100% (v/v) acetone:

$$C_a = 11.75E_{662} - 2.35E_{645}$$

$$C_b = 18.61E_{645} - 3.96E_{662}$$

$$C_{x+c} = \frac{1000E_{470} - 2.27C_a - 81.4C_b}{227}$$

```
1000 REM
     **********************
     **CHLCALC-A PIGMENT**
     *ESTIMATION PROGRAM*
     **********************
1010 REM
     **********************
     **A.R.WELLBURN 1983**
     *H.K.LICHTENTHALER**
     **********************
1020 GOSUB 1640
1030 PRINT "ETHER..........1": PRINT
1040 PRINT "LIGHT PETROLEUM.2": PRINT
1050 PRINT "METHANOL......3": PRINT
1060 PRINT "96% ETHANOL....4": PRINT
1070 PRINT "80% ACETONE....5": PRINT
1080 PRINT "100% ACETONE....6": PRINT
1090 INPUT "CHOOSE...(1-6)? ";X
1100 PRINT : PRINT
1110 IF X < 1 OR X > 6 THEN GOTO 1020
1120 ON INT (X) GOTO 1130,1210,1290,1370,1450,1530
1130 INPUT "O.D. 644        ";B
1140 PRINT
1150 INPUT "O.D. 662        ";A
1160 GOSUB 1880
1170 E = (A * 10.05 - B * 0.766) * D
1180 F = (B * 16.37 - A * 3.14) * D
1190 I = (((1000 * C) - (1.28 * (E / D)) - (56.7 * (F / D))) / 230) * D
1200 G = E + F:H = E / F: GOTO 1610
1210 INPUT "O.D. 642        ";B
1220 PRINT
1230 INPUT "O.D. 660.5      ";A
1240 GOSUB 1880
1250 E = (A * 9.89 - B * 1.02) * D
1260 F = (B * 16.11 - A * 2.82) * D
1270 I = (((1000 * C) - (5.04 * (E / D)) - (25.23 * (F / D))) / 257) * D
1280 GOTO 1600
1290 INPUT "O.D. 653        ";B
1300 PRINT
1310 INPUT "O.D. 666        ";A
1320 GOSUB 1880
1330 E = (A * 15.65 - B * 7.34) * D
1340 F = (B * 27.05 - A * 11.21) * D
1350 I = (((1000 * C) - (2.86 * (E / D)) - (129.2 * (F / D))) / 245) * D
1360 GOTO 1600
1370 INPUT "O.D. 649        ";B
1380 PRINT
1390 INPUT "O.D. 665        ";A
1400 GOSUB 1880
1410 E = (A * 13.95 - B * 6.88) * D
1420 F = (B * 24.96 - A * 7.32) * D
1430 I = (((1000 * C) - (2.05 * (E / D)) - (114.8 * (F / D))) / 245) * D
1440 GOTO 1600
1450 INPUT "O.D. 646        ";B
1460 PRINT
1470 INPUT "O.D. 663        ";A
1480 GOSUB 1880
1490 E = (A * 12.21 - B * 2.81) * D
1500 F = (B * 20.13 - A * 5.03) * D
1510 I = (((1000 * C) - (3.27 * (E / D)) - (104 * (F / D))) / 229) * D
1520 GOTO 1600
1530 INPUT "O.D. 645        ";B
1540 PRINT
1550 INPUT "O.D. 662        ";A
1560 GOSUB 1880
1570 E = (A * 11.75 - B * 2.35) * D
1580 F = (B * 18.61 - A * 3.96) * D
1590 I = (((1000 * C) - (2.27 * (E / D)) - (81.4 * (F / D))) / 227) * D
1600 G = E + F:H = E / F
1610 GOSUB 1800
1620 GOSUB 1700
1630 END
1640 HOME
1650 PRINT "THIS PROGRAM CALCULATES MICROGRAMS ";
1660 PRINT "OF  CHLOROPHYLL AND CAROTENOIDS": PRINT : PRINT
1670 PRINT "DATA REQUIRED:O.D.";
1680 PRINT "MEASUREMENTS IN      SPECIFIED SOLVENT";
1690 PRINT " AND TOTAL VOLUME": PRINT : PRINT : RETURN
1700 INPUT "PRINT RESULTS ? (Y/N) ";B$
1710 IF LEFT$ (B$,1) = "N" THEN GOTO 1770
1720 INPUT "NAME OR NO.OF EXPT ";C$
1730 D$ = CHR$ (4): PRINT D$;"PR#1"
1740 PRINT C$: PRINT
1750 GOSUB 1800
1760 PRINT D$;"PR#0"
1770 INPUT "RUN AGAIN? (Y/N) ";A$
1780 IF LEFT$ (A$,1) = "Y" GOTO 1020: RETURN
1790 IF LEFT$ (A$,1) = "N" THEN END
1800 PRINT "CHLOROPHYLL A       =": TAB( 22); INT (E * 1000 + .5) / 1000
1810 PRINT "CHLOROPHYLL B       =": TAB( 22); INT (F * 1000 + .5) / 1000
1820 PRINT "CHLOROPHYLL(A + B)  =": TAB( 22); INT (G * 1000 + .5) / 1000
1830 PRINT "CHLOROPHYLL A/B     =": TAB( 22); INT (H * 1000 + .5) / 1000
1840 PRINT "TOTAL CAROTENOIDS   =": TAB( 22); INT (I * 1000 + .5) / 1000
1850 PRINT "A + B / X + C       =": TAB( 22); INT (((E + F) / I) * 1000 + .5) / 1000
1860 PRINT : PRINT "64X =";B;"  66Y =";A;"  470 =";C;"     VOL. =";D;"MLS"
1870 PRINT : RETURN
1880 PRINT
1890 PRINT "O.D. 470        ";C
1900 PRINT
1910 INPUT "TOTAL VOLUME    ";D
1920 PRINT : RETURN
```

Table 1. Redetermined specific absorption coefficients (SAC) for chlorophyll a and b at the peak maxima of each (underlined) and at 470nm plus the $E_{1cm}^{1\%}$ values at 470 nm for total leaf carotenoids (xanthophylls = x plus carotenes = c) in different solvents.

| Solvent | Chlorophyll a | | Chlorophyll b | | $E_{1cm}^{1\%}$ |
	nm	SAC	nm	SAC	for x+c
Di-ethyl ether*	<u>662</u>	101†	662	4.72	
	<u>644</u>	19.40	<u>644</u>	62†	
	470	1.20	<u>470</u>	56.70	2300
Methanol*	<u>666</u>	79.29	666	21.51	
	<u>663</u>	32.86	<u>663</u>	45.88	
	470	2.86	<u>470</u>	129.18	2450
96%(v/v) ethanol	<u>665</u>	83.83	665	23.11	
	<u>649</u>	24.57	<u>649</u>	46.84	
	470	2.05	<u>470</u>	114.75	2450
80%(v/v) acetone	<u>663</u>	86.86	663	12.12	
	<u>646</u>	21.73	<u>646</u>	57.70	
	470	3.27	<u>470</u>	104.03	2290
100%(v/v) acetone*	<u>662</u>	88.88	662	11.22	
	<u>645</u>	18.91	<u>645</u>	56.11	
	470	2.27	<u>470</u>	81.36	2270

* Spectroscopically pure solvent.   † accepted from Smith, Benitez (1955)

(Wintermans and de Mots 1965) were also to be found quite accurate whilst those for methanol (Visniac 1957) are particularly poor. There is a considerable difference between the spectroscopic properties of 80% (v/v) acetone and pure dry acetone which is reflected in the two equations we now provide. One of the reasons why the method of Arnon (1949) using 80% (v/v) acetone is prone to inaccuracy is because the specific absorption coefficients were originally taken from Mackinney (1941) who used pure acetone.

The coefficients given in Table 1 represent mean values from measurements performed with two recording spectrophotometers, a Shimadzu UV 200 and a Unicam SP 8000. However there are variations in spectrophotometer response especially in the relative height of the absorption in the blue-green as compared to that in the red region as well as a shift of the peak maxima of chlorophylls of between 0.5 to 1 nm. By using pure chlorophyll a and b solutions prepared in the manner described, the values at 470 nm and at the peak maximum of the other relative to the specific absorption coefficient at the peak maximum given in Table 1 can be re-adjusted, if necessary, for any spectrophotometer.

REFERENCES

Arnon DI (1949) Copper enzymes in chloroplasts. Phenol oxidase in *Beta vulgaris*, Plant Physiol. 24, 1-15.
Gaudillère JP (1974) Amelioration du dosage spectrophotometrique des chlorophylla a et b et des caroténoides totaux dans des extraits foliares, Physiol. Veg. 12, 585-599.
Holden M (1976) Chlorophylls. In Goodwin TW ed. Chemistry and Biochemistry of Plant Pigments, 2nd edn. pp. 462-488, London:Academic Press.
Lichtenthaler HK and Pfister K (1978) Practium der Photosynthese, Heidelberg:Quelle Meyer Verlag.
Mackinney G (1941) Absorption of light by chlorophyll solutions, J. biol. Chem. 140, 315-322.
Vishniac W (1957) Methods for study of the Hill reaction. In Colowick·SP and Kaplan NO, eds. Methods in Enzymology, Vol. 4, pp. 342-355. New York: Academic Press.
Smith JHC and Benitez A (1956) Chlorophylls:Analysis in plant materials. In Paech K and Tracey MV, eds. Modern methods of plant analysis, vol. 4, pp. 142-196, Berlin:Springer-Verlag.
Wintermans JFGM and de Mots A (1965) Spectrophotometric characteristics of chlorophyll a and b and pheophytins in ethanol, Biochim. Biophys. Acta 109, 448-453.
Ziegler R and Egle K (1965) Zur qualitativen Analyse der Chloroplasten-pigmente, Beitr. Biol. Pflanzen 41, 39-63.

Authors addresses: (ARW) Department of Biological Sciences, University of Lancaster, Bailrigg, Lancaster LA1 4YQ U.K. and (HKL) Botanisches Institut der Universität, Kaiserstraße 12, Postfach 6380, Karlsruhe, F.R.G.

UNUSUAL PIGMENTS IN A PRIMITIVE GREEN ALGA

Jeanette S. Brown/ Carnegie Inst. Washington, Stanford, CA 94305, U.S.A

INTRODUCTION

Although chlorophyll a is the essential pigment for photosynthesis in all algae and higher plants, its organization within thylakoid membranes and associations with different antenna pigments vary widely in different groups of plants. In general, most of the chlorophyll is bound to at least two or three different proteins. One of these proteins is closely associated with the reaction center of Photosystem I (PSI), another with PSII, and a third with other antenna pigments such as Chl b, Chl c, fucoxanthin or peridinin. To understand fully the light-absorbing steps in photosynthesis, we need to know more precisely how Chl a is attached to each of these different proteins. By comparing spectral characteristics of Chl-proteins isolated from diverse algal groups, we hope to distinguish the common and perhaps essential nature of the photoreactive pigments.

Currently we are examining a group of primitive, eucaryotic microalgae called Prasinophytes. These green flagellates may be evolutionary predecessors of Chlamydomonas and Dunaliella. Micromonas and Mantoniella are among the most primitive morphologically [Manton & Parke, 1960]. Both are highly motile spheres about 1 micron in diameter; distinctions between these two genera are not clear [Norris, 1980]. In addition to Chls a and b, they contain the protochlorophyll- or Chl c-like pigment, magnesium 2,4 divinylpheoporphyrin a_5 (MgDVP) [Ricketts, 1966a]. Their carotenoid composition is also unusual [Ricketts, 1966b]; the majority of their xanthophylls are in a high state of oxidation and show resemblances to fucoxanthin and peridinin. The most abundant of their carotenoids was named microxanthin.

These primitive algae are important for several reasons. Because they are abundant and widespread in both open and coastal waters, they must contribute significantly to productivity and biomass [Jeffrey, Hallegraeff 1980]. Their phylogenetic position near the origin of the Chlorophyta makes them valuable for comparative developmental and genetic studies of green algae and higher plants. Here, we present some initial studies of Chl-proteins isolated from Mantoniella. Spectra of an antenna pigment fraction show an absorption maximum near 632 nm from MgDVP and an absorption and fluorescence excitation maximum near 520 nm probably from microxanthin.

PROCEDURE

Culture: Micromonas pusilla No. 991 and Mantoniella squamata No. 990 were obtained from the Culture Collection at the University of Texas. Because Mantoniella grows much better than Micromonas in an enriched, natural seawater medium [Starr, 1978], most of our experiments have been performed with this alga. Cells were harvested from cultures grown for 7-9 days with continuous illumination and bubbled with air at about 20-25°C, resuspended in 5 mM $MgCl_2$, 50 mM Tricine, pH 7.3, and forced through a French pressure cell. The broken cell particles were washed with 5 mM EDTA, 2 mM Tris-maleate buffer, pH 7. Chlorophyll concentration was determined according to the method of Jeffrey and Humphrey [1975] for phytoplankton in which Chls a, b and c may be present together.

Preparation of Chl-Protein Complexes: The following detergents at various concentrations have been tried for solubilizing the washed cell membranes:

Sybesma, C. (ed.), Advances in Photosynthesis Research, Vol. II. ISBN 90-247-2943-2.
© *1984 Martinus Nijhoff/Dr W. Junk Publishers, The Hague/Boston/Lancaster.*

sodium deoxycholate (DOC), digitonin, sodium dodecylsulfate (SDS), and
Nonidet P40 (similar to Triton X-100). After incubation of detergent with
algal membranes for varying lengths of time at 20°C, the solubilized
pigment-proteins were separated by sucrose density gradient centrifugation
or by polyacrylamide gel electrophoresis (PAGE). Two different gel systems
have been used; one a 5% slab gel with 0.05% DOC [Picaud et al, 1982] and
the other a 7-15% gradient gel with 0.1% SDS.
Spectroscopy: Absorption and fluorescence spectra were measured with a
Cary spectrophotometer and Perkin Elmer fluorometer at 77 K with computer
interfacing as described in [Brown, Schoch 1982].

RESULTS AND DISCUSSION

Assuming that Chl $\underline{c}$ and MgDVP have similar extinction coefficients, we
found that the total chlorophyll concentration in <u>Mantoniella</u> varied
between 2 and 7 x 10^{-8} ug/cell; the ratios of Chl $\underline{a}$ to $\underline{b}$ between 2.6 and
3.4 and to MgDVP between 8.9 and 11. Probably these variations were caused
by changes in light intensity and temperature during growth and indicate
an inherent ability of this alga to adapt to changes in environment.

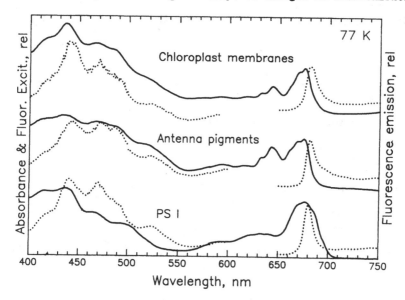

FIGURE 1. Absorption (solid lines), fluorescence excitation (400-600 nm)
and emission (650-750 nm, dotted lines) of <u>Mantoniella</u>. Excitation at
438 nm for emission spectra, sw = 3 nm; emission at 680 nm for excitation
spectra, sw = 4 nm. All fluorescence spectra were corrected for xenon
lamp output and instrument sensitivity.

Spectra of chloroplast membranes shown in Fig. 1 were identical for
either <u>Mantoniella</u> or <u>Micromonas.</u> The red absorption peak of Chl $\underline{b}$ near
643 nm is over 5 nm lower than it appears in other green plants; the band
near 632 nm is normally seen only in Chl $\underline{c}$-containing algae [Mann, Myers
1968] and here may be MgDVP; the band near 520 nm may be from microxanthin.
The fluorescence emission spectrum was also unusual in that there were no
long wavelength maxima (>700 nm) [Brown, 1969]; the absorption spectrum

of the PSI fraction clearly has the long wavelength bands which may fluoresce at longer wavelengths in most plants. The proportion of antenna pigment to PSI chlorophyll is very high in Mantoniella. Possibly most of the fluorescence comes from this antenna, overshadowing PSI emission.

A fluorescence excitation maximum near 520 nm for emission of Chl a at 680 nm strongly suggests resonance energy transfer between microxanthin and Chl a. Presumably there are other peaks of this xanthophyll, but they are masked by Chl b at 470 nm, Chl a near 440 nm, and perhaps MgDVP.

Several procedures were tried for separating Chl-proteins from Mantoniella. The membranes were solubilized by 4% digitonin (Dig:Chl = 40) after stirring at 4°C overnight, by 0.5% Nonidet P40 (NI:Chl = 10), or by 2.5% SDS (SDS:Chl = 65) after incubation at 20°C for 30 min. The pigment-proteins were not as well solubilized by digitonin as by Nonidet or SDS, but the former is a milder detergent and causes less change in their spectroscopic properties. Thus far the best separation of a PSI fraction from an antenna fraction has been after solubilization with Nonidet and centrifugation for 36 hrs at 200,000 x g, 4°C in a 0.1 to 1.0 M linear sucrose gradient. Four distinct colored bands formed in the gradient, containing from the top: 1. yellow carotenoids; 2. yellow-green, solubilized Chl a; 3. dark brown pigments, over 50% of the total chlorophyll, 19% sucrose; 4. green, mostly Chl a, 26% sucrose, Chl/P700 = 87.

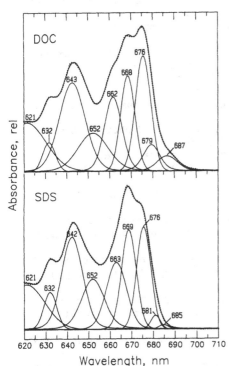

The spectra of bands 3 and 4 are shown in Fig. 1. The antenna pigment fraction (Band 3) was enriched in Chl b and MgDVP, and the PSI fraction (Band 4) was depleted in both. The absorption and fluorescence excitation band near 520 nm observed here is similar to that of fucoxanthin and peridinin seen in diatoms [Mann, Myers 1968] and dinoflagellates [Prezelin, Haxo 1976]. These correlations suggest that microxanthin may be another antenna pigment for photosynthesis.

The fluorescence spectra of the PSI fraction were unexpected because there were no emission bands corresponding to the absorption bands near 685 and 695 nm. Very little Chl b absorption is visible at 643 nm, and yet there is a high fluorescence excitation band at 470 nm. PSI-enriched fractions were also obtained after digitonin, DOC or SDS solubilization, and in no case has any long wavelength emission been found. Probably the fluorescence yield of PSI in this alga is so low that the emission from even a very small amount of contaminating antenna pigment can mask its emission.

FIGURE 2. Absorption spectra at 77 K of Mantoniella antenna Chl-proteins isolated with DOC and treated with 0.1% SDS. The dotted line represents the spectral data and the solid line, the sum of the component curves.

When antenna pigment fractions were obtained after either digitonin, DOC or Nonidet solubilization, the Chl a absorption maximum was at 676 nm, but if SDS was added to either an antenna fraction or chloroplast membranes, this maximum shifted to 668 nm as shown in Fig. 2. Spectra of antenna complexes before and after SDS-action were analysed with the RESOL computer program [Brown, Schoch 1982] using mixed Gaussian-Lorentzian component curves. The effect of SDS-action on the individual components is discussed in detail in Brown, 1983. Although the spectrum of Chl a is altered, CPI and several other pigmented bands can be obtained in SDS gels. All of the separate antenna pigment bands which we have examined spectrally have had the same absorption and fluorescence as shown in Fig. 1. Mantoniella appears to be an excellent alga for the study of pigment formation, evolution and organization.

ACKNOWLEDGEMENTS

I am grateful to Jerome Lapointe for excellent assistance and Shirley Jeffrey for suggesting study of the Prasinophytes. CIW-DPB No. 828.

REFERENCES

Brown JS (1967) Absorption and fluorescence of chlorophyll a in different particle fractions from different plants. Biophys. J. 9, 1542-1552.

Brown JS and Schoch S (1982) Comparison of chlorophyll a spectra in wild-type and mutant barley chloroplasts grown under day or intermittent light. Photosyn. Res. 3, 19-30.

Brown JS (1983) Spectroscopy of chlorophyll in photosynthetic membranes. Carnegie Inst. Year Book 82, in press.

Jeffrey SW and Humphrey GF (1975) New spectrophotometric equations for determining chlorophylls a, b, c_1 and c_2 in higher plants, algae and natural phytoplankton. Biochem. Physiol. Pflanzen 167, 191-194.

Jeffrey SW and Hallegraeff GM (1980) Studies of phytoplankton species and photosynthetic pigments in a warm core eddy of the east Australian current. Mar. Ecol. Prog. Ser. 3, 285-294.

Mann JE and Myers J (1968) On the pigments, growth, and photosynthesis of Phaeodactylum tricornutum. J. Phycol. 4, 349-355.

Manton I and Parke M (1960) Further observations on small green flagellates with special reference to possible relatives of Chromulina pusilla Bucher. J. mar. biol. Ass. U.K. 39, 275-298.

Norris RE (1980) Prasinophytes In Cox ER, ed. Phytoflagellates, pp. 85-145. New York: Elsevier/North Holland.

Picaud A, Acker S and Duranton J (1982) A single step separation of PS 1, PS 2 and chlorophyll-antenna particles from spinach chloroplasts. Photosyn. Res. 203-213.

Prezelin BB and Hazo FT (1976) Purification and characterication of peridinin-chlorophyll a-proteins from the marine dinoflagellates, Glenodinium sp. and Gonyaulax polyedra. Planta 128, 133-141.

Ricketts TR (1966a) Magnesium 2,4-divinylphaeoporphyrin a_5 monomethyl ester, a protochlorophyll-like pigment present in some unicellular flagellates. Phytochemistry 5, 223-229.

Ricketts TR (1966b) The carotenoids of the phytoflagellate, Micromonas pusilla. Phytochemistry 5, 571-580.

Starr RC (1978) The culture collection of algae at the university of Texas at Austin. J. Phycol. 14 suppl. 47-100.

ARTIFICIAL CHLOROPHYLL-PROTEIN COMPLEXES

ELISABETH TOMBÁCZ, Z. VÁRKONYI AND L. SZALAY

1. INTRODUCTION

Chlorophyll-a protein complexes from different plant material (bean, Kaloskas et al; Clenodimium, Prezelin, Boczar; Ricinus, Timko et al; spinach and barley, Anderson et al; etc.) with different molar weights and models of chlorophyll-a protein complexes prepared with different proteins (milk proteins, Widart et al; casein, Meister, Brecht; bovine serum albumin, Tombácz, Vozáry; Zenkevich et al; human serum albumin, Tombácz et al; etc.) are very similar: the Soret-bands and the red bands lie at around 435 and 673 nm, while the fluorescence maxima are at about 685 nm. Since these properties depend on the binding of chlorophylls to amino acids present in proteins, the universality suggests that preferentially binding amino acids should determine the binding sites, while the rest of them practically do not interact. Chlorophyll-a amino acid (tryptophan, tyrosine, histidine, proline, lysine) and poly-pro-lysine polypeptide complexes were therefore prepared and studied spectroscopically.

2. PROCEDURE

$$\lambda_{abs} = 235 \text{ nm} \quad (1)$$

$$\lambda_{abs} = 674 \text{ nm} \quad (2)$$

The preparation of these complexes was similar to that described earlier. Lecithin in chloroform and chlorophyll-a in diethyl ether were mixed to give final concentrations of 10^{-3} M and $2.5 \cdot 10^{-4}$ M, respectively. The solvent was removed under vacuum, and the residue was taken up in Tris buffer (pH-7.2) and sonicated until a liposome system was obtained containing monomeric chlorophylls. In this system amino acids or poly-pro- lysine were dissolved under cautious shaking. The components were separated by gel filtration and the fraction with $E(235)/E(674) \approx$ ≈ 1.8 was used for further studies (Fig. 1). Chromatography reveals the presence of lecithin too in the complex.

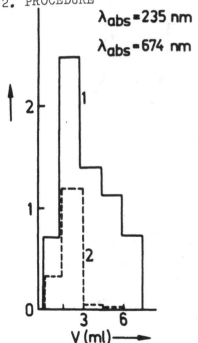

FIGURE 1.

Elution curve

Sybesma, C. (ed.), Advances in Photosynthesis Research, Vol. II. ISBN 90-247-2943-2.
© *1984 Martinus Nijhoff/Dr W. Junk Publishers, The Hague/Boston/Lancaster.*

3. RESULTS AND DISCUSSION

The <u>absorption spectra</u> of chlorophyll-a in complexes with protein polypeptide or amino acid have practically the same shape
in the chlorophyll region (Fig. 2). Non-ionic detergents have
no appreciable effect on these spectra. The <u>fluorescence spectra</u>
are not sensitive to treatment with non-ionic detergents, but
with sodium lauryl sulfate the chlorophyll fluorescence excited
at 235 nm causes a loss of longwave fluorescence (Fig. 2), obviously due to the cessation of energy transfer from the polypeptide. Both the absorption and fluorescence spectra are very
similar to those previously reported. The <u>relative fluorescence
polarization</u> spectra are satisfactorily reproducible (Fig. 3).
They are very sensitive even to mild non-ionic detergent

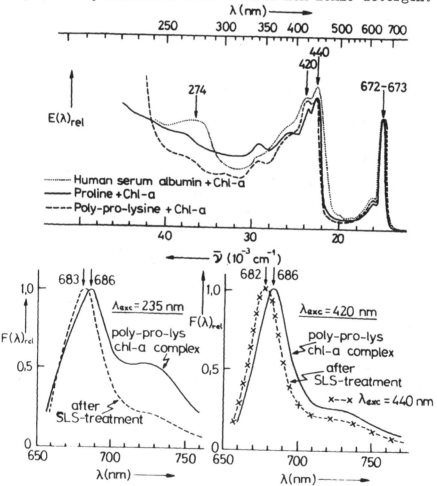

FIGURE 2. Absorption and fluorescence spectra

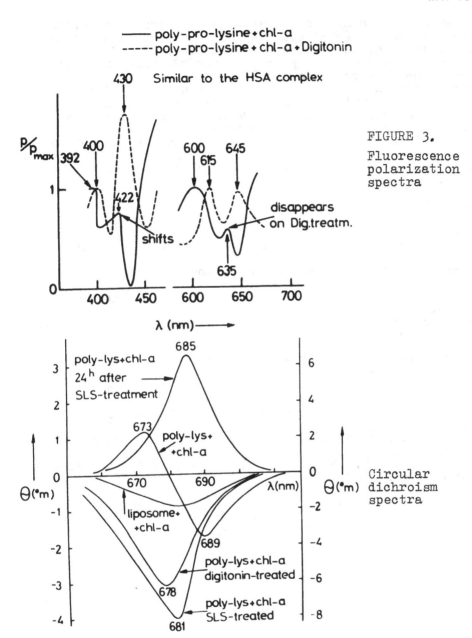

poly-pro-lysine+chl-a
poly-pro-lysine + chl-a + Digitonin

Similar to the HSA complex

FIGURE 3.

Fluorescence
polarization
spectra

Circular
dichroism
spectra

treatment, showing that in both chlorophyll absorption bands
the electronic transitions are affected. This has been found
for the albumin-chlorophyll complex, too. The <u>circular
dichroism spectra</u> show that in the polypeptide complex
chlorophyll dimers are present which disaggregate after detergent treatment (Fig. 3). Sodium lauryl sulfate treatment

leads even to a change in the absolute configuration after a prolonged period (pheophytinization occurs). In order to draw a safe final conclusion a systematic study of more amino acid chlorophyll complexes is needed. However, we can say that the spectroscopic properties of the complexes reported above are practically the same as those of complicated protein-chlorophyll complexes. In addition, we can conclude that even mild detergent treatment leads to an inhibition of the interaction between amino acid and chlorophyll molecules. Though the effect on the absorption and fluorescence is negligible, that on the polarization of fluorescence and on the circular dichroism spectra is very great.

REFERENCES

Anderson JM, Barrett J and Thorne SW (1981) Proc. Vth Int. Photosynthesis Cong. Balaban Int. Sci. Service, Philadelphia, Vol. III, 301.
Kaloskas K, Argyroudi-Akoyunoglou JH and Akoyunoglou G (1981) Ibid. Vol. V, 569.
Meister A and Brecht E (1979) Biochem. Physiol. Pflanzen 174, 305.
Prezelin B and Boczar B (1981) Proc. Vth Int. Photosynthesis Cong. Balaban Int. Sci. Service, Philadelphia, Vol. III, 417.
Timko MP, Triemer RE and Vasconcelos AC (1981) Ibid. Vol. V, 847.
Tombácz E and Vozáry E (1980) Acta Biol. Univ. Szeged, 26, 33.
Tombácz E, Várkonyi Z and Szalay L (1982) Zh. Prikl. Spectr. (Minsk) 36, 64.
Zenkevich EI, Kochubeev GA and Kulba AM (1981) Biofizika (Moscow) 26, 389.
Widart M, Dinant M, Téchy F and Aghion J (1980) Photobiochem. Photobiophys. 1, 103.

Author's address: Elisabeth Tombácz, Institute of Biophysics, József Attila University, Szeged, Egyetem u. 2, H-6722 Hungary.

MODEL SYSTEMS FOR STUDYING CHLOROPHYLL-PROTEIN INTERACTIONS

PETR PANČOŠKA, KAREL VACEK AND LUBOMÍR SKÁLA

1. INTRODUCTION

The chlorophyll-protein complexes - within the membrane framework - bear the fundamental primary functions of the photosynthetic apparatus in vivo. Therefore not only pigment-pigment, but also pigment-protein interactions should be involved in the differentiation and functional specialization of chlorophyll. Any pigment molecule in the protein environment can be influenced by i) physical factors (e.g. variation of dielectric constant); ii) field of the charged groups in the polypeptide side chains (Lys, Glu, Asp); iii) fixation by weak (hydrophobic, $\pi - \pi$) or iv) strong (coordination) interactions. Here the coordination unsaturation of the central Mg atom or the hydrogen-bond capabilities of the chlorophyll carbonyls can be used for the contact with suitable protein groups. All these interactions can modify the chlorophyll physical properties (electronic levels in particular). Such an altered pigment is then involved in the cooperative estabilishment of the final conformation of the complex, resulting in the optimal geometry adjustment of the chlorophyll molecules, necessary for the proper function of the system. In this sense the protein influences also the pigment-pigment interactions. The aim of our work was to study the manifestation of several above mentioned chlorophyll-protein interactions in the pigment properties, using specialy prepared model systems.

2. EXPERIMENTAL

2.1. **Materials.** Chlorophyll a (Chl) was extracted from cyanobacteria Synechococcus elongatus and purified according to Pančoška et al.(1983a). Pheophorbide a was prepared according to Wasielewski (1982). Crosslinked polyhydroxyethylmetacrylate modified by imidazole (HEMA-I, Laboratory Instrument Works, Prague) was prepared according to Čoupek et al.(1979). Chl has been sorbed onto this material from the solution in redistilled n-hexane. Colloid solutions of Chl were prepared by transferring the pigment dissolved in acetone into 0.1 M phosphate buffer, pH=8.5, ammount of acetone no more than 5% of the final sample volume (Kiselev et al.(1979)). Covalent linking of pheophorbide a to ε -aminogroups of lysine side chains in sequential polytripeptide poly(L-lysyl-L-alanyl-L-alanine), m.w. 10 000, was prepared according to Bláha, Pančoška (unpublished).

2.2. **Methods.** Absorption spectra were measured on Specord UV/VIS spectrophotometer (Zeiss, Jena). Circular dichroism (CD) spectra were recorded on Mark III Dichrographe (ISA Jobin Yvon). Quantum chemical calculations were performed with the programme POLY (Skála (1981)), using the porphyrine atom coordinates from Maggiora et al.(1978).

3. RESULTS AND DISCUSSION

3.1. **Model of the conformational origin of Chl spectral forms**. We model the protein by sequential polytripeptide poly(L-Lys-L-Ala-L-Ala) with 35-40 ε -aminogroups in lysine side chains per one molecule. These amino-

Sybesma, C. (ed.), Advances in Photosynthesis Research, Vol. II. ISBN 90-247-2943-2.
© *1984 Martinus Nijhoff/Dr W. Junk Publishers, The Hague/Boston/Lancaster.*

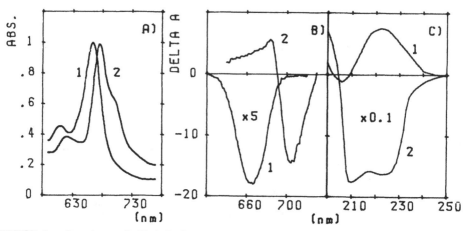

FIGURE 1. Spectra of Model I in trifluoroethanol (1) and dibuthylether
(2); a) longwavelength absorption; b) CD spectra ; c) CD spectra in the
peptide absorption region.

groups are used as the bonding sites for the covalent attachment of
pheophorbide a through the amide bond with its $C17^2$ carboxyl (Model I).
The spectroscopic properties of Model I are presented on Fig.1. The
'monomeric' to 'aggregated form' change in the visible spectra (new
absorption band at 698 nm, red shift from 662 to 673 nm, exciton
splitting of the Cotton effects in main $\pi - \pi$ transitions when going
from trifluoroethanol to dibuthylether) can be explained by the peptide
CD. Random coil to α-helix conformation transition is observed. Thus
(using the standard geometrical parameters for helix and known peptide
primary structure) the distances between two nearest porphyrine macro-
cycles can be estimated as 15 - 22 Å. This allows for the appearance of
pigment-pigment interactions in the studied spectra which are missing in
the irregular, long-distance conformer, formed in the alcohol.

3.2. Model of Chl fixation on the protein. To mimic the possible
naturally occuring noncovalent fixation of Chl to protein we select the
Model II : imidazole - as histidine side chain - chemically immobilized
on the insoluble polymeric support. Results, observed in absorption when
Model II is in contact with mixtures of a) Chl - pheophytine or b)
monomeric - hydrated $(Chl.H_2O)_n$ in n-hexane, are presented on Fig.2. It
can be concluded that the central Mg atom is involved in this fixation:
for mixture a) the pheophytine remains in solution, whereas Chl is
removed with the polymer. In system b) we did not observed the sorption
of 'microcrystalline' (725 nm absorbing) species, in which the Mg atoms
are saturated by $Chl-H_2O-Chl$ crosslinking (Fong (1982)). Using the
absorption measurements, the Chl - Model II complex can be characterised
by the apparent equilibrium constant $K=10^2$ 1/mol, but the value includes
a relatively high error due to the acompanying nonspecific sorption of
the pigment on the polymeric support.

3.3 Models of ionic interactions of chlorophyll with proteins. Ionic
interactions have been studied both on supramolecular and on single
molecular level. Negative charge was observed to stabilize the chloro-

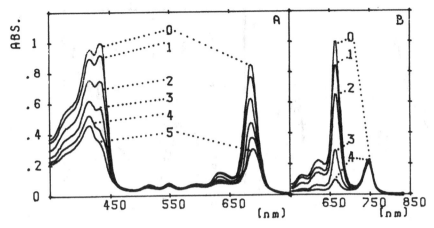

FIGURE 2. Time evolution of absorption spectra of pigment-Model II systems: a) mixture (a) in n-hexane (0 - solution without Model II, 1-7 - Model II added, recorded in 4 min. intervals using solutions after filtering off the (Model II-Chl). b) mixture (b) in n-hexane (0 - without Model II, 1-4 - Model II added, time intervals 1.5 min, recorded as above).

phyll micelles, containing approx. 1000 pigment molecules, which are formed after the introduction of Chl into the water environment (see EXPERIMENTAL). The red absorption maximum of this system peaks at 672-676 nm and is skewed to the longer wavelengths (see Fig. 3a). Positively charged NH_3^+ groups in the lysine side chains of poly(L-Lys-L-Leu-L-Ala) or poly(L-Lys) allows for ionic contact when both partners are present in the solution, resulting in the red shift of the Chl maxima, the effect being more pronounced for poly(L-Lys). The supposed interaction manifests itself also in the disappearance of the colloidal Chl polarographic reduction wave around -1.3 V, when polypeptide is added. We suppose that the flexible polycations 'cover' the Chl micelle surface, prevent the electron exchange with the mercury electrode in this way and their electric field induces - among others - a structural reorganization of pigment molecules, yielding the longwavelength Chl forms. The specifity of the polypeptide action follows from the model experiments: the similar effects can be induced by addition of methylamine hydrochloride (1M) or by increase of H^+ concentration (pH 4) (Fig. 3b,3c). The high concentrations of these substances necessary for observable effect contrast to that of polypeptide (10^{-5} M). Thus, not the average, but the high local ion concentration in polypeptide chains is effectively operative in the Chl system studied.

The effect of the external charges on the porphyrine electronic spectra has been studied theoreticaly (Pančoška et al.(1983b)).We have treated the problem by square-well potential model, FEMO-CI and CNDO/S-CI methods. Irrespectively to the approximations involved, all three approaches predict blue shift of Q_y absorption band (up to 10 nm) when the porphyrin is perturbed by positive charge and a red shift for negative one. Summarizing the results of computations qualitatively: the perturbing charge induces the shift of porphyrin energy levels, the magnitude of which depends both on sign and position (distance) of the charge and

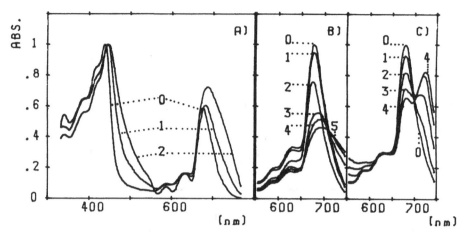

FIGURE 3. a) Absorption spectra of Chl micellar systems (Chl concentration 10^{-4} M, 0.1 M phosphate buffer, pH 8.5): 0 - without polypeptides, 1 - 10^{-5} M of poly(L-Lys-L-Leu-L-Ala) added, 2 - 10^{-5} M of poly(L-Lys) added. b) Time evolution of red absorption: b) 1 M of methylamine hydrochloride added, c) pH changed to 4 (0 - intact system, 1-5 - records in 7 min. intervals

on the energy of the unperturbed level (higher-energy levels are less influenced). Our results are in agreement with experiments of Davis et al.(1981) and are to be taken into account also for the explanation of Chl spectral forms in vivo.

REFERENCES
Čoupek J, Kuhn M and Mohr P (1973) DDR pat. No. 136 269
Davis RC, Ditson SL, Fentinmann AT and Pearlstein RM (1981) Reversible wavelength shifts of chlorophyll induced by a point charge, J. Amer. Chem. Soc. 103, 6823-6826.
Fong FK (1982) Light path of carbon reduction in photosynthesis. In Fong FK, ed. Mol. Biol. Biochem. Biophys. 35, pp. 277-310. Springer-Verlag, Berlin Heidelberg New York.
Kiselev BA Kalasnikova IG (1979) Colloid solutions of Chlorophyll. Electrical charge of particles, Biofizika 24, 811-814.
Petke JP Maggiora GM Shipman LL and Christofersen RE (1978) Stereo-electronic properties of Mg an metal free porphine, J. Mol. Spectr. 71, 64-84.
Pančoška P, Čoupek J and Frydrychová A (1983) Czech. Pat. PV-440-83.
Pančoška P, Kapoun M and Skála L (1983) The effect of external charges on absorption spectra of porphyrins, Photobiochem. Photobiophys. in press.
Skála L (1981) Programme for symmetry clasification in quantum chemical calculations, Comp. Phys. Commun. 24, 135-140.
Wasielewski MR (1982) Synthetic approaches to photoreaction center structure and function. In Fong FK, ed. Mol. Biol. Biochem. Biophys. 35, pp. 234-276. Springer-Verlag, Berlin Heidelberg New York.
Authors adress: Dept. Chem. Phys., Faculty Math. and Phys.,Charles
                University, Ke Karlovu 3, 121 16 Prague 2
                CZECHOSLOVAKIA

COMPARATIVE BIOCHEMISTRY OF CHLOROPHYLL-PROTEIN COMPLEXES

J.P. THORNBER, R.J. COGDELL, R.E.B. SEFTOR, B.K. PIERSON, E.M. TOBIN

The photosynthetic pigments occur in vivo noncovalently bound to protein, forming two functionally and spectrally distinct classes of pigment-protein complexes: the photochemical reaction center and the light-harvesting complexes (Thornber et al. 1978). By using suitable surfactants and standard protein purification techniques, reaction center and the antennae complexes have been independently purified from many photosynthetic organisms (Gingras 1978; Olson, Thornber 1979; Cogdell 1983).

1. PHOTOCHEMICAL REACTION CENTERS OF BACTERIA

The reaction center (RC) has been obtained completely free of antenna pigments and other extraneous material from most purple bacteria (Feher, Okamura 1978; Gingras, 1978) and from one green filamentous bacterium, Chloroflexus aurantiacus which is thought to be the evolutionary precursor of all photosynthetic organisms (cf. Pierson, Thornber 1983). Other green bacteria have not yet yielded their RCs to isolation.

TABLE I A Comparison of the Composition of Some Isolated Photochemical Reaction Centers

Species	Rps. sphaeroides	Rps. viridis	Chloroflexus aurantiacus
Pigment content per P870 or P960 molecule	4 molecules BChl a 2 molecules BPheo a 2 molecules UQ$_{10}$ 1 molecule carotenoid	4 molecules BChl b 2 molecules BPheo b 1 molecule UQ$_{10}$ 1 molecule MQ$_7$ 1 molecule carotenoid	3 molecules BChl a 3 molecules BPheo a No carotenoid
Subunit composition Number per P870 or P960 (and names)	3 (L, M and H)	4 (L, M, H and two c type cytochrome(s)[+])	2 (L and M?)
Apparent size of sub-units from SDS-PAGE (from amino acid analyses)	H - 28 (36) kDa M - 24 (32) kDa L - 21 (28) kDa	H - 35 kDa M - 28 kDa L - 24 kDa cytochrome - 38 kDa	M - 30 kDa L - 28 kDa

[+] 4-5 hemes per P960

The most fully characterized RC is that of Rhodopseudomonas sphaeroides (Feher, Okamura 1978; Clayton 1980). It has a unit containing 4 molecules of BChl a, 2 molecules of BPheo a, 1 molecule of carotenoid, 1-2 molecules of ubiquinone (UQ$_{10}$) and 1 atom of iron. These components are conjugated with one copy each of three polypeptides of MWs 28, 32 and 36 kDa (Sutton et al. 1982) termed L, M and H respectively. A simpler, but less stable, photochemically active unit can be obtained by treating the RC with chaotropic agents which removes the H polypeptide (Feher, Okamura 1978) but none of the pigment molecules. Thus all of the RC pigments are associated with the L or the M subunit or both, but so far in spite of much effort, the primary donor's exact location has not been deduced.

Sybesma, C. (ed.), Advances in Photosynthesis Research, Vol. II. ISBN 90-247-2943-2.
© *1984 Martinus Nijhoff/Dr W. Junk Publishers, The Hague/Boston/Lancaster.*

The quinone(s) has been located on the M subunit (Debus et al. 1982).
The absorption spectrum and the assignment of absorption bands to
chromophores is depicted in Fig. 1.

By using a comparative biochemical and biophysical approach, it becomes
possible to assess those features of an RC's structure and function that
are of general significance. To this end the composition and spectra of
distinct reaction center types are compared in Table I and Fig. 1 respectively. The absorption spectra emphasize most of the major differences
that occur between any one of them and the others. Thus, the Rps. viridis
and Thiocapsca pfennigii (Thornber et al. 1980; Seftor, Thornber 1984)
components have spectral forms absorbing further into the NIR and are
more separated spectrally than those in the other two RCs; this reflects
the presence of BChl and BPheo b rather than a. These two RCs also
contain c-type cytochromes whereas the others shown in Table I and Fig. 1
do not; however, this is not a property of BChl b-containing RCs alone;
Chromatium vinosum RCs also contain cytochrome (Lin, Thornber 1975). The
green bacterial RC contains 3 BPheo a (A_{760}) and 3 BChl a (A_{805}) molecules
(see Fig 1 and Pierson, Thornber, 1983) whereas all the other isolated
RCs have 2 BPheo and 4 BChl (Gingras 1978). All RCs studied so far
contain quinones and, when prepared from wild-type strains of purple
bacteria (Gingras 1978), carotenoids which protect RC pigments from
photodestruction via ^{3}BChl (Cogdell 1983), note however the absence
of carotenoids in the Chloroflexus complex (Fig. 1).

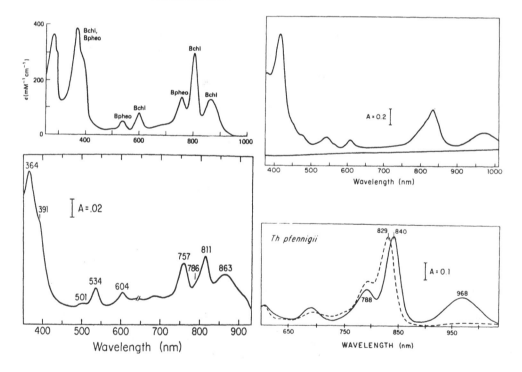

FIGURE 1. Spectra of RCs from Rps. sphaeroides (upper left), Chloroflexus
(lower left), Rps. viridis (upper right) and T. pfennigii (lower right).

Apart from the differences just stressed, other characteristics of the
isolated RCs have proved strikingly similar: Their pigment/primary donor
ratio, their carotenoid and quinone content (but not type) and the sizes
of their protein subunits. It is noteworthy, for studies on all Chl
apoproteins, that the molecular sizes of the Rps. sphaeroides apoproteins,
obtained by SDS-PAGE, have been found to be some 30% lower than their
true molecular weights (Sutton et al. 1982). It remains to be established
whether the two polypeptides in Chloroflexus RCs are equivalent to L and
M, or to M and H, or are not equivalent to those in purple bacterial RCs.
The first choice seems most likely (see Pierson, Thornber 1983) in view
of the LM complex being the simplest active unit of purple bacterial RCs.
The L, M and H subunits have greater hydrophobicities than the higher
plant complexes (see below) which probably indicates that the bacterial
RC polypeptides, particularly L and M, are intrinsic membrane polypeptides
with little of their chains occurring above the membrane surface. The
N-terminal amino acid sequence is known for the first 30 or so residues
of each of the Rps. sphaeroides polypeptides (Sutton et al. 1982) and
the complete sequence of the gene for the M polypeptide should soon be
available from work underway in G. Feher's laboratory.

The mechanism of RC function has come mainly from studies on the Rps.
sphaeroides and Rps. viridis RCs (Clayton 1980; Parson 1982; Cogdell
1983). It is widely believed that during primary change separation,
P^*_{870} (or P^*_{960}) transfers an electron onto one of the BChl a molecules
in a few picoseconds. During the next 200 picoseconds this electron
migrates via a BPheo molecule to a quinone. However, recent data on
T. pfennigii RCs (Seftor, Thornber 1984) indicate that both of its BPheo
molecules are involved in the transfer while its BChl molecules do not
seem to function as intermediary carriers (see also Thornber et al.
1981). This chain of pigmented electron acceptor seems to be an essential feature of RC function in both bacteria and plants (see Cogdell
1983). By rapidly removing the unpaired electron from the vicinity of
$P870^+$ or $P960^+$ wasteful charge recombination is minimized.

The ultimate aim of research on the RC is to obtain its 3-D structure and
its relative orientation in the membrane (Clayton 1980). Until recently,
models for RC organization have come from indirect experimental techniques
due to the lack of availability of crystals for X-ray studies. However,
crystals of the Rps. viridis RC have at last become available (Michel
1982) and therefore the detailed structure of the first RC should soon
be available. If the complete set of RC pigments can be seen, and each
ascribed to one of the six chromophore molecules present, then we will
have an understanding of the mechanism of the primary photochemical event
and a molecular explanation of spectral forms of Chl in the detail
researchers have been seeking for many years.

2. THE LIGHT-HARVESTING PIGMENT-PROTEIN COMPLEXES OF PURPLE BACTERIA

Previously, based on structural and functional data of a few purified
complexes, Thornber et al. (1978) proposed that the antenna complexes
were of two distinct classes: One type is present in all the purple
bacteria, is intimately associated with the RC and apparently occurs in
a fixed stoichiometry relative to the RC (Aagaard, Sistrom 1972; Thornber
et al. 1978). The B890-protein of Rhodospirillum rubrum (Sauer, Austin
1978; Cogdell, Thornber 1979) or B875-protein of Rps. sphaeroides (Broglie

TABLE II Comparison of the Major Antenna Caroteno-Chlorophyll-Proteins in Purple Bacteria

Antenna Type	B890-protein class		B800-850-protein class		
	B890-protein	B875-protein	B800-850-protein Type I	B800-850-protein Type II	B800-820-protein
Examples of bacteria containing antenna type	Rsp. rubrum C. vinosum Rps. acidophila ? Rps. viridis		Rps. palustris Rps. capsulata Rps. sphaeroides Rps. gelatinosa Rps. acidophila 7750 and 7050 (high light grown)	C. vinosum Rps. acidophila 7050 (low light grown)	
BChl a: carotenoid	2:1	2:2	3:1	3:1	3:1
Number of polypeptides in isolated complex	2	2	2 or 3	2	2
Number of amino acid residues in polypeptides	52 and 54	52-58 and 47-48	54-60 and 52	~50-65	~50-65
Intensity of CD spectrum of long wavelength band	Strong	Weak	Strong	Strong	Strong
	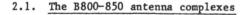 All spectra indicate BChl dimer present in the band				

Note: The quantitative data are based on analysis of complexes isolated from only a few of the species given as examples.

et al. 1978) are good examples of this class (Table II). Many species of purple bacteria contain, in addition, a spectrally and biochemically different class of which the B800-850-proteins of Rps. sphaeroides or Rps. capsulata (Cogdell, Thornber 1979, 1980; Clayton, Clayton 1972; Feick, Drews 1978) are examples. Since this complex varies in amount with respect to the RC (Aagaard, Sistrom 1972), it is a reasonable supposition that this is how bacteria modulate the size of their photosynthetic unit in response to environmental conditions such as changes in incident light intensity (Thornber et al. 1978).

While the functional distinction between the two classes of antenna complex remains valid, it appears, from more recent biochemical analyses of a wider range of purple bacterial light-harvesting complexes, that structurally each of those two classes has at least two subdivisions (Table II).

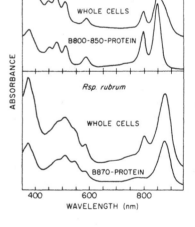

FIGURE 2. Spectra of some typical antenna BChl-proteins.

2.1. The B800-850 antenna complexes

The B800-850 antenna complex from Rps. sphaeroides (Fig. 2) occurs in the photosynthetic membrane as a multimer of a minimum unit (cf. Clayton, Clayton 1972) consisting of 3 molecules of BChl a, 1 molecule of carotenoid and one copy of each of two polypeptides (Sauer, Austin 1978; Cogdell et al. 1980). No one has yet succeeded in isolating this minimum unit in a pigmented form. The smallest pigmented complex isolated has a M_r $\geqslant 100,000$. The polypeptides have low MWs (<10,000 kDa) (Cogdell et al. 1980; Theiler et al. 1982) and are as hydrophobic as the RC apoproteins and more hydrophobic than their plant counterparts. Two BChl a molecules contribute the 850-nm absorption band, while the third gives the 800-nm absorption (Fig. 2) (Sauer, Austin 1978; Cogdell et al. 1980). The carotenoid in this complex, and not that in the B890 or B875 complex undergoes an electrochromic shift during photosynthesis (Holmes et al. 1980).

The B800-850-protein of Rps. capsulata (M_r = 170,000) has been studied with equal intensity to the Rps. sphaeroides complex (e.g. Shiozawa et al. 1982; see also Cogdell, Thornber 1980). It is essentially the same as the Rps. sphaeroides component except for the presence in the complex of an additional colorless M_r = 14,000 polypeptide. The carotenoid molecule and one of the three BChl a molecules are associated with the smallest, and the other two BChl with the middle-sized apoproteins in this B800-850-protein (Webster et al. 1980).

We have termed the above two B800-850 antenna complexes type I B800-850-proteins in Table II. This subtype is characterized by its 850 nm absorption band being about 1.5 times as intense as the 800 nm band (Fig. 2) Alternatively, some other species, C. vinosum and Rps. acidophila (when grown at low light intensity), contain B800-850 antenna complexes in which the 850-nm absorption band is of equal or lower intensity than the 800 nm band (Thornber 1970; Cogdell et al. 1983). We designate these B800-850 antenna complexes as type II complexes. Both have the same composition as the type I complex, but the position of their 850 nm band can vary between 835 and 855 nm, depending on detergent concentration, pH, etc. (Thornber 1970; Thornber et al. 1978; Cogdell, Thornber 1980; Cogdell et al. 1983), and the exact position alters the relative heights of the 800 and 850 nm peaks. It is also noteworthy that, so far, whenever the type II complex occurs in a bacterium there is also a B800-820 antenna complex (Thornber 1970; Cogdell et al. 1983). Their interrelationship is not yet clearly defined, but in Rps. acidophila they appear to have very similar, perhaps identical, polypeptide compositions and in C. vinosum the holocomplexes have very similar sizes and identical pigment contents.

2.2. The B890-protein complex

The B890 complex of R. rubum (Fig. 2, Table II) is the most fully characterized of this class (Cogdell, Thornber 1979; Theiler et al. 1982; Brunisholz et al. 1983). Its minimum unit consists of two molecules of BChl a and one molecule of carotenoid (usually spirilloxanthin) bound to two low MW polypeptides. Both polypeptides have been sequenced and one has 52, the other 54 amino acids (Brunisholz et al. 1983). The 890 nm peak gives an intense CD spectrum, typical of a BChl dimer (Sauer, Austin 1978). The B890-complexes from C. vinosum (Thornber 1970; Cogdell, Thornber 1979) and Rps. acidophia (Cogdell et al. 1983) have these same characteristics.

In contrast, the supposedly analogous B875-protein from some other species (e.g. Rps. sphaeroides) differs (Table II) in that it contains two molecules of BChl a and two molecules of carotenoid per pair of

polypeptides (Broglie et al. 1978), has its NIR maximum at shorter wavelengths than the B890 complex, and the 875-nm absorption band shows only
a weak CD spectrum. The B890 complexes should therefore be segregated
into two subclasses: B890-protein complex and B875-protein complexes
(see Table II).

3. CONCLUDING REMARKS

In spite of the differences between the various antenna types described
above they all are remarkably similar biochemically. Thus, all the
antenna apoproteins are small (<10 kDa), occur in pairs of very nearly
the same size, and are extremely hydrophobic (often being soluble in a
mixture of chloroform and methanol). Secondly, the 3 or 4 pigment molecules are bound to two different polypeptides to form what is believed
to be the minimum building block from which a supermolecular aggregate
$\geqslant 100$ kDa is constructed in situ. Protein sequence data of H. Zuber and
colleagues (e.g. Theiler et al. 1982; Brunisholz et al. 1983) show that
in six organisms both polypeptides of either B800-850 or B890 complex
have three domains: a polar N-terminus located in the cytoplasm of the
bacterium, a hydrophobic core of 20-23 residues which form an α-helix
that crosses the chromatophore membrane, and a polar C-terminus which
protrudes into the periplasm (Fig. 3). In this manner the BChl-protein
molecules are prevented from "flip-flopping" in the membrane. It is
proposed that each α-helix binds one, sometimes two, BChl molecule via a
histidine ligand. Their organization in the membrane resembles somewhat
that proposed by Tobin et al. (1984) for the Chl $\underline{a}$/b-protein of green
plants (Fig. 3). The plant complex is believed to have an α-helix also
near its C-terminus which crosses the thylakoid membrane, and to have a
histidine residue (liganding to Chl) in it which lies close to the inner
membrane surface (Fig. 3). However, it is not phylogenetically homogenous because, like all Chl apoproteins in plants, the polypeptide chain

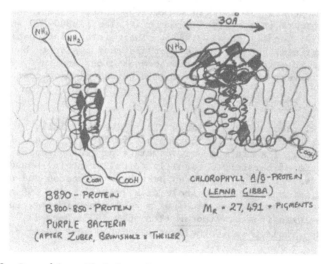

FIGURE 3. Some ideas that have been presented for the organization of
plant and bacterial antenna Chl-proteins (Tobin et al. 1984; Brunisholz
et al. 1983).

is much larger than its bacterial counterparts; furthermore, it probably crosses the membrane at least three times. In addition, the Chl a/b-protein has up to 40% of its polypeptide chain outside the membrane surface on the stroma side within which some of the pigment molecules might be located. These differences are consistent with the less hydro-phobic character of the plant complex. Homogenous features among the antenna complexes of photosynthetic organisms are then likely to be limited to lesser points (e.g. α-helices anchoring the complex to the membrane, histidine liganding Chl to the protein) rather than sequence homology. The bag-like β-pleated sheet structure of the water-soluble BChl a-protein of green bacteria does not, at this stage, seem to be a common feature of Chl-proteins as originally anticipated (Olson, Thornber 1979).

The possibilities of Chl-protein homology between different classes of photosynthetic organisms is greater for the RC complexes. The similarity of the mechanism of the primary event in the purple bacteria and plant PS II could imply their RC components are phylogenetically related. If so, the L and M polypeptides of the bacterial RC should be related to the 43 kDa and/or the 47 kDa proteins of PS II. The equivalence of the H subunit to the so-called plant herbicide-binding protein is already being discussed and tested (cf. Okamura 1984). Obviously, more sequence data and specific antibodies are required to investigate such interrelation-ships. It is more certain that the RC of the proposed progenitor of photosynthetic organisms, Chloroflexus aurantiacus, is closely related to the purple bacterial RC (Table II; see also Pierson, Thornber 1983). Nevertheless, if this is so, we are unable to suggest any evolutionary advantage of replacing one BPheo a in the Chloroflexus RC by a BChl in the purple bacterial RC.

For the future, we particularly need to know more about the antenna holocomplexes. The true molecular size and the number of copies of each of the apoproteins in the holocomplex must be determined. We must obtain the 3-D arrangement of the pigments and the protein which requires crystallization of the pigmented complex, a task which has until recently been approached without confidence because of the belief that detergent-soluble complexes could not be crystallized. Since the more complex photochemical RC has at last been crystallized, similar approaches may yield crystals of the antenna pigment-proteins. We also need to delineate the relative arrangement and amounts of the antenna complexes and the reaction center component in the photosynthetic membrane. Lastly, but not least, we must further our understanding of the biosynthesis of the proteins and how the stoichiometry of polypeptides within and between complexes is controlled.

REFERENCES

Aagaard J and Sistrom WR (1972) Photochem. Photobiol. 15, 209.
Broglie RM, Hunter CN, Delepelaire P, Niederman RA, Chua N-H and Clayton RK (1980) Proc. Natl. Acad. Sci. USA 77, 87-91.
Brunisholz R, Suter F, Zuber H (1983) Eur. J. Biochem. (in press).
Clayton RK (1980) Photosynthesis: Physical Mechanisms and Chemical Patterns, Cambridge Univ. Press.
Clayton RK, Clayton BJ (1972) Biochim. Biophys. Acta 283, 492-504.
Cogdell RJ (1983) Ann. Rev. Plant Physiol. 34, 21-40.

Cogdell RJ, Thornber, JP (1979) Ciba Symp. 61 (new series) Excerpta Medica, Amsterdam, Oxford, New York, pp. 61-79.
Cogdell RJ, Thornber JP (1980) FEBS Lett. 122, 1-8.
Cogdell RJ, Linsay JG, Valentine J and Durant, I (1982) FEBS Lett. 150, 151-154.
Cogdell, RJ, Durant I, Valentine J, Lindsay JG and Schmidt K (1983) Biochim. Biophys. Acta 722, 427.
Cogdell RJ, Lindsay JG, Reid GP and Webster GD (1980) Biochim. Biophys. Acta 591, 312-320.
Debus R, Valkirs G, Okamura MY and Feher G (1982) Biochim. Biophys. Acta 682, 500-503.
Feher G and Okamura MY (1978) In Clayton RK and Sistrom WR, eds. The Photosynthetic Bacteria, pp. 349-386. Plenum Press, New York and London.
Feick R and Drews G (1978) Biochim. Biophys. Acta 501, 499-513.
Gingras G (1978) In Clayton RK and Sistrom WR, eds. The Photosynthetic Bacteria, pp. 119-131. Plenum Press, New York and London.
Holmes NG, Hunter CN, Niederman RA and Crofts AR (1980) FEBS Lett. 115, 43-48
Lin L and Thornber JP (1975) Photochem. Photobiol. 22, 37-40.
Michel H (1983) J. Mol. Biol. 158, 567-572.
Okamura MY (1984) Bacterial chlorophyll-protein complexes. In Hallick R, Staehelin A and Thornber JP, eds., Proc. Symposium on Biosynthesis of Photosynthetic Apparatus. Liss, New York.
Olson JM and Thornber JP (1979) Photosynthetic reaction centers. In Capaldi RA, ed. Membrane Proteins in Energy Transduction, pp. 279-340. Marcell Decker, Inc., New York and Basel.
Parson WW (1982) Ann. Rev. Biophys. Bioeng. 11, 57-80.
Pierson BK and Thornber JP (1983) Proc. Natl. Acad. Sci. USA 80, 80.
Sauer K and Austin LA (1978) Biochemistry 17, 2011-2019.
Seftor REB and Thornber JP (1984) Biochim. Biophys. Acta. (submitted).
Shiozawa JA, Welte W, Hodapp N and Drews G (1982) Arch. Biochim. Biophys. 213, 473-485.
Sutton MR, Rosen D, Feher G and Steiner LA (1982) Biochemistry 21, 3842-3849.
Theiler R, Brunisholz R, Frank G, Suter F and Zuber H (1982) IVth Inter. Symp. Photosynthetic Prokaryotes, Bombannes, France, C-40.
Thornber JP (1970) Biochemistry 9, 2688-2698.
Thornber JP, Cogdell RJ, Seftor REB and Webster GD (1980) Biochim. Biophys. Acta 593, 60-75.
Thornber JP, Seftor, REB and Cogdell RJ (1981) FEBS Lett. 134, 235-239.
Thornber JP, Trosper, TL and Strouse CE (1978) In Clayton RK and Sistrom WR, eds. The Photosynthetic Bacteria, pp. 133-239. Plenum Press, New York and London.
Tobin EM, Neumann, G, Wimpee C, Silverthorne J, Stiekema WJ and Thornber JP (1983) J. Cell. Biochem. (in press).
Webster GD, Cogdell RJ and Lindsay JG (1980) Biochem. Soc. Trans. 8, 184-185.

ACKNOWLEDGEMENTS

We gratefully acknowledge Prof. H. Zuber's generosity in making his protein sequence data available to us. Research was supported by grants from NSF and SRC.

Author's address: Dept. of Biology and Molecular Biology Institute, University of California, Los Angeles, California 90024, USA

FLUORESCENCE PROPERTIES OF CHLOROPHYLL-PROTEIN COMPLEXES FROM SPINACH
CHLOROPLASTS — ON THE SPECIFICITY OF THE EMISSION BANDS.

N. FUAD, K. KINOSITA, Jr., M. MIMURO, Y. INOUE AND Y. FUJITA

1. INTRODUCTION

Interpretation of chlorophyll fluorescence signals from leaves and chloroplasts rely on a number of assumptions regarding fluorescence yields, fluorescence emitters, energy transfer probabilities etc. One of the assumptions
which has not been clearly resolved yet is the assignment of the emission
peaks of chloroplast fluorescence at 77K to specific chlorophyll-protein
complexes. The 685 and 695 nm emissions are usually attributed to PSII
and that in the 710-730 nm region to PSI. The purified light harvesting
complex of PSII (LHCII) has an emission peak at 681 nm but also emissions
at 695 nm and 735 nm. Under certain conditions of isolation, the long
wavelength emission in LHC is of comparable intensity to that of the 695 nm
shoulder (Fuad et al. 1983). Such an LHC, named LHC736, has no apparent
contamination from PSI. The fluorescence properties of chlorophyll-protein
complexes were therefore studied in detail to obtain more information on
the specificity of the emission bands of chloroplasts at 77K. Three approaches were used i) analysis of fluorescence spectra by deconvolution
into gaussian components, ii) measurements of fluorescence lifetimes at 77K
and iii) study of the temperature dependence of fluorescence intensity.
These approaches were applied to thylakoids, and the purified chlorophyll-
protein complexes: PSI, PSII, LHCII, LHC736 and aged LHCII.

2. MATERIAL AND METHODS

Thylakoids and chlorophyll-protein complexes were isolated from spinach
(*Spinacia oleracea*). PSI and PSII were as isolated by Satoh (1982), LHCII
and LHC736 were as isolated by Fuad et al.(1983), aged LHCII was obtained
by leaving LHCII exposed to air and light for approximately 48h at 22°C.
Thylakoids were isolated in the presence of 10 mM $MgCl_2$.

The curve fitting analysis of fluorescence spectra was carried out as
described by Mimuro et al. (1982). Fluorescence spectra were recorded at
77K in the presence of 15% polyethylene glycol and corrected for the spectral
sensitivity of the apparatus (Hitachi MPF6 spectrophotometer).

Measurements of fluorescence lifetimes were carried out using the nanosecond fluorimeter and data analysis described by Kinosita et al. (1981).
Fluorescence was excited at 435 nm from a hydrogen lamp (10 KHz). Emission
wavelengths were set by interference filters and appropriate filter combinations at 684, 698 and 727 nm. The results were analysed assuming one or
two decay components, hence monophasic (M) and biphasic (B) analyses.

For the study of fluorescence intensity as a function of temperature, a
Shimadzu RF 502 spectrofluorimeter was used. Samples were frozen in liquid
nitrogen and after evaporation of the liquid nitrogen, they were warmed up
by blowing warm nitrogen gas into the sample compartment. Interference
from condensation and N_2 vapours was minimised. Temperature was monitored
by a thermocouple and fluorescence was measured in sequence at various
wavelengths as the sample warmed up.

Sybesma, C. (ed.), Advances in Photosynthesis Research, Vol. II. ISBN 90-247-2943-2.
© *1984 Martinus Nijhoff/Dr W. Junk Publishers, The Hague/Boston/Lancaster.*

3. RESULTS AND DISCUSSION

i) Resolution into gaussian bands

The fluorescence spectra at 77K were analysed by deconvolution into gaussian components. The results are shown in *Fig. 1*.

The fluorescence spectra of thylakoids and of the various chlorophyll-protein complexes can be deconvolved into 8-12 bands. Although the emission of chloroplasts at short wavelength is dominated by that of PSII (685 nm), there is also a small contribution of a band near 680 nm presumably due to LHCII. The apparent 695 nm shoulder/peak of chloroplasts, PSII and LHCs results from a combination of two bands located at about 695 nm and 704 nm. The contribution of these two bands varies from sample to sample. In aged LHCII, which has its peak at 702 nm, the contribution of the 704 nm band is much increased relatively to that of the 695 nm band, resulting in the shift of the maximum to 702 nm; concomitantly, the 687 nm band disappeared. The various complexes have many common bands and it is difficult to ascribe one band specifically to one complex, except in the case of the 681 and 685 nm emissions present only in the LHCs and PSII respectively. Beyond 710 nm the deconvolution pattern is complex and rather similar for most samples. In particular, chloroplasts, PSI and LHC736 showed similar deconvolution patterns in this region. Some bands are however absent from LHCII and PSII. It is likely that some of the long wavelength bands may be vibrational bands of the shorter wavelength emissions.

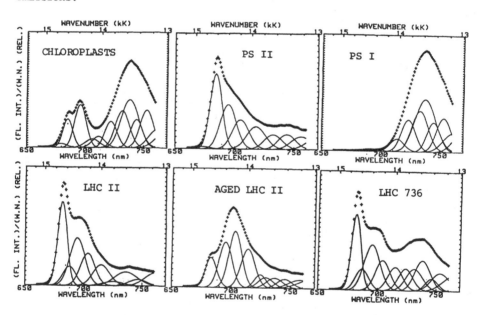

FIGURE 1. *Deconvolution patterns of fluorescence spectra at 77K. Tentative location of bands was determined from the second derivative spectra. Plus signs: real data; solid curves: deconvolution curves.*

ii) Fluorescence lifetimes

TABLE 1. Fluorescence lifetimes at 77K

Wavelength / Sample	685 nm (B)	685 nm (M)	698 nm (B)	698 nm (M)	727 nm (M)	727 nm (M)
CHLOROPLAST	0.72 + 5.56 (94) (6)	1.28	1.84 + 7.90 (94) (6)	2.48	2.97 (100)	2.97
PS I					2.77 (100)	2.77
LHC 736	0.72 + 3.31 (97) (3)	0.87	1.34 + 4.29 (81) (19)	2.26	1.95 + 4.1 (55) (45)	3.22
LHC II	0.67 + 2.89 (97) (3)	0.75	1.17 + 4.29 (79) (21)	2.3		
Aged LHC II	0.21 + 2.03 (97) (3)	0.34	0.62 + 2.84 (80) (20)	1.55		
PS II	0.83 + 5.17 (83) (17)	2.80	1.64 + 5.93 (79) (21)	3.26		
Chl a in 80% acetone	6.64 (M) (100)	6.64 (M)			6.73 (M) (100)	6.73 (M)

For each sample, the first line shows fluorescence lifetimes in ns as obtained by analysis assuming one (M) or two (B) decay components. The numbers in brakets show the contribution in amplitude of the components (in percent).

As expected, purified chlorophyll a showed a monophasic decay for both the main and the vibrational emissions and had a long lifetime (6 ns). The emission viewed at 698 nm was biphasic for all samples examined. The emission at 685 nm was essentially monophasic, but with a very small contribution (3%) of a longer lifetime component, except in the case of PSII where this contribution was greater (17%). The emission at 727 nm was monophasic for chloroplasts and PSI, and with the same lifetime, but biphasic for LHC736 which had different lifetimes. This suggests that the emitter responsible for the long wavelength emission in chloroplasts and PSI is different from that in LHC736. There may be several reasons for the splitting of the 698 nm emission lifetime (and also of that of the 727 nm emission in LHC736) into two components. These include energy transfer from donor(s) decaying with different lifetimes, as well as direct emission from more than one chromophore. In aged LHCII, fluorescence lifetimes were in general very much shortened, suggesting a closer proximity of the chromophores in this sample, may be due to a higher degree of aggregation.

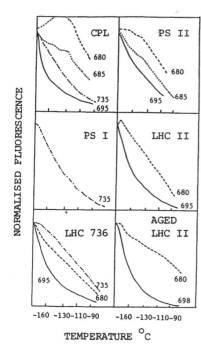

TEMPERATURE °C

FIGURE 2. Temperature dependence of fluorescence. The intensities were normalised to the value at -196°C. CPL: chloroplasts.

iii) <u>Temperature dependence</u>

The profile of the decay of fluorescence intensities at various wavelengths, as the samples were warmed up from -196°C to about -80°C are shown in *Fig. 2*. The 695 nm emission band decreased the most rapidly as the temperature was increased, and the profile of this decrease was similar in all samples. That of the 736 nm emission was similar in PSI and chloroplasts, but quite different in LHC736. These data support the suggestion made earlier that the emitters for the 736 nm peak in LHC736 and in PSI are different.

In conclusion, it appears that only the 681 nm and the 685 nm emissions are specific (to LHCs and PSII respectively). The 695 nm emission which deconvolved into two bands, is present in LHCs as well as in PSII. Although the deconvolution data did not show much differences between PSI and LHC736 in the long wavelength emission, the temperature dependence and the fluorescence lifetime data suggested that the origin of the 736 nm emission in these samples is different. Hence the 736 nm emission does not appear to be an emission specific to PSI only. In chloroplasts isolated in the presence of cations however, the 736 nm emission seems to emanate mainly from PSI.

REFERENCES

Fuad N., Day D.A., Ryrie I.J., Thorne S.W. (1983) A photosystem II light harvesting chlorophyll-protein complex with a high fluorescence emission at 736 nm. Photobiochem. Photobiophys. (in press).
Kinosita K. Jr., Kataoka R., Kimura Y., Gotoh O., Ikegami A. (1981) Dynamic structure of biological membranes as probed by 1,6-Diphenyl-1,3,5-hexatriene: a nanosecond fluorescence depolarization study. Biochemistry 20, 4270-4277.
Mimuro M., Murakami A., Fujita Y. (1982) Studies on spectral characteristics of allophycocyanin isolated from *Anabaena cylindrica*: curve fitting analysis. Arch. Biochem. Biophys. 215, 266-273.
Satoh K. (1982) Fractionation of thylakoid-bound chlorophyll-protein complexes by isolation by isoelectrofocussing. In Methods in Chloroplast Biology. Edman et al. eds. pp 845-861.

ACKNOWLEDGEMENTS

A postdoctoral award to N.F. from the Australian Department of Science and Technology under the Australia-Japan Science and Technology Agreement on Photosynthesis and Crop Productivity is gratefully acknowledged.
Authors address: N. Fuad, K.Kinosita Jr, Y. Inoue, Solar Energy Research Group. RIKEN, Wako, Saitama, 351 Japan.
M. Mimuro, y. Fujita, National Institute for Basic Biology, Okazaki, Aichi, 444 Japan.

SUPRAMOLECULAR STRUCTURE OF PIGMENT-PROTEIN COMPLEXES IN RELATION TO THE
CHLOROPHYLL a FLUORESCENCE OF CHLOROPLASTS AT 25° C OR AT -196° C:
CATION EFFECT, pH EFFECT AND TRYPSINOLYSIS.

J.H. ARGYROUDI-AKOYUNOGLOU AND G. AKOYUNOGLOU/ NRC "DEMOKRITOS", GREECE

1. INTRODUCTION

The oligomeric forms of the thylakoid pigment-protein complexes (LHCP[1],
LHCP[2] and CPIa) are dissociated to their constituents in the presence of
cations (Argyroudi-Akoyunoglou, 1980). Cations have been also found to increase drastically the 77 K F685/F730 ratio of the pigment-protein complexes,
and especially of CPIa, isolated by SDS-sucrose density gradient centrifugation (Argyroudi-Akoyunoglou et al., 1982; Argyroudi-Akoyunoglou, Thomou, 1981)
These results suggested that the increase in the F685/F730 ratio may arise
from changes in the organization of the pigment-protein complexes, i.e. the
dissociation of the complexes to their components. Support to this came also
from the finding that in developing chloroplasts the the fluorescence yield
(Fmax/Chl) at 25° C, follows an initial rise and then a sharp decline to a
plateau (Castorinis et al., 1982); these changes follow closely the pigment-
protein complex monomer to oligomer organization, which occurs during chloroplast development (Kalosakas et al, 1981). Thus high Fmax/Chl is found when
monomers predominate, and low when oligomers get organized.
To check this hypothesis further, we studied whether various conditions,
known to induce or not affect changes in the Chl a fluorescence yield of
chloroplasts at 25° C, or in the F685/F730 ratio at 77 K, can also induce
changes in the pigment-protein complex organization.

2. MATERIALS AND METHODS

The isolation of chloroplasts and thylakoids, the SDS-PAGE for separating
the pigment-protein complexes in the presence or absence of cations, the
isolation of complexes by SDS-sucrose density gradient centrifugation, and
the fluorescence measurements at 25°C or 77 K, were done as described in
(Argyroudi-Akoyunoglou, 1980; Argyroudi-Akoyunoglou et al, 1982; Castorinis
et al, 1982; Argyroudi-Akoyunoglou, Thomou, 1981; Akoyunoglou, 1977). The
pH effect was determined after incubating the thylakoids in various buffers,
and then analyzing the solubilized material by PAGE; or determining the
F685/F730 ratio in liquid Nitrogen of thylakoids or of the isolated CPIa.
Trypsinolysis was done as described (Jennings et al, 1978) in Tricine suspended thylakoids; for PAGE analysis, aliquots removed at various times from
a thylakoid sample (150 ug Chl/ml) incubated with trypsin (30 ul of 1.5 ug/
/ml Tricine without $CaCl_2$) were repeletted and solubilized in SDS(1 mg Chl/
/ml, SDS/Chl=10). For fluorescence measurements at 25° C, the cuvette had
12.5 ug Chl/2.5 ml Tricine, and 7.5 ul trypsin. $MgCl_2$ was added at various
times (5 mM final) and the fluorescence at 685 nm recorded. For 77 K emmission measurements, a spatula tip of trypsin was added to a thylakoid sample
at 25° C (at 5 ug Chl/ml); at various times aliquots were frozen and the
spectra recorded.

3. RESULTS

Effect of pH. Fig. 1 shows the effect of pH on the ratio of the oligomeric
to monomeric forms of the complexes: as shown, oligomers predominate at high
pH; in contrast, low pH induces their dissociation to monomers. Similarly,

Sybesma, C. (ed.), Advances in Photosynthesis Research, Vol. II. ISBN 90-247-2943-2.
© *1984 Martinus Nijhoff/Dr W. Junk Publishers, The Hague/Boston/Lancaster.*

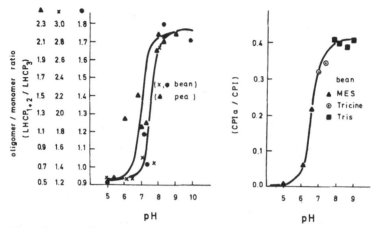

Fig. 1. The oligomer/monomer ratio of pigment-protein complexes as affected by pH. Thylakoids incubated at 0° C for 30 min in 0.05 M buffers (400 ug Chl/ /ml) were repeletted and either directly solubilized in SDS (left, •, 1 mg Chl /ml), or resuspended in the respective buffer and then solubilized (left, x,▲ 580 ug Chl/ml; right, 640 ug Chl/ml). Solubilization in 1% SDS-10% glycerol.

the 77 K fluorescence F685/F730 ratio of thylakoids or of the isolated CPIa complex (Figs.2,3) follows a similar trend: higher F685/F730 is found at low pH, which gradually decreases with increase in pH. The results suggest that not only Mg^{++}, but H^+ as well, is effective in dissociating the oligomers to their constituents; this dissociation parallels the increase in the F685/F730 ratio at 77 K in thylakoids and in the isolated CPIa complex. Effect of Zn^{++}, Cd^{++} or Mg^{++}. Table I compares the effect of these cations on the oligomer/monomer ratio in bean or spinach thylakoids and on their

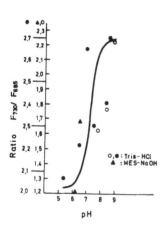

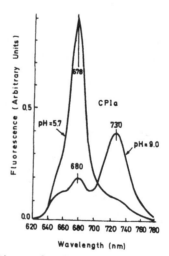

Figure 2. The pH effect on the 77K F730/F685 ratio (pea thylakoids in various pH buffers at 20 ug Chl/ml)

Figure 3. The 77K emmission spectra at pH 9.0 or 5.7 of CPIa, isolated by sucrose density gradient centrifugation

TABLE 1. The effect of Zn^{++}, Cd^{++} or Mg^{++} on the oligomer/monomer ratio in the Chl-protein complexes, and on the 25° C fluorescence characteristics of plastids washed and suspended in 50 mM Tricine-NaOH, pH 7.3. Salts added to solubilization buffer prior to PAGE. For fluorescence measurements cations were incubated with thylakoids (5 ug Chl/ml) for 5 min.

	Cation added (mM)	oligomer/monomer CPIa/CPI	$LHCP^{1+2}/LHCP^3$	Fmax	Fo	Fmax/Fo
A. Spinach	0.0	0.66	0.75	2.8	1.1	2.54
	1.2 Mg^{++}	0.20	0.35	–	–	–
	6.0 Mg^{++}	0.00	0.13	4.8	1.1	4.36
	6.0 Zn^{++}	0.30	0.93	3.1	1.2	2.58
B. Bean	0.0	0.60	1.46			
	7.9 Mg^{++}	0.00	0.25			
	7.9 Zn^{++}	1.00	2.30			
C. Bean	0.0	1.25	1.49			
	3.8 Mg^{++}	0.29	0.57			
	3.8 Cd^{++}	0.53	1.75			

fluorescence parameters at 25° C. In accordance to earlier reports (Li, 1975; Murata et al, 1970) Zn^{++} and Cd^{++}, contrary to Mg^{++} do not increase the Fmax/Fo of Tricine suspended thylakoids. As shown, these cations are also less effective than Mg^{++} in dissociating the oligomers to monomers; on the contrary, they enhance the oligomeric structures, so that in their presence the oligomer/monomer ratio is even higher than that at zero cation. Similarly, the effect of Zn^{++} on the 77 K F685/F730 ratio of the isolated CPIa complex is nil compared to that of Mg^{++} (see Figure 4).

Effect of trypsin treatment. To test further our hypothesis, we studied the effect of trypsinization on the oligomer/monomer ratio. Trypsinized thyla-

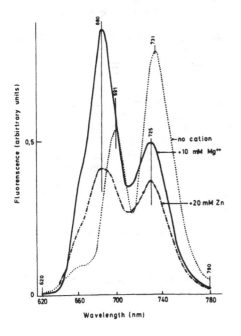

Figure 4. The effect of Mg^{++} or Zn^{++} addition to the CPIa complex, isolated by SDS-sucrose density gradient centrifugation, on the 77 K emmission spectra.

(....): no cation added.
(-·-·): in the presence of Zn^{++}.
(———): in the presence of Mg^{++}.

Table 2. Effect of trypsinolysis on the Mg^{++}-induced increase in Chl a fluorescence yield of spinach chloroplasts, and on the F730/F685 ratio at 77 K of pea thylakoids suspended in Tricine. (Details in Materials and Methods)

Trypsin action (min)	no cation added			Mg^{++} added			Fmax, 77K (nm)	Ratio F730/F685
	Fm	Fo	Fm/Fo	Fm	Fo	Fm/Fo		
0	2.8	1.1	2.54	4.8	1.1	4.36	686, 732	1.50
8	2.8	1.2	2.33	3.4	1.2	2.83	684, 731	1.10
18	2.4	1.2	2.00	2.9	1.2	2.42	684, 731	0.85
control (30 min)							685, 732	1.50

koids, suspended in Tricine, do not show the Mg^{++} effect on the Chl a fluorescence yield increase (Table 2). When analyzed by PAGE, they are deficient in the oligomeric forms of the complexes, and especially CPIa. Their fluorescence parameters at 77 K suggest that the F730/F685 in trypsinized samples is lower than that in controls. The inability of Mg^{++}, therefore, to induce changes in the Chla fluorescence yield in trypsinized thylakoids may be due to extensive trypsinolysis of the oligomers, so that no further dissociation by Mg^{++} is possible, and thus no increase of the fluorescence yield at 25°C.

CONCLUSIONS

Cation or pH-induced changes in the Chl a fluorescence yield at 25° C, and in the 77 K fluorescence emmission spectra of chloroplasts were considered to reflect spillover from PSII to PSI or vice versa (Murata, 1969). The results of this study suggest that these changes can be rather attributed to changes in the organization of the PSI and PSII unit Chl-protein complexes.

REFERENCES

Akoyunoglou G (1977) Development of the Photosystem II unit in plastids of bean leaves greened in periodic light. Arch. Biochem. Biophys. 183, 571-580.
Argyroudi-Akoyunoglou JH (1980) Cation-induced transformation of oligomeric to monomeric forms in pigment-protein complexes of the thylakoid. Photobiochem. Photobiophys. 1, 279-287.
Argyroudi-Akoyunoglou JH Castorinis A and Akoyunoglou G (1982) Cation-induced increase in the low temperature fluorescence F685/F730 ratio in detergent solubilized pigment-protein complexes separated by sucrose density gradient centrifugation. Photobiochem. Photobiophys. 4, 201-210.
Argyroudi-Akoyunoglou JH and Thomou H (1981) Separation of pigment-protein complexes by SDS-sucrose density gradient centrifugation. FEBS Lett. 135, 177
Castorinis A Akoyunoglou G and Argyroudi-Akoyunoglou JH (1982) Correlation between the organization of Chl-protein complexes and the Chla fluorescence yield of chloroplasts during development. Photobiochem. Photobiophys. 4, 283.
Jennings RC et al (1978) Studies on cation-induced thylakoid membrane stacking, fluorescence yield and photochemical efficiency. Plant Physiol. 62,879.
Kalosakas K Argyroudi-Akoyunoglou JH and Akoyunoglou G (1981) In Akoyunoglou G. ed. Photosynthsis Vol V pp 569-580 Phil. Pa: Balaban Intern Sci. Services.
Li YS (1975) Salts and chloroplast fluorescence. Biochim. Biophys. Acta 376, 180-188.
Murata N (1969) Control of excitation transfer in Photosynthesis. Biochim. Biophys. Acta 189, 171-181.

Author's address: Biology Department, Nuclear Research Center "Demokritos" Aghia Paraskevi Attikis, Athens, GREECE.

THE CHLOROPHYLL-PROTEIN COMPLEXES OF PROCHLORON

A.W.D. LARKUM AND R.G. HILLER

1.   INTRODUCTION

Prochloron is a prokaryotic oxygenic organism whose structure shows
affinities with the green algae and higher plants. In particular it
posseses Chl b and stacked thylakoids. However recent studies on its
r-RNA (Seewaldt, Stackebrandt 1982, Van Valen 1982) do not rule out an
acquisition of Chl b independent of that which gave rise to the green
algae.
        A single report has suggested that Prochloron contains light-
harvesting Chl a/b protein with electrophoretic and spectral properties
similar to LHCP of green algae and higher plants (Withers et al. 1978).
Our studies of the Chl-protein complexes of Prochloron indicate that there
are significant differences and further more that there are differences
in the P700-Chl a protein.

2.   MATERIALS AND METHODS

Prochloron was obtained from colonies of the ascidian Diplosoma virens
growing under dead stagshorn coral on the reef flat at Heron and One Tree
Reefs (Capricornia region, Great Barrier Reef). Material was transported
to the laboratory as frozen cells or frozen crude thylakoid perparations
or in living ascidian colonies. Algal cells were pressed out of cut
ascidian colonies into buffered seawater (0.1 M Tris-acetate pH 9.2) and
broken by a single pass through a Yeda (1500 psi) or French (4000 psi)
pressure cell. After filtering through 2 layers of miracloth, unbroken cells
and cell debris were removed by centrifugation at 500g for 10 min.
Thylakoids were obtained by centrifugation at 5000g for 15 min, washed in
0.1 M Tris-acetate (pH 9.2) containing 10 mM EDTA and washed again in Tris-
acetate. After solubilisation in SDS:Chl 15:1, w/w) the samples were
separated by SDS-PAGE (6% acrylamide) at 4°C. Peptides were separated by
the procedure of Laemmli (1970).

3.   RESULTS

In comparison with spinach chloroplasts the Chl-protein complexes of
Prochloron thylakoids showed a number of differences upon SDS-PAGE (Fig 1).
A high M_r species (CPI*) was present and the CPI and CPIa bands were missing.
No band corresponding with LHCP3 was present and the fastest-moving complex
corresponded with CPa of spinach chloroplasts. Pigment analysis and
absorption and fluorescence spectra of the various bands (not shown except
figs 2 and 3) showed that all bands contained Chl b as well as Chl a but
that no band contained Chl a/b ratios approaching the very low ratios of
1.2-1.4 found for LHCP3 of spinach chloroplasts.
        The Chl a/b ratios of whole cells varied from about 3 in the summer to 6
in the winter. Summer material was used in the experiment of Fig 1. Use
of winter material showed that the fastest migrating Chl-protein was really
composed of two complexes with very different Chl a/b ratios. In this
experiment the band designated Chl a-b* was cut from the gel and reelectrophoresed. The Chl a-b* band gave rise to the original complex, some
Chl a-b complex and some free pigment.
        The absorption spectra of the Chl a-b* complexes from summer and winter

Sybesma, C. (ed.), Advances in Photosynthesis Research, Vol. II. ISBN 90-247-2943-2.
© *1984 Martinus Nijhoff/Dr W. Junk Publishers, The Hague/Boston/Lancaster.*

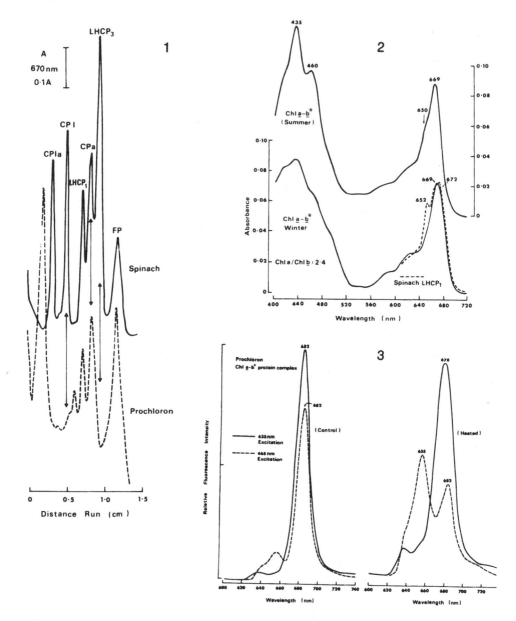

FIGURE 1. Densitometric scans (670nm) of green gels (SDS-PAGE) of _Prochloron_ thylakoids (summer material) and spinach thylakoids.

FIGURE 2. Absorption spectra of the Chl a-b* protein complex obtained by SDS-PAGE for summer and winter material.

FIGURE 3. Low temperature fluorescence emission spectra (77 K) of the Chl a-b* protein complex before and after heating at 60°C.

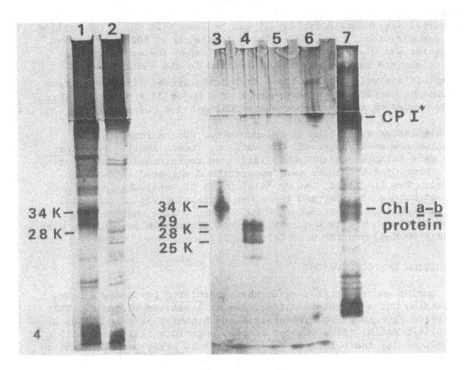

FIGURE 4. Polypeptide analysis of 1, Prochloron thylakoids; 2, spinach thylakoids; 3, Prochloron Chl a-b; 4, spinach LHCP₃; 5, Prochloron Chl a-b*; 6, CPI*; 7, Prochloron thylakoids. Lanes 1 and 2, samples boiled with SDS; lanes 3-7, samples unboiled. Gels were silver stained.

FIGURE 5. Polypeptide analysis of Chl-proteins isolated on a sucrose gradient. 1, spinach thylakoids; 2, Prochloron Chl a-b and Chl a-b*; 3, Prochloron CPI; 4, protein standards. Samples were boiled with SDS and Coomassie Blue stained.

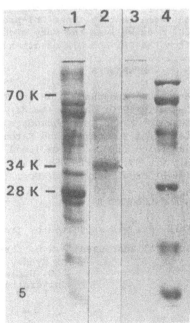

material are shown in Fig 2. The chl $\underline{a/b}$ ratios were about 2.2 and 2.4 respectively. Fluorescent emission spectra for the Chl $\underline{a}$-$\underline{b}$* complex from summer material are shown in Fig 3. It can be seen that Chl $\underline{b}$ (absorption at 465nm) was efficiently coupled to Chl $\underline{a}$ (emission at 682nm). However energy transfer was disrupted by heating (60°C) when the majority of fluorescence excited by 465nm light was emitted in the 655nm band.

Polypeptide analysis of Prochloron thylakoids showed differences from those of spinach, particularly the presence of a major 34 K M_r band. This polypeptide occurs on reelectrophoresis of Chl $\underline{a}$-$\underline{b}$ and Chl $\underline{a}$-$\underline{b}$* bands (Fig 4).

To obtain larger quantities of Chl-proteins, SDS-sucrose gradient centrifugation was used (Argyroudi-Akoyunoglou, Thomou 1981). Prochloron thylakoids were solubilised (SDS:Chl, 15:1) and centrifuged for 16 h at 150,000g. Three distinct zones were apparent and the gradient was sub-fractionated. Two fractions, one containing 62% of total Chl in Chl $\underline{a}$-$\underline{b}$ plus Chl $\underline{a}$-$\underline{b}$* but lacking CPI and the other containing 92% of the total Chl in CPI and CPI* (and enriched in P700 content), were analysed for peptide composition. The 34 K M_r peptide predominated in association with Chl $\underline{a}$-$\underline{b}$ and a 70 K M_r peptide with CPI (Fig 5).

4.   DISCUSSION AND CONCLUSIONS

The major conclusion from this work is that Prochloron possesses a novel light-harvesting Chl $\underline{a}$-$\underline{b}$ protein which forms at least one oligomer. This protein differs from LHCP from green algae and higher plants in having i) a higher Chl $\underline{a/b}$ ratio, ii) a reduced electrophoretic mobility and iii) a higher M_r (34 K) for the principle polypeptide. At present there seems to be sufficient similarity between the two proteins for a common origin to be possible. However fuller confirmation will have to await better character-isation of both proteins.

In its other major Chl-protein, the P700-Chl $\underline{a}$ protein of photosystem I, Prochloron resembles the cyanobacteria in possessing Chl-proteins of very high M_r on SDS-PAGE (Takahashi et al 1982). However it differs from cyano-bacteria in that these Chl-proteins all contain Chl $\underline{b}$. The possession of Chl $\underline{b}$ as well as the very high M_r makes these Chl-proteins quite distinct from the CPI and CPa of green algae and higher plants.

5.   REFERENCES

Argyroudi-Akoyunaglou JH and Thomou H (1981) FEBS Lett. 135, 177-181
Laemmli (1970) Nature (Lond.) 227, 680-685
Seewaldt E and Stackebrandt E (1982) Nature (Lond.) 295, 618-620
Takahashi Y, Koike H and Katoh S (1982) Arch. Biochem. Biophys. 219, 209-218
Van Valen (1982) Nature (Lond.) 298, 493-494
Withers NW, Alberte RS, Lewin RA, Thornber JP, Britton G and Goodwin TW (1978) Proc. Nat. Acad. Sci. (US) 75, 2301-2305

ACKNOWLEDGEMENTS

This work was supported by grants from Macquarie and Sydney Universities.

Authors' addresses: A.W.D. Larkum, School of Biological Sciences, University of Sydney, NSW 2006, Australia
R.G. Hiller, School of Biological Sciences, Macquarie University, N.Ryde, NSW 2113, Australia

PIGMENT-PROTEIN COMPLEXES OF CODIUM, A MARINE GREEN ALGA

JAN M. ANDERSON

Some marine green Siphonales algae have enhanced amounts of chl b, and
siphonaxanthin and/or siphonein as their main xanthophyll(s). Three
pigment-protein complexes of Codium thylakoids were isolated by sucrose
density gradient centrifugation, following fragmentation of thylakoids with
Triton X-100. The main light-harvesting complex of PS II (LHC-II) is a
siphonaxanthin-siphonein-Chl a/b-protein complex (Chl a/b ratio of 0.66)
which has enhanced absorption in the blue-green and green region compared
to the lutein-Chl a/b-protein complex of higher plants and most Chlorophyta.
In Codium LHC-II, siphonaxanthin and siphonein (510 and 538 nm) and Chl b
(470 nm) all efficiently transfer excitation energy to Chl a only. The
fluorescence emission spectrum of Codium LHC-II is identical with lutein-
LHC-II of higher plants (λmax, 682 nm). An undissociated PS I complex has
CP1 (P700-Chl a-protein) and its own antenna complex (LHC-I) which contains
Chl a, Chl b and siphonaxanthin bound to specific apoproteins (25-19 kDa)
different from those of LHC-II. Hence as in higher plants, PS I has an
antenna complex distinct from the main antenna complex, LHC-II. The Codium
PS II complex has CPa (core PS II complex) and some LHC-II. Codium LHC-II
has 30, <u>28</u>, <u>26</u>, 25 kDa apoproteins similar to those of spinach LHC-II, but
also two extra 35 and 34 kDa apoproteins.

1. INTRODUCTION

Siphonous marine green algae contain enhanced amounts of Chl b compared to
freshwater algae and higher plants (Jeffrey 1965; Ogawa et al. 1975). Some
contain siphonaxanthin and its esterified form, siphonein, instead of
lutein, the predominant xanthophyll of most green algae and higher plants.
Siphonaxanthin and siphonein when complexed to protein effectively absorb
light in the blue-green and green region (Yokohama et al. 1977; Kageyama et
al. 1977; Yokohama 1981); the dominant light of deep oceanic or turbid
coastal waters is green (Jerlov 1976).

Recently the pigment-protein complexes of a Codium species were isolated by
mild SDS PAGE and spectrally characterized (Anderson 1983). This study
compares the composition of pigment-protein complexes of Codium which have
been isolated by an alternative procedure, whereby Codium thylakoids are
fragmented with Triton X-100 and resolved by a simple one-step sucrose
density gradient centrifugation (Figure 1).

2. METHODS

Codium was collected from Cronulla NSW, then transported
to Canberra and well-washed thylakoids prepared as described
(Anderson 1983). The Triton X-100 fractionation procedure
was a modified method of Mullet et al. (1980). Thylakoids
were washed with 50 mM sorbitol, 0.75 mM EDTA (pH 7.8),
then 50 mM sorbitol, 5 mM Tricine-$(CH_3)_4$NOH (pH 7.8) and
incubated for 30 min at 4°C in H_2O containing 0.7%
Triton X-100 (0.8 mg Chl/ml). Following centrifugation
at 40,000 x g for 30 min, the supernatant ($\sim$ 86% total
chl) was loaded onto SW 41 tubes containing a continuous
0.1 - 1.0 M sucrose gradient with 0.04% Triton X-100, and

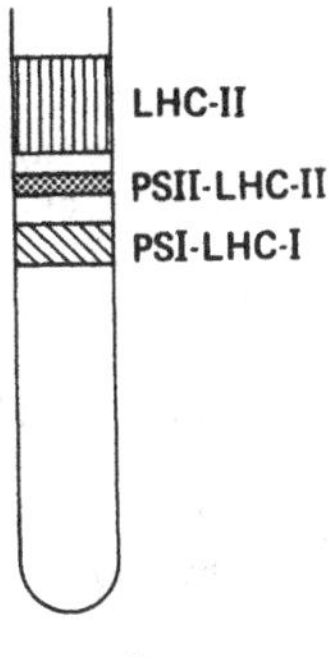

LHC-II

PSII-LHC-II

PSI-LHC-I

Figure 1.

Sybesma, C. (ed.), Advances in Photosynthesis Research, Vol. II. ISBN 90-247-2943-2.
© *1984 Martinus Nijhoff/Dr W. Junk Publishers, The Hague/Boston/Lancaster.*

TABLE 1. Chlorophyll composition of Codium pigment-protein complexes

	Chl a/b ratio	Chl/P700	Green gel bands
Thylakoids	1.45	450	$CPla^{1-3}$, CP1, Cpa, L^1, L^3
PS I-LHC-I	2.34	110	$CPla^{1-3}$, CP1
PS II-LHC-II	2.68	0	CPa, L^1, L^3
LHC-II	0.66	0	L^1, L^3

centrifuged at 200,000 x g for 16-20 h (Fig. 1). LHC-II was precipitated by stirring with 10 mM MgCl for 15 min at 20°C and collected as in Burke et al. (1978). The intermediate and lower bands were diluted with an equal volume of 10 mM Tricine (pH 8.0) and centrifuged at 150,000 x g for 2 h. Mild SDS PAGE (Anderson 1983), SDS slab gels (10-16%) (Anderson 1980), P700 determination and fluorescence spectrometry (Anderson 1983) were as described previously.

3. RESULTS AND DISCUSSION

Pigment composition of pigment-protein complexes of Codium. Codium chloroplasts have a Chl a/b ratio of 1.45, considerably lower than those of extreme shade plants. Following Triton X-100 fragmentation of Codium thylakoids, three major bands were resolved by sucrose density gradient centrifugation (Fig. 1) and analysed for pigments, Chl-proteins and polypeptide content. Recently, eight pigment-proteins were resolved from Codium thylakoids by mild SDS PAGE (Anderson 1983) and characterized by pigment composition and spectral properties as follows : $CPla^1$, $CPla^2$ and CP1 are partly dissociated PS I complexes which in addition to the core reaction centre complex, CP1 contained distinct siphonaxanthin-Chl a/b-proteins of PS I. $LHCP^1$ and $LHCP^3$ are orange-brown Chl a/b-proteins belonging to the main light-harvesting antenna of PS II (LHC-II). The purified LHC-II of Codium thylakoids had a Chl a/Chl b ratio of 0.66 and no P700. The purity of LHC-II was also seen by mild SDS PAGE, which resolved only two green bands, $LHCP^1$ (70% Chl) and $LHCP^3$ (26% Chl) with little loss of pigment. Following mild SDS PAGE, the middle band (Chl a/Chl b ratio of 2.7) contained CPa, $LHCP^1$ and $LHCP^3$ with no Chl-proteins of PS I. In contrast, the lower band (Chl a/b ratio of 2.3) had $CPla^1$, $CPla^2$ and CP1 only. In both cases the amounts of free pigments were very low indeed. The pigment composition of the Triton X-100 complexes of Codium are listed in Table 1.

The absorption spectra of the Codium complexes (Fig. 2a) show very different light-harvesting properties. PS I-LHC-I has an absorption maximum at 679 nm and its enhanced chl b content (compared to that of spinach PS I-LHC-I) is evident at 472 and 650 nm. PS II-LHC-II has a slightly higher chl a/b ratio but Chl b is clearly seen. Codium LHC-II has more Chl b than spinach LHC-II (Fig. 2b) and enhanced light-harvesting in the green region of the spectrum. This is due to the xanthophylls, siphonaxanthin and siponein which absorb at 508 and 538 nm in vivo (Anderson 1983). Since about 60% of the total Chl of Codium thylakoids is accounted for by LHC-II the absorption in the blue-green and green region is enhanced compared to land plants.

The fluorescence emission spectra at 77 K of Codium PS II-LHC-II and PS I-LHC-I (Fig. 3a) are very different and characterestic of PS II and PS I complexes.

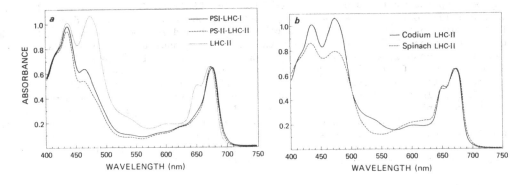

Figure 2. Absorption spectra of a) Codium PS I-LHC-I and PS II-LHC-II and b) Codium LHC-II and spinach LHC-II.

With higher plant thylakoids, the 685 and 695 nm bands are associated with PS II and 735 nm with PS I (Anderson et al. 1981). The far-red emission of Codium chloroplasts, however, as with many other algae occurs at lower wavelengths (710-718 nm) (Anderson 1983). The excitation spectrum for PS II-LHC-II and PS I-LHC-I (not shown) are identical with their respective absorption spectra demonstrating that the antenna complexes with Chl b and siphonaxanthin efficiently transfer excitation energy to Chl a of the core complex only.

The fluorescence emission spectrum of Codium LHC-II was maximal at 681 nm irrespective of whether the fluorescence was excited at 436 nm (Chl a), 468 nm (Chl b) or 510 and 540 nm (siphonaxanthin and siponein) (Fig. 3b) demonstrating that all pigments are integral components of the complex. Moreover, the fluorescence emission spectrum of Codium LHC-II is identical with that of spinach LHC-II which is a lutein-Chl a/b-protein whose Chl a/b ratio is ∿ 1.2 (Fig. 3b). However, there is a dramatic difference in the excitation spectra of Codium LHC-II and spinach LHC-II (not shown) both of which resemble their respective absorption spectra (Fig. 3b).

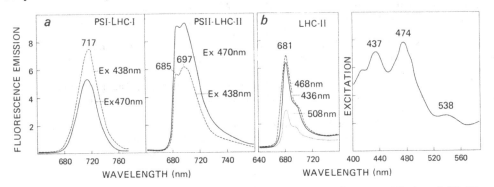

Figure 3. Fluorescence emission spectra at 77 K of a) PS I-LHC I and PS II-LHC-II and b) emission spectrum of LHC-II excited at 436, 468 and 508 nm and excitation spectrum for F681 nm.

The polypeptide profiles of Triton X-100 Codium PS I-LHC-I and PS II-LHC-II were similar to those of spinach complexes (not shown). However, Codium LHC-II in addition to the usual 4 bands of 30, 28, 26 and 24.5 kDa comparable to those of spinach LHC-II, has two additional strong bands at 35 and 33 kDa (Anderson, unpublished result). Although the specific Chl a/b-protein antenna associated with Codium PS I-LHC-I has not been isolated as a separate entity, the polypeptides of LHC-II (36, 34, 30, 28, 26, 24.5 kDa) are not present in PS I-LHC-I. Instead there are a group of 6 polypeptides of 25-19 kDa some of which are implicated.as the apoproteins of LHC-I. In Chlamydmonas reinhardtii, a specific Chl a/b-protein CPO of PS I has been shown to have a 26 kDa apoproteiı (Wollman, Bennoun 1982). In spinach thylakoids, PS I-LHC I (Chl/P700 ratio of 180) there are 6 polypeptides in the 25-19 kDa range some of which are candidates for LHC-I apoproteins (Anderson et al. 1983).

REFERENCES
Anderson JM (1980) P700 content and polypeptide profile of chlorophyll-protein complexes of spinach thylakoids, Biochem. Biophys. Acta 591, 113-126.
Anderson JM, Barrett J and SW Thorne (1981) Chlorophyll-protein complexes of photosynthetic eukaryotes and prokaryotes. In Akoyunoglou G. ed. Photosynthesis III, pp. 301-315. Philadelphia: Balaban International Science Service.
Anderson JM (1983) Chlorophyll-protein complexes of a Codium species. Biochim. Biophys. Acta in press.
Anderson JM, Brown JS, Lam E and Malkin R (1983) Chlorophyll b: an integral component of photosystem I of chloroplasts, Photochem. Photobiol. in press.
Burke JJ, Ditto CL and Arntzen CJ (1978) Involvement of the light-harvesting complex in cation regulation of excitation energy distribution in chloroplasts, Arch. Biochem. Biophys. 187, 252-263.
Jeffrey SW (1965) Pigment composition of siphonales algae in the brain coral, Fabia Biol. Bull. 135, 141-148.
Jerlov NG (1976) Marine Optics, Amsterdam: Elsevier.
Kageyama A, Yokohama Y, Shimura S and Ikawa T (1977) An efficient excitation energy transfer from a carotenoid siphonaxanthin to chlorophyll a observed in a deep water species of Chlophycean seaweed, Plant and Cell Physiol. 18, 477-480.
Mullet JE, Burke JJ and Arntzen CJ (1980) Chlorophyll-proteins of photosystem I, Plant Physiol. 65, 814-822.
Ogawa T, Nakamura K and Shibata K (1975) Algological studies. 14, 37-48.
Wollman FA and Bennoun PA (1982) A new chlorophyll-protein related to PS I in C. reinhardii, Biochim. Biophys. Acta 680, 352-360.
Yokohama Y, Kageyama A, Ikawa T and Shimura S (1977) A carotenoid characteristic of Chlorophycean seaweeds living in deep coastal waters, Bot. Mar. 20, 433-436.
Yokohama Y (1981) Distribution of the green light-absorbing pigments siphonaxanthin and siphonein in marine green algae, Bot. Mar. 24, 637-640.

Authors address: Dr. Jan M. Anderson, CSIRO, Division of Plant Industry
                  GPO Box 1600, Canberra A.C.T. 2601, Australia.

PIGMENT COMPOSITION AND STABILITY IN THE INTERTIDAL ALGA CODIUM FRAGILE

ANDREW H COBB/ERICA E BENSON

1. INTRODUCTION

Codium fragile exhibits several photosynthetic adaptations to the low intensity light fields encountered at high tide. Chloroplast photosynthesis in the alga saturates at 50-100 μE.M^{-2}.s^{-1} P.A.R. (Benson, 1983) and chlorophyll b enrichment is reflected in the low chlorophyll a:b ratios of 1.5-1.8:1 of the fronds (Benson, Cobb, 1981). However when sampled during the summer C.fragile fronds are often bleached and this may suggest that this alga is susceptible to photooxidative damage when the tide recedes. Thus, despite the low light intensity adaptations demonstrated in C.fragile, photoprotective mechanisms must also be operational in this alga. The aim of this study is to investigate light adaptations in C.fragile with respect to the intertidal zone environment. This is to be achieved by characterising the pigment content of the pigment/protein complexes isolated from the alga and investigating the effects of light regimes simulating tidal exposure on frond pigment stability.

2. MATERIALS AND METHODS

C.fragile was sampled from Bembridge, Isle of Wight (U.K.) and maintained as described by Benson et al (1984). Pigment/protein complexes were characterised using SDS/PAGE (Benson, Cobb 1984) and their pigments quantitatively assayed using the methods of Benson, Cobb (1981). For the photostability experiments 50 g samples of vegetative and reproductive C.fragile fronds were exposed to light intensities of 500, 750, 1,000, 1,500 and 2,000 μE.M^{-2}.s^{-1} P.A.R. supplied by an Osram 1000 W solar Colour SON/T Lamp. Exposure times were chosen to simulate various 'natural' exposures (0.5, 1.0, 1.5, 2.0, 3.0, 5.0 and 7 hours) when fronds in the intertidal zone are exposed to daylight as the tide recedes (up to 2,500 μE.M^{-2}.s^{-1} P.A.R.). Storage controls were incorporated into each experiment. All fronds were moistened with seawater to prevent desiccation and maintained at 25-27°C. Pigment analyses were performed as described above and the data processed using the University of Pittsburgh SPSS-10 statistical package. 3 and 2-way analyses of variance (ANOVA's) were used to investigate the effects of light intensity and exposure time on the chlorophyll and carotenoid content of the fronds.

3. RESULTS AND DISCUSSION

The following pigment/proteins were separated from C.fragile thylakoids in order of least mobility on SDS/PAGE:- CP1a, CP1 (P700 complexes) LHCP$_1$ and LHCP$_2$ (dimeric and monomeric light-harvesting complexes). An additional complex tentatively identified as CPa (associated with PSII) was occassionaly resolved between LHCP$_1$ and LHCP$_2$. The P 700 complexes were enriched in chlorophyll a and had Chl.a:b ratios of 1.0-1.4. In contrast, the LHCP complexes were enriched in chlorophyll b (chl.a:b 0.69-0.84). α-carotene was the only carotenoid to be consistently measurable in the P 700 complexes, although variable amounts of siphonein, neoxanthin and siphonoxanthin were also detected. Over 70% of the total pigment content of the complexes was located in the LHCPs. The average carotenoid distribution within

Sybesma, C. (ed.), Advances in Photosynthesis Research, Vol. II. ISBN 90-247-2943-2.
© *1984 Martinus Nijhoff/Dr W. Junk Publishers, The Hague/Boston/Lancaster.*

these complexes was α-carotene (14%), siphonoxanthin (40%), siphonein (19%), and neoxanthin (27%). Siphonein and siphonoxanthin showed the characteristic light-harvesting in vivo absorbance at 540 nm in the LHCPs of C.fragile (Benson, Cobb, 1984; Kageyama, Yokohama, 1978). The above findings suggest that C.fragile may be adapted to the low intensity submarine light fields encountered at high tide. The light-harvesting apparatus of the alga is similar to that observed in shade-adapted higher plants as indicated by the relatively high chlorophyll b concentration (Anderson, et al 1973). In addition, the presence of siphonein and siphonoxanthin suggests that the alga is able to make efficient use of the 'green gap' region of the visible spectrum. This may be a particularly important adaptation as coastal waters can selectively absorb in the red and blue regions of the spectrum (Jerlov 1977).

Pigment stability studies showed that exposure time was not a significant parameter in affecting pigment stability, whereas light intensity significantly affected pigment content, and differences were apparent between reproductive and vegetative fronds. Light intensity significantly affected all the pigments in reproductive (summer-sampled) fronds but only the chlorophylls and ε-carotene in vegetative (winter-sampled) tissue. These findings may suggest that the chlorophyll pigments in C.fragile are particularly unstable at high light intensities. Exposure to increased light intensities did not produce a uniform trend as a decrease in pigment content, rather responses were manifest as an overall trend towards increasingly erratic behaviour in pigment composition. This trend was more erratic as light intensity increased and was very marked in reproductive tissue. This probably reflects the fact that the tissue had been previously exposed to high midsummer light intensities before sampling. However this apparent disruption of pigment stability (as $\mu g.g\ fwt^{-1}$), masked an inherent control of relative pigment composition in both reproductive and vegetative fronds as demonstrated by the stability of pigment ratios and % carotenoid distributions (Table 1). The apparent lack of variability in this data is indicated by the particularly low standard error values. Where n=a maximum of 40 observations, 5 light intensities x 8 exposure times.

The stability of these ratios does indicate that C.fragile chloroplasts may be protected against excessive photooxidation. This may be due in part to a tightly controlled mechanism in which pigment breakdown is coordinated with pigment synthesis. Grumbach, Lichtenthaler (1982) have shown that radish seedlings grown in high intensity light regimes exhibit an increased turnover rate of pigment synthesis compared to seedlings grown under shade conditions and similar phenomenon may be operational in C.fragile.

From these findings in Table 1 midsummer frond bleaching in C.fragile may not be primarily due to photooxidation as midsummer-sampled reproductive tissue still demonstrated pigment stability on exposure to controlled high intensity light regimes. Additional studies using this alga (Benson etal 1984) indicate that summer-bleaching in C.fragile may be partle due to chlorosis induced by nutrient deficiency. Further investigations studying the interaction of light and nutrient availability may elucidate the primary cause of frond bleaching in the alga. However, it is evident from these studies that C.fragile may possess adaptations to both low and high intensity light, both being particularly advantageous in the constantly changing intertidal zone light environment.

TABLE 1: STABILITY OF PIGMENT RATIOS AND % CAROTENOID DISTRIBUTIONS OF
FRONDS EXPOSED TO VARIOUS LIGHT REGIMES.

	Pigment distribution ± S.E.	
	Reproductive	Vegetative
Pigment ratio		
Total xanthophyll:Total carotenes	5.3 ± 0.21	7.0 ± 0.22
Total chlorophyll:α-carotene	25.2 ± 1.2	110.0 ± 3.3
Siphonoxanthin:Siphonein	2.5 ± 0.06	2.2 ± 0.04
Chlorophyll a:b	1.43 ± 0.02	1.51 ± 0.02
% Carotenoid distribution		
α-carotene	15.0 ± 0.6	11.4 ± 0.32
ε-carotene	0.98 ± 0.14	1.2 ± 0.19
Siphonoxanthin	35.2 ± 1.1	43.8 ± 0.5
Siphonein	14.7 ± 0.31	17.8 ± 0.25
Neoxanthin	27.8 ± 0.7	20.0 ± 0.3
Violaxanthin	5.8 ± 0.15	5.5 ± 0.16

REFERENCES

Anderson JM Goodchild DJ and Boardman NK (1973) Composition of the
photosystems and chloroplast structure in extreme shade plants. Biochim.
et Biophys Acta 325, 573-585
Benson EE (1983) Studies on the structure and function of Codium fragile
chloroplasts. Ph.D thesis, Trent Polytechnic, UK.
Benson EE and Cobb AH (1981) The separation, identification and quantitative determination of photopigments from the siphonaceous marine alga
Codium fragile. New Phytol. 88, 627-632.
Benson EE and Cobb AH (1984) Pigment/protein complexes of the intertidal
alga Codium fragile. In press.
Benson EE Rutter JC and Cobb AH (1984) Seasonal variation in frond mor-
phology and chloroplast physiology of the intertidal alga Codium fragile.
In press.
Grumbach KH and Lichtenthaler HK (1982) Chloroplast pigments and their
biosynthesis in relation to light intensity. Photochem and Photobiol
35, 205-212.
Jerlov NG (1977) Classification in terms of quanta irradiance. J. Cons.
Int. Explor. Mer. 37, 281-287.
Kageyama A and Yokohama Y (1978) The function of siphonein in a siphonous
green alga Dichotomosiphon tuberosus. Jap. J. Phycol. 26, 151-155.

ACKNOWLEDGEMENTS

The authors acknowledge the Science and Engineering Research Council of
the U.K. for financial support.

Authors address: Department of Life Sciences, Trent Polytechnic,
                  Nottingham NG1 4BU, U.K.

CHANGES IN CHLOROPHYLL PROTEIN COMPLEXES DURING THE LIFE CYCLE OF SCENEDESMUS

Horst Senger, Karin Krupinska and George Akoyunoglou*

Fachbereich Biologie/Botanik, Universität Marburg, Lahnberge, 355 Marburg,BRD.
*on leave from Nuclear Research Center "Democritos", Athens, Greece.

1. INTRODUCTION

Synchronous cultures of unicellular algae provide an excellent tool to study changes in the photosynthetic apparatus under physiological conditions. In cells of Scenedesmus obliquus, synchronized under a light-dark regime of 14:10 hours, photosynthetic capacity and quantum yield change in parallel. Both have their maximum in the 8 th hour and their minimum in the 16 th hour of the cell cycle (H.Senger, 1970a). More detailed investigations show that PS II activity follows these changes - it has about 50% higher capacity in 8 hour old cells compared to 16 hour cells - while PS I measured as photoreduction remains constant in the light limiting region during synchronous growth (H.Senger, 1970b). In order to gain information about the composition and structure of Scenedesmus thylakoids and their relevance for the described changes in photosynthetic efficiency, in this study the chlorophyll-protein complexes and the effect of cations on low temperature fluorescence of thylakoids derived from synchronous cells were investigated.

Furthermore we examined the possibility of attributing the change in photosynthetic capacity to direct variations in the properties of PS II by calculating the photosynthetic unit size (S.Malkin et al.,1981) and the heterogeneity of PS II (A.Melis, P.H.Homann, 1975; A.Melis, G.Akoyunoglou, 1977). In addition the state I/II-transition during the life cycle of Scenedesmus was studied (J.Lavorel, A.L.Etienne, 1977).

2. MATERIAL AND METHODS

Cultures of the green alga Scenedesmus obliquus, strain D3, in liquid inorganic medium were grown and synchronized under a light-dark regime of 14:10 hours (H.Senger, N.I.Bishop, 1974).

Chloroplast particles from Scenedesmus were prepared by mechanical breaking (V.Mell, H.Senger, 1974). Further purification of the particles was achieved by sucrose density gradient centrifugation (N.H.Chua, P.Bennoun, 1975). After purification the chloroplast particles were used for PAGE (P.Delepelaire, N.H.Chua, 1979 or J.H.Argyroudi-Akoyunoglou, G.Akoyunoglou, 1979).

3. RESULTS AND DICUSSION

3.1. Chlorophyll-protein complexes: Thylakoid membranes of 8 and 16 hour old cells show no distinct differences in composition and quantitative distribution of chlorophyll-protein complexes, when analyzed by the procedure of P.Delepelaire and N.H.Chua (1979).

Preparation and resolution of the complexes according to J.H.Argyroudi-Akoyunoglou and G.Akoyunoglou (1979) however reveals a significant variation in the ratio of oligomers to monomers in thylakoids of different developmental stages (Fig.1 and table 1). Thylakoids derived from 8 hour old cells have a higher portion of CP I-oligomers and of LHCP1+2 compared to 16 hour cells. Assuming a correlation between the low temperature fluorescence ratio F720/F686 and the oligomer/monomer ratio (A.Castorinis et al., 1982), the low temperature fluorescence spectra of 8 and 16 hour old cells (Fig.2), already shown by H. Senger and N.I. Bishop (1971), confirm this result. While the ratio

Sybesma, C. (ed.), Advances in Photosynthesis Research, Vol. II. ISBN 90-247-2943-2.
© *1984 Martinus Nijhoff/Dr W. Junk Publishers, The Hague/Boston/Lancaster.*

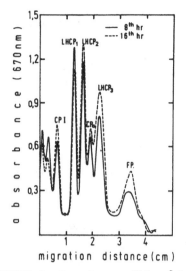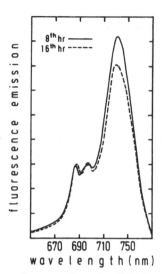

FIGURE 1: Scanning profiles of gels prepared according to J.H.Argyroudi-Akoyunoglou and G. Akoyunoglou (1979). The samples contained triton and LDS in a ratio of 3 triton: 2 LDS : 1 Chl. Gels were scanned at 670 nm.

FIGURE 2: Fluorescence emission spectra at 77°K: comparison of 8 and 16 hour old cells of a synchronous culture of Scenedesmus. The samples contained 15 µg Chl/ml. Exciting light was of 440 nm.

	CPI-olig.	CPI	CPA	LHCP1+2	LHCP3	F.P.	ratio:olig./monomers CPI	LHCP
8 hr	17.1	8.7	9.4	38.1	18.8	7.8	2.0	2.0
16 hr	11.2	9.8	8.4	34.7	23.9	12.0	1.1	1.5

TABLE 1: Distribution of chlorophyll (%) among the chlorophyll-protein complexes and the free pigment (F.P.) in thylakoids of synchronous cells.

F 720/F686 amounts to 3.0 in cells of the 8th hour, it is only 2.5 in 16 hour old cells. The higher oligomer/monomer ratio of thylakoids derived from 8 hour old cells demonstrates that a higher photosynthetic capacity corresponds with a structurally fixed and stable apparatus.

3.2. Magnesium effect on chloroplast particles of synchronous Scenedesmus cells. Further support for the higher oligomer content of thylakoids deriving from 8 hour old cells is given by the variation in magnesium effects on thylakoid particles of synchronous cells. Table 2 shows the differences in F_{max} and F_o of particles prepared in the presence or absence of magnesium ions in a final concentration of 2 mM. The relative change of F_{max} by Mg^{2+} obviously is larger in particles deriving from 8 hour old cells than from 16 hour old cells The more pronounced Mg^{2+}-effect on 8 hour particles may be caused by the higher oligomer content of the corresponding thylakoids (J.H.Argyroudi-Akoyunoglou, 1983).
Under both, high and low salt conditions the fluorescence yield of 16 hour particles is higher than of 8 hour particles (Table 2), as already reported

	8 hr,$-Mg^{2+}$	8 hr,$+Mg^{2+}$	16 hr,$-Mg^{2+}$	16 hr,$+Mg^{2+}$
F_{max}	4.4	6.7	5.9	7.5
F_{max} (%)	100	151	100	128
F_{max}/F_o	1.9	2.5	1.8	2.3
F_{max} 16 hr/8 hr x 100	100	100	133	113
F720/F686	3.2	2.3	2.4	1.9

TABLE 2: The effect of magnesium ions on thylakoid particles of synchronous Scenedesmus cells. The particles were prepared in the absence or prescence of Mg^{2+} (2mM). The samples contained 5 µg chlorophyll/ml and DCMU in a final concentration of $10^{-5}M$.
Room temperature fluorescence was excited with 447 nm light and monitored at 683 nm. The exciting light for the low temperature fluorescence spectra was of 440nm.

for whole cells of Scenedesmus (H. Senger, N.I. Bishop, 1971). These results are confirmed by parallel variations of the low temperature fluorescence of low and high salt chloroplast particles from 8 and 16 hour old cells (Table 2).

3.3. Unit size and heterogeneity of photosystem II: Determination of the unit size by fluorescence induction measurements in the presence of DCMU ($10^{-5}M$) (S.Malkin et al., 1981) results in a 1.5 larger size of the area over the normalized fluorescence induction curve of 8 hour old cells compared to 16 hour old cells. That means that 8 hour old synchronous cells have smaller PS II units than cells of the 16th hour.
Besides the higher stability of thylakoids derived from 8 hour cells their smaller PS II unit size, too, can account for the differences in photosynthetic activity. Since the P700/chlorophyll ratio remains constant during the life cycle of Scenedesmus (V.Mell, H. Senger, 1978) we have to assume a higher PS II/PS I ratio in the stage of maximal photosynthetic capacity (8 hr) compared to the stage of minimal photosynthetic capacity (16 hr).
The calculation of the relative amounts of α- and ß-centers in both cell types, based on the same fluorescence induction curves (A.Melis, P.H.Homann, 1975; A.Melis, G. Akoyunoglou, 1977), yields not significant differences (Table 3). Only the rate constants K_α and Kß are different: both are smaller in the 8th than in the 16th hour samples, measured under the same limiting light intensity and with the same chlorophyll content (Table 3).
The correlation of smaller unit size with smaller value of K_α and Kß also is known from greening systems of higher plants (A.Melis, G.Akoyunoglou, 1977).

3.4. State I/II-transition: Room and low temperature fluorescence measurements were used in order to investigate the possibility to get state I and state II conditions (J.Lavorel,A.L.Etienne, 1977) in synchronous Scenedesmus cells after illumination with light of either 646 nm (II) or 701 nm(I).
Whereas in 16 hour old cells at room temperature Fmax of state I is about 15% higher than Fmax of state II and the low temperature fluorescence ratio F720/F686 is 3.1 for state I and 3.5 for state II, no changes were observed in 8 hour old cells. The amount of the state I/II-transition follows closely the course of quantum requirement (Fig. 3). In the stage of optimal photosynthetic capacity no state I/II transformation can occur while it has its greatest extent around the stage of minimal photosynthetic capacity.
The result that the 8 hr cells show the largest Mg^{2+}-effect but no state I/II-

transition demonstrates clearly that Mg^{2+} and state I-light do not necessarily act in the same way, as it has been assumed (J.Lavorel, A.L. Etienne, 1977).

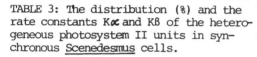

	8 hr	16 hr
ßmax (%)	49	55
Kß x sec^{-1}	2.2	4.9
Kα x sec^{-1}	15.6	21.2

TABLE 3: The distribution (%) and the rate constants Kα and Kß of the heterogeneous photosystem II units in synchronous Scenedesmus cells.

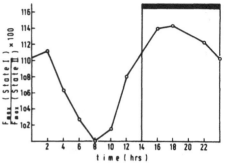

FIGURE 3: The percent change in the F_{max} during state I/state II transformation of Scenedesmus during the synchronous life cycle. The cells were illuminated either with light of 646 nm (state II) or with light of 701 nm (state I) prior to the measurement of fluorescence induction in the presence of DCMU. Room temperature fluorescence was excited with light of 447 nm and monitored at 683 nm.

REFERENCES

Argyroudi-Akoyunoglou JH, Akoyunoglou G (1979) The CPs of the thylakoid in greening plastids of Phaseolus vulgaris, FEBS Letters 104,78-84.
Argyroudi-Akoyunoglou JH (1983) Polypeptide comp. of the pigment-protein complexes of Phaseolus vulg. thylakoids. Cation-induced disaggreg. of oligomeric to monomeric forms correlates with the increase in the ratio F685/F730..., In Akoyunoglou et al. eds., Cell function and differ., II,277-289,Liss,New York
Castorinis A,Akoyunoglou G,Argyroudi-Akoyunoglou JH (1982) Correlation between the organization of pigment-protein complexes and Chl a fluorescence yield of chloroplasts during develop. in Phas.vulg.,Photobiochem.Photobiophys.4,283-291.
Chua NH,Bennoun P (1975) Thylakoid membrane polypeptides of Chlamyd.reinh.: wild-type and mutant strains deficient in PS II RC,PNAS USA 72,2175-2179.
Delepelaire P,Chua NH (1979) LDS/PAGE of thylakoid membranes at 4°C:characterization of 2 additional Chl a protein complexes, PNAS USA 76, 111-115.
Lavorel J, Etienne AL (1977) In vivo chlorophyll fluorescence. In Barber J ed. The intact chloroplast, Vol. 2, pp.206-268, Elsevier, Netherlands.
Malkin S, Armond PA, Mooney HA, Fork DC (1981) PS II photosynthetic unit sizes from fluorescence induction in leaves, Plant Physiol. 67, 570-579.
Melis A, Akoyunoglou G (1977) Development of two PS II units in etiolated bean leaves, Plant Physiol. 59, 1156-1160.
Melis, A, Homann PH (1975) Kinetic analysis of the fluorescence induction in DCMU poisoned chloroplasts, Photochem.Photobiol. 21,431-437.
Mell V, Senger H (1978) Photochemical activities, pigment distribution and photosynthetic unit size of subchloroplast particles isolated from synchronous cells of Scenedesmus obliquus, Planta 143,315-322.
Senger H (1970a) Charakterisierung einer Synchronkultur von Scenedesmus obl., ihrer pot.PS-Leistung und des PS-Quotienten ..., Planta 90,243-266.
Senger H (1970b) Quantenausbeute und unterschiedliches Verhalten der beiden Photosysteme des Photosyntheseapparates während des Entwicklungsablaufes von Scenedesmus obl. in Synchronkulturen, Planta 92, 327-346.
Senger, H, Bishop NI (1971) Changes in fluorescence and absorbance during synchronous growth of Scenedesmus,Proc.Int.Congr.Photosynth.,2nd,677-687.

CHARACTERIZATION OF CHLOROPHYLL-PROTEIN COMPLEXES FROM A CRYPTOPHYCEA:
CRYPTOMONAS RUFESCENS

C. LICHTLÉ and J.C. DUVAL
Laboratoire de Botanique de l'Ecole Normale Supérieure, 24 rue Lhomond,
F-75231 Paris Cedex 05, France

1. INTRODUCTION

Phycoerythrin and chlorophyll c_2 are the accessory pigments for photo-synthesis in the Cryptophycea (Lichtlé *et al*. 1980).
Comparative studies of the separation of chlorophyll-protein complexes either by LiDS-PAGE or digitonin treatment followed by sucrose-density gradient, have led to the isolation of four and three complexes respectively. These complexes have been characterized by their spectral properties and their polypeptides separated. These new data are in agreement with the structural scheme previously proposed.

2. MATERIAL AND METHODS

The culture conditions of *C. rufescens* were previously described (Lichtlé *et al*. 1980).
Thylakoid isolation: Algal cells suspended in 50 mM Hepes-KOH buffer (pH 7,4), 5 mM Mg Cl_2, 2%₀ BSA, were broken at 4°C using a French pressure cell, centrifuged and washed twice to eliminate phycoerythrin.
LiDS-PAGE of chlorophyll-protein complexes: Thylakoids were solubilized in a mixture of LiDS and Deriphat 160, using a detergent-chlorophyll ratio of 10/1 and 20-30/1 (W/W) respectively. Samples containing 10 to 30 µg of chlorophylls were immediatly put on the gel. PAGE were performed on slab gel according to Laemmli (1970) with a 2,5 % stacking gel and a 9 % resolving gel.
Digitonin treatment and sucrose-density gradient: Thylakoidal fractions were treated with 2 % digitonin at a ratio of 80/1 (W/W) and put on a sucrose-density gradient (4-6-8-10-15-20-30-40 % w/w) and centrifuged at 130000 x g for 17 h.
Electrophoresis of Polypeptides from digitonin fractions: Fractions collected on gradient were recentrifuged and the pellet suspended in 0,25 M Trisglycine buffer (pH 8,3) with LiDS (1 g/l) and β-Mercaptoethanol (100 mM) and put on gel.
Absorption, fluorescence emission spectra at -196°C and Gaussian analysis were previously described (Berkaloff *et al*. 1981).

3. RESULTS

3.1. Absorption and fluorescence spectra of digitonin-fractions
A sucrose-density gradient with digitonin-treated thylakoids gave three membrane fractions B_1, B_2, B_3 respectively in 30 %, 20 % and 10 % sucrose layer. Absorption spectra at -196°C (fig. 1) showed that the three bands contained carotenoids and chl c_2 with a sharp 640 nm peak only for B_2. Each fraction exhibited different chl a main peaks, 670 and 676 nm for B_1 and 672 nm for B_2 and B_3.
The fluorescence emission of B_1 (fig. 2) was very low with two emission peaks at 720 nm and 694 nm (F 720/F 694 = 1,6), and a small peak at 640 nm. For B_2, the maximum of emission was near 682 nm with only a small peak near 740 nm (F 740/F 682 = 0,07). No emission at 640 nm was detected. For B_3, the major emission was also near 682 nm, but the ratio F 730/F 682 = 0,1 and there was a small peak at 640 nm. The B_1 presents many similarities

Sybesma, C. (ed.), Advances in Photosynthesis Research, Vol. II. ISBN 90-247-2943-2.
© *1984 Martinus Nijhoff/Dr W. Junk Publishers, The Hague/Boston/Lancaster.*

with the P 700 chlorophyll-protein complex (CPI), while B_2 and B_3 can be related to PS II antenna.

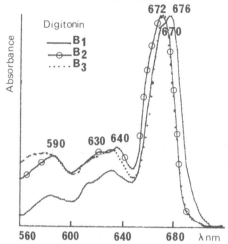

FIG.1.Absorption spectra at -196°C of digitonin fractions

FIG.2.Fluorescence emission spectra at -196°C of digitonin fractions

3.2. Gaussian analysis of absorption spectra at -196°C

Eight elementary Gaussian components were needed for analysis of the absorption spectra (fig. 3). It was necessary to vary slightly maxima and band widthes to allow a good resolution in the three bands.

Three main forms covered about 80 % of the whole absorption : C_a 658, C_a 668, C_a 678. Their relative ratio were roughly the same for the three bands ; in all cases, C_a 668 was the main form as previously seen in *Fucus* and *Cystoseira* (Berkaloff et al. 1981). These formsrepresent probably a light-harvesting system associated with both PS I and PS II.

C_a 687 appeared to be particularly important in B_1, which is the only complex containing C_a 705 ; this last form was clearly associated with a particular abundance in long wavelength forms. Nevertheless, C_a 685 and C_a 695 were also found in B_2 and B_3 but in lower amounts.

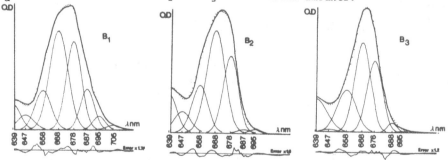

FIG.3. Gaussian analysis of absorption spectra at - 196°C

3.3. Electrophoresis of polypeptides of digitonin fractions

B_1 presented a major polypeptide near 79 KD which appeared bound to chlorophylls and generally 8 lighter polypeptides from 39 KD to 15.8 KD (fig. 4). The number of polypeptidesof the B_2 was always lower than in B_1

with a 39 KD polypeptides and 4 other polypeptides from 26 KD to 18 KD. In the B_3 less polypeptides were evidenced.

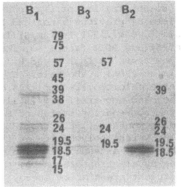

FIG.4. LiDS PAGE of polypeptides of digitonin bands after staining with Comassie blue (molecular weight in KD)

3.4. Electrophoresis of chlorophyll-protein complexes

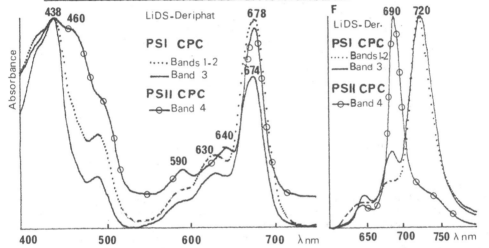

FIG.5. Absorption spectra of chlorophyll-protein complexes (CPC)

FIG.6.Fluorescence emission spectra at -196°C of CPC

By the combination of LiDS and deriphat detergents, we obtained four bands of chlorophyll-protein complexes. The three heavy bright-green complexes contained about 30 % of the total chlorophylls and they presented an identical absorption spectrum (fig. 5). The Chl a maximum absorption was near 678 nm with two additional peaks in the 590 nm and 630 nm regions. The fluorescence emission spectra at -196°C showed a major emission peak at 720 nm and two small peaks near 640 nm and 690 nm; which were slightly different according to the bands (fig. 6). These three bands can be related to the digitonin B_1 and identified to CPI. The lightest complex, yellow-green, represented about 14 % of the total chlorophylls. It showed the shorter chl a wavelength (674 nm), the chl c_2 peaks at 460 nm and 640 nm appeared clearly. The fluorescence emission peak was maximum at 690 nm while the F 740 and F 640 nm emission were very low. This complex appears characteristic of PS II antenna and can be compared to digitonin B_2 and B_3 fractions. However, there was always an important percentage of free chlorophylls despite the numerous methods used to reduce them.

4. DISCUSSION

The solubilization of *Cryptomonas* thylakoids by detergents is always incomplete as previously noticed by many authors working on Chromophytes (Berkaloff *et al.*1981;Ingram,Hiller,1983). The two treatments we used, separate PSI and PSII fractions. The PSI fraction contains always chl a and chl c2 as shown by Ingram and Hiller (1983) in a *Chroomonas* ,while in *Fucus*, there is only chl a (Berkaloff *et al.*1981). This PSI band can be resolved into three chlorophyll-complexes by electrophoresis, while in *Chroomonas* only one band appears(Ingram,Hiller,1983). These bands probably represent different aggregation states of P700-chlorophyll complexes as in Cyanobacteria(Thomas,Mousseau,1981). For PSI, chlorophylls appear linked to a high molecular weight polypeptide as it is the case for higher plants (Anderson 1980).For PSII antenna, we obten two fractions by digitonin treatment, and only one by electrophoresis. By their spectral properties and especially,by the efficient transfer of energy from chl c2 to chl a,shown by fluorescence emission, B_2 is comparable to the light-harvesting chl a/chl c2 obtained on *Chroomonas*(Ingram,Hiller,1983)and on other brown algae (Alberte *et al.*1981). It is difficult to know what polypeptides are linked to the chlorophylls, but as noticed by Mullet *et al.*(1980) on higher plants, we find two light polypeptides (24KD and 26 KD) associated with this fraction.We can also associate the B_3 fraction to PSII antenna, but energy transfer is less efficient than in B_2. We agree with Ingram and Hiller(1983) to assimilate B_2 to the light-harvesting A_0 in the Cryptophyceae photosynthetical scheme that we have previously proposed (Lichtlé *et al.*1980), while B_3 may be a component of the A_2 antenna.

5. REFERENCES

- Anderson JN (1980) P-700 content and polypeptide profile of chlorophyll-protein complexes of spinach and barley thylakoids,Biochim.Biophys.Acta 591,113-126.
- Alberte RS,Friedman AL,Gustafson DL,Rudnick MS and Lyman H(1981) Light-harvesting systems of brown algae and diatoms,isolation and characterization of chlorophyll a/c and chlorophyll a/fucoxanthin pigment-protein complexes, Biochim.Biophys.Acta 635,304-316.
- Berkaloff C,Duval JC,Jupin H,Chrissovergis F and Caron L(1981) Spectroscopic studies of isolated pigment protein complexes of some brown algae thylakoids,In Photosynthesis III.Ed.Akoyunoglou. pp 485-494.
- Ingram K and Hiller R (1983) Isolation and characterization of a major chlorophyll a/c2 light-harvesting protein from a *Chroomonas* species (Cryptophyceae), Biochim.Biophys.Acta 722,310-319.
- Laemmli UK (1970) Cleavage of structural proteins during the assembly of the head of bacteriophage T4, Nature 227, 680-685.
- Lichtlé C,Jupin H and Duval JC (1980) Energy transfers from photosystem II to photosystem I in *Cryptomonas rufescens* (Cryptophyceae), Biochim.Biophys. Acta 591, 104-112.
- Mullet JE,Burke JJ and Arntzen CJ (1980) Chlorophyll-proteins of photosystem I, Plant Physiol. 65,814-822.
- Thomas JC and Mousseau A (1981) Characterization of chlorophyll-protein complexes from Cyanophyceae. The study of their modification in different culture conditions,In Photosynthesis III. Ed.Akoyunoglou.,Balaban Int. Sc.Services, Philadelphia, pp 435-444.

CHLOROPHYLL- AND HEME-BINDING PROTEINS IN <u>PORPHYRIDIUM CRUENTUM</u> THYLAKOIDS

T. REDLINGER/E. GANTT

1. INTRODUCTION

Chl-protein complexes (CP) can be isolated from thylakoids as detergent soluble complexes. In our studies on the relationships of PS I and PS II with the phycobilisomes, we have been isolating and characterizing the Chl-protein complexes of <u>P. cruentum</u>. Previously (Redlinger, Gantt 1981a) we reported that this red alga contained three Chl a-protein complexes, but lacked CP II, the light harvesting Chl <u>a/b</u> protein complex. In the present study, we further characterize the Chl-protein complexes and report that the major complex isolated is equivalent to CP I, and that two lesser complexes (CP III and IV) appear to be part of PS II. In addition we found four heme-binding proteins, which correspond to cytochromes typical of thylakoids.

2. MATERIALS AND METHODS

Thylakoid membranes were isolated from <u>Porphyridium cruentum</u> cells which were washed free of phycobilisomes and other surface binding proteins as described by Redlinger and Gantt (1982b).

Polyacrylamide gel electrophoresis was performed according to Neville (1971) except the separating gel was 7.5 to 15% gradient. Purified thylakoids were solubilized at 100°C for 1 min or at 4°C for 5 min in 2% SDS (w/v), 25 mM Na_2CO_3, 100 mM mercaptoethanol, 20% glycerol (v/v) to a final Chl concentration of 500 µg/ml. The Chl/SDS ratio was 1:40 or diluted accordingly with buffer. In some preparations Triton X-100 was added to a final concentration of 1% (v/v). Electrophoresis was carried out at 4°C for isolation of Chl-protein complexes, or 25°C for resolution of denatured proteins. Gels were stained with Coomassie blue R or with 3,3',5,5'-tetramethylbenzidine (TMBZ).

Samples for spectroscopy were isolated from polyacrylamide gels and dialyzed against 50 mM Na-PO_4 buffer (pH 7.5) and 20% glycerol (v/v) prior to spectral analysis.

P700 was estimated from reversible light-induced absorbance changes at 700 nm using the extinction coefficient and method outlines by Shiozawa et al. (1974). Samples were measured in 50 mM HEPES, pH 7.5, 1 mM $MgCl_2$, and 2 mM sodium ascorbate.

3. RESULTS AND DISCUSSION

In accordance with our earlier results (Redlinger, Gantt 1981a), three Chl-protein complexes (Fig. 1A) were resolved after optimizing the electrophoretic procedure. These Chl proteins are designated as CP I, CP III, and CP IV.

<u>Spectral characterization of the three chlorophyll-protein complexes.</u> The relative Chl distribution estimated by scanning the unstained gel at 675 nm (Fig. 1A) was: CP I = 41%, CP III = 3%, CP IV = 4%, Free = 52%. The three Chl-protein complexes had room temperature absorption maxima: CP III and IV at 675 nm, and CP I at 678 nm (Fig. 2).

Sybesma, C. (ed.), Advances in Photosynthesis Research, Vol. II. ISBN 90-247-2943-2.
© *1984 Martinus Nijhoff/Dr W. Junk Publishers, The Hague/Boston/Lancaster.*

All three had a Soret peak at 435 nm characteristic of Chl a. The calculated M ratio of Chl a/carotenoid was: CP I = 15:1, and CP III and CP IV = 20:1.

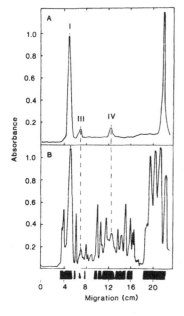

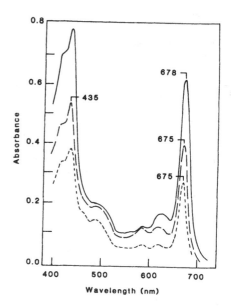

Fig. 1. Electrophoretic analysis of P. cruentum thylakoid membranes: (A) unstained gel scanned at 675 nm to detect Chl and (B) Coomassie blue stained gel scanned at 590 nm to detect protein. Peaks I, III, IV are the Chl-protein complexes. The stained gel is shown at the bottom.

Fig. 2. Absorption spectra (20°C) of the Chl-protein complexes: CP I ( ), CP III ( ), and IV ( ).

By low temperature fluorescence emission analysis CP I had a max at 720 nm and was clearly distinguishable from CP III and CP IV which had maxima at 690 nm (Fig. 3).

Only CP I contained a reversible light-induced absorbance change at 700 nm. The Chl/P700 ratio for CP I was 450:1 which is considerably higher than for washed thylakoids which had a Chl/P700 ratio of 250:1. The lower ratio in CP I can be attributed to extended times in SDS (12 h) required for electrophoretic separation of CP I from CP III.

Protein characterization of the three chlorophyll-protein complexes. The three complexes were located among the other thylakoid proteins by comparing densitometric scans of the unstained gel with those of the Coomassie stained gel (Fig. 1A and B, respectively). When not yet fully denatured the apparent mol wt were: CP I = 120,000; CP III = 92,000; CP IV = 45,000.

The polypeptide composition of the three complexes observed as green bands was determined by isolating these bands from preparative gel slabs and denaturing them by heating (100°C, 1 min). The CP I complex contained a single polypeptide of 68,000 D (Fig. 4B). Since the initial

CP I complex had an apparent mol wt of 120,000 D it was probably a dimer consisting of two 68,000 D polypeptides. CP I contained 10 to 15 Chl/68,000 D before complete denaturation.

On denaturation, the CP III complex (92,000 D) always yielded a major polypeptide at ca. 52,000 D (Fig. 4C) while the CP IV complex yielded three polypeptides at 40,000, 48,000, and 52,000 D (Fig. 4D). The 52,000 D polypeptide corresponded in relative molecular mass to the major component of CP III, indicating a possible relationship between CP III and CP IV. The Chl/protein ratio for the latter two complexes was always similar but varied between preparations.

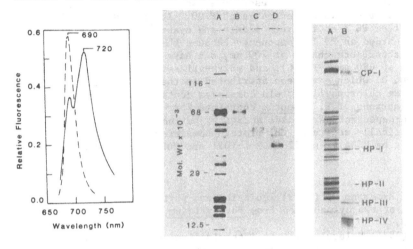

Fig. 3. (left) Fluorescence emission spectra (- 196°C) of CPI (   ) and CPIII/IV (---). Excitation was at 440 nm.

Fig. 4. (center) Polypeptides of the three Chl-protein complexes. The complexes were separated on preparative gels, isolated, denatured and reelectrophoresed. The gel was stained with Coomassie blue: (A) thylakoids, (B) CP I, (C) CP III, (D) CP IV.

Fig. 5. (right) Thylakoid proteins showing heme-dependent peroxidase activity: HPI-IV. Gel stained first with TMBZ (B), and then with Coomassie blue (A).CP I still detectable after TMBZ staining because of Chl.

Analysis of thylakoids for heme-dependent peroxidase activity. To determine whether the Chl-protein complexes were associated with cytochromes, the thylakoid polypeptides of $\underline{P}$. cruentum were separated electrophoretically and analyzed for heme-dependent peroxidase activity. Four TMBZ-oxidation bands were observed and designated heme-binding proteins HP I-IV with mol. wt of 33,000, 18,500, 14,000 and 11,500 D, respectively (Fig. 5). The HP-IV band was observed only when thylakoids were not washed with EDTA prior to isolation, suggesting that it is surface bound and susceptible to removal by EDTA. The four TMBZ bands show the following correspondence: HP-I to cytochrome $\underline{f}$; HP-II to a possible cytochrome $\underline{b_6}$ type; HP-III and HP IV to cytochrome-550 and cytochrome-553, respectively. Our results on cytochrome f and cytochrome-553 are in agreement with those of Evans and Krogmann (1979), who previously reported their isolation from $\underline{P}$. cruentum. We did not detect any heme-binding proteins in association with electrophoretically separated Chl-protein complexes.

4. SUMMARY

Electrophoresis of $\underline{P}$. cruentum thylakoids resolved three green protein bands (CP I, CP III, and CP IV). Evidence for CP I containing the PSI reaction center are its fluorescence emission at F720 nm (- 196°C) and the photoreversible absorbance change at P700. Its mol wt (68,000 D) is characteristic of PS I as in green plants and cyanobacteria. CP III and CP IV are regarded as PS II components because they lack a photoreversible absorbance change at 700 nm, but have a fluorescence emission maximum at 690 nm (-196°C), and polypeptides in the mol wt range (40,000 to 52,000 D) which have been attributed to the PS II reaction center. All three chlorophyll-protein complexes contained carotenoids. The thylakoid membranes of $\underline{P}$. cruentum contained four cytochromes, detected by heme-dependent peroxidase activity, but not associated with the electrophoretically separated Chl-protein complexes.

Supported in part by USDA grant 82-CRCR-1-1051 and DOE AS 05-76-ER-04310.

REFERENCES

Evans DK and Krogmann DW (1979) Isolation and partial characterization of two cytochromes from Porphyridium cruentum. Plant Physiol. 63, S55.
Neville DM (1971) Molecular weight determinations of protein-dodecyl sulfate complexes by gel electrophoresis in a discontinuous buffer system. J. Biol. Chem. 246, 6328-6334.
Redlinger T and Gantt E (1981a) Identification of a 95,000 polypeptide in thylakoids and phycobilisomes of Porphyridium cruentum. In Akoyunoglou G, ed. Photosynthesis III. Structure and Molecular Organization of the Photosynthetic Apparatus, pp 257-262. Philadelphia, PA: Balaban Internatl. Sci. Ser.
Redlinger T and Gantt E (1981b) Phycobilisome structure of Porphyridium cruentum: polypeptide composition. Plant Physiol. 68, 1375-1379.
Redlinger T and Gantt E (1982) A M_r 95,000 polypeptide in Porphyridium cruentum phycobilisomes and thylakoids: Possible function in linkage of phycobilisomes to thylakoids and in energy transfer. Proc. Nat. Acad. Sci. 79, 5543-5546.
Shiozawa JA, Alberte RS and Thornber JP (1974) The P700 chlorophyll protein. Arch. Biochem. Biophys. 165, 388-397.

AUTHORS' ADDRESS: T. Redlinger/E. Gantt, Smithsonian Environmental Research Center, 12441 Parklawn Drive, Rockville, Maryland 20852-1773 U.S.A.

EFFECT OF HYDROLYTIC ENZYMES ON CYANOBACTERIAL PIGMENT-PROTEIN COMPLE-
XES OF PHOTOSYSTEM 1

JIŘÍ HLADÍK, RŮŽENA BÍSKOVÁ, DANUŠE SOFROVÁ

1. INTRODUCTION

Degradation by means of enzymes can provide valuable information on the
structure and spatial arrangement of pigment-protein complexes (Schmidt
1983, Süss et al. 1976, Jennings et al. 1979). A number of studies has
been devoted to the influence of enzymes on thylakoid membranes or
changes in polypeptide composition of the membranes. The results have
shown that a number of peripheral proteins is digested virtually without
residue, while proteins or pigment-protein complexes immersed into the
matrix are degraded little or not at all. The main reason is probably
inaccessibility of integral proteins to hydrophilic enzymes, which are
incapable of penetrating the lipid bilayer. Our study was aimed at
investigating the effect of hydrolytic enzymes on isolated pigment-protein complexes of photosystem 1 of cyanobacteria with the purpose of
obtaining more information on the molecular organisation of this complex.

2. MATERIAL AND METHODS

Pigment-protein complexes (PPC) were obtained by solubilisation of
thylakoid membranes of the cyanobacterium Plectonema boryanum using the
detergent Triton X-100, followed by chromatography on a column of
DEAE-cellulose (Hladík, Sofrová 1981). The chromatographic fractions,
denoted in the order of elution II and III, were then incubated with
hydrolytic enzymes usually in the ratio of 20 mg enzyme per 1 mg
chlorophyll at 10 - 15°C in the dark, under gentle agitation. Incubation
was stopped by rapid cooling to 0 °C and electrophoretic analysis was
immediately carried out or, samples were frozen and stored at -20°C in
the dark. Electrophoresis was performed on 5% polyacrylamide gel in 25 mM
Tris-borate buffer, pH 8.3, containing 0.1% SDS, for 30 min at 4°C. Gels
were dyed with Coomasie Blue R-250.

3. RESULTS AND DISCUSSION

Chromatography of the Triton extract on DEAE-cellulose provided two green
fractions of pigment-protein complexes. These fractions had been earlier
ascribed to photosystem 1 (Hladík, Sofrová 1981, Hladik et al. 1982). One
of the fractions contains the pigment-protein complex denoted III of
M_r 260 000, the second contains pigment-protein complexes IIa and IIb of
M_r 120 000 and 112 000. It was also found that complex III is a higher
structural unit, since it can provide the two complexes IIa and IIb under
suitable conditions without deep-going degradative or structural changes
(Bísková 1983).
Under the influence of the anionic detergent SDS, which causes denaturation of native PPCs when used in higher concentrations, dissociation of
complex III to complexes IIa and IIb is also observed within a relatively
short period up to 60 minutes (Fig. 1).
The course of dissociation of complex III indicates the possibility of
gradual degradation of the M_r 260 000 complex via the M_r 190 000
complex. It is clear, however, from the course of dissociation of the
complex II, that M_r 190 000 is rather an association product than a

Sybesma, C. (ed.), Advances in Photosynthesis Research, Vol. II. ISBN 90-247-2943-2.
© *1984 Martinus Nijhoff/Dr W. Junk Publishers, The Hague/Boston/Lancaster.*

FIGURE 1. Effect of SDS on
stability of PPCs incuba-
ted at 45°C for 45 min
with detergent in a given
SDS/Chl weight ratio.
* - bands contain no pig-
ment, FP - free pigments.
a - control, b - SDS/Chl =
100, c - SDS/Chl = 250,
d - SDS/Chl =1000.

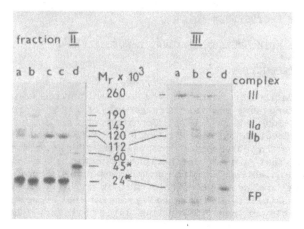

product of degradation, similarly to M_r 145 000. At higher SDS
concentrations, the newly formed complexes IIa and IIb continue to
dissociate, forming a PPC of M_r 60 000 and a polypeptide of M_r
45 000. In this process, sensitivity of complex IIa towards SDS is
considerably higher than that of complex IIb. It is degraded in the
fraction II even at the lowest SDS concentration.
Incubation of isolated thylakoid membranes with the hydrolytic enzymes
pronase and lipase has shown considerable resistivity of PPCs to
treatment by these enzymes (Fig. 2), which fact corroborates earlier
results (Carter, Staehelin 1980, Süss et al. 1976, Laszlo, Gross 1981,
Sofrová et al. 1980).
Complex III remained without substantial changes, only its ammount
relative to the other components of the thylakoid membrane decreased.
While complex IIa (120 000) likewise remained substantially unchanged,
the ammount of complex IIb (112 000) was considerably lower after
treatment with pronase as well as lipase. Non-pigmented proteins –
probably peripheral proteins of the thylakoid membrane – were also
totally degraded by pronase. The results thus demonstrate better accessibility or lower stability of complex IIa with respect to hydrolytic
enzymes. It is difficult to decide, whether higher or lower stability of
PPCs is due to their quarternary structure – stabilised by immersion in
the lipid matrix – or their tertiary structure, which is distinctly
influenced by bonding of chlorophyll (Jennings et al. 1979).
In the following part of our study we, therefore, investigated the
influence of trypsin, pronase and lipase on isolated PPC II and III (Fig.
3).It was found, that substantial changes in the electrophoretic profiles
of the two complexes could not be observed even after treatment with
relatively high enzyme concentrations (up to 80 mg per 1 mg chlorophyll)
over periods of 2 to 3 hours. Changes were recorded only when enzyme
treatment was prolonged, but even these changes were not very distinct.
Trypsin, being a specific protease, was rather ineffective towards
complex III as well as complexes IIa and IIb. The same applies to lipase
even after duration of 7 hours. After 22 hours of treatment none of the
complexes was completely degraded, only the chlorophyll content in the
free pigment band was elevated at the expense of the complexes. In the
case of lipase, complex III dissociated into the two complexes IIa and
IIb in approximately equal ratios.

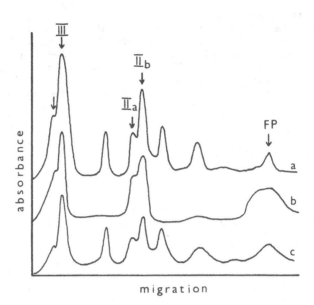

FIGURE 2. Electrophoretic pattern of the PPCs isolated from thylakoid membranes pretreated with pronase (b) and lipase (c), (a) - control. Incubation 60 min. Arrows indicate the PPCs, FP-free pigments.

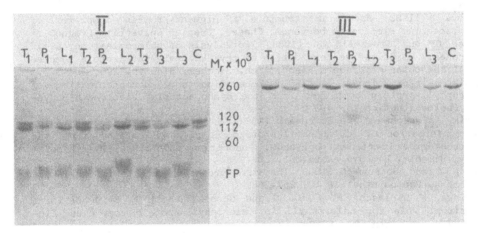

FIGURE 3. Electrophoretic pattern of PPC II and III after treatment with trypsin (T), pronase (P) and lipase (L): incubation for 2 hours (1), 7 hours (2) and 22 hours (3) under the conditions described in METHODS. Equal chlorophyll ammounts were applied to the gel in all cases.

Pronase, which is a non-specific protease, had a more severe effect. After 2 hours of incubation, a degradation product of M_r 60 000 can be observed already and after a longer incubation period the relative amounts of complexes IIa and IIb are considerably lower, complex IIa again being less stable dissapearing from the gel completely after 22 hours of incubation. Similarly, pronase degrades fraction III more quickly, both complexes IIa and IIb being formed. It is remarkable, that while complex IIa is degraded more rapidly during incubation of fraction

II, the degradation product - complex IIb - is hydrolysed more quickly in the case of complex III, which fact apparently relates to the different mechanisms of formation of these complexes. Setting out from the structural model of PPCs in the photosystem 1 of cyanobacteria which we proposed earlier (Hladík, Sofrová 1983) we may assume that complexes IIa and IIb were formed in fraction II by the effect of Triton X-100 on the supramolecular complex III probably in the course of its solubilisation and release from the thylakoid membrane. Their relative ratio is given by their stoichiometry in the complex III (IIa:IIb = 1:2). After enzyme treatment of complex III, complex IIa appears only when complex IIb has been degraded partly or completely. In this case, stoichiometry cannot be maintained and complex IIa, although less stable, is present in greater amount.

The possible role of the detergent Triton X-100, which surrounds PPC in solution, should be kept in mind when assesing the stability of PPCs with respect to hydrolytic enzymes. Inactivation of enzymes by the detergent has been excluded (Bísková 1983). It was found for model systems, that activity of proteases in a solution of 0.2% Triton X-100 actually rises to 105 - 108%, that of lipase decreases to 65 - 70%. In the case of PPC - Triton systems, however, the relatively large Triton molecules can be a steric hindrance for the contact of complexes with the active site of enzymes, thus contributing significantly to resistivity of the pigment-protein complexes towards enzymes.

REFERENCES

Bísková R (1983) Molecular structure of pigment-protein complexes of cyanobacterium Plectonema boryanum, Theses, Charles University Prague.

Carter DP and Staehelin LA (1980) Proteolysis of chloroplast thylakoid membranes. I. Selective degradation of thylakoid pigment-protein complexes at the outer membrane surface, Arch. Biochem. Biophys. 200, 364-371.

Hladík J and Sofrová D (1981) Separation and characterisation of chlorophyll-protein complexes from the cyanobacterium Plectonema boryanum, Photosynthetica 15, 490-503.

Hladík J Pančoška P and Sofrová D (1982) The influence of carotenoids on the conformation of chlorophyll-protein complexes isolated from the cyanobacterium Plectonema boryanum. Absorption and circular dichroism study, Biochim. Biophys. Acta 681, 263-272.

Hladík J and Sofrová D (1983) The structure of cyanobacterial thylakoid membranes, Photosynthetica 17, 267-288.

Jennings RC Garlaschi FM Forti G and Gerola PD (1979) Evidence for a structural role for chlorophyll in chlorophyll-protein complexes, Biochim. Biophys. Acta 581, 87-95.

Laszlo JA and Gross EL (1981) Effect of proteolysis on the photochemical activity and polypeptide composition of the photosystem II core complex isolated from spinach chloroplasts, Plant Physiol. 68, 1008-1013.

Schmidt O (1983) Spatial arrangement of thylakoid membrane proteins analysed by controlled proteolytic digestion, Photosynthetica 17, 69-76.

Sofrová D Čacká H Masojídek J and Leblová S (1980) Functional and structural changes of cyanobacterial thylakoids after treatment with pronase and lipase, Photosynthetica 14, 198-203.

Süss KH Somidt O and Machold O (1976) Biochim. Biophys. Acta 448, 103-113.
Authors address: Dept.Biochem.,Charles Univ.,Albertov 2030
                 128 40 Prague, CZECHOSLOVAKIA

LUMINESCENCE STUDY OF PS 1 PIGMENT-PROTEIN COMPLEXES ISOLATED FROM THE CYANOBACTERIUM PLECTONEMA BORYANUM

PETR PANČOŠKA, MILAN AMBROŽ, *JIŘÍ HLADÍK AND *DANUŠE SOFROVÁ

1. INTRODUCTION

We have recently isolated and characterised the PS 1 pigment-protein complexes (PPCs) from the cyanobacterium Plectonema boryanum (Hladík and Sofrová (1981)). Isolation procedure - extraction of thylakoid membranes by Triton X-100, followed by the chromatography on DEAE cellulose column - yields two fractions; first containing complex of M_r 260 000 (denoted as PPC III), the second containing complex of M_r 118 000 (denoted as PPC II). It was shown (Hladík et al. (1982)) that the quarternary structure of the PPC III protein is - at least partly - conserved : this complex was found to be the oligomer of PPC II. We have proved that specificaly oriented carotenoids (characterised by strong positive circular dichroism bands at 480 and 505 nm) mediate the substantial part of the interactions between the pigment-protein subunits. The presence of protein-fixed, chiraly arranged carotenoids in PPCs seems to be a more general property of these systems: they were found in bacteria (Cogdell et al. (1976)) and also in PPCs from higher plants (Schubin et al. (1981)). The aim of our work was to study the functional consequences of this structural feature of PS 1 complex.

2. EXPERIMENTAL

2.1. Materials. Isolation procedure is described in Hladík et al. (1982). For the luminescence measurements we have used a) the PPCs solutions directly eluted from DEAE-cellulose column; b) gel slices: PPC solutions were applied on 5% polyacrylamide gel (tubes 0.5x8 cm). Electrophoresis was done according to Markwell et al. (1978). Corresponding zones were cut out from the gel and used for measurements. Cells of optical path length 0.01 cm were used, sample absorbances were approximately 0.01.

2.2 Methods. Spectra were measured on low-temperature luminescence apparatus (excitation part : 150W xenon arc lamp, H-20 UV monochromator with the microprocessor controlled wavelength setting, light chopper PAR 125A). Samples were placed in cryostat VSK-3-300 (Leybold Heraeus). The emitted radiation was focused onto the entrance slit of microprocessor controlled HT-20 monochromator and detected by R928 photomultiplier (Hamamatsu). Output signal was processed by synchro-het lock-in amplifier PAR 186A.

3. RESULTS AND DISCUSSION

Fluorescence spectra of PPC III in polyacrylamide gel, excited at main absorption bands and measured at temperatures from 88 K to 261 K are presented in Fig.1. It is seen that the short-wavelength emission (676 nm), observed for excitation in Soret region, is strongly suppressed when carotenoids are excited. The same effect was found for PPC III solution in 0.05 M Tris-HCl buffer, pH 8.0, containing 0.2% Triton X-100. For this sample, the intensity of the 676 nm emission band equals to intensity of

Sybesma, C. (ed.), Advances in Photosynthesis Research, Vol. II. ISBN 90-247-2943-2.
© *1984 Martinus Nijhoff/Dr W. Junk Publishers, The Hague/Boston/Lancaster.*

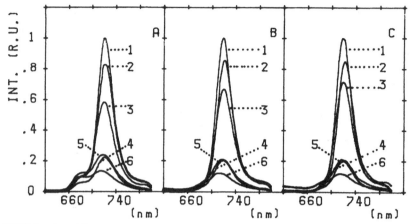

FIGURE 1. Fluorescence spectra of PPC III in polyacrylamide gel.
Temperatures : 1 - 88 K, 2 - 100 K, 3 - 140 K, 4 - 181 K, 5 - 221 K,
6 - 261 K. Excitation wavelengths : A) 434 nm, B) 468 nm, C) 496 nm.

the 676 nm emission band equals to intensity of 727 nm maximum for λ_e
=445nm and decreases to less than 0.2 of the original value for λ_e =470
nm (Ambrož M, Pančoška P, unpublished).
The selectivity in the distribution of excitation energy, absorbed by
carotenoids in PPC III follows from the comparison of fluorescence
excitation spectra of this sample for λ_f =727 nm and λ_f =676 nm (see
Fig. 2a). Similar results were obtained by Searle et al. (1981). We
explain this effect using the scheme shown in Fig. 2c. When the antennae
chlorophylls are excited in Soret as well as in red region, the absorbed
energy is proportionaly transferred to the chlorophyll forms (F676) and
(F727) characterized by the corresponding fluorescence. There is no such
a transfer between the 468-500 nm absorbing carotenoids (C500) and (F676)
chlorophyll form. We have further noticed that between 430 and 550 nm the
maxima in the fluorescence excitation spectrum for λ_f =727 nm coincide
with the positions of the dichroic bands, observed for this sample in CD
curve at 430-550 nm. Relationship between the conformation of carotenoids
and their energy chanelling specifity can be confirmed by comparison of
the above data with results obtained on i) PPC II ; ii) altered PPC III.
i) We have found that the ordered carotenoids are not detectable by CD
spectra, in PPC II, although there are no differences in other - i.e.
chlorophyll - regions of the complex II and III CD curves . The intensity
ratio I(500nm)/I(434nm) in the fluorescence excitation spectra for λ_f
=727 nm can be considered as the measure of energy distribution
selectivity for (C500) carotenoid form (see Fig. 2b). This ratio was
found to be 0.57 for PPC III and 0.24 for PPC II, which correlates well
with the proved higher degree of carotenoid organisation in PPC III.
ii) When carotenoids are extracted from the PPC III and then the attempt
to reconstitute the original form of the complex is performed (addition
of excess of carotenoids in n-heptane), neither the ordering of carotenoids nor the energy transfer between the two pigments is restored (see
Fig. 3a).
The temperature dependence of the PPC III fluorescence excitation spectra
for λ_f =727 nm is presented in Fig. 3b. Within the studied temperature

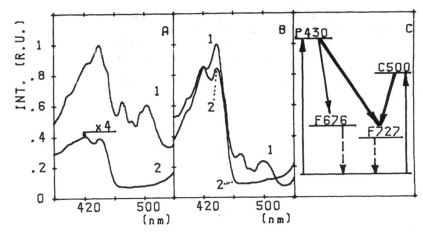

FIGURE 2. Comparison of fluorescence excitation spectra of PPC in polyacrylamide gel; fluorescence maxima : 1 - 721 nm, 2 - 676 nm. A) PPC III, B) PPC II, temperature 88 K. C) Energy level diagram for the considered pigment forms.

interval the selectivity of energy distribution for (C500) carotenoids remains practically unchanged. The nonlinear decrease of the overall fluorescence yield is observed when the temperature is raised from 88 to 261 K, but the positions and relative intensity ratios of maxima remain constant in these spectra.

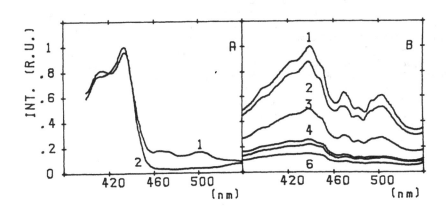

FIGURE 3. A) Comparison of fluorescence excitation spectra for fluorescence maximum at 724 nm - 1 - PPC III in Tris-HCl buffer + 0.2% Triton X-100.; 2 - sample 1 after extraction of carotenoids (heptane) and their back-reconstitution. Temperature - 100 K. B) Temperature dependence of fluorescence excitation spectra of PPC III in polyacrylamide gel; for emission maximum at 721 nm. Temperatures : 1 - 88 K,

From the general point of wiew: the apparent characteristics of C500 carotenoids are similar to the properties of focusing antennae complexes (Beddard and Cogdell (1982), Rubin and Paschenko (1982)). Our results could be considered as evidence that the fixation of photosynthetic pigments by protein in specific, ordered conformation is one of the requirements for ensuring the directed, selective distribution of the absorbed excitation energy.

REFERENCES

Beddard GS and Cogdell RJ (1982) Structure and excitation dynamics of light-harvesting protein complexes, Mol. Biol. Biochem. Biophys. 35, 46-79.

Cogdell RJ Parson WW and Kerr MA (1976) The type, amount, location, and energy transfer properties of the carotenoid in reaction centers from Rhodopseudomonas sph., Biochim. Biophys. Acta 430, 83-93.

Hladík J and Sofrová D (1981) Separation and characterisation of chlorophyll-protein complexes from the cyanobacterium Plectonema boryanum, Photosynthetica 15, 490-503.

Hladík J Pančoška P and Sofrová D (1982) The influence of carotenoids on the conformation of chlorophyll-protein complexes isolated from the cyanobacterium Plectonema boryanum. Absorption and circular dichroism study, Biochim. Biophys. Acta 681, 263-272.

Markwell JP Reinman S and Thornber JP (1978) Chlorophyll-protein complexes from higher plants: a procedure for improved stability and fractionation, Arch. Biochem. Biophys. 190, 136-141.

Rubin LB and Paschenko VZ (1982) Picosecond fluorometry of the exciton difussion process in chloroplast pigment apparatus of higher plants, Izv. Acad. Sci. ESSR 31, 192-199.

Searle GFW van Brakel GH Vermaas WFJ van Hoek A and Schaafsma TJ (1981) Chlorophyll triplets and transfer in isolated PS I chlorophyll proteins from blue-green algae. In Akoyunoglou GA, ed. Photosynthesis, Vol. I, pp. 129-141. Philadelphia, USA : Balaban Int. Sci. Services.

Shubin VV Efimovskaya TV and Karapetyan NV (1981) Structure of photoinduced absorption and CD spectra of P700, Zh. Fiz. Khim. 55,2916-2921.

Authors adress: Dept. Chem. Phys., Fac. Math.&Phys., Charles
                University, Ke Karlovu 3,121 16 Prague
              * Dept. Biochem., Charles Univ., Albertov 2030
                128 40 Prague, CZECHOSLOVAKIA

A 160-KILODALTON PHOTOSYSTEM-I REACTION-CENTER COMPLEX:
LOW-TEMPERATURE DIFFERENCE ABSORPTION, EPR AND FLUORESCENCE
SPECTROSCOPY OF THE PRIMARY ELECTRON ACCEPTORS[*]

I. IKEGAMI[**] and B. KE, Charles F. Kettering Research Laboratory
Yellow Springs, Ohio 45387, U.S.A.

1. INTRODUCTION AND SUMMARY

(1) A photosystem-I reaction-center complex having 10-15 chlorophyll (including $\leqslant 1$ Chl $\underline{b}$), 12 atoms each of Fe and S, and 160 kDalton protein per P700, has been isolated.

(2) Absorbance changes possibly due to A_2 (or X) and A_1 were obtained with the RC-complex by sequential and selective reduction followed by freeze-trapping the reduced acceptors.

(3) Fluorescence emission peaking at 705 nm (F705) at 77^0 K developed to the maximum level when A_2 is reduced, while it gradually disappeared when A_1 becomes reduced. F705 is attributed to luminescence arising from charge recombination between $P700^+$ and A_1^-.

(4) Spectral changes due to reversible (photo)reduction of chlorophyll $\underline{b}$ have been observed; the changes occur at the redox level of A_2.

2. PREPARATION AND CHEMICAL PROPERTIES OF THE PS-I RC-COMPLEX

FRACTIONS	Chl/P700 (mol/mol)	Protein (mg)	Chl (mg)	Chl a/b (w/w)	Fe or S/P700 (atom/mol)		Protein/P700 (g/mol)
Spinach Chloroplast	380	3,000	150	3.75	—	—	2,600,000
Digitonin Fractionation (∼ 5%, dig./Chl = 10 w/w)							
Photosystem-I Particle	160	270	62	8.0	—	—	572,000
Ether Wash (Ether 80% sat'd with H_2O)							
P700-enriched Particle	7-12	200	3.1	15.5	11.0	12.6	512,000
Triton X-100 Treatment (0.3%; Triton/Chl ≏ 40 w/w) (1 hr at 0°C, pH 8.5)							
$(NH_4)_2SO_4$ Fractionation (0.1 g/ml, 0°C) (ppt. dissolved in 10 mM PB, pH 8.5)							
Sucrose-density Gradient Centrifugation (5-50%, at 340 kg for 5 hrs)							
PS-I RC Complex	10-15	—	0.92	16.5	11.9	12.3	160,000
(at 20% layer)							

The complex contains $\leqslant 1$ Chl $\underline{b}$ per P700, although the latter could also be completely removed by further extraction with ether of higher water saturation without appreciable change of the Chl $\underline{a}$/P700 ratio. The results indicate that one Chl $\underline{b}$ (P700) is located very close to, but not in, the photosystem-I reaction center.

Sybesma, C. (ed.), Advances in Photosynthesis Research, Vol. II. ISBN 90-247-2943-2.
© 1984 Martinus Nijhoff/Dr W. Junk Publishers, The Hague/Boston/Lancaster.

3. PHOTOCHEMICAL ACTIVITY (P700) OF THE REACTION-CENTER COMPLEX

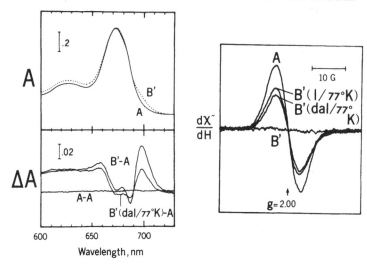

The RC-complex, when appropriately redox-poised, is fully photochemically active. With ascorbate present, almost all P700 could undergo reversible photobleaching. At 77° K, more than half of the P700 was photooxidized; only 10-20% of which was reversible as shown by absorbance (left) and EPR (right) changes.

4. OXIDATION AND (SEQUENTIAL) REDUCTION OF THE PS-I ELECTRON ACCEPTORS

Alphabetical designations for the chemical and photochemical manipulations on the RC-complex at appropriate temperatures to provide oxidation or selective and sequential reduction of the electron carriers

NOTATION	TREATMENT	P	A_1	A_2	B	A
		REDOX STATE CHANGES FROM				
[NA]	no addition, RC-complex as prepared	P^+				
[A]	0.5 mM FeCN, dark or illuminated	P^+				
[B']	10 mM Asc+50 μM DCIP, dark or illuminated at LN_2					
[B]	10 mM $S_2O_4^=$, dark				B^-	A^-
[C]	10 mM $S_2O_4^=$, illuminated near 0°C for 2 min, then frozen in LN_2 in the dark			A_2^-	B^-	A^-
[D]	10 mM $S_2O_4^=$, illuminated near -40°C for up to 20 min, then frozen in LN_2		A_1^-	A_2^-	B^-	A^-
[D']	sample in D above, briefly thawed and refrozen in LN_2			A_2^-	B^-	A^-

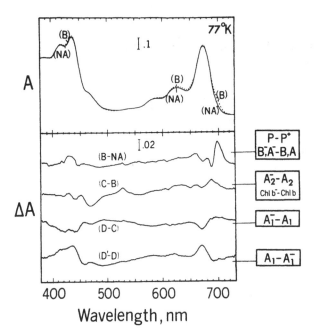

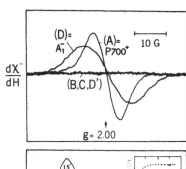

This figure shows the absorption and difference absorption spectra determined under the various redox conditions shown in the Table.

The redox state of the electron carriers under the different conditions are shown in the boxes at right.

$(B-NA) \rightarrow \Delta A(P700-P700^+) + \Delta A(B \cdot A^- - B \cdot A)$

$(C-B) \rightarrow \Delta A(A_2^- - A_2) +$ Chl electrochromic shifts $+ \Delta A$(Chl $\underline{b}$ redox change at 468,527,650 nm)

$(D-C) \rightarrow \Delta A(A_1^- - A_1)$

$(D'-D) \rightarrow \Delta A(A_1 - A_1^-) =$ reverse $\Delta A(A_1^- - A_1) \rightarrow$ (mirror image)

The formation of P700$^+$ and A_1^- in A and D, respectively, are corroborated by EPR results. The EPR signal due to A_1^- disappeared completely in D!

This figure shows the development of the 14 G-wide EPR signal of A_1^- as a function of illumination time under condition D.

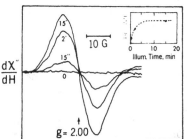

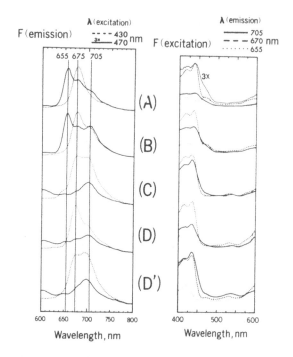

F675 was essentially constant under all redox conditions; its excitation spectrum suggests that F675 originates from the antenna chlorophyll remaining in the complex.

F705 developed to the maximum level in C (A_2 reduced), but completely disappeared in D (A_1 reduced). These results suggest that F705 may originate from charge recombination between $P700^+$ and A_1^-. The excitation spectrum of F705 indicates that its emitter has an absorption band near 455 nm.

F655 disappears from C on (A_2 reduced); its major band at 468 nm in the excitation spectrum indicates that Chl b remains capable of transferring its excitation energy to the emitters of F675 and F705.

REFERENCES

Ikegami, I and Katoh, S (1975) Biochim. Biophys. Acta, 376, 588-592
Shuvalov, VA et al. (1979) Proc. Nat. Acad. Sci.,USA, 76, 770-773
Shuvalov, VA (1976) Biochim. Biophys. Acta, 430, 113-121
Swarthoff, T (1982) FEBS Lett., 146, 129-132
Ikegami, I (1976) Biochim. Biophys. Acta, 449, 245-258
Breton, J (1982) FEBS Lett., 142, 16-20

* Contribution No. 808 from the Charels F. Kettering Research Laboratory; Work supported in part by NSF grants PCM-7900831 and PCM-8211139.

** I. Ikegami
   Laboratory of Chemistry, Faculty of Pharmaceutical Sciences
   Teikyo University, Sagamiko, Kanagawa 199-01 Japan

THE NOVEL CHLOROPHYLL RC I ASSOCIATED WITH PS I

Dieter Dörnemann and Horst Senger
Fachbereich Biologie/Botanik, Universität Marburg, Lahnberge, 3550 Marburg BRD

1. INTRODUCTION

A new chlorophyll, designated chlorophyll RC I (Chl RC I), with absorption and
fluorescence characteristics different from all other known chlorophylls has
been isolated from spinach chloroplasts and Scenedesmus cells (Dörnemann and
Senger 1981b/1982). Its enrichment in PS I-particles and the identity of its
molar ratio with that of P-700 (Dörnemann and Senger 1981a) suggests its
close correlation with the reaction center of PS I, if not its identity with
the chromophore of P-700 itself.
Pigment mutant C-2A' of Scenedesmus has no P-700 when greened in the presence
of chloramphenicol. From those cells no Chl RC I could be isolated (Dörnemann
and Senger 1981b).
Chl RC I is also characterized by its different reaction kinetics with iodine
and its low fluorescence yield compared to Chl a (Dörnemann and Senger, 1982).
Gel filtration on Sephadex LH 20 and plasma desorption mass spectrometry
(PDMS) revealed a higher molecular weight of Chl RC I than Chl a. These data
will be critically revised in this paper.
The fluorescence yield experiments will be supplemented by fluorescence life
time data.
Data on spectral shifts and circular dichroism of the RC I phaeophorbide
compare best with those of a 20-Cl-methylphaeophorbide a, kindly provided
by Dr. Hugo Scheer, Univ. of Munich.
To show a possible insertion of chlorine in the phorphyrin structure of
Chl RC I and its possible position, NMR spectroscopy and neutron activation
experiments were carried out.

2. MATERIALS AND METHODS

Chlorophyll RC I was isolated and purified from the mutant C-6E of the green
alga Scenedesmus obliquus as described earlier (Dörnemann and Senger, 1982).
Preparations of Chl RC I from Anacystis nidulans were performed in cooperation with Dr. Tetzuya Katoh, Univ. of Kyoto, Japan in the same way after
the preparation of photosynthetic membranes from broken cells by a differential centrifugation. For high purification of the chromophor a system for
HPLC was developed: a solvent mixture of acetonitrile and methanol (75/25)
was superimposed by a multilinear gradient of water content in the solvent
starting with 25% H_2O down to 5% after 10 min. Within the next 30 min
the water content is diminished to 0%, followed by a regeneration phase of
10 min to reach the starting point of 25% H_2O again.
For neutron activation analysis highly pure samples were sealed in quartz
ampules after evaporation and exposed to a neutron flux of $4 \cdot 10^{12}$ n/sec cm^2
for 30 min to form ^{38}Cl, if present: samples were a dummy (no content), 20.8μg
methylphaeophorbide a, 22.4 μg 20-Cl-methylphaeophorbide a, 1.5 μg NaCl and
30 μg of Chl RC I. The experiment was carried out in the Forschungsreaktor
of the University in Mainz and kindly supported by Dr. Trautmann and coworkers. The fluorescence life time measurements were carried out by
Dr. Holzwarth and coworkers in the MPI für Strahlenchemie in Mühlheim/Ruhr.
The ^1H-NMR-measurements were performed with a 400 MHz Brucker instrument
by Dr. Berger and coworkers in the Fachbereich Chemie of the University in

Sybesma, C. (ed.), Advances in Photosynthesis Research, Vol. II. ISBN 90-247-2943-2.
© 1984 Martinus Nijhoff/Dr W. Junk Publishers, The Hague/Boston/Lancaster.

Marburg. The solvent was d_6 -acetone (〉99,96% deuterated).

3. RESULTS AND DISCUSSION

From the differences in absorption- and fluorescence data as well as in chemical behavior, published in previous papers (Dörnemann and Senger, 1981a, b; 1982) it is obvious that the differences are due to a different structure of Chl RC I. A first hint to this problem was the fact, that absorption data of methylphaeophorbide RC I, especially the interchanged intensities of the two typical phaeophytin peaks in the region between 500 and 550 nm, corresponded with the data of 20-Cl-methylphaeophorbide a as well as the CD-spectra, both kindly provided by Dr. H. Scheer, Univ. of Munich. A substitution in the 20-position of the chromophore should be clearly seen by the disappearance of the δ(20)-proton signal in the ^{1}H-NMR spectrum. We could clearly demonstrate that both in Chl RC I and in methylphaeophorbide RC I no δ-proton signal appears in the spectrum (Fig. 1). This suggests that the δ(20)-proton should be substituted by a group or atome giving no ^{1}H-NMR signal, for example chlorine. The data are confirmed by the results of H. Scheer (also published in this congress), who could as well show the substitution at C_{20} by a non ^{1}H-NMR active group as also an additional change in the 13^2-position. Scheer proposed from the molecular weight and other data that the second change is the replacement of a proton by an OH-group.

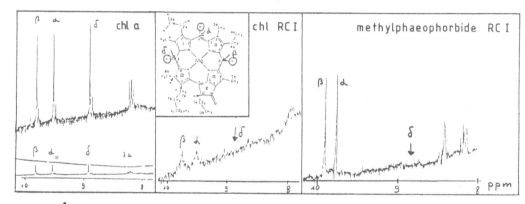

Fig. 1. ^{1}H-NMR-spectra of Chl a, Chl RC I and methylphaeophorbide RC I. Only the low field region is shown, clearly indicating the disapperance of the δ(20)-proton.

We reported earlier (Dörnemann and Senger, 1982) a molecular mass increase of 51-55 units for Chl/RC I compared to Chl a. By repeated experiments it could be shown that Chl RC I did not contain water, but that only Chl a under the conditions of plasma desorption massspectrometry (PDMS) undergoes the addition of a 15 mass unit breakdown product with an M^+-ion of 909 units.
Due to calibration of the PDMS-apparatus which is done by computer extrapolation of the curve from lower masses, molecular weights turned out to be about 2 mass units lower than published earlier (Dörnemann and Senger, 1982). Our results are now getting close to the data obtained by Scheer (pers. comm.) with several methods, who finds a mass increase of 50 units (Fig. 2). This would nicely fit with the findings of the substitution of 2 protons by a chlorine atome and an OH-group.

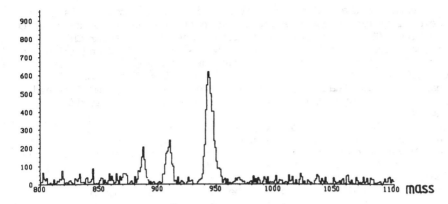

Fig. 2. PDMS-spectrum of enriched Chl RC I (only extended M^+-peak region) showing also peaks of Chl a-add. and phaeo a -add.

To indicate the presence of chlorine in the molecule a neutron activation analysis was performed. The results show clearly that one atom of chlorine per molecule of chromophore is present. For calibration 20- Cl- pyromethyl phaeophorbide a, NaCl, methylphaeophorbide a and a dummy (empty quartz ampoule) were also measured to give 0 and 100% value. (Fig. 3). Labelling experiments with ^{36}Cl, administered as Na ^{36}Cl to chloride-starved cultures of Scenedesmus mutant C-2A' during greening and also just before extraction to the pellet of algae and to chloride-starved C-6E during cultivation in the dark and again to the algae pellet just before extraction did not yield clear results, especially a too low content of label, which might be due to non-optimum experimental conditions.
Further experiments will have to be done to get significant label.

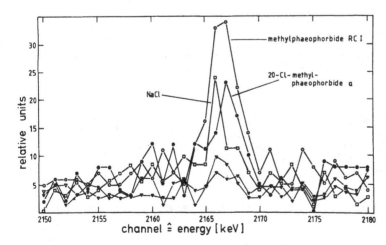

Fig. 3. Neutron activation analysis data: dummy▼—▼, methylphaeophorbide a▽—▽, NaCl□—□, 20-Cl pyromethylphaeophorbide a, ●—● methylphaeophorbide RC I. ○—○

Some more evidence for the presence of chlorine in the C_{20}-position is given by fluorescence life time measurements, kindly performed by Dr. Holzwarth from the MPI at Mühlheim/Ruhr. These data gave life times of 4.3 ns for Chl a, 4.49 ns for phaeophytin a, but 1.85 ns and 1.50 ns for Chl RC I and its phaeophorbide and 2.51 ns for 20-Cl-pyromethylphaephorbide a and thus suggest a more close relationship of the 20-Cl derivative to Chl RC I than to Chl a. All these data point out that the structure of the isolated Chl RC I should be a 20-chloro-13^2-hydroxy-chlorophyll a, as proposed by H. Scheer during this meeting. (Tab. 1)

The presented data, however, cannot exclude that the chlorine enters the molecule during extraction when the natural surroundings of the chromophores are destroyed. Nevertheless Chl RC I must then derive from another chlorophyll species than the bulk Chl a, as it could be shown that increasing amounts of Chl a being present during extraction did not increase the isolated Chl RC I amount. The amount depended only on the type of organism or particle used for the determination. Thus we conclude that Chl RC I can on the one hand be present in vivo in the proposed structural form or on the other hand is an isolation artefact which derives from an in vivo very reactive Chl species in the reaction centre of PS I.

We hope to get more information about the in vivo role of Chl RC I by in vivo and in vitro ENDOR and fluorescence life time measurements.

sample	fluorescence life time(ns)
Chl a	4.30
phaeo a	4.49
20-Cl-pyro- methylphaeo a	2.51
Chl RC I	1.85
phaeo RC I	1.50

Tab. 1. Flurorescence life times of chlorophylls and derivatives (performed by Holzwarth, MPI Mühlheim/Ruhr.)

4. REFERENCES

D. Dörnemann, H. Senger (1981a): proceedings of the Fifth International Photosynthesis Congress, G. Akoyunoglou ed. Vol. V, 223-231
D. Dörnemann, H. Senger (1981b): FEBS Letters 126 (2), 323-327
D. Dörnemann, H. Senger (1982): Photochem. Photobiol. 35, 821-826
H. Scheer et al., this meeting.

This contribution was kindly supported by the Deutsche Forschungsgemeinschaft

STRUCTURE OF A CHLOROPHYLL-RC1

H. Scheer, H. Wieschhoff, Botanisches Institut, Universität
München; W.Sch.aefer, Max-Planck-Institut Biochemie, Martinsried;
E. Cmiel, B. Nitsche, Institut für Physikalische Chemie, TU
München; H.-M. Schiebel, Institut für Organische Chemie, TU
Braunschweig, H.-R. Schulten, Fachhochschule Fresenius, Wiesbaden

INTRODUCTION

The primary donor in the reaction center of photosystem I (P 700)
has a red-shifted absorption maximum as compared to the bulk of
the antenna pigments. P7oo is generally believed to be one or a
pair of chlorophyll a molecule(s). The recent isolation of a new
chlorophyll (chlorophyll-RC1) which has been related quantiatively
to the content of P700 in a variety of preparations (1) has
raised the possibility that a structurally different molecule may
be responsible for this function. Chlorophyll-RCI had originally
been isolated from Scenedesmus, and subsequently also from a
cyanobacterium and from spinach (2). Following a gift of a
chlorophyll-RC1 sample from Scenedesmus to us by Doernemann
and Senger, we have also isolated a similar pigment from a
different cyanobacterium, Spirulina geitleri and converted it to
its methylphophorbide(s). Here, we wish to report its molecular
structure as 13^2(R)-Hydroxy-20-chloro-17(S), 18(S)-Methyl-
pheophorbide a (structure 1), which corresponds structure 2 for
chlorophyll-RC1.

Methylpheophorbide-RC1 Preparation

Spirulina geitleri (100 gms) was extracted with cold methanol,
and the extract demetalated with dilute hydrochloric acid under
nitrogen. The crude pheophytins were chromatographed on silica
60 (Merck) with carbontetrachloride/acetone = 96:4. The fractions
containing pheophytin a were transesterified with methanol/
sulfuric acid under nitrogen. The resulting methylpheophorbides
were chromatographed first on a silica 60 (Merck) column with
methylene chlorid/acetone = 96:4, then on silica (Merck) thin
layer plates with carbontetrachloride/acetone = 90:10, and finally by reverse-phase HPLC on lichrosorb-RP 18 (Merck) with
methanol. Two pure fractions (FI, FII) with essentially
identical absorption spectra (Fig. 1) were obtained in a yield
of 36 and 5 µg, respectively. Both are unstable in light and
convert to two well defined products with higher and lower Rf
values, respectively, than the parent compound. The structural
data given below were obtained with FI of methylpheophorbide -
RC1.

ABSORPTION AND CIRCULAR DICHROISM

The absorption spectrum of methylpheophorbide -RC1 (FI) is
red-shifted with respect to that of methylpheophorbide a. The
only other significant difference is the inverted intensity
ratio of the two minor absorptions in the wavelength range
between 500 and 550 nm. The only group of chlorophylleus pigments
showing these spectral features all have a substituent at the

Sybesma, C. (ed.), Advances in Photosynthesis Research, Vol. II. ISBN 90-247-2943-2.
© *1984 Martinus Nijhoff/Dr W. Junk Publishers, The Hague/Boston/Lancaster.*

C-20 methine bridge. The Cd-spectrum of FI is red-shifted as well
but otherwise similar to that of methylpheophorbide a (structure
3) with respect to the sign of the major bands (fig. 2). This
indicates the same absolute S-configuration at the asymmetric
centers, C-17 and C-18. Also present is the intense negative
band characteristic for the 13^2 - carbomethoxy group in the
13^2 (R) configuration (3). A further significant difference is
the relatively large ellipticity of the red as compared to the
soret-band, which is typical for steric hindrance due to methine
substituents (3).

MASS SPECTRA

The mass spectrum of methylpheophorbide -RC1 (FI) has a molecular
ion at 656 mass units. This corresponds to an increase of 50
mass units as compared to the molecular ion of methylpheophorbide
a. The fragmentation of the two pigments in the electron impact
spectrum is similar. The (M+2) ion is relatively large. Since
hydrogeation-dehydrogenation processes are well known for tetrapyrrole mass spectra (4), a field-desorption spectrum was taken
of the same pigments. The molecular ion of FI at 656 as well as
the intense ion at 658 mass units are observed in these spectra,
too (fig. 3). The observed peak pattern fits well to the
structural formula $C_{36}H_{37}N_4O_6Cl$. (0.7 % error), corresponding to
the addition of one atom of oxygen and one atom of chlorine to
methylpheophorbide a.

PROTON NMR SPECTRA

The proton nmr spectrum of methylpheophorbide -RC1 (FI) was also
quite similar to that of methylpheophorbide a (fig. 4). There
is, however, no signal in the range of δ = 8-9 ppm, where the
C-20 proton of pheophorbides is generally observed (5). Since
there is no additional signal in the low field range, there must
be an nmr inactive substituent at C-20. This is supported by
smaller but distinct shifts of one aromatic methyl signal (probably C-2) and the signals of the C-18 substituents, which are
neigbers to C-20. The only other distinct difference is the high
field shift of the singlet assigned to the proton at C-13^2. A
similar shift has been reported for an allomer of bacteriochlorophyll a bearing a hydroxy group rather than a proton at the 13^2
position (6).

DISCUSSION

Based on the spectroscopic data presented above, the methylpheo-
phorbide-RC1 (FI) is 13^2-hydroxy-20-chloro- 17(S), 18(S)
methylpheophorbide a (structure 1), which corresponds to
structure 2 for chlorophyll -RC1. Absorption and circular
dichroism reflect the new substituent at C-20, which is proved
by the mass spectra to be a chlorine atom. Sign and magnitude of
the nmr shifts in the neighborhood of C-20 are comparable to
those reported by Hynninen et al. for 20-chloro-methylpheophor-
bide a (7). The heavy-atom effect of the chlorine would also
explain the decreased fluorescence observed by Doernemann and
Senger for chlorophyll -RC1 (2) and by us for its methylpheophor-

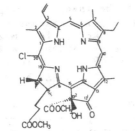

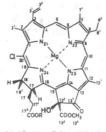

1: METHYLPHEOPHORBIDE - RC1 (F1)

2: CHLOROPYLL - RC1 (F1)

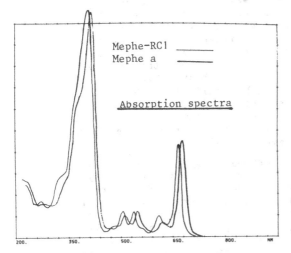

Mephe-RC1 _____
Mephe a _____

Absorption spectra

200.   350.   500.   650.   800.   NM

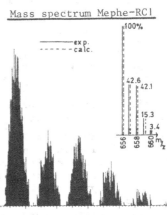

Mass spectrum Mephe-RC1

—— exp.
---- calc.

100%

42.6   42.1

15.3
3.4

656   658   660   m/z

Proton NMR spectrum of Methylpheophorbide-RC1

We thank the Bruker company for letting us use their 500 MHz spectrometer

— CHCl₃

2,7,12,13⁴,17⁶

HDO

8²

8¹

18¹

5,10   Hₓ   Hₐ H_B 13²   18   17   17¹ 17²

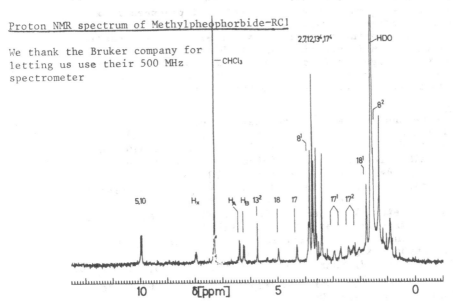

10   δ[ppm]   5   0

bide. The oxygen must then be located at the isocyclic ring, most likely as a hydroxy-substituent at C-13^2. An exchange of the two groups (OH at C-20, Cl at C-13^2) is unlikely, since meso-hydroxyporphyrins are generally present as the oxophlorin-isomers which have quite different absorption and NMR spectra (8). This structure would also explain the second pigment (FII) with an absorption similar to that of FI as the 13^2 (S)-isomer.

Both the substitution with chlorine at C-20 and the oxidation at C-13^2 are well known artifacts in chlorophyll chemistry (6, 7,9). The introduction of a chlorine atom during the conversion of chlorophyll -RC1 to its methylpheophorbide can be ruled out, because the product was obtained as well when any chlorine containing chemicals (chlorinatd hydrocarbons, hydrochloric acid, sodium chloride, etc) were omitted during the entire procedure. It is thus very likely that this substituent is already present in chlorophyll -RC1 in vivo and is responsible for the red-shifted absorption. It should be noted, however, that chloride ions are present in the photosynthetic membrane. Oxidation at C-13^2 is principally possible, too, but unlikely with the precautions taken by us. The ^{252}Cf plasma desorption mass spectrum of chlorophyll -RC1 (2) has a molecular ion indicative of one additional oxygen being present together with one chlorine atom already in the original pigment, too.

The structure 2 suggested for chlorophyll -RC1 is not readily compatible with its function as the primary donor of photosystem 1 (e.g. chemical reactivity, redox behavior (2)). There are at least two possible explanations for this apparent contradiction: firstly, this pigment could have another function e.g. as an electron acceptor or a special antenna pigment. Since all P700 preparations contain rather large amounts of additional (antenna ?) chlorophylls, this possibility cannot be ruled out. Secondly, chlorophyll -RC1 may indeed be an artifact produced from an unusually reactive chlorophyll (a) species (e.g. P700), which is formed even during the extraction under very mild conditions (cold methanol, nitrogen atmosphere). Since the primary photo-reactants are expected to show an unusual chemistry, this possibility cannot be ruled out either. Further work is now in progress to try and clarify these questions.

References
1. D. Dörnemann, H. Senger. FEBS Lett. 126, 323 (1981)
2. D. Dörnemann, H. Senger, Photobiol. 35, 821 (1982)
3. H. Wolf, H. Scheer, Ann. N. Y. Acad. Sci. 206, 549 (1973)
4. H. Budzikiewicz, in D. Dolphin(ed.) "The Porphyrins". Vol. III, chapter 9, Academic Press, New York, 1978
5. H. Scheer, J.J. Katz, in K. M. Smith (ed.) "Porphyrins and Metalloporphyrins", Elsevier, New York, 1975
6. R. G. Brereton, V. Rajanada, T. J. Blake, J.K.M. Sanders, D.H. Williams, Tetrahedron Lett. 21, 1671 (1980)
7. P.H. Hynninen, S. Lötjönen, Tetrahedron Lett. 22, 1845 (1981)
8. J.H. Fuhrhop, Angew. Chem. 86, 363 (1974)
9. R.B. Woodward, V. Skaric, J. Am. Chem. Soc. 83, 4676 (1961)

ACKNOWLEDGEMENTS: This work was supported by the Deutsche Forschungsgemeinschaft, Bonn.

PHOTOSYSTEM I   REACTION CENTER FROM CYANOBACTERIUM GREEN ALGA AND HIGHR
PLANTS.

RACHEL   NECHUSHTAI and NATHAN NELSON/ Department of Biology, Technion –
Israel Institute of Technology, Haifa 32000, Israel.

1. ABSTRACT

Photosystem I reaction center isolated from the thermophilic
cyanobacterium Mastigocladus laminosus was found to resemble that of
green algae, which also contains four different subunits. The molecular
weight, P$_{700}$ content and appearance on sodium dodecyl sulfate gels of
subunits I of photosystem I reaction center from higher plants, green
algae and cyanobacterium were found to be similar . Subunit I of
photosystem I reaction center from these three different sources shows
immunological cross reactivity. This is also true for subunit II of
photosystem I reaction center which is known to be a product of the
cytoplasmic translation system in green alga and higher plants.
The protein composition of leaves was followed during greening of
dark grown plants, using specific antibodies raised against the
individual subunits of some of the chloroplast protein complexes.
Peptides that are not directly involved in a photobiochemical reaction
were found to be present in etiolated leaves and their amount did not
change significantly during the greening process. On the other hand,
the synthesis of the different subunits of photosystem I reaction
center was found to be light induced. Subunit II of the complex is the
first to be synthesized in the light, and the synthesis of the other
subunits then follows sequentially. Subunit I of photosystem I reaction
center was found to be present in etiolated leaves and its amount was
only slightly increased during the first few hours of greening.

2. INTRODUCTION

Photosystem I reaction center has an indespensible role in the
process of photosynthesis. The photobiochemical function of this
complex enables the charge separation and electron transfer which
eventually cause the reduction of ferredoxin. Photosystem I reaction
center from higher plants consists of seven distinctive subunits
(Bengis, Nelson 1975; Bengis, Nelson 1977; Nechushtai et al. 1981),
while the same complex from green algae has only four subunits
(Nechushtai, Nelson 1981; Nechushtai et al. 1983). Recently,
photosystem I reaction center was isolated from cyanobacteria
(Takahashi, Katoh 1982; Nechushtai et al. 1983), and found to contain
four subunits with molecular weights of 70,000 (subunit I), 16,000
(subunit II), 11,000 (subunit III) and 10,000 (subunit IV) daltons
(Nechushtai et al. 1983). Similarity in molecular weight, pattern of
appearance in sodium dodecyl sulfate gels and P$_{700}$ content, was found
between subunits I of photosystem I reaction center isolated from
higher plants, green alga and cyanobacterium. Subunits I of photosystem
I reaction center from all three sources also show immunological cross
reactivity. The same phenomenon was observed for subunit II that is
known to be a product of the cytoplasmic translation system in green
alga and higher plants (Nechushtai, Nelson 1981; Nechushtai et al.
1981). This observation may suggest that several amino acid sequences

Sybesma, C. (ed.), Advances in Photosynthesis Research, Vol. II. ISBN 90-247-2943-2.
© *1984 Martinus Nijhoff/Dr W. Junk Publishers, The Hague/Boston/Lancaster.*

had been preserved upon transfer of the gene coding for subunit II from the prokaryotes to the nucleus of the eukaryotes.

The development of the photosynthetic apparatus had been intensively investigated in green alga and higher plants (Ohad et al. 1967; Bar - Nun et al. 1977; Oelze-Karow, Butler 1971; Baker, Butler 1976). On these investigations it was found that although plastid of etiolated leaves contain many of the chloroplast components, no photosynthetic electron transport and photophosphorylation are carried out (Bradbeer 1981). Recently, we studied the assembly of photosystem I reaction center during greening using specific antibodies raised against individual subunits of this complex (Nechushtai, Nelson 1983).

3. MATERIALS AND METHODS

Analytical Methods

Chlorophyll concentration was determined according to Arnon (1949). Electrophoresis in slab gels containing 12.5% acrylamide or an

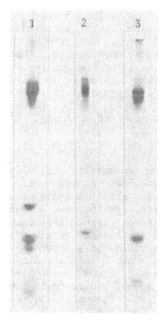

Figure 1: The polypeptide composition of photosystem I reaction center, isolated from Mastigocladus, Chlamydomonas and Swiss chard.

Photosystem I reaction centers of Chlamydomonas, Mastigocladus and Swiss chard were purified as described under Materials and Methods. The preparations were incubated for about 2 hours at room temperature with 2% SDS, and 2% β-mercaptoethanol. Dissociated samples containing about 30 μg of protein were electrophoresed on an exponential 10-15% acrylamide gel. Lane 1, Swiss chard; lane 2, Chlamydomonas; lane 3, Mastigocladus photosystem I reaction centers.

exponential gradient of 10-15% acrylamide was performed according to Douglas and Butow (1976). Proteins were transferred from gels to nitrocellulose papers by electrotransfer as previously described (Rott, Nelson 1981). The nitrocellulose papers were immunodecorated with specific antibodies and ^{125}I-protein A as described in (Rott, Nelson 1981; Nelson 1983). Antibodies against the individual subunits of photosystem I reaction center isolated from spinach or Swiss-chard were raised in rabbits as previously described (Nelson 1983).

Preparations

Photosystem I reaction centers from spinach, Swiss chard Chlamydomonas reinhardii, and Mastigocladus laminosus were purified as described in (Nechushtai et al. 1981; Nechushtai, Nelson 1981) and (Nechushtai et al. 1983) respectively. Greening experiments in dark grown spinach oat and bean plants were performed as previously described (Nechushtai, Nelson 1983).

4. RESULTS

Photosystem I reaction center had been isolated from higher plants (Swiss chard) green alga (Chlamydomonas) and cyanobacterium

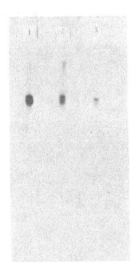

Figure 2: Immunological crossreactivity between subunits I of photosystem I reaction centers from higher plants, green alga and cyanobacterium.

Photosystem I reaction centers of spinach , Chlamydomonas and Mastigocladus were electrophoresed on an exponential 10-15% acrylamide gel. The proteins were electrotransfered to nitrocellulose paper , and the paper was immunodecorated with specific antibodies raised against subunit I of photosystem I reaction center from Swiss chard , and ^{125}I-protein A as described under Materials and Methods. Lane 1, photosystem I reaction center of spinach equivalent to 0.1 µg of chlorophyll; lane 2, photosystem I reaction center of Chlamydomonas equivalent to 0.3 µg of chlorophyll; lane 3, photosystem I reaction center of Mastigocladus equivalent to 0.4 µg of chlorophyll.

(Mastigocladus). The polypeptide composition of the purified reaction centers from these three different sources is shown in Fig. 1. Mastigocladus photosystem I reaction center resembles that of the green algae Chlamydomonas and Dunaliella in several respects. These preparations contain only four subunits, while spinach photosystem I reaction center consists of seven subunits. In all four species the molecular weights of subunits I are 70,000 daltons and they have very similar appearance on sodium dodecyl sulfate gels. The immunological cross reactivity between subunits I of photosystem I reaction center from Mastigocladus, Chlamydomonas and spinach is established in Fig. 2. Fig. 3 shows that the same phenomenon can be detected for subunits II of photosystem I reaction centers isolated from the three different sources.

Subunit II of photosystem I reaction centers from green alga (Nechushtai, Nelson 1981) and higher plants (Nechushtai et al. 1981) is a product of the 80s ribosomal translation system. Fig. 4 demonstrats the effect of chloramphenicol and cycloheximide on the synthesis and assembly of the individual subunits in photosystem I reaction center.

The amount of individual subunits of some of the chloroplast protein complexes was followed during greening of dark grown plants. Using the immunodecoration technique (Rott, Nelson 1981; Nelson 1983) it was found that many of the chloroplast protein complexes were

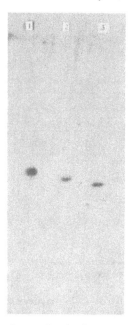

Figure 3: Immunological crossreactivity between subunits II of photosystem I reaction center from spinach, Chlamydomonas and Mastigocladus.

The conditions were as described in the legend to Fig 2. The immunodecoration was carried out with antibodies raised against subunit II of photosystem I reaction center from Swiss chard. Lane 1, spinach; lane 2, Chlamydomonas; lane 3, Mastigocladus.

present in the etiolated leaves, and their amounts did not change
significantly after exposing the plants to light. Fig. 5 shows two
representatives of this category of proteins, the β subunit of the
chloroplast coupling factor and subunit I of the cytochrome b_6-f
complex during greening of oat leaves.The same method was applied to
follow the development of photosystem I reaction center in spinach
leaves. Fig. 6 shows that while subunit I of photosystem I reaction
center is present in etiolated leaves and its amount is only slightly
increased during greening, the synthesis of the other subunits is
induced by light. The synthesis of subunit II starts 2 hours after the
plants were exposed to light and a lag period of 4 hours is observed
between the exposure to light and the detection of subunit III. Subunit
IV may be detected only after 14.5 hours of greening.The quantitative
analysis shown in Fig. 7 reveals that after two days of greening, the
amount of the different subunits of photosystem I reaction center
reaches the level of these subunits in green plants that were grown
under continuous light.

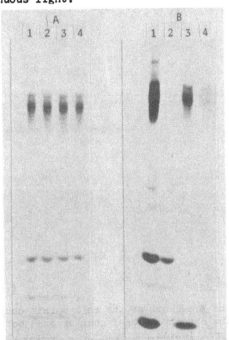

Figure 4: Effects of chloramphenicol and cycloheximide on the
labeling of the different subunits of photosystem I reaction center.
Chlamydomonas cells were labeled with [^{35}S]-sulfate in the
presence or absence of protein-synthesis inhibitors as described
(Nechushtai, Nelson 1981). Photosystem I reaction center was then
isolated and applied on SDS-polyacrylamide gels,that were stained with
coomassie blue (A), or exposed to X-ray film (B). The complex was
isolated from labeled untreated cells (lane 1) and from cells labeled
in the presence of 2mM chloramphenicol (lane 2), 100μg/ml cycloheximide
(lane 3), and 100 μg/ml cycloheximide for 1 hour incubation prior to
the labeling (lane 4).

5. DISCUSSION

Photosystem I reaction center, that functions on the reducing side of the electron transport chain, may have evolved from green photosynthetic bacteria (Olson 1970). Cyanobacteria which are the only free-living prokaryotes containing photosynthetic systems of the higher plants type, played a central role in the evolution of the complex (Padan 1979).

This work shows that photosystem I reaction center isolated from Mastigocladus laminosus resembles that of green algae and higher plants. As in green algae and higher plants, the purified photosystem I reaction center contains about 40 molecules of chlorophyll a per one molecule of P_{700} (Bengis, Nelson 1975; Bengis, Nelson 1977; Nechushtai, Nelson 1981; Nechushtai et al. 1983). The "heart" of the complex is subunit I, which contains the P_{700} in all three isolated photosystem I

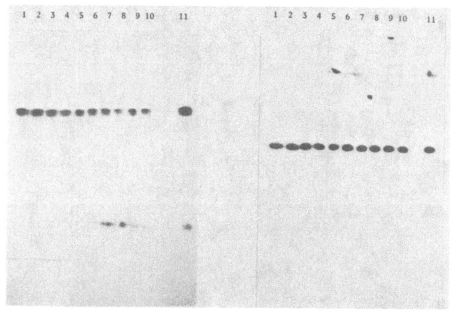

Figure 5: The amount of the β subunit of the chloroplast coupling factor and subunit I of the cytochrome b₆-f complex does not change during greening of etiolated oat leaves.

Oat plants grown in the dark for 7 days were exposed to light. Samples of leaves were taken during the greening process and the proteins were extracted as described under Materials and Methods. Left; immunodecoration with antibodies raised against subunit β of the chloroplast coupling factor. Right; immunodecoratin with antibodies raised against subunit I of the cytochrome b₆-f comlex. Exposure of leaves to light was as follows: 1) 0 min (etiolated leaves), 2) 30 min , 3) 60 min , 4) 105 min , 5) 3.5 h , 6) 5 h , 7) 8 h , 8) 14 h , 9) 22.5 h , and 10) 40 hours. Lane 11 represents immunodecoration of a sample taken from oat plants grown for 7 days in the light.

reaction centers. The stoichiometry of subunit I and the other subunits in the complex from all three sources is 2 to 1. This, and the fact that immunological cross-reactivity between subunits I of photosystem I reaction centers from cyanobacterium, green alga and higher plants exists, implies that essential amino acid sequences in the structure of subunit I were maintained during the evolution from prokaryotes to the chloroplasts of higher plants.

While subunit I of the various photosystem I reaction centers have apparently the same molecular weight, there are variations on that of subunit II. However, immunological cross-reactivity between subunits II from Mastigocladus, Chalmydomonas and spinach can be detected (Fig. 3).As was demonstrated in Fig. 4, this subunit is a product of the cytoplasmic translation system in green alga and higher plants (Nechushtai et al. 1981). Fig. 4 also shows that a one hour inhibition of subunit II synthesis (lanes 4),prevents the assembly of the entire photosystem I reaction center complex. This may suggest that subunit II serves as template for the assembly of the complex (Nechushtai et al. 1981; Nechushtai, Nelson 1983).

During the greening process major changes occur in the chloroplast structure and function (Bradbeer 1981). During this process the amount of different chloroplast protein complexes was followed using specific antibodies. Subunits of the chloroplast protein complexes which are not directly involved in a photobiochemical reaction accumulate in etiolated leaves (Bradbeer 1981; Nechushtai, Nelson 1983), so their amount did not change during greening (Fig. 5). The synthesis of all the subunits (except subunit I) of photosystem I reaction center, which does participate in a photobiochemical reaction, is induced by light.

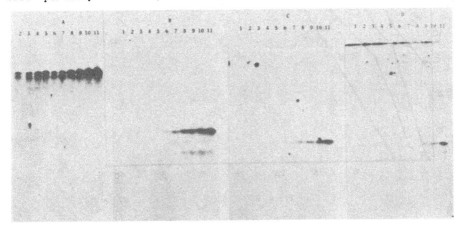

Figure 6: The light induced synthesis of the different subunits of spinach photosystem I reaction center.

Greening of etiolated spinach leaves and protein extraction were performed as described under Materials and Methods. Immunodecoration was carried out with antibodies raised against subunits I (A), II (B), III (C) and IV (D) of photosystem I reaction center. Samples were taken from etiolated leaves (1) and after exposing the leaves to light for 30 min (2), 60 min (3), 90 min (4), 120 min (5), 240 min (6), 6 h (7), 8 h (8), 14.5 h (9), 21.5 h (10) and 40 hours (11).

As pointed out before, subunit I is the "heart" of photosystem I, and therefore its presence in etiolated leaves ensures that electron transport would start as soon as the leaves are exposed to light (Oelje-Karow, Butler 1971; Baker, Butler 1976; Bradbeer 1981; Nechushtai, Nelson 1983). Subunit II synthesis is light induced, and the fact that its appearance can be detected prior to that of the other subunits, may support the hypothesis of subunit II serving as a template for the complex assembly.

The evidence presented in this work may elucidate the pattern of the evolutionary development, biogenesis, and assembly of photosystem I reaction centers.

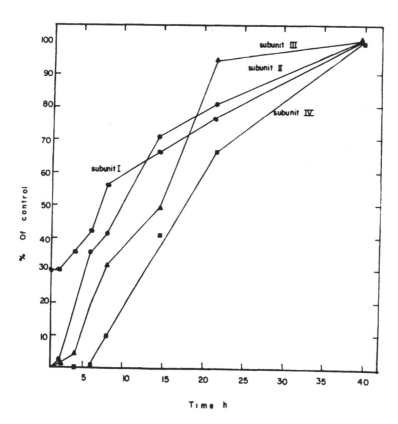

Figure 7: The sequence of appearance of the different subunits of photosystem I reaction center during the greening of etiolated spinach leaves.

Proteins were extracted from the leaves as described under Materials and Methods. Following electrophoresis and immunodecoration the nitrocellulose papers were used for quantitative determination of the different subunits as described in (Nechushtai, Nelson 1983).

6. REFERENCES

Arnon DI (1949) Copper enzymes in chloroplasts. Polyphenol oxidases in Beta vulgaris, Plant Physiol. 24, 1-15.

Baker NR and Butler WL (1976) Development of the primary photochemical apparatus of photosynthesis during of etiolated bean leaves, Plant Physiol. 58,526-529.

Bar-Nun S, Schantz R and Ohad I (1977) Appearance and composition of chlorophyll-protein complex I and II during chloroplast membrane biogenesis in Chlamydomonas reinhardi Y-1, Biochem. Biophys. Acta 459, 451-467.

Bengis C and Nelson N (1975) Purification and properties of the photosystem I reaction center from chloroplasts, J. Biol. Chem. 250, 2783-2788.

Bengis C and Nelson N (1977) Subunit structure fo chloroplast photosystem I reaction center, J. Biol. Chem. 252, 4564-4569.

Bradbeer JW (1981) Development of photosynthetic function during chloroplast biogenesis. In Hatch MD and Boardman NK eds. The biochemistry of plants, Vol. 8, pp 423-472. Academic Press, New-York, London, Toronto, Sydney, San Francisco.

Douglas M and Butow RA (1976) Variant forms of mitochondrial translation products in yeast: evidence for location of determinants on mitochondrial DNA, Proc. Natl. Acad. Sci. USA 73, 1083-1086.

Nechushtai R and Nelson N (1981) Purification properties and biogenesis of chlamydomonas reinhardi photosystem I reaction center, J. Biol. Chem. 256, 11624-11628.

Nechushtai R and Nelson N (1983) Biogenesis of photosystem I reaction center during greening of oat, bean and spinach leaves, Plant Physiol. submitted.

Nechushtai R, Nelson N, Mattoo A and Edelman M (1981) Site of synthesis of subunits to photosystem I reaction center and the proton-ATPase in spirodela, FEBS Letts. 125, 115-119.

Nechushtai R, Muster P,Binder A, Liveanu V and Nelson N (1983) Photosystem I reaction center from the thermophilic cyanobacterium mastigocladus laminosus, Proc. Natl. Acad. Sci. USA 80, 1179-1183.

Nelson N (1983) Structure and synthesis of chloroplast ATPase, Methods Enzymol. 97, 510-523.

Oelze-Karow H and Butler WL (1971) The development of photophosphorylation and photosynthesis in greening bean leaves, Plant Physiol. 48, 621-625.

Ohad I, Siekevitz P and Palade GE (1967) Biogenesis of chloroplast membranes, J. Cell Biol. 35, 521-584.

Olson JM (1970) The evolution of photosynthesis, Science 168, 438-446.

Padan E (1979) Facultative anoxygenic photosynthesis in cyanobacteria. Ann. Rev. Plant Physiol. 30, 27-40.

Rott R and Nelson N (1981) Purification and immunological properties of proton-ATPase complexes from yeast and rat liver mitochondria, J. Biol.Chem. 256, 9224-9228.

Takahashi Y and Katoh S (1982) Functional subunit structure of photosystem I reaction center in synechococcus sp., Arch. Biochem. Biophys. 219, 219-227.

EVIDENCE THAT CP 47 (CPa-1) IS THE REACTION CENTRE OF PHOTOSYSTEM II

BEVERLEY R. GREEN AND EDITH L. CAMM, UNIVERSITY OF BRITISH COLUMBIA, VANCOUVER, B.C., CANADA, V6T 2B1

SUMMARY
Two lines of evidence support the identification of the chlorophyll a-protein complex CPa-1 (CP 47) as the reaction centre core complex of Photosystem II (PS II). When PS II core particles are prepared by octyl glucoside extraction and sucrose gradient sedimentation, maximum PS II activity (DPC $\rightarrow$ DCPIP) coincides with the peak of CPa-1, and does not coincide with that of any other chlorophyll-protein complex. Secondly, the complex CPa-1 is the only complex missing from the corn mutant hcf*-3, which is totally lacking in PS II activity.
   We propose that the other chlorophyll-protein complex associated with reaction centre preparations, CPa-2 (CP 43) may be an internal antenna complex. A model is presented suggesting a possible organization of these complexes in the core of PS II, associated with the minor Chl a+b complex CP 29 and a variable number of units of the light-harvesting complex (LHCP).

1. INTRODUCTION

A number of different chlorophyll-protein complexes have been isolated from higher plants in the last few years. Two of these complexes, CPa-1 (CP 47) and CPa-2 (CP 43) (Camm, Green, 1980) are thought to be part of the core of Photosystem II (Machold et al, 1979; Delepelaire, Chua, 1979). Each of these complexes comprises only a few percent of the total chlorophyll (Chl), contains only Chl a, and has a fluorescence emission maximum red-shifted compared to the Chl a+b complexes of PS II (Green et al, 1982; Camm, Green, 1980). In this paper, we present evidence that one of these complexes, CPa-1, is the reaction centre complex of PS II, and suggest that the other Chl a complex, CPa-2, as well as the Chl a+b complex CP 29, are internal antenna complexes in PS II. A preliminary report of this work is in press (Camm, Green, 1983).

2. MATERIALS AND METHODS

Corn was grown in a greenhouse; spinach was purchased in a local market. The corn mutant hcf*-3 was isolated and characterized in the laboratory of Dr. C.D. Miles, who kindly made the seeds available (Leto, Miles, 1980). Recessive mutant segregants were identified by their high chlorophyll fluorescence under long-wavelength UV light. Chloroplasts were isolated and washed thylakoids prepared as described (Camm, Green, 1980). To display total chlorophyll-protein complexes, corn thylakoids were washed with cold 2 mM Tris-maleate, pH 8.0, solubilized with a mixture of 2% octyl glucoside and 0.5% SDS to give detergent/Chl ratios of 40 and 10 respectively and electrophoresed on 10% polyacrylamide gels in the presence of 0.1% SDS (Camm, Green, 1980).
   For PS II reaction centres: spinach thylakoids were first suspended in 100 mM sorbitol, 50 mM tricine-NaOH, pH 7.6, 10 mM NaCl, 0.5 mM MgCl$_2$, incubated for 15 min, and pelleted at 10,000 x g. The pellet was extracted twice with 30 mM octyl glucoside in 2 mM Tris-maleate, pH 8.0. The first extract was discarded, and the second extract was layered on a

Sybesma, C. (ed.), Advances in Photosynthesis Research, Vol. II. ISBN 90-247-2943-2.
© *1984 Martinus Nijhoff/Dr W. Junk Publishers, The Hague/Boston/Lancaster.*

10-30% sucrose gradient containing 30 mM octyl glucoside, 0.75 mM EDTA
and 2 mM Tris-maleate, pH 8.0. After centrifugation at 110,000 x g for
16 hr at 4°, the gradient was fractionated and fractions assayed for Chl
content and Chl a/b ratio (Arnon, 1949); for PS II activity by measuring
the light-driven reduction of diphenylcarbazide (DPC) by 2,6-dichloro-
phenolindophenol (DCPIP) (Vernon, Shaw, 1969); and for amounts of chlorophyll-protein complexes by SDS-polyacrylamide gel electrophoresis
(Camm, Green, 1980). Unstained gels were scanned at 680 nm with a
Helena R and D densitometer and the amounts of Chl in each complex calculated from the area under the peaks.

3. RESULTS

3.1. Distribution of chlorophyll-protein complexes and PS II activity on sucrose gradients.

The distribution of PS II activity and various complexes across the gradient in a typical experiment is given in Fig. 1. The peak of total Chl
is in the fraction at 11% sucrose, which has a Chla/b ratio of 2.1, characteristic of a high concentration of LHCP (Fig. 2a). The fractions with
high PS II activity have a/b ratios of 5-6, indicating the absence of Chl
b-containing complexes. These fractions contain only the two minor Chl a
complexes CPa-1 and CPa-2 (Fig. 2b). As can be seen in Fig. 1, the peak
of PS II activity coincides with the peak of CPa-1 but not CPa-2. This
strongly suggests that CPa-1 and not CPa-2 is the reaction centre complex
of PS II.

3.2. Photosystem II-defective mutant hcf*-3

In the homozygous recessive condition, the corn mutant hcf*-3 has no
detectable PS II activity, although its PS I activity is near normal
(Leto, Miles, 1980). Seedlings are able to survive on their seed

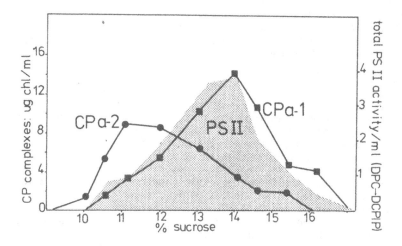

FIGURE 1. Distribution of chorophyll-protein complexes and total PS II
activity along sucrose gradient. PS II activity/ml=µmoles DCPIP
reduced/hr by 1 ml of gradient fraction(Shaded area)

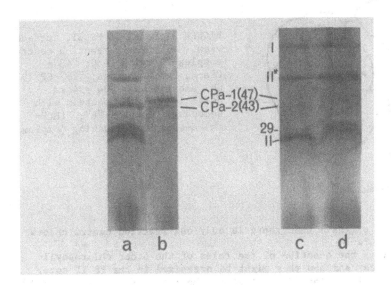

FIGURE 2. SDS gel electrophoresis of chlorophyll-protein complexes (unstained). (a) and (b) fractions from sucrose gradient applied directly to gel: (a) 11% sucrose, (b) 14% sucrose. (c) and (d) thylakoids solubilized with SDS and octyl glucoside: (c) hcf*-3 mutant phenotype (d) normal phenotype.

reserves for about 14 days, during which period they synthesize both Chl a and b, and look quite normal, although growing slightly more slowly than normal siblings (Miles 1980).

Fig. 2c and d show the chlorophyll-protein complexes from mutant and normal segregants of hcf*-3. The normal membranes (d) contain the full complement of complexes: CP I, the reaction centre complex of PS I; CP II* and CP II, the oligomer and monomer forms of the LHCP; CPa-1; CPa-2; and CP 29, the minor Chl a+b complex associated with PS II. Membranes from plants with the mutant phenotype contain all the other complexes but contain little or no CPa-1. It is particularly interesting that they do contain CPa-2, since this complex has also been proposed as part of the reaction centre. This evidence also supports our contention that CPa-1 is the PS II reaction centre complex.

4. DISCUSSION

Both lines of evidence given above support our proposal that CPa-1 (CP 47) is the reaction centre chlorophyll-protein complex of PS II. Only CPa-1 and CPa-2 are found in the PS II-active fractions, and only the distribution of CPa-1 parallels the activity distribution. There is, of course, the possibility that CPa-2 also has PS II activity but has become partially separated from some unknown cofactor on the gradient. This possibility cannot be rejected out-of-hand, but in view of the presence of CPa-2 in hcf*-3, it is not particularly likely. The evidence from

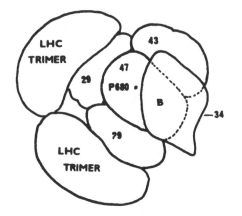

FIGURE 3. Model of PS II, surface view. "47"=CPa-1, reaction centre complex carrying P680; "43"= CPa-2, Chl a antenna; "29"=CP 29, Chl a+b antenna; "34"=34 kd herbicide-binding protein with secondary acceptor "B"; "LHC"= Chl a+b Light-Harvesting Complex.

hcf*-3, in fact suggests that there is only one reaction centre chlorophyll-protein, CPa-1.

This raises the question of the roles of the other chlorophyll-protein complexes, and how they might be organized in the PS II core. For purposes of discussion, we propose the model in Fig. 3, which shows a view looking down on the outside surface of a thylakoid. The O_2-evolving system is hidden behind the plane of the paper, on the inside thylakoid surface. We propose that CP 29 and CPa-2 (43) are antenna complexes associated with the reaction center CPa-1 (47) and represent them in their approximate ratio 2:1:1 (Green et al, 1982). The 34 kd herbicide-binding protein, thought to be the apoprotein of the two-electron carrier "B", partially covers CPa-1 as "proteinaceous shield". The basic LHC unit may be a trimer made up of two CP II_1 and one CPII$_2$ chlorophyll-proteins (Camm, Green, 1982). LHC units are shown toward the outside to allow for flexibility in number of units/core, and next to CP 29 because the two complexes are often extracted together. However, this is <u>not</u> meant to imply that energy transfer is LHC → CP 29 → CPa-1. There is no reason why the arrangement of complexes in three dimensions should not be much more asymmetrical than drawn, possibly putting LHC and CPa-1 in closer proximity.

REFERENCES
Arnon DI (1949) Plant Physiol. 24, 1-15
Camm EL and Green BR (1980) Plant Physiol. 66, 428-432
Camm EL and Green BR (1983) Biochem. Biophys. Acta, in press
Delepelaire P and Chua NH (1979) Proc. Nat. Acad. Sci. U.S.A. 76, 111-115
Green BR and Camm EL (1982) Biochim. Biophys. Acta 681, 256-262.
Green BR, Camm EL, Van Houten J (1982) Biochim. Biophys. Acta 681, 248-255
Leto KJ and Miles CD (1980) Plant Physiol. 66, 18-24
Machold O, Simpson DJ, Moller BL (1979) Carlsberg Res. Comm. 44, 235-254
Miles CD (1980) in San Pietro A, ed. Meth. Enzymol. 69, pp 3-23, Academic, New York
Vernon LP, Shaw ER (1969) Plant Physiol. 44, 1645-1649

CHARACTERIZATION OF THE PHOTOSYSTEM II REACTION CENTER POLYPEPTIDE (CP47)

H.Y. NAKATANI, C.J. ARNTZEN AND Y. INOUE, MSU-DOE PLANT RESEARCH LABORATORY
AND RIKEN

1. INTRODUCTION

Chlorophyll fluorescence has been extensively utilized as an intrinsic
membrane probe to study the photochemistry of photosynthesis. The assignment of the various fluorescence emission bands to specific functional
complexes in the membrane has been attempted in many laboratories (Reviewed in: Breton, 1982, Butler, 1979; Satoh, this symposium). Two chlorophyll fluorescence emission bands at 685 and 695 nm (at 77°K) have
been identified to arise from photosystem II (PS II). In this paper we
provide evidence for the fact that these two emission maxima arise from
separate pigment-proteins within the PS II reaction center complex.
In 1979, Camm and Green introduced a procedure using octyl-glucopyranoside
to solubilize chloroplast membranes; allowing the electrophoretic separation of two PS II reaction center polypeptides designated CP47 and CP43
(corresponding to apparent apoprotein molecular weights of 47,000 and
43,000, respectively). We have extended these observations by examining
the excised CP47 and CP43 pigment-proteins via spectral characterization
and have identified CP47 as the polypeptide having the PS II reaction center components, P680 and pheophytin (Pheo). This assignment was based
upon the detection of absorption changes for these components in the isolated CP47 (Nakatani, 1983; Nakatani, et. al., 1984). In these studies
CP43 was assigned a light-harvesting role for the PS II reaction center.
The data presented herein are further analysis of the spectral properties
of CP47 and CP43.

2. MATERIALS AND METHODS

Photosystem II-enriched fractions were obtained by treatment of spinach
chloroplasts with octyl-β,D-glucopyranoside essentially as described by
Camm, Green, 1980. The detergent extract was then centrifuged at 110,000
xg for 30 min. the supernatant fraction, enriched in the chlorophyll
proteins associated with PS II, was concentrated against solid sucrose
and subjected to non-denaturing lithium-dodecylsulfate polyacrylamide
gel electrophoresis (LDS-PAGE) at 4°C. The two chlorophyll-a binding
proteins at 47 kDa and 43 kDa were usually analyzed directly from the
gel slices containing the pigment proteins. Small amounts of the pigment
proteins were obtained in soluble form by electroelution from the excised
gel slices to examine their low temperature absorption characteristics.
Low temperature absorption measurements were carried out using a Shimadzu
UV 3000 spectrophotometer fitted with a liquid nitrogen dewar. Fluorescence measurements were carried out with a Shimadzu Solar Radiant
Energy Spectrum Analyzer. A PS II core complex was prepared as
previously described (Satoh, et. al., 1983).

3. RESULTS AND DISCUSSION

The low temperature (77°K) fluorescence emission bands of the two
chlorophyll proteins are shown in Fig. 1. CP47 emits maximally at 695
nm whereas CP43 emits maximally at 685 nm. A number of reports had

Sybesma, C. (ed.), Advances in Photosynthesis Research, Vol. II. ISBN 90-247-2943-2.
© *1984 Martinus Nijhoff/Dr W. Junk Publishers, The Hague/Boston/Lancaster.*

previously assigned the 695 nm emission band to the PS II reaction center (Butler, 1979; Cho, Govindjee, 1970). Although the light-harvesting chlorophyll protein complex (LHC) has also under certain circumstances been known to have a fluorescence emission band at this wavelength (Mullet and Arntzen, 1980), the 685 or 695 nm bands in this study cannot be ascribed to the LHC since there is no cross-contamination of LHC in either CP47 or CP43 preparations (see Fig. 2 and Camm, Green, 1980). As shown in the excitation spectra for CP47 and CP43 in Fig. 2, energy transfer to the fluorescence emission bands at 695 and 685 nm, respectively, is observed to occur from carotenoids (likely β-carotene) and pheophytin, in addition to chlorophyll a.

An examination of the fluorescence emission maxima as a function of temperature is shown in Fig. 3. The emission band maximum for the CP43 complex remained essentially constant at 685 nm from 25°C to -196°C, whereas the emission band maximum for the CP47 complex was shifted from 685 to 695 nm from room temperature (25°C) to liquid nitrogen temperature (-196°C). We interpret these data to indicate that CP47 contains both antennae chlorophyll as well as P680 and Pheo. At room temperature, efficient bidirectional energetic coupling occurs among both the trap and antennae chlorophyll, with most of the emission occurring from the latter. As the temperature is lowered, the reaction center components are energetically isolated from the antennae (i.e., energy transfer is preferentially unidirectional to P680) and the fluorescence emission shifts predominantly to 695 nm. This interpretation

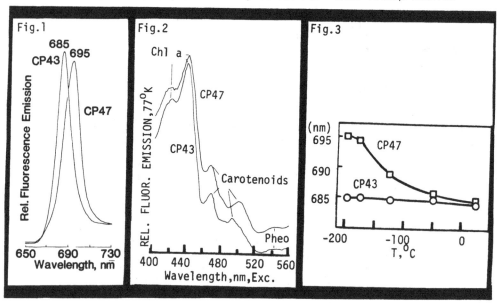

FIGURE 1. 77°K fluorescence emission spectra (4 nm Em. and Exc. Slits), excitation at 440 nm.
FIGURE 2. 77°K fluorescence excitation spectra (4 nm Exc. and Em. Slits).
FIGURE 3. Temperature dependence of maximum fluorescence emission band.

recognizes that CP47 contains the reaction center components, P680 and Pheo (Nakatani, 1983; Nakatani et. al. 1984) and is based upon the concept that it is the PS II trap, per se, which emits at 695 nm. Butler, 1979; Cho, Govindjee, 1970, see also Rijgersberg, et. al., 1979). The presence of a 680 nm absorbing species in CP47 was previously shown (Nakatani et. al., 1983) and is further substantiated by an examination of the 4th derivative of its low temperature (77°K) absorption spectrum; see Fig. 4. In a comparison with a PS II core complex (Satoh, et. al.,

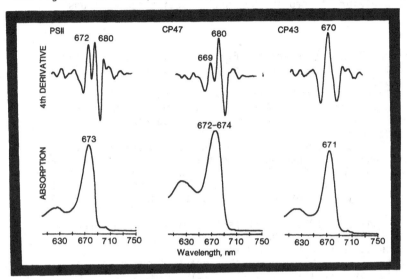

FIGURE 4. 77°K absorption and 4th derivative spectra. 4th derivative obtained at 3.6 nm increments on Shimadzu UV 3000 spectrophotometer.

1983) and CP43, the major difference observed is the presence of a major absorbing species at 680 nm for both the CP47 and PS II core complex. This is consistent with the presence of the reaction center trapping species in both CP47 and the PS II core complex. A small shoulder at 680 nm in the 4th derivative spectrum of CP43 indicates the presence of some antennae chl a-680 in CP43. The absolute amount of P680 cannot be determined by this analysis; however, we do know that the antenna size of the PS II core complex is about 50 chl a (see Satoh et. al., 1983). Delepelaire and Chua, 1979, had previously estimated that the two polypeptides (CP47 and CP43) contained approximately 5 chlorophyll a per polypeptide; however, this would have to be an underestimate since we know that these complexes undergo photobleaching during sample preparation (Nakatani et. al., 1984).

4. SPECULATIONS

The 685 and 695 nm (77°K) fluorescence bands from PS II have been assigned to individual polypeptides, CP43 and CP47, respectively; absorption maxima are at 670 and 680 nm, respectively (77°K). Although the redox state of

Pheo controls th yield of the 695 nm fluorescence emission, Pheo, itself, may not be responsible for the emission band. From an examination of the absorbing species present in PS II, i.e., 670, 680 and 685 nm (Pheo), a longer wavelength emission band may be postulated for Pheo. Recent studies indicate that a longer wavelength fluorescence emission band is correlated with the presence of pheophytin (formed by degradation of the native chl) between 700-710 nm; a 703 nm emission band has been observed, see N. Fuad, this Symp. However, further studies are required to determine the behavior of Pheo in the PS II reaction center traps.

REFERENCES

Breton J (1982) The 695 nm fluorescence (F_{695}) of chloroplasts at low temperature is emitted from the primary acceptor of photosystem II, FEBS Lett. 147, 16-20

Butler WL (1979) Tripartite and bipartite models of the photochemical apparatus of photosynthesis. In CIBA Foundation Symp. 61, Chlorophyll Organization and Energy Transfer in Photosynthesis, pp 237-252. Amsterdam, Excerpta Medica

Camm EL and Green BR (1980) Fractionation of thylakoid membranes with the nonionic detergent octyl-β-D-glucopyranoside, Plant Physiol 66, 428-432

Delepelaire P and Chua NH (1979) Lithium dodecyl sulfate/polyacrylamide gel electrophoresis of thylakoid membranes at 4°C: characterizations of two additional chlorophyll a-protein complexes. Proc Natl Acad Sci USA 76, 111-115

Klimov VV Dolan E and Ke B (1980) EPR properties of an intermediary electron acceptor (pheophytin) in photosystem II reaction centers at cryogenic temperatures, FEBS Lett 112, 97-100

Mullet JE and Arntzen CJ (1980) Simulation of grana stacking in a model membrane - Mediation by a purified light-harvesting pigment protein complex from chloroplasts. Biochim Biophys Acta 589, 100-117

Nakatani HY (1983) Correlation of the low temperature 695 nm fluorescence emission with the reaction center of PS II (CP47). In Inoue Y, ed., Proc Intl Symp on Photosynthetic Water-oxidation and PS II Photochemistry, NY, Acad Press, in press

Nakatani HY Ke B Dolan E and Arntzen CJ (1984) Identity of the photosystem II reaction center polypeptide, submitted

Rijgersberg CP Melis A Amesz J and Swager JA (1979) Quenching of chlorophyll fluorescence and photochemical activity of chloroplasts at low temperature. In CIBA Foundation Symp 61, Chlorophyll Organization and Energy Transfer in Photosynthesis, pp 305-318. Amsterdam, Excerpta Medica

Satoh K Nakatani HY Steinback KE Watson J and Arntzen CJ (1983) Polypeptide composition of a photosystem II core complex: Presence of an herbicide binding protein. Biochim Biophys Acta, in press

ACKNOWLEDGMENTS
The author wishes to acknowledge helpful discussions with Drs. Govindjee and H. Koike and the financial support of the NSF (grant # PCM-8023031), DOE (contract # DE-ACO2-76ERO-1338) and STA, Japan for portions of this work.
Authors address: H.Y. Nakatani, C.J. Arntzen/MSU-DOE Plant Research Laboratory, Michigan State University, E. Lansing, MI, USA 48824 Y. Inoue, RIKEN, The Institute of Physical and Chemical Research, Wako, Saitama, 351, Japan

LIFETIME STUDIES BY PHASE FLUORIMETRY IN ISOLATED
LIGHT-HARVESTING COMPLEXES(LHC) : dependance on the
agregation state .

ISMAEL MOYA and PIERRE TAPIE

The chlorophylle_protein complex LHC has been extensively
studied in-vivo and in-vitro during the last years.
Low temperature fluorescence measurements have shown that
the fluorescence emission spectrum exhibits a more or less
pronounced shoulder near 695 nm , acompaniing the main
emission peak located at 682-685 nm , following the prepa
rations. In order to analyse the elementary components
constituing the emission spectra , we have carried out
measurements under high frequency modulated ligth and the
effect of agregation has been studied at two different
temperatures : 3oo and 77 K.

MATERIEL AND METHODES

LHC particules were prepared according the MULLET's
technique. LHC "agregated"is the resuspended pellet after
Mg-agregation. Further dissolution with digitonine and
Octyl- -Dglucopyranoside provides monomeric LHC: this
particle does not fall down on a sucrose 0.1 M cushion
after 10 minutes centrifugation at 4000 g.
The analysis of the fluorescence emission spectra
under high frequency modulated light provides , in addition
to the time integrated spectrum Fc (i.e. at the zero
frequency),two supplementary spectra : Re and Im (respectively the real and imaginary part of the fluorescence
at the frequency of modulation). If Fc is a linear combination of a basis of elementary components, Re and Im
must be also a linear combination of the same basis. By
repiting the experiment at n different frequencies,2n+1spectra
can be obtained. To evidence lifetime values Re and Im
can be remplaced by two others parametres defined as follows

$$T_p = \frac{1}{2\pi f}\frac{Im}{Re} \quad \text{and} \quad T_m = \frac{1}{2\pi f}\left(\frac{Fc^2}{Im^2+Re^2}-1\right)^{1/2} ; \quad f = \text{frequency of modulation}$$

In the case where the emission is homogenous (i.e. a single
component) we have:
Tp = Tm = τ
In the case of an heterogenous emission we have in general:
Tp$<\tau<$ Tm. Tp and Tm are in this case frequency-dependant.
The phase fluorimeter used in this work has been described
elsewhere (Moya and Garcia , 1983).The new point is the
utilisation of the synchretron radiation of the storage
ring of A.C.O. (LURE-ORSAY) as a source of high frequency
white modulated light.

Sybesma, C. (ed.), Advances in Photosynthesis Research, Vol. II. ISBN 90-247-2943-2.
© 1984 Martinus Nijhoff/Dr W. Junk Publishers, The Hague/Boston/Lancaster.

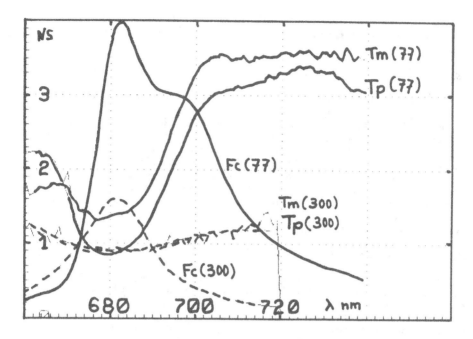

FIGURE 1. AGREGATED LHC
         Continuous line : mesurements made at 77 K.
         Tp:lifetime deduced from the phase-shift.
         Tm:lifetime deduced from the relative modulation.
         Fc:fluorescence intensity.
         Dashed line : room temperature measurements.
         The same sample in the same optical conditions
         has been used. Frequency of modulation:54.47 Mhz

    Figure 1 shows the comparison of the lifetime and
fluorescence intensity spectra of agregated LHC at 300 K
and 77K, in identical conditions. One can see that at
room temperature the $Tp(\lambda)$ and $Tm(\lambda)$ spectra are identical
and almost flats: there is a strong evidence for a quite
homogeneous emission. When the temperature is lowered to
77 K,a shoulder grow-up and the lifetime spectra exhibits
a strong wavelength dependency. However the lifetime near
680 nm remains almost unchanged.

    In figure 2 the comparison between the low and room
temperature spectra have been performed for the monomeric
LHC. In the present case $Tm(\lambda) > Tp(\lambda)$, which is an indication
of an emission heterogeneity. However the wavelength dependency remains small: a small contamination by the agregated
form which has the same emission spectra at room temperature
can explain the difference between $Tp(\lambda)$ and $Tm(\lambda)$. As in
the case of agregated LHC , the lifetime values around 680 nm
remains almost unchanged between 300 K and 77 K.

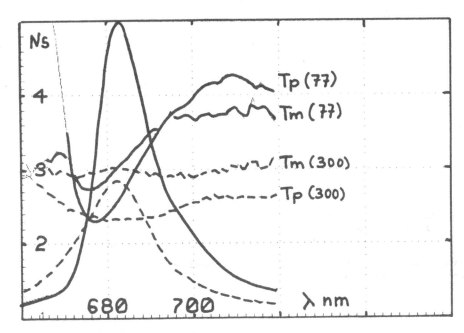

FIGURE 2. MONOMERIC LHC
Continuous line : measurements made at 77 K.
Dashed line : room temperature measurements.
Same abreviations that in Figure 1.

At variance whith the $Fc(\lambda)$ spectra of the agregated form the shoulder in the 700 nm range is very small. However the lifetime spectra exhibits the same strong increase in this wavelength range that in the agregated case. This is in line whith the idea that in the two types of preparations , the effect of the cooling process leads to the growing of the same elementary components, as we show in the following.

THE LORENTZIAN DECOMPOSITION OF LHC
A simultaneous least squares fitting of $Fc(\lambda)$, $Re(\lambda)$ and $Im(\lambda)$ by a model of lorentzian components has been performed on the low temperature spectra of both monomeric and agregated forms of LHC (figures 3 and 4). Three elementary components are at least required to discribe our data. The formulae used are those of the phase fluorimetriy method :

$$L(\sigma_i, \tau_i) = \sigma_i^2/((\lambda - \lambda_i)^2 + \sigma_i^2) \qquad ; \quad i = 3$$

$$Fc(\lambda) = \sum_i F_i(\lambda) = \sum_i A_i \cdot L(\sigma_i, \lambda_i)$$

$$Re(\lambda) = \sum_i F_i(\lambda) \cdot (1 + (2\pi f \tau_i)^2)^{-1}$$

$$Im(\lambda) = \sum_i F_i(\lambda) \cdot (1 + (2\pi f \tau_i)^2)^{-1} (2\pi f \tau_i)$$

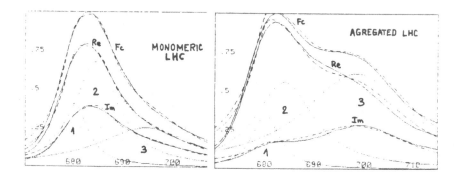

FIGURES 3 and 4. THE THREE LORENTZIAN DECOMPOSITION.
        Continuous line : experimental spectra
        Dashed line : calculated spectra
        Dotted line : elementary lorentzian components

The resulting parameters are listed below:

PREPARATION	AMPLITUDE	PEAK	FWHM	LIFETIME
AGREGATED	0.44	680 nm	7 nm	0.7 ns
LHC	0.52	685	12.5	0.3
	0.62	698.5	25	3.46
MONOMERIC	0.4	680	6.5	1.9
LHC	0.68	684.5	10.5	3.1
	0.26	694.5	23.5	4.35

Although the emission spectra are quite differents for the
two agregation states , the same wavelength positions and
similar widths are found. However in some cases the peak
of $Fc(\lambda)$ is to be found near 685 nm and the lifetimes for
the two first band are respectively 0.1 ns and   3 ns in
the monomeric state and 0.1 ns and 0.7 ns in the agregated
state.

CONCLUSION

        Our data are consistent with the following ideas:
1./ At room temperature high coupling exists between the
    3 components evidenced by the phase fluorimetric analysis
    but energy transfert between monomeric units can only
    take place in the agregated form of LHC.
2./ The fluorescence quantum yield at room temperature
    is to be found 2.5 times lower in the agregated LHC
    than in the monomeric LHC: this can be due to a larger
     collecting antenna pigments transfering the energy to
    eventual traps.
3./ At 77 K the up-hill transfert is preferentially inhi-
    bited : the energy trends to be accumulated in the red
    edge of the spectrum before radiative emission.
4./ This phenomena is strengthened in the agregated LHC by
    energy transfert from F680 and F685 to F695-698 which can
    take place between several monomeric subunits.

Authors address:Ismaël Moya . CNRS Lab PHOTOSYNTHESE 91190
GIF/YVETTE . Pierre Tapie . DB SBph,CEN Saclay 91191 GIF/
YVETTE - FRANCE .

CHLOROPHYLL a/b-PROTEINS IN THEIR RELATION TO THE LIGHT-
HARVESTING COMPLEX

OTTO MACHOLD

1. INTRODUCTION

In 1966 two research groups (Ogawa et al. 1966; Thornber
et al. 1966) were the first to demonstrate that sodium dodecyl sulfate dissociated thylakoid membranes can be separated into three different zones two of which represent
chlorophyll-proteins. Analysis of the two chlorophyll-
protein zones, which were termed components or complexes I
and II, respectively, revealed most, if not all, of the
Chl b associated with complex II whilst complex I contained
mainly Chl a.
Although extensive research work over the past 17 years has
provided a large body of informations, the detailed composition of chlorophyll-protein bands is still subject to
many uncertainties. This mainly concerns the major band previously termed complex II and later on light-harvesting
chlorophyll a/b-protein (Thornber, Highkin 1974). The
following experiments were performed under the aspect of
its further characterization.

2. PROCEDURE

2.1. Material and Methods

2.1.1. Plant material. Vicia faba plants were cultivated in
a growth chamber for 13 days at 20°C under light (10 000 lux;
16 h) and 15°C in darkness (8 h). Seeds of Hordeum vulgare
were germinated in vermiculite moistened with tap water and
grown for 9 days at 20°C in continuous light (1500 lux).
Light was provided by Narva LS 40 lumoflor fluorescent lamps.

2.1.2. Thylakoid membrane isolation and dissociation. Chloroplasts were isolated by two steps of differential centrifugation. Following bursting open of envelopes by hypothonic
shock using a 3 mM NaCl solution, two extensive washings of
the thylakoid membranes were performed to remove stroma com-

Sybesma, C. (ed.), Advances in Photosynthesis Research, Vol. II. ISBN 90-247-2943-2.
© 1984 Martinus Nijhoff/Dr W. Junk Publishers, The Hague/Boston/Lancaster.

ponents. For the buffers applied for membrane dissociation
see the figures. The light-harvesting complex was prepared
according to Burke et al. (1973).

2.1.3. Electrophoresis. Lithium dodecyl sulfate (LDS) electrophoresis was performed at 5°C using gradient slab gels
of 350 mm length. The composition of the buffer systems
applied described previously (Machold et al. 1979).

3. RESULTS

Previous experiments with Vicia faba performed under the
aspect of better resolution of the major chlorophyll a/b-
protein zone yielded two bands termed CP III and CP II'
(Machold, Meister 1979) and later on Chl a/b-P 1 and Chl
a/b-P 2 (Machold et al. 1979). The results were obtained
with gradient gels of 200 mm length. Membrane dissociation
and electrophoresis were carried out at 13°C. Based on the
polypeptide compositions it was concluded the Chl a/b-P 1
band represents a single chlorophyll a/b-protein whilst
Chl a/b-P 2 is a mixture of at least two different chlorophyll a/b-protein species (Machold 1981).

For better resolution of the Chl a/b-P 2 band and for
keeping denaturation of chlorophyll-protein molecules as low
as possible, membrane dissociation and electrophoresis were
performed at 5°C using gels of 350 mm length. The results
obtained under these conditions with Hordeum vulgare are
comparable to those achieved at 13°C with 200 mm slabs. As
shown in Fig. 1, the chlorophyll a/b-protein zone can be resolved into two sharp bands revealing spectral properties
as found previously with Vicia faba (Machold, Meister 1979).

To analyse the polypeptide composition, the two bands were
excised and prepared for re-electrophoresis. Interpretation
of the results of re-electrophoresis has to consider that
chlorophyll-proteins and extraneous membrane polypeptides
may co-migrate and form common bands after the first separation. Since chlorophyll-proteins migrate anomalously in

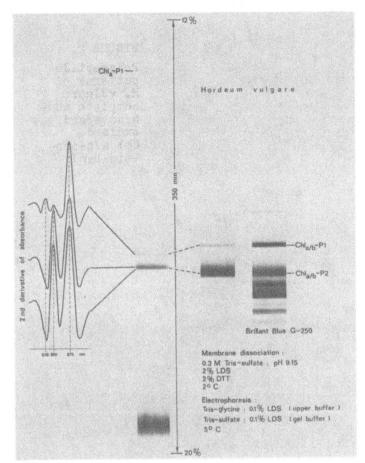

FIGURE 1.

Chlorophyll a/b-protein pattern of H. vulgare

SDS gels (Delepelaire, Chua 1981) and since the pigment de-pleted apoproteins are of increased electrophoretic mobility as compared with their holoproteins, existing "contaminants" can be separated from the apoproteins if re-electrophoresis is performed with denaturing systems. Under this aspect, in-crease of electrophoretic mobility subsequent to chlorophyll-protein denaturation can be used as criterion to identify apoproteins.

As shown in Fig. 2, re-electrophoresis of the Chl a/b-P 1 band yields two polypeptides only one of which has a signifi-cantly increased electrophoretic mobility. There is good evi-dence that the 30 kD polypeptide is not Chl-bearing and that

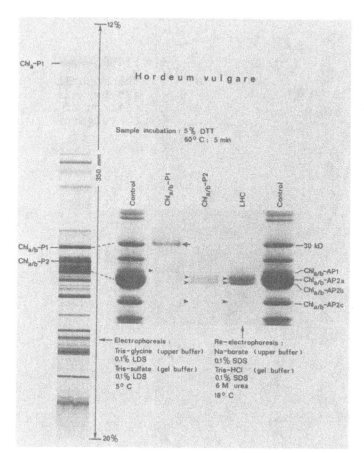

FIGURE 2.

Polypeptide patterns of H. vulgare complete membranes and excised Chl a/b-protein bands

the Chl a/b-Protein 1 band represents only a single chlorophyllprotein (Machold 1931). In contrast, re-electrophoresis of the Chl a/b-P 2 band yields three polypeptides of increased mobility indicating that this band is a mixture of three different chlorophyll a/b-proteins not resolvable in the holoprotein state. The three polypeptides correspond exactly with the constituent polypeptides of the dissociated light-harvesting complex (LHC).

In comparison with Hordeum vulgare, and also as compared with previous results obtained at 18°C using 200 mm gels (Machold 1931), electrophpresis of Vicia faba thylakoid membranes performed at 5°C reveals a completely different chlorophyll a/b-

protein pattern. Under the analytical conditions specified
in Fig. 3, the chlorophyll a/b-protein zone can be resolved
into 5-7 discrete bands all of which contain Chl_a and Chl_b,
even in various ratios. Staining indicates each chlorophyll
band to be associated with a polypeptide band. All chlorophyll bands are located between two major polypeptides of
30 kD and 26.5 kD which can serve as native markers.

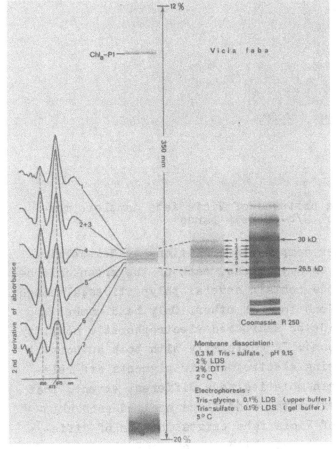

FIGURE 3.

Chlorophyll
a/b-protein
pattern of
V. faba

Re-electrophoresis of the excised chlorophyll a/b-protein
bands under denaturing conditions reveals 6 different polypeptides but, as found with Hordeum vulgare, only 4 are characterized by significantly increased electrophoretic mobilities (Fig. 4). The two major 30 kD and 26.5 kD polypeptides
did not alter their positions and very probably do not bear
pigments.

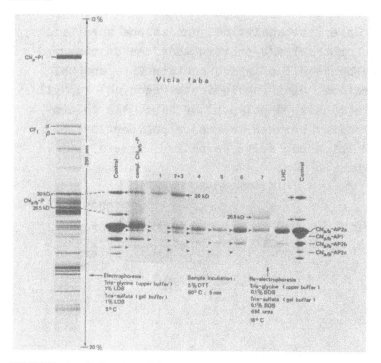

FIGURE 4. Polypeptide patterns of Vicia faba complete membranes and excised Chl a/b-protein bands

As already found under comparable conditions by Delepelaire
and Chua (1981) with Chlamydomonas, each of the green chlorophyll a/b-protein bands contain several polypeptides,although
in different proportions to each other. Only band number 7
reveals one polypeptide of increased electrophoretic mobility
besides the 26.5 kD band. In conformity with both authors it
is concluded that during electrophoresis pigments are released from their protein moieties to a different extent. That
means, under the experimental conditions applied each chlorophyll a/b-protein of Vicia faba exists in form of different states or conformations which are characterized by specific charge to mass ratios and, consequently, by specific
electrophoretic mobilities. This hypothesis could explain
the discrepancy between the number of chlorophyll a/b-protein
bands and the number of apoproteins.
The experiments indicate that, in dependence upon the analytical conditions, the chlorophyll a/b-protein patterns ob-

tained with Hordeum vulgare and Vicia faba reveal differences
which are characteristic when membrane dissociation and electrophoresis are performed at low temperature. Inspite of this
specific electrophoretic behavior, however, the chlorophyll
a/b-proteins of both plant species are characterized by common features. With both species four polypeptides were found,
which on the basis of their electrophoretic behavior must be
attributed to chlorophyll a/b-proteins. Furthermore, all the
four polypeptides are translated by cytoplasmic ribosomes
(Machold 1983) and enriched in the grana fraction (Faludi-
Daniel et al. 1983).

As an additional characteristic feature of both plant species
Chl a/b-P 1 was found to differ in at least three details
from the other chlorophyll a/b-proteins. Chl a/b-P 1 is characterized by a higher Chl a : Chl b ratio and an additional
peak in the position of 637 nm (Machold, Meister 1979). Chl
a/b-P 1 is no constituent of the light-harvesting complex
and was found to be missing in all LHC preparations analysed
(Machold 1981). Chl a/b-P 1 can be selectively removed from
the membrane by mild detergent treatment under conditions,
in which the other chlorophyll a/b-proteins are weakly soluble or insoluble indicating different binding forces or
different sites of localization.

3. CONCLUSIONS

Although characterized by species specific differences, the
experiments performed with Vicia faba and Hordeum vulgare
point to the existence of at least four different chlorophyll a/b-proteins all of which are translated by cytoplasmic
ribosomes and enriched in the grana region of the membranes.
Three chlorophyll a/b-proteins were found to be constituents
of the light-harvesting complex. One chlorophyll a/b-protein
(Chl a/b-P 1) differs in at least three details from the
others.

1. Chl a/b-P 1 is characterized by an increased Ch a : Ch b
   ratio and an additional absorption peak at 637 nm.

2. Chl a/b-P 1 was missing in the light-harvesting complex preparations analysed.

3. Chl a/b-P 1 can be selectively removed from the thylakoid membranes by mild detergent treatment under conditions in which the other chlorophyll a/b-proteins are weakly soluble or insoluble.

From these results the existence of two different chlorophyll a/b-protein populations is inferred.

REFERENCES

Burke JJ, Ditto CL and Arntzen CJ (1978) Arch.Biochem.Biophys. 187, 252-263.

Delepelaire P and Chua NH (1981) J.Biol.Chem. 256, 9300-9307.

Faludi-Daniel A, Schmidt O, Szczepaniak A and Machold O (1983) Eur.J.Biochem. 131, 567-570.

Machold O and Meister A (1979) Biochim.Biophys.Acta 546, 472-480.

Machold O, Simpson DJ and Lindberg Møller B (1979) Carlsberg Res.Comm. 44, 235-254.

Machold O (1981) Biochem.Physiol.Pflanzen 176, 805-827.

Machold O (1983) Biochim.Biophys.Acta 740, 57-63.

Ogawa T, Obata F and Shibata K (1966) Biochim.Biophys.Acta 112, 223-234.

Thornber JP, Smith CA and Baily JL (1966) Biochem.J. 100, 14p-15p.

Thornber JP and Highkin HR (1974) Eur.J.Biochem. 41, 109-116.

ACHNOWLEDGEMENTS

The author gratefully acknowledges technical assistance of Sigrid Riese, Jutta Grosse, and Fritz Bosse. Thanks are due to Dr. Armin Meister for carrying out the spectroscopic measurements.

Authors address: Otto Machold, Zentralinstitut für Genetik und Kulturpflanzenforschung der AdW der DDR 4325 Gatersleben (G.D.R.)

MAGNETIC RESONANCE AND PICOSECOND LASER SPECTROSCOPY OF LIGHT-HARVESTING CHLOROPHYLL A/B-PROTEIN

G.F.W. Searle, J.G.E.M. Fraaije, and T.J. Schaafsma.
Department of Molecular Physics, Agricultural University,
De Dreijen 11, NL-6703 BC, Wageningen, The Netherlands.

ABSTRACT

Chlorophyll (Chl) a/b - protein 2, a light-harvesting protein, isolated in the monomeric form from barley, has unusual spectroscopic properties. The fluorescence detected magnetic resonance spectrum at 4.2 K shows a narrow (< 10 MHz) intense 2E line, which is lost on partial denaturation, whilst fluorescence fading experiments at 4.2 K indicate triplet state spin level decay rate constants which are much slower than for either Chl a or Chl b in vitro. At 293 K the Chl a fluorescence decay cannot be described by a single exponential and is best described by two exponential components (2.5 and 4.7 ns). These observations have led to a model for chlorophyll interactions within the protein in which one or more Chl a is closely associated with a Chl b, whilst remaining Chl a shows much weaker Chl-Chl interactions. Fluorescence spectra of Chl a/b-protein 2 at 293 K and 4.2 K support this hypothesis of a uniquely strong interaction between Chl a and Chl b molecules in Chl a/b-protein 2.

INTRODUCTION

Chl a/b-protein 2 is similar in composition to preparations denoted by CP 2 (CP II) and is the monomeric building block of the light-harvesting complex (LHC). It contains about 4-5 Chl a and 4 Chl b molecules per 32 kD molecular weight and lacks any reaction centre. A second minor protein denoted Chl a/b-protein 1 can be separated electrophoretically from Chl a/b-protein 2. Using the fluorescence detected magnetic resonance (FDMR) technique we have previously shown that proteins lacking reaction centres can give FDMR spectra which give information on Chl ligation by comparing to spectra of Chl in vitro (Searle et al. 1981). This work has been extended and supplemented by studies of fluorescence decay kinetics, using low-intensity repetitive-pulse laser excitation (So as to avoid singlet-singlet and singlet- triplet exciton annihilation artefacts, see Beddard et al. 1979). The aim of this work is to construct a model for Chl-Chl interactions within Chl a/b-protein which can explain the rapid and efficient transfer of singlet excitation energy from this protein to reaction centre proteins in vivo.

EXPERIMENTAL

The major Chl a/b-protein 2 and minor Chl a/b-protein 1 were isolated from barley (Hordeum vulgare L. cv Bonus) free from contamination by other Chl-proteins using sodium dodecyl sulphate (SDS) polyacrylamide gel electrophoresis (Machold et al. 1979). FDMR at 4.2 K was performed as previously described (Searle et al. 1981) except that the 476 nm line of a Coherent Radiation CR3 Ar^+ (CW) laser was used to excite Chl b specifically. The same set-up (detailed in Searle et al., to be published) was used to measure front-surface fluorescence excitation and emission spectra, and also to carry out the fluorescence fading measurements (Avarmaa 1977), all at 4.2 K. Chl fluorescence decay kinetics and fluorescence emission spectra at 293 K were measured using excitation from a mode-locked Coherent Radiation CR18 Ar^+ laser (100 ps FWHM pulses with < 10^{11} photons cm^{-2} $pulse^{-1}$), the pulse train being modulated at 330 kHz.

Sybesma, C. (ed.), Advances in Photosynthesis Research, Vol. II. ISBN 90-247-2943-2.
© *1984 Martinus Nijhoff/Dr W. Junk Publishers, The Hague/Boston/Lancaster.*

Fluorescence was detected by a Philips PM2254 B photomultiplier (S20 photocathode sensitivity) and single photon counting detection.

RESULTS

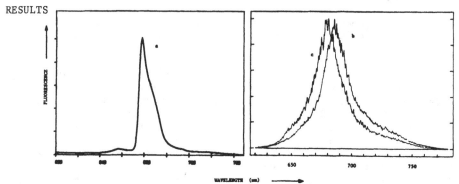

Figure 1. Fluorescence emission spectra of Chl a/b-protein 2. Bandwidth 4 nm for all spectra.

Fig. 1 curve a shows that at 4.2 K singlet energy transfer from Chl b to Chl a is still efficient as Chl b emission at 657 nm is low. The Chl a emission band is heterogeneous with a maximum at 678-680 nm and a district shoulder at 686 nm. Curves b and c show the 293 K emission spectrum of native and partially denatured Chl a/b-protein 2 respectively (the curves are normalized), a shift from 686 nm to 680 nm being observed.

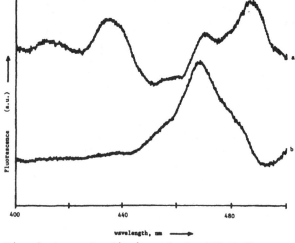

Figure 2. Fluorescence excitation spectra of Chl a/b-protein 2.

Fig. 2. gives the 4.2 K fluorescence excitation spectra for emission at 680 nm (curve a) and 640 nm (curve b), which represent Chl a and Chl b fluorescence respectively (the emission bandwidth was 10 nm) from native protein.

Fig. 3 shows the fitting of the 293 K fluorescence decay kinetics of native protein by one and two components. It is seen that in agreement with Lotshaw et al. 1982 fitting requires two exponentials. On partial denaturation the lifetimes were found to shorten to about 1 and 3 ns. Similar biexponential kinetics are observed for Chl a/b-protein 1, but with slightly different amplitudes of the two components.

The most striking feature of the FDMR spectrum of the native protein (Fig. 4, curve a) is the intense narrow 2E line at 220 MHz which has a positive sign (increase of fluorescence on resonance). The relative intensity of the lines in the spectrum is dependent on excitation intensity (not shown).

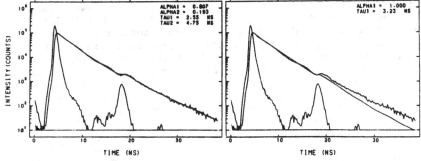

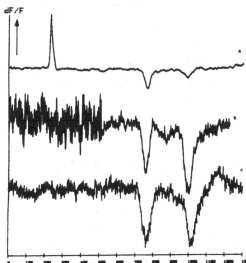

Fig. 3. Fluorescence decay kinetics of Chl a/b-protein 2 at 293 K. The same data is fitted by either two (left hand) or one (right hand) exponential component by an iterative least square fitting procedure. The instrumental response to the excitation pulse is also shown.

Fig.4. Fluorescence detected magnetic resonance spectra of Chl a/b-protein 2 and of Chl a in vitro. For native protein (a) $D = 293 \pm 3$, $E = 39 \pm 3 \times 10^{-4} cm^{-1}$, for denatured protein (b) $D = 291 \pm 3$, $E = 40 \pm 3 \times 10^{-4} cm^{-1}$; and for Chl a in 95% THF/5% ethanol (c) $D = 297 \pm 3$, $E = 42 \pm 3 \times 10^{-4} cm^{-1}$.

On denaturing the protein (100°C, 1 min) the 2E line is lost specifically (curve b) and the spectrum is similar to that for Chl a in tetrahydrofuran (curve c). In table 1 the decay kinetics are given for the Chl triplet states in the protein at 4.2 K as detected by fluorescence fading at 692 nm. It is seen that these kinetics are very slow compared to both Chls a and b, in vitro, for example in MTHF (2-methyltetrahydrofuran).

Table 1. Kinetic constants for the lowest triplet level T_0 at 4.2 K (S^{-1})

	k_1	k_2	k_3
Chl a/b-protein 2	73 ± 2	6 ± 2	–
Chl a.2 MTHF	930 ± 90	310 ± 40	150 ± 25
Chl b.2 MTHF	380 ± 40	195 ± 30	20 ± 10

DISCUSSION

These spectroscopic properties of Chl a/b-protein 2 point to the existence of a strong interaction between Chl a and b molecules. The fluorescence band at 686 nm (Fig. 1) is probably emission from a strongly interacting Chl a/b pair, shifted slightly from the emission maximum of non-interacting Chl a (678-680 nm).

The excitation spectrum of Chl a emission (Fig. 2, curve a) shows bands at 412 and 435 nm (direct absorption by Chl a), and at 470 nm (singlet energy transfer from Chl b). The band at 487 nm does not correspond to either Chl a or b and we suggest that it could represent absorption into the lowest energy level of an interacting Chl a/b pair. The heterogeneity of Chl a molecules within Chl a/b-protein 2 is also seen in the decay kinetics of Fig. 3. The two components probably represent emission from two types of Chl a - that which is interacting strongly with Chl b and that which is relatively "isolated", in agreement with the model postulated. FDMR and fluorescence fading results at 4.2 K indicate the presence of a unique Chl triplet state most likely associated with the closely interacting Chl a/b pair. The 2E line which is not usually detected in FDMR spectra of Chl-proteins, has a linewidth of about 10 MHz compared to about 25 MHz for the D-E and D+E lines. The uniqueness of this triplet state is lost on partial denaturation (Fig.4, curve b) as would be expected if this depends on a highly specific relative orientation of two Chl molecules embedded in the protein matrix. The slow triplet state decay kinetics (Table 1) are also on indication of an unusual type of Chl triplet state. However, the mechanism which leads to these unique properties is not yet understood fully and is the subject of current investigations.

REFERENCES

Avarmaa, R. (1977). Fluorescence detection of triplet state kinetics of chlorophyll, Chem. Phys. Lett. 46, 279-282.

Beddard, G.S., Fleming, G.R., Porter, G., Searle, G.F.W., and Synowiec, J.A. (1979). The fluorescence decay kinetics of in vivo chlorophyll measured using low intensity excitation, Biochim. Biophys. Acta, 545, 165-174.

Lotshaw, W.T., Alberte, R.S., and Fleming, G.R. (1982). Low-intensity sub-nanosecond fluorescence study of the light-harvesting chlorophyll a/b protein, Biochim. Biophys. Acta, 682, 75-85.

Machold, O., Simpson, D.J., and Møller, B.L. (1979). Chlorophyll-proteins of thylakoids from wild-type and mutants of barley (Hordeum vulgare L), Carlsberg Res. Comm. 44, 235-254.

Searle, G.F.W., Koehorst, R.B.M., Schaafsma, T.J., Møller, B.L., and von Wettstein, D. (1981). Fluorescence detected magnetic resonance (FDMR) spectroscopy of chlorophyll-proteins from barley, Carlsberg Res. Comm. 46, 183-194.

ACKNOWLEDGEMENTS

This work was supported by the Commission of the European Communities (contract no. ESD-013-DK(G) of the Solar Energy Programme), and by the Netherlands Foundation for Chemical Research (SON) with financial aid from the Netherlands Organisation for the Advancement of Pure Research (ZWO). We are particularly indebted to Prof. D. von Wettstein and Dr. B. L. Møller for their valuable co-operation. The assistance of Arie van Hoek is gratefully acknowledged, who together with Dr. A.J.W.G. Visser developed the picosecond laser apparatus.

STRUCTURE OF TWO-DIMENSIONAL CRYSTALS OF THE LIGHT-HARVESTING
CHLOROPHYLL a/b-PROTEIN COMPLEX

WERNER KUEHLBRANDT/THOMAS THALER/ERNST WEHRLI

The light-harvesting chlorophyll a/b protein complex (LHC)
isolated from pea chloroplasts contains two polypeptides of
25,000 and 27,000 apparent molecular weight which seem to be
largely homologous (see Thaler et al., this volume). Upon
dialysis against cations the detergent-solubilised complex
forms extensive two-dimensional crystals. The molecular
structure of LHC was investigated by electron microscopy and
image analysis of the crystalline sheets.

Heavy-metal shadowing indicated that crystalline monolayers
were roughly 60 Å thick. The surface pattern of these
crystalline sheets resembled a honeycomb with holes on a
hexagonal lattice (centre-to-centre distance 125 Å) extending
through the sheet. Freeze-etching studies and double replicas
showing complementary fracture faces suggested that the
crystalline sheets were symmetrical, containing a two-fold
symmetry axis in the plane of the sheet. The structure on
both surfaces of the crystalline sheets appeared to be the
same in freeze-etched specimens. Image processing revealed
that complementary fracture faces were structurally
identical, indicating that the sheets fractured into two
equal halves.

Negatively stained, two-dimensional crystals showed hexagonal
arrays of darkly staining regions which corresponded to the
holes seen in the freeze-dried sheets. Image analysis
revealed two units of three-fold symmetry per crystallo-
graphic unit cell apparently of opposite hand in projection.
Edge-on views of stacked sheets gave a thickness of
individual sheets of 54-56 Å (see Kühlbrandt et al., 1983).

A three-dimensional map of LHC was calculated by processing
images of negatively stained, two-dimensional crystals tilted
in the electron microscope. Amplitude and phase data could be
collected to a resolution of 16-18 Å. The three-dimensional
map showed two light-harvesting complexes in the crystallo-
graphic unit cell, each having three-fold symmetry relative
to symmetry axes perpendicular to the plane of the sheet. The
two complexes in the unit cell were related by a two-fold
symmertry axis in the plane of the sheet, parallel to the
edge of the unit cell (space group P321). The orientation of
adjacent complexes thus alternated with respect to the plane
of the crystal. The complex appeared to be a trimer composed
of three identical subunits each of which presumably was a
hetero-dimer of the two polypeptides found by the biochemical
analysis. The complex spanned the thickness of the two-
dimensional crystal, exposing different structural features
and protruding by different amounts on each side. The
distribution of stain-excluding material within the complex

Sybesma, C. (ed.), Advances in Photosynthesis Research, Vol. II. ISBN 90-247-2943-2.
© 1984 Martinus Nijhoff/Dr W. Junk Publishers, The Hague/Boston/Lancaster.

was asymmetrical relative to the plane of the sheet. The main mass was found on the side which exhibited a handed feature resembling three hook-like arms. The arms extended from the position of the three-fold axis in the centre around three darkly staining regions which looked like clefts in the surface of the complex. The other side showed three protrusions which appeared to be continuous with the two ends of the hook-like features on the opposite side of the complex.

REFERENCES
Kühlbrandt W, Thaler T and Wehrli E (1983) The structure of membrane crystals of the light-harvesting chlorophyll a/b-protein complex, J. Cell Biol. 96, 1414-1424.

Author's address: Dr. Werner Kühlbrandt, Institut für
                    Zellbiologie, ETH-Hönggerberg,
                    8093 Zürich, Switzerland.

BIOCHEMICAL ANALYSIS OF POLYPEPTIDE COMPOSITION OF THE LIGHT-HARVESTING
CHLOROPHYLL a/b-PROTEIN COMPLEX FROM PEA CHLOROPLASTS.

T. Thaler, W. Kühlbrandt and K. Mühlethaler

1. INTRODUCTION

The light-harvesting chlorophyll a/b-protein complex (LHC) accounts for
about 50% of protein and chlorophyll in the thylakoids of higher plants.
Apart from its antenna function it appears to be the factor responsible
for stacking of thylakoids into grana. The individual complexes can be
purified from membranes solubilized with non-ionic detergents. The isolated complex precipitates from solution upon addition of mono- and divalent
cations, forming membrane crystals that tend to aggregate into stacks.
Biochemical analysis as well as electron microscopy and image processing
showed three-fold symmetry of the complex. This is consistent with biochemical findings which suggest that the native complex is an oligomer
composed of three identical monomers.

2. CHARACTERISATION OF THE COMPLEX USING SODIUM- (SDS) AND LITHIUM-DODECYL-SULFATE (LDS)-GELS

The oligomer and the monomer were resolved on unstained gels run under
partly dissociating conditions using low concentrations of SDS. SDS gel-
electrophoresis under fully dissociating conditions showed two polypeptides
of 25 kD and 27 kD apparent molecular weight. Two-dimensional electrophoresis of the LHC showed that the oligomer as well the monomer contained the
two polypeptides resolved on the fully dissociating SDS gel. Both polypeptides migrated as a single band upon re-electrophoresis.
When LDS was used instead of SDS, the undenaturated chlorophyll-protein-
complex ran ahead of the two polypeptide bands, which were only visible
after staining the gel. Re-electrophoresis of the pigmented band gave the
characteristic pattern of two polypeptide bands.

3. DIGESTION WITH LOW TRYPSIN CONCENTRATIONS

Treatment of the native, solubilized complex with small amounts of trypsin
led to the disappearance of the 27 kD band on SDS gels, while the 25 kD
band remained visible.

4. PEPTIDE MAPPING

Peptide mapping by limited enzymatic proteolysis of the 25 kD and the 27 kD
polypeptides using trypsin (cleaves at the carbonyl-side of arginine and
lysine) and Staphylococcus aureus V8 protease (cleaves at the carbonyl-
side of aspartic- and glutamic acid) showed significant differences in
the band pattern of the degradation products. The chemical cleaving reagent
N-chlorosuccinimide (cleaves at the carbonyl-side of tryptophan) also
caused a different band pattern of the 25 kD and the 27 kD digestion
products. Treatment of purified LHC with low concentrations of trypsin
(cleaves the 2 kD segment from the 27 kD polypeptide) and subsequent

Sybesma, C. (ed.), Advances in Photosynthesis Research, Vol. II. ISBN 90-247-2943-2.
© *1984 Martinus Nijhoff/Dr W. Junk Publishers, The Hague/Boston/Lancaster.*

peptide mapping with Staphylococcus aureus V8 protease suggested that the
25 kD degradation product of the 27 kD band was not identical to the 25 kD
polypeptide.
The band pattern after trypsin, Staphylococcus aureus V8 protease or N-chlorosuccinimide digestion suggested extensive homology but also showed some
differences in primar structure.

5. RESULTS

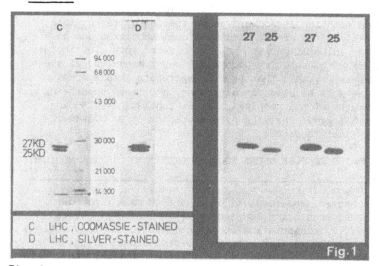

Fig. 1
The two polypeptides were resolved on SDS-page of purified LHC under fully
dissociating conditions. Upon re-electrophoresis on SDS gels, both components
migrated as a single band, demonstrating that they represent different polypeptides rather than different electrophoretic forms of the same protein.

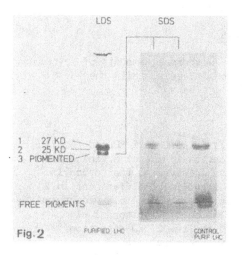

Fig. 2
Polyacrylamide gel electrophoresis
using LDS instead of SDS showed one
pigmented band (band 3) on the unstained gel. After staining the gel,
the 27 kD (band 1) and the 25 kD
(band 2) polypeptide were visible.
Re-electrophoresis of the pigmented
band gave the characteristic pattern
of the two polypeptides.

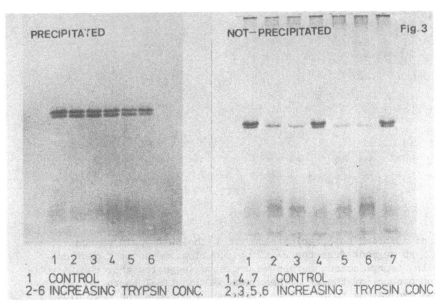

PRECIPITATED   NOT—PRECIPITATED   Fig. 3

1 2 3 4 5 6   1 2 3 4 5 6 7

1 CONTROL   1,4,7 CONTROL
2-6 INCREASING TRYPSIN CONC.   2,3,5,6 INCREASING TRYPSIN CONC.

Fig. 3
Treatment of the native, precipitated LHC with low concentration of trypsin produced no alteration in the polypeptide pattern of the LHC. Treatment of the native solubilized complex with the same amounts of trypsin led to the disappearance of the 27 kD band on SDS gels, while the 25 kD band remained visible.

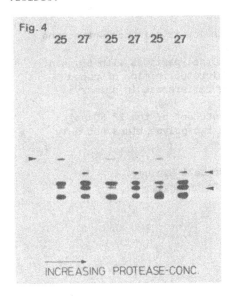

Fig. 4

25 27   25 27   25 27

INCREASING PROTEASE-CONC.

Fig. 4
Peptide mapping with Staphylococcus aureus V8 protease (which cleaves at the carbonyl-side of aspartic- and glutamic acid) of the isolated 27 kD and 25 kD polypeptides showed significant differences in the degradation products.

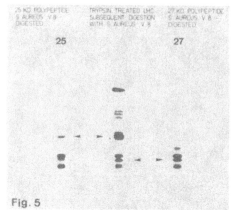

Fig. 5

Fig. 5
Treatment of purified not-precipitated LHC with low trypsin concentrations
cleaves a 2 kD segment from the 27 kD polypeptide. Subsequent peptide
mapping with Staphylococcus aureus V8 protease suggested that the 25 kD
degradation product of the 27 kD band was not identical to the 25 kD
polypeptide indicating that further differences between the two polypeptides exist other than the presence of a 2 kD segment.

6. CONCLUSIONS

a) The purified LHC of peas consists of two polypeptides with apparent
   molecular weights of 25 kD and 27 kD.
b) The 25 kD polypeptide is not a degradation product of the 27 kD polypeptide.
c) The 27 kD polypeptide can be degrated with low trypsin concentrations
   to a 25 kD polypeptide.
d) Peptide mapping of the 27 kD and the 25 kD polypeptides with trypsin,
   Staphylococcus aureus V8 protease or N-chlorosuccinimide produced
   different degradation products indicating differences in primary
   structure.
e) Peptide mapping of the 25 kD degradation product of the 27 kD polypeptide suggested differences between the two polypeptides not only
   in a 2 kD segment.

Authors address: Thomas Thaler, Institute for Cell Biology,
                 Swiss federal Institute of Technology,
                 8093 Zürich, Switzerland.

DETACHED LIGHT-HARVESTING COMPLEX IN CHLOROPLASTS OF COLD-GROWN
TRITICALE SEEDLINGS

BEVERLY ELFMAN/WILLIAM G. HOPKINS

1. INTRODUCTION

In our recent studies we reported that growth of winter-hardy Puma rye
at cold-hardening temperatures resulted in various structural
alterations in the thylakoid membranes. Specifically, we observed
changes in the size distribution of freeze fracture particles on the EF
fracture face (Huner et al, 1983) and changes also in the stability of
the pigment-protein complexes associated with PSII (Elfman et al,
1983). We have extended our investigation to include another
winter-hardy species - OAC wintri - a hexaploid triticale variety.
Preliminary results indicate similar alterations in the stability of
CPIV (CPa) upon detergent solubilization and electrophoresis at room
temperature as was found with Puma rye. In this manauscript we report
changes found in PSII electron transport activity upon growth of the
triticale seedlings at cold-hardening temperatures. We conclude that a
certain population of the light-harvesting chl a/b-protein (LHC) is
functionally detached from the PSII antennae and is correlative of the
apparent deficiency in light-harvesting capabilities at limiting light
intensities.

2. MATERIALS AND METHODS

Triticale seedlings were grown at either 5°C or 20°C as previously
described (Huner, MacDowall 1976). Chloroplasts were isolated according
to the method of Kyle et al, (1982). PSII dependent (DCMU-sensitive)
electron transport was measured as the photoreduction of dichloropheno-
indophenol (DPIP) at 590 nm. The reaction mixture consisted of 200 mM
Na phosphate, pH 7.8, 0.1 M sorbitol, 5 mM $MgCl_2$, 10 mM NaCl, 1 mM
NH_4Cl, 10^{-7} M gramicidin, 33 µM DPIP and chloroplasts at 10 µg chl/ml.
Actinic light was supplied as described (Hopkins et al, 1980).

Room temperature fluorescence induction was detected by a Dumont KM 2433
photomultiplier tube (S.20 surface) blocked with a Corning CS2-64
filter. Excitation of chloroplasts (5 µg chl/ml) in a mixture of 15 mM
tricine-NaOH pH 7.8, 0.1 M sorbitol, 5 mM $MgCl_2$, 10 mM NaCl ± 5 µM DCMU
was accomplished by passing light from an illuminator (see lamp
description above) through a Corning CS4-96 blue filter. The signal
from the photomultiplier tube was stored on a Princeton Applied Research
scan recorder (Model 4101) or recorded directly on a Houston 2000 X-Y
recorder. Low temperature (77K) fluorescence emission spectra were
obtained as described (Kyle, Zalik 1982).

3. RESULTS

Light saturation curves for DCMU-sensitive PSII electron transport
revealed two significant differences between TH and TNH chloroplasts
(Fig. 1). First, TH chloroplast activity saturated at a much higher
light intensity than TNH chloroplasts and, second, the quantum
efficiency of the TH chloroplasts was lower at limiting light

Sybesma, C. (ed.), Advances in Photosynthesis Research, Vol. II. ISBN 90-247-2943-2.
© *1984 Martinus Nijhoff/Dr W. Junk Publishers, The Hague/Boston/Lancaster.*

intensities. From these data alone we might conclude that there is a deficiency in the light harvesting complex in TH chloroplasts. However, an analysis of the membrane components associated with the electron transport system did not support this conclusion as we found no significant differences in the chl a/b ratios (TH = 2.9 ± 0.2, TNH = 3.0 ± 0.1), in the relative amounts of the major chlorophyll-protein complexes or in the thylakoid polypeptide profile obtained by LDS-PAGE (data not presented).

There was also no observable change in the active plastoquinone pool size as measured by room temperature fluorescence induction. However, TH chloroplasts exhibited both a higher initial fluorescence (F_0) and a higher F_{max} in comparison to TNH chloroplasts.

A convenient method for monitoring the interaction of LHC with other pigment proteins is by low temperature (77K) fluorescence emission spectra. Fig. 2 is representative of TH and TNH spectra where the emission has been normalized at 685 nm. Emissions above 700 nm are attributed to PSI whereas those below are attributed to PSII (Homann and Liu, 1981). These spectra show a reduced emission from TH chloroplasts

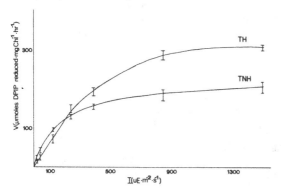

Fig. 1. Light saturation curves for PSII electron transport by TH and TNH chloroplasts.

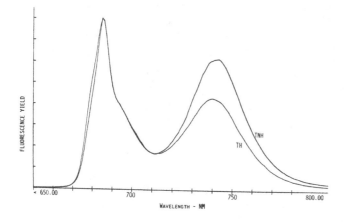

Fig. 2. Low temperature (77K) fluorescence emission spectra of TH and TNH chloroplasts.

in the long-wavelength PSI associated peak relative to the 685 nm peak and also the presence of a shoulder on the 685 nm peak in TH chloroplasts.

4. DISCUSSION

Decreases in light-harvesting efficiency are typically correlated with a smaller quantity of the light-harvesting chl a/b-protein complex. Light-saturation curves similar to that of TH chloroplasts have been reported for a chl b-deficient, light-sensitive mutant of maize (Hopkins et al, 1980) and for intermittent-light grown pea seedlings (Arntzen et al, 1976). In both cases it was reported that the chloroplast membranes were deficient in the light-harvesting chl a/b-protein complex and this resulted in: 1) a higher chl a/b ratio, 2) a smaller photosynthetic unit size and 3) an inability to harvest energy at low light intensities.

However, in contrast, we found that a decrease in light-harvesting efficiency in TH thyalkoids was not accompanied by any apparent reduction in the amount of LHC as judged by gel electrophoresis of chlorophyll-protein complexes and membrane polypeptides. These conflicting observations led us to believe that perhaps there was an alteration in the association of LHC with the PSII reaction center complex. This association can be monitored by both room temperature (20°C) and low temperature (77K) fluorescence. Our examination of PSII associated fluorescence (20°C) revealed an increased F_0 and F_{max} in TH chloroplasts compared to TNH chloroplasts. It was important to note, however, that the ratio of F_{max} to F_0 was the same for both TH and TNH chloroplasts. This result is interpreted as an indication of no change in the photosynthetic unit size (Argyroudi-Akoyunoglou et al, 1982). Of greater importance was the higher initial fluorescence (F_0) found in TH chloroplasts. Since this initial rise component is thought to represent 'inactive' chlorophyll (Butler, 1978) we conclude that there is a proportion of chlorophyll associated with PSII (LHC or reaction center antennae) that is functionally disconnected.

Further evidence for an alteration in the association of LHC with PSII in TH thylakoids was obtained from the low temperature fluorescence emission spectra as shown in Fig. 2. Of significance to this investigation was the presence of a shoulder on the F685 peak which was found to be maximally different from TNH samples at 680 nm (data not shown). Low temperature fluorescence emission at 680 nm has been found with: (1) detergent-solubilized LHC (Burke et al, 1978), (2) senescing chloroplasts of Chlamydomonas (Burke et al, 1978), (3) the PSII activity deficient hcf*-3 mutant of maize (Leto, Arntzen 1981) and (4) a virescens mutant of barley (Kyle, Zalik 1982). In Homann and Liu's (1981) investigation of various tobacco varieties exhibiting structurally different chloroplasts, they concluded that LHC fluorescence (F680) is not significant unless the association of LHC with the PSII pigment-protein complex is disturbed.

We therefore conclude that TH chloroplasts contain a population of LHC which is detached from PSII reaction center complexes and which may explain the reduced light harvesting efficiency seen. This data plus the results of the room temperature fluorescence study lead us to believe that growth of triticale seedlings at cold-hardening temperatures has resulted in the production of chloroplast thylakoids in which the association of LHC with PSII is altered. That this alteration may affect the excitation energy distribution in the membranes is evidenced by the significant difference in the reduced amount of PSI-associated long wavelength fluorescence emission (77k) in TH chloroplasts as compared to TNH chloroplasts (Fig. 2).

REFERENCES
Argyroudi-Akoyunoglou J H, Akoyunoglou A, Kalosakas K and Akoyunoglou G (1982) Reorganization of the photosystem II unit in developing thylakoids of higher plants after transfer to darkness, Plant Physiol. 70, 1242-1248.
Arntzen C J, Armond P A, Briantais J-M, Burke J J, and Novitzky W P (1977) Dynamic interactions among structural components of the chloroplast membrane, Brook Symp. Biol. 28, 316-337.
Burke J J, Steinback K E, Ohad I, and Arntzen C J (1978) in Chloroplast Development, Akoyunoglou G and Argyroudi-Akoyunoglou JH eds, pp. 461-466, Elsevier/North Holland Press, Amsterdam.
Butler W L (1978) Energy distribution in the photochemical apparatus of photosynthesis, Annu. Rev. Plant Physiol. 29, 354-378.
Elfman B, Huner N P A, Griffith M, Krol M, Hopkins W G, Hayden D B (1983) Growth and development at cold-hardening temperatures. Chlorophyll protein complexes and thylakoid membrane polypeptides, Can. J. Bot., in press.
Homann P H and Liu Y-Y (1981) Properties of thylakoids and thylakoid particles derived from structurally different chloroplasts, Biochim. Biophys. Acta 638, 12-21.
Hopkins W G, German J B, Hayden D B (1980) A light-sensitive mutant in maize (Zea mays L.). II. Photosynthetic properties. Z. Pflanzenphysiol. Bd. 100.S., 15-24.
Huner N P A, Elfman B, Krol M, MacIntosh A (1983) Growth and development at cold-hardening temperatures. Chloroplast ultrastructure, pigment content and composition, Can. J. Bot., in press.
Huner N P A and MacDowall F D H (1976a) Chloroplastic proteins of wheat and rye grown at warm and cold-hardening temperatures, Can. J. Biochem. 54, 848-853.
Kyle D J and Zalik S (1982) Photosystem II activity, plastoquinone A levels, and fluorescence characterization of a virescens mutant of barley, Plant Physiol. 70, 1026-1031.
Kyle D J, Haworth P and Arntzen C J (1982) Thylakoid membrane protein phosphorylation leads to a decrease in connectivity between photosystem II reaction centers, Biochim. Biophys. Acta 680, 336-342.
Leto K J, Keresztes A and Arntzen C J (1982) Nuclear involvement in the appearance of a chloroplast-encoded 32,000 dalton thylakoid membrane polypeptide integral to the photosystem II complex, Plant Physiol. 69, 1450-1458.

ACKNOWLEDGEMENTS
This research was supported in part by an operating grant from National Science and Enginerring Research Council.
Authors address: Department of Plant Sciences, University of Western
                 Ontario, London, Ontario N6A 5B7

POLYPEPTIDES OF THE LIGHT-HARVESTING COMPLEX IN A CHLOROPHYLL
b-LESS MUTANT OF CHLAMYDOMONAS REINHARDII

HANS-PETER MICHEL, MATHIAS TELLENBACH AND ARMINIO BOSCHETTI

1. INTRODUCTION

The studies on chlorophyll b-deficient mutants of higher plants
or green algae were of great help in the identification of
the chlorophyll-protein complex II (CP II) as the light-
harvesting chlorophyll a/b-protein complex (LHCP) (Thornber,
Highkin, 1974). This complex is not involved in any direct
photosynthetic reaction. However, it has been found that it
is involved in the stacking of thylakoids and in the energy
distribution between photosystem I and II (Burke et al.,
1979). This energy distribution is regulated by phosphorylation of the LHC-apoproteins (Bennett et al., 1980). The chlorophyll b-less barley mutant chlorina f2 has been studied very
thoroughly. In this mutant, which completely lacks in chlorophyll b and in CP II, the thylakoids are still extensively
stacked (Miller et al., 1976), but energy distribution is reduced (Haworth et al., 1982), indicating that the LHCP is predominantly involved in the regulatory function of energy distribution and less so in the formation of membrane stacks.
However, in this mutant two of the three LHC-apoproteins are
also missing (Bellemare et al., 1982). This renders it difficult to discriminate whether the absence of chlorophyll b or
of the apoproteins is responsible for the behaviour of the
mutant. Therefore, it is interesting to analyse more chlorophyll b-deficient mutants with respect to modified LHC-apoproteins, phosphorylation, photosynthetic activity and morpho-
logy. By screening for pigment mutants of Chlamydomonas reinhardii we selected a new chlorophyll b-deficient strain, pg 113.
The mutant shows a new stacking arrangement of thylakoids.
However, apart from the absence of chlorophyll b, we have not
yet been able to find any difference between the chemical composition of the parent strain and that of the mutant.

2. MATERIAL AND METHODS

Culture conditions were previously described (Boschetti et al.,
1978). All media contained arginine • HCl (100 mg/l). Mutants
were induced by ultraviolet irradiation (7 % survival) of the
parent strain, Chlamydomonas reinhardii arg$_2$mt$^+$. Among colonies
with changed colour, the pigment mutant pg 113, which does not
contain any chlorophyll b, was found.
Oxygen evolution was assayed in a Clark-type electrode containing about 30 μg of total chlorophyll in 4 ml of an oxygen-
depleted buffer of 50 mM sodium phosphate (pH 7.6), 5 mM NaHCO$_3$
and 10 mM KCl.
For in vivo phosphorylation cells were harvested and resuspended in culture medium containing sodium acetate (2 g/l) to
a concentration of 4 x 10^7 cells per ml. After incubation in
the dark for 10 hours, they were labelled with 21 MBq/ml [^{32}P]-
orthophosphate for half an hour either in darkness or in the

Sybesma, C. (ed.), Advances in Photosynthesis Research, Vol. II. ISBN 90-247-2943-2.
© *1984 Martinus Nijhoff/Dr W. Junk Publishers, The Hague/Boston/Lancaster.*

light. In order to avoid dephosphorylation all media used
henceforth contained 20 mM NaMoO$_4$.
Polypeptides were analysed by SDS-gel electrophoresis (Lämmli,
1970) on a polyacrylamide gradient gel. In order to detect
chlorophyll-protein complexes, partially solubilized thylakoids were analysed by LDS-gel electrophoresis (Delepelaire,
Chua, 1979). The isolation of thylakoids was previously described (Boschetti et al., 1978).
For electron microscopy whole cells were embedded in the epoxy
resin of Spurr according to the instructions given by the manu-
facturer (Balzers Union). Thin sections were stained with lead
citrate and observed in a Philips 300 electron microscope.

3. RESULTS AND DISCUSSION

The pigment mutant pg 113 is capable of growing under autotrophic conditions. Its chlorophyll b content is below the
limit of the spectrophotometric detection so that the apparent
chlorophyll a/b ratio is very high (Table 1). The photosynthetic O$_2$-evolution of mutant cells is comparable to that of
the parent strain. Under light saturating conditions, the activity based on the chlorophyll a content is about the same
in the parent strain and in the mutant, indicating that an
equal number of reaction centres is present in both strains.
At low light intensities O$_2$-evolution is almost equal in both
strains when based on a total chlorophyll content, showing
that the mutant still contains light-harvesting device with
similar efficiency as compared with the parent strain.

TABLE 1. Chlorophyll content and photosynthetic activity of
parent strain (ps) and chlorophyll b-less mutant (pg-113) of
Chlamydomonas reinhardii
Chlorophyll a and b content in μg per 10^6 cells, O$_2$-evolu-
tion in μMoles/h $\cdot$ mg of total chlorophyll at 24° C. Low light:
12 J $\cdot$ m^{-2} $\cdot$ s^{-1}; high light: 460 J $\cdot$ m^{-2} $\cdot$ s^{-1}.

	chl a	chl b	a/b	O$_2$-evolution low light	high light
ps	2.4	1.0	2.4	20.8	107.6
pg-113	2.3	0.08	>30	19.3	138.8

The absence of LHCP is demonstrated by the lack of CP II after
electrophoresis of partially solubilized thylakoids on LDS-
polyacrylamide gels (Fig. 1a). In the mutant pg-113 CP II is
completely absent while CP I as well as two minor complexes
running between CP I and CP II, which may be attributed to
reaction centres of photosystem II, are still present.
From electron micrographs of thin sections of light-grown
whole cells (Fig. 1b) we found stacked as well as unstacked
regions of thylakoids both in the parent strain and in the

the mutant. However, in the parent strain the stacked regions were distributed evenly over the chloroplast, whereas in the mutant, they were concentrated near the envelope and in the "fingers" of the chloroplast. In the interior of the mutant chloroplast, i.e. around the pyrenoid, an area with long wavy or small circular unstacked double membranes could be observed. The reason for this morphological modification is not clear.

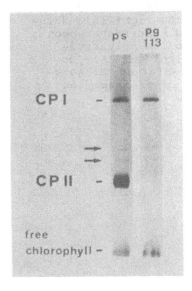

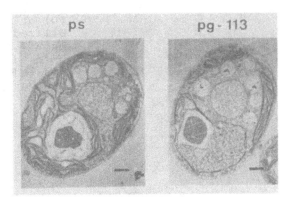

FIGURE 1a (left). LDS-gel electrophoresis of partially solubilized thylakoids. Photograph of unstained gel.

FIGURE 1b (right). Thin sections of whole cells. Bars equal 1 μm. ps: parent strain; pg-113: mutant strain.

It was of interest to find out whether the chlorophyll b-deficiency is the one and only reason for the absence of CP II or whether other components are also affected by the mutation. Primary candidates would be the apoproteins of LHCP, which are the main substrates for phosphorylation. Tje electrophoretic analysis of [^{32}P]-phosphate labelled whole cells shows that all the prominent Coomassie Blue-stained polypeptides of the parent strain are also present in the mutant pg 113 (Fig. 2a) and the phosphorylation pattern is also the same (Fig. 2b).
It can be seen that the three prominent LHC-apoproteins (polypeptides 11, 16 and 17) are also present in the mutant and that their relative ratios are the same in the mutant as compared to the parent strain. Furthermore, light-dependent phosphorylation is not affected by the lack of chlorophyll b. Specially the LHC-apoproteins show the same phosphorylation behaviour.
By using antiserum raised against CP II of the parent strain we have further been able to identify the polypeptides of the mutant pg 113 as the LHC-apoproteins.
From our results with the mutant pg 113 two conclusions which are not available with chlorina f2 must be emphasized. Considerable amounts of LHC-apoproteins can be formed even in

the absence of chlorophyll b. Phosphorylation of LHC-apoproteins is not dependent on the intactness of the LHCP.

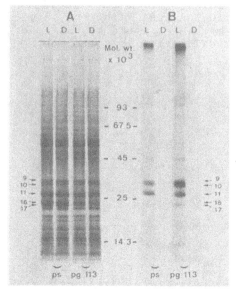

FIGURE 2. Electrophoretic analysis of [^{32}P]-phosphate labelled whole cells. After a
10 hour dark period cells were
labelled either in the dark
(D) or in the light (L), solubilized and electrophoresed
on a 7.5 to 15% polyacrylamide gradient gel. A: Coomassie Blue-stained, B: Auto-
radiograph of the same gel.

4. REFERENCES

Bennett J, Steinback KE and Arntzen CJ (1980) Proc. Natl.
Acad. Sci. USA 77, 5253-5257.
Bellemare G, Bartlett SG and Chua NH (1982) J. Biol. Chem.
257, 7762-7767.
Boschetti A, Sauton-Heiniger E, Schaffner JC and Eichenberger W
(1978) Physiol. Plant. 44, 134-140.
Burke JJ, Steinback KE and Arntzen CJ (1979) Plant. Physiol.
63, 237-243.
Delepelaire P and Chua NH (1979) Proc. Natl. Acad. Sci. USA
76, 111-115.
Haworth, P, Kyle DJ and Arnzten CJ (1982) Arch. Biochem. Bio-
phys. 218, 199-206.
Laemmli UK (1970) Nature 227, 680-685.
Miller KR, Miller JG and McIntyre KR (1976) J. Cell Biol. 71
624-638.
Thornber JP and Highkin HR (1974) Eur. J. Biochem. 41, 109-116.

ACKNOWLEDGEMENT

This work has been supported by the Swiss National Foundation
for Scientific Research.
Authors' Address: H.P. Michel, M. Tellenbach and A. Boschetti
                  Institute of Biochemistry, University of
                  Berne, CH-3012 Berne, Switzerland

LIPID INDUCED TRANSFORMATION OF THE MONOMERIC TO OLIGOMERIC FORM OF THE
LHCP

REMY R.[1], AMBARD-BRETTEVILLE F.[1], DUBACQ J.P.[2] and TREMOLIERES A.[3]
1) Laboratoire de Photosynthèse, CNRS, 91190 Gif sur Yvette, France
2) Laboratoire de Botanique et de Cytophysiologie Végétale (L.A 311)
   ENS, 24 rue Lhomond, 75005 Paris, France
3) Laboratoire du Phytotron, CNRS, 91190 Gif sur Yvette, France

1. INTRODUCTION

It is now clearly established that all the chlorophyll and carotenoids in
the thylakoids of higher plants are non-covalently bound to specific proteins associated in supramolecular complexes. The improvment of solubilization and analysis technics of thylakoid membranes allowed to extract
these supramolecular complexes closer to the "in vivo" state. Oligomeric
forms of P 700-chl a protein and light harvesting chlorophyll a/b protein
were isolated (Rémy et al., 1977, 1978). Recently, the presence of lipids
in these chlorophyll-proteins was reported by different groups (Selstam,
1980 ; Trémolières et al., 1981 ; Ryrie et al., 1980). Rawyler et al.,
1980, Trémolières et al., 1981 have shown that LHCP was rich in phospha-
tidylcerol (PG) containing a high amount of 3-trans-hexadecenoic acid, an
acyl lipid which has been implicated as the LHCP in the formation of the
grana stacks (Dubacq, Trémolières, 1983). Moreover, we reported recently
that this PG enrichment was more significant in an oligomeric form of
LHCP and that it was possible to convert partially monomeric LHCP into
oligomeric LHCP using PG liposomes (Rémy et al., 1982). In the present
paper we intend :
1) to define more precisely than we have done before the different chlorophyll b containing complexes separated in our SDS-PAGE analyses since the
presence of two oligomeric forms (LHCP1 and LHCP2) were described in
other reports (Anderson et al., 1978)
2) to introduce the problem of a lipid specificity in the formation of the
oligomeric LHCP1.

2. MATERIALS AND METHODS

SDS-PAGE of thylakoids, purification of LHCP3 were performed according to
Rémy et al., 1977 .
Lipid analysis of chlorophyll-protein complexes, liposomes preparation
were previously described (Trémolières et al., 1981 ; Rémy et al., 1982).
The reconstitution of the oligomeric form LHCP1 was done by a mixture of
monomeric LHCP3 with liposomes (lipid/LHCP3 chlorophyll ratio = 2) submitted to a vigorous shaking for 15 min. The reconstitution of LHCP1 was
estimated after electrophoresis and quantified by the area of its peak
after scanning the gels at 670 nm.

3. RESULTS

3.1. Characterization of chlorophyll a/b proteins
In Fig.1A is represented a classical separation of chlorophyll-protein
complexes. L1 and L3 correspond respectively to the well described oligomeric LHCP1 and monomeric LHCP3 as shown by their polypeptidic composition
(Fig. 1B and Fig. 1C) resolved by 2D-electrophoresis. Lx or LHCPx is a

Sybesma, C. (ed.), Advances in Photosynthesis Research, Vol. II. ISBN 90-247-2943-2.
© *1984 Martinus Nijhoff/Dr W. Junk Publishers, The Hague/Boston/Lancaster.*

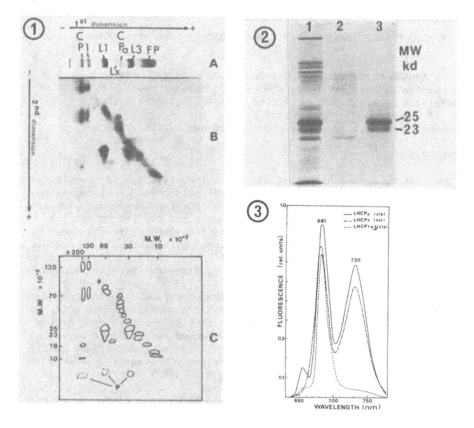

Fig. 1 A) Separation of chlorophyll-protein complexes, CP1 , LHCP1 (L1),
LHCPx (Lx), CPa, LHCP3 (L3) and free pigments (FP). B) 2D-gel showing the
polypeptidic composition of the different chlorophyll-protein complexes
after coomassie blue staining. C) Schematic map of the 2D-separation.
Fig. 2 SDS gels of:thylakoid polypeptides (lane 1), LHCPx (lane 2),
LHCP3 (lane 3).
Fig. 3 Fluorescence emission spectra at 77 K of LHCPx and LHCP1 or LHCP3

complex that we have in previous reports erroneously assimilated to the
LHCP2 described by Anderson et al., 1978. LHCPx possesses a polypeptidic
composition different from LHCP1 and LHCP3 showing only one polypeptide
of about 18-20 kd (Fig. 1B, 1C and Fig. 2). Moreover, it contains twice
less chlorophyll b than LHCP1 and LHCP3. As shown in Fig. 3, LHCPx is
also distinguishable from LHCP1 and 3 by its fluorescence emission spectra showing a strong emission at 730 nm different from that of CP1 which
is at 720 nm. Since the polypeptide 18-20 kd of LHCPx has been found
associated with oligomeric CP1, we think that it may represent part of
the light-harvesting antennae of CP1.

3.2. Reconstitution of the oligomere LHCP1 with liposomes and monomere LHCP3
3.2.1. Kinetics of reconstitution with different kinds of liposomes

In order to check a possible specificity of lipids in the reconstitution of LHCP1, liposomes differing by their polar heads or acyl lipid composition were tested in their ability to reconstitute LHCP1. It is clear in Fig. 4A that with PG liposomes containing trans Δ 3 hexadecenoic acid, the reconstitution of LHCP1 is more rapid than with dipalmitate PG liposomes.

In Fig. 4B are compared the kinetics of reconstitution with other chloroplastic or synthetic lipids. Here again, the best reconstitution is obtained with PG containing trans hexadecenoic acid. Among all the lipids so far examined, only monogalactosyldiacylglycerol and dipalmitate phosphatidylethanolamine failed to reconstitute LHCP1.

3.2.2. Necessity of the integrity of LHCP3 apoproteins for the reconstitutions

A mild trypsin-treament of the LHCP3 which causes the loss of about 1 kd fragments from the two subunits of LHCP3 makes the LHCP1 reconstitution impossible (Fig. 5A and B).

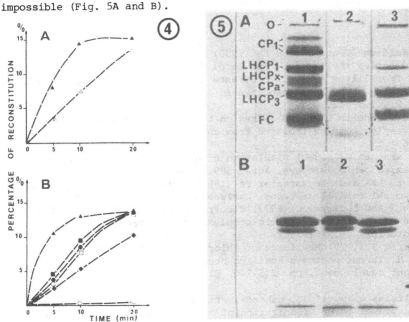

Fig. 4 Comparison of the kinetics of LHCP1 reconstitution with liposomes of:
A) trans hexadecenoic acid PG (▲), dipalmitate PG (Δ),
B) tobacco phosphatidylcholine (□), synthetic dioleate phosphatidylcholine (■), tobacco phosphatidylethanolamine (●), synthetic dipalmitate phosphatidylethanolamine (O), digalactosyldiacylglycerol (◆), tobacco PG as control (▲).

Fig. 5 A) SDS gels of chlorophyll-proteins : lane 1, control with thylakoids; lane 2, isolated LHCP3 ; lane 3, mixture of LHCP3 + PG liposomes = reconstitution of LHCP1 ; lane 4, same as 3 but trypsin-treated LHCP3 = no LHCP1 reconstitution (0.2 μg trypsin/ 15 μg LHCP3 chlorophyll).

B) Coomassie blue stained gels of : lane 1, reconstituted LHCP1; lane 2, remaining LHCP3; lane 3,trypsin-treated LHCP3;lane 4, standards.

4. DISCUSSION AND CONCLUSION

The experiments reported above points out a possible lipid specificity in
the supramolecular organization of the LHCP. The faster kinetics of reconstitution always observed with PG containing trans hexadecenoic acid may be
explained by a better recognition of some anchoring sites between two
LHCP3 molecules. Trypsin-treatments suggest that the same peptidic fragments
are involved in the formation of the oligomere LHCP1 and in experiments
which simulate the grana stacking process with LHCP included in liposomes
(McDonnel, Staehelin, 1980; Mullet, Arntzen, 1980; Ryrie, Fuad, 1982).
Nevertheless, the question of the real significance of the oligomere LHCPs
vizualized by electrophoresis remains asked. Do they represent traces of an
intrathylakoidal organization or rather traces of a linkage between two
adjacent thylakoids ? Independently, no doubt that lipids play fundamental
role in the orientation of the LHCP molecules in the interaction between
these molecules and with others as reaction centers.

REFERENCES

Anderson JM, Waldron JC and Thorne SW (1978) Chlorophyll-protein complexes
of spinach and barley thylakoids, FEBS Lett. 92, 227-233.
Dubacq JP and Trémolières A (1983) Occurence and function of phosphatidyl-
glycerol containing trans-hexadecenoic acid in photosynthetic lamellae,
Physiol. vég. 21, 293-312.
McDonnel A and Staehelin LA (1980) Adhesion between liposomes mediated by
the light-harvesting complex isolated from chloroplast membranes, J. Cell
Biol. 84, 0-56.
Mullet JE and Arntzen CJ (1980) Simulation of grana stacking in a model
membrane system, Biochim. Biophys. Acta, 589, 100-117.
Rawyler A, Henry LEA and Siegenthaler PA (1980) Acyl and pigment lipid com-
position of two chlorophyll-proteins, Carlsberg Res. Commun. 45, 443-451.
Rémy R, Hoarau J and Leclerc JC (1977) Electrophoresis and spectrophotometric studies of chlorophyll-protein complexes from tobacco chloroplast.
Photochem. Photobiol. 26, 151-158.
Rémy R and Hoarau J (1978) New forms of chlorophyll-protein complexes from
thylakoids of different photosynthesizing organisms. In Akoyunoglou et al.
eds. Chloroplast development, pp. 235-240. Elsevier/North-Holland Biomedical Press.
Rémy R, Trémolières A, Duval JC, Ambard-Bretteville F and Dubacq JP (1982)
Study of the supramolecular organization of light-harvesting chlorophyll-
protein , FEBS Lett. 137, 271-275.
Ryrie IR, Anderson JM and Goodchild DJ (1980) The role of the light-harves-
ting chlorophyll a/b-protein complex in chloroplast membrane stacking,
Eur. J. Biochem. 107, 345-354.
Ryrie JR and Fuad N (1982) Membrane adhesion in reconstituted proteolipo-
somes containing the light-harvesting chlorophyll a/b-protein complex :
the role of charged surface groups, Arch. Biochem. Biophys. 214, 475-488.
Selstam E (1981) Lipids of the light-harvesting chlorophyll protein complex.
In Akoyunoglou G, ed. Photosynthesis III, pp. 631-634. Balaban International
Science Service, Philadelphia.
Trémolières A, Dubacq JP, Ambard-Bretteville F and Rémy R (1981) Lipid com-
position of chlorophyll-protein complexes, FEBS Lett. 130, 27-31.

FUNCTION OF CHLOROPHYLLS AND CAROTENOIDS IN THYLAKOID MEMBRANES.
PIGMENT BLEACHING IN RELATION TO PS-I and PS-II ACTIVITY OF SUBCHLORO-
PLAST PARTICLES PREPARED WITH DIGITONIN.

TH.BRAUMANN, TH. GRÖPPER, I.DAMM AND L.H.GRIMME
UNIVERSITY OF BREMEN, FB BIOLOGIE/CHEMIE, 2800 BREMEN 33, FRG.

1. INTRODUCTION

Carotenoids perform two major functions in photosynthesis. They serve as
light-harvesting pigments, and they protect the chlorophylls (chl) from
harmful photodestructive reactions which occur in the presence of oxygen (Krinsky, 1979). Carotenoids are not homogeneously distributed across
the thylakoid membrane but rather specifically bound to pigment-protein
complexes from which each shows a distinct carotenoid pattern (Braumann
et al., 1982). The reason for this specific binding is not readily apparent because any carotenoid with 9 or more conjugated double bonds can
fulfill either light-harvesting or photoprotection. It has been suggested that carotenoids may also be responsible for the proper assembly
of the pigment-protein complexes to the three supramolecular complexes,
PS-I, PS-II, and the light-harvesting complex, of the thylakoid membrane.

We have fractionated the thylakoids into light and heavy particles by
digitonin (Anderson, Boardman, 1966). The heavy particles (D-10) show
good PS-II activity, very little PS-I activity, a low chl a/b ratio,
and are enriched in xanthophylls. The light fraction (D-144) is highly
enriched in P-700, contains essentially no PS-II activity, shows a high
chl a/b ratio, and includes large amounts of ß-carotene. We have illuminated thylakoids, D-10, and D-144 with high light intensities for several hours to analyse the loss in PS-II and PS-I activities and to relate these data with the concomitant bleaching kinetics of chlorophylls
and carotenoids. The results allow a differentiation between the individual carotenoids with respect to their function within the different
particles and within the thylakoid membrane.

2. MATERIAL AND METHODS

Spinach chloroplasts were isolated in 0.05 M Tris-HCl (pH 8), 0.01 M KCl
and 0.3 M sucrose, washed and osmotically shocked in the same buffer
without sucrose. Chl concentration was adjusted to 0.4 mg chl/ml and a
freshly prepared digitonin solution was added to give a final digitonin
content of 0.5 % (w/v). This suspension was incubated in the dark at
$4^{O}C$ for 30 min. Sequential centrifugation (Anderson, Boardman, 1966)
yielded D-10 (sedimented at 10,000g) and D-144 (144,000g). The different
fractions were diluted to give 40 µg chl/ml and illuminated with white
light of 50 klux, the temperature was maintained at $11-14^{O}C$. Samples
were taken from the same suspension to measure pigment composition and
photosynthetic activity.

Pigments were determined by reversed-phase HPLC (Braumann, Grimme,1981).
DCPIP-reduction of thylakoids and D-10 was measured with either water or
diphenylcarbazide (DPC) as electron donor. P-700 was determined chemically and photochemically using an extinction coefficient of 64 $mM^{-1}cm^{-1}$
and an isosbestic point at 725 nm. In the chemical assay, to the sample
were added 0.36 mM ferricyanide, 1 µM phenazinmethosulphate (PMS),60 µg

Sybesma, C. (ed.), Advances in Photosynthesis Research, Vol. II. ISBN 90-247-2943-2.
© *1984 Martinus Nijhoff/Dr W. Junk Publishers, The Hague/Boston/Lancaster.*

chl and 0.1% Triton X-100 (only with thylakoids); in the reference, ferricyanide was substituted by 1.6 mM Na-ascorbate. The light-induced absorption change at 699 nm was measured in sample containing 75 μM methylviologen and 5 μM PMS; to the reference were added 1.6 mM Na-ascorbate
and 5 μM PMS. Measurements were performed with an Aminco DW-2 spectrophotometer equipped with side-illumination accessory. The actinic light
was filtered through a Schott BG 28, and the photomultiplier was protected by a Schott RG 665.
Experiments were repeated three times and gave similar results with respect to the overall behaviour of pigment bleaching and the loss of photosynthetic activities in the different illuminated samples. However,
single values after certain illumination times can vary up to 20% depending on the quality of the starting material.

3. RESULTS AND DISCUSSION

Fig.1 shows the decrease of total chl and total car during illumination
with high light intensities in thylakoids, D-10, and D-144. In all cases
car are bleached faster than chl, as has been reported previously (Ridley, 1977). However, there are important differences with respect to
the bleaching kinetics within the different fractions. In thylakoids,chl
and car show a distinct lag phase of about 150 min before rapid bleaching commences whereas in D-10 only chl is retained for some 30 min, and
car decreases with the onset of illumination. In D-144, chl and car are
immediately degraded with the start of illumination. Fig.2 shows the
corresponding PS-II activities of thylakoids and D-10.

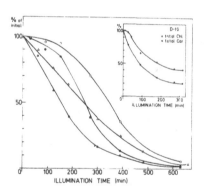 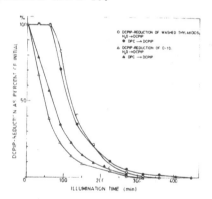

*FIGURE 1. Decrease of total chlorophylls and total carotenoids during
illumination. Open symbols represent total chl, black symbols indicate
total car in thylakoids (circles) and D-144 (triangles) Insert : the
same with D-10.*

*FIGURE 2. DCPIP reduction of thylakoids and D-10 during illumination.
Thylakoids showed an initial activity of 80-120 μmoles DCPIP-red/mg chl
x h with either water or DPC as electron donor. The activity of D-10
was 30-60 μmoles DCPIP-red/mg chl x h with water and 50-80 μmoles DCPIP-
red/mg chl x h with DPC as electron donor, respectively.*

Again, a lag phase is observed for the inhibition of DCPIP reduction in thylakoids. However, this phase is much shorter so that after i.e.135min of illumination only about 40% of the initial activity is retained. At the same time, still about 95% of the chl and 85% of the car are bound to the membranes (Fig.1). In D-10, PS-II activity is rapidly inhibited with the onset of illumination.

Table 1 indicated that the three membrane preparations possess different pigment compositions although the concentration of total chl and total car per ml sample are similar. Therefore, the different bleaching kinetics of chl and car cannot be explained with a lack of carotenoids per se but point to a specific function of each car species.

Table 1. Pigment content (in nmoles of a 4 ml sample) of thylakoids, D-10, and D-144 prior to illumination.

	Thylakoids	D-10	D-144
Neoxanthin	8.7	10.4	5.1
Violaxanthin	18.7	10.7	12.6
Lutein	27.1	27.9	22.8
Chlorophyll b	40.0	50.8	33.2
Chlorophyll a	129.5	104.4	132.6
ß-Carotene	24.8	14.9	29.6
Chlorophyll a/b	3.24	2.06	3.99
Xanthophylls/ß-car.	2.20	3.29	1.37
Chl/car	2.13	2.43	2.37

A closer examination of the breakdown of each carotenoid in thylakoids, D-10, and D-144 (Fig.3) allows a more specific statement to be made. In thylakoids, neoxanthin,violaxanthin, and ß-carotene show essentially the same bleaching kinetic with a significant lag phase whereas lutein degradation starts later and continues slower. Since both PS-II activity (Fig.2) and PS-I activity (data now shown) are inhibited before massive breakdown of pigments occurs, either the bleaching of car is not responsible for the rapid loss in activity, or only very few car and chl molecules are affected during the early phase of the experiment. From the data of Fig. 3, these specific pigments cannot be identified.

The bleaching of D-10 is characterized by a similar behaviour of neoxanthin, violaxanthin, and lutein; ß-carotene however shows a linear breakdown during the first 200 min of the experiment. This breakdwon coincides well with the destruction of chl a (about 40% of the initial amount is lost after 200 min) and with the inhibition of DCPIP reduction (Fig.2). In this case, a close association of ß-carotene and chl a probably in the reaction centre itself is evident so that the rapid ß-carotene bleaching is a manifestation of its photoprotective action with the D-10 fraction.

In D-144, again lutein is the most stable carotenoid showing a short lag phase of 30 min. Neoxanthin is also retained during the first 30 min but is then the most rapidly degraded carotenoid. Violaxanthin and ß-carotene are linearly bleached with the onset of illumination. P-700 measurements in D-144 during illumination have shown that the activity of the reaction centre of PS-I is remarkably stable (data not shown) being destroyed only when less than 5% of chl and car remain present within

D-144. When inhibition of P-700 is initiated, the chl a/P-700-ratio was about 25 and the ß-carotene/P-700-ratio was approximately 1. These results clearly indicate that ß-carotene is essential for the preservation of PS-I activity by photoprotection of the reaction centre-chl a molecules.

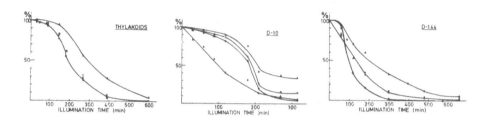

FIGURE 3. Bleaching kinetics of carotenoids (as percent of initial) in a) thylakoids, b) D-10, and c) D-144. Pigment identification: o Violaxanthin, ● Neoxanthin, ▲ Lutein, Δ ß-Carotene.

Since the consistently greater stability of lutein over all other carotenoids is not the result of different physical properties of the chromophore (cf. Krinsky, 1979), we conclude that lutein must be situated in a different environment than the other polyenes. The function of lutein is probably not to protect chl against photodestruction but to stabilize the structure of the pigment-protein complexes.

4. REFERENCES

Anderson JM and Broadman NK (1966) Fractionation of the photochemical systems of photosynthesis.I.Chlorophyll contents and photochemical activities of particles isolated from spinach chloroplasts, Biochim. Biophys. Acta 112, 403-421.
Braumann Th and Grimme LH (1981) Reversed-phase high-performance liquid chromatography of chlorophylls and carotenoids. Biochim.Biophys.Acta637, 8-17.
Braumann Th, Weber G and Grimme LH (1982) Carotenoid and chlorophyll composition of light-harvesting and reaction centre proteins of the thylakoid membrane, Photobiochem. Photobiophys. 4, 1-8.
Krinsky NI (1979) Carotenoid protection against oxidation, Pure Appl. Chem.51, 649-660.
Ridley SM (1977) Interaction of chloroplasts with inhibitors, Plant Physiol. 59, 724-732.

FUNCTION OF CHLOROPHYLLS AND CAROTENOIDS IN THYLAKOID MEMBRANES: CHLORO-
PHYLLS BETWEEEN PIGMENT-PROTEIN COMPLEXES MIGHT FUNCTION BY STABILIZING
THE MEMBRANE STRUCTURE

GRIMME, L. HORST[+]and JEANETTE S. BROWN[++]
[+]University of Bremen, FB Biologie/Chemie, D-2800 Bremen 33, FRG
[++]Carnegie Institution of Washington, Biology Dept., Stanford, CA94305
USA

I. INTRODUCTION

The molecular arrangement of chlorophylls and carotenoids with proteins
and lipids in the thylakoid membrane of chloroplasts determines their
function in light harvesting and energy transfer reactions. Seven different pigment-protein complexes have been isolated from spinach thylakoid membranes which were solubilized by SDS, isolated by PAGE and analyzed with respect to their pigment composition by RPLC (Braumann et al.
1982). All of these complexes contain chlorophyll a (chl a), chlorophyll b (chl b) and carotenoids (car), each in different quantities and
in different proportions: CP 1 and CP a show a high chl a/chl b ratio
and a low chl a/ß-car ratio, the LHCs show a low chl a/chl b ratio and
high amounts of xanthophylls (Menger et al., these proceedings).
Quantitatively most chl a molecules are arranged in LHCs and seem to act
as a funnel through which photons are channelled to few special chl a
molecules in reaction centre (RC) complexes. The broad red absorption
band of chl a observed in photosynthetic membranes has been attributed
to individual and overlapping absorption bands of differently bound
forms of chl a, which have been demonstrated by curve analysis of absorption spectra (630-730 nm) for many photosynthetic organisms. One
objective of our studies with the green alga Chlorella fusca is to gain
more insight into the molecular arrangement of its photosynthetic pigments within the complexity of the thylakoid membrane and into the stability of this arrangement. Especially absorption and fluorescence spectra of thylakoids and of pigment-protein complexes derived from them by
the use of detergents should give a key to see in which way chl molecules are involved in LHCs and RC-complexes and in the interaction
between these components.

II. MATERIALS AND METHODS

Chlorella thylakoid membranes were prepared by passing photoautotrophically grown cells (strain 211-15 of the Algal Culture Collection Göttingen, FRG) through a French press at 15,ooo psi in a medium containing
100 mM sucrose, 10 mM KCl, 1% bovine serum albumin, and 125 mM potassium
phosphate buffer, pH 7.2. Membranes were sedimented at 20,000g after
removal of large particles and whole cells by a low speed centrifugation. The membranes were washed successively with 10 mM KCl, 125 mM
potassium phosphate buffer, pH 7.2, 1mM EDTA, and finally with 50 mM
Tricine-NaOH, pH 8.
The washed membranes were resuspended to 1 mg chl/l in the latter buffer.
An equal volume of 1% SDS in 0.6 M Tris-HCl, pH 8 as solubilization medium was added to give a final weight ratio of SDS:chl = 10:1. The mixture was incubated at 4°C for 30 min in darkness, homogenized and centrifuged at 40,000g for 15 min. The supernatant (solubilizate) was
used for spectral measurements at 77K and for PAGE.

Sybesma, C. (ed.), Advances in Photosynthesis Research, Vol. II. ISBN 90-247-2943-2.
© *1984 Martinus Nijhoff/Dr W. Junk Publishers, The Hague/Boston/Lancaster.*

PAGE was performed according to Brown et al. (1982), absorption and fluorescence spectra at 77K were performed with instruments and by the procedure described by Brown and Schoch (1981).

III. RESULTS AND DISCUSSION

The absorption and fluorescence emission spectra of Chlorella thylakoid membranes before and after solubilization with SDS are shown in Fig.1. It is seen that this detergent affects drastically both, the absorption and the fluorescence emission spectra. The effect of SDS is obvious from the decrease in absorbance at 678 nm and the increase at 670 nm. The fluorescence emission spectrum lost all of the long wavelength fluorescence at 722 nm. Also most of the photochemical activity during solubilization is lost (data not shown).

However, the long wavelength absorption at 678 nm and the long wave-length fluorescence emission at 722 nm are not really lost completely in the SDS solubilizate. The separation by PAGE of this solubilizate reveals seven pigmented bands (band 1-7), of which six (band 1-6) were pigment-protein complexes. The low temperature absorption and fluores-cence emission spectra of all bands are shown in Fig. 2 A-D.

Fig. 2A collects the spectra of band 1,2,6 and 7, which all show only small absorption at 650 nm, indicating a reduced chl b content. Band 2 and to some extent also band 1 have long wavelength absorption, which is reflected by the fluorescence emission at 722 nm for band 2 and at 710 nm for band 1 in Fig. 2B. Also bands 3,4 and 5, which show a pro-nounced peak at 650 nm, indicating a high proportion of chl b in these LHCs, reveal some long wavelength absorption at 678 nm (Fig. 2C) how-ever no long wavelength fluorescence emission (Fig.2D).

The results show that SDS solubilization of thylakoid membranes from Chlorella fusca causes spectral shifts which seem to be based on spe-cific interactions between detergent and membrane components since repeated experiments of solubilization and of separation of pigment-protein complexes lead to the same spectral characters. Three different explanations for the spectral changes in Fig.1 have been given (Fawley et al., 1982): first, a specific and small portion of chl is solubil-ized into SDS micelles where it absorbes at 670 nm and fluoresces at 678 nm. Because the fluorescence yield of solubilized chl is high, a small amount can cause a large change in the emission spectrum; second, the interaction of different pigment-protein complexes are responsible for some of the observed spectral characters (Mullet et al., 1980) and detergents may break these interactions altering the spectral charac-teristics; third, noncovalently bound chl molecules are located in different regions of a protein which leads to different absorbing chlorophyll forms; point charges will cause spectral shifts of adja-cent pigment molecules (Davies et al., 1981) and any detergent in-fluencing such point charges will cause changes in the chl absorption. Probably a combination of these effects is operating. Depending on the chemical nature of the detergent, however, there might be differences. SDS will change absorption and fluorescence emission by specific solu-bilization of certain amounts of chl with concomitant dislocation of chl within the pigment-protein complexes as shown by computer assisted analysis of spectra (Fawley et al., 1982). Digitonin and tauro-desoxy-cholate (T-DOC) do not alter the absorption and, additionally T-DOC does not alter the fluorescence emission spectrum of Chlorella membranes

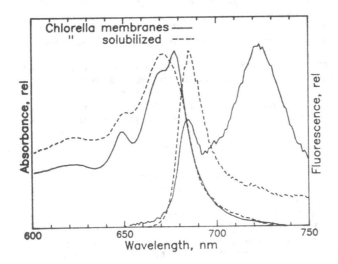

Fig. 1 Absorption and fluorescence emission spectra (77K) of
Chlorella membranes before (solid lines) and after (dashed lines)
treatment with the detergent SDS. Excitation at 438 nm;
slitwidth, 10nm; emission slitwidth, 3 nm.

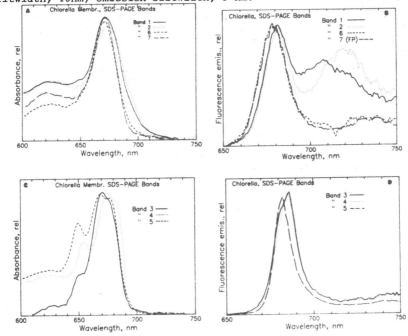

Fig.2 Absorption and fluorescence emission spectra (77K) of pigment-protein complexes from Chlorella fusca. Excitation at 438 nm, slitwidth, 10nm;
emission slitwidth, 3nm. All curves were normalized to similar peakheights.

(unpublished results), but both might shift the spectral components of isolated pigment protein complexes.

It has been suggested that the net bilayer structure of biomembranes may result from a complex interplay between protein and lipids which may not support bilayer structure of themselves (Madden, Cullis 1982). Detergents may have the unexpected effect serving to stabilize rather than destabilize bilayer organization. From our results we might conclude that SDS destabilizes the thylakoid bilayer structure by solubilizing structurally important chl molecules which themselves act as detergents in stabilizing the interaction between pigment-protein complexes within the bilayer structure of the photosynthetic membrane.

IV. REFERENCES

Braumann Th, Weber G and Grimme LH (1982) Carotenoid and chlorophyll composition of light-harvesting and reaction centre proteins of the thylakoid membrane, Photobiochem. Photobiophys. 4, 1-8.

Brown JS, Anderson JM and Grimme LH (1982) Antenna chl a complexes in mutant and developing barley,Photosynth. Res. 3, 279-291.

Brown JS and Schoch S (1981) Spectral analysis of chlorophyll-protein complexes from higher plant chloroplasts, Biochim. Biophys. Acta 636, 201 - 209.

Davies RC, Ditson SL, Fentiman AF and Pearlstein RM (1981) Reversible wavelength shifts of chlorophyll induced by a point charge, J.Amer.Chem.Soc. 103, 6823-6826.

Fawley M, Grimme LH and Brown JS (1982) Spectral analysis of detergent-solubilized photosynthetic membranes. Carnegie Inst. Yearbook 81, 38-40

Madden TD and Cullis PR (1982) Stabilization of bilayer structure for unsaturated phosphatidylethanolamines by detergents. Biochim.Biophys.Acta 684, 149-153.

Mullet JE, Burke JJ and Arntzen CJ (1980) A developmental study of photosystem I peripheral chlorophyll proteins, Plant Physiol.65,823-827.

V. Acknowledgements

We thank Marvin Fawley for skillful technical assistance.
Support from the Deutsche Forschungsgemeinschaft for LHG is gratefully acknowledged.

FUNCTION OF CHLOROPHYLLS AND CAROTENOIDS IN THYLAKOID MEMBRANES: PIGMENT
COMPOSITION OF LHCP-COMPLEXES OF SPINACH AND CHLORELLA FUSCA

W.MENGER, J.KNÖTZEL, Th.BRAUMANN AND L.H.GRIMME
UNIVERSITY OF BREMEN, FB BIOLOGIE/CHEMIE, 2800 BREMEN 33, FRG.

1. INTRODUCTION

Three different groups of pigment-protein complexes (PPC) have been isolated from chloroplast membranes by SDS solubilization and SDS-PAGE:
CP 1 and CP 1a as reaction centre complex of PS I, CP a as reaction centre complex of PS II and a group of light-harvesting chl a/b-proteins
(LHCP) (Anderson et al., 1978).
Since LHCP have similar characteristics, it has been concluded that the
main complex LHCP[3] may represent the monomer of oligomeric forms(Machold
et al., 1979). However, the pigment composition reported for LHCP is
still a matter of controversy, because a higher Chl a/b-ratio than postulated, varying amounts of ß-carotene, and different proportions of xanthophylls have been reported (Lichtenthaler et al., 1982; Siefermann-
Harms, Ninnemann, 1982).
These divergent results are mainly due to (i) polar stationary phases for
pigment separation, which produce artifacts, (ii) incomplete elution of
chl-protein complexes from homogenized gel slices, (iii) prolonged contact of chl-protein complexes with SDS, and (iv) co-migrating SDS-complexed pigments.
To overcome these problems, we used a mild SDS-PAGE (Braumann et al.,
1982) to separate chl-protein complexes from spinach and *Chlorella* thylakoids, developed a rapid reversed electrophoresis to remove co-migrating
pigments and analyzed the pigment content of LHCP by RPLC (Braumann,
Grimme, 1981).

2. MATERIAL AND METHODS

Spinach thylakoids were isolated, solubilized by SDS and separated by
SDS-PAGE according to Braumann et al. (1982).*Chlorella* thylakoids were
isolated from photoautotrophically grown cultures of *C.fusca* (strain
211-15, Algal Culture Collection Göttingen, FRG) by passing them through
a French press (20,000 psi) in a buffer containing 0.05 M potassium
phosphate, 0.01 M KCl and 0.3 M sucrose, pH 7.2. The homogenate was centrifuged at 21.000 g for 10 min. The pellet was incubated with 1 % SDS
and a SDS-Chl-ratio of 5 : 1 for 30 min in the dark at 4° C and then
centrifuged at 21.000 g. SDS-PAGE was performed according to Braumann
et al., 1982.

The LHCP were eluted from the gels by a modified reversed electrophoresis (Šnejdárková, Šnejdarková, 1981), or, for comparison, by the
procedure Braumann et al., 1982, forcing gel slices through a nylon net
and extracting overnight in 0.041 M Tris-boric acid buffer (pH 8.64).

Chlorophyll and carotenoid composition of LHCP from spinach and
Chlorella were determined by reversed-phase HPLC (Braumann, Grimme,
1981).

Sybesma, C. (ed.), Advances in Photosynthesis Research, Vol. II. ISBN 90-247-2943-2.
© *1984 Martinus Nijhoff/Dr W. Junk Publishers, The Hague/Boston/Lancaster.*

3. RESULTS AND DISCUSSION

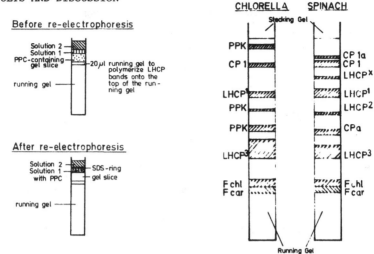

FIGURE 1. Reversed electrophoresis of gel slices.
FIGURE 2. Gel diagramms after SDS-PAGE with SDS-solubilized thylakoids from spinach and Chlorella.
Chl-protein complexes of spinach thylakoids are designated according to Anderson et al., 1978. Three PPC from Chlorella could not assigned unequivocally to the corresponding complex from spinach due to different positions in the gel.

Fig. 1 shows the experimental set up of the reversed electrophoresis. Tubes of 45 mm length (inner diameter 7 mm), filled with running gel (8 % acrylamide/bisacrylamide) about 10-15 mm below the top of the tube were pre-electrophorized for 3 h at 2 mA per tube. Gel slices of LHCP-bands were polymerized onto the top of the gel using 20 µl running gel. This procedure ensures that no air bubbles are trapped between gel slices and running gel. The PPC containing gel slices was overlayered with 0.15 ml solution 1 (0.025 M Tris, 0.075 M Glycin, 60% Glycerin (v/v), pH 8.8) and the tube was filled with solution 2 (2 M NaCl). Using the usual electrophoresis buffers, these gels were run with reversed polarity for 3 h at 3.5 mA per tube. After re-electrophoresis, SDS appeared as a sharp ring at the border between solution 1 and 2, while PPC migrated into solution 1. Solution 1 was carefully removed with a 50 µl-syringe.

Fig. 2 shows the separation pattern for spinach and Chlorella thylakoids. It is seen that least 3 PPC of Chlorella were assigned to known complexes of spinach thylakoids, CP 1, LHCP^1 and LHCP^3. LHCP^3 from Chlorella was analyzed in respect to its pigment composition and compared with the pigment composition of LHCP from spinach.

The results are summarized in Table 1 and Table 2. All LHCP-complexes are characterized by low Chl a/b-ratios and by high amounts of xanthophylls. However, the pigment data indicate that the multiple zoning on electrophoretic pattern is most probably not due to the aggregation of only one basic LHCP to various oligomeric forms. Table 2 indicates two distinct groups of LHCP with respect to their Chl and carotenoid compo-

sition: LHCP[1] and LHCP[3] with Chl a/b-ratio of about 1.5 and little ß-carotene, LHCP[x] and LHCP[2] with a relatively high Chl a/b-ratio and a Xan/
Car-ratio of about 10. The two groups are also distinguishable with respect to their ratios of Chl b/ß-carotene and Chl b/Neo.

TABLE 1. Distribution of chlorophylls and carotenoids in LHCP-complexes
of spinach and Chlorella (%Pigment in chl-protein complexes).

TABLE 2. Ratios between chlorophylls and carotenoids in LHCP-complexes
of spinach and Chlorella thylakoids (mol/mol).

| | | Spinach | | | Chlorella |
	LHCP[x]	LHCP[1]	LHCP[2]	LHCP[3]	LHCP[3]
Neoxanthin	2.35	3.17	2.23	4.17	3.5
Trihydroxy-α-carotene	-	-	-	-	7.1
Violaxanthin	0.43	0.76	1.11	1.23	1.9
Lutein	7.17	10.76	7.92	9.49	5.0
Chl b	28.99	35.68	28.84	30.27	38.8
Chl a	58.07	48.56	55.57	53.23	43.6
ß-Carotene	2.99	1.06	4.33	1.61	0
Chl a/b	1.89	1.52	1.99	1.54	1.3
Chl a/ß-Carotene	9.46	25.70	9.02	19.90	∞
Chl b/Lutein	2.42	2.13	2.35	2.30	7.76
Chl b/ß-Carotene	4.94	17.21	4.55	13.47	∞
Xan/ß-Carotene	2.77	11.83	2.82	9.73	∞
Chl b/Neo	8.83	5.80	7.51	4.56	11.09

Data represent the mean of three experiments

The pigment composition of the LHCP[3] from Chlorella (Tables 1 and 2)
clearly shows that this LHCP belongs to the first group of the LHCP of
spinach. However, there are some important points to be mentioned. This
LHCP[3] obviously does not contain any ß-carotene, as has been proposed for
highly purified LHCP-preparations, and shows a lower Chl a/b-ratio of
1.3. Since LHCP[3] from Chlorella is the most prominent chl-protein complex upon SDS-PAGE containing 60% or more of the Chl, this preparation
is probably in a purer state than the corresponding LHCP[3] from spinach
thylakoids. In other words, the divergent pigment data of LHCP[3] may be
explained by varying amounts of either co-migrating free pigments or
loosely associated chl and carotenoids of the LHCP which are dissociated
to a different extent depending on the experimental conditions.

We have therefore developed a technique which allows the rapid elution
of chl-protein complexes from the gel and which includes a further purification step (see Fig. 1). Upon reversed-electrophoresis, small molecules like SDS and SDS-complexed free pigments migrate fast into solution 1 and collect as a boundary layer between solution 1 and 2. The
slow-migrating chl-protein complexes can then be removed from solution1
without any impurity originating from this layer.However, up to now the
yield of the eluted chl-proteins is too small to characterize further
also the minor LHCPs.

Table 3 collects the pigment data of these purified preparations.
Table 4 compares the pigment pattern with those obtained with the
conventional elution technique.

TABLE 3. *Pigment distribution and ratios in re-electrophotized LHCP[1]
and LHCP[3] complexes from spinach*

| | Spinach | | | Spinach | |
	LHCP[1]	LHCP[3]		LHCP[1]	LHCP[3]
Neoxanthin	4.8	3.1	Chl a/b	1.0	1.0
Violaxanthin	1.3	1.0	Chl a/ß-Car	∞	∞
Lutein	10.2	10.7	Chl b/Lutein	4.1	4.0
Chl b	41.6	42.7	Chl b/ß-Car	∞	∞
Chl a	42.1	42.6	Xan/ß-Car	∞	∞
ß-Carotene	0	0	Chl b/Neo	8.7	13.8

TABLE 4. *Distribution of chlorophylls and carotenoids in LHCP-complexes
of spinach and Chlorella.*

| | Mol pigment/100 Mol Chl a | | | | | | |
| | | Spinach | | | Chlorella | Spinach | |
	LHCP[x]	LHCP[1]	LHCP[2]	LHCP[3]	LHCP[3]	LHCP[1]	LHCP[3]
Neoxanthin	4	7	4	8	8	11	7
Trihydroxy-α-carotene	-	-	-	-	16	-	-
Violaxanthin	1	2	2	2	4	3	2
Lutein	12	22	14	18	12	24	25
Chl b	50	74	52	57	89	99	100
ß-Carotene	5	2	8	3	0	0	0

It is seen that both, LHCP[1] as well as LHCP[3], show a Chl a/b-ratio of
1.0 and essentially no ß-carotene. The question now is whether these results do represent the true amount of Chl and carotenoids within LHCP or
are produced solely by the experimental procedure. We believe the first
to be the case for several reasons: (i) Chl a/b-ratio of 1.0 is completely reproduceable for both complexes also when different amounts of
Chl are applied onto the gels, (ii) LHCP[3] from *Chlorella* shows already
with the conventional elution procedure a Chl a/b-ratio of 1.3 and no
ß-carotene (because LHCP[3] is the major chl-protein complex and is therefore to a lesser extent subject to contaminations). Preliminary results
(data not shown) indicate that also with this complex a Chl a/b-ratio of
1.0 is achieved upon reversed electrophoresis, (iii) even if Chl a/b-ra-
tios above 1.0 and small amounts of ß-carotene are not the result of co-
migrating free pigments but indicate a small fraction of very loosely
bound pigments, the question arises whether these pigments are really
an essential part of what is called the light-harvesting apparatus.
In summary, at least LHCP[3] and its oligomer LHCP[1] in spinach and *Chlorella* thylakoids bind Chl a and Chl b in a stoichiometric relation of
1:1, contain no ß-carotene, and include 7 molecules neoxanthin, 2 molecules violaxanthin and 25 molecules lutein per 100 molecules Chl a and
Chl b. In *Chlorella* lutein is partly substituted by trihydroxy-α-caro-
tene which is the major xanthophyll within LHCP[3] of this green alga.

4.REFERENCES

Anderson JM, Waldron JC and Thorne WS (1978), FEBS Lett.92,227-233.
Braumann Th and Grimme LH (1981),Biochim.Biophys.Acta 637, 8-17.
Braumann Th, Weber G and Grimme LH (1982) Photobiochem.Photobiophys.4,
1-8. Lichtenthaler HK, Prenzel U and Kuhn G(1982),Z.Naturforsch.37c,
10-12. Machold O, Simpson and Møller (1979), Carlsberg Res.Commun.44,
235-254. Siefermann-Harms D and Ninnemann H (1982),Photochem.Photobiol.
35,719-731.Snejdárková (1981),Anal.Biochem.111, 111-114.

β-CAROTENE-PROTEIN AND OTHER CAROTENOID-PROTEINS OF BROWN
ALGAE AND OF CYANOPHYTES.

JACK BARRETT.

1. INTRODUCTION

Reaction centre complexes I and II (RCI,RCII) of green plants
brown algae and cyanophytes show spectral and HPLC evidence of
the presence of carotenoids (cf. Larkum and Barrett, 1983).
Estimated from the absorption at 480-500 nm the amount of carotenoid is 2-3 times greater in RCII than in RCI of green plants:
not all of this is β-carotene (Brauman et al., 1982). In the
RCs of brown algae only β-carotene is found (molar ration P-700:
Chla: β-carotene of 1:38: 4). Attempts to remove the inner antenna β-carotene with detergents causes collapse of the spectral
features of the red absorption of the RCs, and disappearance of
the long wave fluorescence emission peaks but an increase in the
overall fluorescence. The question is raised as to whether the
β-carotene is bound in a single pigment protein complex or is
directly associated with Chla molecules in the same protein.
In experiments in which RCI and RCII of the brown seaweed
Acrocarpia paniculata were exposed to the steroid detergents
cholate or deoxycholate (<0.1% w/v) no shifts of the A_{max}
were seen. In contrast, when either RCI or RCII were exposed
to the same concentration of detergents having long alkyl tails,
Triton X-100,LDAO,Zwittergens or Tweens then shifts to the blue
of certain A_{max} occured (Barrett, unpublished). The A_{max} of
the β-carotene of the RCI was displaced from 503 to 498 nm;
499 to 494 nm for RCII. Difference spectroscopy showed that
the inner antenna Chl680 was converted to a Chl660 form: the
Chl670 form remained unaltered. Hoarau and Remy (1978) found
similar shifts for Chl680 of the LHCP and of P-700 of some
green plants and red algae. These combined findings suggest
that the β-carotene and the Chla giving rise to the 680 nm
absorption lie in a mutual hydrophobic environment. Exposure
of both RCs to Triton X-100 at 1% level, of to deoxycholate,
yielded after density gradient centrifugation a highly buoyant
fraction enriched in β-carotene relative to Chl670. Any 660 nm
Chla-protein present would have been obscured by the excess of
the Chl670 present in this fraction from the gradient.

    In this current study attention has been focused on
the carotenoid-proteins that can be removed from O_2-evolving
preparations from the mesophilic cyanophyte Phormidium luridium
and the thermophile Mastigocladus sp. Two approaches have been
made. Firstly the use of, in glucose-salts buffer, of the
steroid conjugates glycocholate and glycodeoxycholate. Secondly,
the use of these and of alkyl detergents, having as polar group
either a sugar or a zwitterion, in the presence of 0.5 or 1 molar
citrate.

Sybesma, C. (ed.), Advances in Photosynthesis Research, Vol. II. ISBN 90-247-2943-2.
© 1984 Martinus Nijhoff/Dr W. Junk Publishers, The Hague/Boston/Lancaster.

2. EXPERIMENTAL

2.1 M̥ethods. P. Luridium was cultured at 37°C and Mastigocladus at 55°C in standard media, in an atmosphere of air plus 0.5 CO_2. 6 day old cultures were centrifuged at 3k x g and the cells washed twice with fresh media and then with a pH 7.1 buffer containing 0.1% w/v of albumin Fr. V (sucrose 0.2M/NaCl 30mM/MgSo$_4$ and K_2SO_4 10mM/CaCl$_2$ and MnSO$_4$ 1mM/MOPS 10 mM, pH 7.1). Algae from 1 L were suspended in 80 ml of the buffer plus albumin and were passed twice through a French press at 15 lb/sq. inch. The suspension was centriguged at 2k x g and the supernatant was similarly centrifuged. Particles were then obtained from centrifuging the supernatant serially, at 40k x g and at 185k x g. These fractions are termed heavy and light. Particles were washed with the buffer until no more blue tint could be seen and the particles were a bright yellow-green. Even so a small amount of tightly bound phycocyanin and allophycocyanin remain. Electrophoresis on Laemli gels of the particles revealed only slight differences in polypeptides. The two lots of particles, assayed in the buffer at pH 7.1 and 30°C had an O_2 -evolution rate of 300 µmol. mg Chl h^{-1} .

2.2 Experimental and Results

In sucrose-salt buffer.

Heavy particles from P. luridium (A440 nm / A485 nm = 2.4) were treated with glycocholate in buffer, pH 7.2 (0.1% w/v, detergent to Chla, 10/1) for 12 h at 4°C. This was followed by a second extraction. The particles were then treated with glycodeoxycholate in the same manner, but at 0.5% concentration of the detergent. Both solutions were centrifuged at 180k x g for 1 h. The clarified solutions were concentrated using Amicon cone filters (CF25) and the concentrates centrifuged through a 5-25 % sucrose gradient for 24 h at 250k x g. The glycocholate material gave a zone of carotenoid-protein at about 1/8th down the gradient, with some carotenoid at the bottom of the tube. This may be polymerized material. The glycodeoxycholate extract gave a zone a third of the way down the gradient. The carotenoid protein from the first detergent had an absorption spectrum free of Chl a and with Amax 485>465>430(sh) nm.
The carotenoid -protein isolated using glycodeoxycholate still had a trace of Chla. Amax were at 494>465>444 nm. The particles after this double detergent treatment had an Amax 438/Amax495 of 3.8.

The blue supernatant from the 180k x g centrifugation of the O_2-evolving particles was further centrifuged at 180k x g for 48 h. After this a narrow zone of a bright yellow carotenoid-protein was close to the meniscus of the gradient: traces of Chla could be removed by further such centriguation of diluted smaples. Amax were at 458>484>440,520(shs)nm. It should be noted that no detergent had been used to isolate this material. It has been observed that polyvalent ions, eg. sulphate, succinate or citrate, treble of quadruple the O_2 - evolving capacity of isolated spinach chloroplasts (Stewart, 1982). A similar enhancement of H_2O-splitting has been observed for the light and heavy particles of <u>Mastigocladus</u>, during this study. Furthermore the particle preparations which were initially clear when dispersed in sulphate of citrate at 0.5 - 2.0M concentration aggregated into membranous vesicles, which were dispersable in buffer alone. It was thought that the these polyvalent anions might be abstracting water molecules from the surface of the membranes, thus exposing new areas for attack by detergents or weakening some protein interactions. Detergents were thus chosen which because of their polar head group or presence of a polar amino acid would remain in solution in citrate or sulphate at high molarity.

Light particles were suspended in buffer at pH 7.1 containing 0.5M citrate, 1M citrate, and in buffer without any extra salts. A similar series was made with K_2SO_4. Glycocholate at pH 7.1 at 0.2% w/v was added and the mixtures were incubated, with stirring, at $21^\circ C$, for 2 h. After the incubation the suspensions were centrifuged at 185k x g for 18 - 36 h. With buffer only small amounts of the phycocyanins were found in the supernatants, and some residual buffer-soluble carotenoid-protein described above. With 0.5 and 1 molar citrate a yellow carotenoid with a pinkish tinge was found throughout the tubes. At 0.5M the particles were deposited at the bottom of the tube. A dramatic difference was seen with the experiment in 1M citrate. Here there was a narrow zone of yellow carotenoid protein, and below this a broad zone of the pinkish carotenoid. Half way down the tube the major part of the Chl-protein membrane had formed a sharp zone of dark green lamellae, which examination by the light microscope suggested had an ordered structure. This material when dispersed in 1M citrate in buffer was found to have O_2 evolving activity. Evidently there was considerable removal of carotenoid from the particles since the A440 nm / A490 nm had risen from 2.6 to 3.5. Taurocholate gave similar results. The dihydroxyl glycodeoxycholate was sufficiently soluble in 0.5M citrate to release some of the pinkish carotenoid from the membranes but was not effective in the 1M citrate.

Of the alkyl detergents (8 to 14 carbons) used,Zwittergen 14, dodecylmaltose and octylglucose, gave interestingly different results. Dodecylmaltose (DDM) gave a yellow zone of carotenoid-protein, after centrifugation at 185k x g for 36 h, at the top of the tube. While through the middle third of the tube extended a deep green band with a center of concentrated pigment at mid-tube. A blue-green granular deposit contained about equal amounts of both phycocyanins and a small amount of Chla-protein. The Chla-protein from the centre of the green zone had a ratio of A440 nm /A495 nm of 4.4, and was thus the freest of carotenoid-protein. Zwittergen 14 gave similar results, but was less soluble than DDM. With octyl-glucose the particles fragmented differently. Some carotenoid-protein floated to the top third of the tube, with 1M citrate. The Chl-containing membrane formed two zones. The major zone was a third way down the tube, while a minor zone was at two thirds down the tube. The

A438/490 nm was 3.4 and 3.6 respectively. A deposit of phycocyanins and Chl<u>a</u>-protein was at the bottom of the tube.

   Since the Chl-protein membrane, A440/495 of 4.4, prepared by the use of DDM, was free of carotenoid other than β-
carotene, it was exposed to dihydroxyl-steroid detergents to
test whether this remaining carotenoid was integral to the
inner assembly of pigments or was not integrated with the inner
Chl 680, referred to in the Introduction. The DDM ratio 4.4
Chl-membrane was incubated separately, in deoxycholate or glycodeoxycholate, at pH 7.1, 4°C, for 12 h. The deoxycholate
treated membranes gave a colourless supernatant when the incubated material was centrifuged at 185k x g for 6 h.   The
glycodeoxycholate gave a pale yellow supernatant and a light
fluffy deposit overlaying a firm dark green deposit. The supernatant contained β-carotene-protein, while the fluffy deposit
gave a Chl-protein complex which exhibited an altered spectrum:
A436/A495 nm of 2.7, with red peak at 675 nm, and marked
broadening on the lower wavelength side. The hard deposit had
a spectrum similar to the starting material. Clearly in order
to pull out the β-carotene-protein there results a change in the
original inner Chl 680. These findings are in accord with observations made by me with the brown alga <u>A</u>. <u>paniculata</u> (see
Introduction).

3.1 DISCUSSION

   Carotenoid-proteins in the photosynthetic membranes
may have three functions: as light-harvesters, as photopro-
tectants and as structural components which contribute to the
architecture of the PS membranes. In this study attention has
been centered on the O_2-evolving membranes of Cyanobacteria,
since essentially all light-harvesting pigments, other than
carotenoids are easily removed by procedures that do not release
the carotenoids. The glucose-salt buffer soluble carotenoid-
protein released from the PS membrane of <u>Mastigocladus</u> may be
removable because of phase transition changes which occur at
about 45°C, on going from T^{o} of growth to 4°C (cf. Quinn and
Williams, 1983). Holt and Krogmann (1981) have isolated water-
soluble carotenoid-proteins from wild cyanophytes. While some
carotenoids can be removed from the O_2-evolving membrane with
steroids having zwitterionic conjugates and in the presence of
citrate at high molarity (retaining O_2-evolving ability in
the membrane), alkyl-chain detergents which have a highly
water-soluble head are required for complete removal of all
but the inner core antenna β-carotene protein. Again the
structure of the detergent is very important, since the absence of the zwitterionic polar group results in rupture of
the β-carotene-protein Chl 680 relationship.

ACKNOWLEDGEMENTS
Supported by the National Science Foundation, U.S.A.

<u>Authors address</u>: Dr. Jack Barrett, 1230 York Ave., N.Y.C.

PROPOSAL FOR THE NOMENCLATURE OF REACTION CENTRES AND LIGHT-HARVESTING
COMPLEXES FROM THE PURPLE PHOTOSYNTHETIC BACTERIA.

At the pre-meeting workshop in Zürich on "The Molecular Structure and
Function of Light-harvesting Pigment-Protein Complexes and Photosynthetic
Reaction Centres" it was decided to try and standardise the nomenclature
for the reaction centres and antenna complexes from the purple photosynthetic bacteria. This was seen to be necessary as there has been some confusion
in the literature and because it is now necessary to name their constituent
polypeptides. The following scheme was proposed and generally accepted.

Reaction Centres

The term RC is the approved abreviation, with specific RC pigments being
identified as P870 or P800 i.e. with the use of "P". The RC polypeptides
should continue to be called "H", "M" and "L" not "α", "β", "γ".

Antenna complexes

The antenna complexes, when referred to in the context of the pigmented
holocomplex, should be called B800-850, B890 or B875 etc i.e. with the
use of "B" then the wavelength maximum (a) of the Bchl absorption(s) in
the near infra-red. Terms such as LH1, LH2 or LHP etc should be avoided
since they do not convey sufficient information as to which complex is
being referred to.
The antenna polypeptides should be called by greek letters, thus the
B890 from R. rubrum continous B890-α and B890-β as its polypeptides. α is
the group defined by the amino acid sequences with only one conserved His,
while the β chains contain two conserved His residues. Other polypeptides,
such as the colorless 14 KD one from B800-850 from Rps. capsulata should
be given other greek letters i.e. B800-880-γ.
We hope that this simple informative set of rules will serve to clarify
the literature.

Sybesma, C. (ed.), Advances in Photosynthesis Research, Vol. II. ISBN 90-247-2943-2.
© 1984 Martinus Nijhoff/Dr W. Junk Publishers, The Hague/Boston/Lancaster.

STRUCTURE AND FUNCTION OF THE REACTION CENTER FROM RHODOPSEUDOMONAS
SPHAEROIDES[*]

G. FEHER AND M. Y. OKAMURA

1. Introduction
2. Composition of the RC
3. Primary Structure of the Subunits
4. Three Dimensional Structure of RCs
5. Location of the Reactants on the RC
6. Should LM or LMH Be Defined as the RC?; The Role of the H Subunit
7. The Electrogenicity of the Electron Transfer Steps
8. Summary
9. Acknowledgements

1. INTRODUCTION

   A great deal of progress has been made in the elucidation of the
structure and function of RCs during the past decade. This topic has been
reviewed at several conferences on photosynthesis during the last few years.
Looking over these reviews, one discerns a steady buildup of new information and understanding at each successive conference. We hope that you
will agree with us at the end of the talk that this trend is still proceeding at a healthy rate.

   Some of the material to be presented has already been described in the
literature (for reviews, see for example, Refs. 1-4), but will nevertheless
be covered here briefly for the sake of completeness; the unpublished
material will be described in more detail.

2. COMPOSITION OF THE RC

   The RC is a multi-subunit protein to which several prosthetic groups
are attached in the right juxtaposition to provide an efficient transducer
for the conversion of electromagnetic (light) energy into chemical (free)
energy. It is located in the bacterial membrane. Consequently, it is very
hydrophobic and requires a detergent for isolation and to keep it in solution. We shall not describe the various purification procedures that have
been developed (5) but will assume as a starting point a uniform population
of purified RCs. Unless otherwise specified, we shall focus in this talk
on RCs obtained from the carotenoidless mutant R-26 of Rhodopseudomonas
sphaeroides in the presence of the detergent, lauryl-dimethyl amine oxide
(LDAO).

   The subunit structure of RCs from several bacterial species is shown
in Fig. 1. In each case, three subunits designated L, M, and H, are
clearly discernible. A similar pattern has been observed in R. capsulata,
a bacterium on which promising genetic experiments have been carried out (6).
The stoichiometry of the subunits has been determined to be 1:1:1 (1). The
molecular weights of the subunits as estimated from the electrophoretic
mobility (see values in parentheses in Fig. 1) deviate significantly from
their true values, as will be discussed later.

[*] Department of Physics, University of California, San Diego; La Jolla,
California 92093, U.S.A.

Sybesma, C. (ed.), Advances in Photosynthesis Research, Vol. II. ISBN 90-247-2943-2.
© *1984 Martinus Nijhoff/Dr W. Junk Publishers, The Hague/Boston/Lancaster.*

In addition, the following prosthetic groups are attached to the RC protein:

    4   Bacteriochlorophylls
    2   Bacteriopheophytins
    2   Ubiquinones
    1   Fe^{2+}

3. THE PRIMARY STRUCTURE OF THE SUBUNITS

The determination of the amino acid sequence of the RC polypeptides is essential to the detailed elucidation of its structure. The sequences of 25 to 28 amino acid residues of each subunit, determined by automated Edman degradation (7), are as follows:

```
                5              10              15             20             25
L:  H₂N-ALA-LEU-LEU-SER-PHE-GLU-ARG-LYS-TYR-ARG-VAL-PRO-GLY-GLY-THR-LEU-VAL-GLY-GLY-ASN-LEU-PHE-ASP-PHE-HIS)VAL-

M:  H₂N-ALA-GLU-TYR-GLN-ASN-ILE-PHE-SER-GLN-VAL-GLN-VAL-ARG-GLY-PRO-ALA-ASP-LEU-GLY-MET-THR-GLU-ASP-VAL-ASN-LEU-ALA-ASN-

H:  H₂N-MET-VAL-GLY-VAL-THR-ALA-PHE-GLY-ASN-PHE-ASP-LEU-ALA-SER-LEU-ALA-ILE-TYR-SER-PHE-TRP-ILE-PH, LEU-ALA- X -LEU-ILE-
```

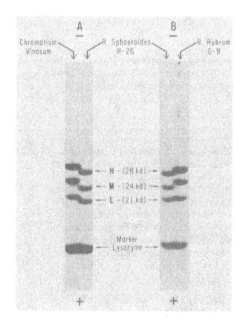

The sequences inside the shaded boxes were used to synthesize oligonucleotide hybridization probes (8). These were used to isolate restriction fragments from the DNA of R. sphaeroides that contain the structural genes of the subunits. The nucleotide sequence of the gene encoding the M-subunit was determined by the dideoxy method (9) using the phage vector Ml3 mp 7, 8, and 9. The sequencing strategy, as well as the restriction

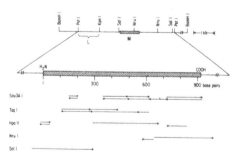

Fig. 1. Comparison of the subunit structure of RCs from several bacterial species by electrophoresis on a split SDS-polyacrylamide gel (from Ref. 1).

Fig. 2. Restriction map of the 13 kb Bam Hl fragment and sequencing strategy for the gene encoding the M-subunit. Arrows indicate the extent and direction of sequencing of subcloned fragments (from Ref. 8).

map of the fragment (Bam
Hl) that hybridized with
both the L and M probe,
is shown in Fig. 2. The
probe for the H subunit
did not hybridize to the
fragment.

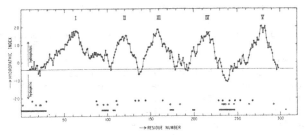

Approximately 94% of
the sequence was deter-
mined on both strands.
Additionally, 24% of the
sequence was confirmed by
amino acid sequence data
obtained from tryptic and
chymotryptic digests. The
amino acid sequence of
the carboxy terminus was
also determined.

*Fig. 3. Hydropathy profile of the sequence of
the M-subunit using a moving window of 19 amino
acids. Positions of basic (Lys, Ag) and acid
(Glu, Asp) are denoted by ⊕ and ⊖ . Solid line
indicates residues obtained by peptide analysis
(from Ref. 8).*

The sequence of the M-subunit was examined for potential membrane spanning regions by plotting the hydropathy value of a moving segment, as shown
in Fig. 3 (8,10). This analysis suggests that there are five hydrophobic
segments (having a hydropathy index > 1.5) that are long enough ($\sim$ 20 residues) to span the membrane in α helix (see peaks I to V of Fig. 3). A
significant homology was found between the amino acid sequence of the M-
subunit and a thylakoid membrane protein (M_r 32,000) from spinach (11) that
has been implicated in herbicide and quinone binding.

4. THE THREE DIMENSIONAL STRUCTURE OF RCS

The first step in obtaining the three dimensional
structure of RCs is to crystallize them. This has
been accomplished by Michel (12) who crystallized RCs
from R. viridis. In view of the smaller size and
better characterization of the RC from R. sphaeroides
we have concentrated on that system. In collaboration with Jim Allen, we have succeeded in crystallizing
RCs by two different methods (13):

A) The method of Michel (12) in which small
amphipathic molecules were included with RCs in LDAO
and ammonium sulfate followed by concentration of the
solution by vapor diffusion.

B) The method of Garavito and Rosenbush (14)
in which the protein is purified in β-octyl glucoside
and concentrated in the presence of polyethylene
glycol (PEG) by vapor diffusion.

*Fig. 4. Crystals of
RCs from R.sphaeroides
obtained by two dif-
ferent procedures. A)
Method of Michel (12).
B) Method of Garavito
and Rosenbush (14).*

Both of these methods gave crystals as shown in
Fig. 4. Interestingly, the growth pattern of the
crystals is very different in the two procedures as
is the size of the unit cell. The unit cell dimen-
sions of the RC crystall from R. sphaeroides were found to be significantly
smaller than those reported for R. viridis (12). This is not too surprising
since RCs from R. viridis have several cytochromes attached to them.

5. LOCATION OF THE REACTANTS ON THE RC

The location of the binding site of the <u>primary quinone, Q_A</u>, was obtained by photoaffinity labeling (15). The analog 2-azido[^{3}H] anthraquinone (AzAQ) was used to substitute for ubiquinone as the primary acceptor in RCs. When the reconstituted RCs were illuminated at 80 K and subsequently warmed, some of the label became covalently attached to the protein. From a gel electrophoresis profile of the labeled RCs, we found that the M-subunit was selectively labeled (Fig. 5a) as compared to control preparations in which the primary quinone binding site was filled by ubiquinone before labeling (Fig. 5b). Thus, the primary quinone binding site is located on or very close (within ~ 5 Å) to the M subunit. Experiments are in progress to determine the amino acid sequence of the peptide to which the labeled AzAQ is covalently bound.

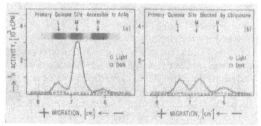

Fig. 5. *SDS-PAGE of RCs labeled with* [^{3}H] *AzAQ. a) RCs depleted of ubiquinone prior to labeling with AzAQ. b) RCs with primary quinone filled by ubiquinone prior to labeling (from Ref. 15).*

The location of the binding site of the <u>secondary quinone, Q_B</u>, could not be obtained by photoaffinity labeling since no suitable analog compound that binds to the site has so far been found. Instead, we determined the location of the Q_B-binding site by measuring the inhibition of the electron transfer from Q_A to Q_B in the presence of affinity purified antibodies directed against the different RC subunits (16).

The electron transfer was analyzed by measuring the charge recombination between Q_A^- or Q_B^- and $(BChl)_2^+$. Since the recombination time between Q_A^- and $(BChl)_2^+$ is an order of magnitude faster ($\tau \approx 100$ msec) than between Q_B^- and $(BChl)_2^+$ (~ 1 sec) (see Fig. 11), a measurement of the slow component yields the fraction of RCs with functional Q_B.

Figure 6 shows that electron transfer to Q_B is inhibited in the presence of anti-M Fab but is unaffected by anti-H Fab and anti-L Fab. From these experiments, as well as studies that showed that anti-M Fab competes with Q_B for the binding site, we conclude that the Q_B binding site is also located on or near the M-subunit (16).

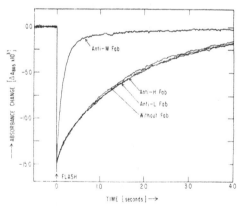

Fig. 6. *Recovery kinetics of absorbance changes at 865 nm of RC in the presence of affinity purified antibodies (from Ref. 16).*

The location of the binding site of <u>cyt c</u> was investigated by several different approaches (17). In one, crosslinking agents (e.g., dithiobispropionimidate, dimethyl suberimidate) were used to link cyt c and cyt c_2

to RCs; cyt c crosslinked to the L and M subunits, whereas cyt c_2 cross-linked only to L. In the other method the inhibition of electron transfer by antibodies against the subunits was investigated as described above (Fig. 6). Fab fragments of antibodies specific against the L and M sub-units blocked electron transfer from both cyt c and cyt c_2, whereas antibodies specific for the H subunit did not block the reaction. We con-clude that the cytochrome binding site is close ($\sim$ 10 Å) to both the L and M subunits.

6. SHOULD LM OR LMH BE DEFINED AS THE RC?; THE ROLE OF THE H-SUBUNIT

Several years ago, we showed that the H subunit can be dissociated from the RC to yield a photochemically active LM complex (18,1). What is then the function of H? To answer that question we have investigated differences in the physical and chemical characteristics of the LM and LMH complexes; in particular, we focused our attention on the electron transfer rate from Q_A to Q_B (19).

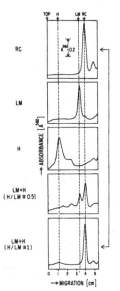

In order to ascertain that the LM complex was not denatured during the dissociation procedure, we developed conditions to reconstitute LM and H to form "native" RCs (i.e., LMH). The reconstitution was assayed by gel electrophoresis under nondenatu-ring conditions, as shown in Fig. 7. The H, LM, and RC are easily distinguished by their electro-phoretic mobility. The bottom frame of Fig. 7 shows that equimolar proportion of LM and H insured complete reconstitution.

To test the functional integrity of the recon-stituted complex, the kinetics of charge recombination following a laser flash was monitored by absorption changes at 865 nm (Fig. 8). The identical kinetics observed in RCs and the recon-stituted LMH complex shows that neither LM nor H were damaged by the dissociation procedure. The characteristic time in the LM unit is faster than that observed in RCs indicating that the recombina-tion had occurred from Q_A rather than from Q_B, as is the case in RCs. This suggests an impairment of the electron transfer from Q_A to Q_B in the LM complex.

Fig. 7. Assay for re-constitution of RCs from LM and H by elec-trophoresis under non-denaturing conditions (from Ref. 19).

The electron transfer from Q_A to Q_B was determined more directly by a method introduced by Parson (20) in which cyt-c oxidation is measured after two successive light flashes. After the first flash, the state DQ_AQ_B is formed which cannot be excited by the second flash since Q_A cannot accept another electron. If, however, the electron is transferred to Q_B, the state DQ_AQ_B is excitable and D^+ can oxidize a cyt c. Thus, the degree of cyt-c oxidation depends on the electron transfer. The results of this experiment are shown in Fig. 9. The rate of electron transfer from Q_A to Q_B is $\sim 10^3$ slower in LM than in RC. Other findings include a decrease in the stability to denaturation by detergents in the absence of the H-subunit and a decrease in the affinity of ubiquinone binding by approximately an order of magnitude in the LM unit (19).

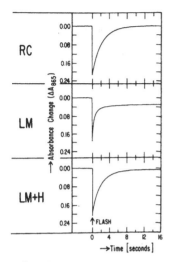

Fig. 8. *Charge recombin-ation between Q_A^- or Q_B^- and D^+ in RCs, LM, and Reconstituted LMH. The lack of complete recov-ery in the LM is attri-buted to a loss of the electron from the qui-none (from Ref. 19).*

We now turn to the question of whether LM or LMH should be called the reaction center. The answer to this question depends on the choice of the definition of an RC. If we wish to define the RC in terms of a unit capable of performing the charge separation between D^+ and Q_A^-, then LM is the appropriate complex. If, on the other hand, we want to include the effective electron transfer from Q_A to Q_B, the H subunit seems to be essential and LMH is the appropriate unit. Thus, the de-finition of the RC depends on one's point of view, as illustrated in the cartoon of Fig. 10.

7.   THE ELECTROGENICITY OF THE ELECTRON TRANSFER STEPS

Following the absorption of a photon, the electron transfers through a series of reactions (Fig. 11) that establish a potential difference across the bacterial membrane. According to the chemiosmotic hypothesis (22), this voltage dif-ference, together with the concomitant proton gradient provides the driving force for ATP production (23). It is, therefore, of interest to determine the contributions of the different electron transfer steps to the potential difference (i.e., the "electrogenicity" of each step). This problem has been addressed by several groups by measuring either electrochromic shifts (24) or electrical signals (25-27). The results to be discussed here were obtained by a tech-nique introduced by Trissl et al. (28). It is based on the measurement of the dis-placement current (or the associated voltage) produced by the photon induced charge separation in RCs deposited on a thin teflon film.

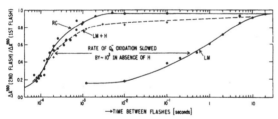

Fig. 9. *The amount of cyt c oxidized on the second flash as a function of the time be-tween flashes in RCs, LM, and reconstituted LMH (from Ref. 19).*

A schematic representation of the experi-mental setup is shown in Fig. 12. A thin teflon film separates two aqueous compart-ments. One compartment contains phospho-lipid vesicles into which RCs were incorporated. The vesicles break up at the air-water interface to form a monolayer (30) which is adsorbed onto the teflon film. Il-lumination of the adsorbed layer evokes an electrical signal. From the time dependence of the signal, the particular transfer

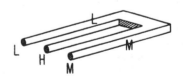

Fig. 10. *Cartoon representa-tion of the question whether LM or LMH is the smallest active unit. It depends on our point of view (i.e., the way we look at it).*

can be identified; from the amplitude the relative electrogenicity of the step can be deduced. To obtain absolute values, the effective orientation of the RCs has to be known. This was determined by comparing the signal amplitude to that obtained from 100% functionally oriented RCs in the presence of cyt c (31). The effective orientation thus obtained was $\sim$ 15% with D preferentially facing the aqueous solution.

The voltage pulse following a laser flash obtained from RCs with one quinone is shown in Fig. 13. The voltage decays

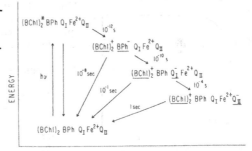

Fig. 11. *Schematic representation of the electron transfer reactions. Transfer times are given for room temperature and rounded to the nearest power of 10 (after Ref. 21).*

with a time characteristic of the recombination rate between Q_A^- and D^+ ($\sim$ 100 ms). When excess quinone was added to the system, the time increased to $\sim$ 1 sec in accord with the known recombination rate between Q_B^- and D^+. No voltage component having a characteristic time of $\sim$ 100 µsec, corresponding to the electron transfer from Q_A to Q_B, was observed. In the presence of cyt c a biphasic recovery was obtained (Fig. 14). The slow component is due to the electron transfer from cyt c^{2+} to D^+ and is in agreement with previous findings (26). When two laser flashes were used in the presence of cyt c^{2+}, a voltage buildup with a characteristic time of $\sim$ 2.5 ms was observed after the second flash. This was attributed to the electron transfer step $Q_A^- Q_B^- \rightarrow Q_A Q_B^=$.

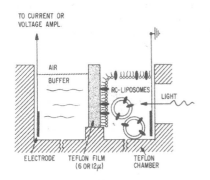

Fig. 12. *Experimental setup (not to scale) used to measure electrogenicity of electron transfer steps (after Ref. 29).*

From the above results, the following relative electrogenicities of the electron transfer step were obtained:

$$DQ_A Q_B \xrightarrow[\text{hv}]{1.0} D^+ Q_A^- Q_B \xrightarrow{<0.03} D^+ Q_A Q_B^- \xrightarrow[\text{cyt c}]{0.4} DQ_A Q_B^-$$

$$DQ_A Q_B^- \xrightarrow[\text{hv}]{1.0} D^+ Q_A^- Q_B^- \xrightarrow{0.12} D^+ Q_A Q_B^=$$

where the numbers above the arrows indicate the relative electrogenicity of the step. In the above scheme the intermediate acceptor (BPh, see Fig. 11) was omitted since the time resolution of our system ($\sim$ 50 ns) made it difficult to observe the recombination reactions between D^+ and

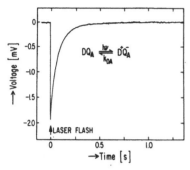

Fig. 13. *Voltage response from RCs with one quinone. The same result was obtained from RCs with 2 quinones in the presence of one electron transfer blocker ($\sim 10^{-4}$ M terbutryne) (from Ref. 29).*

BPh^-. Trissl was able to measure this reaction with a time resolution of 2 ns and found that it contributed 2/3 to the electrogenicity of the D^+Q^- step (27).

8. SUMMARY

In this talk, several topics on the structure and function of RCs were reviewed and some recent findings that seem to us of particular interest were highlighted. In view of the time limitation and the extent of the field, several topics could not be covered (e.g., the mechanisms of electron transfer and action of electron transfer blockers, the interaction of RCs with the bc_1 complex, the question of protonation and proton pumping, the development of a genetic system to study RCs, etc.). Some topics dealing with questions to which no definitive answers have been obtained were omitted. One of these deals with the role of Fe^{2+}, another with the assignment of functions to <u>all</u> six tetrapyrrole pigments.

The determination of the primary structure of RCs from several bacterial species will undoubtedly be completed in the near future. The methodology used in this endeavor (i.e., the isolation of the gene) provides a basis of analyzing and producing site specific mutations, for the study of the regulation of the synthesis of the RC protein, and for the evaluation of evolutionary relationships among photosynthetic organisms.

Now that RCs have been crystallized, their three-dimensional structure should be forthcoming, although in view of the

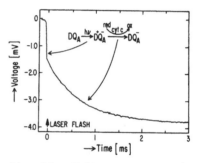

Fig. 14. *Voltage response in the presence of cyt c^{2+} (after Ref. 29).*

size of the protein it may take a few years. Once the coordinates of the reactants become known, advances in the detailed theoretical understanding of the electron transfer mechanisms are expected. Hopefully, some of the problems and questions raised here will be answered at the next (7th) International Conference on Photosynthesis.

9. ACKNOWLEDGEMENTS

The work reviewed here represents the effort and collaboration of a large number of individuals. We are particularly indebted to those whose work has not yet appeared in print: E. Abresch, Y. Blatt, R. Debus, A. Gopher R. A. Isaacson, D. Kleinfeld, M. Montal, M. I. Simon, L. A. Steiner, and J. C. Williams. The work from our laboratory was supported by grants from the National Science Foundation (PCM 82-02811), the National Institutes of Health (GM-13191), and the U. S. Department of Agriculture (GM-00106-05).

REFERENCES

(1) Feher G and Okamura MY (1978). In Clayton RK and Sistrom WR, eds. The photosynthetic bacteria, Chap. 19, pp. 349-386. New York: Plenum Press.

(2) Blankenship RE and Parson WW (1978), Annu. Rev. Biochem. 47, 635-653.

(3) Olson JM and Thornber JP (1979). In Capaldi RA, ed. Membrane proteins in energy transduction, pp. 279-340. New York: M. Dekker.

(4) Okamura MY, Feher G, and Nelson N (1982). In Govindjee, ed. Photosynthesis: energy conversion by plants and bacteria. Vol. 1, Chap. 5, pp. 195-272. New York: Academic Press.

(5) Gingras G (1978). In Clayton RK and Sistrom WR, eds. The photosynthetic bacteria, pp. 191-231. New York: Plenum Press.

(6) Marrs B (1981), J. Bacteriol. 146, 1003-1012.

(7) Sutton MR, Rosen D, Feher G, and Steiner LA (1982), Biochem. 21, 3842-3849.

(8) Williams JC, Feher G, and Simon MI (1983), Biophys. J. 41, 122a. Williams JC, Steiner LA, Ogden R, Simon MI, and Feher G (1983), Proc. Natl. Acad. Scis. (USA) (in press).

(9) Sanger F, Nicklen S, and Coulson AR (1977), Proc. Natl. Acad. Scis. (USA) 74, 5463-6467.

(10) Kyte J and Doolittle RF (1982), J. Mol. Biol. 167, 105-132.

(11) Zurawski G, Bohnert HJ, Whitfeld PR, and Bottomley W (1982), Proc. Natl. Acad. Scis. (USA) 79, 7699-7703.

(12) Michel H (1982), J. Mol. Biol. 158, 567-572.

(13) Allen J and Feher G (unpublished results), presented by Feher G at the Biophysical Society Meeting, San Diego, February 1983 (Biophys. J 41, 1983).

(14) Garavito M and Rosenbush JP (1980), J. Cell Biol. 86, 327-329.

(15) Marinetti TD, Okamura MY, and Feher G (1979), Biochem. 18, 3126-3133.

(16) Debus RJ, Valkirs GE, Okamura MY, and Feher G (1982), Biochim. Biophys. Acta 682, 500-503.

(17) Rosen D, Okamura MY, Abresch EC, Valkirs GE, and Feher G (1983) Biochem. 22, 335-341. Okamura MY and Feher G (1983), Biophys. J. 41, 122a.

(18) Okamura MY, Steiner LA, and Feher G (1974), Biochem. 13, 1394-1403.

(19) Debus RJ, Okamura MY, and Feher G (1981), Biophys. J. 33, 19a. Debus
      RJ, Okamura MY, and Feher G (1983) (unpublished data; manuscript in
      preparation).

(20) Parson WW (1969), Biochim. Biophys. Acta 139, 384-396.

(21) Okamura MY, Debus RJ, Kleinfeld D, and Feher G (1982). In Trumpower BL,
      ed. Function of quinones in energy conserving systems, Chap. V,
      pp. 299-317. New York: Academic Press.

(22) Mitchell P (1968), Glynn Research, Bodmin, England.

(23) For reviews, see: Junge W and Jackson JB (1982) and Ort DR and
      Melandri BA (1982). In Govindjee, ed. Photosynthesis, Vol. 1,
      Chapters 12 and 13. New York: Academic Press.

(24) For a review, see: Cramer WA and Crofts AR (1982). In Govindjee, ed.,
      Photosynthesis, Vol. 1, Chap. 9. New York: Academic Press.

(25) Drachev LA, Frolov AD, Kaulen AA, Kondrashin VD, Samuilov VD, Semenov AY
      and Skulachev VP (1976), Biochim. Biophys. Acta 440, 637-660.

(26) Packham NK, Dutton PL, and Müller P (1982), Biophys. J. 37, 465-473.

(27) Trissl H-W (1983), Proc. Natl. Acad. Scis. (USA), in press.

(28) Trissl H-W, Darszon A, and Montal M (1977), Proc. Natl. Acad. Scis.
      (USA) 74, 207-210.

(29) Blatt Y, Gopher A, Montal M, and Feher G (1983), Biophys. J. 41, 121a.
      Blatt Y, Gopher A, Kleinfeld D, Montal M, and Feher G (1983), manuscript
      in preparation.

(30) Schindler H (1979), Biochim. Biophys. Acta 555, 316-336.

(31) Schonfeld M. Montal M, and Feher G (1979), Proc. Natl. Acad. Scis.
      (USA) 76, 6351-6355.

Authors' address: Department of Physics, B-019
                   University of California, San Diego
                   La Jolla, California 92093 U.S.A.

THE BACTERIAL REACTION CENTER: WHERE IS THE CENTER?

V. WIEMKEN and R. BACHOFEN

1. INTRODUCTION

The reaction center (RC) of phototrophic bacteria is responsible for primary photochemistry. It was originally described as a complex composed of three polypeptides (H, M, and L) with the following cofactors: four bacteriochlorophylls, two bacteriopheophytins, one carotenoid, one or two quinones, and a non heme iron (1). Subunit H can be removed from the RC without loss of photochemical activity (1,2). Recently, the characteristic absorption spectrum and its changes upon photooxidation reflecting the primary charge separation were found also in a preparation showing one single band on SDS-PAGE at the position of subunit L (3). However, this is not sufficient to prove that only one polypeptide is present in the preparation and that the band in the gel really contains subunit L. The preparation was obtained by ultrafiltration in the presence of LDAO and DOC at room temperature in the absence of protease inhibitors. Hence it may well be that breakdown products of the larger subunits comigrated in the gel with subunit L (see our results).

To get more insight what the smallest functional unit of the RC may be we digested isolated RC (H,M,L) progressively with proteinase K. A still functional preparation of the RC was obtained containing two polypeptides. One of them, migrating in PAGE with an apparent molecular weight of 21 kDa was identified by amino terminal sequencing as subunit L. The other is smaller approximately 18kDa and most probably is a fragment of subunit M. The two polypeptides were so intimately associated that they could not be separated without inducing a concomitant loss of function.

2. MATERIALS AND METHODS

Chromatophores and RC of Rhodospirillum rubrum G-9 were prepared according to published methods (4) except that Tris-HCl buffer was replaced by 25mM triethanolamine buffer pH 8, 0.025% LDAO (=TEA-buffer). RC (OD 802nm,1cm =2; OD 280/802nm=1.4) were digested with proteinase K(0.8mg/10ml)(Boehringer, Mannheim, FRG) under the conditions given in the legends to the figures. The activity of proteinase K was stopped by the addition of 4mM phenylmethanesulfonyl fluoride from a 200mM stock solution in ethanol. Proteinase K was separated from the digested RC by gel filtration on Sepharose 6B (column 35x3.2cm) in TEA-buffer. The grey-blue fractions were pooled and further purified by adsorption on DEAE-cellulose (column 20x1.5 cm) followed by washing with 200 ml of the TEA-buffer, 200 ml of the buffer containing 60 mM NaCl, and 50 ml of the buffer with 100 mM NaCl. The pigmented fraction still adsorbed on the top of the column was eluted with 300 mM NaCl in the buffer .It was pooled, dialyzed against the buffer and then used for PAGE and activity tests. Part of the same sample was dialyzed extensively against distilled water for Edman degradation. SDS-PAGE and sample preparation were performed as before (5). Absorption spectra were measured on an Uvicon 810 spectrophotometer. Photochemical activity of the RC was measured as photooxidation of P-870 and of cytochrome c_2. Measurements were performed on an Aminco DW-2 spectrophotometer using the dual wavelength mode 600nm-650nm for photooxidation of P-870 and 540nm-550nm for photooxidation of cytochrome c_2. Samples were illuminated with light of 700nm-900nm (IR-filter, Balzers, Lichtenstein) at an intensity 1.5x10^4 erg/cm^2 x s . Amino-terminal sequences were determined by Edman degradation(6).

Sybesma, C. (ed.), Advances in Photosynthesis Research, Vol. II. ISBN 90-247-2943-2.
© 1984 Martinus Nijhoff/Dr W. Junk Publishers, The Hague/Boston/Lancaster.

3. RESULTS AND DISCUSSION

The near IR spectrum of the RC remained virtually unchanged when the purified
RC (subunit H,M, and L) were treated with the unspecific proteinase K for up
to 50 min (Fig.1). Only after prolonged incubation the absorption at 870nm decreased while the absorption at 770 nm increased indicating a change of the
pigment-protein interactions which paralleles loss of photochemical activity.
The proteinase K treated preparations still having the characteristic absorption spectrum of native RC, however, had a drastically changed polypeptide
pattern as shown by PAGE (Fig.2).

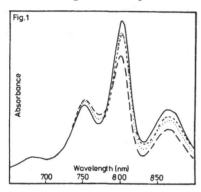
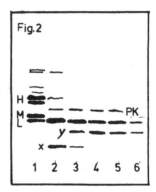

FIGURE 1. Near-IR absorption spectra of RC incubated with proteinase K for 0—
, 30···· ,50 - - -, and 90— —min at 25°C. FIGURE 2. SDS-PAGE of RC (subunit
H,M,L) incubated with proteinase K (PK) for 1) 0, 2) 5, 3) 10, 4) 20, 5) 30,
6) 60min, at 25°C. x=unknown digestion fragment,y=supposed subunit M.

After only a short incubation time subunit H and M disappeared from their original position and at lower molecular weights in region of subunit L the
staining intensity was much increased. A small polypeptide(x) appeared (Fig.2,
slot 2) and after prolonged incubation disappeared again. Finally the band at
the position of subunit L regained the original staining intensity and a new
band of a smaller polypeptide appeared (y) (Fig.2,slot4). Both bands showed a
similar intensity of Coomassie blue staining. A large preparation of the two-
polypeptide RC (Fig.2, slot 4) was prepared under the conditions given in the
legend to Fig.3. The preparation was repurified by gel filtration and ion exchange chromatography as described to remove the protease and the digestion
products of the RC (Fig. 3).

The photochemical activity of the three-subunit RC (control) was compared
with the activity of the digested two-polypeptide RC. The reduction of oxidized P-870 in the dark is slower in the digested preparation than in the control (Fig.4). The reduction rate is in both preparations faster in the presence of ascorbic acid. However, the kinetics of the reversible photooxidation
of cytochrome c_2 are closely similar for the two-polypeptide RC and the control (Fig.5).This suggests that subunit L, the only subunit intact upon digestion, carries the binding site for cytochrome c_2. This is in agreement with
the result that cytochrome c_2 was specifically cross-linked to subunit L (7).
The different kinetics for reversible photooxidation in the digested preparation compared with the control might be due to the fact that subunit M carries
the binding site for quinones (8); since M is reduced in size by the digestion, the quinone dependent reduction of P870 may be impaired.

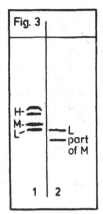

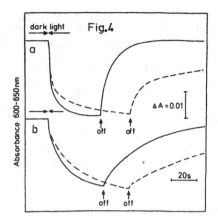

FIGURE 3. SDS-PAGE of the control (1) and of repurified digested RC (2) incubated with proteinase K for 30 min at 25°C. FIGURE 4. Reversible photooxidation of RC (OD 802nm,1cm=0.75) a) control without --- and —— with ascorbic acid (0.5mM) b) two-polypeptide RC without ---; and with —— ascorbic acid(0.5mM).

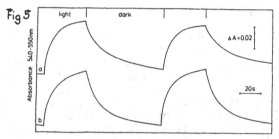

FIGURE 5. Reversible photooxidation of cytochrome c_2 (5μM) with RC(1.5μM) and ascorbic acid 5μM a) control b) two polypeptide RC.

        For an unequivocal identification of the two polypeptides obtained after digestion and purification the amino-terminal sequences in the preparation were determined by Edman degradation. Only two main amino acid sequences were found, namely Ala, Leu, Leu, Ser, Phe, Glu, -, Lys and Gly, Val, Gly, Asp, - , Ala. The former is identical with the amino-terminal sequence of subunit L as found in Rhodopseudomonas sphaeroides (9) and in Rhodospirillum rubrum (G-9) (10). One of the polypeptides in the digested preparation can therefore clearly be attributed to subunit L. The latter sequence can not yet be identified as part of other subunits as it does not correspond with any of the known amino-terminal sequences of the subunits H,M or L. It is not likely to be a fragment of subunit H since this subunit is rapidly digested, even in intact chromatophores without producing fragments which can be recognized on the gel by formation of either new bands in the region of the RC bands or producing intensity change of the M or L bands(11).

Rather unlikely is finally the possibility that subunit L is digested from its C-terminal end producing the band of the smaller peptide while the remainder migrates exactly to the position of subunit L, since treatment of R with carboxypeptidase Y had no effect on any of the subunits as judged from the banding pattern on SDS-PAGE. This indicates . that the C-terminal ends are well protected from protease. Thus the two-polypeptide RC seems to be compose of the intact subunit L and of subunit M shortened by around 60 amino acids o the totally around 230. Since the fragment of the polypeptide lost is quite substantial, one would expect, if subunit M alone would be the pigment carry-ing polypeptide,that the pigment-protein interactions are disturbed.

Attempts to digest selectively and without loss of activity one of the two polypeptides by other specific proteases (trypsin,α-chymotrypsin) failed Also after gel filtration in the presence of 0.025% LDAO and 40mM DOC on Sephacryl S-200 (column 125x2.5cm) the two polypeptides remained together. Interestingly, when chromatophore membranes are treated with proteinase K a cytosolically exposed piece of 16 amino acids is cleaved from subunit L (12) whereas this does not occur when isolated RC are treated. Possibly, in the isolated RC this piece is protected from an attack by proteases by detergent molecules.

From these results we propose that subunit L as well as part of subunit M form the smallest functional unit of the RC.

References

1. Feher G and Okamura MY (1978). In Clayton RK and Sistrom W, eds. The pho-tosynthetic bacteria p. 349-396. New York: Plenum Press. 2.Snozzi M and Bachofen R (1979) Biochim. Biophys. Acta 546, 236-247. 3.Gimenez-Gallego G, Suanzes P and Ramirez JM (1982) . FEBS Lett. 149, 59-64 4.Zurrer H. Snozzi M.Hanselmann K, and Bachofen R (1977) Biochim. Biophys. Acta 460, 273-279. 5. Wiemken V, Theiler R, and Bachofen R (1981) J. Bioenerg. Biomembranes 181-194. 6. Petersen JD, Nehrlich S, Oyer PE, and Steiner DF (1972) J.Biol. Chem 4866-4871. 7. Rosen D, Okamura MY, Abresch GE, Valkirs GE and Feher G (1983) Biochemistry 22,335-341. 8. Marinetti TD, Okamura MY and Feher G (1979) Biochemistry 18, 3126-3133. 9. Sutton MR, Rosen D, Feher G, and Steiner LA (1982) Biochemistry 21, 3842-3849. 10. Theiler R, Cuendet PA, Wiemken V, ar Zuber H (1980) .Fifth International Congress on Photosynthesis. p.567 Halkidi ki, Greece 11. Wiemken V and Bachofen R (1982) Biochim. Biophys. Acta 681, 72-76. 12. Brunisholz R Wiemken V Suter F Bachofen R and Zuber H (1983) Workshop on molecular structure and function of light-harvesting pigment-protein complexes and photosynthetic reaction centers. Zurich.

ACKNOWLEDGEMENT We are grateful to Rolf Theiler and Franz Suter for the amino-terminal sequencing of our preparation. We thank Markus Purro for his excellent technical assistence and Andres Wiemken for help with the manuscript. This work was supported by grant 3.582.79 of the Swiss National Foundation.

Authors address: Vreni Wiemken Institute of plant Biology, University Zurich,Zollikerstr.107 8008 Zurich,Switzerland

STUDIES ON THE IRON-BINDING SITE
OF BACTERIAL REACTION CENTERS

STEFAN BODMER and REINHARD BACHOFEN

1. RHODOSPIRILLUM RUBRUM

1.1 INTRODUCTION

Most reaction center preparations from photosynthetic bacteria are usually isolated as a complex of three polypeptides (H, M and L) and contain, besides other cofactors, nonheme iron in stoichiometric quantities. The exact structural arrangement of the various components of the reaction center is not yet established. However subunit H could be separated from the M-L complex without loss of photochemical activity (Okamura et al., 1974). Marinetti et al. (1979) showed that the primary quinone binding site is located on or close to the M subunit, whereas Gimenez-Gallego et al. (1982) found that even subunit L alone retains the characteristic absorption spectrum and the photochemical activity. We found no report so far on the binding site of the nonheme iron and therefore decided to study the iron binding site of the reaction center of Rhodospirillum rubrum using radioactive labeling and electrophoretic techniques.

1.2 METHODS

Rhodospirillum rubrum G-9 was grown phototrophically on the medium described by Ormerod et al. (1961) without yeast extract and pepton, but supplemented with Iron-59 (Ferric chloride, 1 mCi/280 ml).
For isolation of chromatophore membranes, the cells were washed and resuspended in 20 mM Tris buffer (pH8.0) and passed twice through a French pressure cell at 100 bar. The crude chromatophores in the supernatant were then purified by several successive centrifugations in 10 mM phosphate buffer (pH 7.0) and treatment with 5 mM NaEDTA (pH 7.0) to remove proteins loosely bound to the membranes. Reaction centers were prepared according to the method of Snozzi and Bachofen (1979). All steps were carried out in the dark at 4°C.
Samples for electrophoresis were prepared by solubilization of concentrated reaction center preparations (Abs at 280nm=9) in 10 mM Tris buffer (pH 7.0), 5% sucrose, 1% LiDodSO$_4$ at 50°C for 20 minutes. LDS-polyacrylamid gel electrophoresis was performed as follows: slab-gels (1.5 mm thickness) with a gradient of acrylamide (8 to 16%) were run at constant power in the cold (approx. 4°C) for about eight hours using the buffer system of Laemmli (1970), but with LiDodSO$_4$ instead of NaDodSO$_4$. All gels were prerun for 45 minutes prior to loading of the samples.
Identical lanes on a single gel were either stained for total protein with Coomassie brilliant blue or for c-type cytochromes according to Thomas et al. (1976). From the same gel duplicate tracks were cut out and sliced into 2 mm pieces and the radioactivity in the samples measured in a liquid scintillation counter. Autoradiograms of 59-Fe labeled samples were obtained using Fuji RX-(safety)-4 films, exposed to gels immediately dried after electrophoresis. This seems to be mandatory, since gels which were pretreated by staining and destaining retained no 59-Fe which comigrated with a polypeptide of the reaction center.
The position of the reaction center subunits H, M and L on gels is already well established.

Sybesma, C. (ed.), Advances in Photosynthesis Research, Vol. II. ISBN 90-247-2943-2.
© *1984 Martinus Nijhoff/Dr W. Junk Publishers, The Hague/Boston/Lancaster.*

1.3 RESULTS AND DISCUSSION

We have isolated reaction centers from Rhodospirillum rubrum G-9 cells labeled
with the isotope 59-Fe. On LDS-polyacrylamide electrophoresis the preparation
shows the typical three bands H, M and L and some minor contaminations (Figure
1). The specific staining technique for c-type cytochromes gave no positive
reaction, indicating that there is no such cytochrome among the contaminations
(data not shown).

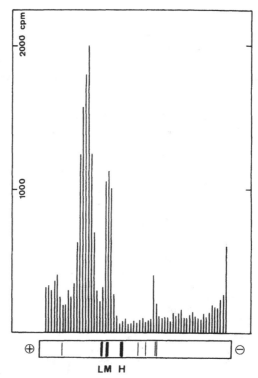

Figure 1. Analysis of bound 59-Fe to the reaction center of Rhodospirillum
rubrum G-9. The exact conditions are stated in the text.

By using mild conditions during denaturation of the reaction center and LDS-
electrophoresis in the cold we were able to separate the reaction center into
its subunits without total loss of bound iron. Still some of the iron bound to
the reaction center before denaturation dissociates from the protein and
migrates as a broad, non specific band below the L subunit (figure 1).
Of the three components of the reaction center only subunit M still has bound
59-Fe. Since the bound nonheme iron is in close association with the primary
acceptor of the reaction center, we conclude from our results that the primary
acceptor site is located on the M subunit, rather than on the L subunit.

2. MASTIGOCLADUS LAMINOSUS

2.1 INTRODUCTION

Nechushtai et al. (1983) reported on a purification of the photosystem 1 reaction center from this thermophilic cyanobacterium and compared its subunit composition, photochemical and immunological properties with those of photosystem 1 reaction centers from green algae and higher plants.
The isolation and characterisation of subunit 1 as a fully active photosystem 1 reaction center by its own is described by Binder et al. (this symposium).
Analogous to the study on the binding site of the nonheme iron in Rhodospirillum rubrum we tried to localize the iron binding site of photosystem 1 reaction center of Mastigocladus laminosus.

2.2 METHODS

Mastigocladus laminosus was grown on medium D of Castenholz (1970) complemented with 59-Fe (Ferric chloride, 1 mCi/1000 ml). Cells were grown for 3 days at 40°C, during which some 28% of the total label was incorporated in the cells (4 g wet weight). The method described by Nechushtai et al. (1983) was followed to prepare membranes and to extract and purify reaction centers. The fractions containing the reaction centers eluting from a DEAE cellulose column (checked by absorption spectrum and SDS-polyacrylamide gel electrophoresis) were pooled and concentrated on another DEAE cellulose column. All steps were carried out in the dark at approx. 4°C.
Samples for electrophoresis were prepared by solubilization of concentrated reaction center preparations with 1% $LiDodSO_4$ at various temperatures and incubation times. LDS-polyacrylamide gel electrophoresis was performed as described in the section 'Rhodospirillum rubrum'.

2.3 RESULTS AND DISCUSSION

We have isolated reaction center 1 of Mastigocladus laminosus cells labeled with the isotope 59-Fe. The exact subunit composition of this reaction center is still a matter of discussion (see Binder et al., this symposium). At room temperature (25°C) the reaction center 1 does not dissociate into the subunits with apparent molecular weight of 70 kDal and 50 kDal, respectively (Figure 2).

By increasing both temperature and incubation time (or detergent to protein ratio) during sample preparation dissociation is observed. Figure 2 shows that the upper of the two high molecular bands (around 100 kDal) in the non-dissociated form disappears with increasing dissociation and at the same time the 70 kDal and 50 kDal bands appear. The lower of the two high molecular bands seems unchanged on the Coomassdie blue stained gel (figure 2 A). Bound chlorophyll can be found on both high molecular bands in the non-dissociated form as well as on the 70 kDal and 50 kDal form of subunit 1, when treated for 20 minutes at 50°C. The autoradiogram (figure 2 B) however shows that neither the 70 kDal nor the 50 kDal form contain radioactive iron in detectable amounts. In the non-dissiciated form (25°C, no incubation with detergent) the upper of the two high molecular bands contains most of the iron, whereas after treatment for 20 minutes at 50°C almost all the iron is found attached to the lower band. Although Binder et al. (this symposium) show a fully active photosystem 1 reaction center with a apparent molecular weight of 50 kDal, we could not detect any bound iron to this subunit after electrophoresis.

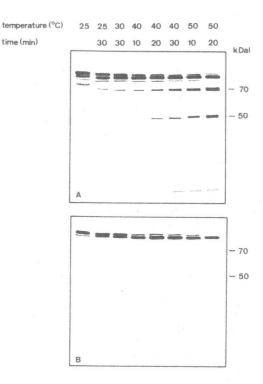

Figure 2. LDS–polyacrylamide gel electrophoresis of Mastigocladus laminosus reaction center 1. Varying temperature and incubation time during sample preparation. Explanations see text.

REFERENCES

Binder A et al. The photosystem 1 reaction center of a thermophilic cyanobacterium, this symposium.
Castenholz RW (1970) Schweiz Z Hydrologie 32, 538–551.
Gimenez–Gallego G et al.(1982) FEBS letters 149, 59–62.
Laemmli UK (1970) Nature 227, 680–685.
Marinetti TD et al. (1979) Biochem. 18, 3126–3133.
Nechushtai R et al. (1983) Proc. Natl. Sci. USA 80, 1179–1183.
Okamura MY et al. (1974) Biochem. 13, 1394–1402.
Ormerod JG et al. (1961) Arch. Biochem. Biophys. 94, 449–463.
Snozzi M and Bachofen R (1979) Biochim. Biophys. Acta 546, 236–247.
Thomas PE et al. (1976) Anal. Biochem. 75, 168–176.

ACKNOWLEDGEMENT

This work was supported by the Swiss National Foundation grant no. 3.582.79

Authors address: Institute of Plant Biology, University of Zurich, Zollikerstrasse 107, CH-8008 Zurich, Switzerland.

LOW TEMPERATURE MAGNETIC RESONANCE AND OPTICAL SPECTROSCOPY OF CHROMATO-
PHORES AND (CRYSTALLINE) REACTION CENTERS OF *RHODOPSEUDOMONAS VIRIDIS*.

F.G.H. van Wijk, T.J. Schaafsma
Department of Molecular Physics, Agricultural University
Wageningen, De Dreijen 11, NL-6703 BC Wageningen,
The Netherlands
and
H. Michel
Max Planck Institüt für Biochemie, D-8033 Martinsried
bei München, W. Germany.

ABSTRACT
The low temperature spectroscopic properties of the reaction centers
(RC's) of *Rps. viridis* in crystalline form have been compared with those
of reaction centers in suspension and chromatophores. At. 4.2 K and 1.5 K
the fluorescence spectrum of the dithionite reduced chromatophores exhibits a band at 1065 $+_1$nm. This emission was used to obtain FDMR spectra
of the RC P^R - state, yielding $|D| = (160 \pm 2) \times 10^{-4} cm^{-1}$, $|E| = (38 \pm 1)$
$\times 10^{-4} cm^{-1}$, in good agreement with published EPR results. The FDMR spectra correspond to an increase of fluorescence, whereas a decrease is predicted for Bchl-b in vitro (Den Blanken, Hoff (Chem. Phys. Lett. (1983)
96, 343). The sign-reversal can be explained by singlet energy transfer
from antenna to RC.
The fluorescence spectrum from the crystalline RC's at low temperature is
identical with that from suspensions of RC's except for the relative intensities of the two bands. This is further support for the nativeness
of the crystalline form.

INTRODUCTION
It is now possible to obtain reaction centers (RC's) in crystalline
form from the Bchl-b containing purple bacterium *Rhodopseudomonas viridis*.
These crystalline RC's showed no substantial bleaching of the primary
donor without addition of ubiquinone. (Michel, 1982). Improved preparations did show bleaching of the primary donor P960 at room temperature
(Zinth et al., 1983). At low temperature however the crystalline structure may be damaged by ice-formation of intra-crystalline water. This may
also damage the nativeness of the RC's. We therefore studied the low-temperature spectroscopic properties of the crystalline RC's.

MATERIAL AND METHODS
Chromatophores were diluted in glycerol/water mixtures at pH 7.0 or in
65% w/w sucrose/water. Crystalline RC's and suspension of RC's were also
brought into 65% sucrose. Prior to all experiments the sample was reduced
in dim green light with an aliquot of solid $Na_2S_2O_4$, followed by rapid
freezing to 77 K and immersion in liquid helium. All experiments were
carried out using 600 nm excitation light from a CW Rhodamine 6G dye
laser (Coherent Radiation 590) pumped by an Ar^+-laser (Coherent Radiation
CR-3 equiped with a CR-4 plasmatube). The FDMR apparatus is described
elsewhere (G.F.W. Searle et al., to be published in Proceedings Symposium
on picosecond Chemistry and Biology, London, 1982), modified by use of
the dyelaser and of a Varian 152A instead of a RCA 31034 photomultiplier
tube, as emission from RC's and chromatophores is in the range 900 -
1100 nm.

Sybesma, C. (ed.), Advances in Photosynthesis Research, Vol. II. ISBN 90-247-2943-2.
© *1984 Martinus Nijhoff/Dr W. Junk Publishers, The Hague/Boston/Lancaster.*

RESULTS AND DISCUSSION

The emission spectrum from the reduced chromatophores shows one peak at 1065 $\pm$ 1 nm (fig. 1), shifting to the blue upon ageing.

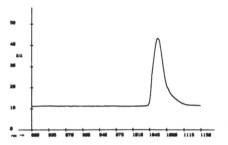

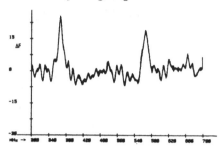

Fig.1. Emission spectrum from dithionite reduced chromatophores at 4.2 K in glycerol/water at pH7. Maximum at 1065 $\pm$ 1 nm. Emission intensity in arbitrary units.

Fig.2. FDMR spectrum from dithionite reduced chromatophores as detected at 1065 nm at 4.2 K. No other peaks were detected outside the range shown. $|D| - |E|$ = 367 MHz $|D| + |E|$ = 593 MHz. Signal intensity in arbitrary units.

This emission is assigned to fluorescence from the antenna-Bchl-b S_1-state. The fluorescence detected magnetic resonance (FDMR) spectrum given in fig. 2 was obtained in this emission band. (For a recent review on FDMR of bacterial photosynthesis see Hoff, 1982). From this spectrum $|D|$ was calculated to be $(160 \pm 2) \times 10^{-4}$ cm^{-1}, and $|E| = (38 \pm 1) \times 10^{-4}$ cm^{-1}. This is in good agreement with previously published results obtained with EPR-spectroscopy (Prince et al., 1976).

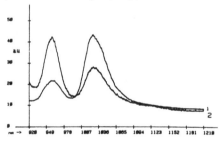

Fig. 3. Emission spectra from dithionite reduced RC's at 1.5 K in 65% sucrose. 1. suspension, maxima at 960 $\pm$ 2 nm and 1026 $\pm$ 2 nm. I_{960}/I_{1026} = 0.4 and
2. crystalline, maxima at 959 $\pm$ 2 nm and 1028 $\pm$ 2 nm. I_{959} I_{1028} = 1.7. a.u. = arbitrary units.

Fig. 3 shows the emission spectrum from crystalline RC's and a suspension of RC's. Although assignment of the two emissions is difficult, both of the bands represent emission from a level higher in energy than the antenna S_1-state at 1065 nm. This permits two ways for radiationless decay from the high energy level created by excitation with 600 nm light as shown in fig. 4.

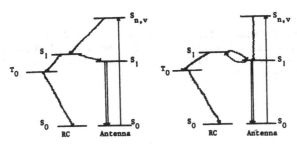

Fig.4. Two models for radiationless decay from a higher excited antenna singlet energy level. Left: decay via the S_1-state of the primary donor in the RC; right: decay directly to the S_1-state of the antenna bchl-b.

From the work of Den Blanken and Hoff (1983) one predicts a decrease of fluorescence in FDMR of Bchl-b in vitro, and also for the case of the lefthand model in fig. 4. Since an increase of fluorescence has been found (fig. 2) one can rule out this model. The sign reversal can be explained by a single singlet energy transfer. This agrees with the sign of FDMR-spectra from crystalline RC's, which show a decrease in fluorescence change (data not shown).

From the emission spectra in fig. 3 it is concluded that the nativeness of the RC's in the crystal is not seriously harmed since both peaks are present and no shifts beyond experimental error are observed. Partial denaturation caused both peaks to disappear without a change in relative intensities.

Several possibilities exist for the assignment of the two emission peaks: vibration progression; fluorescence from the excited dimer and an intermediate state; from the excited dimer and another photosynthetic pigment such as 'voyeur' Bchl-b, or from the reduced and unreduced state of the primary donor. In the last case the difference found in the relative emission intensities in crystalline form and in suspension may be explained by poor accessibility of the crystal to the reducing agent. This will be the subject of further study.

REFERENCES

Den Blanken, H.J., and Hoff, A.J. (1983). Sublevel decay kinetics of the triplet state of bacteriochlorophyll a and b in methyltetrahydrofuran at 1.2 K, Chem. Phys. Lett. 96, 343–347.

Hoff, A.J.(1982). ODMR spectroscopy in photosynthesis II. The reaction center triplet in bacterial photosynthesis in Triplet state ODMR spectroscopy, Clarke R.H., ed. John Wiley & Sons, New York, chapt. 9.

Michel, H. (1982). Three dimensional crystals of a membrane protein complex. The photosynthetic reaction centre from *Rhodopseudomonas viridis*, J. Mol. Biol. 158, 567–572.

Prince, R.C., Leigh, J.S., and Dutton, P.L. (1976). Thermodynamic properties of the reaction center of *Rhodopseudomonas viridis*. In vivo measurement of the reaction center bacteriochlorophyll–primary acceptor intermediary electron carrier, Biochim. Biophys. Acta. 440, 622–636.

Zinth, W., Kaiser, W., and Michel, H. (1983). Efficient photochemical activity and strong dichroism of single crystals of reaction centers from *Rhodopseudomonas viridis*, Biochim. Biophys. Acta, 723, 128–131.

ACKNOWLEDGEMENTS
The authors thank dr. H. Scheer for a generous gift of Bchl-b.
This investigation was in part supported by the Netherlands Foundation
for Chemical Research (SON) with financial aid from the Netherlands
Organization of Pure Research (ZWO).

POLARIZED INFRARED SPECTROSCOPY OF BACTERIAL REACTION CENTERS : THE LMH
AND LM COMPLEXES IN RECONSTITUTED MEMBRANE.

E. NABEDRYK, D. TIEDE, P.L. DUTTON and J. BRETON

INTRODUCTION

The bacterial photosynthetic reaction center (RC) of Rps. sphaeroides is
composed of three protein subunits called L, M, H, having the respective
molecular weight of 28000, 32000, 36000 as determined from the amino
acid compositions (1). After isolation of the RC, the H subunit can be
removed and all of the chromophores remain in the final LM complex which
is still fully functional with respect to the primary light-induced
reactions (1-3).
We have recently used polarized infrared (IR) spectroscopy to examine
the extent and organization of protein structures in membrane associated
proteins (4-6). In particular, with membrane reconstituted RC, we have
determined the net orientation of the α-helical segments with respect
to the membrane plane (6). In this article, we have made a comparison of
the protein structures in the LMH and LM complexes in order to analyze
the composition and the orientation of the secondary structures localized within the LM complex. We have investigated the UV circular dichroism (CD) and polarized IR spectra of the LMH and LM complexes reconstituted in lipid vesicles. Our results show that transmembrane
α-helices are present in both LMH and LM complexes. In addition, we also
discussed some general features of membrane IR dichroism spectra.

2. MATERIALS AND METHODS

The photosynthetic bacteria Rps. sphaeroides R 26 was grown and the
chromatophore membranes prepared as described earlier (2, 6). RCs
were isolated with the detergent LDAO (2, 6). The LM complex was prepared by precipitation of the H subunit (1, 2). Reconstitution of the RC
complexes into phospholipid vesicles was performed as already reported
(2, 6). Orientation of the membrane vesicles for IR dichroism analysis
was obtained by air-drying the suspension onto CaF_2 discs (4-6). The
method for quantitative analysis and calculation of α-helix tilt angles
from the IR dichroism spectra and UVCD spectra has been extensively described in our previous papers (4-6).

3. RESULTS AND DISCUSSION

3.1. Analysis of IR spectra

Polarized IR difference spectra $(A_{//}-A_{\perp})$ in the 1900-1100 cm^{-1} region
for the reconstituted LMH and LM membrane multilayers are compared in
Fig. 1. Both spectra show similar dichroism for the amide I (at 1656
cm^{-1}) and the amide II (at 1545 cm^{-1}) bands attributable to oriented
α-helices. Qualitatively, a positive $A_{//}-A_{\perp}$ dichroism signal is associated with the alignment at less than 55° average from the membrane
normal. The positive signal for the amide I band and the negative signal
for the amide II band are consistent with an alignment of the axes of
the α-helix segments preferentially along the normal to the membrane,

Sybesma, C. (ed.), Advances in Photosynthesis Research, Vol. II. ISBN 90-247-2943-2.
© *1984 Martinus Nijhoff/Dr W. Junk Publishers, The Hague/Boston/Lancaster.*

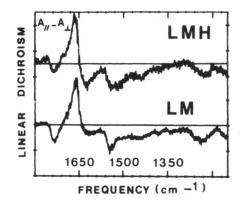

FIGURE 1. IR linear dichroism spectra of air-dried LMH and LM reaction centers reconstituted in phosphatidylcholine vesicles.

since in these structures the amide I transitions tend to be aligned along the helix axis, while the amide II transitions tend to have an orientation perpendicular to this axis (4). In addition, the dichroism of the lipid carbonyl ester (1738 cm^{-1}) and phosphate (1240 cm^{-1}) groups is characteristic of oriented bilayers structures. In these respects, the IR dichroism of the reconstituted LM complex is very similar to that of the LMH complex indicative of a similar orientation of the α-helix axes out of the membrane plane in both cases.

Furthermore, the LMH dichroism spectrum (Fig. 1) shows additional absorptions, in particular a negative dichroism at 1630 cm^{-1} which is absent in the LM complex. This additional feature which has a derivative shape gives the LMH amide I dichroism band an asymmetrical shape. Although this allows for the interpretation of two independent absorptions (α-helix at 1656 cm^{-1} and β-structure at 1630 cm^{-1}) in the amide I region with opposite dichroism, there are other phenomena which can lead to a derivative type distortion appearing in these spectra. Differences of similar amplitudes as those shown on Fig. 2 have been observed for other treatments of the samples. Firstly, we have found that the extent of dehydration of the sample modifies in a complex manner the IR spectra in the amide I and amide II regions (Fig. 3) as has been described for soluble proteins (7). In addition to the loss of free water absorptions, changes in the protein absorption bands were observed, notably in the amide I (1656 cm^{-1} and 1630 cm^{-1}) and amide II (1545 cm^{-1}) regions (Fig. 3). It must be noticed that in all cases, the IR dichroism spectra of the reconstituted membranes were found to have

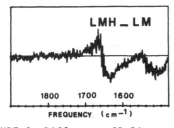

FIGURE 2. Difference IR linear dichroism spectra between LMH and LM reaction centers.

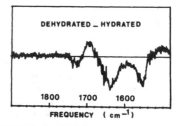

FIGURE 3. Difference IR spectra between dehydrated and hydrated reaction centers.

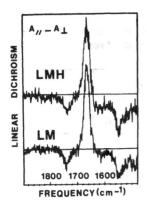

FIGURE 4. IR linear dichroism spectra of air-dried LMH and LM reaction centers. Samples were covered with nujol.

little dependence on the state of hydration.

In addition, optical interference and dispersion arising from multiple reflections occuring at the various dielectric interfaces present in an air-dried sample can perturb the observed dichroism spectra. In order to remove the major source of reflection occuring at the membrane/air interface, we have investigated the IR dichroism spectra after sealing the membrane multilayers with a covering of spectroscopic grade paraffin oil (nujol). The preservation of the linear dichroism of the RC chromophores in the visible region and of the polar lipid groups in the IR region shows that the samples have not been disordered by this treatment. However, the nujol has several effects on the visible and IR spectra including a significant improvement in the baselines of the polarized spectra. Furthermore, the IR dichroism spectra of LMH and LM reconstituted membranes (Fig. 4) become quite similar. The amide I and amide II dichroism signals now are more symmetrical and the negative dichroic signal at 1630 cm^{-1} for the LMH samples is removed. However, the extent of the amide dichroisms due to α-helix orientation is still retained.

3.2. Estimation of the tilt angle of the α-helices

Both random and α-helix secondary protein structures have overlapping amide I and amide II absorptions. For a quantitative determination of α-helix orientation, the experimental dichroic ratio ($D = A_{//}/A_{\perp}$) must be corrected to account for the extent of α-helix contribution to yield the α-helix dichroic ratio, D_α. This extent of α-helix was determined from analysis of UVCD spectra. The UVCD spectra of LMH and LM are quite similar and their fitting with reference spectra (6, 8) yields similar α-helix contents of 48 % for LMH and 50 % for LM. With these fractions of α-helix content, the values of D_α and the tilt angle ϕ_α of the α-helices with respect to the membrane normal can be calculated. These values D, D_α and ϕ_α are summarized in Table I. They are found to be essentially the same for the LMH and LM complexes suggesting that similar values must also exist for the H subunit. In order to include the extremes in the range of possible α-helix contents (4, 6), upper limits for ϕ_α are placed at 35° for both LMH and LM complexes. As discussed previously (4, 6), the likely non-perfect ordering of the multilayers will make the actual value be smaller than this.

TABLE I. Orientation of α-helices in LMH and LM reaction centers. The average was obtained from five different samples for each type of preparation.

% α-helix	LMH 48 $\pm$ 4		LM 50 $\pm$ 9	
	Amide I	Amide II	Amide I	Amide II
D_α	1.15	.89	1.14	.89
D	1.32	.79	1.30	.80
ϕ_α	31°	29°	32°	30°

To summarize, from our data, it can be concluded that in both LMH and LM complexes, all the protein secondary structures are oriented almost to the same extent. This observation suggests a model of the RC in which all three subunits L, M, H, have transmembrane α-helices. The observed differences between LMH and LM in the absence of nujol could be due to differences i) in the state of hydration in LMH and LM, ii) in the optical properties of the multilayers, iii) in oriented secondary structures (ex : β-structure) which are removed with nujol. Similar findings were observed on other oriented air-dried membranes such as chromatophore, thylakoid and purple membrane. We find that distortions of the amide I and amide II absorptions brought about by dehydration and/or optical reflection effects are a general feature of protein infrared spectra. These effects are more evident in samples with small extents of dichroism (eg. air-dried samples of soluble proteins such as cytochrome c or ribonuclease) since these effects are present to the same extent, but largely hidden in samples with large dichroic signals (eg. oriented purple membrane). These results indicate that a more precise determination of membrane protein orientation by polarized IR spectroscopy requires these effects to be taken into account.

REFERENCES

1. Feher G. and Okamura MY (1978) in the Photosynthetic Bacteria. (Clayton RK and Sistrom WR eds.) pp. 344-386. New-York, Plenum Press.
2. Tiede DM and Dutton PL (1981) Biochim. Biophys. Acta 637, 278-290.
3. Blankenship RF and Parson WW (1979) Biochim. Biophys. Acta 545, 429-444.
4. Nabedryk E and Breton J (1981) Biochim. Biophys. Acta 635, 515-524.
5. Nabedryk E, Gingold MP and Breton J (1982) Biophys. J. 38, 243-249.
6. Nabedryk E, Tiede DM, Dutton PL and Breton J (1982) Biochim. Biophys. Acta 682, 273-280.
7. Poole PL and Finney JL (1983) Biopolymers 22, 255-260.
8. Chen YH, Yang JT and Chau KH (1974) Biochemistry 13, 3350-3359.

Authors address : E.Nabedryk, D.Tiede, J.Breton, Service Biophysique, Dept. Biologie, CEN Saclay, 91191 Gif/Yvette Cedex, France.
[+]P.L.Dutton, Dept. Biochem. and Biophys., U. of Penn, Philadelphia PA 19104.

A LIGHT-HARVESTING PIGMENT-PROTEIN COMPLEX ASSOCIATED WITH THE REACTION
CENTER OF GREEN BACTERIA

H. VASMEL, T. SWARTHOFF, H.J.M. KRAMER and J. AMESZ

1. INTRODUCTION

Green photosynthetic sulfur bacteria have an extensive antenna system consisting of 1000 - 2000 bacteriochlorophyll (BChl) molecules per reaction
center. Most of these are BChl c, d or e, contained in the so-called
chlorosomes. BChl a is situated in or close to the cytoplasmic membrane,
which also contains the reaction center (Olson, 1978). Incubation of the
membrane with Triton X-100 produces a pigment-protein complex (PP), that
contains about 80 BChl a per reaction center. Subsequent treatment with
guanidine-HCl removes two molecules of the light-harvesting BChl a protein,
each containing 21 BChl a and produces the RCPP complex (about 35 BChl a;
Swarthoff, Amesz, 1979). Fluorescence emission and excitation and linear
dichroism (LD) spectra indicate that the RCPP complex still contains a
significant amount of BChl a protein in a bound form, in addition to a
'core complex', that is associated with the reaction center (Kramer et
al., 1982; Swarthoff et al., 1980). In this paper we describe the isolation
of this core complex and some of its properties.

2. MATERIALS AND METHODS

The PP and RCPP complexes were prepared from Prosthecochloris aestuarii
as earlier described (Swarthoff, Amesz, 1979). Low-temperature spectra were
measured in a non-crystallizing medium, obtained by addition of sucrose and
glycerol. LD spectra were measured in a polyacrylamide gel, pressed in two
perpendicular directions, so that axially symmetric samples were obtained.
The increase in length along the orientation axis was by a factor of 1.5.

3. RESULTS AND INTERPRETATION

Incubation of the PP complex with 0.2 % lithium dodecyl sulfate (LDS)
during 1 h at room temperature and centrifugation on a sucrose gradient
for 16 h at 200 000 x g gave a blue fraction at 30 % sucrose, which showed
an enhanced absorption in the long-wavelength region. Other fractions showed the absorption spectra of the PP complex and of the BChl a protein.
Fig. 1 shows absorption spectra of the 30 % sucrose fraction. At 77 K the
preparation showed BChl a bands at 836, 813 and 801 nm and a shoulder near
785 nm. The band at 825 nm, typical for the BChl a protein, was lacking.
Electrophoresis on SDS polyacrylamide gel showed the absence of the 40 kD
peptide band, due to subunits of the BChl a protein (Olson, 1978).
The above observations indicate that the 30 % sucrose fraction consists of
the core complex mentioned in the Introduction. From the elution volume on
a Sepharose CL-6B column a particle weight of the complex of 200 ± 50 kD
was calculated.
No reaction center activity could be detected in the core complex. Measurement of P840 oxidation or of triplet formation, either by the flash-induced
absorbance increase at 425 nm or by changes in fluorescence yield induced
by a magnetic field (Kramer et al., 1982) indicated that the amount of
photoactive reaction centers had been reduced to less than 10 % of that
originally present in the PP complex.
Low-temperature emission spectra of the core complex are shown in fig. 2.

Sybesma, C. (ed.), Advances in Photosynthesis Research, Vol. II. ISBN 90-247-2943-2.
© *1984 Martinus Nijhoff/Dr W. Junk Publishers, The Hague/Boston/Lancaster.*

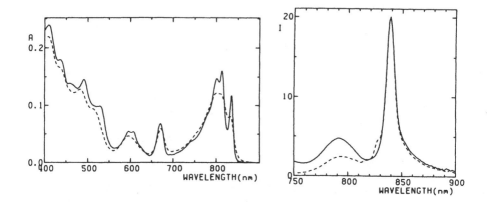

FIGURE 1. Absorption spectra of the core complex at room temperature
(---) and at 77 K (——).

FIGURE 2. Fluorescence emission spectra of the core complex normalized
at 77 K. Excitation at 670 nm (——) or 606 nm (---).

The main emission band was at 839 nm, which band has probably the same
origin as the 838 nm emission from the PP and RCPP complexes (Kramer et
al., 1982; Swarthoff et al., 1981). The band at 828 nm, due to the BChl $\underline{a}$
protein was very weak, and comparison with the fluorescence emission spectra
of the PP complex (Swarthoff et al., 1981) indicated that close to 99 % of
the soluble and bound BChl $\underline{a}$ protein had been removed from the PP complex
by LDS incubation.
The circular dichroism (CD) spectrum at 77 K of the core complex showed a
strong negative band at 810 and positive bands at 794, 820 and 837 nm.
The spectrum is clearly different from that of the soluble BChl $\underline{a}$ protein

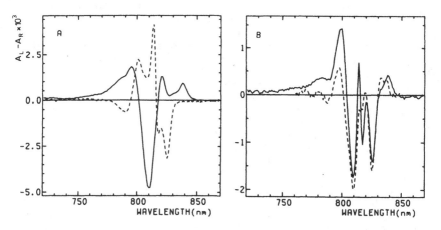

FIGURE 3. CD spectra measured at 77 K. All spectra correspond to $A_{810} = 1.0$
at room temperature. A: Core complex (——) and BChl $\underline{a}$ protein (---). B:
Measured (---) and simulated (——) spectra of the RCPP complex (see text).

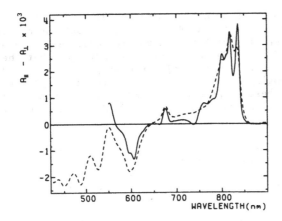

FIGURE 4. LD spectrum ($A_{//} - A_{\perp}$) of the core complex at room temperature (---) and at 77 K (———). A_{810} = 0.04 at room temperature

(broken line of fig. 3A), which agreed with that earlier obtained by Philipson and Sauer (1973), whereas the spectrum of the RCPP complex (fig. 3B, broken line) appears to be a composite of both spectra. In fact, a reasonable simulation of the spectrum of the RCPP complex was obtained by a linear combination of the spectra of the BChl a protein and the core complex (fig. 3B, solid line). This indicates that the structure of the bound BChl a protein, at least with regard to the arrangement of the bacteriochlorophylls, is identical to that of the soluble protein. The results also indicate that the structure of the core complex is left basically unchanged by the LDS treatment.
LD spectra of the core complex are shown in fig. 4. In the near-infrared region three positive bands are visible (at 800, 818 and 836 nm at 77 K) of BChl a transition dipoles that all make an angle smaller than the magic angle ($5\overline{4}.7\,^{\circ}$) with the orientation axis, and consequently with the plane of the membrane (Swarthoff et al., 1980). The band at 818 nm may correspond to the transition that causes the positive band at 820 nm in the CD spectrum. The BChl a Q_x bands at 595 and 606 nm and the weak Q_y band at 785 nm show negative polarization. Carotenoids show a more or less perpendicular orientation, as was also noted for the PP and RCPP complexes (Swarthoff et al., 1980). The negative band at 814.5 nm of the BChl a protein, which is quite prominent in the spectra of the PP and RCPP complexes (Swarthoff et al., 1980), is completely absent. Around 670 nm the LD spectrum of the core complex is also different from that of the PP and RCPP complexes: instead of two or three bands between 650 and 680 nm, only one, positive band at 666 nm is seen.

4. CONCLUSION

We conclude that by treatment of the PP complex with LDS a purified core complex is obtained from which 99 % of the BChl a protein has been removed. The complex retains its structural integrity, but does not show reaction center activity. Related to this lack of activity may be changes near 670 nm in the LD spectrum (fig. 4) and CD spectrum (not shown) as compared to the RCPP complex, and the presence of only one absorption band in the long-wave region at 836 nm, whereas the RCPP complex has two bands at

77 K, at 834 and 836.5 nm (Swarthoff, Amesz, 1979).
The simulated CD spectrum of the RCPP complex (fig. 3B) was obtained by
a combination of the spectra of the core complex and the BChl a protein in
a ratio of about 3 : 2 on a BChl a basis. This indicates that the RCPP
complex contains 2 subunits of the BChl a protein (14 BChl a) and about 20
BChl a molecules belonging to the core complex, in agreement with the
molecular weights of these components. The PP complex contains in addition
2 molecules (6 subunits) of the BChl a protein (Swarthoff, Amesz, 1979).

REFERENCES
Kramer HJM, Kingma H, Swarthoff T and Amesz J (1982) Prompt and delayed
fluorescence in pigment-protein complexes of a green photosynthetic
bacterium, Biochim. Biophys. Acta 681, 359-364.
Olson JM (1978) Bacteriochlorophyll a-proteins from green bacteria. In
Clayton RK and Sistrom WR, eds. The Photosynthetic Bacteria, pp. 161-178.
New York-London: Plenum Press.
Philipson KD and Sauer K (1973) Exciton interaction in a bacteriochloro-
phyll-protein from Chloropseudomonas ethylica. Absorption and circular
dichroism at 77 degrees K, Biochemistry 11, 1880-1885.
Swarthoff T and Amesz J (1979) Photochemically active pigment-protein com-
plexes from the green photosynthetic bacterium Prosthecochloris aestuarii,
Biochim. Biophys. Acta 548, 427-432.
Swarthoff T, de Grooth BG, Meiburg RF, Rijgersberg CP and Amesz J (1980)
Orientation of pigments and pigment-protein complexes in the green photosynthetic bacterium Prosthecochloris aestuarii, Biochim. Biophys. Acta
593, 51-59.
Swarthoff T, Amesz J, Kramer HJM and Rijgersberg CP (1981) The reaction
center and antenna pigments of green photosynthetic bacteria, Israel J.
Chem. 21, 332-337.

ACKNOWLEDGEMENTS
Thanks are due to Mrs. L.M. Blom for help with the preparation of the
complexes, to A.H.M. de Wit for culturing the bacteria, to L.J. de Vos
for assistance with the CD and LD measurements and to H. Kingma for
measurement of the magnetic field effect on fluorescence. The investi-
gation was supported by the Netherlands Foundation for Chemical Research
(SON), financed by the Netherlands Organization for the Advancement of
Pure Research (ZWO).
Authors' address: Department of Biophysics, Huygens Laboratory of the State
University, P.O. Box 9504, 2300 RA Leiden, The Netherlands.

REACTION CENTER COMPONENTS OF THE GREEN PHOTOSYNTHETIC BACTERIUM CHLORO-
FLEXUS AURANTIACUS

HENK VASMEL, RON F. MEIBURG, HERMAN J.M. KRAMER, HUBERT J. den BLANKEN[*],
LEO J. de VOS and JAN AMESZ

1. INTRODUCTION

The thermophilic gliding green bacterium Chloroflexus aurantiacus occupies
an intermediate position between the 'classical' green bacteria and the
purple bacteria. With the green bacteria it shares its antenna organization
in the form of so-called 'chlorosomes', large bodies containing mainly BChl c
and attached to the membrane, while its primary photochemistry resembles that
of purple bacteria. This paper presents the results of an investigation of
the structure of the primary donor P865 and the pigment arrangement in the
reaction center of C. aurantiacus. We have also investigated the electron
acceptor chain to shed more light on the position of Chloroflexus with respect to the other groups of photosynthetic bacteria.

2. MATERIALS AND METHODS

Reaction centers were isolated by a method similar to that of Pierson and
Thornber (1983). Low-temperature absorption, CD and LD experiments were performed as described elsewhere (Vasmel et al., 1983). Absorption-detected
electron spin resonance in zero magnetic field (ADMR) experiments were performed at 1.2 K as described by Den Blanken and Hoff (1982).

3. RESULTS AND DISCUSSION

The absorption spectrum of the reaction center at 4 K (fig. 1) shows bands
at 887, 813, 787 and 607 nm, due to BChl a, while BPheo a bands are located
at 757, 540, 533 and 498 nm. The broad band at 887 nm is the Q_y band of the
primary electron donor P865. The reaction center is devoid of carotenoids.
The fluorescence excitation spectrum detected at 925 nm (not shown) showed
equally efficient energy transfer from all pigments to P865. The splitting

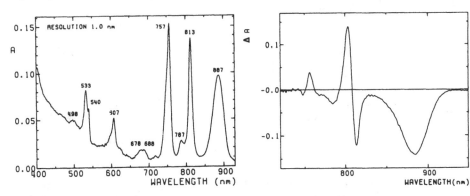

FIGURE 1 (left). Absorption spectrum at 4 K of reaction centers of C.
aurantiacus in non-crystallizing medium.

FIGURE 2 (right). Absorption difference spectrum at 77 K.

Sybesma, C. (ed.), *Advances in Photosynthesis Research, Vol. II. ISBN 90-247-2943-2.*
© *1984 Martinus Nijhoff/Dr W. Junk Publishers, The Hague/Boston/Lancaster.*

of the Q_x band of BPheo a at 540 and 533 nm in the unusual ratio 1 : 2 and the strong Q_y band at 757 nm indicate that 3 BPheo a molecules are present per reaction center instead of 2 as found for purple bacteria.
The BChl a content was determined by comparing the amplitude of the BPheo a Q_x band in acid and neutral extracts of the reaction center (Reed, Peters, 1972). For Chloroflexus their ratio was 2.3 ± 0.3, while control experiments on Rhodopseudomonas sphaeroides R-26 reaction centers yielded the value of 3.0 ± 0.3. Assuming that 3 BPheo a molecules are present, we thus conclude that the reaction center contains 4 molecules of BChl a.
By illumination and cooling to 77 K the reaction center could be frozen in the oxidized form. The absorption difference spectrum thus obtained (fig. 2) showed bleachings of the Q_y band of the primary donor at 887 nm, bleaching of the bands at 790 nm (787 nm at 4 K) and 813 nm and formation of a band at 806 nm. A slight absorption increase of the BPheo a band at 757 nm is observed. The difference spectrum shows a clear resemblance to those obtained for purple bacteria.
The LD and CD spectra of reduced and oxidized reaction centers are shown in figs. 3 and 4, respectively. In the LD spectrum of the reduced form two BChl a transitions are observed, one near 790 nm and the other at 887 nm, both making an angle smaller than the magic angle (54.7 °) with the orientation axis. (The gel was pressed in two perpendicular directions). This seems to rule out the possibility that the 790 nm band is due to exciton splitting of P865. The similar sign of the corresponding CD bands is in agreement with this conclusion. The BChl a transition at 813 nm is oriented close to the magic angle. Upon oxidation the LD spectrum is drastically altered. The Q_y band of P865 disappears and a new LD band is observed at 806 nm, while the 790 nm band vanishes. The LD and CD bands of BPheo a are little affected.
Absorption, CD and LD data, combined with the results of extraction experiments, can be accounted for in the following model: the reaction center of C. aurantiacus consists of 3 BPheo a and 4 BChl a molecules; two of the latter are closely coupled and form the primary donor P865. The other two

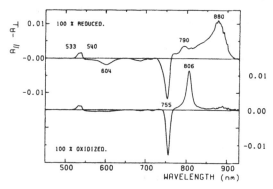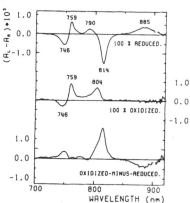

FIGURE 3 (left). LD spectra of reaction centers of C. aurantiacus in pressed polyacrylamide gel at 77 K. A_{865} = 0.04 at room temperature in reduced sample. The spectra were obtained by a linear combination of experimental spectra.

FIGURE 4 (right). CD spectra at 77 K. A_{865} = 0.25 at room temperature. Oxidation by addition of 2 mM potassium ferricyanide.

BChl a molecules are also strongly interacting and cause the absorption
bands near 790 and 813 nm in reduced reaction centers. Upon oxidation of
the primary donor this interaction is broken, leading to spectral shifts of
both bands to 806 nm (fig. 2) and a concomitant reorientation of the transitions as observed with LD (fig. 3), as is to be expected for exciton
coupled transitions. In this approximation the extent of splitting (−163 cm⁻¹)
is compatible with an in-plane head-to-tail arrangement of two chromophores,
absorbing in uncoupled form at 806 nm, with a distance of 10 − 11 Å, making
an angle of approximately 59 ° as indicated by the ratio of their dipole
strengths.

Further support for a dimeric model for the primary donor comes from the
triplet-minus-singlet absorption difference spectrum (T − S spectrum) obtained by ADMR (fig. 5). Both the T − S spectrum and the oxidized-minus-
reduced difference spectrum of P865 (fig. 2) show a bleaching of the Q_y
band at 887 nm. However, the absence of a bleaching of a band at 787 nm
in the T − S spectrum again indicates that this band is not due to the
primary donor. The characteristics of the T − S spectrum are very similar
to those of reaction centers of Rps. sphaeroides R-26 and Rps. viridis
(Den Blanken, Hoff, 1982) and strongly suggest that P865 is a BChl a dimer.
The appearance of a monomer absorption band at 807 nm upon triplet formation
indicates that the triplet state is localized on one of the constituent
pigments of the dimer on an optical time scale.

To investigate the nature of the secondary electron acceptor the reaction
center was illuminated in the presence of reduced N-methylphenazonium methosulfate and ascorbate. Under these conditions the rereduction of P865⁺ was
strongly accelerated while at most wavelengths a second, slower component
became visible. Its absorption difference spectrum, shown in fig. 6, indicated the reduction of menaquinone (vitamin K_2), the only quinone present
in the membrane of C. aurantiacus (Hale et al., 1983). A stoichiometry of
approximately one reduced menaquinone per oxidized P865 was calculated

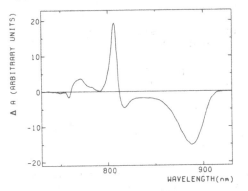

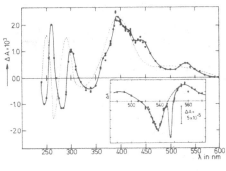

FIGURE 5 (left). Triplet-minus-singlet absorption difference spectrum, re
corded by ADMR at 1.2 K. Frequency of resonant microwaves: 735 MHz.

FIGURE 6 (right). Solid line: absorption difference spectrum measured 10 ms
after a flash, in the presence of 25 μM N-methylphenazonium methosulfate
(PMS) and 5 mM ascorbate. A_{865} = 4.5 x 10⁻². Dashed line: in vitro absorption
difference spectrum of vitamin K_1 (Q⁻ − Q) in methanol (E.J. Land, personal
communication), normalized at 395 nm. Insert: 100 μM PMS, 10 mM dithionite,
300 ms illumination. A_{865} = 2.0 x 10⁻².

(Vasmel, Amesz, 1983). In the presence of dithionite to keep menaquinone reduced the spectrum showed negative bands at 538 and 555 nm (insert), presumably due to reduction of BPheo a and oxidation of cytochrome c-554. This suggests that BPheo a acts as an early electron acceptor, as was also suggested by flash spectroscopy (C. Kirmayer et al., personal communication).

4. CONCLUSIONS

The reaction center of C. aurantiacus contains the following components: the primary donor, P865, is a BChl a dimer while two additional BChl a molecules are present, which are responsible for the absorption bands around 800 nm. Three molecules of BPheo a are present in contrast to two in the reaction center of purple bacteria. The one with Q_x absorption band at 540 nm at 4 K probably acts as an early electron acceptor. Menaquinone (vitamin K_2) is a secondary electron acceptor. The reaction center pigment arrangement and the organization of the electron acceptor chain show clear similarities between the reaction center of C. aurantiacus and those of purple bacteria.

REFERENCES
Den Blanken HJ and Hoff AJ (1982) High-resolution optical absorption-difference spectra of the triplet state of the primary donor in isolated reaction centers of the photosynthetic bacteria Rhodopseudomonas sphaeroides R-26 and Rhodopseudomonas viridis measured with optically detected magnetic resonance at 1.2 K, Biochim. Biophys. Acta 681, 365-374.
Hale MB, Blankenship RE and Fuller RC (1983) Menaquinone is the sole quinone in the facultatively aerobic green photosynthetic bacterium Chloroflexus aurantiacus, Biochim. Biophys. Acta 723, 376-382.
Pierson BK and Thornber JP (1983) Isolation and spectral characterization of photochemical reaction centers from the thermophilic green bacterium Chloroflexus aurantiacus strain J-10-fl, Proc. Natl. Acad. Sci. USA 80, 80-84.
Reed DW and Peters GA (1972) Characterization of the pigments in reaction center preparations from Rhodopseudomonas sphaeroides, J. Biol. Chem. 247, 7148-7152.
Vasmel H and Amesz J (1983) Photoreduction of menaquinone in the reaction center of the green photosynthetic bacterium Chloroflexus aurantiacus, Biochim. Biophys. Acta, in the press.
Vasmel H, Meiburg RF, Kramer HJM, de Vos LJ and Amesz J (1983) Optical properties of the photosynthetic reaction center of Chloroflexus aurantiacus at low temperature, Biochim. Biophys. Acta, in the press.

ACKNOWLEDGEMENT
This investigation was supported by the Netherlands Foundation for Chemical Research (SON), financed by the Netherlands Organization for the Advancement of Pure Research (ZWO).

Authors' address: Department of Biophysics and *Centre for the Study of the Excited States of Molecules, Huygens Laboratory of the State University, P.O. Box 9504, 2300 RA Leiden, The Netherlands.

REGULATION OF PHOTOSYNTHETIC UNIT STRUCTURE IN RHODOSPIRILLUM RUBRUM
WHOLE CELLS

PAUL A. LOACH, PAMELA S. PARKES AND PEGGY BUSTAMANTE

1. INTRODUCTION

The photosynthetic unit (PSU) in Rhodospirillum rubrum consists of
a reaction center (RC) which is about 95 kd (Gingras, 1978) and a light-
harvesting (LH) antenna complex which has a molecular weight of about 150
kd (Picorel et al., 1983). The RC contains 3 polypeptides, 4 bacteriochlorophyll, 2 bacteriopheophytin and several ubiquinone molecules. The
LH complex contains 21 ± 3 bacteriochlorophyll, 10 ± 2 carotenoid and 9 to 12
copies of each of two small polypeptides. Thus, the total size of this
fundamental unit is about 250 kd.

In the in vivo state, Vredenberg and Duysens (1963) showed that these
PSU's behave in a cooperative fashion in that light energy absorbed in the
antenna complex of a particular PSU may be efficiently transferred to several other PSU's if its own RC is closed (already in a charge-separated
state). Similar results have been obtained by Clayton with Rhodopseudomonas sphaeroides (1966).

Recently, in studying alanine uptake by whole cells of R. rubrum, we
found an Mg^{2+}-dependent reversible loss of transport activity which occurred in parallel with a reversible loss in cooperativity among the PSU
(Zebrower and Loach, 1981; Zebrower and Loach, 1982). From this study we
became interested in the molecular events that are responsible for the
change between cooperative and noncooperative states. We felt that Mg^{2+}
was playing a key role, perhaps part of which involved activation of one
or more ATP kinases which then phosphorylated membrane proteins leading
to a change in surface charge distribution similar to events postulated
for regulation of the state 1 to state 2 transitions in chloroplasts
(Horton, 1983). However, in the case of R. rubrum, there are no LH complexes except those intimately associated with the RC. We would propose
that phosphorylation may lead to association to form the cooperative PSU
state. This would perhaps be contrasted to part of the phenomena noted
in chloroplasts where protein phosphorylation parallels a decrease in PS
II-PS II interaction (Kyle and Arntzen, 1983).

In the present paper we wish to report the results of our initial
experiments which were designed to specifically determine whether LHP 1
(the alpha-polypeptide of the LH complex) was phosphorylated to a different extent in cooperative and noncooperative cells. We chose this peptide
because it normally dissolves in chloroform/methanol (1/1), is readily
purified (Tonn et al., 1977), and its primary structure is known (Gogel
et al., 1983).

2. METHODS

Bacteria were grown on modified Hutner's media as described previously (Tonn et al., 1977). Samples were taken for analysis between the second
and fourth days of photosynthetic growth. The scheme used for preparing
the cells in their cooperative and noncooperative states for LHP 1 isolation is shown in Figure 1.

Sybesma, C. (ed.), Advances in Photosynthesis Research, Vol. II. ISBN 90-247-2943-2.
© *1984 Martinus Nijhoff/Dr W. Junk Publishers, The Hague/Boston/Lancaster.*

PROCEDURE FOR ISOLATION OF COOPERATIVE AND
NONCOOPERATIVE CELLS FOR PROTEIN EXTRACTION

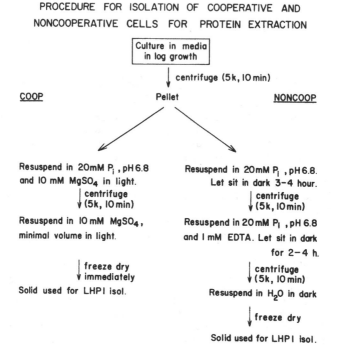

FIGURE 1. Flow diagram for preparation of cooperative and noncooperative
whole cell samples used for LHP 1 isolation.

The cells were initially centrifuged from their growth media at 5,000
RPM for 10 min. at about 15° in the dark. The cells to be maintained in
their cooperative state (Zebrower and Loach, 1981) were purposely illuminated at all times except during centrifugation steps and always resuspended
in phosphate buffer containing 10 mM Mg^{2+}. The cells to be converted to
the noncooperative condition and subsequently studied were resuspended in
phosphate buffer which did not contain Mg^{2+} but did contain 1 mM EDTA.

After the final centrifugation following the wash steps, each sample
of cells was either immediately frozen and lyophilized to dryness or resuspended for activity measurements. LHP 1 was isolated from the lyophilized whole cell samples following the procedures we previously employed
for isolation from washed and lyophilized chromatophores (Tonn et al.,
1977). The procedure makes use of selectively dissolving the protein in
chloroform/methanol (1:1) followed by purification on an LH 60 Sephadex
column. The elution profiles for material extracted from cooperative and
noncooperative cells are shown in Figure 2.

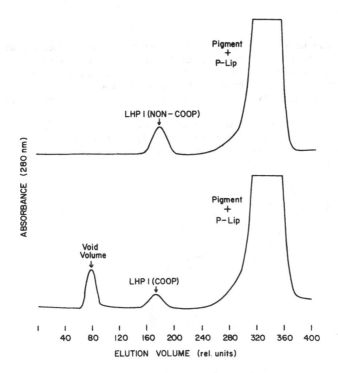

FIGURE 2. Elution profile of CHCl₃/CH₃OH extracts of cooperative and noncooperative whole cells run on a Sephadex LH 60 column.

The extent to which the cells were cooperative was measured in two ways, each utilizing a beam of continuous illumination whose intensity was varied to change the percentage of reaction centers which were in a closed (charge separated) state (Zebrower and Loach, 1981). A second interrogating light pulse from a Xenon flash lamp was adjusted to between 10 and 30% saturating intensity and used to measure the efficiency of light utilization at each selected state of closed reaction centers. The whole cell sample was either measured directly or suspended in 70% glycerol at an absorbance at 880 nm of 1.0 for absorbance change measurement at 792 nm or to an absorbance at 880 nm of 40 for ESR measurements, both of which monitored the oxidation of P865. MgSO₄(10mM) was present in 20 mM phosphate buffer (pH 6.8) for cooperative cells and 1 mM EDTA replaced the MgSO₄ in the noncooperative cell suspensions. Fresh cell samples were used after every 1 to 2 hours.

The phosphate content of the LHP 1 protein was determined in two different ways. In one of these, the cells were grown in the presence of ³³P-inorganic phosphate and the amount of phosphate in LHP 1 evaluated by comparing the cpm in a sample with that of a known quantity of phosphate in the phospholipid fraction. In the second method, phosphate was determined chemically by a modification of the Bartlett procedure (1959).

The concentration of LHP 1 was determined by measurement of the protein's UV spectrum in (2,2,2-)trifluoroethanol, in which solvent we have determined the extinction coefficient to be 10,200 cm⁻¹ at 290 nm.

3. RESULTS

   Whole cells can be converted from a cooperative to a noncooperative state by depleting their Mg^{2+} concentration (Zebrower and Loach, 1981). Examples of data which relate the efficiency of light utilization by such cells are shown in Figure 3. These cells were treated in the same way as those which were subsequently used for LHP 1 isolation for measurement of its phosphate content.

   LHP 1 which had been isolated from cooperative and noncooperative cells grown in the presence of ^{33}P were found to contain significant amounts of radioactivity. The data for an example of such a preparation are shown in Table 1.

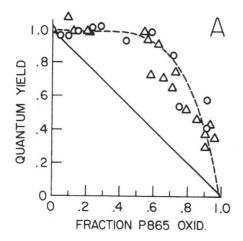

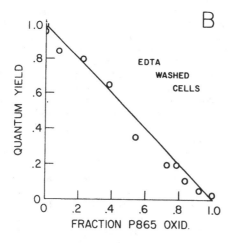

FIGURE 3. Quantum efficiency of P865 oxidation at various fractions of open reaction centers. A, cells maintained in Mg^{2+} and light. B, cells treated with EDTA.

TABLE 1. Determination of ^{33}P in LHP 1 prepared from cooperative and noncooperative whole cells of R. rubrum

SAMPLE	^{33}P (cpm)	mg LHP 1	cpm/mg LHP 1
COOP LHP 1	678	1.9	325
NONCOOP LHP 1	654	3.3	176

Although the cpm in control samples taken before and after the LHP 1 peak on the LH 60 column showed some variation, they were consistently less than LHP 1 and in the range of 70-110 cpm. Interestingly, the ratio of cpm per mg of protein is several fold higher in LHP 1 from cooperative cells than it is from noncooperative cells.

The results of chemical analyses of the phosphate content in LHP 1 preparations are summarized in Table 2. Control samples taken from fractions eluted before and after the LHP 1 peak typically were at background levels (3 to 6 nmoles). Thus, the results again show significant phosphate to be present in the LHP 1 isolated from cooperative cells.

Table 2. Chemical determination of phosphate in LHP 1 prepared from cooperative and noncooperative whole cells of R. rubrum

SAMPLE	Phosphate (nmoles)	Protein (nmoles)	Phosphate/Protein
Experiment 1			
COOP LHP 1	34	123	.28
COOP LHP 1	62	261	.24
NONCOOP LHP 1 *	13	123	.11
Experiment 2			
COOP LHP 1	44	353	.125
NONCOOP LHP 1 **	17	364	.047

* These cells were washed in phosphate buffer, then distilled water before lyophilizing. Note that neither EDTA nor Mg^{2+} were present.

** This sample was prepared using the EDTA wash as shown in the scheme of Figure 1.

The level of phosphate present in the LHP 1 from noncooperative cells is much less. As can further be seen from the data of Table 2, the amount of phosphate per mole protein is between .13 and .28 moles in LHP 1 from cooperative cells. These experiments have been conducted several times and the actual phosphate content seems to vary somewhat in the range reported in Table 2. The amount of phosphate found in LHP 1 isolated from cooperative cells has always been several fold higher than it is from noncooperative cells.

We have also conducted chemical phosphate analysis on samples of the LHP 1 isolated from chromatophores which are normally prepared in Mg^{2+} free buffer. From several such samples examined, very little phosphate was found (less than .05 mole per mole LHP 1). When assayed for efficiency of light utilization, these chromatophore samples are routinely found to be noncooperative.

Another interesting difference in the behavior of cooperative and noncooperative cells is reflected in the amount of LHP 1 that can be isolated from the cells. Typically, 2 to 3 times as much LHP 1 can be extracted and purified from noncooperative cells as from cooperative cells see Table 3). This difference is also readily seen in the comparisons of column elution profiles shown in Figure 2. Furthermore, considering the additional mass of other components that whole cells contain relative to chromatophores, the yield of pure LHP 1 from noncooperative whole cells appears to be as high as that from chromatophores.

Finally, the data presented in Figure 2 may also be used to point out another interesting difference in the material extracted from cooperative and noncooperative whole cells. Note that the peak at the void volume in the extract from noncooperative cells is not found, or when present, is very small. Preliminary SDS PAGE analyses of the material in this peak at the void volume shows that it contains several polypeptides but also much material that is not protein and has little UV absorbance. Interestingly, a significant amount of phosphate is also contained in this band.

During the course of our measurements on the cooperativity of whole cell samples, we wished to stabilize them in their native state because otherwise they can change in complex ways during the course of the measurements and thus contribute some uncertainty to the measurement. As seen in the data presented in Figure 3, 70% glycerol was used for some experiments. Although this turned out to be useful at room temperature because the decay times were significantly slowed allowing easier light saturation of the samples, one of our initial reasons for using glycerol was to enable us to freeze the whole cells to liquid nitrogen temperature as a clear glass and then perform ΔA measurements under conditions where the sample would not change with time. Interestingly, whereas these samples were indeed quite stable and ideal to measure, they gave results that indicated the PSU's were in a noncooperative state, even if the sample before freezing had been in a highly cooperative state. We found this to be true both for samples at low (absorbance at 880 nm = 1) and high (absorbance at 880 nm = 40) sample concentrations.

Obviously, it would be desirable to examine the other polypeptides in the LH and RC complexes for possible sites of phosphorylation. We are continuing these experiments using more highly ^{33}P-labelled cells and examining various polypeptides by SDS PAGE. In addition, the actual group on LHP 1 to be phosphorylated should be determined.

TABLE 3. Yield of LHP 1 isolated from cooperative and noncooperative whole cell and chromatophores of R. rubrum

SAMPLE EXTRACTED FROM	LHP 1 (mg/100 mg dry wt.)
Experiment 1	
COOP WHOLE CELLS	1.05
NONCOOP WHOLE CELLS *	1.56
Experiment 2	
COOP WHOLE CELLS	0.80
NONCOOP WHOLE CELLS **	1.69
CHROMATOPHORES	2.5 to 3.5

* These cells were washed in phosphate buffer, then distilled water before lyophilizing. Note that neither Mg^{2+} nor EDTA were present.

** This sample was prepared using the EDTA wash as shown in the scheme of Figure 1.

4. DISCUSSION

Although additional quantitative experiments need to be conducted, the data presented here clearly show that changes in structure and/or interaction between photosynthetic units in R. rubrum whole cells are dependent upon Mg^{2+} levels which in turn may control the activity of enzymes involved with phosphate metabolism in the cell (e.g., protein kinases). Thus, a change in surface charge due to phosphorylation of the LH complex, which is perhaps located around the perimeter of the reaction center complex, would be an effective way to control PSU-PSU interaction. In this case, increased phosphorylation leads to increased interaction.

Although the amount of phosphate, presumably covalently bound to LHP 1, isolated from cooperative cells is only .13 to .28 mole phosphate per mole of protein, it should be kept in mind that this polypeptide is only 6.1 kd molecular weight and that there are probably 10 to 12 copies in the in vivo LH complex associated with one reaction center. Thus, there would be perhaps between 2 and 4 phosphates bound per LH complex in the cooperative in vivo state. In addition, it is likely that an equivalent amount of a second polypeptide (LHP 2, which is 5.8 kd) is involved in the formation of the LH complex (Gingras, 1983; Cogdell et al., 1982). We have not yet examined this polypeptide for possible phosphorylation.

It is interesting to contemplate the location of the phosphate on the LHP 1 polypeptide. From the primary structure which is known (Gogel et al., 1983), there are two sequences of amino acids near the c-terminal end that are reminiscent of sequences found to be phosphorylation sites in other proteins (Haworth et al., 1982; Shenolikar and Cohen, 1978). These are the sequences Ser-Thr-Lys and Lys-Pro-Val-Gln-Thr-Ser. Future work should establish the location of the bound phosphate in LHP 1.

A question that is interesting to raise is the following: Why would cells possess a regulatory system for controlling the cooperativity of photosynthetic units? In their natural habitat, why would the cell care whether the photosynthetic machinery is in an associated state in the dark?

We have previously suggested (Zebrower and Loach, 1981) that the answer might lie in the possibility that the PSU's are also intimately associated with energy requiring systems (e.g., alanine transport and phosphorylation of ADP) when they exist in their "cooperative" state and that it would indeed be beneficial to disengage such connections in the absence of light so that what could become energy leakage pathways would not exist.

REFERENCES

Bartlett GR (1959) Phosphorus assay in colum chromatography, J. Biol. Chem. 234, 466-468.

Clayton RK (1966) Relations between photochemistry and fluorescence in cells and extracts of photosynthetic bacteria, Photochem. Photobiol. 5, 807-821.

Cogdell R, Lindsay G, Valentine J and Durant I (1982) A further characterization of the B890 light-harvesting pigment-protein complex from Rhodospirillum rubrum strain S1, FEBS Lett. 150, 151-154.

Gingras G (1978) A comparative review of photochemical reaction center preparations from photosynthetic bacteria. In The Photosynthetic Bacteria, Clayton RK and Sistrom W, eds. Plenum Press, NY. pp. 119-131.

Gogel GE, Parkes PS, Loach PA, Brunisholz RA and Zuber H (1983) The primary structure of a light-harvesting bacteriochlorophyll-binding protein of wild-type Rhodospirillum rubrum, Biochim. Biophys. Acta, in press.

Haworth P, Kyle DJ, Horton P and Arntzen CJ (1982) Chloroplast membrane protein phosphorylation, Photochem. Photobiol. 36, 743-748.

Horton P (1983) Control of chloroplast electron transport by phosphorylation of thylakoid proteins, FEBS Lett. 152, 47-52.

Kyle DJ and Arntzen CJ (1983) Thylakoid membrane protein phosphorylation selectively alters the local membrane surface charge near the primary acceptor of Photosystem II, Photobiochem. Photobiophys. 5, 11-25.

Picorel R, Belanger G and Gingras G (1983) Antenna holochrome B880 of Rhodospirillum rubrum S1. Pigment, phospholipid, and polypeptide composition, Biochemistry 22, 2491-2497.

Shenolikar S and Cohen P (1978) The substrate specificity of cyclic AMP-dependent protein kinase, FEBS Lett. 86, 92-98.

Tonn SJ, Gogel GE and Loach PA (1977) Isolation and characterization of an organic solvent soluble polypeptide component from photoreceptor complexes of Rhodospirillum rubrum, Biochemistry 16, 877-885.

Vredenberg WJ and Duysens LNM (1963) Transfer of energy from bacteriochlorophyll to a reaction centre during bacterial photosynthesis, Nature 197, 355-357.

Zebrower M and Loach PA (1981) Role of Mg^{2+} in the cooperative association of photoreceptor complexes in Rhodospirillum rubrum. In Developments in Biochemistry, Vol. 20, Selman BR and Selman-Reimer S, eds., Elsevier/North-Holland, NY. pp. 333-340.

Zebrower M and Loach PA (1982) Efficiency of light-driven metabolite transport in the photosynthetic bacterium Rhodospirillum rubrum, J. Bacteriol. 150, 1322-1328.

ACKNOWLEDGEMENTS
This investigation was supported by research grants from the National
Institutes of Health (GM-11741) and National Science Foundation (PCM-
8110620)
Authors' address: Paul A. Loach, Pamela S. Parkes and Peggy Bustamante
                     Department of Biochemistry, Molecular Biology and Cell
                     Biology, Northwestern University, Evanston, Illinois,
                     60201, USA.

MORE THAN TWO STRUCTURALLY DISTINCT TYPES OF ANTENNA IN RHODOSPIRILLALES

BRUNO ROBERT, ANDRE VERMEGLIO[+] and MARC LUTZ

1. INTRODUCTION, METHODS

Resonance Raman (RR) spectroscopy can be used to yield information about ground state interactions assumed by bacteriochlorophyll a (BChl) in antenna structures of *Rhodospirillales*. This technique permits investigations of the local environment of BChl molecules within the intracytoplasmic membrane and within antenna complexes, and, in particular, of their bonding with their host polypeptides (Lutz, 1981). Using this method, we studied chromatophores and antenna complexes of several species of *Rhodospirillaceae* and *Chromatiaceae* in order to compare, on a structural basis, complexes defined by their biochemical and electronic properties.

TABLE 1. Antenna complexes investigated.

B 880	B 850-800	B 850	B 820-800
Rps sphaeroides 2.4.1. ; R 26 ; R 26.1	Rps sphaeroides 2.4.1.	Rps sphaeroides 2.4.1.	Chromatium vinosum Thiocapsa roseop.
Rsp rubrum S1 ; G9	Rps palustris Thiocapsa roseop.	R 26.1 Rps palustris	
Rps palustris Thiocapsa roseop. Chromatium vinosum			

Complexes studied in their isolated states were extracted using LDAO (*Rhodospirillaceae*) or DOC (*Chromatiaceae*) and purified by ultracentrifugation in sucrose gradients (Robert 1983). B 850 complexes were selectively depleted in BChl B 800 by treatment with LiDS 2%, according to the method described in (Clayton and Clayton, 1981). Raman spectroscopy was conducted as in ref. 4 (Lutz et al., 1982), at 20 K, using 363.8 nm excitation.

2. RESULTS AND DISCUSSION

2.1 Multiplicity of types of antenna complexes

We found that in B 880-type complexes from eight different species of bacteria, the BChl environments are extremely similar. These environments insure the same sets of binding sites for the conjugated carbonyls (Table 2) and, probably, for the Mg atoms, as well as the same local permittivity seen by the dihydrophorbin rings. Hence, the B 880-type complexes of *Rhodospirillales*, as defined by their biochemical and electronic absorption properties, also share nearly identical structures around their dweller BChls.

By contrast, two distinct families are readily distinguished among the B 850-800-type complexes. Indeed, the BChl host sites in B 850-800 complexes from *Rhodospirillaceae* appear very different from those of the B 850-800 complexes from *Chromatiaceae*. In particular, they provide different liganding for the conjugated carbonyls of the BChl molecules (table 2).

Sybesma, C. (ed.), Advances in Photosynthesis Research, Vol. II. ISBN 90-247-2943-2.
© *1984 Martinus Nijhoff/Dr W. Junk Publishers, The Hague/Boston/Lancaster.*

It is concluded that at least three types of structurally distinct antenna complexes must be involved in *Rhodospirillales*, namely B 880, "non sulfur" B 850-800 and "sulfur" B 850-800, and not just two, as proposed earlier (Thornber et al., 1978).

The structures of the host sites of the two B 850 molecules in B 850-800 complexes from various *Rhodospirillaceae* appear quite similar. However, the sites occupied by the B 800-type BChls in complexes from *Rps palustris* and *Rps sphaeroides* are different. For example, it is seen from Table 2 that the 9-keto carbonyl of BChl B 800 of *Rps sphaeroides* 2.4.1. is free from bonding, while in *Rps palustris* this group is bonded. Also the 2-acetyl carbonyls of these molecules are bonded in different sites in the two species. On the other hand, B 850-800 and B 820-800 complexes from *Chromatiaceae* yielded very similar spectra, indicating that they may correspond to a single structural type.

TABLE 2. Observable stretching frequencies of the conjugated carbonyl groups of BChl in antenna complexes.

	B 880							B 800-850			B 800-820			B 850		B 800	
Prosthecochloris aestuarii (+)	2.4.1.	R 26	Rps. pal.	S 1	G 9	Thioc. roseop.	Chrom. vinos.	2.4.1.	Rps. pal.	Rps. caps.	Thioc. roseop.	Thioc. roseop.	Chrom. vinos.	Rps. pal.	2.4.1.	Rps. pal.	2.4.1. (++)
1632								1634	1633	1634				1633			1632
1641 1647	1641	1542	1646	1645	1644	1645	1646	1642	1640	1641				1641	1645		
1657											1656	1657	1659				
1663 ?																	
1668	1668	1668	1667	1667	1667	1668	1664	1667	1664	1668	1671	1671	1667	1670	1666		1665
1677	1676	1678	1678	1674	1674	1677	1679		(?)		1679	1678	1681				
1684 ?											1687	1686					
1692																	
1697																	
1706								1702			1702						1702

(+) : from Lutz et al, 1981 ;   (++) : from Lutz et al, 1983

2.2 Some structural properties shared by antenna complexes of *Rhodospirillales*

Resonance Raman spectroscopy permits the assessment of a number of properties common to the various sites occupied by the B 800, B 820, B 850 and B 880 molecules in antenna structures of *Rhodospirillales* :

    - These sites provide anchorage groups for both keto and acetyl carbonyls of the BChls. None of these groups are Mg atoms of other BChls. B 800 of *Rps sphaeroides* 2.4.1. however retains a free 9C=O group. This is probably true in *Rps capsulata* also.

    - These sites provide anchorage groups for the Mg atoms of the BChls, which generally are singly and not doubly liganded. In the majority of the cases, histidine is the most likely candidate as the Mg ligand. This has been shown by comparison with RR spectra of the soluble complex from *Prosthecochloris aestuarii* and with RR spectra of monomeric BChl complexed to various ligands, including substituted imidazoles. However, the B 800-type molecules may be bound to ligands other than histidines in *Rps palustris* (and in *Rps sphaeroides*), as indicated by specific frequencies (792 cm^{-1}) of a deformation mode of CNC bonds, observed at 796 cm^{-1} for the BChls of other types, in the *P. aestuarii* complex and in model systems involving imidazole.

     -Extraction of the protein complexes from the membrane of *Rsp rubrum* or *Rps palustris* results in no detectable change to the binding sites of the BChls. Hence, the local environments of the polar heads of the BChls most probably are purely proteic and are not protein/ lipid or protein/detergent interfaces. In addition, the B 800 molecule of *Rps sphaeroides* 2.4.1. ocuupies a site which is not readily accessible to small polar molecules such as water.

     - Identical frequencies were observed for almost all of the homologous modes of the BChl dihydrophorbin rings, in the present complexes and in the soluble complex from *P. aestuarii* (Lutz et al 1982). This indicates that all of the BChl binding sites, including those occupied by the B 800-type molecules, probably assume low dielectric permittivities, comparable for example to that of the interior part of the *P. aestuarii* proteic complex.

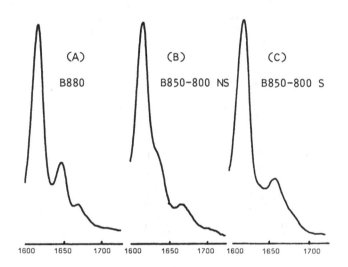

Figure n°1 : Resonance Raman spectra (stretching of the carbonyl groups
                region) of antenna complexes ; A : B 880-type complex of
                Rps palustris ; B : B 850-800-type complex of *Rps palustris* ;
                C : B 850-800-type complex of *Thiocapsa roseopersicina*.

2.3. Antenna composition of the R 26 and R 26.1 strains from *Rps sphaeroides*

Resonance Raman spectra of the intracytoplasmic membranes of the R 26 and R 26.1 strains (Davidson and Cogdell, 1981) permit the antenna complexes contained by those two mutants to be characterized. The original R 26 mutant contains a single B 880-type complex, undistinguishable from that present in the wild type (table 3). By contrast, the R 26.1 strain yielded RR spectra different from those of any purified antenna comple of the wild type. Actually, all of the characteristics of these spectra are intermediate between those of the B 880 and the B 850 (800 depleted) spectra, indicating that the R 26.1 membrane contains both of these types of complexes. Indeed, it is possible to generate a R 26.1 RR spectrum by adding together appropriately weighed spectra of the B 880 and the B 850(800 depleted) complexes of *Rps sphaeroides* 2.4.1. (figure 2). We conclude that the R 26.1 mutant has partially reverted from the original R 26 state and now synthetize both the genuine B 880-type complex and a genuine B 850-type complex devoid of B 800.

TABLE 3. Stretching frequencies of the conjugated carbonyl groups of BChl in antenna complexes of *Rps sphaeroides* 2.4.1., R 26 and R 26.1

2.4.1. B 880			1645	1667	1675	
2.4.1. B 850-800	1634	1642		1667		1702
2.4.1. B 850		1641		1667		
R 26			1645	1667	1678	
R 26.1		1643		1667		

Figure n° 2 : (right): Resonance Raman spectrum (carbonyl stretching region)
        of the R 26.1 mutant of *Rps spheroides*,
        (left): Calculated resonance Raman spectrum,
        $\lceil$ B850(800 depleted) from 2.4.1.$\rfloor$ - 0.22 x $\lceil$B880 from 2.4.1$\rfloor$

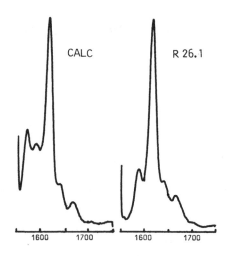

REFERENCES
Clayton R.K. and Clayton B.J. (1981) in Photosynthesis, G Akoyonoglou ed., Balaban, vol. 3, p. 377
Davidson E. and Cogdell R.J. (1981) Biochim. Biphys. Acta 635 : 295-303
Lutz M., Hoff A.J. and Bréhamet L. (1982) Biochim. Biophys. Acta 679 : 331
Lutz M., Robert B., Vermeglio A. and Clayton R.K. (1983) Biophys. J. 41, 316
Robert B. (1983) thèse de troisième cycle, Université Paris VI.
Thornber J.P., Trosper T.L. and Strouse C.E. in "The Photosynthetic Bacteria" Clayton R.K. and Sistrom W.R., eds, Plenum, N.Y., 1978, p.133

Authors' address : B. Robert, M. Lutz, Département de Biologie, CEN Saclay
                            91191 Gif sur Yvette CEDEX, France.
            A. Vermeglio, Département de Biologie, CEN Cadarache,
                       B.P. 1, 13115 St Paul lez Durance, Fr.

PIGMENT ORGANIZATION IN THE LIGHT-HARVESTING B800-850 ANTENNA COMPLEX OF
RHODOPSEUDOMONAS SPHAEROIDES

HERMAN J.M. KRAMER, RIENK van GRONDELLE[+], C. NEIL HUNTER[*] and JAN AMESZ

1. INTRODUCTION

Rhodopseudomonas sphaeroides contains three major pigment-protein complexes,
B800-850, B875 and reaction centers. B800-850 can be isolated by several
procedures and its properties have been characterized rather extensively
(Cogdell, Thornber, 1980; Clayton, Clayton, 1981; Breton et al., 1980;
Van Grondelle et al., 1982, 1983). B800-850 isolated with lauryl dimethyl
amine oxide (B800-850/LDAO) shows a high bacteriochlorophyll (BChl) B800
content and consists of large aggregates of several hundreds of connected
BChl's. However, the complex isolated by means of lithium dodecyl sulfate
(B800-850/LDS) contains very little B800 and is much smaller, containing
about 30 BChl's (Van Grondelle et al., 1983). Upon exposure to LDAO the
B800-850/LDS regains its normal B800 content (Clayton, Clayton, 1981).

Fluorescence polarization and linear dichroism of B800-850/LDAO showed
that the Q_y transition moments (TM) of B850 are circularly degenerate in
a plane, while the Q_x's are perpendicular to this plane. The porphyrin
ring of B800 was found to be almost parallel to the plane formed by the
B850 Q_y's (Breton et al., 1980).

Low-temperature fluorescence measurements showed that the B800-850
dipole-dipole distance is less than 20 Å. 20 - 30 % of the carotenoid was
found to transfer excitations directly to B800 while the remaining 70 -
80 % transfers directly to B850, suggesting two types of carotenoid, one
associated to B800, the other to B850 (Van Grondelle et al., 1982).

This work describes low-temperature fluorescence and fluorescence
polarization measurements in B800-850/LDS and in B800-850/LDS exposed to
LDAO (B800-850/LDS-LDAO). We conclude that the basic unit contains at
least 4 B850, 2 B800 and 3 carotenoid molecules.

2. RESULTS AND DISCUSSION

2.1. Absorption and excitation spectra

Fig. 1 shows the absorption spectrum of B800-850/LDS of Rps. sphaeroides.
The B800 band is small compared to that of B800-850/LDAO and the spectrum
shows a weak band around 890 nm due to some residual B875. The B800 content
varied somewhat for different preparations. Exposure to 0.1 % LDAO resulted
in the total recovery of the B800 band. The B850 band narrowed and shifted
to shorter wavelengths, the B875 band disappeared and an absorption increase
at 588 nm reflected the recovery of B800, while a red shift and an intensity
decrease of the carotenoid absorption bands was observed. Three possible
mechanisms can account for the B800 absorption increase: (1) binding of
'free' BChl absorbing at 790 nm, visible as a slight shoulder on the B800
band; (2) binding of BChl dissolved from the small fraction of B875; (3)
narrowing of the B850 band (Clayton, Clayton, 1981).

Emission spectra of both complexes recorded at 4 K are also shown in
fig. 1. The spectrum of B850 emission peaks at 884 nm for B800-850/LDS and
at 869 nm for B800-850/LDS-LDAO. The latter complex lacks the B875 emission
band. The 4 K excitation spectra for B850 emission are given in fig. 2.
The carotenoid to B850 energy transfer efficiency was calculated to be
between 70 and 80 % in B800-850/LDS. In B800-850/LDS-LDAO this efficiency

Sybesma, C. (ed.), Advances in Photosynthesis Research, Vol. II. ISBN 90-247-2943-2.
© *1984 Martinus Nijhoff/Dr W. Junk Publishers, The Hague/Boston/Lancaster.*

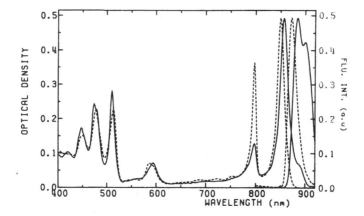

FIGURE 1. Absorption and fluorescence spectra at 4 K of B800-850/LDS (——) and B800-850/LDS-LDAO (---). The spectra were normalized at their maxima.

increased to 90 - 100 %, identical to the values found in B800-850/LDAO. At room temperature similar results were obtained. The value of 70 - 80 % agrees surprisingly well with the fraction of energy transferred directly from carotenoid to B850 and suggests that 20 - 30 % of the carotenoid is directly connected to B800 (Van Grondelle et al., 1982).

2.2. Fluorescence polarization

The polarization (p) of the B850 emission upon excitation in the 850 nm band was 0.14 for all three complexes (table 1). The values for p at 580 and 600 nm show that for the Q_x of B850 p is strongly negative, whereas excitation in the Q_x of B800 gives less negative p-values. For B800-850/LDAO these results agree with those of Breton et al. (1980). Our observations show that the p-values are independent of the size of the complexes. They are also independent of the temperature. This indicates that the

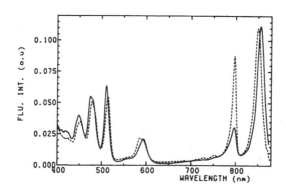

FIGURE 2. Excitation spectra of B800-850/LDS (——) and B800-850/LDS-LDAO (---) at 4 K. Detection at 875 and 880 nm, respectively.

TABLE 1. p-values at various wavelengths and temperatures. The standard errors are ± 0.02 for B850 emission and ± 0.05 for B800 emission.

Complex	λ_{det}	T	Excitation wavelength, nm					
			515	580	588	600	800	850
B800-850/LDS	865	300 K	-0.09	-0.11		-0.22	0.14	0.14
B800-850/LDS-LDAO	865	300 K	-0.075	-0.05			0.14	0.14
B800-850/LDAO	865	300 K	-0.07	-0.03		-0.20	0.14	0.13
B800-850/LDS	880	4 K					0.15	0.14
B800-850/LDS-LDAO	875	4 K					0.15	0.14
B800-850/LDAO	875	4 K			+0.01	-0.20	0.14	0.13
B800-850/LDAO	806	4 K	+0.10		+0.06			

B800-850/LDAO complex must be a highly ordered array of smaller units.

In the carotenoid region the p-value of B800-850/LDS is somewhat more negative than of B800-850/LDS-LDAO. If we assume that the carotenoid trans- ferring excitations to B800 does not contribute to the p-value in the B800-850/LDS complex, this implies an orientation more parallel to the plane of the B850 Q_y's for the carotenoid associated with B800.

The polarization of the B800 emission recorded at 4 K upon excitation in the Q_x region is positive, in contrast to that observed for the B850 emission. Because no back transfer from B850 to B800 occurs at this temper- ature, this indicates that rapid excitation transfer occurs between at least two B800 molecules with approximately perpendicular Q_y and with their porphyrin rings approximately parallel to each other. To obtain sufficient depolarization the rate of excitation transfer between the two B800's must significantly exceed that from B800 to B850 ($k = 3 \times 10^{11}$ s^{-1} at 4 K) (Van Grondelle et al., 1983). Thus taking $k_{800-800} \simeq 3 \times 10^{12}$ s^{-1}, and calculating R_0 from the B800 absorption and emission spectra ($R_0 \simeq 100$ Å; the orientation factor, $\kappa^2 = 1$), we find $R_{800-800} \leq 16$ Å using the well- known Förster equation (Van Grondelle et al., 1983).

The polarization of the B800 emission upon carotenoid excitation again has a small positive value, suggesting that the orientation of the caroten- oid TM associated with B800 is more or less parallel to the porphyrin ring of B800, in agreement with the results given above, and supporting the suggestion that two functionally different carotenoids exist, one associ- ated with B800 and one with B850 in a ratio of 1 : 2-3. The carotenoid molecules seem to be more or less parallel to the planes of the Q_x's of the associated BChl's, which results in a different orientation of the two types of carotenoid with respect to the plane of the Q_y's of B850.

2.3. Circular dichroism (CD)

In fig. 3 the CD spectra at 77 K are given. The CD spectrum of B800-850/LDS is non conservative in the B850 band, indicating interactions with higher excited states. For B800-850/LDAO the band is more nearly conservative, in agreement with Sauer and Austin (1978), possibly due to a positive non conservative contribution from B800-850 interaction, which causes a similar negative band in the 800 nm region. In addition, there seems to be a con- servative B800 exciton signal, in agreement with the B800-B800 interaction proposed above. The CD of B800-850/LDS-LDAO is analogous but not identical to that of B800-850/LDAO. Apparently the native configuration of B800 is

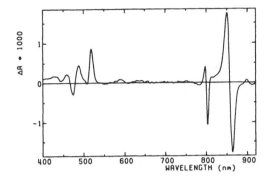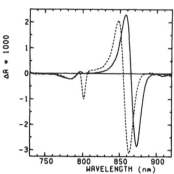

FIGURE 3. CD spectra at 77 K of B800-850/LDAO (left) and of B800-850/LDS (——) and B800-850/LDS-LDAO (---).

not totally recovered, although other experiments seem to indicate so.

On basis of our experiments we propose a basic unit, consisting of 4 polypeptides (2 x 10 kD, 2 x 8 kD). The α-helical section of each polypeptide binds one B850 (R.A. Niederman, personal communication) while the 8 kD subunits each contain an additional B800. All Q_y's are in approximately parallel planes, as are the Q_x's of the B800's, while the Q_x's of the B850 molecules are perpendicular to these planes. The carotenoids associated with B850 are perpendicular and the B800 carotenoid parallel to the plane of the Q_y's.

REFERENCES
Breton J, Vermeglio A, Garrigos M and Paillotin G (1981) Orientation of the chromophores in the antenna system of Rhodopseudomonas sphaeroides. In Akoyunoglou G, ed. Photosynthesis, Vol. 3, pp. 445-459. Philadelphia: Balaban Int. Science Services.
Clayton RK and Clayton BJ (1981) B850 pigment-protein complex of Rhodopseudomonas sphaeroides: Extinction coefficients, circular dichroism, and the reversible binding of bacteriochlorophyll, Proc. Natl. Acad. Sci. U.S.A. 78, 5583-5587.
Cogdell RJ and Thornber JP (1980) Light-harvesting pigment-protein complexes of purple photosynthetic bacteria, FEBS Lett. 122, 1-8.
Sauer K and Austin LA (1978) Bacteriochlorophyll protein complexes from the light-harvesting antenna of photosynthetic bacteria, Biochemistry 17, 2011-2017.
Van Grondelle R, Kramer HJM and Rijgersberg CP (1982) Energy transfer in the B800-850 carotenoid light-harvesting complex of various mutants of Rhodopseudomonas sphaeroides and of Rhodopseudomonas capsulata, Biochim. Biophys. Acta 682, 208-215.
Van Grondelle R, Hunter CN, Bakker JGC and Kramer HJM (1983) Size and structure of antenna complexes of photosynthetic bacteria as studied by singlet-singlet quenching of the bacteriochlorophyll fluorescence yield, Biochim. Biophys. Acta 723, 30-36.

Authors' addresses: Department of Biophysics, Huygens Laboratory of the State University, P.O. Box 9504, 2300 RA Leiden, The Netherlands. †Department of Biophysics, Physics Laboratory of the Free University, De Boelelaan 1081, 1081 HV Amsterdam, The Netherlands. *Department of Biochemistry, University of Bristol Medical School, University Walk, Bristol BS8 1TD, U.K.

LINEAR DICHROISM AND FLUORESCENCE EMISSION OF ANTENNA COMPLEXES DURING
PHOTOSYNTHETIC UNIT ASSEMBLY IN RHODOPSEUDOMONAS SPHAEROIDES.

HERMAN J.M. KRAMER, C. NEIL HUNTER[*], RIENK VAN GRONDELLE[**] and WOUTER BAKKER

1. INTRODUCTION

When aerobically grown, unpigmented cells of Rhodopseudomonas sphaeroides
are subjected to low oxygen growth conditions, an intracytoplasmic
membrane system forms which bears antenna and reaction centre pigment
proteins. The inner regions of the photosynthetic units, which consist
of reaction centres surrounded and interconnected by B875 antenna and
some B800-850 antenna, arise in small invaginations of the cytoplasmic
membrane. Much of the B800-850 is not synthesised until a later stage
of invagination is reached (Niederman et al., 1976); this stepwise mode
of assembly was confirmed in a fluorescence study by Pradel et al. (1978).
Niederman et al. (1979) demonstrated that small and large invaginations
of the cytoplasmic membrane are separable by sucrose density gradient
centrifugation; these are the upper pigmented band (UPB) and intracytoplasmic membrane (ICM) fractions. The UPB is thought to be derived
from new sites of membrane invagination that eventually expand and
assume the sedimentation characteristics of mature ICM. Hunter et al.
(1979) examined the fluorescence properties of the UPB and showed that
despite its relatively low B800-850 content, the UPB membrane in photosynthetically grown cells exhibited relatively high emission at 865nm;
this was attributed to some disconnected B800-850 antenna within this
fraction.
    In this paper, linear dichroism and low temperature fluorescence
properties of UPB and ICM membranes have been investigated at various
stages of repigmentation of aerobically grown cells. We conclude that
the efficiency of energy transfer from B850 to B875 bchls is high
(generally greater than 90%) in both types of membrane and at all stages
of development. During photosynthetic unit assembly, the high degree
of B875 orientation was conserved in both membrane fractions.
Relatively low degrees of B800-850 orientation were observed, especially
in the ICM, which is the main site for expansion of the 'lake' of
B800-850.

2. RESULTS AND DISCUSSION

2:1. Absorbance and fluorescence emission spectra.

    Figure 1 shows the absorbance and emission spectra for UPB and ICM
membranes at various stages of pigmentation. The 0h and 5h UPB fractions
contain very little bacteriochlorophyll (bchl) in relation to the amount
of cytochrome absorbing maximally around 420nm. Although ICM contained
more bchl than UPB membranes at zero time, the ICM fraction was difficult
to detect in sucrose gradients following cell disruption and rate-zone
sedimentation, in agreement with the observations of Niederman et al.
(1976). The material used here was obtained as a result of pooling
and concentrating the appropriate fractions from many gradients. In
most samples fluorescence is emitted from B850 bchls at approximately
870nm, which suggests some inefficient energy transfer between B850 and
B875; however, a calculation of the efficiency of energy transfer between
B850 and B875 bchls (Table 1) reveals no significant differences between
the membranes.
    The results in Table 1 were calculated from excitation spectra

Sybesma, C. (ed.), Advances in Photosynthesis Research, Vol. II. ISBN 90-247-2943-2.
© *1984 Martinus Nijhoff/Dr W. Junk Publishers, The Hague/Boston/Lancaster.*

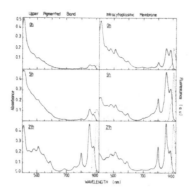

FIGURE 1. Absorbance (————) and emission (-------) spectra of
membranes at 4K; excitation at 590nm. The spectra were normalised at
their near IR maxima.

and were in good agreement with calculations made from the emission
spectra. The room temperature fluorescence data of Hunter et al. (1979)
did not permit any calculation of efficiency and the conclusion that
the UPB possesses some disconnected B800-850 units was based solely on
the observation of extra B800-850 emission at 865nm. This extra emission
is also present in the 21h UPB membranes, when compared with 21h ICM,
and the extent of transfer between B850 and B875 is equally high for
both samples. In the light of this new data, we must conclude that
although UPB and ICM membranes do exhibit small differences in the extent
of B850 to B875 energy transfer this process is extremely efficient even
at very early stages in photosynthetic development.

TABLE 1. The efficiency of energy transfer from B850 to B875 within
the expanding photosynthetic unit.

Membrane preparation	Time (hours)	% efficiency
Upper pigmented band	0	95
	5	85
	21	98
Intracytoplasmic membrane	0	98
	5	100
	21	95

2.2. Linear dichroism (LD)

The linear dichroism of these membranes was examined using the
polyacrylamide gel squeezing technique of Abdourakhmonov et al. (1979).
$A_{11}- A_{\perp}/A$ ratios at various wavelengths are presented in Table 2.
The LD/A values for B875 are high and demonstrate that the Q_y
transition of these bchls is strongly oriented parallel to the stretch

TABLE 2. LD/A ratios for UPB and ICM membranes

Membrane preparation	LD/A		
	800nm	850nm	875nm
UPB – 0h low oxygen growth	0.18	0.31	0.37
5h "	0.25	0.39	0.43
21h "	0.30	0.40	0.45
high light phototrophic growth	0.24	0.47	0.53
low light "	0.18	0.30	0.59
ICM – 0h low oxygen growth	0.14	0.28	0.41
5h "	0.20	0.34	0.51
21h "	0.18	0.25	0.41
high light phototrophic growth	0.10	0.26	0.40
low light "	0.13	0.20	0.35

axis. The degree of orientation of B800-850 was smallest in the highly pigmented intracytoplasmic membranes (low light growth) and reflects the increasing degree of disorder that results from the assembly of a 'lake' of B800-850 composed of many hundreds of bacteriochlorophylls. The fact that LD/A values at 850nm remain higher under almost all circumstances for the UPB membrane than for the ICM membrane, suggests the presence of a small and relatively well ordered arrays of B800-850 and B875 in the immature UPB membrane. The radiolabelling data of Hunter et al. (1982) also demonstrates the presence of a relatively small amount of B800-850 within the UPB that surrounds B875-reaction centre aggregates; the B800-850 'lake' is the predominant feature of the larger ICM invaginations.

In conclusion, the assembly and expansion of the photosynthetic unit results from the synthesis of B875 and some B800-850 antennae in well ordered arrays within the regions of limited invagination that give rise to the UPB fraction in sucrose gradients. Energy tranfer between these antennae is high. Subsequent expansion of the invaginations, and of the unit preserves the high degree of orientation of B875 to some extent but the incorporation of the bulk of the B800-850 in the ICM tends to give more random orientation of this pigment.

REFERENCES

Abdourakhmonov, I.A., Ganago, A.O., Erokhin Yu, E., Soloviev, A.A. and Chugunov, V.A. (1979) Orientation and linear dichroism of the reaction centres from Rhodopseudomonas sphaeroides R-26. Biochim. Biophys. Acta 546, 183-186

Hunter, C.N., van Grondelle, R., Holmes, N.G., Jones, O.T.G. and Niederman, R.A. (1979) In vivo fluorescence yield properties of a membrane fraction enriched in newly synthesised bacteriochlorophyll a protein complexes from Rhodopseudomonas sphaeroides. Photochem. Photobiol. 30, 313-316

Hunter, C.N., Pennoyer, J.D. and Niederman, R.A. (1982) Assembly and structural organisation in pigment protein complexes of Rhodopseudomonas sphaeroides. In: Akoyunoglou et al. eds. Cell Function and Differentiation Part B: Biogenesis of Energy Transducing Membranes and Membrane and Protein Energetics. pp 257-266, New York. N.Y. Alan R. Liss Inc.

Niederman, R.A., Mallon, D.E. and Langan, J.J. (1976) Membranes of
Rhodopseudomonas sphaeroides. IV. Assembly of chromatophores in low
aeration cell suspensions. Biochim. Biophys. Acta 440, 429–447
Niederman, R.A., Mallon, D.E. and Parks, L.C. (1979) Membranes of
Rhodopseudomonas sphaeroides. VI. Isolation of a fraction enriched in
newly synthesised bacteriochlorophyll a protein complexes. Biochim.
Biophys. Acta 555, 210–220
Pradel, J., Lavergne, J. and Moya, I. (1978) Formation and development
of photosynthetic units in repigmenting Rhodopseudomonas sphaeroides
wild-type and 'phofil' mutant strain. Biochim. Biophys. Acta 502,
169–182

Authors addresses: Department of Biophysics, Huygens Laboratory of the
State University, P.O. Box 9504, 2300RA Leiden, The Netherlands.
*Department of Biochemistry, University of Bristol Medical School,
University Walk, Bristol BS8 1TD, UK. ** Department of Biophysics,
Physics Laboratory of the Free University, De Boelelaan 1081, 1081HV
Amsterdam, The Netherlands.

RESONANCE RAMAN SPECTRA OF LIGHT-HARVESTING BACTERIOCHLOROPHYLL
IN PURPLE PHOTOSYNTHETIC BACTERIA

HIDENORI HAYASHI/MITSUO TASUMI

1. INTRODUCTION

Elucidation of the state of the chlorophyll molecules *in vivo* is essential for understanding the primary processes of photosynthesis. Vibrational spectroscopy is a potentially powerful tool for investigating the structure of chlorophyll and interactions between chlorophyll and its environment. In fact, infrared spectroscopy has been shown to be useful for studying chlorophyll-chlorophyll and chlorophyll-solvent interactions *in vitro* (J.J. Katz et al., 1966). Resonance Raman spectroscopy, on the other hand, provides a unique technique for probing the state of chlorophyll *in vivo*, because the Raman bands arising from the chlorophyll moiety can be selectively observed using a laser line in resonance with the electronic absorption of chlorophyll (M. Lutz et al., 1976).

Most bacteriochlorophyll (Bchl) molecules in purple photosynthetic bacteria act as light-harvesting antennae. They show a variety of absorption bands in the 800-900 nm region (due to the Qy transition). The absorption bands of Bchl in chromatophores are typically observed at 800, 850, and 870 nm. These absorption bands are usually called B800, B850, and B870, respectively. This multiplicity of the absorption bands must be related to the inhomogeneity of Bchl states *in vivo*. Actually a variety of Bchl-protein complexes have been prepared from purple photosynthetic bacteria, and they show the near-infrared absorption bands different from each other (R.J. Cogdell and J.P. Thornber, 1982).

Comparison of the resonance Raman spectra of Bchl-protein complexes having different absorption spectra will give new information on the Bchl state *in vivo*. Preliminary data have been reported by Lutz (1981) on complexes from the R-26 mutant of *Rhodopseudomonas sphaeroides* (mainly containing B870 Bchl) and complexes from the wild type of *R. sphaeroides* (mainly containing B800-850 Bchl). In this study, we measured the resonance Raman spectra of Bchl-protein complexes (and chromatophores) from several strains of purple photosynthetic bacteria. These materials are classified into two groups; one showing the B870 absorption band and the other showing the B800 and/or B850 absorption band.

To obtain definite information concerning the Bchl state *in vivo* from the resonance Raman spectra, the assignment of the Raman band should be established. Isotope-substituted molecules give useful information for assigning the Raman bands. The frequency shifts of the Raman bands of Bchl on ^{15}N-substitution have been reported (M. Lutz et al., 1982). To obtain further information for the assignment, we measured the resonance Raman spectra of Bchl molecules in chromatophores substituted with ^{13}C and ^{2}H as well as ^{15}N.

2. MATERIALS AND METHODS

Chromatophores and Bchl-protein complexes were prepared from the carotenoidless mutant of *R. sphaeroides* and the wild types of *R. sphaeroides*, *Rhodospirillum rubrum*, *Rhodopseudomonas palustris*, and *Chromatium vinosum*, according to the procedures reported previously (K. Sauer and L.A. Austin, 1978; H. Hayashi et al., 1982a). Materials were suspended in a Tris buffer containing thioglycolic acid. The resonance Raman spectra were

Sybesma, C. (ed.), *Advances in Photosynthesis Research, Vol. II. ISBN 90-247-2943-2.*
© *1984 Martinus Nijhoff/Dr W. Junk Publishers, The Hague/Boston/Lancaster.*

measured at room temperature with the 363.8 nm line of an Ar⁺ laser (about
2.5 mW at the sample point).   To supress the sample degradation with the
laser irradiation, a rotating cell was used.

The resonance Raman spectra of isotope-substituted Bchls were obtained
with chromatophores and Bchl-protein complexes prepared from the cells of
C. vinosum cultured autotrophically in the medium containing $^{15}NH_4Cl$,
$NaH^{13}CO_2$, or ^2H_2O.

3. RESULTS AND DISCUSSION

3.1. Resonance Raman spectra of Bchl-protein complexes

In Fig. 1 the resonance Raman spectra of the Bchl-protein complexes are
schematically shown.   The upper four spectra are those of the materials
which are classified into the B870-reaction center (RC) complexes.   The
Raman spectra of the carotenoidless mutant of *R. sphaeroides* and the wild
type of *R. rubrum* were measured only with chromatophores.   However, these
chromatophores were practically free from the B800–850 complexes and they
showed an absorption band at 870–880 nm and a weak band at 800 nm.
Accordingly, we classified these two materials into the same category as
the B870–RC complexes.   The resonance Raman spectra of the B870 complexes
which were obtained from the carotenoidless mutant of *R. sphaeroides* and
the wild type of *R. palustris* by the removal of reaction centers were
practically the same as those of the B870–RC complexes, although a slight

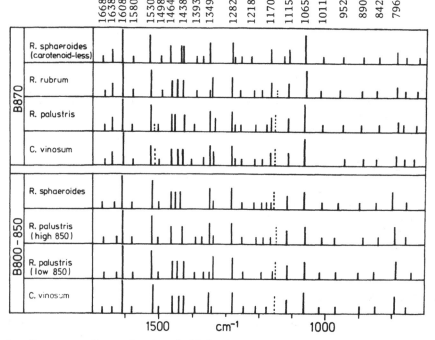

Fig. 1.   Resonance Raman bands of light-harvesting Bchl.
Excited with 363.8 nm, at room temperature.   Frequencies are those observed
for chromatophores of the carotenoidless mutant of *R. sphaeroides*.   Dotted
line; the band superimposed with the Raman band of carotenoids.

intensity decrease of the band at 1580 cm^{-1} (probably attributable to bacteriopheophytin in the reaction center) was observed. Thus, the Raman spectra of the B870-RC complexes should arise from Bchls which give the absorption band at 870 nm.

Lower four spectra are those of the B800-850 complexes. Although these four materials showed the absorption peaks around 800 and 850 nm, their absorption spectra in the 800-900 nm region do not always coincide. For example, the B800-850 complex from R. palustris cultured with a low-intensity light showed much weaker absorption at 850 nm than that from R. palustris cultured with a high-intensity light. The absorption band at 800 nm of the complex from C. vinosum is broader than those of the others. However, in this study, we classified these complexes into the category of the B800-850 complexes, according to their spectroscopic and proteinic properties (H. Hayashi et al., 1982b).

Many Raman bands were commonly observed for all the samples, such as the strong band at 1610 cm^{-1}, and the bands at 1530-20, 1285, 1115, and 790 cm^{-1}. There were some bands whose frequencies and intensities varied slightly from sample to sample, such as two or three overlapped bands at 1350-40 or 1470-30 cm^{-1}, weak bands at 1250-1180 and around 1400 cm^{-1}, and the band superimposed with the Raman band of carotenoids at 1150 cm^{-1} (Fig. 1, dotted lines).

There are some differences between the Raman spectra of B870 Bchl and B800-850 Bchl. All the B870-RC complexes showed a distinct band at 1640 cm^{-1}, while the B800-850 complexes exhibited a band at about 1635 cm^{-1} as a shoulder of the strong band at 1610 cm^{-1}. The B800-850 complex from C. vinosum had a Raman band at 1645 cm^{-1}, but its intensity is relatively low compared with that of the band at 1640 cm^{-1} of the B870-RC complex. The bands at 1530-25 cm^{-1} observed in the spectra of the B870-RC complexes shifted to 1525-20 cm^{-1} in those of the B800-850 complexes. In addition, the relative intensity of the bands at 1060 and 790 cm^{-1} appeared to be different between the B870-RC and B800-850 complexes. Thus, there is a correlation between the Raman spectra and the absorption spectra of Bchl in vivo, reflecting the differences between the environments of the Bchl molecules in the two groups of materials.

3.2. Resonance Raman spectra of isotope-substituted Bchls

In Table 1 the Raman frequencies of the ^{13}C- and ^{15}N-substituted Bchls in chromatophores of C. vinosum are compared with those of normal Bchls. On ^{13}C-substitution most bands shifted to lower frequencies. The bands at 1608 and 1438 cm^{-1} showed shifts of more than 3.5 %, the bands at 1520, 1463, 1338, and 790 cm^{-1} 2.5-3.0 %, and the other bands in the 1400-700 cm^{-1} region about 2.0 %. The strong band at 1608 cm^{-1} shifted by only 0.9 % on ^{2}H-substitution.

On ^{15}N-substitution the bands at 1114 and 790 cm^{-1} showed shifts of 0.8 % and the bands at 1284 and 1062 cm^{-1} 0.5 %, while the other bands showed little shifts. Similar results were obtained previously (M. Lutz et al., 1982).

From the above results it is reasonable to assign the bands at 1608 and 1438 cm^{-1} to the modes which mainly consist of the C=C (or C≡C) stretches. The bands at 1284, 1114, 1062, and 790 cm^{-1} seem to arise from the modes having the C-N (or C≡N) stretching components.

The Raman band at 1640 cm^{-1} observed for the B870-RC complex showed a shift of 2.5 % on ^{13}C-substitution and no shift on ^{15}N-substitution. These experimantal results support the previous assignment by Lutz (1981) that this band is due to the C=O (or C≐O) stretch.   Recently the C=O (or C≐O) stretch of retinal was shown to shift from 1667 cm^{-1} of normal species to 1630 cm^{-1} of ^{13}C-substituted species (R. Mathies 1983, private communication).   This corresponds to a shift of 2.2 % and confirms the above assignment of the 1640 cm^{-1} band of the B870–RC complex to the C≐O stretch.

Table 1.   Raman frequencies of isotope-substituted Bchl[a]

normal		^{13}C		^{15}N	
ν(cm^{-1})		ν(cm^{-1})	$\delta\nu$(%)	ν(cm^{-1})	$\delta\nu$(%)
1666[b]	w	1622[b]	2.6	1664[b]	0.1
1640[b]	m	1599[b]	2.5	1640[b]	0
1608	vs	1550	3.6	1608	0
1520	s	1478	2.8	1524	–
1463	m	1420	2.9	1464	–
1438	m	1385	3.7	1435 ?	
1424 ?		1363 ?		1420 ?	
1392	w	1353 ?		1388	0.3
1378	w	1340 ?		1375	0.2
1350	m	1324	1.9	1348	0.1
1338	m	1300	2.8	1334	0.3
1284	s	1254	2.3	1278	0.5
1206	w	1182	2.0	1204	0.2
1152	m	1126	2.1	1150	0.2
1114	m	1092	2.0	1105	0.8
1062	s	1040	2.1	1057	0.5
968	w	955	1.3	965	0.3
944	w	924	2.3	942	0.2
890	w	872	2.0	893	–
790	m	766	3.0	784	0.8

Measured with a) chromatophores and b) B870–RC complex

REFERENCES
Cogdell RJ and Thornber JP (1980) Light-harvesting pigment-protein complexes, FEBS Lett. 122, 1–8.
Hayashi H, Miyao M and Morita S (1982a) Absorption and fluorescence spectra of bacteriochlorophyll-protein complexes, J. Biochem. 91, 1017–1027.
Hayashi H, Nakano M and Morita S (1982b) Comparative studies of bacterio-chlorophyll-protein complexes, J. Biochem. 92, 1805–1811.
Katz JJ, Dougherty RC and Boucher LJ (1966) Infrared and NMR spectroscopy of chlorophyll, In Vernon LP and Seely GR eds. The chlorophyll, pp. 185–251. Academic Press, New York.
Lutz M, Kleo J and Reiss-Husson F (1976) Resonance Raman scattering of bacteriochlorophyll, Biochem. Biophys. Res. Commun. 69, 711–717.
Lutz M (1981) Bonding interactions on pigments in bacterial antenna, Photosynth., Proc. Int. Congr 3, 461–476.
Lutz M, Hoff AJ and Brehamet L (1982) Bacteriochlorophyll a-protein interactions, Biochim. Biophys. Acta 679, 331–341.
Sauer K and Austin LA (1978) Bacteriochlorophyll-protein complexes from the light-harvesting antenna, Biochemistry 17, 2011–2019.

Authors address: Hidenori Hayashi/Mitsuo Tasumi, Department of Chemistry, Faculty of Science, The University of Tokyo, Hongo, Tokyo 113, Japan.

ASSOCIATIONS OF PIGMENT-PROTEINS AND PHOSPHOLIPIDS INTO SPECIFIC DOMAINS
IN *RHODOPSEUDOMONAS SPHAEROIDES* PHOTOSYNTHETIC MEMBRANES AS DETERMINED BY
LITHIUM DODECYL SULFATE/POLYACRYLAMIDE GEL ELECTROPHORESIS*

C. W. RADCLIFFE, J. D. PENNOYER, R. M. BROGLIE AND R. A. NIEDERMAN

Department of Biochemistry, Bureau of Biological Research, Rutgers
University, P. O. Box 1059, Piscataway, NJ 08854 USA

1. ABSTRACT

When chromatophore vesicles derived from the intracytoplasmic photosynthetic
membranes of *Rhodopseudomonas sphaeroides* are subjected to lithium dodecyl
sulfate/polyacrylamide gel electrophoresis at 4°C, specific domains of the
B800-850 light-harvesting complex can be isolated which include their associated phospholipid components. Electrophoresis of chromatophores from $^{32}P_i$-
labeled cells at various lithium dodecyl sulfate concentrations indicated
that approx. 10-19% of the radioactivity which entered the gel remained
associated near the top with the major B800-850 band with most of the rest at
the gel front. In chromatophores in which cardiolipin was present, B800-850
was enriched 2- to 6-fold in this phospholipid. Up to three phospholipid
molecules were present per complex. Although about 40% of the B800-850 in
the gel was found in complexes which contained B800-850 still associated
with the B875 light-harvesting complex, very little radioactivity was associated with these or with the spectrally homogeneous B875 complex. After
pulse labeling cells 1 min with 2-[^{3}H]glycerol, the specific phospholipid
radioactivity in B800-850 was up to one-third lower than in chromatophores.
This indicates that the exchange of newly synthesized phospholipid into
B800-850 is restricted and that stable phospholipid associations in this
complex may be preserved during electrophoresis. Thus, the major B800-850
band, which is thought to arise from an outer array of this antenna, appears
to associate specifically with phospholipids. The relation of these findings
to models for the organization of the pigment-protein complexes in the membrane is discussed.

2. INTRODUCTION

Spectroscopic studies on the organization of phospholipids in the chromatophore membrane of the photosynthetic bacterium *Rhodopseudomonas sphaeroides*
with fluorescent (Fraley et al., 1978) and spin label (Birrell et al., 1978)
probes have suggested that protein significantly restricts fatty acyl chain
motions within the bilayer and that negatively charged phosphoglycerides
associate preferentially with light-harvesting antennae. Although Sauer and
Austin (1978) reported that up to 15-20% of the dry weight of B800-850 light-
harvesting complexes isolated from *R. sphaeroides* with Triton X-100 was
accounted for by phospholipid, specific associations of phospholipids in
isolated antenna preparations have not been demonstrated.

Broglie et al. (1980) have reported that when *R. sphaeroides* chromatophores are subjected to lithium dodecyl sulfate (LDS)/polyacrylamide gel
electrophoresis at 4°C up to 11 pigment-protein complexes can be resolved.

---

*Supported by U.S. Public Health Service grant GM26248 and U.S. National
Science Foundation grants PCM79-03665 and PCM82-09761 and Fellowships from
the Charles and Johanna Busch Memorial Fund Award to C.W.R., J.D.P. and R.M.B.

Sybesma, C. (ed.), Advances in Photosynthesis Research, Vol. II. ISBN 90-247-2943-2.
© *1984 Martinus Nijhoff/Dr W. Junk Publishers, The Hague/Boston/Lancaster.*

The largest and most abundant band consisted mostly of the B800-850 light-harvesting pigment-protein complex. The smallest light-harvesting complex contained only B875 which migrated slightly behind a reaction center particle The remaining complexes migrated to intermediate positions in the gel and contained B800-850 and B875 still associated functionally in various proportions (Hunter et al., 1981). It is reported here that the majority of the phospholipid isolated with these pigment-protein complexes is present in the major B800-850 band; up to three molecules of phospholipid per complex are present which may in part represent stable associations that existed within the membrane.

3. RESULTS AND DISCUSSION

In Fig. 1, the distribution of radioactivity after preparative LDS/PAGE of chromatophores from *R. sphaeroides* labeled with $^{32}P_i$ is shown. Although the majority of the radioactivity migrates with the free pigment at the gel front, a significant and reproducible quantity is found with the major B800-850 band near the top of the gel. It is also shown in Fig. 1, and after LDS/PAGE on a polyacrylamide gradient gel (Fig. 2), that very little radioactivity was present in the other pigment-protein complexes. After PAGE

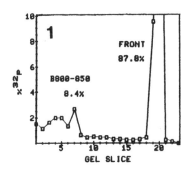 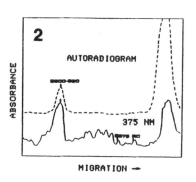

FIGURE 1. Distribution of ^{32}P after LDS/PAGE. Chromatophores were isolated from strain 8253 grown phototrophically to late exponential phase as in Broglie et al. (1980), that was labeled uniformly with $^{32}P_i$ (1.0 µCi/ml). The isolated chromatophores were treated with LDS and subjected to electrophoresis at 4°C on a 6-mm preparative gel prepared with 7.5% polyacrylamide. A lane from the gel slab was cut into 5-mm slices that were digested overnight with 0.5 ml 30% H_2O_2 at 60°C. After addition of 0.5 ml of water and Aqualyte Plus scintillant, radioactivity was determined by liquid scintillation counting.

FIGURE 2. Comparison of distribution of radioactivity and pigment after LDS/PAGE at 4°C. The ^{32}P-labeled chromatophores described in Fig. 1 were subjected to LDS/PAGE on a 1.5-mm analytical gel prepared with a 5-10% polyacrylamide gradient. The gel was scanned at 375 nm on an Ortec model 4310 scanning densitometer to determine the distribution of Bchl and subjected to radioautography. The scan of the autoradiogram is also shown.

over a 10-fold range of LDS concentrations, 12-19% of the ^{32}P-labeled material which entered the gel remained associated with the major band (not shown).

Typically 90% of the ^{32}P in chromatophores isolated from *R. sphaeroides* NCIB 8253 could be extracted into chloroform-methanol and recovered as phospholipid. Essentially the same result was obtained with isolated B800-850, but the polypeptides of this complex are soluble in organic solvents (Theiler et al., 1982). To test whether any radioactivity could be accounted for by phosphorylation of the B800-850 proteins, chromato-phores and isolated B800-850 were heated for 1 min at 100°C in SDS and subjected to PAGE in the presence of this detergent as in Broglie et al. (1980) (Fig. 3). Although a small shoulder of radioactivity overlapped with the stained B800-850 protein, again the majority of ^{32}P was found at the gel front. With regard to the phospholipid composition of B800-850 (determined as in Onishi, Niederman, 1982), when chromatophores were exam-ined in which cardiolipin was present, this antenna complex was enriched 2- to 6-fold in this phosphoglyceride; however, no consistent specificity in the association of the major phospholipid species with B800-850 has yet been observed.

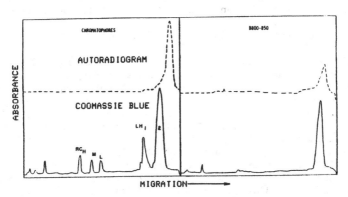

FIGURE 3. Distribution of ^{32}P after SDS/PAGE of chromatophores and the isolated B800-850 complex. Details are presented in the text.

In Table 1, it is shown that about three molecules of phospholipid are isolated per B800-850 complex. The phospholipid content of the B800-850 isolated by LDS/PAGE is somewhat higher than that obtained by an LDAO/Triton procedure. It is also higher than the phospholipid content of the antenna holochrome B880 of *Rhodospirillum rubrum* in which about 2 phospho-lipid molecules per 23,000 g of protein were reported by Picorel et al. (1983). The pigment composition of the isolated B800-850 complexes (Table 1) confirms that of Broglie et al. (1980).

To examine the stability of phospholipid associations with the B800-850 complex, phototrophically growing cells were equilibrated to 15°C to limit membrane diffusion and pulse labeled for 1 min with 2-[^{3}H]glycerol (Table 2). In the NCIB 8253 strain, the specific phospholipid radioactivity in the B800-850 complex was significantly lower than that of chromatophores. This is consistent with the possibility that the exchange of newly synthe-sized phospholipid into the B800-850 membrane domain is partially restricted. It also suggests that stable phospholipid associations may be preserved

TABLE 1. Phospholipid and pigment content of isolated B800-850 complexes

Molar ratio	Preparation	
	LDS/PAGE[a,f]	LDAO/Triton[b]
Phospholipid[c]/complex[d]	3.3	2.6
Bchl[e]/complex	3.2	3.8
Carotenoid[e]/complex	1.5	1.8

[a] Performed by procedure of Broglie et al. (1980).
[b] Isolated by the procedure of Clayton, Clayton (1972), LDAO, lauryl dimethylamine oxide.
[c] Lipid phosphorus was determined after extraction with chloroform-methanol (1:2, vol/vol). A mol. wt. of 750 was used for phospholipid.
[d] Protein was determined by the method of Bradford (1976). A mol. wt. of 18,000 was used for B800-850 (Broglie et al., 1980). Sequencing studies of Theiler et al. (1982) suggest a mol. wt. about 20% lower than this.
[e] Procedures described in Niederman et al. (1976). Bchl, bacteriochlorophyll a.
[f] Lithium dodecyl sulfate/polyacrylamide gel electrophoresis.

during the isolation procedure. The difference between B800-850 and chromatophore specific radioactivities were not as great in the 2.4.1 strain, but this experiment was complicated by significant polyphosphate production in the low phosphate medium employed.

These results support previous suggestions (Broglie et al., 1980) that the major B800-850 band isolated by LDS/PAGE arises from an outer array of this antenna. Furthermore, they suggest that this complex associates specifically with a significant quantity of phospholipid. This may either reflect phospholipid associations at the outer surfaces of the photosynthetic units or separation of the outer B800-850 array from the B875 associated B800-850; this was isolated as intermediate complexes in which no phospholipids were detected. A critical test of these alternatives is underway in which the phospholipid content of chromatophores is increased by fusion to liposomes.

TABLE 2. Stability of phospholipid association with the B800-850 light-harvesting complex.

Preparation	Experiment	
	1[a] $[^3H]/Pi$	2[b] $[^3H]/^{32}P$
	DPM/mg	CPM/CPM
Chromatophores	83,350	0.39
B800-850	56,020	0.34
Gel front	86,000	0.36

[a] Strain NCIB 8253 grown as in Broglie et al. (1980) to mid-exponential phase was brought to 15°C and pulse labeled 1 min with 2-$[^3H]$glycerol (1.3 µCi/ml). Lipid phosphorus was determined as in Table 1.

[b]Strain 2.4.1 was steady-state labeled with 0.3 μCi/ml of $^{32}P_i$ in a low-phosphate succinate medium (Fraley et al., 1979) at 4,200 lux and treated as above.

REFERENCES

Birrell GB, Sistrom WR and Griffith OH (1978) Biochemistry 18, 3768-3773.
Bradford M (1976) Anal. Biochem. 72, 248-254.
Broglie RM, Hunter CN, Delepelaire P, Niederman RA, Chua NH and Clayton RK (1980) Proc. Natl. Acad. Sci. USA 77, 87-91.
Clayton RK and Clayton BJ (1972) Biochim. Biophys. Acta 283, 492-504.
Fraley RT, Jameson DM and Kaplan S (1978) Biochim. Biophys. Acta 511, 52-69.
Fraley RT, Lueking DR and Kaplan S (1979) J. Biol. Chem. 254, 1980-1986.
Hunter CN, Niederman RA and Clayton RK (1981). In Akoyunoglou G, ed. Photosynthesis, Vol. III, Structure and molecular organization of the photosynthetic apparatus, pp. 539-546. Balaban International Science Services, Philadelphia, Pennsylvania.
Niederman RA, Mallon DE and Langan JJ (1976) Biochim. Biophys. Acta 440, 429-447.
Onishi JC and Niederman RA (1982) J. Bacteriol. 149, 831-839.
Picorel R, Bélanger G and Gingras G (1983) Biochemistry 22, 2491-2497.
Sauer K and Austin LA (1978) Biochemistry 17, 2011-2019.
Theiler R, Brunisholz, R, Grank G, Suter F and Zuber H (1982) Abstr. IV Int. Symp. Photosynth. Prokaryotes. C40.

R. M. Broglie's present address is : The Rockefeller University, 1230 York Avenue, New York, N.Y. 10021 USA.

REVERSIBLE PH INDUCED ABSORPTION CHANGE IN THE BCHL B-LH-PROTEIN OF THE ALKALOPHILE ECTOTHIORHODOSPIRA HALOCHLORIS.

R. Steiner, H. Scheer

Botanisches Institut der Universität, München, FRG.

Introduction: Ectothiorhodospira halochloris and E. abdelmalekii are extremely halophilic and alcalophilic photosynthetic bacteria (1). They belong to the few species containing bacteriochlorophyll b (bchl b). In both Ectothiorhodospira species, it is esterified with the unusual alcohol 2,10-phytadienol (2). Like other bchl b-containing species, e.g. Rhodopseudomonas viridis, it has an infrared absorption maximum around 1020 nm. Unlike Rp. viridis, however, the Ectothiorhodospira species contain a second, structured infrared band peaking at 800 and 830 nm. It had originally been suspected (1) to originate from the presence of bchl a besides bchl b, which is, however, disproved by its pigment analysis (2). The 800/830 nm bands must then originate as well from bchl b. Here, we wish to report the isolation of a bchl b-containing antenna complex (B1020/800) from E.halochloris, and a reversible, pH-induced absorption change of the 1020 nm component.

Isolation of B1020/800: E.halochloris was grown anaerobically at 35°C in a medium modified from IMHOFF and TRÜPER (1). Rhodopseudomonas viridis was grown anaerobically at 28°C in HUNTERS medium (3). After disruption by french press treatment of the cells chromatophores were isolated by the method of OKAMURA and FEHER (4). For the fractionation of the antenna pigments, chromatophores ($A_{1020} = 50$) were treated with 1 % Triton X-100, dialysed overnight against 0.1 % Triton and fractionated on DEAE cellulose.
According to SDS gel electrophoresis, the B1020/800 antenna pigment contains two major bands corresponding to a MW of 13500 and 13000 Dalton, if calibrated with a standard set of globular, hydrophilic proteins. When compared with the two antenna complexes of Rp. spheroides 2.4.1, the two bands move between the 6800 D of B875 and the 5400 D band of B800/850. The electrophoresis also shows a weakly stained very low molecular weight band at 4500 D, as well as two higher MW bands at 15500 and 27500 D, which may be due to impurities.

ABSORPTION SPECTRA: When whole cells, spheroplasts or chromatophores of E. halochloris or E.abdelmalekii (fig. 1a) or, isolated antenna fractions of E.halochloris are titrated with acid in the pH range between 7.5 and 5.7, the 1020 nm absorption is replaced by a new absorption peaking at 960 nm. The 830/800 nm band is essentially uneffected by this treatment. The 960 nm band is unstable and decreases again with time and especially upon addition of more acid. It is then replaced by rather broad and unstructured absorptions below 900 nm. At pH 5.7, the species absorbing at 960 nm has a half life of appx. 1 hr. If the decay of the "low pH" form is avoided by an immediate back titration with base, the 1020 nm form is at least partially restored. (fig. 1b). The losses are probably due to the instability of the 960 nm form during the titration. Neither of the two forms, nor their interconversion is effected by α-bromopropionate or its methyl-ester, or by high salt.

FLUORESCENCE SPECTRA: The natural "high pH" form has a single emission band peaking at 1018 nm (ambient temperature) or 1000 nm (1.7K). The acid induced "low pH" form has its emission maximum blue-shifted to 975 nm (992 nm at 1.7K), but additional emission bands are also observed at 803 nm (298K) or 824 nm (1.7K). The short wavelength emission of the "low pH" form is much more pronounced at low temperature. This data indicate an energy transfer from the chromophores absorbing at 800 nm to the ones absorbing at 1020 nm in the native "high pH" form, which is uncoupled in the "low pH" form.

CIRCULAR DICHROISM SPECTRA: The CD spectrum of the natural "high pH" form shows a complex pattern at the 1020 (limited in our machine to 1010 nm) and the 800/830 nm bands (fig. 2), which is indicative of exiton coupling between more than two chromophores. The spectrum of the "low pH" form is less complex, with only a doublet in the region of the 960 nm band, and a strongly decreased optical activity in the 800/830 nm band region.

DISCUSSION: A bchl b-containing antenna complex (B1020/800) has been isolated from Ectothiorhodospira halochloris. These absorptions correspond to the ones seen in whole cells. This native form is stable at pH 8, but is reversibly transformed to a form absorbing at 960 and 800 (B960/800) at lower pH, with an apparent pK of 6.3. This value is close to the pK of histidine (6.8), or possibly carboxylic acid side chains (pK $\approx$ 4.7). Attempts to modify histidine residue(s) with α-bromopropionate or its methyl ester did neither affect the 1020 nor the 960 nm form, which argues against the participation of histidine. High salt is also ineffective. Since the changes are observed in whole cells, spheroplasts (= "outside out" vesicles) and chromatophores (="inside out" vesicles), whichever group is affected is accessible from either side of the membrane, or the membrane is readily permeable to protons.
The fluorescence data indicate for the natural, "high pH" form a good energy transfer from the chromophores absorbing at 800/830 nm to ones absorbing at 1020 nm. This transfer is (partly) disrupted in the "low pH" form, as evidenced by the second emission band in the 800 nm region. The CD spectra would also support a transformation from a tightly coupled to a weakly coupled state. One possible explanation would be the assignment of the 1020 and the 800/830 nm band to two bchl b-polypeptides, which are in close association in the "high pH" form. There are at two strongly interacting chromophores absorbing in the 800 nm region, and more than two in the 1020 nm region, if judged from the cd data.

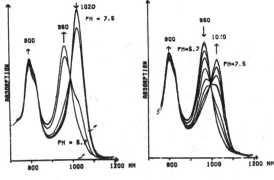

<u>Titrattion of the B800/1020</u>
antenna complex of E.halochloris
with acid (left and base (right
Absorption spectra in the near
infrared spectral range.

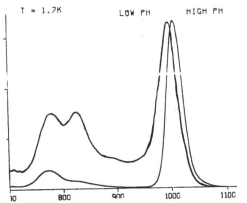

<u>Low-temperature fluorescence</u>
<u>emission spectrum</u> of the
low pH and high pH forms of
the B800/1020 antenna complex.
Excitation at 599 nm. We thank
K. Angerhofer in the group of
Prof. H.Wolf (Stuttgart) for
providing these spectra to us.

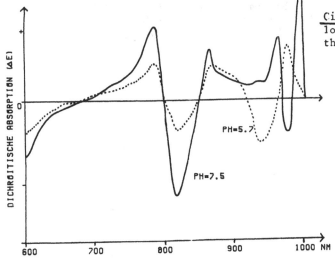

<u>Circular dichroism</u> of the
low pH and high pH forms of
the antenna complex

Bchl <u>a</u>-containing bacteria contain an antenna fraction B800/850.
It has also two absorption maxima, which can be shifted by
various treatments. (5). The B1020/800 antenna complex might then
be the bchl <u>b</u>-containing analogue. However, the B800/850 contains
carotenoids, and does not show such pronounced CD-couplings. We
have also shown, that the absorption maxima (although not the
relative absorption) of the two bands are acid-insensitive down
to pH 1.7. It has also a similar polypeptide composition, if one
disregards the very low molecular weight components present in
<u>E. halochloris</u>. It should be noted, however, that a similar
peptide has been found in another bchl <u>b</u>-containing organism, e.g.
<u>Rp. viridis.</u>
The increased sensitivity of <u>E.halochloris</u> to positive charges
may reflect a specific adaptation to the high pH ($\simeq$11) prevailing
in its natural environment.

ACKNOWLEDGEMENTS: This work was supported by the Deutsche
Forschungsgemeinschaft, Bonn. We thank W. Angerhofer in the group
of Prof. H.C. Wolf (Stuttgart) for the fluorescence measurements.

REFERENCES:

1) J. Imhoff, H. Trüper, Arch. Microbiol. <u>114</u>, 115 (1977)
2) R. Steiner, W. Schäfer, H. Wieschhoff, H. Scheer, Z. Naturforsch. <u>36c</u>, 417 (1981)
3) G. Cohen-Bazire, W.R. Sistrom, R.Y. Stanier, J.Cell.Comp.
   Physiol. <u>49</u>, 25 (1957)
4) M.Y. Okamura, G. Feher in "The Photosynthetic Bacteria" (R.K.
   Clayton, W.R. Sistrom, eds) Academic Press, New York, 1978
5) R.J. Cogdell, J.P. Thornber, FEBS Lett. <u>122</u>, 1 (1980)

ON THE EFFECT OF PROTEINASES,PHOSPHOLIPASES AND GROUP-SPECIFIC REACTANTS
ON CAROTENOID SPECTRA AND ELECTROCHROMIC RESPONSE OF MEMBRANE PREPARATIONS
OF PHOTOSYNTHETIC PURPLE BACTERIA.

C. Swysen, M. Symons[x] and C. Sybesma, Biophysics Laboratory, Vrije
Universiteit Brussel, 1050 Brussels .

[x]present address : Membrane Research, Weizmann Institute for Sciences,
76100 Rehovot, Israel .

INTRODUCTION

In photosynthetic purple bacteria <u>Rhodopseudomonas capsulata</u> and <u>sphaeroides</u>
it has been shown that the carotenoids consist of two spectrally but not
chemically different pools (1,2). In vesicle preparations of these bacteria, carotenoid absorbance changes can be induced by illumination. It
is generally believed that the mechanism of the carotenoid bandshift has
an electrochromic character (3).
Only the smallest carotenoid pool containing 20-30% of the total carotenoid
content shows a substantial spectral shift in response to the generation
of transmembrane potentials. This field-sensitive pool is associated with
the light-harvesting complex LH II which shows bacteriochlorophyll absorption bands at 800 and 850 nm (4). The field-insensitive carotenoids, accounting for 75 % of total carotenoid content, are distributed between both
light-harvesting complexes : 1/3 is associated with the LH I complex
(Bchl absorption at 870 nm) and 2/3 with the LH II complex (4).
Although the protein composition and the organisation of the LH complexes
are known, their location in the membranes is still not well documented.
In a previous study, we incubated chromatophores of Rps.caps. with a specific pronase (type III, Sigma) from Streptomyces griseus (5). The effect
of pronase digestion on the spectral characteristics of both the field-
sensitive and the field-insensitive carotenoid pools was monitored. It
was shown that the field-responding carotenoids are more sensitive to
pronase treatment than the non-responding pool. In the present study,
these experiments were extended to other digestive treatments of chromatophores of Rhodopseudomonas capsulata as well as sphaeroides.

MATERIALS AND METHODS

Chromatophores of Rps. caps. and sphaeroides were incubated at 36°C with
different types of enzymes at 60 μg Bchl per ml in Tris 10mM buffer with
Ca^{++} 1mM at pH = 7 (see ref. 5). Digested samples were taken at different
incubation times and diluted tenfold with Tris 10mM, pH = 7 for pronase
studies and Tris 10mM, EDTA 10mM, pH = 7 for lipase studies. Measurements
at different temperatures allowed us to monitor fluidity effects.
The following enzymes have been used :
proteinases : pronase K, a non-specific pronase from Tritirachium album,
    trypsine and pronase type VI from Streptomyces griseus.

phospholipases : phospholipase A$_2$ (from Naja Naja venom) which catalyzes
    the specific hydrolysis of the acyl group from the 2-
    position of glycerophospholipids.
    phospholipase C (from Clostridium perfringens) which
    cuts off the complete polar head group - including the
    phosphorus - of a phospholipid resulting in a diglyceride.
    phospholipase D (from peanut) converts glycerophospholipids
    into phosphatidic acid.

Sybesma, C. (ed.), Advances in Photosynthesis Research, Vol. II. ISBN 90-247-2943-2.
© 1984 Martinus Nijhoff/Dr W. Junk Publishers, The Hague/Boston/Lancaster.

Group specific reaction with dithiotreitol (10mM) and by N-ethylmaleimide (10mM) was also investigated.
Methods to measure λ^i, the position of the isosbestic point of the long-
wavelength carotenoid absorption band, λ°, the band-center of that band,
ΔA, the light-induced absorbance changes and the Bchl spectrum are described in ref (5).

RESULTS

In Rps. caps., our findings reported for incubation with pronase type VI
(5), were corroborated by digestion with pronase K, a proteinase with a
broad action spectrum : during the first 5 min. of digestion, the isos-
bestic point λ^i shifts much faster to the blue than the top of the band λ°.
An indication of the pool-size of carotenoids affected by the pronase
digestion is given by the parameter $\frac{\Delta\lambda^\circ}{\Delta\lambda^i}$ (t_i), where $\Delta\lambda^\circ$ and $\Delta\lambda^i$ are the
shifts of λ°, resp. λ^i during the time interval ($t_i - t_{i-1}$).
It appears that the pool size affected by 4 min. digestion with pronase
K up to $10\mu g/60\mu g$ Bchl (see table 1) or lipase A_2, 1 unit/60 μg Bchl, is
20-30 % of total carotenoids, a fraction very similar to the estimated
pool size of the field-sensitive carotenoids in this strain. At $40\mu g/$
$60\mu g$ Bchl, more carotenoids are affected (40-50%) by pronase, even in the
first 4 min. It was seen that with $18\mu g/60\mu g$ Bchl, the pool parameter
was 0,22 after the first 2 min and increased to 0,72 in the next 3 min.,
showing that gradually both carotenoid pools are affected.
The kinetic dependence of the shifts on enzyme concentration indicated
that the average rate of the λ^i shift during the first 15 min increased
more than the rate of λ° shift (after 15 min the rate slowed down until
the response became saturated).
The blue shift of the isosbestic point and the degradation of the Bchl
. band at 800 nm showed similar kinetics in Rps. caps. as well as in sphaeroide
Rps. sphaeroides, when briefly digested with pronase type VI, showed no
measurable shift of λ° while λ^i was shifted 0,6 nm to the blue. The same
phenomenon was apparent during short incubation with lipase A_2.
The similar responses seen on incubation with different enzymes, seem to
indicate that the field-sensitive carotenoids are located closer to the
chromatophore surface than the field-insensitive pool in both bacteria.
Moreover, contrary to reactions limited by surface access of the bulky
enzymes, reactions with a small membrane penetrating and environment
changing molecule, whereby all carotenoids are supposedly equally accessible, long incubations of control samples at 36°C or temperature and
fluidity variations of non-incubated samples affected both λ^i and λ°
simultaneously and to the same extent.
In Rhodopseudomonas capsulata, dithiotreitol (10mM) induced a small red
shift (+ 0,3 nm) and N-ethylmaleimide (10mM) a small blue shift of both
λ^i and λ°.
Measurements of non-incubated control samples at 26°C shifted both λ^i
and λ° 0,7-1 nm to the blue compared to measurements at 10°C.
Long incubations of control samples in buffer at 36°C shifted λ^i and λ°
gradually and about equally to the blue.

The two carotenoid populations still showed significant wavelength shift
with respect to one another, even after long digestion (1-2 hrs) with
all enzymes tested, although their absorption bands "approached" each
other : the field-responding pool in both Rhodopseudomonas capsulata and
sphaeroides was more blue shifted than the field-insensitive pool when
the response became saturated.

In Rhodopseudomonas capsulata the maximal blue shifts were :

$$\Delta\lambda^i \; : \; 7 \pm 0,5 \; nm$$
$$\Delta\lambda^o \; : \; 4,3 \pm 0,5 \; nm$$

In Rhodopseudomonas sphaeroides the maximal blue shifts were :

$$\Delta\lambda^i \; : \; 2 - 6 \; nm$$
$$\Delta\lambda^o \; : \; 1,6-2,2 \; nm$$

Two corrections have to be made to these experimentally measured maximal
shifts $\Delta\lambda^o$ to calculate the exact shift of the field-insensitive population only (see also ref. 5) : the shift $\Delta\lambda^o$ comprises a contribution of
the 25 % of field-sensitive carotenoids - the maximal shift $\Delta\lambda^i$ of which
is determined experimentally - and of the 25 % of field-insensitive carotenoids, associated with B870, whose spectrum is most probably not affected
by digestive treatments. Even when appropriate corrections were made,
the maximal shifts $\Delta\lambda^o$ were always smaller than $\Delta\lambda^i$.

The fact that the 2 populations remain spectrally different implies that
- even at this stage of extensive digestion - solvatochromism as well as
specific interactions with the proteins determine the spectral characteristics of the carotenoid pools (see the model of Kakitani et al (6)).

Phospholipase D incubation induced large spectral changes in Rps. sphaeroides, but had no effect in Rps. caps. In both bacteria no measurable
effects were seen after prolonged phospholipase C incubation although
some phosphorus could be detected in the supernatant of the digested
samples.

It remains to be seen, whether these observations may indicate some
specific lipid - LH complex associations.

In conclusion, the facts that temperature variations, long incubations
at 36°C or reactions with small membrane penetrating molecules, affect
both carotenoid populations simultaneously and to the same extent, while
enzymatic digestions with different types of proteinases and with phospholipase A_2 affect the two populations differentially, the field-sensitive carotenoids being the most sensitive ones ,seem to indicate that in
chromatophores of Rhodopseudomonas capsulata and sphaeroides the field-
sensitive carotenoids are located closer to the surface than the field-
insensitive pool.

It was shown that the blue shift of the isosbestic point and the loss of
electrochromic response are linked kinetically to the degradation of the
Bchl absorption band at 800 nm in Rhodopseudomonas sphaeroides, in agreement with the observations made by Webster et al (4) in Rps. capsulata.

TABLE 1   Effect of pronase K incubation (4 min) on carotenoid spectral
characteristics in Rhodopseudomonas capsulata chromatophores,
measured at 10°C.

Additions	λ° (nm)	λ^{i} (nm)	$\dfrac{\Delta\lambda^{\circ}}{\Delta\lambda^{i}}$
–	$510,7 \pm 0,1$	$519,55 \pm 0,05$	–
5 μg pronase/60μg Bchl	$510,6 \pm 0,1$	$519,05 \pm 0,05$	0,2
10μg pronase/60μg Bchl	$510,5 \pm 0,1$	$518,70 \pm 0,05$	0,2
40μg pronase/60μg Bchl	$510,1 \pm 0,1$	$518,30 \pm 0,05$	0,5

REFERENCES

(1) Symons M., Swysen C. and Sybesma C. (1977), Biochim.Biophys.Acta,
462, 706-717.
(2) Symons M., Swysen C. and Sybesma C. (1977) in Bioenergetics of
membranes (Packer L., Papageorgiou G.C. and Trebst A., eds.),
p. 477-483, Elsevier, North-Holland, Biomedical Press Amsterdam.
(3) Junge, W. (1977) Ann. Rev. Plant Physiol., 28, 503-538.
(4) Webster G.D., Cogdell R.J., Lindsay J.G. (1980) Biochim. Biophys.
Acta, 591, 321-330.
(5) Symons M., Swysen C., (1983) BBA, 723, (3) 454-457.
(6) Kakitani T., Honig B., and Crofts A.R. (1982), Biophys. J., 39
57-63.

AKNOWLEDGEMENTS

We thank Dr. L. Slooten for helpfull discussions and Mr. A. Nuyten for
expert technical assistance.

RADICAL FORMATION BY DARK OXIDATION OF ANTENNA BACTERIOCHLOROPHYLL IN
RHODOPSEUDOMONAS

ISABEL GOMEZ[*] and FRANCISCA F. DEL CAMPO[**]/INSTITUTO DE BIOLOGIA CELULAR
CSIC[*]AND DEPARTAMENTO DE BOTANICA Y FISIOLOGIA VEGETAL, UAM[**]. SPAIN.

1. INTRODUCTION

Bacteriochlorophyll (Bchl) is distributed in two fractions, quite distinct
functionally. A small fraction, about 3% of the total, forms part of the
photoreaction center (RC), which carries out the photochemical reaction. The
rest of the Bchl belongs to the light harvesting (LH) antenna, which absorbs
radiant energy and transfers it to the RC. In any case, Bchl is always complexed to proteins and other pigments. Two types of LH complexes have been
described in purple bacteria, B870 and B800-850, their designation referring
to the approximate maxima in the near infrared (NIR) absorption spectrum (for
a revision see Cogdell and Thornber 1980).All purple species studied so far
contain complexes of the two types, except *Rhodospirillum rubrum* which
possesses only the B870 class.

Energy transfer among the LH Bchl molecules and from them to the RC occurs
without any chemical transformation, whereas the photochemical reaction
implies the oxidation of the RC Bchl P870, whose midpoint redox potential is
about 440 mV (Kuntz et al. 1964). P870 oxidation gives rise to a radical
which exhibits a characteristic electron spin resonance (ESR) signal, with a
gaussian shape, a g factor of 2.0025 and a peak to peak linewidth (ΔHpp) of
9.5 gauss (McElroy et al. 1969). The oxidation is also accompanied by an ab
sorbance increase at about 1250 nm (Clayton, 1962).

We have reported recently that oxidation of antenna Bchl, both from RC-lack
ing and wild-type strains of *R.rubrum*, originates a radical detectable by
ESR and NIR spectrophotometry (Gómez et al. 1982b). This radical differs
from that of oxidized RC Bchl in its midpoint redox potential (555 mV) and
in its spectroscopic properties: it shows a larger and narrower (ΔHpp=3.8 G)
ESR signal and a band at 1230 nm (Gómez et al. 1982a, 1982b). The estimation
of the Bchl oxidized and that of the total spins allowed to suggest, first
that the antenna Bchl responsible for both the ESR signal and the 1230 nm
band is about 30% of the total Bchl complement, and second that the unpaired
electron of the radical formed is shared by two or more Bchl molecules.

In order to investigate whether the described spectroscopic changes were
restricted to the LH Bchl of *R.rubrum*, we set out to study the oxidation of
this Bchl in other Rhodospirillaceae. The results obtained with different
Rhodopseudomonas strains are described here. They show, first that chemical
oxidation of LH Bchl from *Rps.sphaeroides* and *capsulata* induces the formation of a radical which exhibits an ESR signal (3.6 G) and a NIR absorption
band (1235 nm) similar to *Rds.rubrum* LH Bchl, and second, that such a radical is probably due exclusively to the B870 Bchl. These results indicate that
the structure of the B870 complexes of different Rhodospirillaceae has some
common features.

2. MATERIALS AND METHODS

The bacterial species used and their phenotypes were: *Rps.sphaeroides*,
wild-type; *Rps.capsulata*, wild-type (strain SB1003) and their LH-defective
derivative strains Y142 and MW442, lacking respectively complex B870 or

B800-850 and RC (Drews et al. 1976 and Zannoni et al. 1981). *Rps.capsulata* strains were generously provided by Dr. B.L.Marrs. Cells of all strains were grown in the dark under low oxygen tension as described before (del Valle-Tascón et al. 1975). Wild-type cells were also grown anaerobically in the light. The induction by ferricyanide of spectroscopic changes was performed as reported earlier (Gómez et al. 1982a, 1982b).

3. RESULTS AND DISCUSSION

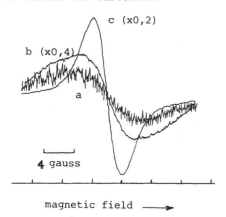

FIGURE 1. ESR signals of chromato-phores from *Rps.sphaeroides* at different redox potentials: a, 390; b, 420; and c, 560 mV. Initial A_{850}= 30.0

Treatment of *Rps.sphaeroides* chromato-phores with potassium ferricyanide at potentials above 380 mV caused the appearance of ESR signals with nearly a gaussian shape, a g value of 2,0025, and with a variable magnitude and line-width, depending on the redox potential. Fig. 1 shows the first derivative of the signals corresponding to a chromatophore preparation poised at 385, 420 and 560 mV. The ΔHpp changed from about 9 to 3.8 G at increasing potentials, the greatest amplitude corresponding to the highest potential. All the chromatophore preparations assayed behaved in a simi-lar manner although, due to their diver-se reductive capacities, they exhibited at the same redox potential ESR signals with slightly different ΔHpp and magni-tude. The ESR signals disappeared in all instances upon the addition of ferro-cyanide to bring the redox potential under 370 mV.

It seems clear that the 9G ESR signal observed in Fig. 1 at the low potential corresponds to oxidized RC Bchl (McElroy et al. 1969). We think that the narrower signals obtained at increasing potentials are mixtures of that sig-nal and a narrower one (3.8 G) due to oxidized antenna Bchl. The different width would result from a bigger or lesser contribution of each signal type, depending on the redox potential, in such a way that the contribution of the 9 G signal is insignificant at 560 mV. We assign the narrow signal to oxidiz-ed LH Bchl, for its g value is specific for a Bchl radical,and its maximum amplitude is rather bigger than that corresponding to the RC present in the chromatophore preparation.

The above interpretation is also supported by two reversible spectral chan-ges in the NIR induced by ferricyanide at potentials higher than those need-ed for a complete oxidation of RC Bchl: the appearance of a band at 1235 nm and an absorbance decrease at about 880 nm. Thus, Fig. 2 shows the spectral changes in the 1150-1300 nm range at different redox potentials. In every case a new band appears, but the wavelength of its maximum absorbance (λmax), as well as its magnitude, varies according to the potential: the λmax is about 1250 and 1240 nm at 395 and 500 mV respectively.The potentials at which the λmax is 1250 nm are similar to those which induce the 9 G ESR signal. This is in agreement with the fact that the 1250 nm band is due to oxidized Bchl. The 1240 nm band should correspond, therefore, to oxidized Bchl both from RC and antenna. The λmax of the band due to the antenna

Bchl is approximately 1235 nm, as calculated by the difference between the spectra obtained at 395 and 560 mV (Fig. 2). This band suggests, furthermore, that the oxidized LH Bchl is oligomeric, for only nonmonomeric Bchl seems to be able to originate bands in this spectral region (Fajer et al. 1975).

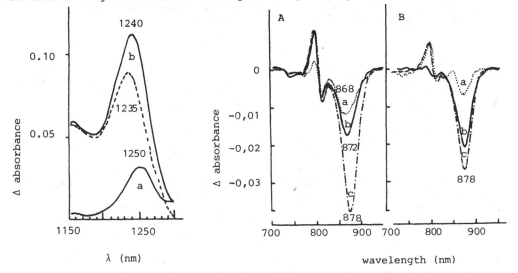

FIGURE 2. Oxidized minus reduced diffe__rence spectra in the 1150-1300 nm range of chromatophores from *Rds.sphaeroi_des*. The potential(mV)of the oxidized__ samples was: a, 395; b, 560; and that of the reduced samples 200. Initial A_{850}=30.0

FIGURE 3. Oxidized minus reduced difference spectra in the 700-900 nm range of chromatophores from *Rps. sphaeroïdes* A) The potential(mV)of the oxidized sample was: a, 395; b, 460; c, 500; and that of the reduced sample 200 B) a, 460-395; b, 500-460; c, 500-395. Initial A_{850}=1.0

Fig 3 A and B illustrates the reversible absorption changes in the 800-900 nm range elicited by ferricyanide in *Rps.sphaeroïdes* chromatophores. At 395 mV an oxidized minus reduced spectrum similar to that of purified RCs can be observed, with an absorbance decrease centered at 868 nm. At higher potentials the decrease grows and is centered at longer wavelengths (Fig. 3A). The differences between the spectra obtained at various potentials (Fig. 3B) show that after P870 is oxidized a band at about 878 nm is bleached. This indicates that the oxidized antenna Bchl responsible for both the ESR and NIR changes observed at high potential belongs to the B870 complex.

The data presented hitherto refer to chromatophores isolated from light-grown cells. Similar results were obtained with cells grown at low oxygen tension in the dark (data not shown). On the other side, parallel experiments were performed with chromatophores both from dark and light-grown cells of *Rps.capsulata* and analogous results were also obtained. Preliminary experiments have been performed with mutants of *Rps.capsulata* defective in the B870 or B800-850 complexes, and the data obtained (not shown) corrobora__te the conclusion drawn above that just the B870 complexes are responsible for both the 3.8 G ESR signal and the 1235 nm band.

The herein results, along with those previously reported for *Rds.rubrum* (Gómez et al. 1982b) strongly indicate that in Rhodospirillaceae a fraction of the LH Bchl B870 complexes can be oxidized reversibly. The oxidation product exhibits some properties very similar to those of oxidized RC Bchl suggesting, therefore, a certain structural similarity between both Bchl po̱pulations. (Work aided by a grant from the CAICYT).

REFERENCES

Clayton RK (1962) Primary reactions in bacterial photosynthesis I. The nature of light-induced absorbancy changes in chromatophores; evidence for a special bacteriochlorophyll component, Photochem. Photobiol., 1, 201-210.
Cogdell RJ and Thornber JP (1980) Light-harvesting pigment-protein complexes of purple photosynthetic bacteria, FEBS Lett., 122, 1-8.
Drews G, Dierstein R and Schumacher A (1976) Genetic transfer of the capacity to form bacteriochlorophyll-protein complexes in *Rhodopseudomonas capsulata*, FEBS Lett. 68, 132-136.
Fajer J, Brune DC, Dris MS, Forman A and Spaulding ND (1975) Primary charge separation in bacterial photosynthesis:oxidized chlorophylls and reduced pheophytin. Proc. Natl. Acad. Sci. USA, 72, 4956-4960.
Gómez I, Picorel R, Ramírez JM, Pérez R and del Campo FF (1982a) Reversible oxidation of antenna bacteriochlorophyll in two photoreaction centerless mutants of *Rhodospirillum rubrum*, Photochem. Photobiol. 35, 399-403.
Gómez I, Sieiro C, Ramírez JM, Gómez-Amores S and del Campo FF (1982b). The antenna system of *Rhodospirillum rubrum*: radical formation upon dark oxidation of bulk bacteriochlorophyll, FEBS Lett. 144, 117-120.
Kuntz ID jr, Loach PL and Calvin M (1964) Absorption changes in bacterial chromatophores, Biophys. J., 4, 227-249.
McElroy JD, Feher G and Mauzerall DC (1969) Characterization of primary reactants in bacterial photosynthesis, Biochim. Biophys. Acta, 172, 180-183.
del Valle-Tascón S, Giménez-Gallego G and Ramírez JM (1975) Light-dependent ATP formation in a non-phototrophic mutant of *Rhodospirillum rubrum* deficient in oxygen photoreduction, Biochem. Biophys. Res. Commun., 66, 514-519.
Zannoni D, Scolnik PA and Marrs BL (1981) The lack of carotenoid band-shifts in a mutant of *R.capsulata* deficient in light-harvesting II. In Proc. of the V Intern. Photos. Conq. I, pp. 525-533. Akoyunoglou G, ed. Philadelphia. Balaban International Science Services.

Authors' addresses: [*]Institutc de Biología Celular, C.S.I.C., Velázquez,144 Madrid-6. Spain.
[**]Departamento de Botánica y Fisiología Vegetal, U.A.M., Cantoblanco, Madrid. Spain.

IS THE TRANSMEMBRANE ELECTROCHEMICAL POTENTIAL A COMPETENT INTERMEDIATE
IN MEMBRANE ASSOCIATED ATP SYNTHESIS?

BRUNO ANDREA MELANDRI[*],GIOVANNI VENTUROLI[*],RITA CASADIO[*],GIOVANNI F.AZZONE[+],
DOUGLAS B.KELL[']HANS V.WESTERHOFF["]/[*]INST.OF BOTANY,UNIV.OF BOLOGNA,[+]CNR
UNIT FOR THE STUDY OF PHYSIOLOGY OF MITOCHONDRIA,PADOVA,[']DEPT.OF BOTANY-MI-
CROBIOLOGY,UNIV.OF WALES,["]INST.OF BIOCHEMISTRY,UNIV.OF AMSTERDAM.

There is no doubt that the chemiosmotic hypothesis for oxidative and photosynthetic phosphorylation(Mitchell,1966)has given a powerful impulse to
the study and the understanding of membrane associated electron transfer
and its coupling to ATP formation.The hypothesis has been of particular
importance in emphasizing the vectorial aspects of oxidoreduction reactions
in energy conserving membranes and the relevance of protonic activity and
electrostatic potential differences for the coupling of energy transducing
enzymes.Many basic features of the hypothesis have been   experimentally
substantiated;such features include the asymmetric distribution of redox
carriers across the membranes,the formation of transmembrane differences
in pH and/or in electrostatic potential by the various redox complexes
and by the ATPase,the possibility of driving ATP synthesis by artificially
imposed ΔpH or diffusion potentials of ions,or by a combination of these
two forces,and the possibility of coupling chemiosmotically in reconstituted vesicles heterologous energy transducing complexes(Boyer et al.,1977;
Ferguson,Sorgato,1982).
In the last decade the quantitative aspects of the hypothesis have been
subjected to an intensive scrutiny.These studies,aimed to verify whether
the thermodynamic and kinetic expectations of chemiosmotic coupling were
met in different systems,have generally been much less in agreement with
chemiosmosis than the qualitative enzymatic and structural studies.Studies
in various systems have often disclosed severe quantitative deviations
(with the noticeable exception of higher plant chloroplasts in continuous
light (Portis, McCarty, 1974))from the expected behaviour of a chemiosmotic
system.It is the purpose of this paper to briefly review these observations,
grouping in three main types of anomalies,and to try to derive a minimum
model for a modified chemiosmotic coupling mechanism which would be capable
of explaining the kinetic and thermodynamic behaviour of energy transducing
membranes.
The simplest possible version of the chemiosmotic model visualizes the
membrane in a fluid moasaic structure in which the various enzyme complexes
are freely diffusable and energetically coupled through the circulation
of protons. No barrier for the diffusion of protons is assumed between
the aqueous bulk compartments,facing both sides of the membrane,and the
proton releasing or accepting sites of the proton translocating enzymes.
Thus these sites must be considered in rapid equilibrium with the proton
activity of the aqueous phase facing them. Quantitative considerations of
the diffusion rate of protons in aqueous enviroments in fact demonstrate

Sybesma, C. (ed.), Advances in Photosynthesis Research, Vol. II. ISBN 90-247-2943-2.
© *1984 Martinus Nijhoff/Dr W. Junk Publishers, The Hague/Boston/Lancaster.*

conclusively, that, in the time range for the occurrence of the redox or/of
the ATP forming reactions, no substantial proton activity difference can
be maintained between points in space at a distance corresponding to the
size of a bacterium or an intracytoplasmic organelle(Mitchell,1981).It
follows that the thermodynamic force seen by the energy transducing
enzymes and driving ,for example ATP synthesis or reversed electron transfer, is measured by the bulk-bulk electrochemical potential difference of
protons. For ATP synthesis therefore, the maximal(positive)free energy
change measured when no further net ATP formation can take place (state 4)
can never exceed the value of $\Delta\bar{\mu}_{H}+$ multiplied by a stoichiometric coefficient, which is related to the number of protons translocated per molecule
of ATP formed:

$$\text{eq.1)} \quad \Delta G_{ATP} \leqslant n_{Hp} \Delta\bar{\mu}_{H}+$$

In eq.1 the "less than" symbol is meant to indicate that the ATP synthetase
enzyme can be intrinsically uncoupled and that some free energy loss can
occur in the coupling of $\Delta\bar{\mu}_{H}+$ to ATP formation within the complex itself
(Baccarini Melandri et al.,1977;Rottenberg,1973).These losses should be
however relatively constant so that the force$(\Delta G_{ATP}/\Delta\bar{\mu}_{H}+$ in state 4) should
be approximatively constant when $\Delta\bar{\mu}_{H}+$ is varied.
From a kinetic stand point, both the redox complexes and the ATP synthase
are considered as independent units in the coupling membrane,whose interaction is only mediated by the protonic activity;since proton concentrations are obviously involved in the kinetics of any proton translocating
system, the rate of all energy transducing complexes will be a function,inter
alia,of the proton activity on the two sides of the membrane.In other
words,for a given set of experimental conditions(external pH constant,
ionic and substrate concentrations constant)the rate of catalysis should
be a single-valued function of the proton electrochemical potential difference,being the state of all the other enzyme complexes in the membrane
immaterial.Thus,for example,the rate of ATP synthesis should be determined by the ATP,ADP ,P_i and Mg^{++} concentrations,and by the extent of the
proton gradient(or the rate of respiration by the substrate and oxygen
concentrations and by $\Delta\bar{\mu}_{H}+$):

$$\text{eq.2)} \quad V_{ATP} = f (\Delta\bar{\mu}_{H}+,\Delta G_{ATP});\text{ at constant external pH}$$

These quantitative expectations stem out from the simplest version of the
chemiosmotic model.We will refer to it as to the "delocalized"chemiosmotic
coupling since the two bulk phases are considered in electrochemical
equilibrium for protons and high potential protons are supposed to be
available with an identical probability to all sites facing the same bulk
phase.
The simultaneous quantitative evaluation of the thermodynamic and kinetic
parameters of photosynthetic or oxidative phosphorylation have demonstrated
remarkably large deviations from this expected behaviour. These anomalies
can be order into three groups and will be discussed briefly below.

Anomaly 1:The Force Ratio in State 4 is not Constant Different Values of $\Delta\bar{\mu}_H{}^+$.

If an energy transducing system is allowed to synthetize ATP until no net ATP formation is observed,the maximal free energy change for ATP formation is can be evaluated and compared with the extent of $\Delta\bar{\mu}_H{}^+$.Under these conditions ,according to eq.1,the force ratio should be the minimal estimate for the H^+/ATP stoichiometry.In early experiments in mitochondria these ratios were generally found to be higher than the value of 2 proposed originally by Mitchell(Azzone et al.,1977;Van Dam et al.,1978);these discrepancy is not related to the polarity of the energy transducing membrane utilized in the experiments,since it was observed both in mitochondria,where the H^+-P_i symport mechanism or the electrogenic ADP-ATP exchange proposed for the translocation of phosphorylation substrates could cause complications in the correct analysis of these parameters,and in chloroplasts(Avron,1979)or bacterial membrane fragments(Kell et al.,1978;Baccarini Melandri et al.,1977)where F_1 faces the external phase of the vesicles.

The most obvious inconsistency with eq.1,however,was the observation that $\Delta G_{ATP}/\Delta\bar{\mu}_H{}^+$ varied with different values of $\Delta\bar{\mu}_H{}^+$ and increased with decreasing $\Delta\bar{\mu}_H{}^+$.Moreover,the force ratio was found to change when conditions where set to alter the electrostatic versus the concetration components of $\Delta\bar{\mu}_H{}^+$ (Wilson, Forman, 1982).' In this type of experiments possible interference by the adenylate kinase equilibrium should be carefully taken into account,especially when low values of ΔG_{ATP} are measured.There is no doubt,nevertheless,that the bulk to bulk $\Delta\bar{\mu}_H{}^+$ is not a single-valued relation with ΔG_{ATP} as expected for a fully reversible ATPase in equilibrium with aqueous protons.

Anomaly 2:There is a Limited Correlation between $\Delta\bar{\mu}_H{}^+$ and the Rates of Electron Transfer or of ATP Synthesis.

The kinetic behaviour of ATP synthase should be a single-valued function of $\Delta\bar{\mu}_H{}^+$(at least at constant pH for the outside compartment).In many systems,however,(except for chloroplast in continuos light)it was observed that large variations of the rate of ATP formation could be caused by the inhibition of the rate of electron transfer without a corresponding significant decrease in the value of $\Delta\bar{\mu}_H{}^+$.This observation,originally reported for chromatophores of photosynthetic bacteria(Baccarini et al.,1977) has been extended to many bacterial and mammalian respiratory systems(Zoratti et al.,1982;Mandolino et al.,1983; Decker, Lang, 1978).As a complementary aspect of this phenomenon,it was observed in a pioneer work by Padan and Rottenberg(1973)that the stimulation of respiration by ADP and P_i was accompanied by a decrease in $\Delta\bar{\mu}_H{}^+$ markedly smaller than that needed to promote a comparable stimulation by uncouplers(Zoratti et al., 1983).In general this type of observation indicates a coupling between the redox and the ATP forming enzymes tighter than that existing between these energy transducers and the bulk-to-bulk $\Delta\bar{\mu}_H^+$. This kinetic situation can be interpreted as evidence for a coupling mechanism alternative, or at least parallel, to the "delocalized" chemiosmotic one, or for the existence of mutual regulatory controls between redox and ATP forming complexes. This second interpretation, on the other hand, appears unreconcileable with the observation discussed in Anomaly 1, since kinetic controls cannot affect, in principle, equilibrium states.

In bacterial photosynthesis the light-induced and cyclic nature of the electron transfer chain allows for control of the rate of redox reactions by using trains of single turnover flashes fired at variable frequency. Coupling this technique with the sensitive luciferin-luciferase assay for ATP, Venturoli and Melandri(1982) were able to demonstrate that the amount of ATP formed per flash was strictly proportional to the number of photo-synthetic units still operative when electron transfer was progressively inhibited by antimycin A. At high ADP/ATP concentration ratios, ATP could be formed, although with a yield per flash of about half of that maximally observed,already after a single turnover of the electron transfer chain, when no ΔpH was yet formed and a membrane potential of about 70 mV was produced (Melandri et al.,1980). This observation is in agreement with the high force ratio observed at low $\Delta\bar{\mu}_H^+$ in respiratory systems (Anomaly 1). Moreover, it could be demonstrated that no ATP formation took place in preilluminated membranes which maintained a high and slowly decreasing $\Delta\psi$, unless one additional turnover of the electron transfer chain was elicited by a single turnover flash. Under those specific conditions, one flash alone was unable to drive ATP formation per se and the photophospho-rylation was dependent upon the preenergization of the membrane. Under these conditions the decrease in the ATP yield accurately followed the decay of the membrane potential; both parameters were destabilized by K^+ and valino-mycin (Melandri et al.,1980). This observation, if one compares the single turnover behaviour of bacterial chromatophores to that in steady state of respiratory systems, indicate that both a competent $\Delta\bar{\mu}_H^+$ and electron transport are conditions required for ATP formation.

Anomaly 3: The Response of Energy Transducing Systems to Double Inhibition of ATPase and of the Electron Transfer Chain.

In their work on oxidative phosphorylation in mitochondria, Baum et al.(1971) pioneered the use of double inhibitor titration of electron transfer reactions and of ATPase as an approach for the study of the interaction between energy transducers. In the chemiosmotic model, the coupling between two enzyme complexes is mediated by the "delocalized" protons, so that if one of the two transducers is kinetically limiting the overall rate, the inhibition of the other complex should not influence the overall velocity of the process. It was, on the contrary found that this was not the case: when, for example, NAD^+ reduction from succinate was driven by ATP hydrolysis, the sensitivity to rotenone inhibition was increased rather then decreased following partial

inhibition of ATPase with oligomycin. This observation was subsequently
generalized to many direct or reverse processes of mitochondrial (Westerhoff
et al.,1983; Baum et al.,1971) or bacterial respiration (Kell et al.,1979;
Parsonage, Ferguson, 1982). In photosynthetic systems the double inhibition
approach has been extensively utilized in chromatophores by Hitchens and
Kell (1982). Using many combinations of inhibitors, they also obtained
evidence for a direct kinetic interaction between ATP synthase and electron
transport. With an analogous rationale Venturoli and Melandri (1982) observed
that the degree of DCCD inhibition of ATP synthesis was not affected by
varying the spacing in time of single turnover flashes by two orders of
magnitude. The synergistic inhibition generally observed in dual inhibitor
titrations again points to a direct functional interaction between energy
transducing complexes not mediated by a "delocalized" intermediate. This
approach in principle does not require the evaluation of $\Delta\bar{\mu}_{H}+$, a measure
always open to theoretical and experimental criticisms. The possible nonlinear behaviour of the kinetics of the energy transducers versus the driving
thermodynamic forces could complicate the interpretation of such results.
An other possible source of uncertainty is a possible rapid exchange of
the inhibitor between the complexes and / or an energy dependent binding,
which could delocalize, or respectively alter the effectiveness of the
inhibitor. In the authors' opinion nevertheless, this experimental approach,
only recently reintroduced in the study of energy transduction, should be
pursued further and preferibly substantiated by parallel measurements of the
proton gradient.

Devising a Minimal Hypothesis

The observations discussed in the preceeding sessions are irreconciliable
with a fully "delocalized" chemiosmotic model. Any model capable of predicting
the features of coupling in energy transducing membranes should therefore
incorporate Anomalies 1-3 as well as the data supporting the "delocalized"
hypothesis. We shall try to propose such a model and to discuss briefly
how it can explain the "localized" coupling phenomena described above.
In the classic chemiosmotic hypothesis the various energy transducing
complexes are considered to be freely diffusable: no restriction is imposed
upon the random distribution of these complexes in the membrane except the
obvious one related to the parallel orientation of all complexes which are
perpendicular to the membrane plane (as far as proton translocation is
concerned).
We shall propose, on the contrary, that the various complexes are laterally
ordered structurally or functionally so that a close protonic interaction
between them can occur. We propose, moreover, that the proton domain in
which the coupling takes place is separated from the bulk phases by diffusion
barriers,so that the leakage of protons from the domain into the aqueous
phase is a rare event,at least in the time range in which the redox or
ATP-forming reactions occur. The coupling membrane is considered to be
composed of single coupling units, e.g. including a single electron transfer

chain and an associated ATPase, each one at least partially isolated from
the others and from the bulk phases by diffusion barriers for protons. In
this way the protonic coupling will not occur by means of the circulation
of protons through the bulk phases, but rather preferentially within the
protonic domain of a single coupling unit. In this sense the coupling is
still chemiosmotic, but the phosphorylating unit no longer coincides with
the whole coupling membrane. The energy transducing membrane is composed
by a mosaic of coupling units which are at least partially independent.
In the model any dissipation is considered to occur both through leaks
within a single coupling unit (this kind of energy dissipation within a
protonic pump has been proposed previously as a slip of the pump (Pietrobon
et al. 1981,1982)) and through leaks between the two bulk phases. According
to this model the bulk-to-bulk $\Delta\bar{\mu}_{H^+}$ results from the slow diffusion of
protons from the localized protonic domain to the bulk phases. Analogously,
the charge separation phenomena within the various coupling units will
form an array of dipoles oriented across the membrane (Skulachev 1982).
The resulting electric field will not be completely equipotential along
the planes parallel to the membrane, but will be slowly delocalized by the
diffusion of protons in the bulk phases and compensated by ion redistribution.
The resulting voltage profile will be a function of the physical spacing
between the coupling units, the electrical conductivity and the exact
location of the proton diffusion barriers,and the conductivity of the
bulk-to-bulk leaks (cf. Zimanyi,Garab 1982).
The proton electrochemical potential difference within the domain of a
single coupling unit (in the following indicated as $\Delta\lambda_{Hi}$) is not in
equilibrium and is, in general, greater than the bulk phase potential
difference $\Delta\bar{\mu}_{H^+}$. Again, the actual difference between these two parameters
will depend upon the conductivity of the diffusion barriers as compared
with membrane leaks. The more the conductivity of the membrane leaks
exceeds that of the diffusion barrier, the more the different coupling
units will behave independently from one another. The meaning of the model
therefore becomes clear: a proton released by the primary pump into the
domain will have a much greater probability of being utilized by the
secondary pump associated with that domain, than to diffuse into the bulk
phase and to be utilized by other secondary pumps or dissipated through
the bulk-to-bulk leaks.
Using these assumptions the anomalies can be qualitatively interpreted
(for a quantitative interpretation cf. Westerhoff et al. in preparation).
In double inhibitor titration experiments (Anomaly 3) the inhibition of
either the primary or the secondary pump of one unit will result in the
effective block of the entire unit, since the efficiency of proton utilization
by the other units will be much smaller. For these experiments the most
clear cut results have to be expected with an "all or none" inhibitor, i.e.
for inhibitors whose residence time on the accepting site is at least as
long as the turnover time of the target complex. Likewise, when the number
of primary pumps is decreased by an electron transfer inhibitor (Anomaly 2),

the rate of ATP synthesis will follow the pattern of that inhibition, while the value of $\Delta\bar{\mu}_H^+$ will have only a limited relevance for the actual rate of phosphorylation. More complex is the analysis of the behaviour of the model as far as $\Delta G_{ATP}/\Delta\bar{\mu}_H^+$ vs $\Delta\bar{\mu}_H^+$ is concerned (Anomaly 1). The inhibition of a primary pump will not markedly alter the value of $\Delta\Lambda_{Hi}$ in the un-inhibited units, but will affect the value of $\Delta\bar{\mu}_H^+$ by altering the ratio between the active units and the bulk-to-bulk leaks. A similar effect will be obtained by increasing the leaks with a protonophore (which could, however, also affect the conductivity of the diffusion barriers of the coupling units and therefore perturb the $\Delta\Lambda_{Hi}$ vs $\Delta\bar{\mu}_H^+$ relationship). The result will be a preferential decrease in $\Delta\bar{\mu}_H^+$ accompanied by a less marked one of $\Delta\Lambda_{Hi}$, and consequently of ΔG_{ATP} which is in equilibrium with the latter. The value of $\Delta G_{ATP}/\Delta\bar{\mu}_H^+$ will therefore be higher at low $\Delta\bar{\mu}_H^+$s as experimentally observed.

The exact nature of the diffusion barrier has been left intentionally undefined in the model. We shall notice, however, that any hydrophilic structure in physical contact with the aqueous phases (e.g. cf. Kell 1979) seems to be kinetically inadeguate to delay the diffusion of protons for a time interval of the order of magnitude of the turnover of the energy transducer complex (several milliseconds). The barrier must be therefore part of more complex structures of the coupling membrane. Experimental evidence for the existence of such structures and of their lability during "in vitro" storage or following mechanical or themal stress is beginning to emerge (Ausländer, Junge, 1974; Theg et al.,1982; Hong, Junge, 1983).

REFERENCES

Auslaender W and Junge W (1974) Biochim. Biophys. Acta 357,285-298
Avron M (1979) In Transport by Proteins (Blauer G and Sundi H eds.)pp.151-161, Walter de Gruyter, Berlin
Azzone GF, Massari S and Pozzan T (1977) Mol. Cell. Biochem. 17,101-112
Baccarini-Melandri a, Casadio R and Melandri BA (1977) Eur. J. Biochem. 78, 389-402
Baum H, Hall GS, Nalder J and Beechey RB (1971) In Energy Transduction in Respiration and Photosynthesis (Quagliariello E, Papa S and Rossi GS eds.) pp.747-755, Adriatica Ed., Bari
Boyer PD, Chance B, Ernster L, Mitchell P, Racker E. and Slater EC (1977) Ann. Rev. Biochem. 46,955-1026
Decker SJ and Lang DR (1978) J. Biol. Chem. 253,6738-6743
Ferguson SJ and Sorgato MC (1982) Ann. Rev. Biochem. 51,185-217
Hitchens GD and Kell DB (1982) Biochem. J. 206,351-357
Hitchens GD and Kell DB (1982) Biosc. Rep. 2,743-749
Hong YQ and Junge W (1983) Biochim. Biophys. Acta 722,197-208
Kell DB (1979) Biochim. Biophys. Acta 549,55-99
Kell DB, John P and Ferguson SJ (1978) Biochem. J. 174,257-266

Kell DB, Ferguson SJ and John P (1979) Biochem. Soc. Trans. 6,1292-1295
Mandolino G, De Santis A and Melandri BA (1983) Biochim. Biophys. Acta 723,428-439
Melandri BA, Venturoli G, De Santis A and Baccarini-Melandri A (1980) Biochim. Biophys. Acta 592,38-52
Mitchell P (1966) Biol. Rev. 41,445-502
Mitchell P (1981) in"Of Oxygen,Fuels and Living MAtter",Part 1 (Semenza G,ed.) pp. 1-160, John Wiley & Sons, New York
Padan E and Rottenberg H (1973) Eur. J. Biochem. 40, 431-437
Parsonage D and Ferguson SJ (1982) Biochem. Soc. Trans. 10,257-258
Pietrobon D, Azzone GF and Walz D (1981) Eur. J. Biochem. 117,389-394
Pietrobon D, Zoratti M, Azzone GF, Stücki JW and Walz D (1982) Eur. J. Biochem. 127,483-494
Portis AR and McCarty RE (1974) J. Biol. Chem. 249,6250-6254
Rottenberg H (1973) Biophys. J. 13, 503-511
Skulachev VP (1982) FEBS Lett. 146,1-4
Theg SM, Johnson JD and Homann PH (1982) FEBS Lett. 145, 25-29
Van Dam K, Wichmann AHCA, Hellingwerf KJ, Arens JC and Westerhoff HV (1978) Fed. Eur. Biochem. Soc. Symp. 45, 121-132
Venturoli G and Melandri BA (1982) Biochim.Biophys.Acta 680, 8-16
Westerhoff HV, Colen AM and Van Dam K (1983) Biochem. Soc. Trans.11, 81-85
Westerhoff HV,Melandri BA, Venturoli G, Azzone GF and Kell DB, in preparation
Wilson DF and Forman NG (1982) Biochemistry 21, 1438-1444
Zimanyi L and Garab G (1982) J. Theor. Biol. 95, 811-821
Zoratti M, Pietrobon D and Azzone GF (1982) Eur. J. Biochem. 126, 443-451
Zoratti M, Pietrobon D and Azzone GF (1983) Biochim. Biophis. Acta in press

ACKNOWLEDGEMENTS

This work has been partially supported by the grant n° 81.00278.04 to BAM, RC and GV from Consiglio Nazionale delle Ricerche of Italy.

First Author's Address: Bruno Andrea Melandri, Institute of Botany, Via Irnerio 42, 40126 Bologna, Italy.

ELECTRICAL EVENTS AND P515 RESPONSE IN THYLAKOID MEMBRANES

WIM J. VREDENBERG, OLAF VAN KOOTEN and ROBERT L.A. PETERS

1. INTRODUCTION

Since its discovery (Duysens, 1954) the light-induced absorbance change
in the 480-550 nm wavelength region, with its characteristic difference
spectrum, has been studied in much detail. Ample evidence has been presented (e.g. Junge, 1977) that this so-called P515 absorbance change is due
to an electrochromic band shift of a pigment moiety with a maximum around
515 nm. This has made it a useful intrinsic probe for studying the magnitude
and kinetics of formation and decay of the transmembrane electric potential
of the inner chloroplast (thylakoids) membranes. There has been found
reasonable good agreement on the value of the initial potential generated
by saturating single turnover light flashes, measured directly by microelectrodes and calculated by calibrating the absorbance change caused by
salt-induced diffusion potentials (e.g. Vredenberg, 1981). The kinetics
of P515 responses in continuous light as well as in single turnover flashes
are however at variance with the responses measured with microelectrodes.
There is now general agreement, based on observations in intact cells as
well as in intact and broken chloroplasts that the P515 absorbance change
in single turnover flashes is a multi-phasic phenomenon, reflected by a
biphasic rise and a decay which consists of at least three different
single exponentials. By using double flashes Schapendonk et al. have
shown the reliability of the deconvolution of the single flash response
into three different responses: reaction I, II and an as yet unidentified
component with an extremely slow decay rate, called phase d, respectively
(Schapendonk et al., 1979). For the sake of a unified characterization
we will denote the third component of the P515 response, i.e. the one with
the slow phase d decay component, by reaction III.

This paper aims to give a short survey of new information about the characteristics of the three distinguished reaction components I, II and
III of P515, in particular in relation to the functioning of the thylakoid
membrane during energization.

2. MATERIAL AND METHODS

Plant material (Spinach and Peperomia metallica) was obtained from
laboratory cultures. Chloroplast isolation and preparation were done
following procedures, described before (Schapendonk, 1980). Absorbance
difference measurements were done with an instrumentation and computer-
assisted processing as described elsewhere (van Kooten et al., 1983).
Potential measurement in Peperomia chloroplasts by implanted micro-capillary
electrodes, were also as described before (Schapendonk, 1980).

3. RESULTS AND INTERPRETATION

We will give the result by presenting new data on the separate components
of the P515 response in single turnover saturating light flashes.

3.1 Reaction I

Reaction I, characterized by a fast (ns) rise and a single exponential

Sybesma, C. (ed.), Advances in Photosynthesis Research, Vol. II. ISBN 90-247-2943-2.
© *1984 Martinus Nijhoff/Dr W. Junk Publishers, The Hague/Boston/Lancaster.*

decay ($k \sim 10 \, s^{-1}$), is observed without interference from the other reaction components in aged chloroplasts, or in broken chloroplasts which have been rethawed from a frozen preparation. Reaction I is found to be the main, if not exclusive, component of the P515 overall response in leaves and chloroplasts isolated from plants (spinach) which have been grown at a reduced light intensity ($< 10 \, W.m^{-2}$). Leaves and chloroplasts of the same plant variety grown at a higher intensity show a response which in addition to Reaction I has a large component of Reaction II (R.L.A. Peters, unpublished results). Reaction I is also the nearly exclusive responding component of P515 in freshly prepared broken chloroplasts from normal grown leaves which after osmotic shock in 1 mM $MgCl_2$ have been suspended in an ion-free TRIS-MES medium (see R.L.A. Peters et al., this symposium). Finally Reaction I is exclusively observed when the membrane is pre-energized either by a few light flashes, or by dark ATP-hydrolysis (R.L.A. Peters, et al., 1983). In all cases where Reaction I is the exclusive component the first order rate constant of the decay titrates linearly with the concentration of permeant ions in the absence and presence of ionophores (R.L.A. Peters et al., this symposium). From these experiments a relatively high permeability of Cl^- in the thylakoid membrane can be concluded. Our new data confirm earlier suggestions (e.g. Vredenberg, 1981) that Reaction I reflects the generation and decay of the delocalized transmembrane electric field induced by the light-induced charge separation in PS 1 and PS 2.

3.2 Reaction II

In contrast to Reaction I and Reaction III (see below) Reaction II is never observed in the light in the absence of the other reactions. Consequently, the response of Reaction II can only be obtained by subtracting the other components from the overall P515 response. Methods for measuring these components have been described elsewhere (Schapendonk et al., 1979). Reaction II is characterised by a relatively slow rise, occurring in 20-100 ms and a decay with a first order rate constant of the order of 1 s^{-1}, i.e. a morefold of the rate constant of Reaction I. Its occurrence is strictly dependent on the integrity of the membrane, i.e. it disappears upon ageing and is absent in plant material grown under low light intensities. Its disappearance in the presence of DBMIB (e.g. Vredenberg, 1981), its reappearance in the presence of DCMU with H-donors to the Fe-S cyt.b.-f protein complex (Selak, Whitmarsh, 1982), and its presence in PS I vesicles (A.L.J. Peters et al., this symposium), confirm earlier evidence (Schapendonk, 1980) that it is associated with an energetic process driven by PS I.

It has been proposed that Reaction II is caused by a lateral and transversal delocalization on inner membrane electric fields associated with the liberation of protons in inner membranes domains near the Fe-S cyt b-f protein complex (Westerhof et al., 1983). These domains might be connected via lateral H-conductive channels with other membranes domains i.e. the ATP synthetase. In this respect it is of interest to mention that, in confirmation with result of others (Schuurmans et al., 1981; Schreiber, Rienits, 1982), Reaction II can also be induced in the dark towards its saturation level by ATP driven reversed electron flow. The light-induced response of Reaction II, absent during ATP hydrolysis, reappears after the ATPase has become inactivated or, with broken chloroplasts in the presence of DTE, when ATP has been consumed (R.L.A. Peters, this symposium). Whether the existence of the membrane domains and their highly efficient and low energy requiring saturable H^+-stabilising function (Reaction II is saturated in well coupled chloroplasts by one or two saturating flashes is dependent on the synthesis and presence of mono-galactolipids which, according to a proposed hypothesis (Murphy, 1982) would preferentially orient in inverted micelles, is under investigation. The absence of Reaction II in low light grown plants (winter-spinach) suggests this plant material to be well suited material for further functional and structual studies as compared to normally grown plant material. In preliminary experiments with two different spinach varieties (R.L.A. Peters et al., unpublished) it was found that the ratio between mono- and di-galactolipids (galacto-diacylglycerol) in low light (6 $W.m^{-2}$) grown plants is approx 35% lower as compared to that in plants grown at high (60 $W.m^{-2}$) light intensity.

It has been suggested by others (e.g. Vredenberg, 1981) that the slow rise in the flash-induced P515 response, i.e. the component caused by Reaction II, reflects a transmembrane electric field generated by electron transfer in a Q-cycle. An association with cyt b_6 reoxidation has been suggested (Selak, Whitmarsh, 1982). However, the evidence is based on data, which does not account for the fact that the field generated by this secondary electron transfer decays with a half time which is considerably higher than the decay of the potential (Reaction I) generated by electron transfer through the reaction centers. Moreover, it has been shown (van Kooten et al., 1983) that the seeming correspondence between the potential associated with this secondary electron transport and a turnover of cyt b_6 only holds for a single flash. A second turnover of cyt b_6 induced by a second flash, given 100 ms after a first one does not, or at least much less, cause a Reaction II response.

So far, Reaction II type kinetics were not observed when transmembrane potentials were measured with microelectrodes. However, recently van Kooten et al. (this symposium) have shown that the transmembrane potential generated by flashes has a slow rising phase. Their results have been obtained with Peperomia plants which were grown under different nutritional conditions. The results suggest evidence that a transversal delocalisation of inner-membrane localized fields occurs under, as yet unknown, conditions which promote the formation of hydrophylic inner membrane micro-compartments. Protons may be stabilized in these compartments and are hindered in their transversal and lateral diffusion. The consequence of hindered diffusion as suggested by others (Hong, Junge, 1983) on the formation of isopotentials under continuous illumination have been similated in a model (van Kooten et al., this symposium). There appears a reasonable agreement with the experimental data and the simulation model. Thus, with respect to Reaction II there is accumulating evidence that it has its origin in innermembrane phenomena which are dependent on the existence of domains which facilitate proton-interaction in the lateral plane of the membrane and may give rise to localized chemiosmotic coupling mechanisms, and electrochromic effects on the native P515 pigment complex.

3.3 Reaction III

It has been reported (Schapendonk et al., 1979) that the relatively small and ultra-slow decay component of the P515 absorbance change, originally denoted as phase d, but here referred to as Reaction III, has a spectrum different from that of P515. Reaction III is insensitive to ionophores which at low concentration abolish Reaction II (Schapendonk, 1980) and at higher concentration enhance the decay of Reaction I manifold. Thus

in the presence of high concentration of valinomycin (plus KCl) or gramicidin Reaction III is observed solely. It appears that upon repetitive flashes
the absorbance change of Reaction III is cumulative, at least in a flash
train of 10 flashes given at a frequency of 1 Hz. The spectrum of Reaction
III is shown in Fig. 1. It has a broad maximum around 535-540 nm and is
positive in the wavelength region from 495 to above 575 nm. The spectrum
suggests that the absorbance change is due to scattering changes, when
compared to published spectra of scattering changes in continuous light.
Schreiber (personal communication) has found a periodicity of two in the
magnitude of Reaction III in consecutive flashes with a higher response
in uneven flashes. This periodicity is illustrated in Fig. 2, although it
is somewhat weak in the first two flashes. In the presence of 3 μM DCMU
the response in the first two flashes is unaffected, whereas in the third
and following flashes inhibition occurs. This leads to the conclusion

FIGURE 1. Spectrum of the phase
d component (reaction III) of the
flash-induced P515 response in
broken chloroplasts in the presence of gramicidin (10 μM). The
cumulative $\Delta I/I$ of 10 flashes
with dark intervals of 500 ms
is plotted. 1 a.u. corresponds
to $\Delta I/I \sim 7.3 \cdot 10^{-3}$.

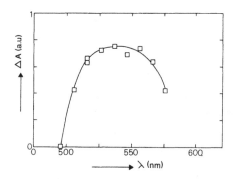

FIGURE 2. Changes in absorbance
in dark-adapted broken chloroplasts at 540 nm induced by 4
flashes in the presence of gramicidin in the absence (upper traces) and presence of 3 μM DCMU.
The smoothed traces are the
average of 3 measurements each
with fresh samples. Temperature
10°C.

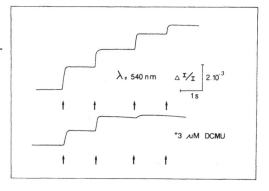

that Reaction III is mainly, if not exclusively, activated by PS 2. The apparent difference between the spectrum of Reaction II with that of P515 suggests that it is not caused by an electrochromic effect. We presume that it senses a slow reversible conformational effect at the outer thylakoid membrane surface caused by the formation of Q^- and I^- (in the presence of DCMU). It is suggestive to think that these negative charges cause an altered exposure of negative charges of a protein at the membrane surface which relaxes slowly. This exposure may cause changes in scattering.

4. CONCLUSION

The distinct kinetic components of the P515 response in intact cells, and chloroplasts are a useful tool for the study of trans-, inner- and surface membrane electrical phenomena which accompany the energization of the membrane. The fact that one deals with the response of a native pigment makes it in particular of relevance because the system can be studied under physiological conditions without possible artificial effects on membrane conformation by external probes.

REFERENCES

Duysens LNM (1954) Reversible changes in the absorption spectrum of chlorella upon irradiation, Science 120, 353-354.
Hong YQ, Junge W (1983) Localized or delocalized protons in phosphorylation. On the accessibility of the thylakoid lumen for ions and buffers, Biochim. Biophys. Acta 722, 197-208.
Junge W (1977) Membrane potentials in photosynthesis, Annu. Rev. Plant Physiol. 28, 503-536.
van Kooten O, Gloudemans AGM and Vredenberg WJ (1983) On the slow component of P515 and the flash induced reduction of cyt b563 in chloroplasts, Photobiochem. Photobiophys. 6(1), 9-14.
Murphy DJ (1982) The importance of non-planar bilayer regions in photosynthetic membranes and their stabilisation by galactolipids, FEBS Letters 150(1), 19-26.
Peters RLA, Bossen M, van Kooten O and Vredenberg WJ (1983) On the correlation between the activity of the ATP-hydrolase and the kinetics of the flash-induced P515 electrochromic bandshift in chloroplasts, J. Bioenerg. Biomembr. (in press).
Schapendonk AHCM, Vredenberg WJ and Tonk WJM (1979) Studies on the kinetics of the 515 nm absorbance changes in chloroplasts, FEBS Letters 100, 325-330.
Schapendonk AHCM (1980) Electrical events associated with primary photosynthetic reactions in chloroplast membranes, Agricultural Research Reports 905, ISBN 90 220 0756 1, Wageningen.
Schreiber U and Rienits KG (1982) Complimentarity of ATP-induced absorbance changes around 515 nm, Biochim. Biophys. Acta 682, 115-213.
Schuurmans JJ, Peters ALJ, Leeuwerik FJ and Kraayenhof R (1981) On the association of electrical events with the synthesis and hydrolysis of ATP in photosynthetic membranes. In Palmiere F et al., ed. Vectorial reactions in electron and ion transport in mitochondria and bacteria, pp. 359-369, Elsevier, Amsterdam.
Selak MA and Whitmarsh J (1982) Kinetics of the electrogenic step and cytochrome b_6 and f redox changes in chloroplasts, FEBS Letters 150, 286-292.

Vredenberg WJ (1981) P515: a monitor of photosynthetic energization in chloroplast membranes, Physiol. Plant. 53, 598-602.
Westerhof HV, Helgerson SL, Theg SM, van Kooten O, Wikstrom MFK, Skulachev VP, Dancshazy Zs (1983) The present state of the chemiosmotic theory, Acta Biol. Acad. Sci. Hung. (in press).

Authors address: Lab. of Plant Physiological Research, Agricultural University, Gen. Foulkesweg 72, 6703 BW Wageningen, The Netherlands.

PROTOLYTIC REACTIONS IN THE THYLAKOID INTERIOR

W. JUNGE, YQ. HONG, S. THEG, LP. QIAN & A. VIALE / UNIVERSITAET OSNABRUECK

The principles of electric potential generation and proton pumping in thylakoids are understood. There is one electrogenic reaction in each of the two photosystems and a third one associated with cyclic electron transfer through the cytochrome-b_6-segment of the electron transfer chain. Proton deposition into thylakoids occurs at the level of water oxidation and of plastohydroquinone oxidation. Proton uptake from outside is initiated when plastoquinone is reduced by photosystem II (or during cyclic electron transfer) and upon reduction of the terminal electron acceptor. Broadly speaking proton pumping in green plant photosynthesis seems to conform with the concept of alternating electron-hydrogen transport across the membrane. This was postulated by Mitchell (1961,1966). We have reviewed these general features recently (Junge, Jackson (1982)). In this manuscript we focus on unorthodox aspects of proton transport: transient release and rebinding of protons by proteinaceous groups in response to electron transfer (membrane Bohr effect proton pump?), the question of localized or delocalized pathways for protons when they cycle between pumps and ATP-synthases and the entry of protons into CFO-CF1.

Digression on measuring techniques. Our work was based on two spectrophotometric techniques, namely, the measurement of the transmembrane electric field via electrochromic absorption changes (Junge, Witt (1968)) and the measurement of pH-transients via pH-indicating dyes, in particular via neutral red for the internal phase (Auslaender, Junge (1975) - Junge et al.(1979) - Hong, Junge (1983)). While the former technique seems to be generally accepted, the latter, although adopted by other laboratories (Saphon, Crofts (1977) - Hope, Morland (1979) - Velthuys (1980)), may require a short explication. In a chloroplast suspension at 10uM chlorophyll the external volume of thylakoids is by three orders of magnitude larger than the internal one. Hence, hydrophilic pH indicators are practically selective for pH-transients in the external phase (see Fig.2 in Junge et al.(1978)). In contrast, neutral red, which binds to a neutral membrane with a distribution coefficient of 3×10^3 in favour of the membrane (Hong, Junge (1983)), reports pH-transients in both aqueous phases. Under selective buffering of the external phase it is selective for pH-transients in the internal space (Auslaender, Junge (1975)). Intrinsically, neutral red monitors transients of the surface pH. This became experimentally apparent via the dependence of its apparent pK on the surface potential at the inner side of the thylakoid membrane (see Figs. 1&3 in Hong, Junge (1983)). Due to the very rapid equilibration of protons between bulk water and interface, neutral red is also a kinetically competent indicator of pH tansients in bulk water (for model studies with this dye in detergent micelles see Gutman et al.(1981)). For the inner phase of highly swollen thylakoids this was experimentally evident from the effect of very hydrophilic buffers on the pH_{in}-indicating absorption changes of neutralred (see Figs. 3&4 in Junge et al.(1979)). In thylakoids we did not observe any "non-pH-response" of neutral red (see Fig.2 in Junge et al.(1979)) and, in particular, no redox response as suggested by Prince et al. (1981). The extent of the pH_{in}-indicating absorption changes of neutral red could be calibrated (Junge et al.(1979)). Under excitation of both photosystems by one single-turnover flash we found that the pH in the thylakoid interior was decreased by 0.06 units.

Sybesma, C. (ed.), Advances in Photosynthesis Research, Vol. II. ISBN 90-247-2943-2.
© 1984 Martinus Nijhoff/Dr W. Junk Publishers, The Hague/Boston/Lancaster.

SHORT REVIEW OF COMPLICATING DETAILS OF PROTON PUMPING

Proton deposition into the thylakoid interior occurred with complex kinetics. Half of the flash-induced extent, with half-rise times of 100μs to 1ms was attributed to water oxidation by photosystem II and the other half with half-rise time of approximately 20ms to plastohydroquinone oxidation by photosystem I (Auslaender, Junge (1975)). Proton release during water oxidation was kinetically complex in itself. When dark-adapted material was excited by a series of flashes we observed that most of the kinetically distinct components could be associated with certain transitions between the four oxidation states, $\{S_i\}$, of the water-oxidizing enzyme (Foerster et al., 1981). However, one very rapid component did not fit into this scheme. Possibly it was due to the protolytic reaction which was observed after DCMU treatment (Hong et al. (1981). Tris treatment confirmed a second proton-releasing site at the donor side of photosystem II apart from water oxidation (Foerster, Junge, these proceedings); from the time course of proton release and rebinding it was evident that it had to be correlated with the oxidation of the intermediate electron carrier Z(=ESR (IIf)) in photosystem II. Probably the "push and pull" of protons was done by the protein which contained Z and which responded to redox transients of Z with pK shift of amino acid residues (Foerster, Junge, these proceedings). It is evident that transient proton release and proton uptake which occurs at the same side of the thylakoid membrane does not contribute to net proton pumping. In conclusion, net proton transport in thylakoids is caused by alternating electron hydrogen transfer. There is at least one additional protolytic reaction of the Bohr effect type which, however, does not contribute to net proton pumping.

An intermediate protolytic reaction of an apoprotein seems to occur also at the reducing side of photosystem II. There was no evidence for direct protonation of the singly reduced bound quinone acceptors Q (Pulles et al. (1976)) or R (Diner (1977)). On the other hand, very little oscillation of proton uptake was observed when this acceptor system was expected to oscillate through the singly and doubly reduced state (Velthuys (1978) - Hope, Morland (1979) - Foerster et al. (1981)). We proposed that a proton was transiently taken up by the host protein(s) of Q&R, where it served as specific counter ion to the semiquinone anion (Foerster et al.,1981). For unknown reason direct protonation of the quinone seemingly occurred only in the doubly reduced state.

With the kinetics of proton deposition at hand we asked whether any of the protolytic reactions at the oxidizing side of photosystem II contributed to the transmembrane electric potential. Under high time resolution of the electrochromic absorption changes (2ns), under linear electron transport and at room temperature we found no component which rose more slowly than instrument limited. At lower temperature ($-30°C$) a slower component appeared (extent appr. 50%, half-rise 30ns), which could be attributed to the electric field which was generated in photosystem I. We interpreted this by hindered delocalization of this field from the stroma into the grana domains where the electrochromic pigments resided in vicinity of photosystems II (Klose et al., unpublished). In conclusion, none of the protolytic reactions at the inner side of the thylakoid membrane seems to be electrogenic, i.e. these reactions occur in an environment of high dielectric constant and/or the path from the proton storage sites to the membrane water interface is very short. It is noteworthy, that translocation of charge across the membrane-water interface is expected to contribute comparatively little to the transmembrane voltage due to the high specific capacitance of a diffuse ionic double layer (some $100\mu F/cm^2$) as compared to the capacitance of the membrane (some $1\mu F/cm^2$) (see Feldberg, Delgado (1978)).

In view of the extensive and controversial literature on the question whether or not protons pass through aqueous bulk phases when they cycle between pumps and users (ATP-synthases) we put this alternative on a test. We started with chloroplasts which were stored under liquid nitrogen under cryoprotection by DMSO. After thawing, these chloroplasts were suspended in hypotonic buffer. With such material we observed the following: 1.) The pH_{in}-indicating absorption changes of neutral red were quenched by any buffer, even by extremely hydrophilic ones like pyrophosphate. The degree of quenching was proportional to the buffering capacity which we calculated under the assumption that these buffers were in aqueous environment (Junge et al. (1979)). 2.) The apparent pK of neutral red which was bound to the inside of thylakoids varied as function of added cations as expected on basis of the Gouy-Chapman theory (Hong, Junge (1983)). This required the existence of an extended ionic double layer. Both observations were made even for the most rapid components of proton deposition. In conclusion, in freeze-thawed and osmotically swollen thylakoids protons are deposited into an internal aqueous bulk phase. We carried out similar studies on thylakoids which originated from freshly prepared class I chloroplasts which were briefly exposed to osmotic shock. Here we found that hydrophilic buffers did not act on the pH_{in}-indicating absorption changes of neutral red, neither was the apparent pK of neutral red influenced by salts (Hong, Junge (1983)). Since the complex kinetics of the flash-induced absorption changes of neutral red were similar in both cases (Foerster, unpublished) we concluded that neutral red still indicated proton deposition, however, the narrow interior of fresh thylakoids lacked properties of an extended aqueous phase. We tentatively interpreted the lacking influence of salts on the apparent pK of neutral red at the inner side of the membrane as indication that opposing membranes of one thylakoid were so tightly appressed that the access of "free" ions together with their hydration shells was limited (Hong, Junge (1983)). Such effects were reported for highly compressed lipid multilayers (Lis et al.(1981)). The applicability of this model to thylakoids gained circumstantial support by the tendency to aggregate of inside-out thylakoid membranes (Albertson (1982)). The fact that the interior of "good thylakoids" did not behave like an aqueous bulk phase did not argue against the "chemi-osmotic" concept, as freeze-thawing and hypoosmolar suspension, which produced a behaviour which was characteristic for an internal aqueous bulk phase, did hardly impair the ability of these thylakoids to phosphorylate at high rate.

The unusual properties of the internal phase in "good" thylakoids induced us to reinvestigate proton flow through the partition regions at the outer side of stacked thylakoids. Previously we observed that proton uptake from the external phase and at the reducing sides of both photosystems was "seen" by water soluble pH-indicating dyes only at considerable delay (some 50ms - Auslaender, Junge (1974)). We postulated a diffusion barrier for protons. In extension of this work we observed that this barrier was lowered if thylakoids were unstacked (Polle&Junge, these proceedings). We have proposed that the apparent diffusion barrier for the flux of protons from the external aqueous bulk phase to the uptake sites, say at photosystem II, is caused by multiple binding-debinding of protons to tightly packed proteins in the partition regions (Hong, Junge (1983)). With reasonable estimates for the buffering capacity in these regions and applying theory of diffusion (Crank (1979)) we calculated that proton diffusion along the partition regions should be slowed by four orders of magnitude as compared to diffusion in water (Hong, Junge (1983)). We found that the partition regions were only badly accessible to bovine serum albumin, however, in contrast to the internal phase, they were accessible to phosphate buffer.

TRANSIENT TRAPPING OF PROTONS FROM WATER OXIDATION BY SUBCOMPARTMENT WITH
PREEXISTING pH-DIFFERENCE AGAINST EXTERNAL PHASE (Theg, Junge (1983)).

The following experiments were stimulated by reports from the laboratories
of Dilley and Homann (Prochaska, Dilley (1978) - Baker et al. (1981) - Theg,
Homann (1982) - Theg et al. (1982)) who showed that thylakoids sustain pH
gradients even in the dark and that there may be special pathways for protons
from water oxidation as contrasted to protons from plastoquinone oxidation.
Most of our experiments were carried out with freshly prepared pea chloroplasts
or with chloroplasts which were stored under liquid nitrogen under
cryoprotection with ethylene glycol (Farkas, Malkin (1979)). We measured the
extent of the pH_{in}-indicating absorption changes of neutral red in the absence
and in the presence of nanomolar concentrations of gramicidin (Fig.1). The
traces in the absence of gramicidin (left) show the above mentioned rapid
proton deposition by water oxidation and the slower one by plastohydroquinone
oxidation. It was obvious that small concentrations of gramicidin made proton
deposition by water oxidation undetectable by neutral red.

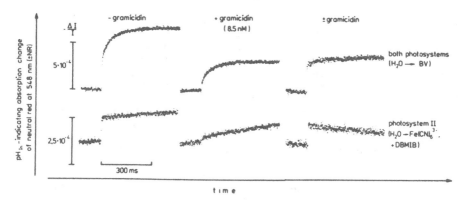

FIGURE 1. Effect of gramicidin on flash induced proton release from both
photosystems (above) and from photosystem II alone (below). Right traces
resulted from subtraction of left traces (-gramicidin) from middle traces
(+gramicidin). pH8.

We asked whether or not non-detected protons were released. For this we
measured the pH-changes in the outer phase. The result is shown in Fig.2.

FIGURE 2. Effect of
gramicidin on pH-
changes in external
(above) and internal
phase (below) as
measured with cresol
red and neutralred,
respectively. pH 7.7.

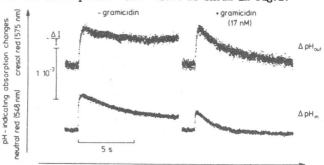

It is evident that protons, which were not to be detected in the internal phase later compensated the external alkalinization to zero level. This proved that they had been produced and that they had passed across the membrane. By measuring the electrochromic absorption changes we checked whether transiently disappeared protons had rapidly crossed the membrane dielectric (see Fig.6 in Theg, Junge (1983)). Addition of gramicidin in nanomolar concentration did not affect the extent but it accellerated the decay of the electric field (typical half-decay time 2ms). However, the accelleration was too low to account for the disappearence of the most rapidly deposited protons from water oxidation (half-time of internal acidification 100μs). We concluded that there exist special domains which, when activated, transiently trap protons which are released during water oxidation. These domains could be activated by nanomolar concentrations of gramicidin or nigericin and by micromolar concentrations of FCCP or A23187 but not by the potassium specific ionophore valinomycin. pH titration of this effect showed, that ionophores activated the extra buffering power of the domain only if the medium pH was above 7.5. The experimental titration curve resembled the one of a divalent acid ($n=2$). Since disappearence of protons from water oxidation required protonophores and alkaline pH we concluded that the special domains are normally saturated with protons, i.e. they are not in equilibrium with the medium pH. Protonophores and high pH of the medium deplete these domains from bound protons so that they are ready to trap incoming protons from water oxidation. We do not know whether the retention of protons by these domains in the absence of protonophores reflects an impermeable barrier or, alternatively, the presence of a slow proton pump. We titrated the buffering capacity of the special domain by firing a series of flashes. The lost observability of protons from water oxidation recovered with increasing flash number (see Fig.8 in Theg, Junge(1983)). 90% recovery was observed after six flashes. We have interpreted these observations as follows: Protons from water oxidation are intrinsically deposited into special domains. These have limited buffering capacity (4-6H /PSII). If their proton binding sites are empty (at alkaline pH and with added protonophore) protons from water oxidation transiently disappear. If the sites are occupied protons rapidly overflow into the thylakoid interior. We cannot tell up to now whether or not there is a separate domain for each of the reaction centers II. Upon superficial inspection the existence of these domains seems to argue in favour of postulated concepts of localized pathways for protons from pump sites into the ATP-synthases (see Williams (1961) - Kell (1979) - Dilley et al. (1981)). Closer inspection reveals that the existence of these special proton trapping domains scarely disturbs the orthodox chemiosmotic concept since they saturate after few turnovers of the proton pumps.

TRANSIENT TRAPPING OF PROTONS FROM WATER OXIDATION BY CFO AFTER EXTRACTION OF FOUR-SUBUNIT CF1 (Junge, Qian, Hong, Viale, unpublished).

This study aimed at proton conduction through CFO, the integral membrane portion of the ATP-synthase. When CF1, the catalytic portion, was detached from CFO a proton conducting channel appeared. This could be closed by addition of CF1 or by DCCD (for reviews see McCarty (1979) - Nelson (1981)). Usual extraction procedures of CF1 induced large side conductances which masked enhanced proton flow through CFO (e.g. Schmid et al. (1976)). To avoid this, in the present study we chose mild extraction conditions which removed only 5-10% of CF1 from the membrane. When chloroplasts were prepared

in the absence of magnesium ions relatively low concentrations of EDTA
(20-40μM) were sufficient to induce extraction of 5-10% of CF1. Chloroplasts
were incubated for 10 min in EDTA containing medium and then the extraction
was stopped by addition of magnesium 1-2mM. We determined the extracted
amount of CF1 by immuno-electrodiffusion (Laurell (1966)). After
immunoprecipitation we analysed the subunit composition of extracted CF1 by
SDS-PAGE (Laemmli (1970)). Gels were stained by Coomassie blue. Inspection
of the photometric scan showed that the contribution of subunit delta was
drastically reduced under mild extraction conditions. This was corroborated
by quantitative analysis. By cutting the gels and photometric assay of the
extract we calculated the proportion between subunits under consideration of
the Coomassie binding factors as determined by Binder et al. (1978). When
normalizing the content of gamma subunit to 1 we obtained the following:
under extraction at 25μM EDTA the ratios of alpha to epsilon were
1.6/2.0/1.0/0.2/1.4. Under strong extraction at 750μM EDTA the ratios were
1.8/2.0/1.0/0.9/1.3. Low EDTA favoured the extraction of only four subunits.
We measured the electric potential difference via electrochromic absorption
changes and proton deposition inside after excitation of chloroplasts with
xenon flashes. The result is documented in FIG.3.

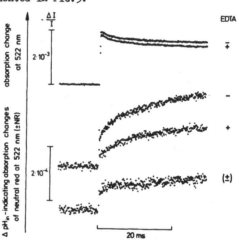

FIGURE 3. Electrochromic
absorption changes (top) and
pH_{in}-indicating absorption
changes of neutral red (middle)
in mildly extracted spinach
chloroplasts (+EDTA, 20μM) and
in controls (-EDTA). The bottom
trace represents the arithmetic
difference of the two traces
in the middle.

The most conspicuous effect of mild EDTA-treatment was the disappearance of
rapid proton deposition from water oxidation (Fig.3, middle and most obvious
in the bottom trace, which represents the missing component). The extent of
the electric field was slightly altered and its decay was only negligibly
accelerated. Again we checked whether or not the undetectable protons were
not produced. We measured pH-transients in the external phase (same
procedure as documented in Fig.2 in other context). We found that "lost"
protons later reappeared in the external phase. Therefore we had to conclude
that mild extraction of four-subunit-CF1 opens an additional buffering
capacity which rapidly and transiently traps protons which are released
during water oxidation. Transiently trapped protons leak out in the time
domain of some 100ms.

We asked whether this transient trapping was caused by CFO. For this we
added DCCD (dicyclohexylcarbodiimide, 20μM) to chloroplasts which were mildly
extracted by EDTA-treatment (20μM). This reversed both the undetectability
of protons from water oxidation and the small decrease of the extent of the
electric potential difference. Since one molecule of DCCD on six copies of

subunit III of CFO is known to block the proton channel (Sigrist-Nelson et al. (1978)) it was probable that the extra buffering power was caused by CFO. We titrated this extra buffering power by variation of the medium pH and measurement of the relative extent of disappearing protons from water oxidation. The result is documented in Fig.4.

FIGURE 4. Relative extent of proton uptake which became undetectible upon mild extraction of four-subunit CF1 by EDTA as function of the medium pH. Circles (spinach) and squares (pea chloroplasts) experimental, curves calculated for a hexacooperative buffer according to the following equation:

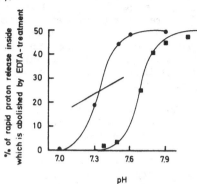

$$[A^{n-}]/[A_{total}] = (1-10^{n(pK-pH)})^{-1}$$

where $[A^{n-}]$ denotes the concentration of the unprotonated buffer and n=6.

It was to our surprise that the titration curves were so steep. Our calculation showed that the extra buffering capacity behaved as if it buffered six protons at once in strict cooperativity. Although, cooperativity of six was already suggested by studies on the deactivation of the proton channel by DCCD (Sigrist-Nelson et al. (1978)), here, the degree of cooperativity seemed too high if one assumed that the channel was formed by only six copies of the 8 kD subunit III. We had rather expected that electrostatic repulsion of protons had reduced the degree of cooperativity when they were bound to the same channel. Such anticooperative effects are known to occur in dicarboxylic acids (Bell (1973)). If the high cooperativity documented in Fig.4 stands further experimental test (e.g. competition between DCCD and protons) we will have to conclude that the proton channel through CFO is formed by more than six copies of the 8kD protein.

As documented in Figs.3 and 4 the transient disappearence of protons from water oxidation was accompanied by a small decrease of the extent of the electric potential difference. This effect was also reversed upon addition of DCCD. It is known from previous work that the extent of the electrochromic absorption changes is pseudolinear as function of translocated charges (for reviews see Witt (1971) - Junge (1977)). The extent induced by a saturating flash of light with methyl viologen as terminal electron acceptor corresponds to two elementary charges per 500-600 chlorophyll molecules. From traces as documented in Fig.3 we calculated the lowering of the initial extent and we found that it corresponded to 0.2 charges (per 500-600 chl) while 0.5 protons had disappeared in the response of neutralred. We concluded that water oxidation protons which were transiently buffered had crossed less than one half of the membrane dielectric within less than 20us, while further passage towards the external surface was inhibited. Only at higher degrees of extraction we observed a more pronounced acceleration of the electric field decay. However, this occurred concomitant with increasing extraction of delta subunit. It is probable that the delta subunit of CF1 is the plug which regulates passage of protons across CFO and into CF1. This proposal is consistent with results by Andreo et al. (1982) who found that the delta subunit was not necessary for binding of CF1 to CFO and of Roos and Berzborn (1982) who found that delta alone can rebind to CFO.

We observed specificity of the extra buffering capacity which was related to CFO (plus delta) for protons which were released during water oxidation. For well stacked thylakoids it is established that photosystem II with the water-oxidizing enzyme is located in the stacked domains while CF1 resides in stroma lamellae (review Anderson (1981)). There are two ways of visualizing this specificity: 1.) After extraction of bulky and hydrophilic CF1, CFO is free to move into stacked domains where it forms associates with photosystem II. 2.) Protons which are liberated during water oxidation travel via special conducting channels into CFO. The second model was previously proposed by Prochaska and Dilley (1978). The first concept has the advantage that it can be experimentally verified or disproved by known experimental techniques.

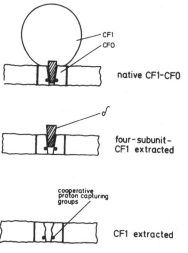

FIGURE 5. Speculative scheme which relates the experimental results to the structure of CFO-CF1. CFO, CF1, the delta subunit of CF1 and hexacooperative buffering groups in CFO are indicated. If CF1 is attached to CFO (top) delta is plugged into CFO so that the extra buffering groups are shielded and inside deposited protons can be detected by neutral red. If four-subunit-CF1 is extracted (middle) buffering groups are exposed but passage of protons across the membrane is still blocked. That the buffering groups are only accessible for protons but not for other cations (as evident from lowered extent of electric field) supports Mitchell's concept of a proton well which rationalizes the equivalence of pH and electric potential for photophosphorylation. If all subunits of CF1 are extracted very rapid buffering occurs together with increased electric conductance of the membrane.

REFERENCES

Albertsson PA (1982) FEBS Lett.149,186-190
Anderson JM (1981) FEBS Lett.124,1-10
Andreo C, Patrie W and McCarty RE (1982) J.Biol.Chem. 9968-9975
Auslaender W and Junge W (1974) Biochim.Biophys.Acta 357,285-298
Auslaender W and Junge W (1975) FEBS Lett.59,310-315
Baker GM, Bhatnager D and Dilley RA (1981) Biochemistry 20,2307-2315
Bell RP (1973) The Proton in Chemistry, ch.6, Chapman&Hall, London
Binder A, Jagendorf A and Ngo E (1978) J.Biol.Chem.253,3094-3100
Crank J (1979) The Mathematics of Diffusion, ch.14, Clarendon, Oxford
Dilley RA, Baker GM, Bhatnager D, Millner J and Laszlo J (1981) in: Energy
        Coupling in Photosynthesis (Selman&Selman-Reimer,eds) p.47-58
        Elsevier, Amsterdam
Diner BA (1977) Biochim.Biophys.Acta 460,247-258
Farkas DL and Malkin S (1979) Plant Physiol. 64,942-947
Feldberg SW and Delgado AB (1978) Biophys.J. 21,71-86
Foerster VG, Hong YQ and Junge W (1981) Biochim.Biophys.Acta 638,141-152
Gutman M, Huppert D, Pines E and Nachliel E (1981) Biochim.Biophys.Acta
        642,15-26
Hong YQ and Junge W (1983) Biochim.Biophys.Acta 722,197-208
Hope AB and Morland A (1979) Aust.J.Plant Physiol.6,1-16

Junge W (1977) Ann.Rev.Plant Physiol.28,503-536
Junge W and Jackson JB (1982) in: Photosynthesis (Govindjee, ed.) vol.1,
    pp.589-646, Academic Press, New York
Junge W and Witt HT (1968) Z.Naturforsch.23B,244-254
Junge W, McGeer AJ, Auslaender W and Kollia J (1978) in: Energy
    Conservation in Biological Membranes (Schaefer&Klingenberg, eds.)
    pp.113-127, Springer Verlag, Berlin
Junge W, Auslaender W, McGeer AJ and Runge T (1979) Biochim.Biophys.Acta
    546,121-141
Kell DB (1979) Biochim.Biophys.Acta 549,55-99
Laemmli UK (1970) Nature 227,680-685
Laurell CB (1966) Analyt.Biochem. 15,45-52
Lis LJ, Lis WT, Parsegian VA and Rand RP (1981) Biochemistry 20,1771-1777
McCarty RE (1979) Ann.Rev.Plant Physiol. 30,79-104
Mitchell P (1961) Nature 191,144-148
Mitchell P (1966) Biol.Rev.Cambridge Philos.Soc. 41,445-502
Nelson N (1981) Curr.Top.Bioenergetics 11,1-33
Prince RC, Linkletter SJG and Dutton PL (1981) Biochim.Biophys.Acta
    635,132-148
Prochaska LJ and Dilley RA (1978) Arch.Biochem.Biophys.187,61-71
Pulles MPJ, vanGorkum HJ and Willemsen JG (1976) Biochim.Biophys.Acta
    449,536-540
Roos P and Berzborn RJ (1082) Proc.2nd Europ.Bioenergetics Conf. pp.99-100
    LBTM-CNRS, Villeurbanne
Saphon S and Crofts AR (1977) Z.Naturforsch.32C,617-626
Sigrist-Nelson K, Sigrist H and Azzi A (1978) Europ.J.Biochem.92,9-14
Schmid R, Shavit N and Junge W (1976) Biochim.Biophys.Acta 430,145-153
Theg SM and Homann PH (1982) Biochim.Biophys.Acta 679,221-234
Theg SM and Junge W (1983) Biochim.Biophys.Acta 723,294-307
Theg SM, Johnson JD and Homann PH (1982) FEBS Lett.145,25-29
Velthuys BR (1978) Proc.Natl.Acad.Sci USA 76,2765-2769
Velthuys BR (1980) FEBS Lett.115,167-170
Williams RPJ (1961) J.Theor.Biol. 1,1-13
Witt HT (1971) Quart.Rev.Biophys. 4,365-477

ACKNOWLEDGEMENTS

We are very grateful to our collegues V.Foerster, U.Klose, Dr.U.Kunze,
M.Offermann, A.Polle, N.Spreckelmeyer and Dr.HW.Trissl for collaboration.
Different parts of this work were financially supported by the Deutsche
Forschungsgemeinschaft, the European Commission and the Niedersaechsisches
Ministerium fuer Wissenschaft und Kunst. The stay of YQH in Osnabrueck was
sponsored by the Chinese Ministery of Education and the stay of WJ in Rosario
by the Consejo Nacional de Investigaciones Cientificas y Technicas, Argentina
and the Deutscher Akademischer Austauschdienst.

Authors Address:
WJ   Biophysik, Fachbereich Biologie/Chemie, Universitaet Osnabrueck,
        Postfach 4469, D-4500 Osnabrueck, FR-Germany
YQH&LPQ Institute of Plant Physiology, Academia Sinica, 300 Fongling Road,
        Shanghai, China
ST   Dept. of Biological Sciences, Purdue University, West Lafayette,
        IN 47907, USA
AV   Centro de Estudios Fotosinteticos y Bioquimicos, CONICET, Fundacion
        Miguel Lillo, Universidad de Rosario, Suipacha 536, 2000 Rosario,
        Argentina

THE INNER- AND INTER-THYLAKOIDAL ELECTRIC POTENTIAL OF ISOLATED
CHLOROPLASTS IN THE DARK

DIETER WALZ/BIOZENTRUM (UNIVERSITY OF BASEL)

1. RATIONALE

A thermodynamic assessment of light-driven H^+-pumps requires the estimation
of the conjugate force of the flow of H^+-ions which is the difference in
electrochemical potential of these ions, $\Delta\tilde{\mu}_H$, between the inner thylakoidal
space and the suspending medium. The partitioning of suitable markers is
used to determine the two constituents of this quantity, e.g. methylamine
for the chemical part and thiocyanate for the electric part, respectively
(Rottenberg et al., 1972). The analysis of such data is based on the
assumption that the concentration of markers in the whole space outside
the thylakoids is constant and equal to that in the suspending medium.
The amount of marker found in the pellet of thylakoids is corrected
accordingly and the concentration in the inner space calculated by means
of the internal volume.

The surfaces of thylakoid membranes carry charges associated with the
proteins. Hence, diffuse double layers of ions on the membrane surfaces
exist in which the concentration of ions deviates from that in the bulk
phase. This phenomenon is particularly relevant to the interthylakoidal
space in grana stacks since the spacing between thylakoids is of the
order of the Debye-Hückel length. Moreover, it becomes the more prominent
the lower the concentration of salt.

2. MODEL FOR CALCULATING ELECTRIC POTENTIAL PROFILES

The thylakoid membranes are represented by plane sheets of a dielectric
with constant D_m. The charges are considered as smeared over the surface
thus yielding a charge density σ. The space in and outside the thylakoids
contains an aqueous solution (dielectric constant D_w) of a 1,1-electrolyte.
With a reference point for the electric potential Ψ^w in the bulk phase,

$$\Psi(x) = -(RT/F) \ln \{c_+(x)/c_\infty\} = (RT/F) \ln \{c_-(x)/c_\infty\} \qquad (1)$$

where x denotes the space coordinate in a system as shown in Fig.1; c_+,
c_- and c_∞ refer to the concentration of positive ions, negative ions and
of salt in the bulk phase, respectively. The constant field approximation
(no mobile charges in the membrane) yields

$$\Psi(x) = ax + b \qquad \text{in the membrane.} \qquad (2)$$

Integration of the Poisson equation pertinent to the aqueous phases requires complex procedures in the general case (Duniec, Thorne, 1983) but
for a univalent electrolyte in a one-dimensional system

$$c_s(x)/c_\infty = \lambda sn^2 \{K(\lambda^2) + (x-x_o)/(2\lambda^{1/2}\delta)|\lambda^2\} \quad \text{with } s = sgn(\sigma) \qquad (3)$$

where sn and K denote, respectively, a Jacobian elliptic function with
parameter λ^2 and the complete elliptic integral of the first kind; x_o
is the coordinate where Ψ has a minimum, and δ is the Debye-Hückel length

$$\delta = \{RT\varepsilon_o D_w/(2F^2 c_\infty)\}^{1/2}. \qquad (4)$$

Sybesma, C. (ed.), Advances in Photosynthesis Research, Vol. II. ISBN 90-247-2943-2.
© *1984 Martinus Nijhoff/Dr W. Junk Publishers, The Hague/Boston/Lancaster.*

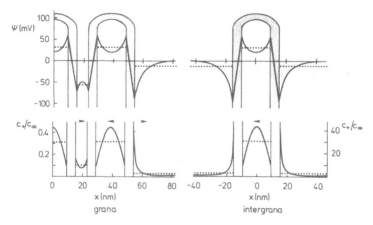

FIGURE 1. Profiles for the electric potential (top) and the concentration of positive ions (bottom) in stacked (left) and unstacked (right) thylakoids. Dotted lines indicate average values and arrows point to the pertinent scale. Membranes have a charge density σ (C/m^2) of 0.009 on inner surface, -0.01 on outer surface in stacks and -0.02 on surfaces of stacks and unstacked membranes. D_m = 5, D_w = 80, c_∞ = 2.5 mM and T = 20°C.

The coefficients a, b and λ are adjusted to fit the boundary conditions at the membrane/water interfaces with coordinates x_i

$$\varepsilon_{x < x_i} (d\Psi/dx)_{x \to x_i} - \varepsilon_{x > x_i} (d\Psi/dx)_{x \leftarrow x_i} = \sigma, \ \Psi_{x \to x_i} = \Psi_{x \leftarrow x_i} \qquad (5)$$

where ε is the permittivity (absolute permittivity ε_o times dielectric constant). Fig.1 shows an example of the electric potential profile together with the pertinent concentration ratio c_+/c_∞ calculated with Eqns.1-5.

3. PARTITIONING OF MARKERS

3.1. Experimental

Chloroplasts were isolated as described previously (Walz et al., 1974), final washing and resuspending was in 100 mM Sucrose, 5 mM Hepes, pH 7.5. Partitioning of markers ([^{14}C]-methylamine, -thiocyanate and -carboxyinulin) was assayed with the silicon oil centrifugation technique using [^{3}H]-H$_2$O and [^{14}C]-methoxyinulin for estimating the total and interstitial volume, respectively. The reaction medium contained 100 mM sucrose and an appropriate Good buffer whose concentration was chosen such that, after adjusting the pH with KOH, K$^+$ concentration amounted to 2.5 mM. Centrifugation was performed for 2 min after 10 min preincubation, both steps were carried out in the dark and at 20°C.

3.2. Apparent $\Delta\tilde{\mu}_H$

The analysis of the data for partitioning of methylamine and thiocyanate by the conventional procedure (Rottenberg et al., 1972) yielded values for $\Delta\tilde{\mu}_H$ such as shown in Table 1. For all pH-values used, there was a significant deviation of this quantity from the expected equilibrium value of

TABLE 1. Chemical and electric part of $\Delta\tilde{\mu}_H$. The ratio of total concentration of marker in the inner space and in the bulk phase is denoted by q with index M and T for methylamine ($pK_M = 10$) and thiocyanate, respectively.

pH	q_M	q_T	$\Delta\mu_H$ (kJ/mol)	$F\Delta\Psi$ (kJ/mol)	$\Delta\tilde{\mu}_H$ (kJ/mol)
5.2	4.35+0.14	3.24+0.13	3.62+0.08	2.89+0.11	6.51+0.19
6.0	8.01+0.69	0.84+0.05	5.06+0.21	-0.41+0.15	4.65+0.36
8.6	15.5 +0.6	0.40+0.07	6.76+0.10	-2.24+0.22	4.52+0.32
5.2	4.26*	3.39*	3.53	2.98	6.51

* Values calculated from the data shown in Fig.1 for a stack of 10 thylakoids without intergrana membranes.

zero. Possible errors in the estimation because of binding of markers to the membrane or due to a persisting non-equilibrium of H^+-ions could be excluded since the presence of unlabelled markers in excess and addition of valinomycin and/or CCCP did not change the results significantly. The non-vanishing $\Delta\tilde{\mu}_H$-values then indicate substantial contributions from the electric potential in the interthylakoidal space (cf. Fig.1).

3.3. Inner- and interthylakoidal electric potential

Carboxyinulin is a negatively charged marker which should penetrate into the inter- but not the inner-thylakoidal space. Hence, its partitioning should report on the electric potential in the former space and in the surface layers. It can then be derived that

$$<\Psi_{in}> = (RT/F)\ln\gamma_{in}^- \quad \text{with} \quad \gamma_{in}^- = q_T - q_C(1+p_C) \tag{6}$$

$$<\Psi_{is}> = -(RT/F)\ln\gamma_{is}^+ \quad \text{with} \quad \gamma_{is}^+ = \{1/\gamma_{in}^- - q_M(1+p_M) + p_M\}/\{q_C(1+p_C)\} \tag{7}$$

$$\text{and} \quad \bar{v}_{is} = \bar{v}_{in}q_C(1+p_C) \ \gamma_{is}^+/(1-\gamma_{is}^+) \tag{8}$$

$$\text{where} \quad p_C = \exp\{\ln10(pK_C-pH)\} \quad \text{and} \quad p_M = \exp\{\ln10(pH-pK_M)\} \tag{9}$$

Herein, γ represents an average concentration ratio of ions with valence as indicated and the index "is" shall indicate that the pertinent quantity relates to both interthylakoidal space and surface layer of membranes. For the meaning of q see Table 1 (index C for carboxyinulin, $pK_C = 4.5$). Fig.2 then presents results obtained by means of Eqns.6-9.

4. DISCUSSION

The resolution of the data for partitioning of markers in non-energized thylakoids reveals a positive electric potential in the internal space at low pH-values which decreases to slightly negative values with increasing pH. This is in line with an H^+-binding group at the inner membrane surface with $pK \simeq 5.2$ (Walz et al., 1974) whose protonated species is positively charged. The negative values of $<\Psi_{in}>$ in the pH-range, where the positive charge inside vanishes, reflect the negative charges on the outer membrane surface as verified by simulations with the model.

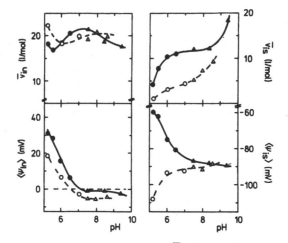

FIGURE 2. Dependence of specific volume $\bar{v}$ and average electric potential $<\Psi>$ on pH for the inner space (left) and for the combined space comprising the interthylakoidal space and the surface layers (right). Different symbols refer to different chloroplast preparations.

A straightforward interpretation of the average electric potential $<\Psi_{is}>$ is not possible; this quantity comprises contributions from the interthylakoidal space and the surface layers whose relative proportions depend on the ratio of stacked to unstacked membranes. Increasing pH-values give rise to more negative charges on the membrane surface which causes a stronger repulsion of adjacent thylakoids in the stacks (Barber, Chow, 1979) and, as a consequence, the interthylakoidal space should increase. This phenomenon is documented by the specific volume $\bar{v}_{is}$ although it should be noted that this quantity too is partially determined by the surface layers. However, $<\Psi_{is}>$ can increase or decrease with increasing pH in the medium since the profile of Ψ strongly depends on the distance between thylakoids. This and other aspects including the effect of salt concentration or energization of thylakoids will be discussed elsewhere (D. Walz, in preparation).

REFERENCES

Barber J and Chow WS (1979) A mechanism for controlling the stacking and unstacking of chloroplast thylakoid membranes, FEBS Lett. 105, 5-10.

Duniec JT and Thorne SW (1983) Electrostatic potentials in membrane systems, Bull. Math. Biol. 45, 69-90.

Rottenberg H, Grunwald T and Avron M (1972) Determination of ΔpH in chloroplasts. 1. Distribution of $[^{14}C]$methylamine, Eur. J. Biochem. 25, 54-63.

Walz D, Goldstein L and Avron M (1974) Determination and analysis of the buffer capacity of isolated chloroplasts in the light and in the dark, Eur. J. Biochem. 47, 403-407.

Author's address: Dieter Walz/Biozentrum, University of Basel, Klingelbergstr. 70, CH-4056 Basel, Switzerland

DIFFUSION BARRIERS FOR PROTONS AT THE EXTERNAL SURFACE OF THYLAKOIDS

ANDREA POLLE AND WOLFGANG JUNGE / UNIVERSITAET OSNABRUECK

INTRODUCTION

Light induced electron transfer across photosynthetic membranes of higher plants is coupled with proton translocation from the outer phase into the lumen of thylakoids. The primary photochemical charge separation in PSII transfers one electron to special plastoquinone molecules Q and R which act in series (H.H. Stiehl, H.T. Witt,1968,1969; B.A. Diner ,1977; B.R. Velthuys, J. Amesz ,1974 ; B. Bouges-Bocquet ,1973). R stores two electrons and mediates electron transfer into the plastoquinone pool (B.R.Velthuys, J. Amesz ,1974)

It was shown that under repetitive flash excitation the reduction of the plastoquinone acceptors was followed by the uptake of $1H^+/e^-$ (W.Schliephake et al., 1968; W. Junge, W.Auslaender ,1973; S.Saphon, A.R.Crofts ,1977). However, kinetic studies on spinach chloroplasts with the water soluble dye cresol red revealed that the uptake of protons seemed to be retarded by two orders of magnitude ($t_{1/2}$= 60ms) (W.Auslaender, W.Junge,1974) in comparison with the reduction of R ($t_{1/2}$= 600µs) (H.H.Stiehl, H.T.Witt ,1968,1969). The retardation could be partially removed by mechanical desintegration of thylakoids and by addition of protonophores (W.Auslaender, W.Junge ,1974). By these treatments the proton uptake could be accelerated up to $t_{1/2}$= 2ms. This half rise time coincided with a lag phase of 2ms prior to the reduction of P_{700}^+ by plastoquinol (W.Haehnel,1976). From this a diffusion barrier for protons was inferred (W.Auslaender, W.Junge ,1974).

When outer membrane proteins were modified by trypsin treatment, the plastoquinone acceptor Q became accessible to externally added ferricyanide (G.Renger, 1976). From this a "proteinaceous shield" covering the reducing side of PSII was apparent. Since tryptic attack predominantly modified the DCMU-binding 32 kD protein, Mattoo et al.(1981) speculated that this protein might fullfill the requirements of the diffusion barrier. However, proton uptake was not observed in trypsinated chloroplasts (G.Renger , R.Tiemann,1979). It remained to be documented whether the diffusion barrier for ferricyanide was identical with the diffusion barrier for proton uptake or not.

Since PSI and PSII are at different locations in stacked thylakoids, it was surprising that proton uptake at both sites showed the same slow kinetics. In presence of methyl viologen (MV) proton uptake at PSI is caused by the Mehler reaction:

$$2 MV^{\cdot-} + 2O_2 \longrightarrow 2MV + 2O_2^{\cdot-}$$

$$2 O_2^{\cdot-} + 2H^+ \longrightarrow H_2O_2 + O_2 \text{ (Mehler reaction)}$$

We asked whether proton uptake induced by PSI was rate limited by the residual amount of superoxide dismutase (SOD), which catalyses the Mehler reaction.

Then we investigated whether or not multiple binding and debinding of protons in grana stacks delayed the access of protons to the uptake sites at PSII as proposed recently (Y.Q.Hong, W.Junge, 1983).

Sybesma, C. (ed.), Advances in Photosynthesis Research, Vol. II. ISBN 90-247-2943-2.
© *1984 Martinus Nijhoff/Dr W. Junk Publishers, The Hague/Boston/Lancaster.*

MATERIALS AND METHODS:

Broken chloroplasts were prepared from 14-18 days old pea seedlings according to Foerster et.al. (1981). Chloroplasts were stored frozen under liquid nitrogen with 30% ethylen glycol as cryoprotective. Chloroplasts equivalent to 10µM chlorophyll were added to an assay medium containing 25mM KCl, 3mM $MgCl_2$ (stacked thylakoids) or were incubated for 10min in 10mM NaCl (destacked thylakoids) (J.Barber,1980). To induce restacking 25mM KCl and 3mM $MgCl_2$ were added to destacked thylakoids and the suspension was incubated for another 5min. The pH was adjusted to pH7.9 . Other additions are given in the legends to figures.
Absorption changes were measured as described elsewhere (V.Foerster et al.,1981). The pH_{out}-indicating absorption changes of cresol red at 575nm were obtained by subtraction of signals recorded in the absence from those recorded in the presence of the dye (15µM). The samples were excited by saturating flashes (λ>610nm, 15us, $1mJ/cm^2$). Typically 20 or 40 samples were averaged, repetition rate of flashes 0.1Hz.

RESULTS AND DISCUSSION:

Rate of proton uptake from the external bulk phase by PSI is limited by superoxide dismutase. Fig.1 shows the pH_{out}-indicating absorption changes of cresol red at 575nm. In the presence of MV proton uptake by PSII and by the Mehler reaction induced by PSI had a half rise time of approximate 120ms (upper trace, left). The rise of the signal was not monoexponential. A detailed analysis of the data will be presented elsewhere. Here we point out qualititative effects.

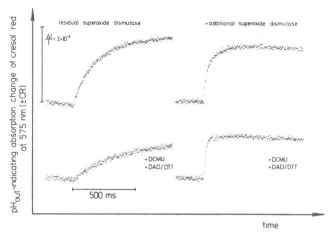

Figure 1: Proton uptake from the external bulk phase by PSI and PSII (upper traces) and by the Mehler reaction of PSI only (lower traces). The assay medium contained MV 10µM and where indicated DCMU 5µM, DAD 10µM, DTT 40µM, SOD 5µg/ml. Time resolution 1ms/point

When SOD was added (upper trace, right) the rise of the pH_{out}-indicating signal was accelerated. It showed a fast and slower phase. Obviously one part of the alkalinization of the external phase was rate limited by the activity of SOD. We inhibited PSII by DCMU and restored PSI activity by addition of the donor couple diaminodurene (DAD) and dithiothreitol (DTT).

We measured the pH$_{out}$-indicating absorption change of cresol red (fig.1, lower traces), which was caused by PSI activity only. The amplitude of the signal was about 50% of the control and the rise was slow (Fig.1 lower trace, left). When SOD was added, only the fast phase of the control signal reappeared (fig.1, lower trace,right). It is apparent, that the rise of the pH$_{out}$-indicating absorption change caused by the Mehler reaction was rate limited by SOD. In contrast proton uptake at the reducing side of photosystem II was still retarded by a diffusion barrier.

Rate of proton uptake from the external bulk phase by PSII is predominantly limited by multiple binding and debinding of protons within the grana stacks. We investigated the effect of destacking on the diffusion barrier for proton uptake by the reaction centers of PSII which are predominantly located in the grana stacks (B.Andersson,W.Haehnel, 1982). Figure 2 shows proton uptake at the reducing side of PSII only, with ferricyanide used as electron acceptor for PSI. The upper trace shows that the pH$_{out}$- indicating absorption change of cresol red had a $t_{1/2} \approx 60$ms, when pea thylakoids were stacked. When thylakoids were destacked , the rise of alkalinization was biphasic (figure 2, middle trace). Obviously destacking accelerated proton uptake from the external bulk phase at the reducing side of PSII. In contrast shielding of the reducing side of PSII against the access of ferricyanide still existed, since the amplitude of the pH$_{out}$-indicating signal was not markedly reduced. The biphasic rise might indicate that destacking of grana was incomplete.

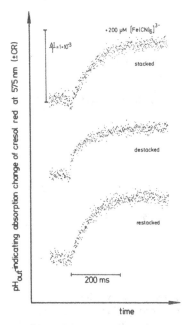

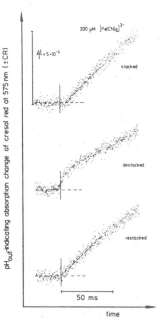

Figure 2: Proton uptake from the external bulk phase at the reducing side of PSII in stacked (upper trace), destacked (middle trace) and restacked thylakoids (lower trace). Time resolution 1ms/point

Figure 3: Proton uptake from the external bulk phase at the reducing side of PSII. Same assay conditions as in figure 2, time resolution 200 μs/point.

Fig.2 (lower trace) shows, that the fast rise of the pH-signal completely disappeared when membranes were restacked. These findings support our previous proposal, that proton uptake from the external bulk phase to the sinks at the reducing side of PSII is retarded by multiple binding and debinding of protons which diffuse through the grana stacks (Y.Q.Hong, W.Junge, 1983).

<u>Proton uptake by PSII shows a lag phase (proteinaceous barrier?).</u>   We investigated proton uptake at the reducing side of PSII at higher time resolution (200 μs/point). Fig.3 (upper trace) shows the pH_{out}-indicating absorption change of cresol red in the external phase of stacked pea thylakoids under repetitive flash excitation. It is apparent, that the absorption of cresol red changed only after a distinct lag phase of 6-7ms. Obviously this lag phase did not coincide with the lag phase which was observed before the reduction of P_{700}^{+} by plastoquinol (W.Haehnel, 1976). The middle trace of fig.3 shows that the lag was abolished, when the membranes were destacked. When membranes were restacked , the original rise of the signal reappeared, but not the lag phase.

To find out whether the lag could be attributed to a shielding protein, which was lost during destacking and did not rebind, we separated proteins of the supernatant of stacked, de-and restacked thylakoids by SDS gel electrophoresis. We found that two proteins (~55 kD,~30 kD) which were lost under destacking conditions, did not rebind, when thylakoids were restacked again (not shown).

REFERENCES:
Barber J, (1980) Biochim.Boiphys.Acta 594,253-308
Andersson B, Haehnel W (1982) FEBS Lett.,146,13-17
Auslaender W, Junge W (1974) Biochim.Biophys.Acta 357,285-298
Bouges-Bocquet B (1972) Biochim.Biophys.Acta 92,772-785
Diner BA (1977) Biochim.Biophys.Acta 460,247-258
Foerster V, Hong YQ, Junge W (1981) Biochim.Biophys.Acta 638,141-152
Haehnel W (1976) Biochim.Biophys.Acta 440,506-521
Hong YQ, Junge W (1983) Biochim.Biophys.Acta 722,197-208
Junge W, Auslaender W (1973) Biochim.Biophys.Acta 333,59-70
Mattoo AK, Pick U, Hoffman-Falk H, Edelmann M (1981)
                    Proc.Natl.Acad.Sci.USA,78,1572-1576
Renger G (1976) Biochim.Biophys.Acta 440,287-300
Renger G, Tiemann R (1979) Biochim.Biophys.Acta 545,316-324
Saphon S, Crofts AR (1977) Z.Naturforsch.,32c,810-816
Schliephake W, Jumge W, Witt HT (1968) Z. Naturforsch.,23,1571-1578
Stiehl HH, Witt HT (1968) FEBS Lett.,39,205-208
Stiehl HH, Witt HT (1969) Z.Naturforsch.,24b,1588-1598
Velthuys BR, Amesz J (1974) Biochim.Biophys.Acta 333,85-94

ACKNOWLEDGEMENTS:
We wish to thank H. Kenneweg for photographs, V. Foerster for valuable discussion, Dr. U. Kunze and H. Lill for computer programms. Financial support by the European Commission (Solar Energy Programm) is gratefully acknowledged.

<u>Authors adress:</u>   Andrea Polle and Wolfgang Junge /Biophysik,
                    Fachbereich Biologie/Chemie
                    Universitaet Osnabrueck, Postfach 4469
                    D-4500 Osnabrueck, Germany (FRG)

INDICATIONS FOR THE CHLOROPLAST AS A TRI-COMPARTMENT SYSTEM:
MICRO-ELECTRODE AND P515 MEASUREMENTS IMPLY SEMI-LOCALIZED CHEMIOSMOSIS

OLAF VAN KOOTEN/FRANS A.M. LEERMAKERS/ROBERT L.A. PETERS/WIM J. VREDENBERG

1. INTRODUCTION

The electric response of a thylakoid membrane to a single turnover saturating light flash differed if measured by the electrochromic absorbance
change (P515) from the response as measured with the micro-electrode. The
slower kinetic components in the P515 response being absent in the micro-
electrode measurements (Vredenberg 1981). These slower kinetic components,
called reaction II (Schapendonk et al. 1979), in the P515 response were
shown not to be a response to a secondary charge separation, known as the
Q-cycle (Van Kooten et al. 1983). Because of the difference between micro-
electrode and P515 measurements, it was postulated that reaction II was
an intramembraneous electric event (Westerhoff et al. 1983).

However we have now been able to measure a reaction II type response with
the micro-electrode in *Peperomia metallica* chloroplasts. We correlated
micro-electrode measurements from flash-induced responses with long light
pulse responses and with P515 flash-induced responses. From this we conclude
that the chloroplast membrane responds heterogenously to a charge separation. Parts of the chloroplast lumen seem to respond "classically" in
that the dissipation of the field is through passive ion fluxes across the
membrane (Vredenberg 1976). These parts are described by the Goldman-
Hodgkin-Katz equation for diffusion. Other parts of the lumen seem to
hinder ion and proton movement. This leads to a different electric response
due to local charge separations. A physical explanation is given how the
lumenal structure can virtually decrease the diffusion coefficient of free
protons.

2. MATERIAL AND METHODS

Peperomia metallica was grown as described before (Bulychev et al. 1973),
except for the growing conditions, which are slightly altered with
respect to the fertility of the soil. Micro-electrode and P515 absorbance
change measurements in leaf cuts were done at room temperature as described
elsewhere (Schapendonk 1980). Calculations of the transmembrane potential
in a 2-compartment system are based on the Goldman equation (Schultz 1980).
The parameters for these calculations were taken from Vredenberg (1976).
Calculations for proton concentration profiles in a plane containing mobile
or immobilized buffer groups were derived from Crank (1975).

3. RESULTS AND INTERPRETATION

After discovering that the occurrence of reaction II in P515 flash-induced
responses is correlated with optimal growing conditions of spinach plants
(Peters, unpublished results), we optimized the growing conditions of our
Peperomia metallica culture. In fig. 1 it is seen that chloroplasts from
leaves of these cultures have a reaction II type response, when activated
by a single turnover light flash or by a short (10 ms) light pulse. A
second flash given a 100 ms after the first shows only a reaction I type
response. This is very similar to what is seen in spinach leaves (Van Kooten
et al. 1983). Fig. 1 shows that micro-electrode and P515 measurements duplicate one another reasonably. This implies that reaction II is either a
transmembrane phenomenon or an intramembrane event, which translates into
a transmembrane phenomenon (Zimanyi, Garab 1982). Because of the insensitivity

Sybesma, C. (ed.), Advances in Photosynthesis Research, Vol. II. ISBN 90-247-2943-2.
© *1984 Martinus Nijhoff/Dr W. Junk Publishers, The Hague/Boston/Lancaster.*

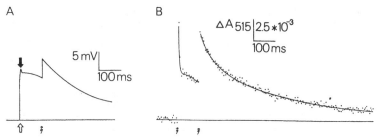

FIGURE 1. (A) Electric response of a *Peveromia metallica* chloroplast to a 10 ms light pulse followed by a single-turnover saturating flash, measured with a glass capilary micro-electrode inserted into the thylakoid lumen. The open arrow indicates the light pulse and the zig-zag arrow indicates the flash. (B) Electrochromic absorbance change measured at 515 nm from a *P. metallica* leaf. The arrows indicate single-turnover saturating light flashes. The spike in the response to the first flash is not an artefact, it appears to have a periodicity of 2 in multiple flash experiments (data not shown). This spike cannot be measured with the micro-electrode, because of electro-magnetic disturbances from the Xenon flash.

of the micro-electrode to light scattering changes, as opposed in the P515 measuring technique, it is now possible to correlate long light pulse responses to flash-induced ones.

In fig. 2A the electric response to a single turnover flash and to a 100 ms light pulse in the same chloroplasts can be seen. The slow decay after the flash (fig. 2B) corresponds with the slow decay after the 100 ms pulse. The fast decay after the pulse corresponds with a reaction I type response and is consistent with calculations of a 2-compartment system (fig. 2C). From these measurements it is clear that the single chloroplast reacts heterogenously to the charge separation.

Part of the system reacts "classically" and can be described by the Goldman-Hodgkin-Katz equation for transmembrane diffusion of ions. We believe that our earlier measurements were done on chloroplasts which completely fit such a system description (fig. 3A). Such chloroplasts revealed only monophasic decay kinetics (reaction 1) after the light was turned off. Recent experiments of others (Remish et al. 1981) reveal the same type of kinetics. Calculations based on passive diffusion of ions show

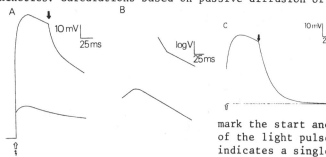

FIGURE 2. (A) Electric response of a *Peperomia metallica* chloroplast to a 100 ms saturating light pulse. The open and closed arrows mark the start and end, respectively of the light pulse. The zig-zag arrow indicates a single turnover saturating light flash, measured in the same chloroplast. (B) Semi-logarithmic plots of the electric responses after the light induced potential changes. (C) Calculated response of a two compartment system to a 100 ms saturating light pulse. The calculation is based on a passive ion redistribution.

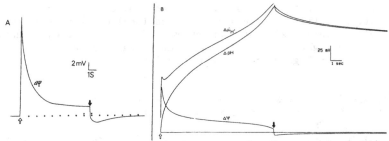

FIGURE 3. (A) "Classical" electric response of a *Peperomia metallica* chloroplast to an 8 second saturating light pulse. (B) Calculated response of a 2-compartment system like a chloroplast to a 10 second saturating light pulse. The calculation is based on the Goldmann-Hodgkin-Katz equation. Back pressure on the proton pump is assumed by the Nernst potential of protons. The majority of the buffer groups in the lumen are assumed to be at pK 5.2.

that such chloroplasts need seconds of saturating illumination in order for the electrochemical potential to supersede the phosphorylation potential (fig. 3B). It has been shown (Galmiche, Girault 1982) that ATP production starts within milliseconds after the light is turned on. In order to explain this discrepancy we postulate a third compartment in the chloroplast. This third compartment can be viewed either as lumenal areas or intramembraneous cavities, which are not at an isopotential with the bulk lumen nor with the stroma. The possibility of such a compartment was first hinted at by Hong and Junge (1983).

We have calculated the spreading of a proton concentration profile in an idealized situation. We consider a strongly appressed lumenal area (distance between membranes < 5 nm) with a high concentration of immobilized buffer groups. Into this area protons are pumped by a constant point source flux. The result is seen in fig. 4. It is seen that a high concentration of buffer groups, when immobilized, can keep the local concentration of free protons.

FIGURE 4. Calculated proton concentrations profiles in a plane as a function of distance to the source flux and of time. The source is considered to be a constant point flux of 10^{-12} mol/s. The diffusion coefficient for protons is taken to be 10^{-5} cm^2/s. The profiles are calculated for 2 situations:

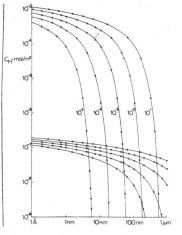

1. Profiles with a large amplitude are calculated on the assumption that the buffering groups are immobilized (e.g. protein residues).
2. Profiles with a small amplitude are calculated on the assumption that the buffering groups are about as mobile as protons in an aqueous phase.

It is assumed that the rate of the buffering reaction is much faster than the speed of diffusion of free protons. The buffering groups have an infinit buffering capacity. Of the total amount of protons pumped into the plane only a fraction 10^{-4} is left as free protons. The inset numbers give the time in seconds of the profile after the onset of the pump.

high. While when the buffer groups are about as mobile as the free protons the concentration profile is spread out and the local concentration is low. This could explain a kind of micro-chemiosmosis as proposed by Haraux et al. (1983), and might be consistent with multiphasic kinetics of the transmembrane potential generated by light activation.

REFERENCES

Bulychev AA, Andrianov VK, Kurella GA, Litvin FF (1973) in Biophysics of membranes, Proc. Symp. on Molecular Mechanisms of Permeability of Membrane Structures, Kaunas, USSR, 104-113.
Crank J (1975) The mathematics of diffusion 2nd ed. Clarendon Press, Oxford.
Galmiche JM, Girault G (1982) Synthesis of ATP induced in pea chloroplasts by single turnover flashes. FEBS Lett. 146, 123-128.
Heraux F, Sigalat C, Morau A, de Kouchkovsky Y (1983) The efficiency of energized protons for ATP synthesis depends on the membrane topography in thylakoids. FEBS Lett. 155, 248-252.
Hong YG, Junge W (1983) Localized or delocalized protons in phosphorylation? On the accessibility of the thylakoid lumen for ions and buffers. Biochim. Biophys. Acta 722, 197-208.
Remish D, Bulychev AA, Kurella GA (1981) Light induced changes of the electrical potential in chloroplasts associated with the activity of photosystem I and photosystem II. J. Exp. Bot. 32(130), 979-987.
Schapendonk AHCM, Vredenberg WJ, Tonk WJM (1979) Studies on the kinetics of the 515 nm absorbance changes in chloroplasts. FEBS Lett. 100, 325-330.
Schapendonk AHCM (1980) Electrical events associated with primary photosynthetic reactions in chloroplast membranes. Thesis Landbouwhogeschool Wageningen, ISBN 90 220 07561, Pudoc Wageningen.
Schultz SG (1980) Basic principals of membrane transport. Cambridge Univ. Press, Cambridge.
Van Kooten O, Gloudemans AGM, Vredenberg WJ (1983) On the slow component of P515 and the flash-induced reduction of cytochrome b563 in chloroplast membranes. Photobiochem. Photobiophys. 6(1), 9-14.
Vredenberg WJ (1976) Electrical interactions and gradients between chloroplast compartments and cytoplasm. In: The Intact Chloroplast, ed. Barber J. Elsevier Biomedical Press North Holland, 53-88.
Vredenberg WJ (1981) P515: A monitor of photosynthetic energization in chloroplast membranes. Physiol. Plant. 53, 598-602.
Westerhoff HV, Helgerson SL, Theg SM, Van Kooten O, Wikstrom MKF, Skulachev VP, Dancshazy Zs (1983) The present state of the chemiosmotic theory. Acta Biol. Acad. Sci. Hung.(in press).
Zimanyi L, Garab Gy (1982) Configuration of the light induced electric field in thylakoid and its possible role in the kinetics of the 515 nm absorbance change. J. Theor. Biol. 95, 811-821.

AKNOWLEDGEMENTS

We would like to thank W.J.M. Tonk for invaluable assistence with the micro-electrode measurements. This research was supported by the Dutch Foundation of Biophysics financed by the Netherlands Organization for the Advancement of Pure Research (Z.W.O.).

Authors address: Lab. of Plant Physiological Research, Agricultural University, Gen. Foulkesweg 72, 6703 BW Wageningen, The Netherlands.

THE KINETICS OF THE FLASH-INDUCED P515 ELECTROCHROMIC BANDSHIFT IN
DEPENDENCE OF THE ENERGETIC STATE OF THE THYLAKOID MEMBRANE AND THE ION
CONCENTRATION IN THE OUTER AQEOUS PHASE

ROBERT L.A. PETERS/OLAF VAN KOOTEN/WIM J. VREDENBERG

1. INTRODUCTION

The P515 electrochromic bandshift following a single saturating light
flash in dark-adapted and well preserved chloroplasts, shows multi-
phasic rise and decay kinetics (A.H.C.M. Schapendonk et al. 1979). These
kinetics cannot be explained in terms of generation and decay of a
transmembrane electric field. Such a field would be expected to follow
a single exponential dark decay with a rate constant determined by the
membrane capacitance and the membrane conductance, provided these are
field-independent (A.A. Bulychev, W.J. Vredenberg 1976). According to
Schapendonk and Vredenberg (A.H.C.M. Schapendonk et al. 1979), the P515
bandshift has components contributed by two different reactions called
reaction 1 and reaction 2. In their interpretation, reaction 1,
characterised by a fast rise (μs) and a subsequent single exponential
dark decay, is the reflection of the flash-induced formation of a
transmembrane electric field followed by a field dissipation in the
dark by a passive flux of ions. Reaction 2, however, might be related
to intramembranal electric phenomena. In this communication we present
results that indicate a different sensitivity of the reaction 1 and the
reaction 2 components of the overall flash-induced P515 bandshift
towards a) the extend of energization of the thylakoid membrane, and
b) the ion concentration in the outer aqeous phase.

2. MATERIAL AND METHODS

Freshly-grown spinach (Spinacia Oleracea) was used for all experiments.
The plants were grown as described previously (R.L.A. Peters et al. 1983).
Intact chloroplasts routinely were isolated according to a modified
method of Walker (W. Cockburn, D.A. Walker 1968) as described by
Schapendonk (A.H.C.M. Schapendonk 1980). Broken chloroplasts were
obtained by a 60 sec osmotic shock on ice, as described previously
(R.L.A. Peters et al. 1983). Determination of a) Hill reaction rate,
b) chlorophyll content, c) absorbance changes at 518 nm and d) ATP
hydrolysis in chloroplasts was performed as described previously (R.L.A.
Peters et al. 1983). For the determination of the flash-induced P515
response in chloroplasts at low ion concentrations, chloroplasts were
isolated in a medium containing: Tris/Mes 5 mM pH 7.8, $Mg(Ac)_2$ 1 mM and
Sorbitol 330 mM. Broken chloroplasts in this case were obtained by
osmotic shock in a medium containing 5 mM Tris/Mes pH 7.8 and 1 mM $Mg(Ac)_2$
and subsequent addition of an equal volume of a medium containing
Tris/Mes 5 mM pH 7.8, $Mg(Ac)_2$ 1 mM and sorbitol 660 mM.

3. RESULTS AND INTERPRETATION

The P515 bandshift following a saturating single turnover light flash
in dark-adapted chloroplasts in the presence of 50 mM Hepes/KOH,
2 mM $MgCl_2$ and 2 mM $MnCl_2$ is presented in Fig. 1 (dashed curve). From
this figure it can be seen that under these conditions the P515
bandshift shows multi-phasic rise and decay kinetics. According to

Sybesma, C. (ed.), Advances in Photosynthesis Research, Vol. II. ISBN 90-247-2943-2.
© *1984 Martinus Nijhoff/Dr W. Junk Publishers, The Hague/Boston/Lancaster.*

Schapendonk and Vredenberg (A.H.C.M. Schapendonk et al. 1979), this P515
bandshift has components contributed by two different reactions called
reaction 1 and reaction 2 (insert Fig. 1). Reaction 1 is characterised
by a fast (μs) rise and a subsequent single exponential dark decay with
a half life of about 75 ms (under these conditions). The response obtained
by subtracting reaction 1 from the overall P515 response has been
attributed to that of reaction 2. The rise kinetics of reaction 2 show
a slow absorbance increase within 150 ms after the flash and a subsequent
dark decay with a first order rate constant of the order of 1 s^{-1}.

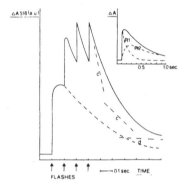 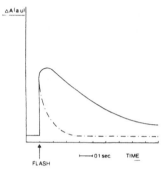

Fig.1: The flash-induced P515 bandshift in dark-adapted broken chloroplasts determined at pH: 7.8 in
the presence of 50 mM Hepes/KOH,
2 mM MgCl$_2$, 2 mM MnCl$_2$ and 330 mM
sorbitol.

Fig.2: The flash-induced P515 bandshift in dark-adapted broken chloroplasts (———) and in broken
chloroplasts after 40 sec of illumination (–·–·–·–) in the presence
of ATP and DTE.

In contrast with reaction 1, reaction 2 appears to be extremely sensitive
towards a pre-established energetic state of the thylakoid membrane.
This pre-energization may result either from light driven electron-
transport (i.e. the second slow rise is largely absent in the P515
response of a single flash following two preceding flashes separated by
100 ms dark time (Fig. 1 solid curve) or from ATP hydrolysis in the dark.
Activation of the chloroplast ATPase by preilluminating chloroplasts in
the presence of ATP and DTE, results in a complete suppression of
reaction 2 from the flash-induced P515 response determined 5 seconds
after the preillumination period (Fig. 2). Moreover, this suppression
appears to be temporary. The duration of the suppression period is
dependent on the amount of ATP present in the sample (R.L.A. Peters et
al. 1983). Under conditions of dark ATP hydrolysis, a light flash causes
only a reaction 1 type electrochromic shift (Fig. 2 dashed curve). The
half life of the dark decay is found to be insensitive to either an
increased number of flashes or to the length of the preillumination
period.

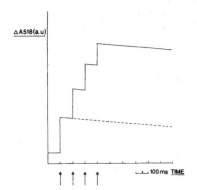

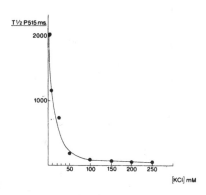

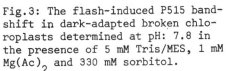

Fig.3: The flash-induced P515 band-shift in dark-adapted broken chloroplasts determined at pH: 7.8 in the presence of 5 mM Tris/MES, 1 mM Mg(Ac)$_2$ and 330 mM sorbitol.

Fig.4: The half life of the dark decay of the flash-induced P515 bandshift in dependence of the concentration KCl in the reaction medium.

A typical example of the flash-induced P515 bandshift in dark-adapted chloroplasts determined at low ion concentration (Tris/Mes 5 mM, Mg(Ac)$_2$ 1 mM) is presented in Fig. 3 (dashed curve). From this figure it can be seen that under these conditions the P515 bandshift is characterised by a fast rise (µs) followed by a single exponential dark decay with a half life of about 2000 ms. This half life of the dark decay could not be accelerated by illuminating the chloroplasts with a series of four flashes fired at time intervals of 100 ms (Fig. 3), solid curve). Also, under these conditions, the second slow rise typically for the presence of the reaction 2 component is absent. As expected, the half life of the dark decay under ion limited conditions appears to be unsensitive towards ionophores like valinomycin and nigericine, and only shows a reduced sensitivity towards gramicidine. On the other hand, the decay rate can be effectively accelerated by the addition of a protonophore like FCCP (data not shown).

From these experiments we conclude that the flash-induced P515 bandshift measured under ion limitation is in accordance with the reaction 1 type electrochromic shift as defined by Vredenberg, and can be interpreted as a flash-induced formation of a transmembrane electric field followed by a field dissipation in the dark by a passive flux of ions. The decay rate can be effectively accelerated by the addition of KCl (Fig. 4). Chloride appears to be the effective permeant ion. Acceleration of the dark decay was much less with KAc (data not shown).

Until now, we cannot explain the obvious absence of reaction 2 at low ion concentration. Surprisingly we found a remarkable alteration in the kinetics of the response after the addition of manganese. This is shown in Fig. 5. The magnitude of the initial rise in the presence of manganese is virtual because of the rapidity of the decay and the 15 ms time

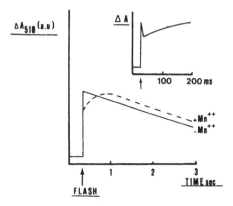

Fig.5: The flash-induced P515 band-shift in dark-adapted broken chloroplasts in the presence (– – – –) and absence (————) of Mn^{++}, at low ion concentration.

resolution at which it has been measured. The insert of the figure shows the first 200 ms of the response at a high (1 ms) resolution. From this figure it can be seen that the overall response in the presence of manganese is characterised by a fast rise (µs) followed by a fast decrease in the next 10 ms and a subsequent relatively slow increase in absorbance during the first 700 ms after the flash. The ultimate decay rate doesn't seem to be altered significantly. From these observations, it is clear that the kinetics of the P515 response determined at low ion concentration in the presence of manganese is essentially different from the kinetics determined under standard conditions (see Fig. 1). Moreover, the kinetics of the overall response under these conditions appeared to be unsensitive towards the extend of energization of the thylakoid membrane i.e. could not be altered significantly by a series of flashes fired at time intervals of 100 ms. As yet, the alterations of the kinetics of the P515 response upon the addition of Mn^{2+} cannot be understood. We presume that the slow rise induced by Mn^{2+} addition is associated with charge stabilization effects near the oxidizing site of PS 2. It might be that the presence of Mn is necessary in order to create the specific conditions in the membrane needed to stabilize intramembranal charged groups.

REFERENCES

Bulychev AA and Vredenberg WJ (1976) Effect of ionophores A23187 and nigericin on the light-induced redistribution of Mg^{++}, K^+ and H^+ across the thylakoid membrane, Biochim. Biophys. Acta 449, 48-58.
Cockburn W and Walker DA (1968) The isolation of spinach chloroplasts in pyrophosphate media, Plant Physiol. 43, 1414-1418.
Peters RLA, Bossen M, van Kooten O and Vredenberg WJ (1983) On the correlation between the activity of the ATP-hydrolase and the kinetics of the flash-induced P515 electrochromic bandshift in spinach chloroplasts, Journal of Bioenergetics and Biomembranes (in press).
Schapendonk AHCM, Vredenberg WJ and Tonk WJM (1979) Studies on the kinetics of the 515 nm absorbance changes in chloroplasts, FEBS Lett. 100, 325-330.
Schapendonk AHCM (1980) Electrical events associated with primary photosynthetic reactions in chloroplast membranes, Doctoral Thesis, Agricultural University Wageningen. ISBN 90 220 07561.

Authors address: Lab. of Plant Physiological Research, Agricultural University, Gen. Foulkesweg 72, Wageningen, The Netherlands

REDOX TITRATIONS OF THE FAST AND SLOW PHASE OF THE 515NM ELECTROCHROMIC ABSORPTION CHANGE IN CHLOROPLASTS

LINDA GIORGI, NIGEL PACKHAM AND JAMES BARBER

ARC Photosynthesis Research Group, Department of Pure and Applied Biology, Imperial College, London SW7 2BB (England).

1. INTRODUCTION

The fast phase (phase a) of the 515 nm electrochromic absorption change, ΔA_{515}, occurs as a result of the transmembrane electric field set up at the photochemical reaction centres upon flash activation (Witt 1979). An additional slow increase in the ΔA_{515} is thought to be due to a sub-sequent electrogenic step occurring during electron transport (Crowther, Hind 1980). Redox titrations of phase a have shown that both reaction centres contribute to the extent of the electric field (Malkin 1978; Diner, Delosme 1983), however the redox component responsible for phase b is less well characterized. Using chloroplast thylakoid membranes, redox titrations of both phase a and phase b have been carried out in order to investigate the various contributions to phase a under different salt conditions and to determine the component responsible for phase b.

2. MATERIALS AND METHODS

Redox titrations of the ΔA_{515} were carried out at $20^{\circ}C$, using the stand-ard criteria required for redox potentiometry (Dutton, Wilson 1971). The sample contained pea chloroplasts osmotically shocked on ice for. 30 sec-onds in 1.5 ml distilled water, in the presence of $MgCl_2$, and then an equal volume of double strength medium added to give final concentrations of 0.33 M sorbitol, 50 mM HEPES/KOH (pH 7.5), 5 mM $MgCl_2$ and 50 µg/ml chlorophyll. For the low salt condition, osmotic shocking was carried out in the presence of KCl to give a final concentration of 5 mM K^+. Each titration point represents the extent of the ΔA_{515}, measured using the apparatus described in Olsen and Barber (1980), resulting from an average of 8 flashes at a particular potential. Both reductive and oxi-dative titrations were carried out using, as titrants, solutions of 50 mM sodium dithionite, freshly prepared before each titration in 0.1 M Tris/HCl pH 9.0, and 50 mM ferricyanide, respectively. The following redox mediators were used, the values in parentheses being their $E_{m,7}$ in mV: ferricyanide (+430), DAD (+220), 1,2-naphtoquinone (+143), 1,4-naphto-quinone (+36), duroquinone (+5), 2,5-dihydroxybenzoquinone (-60), 2-hy-droxy-1,4-naphtoquinone (-139), anthroquinone-2,6-disulphate (-185) and anthroquinone-2-sulphonate (-225), all at conc. 25 µM; phenazine metho-sulphate (+80), at conc. 5 µM.

3. RESULTS

Fig. 1(a) shows a 515 nm electrochromic absorption change in which two phases (a and b) can be distinguished. On addition of 2,5-dibromo-3-methyl-6-isopropyl-p-benzoquinone (DBMIB), phase b is inhibited (Fig. 1(b)) indicating that the electron transfer step responsible for phase b is related to the oxidation of quinol by the Rieske FeS centre located within the cytochrome b-f complex. Subtraction of signals ± DBMIB reveals

Sybesma, C. (ed.), Advances in Photosynthesis Research, Vol. II. ISBN 90-247-2943-2.
© 1984 Martinus Nijhoff/Dr W. Junk Publishers, The Hague/Boston/Lancaster.

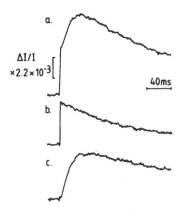

Fig. 1 Flash-induced 515 nm absorption change of stacked chloroplasts, prepared as described in Materials and Methods but in the absence of redox mediators and in the presence of 25 µM DCMU + 1 mM dithionite. (a) Control; (b) control + 15 µM DBMIB; (c) subtraction of trace (b) from trace (a).

the full extent of phase b and its kinetics (Fig. 1(c)).

The redox titrations of the extent of the fast phase a, in the presence and absence of 3-(3,4-dichlorophenyl)-1,1-dimethylurea (DCMU), are depicted in Fig. 2. Fig. 2(a) is for high salt conditions when the thylakoid membranes are stacked. In the absence of DCMU this titration curve follows two waves as the redox potential is lowered. Each wave can be fitted to an $n = 1$ theoretical Nernst curve and contributes a third to the total extent of the ΔA_{515}. The midpoint potentials, $E_{m,7.5}$, of each wave are +60 mV and -195 mV. A residual 515 nm change at $E_h < -400$ mV constitutes a third, non-titratable component. The +60 mV component is inhibited by DCMU, suggesting that it results from the PS2 contribution to the ΔA_{515}. The full extent of the -190 mV component is still observed in the presence of DCMU, but it is replaced by a non-titratable component in the presence of DCMU + methyl viologen (MV) (results not shown). Both these results suggest that this component is possibly associated with PS1. The low

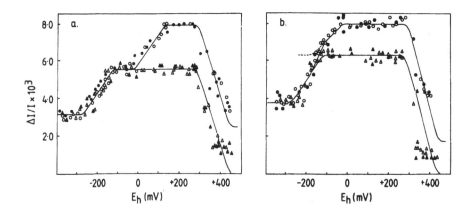

Fig. 2 Redox titrations of the initial (3 msec) extent of the flash-induced ΔA_{515} in (a) stacked and (b) unstacked chloroplasts. No additions, circle symbols; + 25 µM DCMU, triangle symbols. Reductive titrations, open symbols; oxidative titrations, closed symbols.

potential non-titratable wave is observed ± DCMU and is probably due to
a PS1 electron acceptor. At potentials >+300 mV, the ΔA_{515} is made
smaller, presumably due to the chemical oxidation of the PS1 electron
donor. This component contributes more than one third to the total ΔA_{515}
(-DCMU) and follows a single n = 1 Nernst curve with an apparent $E_{m,7.5}$
of +370 mV. The extent of this component remains unaltered in the presence of DCMU.

Three main features are apparent when similar redox titrations are carried out in the presence of low salt (Fig. 2(b)) at concentrations which
reduce the screening of the negative surface charges and hence results in
the unstacking of the thylakoid membranes: (1) There is an apparent increase in the extent of the low potential, DCMU insensitive, non-titratable component, indicating an increase in the PS1 contribution. (2) There
is a concomitant decrease in the DCMU sensitive component, suggesting a
decrease in the PS2 contribution. (3) There is a shift in the $E_{m,7.5}$ of
this latter component, from +60 mV (in Fig. 2(a)) to -100 mV. Although the
contributions of these different components has now been altered, n = 1
Nernst curves can still be fitted. The -195 mV component, however, remains unchanged both in its $E_{m,7.5}$ and extent, and it still exhibits DCMU
insensitivity. The increase in the PS1 contribution under these low salt
conditions, agrees with the observations in Olsen and Barber (1981), and
is also manifested by the increase in the extent of the wave at high potentials. An n = 1 Nernst curve can be fitted to this wave with an apparent $E_{m,7.5}$ = +360 mV and it is DCMU insensitive.

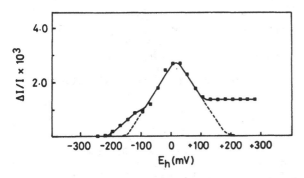

Fig. 3 Redox titration
curve for phase b using
stacked chloroplasts, ob-
tained by subtracting the
redox titration curve for
the extent of the ΔA_{515}
at 10 msec in the presence
of 25 µM DCMU + 15 µM
DBMIB from that obtained
with only 25 µM DCMU.

Fig. 3 shows the redox titration for the extent of phase b in the presence of DCMU. This is the result of subtracting the redox titration
curves obtained for the extent of the ΔA_{515} at 10 msec, in the absence
and presence of DBMIB. The main contribution to phase b appears with an
$E_{m,7.5}$ = +105 mV and it disappears with an $E_{m,7.5}$ = -75 mV. These components both fit n = 1 Nernst curves and follow a 60 mV/pH dependency.
These $E_{m,7.5}$ values correlate with the reduction of the plastoquinone
pool and that of cytochrome b-563, respectively. The shoulder obtained
at low potentials is believed to be the result of a contamination from
the -195 mV component while the contribution to phase b at high potentials
is possibly real. A similar titration curve to that in Fig. 3 is obtained for the phase b observed in the presence of DCMU + MV. This result indicates that cyclic electron flow is not an essential requirement
for the occurrence of phase b. Under low salt conditions, the titration
curve of Fig. 3 is again repeated, but with a smaller maximum extent, in

accordance with the observation of Olsen and Barber (1981).

4. DISCUSSION

The redox titrations under high salt conditions show that the contributions to the transmembrane electric field are equally shared between three components: one associated with PS2, $E_{m,7.5} = +60$ mV; one associated with PS1, $E_{m,7.5} < -400$ mV; and a component with an $E_{m,7.5} = -195$ mV, of unknown origin and exhibiting DCMU insensitivity, which we suggest to be associated with PS1. The extent of the high potential, $E_{m,7.5} = +370$ mV, wave confirms our interpretation of two types of PS1 acceptors. Assuming a linear relationship between absorption change and reaction centre concentration, the number of PS1 reaction centres either exceeds those of PS2 reaction centres or that the oxidation of P700 can lead to the translocation of two charges across the membrane involving a primary event followed by a rapid secondary electrogenic step, as suggested by Diner and Delosme (1983). Lowering the salt level has the effect of causing a shift in the $E_{m,7.5}$ of the PS2 component and also attenuates the extent of the PS2 contribution with a concomittant increase in the extent of the PS1 contribution. The cause of this effect is unclear, but may indicate that the primary PS2 electron acceptor, titrated here, is more susceptible to surface potential changes than the stromal PS1 electron acceptor. Alternatively, it is the gross conformational changes which occur (unstacking and randomisation of PS2) on lowering the salt level that affect the redox properties of the acceptor or the equilibrium properties of the mediators.

Redox titrations of phase b show that its appearance occurs with an $E_{m,7.5} = +105$ mV and that its disappearance occurs with an $E_{m,7.5} = -75$ mV. This indicates that phase b can be readily detected under redox potential conditions when the plastoquinone pool is reduced and cytochrome b-563 is oxidized prior to flash excitation. This, and other unpublished results, implicate phase b with the re-oxidation of the flash-reduced cytochrome b-563.

REFERENCES

Crowther D and Hind G (1980) Partial characterization of cyclic electron transport in intact chloroplasts, Arch. Biochem. Biophys. 204, 568-577.
Diner BA and Delosme R (1983) Oxidation-reduction properties of the electron acceptors of Photosystem II, Biochim. Biophys. Acta 722, 443-451.
Dutton PL and Wilson DF (1974) Redox potentiometry in mitochondrial and photosynthetic bioenergetics, Biochim. Biophys. Acta 346, 165-212.
Malkin R (1978) Oxidation-reduction potential dependence of the flash-induced 518 nm absorbance change in chloroplasts, FEBS Lett. 87, 329-333
Olsen LF and Barber J (1981) Origin of the slow component of the electrochromic shift: a charge delocalisation model, FEBS Lett. 123, 90-94.
Witt HT (1979) Energy conversion in the functional membrane of photosynthesis, Biochim. Biophys. Acta 505, 355-427.

REDOX TITRATION OF THE SLOW ELECTROCHROMIC PHASE IN CHLOROPLASTS

M. Girvin and W. A. Cramer, Dept. of Biological Sciences, Purdue University, W. Lafayette, IN 47907

SUMMARY

The amplitude of the slow phase of the electrochromic bandshift has been measured as a function of redox potential between +200 and -200 mV. Reduction of a quinone-like donor with an E_m=+100 mV is required for generation of the slow phase. 80-100% of the amplitude of this signal with a $t_{\frac{1}{2}}$ = 4-7 msec is observed at -200 mV, where cyt $\underline{b}_6$, as well as any electron acceptor for it, would be fully reduced. At this potential, there is a small, slower absorbance change that might correspond to $\underline{b}_6$ oxidation, and has a $t_{\frac{1}{2}}$ = 10-15 msec. It is concluded that the slow phase of the electrochromic bandshift does not arise from the operation of a mechanism involving turnover of cyt $\underline{b}_6$.

INTRODUCTION

A slow (msec) component of the electrochromic bandshift was observed in algae by Joliot and Delosme (1974) and shown to be associated with PS I electron transfer. Several mechanisms have been proposed to explain the existence of the slow phase of the bandshift. Most of the models are variations of Mitchell's Q cycle hypothesis (Mitchell, 1975) in which the slow phase arises from one of the two electrons released upon PQH_2 oxidation recrossing the membrane and generating a field (Velthuys, 1978; Bouges-Bocquet, 1981; Crowther and Hind, 1982; and Selak and Whitmarsh, 1982). In most of these models this path of transmembrane electron transfer is thought to include cyt $\underline{b}_6$. For generation of the slow phase, one would then need reduced donor, oxidized acceptor, and a hole in the Rieske FeS/cyt $\underline{f}$ region generated by PS I turnover. We undertook a redox study of the slow phase to characterize the donor and acceptor of this proposed cycle.

METHODS

Broken chloroplasts were prepared from spinach. Absorption changes were measured using a single beam spectrophotometer with a dye laser providing the actinic flashes. All traces are the average of 10 runs. The flash frequency was 0.1 Hz. The slow phase was measured as the difference between the signals obtained in the presence and absence of 4 μM DBMIB, both recorded at 515 nm. Cyt $\underline{b}_6$ turnover was measured as the difference between traces recorded at 563 and 570 nm. Redox poise was measured using a Metrohm EA 259 Ag/AgCl combination electrode in the presence of mediators: 1,2 naphthoquinone, 1,4 naphthoquinone, menadione, anthraquinone-2,6-disulfonate and anthraquinone-2-sulfonate.

RESULTS

Figure 1 shows a redox titration of the amplitude of the slow phase of the electrochromic bandshift at pH 7. At high potentials the amplitude is very small and increases to a maximum with a midpoint of +100 mV. The midpoint and slope of the donor are consistent with its being a PQ species. Figure 1 also shows that the slow phase persists at low potentials, well below the E_m (0 to -100 mV) of cyt $\underline{b}_6$. Most of the amplitude is present

Sybesma, C. (ed.), Advances in Photosynthesis Research, Vol. II. ISBN 90-247-2943-2.
© 1984 Martinus Nijhoff/Dr W. Junk Publishers, The Hague/Boston/Lancaster.

even at −200 mV. Figure 2 shows a comparison of the slow phase at +30 mV
and −218 mV. Both traces have similar kinetics, although the $t\frac{1}{2}$ at −218 mV
may be 1–2 msec longer than at +30 mV ($t\frac{1}{2}$ = 4 msec). The amplitude and
kinetics of the 515 nm signal on the first and last flash of the series
were the same. A spectrum of the slow phase at −130 mV is similar to
published spectra for the electrochromic bandshift (data not shown).

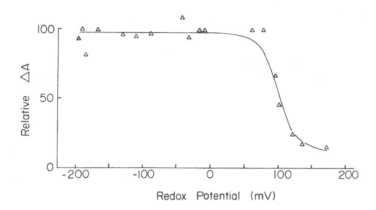

Figure 1. Amplitude of the slow phase of the 515 nm absorption change
as a function of redox potential at pH 7.

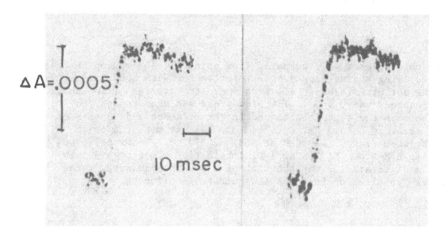

Figure 2. Time course of the slow phase of the electrochromic bandshift
measured at +30 mV (left) and −218 mV (right).

Cyt $\underline{b}_6$ turnover was also measured to determine if the cytochrome was being oxidized by the flash at negative potentials or if it were inaccessible to the redox dyes and was being photoreduced even at these potentials. Figure 3 shows cyt $\underline{b}_6$ turnover at +10 mV and −140 mV. At +10 mV 0.4 molecules $\underline{b}_6$/600 chl were reduced by a flash and full amplitude of the slow phase was seen. At −140 mV, negligible $\underline{b}_6$ reduction was observed, an absorbance change corresponding to photooxidation of no more than 0.3 molecules $\underline{b}_6$/600 chl with a $t^{\frac{1}{2}}$ = 10–15 msec was observed, and 80 to 100% of the maximum amplitude of the slow phase was still present (Figs. 1 and 2).

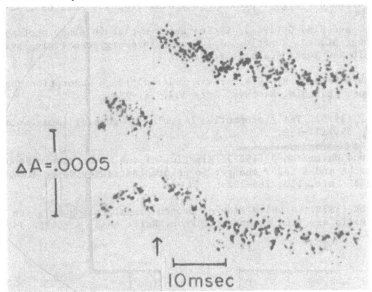

Figure 3. Cytochrome $\underline{b}_6$ turnover (563–570 nm) measured at +10 mV (above) and −140 mV (below).

DISCUSSION

The reduction of a quinone-like species with an E_{m7} of +100 mV is required for the generation of the slow phase of the electrochromic bandshift in spinach chloroplasts. This species, 'U' of Bouges-Bocquet (1981) would be analogous to Q_z in the photosynthetic bacteria. The slow phase is still observed with a halftime of 5–7 msec under conditions ($E_h \simeq$ −200 mV) where cyt $\underline{b}_6$ is reduced before the flash. If the $\underline{b}_6$ is turning over, it shows only a slow ($t^{\frac{1}{2}}$=10–15 ms) oxidation following the flash. At these low potentials any acceptor species thermodynamically capable of oxidizing the semiquinone formed by reduction of the Rieske center should be reduced before the flash. In particular, cyt $\underline{b}_6$ as well as any electron acceptor of this cytochrome, should be initially reduced. These data would indicate that the slow phase is not the result of Q cycle activity. The only component on the donor side of PS I whose turn-over after a single flash would require a reduced donor, and which would continue to turn over at the low redox potentials of this experiment, would be the Rieske FeS center. Hence, we would propose that the reduction of this component, which could be linked to an H^+ pump, is responsible for formation of the slow phase of the bandshift.

ACKNOWLEDGMENT

This research was supported by NSF grant PCM-8022807. We thank Ruth Rafferty for help in preparation of the manuscript.

REFERENCES

Bouges-Bocquet B (1981) Factors regulating the slow electrogenic phase in green algae and higher plants. Biochim. Biophys. Acta 635, 337-340.

Crowther D and Hind G (1982) Cycles and Q-cycles in plant photosynthesis In Trumpower BL, ed. Function of quinones in energy conserving systems, pp. 499-510 New York: Academic Press.

Joliot P and Delosme R (1974) Flash-induced 519 nm absorption change in green algae. Biochim. Biophys. Acta 357, 267-284.

Mitchell P (1975) The protonmotive Q cycle: a general formulation. FEBS Lett. 59, 137-139.

Selak MA and Whitmarsh J (1982) Kinetics of the electrogenic step and cytochrome b_6 and f redox changes in chloroplasts: evidence for a Q cycle. FEBS Lett. 150, 286-292.

Velthuys BR (1978) A third side of proton translocation in green plant photosynthetic electron transport. Proc. Natl. Acad. Sci. USA 75, 6031-6034.

ON THE ORIGIN OF THE SLOW ELECTRIC POTENTIAL COMPONENT INDUCED BY
FERREDOXIN-MEDIATED CYCLIC ELECTRON TRANSFER IN PHOTOSYSTEM I-ENRICHED
SUBCHLOROPLAST VESICLES

FONS A.L.J. PETERS, GUUS A.B. SMIT AND RUUD KRAAYENHOF

1. INTRODUCTION

Cyclic electron transfer around photosystem I (PSI) can be considered as
an important energy-conserving pathway in plant photosynthesis. Its
operation requires ferredoxin as a cofactor while its activity is
strongly dependent on a proper redox balance of the electron carriers.
The arrangement of electron carriers has been only partly characterized
(Crowter, Hind, 1980; Chain, 1982) and is still under investigation. In
particular, schemes of non-serial transfer of electrons from ferredoxin
to P-700 (i.e. Q- and b-cycles) have been proposed to account for an
extra electrogenic step as monitored by the electrochromic carotenoid
absorbance changes. We have studied ferredoxin-mediated cyclic electron
transfer and electric potential generation in PSI-enriched subchloroplast
vesicles that contain all components involved in the native system except
ferredoxin, in amounts originally present in the stroma lamellae (Peters
et al., 1983a). In the experimental system of PSI vesicles, supplemented
with ferredoxin, the activity of cyclic electron transfer can be
adjusted by varying the concentrations of NADPH, $NADP^+$ and O_2 (Peters
et al., 1983b). In this way we studied the flash-induced slow carotenoid
response in relation to redox changes of cytochromes b-563 and c-554.
We present evidence for a reductant-induced oxidation of cyt b-563, but
the slow potential component has its origin before the reoxidation of
b-563.

2. MATERIALS AND METHODS

PSI vesicles were prepared as described before (Peters et al., 1983a).
Flash experiments at 20°C were carried out as described previously
(Peters et al., 1983b); signals of 8-16 flashes fired at a frequency of
0.125 Hz were averaged. The reaction mixture contained 5 mM TES buffer
(pH 7.8), 2.5 mM KH_2PO_4, 20 mM KCl, 20 mM NaCl, 5 mM $MgCl_2$, 5 µM
ferredoxin, 300 µM NADPH and chlorophyll at a final concentration of
50 µg/ml. O_2 concentration was maintained at about 100 µM by periodical
stirring.

3. RESULTS AND DISCUSSION

Under control for optimal redox poising of the ferredoxin-mediated cyclic
system a very prominent slow-rising component, $\Delta A518$(slow), is present
in the flash-induced carotenoid response (Fig. 1A). In this reconstituted
system the carotenoid transient is qualitatively similar to that in
intact chloroplasts (Crowther, Hind, 1980) and whole algae (Bouges-
Bocquet, 1981) and remains identical for at least 30 min. To investigate
the function of cytochromes in the cyclic system spectra of flash-
induced absorbance changes were obtained throughout the cytochrome
α-band region, a representative serie of which is shown in Fig. 1B. The
extents of absorbance changes relative to the initial dark level, at
selected time intervals after the flash, are plotted against wavelength
of measurement in Fig. 2. Whereas in particular the fast carotenoid

Sybesma, C. (ed.), Advances in Photosynthesis Research, Vol. II. ISBN 90-247-2943-2.
© *1984 Martinus Nijhoff/Dr W. Junk Publishers, The Hague/Boston/Lancaster.*

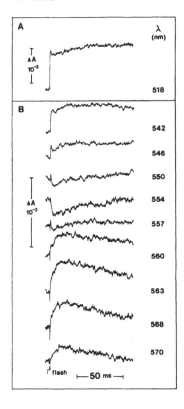

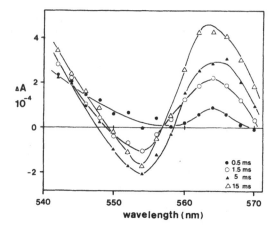

FIGURE 2. Spectra of absorbance changes in the cytochrome α-band region at selected time-intervals after the flash. The extents of absorbance changes, relative to the dark level, are plotted against wavelength of measurement.

FIGURE 1. Flash-induced absorbance changes in PSI vesicles at 518 nm (A) and in the cytochrome α-band region (B).

response disturbs the cytochrome responses below 556 nm, it is obvious from these spectra that transients following 0.5 ms after the flash are primarily due to redox changes of cyt c-554 and b-563. The reduction of b-563 (Fig. 1B) appears to have a biphasic character which is also observed in the reoxidation kinetics (computer-deconvoluted, not shown). Cyt c-554 rereduction kinetics, after the rapid flash-induced oxidation, are single exponential. The biphasic increment in absorbance at 563 nm was also observed in intact chloroplasts, the fast component being interpretated as a result of redox changes of plastocyanin (Crowther, Hind, 1980). The spectrum of absorbance changes in this wavelength region (Fig. 2) indicates that in PSI vesicles the absorbance changes between 0.5 and 1.5 ms are mainly due to a fast reduction of b-563, but some interference from spectral changes of other origin we cannot exclude. Biphasic kinetics in the reoxidation of b-563 were also observed in chloroplasts (Selak, Whitmarsh, 1982).

Fig. 2 (cf. Fig. 1B) shows that the sequence of redox changes after the flash is (1) fast reduction of b-563 ($t_{0.5}$ = 0.5 ms) (2) oxidation of c-554 ($t_{0.5}$ = 2 ms) (3) slower reduction of b-563 ($t_{0.5}$ = 4 ms) (4) rereduction of c-554 ($t_{0.5}$ = 30 ms) and (5) reoxidation of b-563 ($t_{0.5}$ = 90 ms).

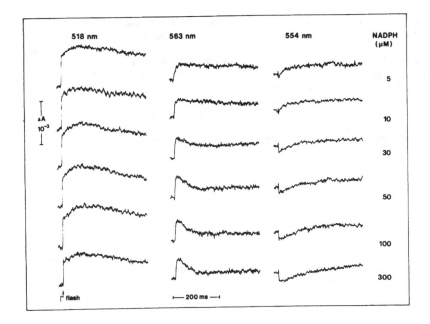

FIGURE 3. Flash-induced absorbance changes at 518, 563 and
554 nm as a function of the concentration of NADPH.

Remarkably, the fast reduction of b-563 appears to preceed the flash-
induced oxidation of c-554. Reduced ferredoxin or another reductant on
the acceptor side of PSI is probably responsible for the fast reduction
of b-563. Relatively little cytochrome b-563 and c-554 is seen to
turnover, about 25% of the content of each; the mutual ratio is about
2 (measured 14 times in different preparations), indicating that for
one c-554 two b-563 cytochromes turnover. The slow-rising component in
the carotenoid response has a $t_{0.5}$ of about 35 ms (Fig. 1), about
similar to $t_{0.5}$ of the rereduction of c-554 (30 ms), but significantly
smaller than $t_{0.5}$ of the reoxidation of b-563 (90 ms). So, it seems
unlikely that the reoxidation of b-563 is responsible for ΔA518 (slow)
as proposed for chloroplasts by Selak and Whitmarsh (1980).
As previously shown (Peters et al., 1983b), the electron pressure on
(i.e. redox balance of) the cyclic chain in PSI vesicles can be
manipulated by altering the input and output of electrons at the level
of ferredoxin, PSII activity being absent. This is achieved by varying
the concentrations of NADPH, $NADP^+$ and O_2. Fig. 3 shows the effect of
increasing NADPH concentrations on the flash-induced ΔA518 and cyt-b563
and c-554. The enhancement of b-563 reoxidation at higher NADPH
concentrations is very prominent, whereas the reduction level of b-563
is independent of NADPH. The flash-induced oxidation of c-554 is larger
at higher NADPH concentrations, suggesting incomplete reduction of c-554
in the dark. However, the rereduction of c-554 is very fast at lower
NADPH concentrations which suggest that the oxidation of c-554 is
masked by its fast rereduction. The redox state of P-700 in the dark
seems similar in all experiments, as indicated by the constant PSI

charge seperation-sensing fast carotenoid absorpbance change.(Fig. 3).
These results are not in harmony with a serial transfer of electrons
from b-563 to c-554 and further on to P-700. At the lower concentrations
of NADPH, a pool of electrons seems to exist on cyt b-563 which is not
easily transferred to following electron carriers. So, in contrast to a
serial transfer of electrons from b-563 to c-554, as predicted by the
Z-scheme, oxidizing equivalents generated by PSI do not drive the
oxidation of b-563, in accordance with findings in chloroplasts (Chain,
1982). We suggest that both the fast and slow reoxidation of b-563 are
strongly dependent on the dark reduction state of some component,
presumable a semiplastoquinone. This "reductant-induced oxidation" of
cyt b-563 points to the existence of a Q- or b-cycle of electrons during
cyclic electron transfer (Mitchell, 1976; Wikström, Krab, 1981).
However, the oxidant-induced reduction of b-563, which would complete
the Q- or b-cycle, cannot be derived from the observed sequence of
events. As is shown by Figs. 1B and 2, the slow reduction of b-563
($t_{0.5}$ = 4 ms) is faster than the rereduction of c-554 ($t_{0.5}$ = 30 ms)
as is also observed by others in chloroplasts (Crowther, Hind, 1980;
Selak, Whitmarsh, 1982). So, modification of a Q- or b-cycle is required,
as stated by Crowther and Hind (1980) (b-563 functions only as a complex
capable of storing electrons from semiplastoquinone) or by Selak and
Whitmarsh (1982) ($PQH_2/PQH^{\cdot}$ couple reduces b-563 and $PQH^{\cdot}/PQ$ couple
reduces the FeS-center and c-554). As Fig. 3 shows (confirming
conclusions from Fig. 1) $\Delta A518$ (slow) does not correlate with the
reoxidation of b-563, because the carotenoid response is rather constant
at variable rates of b-563 reoxidations. Therefore we propose a
generation site concomitant with the slow reduction of b-563.

REFERENCES

Bouges-Bocquet B (1981) Factors regulating the slow electrogenic phase
in green algae and higher plants, Biochim. Biophys. Acta 653, 327-340.
Chain RK (1982) Evidence for a reductant-induced oxidation of chloroplast cytochrome b-563, FEBS Lett. 143, 273-278.
Crowther D and Hind G (1980) Partial characterization of cyclic electron
transport in intact chloroplasts, Arch. Biochem. Biophys. 204, 568-577.
Mitchell P (1976) Possible molecular mechanisms of the protonmotive
function of cytochrome systems, J. Theor. Biol. 62, 327-362.
Peters FALJ, Van Wielink JE, Wong Fong Sang HW, De Vries S and
Kraayenhof R (1983) Studies on well coupled photosystem I-enriched
subchloroplast vesicles. Content and redox properties of electron-transfer
components, Biochim. Biophys. Acta 722, 460-470.
Peters FALJ, Van Spanning R and Kraayenhof R (1983) Studies on well
coupled photosystem I-enriched subchloroplast vesicles. Optimization of
ferredoxin-mediated cyclic photophosphorylation and electric potential
generation, Biochim. Biophys. Acta 724, 159-165.
Selak MA and Whitmarsh J (1982) Kinetics of the electrogenic step and
cytochrome b_6 and f redox changes in chloroplasts, FEBS Lett. 150,
286-292.
Wikström M and Krab K (1980) Respiration-linked H^+ translocation in
mitochondria: stoichiometry and mechanism, Curr. Top. Bioenerg. 10, 51-101.

Authors address: Biological Laboratory, Vrije Universiteit, De Boelelaan
1087, 1081 HV Amsterdam, The Netherlands.

CATION BINDING TO THYLAKOID MEMBRANES : ASSOCIATION CONSTANT AND EFFECT
ON TRANSMEMBRANE POTENTIALS

MAURO S., BOTTE P., LANNOYE R. and HURWITZ H.
Laboratoire de Physiologie végétale, Université Libre de Bruxelles,
28, av. Paul Heger, 1050 Bruxelles, Belgique.

1. INTRODUCTION

Studies with model membranes clearly indicate that cationic interactions
with surface negative charges have to be understood through both an
unspecific screening effect in the aqueous diffuse layer adjacent to the
surface and a specific adsorption to the charged molecules (Mc Laughlin
1977). This dual action has been shown to be well quantitatively described
by a Gouy-Chapman-Stern analysis (Lau et al. 1981).
In contrast, with thylakoid membranes, where a large number of processes
are controlled by surface electrical charges (Barber 1982), specific
adsorption has not been taken into account. This assumption was supported
by the lack of specificity between cations of the same valency group in
bringing about stacking and the high fluorescence state of thylakoid
membranes.
We have approached the problem by compairing the effect of mono and divalent
cations on 9-aminoacridine fluorescence. We present experiments which
investigate the cationic binding and propose a method for the evaluation of
the association constant.

2. MATERIALS AND METHODS

Pea chloroplasts were isolated as described previously (Chow, Barber 1980)
Fluorescence from 9-aminoacridine was observed with a Farrand MKII fluorescence spectrophotometer on excitation at 390 nm and with the emission
monitored at 450 nm at 90° to the excitation beam.

3. RESULTS

9-aminoacridine (9AA) is a fluorescent probe of the diffuse layer adjacent
to the negatively charged thylakoid membranes (Searle et al. 1977). This
dye carries a positive charge at neutral pH. Its accumulation at the
membrane surface is paralleled by a decrease of its fluorescence yield
Cations displace 9AA from the surface and stimulate the sample fluorescence
Results of Fig. 1 are taken as an evidence of the additional ability of
divalent metal cations to bind to negative groups and consequently to be
more efficient than bulkier organic cations like methylviologen in
stimulating 9AA fluorescence.
The order of effectiveness being $Ca^{++}>Ba^{++}>Sr^{++}>Mg^{++}$

One of us (P.Botte) received a fellowship from the Institut pour
l'Encouragement de la Recherche Scientifique dans l'Industrie et
l'Agriculture.

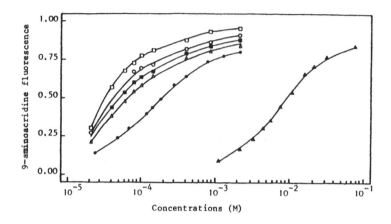

Fig. 1 - Release of 9-aminoacridine fluorescence quenching by cations chloroplasts were suspended in 0.1 M Sorbitol, 1 mM Hepes, 20 μM 9AA, pH 7.5 (0.7 mM KOH). 9AA fluorescence is normalized to the maximum of fluorescence induced by 20 mM CaCl₂.
▲ KCl ; ● methylviologen ; △ MgCl₂ ; ■ SrCl₂ ; ○ BaCl₂ ; ☐ CaCl₂.

Table I shows that Mg^{++} effect is more temperature dependent than Ca^{++} and methylviologen. Methylviologen being used as a reference which reflects the increased diffuseness of the ionic atmosphere near the membrane at higher temperatures.

Table I - Temperature dependence of cation induced release of 9AA fluorescence quenching.

	$CaCl_2$ (M)		$MgCl_2$ (M)		Methylviologen (M)	
	10^{-4}	$6\ 10^{-4}$	10^{-4}	$6\ 10^{-4}$	$1.2\ 10^{-4}$	$7.2\ 10^{-4}$
4°	68.12	83.68	56.33	74.33	60	73
38°	75.47	83.96	71	84.33	64.67	73
	+ 10%	+ 1%	+ 26%	+ 13%	+ 6%	0%

4. DISCUSSION

1°/ Basis for surface charge density and association constant evaluation

In the Gouy Chapman theory, the surface potential (ψ_o) is related to the surface charge density (σ) by :

$$\sigma = (8 \ RT \varepsilon_r \varepsilon_o C_\infty)^{\frac{1}{2}} \ \sinh \ (\frac{F\psi_o}{2RT}) \qquad \ldots \ 1$$

where $R, T, \varepsilon_r, \varepsilon_o$ have their usual meaning. C_∞ is the bulk concentration of 1:1 salt.

The concentrations of divalent cations at the membrane surface (c_o^{2+}) and in the bulk are related through the Boltzmann equation :

$$c_\infty^{2+} = c_o^{2+} \ \exp \ (- \ 2 \ F\psi_o \ / \ RT) \qquad \ldots \ 2$$

Assuming that divalent cations are forming 1:1 complexes with the surface negative groups (P^-), the intrinsinc association constant (K_2) is given by :

$$K_2 = \frac{\left[c^{2+}, P^- \right]}{\left[P^- \right]\left[c_o^{2+} \right]} \qquad \ldots \ 3$$

where P^- and $\left[c^{2+}, P^- \right]$ are the surface concentration of free and bound negative groups.

As explained above 9AA fluorescence is a reflection of the surface potential. Chow and Barber (1980) assumed that ψ_o was identical for an equal release of 9AA fluorescence quenching induced either by a mono (c_∞^+) or a divalent cation (c_∞^{2+}). Under these conditions and providing that only diffusible cations are used, σ_i is given by :

$$\sigma_i = - \ \{2 \ \varepsilon_o \varepsilon_r RT \ \left| (c_\infty^+)^2 - 4 \ c_\infty^+ . c_\infty^{2+} / c_\infty^{2+} \right| \}^{\frac{1}{2}} \qquad \ldots \ 4$$

The value of σ_i for $F/F_{Max} = 0.5$ is equivalent to $\simeq 40 \ 10^{-3} \ C/m^2$.

Knowing σ_i (Eq 3) it is possible to calculate ψ_o for any concentration of methylviologen (Eq 1) and consequently to calculate the surface charge density (σ_b) associated with the concentration of the binding species giving the same release of fluorescence quenching. Thus by combining Eq 1 to 3, it is possible to calculate :

$$K_2 = 2.3 \ M^{-1} \ \text{for} \ Mg^{2+} \ \text{and} \ K_2 = 3.9 \ M^{-1} \ \text{for} \ Ca^{2+}.$$

In our calculations we have ignored the possible binding of K^+ which would probably affect essentially the value of σ_i since our basic medium for binding measurements contained very low levels of K^+ ($\simeq 0.7$ mM).
However calculated values are in a fair agreement with published values for association constant of Mg^{2+}/Ca^{2+} with phosphatidylglycerol liposomes

$$(Mg^{2+} : 6.0 \ M^{-1} ; \ Ca^{2+} : 11.5 \ M^{-1}) \ \text{(Lau et al. 1981)}.$$

2°/ Possible mechanism of ionic binding to thylakoid membranes

The association constant is rather insensitive to the nature of the charged surface. One possible explanation lies in the fact that divalent cations form outer sphere complexes (Lau et al. 1981).
In that case, the ionic sequence illustrated in Fig. 1 and Table 1 would reflect the relative ability of each ionic species to accomodate its hydraticn shell with the structured water layers at the membrane surface.

5. CONCLUSION

Our main conclusion is that cations complexe to a significant amount the negative groups exposed on the outer membrane surface of thylakoids. Strong ionic binding would also greatly reduced the development of any surface transmembrane potential in either dark or light conditions.

REFERENCES

Barber J. (1982) Influence of surface charges on thylakoid structure and function. Ann. Rev. Plant. Physiol. 33 : 261-95
Basolo F., Pearson R.G. (1967) Mechanism of inorganic reactions. John Wiley and Sons, Inc.
Chow W.S., Barber J. (1980) Salt dependent changes of 9-aminoacridine fluorescence as a measure of charge densities of membranes surfaces. J. Biochem. Biophys. Methods 3 : 173-185.
Lau A., Mc Laughlin A., Mc Laughlin S. (1981) The adsorption of divalent cations to phosphatidylglycerol bilayer membranes. Biochim. Biophys. Acta 645 : 279-292.
Mc Laughlin (1977) Electrostatic potentials at membrane-solution interfaces. Curr. Top. Membr. Transp. 9 : 71-155.
Searle G.F.M., Barber J., Mills J.D. (1977) 9-aminoacridine as a probe of the electrical double layer associated with the chloroplast thylakoid membrane. Biochim. Biophys. Acta 461 : 413-425.

TEMPERATURE DEPENDENCE OF ÉLECTRIC EVENTS IN THYLAKOID MEMBRANES: COMPARISON
OF MEMBRANE POTENTIAL, SURFACE POTENTIAL AND AMINOACRIDINE BINDING

RUUD KRAAYENHOF, JOSÉ M.G. TORRES-PEREIRA, FONS A.L.J. PETERS AND
HARRO W. WONG FONG SANG

1. INTRODUCTION

Various methods are available for monitoring electric potential generation in
illuminated photosynthetic membranes. These include the use of electrodes,
free-flow electrophoresis, distribution or binding of amphiphilic dyes and
optical changes of intrinsic or added field-sensitive probes. Some of these
events seem to be correlated kinetically but without certainty about their
precise mechanism and origin they remain difficult to accomodate in a single
physical model. As part of such correlation studies we have determined the
temperature dependence of some of these electric events that are thought to
reflect thylakoid membrane potentials or interfacial electric potentials.

2. MATERIALS AND METHODS

For the preparation of broken spinach chloroplasts and PS I subchloroplast
vesicles and the techniques and materials used here one is refered to Schapendonk et al. (1980), Schuurmans et al. (1981) and Peters et al. (1983).
Optical measurements were done with a thermoelectrically regulated multipurpose cuvette described elsewhere (Kraayenhof et al., 1982). Further details
are given in the legends to the Figures.

3. RESULTS AND DISCUSSION

The flash-induced slow-rising carotenoid absorbance transient at 518 nm is
thought to reflect electrogenic proton displacement by the plastoquinone-
cytochrome b-563.c-554 redox complex (Velthuys, 1979). The extrinsic probe
oxonol VI was found to behave similarly (Peters et al., 1983). Fig. 1 shows
that these responses have a rather broad temperature optimum between 10 and
25 °C in PS I vesicles; in broken chloro-
plasts similar results were obtained. Fig.
1 also shows the decrease of the oxonol VI
half-rise time with increasing temperature.
Both rise and decay kinetics show major
temperature transitions around 19 °C.

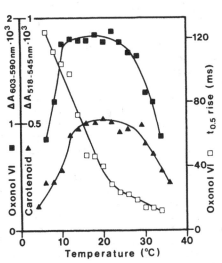

FIGURE 1. Temperature dependence of membrane potential-indicating flash-induced
oxonol VI and slow carotenoid absorbance
changes in PS I vesicles. The reaction
medium contained 20 mM NaCl, 20 mM KCl,
2.5 mM KH_2PO_4, 5 mM $MgCl_2$, 5 mM TES buffer,
pH 8.0, 5 μM ferredoxin, 0.5 mM NADPH, 0.1
mM O_2 and 50 μg Chl/ml. Transients of 8
flashes, fired at 0.125 Hz, were averaged.

Sybesma, C. (ed.), Advances in Photosynthesis Research, Vol. II. ISBN 90-247-2943-2.
© *1984 Martinus Nijhoff/Dr W. Junk Publishers, The Hague/Boston/Lancaster.*

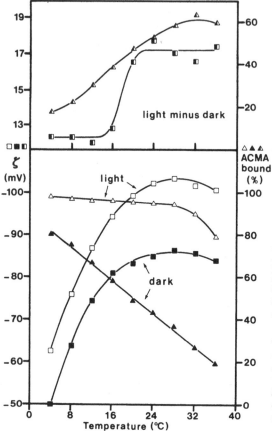

FIGURE 2. Temperature dependence of electrokinetic zeta potential and ACMA binding in broken chloroplasts and the effect of energization. The free-flow electrophoresis medium contained 345 mM sorbitol, 2 mM Tricine buffer, pH 8.0, 20 μM methylviologen and 10 μg Chl/ml. ACMA binding was determined by fluorescence measurements in the same medium with 40 μg Chl/ml and 5 μM ACMA.

The negative electrokinetic zeta potential (ζ) at the hydrodynamic plane of shear of chloroplasts, measured by free-flow electrophoresis, has a similar temperature optimum as the membrane potential (Fig. 1) and shows a pronounced decline below about 20 °C, as is illustrated by Fig. 2. Its light-induced increment shows a transition at about 18 °C. Increasing Mg^{2+} concentrations reduce the oxonol and slow carotenoid changes and zeta potential in the same way (not shown). At the lower temperatures the ionic double layer will become more compact due to the slower thermal motion, the result being equivalent to an increased concentration of counterions in the medium (Mysels, 1959). Apparently, the more effective surface charge screening at lower temperatures is sensed by the field-sensitive probes in the membrane and by net charge density at the plane of shear in the same way.

Fig. 2 also shows the binding behavior of the fluorescent cationic 9-amino-6-chloro-2-methoxyacridine (ACMA) in dark and light as function of temperature. In the dark, ACMA binding linearly increases at decreasing temperature. While at fixed temperature the electrostatic interaction of this type of probe can be taken as a measure of surface charge density (Kraayenhof, 1977) its increased binding at low temperature does not reflect an increased membrane potential (Fig. 1) or ζ potential (Fig. 2). We interpret this as participation of ACMA in the more effective screening of surface anionic groups at lower temperature. In the light, almost complete binding of the added ACMA occurs between 4 and 28 °C. Consequently, the energy-linked extra binding increases with temperature as well. The kinetics of the light-induced ACMA binding and of its dark-release show temperature transitions around 19 °C, as earlier observed for atebrin (Kraayenhof et al., 1971). Binding titrations with the uncoupling aminoacridine atebrin in illuminated chloroplasts (Fig. 3A) clearly show that the number of atebrin binding sites (n) increases and that binding becomes positively cooperative at lower temperatures. Fig. 3B shows that both n, the binding constant

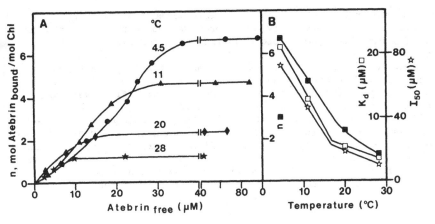

FIGURE 3. Light-induced atebrin binding and its uncoupling effect at different temperatures in broken chloroplasts. The medium contained 50 mM NaCl, 50 mM KCl, 3 mM $MgCl_2$, 1 mM ADP, 3 mM KH_2PO_4, 2 mM Tricine buffer, pH 8.0, 20 μM pyocyanine and 30 μg Chl/ml.

(K_d) and the atebrin concentration giving half-maximal inhibition of ATP synthesis (I_{50}) decrease in parallel with increasing temperature, again with possible transitions at about 18 °C. The increased and positively cooperative binding suggests that another process in addition to electrostatic screening contributes to the binding of aminoacridines to thylakoid membranes. In line with earlier studies (Dell'Antone et al., 1972; Kraayenhof, 1977) we suggest that this process is concentration-dependent aggregation, well-known for this family of dyes. Fig. 4 demonstrates that this assumption is correct for the metachromatic dye acridine orange, that shows an increase of its characteristic dimer fluorescence, both upon energization and at lower temperatures. The monomer fluorescence is quenched under these conditions as with the other aminoacridines. ACMA also forms aggregates (presumably dimers) upon concentration, indicated by an absorbance decrease at 409 nm, giving way to the 392 nm and 427 nm peaks, analogous to the behavior of 3,6-diaminoacridine (Levshin, 1955a; Haugen, Melhuish, 1964), 9-aminoacridine and other acridines (Levshin 1955b).

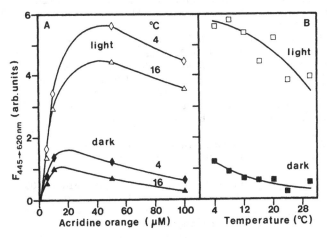

Thus accumulation of aminoacridine dyes in the light (by increase of net negative surface charge) or at lower

FIGURE 4. Light and temperature dependence of acridine orange dimerization in broken chloroplasts. Reaction conditions were as in Fig. 2 (40 μg Chl/ml). A: concentration dependence; B: temperature dependence at 50 μM acridine orange.

temperatures (by slower thermal motion, i.e. more effective screening) is accompanied by aggregation. Obviously, this process should be taken into consideration when using this type of probe as indicator for ΔpH or electric surface potential changes.

In conclusion, at lower temperatures both membrane potential (monitored by the carotenoid and oxonol VI changes) and electrokinetic zeta potential (measured by free-flow electrophoresis) are decreased due to a more effective charge screening in a more compact and less diffuse double layer. The "binding" of the cationic aminoacridines is stimulated under these conditions. Energization causes an increase of both membrane potential and fixed surface charge density in thylakoid membranes. This induces additional aminoacridine binding which is cooperatively stimulated at lower temperatures by dye aggregation. This may explain the anomalous behavior of aminoacridine probes with respect to the actually measured electric potentials and ΔpH as function of temperature.

REFERENCES

Dell'Antone P, Colonna R and Azzone GF (1972) The membrane structure studied with cationic dyes, Eur. J. Biochem. 24, 553-576.
Haugen GR and Melhuish WH (1964) Association and self-quenching of proflavine in water, Trans. Farad. Soc. 60, 386-394.
Kraayenhof R, Katan MB and Grunwald T (1971) The effect of temperature on energy-linked functions in chloroplasts, FEBS Lett. 19, 5-10.
Kraayenhof R (1977) Energy-linked change of electrical surface charges on the membrane and on the ATPase molecule of chloroplasts. In Van Dam K and Van Gelder BF, eds. Structure and function of energy-transducing membranes, pp. 223-236. Amsterdam: Elsevier.
Kraayenhof R, Schuurmans JJ, Valkier LJ, Veen JPC, Van Marum D and Jasper CGG (1982) A thermoelectrically regulated multipurpose cuvette for simultaneous time-dependent measurements, Anal. Biochem. 127, 93-99.
Levshin LV (1955a) The effect of concentration on the optical properties of solutions of 3,6-diaminoacridine, Soviet Physics-JETP 1, 244-253.
Levshin LV (1955b) The effect of concentration on the optical properties of solutions of acridine compounds, Soviet Physics-JETP 1, 235-243.
Mysels KJ (1959) Introduction to colloid chemistry. New York: Interscience Publishers Inc.
Peters FALJ, Van Spanning R and Kraayenhof R (1983) Studies on well coupled photosystem I-enriched subchloroplast vesicles. Optimization of ferredoxin-mediated cyclic photophosphorylation and electric potential generation, Biochim. Biophys. Acta 724, 159-165.
Schapendonk AHCM, Hemrika-Wagner AM, Theuvenet APR, Wong Fong Sang HW, Vredenberg WJ and Kraayenhof R (1980) Energy-dependent changes of the electrokinetic properties of chloroplasts, Biochemistry 19, 1922-1927.
Schuurmans JJ, Peters FALJ, Leeuwerik FJ and Kraayenhof R (1981) On the association of electrical events with the synthesis and hydrolysis of ATP in photosynthetic membranes. In Palmieri F et al., eds. Vectorial reactions in electron and ion transport in mitochondria and bacteria, pp. 359-369. Amsterdam: Elsevier.
Velthuys BR (1979) Electron flow through plastoquinone and cytochromes b_6 and f in chloroplasts, Proc. Natl. Acad. Sci. USA 76, 2765-2769.

Authors address: Biological Laboratory, Vrije Universiteit, De Boelelaan 1087, 1081 HV Amsterdam, The Netherlands. J.M.G.T.-P. was on leave from the Plant Physiology Laboratory, Instituto Universitario, P.O.Box 202, 5001 Vila Real, Portugal.

MICROCHEMIOSMOTIC COUPLING IN THYLAKOIDS : I. DEPENDENCE OF LOCAL $\Delta\mu H$ ON THE TOPOGRAPHY OF THE H^+ ENTRY AND EXIT POINTS

FRANCIS HARAUX/CLAUDE SIGALAT/ANNIE MOREAU/YAROSLAV DE KOUCHKOVSKY

1. INTRODUCTION

It is still debated whether the proton gradient $\Delta\mu H$ is delocalized or not (Storey, Lee ; Wikström, 1981 ; Skulachev, 1982 ; Haraux, de Kouchkovsky, 1983). In thylakoids, we have previously correlated the rate of ATP synthesis — sensitive to $\Delta\mu H$ across the coupling factors CF — to the macroscopic $\Delta\mu H$, varied by manipulation of possible local protonic resistances. In a recent work (Haraux et al., 1983) we have particularly investigated the dependency of photophosphorylation on ΔpH depending on its generation site : PSII-water splitting, PQH_2 reoxidation by PSI, or artificial proton loop around PSI. PSII is concentrated in grana stacks, that is far from CF which, as PSI, are in stromal lamellae (Andersson, Anderson, 1980). A given measured, i.e. mean, ΔpH thus formed more ATP if internal protons were emitted near CF. In the present work, we have extended this study by changing the points of H^+ input with PSII/PSI manipulation and those of H^+ output by nigericin. First, we have verified that our previous results were not due to an unequal $\Delta\Psi$ generation by PSII and PSI (Bulychev et al., 1983). Then, we have compared the effects of nigericin addition and light attenuation on $\Delta\mu H$ and ATP synthesis with PSII and PSI proton circuits. The results suggest that even when redox and phosphorylating H^+ pumps are far apart, they are connected by a proton pathway inaccessible to nigericin.

2. METHODOLOGY

Envelope-free chloroplasts were prepared and essayed as previously (de Kouchkovsky et al., 1982) for the ΔpH (9-aminoacridine method) and ATP synthesis (ferricyanide photoreduction). $\Delta\Psi$ was related to the electrochromic absorbance variation at 520 nm measured on the same sample than the other phenomena, just shifting the analytic wavelength and using appropriate filters. The assay medium was : sorbitol 0.2 M + Tricine 0.01 M + Hepes 0.01 M + KCl 0.01 M + $MgCl_2$ 6 mM + K_2HPO_4 2 mM + ADP 0.5 mM + 9-aminoacridine 4 µM. Dimethylquinone (0.5 mM) or pyocyanin (50 µM) were added when indicated. Chlorophyll concentration was 15 µM. Aerobic conditions.

3. RESULTS AND DISCUSSION

3.1. ATP synthesis versus $\Delta\mu H$ driven by PSII or PSI

In a previous work (Haraux et al., 1983) we have interpreted in microchemiosmotic terms the fact that a given ΔpH produces more ATP when it is generated near CF by the PSI loop, catalyzed by pyocyanin, than when it is formed by the linear chain $H_2O \longrightarrow PSII \longrightarrow$ dimethylquinone, which liberates internal protons far from CF. Since this could have been partly due to a more pronounced $\Delta\Psi$ with PSI chain (Bulychev et al., 1983), we have now also measured the phosphorylation rate versus $\Delta\Psi$ with these two electron-proton pathways. Fig. 1a shows the time-course of the electrochromic variation of absorbance at 520 nm (ΔA 520) related to $\Delta\Psi$: only the fast part of the biphasic dark-decay (top) was considered to be linked to the bulk $\Delta\Psi$, because it was not obscured by optical artefacts

Sybesma, C. (ed.), Advances in Photosynthesis Research, Vol. II. ISBN 90-247-2943-2.
© *1984 Martinus Nijhoff/Dr W. Junk Publishers, The Hague/Boston/Lancaster.*

and is sensitive to valinomycin, contrarily to the slow phase. Fig. 1b (where previous ΔpH data are recalled in insert) shows that ΔpH shift between PSII and PSI curves cannot be cancelled by an opposite shift in ΔΨ : indeed, as for ΔpH, a given phosphorylation rate requires a higher ΔA 520 with the $H_2O \longrightarrow$ dimethylquinone chain compared to the pyocyanin cycle. Our interpretation remains therefore valid : for a given macroscopic ΔμH, the local ΔμH at CF level, responsible for ATP synthesis, is higher when the proton input is in the CF vicinity. This is true even if ΔΨ and ΔpH are only qualitative values or if 9-aminoacridine probes, as neutral red does (Hong, Junge, 1983), interfacial instead of bulk pH.

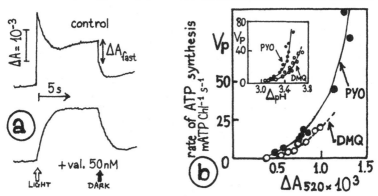

FIGURE 1. a) Time-course of the light-induced ΔA520 without (top) and with (bottom) 50 nM valinomycin. Conditions as in Methodology. b) Rate of ATP synthesis (V_p) versus the fast dark-decay of ΔA520 with $H_2O \longrightarrow$ dimethylquinone (DMQ) or pyocyanin (PYO) chains. Insert : recall of previous data (Haraux et al., 1983) on the ΔpH-dependency of the ATP synthesis with PYO and DMQ.

3.2. Resistance to nigericin of local ΔμH driving ATP synthesis with PSII and PSI chains

It was shown (Casadio et al., 1978) that in bacterial chromatophores, a protonophore addition depresses only slightly ATP synthesis despite a strong reduction of ΔμH. It was proposed that protonophores affect more the macroscopic ΔμH than its local value at the CF level. In the absence of any special structure for lateral H^+ movements, this should depend on the distance between the electron-driven and phosphorylating proton pumps, which we may modulate by using granal (PSII) or stromal (PSI) proton sources as above. Fig. 2 shows that, for PSI (a) as for PSII (b) chains, nigericin abolishes the ATP synthesis less than expected from the ATP versus ΔpH relationship observed at variable light intensity. Even if the proton source (water-splitting enzyme) is far from the coupling factor, all is as if nigericin could not intercept H^+ between their sites of input and of utilization. This conclusion is not explainable in terms of a ΔΨ effect. Indeed, in Fig. 3, the PSII experiment was repeated with 50 nM valinomycin, and the basic results of Fig. 3a were maintained. Our observation — in steady-state and not at equilibrium — seems at variance with those of Baker et al. (1981) and Theg and Junge (1983), which suggest a proton-compartment distinct from the internal bulk phase, but which is opened

by protonophore addition : our lateral proton pathway between redox carriers and ATP-synthetases exhibits different properties. It is likely
that our experiments did not necessarily involve an intramembraneous
domain ; they may have revealed surface conductive polypeptide chains or
locally structured water, slowly exchanging protons with the adjacent
medium. In any event, our view remains different from the direct coupling
hypothesis, since even in such a particular structure, a pH drop, suggested by comparison of PSII and PSI chains, prevents the ΔμH equalization
between the redox carriers and the coupling factors.

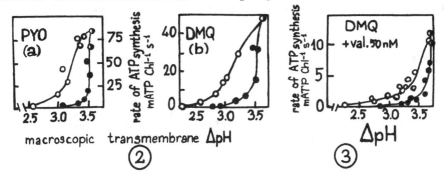

FIGURE 2. Rate of ATP synthesis versus ΔpH at variable light intensity
(●) or with variable nigericin (o). a : pyocyanin chain ; b : dimethylquinone chain. Conditions as in Methodology.

FIGURE 3. Rate of ATP synthesis versus ΔpH at variable light intensity
(●) or with variable nigericin (o). $H_2O \longrightarrow$ dimethylquinone chain.
Conditions as in Fig. 2 + valinomycin (50nM) to cancel ΔΨ.

REFERENCES
Andersson B and Anderson JM (1980) Lateral heterogeneity in the distribution of chlorophyll-protein complexes of the thylakoid membranes of
spinach chloroplasts, Biochim. Biophys. Acta 593, 427-440.
Baker GM, Bhatnagar D and Dilley RA (1981) Proton release in photosynthetic water oxidation : evidence for proton movements in a restricted domain,
Biochemistry 20, 2307-2315.
Bulychev AA, Remish D and Kurella GA (1983) Comparative efficiencies of
photosystems I and II as generators of electrical potential across the
chloroplast thylakoid membrane, Plant Sci. Lett. 29, 73-79.
Casadio R, Baccarini-Melandri A and Melandri BA (1978) Limited cooperativity in the coupling between electron flow and photosynthetic ATP
synthesis. A comparative study in chromatophores phosphorylating at very
different rates, FEBS Lett. 87, 323-328.
Haraux F, Sigalat C, Moreau A and de Kouchkovsky Y (1983) The efficiency
of energized protons for ATP synthesis depends on the membrane topography
in thylakoids, FEBS Lett. 155, 248-252.
Haraux F and de Kouchkovsky Y (1983) The energy transduction theories :
a microchemiosmotic approach in thylakoids, Physiol. Vég., in press.
Hong YQ and Junge W (1983) Localized or delocalized protons in photo-
phorylation ? On the accessibility of the thylakoid lumen for ions and
buffers, Biochim. Biophys. Acta 722, 197-208.
de Kouchkovsky Y, Haraux F and Sigalat C (1982) Effect of hydrogen-

deuterium exchange on energy-coupled processes in thylakoids. A new
illustration of local proton gradients with the energy-transducing biomembranes, FEBS Lett. 139, 245-249.

Skulachev VP (1982) The localized $\Delta\mu H^+$ problem. The possible role of the
local electric field in ATP synthesis, FEBS Lett. 146, 1-4.

Storey BT, Lee CP and Wikström M (1981) Is transmembrane $\Delta\mu H^+$ essential
to mitochondrial energy coupling ? Trends Biochem. Sci. 6, 166-170.

Theg SM and Junge W (1982) The effect of low concentrations of uncouplers
on the detectability of proton deposition in thylakoids. Evidence for
subcompartmentation and preexisting pH differences in the dark, Biochim.
Biophys. Acta 723, 294-307.

ACKNOWLEDGEMENTS
This work was supported by a CNRS-ATP grant.

ABBREVIATIONS
PS : photosystem ; CF : coupling factor ; $\Delta A520$: light-induced absorbance
variation at 520 nm ; PQ, PQH_2 : plastoquinone, plastoquinol ; $\Delta\mu H$:
transmembrane electrochemical potential difference for proton ; ΔpH :
transmembrane pH difference ; $\Delta\Psi$: transmembrane electrical potential
difference.

Authors address : Francis Haraux/Claude Sigalat/Annie Moreau/Yaroslav
                  de Kouchkovsky
                  Laboratoire de Photosynthèse, CNRS, 91190 Gif sur Yvette,
                  France.

MICROCHEMIOSMOTIC COUPLING IN THYLAKOIDS : II. DEPENDENCE OF LOCAL $\Delta\mu H$
ON THE MEMBRANE CONDUCTANCE

CLAUDE SIGALAT/YAROSLAV DE KOUCHKOVSKY/ANNIE MOREAU/FRANCIS HARAUX

1. INTRODUCTION

In bacterial chromatophores, a protonotophore addition depresses ATP synthesis less than expected from the $\Delta\mu H$ diminution (Casadio et al., 1978).
This result, and many others, led to the concept of "microchemiosmotic
coupling", according to which $\Delta\mu H$ at the redox and phosphorylating proton
pumps level, although equal, differ from the average measured $\Delta\mu H$ (Van Dam
et al., 1978). On thylakoids, various experimental approaches have suggested the existence of H^+ subcompartments distinct from the adjacent bulk
phases (Baker et al., 1981 ; Johnson et al., 1982 ; Hong , Junge, 1983).
We have previously reported that in thylakoids, correlations between electron flow control, $\Delta\mu H$ and ATP synthesis could depend on local H^+ resistances modulated in magnitude by isotope exchange (de Kouchkovsky,
Haraux, 1981 ; de Kouchkovsky et al., 1982) or in localization by variable
PSII/PSI contribution to the proton transport (Haraux et al., 1983). In
the present work, the relationships between steady-state thermodynamic
fluxes and forces were varied by the membrane H^+ conductivity. The results
support a microchimiosmotic view where the $\Delta\mu H$ across the coupling factors,
different from the bulk one, does not necessarily equalize that locally
generated by the H^+ redox carriers, at variance with Van Dam et al.
(1978) concept.

2. METHODOLOGY

Envelope-free chloroplasts were prepared and essayed as previously (de
Kouchkovsky et al., 1982) for the ΔpH (9-aminoacridine method), the electron flow (ferricyanide photoreduction) and ATP synthesis (luciferin-
luciferase technique). The assay medium was : sorbitol 0.2 M + Tricine
0.01 M + Hepes 0.01 M + KCl 0.01 M + $MgCl_2$ 6 mM + K_2HPO_4 2 mM + ADP (0.1
or 0.5 mM) + 9-aminoacridine 4 µM + $K_3Fe(CN)_6$ 0.8 mM. 20° C, pH = 7.8.
Chlorophyll concentration was 15 µM. Aerobic conditions.

3. RESULTS AND DISCUSSION

3.1. ΔpH, ATP synthesis and redox chain control modulated by light intensity or nigericin

Fig. 1 shows that a given steady-state ΔpH synthetizes more ATP when
modulated by nigericin than by light intensity, which confirms the results
of others on chromatophores (Casadio et al., 1978) and suggests localized
ΔpH due to protonic resistances between the sites of H^+ translocation and
the basal membrane (Van Dam et al., 1978). This view is strenghtened by
Fig. 2, where the parameter plotted versus ΔpH is the control of the electron flow as expressed by the ratio : fully uncoupled (1 µM nigericin)/
ΔpH-controlled (no or slight nigericin) electron transfer rates. This
control ratio should reflect the actual ΔpH exerted on the plastoquinone
pool : as for that across CF, this local ΔpH is, according to Fig. 2,
higher for a given macroscopic ΔpH when it is varied by nigericin instead
of light intensity. To exclude the contribution of the electrical component $\Delta\Psi$ of $\Delta\mu H$, we have repeated the experiment of Fig. 1 with 50 nM

Sybesma, C. (ed.), Advances in Photosynthesis Research, Vol. II. ISBN 90-247-2943-2.
© *1984 Martinus Nijhoff/Dr W. Junk Publishers, The Hague/Boston/Lancaster.*

valinomycin : Fig. 3. Though less, the curves are still apart. Since an activation of CF1 by PSI acceptors (Mills , Mitchell, 1982) could not occur in our conditions, owing to the lack of the required endogenous electron carriers, the microchemiosmotic mechanism remains the most economical interpretation.

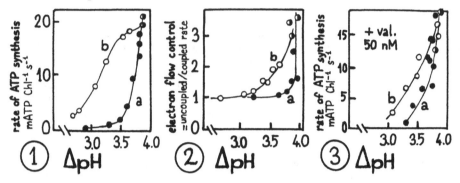

FIGURE 1. Rate of ATP synthesis as a function of macroscopic ΔpH. Conditions as in Methodology. a) Variable light intensity (0.5 % to 100 % of the control) ; b) variable nigericin (10 to 100 nM). ADP 100 µM.

FIGURE 2. Control of the redox chain as a function of ΔpH. Conditions as in Fig. 1 except that ADP = 0. a) Variable light ; b) variable nigericin.

FIGURE 3. Rate of ATP synthesis versus ΔpH. Conditions as Fig. 1 except that ADP = 500 µM and valinomycin 50 nM was present. a) Variable light ; b) variable nigericin.

3.2. Direct coupling versus "generalized microchemiosmosis"

The relative insensitivity of phosphorylation to nigericin should indicate a rather direct H^+ transfer from the redox carriers to the ATP-synthetases. We have however proposed (de Kouchkovsky et al., 1982) that a kinetic barrier for protons could prevent the $\Delta\mu H$ equalization between the redox and phosphorylating H^+ pumps. This picture, called by us "generalized microchemiosmosis" (Haraux , de Kouchkovsky, 1983), was in particular suggested by the dependency on the CF H^+ conductance (modulated by ADP) of the relationship between redox chain control and macroscopic ΔpH. Such a dependency is expected only if a significant protonic resistance separates the site of PQH_2 reoxidation and the CF0-CF1 (Haraux et al., 1982). We have since then reexamined this CF conductance effect with two CF inhibitors, phlorizin and tri-n-butyltin. These two reagents were actually found to accelerate the H^+ efflux, but preliminary experiments revealed that despite the resulting severely depressed ΔpH, the redox chain remains rather highly controlled. If these molecules act specifically on CF and not on the basal membrane, these results should confirm the weight of CF conductance in the relationship between macroscopic and local ΔpH at the redox carriers level : this would be in favour of "generalized microchemiosmosis".

3.3. Specific effects of protonophores

Fig. 4 illustrates an experiment similar to Fig. 1 but with gramicidin D
instead of nigericin. Surprisingly, the result is in the opposite

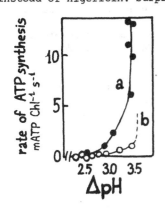

direction : ATP synthesis is much more depressed there than the measured ΔpH (reference : variable light intensity). A simple
$\Delta\Psi$ suppression by this ionophore is not
sufficient to explain this observation. One
could have supposed that gramicidin simply
shunts the phosphorylating H^+ current between the redox generators and the coupling
factors but, compared to nigericin, it was
used at ten times lower concentration. Some
data (P. Gräber, personal communication)
suggest an effect of gramicidin on the coupling factor itself. Moreover, one cannot
exclude the contribution of special H^+ micro-
spaces which would communicate with the
bulk phases only when protonophores are
added (Baker et al., 1981). However, gramicidin being required for such an effect at
only one order of magnitude below nigericin
(Theg , Junge, 1983), one should have
expected comparable behaviours in our Figs.
1 and 4. This problem therefore deserves
additional experiments to be solved.

FIGURE 4. Rate of ATP synthesis versus ΔpH. a) With
variable light;b) variable
gramicidin (1 to 10 nM).ADP =
100 μM.

REFERENCES

Baker GM, Bhatnagar D and Dilley RA (1981) Proton release in photosynthetic water oxidation : evidence for proton movements in a restricted
domain, Biochemistry 20, 2307-2315.
Casadio R, Baccarini-Melandri A and Melandri BA (1978) Limited cooperativity in the coupling between electron flow and photosynthetic ATP synthesis.
A comparative study in chromatophores phosphorylating at very different
rates, FEBS Lett. 87, 323-328.
Haraux F, Sigalat C and de Kouchkovsky Y (1982) Localized pH gradients at
the thylakoid membrane level. In Second european bioenergetic conference
reports, LBTM-CNRS, Villeurbanne, 373-374.
Haraux F, Sigalat C, Moreau A and de Kouchkovsky Y (1983) The efficiency
of energized protons for ATP synthesis depends on the membrane topography
in thylakoids, FEBS Lett. 155, 248-252.
Haraux F and de Kouchkovsky Y (1983) The energy transduction theories : a
microchemiosmotic approach in thylakoids, Physiol. Vég. in the press.
Hong YQ and Junge W (1983) Localized or delocalized protons in photophosphorylation ? On the accessibility of the thylakoid lumen for ions
and buffers, Biochim. Biophys. Acta 722, 197-208.
Johnson JD, Pfister VR and Homann PH (1983) Metastable proton pools in
thylakoids and their importance for the stability of photosystem II,
Biochim. Biophys. Acta 723, 256-265.
de Kouchkovsky Y and Haraux F (1981) ^2H_2O effect on the electron and
proton flow in isolated chloroplasts. An indication for lateral heterogeneity of membrane pH, Biochem. Biophys. Res. Comm. 99, 205-212.
de Kouchkovsky Y, Haraux F and Sigalat C (1982) Effect of hydrogen-

deuterium exchange on energy-coupled processes in thylakoids. A new illustration of local proton gradients with the energy-transducing biomembranes.
FEBS Lett. 139, 245-249.

Mills JD and Mitchell P (1982) Modulation of coupling factor ATPase
activity in intact chloroplasts. Reversal of thiol modulation in the dark,
Biochim. Biophys. Acta 679, 75-82.

Theg SM and Junge W (1982) The effect of low concentrations of uncouplers
on the detectability of proton deposition in thylakoids. Evidence for
subcomportmentation and preexisting pH differences in the dark. Biochim.
Biophys. Acta 723, 294-307.

Van Dam K, Wiechmann AHCA, Hellingwerf KJ, Arents JC and Westerhoff HV
(1978) Functioning of the mitochondrial ATP-synthesizing machinery. In
Nicholls P et al.,eds. Membrane proteins, pp. 121-132. Pergamon Press,
Oxford.

ACKNOWLEDGEMENTS
This work was supported by a CNRS-ATP grant.

ABBREVIATIONS
PQ, PQH_2 : plastoquinone, plastoquinol ; CF0 : proton channel of coupling
factor ; CF1 : catalytic site of coupling factor ; $\Delta\mu H$: transmembrane
electrochemical potential difference for protons ; ΔpH : transmembrane
pH difference ; $\Delta\Psi$: transmembrane electrical potential difference.

Authors address : Claude Sigalat/Yaroslav de Kouchkovsky/Annie Moreau/
                  Francis Haraux,
                  Laboratoire de Photosynthèse, CNRS, 91190 Gif sur Yvette,
                  France.

REGULATION OF THE ENERGIZATION OF THYLAKOIDS IN LEAVES AND
IN ISOLATED CHLOROPLASTS

GYŐZŐ GARAB, KLÁRA BARABÁS, LÁSZLÓ ZIMÁNYI, JACK FARINEAU[*]
BIOLOGICAL RESEARCH CENTER, SZEGED; *CENTRE D ÉTUDES
NUCLEAIRES, SACLAY

1. INTRODUCTION

The flash-induced electrochromic absorbance change, ΔA_{515},
of thylakoids is characteristic to the build-up of the transmembrane electrical field and the consecutive utilization or
dissipation of the electrical field energy (Junge 1979).
Hence ΔA_{515} has been proved to be a powerful tool in studying the energization of chloroplast membranes. Kinetics of
ΔA_{515} has been shown to be regulated by various factors
(Vredenberg 1982). Recently it has been suggested that in
isolated chloroplasts Q-cycle (Selak, Whithmarsh, 1983) or
cycle around photosystem 1 (Crowther, Hind 1980) as well as
ATP-hydrolysis (Schreiber, Rienits 1982) play an important
role in the energization process. We have shown that in
algal cells and in leaves of higher plants CO_2 regulates
both the build up of the transmembrane electrical field and
the ATP synthesis (Garab et al. 1983).
It has not yet been, however, investigated what are the rate limitations of the energization of thylakoids. In the work presented here we determined the rate limiting turnover times
of the photosynthetic electron transport in leaves and in
isolated chloroplasts.

2. MATERIALS AND METHODS

Intact chloroplasts were isolated from 2 weeks old pea leaves
grown in the greenhouse. The standard reaction medium contained chloroplasts with a chlorophyll concentration of about 50
μM in 0.35 M sorbitol, 20 mM tricine and 5 mM $MgCO_3$, pH=7.7.
Freshly harvested 2-3 weeks old barley leaves were infiltrated
by water under gentle vacuum for about 5 min and the leaves
were supplied with CO_2 for 5-10 min (Garab et al. 1983). The
absorbance changes between 460 and 580 nm were carried out in
a single beam kinetic spectrophotometer (Horváth et al. 1979)
equipped with a side illumination attachment. Kinetics of
ΔA_{515} could be satisfactorily fitted by linear combinations
of two exponentials: $\Delta A(t) = a_1 \exp(-k_1 t) - a_2 \exp(-k_2 t)$

3. PRINCIPLE OF DETERMINATION OF THE RATE-LIMITING TURNOVER
TIME

The amplitude of ΔA_{515} upon a saturating flash excitation is
proportional to the density of the open reaction centers
(Junge 1979). With continous background illumination of the
sample density of the open reaction centers and consequently
the amplitude of the flash-induced ΔA_{515} decreases Let S_0

($cm^{-2}s^{-1}$) be the incident photon flux of the background light;
A, the absorbance of the sample at the given wavelength of
the side-illumination; x, (cm) pathlength of the background
beam; n_O (cm^{-3}) density of the reaction centers, n_{st}, (cm^{-3})
density of the open reaction centers in steady state after
the onset of the background light; hence $(n_{st}/n_O) = \Delta = 1$-inhi-
bition. It can be shown that the overall rate limiting turnover time is:

$$\tau_c = \frac{n_O}{S_O} \frac{x}{2.3A} \frac{10^A - 10^{A\Delta}}{10^{A\Delta} - 1}$$

4. RESULTS AND DISCUSSION

Intensity-dependence of the photoinhibition by 650 nm light
is shown in Fig. 1, and variation of the photoinhibition
upon various treatments in Table 1. (It is to be emphasized
that the photoinhibition in our experiments is fully reversible furthermore it is not the consequence of a decrease in the "sensitivity" of field-detecting pigments.)
Our results (also the wavelength dependence, not given here)
show that the photoinhibition is due to a very slow turnover
rate of the photosystem 1 - supposedly operating in a cyclic
regime as suggested by Crowther and Hind (1980). The photoinhibition is initiated in photosystem 1 (cf. also Satoh,
Fork 1982), however, it extends rapidly also to photosystem 2.
In leaves with 650 nm background light the onset of photoinhibition is similar to that observed in isolated chloroplasts, i.e. amplitude of ΔA_{515} decreases rapidly (~1s).
However, somewhat later about 10s-100s after the onset of
illumination ΔA_{515} increases again and in steady state the
photoinhibition is small. In steady state the overall turnover rate in leaves is as low as the still persisting slow
electrochromic rise, i.e. 5-10 ms (Fig. 2). This shows that
in leaves photosystem 2 may considerably ease the photoinhibition initiated in photosystem 1.

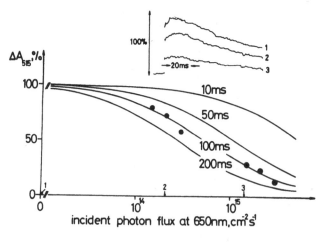

FIGURE 1. Intensity
dependence of the
photoinhibition in
isolated intact
chloroplasts;•• ,
measured values;
—, computed curves.
Typical kinetics are
shown in inset.

TABLE 1. Photoinhibition by 650 nm background light ($S_0 = 6.6 \cdot 10^{14}$) and the corresponding rate limiting turnover times in isolated intact chloroplasts

addition / kinetic component	initial amplitudes in dark	amplitudes of ΔA_{515} relative to the "dark" values,% (in brackets: τ_c, ms)		
		total	initial	slow rise
$\emptyset$	100	48(122)	46(133)	47(127)
2 µM nigericin + 10 mM KCl	95	64 (63)	49(117)	98 (2)
330 µM ADP+P$_i$	105	61 (72)	57 (85)	75 (37)
330 µM ATP	89	52(104)	52(104)	42(156)
ATP+nigericin	85	57 (85)	48(122)	76 (35)
10 mM Na-dithionite	57	63 (66)	48(122)	72 (43)
dithionite+nigericin	48	66 (58)	50(113)	77 (33)
17 µM antimycin A	82	-	74 (39)	-
40 µM PMS	115	-	86 (18)	-

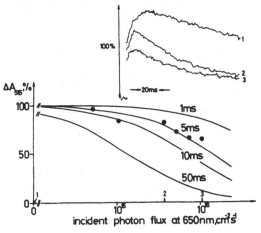

FIGURE 2. Intensity dependence of the photoinhibition in barley leaves ••, measured values, —, computed curves. Typical kinetics are shown in inset.

Concerning the slow rise it must be pointed out that generally it does not correspond to the terminating step of the energization process. Moreover, in some experimental conditions it can be completely "detached" (uncoupled) from the very slow turnover rate processes (cf. Table 1). It was also interesting to observe that the slow rise in uncoupled chloroplasts was not inhibited considerably even in the presence of ATP either in dark or in 650 nm light. No direct correlation was found between the slow rise and the rate limiting turnover rate believed to originate from photosystem 1-driven cyclic electron flow. The same conclusion was reached also from our recent experiments with antimycin A (Garab, Farineau 1983). Our results are consistent with an interpretation correlating the slow electrochromic rise with 5-10 ms translocations of the photoinduced charges (Zimányi, Garab 1982). We conclude that cyclic electron transport around photosystem 1 not only takes part in but also regulates the energization of thylakoids.

REFERENCES

Crowther D and Hind G (1980) Partial characterization of cyclic electron transport in intact chloroplasts, Arch. Biochem. Biophys. 204, 568-577.
Garab Gy and Farineau J (1983) Is antimycin A a specific inhibitor of the slow rise of the electrochromic absorbance change in intact chloroplasts? Biochem. Biophys. Res. Commun. 111, 619-623.
Garab Gy, Sanchez Burgos AA, Zimányi L and Faludi-Dániel Á (1983) Effect of CO_2 on the energization of thylakoids in leaves of higher plants, FEBS Lett., 154, 323-327.
Horváth G, Niemi HA, Droppa M and Faludi-Dániel Á (1979) Characteristics of the flash-induced 515 nanometer absorbance change of intact isolated chloroplasts, Plant Physiol. 63, 778-782.
Junge W (1977) Membrane potentials in photosynthesis, Ann. Rev. Plant Physiol. 28, 503-536.
Satoh K and Fork CD (1982) Photoinhibition of reaction centers of photosystem I and II in intact Bryopsis chloroplasts under anaerobic conditions, Plant Physiol. 70, 1004--1008.
Schreiber U and Rienits KG (1982) Complementarity of ATP-induced and light-induced absorbance changes around 515 nm, Biochim. Biophys. Acta. 682, 115-123.
Selak MA and Whitmarsh J (1982) Kinetics of the electrogenic step and cytochrome b_6 and f redox changes in chloroplasts. Evidence for Q cycle, FEBS Lett. 150, 286-292.
Vredenberg WJ (1981) P515: a monitor of photosynthetic energization in chloroplast membranes, Physiol. Plant. 53, 598-602.
Zimányi L and Garab Gy (1982) Configuration of the light induced electric field in thylakoid and its possible role in the kinetics of the 515 nm absorbance change, J. theor. Biol. 95, 811-821.

Authors' address: Biological Research Center, Szeged, P.O.B. 521, Hungary and *Service de Biophysique CEA/SACLAY 91191 Gif/Yvette, France

PROTOLYTIC REACTIONS AT THE DONOR SIDE OF PSII:
PROTON RELEASE IN TRIS-WASHED CHLOROPLASTS WITH $t_{1/2} \cong 100 \mu s$.
IMPLICATIONS FOR THE INTERPRETATION OF THE
PROTON RELEASE PATTERN IN UNTREATED, OXYGEN-EVOLVING CHLOROPLASTS

VERENA FÖRSTER/WOLFGANG JUNGE

1. INTRODUCTION

Oxidation of H_2O to O_2 by PSII at the expense of 4 quanta of red light
has mainly been studied by observation of the products, O_2 and protons.
In dark-adapted chloroplasts which are submitted to a series of
excitations the PSII population carries out the four-step water
oxidation roughly in phase; protons are released into the internal phase
of thylakoids in a stoichiometric sequence which is supposed to be
determined by the mechanism of water oxidation. The stoichiometric
pattern observed on strictly dark-adapted chloroplasts (>>30min) had
been interpreted according to the scheme of Kok et al. (B.Kok et al.,
1970) as follows:

Time-resolved measurements under these conditions largely confirmed this
interpretation. However, a rapid pH change following the second flash
was not unequivocally explicable under the simple assumption of
single-exponential H^+ release during each transition $S_i \to S_{i+1}$ (V.Förster
et al., 1981). One possible reason for this could be additional,
transient proton release apart from water oxidation.
From proton release measurements in the presence of DCMU (water
oxidation abolished) we got a first hint to a proton-releasing site
associated with the intermediate electron carrier Z between the
water-oxidizing enzyme and P680 (Y.Q.Hong et al., 1981). We have tried
to confirm this finding by the investigation of Tris-treated
chloroplasts where water oxidation is also abolished and proton release
originating from the formation of water-oxidation intermediates is not
expected to occur. Side effects of Tris treatment are the loss of
proteins from the membrane (Y. Yamamoto et al., 1981) and the
alteration of the reaction kinetics at the remaining part of the donor
side of PSII. The half time of $P680^+$ reduction by its immediate donor Z
(also referred to as D_1) is altered from 30ns in untreated chloroplasts
to 5-10µs (at pH7) in Tris-treated chloroplasts; the 5-10µs phase
accounts for ~60% of the reaction centers (Conjeaud et al., 1979;
M.Boska et al.,1983). Z is rereduced in the millisecond range by
unknown donor(s) (G.T.Babcock, K.Sauer, 1976). Renger and Voelker
recently found proton release by Tris-treated inside-out thylakoids;
$0.85H^+/600$Chl were detected in the external aqueous bulk phase by a
water-soluble pH indicator (bromo-cresol purple) with a half rise time
of 1ms. The authors suggested that this pH change was due to a
deprotonation of Z^+ (G.Renger, M.Voelker, 1982). Here, we report on
proton release in rightside-out Tris-treated thylakoids which is

Sybesma, C. (ed.), Advances in Photosynthesis Research, Vol. II. ISBN 90-247-2943-2.
© *1984 Martinus Nijhoff/Dr W. Junk Publishers, The Hague/Boston/Lancaster.*

monitored by a membrane-soluble pH indicator (neutral red) with $t_{1/2} \cong 100 \mu s$.

2. MATERIALS AND METHODS

Tris-treated chloroplasts were prepared from broken pea chloroplasts according to standard procedures. pH changes in the internal phase of thylakoids were monitored flash photometrically via absorption changes of the membrane-soluble pH-indicator dye neutral red at 548nm (for details see V.Förster et al., 1981). Proton release due to plastohydroquinone oxidation by PSI was inhibited by DBMIB (2,5-dibromo-3-methyl- 6-isopropyl- 1,4-benzoquinone).

3. RESULTS AND DISCUSSION

3.1. Proton release in Tris-treated chloroplasts

Fig.1 shows the pH_{in}-indicating absorption change of neutral red in repetitively excited Tris-treated thylakoids when the PSII acceptor couple hexacyanoferrat(III)/DBMIB is used and no artificial donor is added. The signal amplitude corresponds to the liberation of $\sim 0.6 H^+$/P680 (+0.1). The absorption change rises with $t_{1/2} \cong 100 \mu s$ (80-130μs are obtained in different preparations).

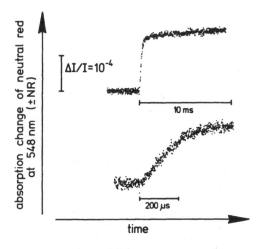

FIGURE 1.
pH_{in}-indicating absorption change of neutral red in Tris-treated chloroplasts, measured at low and high time resolution;
above: 20μs/point
below: 1μs/point
Each trace is an average of 200 scans.

The stoichiometry of $0.6 H^+$/P680 suggests that the proton release is a consequence of the oxidation of Z. In this case, proton detection would be retarded with respect to the redox reaction by one order of magnitude; for methodical reasons a lag phase of 5-10μs corresponding to the oxidation of Z is not easily to be proved. However, a strong argument for the link between Z oxidation and H^+ release is obtained from the relaxation of the pH signal which is shown in the upper trace of Fig.2; two kinetic phases can be distinguished in the decay, a rapid phase ($t_{1/2}$=140ms, 60%) and a slow phase ($t_{1/2}$=1.2s, 40%). We explain them by two processes: rebinding to the proton-releasing site (140ms) and leaking out of protons through the membrane (1.2s). The 140ms half time is on the order of what is commonly observed for the reduction of Z^+ in Tris-treated thylakoids (G.T.Babcock, K. Sauer, 1976). Its most likely explanation is reduction of Z^+ by a non-proton-releasing reductant. We assumed that the slow phase of the decay was due to the leakiness of the membrane because this usually caused proton efflux in the second time range. To support this we induced a pH difference across the thylakoid membrane by photooxidation of hydroquinone/ascorbate; it had to be expected that protons which were released upon such an "irreversible" oxidation would not be taken up at a special site at the inner membrane surface, but would leak out through the membrane. The result is shown in the lower trace of Fig.2; we obtained a decay time of 1.35s. This is very close to the slow decay phase of the pH signal in question.

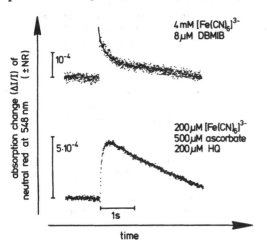

FIGURE 2.
pH_{in}-indicating absorption change of neutral red in Tris-treated chloroplasts;
above: without added artificial donor (as in Fig.1; the solid line corresponds to data given in the text)
below: hydroquinone/ascorbate added as indicated (no DBMIB present)

We propose that the proton-releasing species is a proteinaceous group close to the prosthetic group, Z, which suffers a pK shift when Z is oxidized and consequently deprotonates (V.Förster, W.Junge, in preparation).
Why is proton release retarded with respect to the oxidation of Z? We have found that the apparent pK of the proton-releasing group is ≈5 (not shown). An acid of pK≈5 which is in contact with water is expected to deprotonate much faster ($t_{1/2}$≈1µs, see M.Eigen, 1963) than we observe a proton (100µs). Since protonation of neutral red is not rate-limiting (V.Förster, unpublished) we conclude that the proton has to diffuse through the reaction center protein before it is detected by neutral red.
Despite a striking difference in the rise kinetics we have to assume that our observations are due to the same event as those of Renger and Voelker (G.Renger, M.Voelker, 1982). The slow rise (1ms) found by these authors is most likely explained by H^+ diffusion between stacked membrane layers to the water-soluble indicator, which monitors the bulk pH (multiple binding and debinding at buffering groups; see Y.Q.Hong, W.Junge, 1983 and A.Polle, W.Junge, these proceedings).

3.2. <u>Implications of H^+ release upon Z oxidation for the interpretation of the H^+ release pattern observed in unmodified water-oxidizing thylakoids.</u> Although Tris treatment affects the donor side of PSII considerably (altering of reaction kinetics, loss of proteins,...)

transient proton release linked to Z oxidation is a very likely explanation for unexpected rapid proton release following the second flash after a dark period in unmodified chloroplasts.

Proton release observed after the second flash given to strictly dark-adapted chloroplasts had been expected to be due nearly exclusively to the liberation of one proton during the $S_2 \rightarrow S_3$ transition. Surprisingly, it appeared biphasic. The rapid phase ($t_{1/2}$ =100µs in spinach chloroplasts, 200µs in pea chloroplasts) was faster than the foregoing electron abstraction (~400µs in spinach, see Babcock et al., 1976 and Förster et al.,1981). Transient proton release coupled to the redox change of the intermediate electron carrier Z would be irrelevant for the stoichiometric pattern, but explain the observed kinetics. We suggest that the rapid rise is due to transient proton release in the state S_2Z^+P680 as can be seen from the scheme in Fig.3.

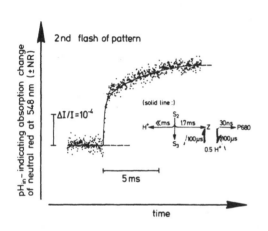

FIGURE 3.
pH_{in}-indicating absorption change of neutral red in unmodified broken pea chloroplasts observed after the second flash given to strictly dark-adapted chloroplasts.
solid line: curve calculated under the assumption of transient proton release upon oxidation of part of the intermediate electron carrier Z.

REFERENCES

Babcock GT, Blankenship R, Sauer K (1976) FEBS Lett.61, 286-289
Babcock GT and Sauer K (1976) Biochim. Biophys. Acta 396, 48-62
Boska M, Sauer K, Buttner W, Babcock GT (1983)
        Biochim.Biophys.Acta 722, 327-330
Conjeaud H, Mathis P, Paillotin G (1979) Biochim.Biophys.Acta 546,280-291
Eigen M (1963) Angew. Chem. 75, 489-588
Förster V, Hong YQ, Junge W (1981) Biochim. Biophys. Acta 638, 141-152
Hong YQ, Förster V, Junge W (1981) FEBS Lett. 132, 247-251
Kok B, Forbush B, McGloin M (1970) Photochem. Photobiol.11, 457-475
Renger G and Voelker V (1982) FEBS Lett. 149, 203-207
Yamamoto Y, Doi M, Tamura N, Nishimura M, (1981) FEBS Lett.133, 265-268

ACKNOLEDGEMENTS
We wish to thank Hella Kenneweg, Dr.Ulrich Kunze, Holger Lill, Andrea Polle, Norbert Spreckelmeyer and Dr.H.W.Trissl for valuable help (discussion, electronics, computer programs and photographs). Financial support from the Commission of the European Communities is gratefully acknoledged.
Authors address: Verena Förster/Wolfgang Junge, Universität Osnabrück/
                Biophysik, Postfach 4469, D-4500 Osnabrück, F.R.G.

INFLUENCE OF A PROTON GRADIENT ON THE ACTIVITY OF THE RECONSTITUTED CHLOROPLAST PHOSPHATE TRANSLOCATOR

U. I. FLÜGGE AND H. W. HELDT, Lehrstuhl für Biochemie der Pflanze, Untere Karspüle 2, 3400 Göttingen, F.R.Germany

Introduction

The phosphate translocator of the chloroplasts is located in the inner envelope membrane and catalyzes a counter exchange of phosphate (P_i), triose phosphate (dihydroxyacetone phosphate (DHAP) and glyceraldehyde phosphate) and 3-phosphoglycerate (3-PGA) (Fliege et al. 1978). In the dark, these substrates are transported equally well into either direction. In the illuminated state, however, the phosphate translocator catalyzes the preferential export of triose phosphate in exchange with phosphate, even though its level in the chloroplast stroma is much lower than that of 3-PGA (Lilley et al. 1977). In the cytosol, triose phosphate serves as the substrate for sucrose biosynthesis. Furthermore, by an export of triose phosphate in exchange with 3-PGA, such an exchange provides the cytosol with ATP and reducing equivalents (Heber, Santarius, 1970). During illumination the functioning of these two shuttles catalyzed by the phosphate translocator implies a preferential export of triose phosphate into the cytosol and a preferential import of 3-PGA into the chloroplast stroma. The influence of light on the preferential export of triose phosphate and import of 3-PGA, respectively, is supposed to be related to the proton gradient across the envelope, resulting from an alkalization of the stroma due to the light driven proton transport into the thylakoid space (Heldt et al. 1978). In the present study we examine the influence of a proton gradient on the activity of the phosphate translocator. For this, the purified phosphate translocator was reconstituted into liposomes in a functional state under well defined conditions (Flügge and Heldt 1978, Flügge et al. 1983) and the influence of a proton gradient generated across the liposome membrane on the substrate transport was studied.

Method

If an inward directed proton gradient across the liposome membrane was to be applied, liposomes with an internal pH of 7.9 were used and buffer was added to change the external medium to pH 6.8. The gradient was maintained for several minutes and could be dissipated by the subsequent addition of nigericin which induced an H^+/K^+ exchange allowing the internal pH of the liposome to equilibrate with the pH of the external medium. For monitoring the proton gradient, liposomes were prepared in the presence of the impermeable dye phenol red, and the absorbancy was measured using a dual wavelength spectrophotometer (Fig 1). When an outward directed proton gradient was to be applied, the liposomes were prepared in a medium of pH 6.8 and the proton gradient was adjusted by adding alkaline buffer solution which changes the external pH to 7.8. Here again, the proton gradient was maintained for several minutes (not shown).

Results and Discussion

Influence of a proton gradient on the uptake of substrates

The present paper investigates the effect of an applied a proton gradient on the transport. For the sake of simplicity, here only transport in the heteroexchange mode, i.e. the exchange of 3-PGA in the internal liposome volume with DHAP or

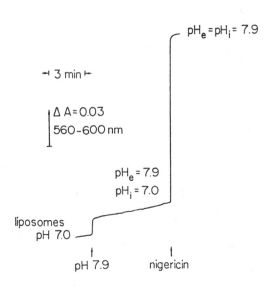

FIGURE 1 Measurement of the internal liposomal pH using phenol red

vice versa will be dealt with. In the latter case, the liposomes are loaded with Pi instead of DHAP, to save costs and because the transport of DHAP behaved like the transport of Pi under the conditions of our experiments. The concentration dependence was studied by varying the external substrate concentrations, and the resulting Km and Vmax were evaluated from double reciprocal plots.

From the results in Table 1, the following conclusions can be drawn:
1) In the case of DHAP transport an inward directed proton gradient results in an inhibition due to an increase in the apparent Km for DHAP with the Vmax being unaltered, an outward directed H^+ gradient having an opposite effect in decreasing the apparent Km for DHAP.
2) In the case of 3-PGA transport an inward directed proton gradient leads to a stimulation associated with an increase of Vmax. Again an outward directed proton has the opposite effect in decreasing the Vmax of 3-PGA.

Obviously, the transport of DHAP and 3-PGA is influenced by a proton gradient in different manner. In the case of DHAP and phosphate transport the observed effect are associated with changes in the apparent Km values whereas in the case of 3-PC transport, the application of a proton gradient results in a change of Vmax. Although there is no doubt, that both substrates are transported by the same carrier (Fliege et al. 1978), these observations suggest, that these two substrates are recognized and bound to the carrier differently. The fact that both substrates are transported as twice negatively charged anions (Fliege et al. 1978) implies that the phosphate ion in the transported DHAP and phosphate is charged differentl than in the 3-PGA. In the case of DHAP and phosphate it contains two negative charges but only one in the case of 3-PGA, where the other negative charge is her

TABLE I   Influence of the H^+ gradient on Km- and Vmax (heteroexchange mode). Km values are in mM, Vmax values are in μmol/mg protein per min.

	external pH	Km without proton gradient	Vmax	Km with proton gradient	Vmax
Transport of DHAP					
inward directed proton gradient	6.8	1.30	9.7	1.93	9.7
outward directed proton gradient	7.8	1.18	9.7	0.74	9.7
Transport of 3-PGA					
inward directed proton gradient	6.8	0.72	10.0	0.72	14.9
outward directed proton gradient	7.8	2.17	10.0	2.17	7.5

contributed by the carboxylic group (pK = 3.42). Since two negative charges appear to be involved in the binding to the carrier, the differences in the charge of the phosphate moiety may partly account for the different responses to an applied proton gradient observed between DHAP and 3-PGA.

Physiological significance of the regulation of the phosphate translocator by a proton gradient

In intact chloroplasts, illumination causes a large increase of the stromal pH leading to a proton gradient across the envelope (inside alkaline). Under these conditions, the efflux of substrates out of the stroma into the cytosol is directed against the proton gradient and corresponds to the experiments presented here in which an outward directed proton gradient was applied across the liposome membran The influx of substrates into illuminated chloroplasts, however, is directed with the proton gradient and is analogous to the transport of substrates under the influence of an inward directed proton gradient. The question arises, whether the effects of a proton gradient observed in the reconstituted liposome system can explain the considerable changes in the transport of DHAP and 3-PGA between illumination and darkness. In experiments with illuminated protoplasts from spinach (Stitt et al. 1980) we have found stromal concentrations of 7.0 mM Pi, 2.1 mM 3-PGA and 0.6 mM DHAP. The cytosolic concentrations were 2.3 mM 3-PGA and 2.8 mM DHAP; Pi may be estimated as 10 mM. Using these concentrations

TABLE II   Model calculation of the influence of the light induced H^+ gradient on the fluxes of triose phosphate and 3-phosphoglycerate

The kinetic constants have been derived from Table I.

	Transport DHAP/3-PGA outward	inward	outward/inward
Light	1.2	0.21	5.9
Dark	0.15	0.62	0.24

and the values for the apparent Km and Vmax of Table I and assuming a simple Michaelis Menten characteristic for the transport, the transport rates for each single metabolite can be calculated. A model calculation presented in Table II shows, that in the light the ratio of DHAP/3-PGA transported is more than 8 fold higher for the outward transport and 3 fold lower for the inward transport than the corresponding ratios obtained in the dark. In the dark, the species transported out would be mainly 3-PGA with DHAP transported mainly into the chloroplasts. These fluxes are reversed under the influence of a proton gradient (light conditions). So, the observed effects of a proton gradient result in a considerable change in the ratio of the substrates transported by the phosphate translocator. The increase of the stromal pH during illumination leading to a proton gradient across the envelope indeed enhances the export of DHAP and the import of 3-PGA.

However, the presented effects of a proton gradient on the selectivity of the substrate transported by the phosphate translocator can not fully explain the observed DHAP/PGA gradients between illumination and darkness as measured in spinach leaves (Heldt et al. these proceedings) or isolated chloroplasts (Heldt et al. 1978) One additional factor might be the increasing stromal Mg^{++} concentration in the light, which results in an inhibition of 3-PGA transport and a stimulation of Pi and DHAP transport (Flügge, unpubl. results).

Acknowledgement This work was supported by the Deutsche Forschungsgemeinschaft

References

Fliege R, Flügge UI, Werdan K and Heldt, HW (1978) Biochim. Biophys. Acta 205, 232 - 247.

Flügge UI and Heldt HW (1981) Biochem. Biophys. Acta 638, 296 - 304

Flügge UI, Gerber J and Heldt HW (1983) Biochem. Biophys. Acta in press

Heber, U and Santarius KA (1970) Z. Naturforsch. 25b, 718 - 728.

Heldt HW et al. (1983) these proceedings

Heldt HW, Flügge UI and Fliege R (1978) in: Mechanism of proton and calcium pumps (M. Avron et al. eds) Elsevier/North Holland, pp 105 - 114

Lilley RMcC, Chon CJ, Mosbach, A and Heldt HW (1977) Biochem. Biophys. Acta 460, 259 - 272.

Stitt M, Wirtz W and Heldt HW (1980) Biochim. Biophys. Acta 593, 85 - 102.

COMPLEX KINETICS OF PROTON DEPOSITION INSIDE CHLOROPLAST THYLAKOIDS

A.B. HOPE and D.B. MATTHEWS

1.   INTRODUCTION

The acidification of the intrathylakoid spaces following single-turnover flashes was first studied by Ausländer and Junge (1975) using neutral red with a non-permeating buffer to prevent response to external pH changes. The signal consisted of half "fast" ($t_{\frac{1}{2}} <$ lms) and half "slow" ($t_{\frac{1}{2}} \simeq$ 20ms) components, attributed to H_2O and plastoquinol oxidation respectively producing 1 H^+ each, per e^-, during non-cyclic electron flow.

This simple picture is frequently inadequate (eg. see Velthuys 1980; Hong, Junge 1983) giving rise to unresolved, contradictory statements about the H^+/e^- ratio, for example.

We have re-examined the kinetics of proton deposition, following single-turnover flashes to dark-adapted, class C pea chloroplasts, in the time range 1-1000 ms under a variety of conditions, with adequate signal-to-noise and a response time-constant of $100\mu s$.

2.   RESULTS

During non-cyclic electron flow, with H_2O the donor and ferricyanide or methylviologen the acceptor, three components have regularly been resolved from a semi-logarithmic plot of the neutral red acidification signal (Fig. 1(a) and (b)). These components could be buffered away by the penetrating buffers imidazole or hydroxyethyl morpholine, but higher concentrations, up to 15mM instead of 4mM, were necessary.

A hitherto unrecorded slow component ($t_{\frac{1}{2}} \simeq$ 100ms) amounted to 25-30% of the total, the remainder being the "fast" (< lms) and intermediate ($t_{\frac{1}{2}} =$ 12-20ms) components already known. The significant difference between our experiments and earlier ones is probably that now only 10 flashes suffice to give a satisfactory kinetic resolution, instead of 50-250.

The slow component was:
(a) sensitive to DCMU, but merely delayed and decreased by DBMIB,
(b) eliminated by 20-40 fast (10-20Hz) preflashes given about ls before test flashes (Figure 2 shows the classical composition of half fast, plus half 20ms components, after this treatment), (c) preferentially decreased and accelerated by 2-4mM hydroxyethyl morpholine, (d) absent during cyclic electron flow (+DCMU + NADPH + ferredoxin, anaerobic), (e) present in spinach as well as pea chloroplasts, in frozen/thawed chloroplasts, and in preparations from mature or immature pea leaves.

The intermediate component:
(a) was equal in size for non-cyclic and cyclic electron flow, (b) had a $t_{\frac{1}{2}}$ of 3ms for cyclic, and for non-cyclic conditions with no added electron acceptor, (c) was sensitive to DBMIB, (d) displayed clear binary periodicity in size (Fig. 3) during the first few flashes to

Sybesma, C. (ed.), Advances in Photosynthesis Research, Vol. II. ISBN 90-247-2943-2.
© *1984 Martinus Nijhoff/Dr W. Junk Publishers, The Hague/Boston/Lancaster.*

dark-adapted suspensions.

The fast component:
(a) developed in size and did not reach a steady level (even allowing
for unequal H^+ releases during S-state transitions) for many turnovers,
during a train of flashes to dark-adapted suspensions, (b) was absent
(+NADPH + ferredoxin, anaerobic) despite the absence of DCMU, but could
be induced by rapid preflashes.

3.   CONCLUSIONS

The slow component of flash-induced proton release is tentatively placed
near PSII water oxidation. Why is it so slow? Three partial explanations could be considered:
(a) some PSII (Q) may be disconnected from electron transport chains,
(b) proton release into the intrathylakoid spaces may be indirect
(Prochaska, Dilley 1978), very pH_1-sensitive, but accelerated under more
acid (rapid pre-flashes) or more alkaline conditions (+ hydroxyethyl
morpholine buffer) than prevailing during test flashes at 0.5Hz, (c)
electron transfer out of PSII may depend on the redox state of the PQ
pool. If prevented, a slow electron cycle may occur, with slow H^+
release. Preflashes oxidise the pool and allow "normal" non-cyclic flow.

The intermediate component, whether of $t_{\frac{1}{2}}$ = 3ms or 15-25ms, originates
from PQH_2 oxidation; the smaller $t_{\frac{1}{2}}$ depends on a reduced pool. The
binary periodicity observed during non-cyclic flow (Fig. 3) matches
closely that for proton uptake in similar preparations (Hope, Matthews
1983). It implies that electrons and protons both attach to and leave
PQH_2 two at a time on alternate flashes, and speaks against a Q-cycle
(Velthuys 1978); Crowther, Hind 1980). When the PQ pool is reduced
('cyclic" conditions) the intermediate component is the same size as the
average for 10 flashes during non-cyclic flow, which tends to eliminate
a Q-cycle under reducing conditions as well.

The fast component, because not fully patent for many turnovers (Fig. 4),
may yield misleading information about H^+ release during S-state
transitions. Conflicting findings are discussed by Förster et al (1981).
Nevertheless the fast component appears to imply a 0112 pattern.

Generally speaking, kinetic resolution of proton deposition into "fast"
and "slow" components, and attribution to PSII and I respectively cannot
be done by eye. The ratio of "slow":"fast" so obtained, though greater
than 1, does not signify a Q-cycle (though it may indicate some electron
donation not from PSII) because some of the "slow" could originate from
PSII water oxidation.

REFERENCES

Ausländer W and Junge W (1975) Neutral red, a rapid indicator for pH
changes in the inner phase of chloroplasts, FEBS Lett. 59, 310-15.
Crowther D and Hind G (1980) Partial characterisation of cyclic electron
transfer around photosystem I in intact chloroplasts, FEBS Lett. 98,
386-90.

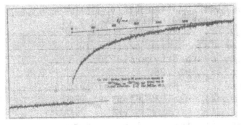

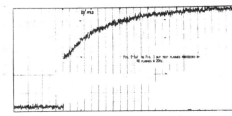

Fig. 2(a2) as Fig. 1 but test flashes preceded by 40 flashes @ 20Hz.

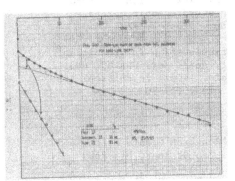

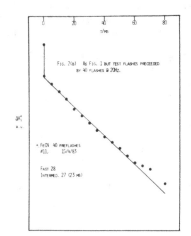

Fig. 2(b)   As Fig. 1 but test flashes preceded by 40 flashes @ 20Hz.

+ FeCN 40 PREFLASHES
#11,   15/4/83

Fast 28
Intermed. 27 (23 ms)

Fig. 3 The magnitude of the fast, intermediate and slow components of proton deposition during the first 7-9 flashes to dark-adapted suspensions (10 such runs accumulated).

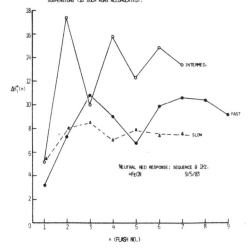

Neutral red response; sequence @ 1Hz.
+FeCN   9/5/83

Fig. 4 The mean of the fast component for flashes 1-4, 5-8, 9-12, etc, compared with those predicted (dashed line) from proton release in a 0,1,1,2 pattern during S-state transitions, normalised to 5 units for flashes 1-4. ●:Flashes given @ 10Hz, ○:1Hz.

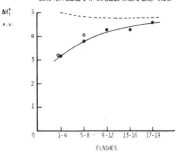

Förster V, Hong Yu-Qun and Junge W (1981) Electron transfer and proton pumping under excitation of dark adapted chloroplasts with flashes of light, Biochim. Biophys. Acta. 638, 141-52.

Hong Yu-Qun and Junge W (1983) Localised or delocalised protons in photophosphorylation? On the accessibility of the thylakoid lumen for ions and buffers, Biochim. Biophys. Acta. 722, 197-208.

Hope AB and Matthews DB (1983) Further studies of proton translocations in chloroplasts after single-turnover flashes, Aust. J. Plant Physiol. 10, in press.

Prochaska LJ and Dilley RA (1978) Site specific interaction of protons liberated from photosystem II oxidation with a hydrophobic membrane component of the chloroplast membrane, Biochem. Biophys. Res. Comm. 83, 664-72.

Velthuys BR (1978) A third site of proton translocation in green plant photosynthetic electron transport, Proc. Nat. Acad. Sci. USA 75, 6031-4.

Velthuys BR (1980) Electron and proton transfer events in chloroplasts during a short series of flashes, FEBS Lett. 115, 167-70.

ACKNOWLEDGEMENTS

The project was supported by the Australian Research Grants Scheme and Flinders University. The assistance of Mrs. Pam Stainer, Mr. Rob Scholten and Mr. West Hiscock is gratefully acknowledged.
Authors address: Flinders University, Bedford Park, South Australia,
                       5042, Australia.

PERMEABILITY PROPERTIES OF THE CHLOROPLAST ENVELOPE FOR MONOVALENT IN-
ORGANIC CATIONS AND THEIR DISTRIBUTION BETWEEN THE INTACT CHLOROPLAST
AND THE EXTERNAL MEDIUM IN THE LIGHT AND DARK.

BARBARA DEMMIG, HARTMUT GIMMLER

1. INTRODUCTION

The chloroplast envelope is known to be a site for regulation of the
intrachloroplastic pools of metabolites. Howewer, it is not known
whether regulatory processes which determine the distribution of inorganic ions and protons take place at the level of the chloroplast
envelope. Mg^{2+}-dependent ATPase activity associated with the chloroplast envelope (Joyard, Douce 1975) has recently been postulated to
perform $K^+(Na^+)/H^+$ antiport (Maury et al., 1981). We present evidence
that K^+ is not transported in a primary active process but that light-
induced cation uptake into isolated intact chloroplasts is driven by
an electrical potential gradient. Furthermore, our data suggest that
different capacities of isolated intact chloroplasts to preserve their
native inorganic ion composition might be one reason for the variation
in CO_2-fixation activity and stability which is being observed with
isolated chloroplasts.

2. MATERIALS AND METHODS are described in the legends of the figures
and elsewhere (Demmig, Gimmler, 1983 ; Gimmler et al. 1981). "Low-
salt chloroplasts" were isolated in "low-salt medium" (0.33 M sorbitol,
25 mM Hepes-Tris, pH 7.6) and stored in the same medium for 1 hour on
ice. "Sorbitol/KCl media" were composed of 0-100 mM KCl, 0.16-0.33 M
sorbitol, 25 mM Hepes-Tris, pH 7.6. All internal concentrations were
calculated using an osmotic volume of the intact chloroplast of 30 ul
mg^{-1} Chl. In uptake studies which were performed in the light the media
included either 0.1 mM methylviologen and catalase or oxaloacetate to
support electron transport.

3. RESULTS AND DISCUSSION
Internal K^+ and Cl^- concentrations of intact spinach chloroplasts.
Intact chloroplasts obtained by gentle rupture of isolated protoplasts
exhibited K^+ concentrations which were as high as and Cl^- concentrations which were twofold higher than the respective concentrations of
leaf sap from the corresponding leaf tissue (FIG 1, A). Internal K^+
concentrations of chloroplasts isolated mechanically, howewer, were
markedly lower (FIG 1, B). Rapid loss of K^+ but not of Cl^- was observed with these chloroplasts suspended in low-salt medium (FIG 1, A,
B). Cl^- concentrations were the same in both chloroplasts obtained
from protoplasts and those isolated mechanically (FIG 1, A,B). K^+
efflux from intact chloroplasts (isolated mechanically) is at least
partly balanced by an uptake of protons which results in a decrease of
stroma pH both in the light and dark (Demmig, Gimmler, 1983). Intactness as determined by the ferricyanide method (Heber, Santarius, 1970)
was 95-100% for chloroplasts obtained from protoplasts and 80-90% for
chloroplasts isolated mechanically. This difference might be important
for the permeability properties of the envelope.
Effect of external KCl on photosynthetic activity of isolated chloroplasts. A decrease of stroma pH as is observed in low-salt chloroplasts
is correlated with a reduced activity of Calvin-cycle enzymes. CO_2

Sybesma, C. (ed.), Advances in Photosynthesis Research, Vol. II. ISBN 90-247-2943-2.
© *1984 Martinus Nijhoff/Dr W. Junk Publishers, The Hague/Boston/Lancaster.*

fixation of low-salt chloroplasts was efficiently stimulated by addition of monovalent cations such as K^+, Na^+ or even Rb^+ (Demmig,
Gimmler, 1979 ; Kaiser et al., 1980) but was inhibited by choline
chloride (Demmig, unpublished). Addition of monovalent cations to the
standard medium affects CO_2 fixation and stability of isolated chloroplasts in a manner dependent on the integrity of the chloroplasts. FIG
2 shows three chloroplast preparations which were treated identically
but exhibited different properties. CO_2 fixation of chloroplasts with
low initial photosynthetic activity (FIG 2, I) was markedly stimulated by addition of KCl to the storage medium whereas chloroplasts with
high photosynthetic activity were not stimulated (FIG 2, III).Intact-
ness of preparation I, II and III was 78, 86 and 93% respectively.
These differences may reflect a different capacity of the three chloroplast preparations for retaining the high native K^+ content.
<u>Light-stimulated K^+/H^+ exchange by intact chloroplasts</u>. Uptake of K^+
and release of protons observed with intact chloroplasts is stimulated
by light (Gimmler et al., 1974). Valinomycin as expected increased
the velocity of K^+ uptake but did not abolish the light-dependent K^+
uptake (FIG 3). K^+ uptake in the light minus K^+ uptake in the dark
reached a maximum between 20 and 40 mM external KCl (FIG 4). Increasing the external KCl concentration beyond 20 mM markedly decreased the
light-induced uptake. Added Tetraphenylphosphonium$^+$ was accumulated in
the dark by intact chloroplasts suspended in low-salt media (FIG 5)
indicating an internally negative electrical potential which is likely
to be a Gibbs Donnan and/or a K^+ diffusion potential. In the light TPP^+
accumulation increased suggesting that this electrical potential becomes more negative in the light (FIG 5). ΔE values were calculated
using the Nernst Equation for TPP^+ steady state distribution which is
reached within 10 min. The light-induced increase of K^+ uptake and TPP^+
accumulation both in the light and dark concomitantly become very small
at high external concentrations of KCl (FIG 3, 5) or other salts (not
shown). The above observations are not consistent with a direct linkage
of light-dependent ion movements via a strict K^+/H^+ antiport. Electrogenic proton pumping, nevertheless, would be consistent with the observed light-induced increase of an electrical potential difference. In
this case K^+ or another cation would passively follow the electrical
gradient. However, the resistance of light-induced uptake of K^+ against
Dio-9 (Gimmler et al., 1981) or Tentoxin (Demmig, unpublished) argues against the involvement of an active ATP-fueled process. Both
stroma pH in the light and light-induced K^+ uptake are even slightly
higher in the presence of energy transfer inhibitors while ATP pool is
significantly decreased (Gimmler et al., 1981). Assuming the involvement of an envelope ATPase one would have to postulate either an unusually high affinity for ATP or a factor other than ATP being the
trigger for the putative pump.

FIGURE 1. Internal K$^+$ (O) and Cl$^-$ (△) concentrations in intact chloroplasts obtained by rupture of protoplasts (A) or isolated mechanically (B). (●) K$^+$ concentrations and (▲) Cl$^-$ concentrations of leaf sap. (A) Chloroplasts equivalent to 300 ug Chlorophyll obtained from protoplasts were 2 to 4 times (2x, 3x, 4x) spun down and resuspended in 10 ml low-salt medium. (B) Leaves were blended and chloroplasts washed 3 times as described above in low-salt medium. K$^+$ was determined by flamephotometry and Cl$^-$ by potentiometry.

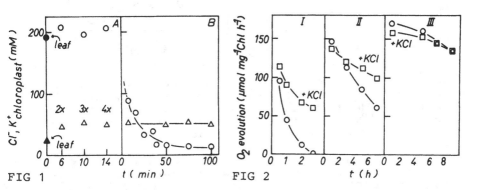

FIG 1   FIG 2

FIGURE 2. KHCO$_3$-dependent O$_2$ evolution by different preparations (I,II, III) of intact chlorplasts at various times after isolation and storage in standard medium (O) and standard medium +50 mM KCl (□). The measurement of CO$_2$-dependent O$_2$ evolution was in each case performed in the standard medium.

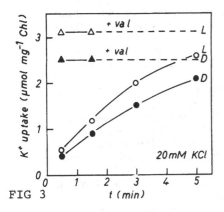

FIG 3

FIGURE 3. Kinetics of K$^+$ uptake in light (L) and dark (D) as influenced by valinomycin (△,▲). Low-salt chloroplasts were added to a sorbitol/KCl medium containing 20 mM KCl (^{86}RbCl as a tracer) and valinomycin (5x10^{-7} M). Illumination was started simultaneously. The reaction was stopped by centrifugation through silicone oil after different periods of time.

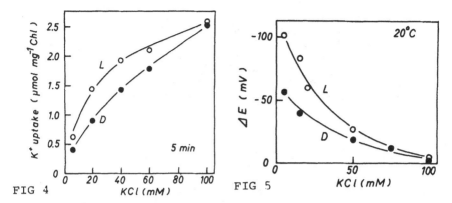

FIG 4

FIG 5

FIGURE 4. K^+ uptake by intact chloroplasts at different external KCl concentrations in the light (L) and in the dark (D). Low-salt chloroplasts were added to sorbitol/KCl media containing different KCl concentrations and ^{86}RbCl as a tracer. Illumination was started at the same time. Reaction was stopped by centrifugation.

FIGURE 5. ΔE_{TPP}^+ for intact chloroplasts in light (O) and dark (●) at various external KCl concentrations. Low-salt chloroplasts were incubated in sorbitol/KCl media with ^{14}C-TPP$^+$ for 10 min at 20°C in dark and light.

ACKNOWLEDGEMENT
This work was supported by the DEUTSCHE FORSCHUNGSGEMEINSCHAFT.

REFERENCES
Demmig B., Gimmler H. (1983) Properties of the isolated intact chloroplast at cytoplasmic K^+ concentrations. I Light-induced cation uptake into intact chloroplasts is driven by an electrical potential difference, Plant Physiol, in press
Demmig B., Gimmler H. (1979) Effect of divalent cation fluxes across the chloroplast envelope and on photosynthesis of intact chloroplasts, Z Naturforsch 34, 233-241
Gimmler H., Demmig B., Kaiser W.M. (1981) The role of K^+ and H^+-fluxes across the chloroplast envelope for photosynthetic CO_2 fixation, in: G. Akoyunoglou, eds, Proc 4th Int Congr on Photosynthesis Research, vol 4, Balaban Int. Sciences Services, Philadelphia pp 599-608
Gimmler H., Schäfer G., Heber U. (1974) Low permeability of the chloroplast envelope towards cations, in: M. Avron, eds, Proc 3rd Int Congr on Photosynthesis Research, vol 3, Elsevier, Amsterdam pp 1381-1892
Heber U., Santarius K,A. (1970) Direct and indirect transfer of ATP and ADP across the chloroplast envelope, Z Naturforsch 25b, 708-728
Joyard J., Douce R. (1975) Mg^{2+}-dependent ATPase of the envelope of spinach chloroplasts, FEBS Lettters 51, 335-340
Kaiser W.M., Urbach W., Gimmler H. (1980) The role of monovalent cations for photosynthesis of isolated intact chloroplasts, Planta 149, 170-175
Maury W.J., Huber S.C., Moreland D.E. (1981) Effects of Mg^{2+} on intact chloroplasts. II Cation specifity and involvement of the envelope ATPase in (sodium)potassium/proton exchange across the envelope, Plant Physiol 68, 1257-1263

Authors address: Barbara Demmig, Lehrstuhl Botanik I der Universität
                 Mittlerer Dallenbergweg 64 , D-8700 Würzburg

FLASH-INDUCED pH CHANGES IN PHOTOSYSTEM I-ENRICHED VESICLES MONITORED
WITH NEUTRAL RED AND CRESOL RED

FRITS A. DE WOLF, LEO P.A. VAN HOUTE, FONS A.L.J. PETERS AND
RUUD KRAAYENHOF

1. INTRODUCTION

Proton concentration changes occurring on either side of photosynthetic
membranes, in the aqueous phases or at the interfaces of the membranes,
are important phenomena accompanying photosynthetic energy conservation.
In single turnover experiments, the relevant changes are transient
and occur within 0.1 second or less. They should be monitored by means
of artificial pH indicators (Grünhagen, Witt, 1970; Hong, Junge, 1983),
since electrodes respond too slowly. We have used photosystem I-
enriched (PSI) vesicles, derived from spinach thylakoids by mild
digitonin treatment, which are capable of ferredoxin-mediated cyclic
electron transfer (Peters et al., 1983a; Peters et al., 1983b). They
provide a less complicated system for the study of proton translocation
than broken chloroplasts (no stacking-destacking, no distinction
between grana and stroma lamellae). Neutral red (NR) or cresol red (CR)
were added just prior to experimentation but we have also tried to
trap these dyes inside the vesicles during their preparation.

2. MATERIALS AND METHODS

PSI vesicles were prepared according to Peters et al. (1983a) and
stored under liquid nitrogen in a medium containing 1 mM Hepes (pH 7.2),
250 mM sorbitol and 5 mM $MgCl_2$. NR or CR (Sigma) were added just
before experimentation at a final concentration of 20 μM. We tried
to trap NR and CR by adding them to the medium during digitonin
treatment, at 50 μM and 200 μM respectively. Single turnover flash
experiments were carried out as described elsewhere (Peters et al.,
1983b) at 20°C in a medium identical to that for storage, except for
additions as indicated in the legends to the figures. Binding of
dyes was quantified by comparing the absorption at 524 nm (NR) or
572 nm (CR) in the dark before and after precipitation of the vesicles
(1 hr at 129,000 g). The millimolar absorption coëfficients were
determined as 10.5 $mM^{-1}cm^{-1}$ (NR) and 5.1 $mM^{-1}cm^{-1}$ (CR).

3. RESULTS AND DISCUSSION

3.1. Binding of externally added dyes

Figure 1 shows that about equal amounts of NR are bound electrostatically
and bound (or accumulated) due to hydrophobicity of NR respectively.
Electrostatic binding is abolished by 20 mM $MgCl_2$. Binding of NR
to PSI vesicles is further accompanied by a shift of the pK_a of NR
from 6.71 (free NR) to 6.81 (in a suspension of vesicles containing
50 μg Chl/ml and with 20 μM NR) and to 6.93 (same suspension under
continuous illumination). From the reversal of the electrostatic
binding by externally added $MgCl_2$, we conclude that NR is at least
partly bound at the outside of the PSI vesicles. In contrast, CR is
hardly bound and its binding is not affected by $MgCl_2$. Others have
described the binding of NR inside thylakoids of freeze-thawed broken

Sybesma, C. (ed.), Advances in Photosynthesis Research, Vol. II. ISBN 90-247-2943-2.
© *1984 Martinus Nijhoff/Dr W. Junk Publishers, The Hague/Boston/Lancaster.*

chloroplasts, resulting in a pK$_a$ shift of NR to 7.25 (Junge et al., 1979), or even to 7.8 (Hong, Junge, 1983). This binding was also reversed by MgCl$_2$ or KCl. However, in our opinion it was not convincingly shown that NR was present at the inside of the thylakoids, the main argument for this being that external bovine serum albumin (BSA) at 1.3 mg/ml could abolish the observed NR response, due to its buffer capacity (± 0.15 mM, derived from the next cited reference) (Ausländer, Junge, 1975).

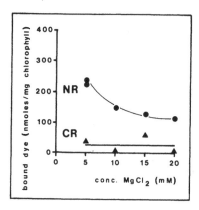

FIGURE 1. The binding of neutral red (NR) and cresol red (CR) as a function of MgCl$_2$ concentration. The medium contained 20 μM NR or CR and PSI vesicles containing 50 μg Chl/ml. Experiments were carried out in the dark.

3.2 Flash-induced responses of externally added dyes

In Fig. 2A, flash-induced absorption changes at 548 nm, where no interference of carotenoid responses occurs, are shown (a). The NR response (trace c) is obtained by subtracting the signal obtained in the absence of NR (trace a) from that obtained with externally added NR (trace b). In trace c, first a fast absorption decrease occurs. This component of the signal has not been reported by Ausländer and Junge (1975) for the case of broken chloroplasts. Possibly, it represents fast binding of NR to PSI vesicles. The slower absorption increase presumably represents NR-monitored acidification because (1) a pH decrease causes an increase of absorption by free NR at this wavelength (in the absence of vesicles) and (2) the observed slow increase of the flash-induced NR absorption (trace d) can be abolished by the buffer Tricine. (Fig. 2B). The extent of the slow NR component is halved by the addition of 80 nM valinomycin and totally abolished by the addition of 3 μM of the uncoupler SF6847 (not shown; higher concentrations not tested);   the fast component of the NR transient is hardly influenced by the agents at the tested concentrations. Since Tricine will not be preferentially present inside the vesicles, and since Tricine is a relatively impermeant buffer, we conclude that NR is mainly present in the outer membrane interface of the vesicles. The influence of external MgCl$_2$ (Fig. 1) supports this conclusion. Thus, NR seems to monitor an acidification at the outer surface of the vesicle membrane, in single turnover experiments. Continuous illumination induced, electrode-monitored alkalinization of the outside medium (not shown) and other data (Peters et al., 1983b) indicate that these vesicles are predominantly oriented right side out.

In contrast to NR, CR does not show any flash-induced absorption change. This is presumably due to the negative charge of CR, being repulsed by the negative surface charge of the vesicles. Thus, CR would be too far from the membrane to monitor small, local pH changes occurring close to the membrane.

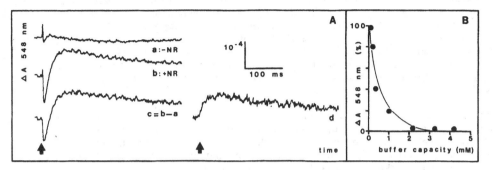

FIGURE 2. Flash-induced absorption changes of neutral red (NR) in the presence of PSI vesicles. A: NR response after correction for nonspecific absorption changes (trace c); pH-monitoring component of the NR response (trace d), obtained by subtracting the response occuring in the presence of 20 mM Tricine (not shown) from the NR response as shown in trace c. B: Amplitude of the pH-monitoring NR response (trace d) as a function of the buffer capacity imposed by external Tricine. (A buffer capacity of 4 mM corresponds to a Tricine concentration of 20 mM.) The medium contained in addition 5 μM ferredoxin, 300 μM NADPH, 100 μM O_2, 20 μM NR and PSI resides containing 50 μg Chl/ml. The transients which followed 16 flashes, fired at 0.125 Hz, were averaged.

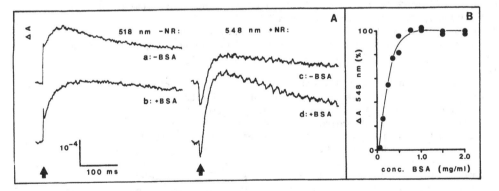

FIGURE 3. The influence of bovine serum albumin (BSA) on the flash-induced carotenoid and neutral red (NR) responses. A: carotenoid (518 nm) and NR (548 nm) responses in the absence and presence of 1 mg/ml BSA. B: Extent of the enhancement of the NR absorbance increase by increasing BSA concentrations, as shown in trace d versus c for 1 mg/ml BSA. The medium and the flashing conditions were as described under Fig. 2.

We think that BSA influences the structure of the vesicle membranes,
resulting in a enhanced NR response (Fig. 3, trace d versus c), since BSA
also influences the slow carotenoid absorption increase at 518 nm in the
absence of NR (Fig. 3, trace b versus a). BSA influence on membrane
systems has been described elsewhere (Friedlander, Neumann, 1968;
Robinson et al., 1980). The enhancement of the slow component of the NR
response by 1 mg/ml BSA, shown in Fig. 3B, is presumably not due to the BSA
buffer capacity: this concentration of BSA corresponds to a buffer
capacity of 0.39 mM. A comparable buffer capacity imposed by Tricine only
induced a 40% decrease of the slow NR absorption increase. Our findings
are at variance with those of Ausländer and Junge (1975), who interpreted
their results as binding of NR inside spinach thylakoids because NR was
supposed to monitor an acidification exclusively on the inside, after
addition of BSA.

Sofar, we did not succeed in trapping NR or CR inside the PSI vesicles
during digitonin treatment, when the vesicles are formed from the
thylakoids and when the interior of the thylakoids is thought to come
into contact with the medium. However, continuous illumination experiments indicated that externally added buffer could be trapped inside
the vesicles. We will try to find a way to trap pH probes inside our
vesicles, since such trapping may seriously contribute to the elucidation
of the mechanism of flash-induced proton translocation.

REFERENCES

Ausländer W and Junge W (1975) Neutral red, a rapid indicator for pH
changes in the inner phase of thylakoids, FEBS Lett. 59, 310-315.
Friedlander M and Neumann J (1968) Stimulation of photoreactions of
isolated chloroplasts by serum albumin, Plant Physiol. 43, 1244-1254.
Grünhagen HH and Witt HT (1970) Umbelliferone as indicator for pH
changes in one turn-over, Z. Naturforsch. 25b, 373-386.
Hong YQ and Junge W (1983) Localized or delocalized protons in photophosphorylation? On the accessibility of the thylakoid lumen for ions
and buffers, Biochim. Biophys. Acta 722, 197-208.
Junge W, Ausländer W, McGeer AJ and Runge Th (1979) The buffering
capacity of the internal phase of thylakoids and the magnitude of pH
changes inside under flashing light, Biochim. Biophys. Acta 546,
121-141.
Peters FALJ, Van Wielink JH, Wong Fong Sang HW, De Vries S and
Kraayenhof R (1983a) Studies on well coupled photosystem I-enriched
subchloroplast vesicles. Content and redox properties of electron
transfer components, Biochim. Biophys. Acta 722, 460-470.
Peters FALJ, Van Spanning R and Kraayenhof R (1983b) Studies on well
coupled photosystem I-enriched subchloroplast vesicles. Optimization
of ferredoxin-mediated cyclic photophosphorylation and electric
potential generation, Biochim. Biophys. Acta 724, 159-165.
Ronbinson HH, Guikema JA and Yocum CF (1980) Reversal of dibromothymo-
quinone inhibition of photosynthetic electron transport by bovine
serum-albumin, Arch. Biochem. Biophys. 203, 681-690.

Authors address: Biological Laboratory, Vrije Universiteit, de
                 Boelelaan 1087, 1081 HV Amsterdam, The Netherlands.

THE EFFECT OF TRYPSIN AND CHYMOTRYPSIN ON THE 519 NM FIELD-INDICATING
ABSORPTION CHANGE IN ISOLATED CHLOROPLASTS

C.A. RAINES/DR. M.F. HIPKINS   GLASGOW UNIVERSITY

INTRODUCTION

Proteolytic enzymes have been used to investigate the location and
function of some of the proteins associated with the thylakoid membrane,
particularly the light-harvesting pigment proteins (Steinbeck *et al.*,
1979), the coupling factor (CF_1) (Moroney, McCarty, 1982) and the
shield over the acceptor side of photosystem II (Renger *et al.*, 1976).
For example, trypsin has been shown to induce an increase in the rate of
light-dependent electron transport which has been associated with
uncoupling of photophosphorylation due to partial digestion of CF_1.   In
addition trypsin, like dithiothreitol (DTT), stimulates the latent ATP-
hydrolase activity of CF_1.
Trypsin and chymotrypsin attack bonds between different amino acids:
in this paper we have compared the effects of digesting thylakoid
membranes with trypsin and chymotrypsin in an attempt to distinguish
differential action of the two enzymes.

MATERIALS AND METHODS

Chloroplasts were isolated from 14-day-old peas by grinding leaf tissue
in the medium of Stokes and Walker (1971). and resuspended in the medium
of Renger *et al.* (1976).   Incubations with trypsin or chymotrypsin (Sigma
types XIII, TCPK & VII TLCK respectively) were at pH 7.2, temp. 23°C and were
stopped by addition of Soya Bean trypsin inhibitor (Sigma).
Steady-state oxygen evolution was measured using a conventional Clarke-
type oxygen electrode.   Field-indicating flash-induced absorbance
changes at 519 nm (Δ519) were measured essentially as described
previously (Hipkins, Hillman, 1981) whilst photophosphorylation and ATP
hydrolysis were measured using the pH method of Nishimura *et al.* (1962).

RESULTS AND DISCUSSION

Enzyme digestion of thylakoid membranes, caused an increase in the rate
of electron transport from water to potassium ferricyanide (Fig. 1), an
increase in the rate of decay of Δ519 (Fig. 2, control, transients A and
C), and a decrease in the rate of phosphorylation (Table 1).   These
results lend support to the proposal (Moroney, McCarty, 1982) that
trypsin causes uncoupling of electron transport due to partial digestion
of CF_1, resulting in an increase in membrane permeability. Chymotrypsin
acts in a similar way, but the effects are less marked.
A significant increase is seen in the rate of decay of Δ519 on addition
of ATP to samples pretreated with D.T.T. or chymotrypsin in the light
(Fig. 2); trypsin acts similarly (data not shown).   Stimulation in the
rate of ATP hydrolysis is seen in membranes treated with trypsin or
chymotrypsin in the light (Table 2).   It is known that trypsin and D.T.T.
treatment in the light can stimulate ATP hydrolysis in chloroplast
membranes (Bakker-Grunwald, Van Dam, 1974).   From our results it would
appear that a correlation exists between ATP hydrolysis and an increase
in the rate of decay of Δ519.   This effect on the band shift may be

Sybesma, C. (ed.), Advances in Photosynthesis Research, Vol. II. ISBN 90-247-2943-2.
© *1984 Martinus Nijhoff/Dr W. Junk Publishers, The Hague/Boston/Lancaster.*

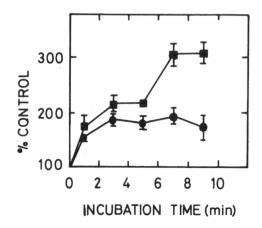

FIGURE 1.   Effect of trypsin and chymotrypsin on light induced electron transport to ferricyanide.   Changes in light induced electron transport in chloroplasts (50 μg ml^{-1}) after treatment with (■) trypsin (4 μg ml^{-1}) (●) chymotrypsin (30 μg ml^{-1}) at pH 7.2, temperature 23°C.

TABLE 1.   Effect of Trypsin and Chymotrypsin on the Rate of Phosphorylation

Enzyme Treatment	ATP min^{-1} (mg Chl)$^{-1}$		
	Control	1 min light	3 min light
Trypsin	1,760	720	0
Chymotrypsin	1,650	700	500

TABLE 2.   Effect of Trypsin and Chymotrypsin on the Rate of ATP Hydrolysis

Enzyme Treatment	Δ H$^+$ ng min-1 (mg Chl)-1		% Stimulation
	-ATP	+ATP	
Control 1 min	382.5	624.5	163
Trypsin Light	307.5	1,142.5	371
Control 3 mins	284.5	482.5	169
Chymotrypsin Light	297.5	1,500.0	505

TABLES 1 and 2:

Phosphorylation and hydrolysis were measured after chloroplasts (40   μg ml^{-1} were pretreated with trypsin (4 μg ml^{-1}), chymotrypsin (30 μg ml^{-1}), pH 7.8, temperature 23°C.   In the subsequent dark period additions were 100 μl Soy Bean trypsin inhibitor, methyl viologen 1.0 μM, ADP or ATP 0.5 mM.

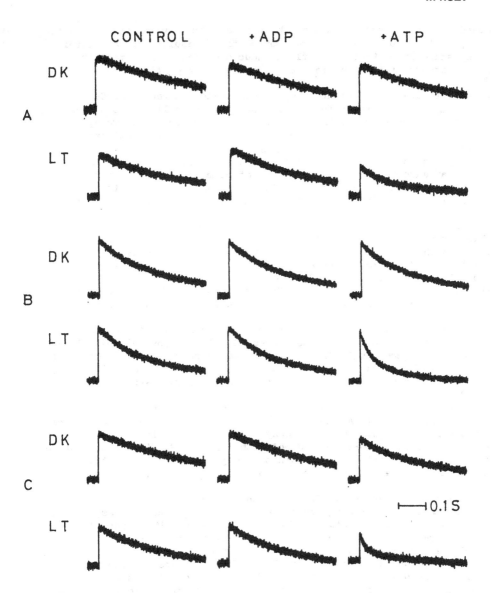

FIGURE 2. Effect of D.T.T. and chymotrypsin on Δ519, chloroplasts
(25 μg ml⁻¹) were incubated at 23°C, pH 7.2 in either (L) or (D).
ATP (0.5 mM) or ADP (0.5 mM) were added on completion of treatment with
potassium ferricyanide (7.5 μM). A. Control, no addition, B. D.T.T.
(5.0 mM) 1 min C. Chymotrypsin (15 μg ml⁻¹) 3 min. In the case of
C, 100 μl of Soya Bean trypsin inhibitor was added prior to
measurements.

caused by reverse proton pumping, brought about by hydrolysis (Carmeli, 1970), building up a large proton motive force resulting in a greater steady state $\Delta\Psi$. Hence, the flash-induced $\Delta\Psi$ decays more rapidly to the new steady state: the rate of decay being proportional to the total $\Delta\Psi$.

Our results are in agreement with the conclusion, Girault and Galmiche (1977), that the increase in the $\Delta519$ decay rate is linked to a change in proton conductivity of the membrane resulting from dark ATP hydrolytic activity. Schuurmans *et al.* (1981) show similar results but are hesitant in accepting this explanation

The only consistent difference observed between the effects of the two enzymes appears to be a smaller effect of chymotrypsin on electron-transport and phosphorylation. This difference may be due to a lack of availability of the specific amino acid bonds which chymotrypsin attacks.

REFERENCES

Bakker-Grunwald T and Van Dam K (1974) Mechanism of activation of the ATPase in chloroplasts, Biochim. Biophys. Acta. 347, 290-298.

Carmeli C (1970) Proton translocation induced by ATPase activity in chloroplasts, FEBS. Lett. 7, 297-300.

Girault G and Galmiche JM (1978) Effect of nucleotides on potential and pH changes across the thylakoid membrane. cf spinach chloroplasts, Biochim. Biophys. Acta. 502, 430-444.

Hipkins MF and Hillman JR (1981) Abscisic acid and ion fluxes through photosynthetic and artificial membranes, Z. Pflanzenphysiol 104, 217-224.

Moroney JV and McCarty RE (1982) Light dependent cleavage of the δ sub-unit of CF_1, by trypsin causes activation of Mg^{2+} - ATPase activity and uncoupling of photophosphorylation in spinach chloroplasts, J. Biol. Chem. Vol. 257, 5915, 5920.

Nishimura M Ito T and Chance B (1962) Studies on bacterial phosphorylation. III A sensitive and rapid method of determination of photophosphorylation, Biochim. Biophys. Acta, 59, 177-182

Renger G Erixon K Doring G and Wolff Ch (1976) Studies on the nature of the inhibitory effect of trypsin on the photosynthetic electron transport of system II in spinach chloroplasts, Biochim. Biophys. Acta. 440, 278-286.

Schuurmans JJ Leevwenk FJ Siu Oen B and Kraayenhof R (1981) Deconvolution of the flash-induced carotenoid and ox nol VI response in broken spinach chloroplasts. In Akoyunoglou G, ed. Photosynthesis I Photophysical processes - membrane energization, pp. 543-552, Philadelphia P.A: Balaban Int. Science Services.

Steinbeck KE Burke JJ and Arntzen CJ (1979) Evidence for the role of surface-exposed segments of the light harvesting complex in cation-mediated control of chloroplast structure and function, Arch. Biochem. Biophys. 195, 546-557.

Stokes DM and Walker DA (1971) Phosphoglycerate as a Hill oxidant in a reconstituted chloroplast system, Plant. Physiol. 48, 163-165.

ACKNOWLEDGEMENTS

We would like to thank the U.K. SERC for support.

Authors Address: Botany Department, University of Glasgow, Glasgow G12 8QQ, U.K.

THE EFFECTS OF AN ELECTRICAL FIELD ON THE PRIMARY REACTIONS OF SYSTEM II

H.J. van GORKOM, R.F. MEIBURG and R.J. van DORSSEN

1. INTRODUCTION

When osmotically swollen chloroplasts (blebs) are subjected to an externally applied electrical field pulse, a greatly enhanced local field is generated in the membrane. In this way a fraction of system II can be exposed to membrane potentials of up to 1 V with sub-ms time resolution. It was reported before that the charge recombination after illumination of system II in the presence of DCMU could be accelerated by many orders of magnitude (De Grooth, Van Gorkom, 1981). We now report field-induced changes of the chlorophyll fluorescence yield which suggest that the field-sensitive reaction is the electron transfer between the 'primary' acceptor Q and the 'intermediary' acceptor Pheophytin (at least in those system II centers that are membrane potential-sensitive at all, see accompanying paper).

2. METHODS

Blebs were prepared from spinach chloroplasts by 200-fold dilution. DCMU and an electron donor were added just before measurement. Fluorescence was excited by a high intensity, blue light modulated at 100 kHz. The photo-multiplier signal was fed into a lock-in amplifier, transient recorder and signal averager.

3. RESULTS

During a membrane potential pulse, the chlorophyll fluorescence yield was either enhanced or decreased, depending on the redox state of Q before the pulse (see fig. 1). Just after opening the excitation beam, when the blebs were still in the dark-adapted, low fluorescent state, the membrane potential enhanced the fluorescence yield. After longer illumination times in the presence of DCMU and an electron donor, when Q was in the reduced state, the membrane potential was accompanied by a quenching of fluorescence. If the artificial electron donor is omitted an endogenous donor is oxidized when Q is reduced, and the charge pair recombines when an electrical field is applied (De Grooth, Van Gorkom, 1981). The fluorescence yield decreases correspondingly and remains low after the pulse until Q is photoreduced again. As illustrated in fig. 2 this fluorescence decrease cannot be repeated by a second pulse of the same polarity: it corresponds to the total variable fluorescence of all system II exposed to a large membrane potential of the right polarity. In addition to the charge recombination at the beginning of the first pulse, both pulses in fig. 2 are seen to induce a reversible fluor-escence increase of the same amplitude, as if the sample were dark-adapted. Since Q was in the oxidized state only in those centers where charge recom-bination had taken place, it appears that the effects described in fig. 1 are due to the same fraction of system II in the sample and also correspond to the total variable fluorescence of this fraction.
At maximum membrane potential the fluorescence reached the same level as before the charge separation had taken place, i.e. the photochemical quenching by the reopened centers disappeared completely. We conclude that a membrane potential of 1 V is large enough to prevent electron transfer from excited P680 to Q.

Sybesma, C. (ed.), Advances in Photosynthesis Research, Vol. II. ISBN 90-247-2943-2.
© 1984 Martinus Nijhoff/Dr W. Junk Publishers, The Hague/Boston/Lancaster.

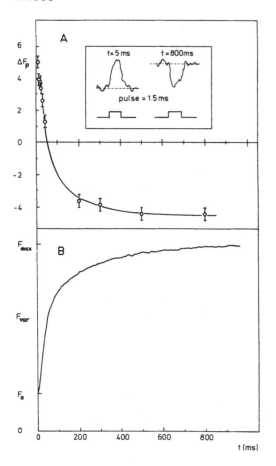

FIGURE 1.
A. Fluorescence change induced by a field pulse of 1100 V/cm, in % of the maximum variable fluorescence, versus illumination time in the presence of 10 μM DCMU and 5 μM Tetraphenylboron. Inset: example of measured traces. B. Fluorescence induction curve under the same conditions.

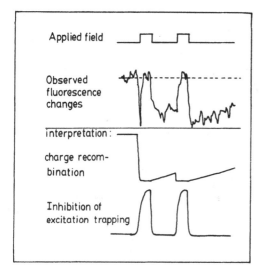

FIGURE 2.
Fluorescence kinetics induced by a pair of pulses of 1100 V/cm of the same polarity, measured 40 ms after a saturating flash in the presence of 5 μM DCMU, but in the absence of an artificial electron donor.

We ascribe the field-induced quenching observed when both Q and the oxygen evolving apparatus were in the reduced state to 'reversed' electron transport from Q$^-$ to the intermediary acceptor, Pheophytin. The fluorescence is low when Phe is in the reduced state (Klimov et al., 1977), probably due to excitation transfer towards Phe$^-$, followed by rapid internal conversion to the ground state (if the electron is actually pushed on towards P680, the same quenching mechanism is expected). By lack of a positive charge to recombine with, the electron returns to Q as soon as the membrane potential decreases.

From the field strength dependence of the quenching (fig. 3) and the bleb size distribution it was calculated that half saturation occurred at a membrane potential of 330 mV, so the free energy difference between the quenching and the non-quenching state is at most 330 mV. This value corresponds to the midpoint potential difference between Phe and Q (cf. Klimov, Krasnovsky, 1981, and Thielen, Van Gorkom, 1981a) and to the activation energy of the charge recombination of P680$^+$Phe Q$^-$ (Döring, 1975). Our interpretation of the field-induced quenching thus requires that electron transfer between Phe and Q is exposed to nearly the total potential difference across the membrane, implying that electron transfer between P680 and Phe is not electrogenic.

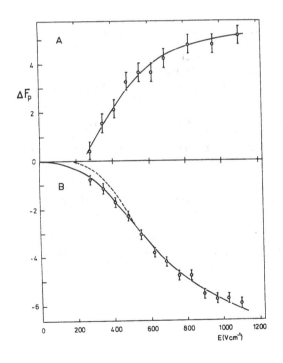

FIGURE 3.
Electric field strength dependence of the field-induced fluorescence changes of Fig. 1.

It seems likely then, that neither our artificial membrane potentials, nor the presence of Q^- inhibit $P680^+Phe^-$ formation, and that F_{var} is really delayed fluorescence, as first suggested by Klimov et al. (1977). The recombination of $P680^+Phe^-$ does not proceed via the triplet state in chloroplasts at room temperature (Kramer, Mathis, 1980; Klimov et al., 1981; Thielen, Van Gorkom, 1981b). Recombination to the singlet state of P680 and energy transfer to antenna chlorophylls is probably much faster than the nanosecond process of electron spin dephasing in $P680^+Phe^-$, which initially is a quadratic function of time (Hoff, 1981). Therefore resonance with singlet excited P680 may effectively preserve the singlet character of the charge pair $P680^+Phe^-$.

REFERENCES

De Grooth BG and van Gorkom HJ (1981) External electric field effects on prompt and delayed fluorescence in chloroplasts, Biochim. Biophys. Acta 635, 445-456.
Döring G (1975) Further results on the photoactive chlorophyll aII in photosynthesis, Biochim. Biophys. Acta 376, 274-284.
Hoff AJ (1981) Magnetic field effects on photosynthetic reactions, Quart. Rev. Biophys. 14, 599-665.
Klimov VV, Ke B and Dolan E (1981) Effect of photoreduction of the photosystem-II intermediary electron acceptor (pheophytin) on triplet state of carotenoids, FEBS Lett. 118, 123-126.
Klimov VV, Klevanik AV, Shuvalov VA and Krasnovsky AA (1977) Reduction of pheophytin in the primary light reaction of photosystem II, FEBS Lett. 82, 183-186.
Klimov VV and Krasnovsky AA (1981) Pheophytin as the primary electron acceptor in photosystem 2 reaction centers, Photosynthetica 15, 592-609.
Kramer H and Mathis P (1980) Quantum yield and rate of formation of the carotenoid triplet state in photosynthetic structures, Biochim. Biophys. Acta 593, 319-329.
Thielen APGM and van Gorkom HJ (1981a) Redox potentials of electron acceptors in photosystem IIα and IIβ, FEBS Lett. 129, 205-209.
Thielen APGM and van Gorkom HJ (1981b) Energy transfer and quantum yield in photosystem II, Biochim. Biophys. Acta 637, 439-446.

ACKNOWLEDGEMENT
This study was supported by the Netherlands Foundation for Chemical Research (SON), financed by the Netherlands Organization for the Advancement of Pure Research (ZWO).
Authors' address: Department of Biophysics, Huygens Laboratory of the State University, P.O. Box 9504, 2300 RA   Leiden, The Netherlands

ELECTRIC MEASUREMENTS OF THE KINETICS OF PHOTOSYNTHETIC EVENTS

PETER GRÄBER and HANS BAUERMEISTER

1. INTRODUCTION

When a suspension of chloroplasts is illuminated with a non-saturating flash a transient electric potential difference can be measured between two electrodes inserted into the suspension if the electrodes are separated along the direction of the light path(Fowler and Kok,1974;Witt and Zickler 1973).This effect has been interpreted to reflect the transient generation of a transvesicular electric potential difference. Due to the partial light absorption a higher number of charge separations occur at that half of the vesicle directed to the light entrance side than at the light exit side.(See scheme in Fig. 1.)

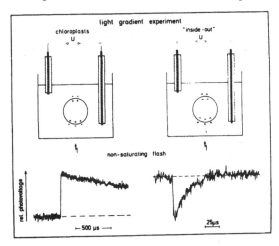

Fig. 1: Schematic explanation of the light gradient experiment. The signals from chloroplasts and "inside-out"-thylakoids have been obtained by excitation with a blue 1 µs flash. Reaction medium: 2 M sucrose, 1mM MES pH 6.5,5 µM chlorophyll, 1o µM benzylviologen for chloroplasts, o.1 mM phenyl-p-benzoquinone for "inside-out"-thylakoids

From the polarity of the photovoltage the direction of the primary photosynthetic electron transport can be determined. It has been shown by this method that "inside-out"-thylakoids show the opposite polarity as chloroplasts (Gräber et al. 1978) This phenomenon is shown with a technically improved measurement in Fig.1.

With the light gradient method also double flash experiments have been carried out which give informations not only about electrogenic steps but also about kinetics of non-electrogenic steps of photosynthetic reactions.

2. MATERIALS AND METHODS

Spinach chloroplasts were prepared according to standard procedures. Tris washing was carried out by resuspending the chloroplasts in o.8M Tris pH 8 and incubation at o°C for 2o min. Chloroplasts and tris-washed chloroplasts were stored under liquid nitrogen. The measuring cell (inner diameter 12 mm) was made from trovidur with windows at two sides. The photovoltage was picked up by two Ag/AgCl electrodes inserted in 3.5 M KCl

Sybesma, C. (ed.), Advances in Photosynthesis Research, Vol. II. ISBN 90-247-2943-2.
© 1984 Martinus Nijhoff/Dr W. Junk Publishers, The Hague/Boston/Lancaster.

and connected via a ceramic diaphragm to the chloroplast suspension (distance between electrodes 2o mm). The photovoltage was amplified (EMV 8o, M&S Elektronik and 1 A 5,Tektronix) stored in a transient recorder (Biomation 81oo) and then transferred to an averager (Nicolet 1o72). One electrode was mounted directly at the input of the EMV 8o amplifier and the cuvette with the amplifier was placed in a Faraday cage. Photosynthesis was excited by two flash lamps (15 μs duration) using fibre optics. The saturating flash was white, the non-saturating flash was filtered through a Schott DAL 673 nm interference filter. The dark time t_d between sucessive saturating flashes was 1 or 1o s. The dark time t_v between saturating and non-saturating was varied between 2o μs and 3 s. The signals from 8 – 64 flashes have been averaged. The reaction medium contained o.4 M sucrose, 2 mM tricine pH 7.4, 1 mM $K_3(Fe(CN)_6)$ and 5 μM chlorophyll.

3. RESULTS

When the photovoltage is measured at increasing times after suspending the chloroplasts in the reaction medium a continuous decrease of its amplitude (about 2o% in the first 3o min) was observed. In order to obtain reproducible results the photovoltage was measured always at the same time after addition of the chloroplasts.Fig. 2 shows a typical experiment: at t= o a saturating

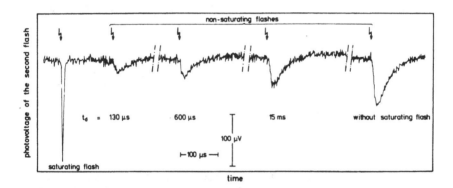

Fig. 2: Photovoltage evoked by a non-saturating flash at different dark times after a saturating flash. The flash profile of the saturating flash is schematically shown.

flash was fired(schematically depicted);at different dark times the signal evoked by the non-saturating flash was measured. The experimental protocol was as follows: the photovoltage of the non-saturating flash was measured three times: without saturating flash, then with saturating flash and then again without saturating flash. The photovoltage with saturating flash $U(t_v)$ was then divided by the arithmetic mean of the photovoltages without saturating flash $U(t_d)$. In the following the dependence of this relative photovoltage of the second (non-saturating) flash on the dark time t_v has been investigated under different conditions.

Fig.3 left shows the relative photovoltage of the second flash in the time range between 2o μs and 3 s in tris- washed chloroplasts. Fig.3 right shows a logarithmic plot of the data. The kinetics in the long time range can be described by a first order reaction with a half life time of 1.2 s (bottom).

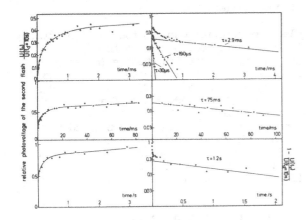

Fig. 3:Left:Relative photovoltage of the second
flash as a function of the
dark time. Three different
time scales are shown.
Right: Logarithmic plot of
the data. Details see text.

This kinetics is extrapolated to the short time range and the difference
between data and extrapolated kinetics is plotted (center) giving a first
order kinetics with a half life time of 75 ms. If this process is repeated
further kinetics with half life times of 2.5 ms, 19o µs and 3o µs are
obtained. The data of this evaluation (half life times and relative amplitudes of the different phases) are collected in Table 1 together with data
obtained from similar experiments with normal chloroplasts.

TABLE 1

normal chloroplasts		tris-washed chloroplasts	
half life time/ms	rel. amplitude	half life time/ms	rel. amplitude
22	o.21o	12oo	o.275
4.2	o.215	74	o.195
o.47	o.4oo	2.9	o.195
o.o95	o.15o	o.19	o.2o5
		o.o3	o.13o

Table 1: Analysis of the data shown in Fig.3 and additional experiments
assuming independent first order reactions. Details see text.

4. DISCUSSION

In normal chloroplasts about half of the photovoltage arises from photosystem I and the other half from photosystem II (Fowler and Kok, 1974).
Therefore, the double flash experiment reveals kinetics from both photosystems. The 22 ms kinetics can be attributed to kinetics of photosystem I
since the Chl a^+ reduction shows this kinetics (Stiehl and Witt, 1969).
The 47o µs kinetics can be ascribed to photosystem II and might represent
the kinetics of the reaction $PQ_A^- + PQ_B^- \longrightarrow PQ_A + PQ_B^{2-}$ (Robinson and Crofts,
1983; Weiss and Renger, 1983).Tentatively, the 4.2 ms kinetics is attri-
buted to photosystem I and the 95 µs kinetics to photosystem II. The latter
might reflect the process $PQ_A^- + PQ_B \longrightarrow PQ_A + PQ_B^-$ (Robinson and Crofts,1983;
Weiss and Renger 1983). In an earlier investigation of the photovoltage in
double flash experiments this fast kinetics has not been observed due to

the limited time resolution (Witt and Zickler, 1974).

In tris washed chloroplasts the electron transfer between the water oxidizing enzyme system,Y, and the donor,D, is interrupted. Measurements of the chl a_{II}-, the PQ_A - and the electrochromic absorption changes (Havemann and Mathis, 1976; Renger and Wolff, 1978, Renger 1979) have revealed under repetitive flash excitation a back reaction chl a_{II}^+ + PQ_A^- →chl a_{II} + PQ_A with 1oo - 2oo µs kinetics.However, the photovoltage double flash experiment reveals a multiphasic kinetics (see Fig. 3 and table 1). Practically no effects (<1o%) arising from photosystem I reactions are observed here as inferred from a measurement of the chl a_I absorption change under our reaction conditions. Therefore, all the different kinetics listed in table 1 right should reflect photosystem II reactions. Fig. 4 shows a tentative assignment of the observed kinetics to different reactions in photosystem II.

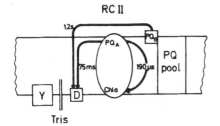

Fig.4: Tentative assignment of the kinetics found in the photovoltage experiment to reactions in photosystem II.

The 1.2 s kinetics might correspond to a back reaction PQ_B^- + D^+→PQ_B + D the 74 ms kinetics might correspond to the PQ_A^- + D^+→PQ_A + D back reaction which might occur in non B-type centers (Lavergne, 1982). This interpretation is in accordance with the interpretation of data of the UV- absorption changes (Weiss and Renger, 1983). The 191 µs kinetics corresponds to the back reaction in the reaction center (see above). The 3o µs kinetics might reflect a second photoreaction of the chl a_{II} (PQ_A being still reduced) which should be possible if a further electron acceptor between chl a_{II} and PQ_A exists (Eckert et al. 1979). The 2.9 ms kinetics cannot be assigned as yet.
Using laser double flash experiments this method will also give informations about the relative localisation of different electron acceptors in the membrane.

5. REFERENCES

Eckert HJ,Buchwald HE and Renger G (1977) Febs Lett. 1o3,291-295
Fowler CF and Kok B (1974) Biochim. Biophys. Acta 357,3o8-318
Gräber P, Zickler A and Åkerlund HE, Febs Lett. 96,233-237
Havemann J and Mathis P (1976) Biochim. Biophys. Acta 44o,346-355
Lavergne J (1982) Photobiochem.Photobiophys. 3,257-285
Renger G (1979) Biochim. Biophys. Acta 547,1o3-116
Renger G and Wolff C (1976) Biochim. Biophys. Acta 423,61o-614
Robinson HH and Crofts AR (1983) Febs Lett. 153, 221-226
Stiehl HH and Witt HT (1969) Z. Naturforsch. 24b, 1588-1598
Weiss W and Renger G (1983) these proceedings
Witt HT and Zickler A (1973) Febs Lett. 37,3o7-3o9
Witt HT and Zickler A (1974) Febs Lett. 39,2o5-2o8

Authors Address: Max-Volmer-Institut für Biophysikalische und Physikalische Chem Technische Universität Berlin, Str. des 17. Juni 135, 1 Berlin

MEMBRANE POTENTIAL MEASUREMENTS IN C_3 PROTOPLASTS

K.L.M. VALLES, M.O. PROUDLOVE, R.B. BEECHEY[1], A.L. MOORE

1. INTRODUCTION

Membrane potentials have been assessed in isolated organelles using the lipophilic cation methyltriphenyl phosphonium (TPMP$^+$). This distributes across membranes in response to their electrical potential (Scott, Nicholls, 1980). There are few data available, however, on membrane potentials in the intact cell. Microelectrode studies probably reflecting the potential between the vacuole and the external media rather than those generated by energy-conserving cytoplasmic organelles. The present study was undertaken to investigate membrane potentials under various metabolic conditions; in particular, the extent to which ATP generated by the chloroplast in the light may control mitochondrial respiratory activity. Respiration is thought to be under adenylate control (Moore, 1978) so, under photosynthetic conditions, any limitation on mitochondrial electron transport would be reflected by an increased membrane potential in this organelle.

2. MATERIALS AND METHODS

Protoplasts were isolated from 10 day-old wheat (Triticum aestivum cv. Maris Huntsman) (Edwards et al., 1978) and finally resuspended in 3-5cm^3 of 0.41M sorbitol, 1mM CaCl$_2$, 10mM NaHCO$_3$ and 30mM TRICINE-TRIS, pH 7.6. Uptake of ^{86}Rb$^+$ and [^{3}H]-TPMP$^+$ by protoplasts (at 50μg chlorophyll cm^{-3}) was measured by the silicone oil centrifugal filtration technique (Hampp, 1980) using a mixture of AR200:AR20 (Wacker-Chemie, Munich, W. Germany), 5:1 (v/v) and a centrifugation time of 50-60s. Osmotic spaces were assessed with [^{14}C]-sorbitol and [^{3}H]$_2$O. Cell compartment fractional volumes were measured by conventional stereological methods (Weibel, 1969).

3. RESULTS AND DISCUSSION

Using wheat protoplasts we have observed greater net accumulation of TPMP$^+$ than of Rb$^+$. Since the Rb$^+$ uptake reflects the plasma membrane potential these differences can only be explained in terms of TPMP$^+$ distributing throughout the cell. Inclusion of greater than equimolar concentrations of tetraphenylboron (TPB$^-$) were found to greatly increase both the rate and the optimum net accumulation ratio of TPMP$^+$, the final ratio being dependent on [TPB$^-$]. Equilibrium was reached within 10-15min, compared to 2-4h without TPB$^-$. The role of TPB$^-$ is at present poorly understood. The variation in TPMP$^+$(+TPB$^-$) accumulation ratios (50-130) may be explained by differences in plasma and intracellular membrane potentials, cell compartment volumes and TPMP$^+$ activity coefficients in different cell compartments. A model system, Equation 1, has been developed to account for the way in which each of these factors may affect the accumulation of TPMP$^+$(+TPB$^-$). Using this we have attempted to assess how the activity coefficients for each compartment ($a_1 \ldots a_n$) and their measured fractional cell volumes ($V_1 \ldots V_n$), see Table 1, may reflect or account for the TPMP$^+$(+TPB$^-$) accumulation ratios. Over the measured range of plasma membrane potentials (-50 to -63mV) it could be shown that only variations in the mitochondrial

Sybesma, C. (ed.), Advances in Photosynthesis Research, Vol. II. ISBN 90-247-2943-2.
© 1984 Martinus Nijhoff/Dr W. Junk Publishers, The Hague/Boston/Lancaster.

EQUATION 1. Net accumulation ratios of TPMP$^+$ in terms of plasma membrane potential and the volumes, activity coefficients and membrane potentials for cell compartments.

$$\text{ARAT} = \frac{[\text{TPMP}^+]_{in}}{[\text{TPMP}^+]_{out}} = 10^{-\Delta\Psi_{pl}/60} \left[\frac{V_{cyt}}{a_{cyt}} + \frac{V_1}{a_1}\cdot 10^{-\Delta\Psi_1/60} + \ldots + \frac{V_n}{a_n}\cdot 10^{-\Delta\Psi_n/60} \right]$$

where ARAT = accumulation ratio; pl = plasma membrane; cyt = cytosol; V = compartment volume; a = compartment activity coefficient; 1...n = number of cell compartments.

activity coefficients resulted in significant variations in the predicted accumulation ratios. This results from the small fractional volume of the mitochondria (Table 1).

TABLE 1. Percent fractional cell volumes of wheat protoplasts. The areas occupied by each compartment were measured from electron micrographs (x7975) using a MOP image analyser and are expressed as mean ± s.e..

Cell compartment

vacuole	cytosol	chloroplast	mitochondria	microbodies	nucleus
56.6±1.5	13.2±1.15	27.1±2.3	1.2±0.1	0.1±0.05	1.8±0.5

Using experimentally determined values of the TPMP$^+$(+TPB$^-$) accumulation ratios and the plasma membrane potentials, Figure 1 gave values of −120 to −163mV for the mitochondrial membrane potential. These are typical of energized mitochondrial membranes (Moore, 1978).

FIGURE 1. Graph of predicted mitochondrial membrane potentials using measured values for plasma membrane potential and accumulation ratio of TPMP$^+$(+TPB$^-$).

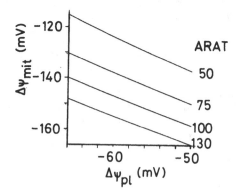

The effects of light and dark on $TPMP^+(+TPB^-)$ and Rb^+ accumulation ratios are presented in Figure 2(a). In the light there was a significant increase in the $TPMP^+(+TPB^-)$ accumulation ratio. After 12min illumination, however, the value had fallen to that found in the dark. This diminution may be due to an adverse effect of TPB^- on chloroplast metabolism since TPB^- inhibits oxygen evolution by these protoplasts. The Rb^+ accumulation ratio did not change appreciably with illumination over this period.

The data presented in Figure 2(b) has been obtained by entering the optimum measured ratios from Figure 2(a) into Equation 1. If a value of +4mV is assumed (Enser, Heber, 1980) for the chloroplast transenvelope potential in non-illuminated protoplasts, the maximum $TPMP^+(+TPB^-)$ accumulation ratio (50, in the dark) corresponds to a mitochondrial membrane potential of -132mV. The same value for the mitochondrial membrane potential in the light would require a chloroplast transenvelope potential of -72mV. This predicted value is not consistent with published values, -20mV (Enser, Heber, 1980). Using the latter value and the maximum $TPMP^+(+TPB^-)$ accumulation ratio of 114 (see Figure 2(a)), the predicted mitochondrial potential is -157mV. These calculations indicate that the potential across the mitochondrial membrane is greater in the light than in the dark. This increase in membrane energization is consistent with an increased cytosolic phosphorylation potential in the light (Moore, 1978) which would result in a diminution of the rate of respiration, corresponding to a state 3 to state 4 transition. The implication of such is that the metabolism of respiratory NADH, produced by the action of glycine decarboxylase for example, must vary with the degree of illumination.

FIGURE 2. Effect of light and dark on the time dependent accumulation of $TPMP^+(+TPB^-)$ and Rb^+ by wheat protoplasts (a) and their effect on mito-chondrial and chloroplast membrane potentials, derived from Equation 1 (b).

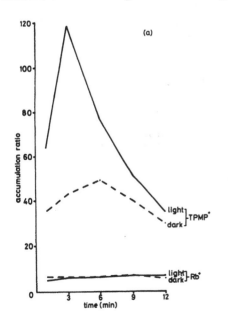
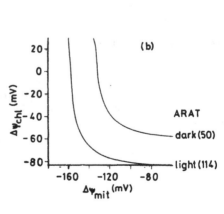

4. CONCLUSIONS

We have measured the fractional cell volumes of the various compartments in isolated wheat protoplasts and developed a mathematical model of the cell which relates the values of the potentials across cellular membranes to the distribution of permeant ions in these compartments. This inform- ation has been used to predict how, or if, the mitochondrial membrane potential changes with illumination. The results show that it increases in value, implying a greater limitation on mitochondrial respiration in the light. Whether the increase in membrane potential is due to adenylate control remains to be clarified. These results have considerable implication on the mechanism of oxidation of NADH produced by glycine decarboxylase.

5. REFERENCES

Edwards GE, Robinson SP, Tyler NJC and Walker DA (1978) Photosynthesis by isolated protoplasts, protoplast extracts and chloroplasts of wheat, Plant Physiol. 62, 313-319.
Enser U and Heber U (1980) Metabolic regulation by pH gradients, Biochim. Biophy. Acta 592, 577-591.
Hampp R (1980) Rapid separation of the plastid, mitochondrial and cytoplasmic fractions from intact leaf protoplasts of Avena, Planta 150, 291-298.
Moore AL (1978) The electrochemical gradient of protons as an intermediate in energy transduction in plant mitochondria. In Ducet G and Lance C eds. Plant Mitochondria, pp 85-92. Amsterdam:Elsevier/North Holland Biomedical Press.
Scott ID and Nicholls DG (1980) Energy transduction in intact synaptosomes, Biochem. J. 186, 21-33.
Weibel ER (1969) Stereological techniques for electron microscopic morph- ometry, Int. Rev. Cytol. 26, 235-296.

ACKNOWLEDGEMENTS

This research was supported by a grant from the ARC (ALM) and a CASE award (KLMV).

Authors addresses: Department of Biochemistry, University of Sussex, Falmer, Brighton, BN1 9QG, U.K. and [1]Shell Research Ltd., Sittingbourne, ME9 8AG, U.K..

THE IONIC CONDUCTANCE BUT NOT THE ELECTRIC CAPACITANCE OF MEMBRANES FROM
PHOTOSYNTHETIC BACTERIA DEPENDS ON THE MEMBRANE POTENTIAL

J B JACKSON AND A J CLARK, DEPARTMENT OF BIOCHEMISTRY, UNIVERSITY OF
BIRMINGHAM, P O BOX 363, BIRMINGHAM B15 2TT, ENGLAND.

1. INTRODUCTION

A proton current, through an ATP synthase, down a proton electrochemical
gradient (Δp), maintained by electron transport, is believed to be
responsible for ATP synthesis in chloroplast and chromatophore membranes
(Mitchell, 1966). This state of affairs can be described by an analogue
electrical circuit, as shown in Fig.1 ($\Delta p = \Delta \psi$ when $\Delta pH = 0$). In steady-
state light, at constant rates of ATP synthesis, the value of the membrane
capacitance is unimportant: the rate of ATP synthesis is fixed by the rate
of the primary H^+-translocation and by the relative rates of H^+ translocation
through the ATP synthase and through leakage pathways (including other metabolic H^+ translocators).

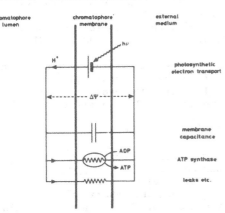

Fig 1. Electrical circuit analogue
of proton currents across the
chromatophore membrane.

Subject to the caveat discussed below, it has been found that the Δp-
driven H^+ efflux from chromatophores has a distinctly non-linear dependence
upon the value of Δp: the H^+ conductance increases steeply as Δp is
increased (Jackson, 1982; Clark et al. 1983a,b; see also Nicholls, 1974).

From a study of the effects of inhibitors such as venturicidin, oligomycin and
dicyclohexylcarbodi-imide and by comparison with the dependence of the rate
of ATP synthesis on Δp we have discovered that the remarkable conductance
properties of the chromatophore membrane arise mainly from "non-Ohmic"
conductance through the F_0 component of the ATP synthase (Clark et al. 1983b).
Three metabolically-important conclusions follow from this (i) the rate of
electron-transport-phosphorylation can be effectively controlled over a very
narrow range of Δp (see also Sorgato, Ferguson, 1979) (ii) a moderately
large Δp can be maintained even at very low rates of electron transport
(Clark et al. 1983a) (iii) low actinic light intensities are sufficient to
abolish respiration in phototrophic cultures (Cotton et al. 1983).

The procedure by which the dependence of H^+ current upon Δp was measured
(Jackson, 1982) relies on the rapid response of the electrochromic carotenoid
pigments to $\Delta \psi$ (see Junge, Jackson, 1982). It is assumed that the initial
rate of decay of $\Delta \psi$ upon promptly darkening the chromatophore suspension is a
measure of the total outward H^+ current. In terms of Fig.1 this is visualised

Sybesma, C. (ed.), Advances in Photosynthesis Research, Vol. II. ISBN 90-247-2943-2.
© *1984 Martinus Nijhoff/Dr W. Junk Publishers, The Hague/Boston/Lancaster.*

as a discharge of the membrane capacitance through the ATP synthase and ionic leakage conductances when the source of the light driven Δp is "disconnected". As explained by Schmid, Junge (1975) the dissipative ionic current (J_{DIS}) is given by

$$J_{DIS} = C. \frac{d(\Delta\psi)}{dt} \qquad \text{(Equation i)}$$

where C is the value of the membrane capacitance. In our experiments we have chosen to operate from the steady-states ($\Delta\psi$ = constant) achieved at progressively diminished light intensities. Central to our analysis is the assumption that the value of C is independent of $\Delta\psi$. In the past we have interpreted our finding that $d(\Delta\psi)/dt$ does not vary proportionately with $\Delta\psi$ by concluding that the membrane conductance is non-Ohmic, having reasoned that it is unlikely that the membrane capacitance is $\Delta\psi$-dependent. In the present experiments we present evidence that C is indeed a constant.

2. EXPERIMENTAL PROCEDURE

Cultures of Rhodopseudomonas capsulata strain N22 were grown and chromatophores were prepared as previously described (Clark et al. 1983b). In experiments with imposed K^+-diffusion potentials, the chromatophores were washed twice in 10% sucrose, 50 mM Na^+-tricine, 8 mM $MgCl_2$, 100 mM NaCl, pH 7.4.
The dependence of J_{DIS} on $\Delta\psi$ was measured by our standard procedure (Jackson, 1982). Diffusion potentials were measured from the carotenoid absorption change with a Perkin-Elmer 356 chopped, dual-wavelength spectrophotometer and flash-induced absorption changes were measured with a home-built, crossed beam dual wavelength instrument.

3. RESULTS

In the following experiments, the strategy was first to apply a K^+-diffusion potential across the chromatophore membranes in the dark. Then, with a pulse of photosynthetic light, a fixed quantity of electronic charge was driven through the reaction centre complexes across the membranes and the ensuing change in $\Delta\psi$ was measured by electrochromism. This change, resulting from the displacement of a fixed quantity of charge is inversely proportional to the membrane capacitance, since by definition,

$$C = \frac{dq}{d(\Delta\psi)} \qquad \text{(Equation ii)}$$

By repeating the experiment at different values of the diffusion potential, changes in the value of membrane capacitance could be evaluated.
After short flashes of light, charge displacement in the direction normal to the membrane takes place in three distinct processes (Jackson, Dutton 1973). These have been identified as (I) P870 $\rightarrow$ UQ_A (II) cytochrome c $\rightarrow$ $P870^+$ (Jackson, Dutton 1973) (III) cytochrome $\underline{b}_{561}$ $\rightarrow$ UQ_C (Crofts et al. 1983). We have chosen to work with only the first of these charge displacements: antimycin A was added to eliminate reaction III and the experiments were carried out at an equilibrium redox potential of about E_h = +400 mV to eliminate reaction II. There were three reasons for this choice of experimental conditions. First because reaction I is exceedingly fast, the extent of the accompanying change in $\Delta\psi$ can be measured without interference from dissipative ionic fluxes. Second, the driving force of reaction I (1.4 <> 0.37 V, see Clayton, 1978) greatly exceeds the values of the K^+-diffusion potential that can be attained (0.14 v, see below) so

that its extent will not be thermodynamically restricted.   Third, at this redox potential poise, cytochrome c is chemically oxidised so that only a single turnover of the photosynthetic reaction centre is possible during the lifetime of the short flash.   These conditions therefore ensure a fixed charge displacement following flash activation at each value of the diffusion potential.

The experimental protocol is shown in Fig.2.

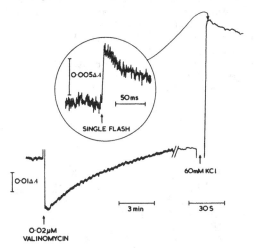

Fig.2.   Experimental protocol for the examination of the dependence of electric capacitance on $\Delta\psi$.   The lower trace shows a recording of the carotenoid absorbance changes in a darkened, K^+-free chromatophore suspension upon addition of valino-mycin and then KCl.   Note the change in recording speed at the break.   The inset shows the flash-induced absorbance change in a duplicate experiment at the time indicated by the arrow.   Experimental medium: 10% sucrose, 20 mM Na^+-tricine, 8 mM $MgCl_2$, 100 mM NaCl, 41 μM Na^+ ferri- and 140 μM Na+ ferrocyanide, 2.5 μM antimycin A, bacteriochlorophyll 10 μM E_h = 400-410 mV, pH = 7.4.

A series of experiments was first carried out to determine the response of the carotenoid band shift to K+-diffusion potentials.   The procedure was similar to that employed on earlier occasions (Jackson, Crofts, 1969; Clark, Jackson, 1981) except that sodium ferri- and ferrocyanide were present as a redox buffer.   The treatment of the chromatophores with valinomycin was followed by a period of 10 min to allow ionic equilibration across the membrane.   The extent of the absorbance change corresponding to the carotenoid band shift, resulting from a subsequent KCl addition was plotted as a function of the final KCl concentration as shown in Fig.3.

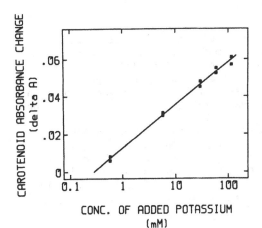

Fig.3.   Dependence of the carotenoid absorbance change on the log of the added K^+-concentration.   Data taken from a series of experiments similar to those described in Fig.2.

The straight line relation between the carotenoid absorbance change and the log of the added $[K^+]$ confirms earlier results (Jackson, Crofts, 1969). The intercept of the line with the x-axis was taken as the internal $[K^+]$ of the vesicles prior to KCl addition and the magnitude of each diffusion potential was then calculated from the Nernst equation.

In a separate series of experiments on the same chromatophore preparation, the above protocol was followed up to and including the KCl addition in the cuvette of a rapidly responding flash spectrophotometer. Immediately after the KCl addition, the spectrophotometer was triggered and the flash-induced carotenoid absorbance change was recorded (see Fig.2). The flash was fired between 6-7 sec after the KCl pulse. This time was sufficient to permit full generation of the diffusion potential but restricted its decay to less than 5% of the total change. In Fig.4 the extent of the flash-induced absorbance change (proportional to 1/C) was plotted against the magnitude of the K^+-diffusion potential. No change in the value of the membrane capacitance was detected in the range 0 - 140 mV.

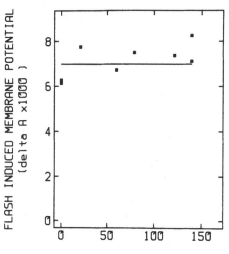

Fig.4. The size of the K^+-diffusion potential does not affect the extent of the carotenoid absorbance change elicited by a single flash. Data taken from a series of experiments similar to those described in the inset to Fig.2. The extent of the flash-induced carotenoid absorbance change is plotted against the value of the diffusion potential reached during the preceding KCl pulse.

The dependence of the initial rate of decay of the carotenoid shift upon darkening an illuminated suspension of chromatophores was then measured as a function of the extent of the carotenoid shift achieved during illumination (see Clark et al. 1983b). Briefly, an anaerobic chromatophore suspension was illuminated (for about 1 sec) until the carotenoid absorbance change reached a constant value. This value ($\Delta A_{503-487 \text{ nm}}$) is proportional to membrane potential. The actinic illumination was then extinguished and the initial rate of decay of the carotenoid absorbance change d(ΔA)/dt, which is proportional to d($\Delta\psi$)/dt was measured. The experiment was repeated at a series of lower actinic light intensities and the results are shown in Fig.5. In confirmation of earlier results the dependence of d($\Delta\psi$)/dt upon $\Delta\psi$ was non-linear. The top axis of Fig.5 is given in mV units using the diffusion potential calibration procedure.

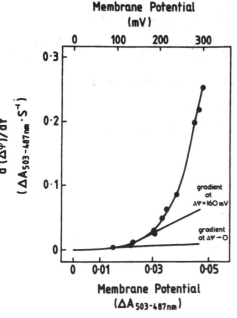

Fig.5 The dependence of the initial rate of decay of the carotenoid absorbance change on the steady-state extent. See Clark et al. (1983b) for details. The chromatophores (10 μM bacteriochlorophyll) were suspended anaerobically in 10% sucrose, 50 mM K^+-tricine, 100 mM KCl, 8 mM $MgCl_2$, 0.5 mM NADH, 0.5 mM succinate pH 7.6. They were illuminated until the membrane potential (measured from the carotenoid absorption change) reached a constant value $(\Delta\psi)$ and then the initial rate of decay of ΔA upon darkening $(d(\Delta\psi)/dt)$ was recorded. The data were collected at a series of reduced light intensities.

4. DISCUSSION

The pronounced upward curvature in Fig.5 could in principle be the result of either a $\Delta\psi$-dependent increase in membrane capacitance (see equation i), or a $\Delta\psi$-dependent increase in membrane conductance or a combination of both. In Fig.4 however it is shown that the membrane capacitance is constant at least in the range 0-140 mV. Because the $\Delta\psi$ generated during the flash amounts to another ~20 mV, the range can be justifiably extended up to 160 mV. In this range of $\Delta\psi$, the slope of the line in Fig.5 changes by a factor of approximately five-fold. This must be due almost entirely to an increase in the value of the membrane conductance. These results, therefore, confirm the assumption that we have made in earlier work (Jackson, 1982; Clark et al. 1983a,b; Cotton et al. 1983) and establish that the ionic conductance of the chromatophore membrane increases steeply with the value of $\Delta\psi$.

5. ACKNOWLEDGEMENTS

This work was supported by a grant from the Science and Engineering Research Council.

6. REFERENCES

Clark AJ, Cotton NPJ and Jackson JB (1983a) Eur. J. Biochem. 130, 575-580
Clark AJ, Cotton NPJ and Jackson JB (1983b) Biochim.Biophys.Acta, 723, 440-453
Clark AJ and Jackson JB (1981) Biochem.J. 200, 389-397
Clayton RK (1978) In: 'Photosynthetic Bacteria' Clayton RK and Sistrom WR, eds) pp 387-396, Plenum Press, New York and London
Cotton, NPJ, Clark AJ and Jackson JB (1983) Eur. J. Biochem. 130, 581-587
Crofts AR, Meinhardt SW, Jones KR and Snozzi M (1983) Biochim.Biophys.Acta 723, 202-218

Jackson JB (1982) FEBS Lett. 139, 139-143

Jackson JB and Dutton PL (1973) Biochim.Biophys.Acta, 325,102-113

Jackson JB and Crofts AR (1969) FEBS Lett. 4, 185-189

Junge W and Jackson JB (1982) In: 'Photosynthesis: Energy Conversion by Plants and Bacteria. Vol.1' (Govindjee, ed.) pp. 589-646, Academic Press, New York.

Michell P (1966) In: 'Chemiosmotic Coupling in Oxidative and Photosynthetic Phosphorylation' Glynn Res. Publ. Bodmin, Cornwall, England

Nicholls DG (1974) Eur. J. Biochem. 50, 305-315

Schmid, R and Junge W (1971) J. Membr. Biol. 4, 179-192

Sorgato MC and Ferguson SJ (1979) Biochemistry, 18, 5737-5742

THE ELECTROCHEMICAL PROTON GRADIENT AND SOLUTE TRANSPORT IN RHODOPSEUDO-
MONAS SPHAEROIDES

WIL N. KONINGS, MARIEKE G.L. ELFERINK, K.J. HELLINGWERF

Rhodopseudomonas sphaeroides is a Gram-negative phototrophic bacterium
which can grow either aerobically in the dark or anaerobically in the
light. The bacteria are surrounded by a cell-envelope which consists of
the outer membrane, the peptidoglycan layer and the cytoplasmic membrane.
This cell envelope functions as a barrier between the cytoplasm and the
environment. The outer membrane is a hydrophobic layer in which proteins
form non-specific hydrophilic pores through which solutes with a molecular
weight of up to about 600 Daltons can penetrate. The peptidoglycan layer
consists of a network of polysaccharides cross-linked by short peptides.
This layer, which is freely permeable even for large molecules, determines
the volume and shape of the cell.

The main barrier for solutes is the cytoplasmic membrane which consists
of a liquid-crystalline bilayer of phospholipids in which proteins are
embedded. The selective movement of solutes through this membrane is
catalyzed by specific carrier proteins. In addition several other cellular
functions are located in the cytoplasmic membrane such as the energy
transducing cyclic and linear electron transfer systems and the $Ca^{2+} Mg^{2+}$-
dependent ATPase complex.

Rps. sphaeroides can also form intracytoplasmic membranes which are vesicular structures associated with the cytoplasmic membrane. Cells grown
anaerobically in the light show an increased synthesis of the components
of the photosynthetic apparatus which are incorporated into the cytoplasmic
membrane. This leads to enlargement of the membrane surface which gives
rise to invaginations (Oelze, Drews, 1972).

Energy-transducing processes in the cytoplasmic membrane of Rps. sphaeroides
such as electron-transfer-linked proton translocation, and solute transport
can be studied in whole cells or in isolated membrane vesicles which have
retained their functional properties. In these latter systems the membrane-
bound activities can be separated from activities in the cytoplasm. Two
procedures have been described for the isolation of membrane vesicles.
Mechanical disruption of the cells by sonication, French-press treatment
or grinding with aluminium yields so-called chromatophores which are
mainly derived from the intracytoplasmic membranes (Oelze, Drews, 1972).
The second procedure implies osmotic lysis of lysozyme-EDTA treated cells.
This procedure leads to the isolation of so-called membrane vesicles
(Hellingwerf et al., 1975).

The orientation of these membrane preparations has been studied by freeze-
etch electronmicroscopy (Michels, Konings, 1978), the localization of
cytochrome c_2 (Michels, Konings, 1978) and by crossed immunoelectrophoresis (Elferink et al., 1979). These studies demonstrated that more than
90% of the chromatophores have an inside-out orientation and that more
than 80% of the membrane vesicles are right-side out orientated.

The chemical composition of the chromatophores and the membrane vesicles
have been compared. Both preparations have qualitatively a rather similar
composition but membrane vesicles contain less protein, carotenoids and
bacteriochlorophyll and more lipids than chromatophores (Michels, Konings,
1978).

Sybesma, C. (ed.), Advances in Photosynthesis Research, Vol. II. ISBN 90-247-2943-2.
© *1984 Martinus Nijhoff/Dr W. Junk Publishers, The Hague/Boston/Lancaster.*

Electron transfer in membrane preparations

Rps. sphaeroides can contain two distinct electron transfer systems: a
respiratory chain in which electrons from electron donors like NADH or
succinate are transferred via cytochrome b, c and a to the electron acceptor oxygen and a cyclic electron transfer system in which light energy
stimulates the oxidation of bacteriochlorophyll (P-870) and the energized
electrons are transferred via bacteriophaeophytin, ubiquinone, the cytochromes b and c back to bacteriochlorophyll.

Membrane vesicles isolated from cells grown   aerobically in the dark
contain a functional respiratory chain which can oxidize the electron
donors NADH, ascorbate-TMPD, malate-TMPD and succinate-TMPD (TMPD =
N,N,N',N'-tetramethyl-1,4-phenyldiamine dihydrochloride (Hellingwerf et al.,
1975). Membrane vesicles and chromatophores isolated from cells grown anaerobically in the light contain a functional cyclic electron transfer
system (Michels, Konings, 1978). Upon illumination of these membrane preparations a proton motive force is formed and energy is supplied for
solute transport and ATP-synthesis (see below).

The generation of an electrochemical proton gradient by cyclic electron
transfer

The generation of a proton motive force by light induced cyclic electron
transfer has been studied in whole cells, membrane vesicles and chromatophores. Several methods have been used for the quantitation of the components of the proton motive force. The membrane potential has been calculated
from the distribution across the membrane of lipophilic cations or anions
and from absorbance changes of membrane-bound carotenoids. The pH gradients have been determined from pH-changes of the external medium, from
the distribution of weak acids or bases and from ^{31}P nuclear magnetic
resonance studies. The membrane potential in right-side out preparations
(whole cells and membrane vesicles) in which the proton motive force
generated by electron transfer will generally be inside negative and
alkaline, can be estimated from the uptake of the lipophilic cation
tetraphenylphosphonium. In inside-out preparations (chromatophores) the
lipophilic anion thiocynate is used as a probe for the membrane potential
(Michels, Konings, 1978; Hellingwerf et al., 1982; Lolkema et al., 1982,
1983; Michels et al., 1981; Elferink et al., 1983a,b). The distribution
of these probes can be determined from the uptake of radioactively labeled
probes or by external concentration measurements with ion selective electrodes. A complicating factor in the exact quantitation of the electrical
potential is the binding of the probe to cellular components. Procedures
have been described to correct for this binding (Lolkema et al., 1982,
1983). The electrical potential values obtained with the ion distribution
method and the carotenoid absorbance shift method have been compared.
The values obtained with the latter method are significantly higher than
those obtained with the distribution method (Michels, Konings, 1978;
Michels et al., 1981; Dijkema et al., 1980). Most likely the absorbance
changes of the carotenoid record not only transmembrane gradients but
also energy dependent changes inside the membrane.

A comparison has also been made of the different methods to measure the
transmembrane pH gradients. In inside-out preparations the fluorescence
quenching of 9-aminoacridine has been used as an indication of the pH
gradients. However, it has been demonstrated that the fluorescence quenching of these compounds cannot be ascribed to accumulation of 9-amino-
acridine in the intravesicular space. A calculation of the ΔpH from the
quenching of 9-aminoacridine is therefore not possible (Kraayenhof,

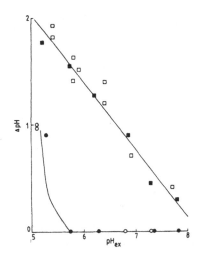

FIGURE 1. Comparison of the ΔpH in <u>Rps. sphaeroides</u> as measured with automated flow-dialysis and ^{31}P-NMR. In the flow dialysis experiments light saturation was ascertained through a series of experiments at various light intensities and cell densities. In the flow-dialysis experiments (^{14}C)-benzoic acid (30 µM) was used as ΔpH probe. Open symbols: ^{31}P-NMR; closed symbols: automated flow dialysis; squares: medium: glycyl-glycine (20 mM), 20 mM N-2-hydroxy ethylpiperazine-N-2-ethane sulphonic acid, 20 mM 2-(N-morpholino)-ethane sulphonic acid, 2 mM EDTA, 5 mM MgSO$_4$, 10 mg chloramphenicol/l, 100 mM KCl brought to the desired pH value with KOH; dots, same medium except that 67 mM Na$_2$SO$_4$ replaced KCl and NaOH was used to adjust the pH of the buffer. (Figure taken from Nicolay et al., 1981).

Arents, 1977; Elema et al., 1978). Recently, the distribution method for determining the ΔpH has been compared with the ^{31}P-NMR method (Nicolay et al., 1981). The values obtained under identical conditions with both methods are identical (Fig. 1), indicating that a reliable ΔpH can be recorded with both methods.
Upon illumination a proton motive force is generated in whole cells, chromatophores and membrane vesicles. In membrane vesicles a proton motive force is generated which is inside alkaline and negative. The ΔpH, determined from the distribution of acetate, depends on the composition of the medium and is maximally 0.6 at an external pH of 7; the $\Delta\psi$, measured from the distribution of triphenylmethyl phosphonium is about -70 mV. The proton motive force in membrane vesicles can thus reach values of around -110 mV (Michels, Konings, 1978).
Illumination of chromatophores results in a ΔpH, inside acid and a $\Delta\psi$ inside positive (Fig. 2). The ΔpH calculated from the distribution of methylamine depends on the external pH and increases from 0.98 at pH 6 to 1.36 at pH 8. The $\Delta\psi$ is not affected and remains about 50-75 mV. The proton motive force in chromatophores can thus reach values of +155 mV (Michels, Konings, 1978; Michels et al., 1981). ^{31}P-NMR has also been used to study the light-dependent generation of the pH-gradient in chromatophores (Hellingwerf et al., 1981). These studies demonstrated that a large degree of heterogeneity exists in the intrachromatophore pH in the

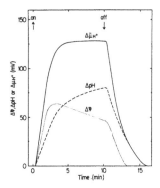

FIGURE 2. The time dependence of the light-induced generation of the electrochemical proton gradient in chromatophores from <u>Rps. sphaeroides</u>. The curves were deduced from the uptake data of methylamine and thiocyanate (.....) membrane potential, $\Delta\psi$, (-----) transmembrane pH gradient, ΔpH; (———) total electrochemical proton gradient, $\Delta\tilde{\mu}_H{}^+$). (Data from Michels, Konings, 1978).

light. This heterogeneity can be largely abolished by the addition of valinomycin.
In whole cells the proton motive force has been studied as a function of light intensity and the external pH. In an Na^+-containing medium at pH 6.4 under saturating light intensity an apparent $\Delta\psi$ (Michels et al., 1981) of about -100 mV and a ΔpH of -90 mV (1.5 pH units) is formed. The magnitude of the components of the proton motive force is strongly affected by the components in the external medium (Michels et al., 1981; Nicolay et al., 1981). In the dark a significant $\Delta\psi$ exists which is sensitive to uncouplers and valinomycin (Elferink et al., 1983). Upon illumination this proton motive force increases rapidly with the light intensity. Under conditions that the proton motive force is composed out of a $\Delta\psi$ only (pH 8, in the presence of nigericin) a maximum $\Delta\psi$ is already reached at 10-15% of the saturating light intensity for solute transport (Elferink et al., 1983a,b).

Solute transport
Solute transport has been studied in all three membrane preparations. In membrane vesicles secondary transport of amino acids (Hellingwerf et al., 1975) occurs upon illumination and this transport is inhibited by inhibitors of electron transfer and uncouplers. Membrane vesicles isolated from aerobically-grown cells accumulate amino acids upon electron transfer in the respiratory chain (Hellingwerf et al., 1975). Upon illumination chromatophores accumulate Ca^{2+} (Michels, Konings, 1978) and Na^+ (Hellingwerf et al., 1982). Sodium is transported via an electro-neutral exchange system against protons. The system functions optimally at pH 8 and is inactive below pH 7.
Direct measurements of the effects of different cations and anions on the components of the proton motive force in whole cells supplied information about the relative permeability of the different ions. In general, cations are less permeant than anions. The anions $SO_4{}^{2-}$, Cl^-, $NO_3{}^-$ and $ClO_4{}^-$ cross the membrane by passive diffusion. The relative permeability of the anions increases in the sequence $SO_4{}^{2-} < Cl^- < NO_3{}^- < ClO_4{}^-$

(Michels et al., 1981).
Whole cells have been shown to accumulate amino acids, phosphate and K^+ upon illumination (Hellingwerf et al., 1975; 1982). The driving force for phosphate transport varied with the external pH. At pH 8 Pi transport is exclusively dependent on $\Delta\psi$, whereas at pH 6 only the ΔpH component is a driving force. Potassium transport is an electrogenic process (Hellingwerf et al., 1982).

The roles of the proton motive force and electron flow in solute transport

The relationship between the proton motive force and secondary transport of alanine has been studied in detail in membrane vesicles (Friedberg et al., 1980) and whole cells (Elferink et al., 1983a,b) of Rps. sphaeroides. The conditions chosen were such that the proton motive force was composed only of a $\Delta\psi$ (pH 8, in the presence of nigericin).
In whole cells a considerable $\Delta\psi$ exists in the dark (-40 to -100 mV) but no uptake of alanine takes place. Upon illumination with low light intensities an increase of the $\Delta\psi$ is observed but uptake of alanine still does not occur. Alanine uptake is only observed at higher light intensities. The rate of uptake of alanine increases with the light intensity even under conditions that no increase and even a decrease of the $\Delta\psi$ is observed. These observations indicate that secondary transport of alanine requires both a proton motive force and electron flow.
The role of the proton motive force and electron transfer has been investigated in more detail under conditions that only one parameter is varied while the other parameter is kept constant (Fig. 3).
Under conditions of constant light intensity the uptake rate of alanine increased sigmoidally with the $\Delta\psi$ in a similar way as has been found for lactose and proline transport in Escherichia coli membrane vesicles (Robertson et al., 1980).
Under conditions of a constant $\Delta\psi$ the rate of uptake of alanine increases linearly with the light intensity. A threshold light intensity is required before alanine uptake occurs. These results strongly indicate a direct interaction between the alanine carrier and the electron transfer chain.
Such an interaction between the electron transfer system and solute transport carriers is not specific for Rps. sphaeroides. In a strain of Rps. sphaeroides in which the E. coli transport protein for lactose (the M-protein) was incorporated by genetic transformation a similar interaction between the rate of cyclic electron transfer and lactose transport was observed (Elferink et al., 1983). Kinetic analysis of the changes in the initial rate of lactose uptake indicates that the regulation is due to a light-dependent change in the number of active carrier molecules in the membrane.
As soon as illumination is stopped the carrier molecules become rapidly deactivated since the rate of solute uptake decreases faster than the dissipation of the $\Delta\psi$.
Recent studies in our laboratory demonstrated that similar interactions between the transport carrier proteins and the linear electron transfer chain exists in Rps. sphaeroides and also in E. coli. This indicates that a regulation by electron transfer is a general property of solute transport systems in bacteria.

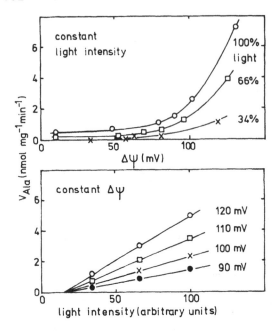

FIGURE 3. The relation between V_{ala} and $\Delta\psi$ at constant light intensity and between V_{ala} and light intensity at constant $\Delta\psi$ in cells of Rps. sphaeroides. Alanine uptake and $\Delta\psi$ were measured simultaneously in cells treated with EDTA and nigericin (Elferink et al., 1983a) at different light intensities. The incubation medium contained: 50 mM potassium phosphate pH 8, 5 mM MgSO$_4$, 50 μM ^{14}C alanine, 4 μM Tetraphenylphosphonium, cells at 0.64 mg protein/ml and various concentrations of valinomycin. Upper panel: The alanine uptake rate (V_{ala}) as a function of $\Delta\psi$ at constant light intensity. Lower panel: V_{ala} as a function of light intensity at constant $\Delta\psi$ values. Data from the upper panel were replotted.

The molecular mechanism of this regulation is not known at this moment. Recently, it was demonstrated that the transport carrier proteins for proline and lactose in E. coli contain redox sensitive dithiol groups. The redox state of these groups determines the affinity of these carriers for their solutes (Konings, Robillard, 1982). The possibility exists that the redox state of these carrier proteins is controlled via an interaction with one of the components of the electron transfer system. A similar regulation by the rate of electron transfer has also been observed for photophosphorylation in Rhodopseudomonas capsulata (Baccarini-Melandri et al., 1977; Casadio et al., 1978) and Rps. sphaeroides (unpublished results). This regulation gives rise to homeostasis of the magnitude of the proton motive force. Large variations in the rates of proton motive force consuming processes are possible, with only minor variations in the magnitude of the proton motive force.

REFERENCES

Baccarini-Melandri A, Casadio R and Melandri BA (1977) Thermodynamics and kinetics of photophosphorylation in bacterial chromatophores and their relation with the transmembrane electrochemical potential difference of protons, Eur.J.Biochem. 78, 389-402.

Casadio R, Baccarini-Melandri A and Melandri BA (1978) Limited cooperativity in the coupling between electron flow and photosynthetic ATP-synthesis: a comparative study in chromatophores phosphorylating at very different rates, Febs Letters 87, 323-328.

Dijkema C, Michels PAM and Konings WN (1980) Light-induced spectral changes of carotenoids in chromatophores of Rhodopseudomonas sphaeroides, Arch.Biochem.Biophys. 201, 403-410.

Elema RP, Michels PAM and Konings WN (1978) Response of 9-amino-acridine fluorescence to transmembrane pH-gradients in chromatophores from Rhodopseudomonas sphaeroides, Eur.J.Biochem. 92, 381-387.

Elferink MGL, Hellingwerf KJ, Michels PAM, Seijen HG and Konings WN (1979) Immunochemical analysis of membrane vesicles and chromatophores of Rhodopseudomonas sphaeroides by crossed immunoelectrophoresis, Febs Letters 107, 300-307.

Elferink MGL, Friedberg I, Hellingwerf KJ and Konings WN (1983) The role of the proton motive force and electron flow in light-driven solute transport in Rhodopseudomonas sphaeroides, Eur.J.Biochem. 129, 583-587.

Elferink MGL, Hellingwerf KJ, Nano FE, Kaplan S and Konings WN (1983) The lactose carrier of Escherichia coli functionally incorporated in Rhodopseudomonas sphaeroides obeys the regulatory conditions of the phototrophic bacterium, submitted.

Friedberg I, Hellingwerf KJ and Konings WN (1980) Regulation of alanine transport by a membrane potential dependent gating effect in Rhodopseudomonas sphaeroides, 13th FEBS Meeting, Jerusalem, S5-P77, p. 120.

Hellingwerf KJ, Michels PAM, Dorpema JW and Konings WN (1975) Transport of amino acids in membrane vesicles of Rhodopseudomonas sphaeroides energized by respiratory and cyclic electron flow, Eur.J.Biochem. 55, 397-406.

Hellingwerf KJ, Konings WN, Nicolay K and Kaptein R (1981) The light-induced pH gradients in chromatophores of Rhodopseudomonas sphaeroides as visualized by ^{31}P NMR, Photobiochem.Photobiophys. 2, 311-319.

Hellingwerf KJ, Friedberg I, Lolkema JS, Michels PAM and Konings WN (1982) Energy coupling of facilitated transport of inorganic ions in Rhodopseudomonas sphaeroides, J.Bacteriol. 150, 1183-1191.

Lolkema JS, Hellingwerf KJ and Konings WN (1982) The effect of "probe-binding" on the quantitative determination of the proton motive force in bacteria, Biochim.Biophys.Acta 681, 85-94.

Lolkema JS, Abbing A, Hellingwerf KJ and Konings WN (1983) The transmembrane electrical potential in Rhodopseudmonas sphaeroides determined from the distribution of tetraphenylphosphonium after correction for its binding to cell components, Eur.J.Biochem. 130, 287-292.

Konings WN and Robillard GT (1982) Physical mechanism for regulation of proton solute symport in Escherichia coli, Proc.Natl.Acad.Sci. USA 79, 5480-5484.

Kraayenhof R and Arents JC (1977)Fluorescent probes for the chloroplast energized state. Energy linked change of membrane surface charge. In Roux E, ed. Electrical phenomena at the biological membrane level, pp. 493-505. Amsterdam: Elsevier.

Michels PAM and Konings WN (1978) Structural and functional properties of chromatophores and membrane vesicles from Rhodopseudomonas sphaeroides, Biochim.Biophys.Acta 507, 353-368.

Michels PAM and Konings WN (1978) The electrochemical proton gradient generated by light in membrane vesicles and chromatophores from Rhodopseudomonas sphaeroides, Eur.J.Biochem. 85, 147-155.

Michels PAM, Hellingwerf KJ, Lolkema JS, Friedberg I and Konings WN (1981) Effects of the medium composition on the components of the electrochemical proton gradient in Rhodopseudomonas sphaeroides, Arch.Microbiol. 130, 357-361.

Nicolay K, Lolkema JS, Hellingwerf KJ, Kaptein R and Konings WN (1981) Quantitative agreement between the values of the light induced ΔpH in Rhodopseudomonas sphaeroides measured with flow-dialysis and ^{31}P NMR, Febs Letters 123, 319-323.

Oelze J and Drews G (1972) Membranes of photosynthetic bacteria, Biochim. Biophys.Acta 265, 209-239.

Robertson DE, Kaszorowski GJ, Garcia M and Kaback HR (1980) Active transport in membrane vesicles from Escherichia coli: the electrochemical proton gradients alters the distribution of the lac carrier between two different kinetic states, Biochemistry 19, 5692-5702.

ACKNOWLEDGEMENTS
The studies were supported by the Netherlands Foundation for Chemical Research (SON) which is subsidized by the Netherlands Organization for the Advancement of Pure Research (ZWO).

Authors address: Wil N. Konings, Marieke G.L. Elferink, Klaas J. Hellingwerf, Laboratorium voor Microbiologie, Kerklaan 30, 9751 NN HAREN, The Netherlands.

SIMULTANEOUS MEASUREMENTS OF ELECTRICAL FIELD CHANGES IN THE OUTER
SURFACE AND CENTRAL REGIONS OF CHROMATOPHORE MEMBRANES OF *Rhodopseudomonas
sphaeroides* BY RESPONSES OF MEROCYANINE DYES AND CAROTENOIDS

Shigeru Itoh/National Institute for Basic Biology

1. INTRODUCTION

The changes in the ionic conditions or pH of the outer medium were shown
to change membrane surface potential and to induce changes of the electrical
field within the membrane under certain conditions (K. Matsuura et al.
1980, S. Itoh 1982). This induces a change of the redox state of the
membrane component depending on its intramembrane localization (S. Itoh
1980a, K. Matsuura et al. 1980). Thus surface potential changes are expected
to affect the movement of electrons within the membrane. On the other
hand, intramembrane electron transport reactions seem to induce only a
small change directly on the surface potential (S. Itoh 1980a, K. Masamoto
et al. 1981). We have been using a series of merocyanine dyes to study
the role of surface potential. The dyes change their distributions at
the outer boundary region of membranes by responding to changes of the
electrical potential at this region with respect to the outer aqueous
phase (K. Masamoto et al. 1981). Movements of the dyes across the membrane
were essentially negligible (S. Itoh 1980b).

2. PROCEDURE

2.1. *Materials and Methods*

Chromatophores were prepared from cells of *Rhodopseudomonas sphaeroides*
green mutant as described in S. Itoh (1982). Absorption changes of
carotenoids and merocyanine dye (NK2274, see Fig. 1 for the structural
formulae) were measured with Hitachi 557 dual wavelength spectrophotometer or with a split beam flash spectrophotometer constructed in
the Institute. A merocyanine dye was purchased from Nippon Kanko
Shikiso Laboratory, Okayama.

2.2. *Results and Discussion*

Fig. 1 shows the response of a merocyanine dye (NK2274) to the
change of surface potential induced by salt addition in the suspension
of chromatophores of *R. sphaeroides*. The scale for the surface potential
on the right-hand ordinate was calculated from the known surface charge
density of -2.5 $\mu C/cm^2$ for this membrane (K. Matsuura et al. 1980) and
concentration of KCl. At each concentration of KCl, illumination of
chromatophores induced the additional absorbance change in response to the
membrane potential change. The broken line indicates the sum of these
light and salt-induced absorbance changes. We can estimate the light-
induced change of surface potential (actually the change of electrical
field within the boundary region of the membrane on the outer surface) from
the right-hand scale. From the difference between the broken and solid lines
in the figure, we can estimate light-induced surface potential changes of
+20-+5 mV at KCl concentrations of 2-200 mM. Under these conditions a light-
induced membrane potential of 260 mV (inside positive) was estimated.

Sybesma, C. (ed.), Advances in Photosynthesis Research, Vol. II. ISBN 90-247-2943-2.
© *1984 Martinus Nijhoff/Dr W. Junk Publishers, The Hague/Boston/Lancaster.*

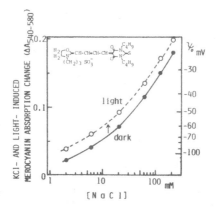

FIGURE 1. *Absorbance change of merocyanine dye induced by NaCl addition (solid line) and sum of light-induced and NaCl-induced absorbance changes (broken line). Scale for the surface potential on the right-side ordinate is calculated from the q value of -2.5 $\mu C/cm^2$ using the Gouy-Chapman equation (S. Itoh 1980a). Reaction mixture contained 0.01 mM $MgSO_4$, 1 mM tricine buffer (pH 7.8), chromatophores corresponding to 20 μM bacteriochlorophyll and varied concentrations of NaCl.*

from the absorbance change of the intrinsic carotenoids. These results suggest that this merocyanine dye senses 2-8% of the field change imposed on the total low-dielectric part of the membrane, probably sensing the field change in the outer boundary region of the membrane, and thus, confirm the results obtained with another merocyanine dye (K. Masamoto et al. 1981).

Fig. 2 shows the effects of a hydrophobic anion, tetraphenylborate (TPB^-) on the response of merocyanine and the intrinsic carotenoids during illumination. In the presence of low concentration of TPB^-, the response of the carotenoid decreased, while that of merocyanine increased.

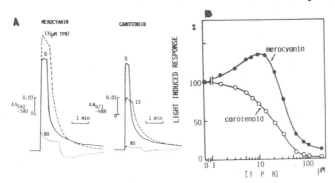

FIGURE 2A. *Effects of TPB^- on the light-induced responses of merocyanine and carotenoid in R. sphaeroides chromatophores. B. Dependences of the extents of absorbance changes of merocyanine and carotenoid on the TPB^- concentration. Reactions were performed at 200 mM NaCl in a medium similar to those in Fig. 1.*

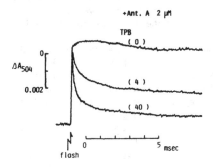
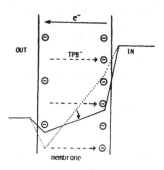

FIGURE 3. *Effect of TPB⁻ on the flash-induced absorbance change of carotenoid. Reaction mixture contained 50 mM Tris buffer (pH 7.8), 50 mM NaCl, 2 μM antimycin A and 15 μM bacteriochlorophyll.*

FIGURE 4. *Schematice model for the movement of TPB⁻ and change of electrical potential profile across the membrane under continuous illumination. Broken line, potential profile in the absence of TPB⁻ movement inside the membrane. Solid line, after the intramembrane movement of TPB⁻ from outer to inner potential energy well. Modified from Fig. 2 of O.S. Anderson et al (1978).*

No such increse of the merocyanine response was induced by use of the valinomycin-K⁺ complex, gramicidin or CCCP (not shown). It is suggested that at low concentrations of TPB⁻, movements of TPB⁻ within the membrane increase the field change on the outer surface of the membrane but decrease it in the center of the membrane where carotenoids sense the field change. Under these conditions, the time response of the merocyanine (half time of about 5 ms) was fast enough to follow the potential change under illumination.

   With measurements of the flash-induced absorbance change of carotenoids, the idea above was further confirmed. TPB⁻ induced a very fast decay phase within a few ms after the flash excitation (i. e., in a time range in which merocyanine can not respond) (Fig. 3). However, the following decay rate was only slightly affected. Higher concentrations of TPB⁻ gave a larger rapid decay phase of the carotenoids with only slight effects on the decay half times of this phase. Similar biphasic decays can be induced with other hydrophobic anions such as dipiclylamine (data not shown) or SCN⁻ (S. Itoh and S. Morita 1982).

   These effects of TPB⁻ are explained by postulating the rapid movement of TPB⁻ within the membrane according to the two-intramembrane-potential-well model proposed by Anderson et al. (1978) in the study of black lipid membranes. According to this model, TPB⁻ moves rapidly from one potential well, which exists just beneath the outer surface, to the other well close to the inner surface upon application of inside positive membrane potential (Fig. 4) and rapidly dissipates the electrical field in the center of the membrane. Equilibration of TPB⁻ concentrations between the well and the outer aqueous phase on each side of the membrane is estimated to be rather slow due to the cooperative interaction between TPB⁻ molecules. Biphasic decay kinetics of the carotenoid can be explained by the response of carotenoids to the field change in the center of the membrane. Thus the change of the electrical field sensed by merocyanine in the boundary region is expected to become larger after movement of TPB⁻ between the wells, as long as the membrane potential is maintained at the same level under continous light.

Similar biphasic decay kinetics of the carotenoid has been reported under phosphorylating conditions in chromatophores (S. Saphon et al. 1975) or in chloroplasts (S. Itoh and S. Morita 1982). In chloroplasts the faster decay phase under phosphorylating conditions was shown to be dependent on the activated state of the ATP-synthetase. Potential wells for protons inside the CF$_o$ moiety of ATP-synthtase are postulated to be analogy with those proposed by hydrophobic anions (S. Itoh and S. Morita 1982). The modulation of the intramembrane field by TPB⁻ shown in the present study seems to provide a new probe for the study of the mechanism of ATP synthesis since we can now selectively change the field strength at different depths inside the membrane.

REFERENCES

Anderson OS, Feldberg S, Nakadomari H, Levy S and McLaughlin S (1978) Electrostatic interactions among hydrophobic ions in lipid bilayer membranes Biophys. J. 21, 35-70.

Itoh S (1980a) Effects of surface potential and membrane potential on the midpoint potential of cytochrome c-555 bound to the chromatophore membrane of Chromatium vinosum, Biochim. Biophys. Acta 591, 346-355.

Itoh S (1980b) Stydy of electrogenic electron transfer steps in chromatophore membrane of Chromatium vinosum by the response of merocyanine dye, Biochim. Biophys. Acta 593, 212-223

Itoh S (1982) Movement of ions and change in the electrical potential profile across the membrane of the Rhodopseudomonas sphaeroides chromatophore, Plant Cell Physiol. 23, 595-605.

Itoh S and Morita S (1982) Decay of membrane potential under phosphorylating conditions in chloroplasts with in vivo activated ATPase, Biochim. Biophys. Acta 682, 413-419.

Masamoto K, Matsuura K, Itoh S and Nishimura M (1981) Membrane potential and surface-potential-induced absorbance changes of merocyanine dyes added to chromatophores from Rhodopseudomonas sphaeroides, Biochim. Biophys. Acta 638, 108-115

Matsuura K, Masamoto K, Itoh S and Nishimura M (1980) Effects of surface potential on the intramembrane electrical field measured with carotenoid spectral shift in chromatophores from Rhosopseudomonas sphaeroides, Biochim. Biophys Acta 547, 91-102

Matsuura K, Takamiya K-I, Itoh S and Nishimura M (1980) Effects of surface potential on the equilibrium and kinetics of redox reactions of membrane components with external reagents in chromatophores from Rhosopseudomonas sphaeroides, J. Biochem. 87, 1431-1437.

Saphon S, Jackson JB, Lerbs V and Witt HT (1975) The functional unit of electrical events and phosphorylation in chromatophores from Rhodopseudomonas sphaeroides, Biochim. Biophys. Acta 408, 58-66.

Autours adress: Shigeru Itoh/National Institute for Basic Biology, Nishigona Okazaki 444, Japan

MEASUREMENTS OF SURFACE POTENTIALS IN CHROMATOPHORE PREPARATIONS OF PHOTOSYNTHETIC BACTERIA IN AGREEMENT WITH THE GOUY–CHAPMAN THEORY; ESTIMATION OF BINDING CONSTANTS

BOTTE P.[*], SYMONS M.[●], SWYSEN C.[Δ], MAURO S.[*], HURWITZ H.[▲], LANNOYE R.[*], and SYBESMA C[Δ].

INTRODUCTION

We have investigated an optimal experimental approach necessary in order to allow an accurate evaluation of the electrochemical properties (surface charges, surface potential and transmembrane potential) of chromatophores of photosynthetic bacteria (*Rps. capsulata*) on the basis of the Gouy-Chapman theory.

The electrochromic shift of the carotenoids was used as internal probe of the surface potential changes induced by addition of non-permeable ions. Experimental data could be adjusted to theoretical curves by introducing three experimental improvements:

- by accounting for the background concentrations in the formulas of Gouy-Chapman (our preparations are more stable in the presence of a small divalent background);

- by introducing the Val-K$^+$ complex in our experimental medium (Matsuura et al. 1979 and Swysen et al. 1980 used CCCP-H$^+$). It is evident that in order to observe the complete surface potential change with the carotenoid shift (which, as internal probe, measures only the potential changes in the membrane), the bulk potentials inside and outside should be inaltered by the surface potential. We showed (Symons and Botte, in preparation), that in the presence of CCCP, this is not the case, making a satisfactory fitting with the Gouy-Chapman theory impossible. The Val-K$^+$ complex fulfils this requirement much better;

- by introducing two "non binding" cations (methyl viologen and N-methyl pyridinium) to estimate the surface potentials and the surface charges. These two organic ions have the particularity of having similar structure (see figure 1): the methyl group hinders binding to the surface.

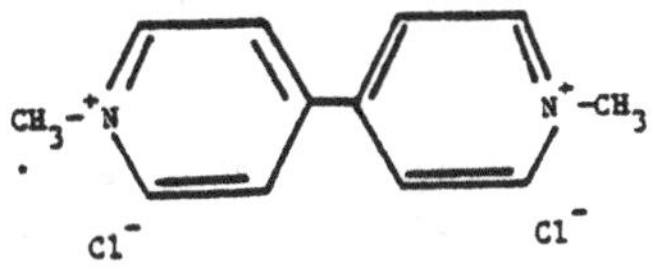

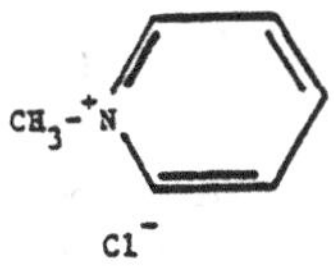

Fig. 1.

Data obtained in presence of methyl viologen and methyl pyridinium are compared with different divalent and monovalent metal cations and allow us to calculate their binding constants (assuming that the difference is only due to binding).

AUTHORS ADDRESS
* U.L.B. Physiologie végétale   Av. P. HEGER   1050 Bruxelles   BELGIUM
Δ V.U.B. Biofysica   Pleinlaan 2   1050 Brussel   BELGIUM
▲ U.L.B. Electrochimie   Av P. HEGER   1050 Bruxelles   BELGIUM
● Present address: WEIZMANN INSTITUTE OF SCIENCE   Membranes   76100 Rehovot   ISRAEL

Sybesma, C. (ed.), Advances in Photosynthesis Research, Vol. II. ISBN 90-247-2943-2.
© 1984 Martinus Nijhoff/Dr W. Junk Publishers, The Hague/Boston/Lancaster.

THEORETICAL METHODS

For a charged surface, such as the negatively charged photosynthetic membranes, the Gouy-Chapman theory predicts that the salt effects should be related only to the valency of the cation and be essentially independent of the associated anion.

Previous authors (Matsuura et al. 1979 and Swysen et al. 1980), in their calculations, did not take the background concentrations into account. The presence of these background concentrations shifts both monovalent and divalent curves towards larger concentrations values, so that comparisons of experimental with calculated curves is not possible without taking these background concentrations into account.

In our preparations, we always use a mixed solution of KOH, KCl and of a divalent cation as background. The Gouy-Chapman equation (Barber 1980) giving the surface charge as a function of the bulk concentrations and the surface potential is:

$$\sigma = - \left[2 \epsilon_o \epsilon_r R T \sum C_i \{ \exp (- \frac{z_i F \psi_o}{R T}) - 1 \} \right]^{\frac{1}{2}}$$ /1/

in which σ is the surface charge, $C_{i \infty}$ the bulk concentration of cation i in the outside medium, ψ_0 the surface potential, ϵ_o, ϵ_r the relative permittivity of a vacuum and of the solution, and other symbols have their usual meaning.

If we assume that the divalent cations are part of symmetrical salts, we can transform equation /1/ to get after adding respectively a monovalent ion concentration C_1^{ad} or a divalent ion concentration C_2^{ad} (in the presence of C_1^{bg} moles of monovalent ions and C_2^{bg} moles of divalent ones):

$$\sigma = - (4 \epsilon_o \epsilon_r R T)^{\frac{1}{2}} \left[(C_1^{ad} + C_1^{bg}) \left\{ \cosh (\frac{F \psi_o}{R T}) - 1 \right\} + 2 (C_2^{ad} + C_2^{bg}) \left\{ \cosh^2 (\frac{F \psi_o}{R T}) - 1 \right\} \right]^{\frac{1}{2}}$$ /2/

Equation /3/ calculates the surface charge, from the added concentrations of monovalent and divalent ion concentrations that give the same surface potential, taking the background concentrations into account:

$$\sigma = - (4 \epsilon_o \epsilon_r R T)^{\frac{1}{2}} \left\{ (C_1^{ad} + C_1^{bg}) (\frac{C_1^{ad} - 4 C_2^{ad}}{2 C_2^{ad}}) + 2 C_2^{bg} \left\{ (\frac{C_1^{ad} - 2 C_2^{ad}}{2 C_2^{ad}})^2 - 1 \right\} \right\}^{\frac{1}{2}}$$ /3/

If we assume that divalent and monovalent cations form only 1:1 complexes with the membrane surface charges, algebraic manipulation of the Langmuir adsorption isotherm and equation /1/ gives:

$$\sigma_{bdg} = \frac{\sigma_{init} \left\{ 1 - (K_2^{bg} C_2^{bg} + K_2^{ad} C_2^{ad}) \exp (- \frac{2 F \psi_o}{R T}) \right\}}{1 + \left\{ (K_1^{bg} C_1^{bg} + K_1^{ad} C_1^{ad}) \exp (- \frac{F \psi_o}{R T}) + (K_2^{bg} C_2^{bg} + K_2^{ad} C_2^{ad}) \exp (- \frac{2 F \psi_o}{R T}) \right\}}$$ /4/

where K are intrinsic association constants, and σ init, σ bdg are the initial surface charge without binding and the modified surface charge in presence of binding cations respectively. From the added concentrations of mono and divalent non-binding cations, we can calculate the initial surface charge with eq./3/. For the same surface potential, the surface charge modified by binding (σ bdg) is calculated with eq. /2/. The binding constants are obtained by replacing the values of respectively σinit and σbdg in eq. /4/.

RESULTS

The initial surface charge is calculated with eq. /3/ using values obtained at low and at high screening on our experimental "non-binding" curves, and the mean value for σ is determined. This value is introduced in a computer program to fix the concentration domains where the error is minimal.

The value of σ^{int} is calculated with eq. /3/ in the concentration range with minimal error. At these concentrations, the surface potential is determined and is used for the determination of σ bdg.

The binding constants are calculated with eq. /4/. Calculated curves are drawn by doing an iterative procedure on combining eq. /2/ and /4/.

The experiments have been performed on two different chromatophore preparations (see table I). Each preparation contained two batches of different background conditions:

- one batch with 20 μM MV^{++} and 2 mM K^+; (MV^{++} = Methyl viologen)
- the other batch with 20 μM Mg^{++} and 2 mM K^+.

The surface charges and the binding constants were estimated from the first batch and calculated curves were made with these values for the two batches. Fig 2 represents the results of preparation 1..

TABLE I

	σ min error	$\Delta\sigma$ / σmin error	K(Mg)	K(Li)
Prep. 1	0.0315 C/m^2	6 %	2.9 M^{-1}	0.3 M^{-1}
Prep. 2	0.0315 C/m^2	10 %	2.7 M^{-1}	0.3 M^{-1}

where σ min error represents the value of surface charges calculated in the concentration range of minimal error, $\Delta\sigma$ is the difference of surface charges measured at low screening and at high screening, and K are binding constants.

DISCUSSION.

- We have introduced a method allowing a quantitative correlation of experimental results with the Gouy-Chapman theory. Previous authors (Matsuura et al. 1979 and Swysen et al. 1980) only obtained qualitative correlation.

- The results allow the calculations of binding constants of some metal cations.

- We show that neglecting the binding effects of metal cations introduces an error on the estimation of the surface potential and of the surface charges (the utilisation of Mg and Li as reference ions, introduces an over-estimate of 1.7 for the surface charges and of 1.3 for the surface potential).

ACKNOWLEDGEMENTS

We appreciated the hospitality and help given by Professor Pecher and his staff during the synthesis of methyl pyridinium. One of us (Botte P.) received a fellowship from the Institut pour l'Encouragement de la Recherche Scientifique dans l'Industrie et l'Agriculture. The many fruitful discussions with Dr. L. Slooten of our laboratory are much appreciated. We thank Mr. A. Nuyten for technical assistance.

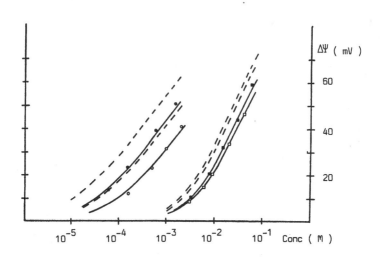

Fig 2.: Surface potential changes obtained by addition of different cations.

experimental conditions: Background: MOPS   2 mM
                                            Potassium   2 mM
                                            Magnesium   20 µM

                            Added ions: O   Methyl viologen
                                         ●   Magnesium
                                         ■   Lithium
                                         □   N-Methyl pyridinium

Drawn lines represent theoretical curves calculated with the data (σ, K) obtained on the first batch containing methyl viologen as background cation. Dashed lines represent theoretical curves calculated on the first batch (containing a background of non-binding cations).

REFERENCES

Barber J.
    B.B.A., 594 (1980) 253-308
Lau A, McLaughlin A and McLaughlin S.
    B.B.A., 645 (1981) 279-292
Matsuura K, Masamoto K, Itoh S and Nishimura M.
    B.B.A., 547 (1979) 91-102
Swysen C., Symons M., Nuyten A., Sybesma C.
    Bioelectrochemistry and Bioenergetics, 7 (1980) 575-585

PROTON PUMPING IN VESICLE-RECONSTITUTED BACTERIAL REACTION CENTRE/bc_1 AND PHOTOSYSTEM I/bf SYSTEMS

P.R. RICH AND P. HEATHCOTE

1. INTRODUCTION

Two light-activated cyclic electron transfer systems have been reincorporated into lipid vesicles in such a way that proton pumping across the membranes may be observed under appropriate conditions. The first of these has been constructed from mammalian cytochrome bc_1 complex and reaction centres isolated from Rhodopseudomonas sphaeroides (RCbc vesicles), a combination used previously by Packham et al. (1980) for single turnover studies in solution. In order to maintain adequate multiple turnover electron flux under our conditions, it was necessary to add both cytochrome c and ubiquinone-2. In the presence of valinomycin, light activation caused the translocation of four protons outwards across the vesicles for each pair of electrons completing a cycle, although this ratio appeared to fall to two after a significant ΔpH had built up.

A second vesicle system was constructed from chloroplast photosystem I and cytochrome bf complex (PSbf vesicles). When supplemented with plastocyanin and plastoquinone-2, light activation promoted a multiple turnover cyclic electron transfer and a more limited proton translocation outward across the vesicles. We report here the preliminary characterisation of these preparations.

2. MATERIALS AND METHODS

Bacterial reaction centres were isolated from the carotenoidless mutant R-26 of Rhodopseudomonas sphaeroides, cytochrome bc_1 complex from beef heart, cytochrome bf complex from lettuce and plastocyanin from parsley. All were prepared by standard methods. Photosystem I was isolated from spinach in such a way that it retained its site of plastocyanin oxidation (Mullet et al. 1980). Ubiquinone-2 was the kind gift of Eisai Co., Ltd. and plastoquinone-2 was synthesised from geraniol and 2,3 dimethylquinol.

Cytochrome c redox changes were monitored at 550nm with a reduced minus oxidised extinction coefficient of 19 and plastocyanin at 575nm with an extinction coefficient of 4.3. The appearance of protons in the external medium was monitored using phenol red spectral changes at either 542nm or 550nm as appropriate. Membrane potential was monitored with the dye safranine (Akerman, Saris 1976). Measurements were made with a single beam instrument and where necessary a reference trace at an appropriate wavelength was computer subtracted from the experimental data. Illumination was provided with light filtered through an RG630 filter and the photo-multiplier was protected with a $CuSO_4$ solution, a Wratten green filter and 580nm cut-off filter.

3. RESULTS

3.1 Vesicle preparation and component orientation

RCbc vesicles were prepared by the technique of cholate dialysis (Kagawa, Racker 1971). A ratio of soya bean lecichin:cholate of 2:1 was employed. This technique produced a preparation with high electron transfer activities

Sybesma, C. (ed.), Advances in Photosynthesis Research, Vol. II. ISBN 90-247-2943-2.
© 1984 Martinus Nijhoff/Dr W. Junk Publishers, The Hague/Boston/Lancaster.

and with the majority of components having their cytochrome c reaction sites
oriented towards the outer aqueous phase. The same technique, however, was
not suitable for the preparation of PSbf vesicles. On the basis of digitonin
stimulation of plastocyanin oxidation it appeared that most of the photosystem I oriented with its plastocyanin oxidation sites facing the vesicle
interior, whereas most of the cytochrome bf complexes had plastocyanin sites
oriented outwards. A technique involving sonication followed by passage
through sephadex G-25 was therefore employed instead. This produced vesicles
with more random component orientation and those components ·with the
desired orientation could be selectively activated by adding plastocyanin
to the outer aqueous phase after the vesicles had been formed.

3.2   Light activated cyclic electron transfer

In the presence of appropriate redox connectors (ubiquinone-2 and cytochrome
c for RCbc vesicles and plastoquinone-2 and plastocyanin for PSbf vesicles)
light activation caused cyclic electron transfer to proceed in both cases.

A RCbc Vesicles   B PSbf Vesicles

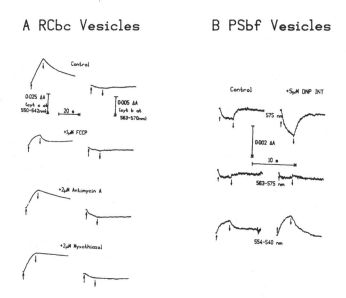

FIGURE 1.   Reaction medium was 160mM sucrose, 40mM KCl and 0.5mM EDTA at
25 C in both cases. In A, vesicles were added to a final bc_1 conc. .of
0.1μM, cytochrome c to 19μM and ubiquinol-2 to 13μM and the pH was 7.1.
In B, vesicles were added to a final bf conc. of 300nM, plastocyanin to
1.1μM, plastoquinol-2 to 17.5μM and valinomycin to 0.1μg/ml and the pH
was 6.65.

The rate of cycling and the redox poise of components in the steady state was dependent on the relative amounts of quinone and cytochrome c (plastocyanin) added and on the number of electrons present in the system. The pH was also critical: a pH too high caused significant non-enzymatic electron transfers whereas a pH too low decreased bc complex activities. Optimal pH values were around 7.2 for RCbc vesicles and around 6.5 for PSbf vesicles. Under appropriate conditions, redox changes of cytochrome b could also be observed. Antimycin A inhibited cytochrome b reoxidation and myxothiazol inhibited donation into the bc complex in the RCbc system, but both were without effect of the PSbf system. Instead 2-n-heptyl-4-hydroxyquinoline-N-oxide could be used to inhibit cytochrome b reoxidation and DNP-INT could be used to inhibit donation into the bf complex. Figure 1 illustrates some typical results.

3.3 Light activated proton translocation

By addition of phenol red, it became possible to monitor proton changes in the external medium. In the presence of appropriate mediators and valinomycin, both systems exhibited proton translocation outwards on light activation (figure 2). Four phases can be seen in these profiles:1) a rapid net alkalinisation on illumination, caused by reduction of quinone to quinol;2) a subsequent acidification as electron transfer proceeds and protons are translocated outwards across the vesicles;3) a rapid further acidification when the light is switched off and quinol is reoxidised; 4) relaxation to the initial pH as protons leak back into the vesicles.

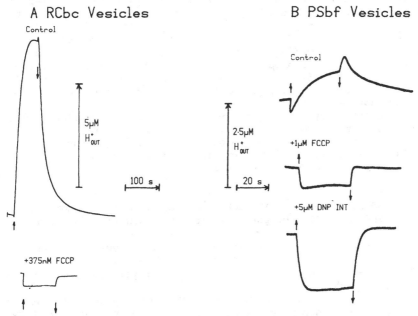

FIGURE 2. Reaction medium as for fig. 1 with 0.1µg/ml valinomycin and 100µM phenol red also added. In A, other additions were KCN, 1mM; vesicles, 53nM in bc_1; cytochrome c, 14.5µM (5% reduced); ubiquinone-2, 4µM; pH=7.2. In B, other additions were: vesicles to 165nM in bf; plastocyanin, 3.6µM; plastoquinol-2, 96µM; catalase, 200U/ml; pH=6.75.

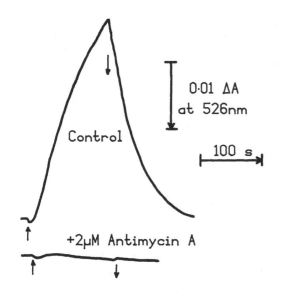

0·01 ΔA
at 526nm

Control

100 s

+2μM Antimycin A

FIGURE 3. RCbc vesicles were suspended to 84nM bc_1 in 0.1M choline chloride and 0.2mM KCl at pH 7.1. 19μM cytochrome c, 13μM ubiquinol-2 and 40μM safranine were also added. Upward deflection indicates absorbance decrease at 526nm and is consistent with the formation of a membrane potential across the vesicles, positive outside.

FCCP abolished the proton translocation and left only those changes associated with change of redox state of the quinone. Addition of bc complex inhibitors caused this net quinone pool change to increase.

Computer analysis of such profiles provided estimates of rates of proton translocation. These could be compared with the rates of electron cycling, obtained by monitoring the rates of cytochrome c (plastocyanin) rereduction on switching off the light after varying periods of illumination. For the RCbc system an $H^+/2e^-$ ratio of four was found soon after the light was switched on, but this dropped to two in the steady state condition with a high ΔpH. A similar calculation for the PSbf vesicles gave $H^+/2e^-$ values of only around 0.6-1.5, probably as a result of the poorer reincorporation of the photosystem I.

$\Delta\psi$ generation may also be observed in the RCbc preparations. Fig. 3 shows the response of RCbc vesicles plus safranine to light activation. A comparison of this change to a change brought about by potassium diffusion potentials in the presence of valinomycin indicated that potentials in excess of 100mV could be generated by light activation.

REFERENCES
Akerman KE and Saris N-E (1976) Stacking of safranine in liposomes during valinomycin-induced efflux of potassium ions Biochim. Biophys. Acta 426, 624-629
Kagawa Y and Racker E (1971) Partial resolution of enzymes catalysing oxidative phosphorylation J. Biol. Chem. 246, 5477-5487
Mullet J., Burke, JJ and Arntzen CJ (1980) Chlorophyll proteins of photosystem I, Plant Physiol. 65, 814-822
Packham NK, Tiede DM, Mueller P and Dutton PL (1980) Construction of a flash-activated electron transport system. Proc. Nat. Acad. Sci. USA 77, 6339-6343

ACKNOWLEDGEMENTS
This work is funded by the Venture Research Unit of British Petroleum Company p.l.c. and was technically assisted by Ms. S.D. Clarke
Authors address: Dept. Biochemistry, University of Cambridge, Cambridge CB2 1 and Dept. Botany & Biochemistry, Westfield College, London NW3 7ST

REACTION CENTERS FROM RHODOPSEUDOMONAS SPHAEROIDES IN RECONSTITUTED
PHOSPHOLIPID VESICLES. STRUCTURAL PROPERTIES AND LIGHT-DEPENDENT PROTON
TRANSLOCATION

KLAAS J. HELLINGWERF

1. INTRODUCTION

Reaction Centers (RC's) from phototrophic bacteria catalyze light-driven
transmembrane electron transfer as a first step in the (cyclic) electron
transfer chain of such bacteria (for a review see: Okamura et al., 1983
and Dutton et al., 1982). Many of the structural and functional features
of RC's have already been elucidated; the remaining questions mainly
focus on (i) the effects of transmembrane gradients (of redox potential
and electrochemical potential of protons) on the reactions catalyzed
by the RC's and (ii) the interactions between RC's and physiological and
artificial electron donors and acceptors. Many of the unsolved aspects
can be optimally investigated under conditions, in which the RC's have
been reconstituted into artificial membranes; either in planar (Schönfeld
et al., 1979) or vesicular form (Crofts et al., 1977; Pachence et al.,
1979). Here I report on the structure of reconstituted RC vesicles and
light-dependent unidirectional proton translocation catalyzed by these
vesicles.

2. MATERIALS AND METHODS

2.1. Materials

Horse heart cytochrome c, L-α-lecithin from soybean and valinomycin were
from Sigma Chemicals, St. Louis, MO; 2,3-dimethoxy-5-methyl-1,4-benzo-
quinone (UQ_O) from Trans World Chemicals, Washington and lauryldimethyl-
ammonium-N-oxide (LDAO) from Onyx Chemical Comp., Jersey City, NJ.
RC's from Rhodopseudomonas sphaeroides R-26 were isolated and stored as
described previously (Feher, Okamura, 1978). These will be referred to as
LDAO-containing RC's. Part of the RC's were treated with Bio-Beads SM-2
(25°C, 3 h, 10% w/v). This treatment removes 66% of the LDAO, originally
present in the suspension (as measured with [^{14}C]LDAO) of which more than
90% was bound to the RC's. These RC's were stored in 200 mM KCl and will
be referred to as LDAO-depleted RC's. Asolectin was isolated from L-α-
lecithin from soybean and stored in an inert atmosphere at -40°C.

2.2. Methods

(i) Reconstitution of RC vesicles: 0.2 ml of the RC suspension plus 0.2 ml
asolectin (40 mg/ml in 200 mM KCl, suspended by vortexing for 15 min) were
co-sonicated in a bath-type sonicator (1700 W) in the dark under argon.
(ii) Sucrose-density gradients were run at 20°C at 216.000 x g for 60 h in
a Beckman Model L2-65B ultracentrifuge. (iii) Proton translocation was
measured in 200 mM KCl in the presence of 2.5 μM valinomycin with a combination pH-electrode connected to a Model 600B Keithley electrometer, essentially as described for bacteriorhodopsin vesicles (Hellingwerf et al., 1979).
Illumination was provided by an Ealing model 22-0004 Fiber light source
after filtering through a long-pass cut-off filter (I_{50} = 660 nm) at
40 mW/cm^2. (iv) The orientation of the reconstituted RC's was measured

Sybesma, C. (ed.), Advances in Photosynthesis Research, Vol. II. ISBN 90-247-2943-2.
© 1984 Martinus Nijhoff/Dr W. Junk Publishers, The Hague/Boston/Lancaster.

via kinetic distinction of the rate of rereduction of the bacteriochlorophyll dimer after photo-oxidation, by either extravesicular cytochrome
c or the endogenous secondary quinone. For these measurements a kinetic
spectrophotometer of local design, having a time resolution of ≥ 1 µs,
was used. An actinic flash was provided by a phase-R DL 2100 C pulsed
dye laser.

3. RESULTS AND DISCUSSION

3.1. Structure of the reconstituted vesicles

Fig. 1 shows an analysis of the structure of sonicated RC vesicles. In
panel A the average density of the various preparations (as measured with
sucrose-density gradient centrifugation (open symbols) is plotted against
the lipid/RC ratio during reconstitution. The data agree closely with
densities calculated from the measured density of lipid vesicles and isolated RC's, assuming ideal mixing of these components during reconstitution (closed symbols). From this we conclude that RC's and lipid associate
at random during reconstitution, in clear contrast with another light-
driven proton pump: bacteriorhodopsin (Hellingwerf, 1979). This conclusion
is fully supported by the measurements of proton translocation in vesicles
reconstituted from asolectin and LDAO-containing RC's (Fig. 1C).
Fig. 1B shows that in all preparations an excess RC's exists having their
cytochrome c binding site oriented towards the exterior of the vesicles,
particularly in vesicles prepared from LDAO-depleted RC's. However, a
complete unidirectional orientation is not obtained under these conditions
(contrast Pachence et al., 1979). This becomes more evident when one rea-
lizes that at the two lowest lipid/RC ratio's the amount of RC's that are
not incorporated into a bilayer, increases significantly. Under these conditions maximally a ratio of 70 to 30 of RC's with opposite orientation
occurs. A higher net-orientation can be obtained by lowering the salt
concentration.
Upon fractionation of vesicles reconstituted from LDAO-depleted RC's at
a molar lipid/RC ratio of 560 on Biogel A-50 it turns out that the sizes
of the vesicles vary from 1200 to 500 nm (as measured with light scattering).
In these fractions the net-orientation of the RC's decreases with vesicle
size from 80/20 to 60/40. This suggests that the RC's spontaneously orient
themselves with their cytochrome c binding site on the outside of large
vesicles, whereas an increased curvature of the bilayer leads to an increased scrambling of the RC's, possibly due to increased steric repulsion.

3.2. Light-dependent proton translocation in reconstituted RC vesicles

In the studies on proton translocation in a suspension of RC vesicles,
reported sofar (Crofts et al., 1977; Darzon et al., 1980; Orlich, Hauska,
1980) use was made of a set of redox mediators such that the unidirection-
ality of proton translocation was lost. In contrast, we have used extravesicular cytochrome c and UQ_0 only. This does lead to unidirectional
translocation of protons upon illumination. However, under some conditions
the process of proton translocation is obscured by changes in the state
of reduction of cytochrome c (Fig. 2A). At the lower pH value (pH = 8.5)
the overall rate of proton translocation is limited by the rereduction of
cytochrome c by UQ_0H_2. Consequently this leads to a complete oxidation of
cytochrome c in the light and since the protonation state of the electron

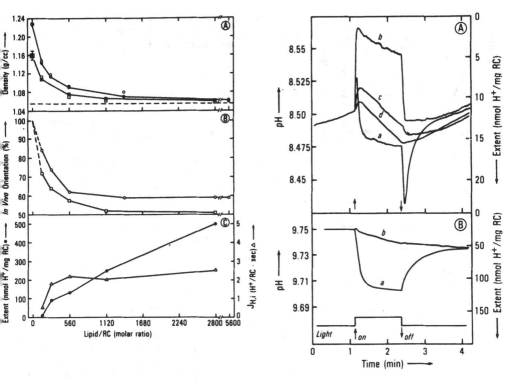

FIGURE 1. Density (A), RC-orientation (B) and Proton translocation (C) in RC vesicles of various lipid/RC ratio's. Squares: LDAO-containing RC's, circles: LDAO-depleted RC's.
FIGURE 2. Light-dependent proton translocation in a suspension of RC vesicles. a: no additions; b: plus 25 μM CCCP; c and d: plus 7.5 and 15 mM o-phenanthroline.

in cytochrome $\underline{c}$ and UQ_oH_2 differs this leads to an -uncoupler insensitive-alkalinization. This rate limitation does not occur at high pH and consequently under those conditions the characteristic pattern of a light-dependent proton pump is observed (Fig. 2B). The titrations with UQ_o (Fig. 3) confirm this interpretation. The amount of UQ_o required for the maximal rate of proton extrusion is strongly pH-dependent: it saturates at a much lower concentration at the highest external pH. The rate of proton translocation at very high pH (> 10.5) is limited by electron transfer from UQ_A to UQ_B within the reaction centers.
Analysis of the light-intensity dependence of the initial rate and extent of proton translocation indicates that the two processes are not saturated at 40 mW/cm^2. From doubly reciprocal plots a degree of saturation of 33 and 69%, respectively, is calculated. These data, together with the light intensity and the absorption spectrum of the RC's allows an estimate of the rate of oxidation of UQ_BH_2 by UQ_o: a value of 44 s^{-1} is obtained at 250 μM UQ_o and 1 μM RC's.

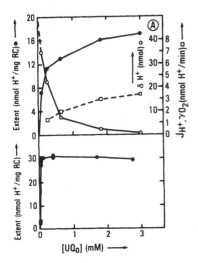

FIGURE 3. Dependence of the extent of proton translocation on the UQ_0 concentration. Upper part: pH 8.5; lower part: pH 9.75.

4. CONCLUSIONS

(i) Reconstituted RC vesicles are homogeneous in lipid/RC ratio and heterogeneous in size; (ii) The largest vesicles contain the highest degree of net orientation of RC's; (iii) Unidirectional proton translocation can be observed upon addition of UQ_0 and cytochrome $\underline{c}$ in the light at high pH; (iv) The reactions that limit the rate of this proton pump have been partly resolved.

REFERENCES

Crofts AR, Crowther D, Celis H, Celis SA and Tierney G (1977) Proton pumps in bacterial photosynthesis, Biochem.Soc.Transactions 5, 491-495.

Darzon A, Vandenberg CA, Schönfeld M, Ellisman MH, Spitzer NC and Montal M (1980) Reassembly of protein-lipid complexes into large bilayer vesicles: perspectives for membrane reconstitution, Proc.Natl.Acad.Sci USA 77, 239-243.

Dutton PL, Mueller P, O'Keefe DP, Packham NK, Prince RC and Tiede DM (1982) Electrogenic reactions of the photochemical reaction center and the ubiquinone-cytochrome b/c_2 oxidoreductase, Curr.Top.Membr.Transp. 16, 323-343.

Feher G and Okamura MY (1978) Chemical composition and properties of reaction centers. In Clayton RK and Sistrom WR, ed. The photosynthetic bacteria, pp. 349-386. New York: Plenum Press.

Hellingwerf KJ (1979) Structural and functional studies on lipid vesicles containing bacteriorhodopsin. PhD Thesis, University of Amsterdam.

Hellingwerf KJ, Arents JC, Scholte BJ and Westerhoff HV (1979) Bacteriorhodopsin in liposomes. II. Experimental evidence in support of a theoretical model, Biochim.Biophys.Acta 547, 561-582.

Okamura MY, Feher G and Nelson N (1983) Reaction centers in bacteria and green plants. In Govindjee R, ed. Integrated approach to plant and bacterial photosynthesis, Ch. 3. New York: Academic Press, in press.

Orlich G and Hauska G (1980) Reconstitution of photosynthetic energy conservation, Eur.J.Biochem. 111, 525-533.

Pachence JM, Dutton PL and Blasie JK (1979) Structural studies on reconstituted reaction center-phosphatidylcholine membranes. BBA 548, 348-373.

Schönfeld M, Montal M and Feher G (1979) Functional reconstitution of photosynthetic reaction centers in planar lipid bilayers, Proc.Natl.Acad. Sci. USA 76, 6351-6355.

ACKNOWLEDGEMENTS

The work described in this communication was performed at the University of California at San Diego in collaboration with George Feher and Mauricio Montal. It will be described in more detail elsewhere. I thank E. Abresch for the isolation of RC's and Z. Kam for performing the light scattering measurements. This study was supported by a grant from the Netherlands Organization for the Advancement of Pure Research (Z.W.O.).

Authors address: K.J. Hellingwerf, Department of Microbiology, University
            of Groningen, Kerklaan 30, 9751 NN HAREN, The Netherlands.

PROTONS, THIOLS AND COUPLING FACTOR 1 IN RELATION TO PHOTOPHOSPHORYLATION

RICHARD E. MC CARTY, JAMES W. DAVENPORT, STUART R. KETCHAM, JAMES V. MORONEY, CARLO M. NALIN, WILLIAM J. PATRIE AND KURT WARNCKE

1. INTRODUCTION

The H^+-ATPase of thylakoids catalyzes ATP synthesis and, in some conditions, hydrolysis in a manner that is linked to proton fluxes. The H^+-ATPase consists of two readily separable parts: coupling factor 1 (CF_1), a hydrophilic, multisubunit enzyme that is extrinsic to the membrane, and F_0, a collection of more hydrophobic proteins that are intrinsic to the membrane. CF_1 contains the active sites of the H^+-ATPase and is composed of five different subunits, labeled α-ε in order of decreasing molecular weight. F_0 constitutes a proton channel and provides specific sites for the attachment of CF_1 to the membrane.

In this communication, we will summarize some of our recent work on CF_1 from spinach chloroplasts. Particular attention will be paid to the structure of the enzyme, proton fluxes in relation to ATP synthesis and hydrolysis, regulation of expression of ATPase activity and functions of the smaller CF_1 subunits.

2. STRUCTURE OF CF_1

The molecular weight of CF_1 was previously accepted to be 325,000 (Farron, 1970), a value most consistent with an $\alpha_2\beta_2\gamma\delta\varepsilon_2$ stoichiometry. There is good evidence that F_1 from other coupling membranes has a higher molecular weight and a subunit composition of $\alpha_3\beta_3\gamma\delta\varepsilon$ (Senior, Wise, 1983). For a number of reasons, we decided to redetermine the molecular weight of CF_1 by two independent methods (Moroney et al., 1983). In Gordon Hammes' laboratory, laser light scattering was used, while we used sedimentation equilibrium. Values of 407,000 ± 20,000 and 400,000 ± 25,000 respectively, were obtained, close to that obtained by Merchant et al. (1983) for CF_1 from Chlamydomonas thylakoids. The higher molecular weight is consistent with an $\alpha_3\beta_3\gamma\delta\varepsilon$ composition, although the data are not of sufficient precision to allow an unambiguous assignment of the copy number for δ and ε. We recently reported (Béliveau et al., 1982) that CF_1 contains 3.5 tryptophans per 325,000 dalton. Since the ε subunit contains two tryptophans and is the only CF_1 subunit to contain tryptophan, this indicated that there are $2\varepsilon/CF_1$. Even though the enzyme used in these studies was highly purified, some ribulose 1,5-bisphosphate carboxylase was likely to have been present. A 1.4% contamination of CF_1 by the carboxylase is sufficient to raise the cysteine and tryptophan contents of a CF_1 preparation by one per CF_1. Passage of CF_1 through an agarose column to which antibodies against ribulose 1,5-bisphosphate carboxylase were covalently attached reduced the tryptophan content of a CF_1 preparation to slightly over $2/CF_1$, based on a CF_1 molecular weight of 400,000. If this procedure does not cause dissociation of the ε subunit, this result indicates that CF_1 contains one ε subunit. This conclusion is in line with our titrations (J. Moroney, unpublished) of the SH content of CF_1 subunits.

Sybesma, C. (ed.), Advances in Photosynthesis Research, Vol. II. ISBN 90-247-2943-2.
© *1984 Martinus Nijhoff/Dr W. Junk Publishers, The Hague/Boston/Lancaster.*

C.W. Akey in S.J. Edelstein's laboratory in this Section has examined CF_1 lacking the δ subunit or lacking both the δ and ϵ subunits by electron microscopy of close-packed monolayers (Akey et al., 1983). By computer techniques, he obtained averaged reconstructions of individual CF_1 molecules. These reconstructed images showed strong hexagonal symmetry (Fig. 1), consistent with an $\alpha_3\beta_3\gamma$ stoichiometry. The reconstructed images of E. coli F_1 were very similar, except that the dimensions of CF_1 were slightly larger.

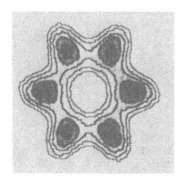

Figure 1. Reconstruction of CF_1 lacking the δ and ϵ subunits. An average of 20 molecules from electron micrographs of three subunit CF_1 was used in the reconstruction. From Akey et al. (1983).

3. PROTON FLUXES IN RELATION TO ATP SYNTHESIS

Our previous analysis (McCarty, Portis, 1976; Davenport, 1983) of ATP synthesis and hydrolysis in relation to proton fluxes provided evidence that the common link between electron transport and ATP synthesis in the steady state is a proton activity gradient (ΔpH). From this analysis, we also derived a maximal phosphorylation efficiency $(P/e_2)_{max}$ of close to $4/3$. The $(P/e_2)_{max}$ is equivalent to $2(H^+/e^-)/(H^+/ATP)$, where H^+/e^- is the ratio of protons appearing in the thylakoid interior to electrons transferred and H^+/ATP is the ratio of protons translocated to ATP synthesized. Thus, if the (H^+/e^-) is 2, (H^+/ATP) is three. (H^+/ATP) ratios of three have been obtained by other methods (Avron, 1978; Handgarter, Good, 1982).

More recently it has been questioned whether protons deposited inside thylakoids by electron flow through photosystem II are equivalent to those deposited by electron flow through photosystem I with respect to their effects on electron flow and ATP synthesis in the steady state (Haraux et al., 1983). We have consistently observed, however, that electron flow and phosphorylation respond to protons from the two photosystems in an identical manner (Portis, McCarty, 1974; Davenport, 1983). An example of this is shown in Table 1. In this experiment, phosphorylation and ΔpH were assayed under the same conditions as a function of light intensity. ΔpH was calculated from the extent of $[^{14}C]$hexylamine uptake determined by a silicone oil microcentrifugation method (Portis, McCarty, 1976). Phosphorylation coupled to electron flow through photosystem II alone responded to ΔpH in a very similar way as that coupled to electron flow through both photosystems. Moreover, in the steady state under nonphosphorylating conditons it was shown that the rate of electron flow in the light (Portis et al., 1975) or ATP hydrolysis in the dark

(Davenport, McCarty, 1981) is proportional to the internal proton concentrations. The relationship between $[H^+]_{in}$ and electron flow was the same when light intensity was used to vary these parameters as when the electron transport inhibitor 3-(3,4-dichlorophenyl)-1,1-dimethyl urea was used (Davenport, 1983). Similarly, the proportionality between rate of ATP hydrolysis in the dark by thylakoids and $[H^+]_{in}$ was the same whether the rates of hydrolysis were changed by increasing concentrations of the ATPase inhibitor, N,N'-dicyclohexylcarbodiimide, or by varying the extent of ATPase activation (Davenport, McCarty, 1981). These results suggest that protons derived from electron flow through either of the two photosystems. Since ATP-P_i exchange, which actually is ATP synthesis from medium ADP and P_i, shows a similar dependence on ΔpH as does photophosphorylation, protons deposited inside thylakoids by ATP hydrolysis by CF_1 are also energetically equivalent to those deposited inside by either photosystem. The discrepancy between our results and those of others may be explained by the difference in methods used to estimate ΔpH. Although it is clear that 9-aminoacridine fluorescence responds to ΔpH, its use as a quantitative indicator of ΔpH has been questioned.

TABLE 1. Response of photophosphorylation coupled to electron flow through photosystems I and II and photosystem II alone to ΔpH

Parameter[a]	Electron Flow System	
	H_2O to $Fe(CN)_6^{3-}$	H_2O to dimethylquinone
Slope	2.26	2.24
Y intercept	-4.50	-4.39
X intercept	1.99	1.95
Linear correlation coefficient	0.995	0.994

[a] Derived from graphs of log phosphorylation rate vs ΔpH. Phosphorylation and ΔpH were determined at five light intensities, ranging from 0.1 to 1.0×10^5 ergs $cm^{-2}s^{-1}$. Rates of phosphorylation were from 6 to 100 μmol h^{-1} mg chlorophyll^{-1} and ΔpH ranged from 2.3 to 2.88. Nitrofluorfen (40 μM) was used to block electron flow between the photosystems when dimethyl quinone was used as the electron acceptor.

4. REGULATION OF EXPRESSION OF ATPase ACTIVITY

Washed chloroplast thylakoids and isolated CF_1 have low ATPase activities. Illumination of thylakoids in the presence of dithiothreitol allows sustained ATP hydrolysis in the dark. ATP hydrolysis is stimulated by uncouplers, is sensitive to phosphorylation inhibitors and is coupled to proton influx. The ATPase activity of soluble CF_1 is also enhanced dramatically by treatment of the enzyme with high concentrations of dithiothreitol.

Reduction of a disulfide bond in the γ subunit is the key to activation by dithiothreitol of the ATPase of both the soluble enzyme and CF_1 in thylakoids. In contrast to previous results (Arana, Vallejos, 1982) we find only one disulfide in CF_1. The activation of soluble CF_1 by dithiothreitol is correlated to the incorporation of anilinonapthylmaleimide specifically into the γ subunit. The fluorescence of proteins modified with this reagent may be readily detected on polyacrylamide gels (Nalin et al., 1983). Incubation of the reduced enzyme with iodosobenzoate causes reformation of the disulfide bond and reverses the activation (Arana, Vallejos, 1982; Nalin, unpublished). If, however, the reduced enzyme is reacted with N-ethylmaleimide prior to exposure to iodosobenzoate, little reversal of activation is seen. This result suggests that at least one of the two SH generated by reduction of the disulfide is accessible to N-ethylmaleimide. Although its reaction with maleimide has little effect on ATPase activity, alkylation of the SH group protects the enzyme from reoxidation (C. Nalin, unpublished).

An analogous situation occurs in membrane bound CF_1. Illumination of thylakoids in the presence of dithiothreitol causes the exposure of an additional SH group(s) in the γ subunit to reaction with N-ethylmaleimide (Table 2). Moreover, the treatment of light- and dithiothreitol-treated thylakoids with N-ethylmaleimide in the dark stabilizes the ability of the thylakoids to catalyze ATP hydrolysis after a light trigger (Table 3) (S. Ketcham, K. Warncke, unpublished).

TABLE 2. Incorporation of N-ethylmaleimide into CF_1 in thylakoids in the dark following dithiothreitol treatment

CF_1 Subunit	Treatment[a]	
	Dithiothreitol in dark	Dithiothreitol in light
	mol N-ethylmaleimide/mol CF_1	
α	0.01	0.03
β	0.02	0.04
γ	0.23	0.92
δ	<0.01	0.01
ϵ	<0.01	0.01

[a] Thylakoids, previously incubated with 2 mM N-ethylmaleimide in the dark to react accessible groups, were incubated in the light or dark with 5 mM dithiothreitol. After removal of the dithiothreitol by washing, 0.25 mM [^{3}H]N-ethylmaleimide (6 x 10^4 cpm $nmol^{-1}$) was added. After 5 min dithiothreitol was added to quench the unreacted N-ethylmaleimide. CF_1 was purified and 9 µg samples subjected to polyacrylamide gel electrophoresis in the presence of sodium dodecyl sulfate.

Reduced CF_1 is capable of sustained ATP hydrolysis both in solution and when membrane-bound. In addition, the reduced enzyme is more active in photophosphorylation at a given ΔpH (Davenport, McCarty, 1981; Mills, Mitchell, 1982) and it is apparent that the reduced form of the enzyme probably is the form the enzyme assumes in illuminated leaves or intact chloroplasts. This raises the interesting question of how ATPase activity is regulated _in vivo_ in the dark.

The γ subunit of CF_1 contains four SH groups. Two of these cysteinyl residues are involved in a disulfide bond whereas the other two may be distinquished on the basis of their reactivity. One SH on the γ subunit of CF_1 in thylakoids reacts rapidly with N-ethylmaleimide in the dark. The alkylation of this accessible sulfhydryl has no effect on either ATP synthesis or hydrolysis. Energy-dependent changes in the conformation of CF_1 in thylakoids exposes a second free SH group in the γ subunit to reaction with N-ethylmaleimide. This SH group is required in some way for photophosphorylation. Analysis by high performance liquid chromatography of tryptic peptides of the γ subunit of CF_1 with the cysteinyl residues specifically labeled with alkylating reagents has shown that each of the four cysteine residues in the γ subunit is distinct (J. Moroney, C. Fullmer, unpublished). A model for the effects of oxidation and alkylation of the sulfhydryl groups of the γ subunit on ATPase activity and phosphorylation is given in Figure 2.

TABLE 3.   Protection of the ATPase activity of thylakoids from iodoso-
           benzoate inhibition.

Thylakoids[a]	ATPase Rate	
	-iodosobenzoate	+ 1 mM iodosobenzoate
Control	341	7
N-ethylmaleimide-treated	288	254

[a] Thylakoids were illuminated in the presence of 5 mM dithiothreitol and were washed to remove the dithiothreitol. Aliquots were incubated for 5 min at room temperature in the presence or absence of 0.25 mM N-ethylmaleimide. The remaining maleimide was reacted with an equivalent amount of dithiothreitol and samples illuminated for 2 min in the presence or absence of iodosobenzoate prior to incubation in the dark at 37°C in the presence of Mg^{2+}-ATP. Rates are given as μmol P_i formed h^{-1} mg chlorophyll^{-1}.

These experiments and others carried out with protease activation (Moroney, McCarty, 1982) point to a central role for the γ subunit in the conversion of CF_1 to more active states. There are also indications (Moroney et al., 1982) that the structure of the γ subunit can influence the proton permeability of the membrane. There may be, then, a direct relationship between the conversion of CF_1 to an active conformation and the opening of a proton gate in CF_1. It is of interest to note that oxidized CF_1 in thylakoids is active in ATP synthesis in the light, but not in ATP hydrolysis in the dark, at least for more than a few seconds. Thus, it seems likely that the γ subunit of oxidized CF_1 in illuminated thylakoids assumes an active conformation that resembles that of the reduced enzyme. In the dark, the structure of the oxidized and reduced enzymes are likely to be different.

5.   ROLE OF THE δ AND ε SUBUNITS

The δ and ε subunits of F_1 from E. coli plasma membranes are required for the attachment of F_1 to the membrane, but not for ATPase activity

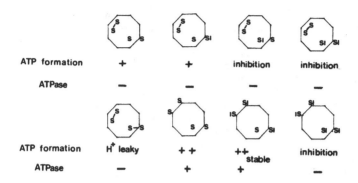

ATP formation	+	+	inhibition	inhibition
ATPase	−	−	−	−

ATP formation	H⁺ leaky	+ +	+ + stable	inhibition
ATPase	−	+	+	−

<u>Figure 2</u>. Effects of oxidation and alkylation of cysteinyl residues in
the γ subunit on ATP synthesis and hydrolysis by thylakoids. In this
model, the octagons represent the γ subunit of CF_1, S, free SH groups,
SI, SH groups reacted with N-ethylmaleimide, and S-S, a disulfide bond.
Activation is correlated with the reduction of a disulfide, whereas
inhibition is correlated with the reaction of an SH group that is exposed
only under energized conditions. The accessible SH on the γ subunit
appears to be nonessential. Alkylation of the thiols formed by reduction stabilizes the enzyme from reoxidation. A similar model can be
generated for the relation between SH groups, S-S bonds and the ATPase
activity of CF_1 in solution.

TABLE 4. Binding of CF_1 to NaBr-extracted thylakoids

CF_1 Preparation	ATPase Activity	ATP Synthesis
Control	243	34
Heat-treated	244	2
$-\delta$	224	3
$-\epsilon$	251	2
Trypsin-treated	54	2
None added	41	1-2

NaBr particles (50 µg of chlorophyll) were incubated in 0.15 ml for 15
min on ice with 125 µg of the CF_1 preparations in the presence of 3 mM
$MgCl_2$. The particles were collected by centrifugation and washed with
0.15 ml of a buffered sucrose solution. Photophosphorylation with N-
methylphenazonium methosulfate as the mediator and ATPase activity in
the presence of octylglucoside were assayed. Rates are given as µmol P_i
formed or esterified h^{-1} mg chlorophyll^{-1}.

(Dunn, Futai, 1981). We were surprised to find (Andreo et al., 1982) that δ-deficient CF_1 binds well to thylakoids freed of CF_1 by NaBr extraction (Nelson, Eytan, 1979). More recently, we have been able to prepare CF_1 lacking the ε subunit (W. Patrie, unpublished). This is accomplished by heating CF_1 in the presence of ATP and a low concentration of Triton X-100 and passing the enzyme solution through a column of hydroxylapatite. This ε-deficient CF_1 preparation binds well to NaBr-extracted thylakoids. However, neither the δ-deficient nor the ε-deficient enzyme restores photophosphorylation, probably because they fail to block the proton channel through F_1 (Table 4). The binding of the CF_1 to NaBr particles appears to be specific. The binding requires cations, is inhibited by trypsin treatment of either the membranes or the CF_1 and is inhibited by treatment of the membranes with phenyl-glyoxal. Moreover, the α subunit of bound CF_1 is protected from cleavage by trypsin. These experiments point to a role for the larger CF_1 subunits in the attachment of CF_1 to F_0 and reinforce the suggestion that a given subunit in F_1 from one coupling membrane need not fulfill the same role in F_1 from another source.

REFERENCES

Akey CW, Crepeau RH, Dunn SD, McCarty RE and Edelstein SJ (1983) Electron microscopy and single molecule averaging of subunit-deficient F_1-ATPases from Escherichia coli and spinach chloroplasts, EMBO Jour, in press.

Andreo CS, Patrie WJ and McCarty RE (1982) Effect of ATPase activation and the δ subunit of coupling factor 1 on reconstitution of photophosphorylation, J. Biol. Chem. 257, 9968-9975.

Arana JL and Vallejos RH (1982) Involvement of sulfhydryl groups in the activation mechanism of the ATPase activity of chloroplast coupling factor 1, J. Biol. Chem. 257, 1125-1127.

Avron M (1978) Energy transduction in photophosphorylation, FEBS Lett. 96, 225-232.

Béliveau R, Moroney JV and McCarty RE (1982) Endogenous fluorescence of coupling factor 1 from spinach chloroplasts, Arch. Biochem. Biophys. 214, 668-674.

Davenport JW (1983) The coupling of proton translocation to ATP synthesis and hydrolysis by chloroplast thylakoid membranes, Ph.D. Thesis, Cornell University.

Davenport JW and McCarty RE (1981) Quantitative aspects of adenosine triphosphate-driven proton translocation in spinach chloroplast thylakoids, J. Biol. Chem. 256, 8947-8954.

Dunn SD and Futai M (1980) Reconstitution of a functional coupling factor from the isolated subunits of Escherichia coli F_1 ATPase, J. Biol. Chem. 255, 113-118.

Farron F (1970) Isolation and properties of a chloroplast coupling factor and heat-activated adenosine triphosphatase, Biochemistry 9, 3823-2828.

Handgarter RP and Good NE (1982) Energy thresholds for ATP synthesis in chloroplasts, Biochim. Biophys. Acta 681, 397-404.

Haraux F, Sigalat C, Moreau A and de Kochkovsky Y (1983) The efficiency of energized protons for ATP synthesis depends on the membrane topography in thylakoids, FEBS Lett. 155, 248-251.

McCarty RE and Portis AR Jr (1976) A simple, quantitative approach to the coupling of photophosphorylation in terms of proton fluxes, Biochemistry 15, 5110-5114.

Merchant S, Shaner SL and Selman BR (1983) Molecular weight and subunit stoichiometry of the chloroplast coupling factor one from Chlamydomonas reinhardii, J. Biol. Chem. 258, 1026-1031.

Mills JD and Mitchell P (1982) Thiol modulation of CF_0-CF_1 stimulates acid/base-dependent phosphorylation of ADP by broken pea chloroplasts, FEBS Lett. 144, 63-67.

Moroney JV, Lopresti, L, McEwen BF, McCarty RE and Hammes GG (1983) The molecular weight of chloroplast coupling factor 1, FEBS Lett., in press.

Moroney JV and McCarty RE (1982) Light-dependent cleavage of the γ subunit of coupling factor 1 by trypsin causes activation of Mg^{2+}-ATPase activity and uncoupling of photophosphorylation in spinach chloroplasts, J. Biol. Chem. 257, 5915-5920.

Moroney JV, Warncke K and McCarty RE (1982) The distance between thiol groups in the γ subunit of coupling factor 1 influences the proton permeability of thylakoid membranes, J. Bioenerg. Biomembr. 14, 347-359.

Nalin CM, Béliveau R and McCarty RE (1983) Selective modification of coupling factor 1 in spinach chloroplast thylakoids by a fluorescent maleimide, J. Biol. Chem. 258, 3376-3381.

Nelson N and Eytan E (1979) Approach to the membrane sector of the chloroplast coupling device. In Mukohata Y and Packer L, eds. Cation flux across biomembranes, pp 409-415. New York: Academic Press.

Portis AR Jr, Magnusson RP and McCarty RE (1975) Conformational changes in coupling factor 1 may control the rate of electron flow in spinach chloroplasts. Biochem. Biophys. Res. Commun., 64, 877-884.

Portis AR Jr and McCarty RE (1974) Effects of adenine nucleotides and of photophosphorylation on H^+ uptake and the magnitude of the H^+ gradient in illuminated chloroplasts, J. Biol. Chem. 249, 6250-6254.

Senior AE and Wise JG (1983) The proton-ATPase of bacteria and mitochondria, J. Membr. Biol. 73, 105-124.

Author' Address: Section of Biochemistry, Molecular and Cell Biology, Cornell University, Ithaca, NY 14853 USA

FLASH-INDUCED ATP SYNTHESIS IN PEA CHLOROPLASTS

J.M. GALMICHE, G. GIRAULT, C. LEMAIRE

1. INTRODUCTION

The onset of the ATP synthesis in chloroplasts depends on the activation of
the ATPase (Mills, J.D. et al., 1980) and on the energy balance (Gräber, P.,
Schlodder, E., (1980). Girault, G., Galmiche, J.M. (1978) have prepared, from
preilluminated leaves, chloroplasts which hydrolyze ATP in the dark and keep
this property quite a long time. We propose to follow the synthesis of ATP
induced by single turnover flashes in such chloroplasts with a fully active
ATPase system as a function of the number of flashes already fired. In those
chloroplasts the synthesis of ATP, induced by single turn-over flashes, is
controlled by the pH difference across the thylakoid membrane and by kinetic
factors which do not seen, to be directly related either to the proton motive
force across the membrane or to the hydrolytic activity of the ATPase system.

2. MATERIAL AND METHOD

Chloroplasts were prepared as described by Girault, G., Galmiche, J.M. (1978)
from 2 week-old pea leaves which were illuminated for 30 min in ice-cold water
before grinding.

ATP was directly measured by recording the luciferin-luciferase luminescence,
ΔpH by the quenching of the atebrin fluorescence and absorbance changes,
ΔA_{515}, as described by Galmiche, J.M., Girault, G. (1982). But, for recording the absorbance at 515 nm the measuring beam passed through the sample only
during the time of the measurement, 1 s, and in the intervals, 10 s, the photomultiplier was lighted by a photodiode to avoid any recovery time in its
response.

3. RESULTS AND DISCUSSION

3.1. Photophosphorylations induced by flashes at 0.1 Hz frequency

   a - The onset of photophosphorylation :

The chloroplasts prepared from preilluminated leaves have an active ATPase.
They hydrolyze ATP in the dark and keep this property for several hours when
stored in 0.4 M sucrose at 0°C. A pH difference of $\sim$ 1.4 was measured across
the thylakoid membrane in the dark from the distribution of the (^{14}C)methyl-
amine (Table 1). ATP hydrolysis in the dark is coupled to an inward transport
of H^+ (Table 1) and to a concomitant change of the transmembrane electric
potential difference (Galmiche, J.M., Girault, G. 1980).

The pH difference across the thylakoid membrane, ΔpH, measured by quenching of
the atebrin fluorescence increased progressively with the number of flashes
fired and reached a stable level 0.1 pH unit above the value measured in the
dark before starting the flash-illumination. The ATP yield/flash increased
with the value of the ΔpH (Figure 1) and became constant above a limit value
$\sim$ 1.4-1.5 of the ΔpH.

Sybesma, C. (ed.), Advances in Photosynthesis Research, Vol. II. ISBN 90-247-2943-2.
© 1984 Martinus Nijhoff/Dr W. Junk Publishers, The Hague/Boston/Lancaster.

TABLE 1. ΔpH ACROSS THE MEMBRANE INDUCED BY 300 μM ATP ADDITION			TABLE 2. ΔpH AND ATP YIELD/FLASH IN THE PRESENCE OF 10 nM NIGERICIN		
Expt. N°	Without ATP	With ATP		CONTROL	+ NIGERICIN
1	1.34	1.51	ATP/flash. 1000 Chl.	0.4	0
2	1.43	1.55	ΔpH initial in dark	1.43	1.18
3	1.06	1.5	ΔpH in dark after 50μM ATP addition	1.65	1.52

Conditions as in Figure 1 but in Table 1 chroroplasts equiv. 125 μg/ml.
ΔpH measured by (^{14}C)methylamine, Table 1 and atebrin fluorescence, Table 2.

We interpret the critical value of the ΔpH as evidence for a critical energy threshold which is expected for ATP synthesis following the chemiosmotic theory. At variance with those results, Schreiber, U., Del Valle-Tascon, S. (1982) do not find any threshold light energy in flash-induced ATP synthesis (flashes of 20 μs duration) in the same type of chloroplasts.

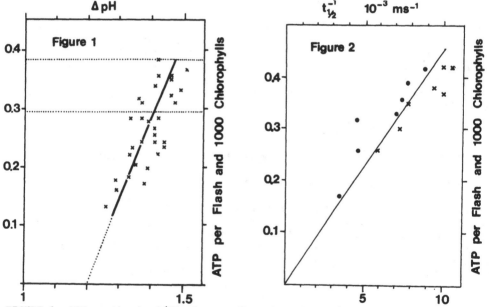

FIGURE 1. ATP synthesized/flash as a function of the ΔpH across the membrane. Chloroplasts (equiv. 40μg Chl.) were suspended in 2 ml reaction medium (40 mM tricine pH 8, 10 mM MgCO$_3$, 2 mM K$_2$HPO$_4$, 20μM ADP, 5μM diadenosylpentaphosphate, luciferin-luciferase from LKB, at 10°C). For pH measurement atebrin was 2.5μM. Single flashes (3μs duration) or groups of 2 flashes (5 to 5000 ms apart) were fired at 0.1 Hz.

FIGURE 2. Flash-induced ATP synthesis and 515 nm absorbance changes. Conditions as in Figure 1 but groups of 2 flashes (10 ms apart).

b - Effect of nigericin :

Nigericin in the presence of K^+ is a potent uncoupler of the steady-state photophosphorylation in chloroplasts. It is generally thought that in flashing light or in the initial period of illumination photophosphorylation is insensitive to nigericin. We observed that even very low concentrations of nigericin (10 nM) inhibited flash-induced ATP formation but did not suppress the dark hydrolysis of ATP and the concomitant increase of the ΔpH (Table 2). That confirms the necessity of a critical value of this ΔpH to observe the ATP synthesis.

c - Onset of photophosphorylation and 515 nm absorbance changes :

The flash-induced electric potential difference across the membrane can be measured by the absorbance changes at 515 nm, ΔA_{515}. When the ATP yield/flash increased with the number of groups of 2 flashes (10 ms apart) already fired, we observed acceleration of the corresponding ΔA_{515} dark decay. ATP yield/flash increased linearly with the reverse of the lifetime of the ΔA_{515}, which gave a crude estimate of the proton flux across the membrane (Figure 2). That seems in good accordance with the chemiosmotic hypothesis (Gräber, P., Schlodder, E., 1981).

3.2. ATP synthesis by groups of two flashes as a function of the dark period, t_D, between the flashes.

Assuming that the first flash of the group was as effective as a single flash given at the same frequency we calculated by difference the ATP yield/2 nd flash of the group.

This ATP yield/2 nd flash was always higher than that for a single flash except at external pH 8 when $t_D < 10$ ms (Figure 3). The maximum value was observed for $t_D \sim 1000$ ms at pH 8 and $t_D \sim 500$ ms at pH 7.

The lower efficacity of the 2 nd flash at external pH 8 when $t_D < 10$ ms could be taken into account by the recovery time of electron transport chain.

At any values of t_D, the ΔpH extent observed with a group of 2 flashes was more or less identical. In the same conditions the transmembrane electric potential difference, recorded by ΔA_{515}, was maximum for $t_D \sim 5$-10 ms and decreased sharply for $t_D > 200$ ms. So the change of the ATP yield/2 nd flash (maximum for $t_D = 1000$ ms at pH 8) were independent of the ΔpH and ΔA_{515}, or more precisely of the proton motive force, $\Delta \mu$ H.

In chloroplasts, the ATP yield/flash depends on a specific state of the membrane which is very sensitive to the value of the external pH. This state takes a long time (500 to 1000 ms) to develop after a single turnover flash. After a 10 s dark period it has almost completely decomposed at pH 7 and only partly at pH 8 and the ATP yield/1 st flash is at pH 8 twice that at pH 7. Whatever is the external pH the full extent of the active state is formed after the first flash with a shorter lifetime at pH 7 than at pH 8.

Because we cannot see any changes in the hydrolytic activity of the ATPase system after a flash or a group of 2 flashes (as that has been also reported by Schreiber, U., Del Valle-Tascon, S. 1982) this active state should act on a short-lived ATP synthetase activity of the enzyme. We cannot decide whether that state, tentatively identified to a red-ox state of some species, sensitive to the external pH, conditions the effects of the flash on the charges transfer and on some local H^+ deposit or on a short-lived activation of the ATPase system increasing its temporary ATP synthetase activity.

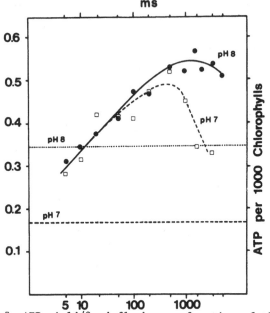

FIGURE 3. ATP yield/2 nd flash as a function of the dark time t_D between the 2 flashes of the group.
Conditions as in Figure 1. Interrupted line : ATP yield/1 st flash at pH 8. Dotted line : ATP yield/1 st flash at pH 7. Dark circles : ATP yield/2nd flash at pH 8. White squares : ATP yield/2nd flash at pH 7.

REFERENCES
GALMICHE J.M., GIRAULT G. (1980) ATP hydrolysis and membrane potential in spinach chloroplasts. FEBS Lett. 118, 72-76.
GALMICHE J.M., GIRAULT G. (1982) Synthesis of ATP induced in pea chloroplast: by single turnover flashes. FEBS Lett. 146, 123-128.
GIRAULT G., GALMICHE J.M. (1978) Effect of nucleotides on potential and pH changes across the thylakoid membrane of spinach chloroplasts. Biochim. Biophys.Acta. 502, 430-444.
GRABER P., SCHLODDER E. (1981) Regulation of the rate of ATP synthesis/hydrolysis by ΔpH and $\Delta\Psi$. in Photosynthesis II (Akoyunoglou G. ed.) Balaban Intern.Sci.Serv. Philadelphia, Pa. pp.867-879.
MILLS J.D., MITCHELL P., SCHURMANN P. (1980) Modulation of coupling factor ATPase activity in intact chloroplasts. FEBS Lett. 112, 173-177.
SCHREIBER U., RIENITS K.G. (1982) Complementarity of ATP-induced and light-induced absorbance changes around 515 nm. Biochim.Biophys.Acta. 682, 115-123.
SCHREIBER U., Del VALLE-TASCON S. (1982) ATP synthesis with single turnover flashes in spinach chloroplasts. FEBS Lett. 150, 32-37.
Authors address : Service de Biophysique-CEN-Saclay 91191 Gif sur Yvette Cede France.

THE DYNAMICS OF MILLISECOND DELAYED LIGHT EMISSION AND ITS RELATION
WITH PHOTOPHOSPHORYLATION

Y. K. SHEN, C. H. XU, J. M. WEI, D. Y. LI, Y. FENG, Z. H. HUANG

Since the discovery (Mayne, 1966) of the phenomenon that the millisecond
delayed light emission (ms-DLE) was closely related to the high energy
state of photophosphorylation (PSP), many works have been done to study
the relation between the electropotential and proton gradient across the
thylakoid membrane and the ms-DLE. However, there are few experiments
which have scrutinized the effects of various factors on the activities
of PSP and the dynamics of ms-DLE simultaneously, though this is of much
importance for the understanding of the relationship between them. We
have tried to investigate this problem in some detail and got the following
results.

1. MATERIAL AND METHODS

Chloroplasts were prepared from spinach leaves with STN solution (sucrose
0.4 M, tricine 20 mM, pH 7.4, NaCl 0.01 M) as before (Wei, 1980). ms-DLE
were measured with a Becquerel phosphoroscope. The dark intervals used
between the 1 ms illumination on the chloroplast suspension and the 1 ms
DLE measurement were 1.8 ms. The flashing light intensity from a mirror
incandecent lamp through the hole of the rotating disk at the surface of
the cuvette of chloroplast suspension was 5×10^6 erg·cm^{-2}·s^{-1}, which was
strong enough to saturate the rate of Hill reaction.
The methods for measuring the PSP and Hill reaction rate and the chlorophyll content were the same as in previous paper (Shen, Shen, 1962).
Chloroplasts with chlorophyll content 10-15 µg were added to 1 ml solution containing 0.05 M Tricine, (pH 7.8), 2 mM $MgCl_2$, 2 mM Pi, 1 mM ADP
and electron acceptor or mediator. The concentrations for FeCy, MV, PMS
or DAD (+Ascorbic acid) was 0.5 mM, 0.1 mM, 10 µM, 0.25 mM respectively.
The concentrations for the other reagent, when added, was as follows:
1×10^{-7} M valinomycin, 10 mM KCl, 1 mM NH_4Cl, 5×10^{-8} M nigericin, 1 µM
DCMU, 1×10^{-7} M DBMIB, 0.1 mM phlorizin, 20 µM DCCD, 20 ppm polymyxin,
3 ppm aureomycin, 30 mM citrate. The activities of PSP and Hill reaction
in the figures expressed in µmoles ATP mgchl^{-1}·hr^{-1} and µmoles FeCy
mgchl^{-1}·hr^{-1}. (PSP* were measured in continuous light with the same
chloroplast preparation).

2. RESULTS AND DISCUSSIONS

2.1 Effects of uncouplers and electron transport inhibitors on the dynamics of ms-DLE

Valinomycin, an ionophore for abolishing the membrane potential, decreased the fast phase ms-DLE significantly and the slow phase (turning
to steady state) ms-DLE slightly, and also had limited effect on the PSP
which was measured under the same flashing light condition for 30 seconds
and thus mainly reflected the steady state PSP (Fig. 1). These results
were in accordance with those reported in the literature in which the
effect of valinomycin on ms-DLE were measured in weak flashing light
(Wraight, Crofts, 1971) and its effect on PSP in continuous strong light
(McCarty, 1969). NH_4Cl and nigericin are known as uncouplers for

Sybesma, C. (ed.), Advances in Photosynthesis Research, Vol. II. ISBN 90-247-2943-2.
© *1984 Martinus Nijhoff/Dr W. Junk Publishers, The Hague/Boston/Lancaster.*

diminishing the proton gradient. Under flashing light condition they also decreased the PSP. As to their effects on ms-DLE, both the fast phase and

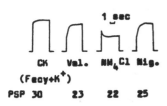

CK   Vol.   NH₄Cl Nig.
(Fecy+K⁺)
PSP 30   23   22   25

Fig.1. Effects of uncouplers
on PSP and ms-DLE

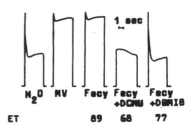

H₂O   MV   Fecy Fecy   Fecy
                 +DCMU +DBMIB
ET   89   68   77

Fig.2. Effects of DCMU and
DBMIB on ms-DLE

the slow phase ms-DLE were lowered by them (Fig. 1). These results were somewhat different from that obtained under weak flashing light condition (Wraight, Crofts, 1971). In the latter case the fast phase ms-DLE was less prominent and it was said that nigericin had no effect on the fast phase ms-DLE. Comparing the effects of NH₄Cl and nigericin in our experiments, it could be seen that nigericin was more efficient in lowering the fast phase ms-DLE than NH₄Cl at concentrations that their effects on PSP were similar. In view of the fact that nigericin is an ionophore which is inserted in the membrane and is to lower the proton gradient by accelerating the exchange between H⁺ and K⁺, while NH₄Cl shows its uncoupling effect by combining of NH₃ with H⁺ in the lumen of the thylakoid (Jagendorf, 1977), the above results perhaps indicate that the proton gradient which affects the fast phase ms-DLE may be closely connected with the membrane which is in accordance with our previous suggestion that there may be different forms of existence of high energy state in thylakoid and one of them was more or less localized to the membrane (Shen et al. 1963, Yin et al. 1979, Wei 1980).

DCMU, which inhibits electron flow near the PS II, lowered both the fast and slow phase ms-DLE, while DBMIB, which inhibits the oxidation of reduced plastoquinone only lowered the slow phase ms-DLE and showed no effect on the fast one (Fig. 2). From this it is inferred that the proton released from the oxidation of water and not from the oxidation of reduced plastoquinone is responsible for the enhancement effect on the fast phase ms-DLE.

2.2. Effects of the high energy state formed from electron flow involving only PS I on the ms-DLE

The ms-DLE and PSP of chloroplasts in the presence of DCMU was significantly increased when PMS or DAD+ascorbate was added to the reaction mixture (Fig. 3) and this enhancement of ms-DLE could be abolished by uncouplers (not shown). The spectrum of this enhanced ms-DLE was identical to that of the ms-DLE under the condition when Hill reaction was going on (with MV alone) (Fig. 4). This points out that even in the former case the ms-DLE is still emitted from the recombination process in PS II.

However, the dynamics of ms-DLE were quite different in these two cases. In the former case the slow phase ms-DLE decreased gradually and became

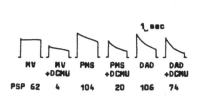

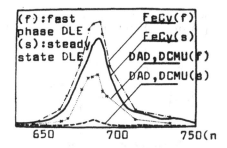

Fig.3. The dynamics of ms-DLE in the presence of MV, DAD or PMS

Fig.4. The spectra of ms-DLE in the presence of FeCy and DAD+DCMU

very low finally, but in the latter case the slow phase ms-DLE quickly stabilized at a higher level, though the PSP was much higher in the former case. This difference seems to be related to the fact that there were only limited PS II reaction in the former case because when no DCMU was added to the PMS or DAD system (Fig. 3) the decrease of ms-DLE after the fast phase was less steep.

2.3. Effects of energy transfer inhibitors and coupling efficiency improving agents on ms-DLE and PSP

Phlorizin, DCCD and antibody against CF_1 were known as energy transfer inhibitors which react with coupling factor and thus prevent ATP synthesis and the extra electron flow induced by PSP (Jagendorf, 1977). In our experiments, they all inhibited PSP and enhanced ms-DLE (Fig. 5, 6).

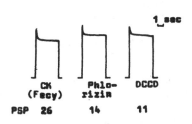

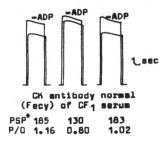

Fig.5. Effects of phlorizin and DCCD on PSP activity and ms-DLE

Fig.6. The effect of CF_1 antibody on PSP activity and ms-DLE

Coupling efficiency improving agents (aureomycin, polymyxin, polybasic acid, etc.) as previously found in our laboratory, are a category of reagents which can induce comformational change of the coupling factor and make it utilize the high energy state more efficiently to synthesize

ATP and thus raise the P/O ratio (Yin et al. 1979, Feng, Shen, 1983).
They all showed more or less enhancement effects on ms-DLE and PSP
simultaneously (Fig. 7,8).

From the above results it appeared that ms-DLE can give much useful information of high energy state in relation to PSP.

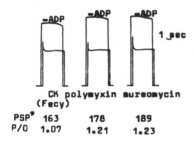

Fig.7. Effects of polymyxin and   Fig.8. Effects of citrate on
       aureomycin on PSP activity   PSP activity and ms-DLE
       and ms-DLE

REFERENCES

Feng Y, Shen YK (1983) Cooperative effect of polybasic acid and
adenylate on electron transport and photophosphorylation, Acta Phytophysiologia Sinica 9(4).
Jagendorf AT (1977) Photophosphorylation, Encyclo. Plant Physiol. V. 5,
Trebst & Avron (ed.), pp. 307-337, Springer Verlag Heidelberg, New York.
Mayne BC et al. (1966) Luminescence of chlorophyll in spinach chloroplasts induced by acid-base transition, Proc. Nat. Acad. Sci. U.S.A.
55:494-497.
McCarty RE (1969) The uncoupling of photophosphorylation by valinomycin
and ammonium chloride, J. Biol. Chem. 244:2492-2498.
Shen YK et al (1963) On the nature of the intermediate accumulated in the
chloroplasts during illumination, Acta Biochimica et Biophysica Sinica
3:278-292.
Shen YK, Shen GM (1962) Studies on photophosphorylation II. The "light
intensity effect" and the intermediate steps of photophosphorylation,
Sci. Sinica 11:1097-1106.
Wei JM et al (1980) Studies on the coupling mechanism of photophosphoryla-
tion, III. The stimulation effect of photophosphorylation by amine, Acta
Phytophysiologia Sinica 6(4): 393-398.
Wraight CA, Crofts AR (1971) Delayed light emission and high energy state
of chloroplasts, Eur. J. Biochem. 19:386-397.
Yin HC et al (1979) Studying the photosynthetic apparatus in operation,
Acta Phytophysiologia Sinica 5(3):295-317.

Authors address : Shanghai Institute of Plant Physiology, Academia
                  Sinica, 300 Fenglin Road, Shanghai 200032, China

INHIBITION OF PSI AND PSII-DEPENDENT FLASH-INDUCED ATP SYNTHESIS
BY TRIPHENYLTIN CHLORIDE AND DCCD

SUSAN FLORES AND DONALD R. ORT

INTRODUCTION

Energy transfer inhibitors are useful probes for investigating the role of coupling
factor accessibility in phosphorylation. Triphenyltin chloride and DCCD both act
as energy transfer inhibitors in chloroplasts and are believed to block ATP
synthesis by preventing proton flux through the CF_0 portion of the coupling
factor. When experimental conditions put great demand on the ATP synthesizing
machinery, as in saturating continuous light, removal of any portion of the pool of
coupling factors would be expected to have an inhibitory effect on
phosphorylation. However, in flashing light or in low light, when ATP synthesis is
proceeding at a slower average rate, it is possible that the availability of coupling
factors exceeds the demand and that removal of a fraction of the pool will have
less effect. If all coupling factors are equally accessible to the protons released
from a particular electron transport complex, then one would expect energy
transfer inhibitors to be less inhibitory in flashing light, since each coupling factor
may turn over several times in the dark period between flashes. But for the
extreme situation of totally localized coupling, removal of each coupling factor
would result in the loss of ATP synthesis coupled to electron flow through the
associated redox complex, regardless of the overall rate of phosphorylation. It
has been reported that in photosynthetic bacteria, inhibition by DCCD is
independent of flash frequency, and a localized coupling was suggested (Venturoli,
Melandri 1982). Additionally, if there are interactions between the electron
transport chain and the coupling factor which are not mediated by a delocalized
protonmotive force, one would predict that only one electron transport complex
would be involved, since there is less than one coupling factor present in
thylakoids per complete electron transport "chain" (Strotmann et al. 1973).
Phosphorylation associated with that site would then be differentially sensitive to
the presence of energy transfer inhibitors.

We found that in chloroplasts, both triphenyltin (TPT) and DCCD are less
inhibitory of flash-induced ATP synthesis than of phosphorylation in continuous
light. Also, the pattern of inhibition by triphenyltin of ATP synthesis induced by
increasing numbers of flashes was very similar for phosphorylation associated with
PSI, PSII, or whole chain electron transport. These results indicate that in
chloroplasts, there is probably not a special association between a particular
electron transport complex and a specific coupling factor. Experimental
procedures have been described previously (Graan, Ort 1982) or are provided in
the figure legends.

RESULTS AND DISCUSSION

The inhibition by triphenyltin of flash-induced ATP synthesis associated with
electron transport from water to methylviologen is shown in Figure 1, as a
function of flash number. The concentration of triphenyltin chloride was 0.9 μM
when present, giving a molar ratio of 37 Chl/1.0 triphenyltin. This concentration
of triphenyltin inhibited ATP synthesis by about 50% in continuous light (Figure
2). At 50 flashes, the yield of ATP per flash for uninhibited samples corresponded
to 0.54 ATP/CF_1, given 1 CF_1/860 Chl (Strotmann et al. 1973). ATP
synthesis after only 3 flashes was inhibited 65% by TPT, however the inhibition
decreased with increasing flash number and was only 35% after 50 flashes.

Sybesma, C. (ed.), Advances in Photosynthesis Research, Vol. II. ISBN 90-247-2943-2.
© *1984 Martinus Nijhoff/Dr W. Junk Publishers, The Hague/Boston/Lancaster.*

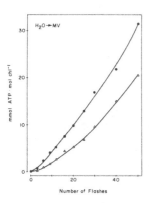

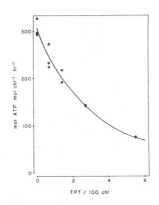

FIGURE 1. Inhibition by triphenyltin chloride of flash-induced ATP synthesis associated with electron transport from water to methylviologen. Reactions were carried out at 4°C in a medium containing 150 mM sorbitol, 10 mM Tricine-KOH (pH 7.8), 3 mM Mg Acetate, 2 mM KCl, 0.1 mM methylviologen, 0.1 mM ADP, 0.5 mM $^{32}P_i$, plus chloroplasts equivalent to 33 μM chl. The concentration of triphenyltin chloride was 0.9 μM when present (Δ-Δ) giving a molar ratio of 37 chl/TPT. The saturating 6 μs flashes were filtered through 3-71 yellow Corning glass filters and delivered at a frequency of 5 Hz.

FIGURE 2. Concentration dependence of triphenyltin inhibition of ATP synthesis in continuous light. Reactions were carried out at 20° in a medium containing 100 mM sorbitol, 25 mM Tricine-KOH (pH 7.8), 3 mM Mg Acetate, 0.5 mM $^{32}P_i$, 1 mM ADP, 5 mM $^{32}P_i$, 0.1 mM methylviologen, chloroplasts equivalent to 14 μM chl, and triphenyltin chloride as indicated. The illumination period was 1 min, and the light was filtered through infra-red blocking filters and a 2-62 red Corning glass filter.

Similar patterns of inhibition were seen for phosphorylation associated with PSI mediated electron transport from durohydroquinone to methylviologen (Fig. 3) and PSII mediated electron transport from water to dimethylquinone (Fig. 4). This was true in spite of large differences in the amount of ATP made per flash : for PSII associated phosphorylation the yield of ATP in the absence of inhibitor was equivalent to 0.11 ATP per CF_1 per flash after 75 flashes. For both partial reactions assayed the degree of inhibition decreased from 4 to 75 flashes, from 65% to 35% for PSI (Fig. 3), and from 85% to 35% for PSII (Fig. 4). The apparent decrease in inhibitory effect with increasing flash number may be an indication of a partial limitation of coupling factor activation during the first few flashes, however this should be interpreted with caution, since there may be an additional uncoupling effect of triphenyltin. This compound was reported to act as an hydroxyl-chloride antiporter, although at higher concentrations of chloride and triphenyltin than were used here (Watling-Payne, Selwyn 1974). The second interpretation is favored by the results from an experiment using DCCD to inhibit ATP synthesis associated with electron transport from water to ferricyanide (Fig. 5), since the extent of inhibition did not decrease at increasing flash numbers. After 4 flashes, ATP synthesis was inhibited only 20% by DCCD at a ratio of 0.05 DCCD/Chl, and at 40 flashes the inhibition was 25%, practically the same.

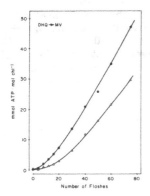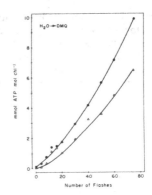

FIGURE 3. Inhibition by triphenyltin chloride of flash-induced ATP synthesis associated with electron transport from durohydroquinone to methylviologen. Reaction conditions were as in Fig. 1, however the reaction medium contained, in addition, 0.5 mM durohydroquinone and 5.0 μM DCMU.

FIGURE 4. Inhibition by triphenyltin chloride of flash-induced ATP synthesis associated with electron transport from water to dimethylquinone. Reaction conditions were as in Fig. 1, however, instead of 0.1 mM methylviologen the reaction medium contained 0.2 mM 2,6-dimethylquinone, 0.1 mM potassium ferricyanide, and 1.0 μM DBMIB.

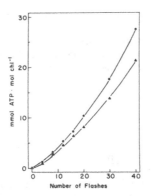

FIGURE 5. Inhibition by DCCD of flash-induced ATP synthesis associated wth electron transport from water to ferricyanide. Chloroplasts were incubated with either DCCD (●-●) at a molar ratio of 0.05 DCCD/chl or an equivalent volume (0.8%) of ethanol (Δ-Δ) for 90 m prior to measurement of phosphorylation. Reaction conditions were as in Fig. 1, however instead of 0.1 mM methylviologen the reaction medium contained 0.1 mM potassium ferricyanide.

DCCD, like triphenyltin after a larger number of flashes, was less inhibitory of flash-induced ATP synthesis than of phosphorylation in continuous light. One problem in comparing the effects of DCCD in flashing vs continuous light is the

TABLE 1. Comparison of DCCD inhibition of ATP synthesis in flashing vs continuous light. The series of 30 single turnover flashes was preceded by a 1 s illumination followed by a 30 s dark period prior to the addition of $^{32}P_i$. The reaction medium contained 100 mM sorbitol, 25 mM KCl, 10 mM Tricine-KOH (pH 7.8), 3 mM $MgCl_2$, 0.2 mM potassium ferricyanide, 0.5 mM $^{32}P_i$, 0.1 mM ADP, and chloroplasts equivalent to 33 μM chl. Other conditions were as in Fig. 1.

	Yield of ATP (mMol/mol Chl)		
Reaction Conditions	-DCCD	+DCCD	% Inhibition
1 s illumination	23.0	9.69	58%
30 flashes	22.9	18.3	20%

increase in reactivity of the compound upon illumination of the chloroplasts (Uribe 1972). However, when this problem was circumvented by providing a 1 s preillumination, prior to the assay of phosphorylation, the subsequent flash-induced ATP synthesis was still inhibited only 20% by a concentration of DCCD which caused 58% inhibition when phosphorylation was assayed during the 1 s of continuous light (Table 1).

CONCLUSIONS

DCCD and triphenyltin chloride are less inhibitory of flash-induced ATP synthesis than of ATP synthesis driven by continuous illumination, which implies that the protonmotive force cannot be localized entirely between particular sites of proton generation and consumption. This idea is supported by the observation that flash-induced phosphorylation associated with different electron transport pathways was equally sensitive to subsaturaing amounts of triphenyltin, since there is less than one CF_1 per complete electron transport "chain", and coupling factors are apparently confined to non-appressed regions of the thylakoid membrane.

REFERENCES

Venturole G and Melandri BA (1982) The localized coupling of bacterial photophosphorylation. Effect of Antimycin a and N,N-dicyclohexylcarbodiimide in chromatophores from Rhodopseudomonas sphaeroides, Ga, studied by single turnover event analysis, Biochim. Biophys. Acta 637, 447-456.
Strotmann H, Hesse H and Edelmann K (1973) Quantitative determination of coupling factor CF_1 of chloroplasts, Biochim. Biophys. Acta 314, 202-210.
Graan T and Ort DR (1982) Photophosphorylation associated with synchronous turnovers of the electron-transport carriers in chloroplasts, Biochim. Biophys. Acta 682, 395-403.
Watling-Payne AS and Selwyn MJ (1974) Inhibition and uncoupling of photophosphorylation in isolated chloroplasts by organotin, organomercury and diphenyleneiodonium compounds, Biochem. J. 142, 65-74.
Uribe EG (1972) The interaction of N,N'-dicyclohexylcarbodiimide with the energy conservation systems of the spinach chloroplast, Biochemistry 11, 4228-4235.

Authors' address: Department of Plant Biology, USDA/ARS, 289 Morrill Hall, University of Illinois, 505 S. Goodwin Avenue, Urbana, Illinois 61801, U.S.A.

CHEMICAL MODELS OF PYROPHOSPHATE BOND SYNTHESIS AND ION-RADICAL MECHANISM OF PHOTOPHOSPHORILATION

A.A.YASNIKOV, N.V.VOLKOVA, N.P.KANIVETS, L.I.VASILENOK
INSTITUTE OF ORGANIC CHEMISTRY OF THE UKRAINIAN ACADEMY OF
SCIENCES, KIEV USSR

1. INTRODUCTION

The molecular mechanism of a final stage of the ATP synthesis
in chloroplasts is a subject of discussion. We used a method
of chemical models in this work for exposing some peculiarities
of the process of photophosphorilation. The ATP synthesis was
done when illuminating water-oil emulsion, containing the ADP,
inorganic phosphate (Pi), phenasinemetosulphate (PMS), askor-
bate (ask.) and Chlorophyll a (chl.a) with visible light. The
reaction of ATP synthesis, found by us, is considered to be
a chemical model of ion-radical mechanism of conjugation of
electron transport and phosphorilation (Yasnikov et al).

2. PROCEDURE

2.1. Material and methods

Refined sunflower-seed oil was used in our experiments. The
experiment was carried out in two ways: a. Water solution
/3.75cm^3/, containing /the quantity in μmol/: ADP - 25,
KH_2PO_4 - 25, PMS - 25, ask. - 12.5, $MgCl_2$ - 25 was shaken
on the vibrator when illuminating it with visible light for
3 - 30 min. The intensity of illumination at the working
place was 64.600wm^{-2}. Water layer was isolated from oil by
centrifuging /15min. at 700g /. To extract oil we used
washing with ether. The solution was concentrated by evaporation in vacuum and nucleotides were defined in the residue
with the help of horisontal electrophoresis (Stransky, 1963).
b. The same emulsion but without ADP and Pi was shaken up
when illuminating it for 3 - 10min. Pi and ADP was added in
the darkness after switching off the light. Shaking-up was
continued for 30min. and then all the operations were done
as per method a.

2.2. Results

The results of two series of experiments are given in table 1
/version a/ and in table 2 /version b/.

Table 1. The dinamics of nucleotides and Pi in the experiments as per method a.

Sybesma, C. (ed.), Advances in Photosynthesis Research, Vol. II. ISBN 90-247-2943-2.
© *1984 Martinus Nijhoff/Dr W. Junk Publishers, The Hague/Boston/Lancaster.*

Time of illumination min.	Pi and nucleotides	Quantity of nucleotides in μmol after the experiment	Total quantity of nucleotides in μmol	% of nucleotides in the mixture	Δ%
3	Pi	30.8			
	AMP	0.8		2.95	+0.51
	ADP	25.0	26.1	95.7	-0.79
	ATP	0.3		1.34	+0.29
10	Pi	29.4			
	AMP	1.4		5.3	+2.8
	ADP	22.5	26.8	83.7	-12.8
	ATP	2.9		11.0	+9.9
30	Pi				
	AMP	3.1		12.1	+9.6
	ADP	18.1	26.1	69.4	-27.11
	ATP	4.9		18.6	+17.51
without illumi- nation 30	Pi	31.0			
	AMP	0.6		2.4	
	ADP	25.6	25.5	96.5	
	ATP	0.3		1.1	

Table 2. The dinamics of nucleotides depending on the preliminary illumination of chl.a - PMS - ask. system

Exposition in min.	Nucleotides	Quantity of nucleotides in μmol
3	AMP	0.9
	ADP	21.9
	ATP	0.3
10	AMP	2.4
	ADP	17.5
	ATP	2.0

The ATP yield fluctuates depending on the time and mode of illumination. Tris-buffer added into the reaction mix sharply reduces the yield but not the Pi consumption. Probably tris phosphorilates here but the product of the reaction was not isolated. Chlorophyll b does not give rise to the ATP synthesis and can't be used as a sensitizer of a model reaction. When added into the system together with chl.a it sharply inhibits the ATP synthesis. No change in the composition of nucleotides and Pi occurs if the system lacks PMS and ask. The ATP yield is considerably reduced without $MgCl_2$ additions. The ATP yield is lower in the reaction carried out as per method b than in the experiments when the whole phosphorilating system is illuminated. Alongside with ATP there appears AMP in the system. The results of the experiments given in table 2 show that a part of ADP is formed by phosphate group transfer from one ADP molecule on the other. In this case no change in nucleotides composition is observed without preliminary illumination of the mix. Thus only the products of a photochemical reaction give rise to phosphorilation of one ADP molecule by the other. The ATP synthesis is not observed at all if the ADP and Pi are added not immediately but 5min. later after the emulsion illumination. The share of this process is small as compared with the ADP phosphorilation at the expence of Pi.

2.3. Discussion

The likely product of a photochemical reaction is a cation-radical PMS (PMS$^+$). According to the hypothesis about the ion-radical mechanism of phosphorilation Pi transformation into the metaphosphate in two one-electron stages - an electron transfer from Pi on an appropriate acceptor with synchronous OH-radical isolation with the ask. help - is an essential stage in photophosphorilation. The PMS can be an electron acceptor and ask. - a hydrogen atom donor in the system we investigated. The PMS and Pi are likely to form at first a complex with charge transfer, which then produces metaphosphate when isolating OH-radical in accordance with the mechanism:

$$\text{PMS}^+ \; \overset{\cdots}{\text{O}}-\overset{\overset{\text{O}}{\|}}{\underset{\underset{\text{HO} \longleftarrow \boxed{\text{H}} \text{ask.}}{|}}{\text{P}}}-\text{O} \longrightarrow \text{PMS}^- + \text{O}\!=\!\!\overset{\overset{\text{O}}{\|}}{\text{P}}\!-\!\bar{\text{O}}; \quad \text{O}\!=\!\!\overset{\overset{\text{O}}{\|}}{\text{P}}\!-\!\bar{\text{O}} + \text{ADP} \longrightarrow \text{ATP}$$

It is interesting to note that Hauska et al (1980) observed photophosphorilation in lyposomes containing chl.a, PMS and a conjugating factor. Low concentrations of PMS by which the ATP synthesis is not observed in model reactions without any ferment were used in these experiments. In accordance with Mitchell theory the act of phosphorilation is considered to be a reaction of nucleophilic substitution of OH-group by the ADP ion in phosphorus atom (V.P.Skulachev, I.A.Kozlov, 1977). Departing OH$^-$-group goes away with the help of a proton transferred from the thylakoid to the outside. Two-stage ion-radical mechanism of synthesis of pyrophosphate bond when photophosphorilation according to which Pi transforms itself into metaphosphate in two one-electron reactions -an electron transfer from Pi to a suitable acceptor and simultaneous OH-radical isolation - was considered on the basis of analysis of chemical models within Mitchell theory. The proton pump is intended for continuous supporting of sufficient concentration of H-atom donors in chloroplasts. The PMS$^+$ is an electron acceptor. They reduce it originally with the help of ask. into the PMSH which is transferred from water phase into the chloroplast membrane. The PMSH transforms itself into PMS when oxidating the electron transport chain with intermediates.

Scheme of the mechanism:

PMS + ask. ───────▶PMSH + desoxiaskorbate (DASK)
<u>electron transportation chain</u>

REFERENCES
Hauska G et al. (1980) Reconstitution of photosynthetic
energy conservation, Enr. J.Biochem, vol.111, N2,pp.535-545.
Stransky Z (1963) Determination of adenine nucleotides by
paper electrophoresis, J. Chromatogr., vol. 10, N4, pp.456-
462.
Skulachev VP, Kozlov IA (1975) Proton adenozinetriphospho-
tases, Moscow Nauka - 170p
Yasnikov AA et al. (1979) Ion-radical mechanisms of phosphorilation and light-dependent transport of protons in
chloroplasts, Photosynthetica, vol. 13, N4, pp.439 - 445.

CHARACTERIZATION OF ATP FORMATION FROM ACETYL PHOSPHATE AND ADP BY PHOTOSYNTHETIC MEMBRANES

HIDEHIRO SAKURAI, KENJI SHINOHARA and RINYA YAMAGISHI

1. ABSTRACT

ATP formation from acetyl phosphate and ADP by illuminated photosynthetic membranes from spinach, A. nidulans and R. rubrum was studied. Studies on the time course of ATP formation revealed that the activity of spinach chloroplasts was usually not so high as previously reported, ranging less than 3-8 % of the activity with Pi + ADP compared at the same concentrations (0.05-0.4 mM) of phosphate and acetyl phosphate. The relative activity of A. nidulans thylakoids was about the same level as that of spinach, and that of R. rubrum chromatophores was very low, if at all. By studying the changes in adenine nucleotide concentrations under phosphorylation conditions using spinach chloroplasts, it was shown that adenylate kinase activity alone cannot explane the ATP formation from acetyl phosphate and ADP.

ABBREVIATIONS: CF1, chloroplast coupling factor 1; DPIP, 2,6-dichlorophenol indophenol; HPLC, high-pressure liquid chromatography; PMS, phenazine methosulfate.

2. INTRODUCTION

β-Naphthyl phosphate, phenyl phosphate, pyridoxal 5-phosphate and acetyl phosphate inhibit photophosphorylation competitively with Pi (Shinohara, Sakurai, 1980, 1981). Except for acetyl phosphate, these organic phosphates behave as typical energy-transfer inhibitors of photophosphorylation, inhibiting photophosphorylation and coupled electron transport, but not affecting the basal rate of electron transport. Interestingly, these organic phosphates inhibit ATPase activity of isolated CF1, which is enhanced by the presence of organic solvents in the reaction mixture (Sakurai et al., 1981), noncompetitively with ATP. In the subsequent experiments, we found that chloroplasts could catalyze light-induced ATP formation from acetyl phosphate and ADP, and proposed the following mechanism for the reaction (Shinohara, Sakurai, 1982);

$$\text{acetyl phosphate + ADP} \longrightarrow \text{ATP + acetate.}$$

The above conclusion was based on the following experiments; 1) the ATP increase paralleled the acetyl phosphate decrease, 2) acetyl phosphate inhibited ATP formation in the presence of high concentration (1 mM) of Pi, but accelerated it in the presence of low concentration (0.2 mM) of Pi.

In order to further characterize the acetyl phosphate-dependent ATP formation, we studied the changes in the levels of adenine nucleotides upon illumination by means of HPLC. We also studied if thylakoids from a blue-green alga, Anacystis nidulans, and from a photosynthetic bacterium, Rhodospirillum rubrum, catalyze ATP formation from acetyl phosphate and ADP.

3. MATERIALS AND METHODS

Preparation of broken chloroplasts from spinach, the reaction conditions, ATP determination by the luciferase assay, and acetyl phosphate determination by the hydroxamate method, were as in Shinohara and Sakurai (1982) unless otherwise stated. Adenine nucleotides were analyzed by HPLC (Waters model QA-1): column, IEX-540 (Toyo Soda Chemical Co., Tokyo); elution medium 400 mM K-phosphate buffer (pH 4.6)-20%(v/v) acetonitrile; flow rate, 0.5 ml/min. Preparation of thylakoids from A. nidulans and of chromatophores from R. rubrum will be described elsewhere (Shinohara et al., in preparation).

4. RESULTS AND DISCUSSION

Sybesma, C. (ed.), Advances in Photosynthesis Research, Vol. II. ISBN 90-247-2943-2.
© *1984 Martinus Nijhoff/Dr W. Junk Publishers, The Hague/Boston/Lancaster.*

We found that <u>A</u>. <u>nidulans</u> thylakoids catalyzed ATP formation, but their activity was not so large. The activity of <u>R</u>. <u>rubrum</u> chromatophores was very low, if at all (Fig. 1).

It was also noted that Pi contaminating acetyl phosphate significantly influenced the apparent activity of acetyl phosphate-dependent ATP formation. We reexamined the steady activity of spinach chloroplasts for ATP formation from acetyl phosphate and ADP, and found that its ratio to the ordinary photophosphorylation (Pi + ADP) varied from preparation to preparation (Table 1). We studied the changes in the levels of each adenine nucleotide upon illumination by means of HPLC (Fig. 2). When Pi was added, almost all the ADP present in the reaction mixture was converted to ATP in 2 min. The rate with 0.4 mM acetyl phosphate was high in the first 2 min, followed by a lower but steady rate of ATP formation (15 μmol/mg chl/hr). The steady rate was not so high as that in the previous report, but it was significantly higher than the control (+ADP only) rate (almost no ATP formation after 2 min, data not shown). The initial 2 minutes' rate may be ascribed to the Pi present in the acetyl phosphate preparation or in ADP. Pi contamination seems to be a little higher in this experiment than in the previous one. Fig. 2 also shows that in the

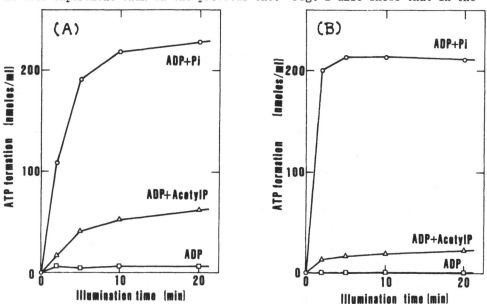

Fig. 1. Time course of ATP formation from acetyl phosphate and ADP by <u>A</u>. <u>nidulans</u> thylakoirs (A) and by <u>R</u>. <u>rubrum</u> chromatophores (B).

The reaction mixture (0.5 ml) for <u>A.n.</u> thylakoids (7.5 μg chl) was 20 mM Tricine-NaOH (pH 8.0), 10 mM NaCl, 0.6 M sucrose, 50 μM PMS, 26 μM DPIP, 20 mM ascorbate, 20 mM glucose, 7 units of hexokinase, 0.2 mM acetyl phosphate or Pi. The reaction mixture (0.5 ml) for <u>R.r.</u> chromatophores (20 μg Bchl) was 20 mM Tricine-NaOH (pH 8.0), 5 mM MgCl2, 2 mM ascorbate, 0.4 mM PMS, 20 mM glucose, 7 units of hexokinase, 0.4 mM acetyl phosphate or Pi. After termination of the reaction by adding 50 μl of 2.5 M perchloric acid, PMS in the acidified reaction mixture was removed by extracting with isobutanol at alkaline pH (10-11), and glucose 6-phosphate was analyzed enzymatically by Lang and Michal (1974).

Table 1. Relative rate of ATP formation a): μmol ATP/mg chl/hr

		Pi/AceP M	ADP + Pi % (rate)	ADP + Acetyl P steady rate %	initial 2 min rate %
S. oleracea	(Exp. 1)	0.2	100 (228) a)	20	26
	(Exp. 2)	2	100 (> 86)	<8	<23
	(Exp. 3)	0.4	100 (>524)	<2.8	<22
	(Exp. 4)	0.05	100 (>100)	<7.7	<37
A. nidulans		0.2	100 (110)	2.5	15
R. rubrum		0.2	100 (> 75)	<0.4	<7

presence of ADP + acetyl phosphate, AMP slightly increased upon illumination. However, the steady rate of ATP formation from 0.4 mM acetyl phosphate + ADP was significantly higher than the rate of AMP formation. Thus, the ATP formation cannot be explained by adenylate kinase activity alone. Considering the Km of Pi (0.25 mM) and of acetyl phosphate (2.4 mM and 0.81 mM obtained from competition with Pi and from acetyl phosphate-dependent ATP formation, respectively) (Shinohara, Sakurai, 1982), a use of higher concentrations of acetyl phosphate may be recommended, but the presence of Pi contaminating the acetyl phosphate preparation made us hesitate in undertaking such experiments.

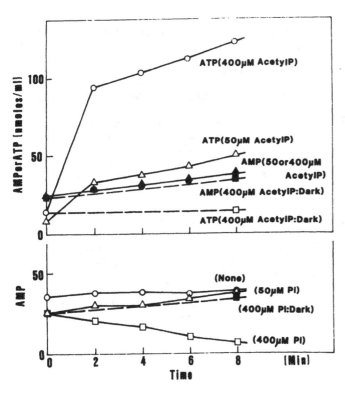

Fig. 2. Time course of ATP formation by illuminated spinach chloroplasts.

Chloroplasts were illuminated unless indicated as "Dark". The reaction mixture contained 20 mM Tricine-NaOH (pH 8.0), 10 mM MgCl2, 50 μM PMS, 0.4 mM ADP and broken chloroplasts containing 20 μg chl/ml. Where indicated, 0.05 or 0.4 mM Pi or acetyl phosphate was also included. Adenine nucleotides were determined by HPLC.

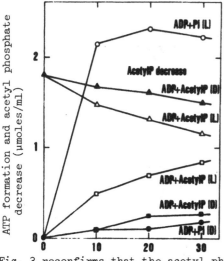

Fig. 3. ATP formation and acetyl phosphate decrease.

The reaction mixture (2.0 ml) contained 20 mM Tricine-NaOH (pH 8.0), 10 mM MgCl2, 50 μM PMS, 0.1 M sucrose, 2 mM ADP, 1.8 mM acetyl phosphate or 2 mM Pi and broken chloroplasts containing 300 μg chl. The reaction was carried out at 18°C. ATP was determined by luciferin-luciferase method. Acetyl phosphate was determined by the hydroxamate method. (L): light, (D): dark.

Fig. 3 reconfirms that the acetyl phosphate decrease and the ATP increase were more pronounced in light than in dark. So, although the rate of ATP formation from acetyl phosphate and ADP by illuminated chloroplasts is usually not so high as previously thought, it seems that the activity does exist. A possible contribution of phosphatase activity to the apparent ATP formation remains to be investigated.

REFERENCES

Shinohara K and Sakurai H (1980) β-Naphthyl oligophosphate: Inhibitors of photophosphorylation and H+-ATPase of spinach chloroplasts, Plant Cell Physiol 21, 75-84; (1981) Pyridoxal 5-phosphate, phenyl phosphate and acetyl phosphate as inhibitors of photophosphorylation competitive with phosphate, Plant Cell Physiol. 22, 1447-1457; (1982) Light-induced ATP formation from acetyl phosphate and ADP by broken spinach chloroplasts, Plant Cell Physiol. 23, 59-66.
Sakurai H, Shinohara K, Hisabori T and Shinohara K (1981) Enhancement of adenosine triphosphatase activity of purified chloroplast coupling factor 1 in an aqueous organic solvent. J. Biochem. 90, 95-102.
Lang G and Michal G (1974) D-Glucose-6-phosphate and D-fructose-6-phosphate. in Bergmeyer HU, ed. Methods of enzymatic analysis (2nd English edn.), pp. 1238-1242. Weinheim. Verlag Chemie.

ACKNOWLEDGEMENTS
The authors wish to express their gratitude to Mr. Toru Hisabori for his instruction in HPLC.

Authors address: Hidehiro Sakurai, Kenji Shinohara and Rinya Yamagishi/ Department of Biology, School of Education, Waseda University, Nishiwaseda, Shinjuku, Tokyo 160, Japan.

This work was supported in part by Grants-in-Aid for Scientific Research from the Ministry of Education, Science and Culture, Japan to H.S.

ENZYME KINETICS OF ATP-ASE AND ENERGY BALANCE IN THYLAKOIDS

LOTHAR JAHN; RETO J. STRASSER

Department of Bioenergetics, Inst. of Biology, Univ. of Stuttgart
Ulmerstrasse 227, 7000 Stuttgart 60, FRG

1. INTRODUCTION

The scientific goals of these investigations are: 1) To measure
and analyse the enzyme kinetics of the ATP-ase which is still
attached to the membranes of chloroplasts or mitochondria.
2) To measure and analyse the energy-flux-balance EB in biological
electron transport systems. The energy-flux-balance is the ratio
of the rate of ATP synthesis (dATP/dt) and the rate of electron
transport (eg. the production of reducing equivalents dRH_2/dt)

$$EB = \frac{dATP/dt}{dRH_2/dt}$$

2. METHODS

The principles of the multi-parameter-analysis (MPA) were used.
- Signal detection: Measurement of the photosynthetic electron
  transport in pea chloroplasts with an oxygen electrode under
  phosphorylating conditions, $H_2O \rightarrow FeCy$ or $H_2O \rightarrow MV$
- Data acquisition: The continuous signal is digitized on line
  with a microcomputer and stored in core memory in digital form.
- Data handling: The original data are converted in such a way by
  the computer that they can be plotted according to the usual
  plots of enzyme kinetics (Fig. 1 and 2).
The experiments were based on the well known phenomenon that the
photosynthetic electron transport gets stimulated when ADP is
added to the suspension thus allowing the membranes to form ATP.
The stimulated electron transport rate decreases again as soon as
the added ADP is consumed. Therefore the membranes are exposed to
a substrate concentration (ADP) between 0 and the concentration
at the beginning of the experiment. It can be shown that the
stimulation of the electron transport is proportional to the actual phosphorylation activity. This makes it possible to determine
any time the ADP concentration in the cuvette, the rate of ATP
synthesis and the rate of total electron transport. On this basis
we have enough information to analyse the experimental reactions
according to the classical enzyme kinetics as shown in Fig. 1.

3. RESULTS

Pea chloroplasts were used and the electron transport activity
was measured with an oxygen electrode measuring oxygen evolution
(the reaction $H_2O \rightarrow FeCy$) or oxygen consumption (the reaction
$H_2O \rightarrow MV$). All experimental conditions were as reported elsewhere.
Fig. 2 and 3 are similar experiments using FeCy (Fig. 2) or
Methylviologen (Fig. 3) as electron acceptor. The analysis of the
experimental curve which is stored as 1000 digitized values with

Sybesma, C. (ed.), Advances in Photosynthesis Research, Vol. II. ISBN 90-247-2943-2.
© *1984 Martinus Nijhoff/Dr W. Junk Publishers, The Hague/Boston/Lancaster.*

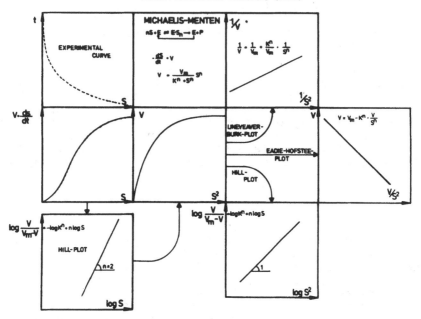

Fig. 1 Multi-parameter-analysis (MPA) of an experimental
curve according to the Michaelis Menten reaction type.
The curves are computer simulations.

12 bit resolution in the computer shows that the number of active
sites of the ATP-ase in situ is 1. K_m and V_{max} can be measured
from the same experiment. In Fig. 4 the same type of experiment
as shown in Fig. 2 and 3 is used. Here we are interested in knowing the energy balance (as defined above) as a function of the
local ADP concentration. For this reason the function of the rate
of ATP synthesis has to be devided by the total electron transport
rate.

It can be shown analytically that the energy balance is a hyperbolic function of the ADP concentration. The constants of the
hyperbolic function are defined by the experimentally measureable
terms P/O (added ADP per parallaxe on the oxygen curve measured
as mole per mole), PC (photosynthetic control which is the ratio
of the electron transport after addition of ADP and after the
consumption of the added ADP) an the Michaelis Menten constant K_m.
The energy balance EB can now be formulated as:

$$EB = \frac{P/O}{\dfrac{K}{PC} + ADP} \cdot ADP$$

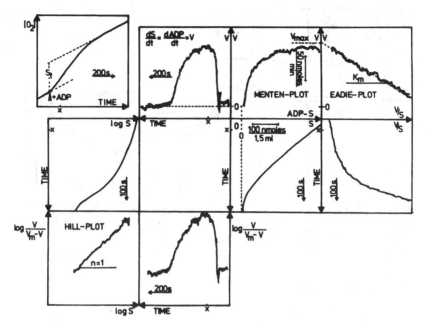

Fig. 2

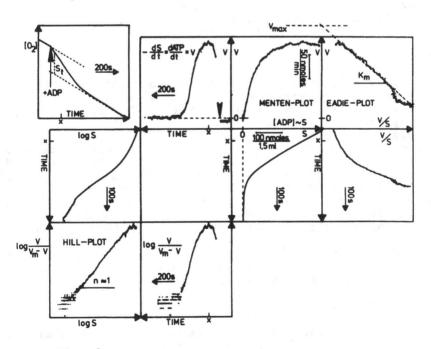

Fig. 3

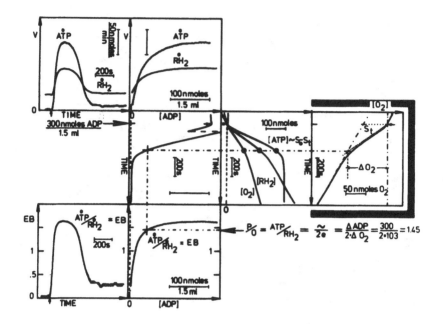

Fig. 4 shows that the EB vs ADP is indeed a hyperbolic function
which can be measured on thylakoids directly. The conventionally
measured P/O value corresponds to the maximal possible value of
the energy balance for a given sample.

3. CONCLUSIONS

- The ATP-ase, bound to biological membranes, behaves like a
  Michaelis type of enzyme with one active site.
- The energy balance in thylakoids versus the ADP concentration
  is a hyperbolic function.
- The classical P/O ratio represents:
  - theoretically the maximal capacity of the energy balance EB
  - experimentally a value sightly lower than the maximal EB
    (because the signal is integrated over the whole range of
    ADP concentrations).
- These techniques show a way to analyse the regulation of ATP-a
  in situ of biological membranes under various conditions.
- The techniques allow to estimate which metabolic pathways domi
  nate in the organells by a given energy balance.
- We hope that these techniques will be a great help in distinguishing the different types of coupling and uncoupling of the
  ATP-ase in thylakoids or mitochondria.

4. REFERENCE

Hall, D. O. (1976) in: The intact Chloroplast (Barber, J. ed.
Elsevier Amsterdam 146-147

ENHANCEMENT OF THE RATE OF PHOTOPHOSPHORYLATION IN SPINACH CHLOROPLASTS
BY LOW CONCENTRATIONS OF CARBOXYLIC IONOPHORES AND AMINES

CHRISTOPH GIERSCH, BOTANISCHES INSTITUT DER UNIVERSITÄT DÜSSELDORF

1. INTRODUCTION

Phosphorylation in chloroplasts is generally accepted to be driven by the
electrochemical proton gradient across the thylakoid membrane as proposed by
the chemiosmotic theory. However, the relation between the magnitude of the
proton motive force (pmf) and the rate of phosphorylation cannot be predicted
by chemiosmosis as mechanistic details of energy transduction are required
for this purpose. Experiments indicate that both the phosphorylation rate and
the phosphorylation potential, $(ATP)/(ADP) \cdot (P_i)$, increase with increasing the
pmf under a variety of conditions (e.g. Avron, 1978). However, other studies
indicate that there may not exist an one-to-one relation between the rate of
phosphorylation and the magnitude of the pmf. Thus, it was reported that the
rate of ATP-synthesis in submitochondrial particles decreases at a constant
pmf (Sorgato et al., 1980). More drastic, it was shown that the phosphorylation
rate in chloroplasts increased at a diminished pmf (Giersch, 1981). A number
of related observations can be found in the literature (e.g., Degani, Shavit,
1972; Wei et al., 1980). However, generally the mere observation was reported,
and the findings were not discussed in terms of energy transduction. Stimulation of phosphorylation under conditions where the pmf was expected to be decreased was observed repeatedly but apparently never studied in detail. In this
contribution it is reported that amines and carboxylic ionophores at low concentrations stimulate the rate of phosphorylation though the pmf and the phosphorylation potential are decreased under these conditions. Consequences of
these observations for the mechanism of energy transduction in chloroplasts
are discussed.

2. RESULTS

2.1 Photophosphorylation and electron transport

Intact isolated chloroplasts suspended in a hypotonic assay medium are capable
of photophosphorylating exogenous ADP in the presence of added P_i. Rates of
ATP formation in a linear electron transport system ($H_2O \longrightarrow$ ferricyanide or
methyl viologen) are generally about 300 µmol/mg chl·h). Addition of 10 mM
NH_4Cl or 2 µM nigericin stimulates electron transport and abolishes phosphorylation; however, low concentrations of these uncouplers do not reduce but
rather increase phosphorylation rates. Figs. 1a und 1b depict the dependence
of electron transport, the phosphorylation rate and of the ATP/e_2 ratio on
the concentrations of methylamine and nigericin, respectively. Phosphorylation
rates are stimulated by concentrations of methylamine and nigericin lower than
1.2 mM and 40 nM, respectively. Uncoupler concentrations causing maximum
stimulation vary somewhat with the preparation. The rate of electron transport
is only slightly increased by addition of uncouplers at concentrations that
stimulate the phosphorylation rate; consequently, the ATP/e_2 ratio is not decreased by low uncoupler concentrations but remains constant or is even
slightly increased as shown in Fig. 1. Data on the stimulation of the rate of
photophosphorylation by these and other uncouplers are given in Table 1. The
effect of monensin (which catalyzes H^+/Na^+ exchange) is comparable to that of
nigericin except that stimulating concentrations of the former ionophore exceed
that of nigericin by a factor of about 10. The extent of stimulation of the
phosphorylation rate induced by a fixed uncoupler concentration varies considerably with the preparation; rates of ATP production in the presence of uncouplers exceed control rates by a factor of more than 1.6 in some cases.

Sybesma, C. (ed.), Advances in Photosynthesis Research, Vol. II. ISBN 90-247-2943-2.
© *1984 Martinus Nijhoff/Dr W. Junk Publishers, The Hague/Boston/Lancaster.*

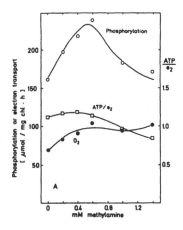
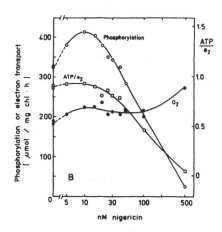

FIGURE 1. Dependence on uncoupler concentration of the rate of electron transport (O_2), phosphorylation rate and the ATP/e_2 ratio. Isolated intact chloroplasts were suspended in an assay medium containing O.1 M sorbitol, 5 mM $MgCl_2$, 10 mM NaCl, 10 mM KCl, 1 mM KH_2PO_4, 1 mM ADP, 40 mM (A) or 10 mM (B) HEPES, and 1 mM ferricyanide (A) or 25 µM methylviologen/1 mM KCN (B) as electron acceptors. Phosphorylation was determined enzymically (A) or by the pH-method (B), electron transport by means of a Clark-type electrode.

A number of other uncouplers like FCCP, DNP, gramicidin or valinomycin was tested but failed to show comparable effects. Also combinations of FCCP and valinomycin (that should mimic nigericin-induced H^+/K^+ exchange) did not induce enhancement of the phosphorylation rate. Nigericin-induced stimulation of phosphorylation is observed only in the presence of K^+ concentrations exceeding about 1 mM and is saturated at about 5 mM (Ch. Giersch, unpublished). This observation suggests that stimulation is related to H^+/K^+ exchange rather than to side-effects of the ionophore. The effects of methylamine and nigericin on the phosphorylation rate failed to be additive: when suboptimal concentrations of nigericin were added to the assay in the presence of suboptimal concentrations of methylamine, stimulation of phosphorylation did not exceed that observed upon separate addition of each of the compounds (Ch. Giersch, unpublished). This suggests that the molecular events leading to stimulation by one type of uncouplers interfere with those induced by the other one.

TABLE 1.   Uncoupler-induced stimulation of phosphorylation. Data are mean values of 5-15 experiments

uncoupler	stimulating concentration	% stimulation by standard concentration	standard concentration
NH_4Cl	0.2 - 0.5 mM	18	0.3 mM
CH_3NH_3Cl	0.2 - 1 mM	39	0.5 mM
nigericin	5 - 15 nM	32	10 nM
monensin	100 - 200 nM	27	150 nM

2.2 Proton motive force

The effect of uncouplers at low concentrations on the magnitude of the proton motive force was studied by means of 9-aminoacridine fluorescence quenching and the electrochromic shift, that monitor ΔpH and $\Delta \psi$, respectively. The proton gradient is decreased by ammonia and methylamine even at low, stimulating concentrations (Giersch, 1981). Also nigericin lowers the proton gradient; the extent of 9-aminoacridine fluorescence quenching increases almost linearly with the ionophore concentration (Fig. 2). It is evident that the proton gradient is lowered even by those nigericin concentrations that stimulate the rate of phosphorylation.

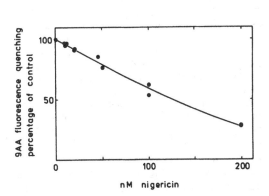

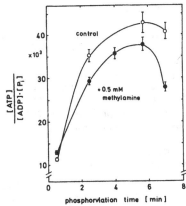

FIGURE 2. Dependence of light-induced 9-aminoacridine (9AA) fluorescence quenching by chloroplasts on the concentration of nigericin

FIGURE 3. Phosphorylation potential of a suspension of illuminated chloroplasts in the absence and presence of 0.5 mM methylamine. Adenylates were separated by HPLC and quantified by an online β-monitor.

The steady state membrane potential is increased by low concentrations of methylamine (Giersch, 1981) and practically not affected by 10 nM nigericin (Ch. Giersch, unpublished). From the relative magnitude of the alterations of ΔpH and $\Delta \psi$ it is evident that the pmf is decreased by low concentrations of amines. The decrease of the pmf is small with both types of uncouplers and does not exceed about 3-5 mV in the presence of 10 nM nigericin or 0.5 mM methylamine. It is concluded that the stimulation of the rate of phosphorylation occurs at a diminished pmf. This conclusion is confirmed by the observation that lowering of the steady-state membrane potential by valinomycin/K[+] does not affect the methylamine induced stimulation of phosphorylation (Giersch, 1981).

2.3 Phosphorylation Potential (PP)

The rate of photophosphorylation depends on a number of factors different from the pmf; the observation that the phosphorylation rate is increased in the presence of a lowered pmf is therefore not necessarily in contradiction with the chemiosmotic concept. On the other hand, the PP is assumed to be determined only by the magnitude of the pmf. Therefore, maximum values of the PP can be used to monitor the pmf. Fig. 3 shows the PP attained by a suspension of chloroplasts in the absence and presence of 0.5 mM methylamine. Maximum values in the absence of the amine are about 43.000'/M ($\Delta G'$(ATP) = 57.78 kJ/mol) in agreement with values obtained by Kraayenhof (1969). In the presence of the amine, the maximum value is reduced to 38.000 /M ($\Delta G'$(ATP) = 57.48 kJ/mol). Likewise, addition of 10 nM nigericin decreases the maximum phosphorylation potential from 45.000 to 33.000 /M (Ch. Giersch, unpublished). From these experiments it is apparent that the PP is decreased by low concentrations of nigericin and methylamine.

3. CONCLUSION

Equilibrium thermodynamics (as well as the non-equilibrium treatment) predict proportionality between the pmf and the maximum attainable PP. It can be calculated that under equilibrium conditions a decrease of e.g. 2% of a pmf of 180 mV leads to decrease in the PP by a factor of about 1.5. The data of Fig. 3 is therefore in agreement with the observation that uncouplers at low concentrations lower the pmf. The simultaneous decrease of the PP and the pmf supports the view that photosynthetic ATP production is driven by the pmf. That the rate of phosphorylation is increased under these conditions illustrates that the kinetics of phosphorylation is governed by factors different from the pmf. At present, the mechanism of uncoupler-induced stimulation of photophosphorylation is unknown. However, preliminary experiments indicate that stimulation may be related to enhanced activation of the CF_1 caused by an uncoupler-induced transient increase of $\Delta\psi$ during the first few seconds after the onset of illumination (Ch. Giersch, unpublished).

It should be noted that the ATP/e_2 ratio is not decreased by low uncoupler concentrations (cf. Fig. 1), though the rate of H^+ leakage across the thylakoid membrane is increased under these conditions (Giersch, 1982). Assuming that neither the number of protons pumped across the membrane per e^- nor the H^+/AT stoichiometry of ATP formation by the CF_0/CF_1 are altered by the uncouplers, constant ATP/e_2 ratio in the presence of increased membrane leakage suggests that protons channeled through the coupling factor are not necessarily in rapid communication with those leaking across the membrane. That phosphorylation can be driven by protons which are not equilibrated throughout the bulk thylakoid lumen is at variance with the classical chemiosmotic concept. The observation that the ATP/e_2 ratio is not decreased in spite of increased proton leakage could mean that localized proton fluxes are involved in energy transformation.

REFERENCES

Avron M (1978) FEBS Letters 96, 225-232
Degani H and Shavit N (1972) Arch. Biochem. Biophys. 152, 339-346
Giersch CH (1981) Biochem. Biophys. Res. Comm. 100, 666-674
Giersch CH (1982) Z. Naturforsch. 37c, 242-250
Kraayenhof R (1969) Biochim. Biophys. Acta 180, 213-215
Sorgato MC, Branca D and Ferguson SJ (1980) Biochem. J. 188, 945-948
Wei JM, Shen YG, Li DY and Zhang XX (1980) Acta Phytophysiol. Sin. 6, 393-398

ACKNOWLEDGEMENTS

I wish to express my gratitude to Prof. N. Shavit for stimulating discussions and to Rita Reidegeld for typing this manuscript.
Authors address: Christoph Giersch, Botanisches Institut der Universität
                 Düsseldorf, D-4000 Düsseldorf, F.R.G.

DIFFERENTIAL INHIBITION OF P_i-ATP EXCHANGE IN RELATION TO ATP SYNTHESIS AND HYDROLYSIS BY GLUTARALDEHYDE MODIFICATION OF CHLOROPLAST MEMBRANES

D. BAR-ZVI and N. SHAVIT/BEN-GURION UNIVERSITY OF THE NEGEV

1. INTRODUCTION

In isolated thylakoid membranes, the chloroplast ATP synthetase catalyses high rates of ATP synthesis, but rather low rates of the reverse reactions. However, higher rates of ATP hydrolysis and P_i-ATP exchange can be obtained after activation by energization of the membranes in the presence of thiol reagents (Shavit, 1980). The participation of different enzyme conformations or altered catalytic site(s) in the catalysis of ADP and ATP utilizing reactions has been suggested (Shavit, 1980; Pedersen, 1978). Chemical modification of the isolated and membrane-bound CF_1 has been used to study its mechanism of action (Vallejos, 1981).

In the present work, we used glutaraldehyde (GA) as a modification agent and studied its effects on several reactions catalysed by the ATP synthetase, using very low reagent concentration and short exposure times. Extensive modification inactivated all reactions tested but did not affect electron transport. However, less extensive (limited) modification inactivated the P_i-ATP exchange reaction with much less effect on the other activities. A mechanism in which GA modification affects only one rate constant can account for the differential inhibition of P_i-ATP exchange in relation to ATP hydrolysis and synthesis.

2. EXPERIMENTAL

Chloroplast thylakoids were prepared from fresh lettuce leaves (Bar-Zvi, Shavit, 1983), washed with cold 50 mM NaCl/ 2 mM tricine-NaOH (pH 8.0), resuspended in the same medium, and used immediately. Thylakoid membranes (100-140 μg chlorophyll/ml) were modified with GA in the dark at $20°C$ in 20 mM NaCl/ 10 mM $MgCl_2$/ 2.5 mM DTT. Modification was stopped by adding $NaBH_4$ to a final concentration of 10 mM.

P_i-ATP exchange and ATP hydrolysis were assayed simultaniously under identical conditions (Bar-Zvi, Shavit, 1983). P_i and ATP were added 5 s after activation of the thylakoids. Substrate mixes contained $^{32}P_i$ or $[\gamma-^{32}P]ATP$ for the assay of P_i-ATP exchange or ATP hydrolysis, respectively. Net rates of ATP hydrolysis were calculated by subtracting the rate of P_i-ATP exchange from that of $^{32}P_i$ released from $[\gamma-^{32}P]ATP$. Other assays are described elsewhere (Bar-Zvi, Shavit, 1983).

3. RESULTS

Inactivation of photophosphorylation by GA modification seems to be biphasic: a first slow phase is followed by a faster phase (Fig. 1). The biphasicity is more apparent if photophosphorylation by the modified thylakoids is assayed under low light intensities. The rate of the fast phase is not affected by the energization state of the membrane, but is increased if DTT and Mg^{2+} are present during modification (not shown). The basal rate of electron transport (H_2O - FeCN) is not affected by GA modification, but FeCN reduction coupled to ATP formation is inhibited (Table 1), suggesting that modification with glutaraldehyde affects the ATP synthetase.

Sybesma, C. (ed.), Advances in Photosynthesis Research, Vol. II. ISBN 90-247-2943-2.
© *1984 Martinus Nijhoff/Dr W. Junk Publishers, The Hague/Boston/Lancaster.*

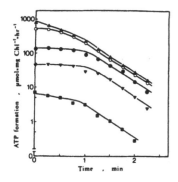

FIGURE 1. Inactivation of photophosphorylation by GA modification. Thylakoids
were modified in the dark with 0.05% GA
for the indicated times, and then
assayed for photophosphorylation. Light
intensities during assay were: $\blacktriangle$, 0.23;
$\bigcirc$, 0.15; $\bullet$, 0.072; $\blacktriangledown$, 0.041 and
$\blacksquare$, 0.021 J x cm^{-2} x s^{-1}.

Fig. 2 shows the effect of GA on ATP hydrolysis and P_i-ATP exchange
(assayed in the dark under identical conditions). Like photophosphorylation, ATP hydrolysis is inactivated in a biphasic time course, whereas
inactivation of P_i-ATP exchange is monophasic. Moreover, P_i-ATP exchange
is much more sensitive to the modification than ATP synthesis or hydrolysis. Thus, limited modification results in a thylakoid preparation that
has lost most of its P_i-ATP exchange activity but little of its capacity
for ATP hydrolysis or synthesis.

The greater sensitivity of the P_i-ATP exchange does not result from
an effect on the activation step, since this activity is more sensitive to
modification even when GA is added together with the substrates to preactivated thylakoids (not shown). The ΔpH supported by ATP hydrolysis
also is not markedly affected by limited modification with GA (Fig. 3)
which suggests that the differential inhibition of P_i-ATP exchange is
probably not due to uncoupling of ATP hydrolysis. Limited modification of
thylakoid membranes with GA also does not affect the tight binding of ADP
to membrane-bound CF_1 (not shown), which was suggested to play a role in
the modulation of the ATP utilizing activities (Bar-Zvi, Shavit, 1980).
Limited modification also does not change the thylakoid polypeptide
pattern, as analysed by polyacrylamide electrophoresis in the presence of
SDS (not shown). However, more extensive modification results in a
decrease in the amount of some polypeptides (including the α and β
subunits of CF_1) and the appearance of crosslinked products (not shown).

TABLE 1. Modification by glutaraldehyde: Effect on electron transport and
photophosphorylation. Thylakoids were modified with 0.05% GA, washed to
remove the DTT and NaBH$_4$, and assayed.

Modification time (min)	Ferricyanide reduction		Photophosphorylation	P/2e
	μmol/mg chlorophyll/hr			
	−ADP	+ADP		
0	288	438	117.6	0.54
1	283	272	41.9	0.31
2	284	263	3.1	0.02

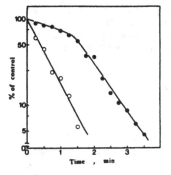 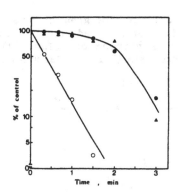

FIGURE 2. (Left) Inactivation of P_i-ATP exchange and ATP hydrolysis by GA modification. Thylakoids were modified with 0.01% GA for the indicated times, then activated and assayed for P_i-ATP exchange (O) or ATP hydrolysis (●).

FIGURE 3. (Right) Modification with glutaraldehyde: Effect on pH supported by ATP hydrolysis. Thylakoids were modified with 0.025% GA for the indicated times. ▲, [H⁺] ratio (in/out); O, ●, rates of P_i-ATP exchange and ATP hydrolysis, respectively.

Limited modification has little effect on the V_{max} for ATP hydrolysis, but increases the apparent K_m for ATP (Table 2). Such modification also increases the apparent K_m values for both ATP and P_i in P_i-ATP exchange, concomitant with a decrease in the V_{max} for this activity. The K_i for ADP as a competitive inhibitor in ATP hydrolysis is not affected.

4. DISCUSSION

GA seems to modify groups outside the catalytic site of the ATP synthetase, resulting in increased apparent K_m values for ATP and P_i in the ATP utilizing reactions, and greater sensitivity of the P_i-ATP exchange relative to the other enzyme activities. Since no crosslinking occurrs upon limited modification of the membranes under these conditions GA either acts as a monofunctional reagent or causes intrasubunit

TABLE 2. Effect of limited GA modification on kinetic parameters for the ATP utilizing reactions.

Activity	Substrate	Apparent K_m (µM)		V_{max} of modified thylakoids (% of control)
		Control	Modified thylakoids	
ATP hydrolysis	ATP	48	83	92[a]
P_i-ATP exchange	ATP	78	180	37
	P_i	1230	2350	50

a P_i-ATP exchange activity of this preparation was completely inhibited.

crosslinks. The greater sensitivity of P_i-ATP exchange may result from an effect of limited GA modification on the catalytic steps rather than on electron transport (Table 1), enzyme activation, or uncoupling of ATP hydrolysis (Fig. 3). Assuming a Uni Bi mechanism for ATP hydrolysis:

$$E \underset{k_{-1}}{\overset{k_1}{\rightleftharpoons}} E \cdot ATP \underset{k_{-2}}{\overset{k_2}{\rightleftharpoons}} E : \begin{matrix} ADP \\ P_i \end{matrix} \underset{k_{-3}}{\overset{k_3}{\rightleftharpoons}} E \cdot ADP \underset{k_{-4}}{\overset{k_4}{\rightleftharpoons}} E$$

we derived rate equations for ATP hydrolysis and P_i-ATP exchange (Cleland, 1975). If limited GA modification of the membranes affects only one rate constant, this would be k_{-2}. A decrease in this rate constant would affect the V_{max} of P_i-ATP exchange more, with less effect on the hydrolysis rate (Fig. 2), would increase the apparent K_m values for both ATP and P_i (Table 2), and would not affect the K_i for ADP as a competitor in ATP hydrolysis. A change of k_{-1} or k_{-3} would also affect the rate of P_i-ATP exchange more than that of ATP hydrolysis, but would not have the right effects on the Michaelis constants. However, if ATP hydrolysis is, as we believe, the reverse of ATP synthesis, a decrease in k_{-2} should also result in the inactivation of photophosphorylation, which does not occur upon limited modification. This discrepancy may be explained by the different types of energization during photophosphorylation (light) and ATP utilizing reaction (dark), which affects the values of some kinetic constant. Indeed, both V_{max} and the K_m for ADP in photophosphorylation depend on the energization level (Bickel-Sandkötter, Strotmann, 1981).

REFERENCES

Bar-Zvi D and Shavit N (1980) Role of tight nucleotide binding sites in the regulation of the chloroplast ATP synthetase activities, FEBS Lett. 119, 68-72.
Bar-Zvi D and Shavit N (1983) Differential inhibition of P_i-ATP exchange in relation to ATP synthesis and hydrolysis by modification of chloroplast thylakoid membranes with glutaraldehyde, Biochim. Biophys. Acta, in press.
Bickel-Sandkötter S and Strotmann H (1981) Nucleotide binding and regulation of chloroplast ATP synthase, FEBS Lett. 125, 188-192.
Cleland WW (1975) Partition analysis and the concept of net rate constants as tools in enzyme kinetics, Biochemistry 14, 3220-3224.
Pedersen PL, Amzel LM, Soper JW, Cintron N and Hullihen J (1978) Structure, function, and regulation of the mitochondrial adenosine triphosphatase complex of rat liver - a progress report, in Energy Conservation in Biological Membranes, Schäfer G and Klingenberg M eds. pp. 159-194. Springer-Verlag, Berlin and Heidelberg.
Shavit N (1980) Energy transduction in chloroplasts: Structure and function of the ATPase complex, Annu. Rev. Biochem. 49, 111-138.
Vallejos RH (1981) Chemical modification of chloroplast coupling factor 1, in Energy Coupling in Photosynthesis, Selman BR and Selman-Reimer S, eds. pp. 129-139. Elsevier/North Holland.

ACKNOWLEDGMENTS

We are grateful to Mrs. Z. Conrad for excellent technical assistance, and to Dr. M. A. Tiefert for help in the preparation of this manuscript.

Authors address: Department of Biology, Ben-Gurion University of the Negev, Beer-Sheva 84105, Israel.

DIFFERENTIAL EFFECTS OF SHORT TIME GLUTARALDEHYDE TREATMENT ON LIGHT INDUCED THYLAKOID MEMBRANE CONFORMATIONAL CHANGES, PROTON PUMPING AND ELECTRON TRANSPORT PROPERTIES.

S.J.Coughlan & U.Schreiber. Biology Department,Brookhaven National Laboratory,Upton,New York 11973,USA & Botanisches Institut,Universität Würzburg,D-8700 Würzburg,West Germany.

1. INTRODUCTION

Chemical cross linking reagents have found much use in probing membrane structure,both as specific modifying agents of the protein or lipid moieties of the membrane;and to stabilise membranes for structural analysis.One of the most widely used of these compounds is glutaraldehyde,which reacts primarily with free amine groups to form a Schiffs base,although there are complicating side reactions.

This reagent has been extensively used in modifying thylakoid structure both in situ and in the isolated membrane.Long incubation times (c.5min) have generally been used in these studies.In general,the thylakoid functions most affected by glutaraldehyde fixation are the coupling factor complex and light induced conformational changes.Electron transport and the associated proton pumping are less affected.

In this report we have studied the short term affects of glutaraldehyde fixation on coupling factor activity and light induced conformational changes.Adopting the protocoll of Dilley and coworkers,we have terminated the glutaraldehyde treatment by the addition of an excess of free amine buffer.By this procedure treatment times as short as 5s could be reproducibly carried out.

2. MATERIALS AND METHODS

All experimental procedures are as in Coughlan & Schreiber (1983). Glutaraldehyde treatment of thylakoids was carried out following the recommendations of Papageorgiou (1982).

3. RESULTS

The first figure shows the effect of increasing time treatments of glutaraldehyde on various photosynthetic parameters at a constant glutaraldehyde/thylakoid ratio of 60 μmoles mg Chl^{-1}.Under these conditions there was a rapid ($t_{\frac{1}{2}}$=5s) inactivation of PMS mediated cyclic photophosphorylation.Concommittant with this was a removal of most of the phosphorylation induced stimulation of electron transport as measured by ferricyanide reduction,and an increase in 9AA fluorescence quenching to the basal level.NADP$^+$ reduction in a thylakoid system reconstituted with 0.1mM ferredoxin was also progressively inhibited with the same kinetics as photophosphorylation.However,over the time period studied (up to 10 min) there was no affect on basal electron transport ($H_2O \rightarrow MV/FeC$) Interestingly, there was a loss of between 20-30% of uncoupled electron transport parallel to the loss of photophosphorylation.The residual 70-80% of uncoupled electron flow remained constant over the remainder of the experimental period.

Sybesma, C. (ed.), Advances in Photosynthesis Research, Vol. II. ISBN 90-247-2943-2.
© 1984 Martinus Nijhoff/Dr W. Junk Publishers, The Hague/Boston/Lancaster.

FIGURE 1

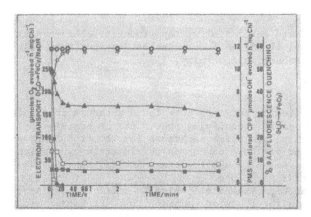

The effect of short term glutaraldehyde treatment on electron trans-
port,cyclic photophosphorylation and 9AA fluorescence quenching.
(A). Electron transport.
The medium (1ml) contained 100mM sorbitol,10mM $MgCl_2$,10mM tricine/10mM
tris pH 7.6,2mM FeCy (H_2O --> FeCy),or 0.1mM ferredoxin,2mM NADP
(H_2O --> NADP) for basal electron transport,plus 10mM KH_2PO_4 and 0.5mM
ADP for phosphorylating conditions,plus 5mM NH_4Cl for uncoupled condit-
ions. $20°C$,50 µg Chl ml^{-1},actinic light (610-800nm) intensity 200 W m^{-2}
(B). Cyclic photophosphorylation.
The medium (1ml) contained 100mM sorbitol,10mM $MgCl_2$,1mM tricine/KOH
pH 7.8,1mM KH_2PO_4,0.5mM ADP,50 µg Chl,$20°C$.Actinic light intensity
(610-800nm) was 600 W m^{-2}
(C). 9AA fluorescence quenching.
The medium (1.5ml) was the same as for basal electron transport with
the addition of 1 uM pAA,15 µg Chl ml^{-1},$20°C$.Fluorescence was excited
at 400nm (1nm slit width,2 W m^{-2}) and emitted light measured at 480nm
(40nm slit width).Actinic light intensity (610-800nm) was 40 W m^{-2}

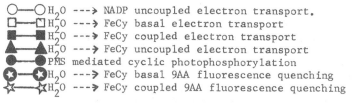

H_2O ---> NADP uncoupled electron transport.
H_2O ---> FeCy basal electron transport
H_2O ---> FeCy coupled electron transport
H_2O ---> FeCy uncoupled electron transport
PMS mediated cyclic photophosphorylation
H_2O ---> FeCy basal 9AA fluorescence quenching
H_2O ---> FeCy coupled 9AA fluorescence quenching

The second figure explores the effect of glutaraldehyde treatment on
light induced scattering changes measured firstly in a laboratory built
instrument (Schreiber & Reinits,1983) which measures the total amount
of scattered light (figure 2A,2B),and secondly the $180°$ (transmittance)
and narrow angle $90°$ components measured on an Aminco DW2A,and anAminco
SPF 500 spectrofluorometer respectively.The slow ($t_{\frac{1}{2}}$ 15-20s) light
induced transmittance changes were completely abolished by short time
glutaraldehyde treatments.The time course of inhibition ($t_{\frac{1}{2}}$ < 5s) was
even more rapid than for photophosphorylation.In contrast,the fast rise

FIGURE 2

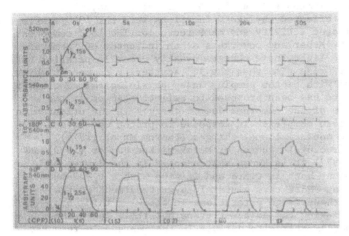

The effect of short term glutaraldehyde treatment on scattering changes
under continuous illumination.
Experimental conditions as in Figure 1 except 0.1mM MV for 2mM FeCy,
15 ug Chl ml^{-1},20°C.Actinic light (630-800nm) was 40 Wm^{-2}.

component of the absorbance change appeared relatively unaffected.
The 90° scattering component was also progressively inhibited by
glutaraldehyde (t$_{\frac{1}{2}}$=20-30s),but the time course of inactivation was
one order of magnitude greater than for the 180° scattering component.
Thus,after 5s glutaraldehyde treatment the 180° component was completely
abolished whereas the 90° component was only 20% inhibited.

4.DISCUSSION

The most widely accepted models of thylakoid electron transport
involve a vectorial arrangement of the electron carriers such that
light driven electron transport results in the inward translocation
of protons into the interthylakoid space at two points,the water
splitting complex and the plastoquinone pool.Concommittant with this
light driven proton uptake is a counter exchange of ions,predominantly
Mg^{2+},to maintain electroneutrality,and scattering changes which are
considered to represent microconformational changes in membrane
structure.These light dependent scattering changes were first reported
by Packer (1963) and have been extensively investigated first on
thylakoids (Dilley,1971),and later using intact chloroplasts (Krause,
1973) and whole leaves (Heber,1969).
Packer et al (1968) were the first to use glutaraldehyde fixation
of thylakoid membranes to inhibit these scattering changes whilst
leaving proton pumping relatively unaffected.In their report,and
subsequent studies,notably by Papageorgiou and coworkers,long treatment
times (mins) were used.Our initial observation was that short (10s)
treatment times selectively inhibited the 180° scattering component

This prompted us to investigate (a) the kinetics of glutaraldehyde
inactivation of various photosynthetic processes in detail, and
(b) to undertake a series of comparative investigations of the 180°
and 90° scattering components (Coughlan & Schreiber, 1983b).
  The most rapidly affected thylakoid functions were the 180° scattering
changes, coupling factor activity and NADP reduction. The 180° component
has been utilised by Heber and coworkers as an indicator of "membrane
energisation" in the light. These changes usually occur in parallel to,
but with a slower $t_{\frac{1}{2}}$ rise time than the 90° scattering component which
has been correlated with proton pumping under physiological conditions
(Hind et al, 1973); and has been proposed to reflect changes in the
selective dispersion of the membrane due to changes in the orientation
of the pigment protein complexes in response to proton uptake and the
associated membrane conformational changes (Thorne et al., 1975).
  From the ultrastructural studies of Murakami and Packer (19 0) and
Miller and Nobel (1972), it is known that changes in the degree of
granal thickness occur both in vitro and in vivo with a $t_{\frac{1}{2}}$=30-60s.
We would propose that short term glutaraldehyde treatment of the
thylakoid prevents these large alterations in stacking thickness,
possibly by cross linking the adjacent grana, but still leave the
pigment protein complexes free to diffuse in the mebrane in response
to light and the associated proton pumping.

5. LITERATURE
Coughlan SJ and Schreiber U (1983) The differential effects of short time
glutaraldehyde treatments on light induced thylakoid membrane conformational changes, proton pumping and electron transport properties.
Biochim. Biophys. Acta (submitted)
Coughlan SJ and Schreiber U (1983b) Light induced changes in the confor-
mation of spinach thylakoid membranes as monitored by 90° and 180°
scattering changes: a comparative study. Biochim. Biophys. Acta (submitted)
Hind G, Nakatani HY and Izawa S (1974) Light-dependent redistribution of
ions in suspensions of chloroplast thylakoid membranes. Proc. Nat. Acad.
Sci. USA 71, 1484-1488.
Krause GH (1973) The high energy state of the thylakoid system as
indicated by chlorophyll fluorescence and chloroplast shrinkage
Biochim. Biophys. Acta 292, 715-728.
Miller MM and Nobel PS (1972) Light induced changes in the ultrastructure
of pea chloroplasts in vivo Plant Physiol. 49, 535-541.
Murakami S and Packer L (1970) Protonation and chloroplast membrane
structure J. Cell Biol. 47, 332-351.
Packer L, Allen JM and Starks M (1968) Light induced ion transport in
glutaraldehyde treated chloroplasts: studies with nigericin.
Arch. Biochem. Biophys. 128, 142-152.
Papageorgiou GC and Isaakidou J (1982) The pH dependence of photosynthet-
ic electron transport in glutaraldehyde treated thylakoids.
FEBS Letts. 138, 19-24.
Schreiber U and Rienits K (1982) Complementarity of ATP-induced and
light induced absorbance changes around 515nm. Biochim. Biophys.
Acta 682, 115-123.
Thorne SW, Horvath G, Kahn A and Boardman K (1975) Light-dependent
absorption and selective scattering changes at 518nm in chloroplast
thylakoid membranes. Proc. Nat. Acad. Sci. USA 72, 3858-3862.

TWO PHOSPHORYLATING CYCLES ASSOCIATED WITH PHOTOSYSTEM I

JONATHAN P. HOSLER AND CHARLES F. YOCUM

INTRODUCTION

Several laboratories have noted that antimycin A inhibits reactions associated with PSI, including a decline in the P/O values of ferredoxin (Fd)-mediated electron transport (Robinson, 1980). We have previously noted that antimycin A has no effect on the P/O value of electron transport to Fd/NADP (Yocum, Hosler, 1981). In this paper we present the results of photophosphorylation experiments on electron transport from H_2O to Fd/O_2 or Fd/NADP in thylakoid membranes, and propose a model for two cyclic pathways around PSI.

MATERIALS AND METHODS

Spinach thylakoids and Fd were prepared as described by Robinson and Yocum (1980). Oxygen evolution (NADP reduction) and oxygen uptake (oxygen reduction) were measured polarigraphically, using saturating red light. ATP synthesis was measured by the esterification of ^{32}Pi to ADP. The following concentrations were used unless otherwise noted: Fd alone, 21μM; Fd/NADP, 10μM/1mM; methylviologen, 67μM; chl, 25-30 μg/ml.

RESULTS AND DISCUSSION

The rate of electron transport to Fd/O_2 is limited by the autoxidation of reduced Fd. Reduced Fd accumulates, and the interphotosystem chain is kept in a relatively reduced condition during steady-state electron flow (data not shown). The rate of electron transport to Fd/NADP is rapid, reduced Fd does not accumulate, and the interphotosystem chain assumes an oxidized condition. A low rate of Fd-mediated oxygen reduction accompanies NADP reduction (data not shown); this rate varies from 5-15% of the total electron transport rate depending upon the concentration of Fd. This rate of oxygen reduction is included in the calculations of the P/O value of NADP reduction in order to obtain the real, rather than an inflated, P/O (see Yocum, Hosler, 1981).

Table 1. Photophosphorylation as a function of Fd concentration

Fd(μM)	H_2O-->Fd-->O_2			H_2O-->Fd-->NADP		
	O_2 rate*	ATP rate*	P/O	O_2 rate*	ATP rate*	P/O
5	51	131	1.28	137	422	1.54
10	74	218	1.47	162	516	1.59
20	96	308	1.60	165	518	1.57
55	120	381	1.59	162	517	1.60

*μmol/hr · mg chl; O_2 rate equals O_2 reduction plus NADP reduction

Sybesma, C. (ed.), Advances in Photosynthesis Research, Vol. II. ISBN 90-247-2943-2.
© *1984 Martinus Nijhoff/Dr W. Junk Publishers, The Hague/Boston/Lancaster.*

Both Fd-mediated oxygen reduction and NADP reduction phosphorylate ADP with P/O values ranging from 1.5-1.6 (Table 1); these values, when compared to the non-cyclic value of 1.25 obtained with methylviologen, indicate concurrent cyclic and non-cyclic photophosphorylation. The rates of oxygen reduction and ATP synthesis in the absence of NADP are dependent upon the concentration of Fd up to 55μM; the P/O value saturates at 20μM (Table 1). Electron transport and photophosphorylation supported by Fd/NADP saturate at 10μM Fd, but the P/O value is high at 5μM Fd; the cycle is fully operative at concentrations of Fd which are subsaturating for NADP reduction. While the cycle observed in the absence of NADP is associated with the accumulation of reduced Fd, it is clear from these data that cyclic photophosphorylation in the presence of NADP does not require a high concentration of reduced Fd, and may only depend upon the turnover of FNR.

Antimycin A inhibits the cycle associated with electron transport to Fd/O_2 but has no effect on cyclic photophosphorylation associated with NADP reduction (Table 2). Antimycin A has been reported to inhibit the PSI cycle during NADP reduction (Arnon, Chain, 1977), but we were unable to repeat this result under a wide variety of reaction conditions. Heparin has been found by us to be a partial competitive inhibitor of Fd reactions at the Fd binding site on FNR (data not shown); this inhibition is presumed to result from the ionic similarity between negatively charged Fd and heparin. Table 2 shows that heparin selectively inhibits the cycle associated with oxygen reduction, and has no effect on non-cyclic photophosphorylation ($H_2O \rightarrow$ Mev). The cycle associated with NADP reduction is only affected at much higher heparin concentrations.

Table 2. The effect of inhibitors on Fd-mediated photophosphorylation

Reaction	P/O		
	Fd/O_2	Fd/NADP	MeV
Control	1.54	1.60	1.25
+1.7μM antimycin A	1.28	1.60	1.25
+10μM heparin	1.27	1.55	1.26
+100μM heparin	1.25	1.34	1.21
+100μM heparin, +1.7μM antimycin A	1.23	1.27	-

The results presented above are consistent with a model of two phosphory-lating cyclic pathways associated with PSI (Figure 1). Q_1 and Q_2 are putative quinone binding sites located near the inside and outside of the thylakoid membrane. The antimycin A-sensitive pathway includes transmembrane electron transfer through cytochrome b-563; this reaction is initiated by plastosemiquinone produced at Q_1 by the one-electron oxidation of reduced PQ. The oxidation of reduced b-563 at Q_2 may depend upon the initial reduction of PQ to PQ^- or $PQH^{\cdot}$ by reduced Fd (through cytochrome b-559 LP). Additional evidence for this cyclic pathway includes the observations (data not shown) that 1) b-559 LP is rapidly reduced by Fd plus NADPH, 2) antimycin A stimulates the photoreduction of b-563 in

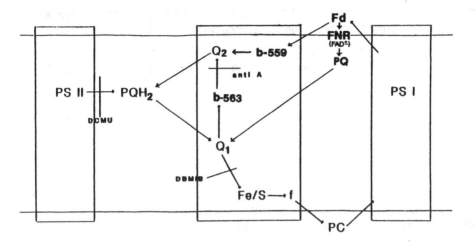

Figure 1. A model for two phosphorylating cycles associated with PSI

dark-adapted chloroplasts, and 3) both DBMIB and trifluralin stimulate the photooxidation of b-563 which was previously reduced by Fd plus NADPH. Heparin inhibition of this cycle may arise from the inhibition of the Fd to b-559 interaction. Using thylakoid membranes in which FNR activity was inhibited up to 90% by treatment with p-chloromercuribenzene sulfonate, we have found that the P/O values obtained for Fd-mediated oxygen reduction are still sensitive to antimycin A (data not shown). This indicates that FNR is not involved in the antimycin Asensitive cycle, however the antimycin A-insensitive cyclic pathway is proposed to flow through FNR, since this cycle is only observed during FNR turnover, and the flavin semiquinone is a potential reductant of PQ. This reaction may involve complete reduction of PQ to PQH_2, however alternative pathways involving a one-electron reduction of PQ to PQH$\cdot$, or PQH$\cdot$ to PQH_2, by the flavin semiquinone are also possible. Electron flow through cytochrome b-563 would be bypassed in favor of the sequential oxidation of PQH_2 by the Fe/S center.

The flow of electrons through these cyclic pathways may be regulated by the redox condition of the interphotosystem chain, and by the pool of reduced Fd available to the thylakoids. The accumulation of reduced Fd during slow electron transport to Fd/O_2 may keep both FNR and the interchain electron carriers reduced, thus inhibiting the formation of FADH$\cdot$, and promoting the reduction of b-563 by PQ$\cdot$ (at Q_1). The net effect would be the inhibition of the antimycin A-insensitive cyclic pathway, and the promotion of the PSI cycle through b-563. Alternatively, the oxidizing conditions created by rapid electron flow to Fd/NADP may promote the antimycin A-insensitive pathway by favoring the sequential oxidation of both electrons of PQH_2 by the Fe/S center; furthermore, if the oxidation of b-563 at Q_2 depends upon the initial reduction of PQ by reduced Fd, this reaction may be inhibited by the low concentration of reduced Fd present during NADP reduction.

Table 3. Fd-mediated photophosphorylation with increasing NADPH

%NADPH*	O_2 reduction**	NADP reduction**	ATP Synthesis**	P/O (-anti. A)	P/O*** (+anti. A)
0	17	159	528	1.50	1.53
50	43	107	475	1.58	1.55
80	45	69	382	1.68	1.63
90	41	38	322	2.04	1.76
95	38	18	278	2.48	1.86

*[NADP + NADPH] = 2mM
**Rates in μmol O_2/hr·mg chl or μmol ATP/hr·mg chl
***Electron transport and photophosphorylation rates not shown

An intermediate redox poise can be created by increasing the concentration of NADPH during electron flow to Fd/NADP. With 95% NADPH and 5% NADP, a slow rate of NADP reduction is evident (Table 3), and the Fd pool is 70% reduced (data not shown). Under these conditions an elevated P/O value is observed, which is partially affected by antimycin A. This is evidence for the concurrent operation of both the antimycin A-sensitive and antimycin A-insensitive cyclic pathways, as predicted by the model.

REFERENCES

Arnon DI and Chain RK (1977) Ferredoxin-catalyzed photophosphorylations: Concurrence, stoichiometry, regulation and quantum efficiency. Plant and Cell Physiol. Special issue: Photosynthetic organelles, pp. 129-147.
Robinson HH (1980) Electron transport and photophosphorylation in chloroplast membranes: Studies on DBMIB inhibition of plastoquinone function. Thesis, University Microfilms, Ann Arbor, Michigan.
Robinson HH and Yocum CF (1980) Cyclic photophosphorylation reactions catalyzed by ferredoxin, methyl viologen and anthraquinone sulfonate, Biochim Biophys. Acta 590, 97-106.
Yocum CF and Hosler JP (1981) Cyclic electron transfer coupled to phosphorylation. In Selman BR and Selman-Reimer S, eds. Energy coupling in photosynthesis, pp. 35-45. Elsevier North Holland, Inc.

ACKNOWLEDGEMENTS

This research was supported by a grant (PCM8214240) to C.F.Y. from the National Science Foundation.
Authors address: Division of Biological Sciences, The University of Michigan Ann Arbor, MI 48109 USA (CFY); Department of Botany, Duke University, Durham, NC 27706 USA (JPH)

COUPLING BETWEEN REDOX AND ACID-BASE ENERGIES BY CHLOROPLAST CYTO-
CHROME b-559

M. HERVAS, F.F. DE LA ROSA, M.A. DE LA ROSA and M. LOSADA

1. INTRODUCTION

It has been previously proposed (Losada 1978; De la Rosa "et al."
1981) that there are two types of energy-transducing redox photosystems: those which decrease their midpoint redox potential upon
light absorption by the reduced form of the pair, like chlorophyll, and those which increase their midpoint redox potential
upon light absorption by the oxidized form of the pair, like flavins. Thus, these photosystems can promote an uphill transfer of
electrons from appropriate electron donors such as H_2O, in the case of chlorophyll, or EDTA, in the case of flavins (Fontes "et
al." 1981). Similarly, there are two types of acid-base energy-
-transducing photosystems which operate by changing their pKa, rather than their redox potential, upon light absorption either by
the acid or by the basic form of the pair. For example, bacteriorhodopsin, the pigment of halobacteria, decreases its pKa upon
absorption of light by the protonated form of the pair (Stoeckenius, Bogomolni 1982), whereas flavins increase their pKa upon
photoactivation of the unprotonated form of the pair (Schreiner
"et al." 1975).
These electron- or proton-affinity changes seem not to be exclusive of energization by light, since similar changes are also observed in systems such as the respiratory (Wilson, Dutton 1970a;
1970b) and photosynthetic (Wada, Arnon 1971) cytochromes, for
which two redox potential values (high and low) have been determined.
On the basis of the preceeding results and specially of those reported by our group (De la Rosa "et al." 1981; Galván "et al."
1983), according to which, a) the photosynthetic cytochrome b-559
is relatively unstable in its high-potential form, being therefore
easily converted into its low-potential form, and b) such a conversion process is dependent on pH and reversible in the presence
of the uncoupler CCCP at low concentrations, our idea is that this
cytochrome, localized between the two photosystems, is the species
responsible for transducing redox energy into acid-base energy and
eventually into phosphate-bond energy, or ATP, during non-cyclic
photophosphorylation.

2. MATERIAL AND METHODS

Chloroplast suspensions were prepared from spinach leaves by the
method of Arnon and Chain (1977). Cytochrome absorbance measurements were carried out in an AMINCO DW-2a dual wavelength spectrophotometer at 570 and 559 nm as reference and measurement wavelengths, respectively. Redox potentials were determined with a ME-
TROHM-HERISAU E512 potentiometer equipped with a combined
Pt-Ag/AgCl INGOLD electrode, previously calibrated against a saturated solution of quinhydrone (E_O', pH 7 = +280 mV). The titrations
were carried out in a 3-ml cell thermostatized at $20°$.

Sybesma, C. (ed.), Advances in Photosynthesis Research, Vol. II. ISBN 90-247-2943-2.
© *1984 Martinus Nijhoff/Dr W. Junk Publishers, The Hague/Boston/Lancaster.*

3. RESULTS AND DISCUSSION

As recently reported (De la Rosa "et al." 1981), fresh chloroplasts, capable of phosphorylating, contain about 2/3 of their cytochrome b-559 in its high-potential form (E_0', pH 7 = +340 mV),
which is reducible by hydroquinone and dithionite, while the remaining 1/3 occurs in its low-potential form (E_0', pH 7 = +130 mV),
which is reducible only by dithionite. However, if chloroplasts
are heated at 50° for 5 min or slightly sonicated -treatments that
produce the total conversion of cytochrome b-559 into its low-potential form-, they are unable to phosphorylate any longer. Furthermore, CCCP, a well-known uncoupler of photophosphorylation,
produces, at high concentrations (30 μM), irreversible conversion
of the high-potential form into the low-potential one. However,
CCCP, at lower concentrations (10 μM), promotes the reversible
conversion between both forms. These facts seem to be, in principle, significant enough so that cytochrome b-559 should be directly implicated in the photophosphorylation process. It was also
found that the midpoint potential of the two forms of cytochrome
b-559 differently depends on pH, and so whereas the high-potential
form presents the same midpoint potential value between pH 6.5 and
pH 8.5, that of the low-potential form changes at the rate of
about -60 mV per unit of pH (De la Rosa "et al." 1981).

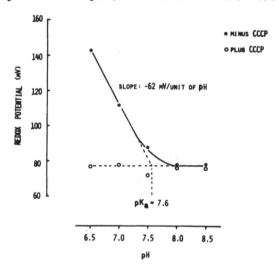

Fig. 1. **Effect of CCCP on the pH-dependence of the low midpoint
potential of cytochrome b-559 from spinach chloroplasts.** Midpoint
redox potentials of several chloroplast suspensions (50 μg chloro-
phyll/ml) were determined at different pH values by titration with
potassium ferricyanide and dithionite as oxidant and reductant,
respectively. The following redox mediators were used: 20 μM
2,3,5,6-tetramethyl-p-phenylenediamine, 20 μM 1,2-naphthoquinone,
and 20 μM duroquinone. Buffers were potassium phosphate (pH 6.5,
and 7) and 50 mM tricine-KOH (pH 7.5, 8.0, and 8.5). Chloroplast
suspensions were supplemented, when indicated, with 33 μM CCCP.

In this paper we report the results obtained when such a dependence on pH of the low midpoint potential was studied in the presence of the uncoupler CCCP. As can be seen in Fig. 1, the midpoint potential value of the low-potential form becomes then independent of pH, thus suggesting that cytochrome b-559 can transduce redox energy into phosphate-bond energy through an intermediate proton gradient. Ammonium, however, another very efficient uncoupler of photophosphorylation, does not affect, in contrast to CCCP, either the interconversion between both forms or the pH-dependence proper to the low-potential form (data not shown).

In consequence, it is likely that CCCP acts as an uncoupler by avoiding protonation of the cytochrome and thus impeding formation of a proton gradient across the thylakoid membrane. In contrast, ammonium would have a different mechanism of action, presumably exerting its uncoupling effect on the F_0F_1-ATPase itself.

The above results reinforce our present point of view according to which cytochrome b-559 operates not only as a redox energy-transducing system at two alternate potentials but also as an acid-base energy-transducing system at two alternate pKa's. In agreement with the proposed mechanism (Fig. 2), the two redox pairs of cyto-

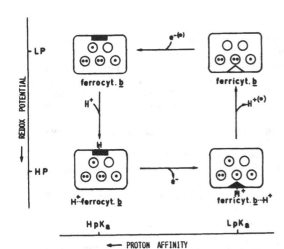

Fig. 2. **Diagrammatic representation of the reversible transduction of redox energy into acid-base energy through electronic energy by the cytochrome b-559 system.** (For details see the text).

chrome b-559 (high and low potentials) share the same reduced form but present two different oxidized forms; namely, the one, corresponding to the low-potential pair, unenergized and unprotonated, and the other, corresponding to the high-potential pair, electronically energized and protonated. Bassically, this system may function as follows: in its low-potential (LP) state, the oxidized and unenergized cytochrome (ferricyt.b) accepts isopotentially from its donor one electron (e*) in a high-energy orbital, and concomitantly accepts a proton at high pKa (H^+), thus becoming not only reduced but also protonated (H^+-ferrocyt.b). In its high-potential (HP) state, the reduced and protonated cytochrome donates isopotentially one electron (e) from a low-energy orbital to its acceptor, thus remaining protonated and becoming oxidized and electronically energized (ferricyt.b*-H$^+$). Oxidation at high potential of the protonated and reduced form implies proton translocation from the acid-base group of the reduced form (high pKa), which loses affinity upon oxidation, to the acid-base group of the resulting energy-rich oxidized form, which concomitantly gains affinity (high pKa*, not shown). The energy-rich oxidized and protonated cytochrome may dissociate afterwards its proton (H$^+$ $^{(*)}$) at low pKa to its basal oxidized form, thus becoming again deenergized and deprotonated and closing the cycle. This proton transfer at low pH is accompanied by the electron fall from the high- to the low-energy orbital. The ferricytochrome $\underline{d}$ electrons have been provisionally located -within the most stable arrangement- in two energy levels separated by the ligand-field splitting energy.

It is important to stress that the crucial requirement in the coupling mechanism is for a common intermediate between the acid-base and phosphate-bond systems which has to be able to occur in both an electronically energized state and a stabilized basal one, the energy difference between these two states being actually the energy transduced by the integrated system.

REFERENCES

Arnon DI and Chain RK (1977) in Photosynthetic Organelles, Spec. iss. of Plant Cell Physiol., pp. 129-147.

De la Rosa FF, Galván F and Losada M (1981). In Akoyunoglou GA, ed. Proc. 5th Int. Congr. on Photosynthesis, Vol. 2, pp. 531-541. Balaban International Science Services, Philadelphia.

Fontes AG, De la Rosa FF and Gómez-Moreno C (1981) Photobiochem. Photobiophys. 2, 355-364.

Galván F, De la Rosa FF, Hervás M and Losada M (1983) Bioelectrochem. Bioenerget. (in press).

Losada M (1978) Bioelectrochem. Bioenerget. 5, 296-310.

Schreiner S, Steiner U and Kramer HEA (1975) Photochem. Photobiol. 21, 81-86.

Stoeckenius W and Bogomolni RA (1982) Ann. Rev. Biochem. 52, 587-616.

Wada K and Arnon DI (1971) Proc. Nat. Acad. Sci. USA 68, 3064-3068.

Wilson DF and Dutton PL (1970a) Biochem. Biophys. Res. Comm. 39, 59-64.

Wilson DF and Dutton PL (1970b) Arch. Biochem. Biophys. 136, 583-584.

ACKNOWLEDGEMENTS

This work was supported by grants from INSALUD, Comisión Asesora de Investigación, Ministerio de Industria and Philips Laboratories.

Authors address: Departamento de Bioquímica, Facultad de Biología y C.S.I.C., Universidad de Sevilla, Aptdo. 1095, Sevilla, Spain.

THE EFFECT OF FLASH-INDUCED ATP-HYDROLYSIS ON THE 515 ABSORBANCE CHANGE

ULRICH SIGGEL/TU BERLIN

1. INTRODUCTION

A specific acceleration of the dark decay of the 515 nm absorbance change
has been reported to occur by addition of ADP plus phosphate after
excitation with one flash of 8 ms (Rumberg, Siggel 1968) as well as with
a group of 4 flashes of 15 us duration (Junge et al. 1970). Later a similar
acceleration by ATP was found (Girault, Galmiche 1976). Accordingly two
alternative interpretations have been given. Either the acceleration
reflects the additional proton flux through the ATPase, coupled to the
reaction of phosphorylation. Or it is caused by a general increase of
proton conductivity of the membrane due to the enhanced internal proton
concentration, brought about by the hydrolysis of added or newly formed ATP.
(Girault, Galmiche 1980). In this paper evidence is presented that the
acceleration has different reasons for ATP and ADP plus Pi.

2. MATERIALS AND METHODS

Freshly prepared thylakoids from market spinach were used with very low
ATPase activity in the dark.
The 515 nm absorbance change was measured with a single beam absorption
spectrophotometer without signal averaging.
Thylakoids were suspended in a medium containing 10 mM tricine (pH 8 or 8.5),
3 mM $MgCl_2$ (eventually plus 20 mM KCl) and 15 uM benzylviologen (or 0.5 mM
ferricyanide) to a final concentration of 10 uM. ADP, ATP and phoshpate
were added when indicated.

3. RESULTS

3.1. Comparison of the effects of ATP and ADP plus Pi

In tightly coupled thylakoids the dark decay of the 515 nm absorbance
change after the first excitation with 10 to 200 ms of saturating red
light is rather peculiar in the presence of some 0.1 mM ATP. The signal
is only partly reversible. The halftime of the multiphasic reversible
portion is smaller than that of the total signal without ATP. After the
second flash the pattern is different. The irreversible phase is no more
present. As a result the signal is smaller and looks as if accelerated
(fig. 1, top). In the ideal case the reversible portions are identical
for the first and second flash. It was not possible to regain the initial
dark state by dark times up to 15 minutes.
This behaviour is not paralleled under conditions optimal for phosphorylation.
The dark decay is somewhat slower than with ATP (in the 2. flash). But the
acceleration relative to the decay without phosphate is already present
in the first flash of 10 ms and does not change in consecutive flashes,
if the darktime between the flashes is sufficiently large to prevent the
formation of an appreciable pH-gradient (fig. 1, bottom).
This difference between the two substrates has been overlooked hitherto
as a consequence of signal averaging.

Sybesma, C. (ed.), Advances in Photosynthesis Research, Vol. II. ISBN 90-247-2943-2.
© *1984 Martinus Nijhoff/Dr W. Junk Publishers, The Hague/Boston/Lancaster.*

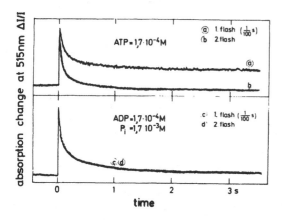

Fig. 1.
Time course of the 515 nm
absorption change. Excitation
with saturating flashes of
10 ms duration. pH of the
medium is 8.5

3.2. More details of the ATP effect

1. If the first flash is given in the absence of ATP, this has no effect on
   the kinetics of the next flash after addition of ATP.
2. The magnitude of the irreversible phase is dependent on the peak value of
   the light-induced potential change, but not on the duration of excitation.
   It does not increase if the flash is prolonged from 20 to 200 ms.
3. The irreversible phase has the normal difference spectrum of the electrochromic absorption change.
4. The ATP effect is eliminated by the energy transfer inhibitor phloridzin.
5. Magnesium ions are obligatory for the ATP effect. Phosphate and especially
   ADP are inhibitory.
6. Ageing of the thylakoids increases the reversibility of the first 515-
   signal and decreases the acceleration of the second.
7. Preillumination (2-5 s) of the thylakoid suspension prior to measurement
   enhances the acceleration of the dark decay. This is also true for slightly
   aged thylakoids, for which the difference between the first and second
   flash may no more be visible.

3.3. Kinetic analysis of the dark decay

The dark decay of the 515 absorbance change after a photoshutter flash
may be resolved into 3 exponential phases, a slow, a fast and a very fast
one. Table 1 shows how ATP and ADP plus Pi change the kinetics of dark
relaxation in the absence of nucleotides.

phase / addition	-	ATP	ADP+Pi
slow	600-700	350-400	500
fast	100	40-50	60-80
very fast	15	10-15	15-20

Table 1.
Halftimes/ms of the 3
exponential phases as
dependent on addition
of substrates

The slow and the fast phase are more effectively accelerated by ATP than by
ADP plus Pi. ATP reduces the magnitude of the very fast phase.

3.4. Analysis of the effects of ADP

A tentative assignment of the different kinetic phases seems possible on the basis of some older experiments on the ADP effects (Siggel, Spetsai Symposium 1977), which were later confirmed in part in flash group experiments (Gräber et al. 1981). ADP has 2 effects on the dark decay of the 515 change. It decelerates at low (phosphate is not necessary) and accelerates at higher concentration (Pi is necessary). These effects are known since a long time for the stationary electron transport (Avron et al.,1958).

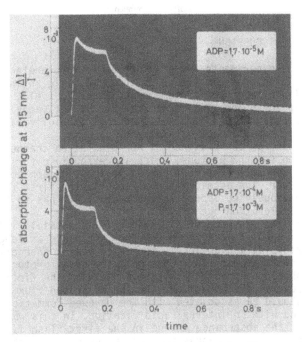

Fig. 2.
Time course of the 515 nm absorption change in the absence (top) and presence (bottom) of phosphorylation. pH=8.4 (KCl present)

Fig. 2 shows the slowest and the fastest dark relaxation after excitation with a flash of some 150 ms. It may be resolved into mainly 2 kinetic phases, the very fast phase being rather small in the thylakoid preparations used. The variation with the ADP concentration of the kinetic constant k_s of the fast phase is strictly parallel to that of the electron transport rate (figs. 3 and 4). As stationary electron transport is determined by the proton efflux, the fast phase should also be correlated with proton fluxes. The slow phase should then correspond to the fluxes of ions other than protons.

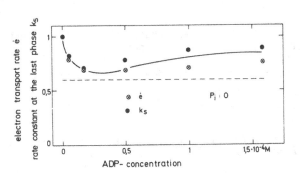

ADP- concentration

Fig. 3.
Variation of stationary electron transport and kinetic constant of the fast phase with ADP in the absence of phosphate.

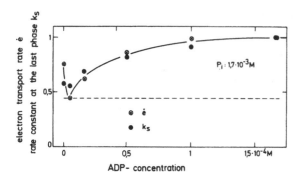

Fig. 4.
Variation of stationary electron transport and kinetic constant of the fast phase of the 515 dark decay with ADP in th presence of phosphate.

4. DISCUSSION

The experiments reported above are interpreted in the following way. The chloroplast ATPase is known to be in an inactive state in the dark. The rise of the transmembrane electric field, as revealed by the 515 change, up to the peak value of some 200 mV activates and energizes the ATPases partly. It has to be assumed that the activation persists some time after the cessation of excitation.

Then, in the presence of ADP plus Pi, a channel through the ATPase is open for the protons, which is responsible for the accelerated dark decay (especially the fast phase). Without phosphate the proton permeability is lower. The acceleration already in the first flash indicates that obviously no appreciable amount of ATP has to be formed before the acceleration occurs.

In the presence of ATP hydolysis is possible in principle for the activated ATPases. It can, however, not take place as long as the electric field stays at a high level, the latter being energetically opposed to an inward proton flux. Thus only the basal proton flux out of the thylakoid should be possible. With the decay of the field the proton flux should reverse its direction. The inward proton flux due to ATP hydrolysis should be associated with the renewed formation of the electric field and with an internal acidification. In this way the irreversible portion of the absorbance change in the first flash is understood. For the consecutive flashes the initial conditions are different because of the enhanced internal proton concentration and some persisting electric field.

5. CONCLUSION

The acceleration of the dark decay of the 515 absorbance change has differen reasons according to the reaction conditions. In the presence of ADP+Pi it reflects the additional proton flux through the ATPase. In the presence of ATP the enhanced internal proton concentration due to ATP hydrolysis especially accelerates the proton efflux in consecutive flashes.

REFERENCES

Authors address: Ulrich Siggel/Max-Volmer-Institut, Technische Universität, Straße des 17 Juni 135, D-1000 Berlin 12, FRG

ATP SYNTHESIS CATALYZED BY RECONSTITUTED CF_oF_1 LIPOSOMES DRIVEN BY AN ARTIFICIALLY GENERATED $\Delta\Psi$ AND ΔpH

PETER GRÄBER, MATTHIAS RÖGNER, DIETRICH SAMORAY[+] and GÜNTHER HAUSKA[+]

1. INTRODUCTION

The mechanism of coupling between proton transport and ATP synthesis is still not known. One approach to study this mechanism is to simplify the energy transducing apparatus of biological membranes and to use biochemically well-defined systems for analysis. The chloroplast ATPase CF_oF_1 can be isolated, purified and reconstituted into liposomes (Pick and Racker, 1979). With such a system the kinetic analysis of proton transport coupled ATP synthesis is possible.

2. MATERIALS AND METHODS

The chloroplast coupling factor CF_oF_1 was prepared as described by Pick and Racker, 1979. Fig. 1 shows the SDS gel electrophoresis pattern (12.5% acrylamide) of the preparation (Samoray and Hauska, unpublished). This protein was reconstituted into asolectin liposomes according to the procedure of Sone et al., 1977. The reconstituted liposomes were energized by an external electric field pulse (20 kV cm^{-1}, duration 200 μs, reaction medium: 1 mM Na_2HPO_4, 1 mM $MgCl_2$, 0.1 mM ADP, 0.3 mM Tricine-NaOH, pH 8, liposomes giving a final protein concentration of 100 μg ml^{-1})(Gräber et al., 1982). The reconstituted liposomes have been energized also by an acid-base transition supplemented by a K^+/valinomycin diffusion potential with a rapid mixing quenched flow device (Schatz et al., 1981; Junesch and Gräber, 1983). Reaction medium I contained 7.5 mM Na-succinat, pH 4.86, 5 mM $NaHPO_4$, 2 mM $MgCl_2$, 1 μM valinomycin, and liposomes giving a protein concentration of 50 μg ml^{-1}. This was rapidly mixed with reaction medium II giving final concentrations of 100 mM tricine-KOH, pH 8.2, 5 mM NaH_2PO_4, 2 mM $MgCl_2$, 0.1 mM ADP and 1 μM valinomycin. After the reaction period (0.1–10 s) the reaction was terminated by mixing with 4% TCA. The ATP concentration was measured with the luciferin/luciferase technique (LKB kit).

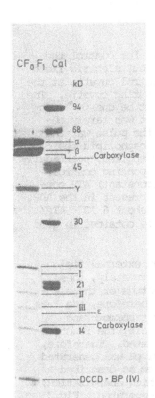

Fig. 1:

SDS gel electrophoresis pattern of the CF_oF_1 preparation and calibration proteins (Cal). A small amount of carboxylase can be seen. $\alpha,\beta,\gamma,\delta,\epsilon$ refer to CF_1 subunits; I,II,III,IV to CF_o subunits.

Sybesma, C. (ed.), Advances in Photosynthesis Research, Vol. II. ISBN 90-247-2943-2.
© *1984 Martinus Nijhoff/Dr W. Junk Publishers, The Hague/Boston/Lancaster.*

3. RESULTS

Fig. 2 shows the ATP concentration in a suspension of liposomes reconstituted with CF_OF_1 as a function of the number of external field pulses. The ATP concentration was measured by taking 10 μl of the suspension and analyzing for ATP with luciferin/luciferase. The time between successive field pulses was about 1 min. The ATP level, when no pulse was applied, corresponds to the ATP background in the ADP and in the CF_OF_1 preparation.

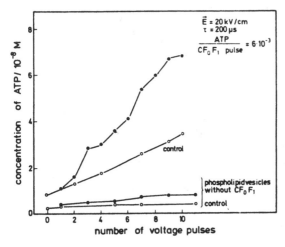

Fig. 2:

ATP concentration in a sample of reconstituted CF_OF_1 liposomes subjected to external electric field pulses (full circles). Open circles: control sample without field pulses. The lower curves show the same experiments using liposomes without CF_OF_1.

Each pulse produces an increase in ATP concentration. For control the ATP level was measured in an identical sample but without electric field pulses. Here the samples have been taken from the control cuvette at the same time as the voltage pulse was applied to the measuring cuvette. The increase of the ATP level without electric pulses might be due to a contamination of this preparation with adenylatekinase but was rarely observed in our preparations. The amount generated by the pulse was the difference between pulsed sample and control. When phospholipid vesicles without CF_OF_1 were prepared and electric field pulses applied, practically no increase of the ATP level was observed. The corresponding control sample (without pulsing) shows practically the same (constant) ATP concentration which here represents the background level present in the ADP. The average ATP yield per pulse in this experiment is about $6 \cdot 10^{-3}$ ATP/ $CF_OF_1 \cdot$ pulse and values up to 0.1 ATP/$CF_OF_1 \cdot$ pulse can be obtained in good preparations.

These experiments show that ATP can be generated by an external field pulse applied to reconstituted CF_OF_1 liposomes. Unfortunately, these data cannot be analyzed quantitatively: (1) the magnitude of the electric field is not constant but decays exponentially; therefore, the rate of ATP synthesis cannot be calculated. (2) Since the transmembrane electric field varies from pole to equator of the vesicle, the functional dependence of the rate on $\Delta\Psi$ also cannot be evaluated. Therefore, ATP synthesis driven by ΔpH was investigated. The Δ pH was generated by an acid-base transition using a quenched flow system as described earlier (Schatz et al., 1981). In order to obtain a high energization, an additional K^+/valinomycin diffusion potential was generated. Fig.3

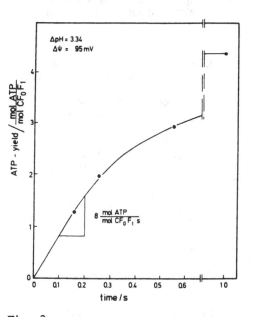

Fig. 3:

ATP yield in an acid-base transition supplemented by a K^+/valinomycin diffusion potential using reconstituted CF_oF_1 liposomes as a function of reaction time.

shows the amount of ATP synthesized by the reconstituted CF_oF_1 liposomes as function of the reaction time. If all protein used in the reconstitution procedure is active for ATP synthesis, each ATPase should carry out 4 turnovers in this experiment (ATP yield after 10 s). From the slope at small reaction times a rate of 8 ATP/CF_oF_1·s can be calculated. This rate is small compared to the rate obtained under similar conditions with chloroplasts which amounts to approximately 327 ATP/CF_oF_1·s (Junesch and Gräber, 1983). However, the rate measured here is by a factor of at least 20 higher than the rate measured in other reconstituted systems (Hauska et al., 1980; Dewey and Hammes, 1981).

It can be concluded from the above results that (1) no proteins other than the coupling factor complex are necessary to catalyze proton transport coupled ATP synthesis and (2) the existence of special proton pathways in the natural thylakoid membrane (Williams, 1961, Kell, 1979) which might connect the electron transport chain with the ATPase is not essential for the coupling mechanism - at least not in this model system. The latter statement, however, is experimentally verified at present only for rates up to 2.5 per cent of the maximal rate. The higher rates observed with thylakoids are assumed to be due to a partial desactivation of the ATPase during the isolation procedure and to a non-optimal reconstitution procedure. However, it also might be taken into consideration that the existence of special proton pathways at or in the thylakoid membrane allows a faster proton supply into the "mouth" of the CF_O part of the ATPase than in the reconstituted system where the protons come from the aqueous inner bulk phase and must cross the water-membrane interface. This latter process might impose strong kinetic limitations on the ATP synthesis rate. Before final conclusions can be drawn about this point, the isolation and reconstitution procedures must be optimized. However, using this simplified model system, the kinetics of ATP synthesis/hydrolysis coupled with proton translocation can be investigated with a biochemically well characterized system und well-defined experimental conditions. If, for example, the reconstituted system is energized by an external electric field or an acid-base transition, the system is "forced" to behave chemiosmotically in an orthodox sense (Mitchell, 1961): the proton translocation must occur from the inner aqueous bulk phase to the outer one. Since other parameters are exactly known (pH_{in}, pH_{out}, ΔG_p, etc.) analysis of energetics and kinetics of coupled ATP synthesis is possible.

4. REFERENCES

Dewey TG and Hammes GG (1981) Steady state kinetics of ATP synthesis and hydrolysis catalyzed by reconstituted chloroplast coupling factor, J. Biol. Chem. 256, 8941–8946

Gräber P, Rögner M, Buchwald HE, Samoray D and Hauska G (1982) Field-driven ATP synthesis by the chloroplast coupling factor complex reconstituted into liposomes, FEBS Lett. 145, 35–40

Hauska G, Samoray D, Orlich G and Nelson, N (1980) Reconstitution of the photosynthetic energy conservation, Eur. J. Biochem. 111, 535–543

Junesch U and Gräber P (1983) The rate of ATP synthesis as function of ΔpH and $\Delta \Psi$ in preactivated and non-preactivated chloroplasts. Proc. VIth Intern. Congr. Photosynth., M. Nijhoff/Dr. W. Junk Publ., The Hague

Kell DB (1979) On the functional proton current pathway of electron transport phosphorylation. An electrodic view. Biochim. Biophys. Acta 549, 55–99

Mitchell P (1961) Coupling of Phosphorylation to electron and hydrogen transfer by a chemiosmotic type of mechanism, Nature 191, 144–148

Pick U and Racker E (1979) Purification and reconstitution of the N,N'-Dicyclohexylcarbodiimide sensitive ATPase complex from spinach chloroplasts, J. Biol. Chem. 254, 2793–2799

Schatz GH, Schlodder E and Gräber P (1981) Kinetics of ATP synthesis and AdN exchange in chloroplasts measured by rapid acid-base transitions. In: Proc. 5th Intern. Congr. Photosynth., Kallithea (Akoyunoglou, G., ed.), Vol. II, Balaban Intern. Sci. Serv., Philadelphia, pp. 945–954

Sone N, Yoshida M, Hirata H and Kagawa Y (1977) Reconstitution of vesicles capable of energy transformation from phospholipids and adenosine triphosphatase of a thermophilic bacterium, J. Biochem. 81, 519–528

Williams RJP (1961) Possible functions of chains of catalysts, J. Theor. Biol. 1, 1–13

5. ACKNOWLEDGEMENTS

This work was supported by the Deutsche Forschungsgemeinschaft.

Authors Address: Max-Volmer-Institut für Biophysikalische und Physikalische Chemie, Technische Universität Berlin, Strasse des 17. Juni 135, 1000 Berlin 12

+ Lehrstuhl für Botanik I, Universität Regensburg, Universitätsstr. 31, D-8400 Regensburg

THE RATE OF ATP-SYNTHESIS AS FUNCTION OF ΔpH AND $\Delta\Psi$ IN PREACTIVATED AND NON-PREACTIVATED CHLOROPLASTS

ULRIKE JUNESCH and PETER GRÄBER

1. INTRODUCTION

ATP can be synthesized in chloroplasts without electron transport when the membranes are energized either by an acid-base transition (Jagendorf and Uribe, 1966) or an external electric field pulse (Witt et al. 1976). For kinetic investigations of the coupling of ATP synthesis with proton translocation the energization with an acid-base transition and limitation of the reaction time to the millisecond range has several advantages:
1. The transmembrane ΔpH is exactly known and does not change during the reaction period.
2. The transmembrane $\Delta\Psi$ can be made zero or adjusted to defined values during the short reaction period with a K^+/valinomycin diffusion potential.
3. Due to the short reaction period, the concentrations of the substrates ADP and P_i remain constant at their initial values. If the ATP concentration is measured also the phosphate potential during the reaction is known.

With this method the reaction kinetics can be investigated under well defined experimental conditions. It has been proposed that the chloroplast ATPase is in an inactive state under non-energized conditions and is converted into an active state upon energization (Gräber et al., 1977). If this is true, the dependence of the rate of ATP synthesis on ΔpH reflects the dependence of the fraction of active ATPases on ΔpH. If the ATPases are activated (e.g., by illumination) and then modified by reaction with DTT so that the activated state is maintained for a longer time, the "true" dependence of the rate of phosphorylation should be observed.
The aim of this work is
(a) to investigate the rate of ATP synthesis as a function of ΔpH and $\Delta\Psi$;
(b) to measure the influence of a preactivation and modification of the ATPase with DTT on the functional dependence of the rate on ΔpH.

2. MATERIALS AND METHODS

The acid-base transition experiments were carried out with a Durrum D133 quenched flow apparatus as described earlier (Schatz et al., 1981). Spinach chloroplasts were isolated according to standard procedures. They were incubated for 25 sec in reaction medium I ($\Delta\Psi$:6 mV: 6.7 mM succinic acid, 3-13 mM KOH pH 4-6, 77-67 mM KCl, 3.3 mM NaCl, 5 mM NaH$_2$PO$_4$, 2 mM MgCl$_2$, 40 µM Chl, 10 µM DCMU, 0.7 µM valinomycin; $\Delta\Psi$:55 mV: 6.7 mM succ., 3-13 mM NaOH pH 4-6, 6 mM KCl, 3.3 mM NaCl, 5 mM NaH$_2$PO$_4$, 2 mM MgCl$_2$, 40 µM Chl, 10 µM DCMU, 0.7 µM valinomycin; $\Delta\Psi$:85 mV: 6.7 mM succ., 3-17 mM NaOH pH 4-7, 6 mM KCl, 3.3 mM NaCl, 5 mM NaH$_2$PO$_4$, 2 mM MgCl$_2$, 40 µM Chl, 10 µM DCMU, 0.7 µM valinomycin) and then rapidly (2 ms) mixed with reaction medium II (200 mM Tricine, 120 mM KOH pH 8.2, 5 mM NaH$_2$PO$_4$, 2 mM MgCl$_2$, 10 µM DCMU). The reaction was allowed to proceed in the age mode for 50 ms - 10 s and was terminated by rapidly mixing

Sybesma, C. (ed.), Advances in Photosynthesis Research, Vol. II. ISBN 90-247-2943-2.
© *1984 Martinus Nijhoff/Dr W. Junk Publishers, The Hague/Boston/Lancaster.*

(2 ms) with 4% TCA. The ATP concentration in the supernatant was then measured with Luciferin/Luciferase (LKB Instruments) using an internal ATP standard.

3. RESULTS AND DISCUSSION

Fig. 1 shows the time course of ATP synthesis in a Δ pH pulse experiment with the transmembrane Δ pH of 3.2 and additional diffusion potentials of Δ Ψ = 6 or 55 or 85 mV. The slopes of these curves directly give the rate of ATP synthesis and within the first 100-200 ms the rate is constant. This corresponds to a constant energization of the membrane. After this time range the energization decreases and after a long time (10 s) only the ATP yield can be measured.

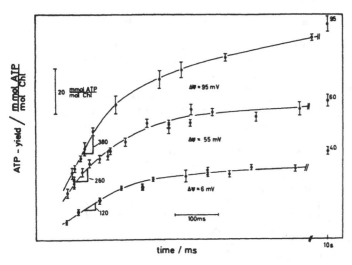

Fig. 1:

ATP yield in an acid-base transition supplemented with K$^+$/valinomycin diffusion potential as a function of reaction time. The numbers give the rate in mMol ATP/Mol Chls.

The following investigations are carried out in the range where the rate of ATP synthesis (and the energization) was constant. Fig. 2 shows the

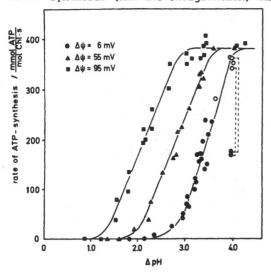

Fig. 2:

Rate of ATP synthesis as function of Δ pH and Δ Ψ as calculated from the linear parts of curves in Fig. 1 and additional sets of experiments. Open symbols refer to corrected measurements, all full symbols are direct measurements.

rate of ATP synthesis (determined in the linear range from the slopes
of the curves shown in Fig. 1 and similar sets of experiments) as a
function of Δ pH. The pH$_{out}$ was kept constant at 8.2. When chloroplasts
are incubated at pH < 4.5 a partial denaturation has been observed. By
appropriate control measurements this effect can be corrected (Gräber et
al., 1983). The open circles refer to corrected data, the dashed lines
connect original data with the corrected ones.

The magnitude of the diffusion potentials was calculated according to
the Goldmann-Hodgkin-Katz equation using literature data of the permeability coefficients (Barber, 1972; Vredenberg, 1976; Muhle, 1973). The
curves have a sigmoidal shape showing a maximal rate of 380 mM ATP/Mol
Chl s. According to the chemiosmotic theory (Mitchell, 1966) Δ pH and
$\Delta \Psi$ should be additive so that the driving force for ATP synthesis is
the electrochemical potential difference of protons $\Delta \tilde{\mu}_H^+ = -(2.3\ RT \Delta pH +
F \Delta \Psi)$, i.e., the same rate of ATP synthesis should be obtained regardless
of the relative contributions of Δ pH or $\Delta \Psi$.

Fig. 3 shows a plot of the rate of ATP synthesis as a function of $\Delta \tilde{\mu}_H^+$
and within the experimental error all data can be described by a single
curve.

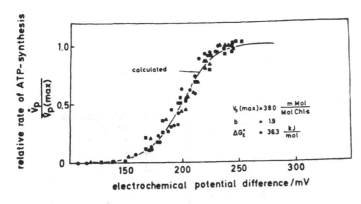

Fig. 3:

Rate of ATP syn-
thesis as a
function of $\Delta \tilde{\mu}_H^+$
as calculated
from data in
Fig. 2. The
solid line is
calculated as
described in the
text.

It has been proposed (Gräber et al., 1977) that the rate of ATP synthesis
is regulated by the fraction of active ATPases and that this fraction depends on the energization of the membrane. A quantitative formulation
based on a simple equilibrium model gives (Gräber and Schlodder, 1981)

$$E_i + b\ H_{in}^+ \rightleftharpoons E_a + b\ H_{out}^+ \tag{1}$$

$$\frac{E_a}{E_t} = \frac{e^x}{1+e^x} \ ; \ x = \ln K_E^O - \frac{b \Delta \tilde{\mu}_H^+}{RT} \tag{2}$$

E_a = active ATPase; E_t = total number of ATPases; K_E^O = equilibrium constant of reaction (1) at $\Delta \tilde{\mu}_H^+ = O$; b = number of protons involved in
the activation process. The solid line in Fig. 3 is calculated from
equation (2) with the parameters b = 1.9 and $K_E^O = 3.7\ 10^{-7}$. This

implies that under non-energized conditions approximately $4 \cdot 10^{-5}$ per cent of the ATPases are in the activated state. Also, literature data of light-induced phosphorylation as function of Δ pH can be fitted by this model using only two parameters (Schlodder et al., 1982), and we conclude that usually the functional dependence of the activation process on membrane energization is measured when the rate of ATP synthesis is measured as a function of Δ pH (Pick et al., 1974; Portis and McCarty, 1974; Gräber and Witt, 1974; Rumberg and Heinze, 1980).

If similar experiments are carried out with chloroplasts where the ATPases are already activated, a different functional dependence of the rate of phosphorylation on Δ pH is expected (Gräber and Schlodder, 1981). Such an activation can be carried out by preillumination in the presence of DTT. Under these conditions the ATPase is activated and modified in such a way that the lifetime of the activated state is increased (Carmeli, 1970; Carmeli and Lifshitz, 1972; Bakker-Grunwald and van Dam, 1973,1974). Qualitative evidence for such an effect has been obtained from measurement of the flash-induced phosphorylation (Harris and Crofts, 1978) and from measurements of the ATP yield in an acid-base transition with preactivated and with non-preactivated chloroplasts (Mills and Mitchell, 1982).

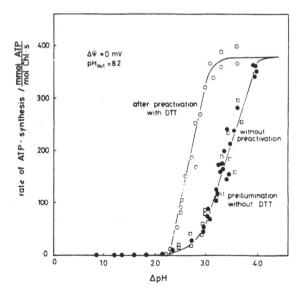

Fig. 4:

Rate of ATP synthesis as function of Δ pH. Full circles: same data as in Fig. 2. Open circles: rates obtained after pre-illumination in the presence of DTT; open squares: rates obtained after preillumination without DTT.

Fig. 4 shows the functional dependence of the rate of ATP synthesis on Δ pH obtained with preactivated and non-preactivated chloroplasts. A preillumination without DTT does not influence the rate of ATP synthesis measured after the preillumination in an acid-base transition (compare open and closed symbols). Whereas, the same maximal rate of ATP synthesis is found with or without preactivation, the curve after preactivation is shifted to lower Δ pH values and the slope of the curve is steeper. Especially the slope at small rates is very steep and constant in contrast to the non-preactivated curve. The crossing point with the zero line represents the equilibrium between protonmotive force and the phosphate potential. At lower Δ pH, ATP hydrolysis occurs. However, due to technical difficulties (the problem is to measure coupled ATP hydrolysis at a low ATP concentration in a short reaction period so that the experimentally indicated Δ pH remains constant), at present we can say only that ATP hydrolysis occurs but we cannot experimentally define the Δ pH to which this rate belongs. At this equilibrium point a slight displacement to higher or lower Δ pH drives the reaction either to synthesis or hydrolysis. This allows a simple determination of the energetics and it results:

$$\Delta G_p = 34.4 \pm 0.4 \text{ kJ/Mol; } pH_{eq} = 2.25; \quad n = \frac{\Delta G_p}{2.3 \text{ RT } \Delta pH} = 2.7$$

This is in accordance with earlier results obtained from ATP yield measurements in acid-base transitions (Hangarter and Good, 1982). Extrapolation of the non-preactivated curve to zero rate causes some ambiguities because of the strong non-linearity at small rates. In this respect the preactivated curve behaves as commonly expected from non-equilibrium thermodynamics, i.e., there exists a linear relation between ATP synthesis rate and driving force ($\Delta G_p + n$ RT $\Delta \widetilde{\mu}_H +$) near equilibrium.

4. REFERENCES

Bakker-Grunwald T and van Dam K (1973) The energy level associated with the light-triggered Mg^{2+}-dependent ATPase in spinach chloroplasts, Biochim. Biophys. Acta 292, 808–814

Bakker-Grunwald T and van Dam K (1974) On the mechanism of activation of the ATPase in chloroplasts, Biochim. Biophys. Acta 347, 290–298

Barber J (1972) Stimulation of millisecond delayed light emission by KCl and NaCl gradients as means of investigating the ionic permeability properties of the thylakoid membranes, Biochim. Biophys. Acta 275, 105–116

Carmeli C (1970) Proton translocation induced by ATPase activity in chloroplasts, FEBS Lett. 7, 297–300

Carmeli C and Lifshitz Y (1972) Effects of P_i and ADP on ATPase activity in chloroplasts, Biochim. Biophys. Acta 267, 86–95

Gräber P and Witt HT (1974) On the relations between the transmembrane electrical potential difference, pH gradient and ATP formation in photosynthesis. In: Proc. 3rd Int. Congress Photosynth. Res. (M. Avron, ed., Rehovot, Israel) Vol I, pp. 427–436

Gräber P and Schlodder E (1981) Regulation of the rate of ATP synthesis/hydrolysis by Δ pH and $\Delta \psi$. In: Photosynthesis II (Akoyunoglou, E., ed.) Balaban Intern. Sci. Serv., Philadelphia, pp. 867–879

Gräber P, Schlodder E and Witt HT (1977) Conformational change of the chloroplast ATPase induced by a transmembrane electric field and its correlation to phosphorylation, Biochim. Biophys. Acta 461, 426–440

Gräber P, Junesch U and Schatz GH (1983) Coupling of ATP synthesis and proton translocation in chloroplasts, Ber. Bunsenges. Physik. Chem., in press

Hangarter RP and Good NE (1982) Energy thresholds for ATP synthesis in chloroplasts, Biochim. Biophys. Acta 681, 397–404

Harris DA and Crofts AR (1978) The initial stages of photophosphorylation studies using excitation by saturating short flashes of light,

Biochim. Biophys. Acta 502, 87–102

Jagendorf AT and Uribe E (1966) ATP formation caused by acid–base transition of spinach chloroplasts, Proc. Natl. Acad. Sci. USA 55, 170–177

Mills JD and Mitchell P (1982) Thiol modulation of CF_o–CF_1 stimulates acid–base dependent phosphorylation of ADP by broken pea chloroplasts, FEBS Lett. 144, 63–67

Mitchell P (1966) Chemiosmotic coupling in oxidative and photosynthetic phosphorylation, Biol. Rev. 41, 445–502

Muhle H (1973) Thesis: Untersuchungen über den Ionentransport bei der Photosynthese, Techn. Univ. Berlin

Pick U, Rottenberg H and Avron M (1974) The dependence of photophosphorylation in chloroplasts on Δ pH and external pH, FEBS Lett. 48, 32–36

Portis AR, McCarty RE (1974) Effects of adenine nucleotids and of photophosphorylation on H^+ uptake and the magnitude of the H^+ gradient in illuminated chloroplasts, J. Biol. Chem. 249, 6250–6254

Rumberg B and Heinze T (1980) Kinetic analysis of proton transport coupled ATP synthesis in chloroplasts, Ber. Bunsenges. Phys. Chem. 84, 1055–1059

Schatz GH, Schlodder E and Gräber P (1981) Kinetics of ATP synthesis and AdN exchange in chloroplasts measured by rapid mixing acid–base transitions. In: Proc. of the 5th Intern. Congr. of Photosynthesis, Kallithea, Greece (Akoyunoglou, G., ed.) Vol. II, Intern. Sci. Serv., Philadelphia, pp. 945–954

Schlodder E, Gräber P and Witt HT (1982) Mechanism of photophosphorylation in chloroplasts. In: Electron transport and photophosphorylation (Elsevier Biomed. Press, Barber, J., ed.), pp. 107–175

Vredenberg WJ (1976) Electrical Interactions and gradients between chloroplasts compartments and cytoplasm. In: The intact chloroplast (Elsevier Biomed. Press, Barber, J., ed.), pp. 53–88

Witt HT, Schlodder E and Gräber P (1976) Membrane-bound ATPase synthesis by an external electrical field, FEBS Lett. 69, 272–276

5. ACKNOWLEDGEMENTS

We thank Dr. G.H. Schatz for helpful discussions. The work was supported by the Deutsche Forschungsgemeinschaft.
Authors Address: Max-Volmer-Institut für Biophysikalische und Physikalische Chemie, Technische Universität Berlin, Strasse des 17. Juni 135, 1000 Berlin 12

ROLE OF THE HYDROGEN ATOMS IN THE FIRST CHEMICAL STEPS OF THE PHOTOSYNTHESIS.

E. Tyszkiewicz and E. Roux

INTRODUCTION.

It is well known, since the work of Neumann J, Jagendorf AT (1964), that the illumination of a suspension of chloroplasts in an unbuffered.medium induces an increase of the pH of the suspension. When the light is switched off the pH returns to its initial value (plain curve, fig. 2). This light induced pH variation is completely reversible in darkness.
Hind·G, Jagendorf AT (1963) have shown also that preilluminated chloroplasts at pH 6 become able to synthesize ATP in darkness at pH 8, in the presence of ADP and orthophosphate. This light induced ability to synthesize ATP in darkness is called X_e and the overall process the two step-phosphorylation. The number of ATP molecules which are formed in those conditions is proportional to the number of protons which are responsible of the observed pH increase, Galmiche et al. (1967).
Many hypothesis have been emitted concerning the fate of the protons during this pH variation and its relationship with the osmotic theory of P. Mitchell (1979), but the molecular mechanism of this ATP synthesis remains unknown.
To check the possible role of the hydrogen atoms in the first chemical reactions of the photosynthesis we have studied the action of a specific hydrogen atom scavenger the 2 propanol, on :
a) the speed of the dichlorophenol-indophenol photoreduction
b) the light induced 515 mm absorption change
c) the light induced pH increase and its dark decrease
d) the different types of photophosphorylations
According to Allan J, Scholes G (1960) and Feitelson J (1971) the 2 propanol scavenges the hydrogen atom following the reaction (reaction 1)

$$\cdot H + H- \underset{\underset{CH_3}{|}}{\overset{\overset{CH_3}{|}}{C}}-OH \longrightarrow H_2 + \cdot \underset{\underset{CH_3}{|}}{\overset{\overset{CH_3}{|}}{C}}-OH \quad \text{then} \quad 2 \cdot \underset{\underset{CH_3}{|}}{\overset{\overset{CH_3}{|}}{C}}-CH_3 \longrightarrow \underset{\underset{CH_3}{|}}{\overset{\overset{CH_3}{|}}{C}}=O + H- \underset{\underset{CH_3}{|}}{\overset{\overset{CH_3}{|}}{C}}-OH \quad (1)$$

In this work, we show that the 2 propanol inhibits these photoreactions. These results lead us to conclude that the 2 propanol acts at two levels :
1) The 2 propanol modifies the interaction of the components of the thylakoid membranes and consequently we observe a decrease of the rate of the photoinduced electron transfer as shown by the lowering of the speed of DCPIP reduction and of the amplitude of the 515 nm absorption change.
2) The 2 propanol acts also as an hydrogen atom scavenger :
a) it suppresses the reversibility in darkness of the photoinduced pH increase.
b) it inhibits the continuous photophosphorylations as well as the two step-phosphorylations.
c) in the two photoreactions (light induced pH increase and cyclic photophosphorylations) where the chloroplast content is sufficient, the formation of acetone could be measured when the 2 propanol was present. According to the scavenging reaction (1), this is the proof that hydrogen atoms are photogenerated in the first chemical step of photosynthesis.

Sybesma, C. (ed.), Advances in Photosynthesis Research, Vol. II. ISBN 90-247-2943-2.
© *1984 Martinus Nijhoff/Dr W. Junk Publishers, The Hague/Boston/Lancaster.*

EXPERIMENTAL RESULTS
1) The 2 propanol slows down the speed of the photoreduction of the dichloro-
phenol indophenol, whereas ethanol has no effect (Table 1)

TABLE 1 . Rate of the reduction of dichlorophenol-indophenol

addition to chloroplast suspension medium	um DCPIP mg^{-1} chlorophyll, h^{-1}
no addition	12.6
+ 2 propanol 1 M	7.8
+ ethanol 1.2 M	13.

The reacting medium is : Tricine pH 7.5 : 4.10^{-2} M ; dichlorophenol-
indophenol 5.10^{-5} M ; chloroplasts : 4 µg chlorophyll ; total volume :
2.5 ml. The rate of the reduction of DCPIP is measured as described by Remy R
(1973).

2) The 2 propanol causes a decrease in the amplitude of the light induced
515 nm absorption change, eliminating its slow generation phase and accelerates its decay in darkness (fig. 1).

3) The 2 propanol causes a decrease in the amplitude of the light induced pH
increase and suppresses its reversibility in darkness (fig. 2).

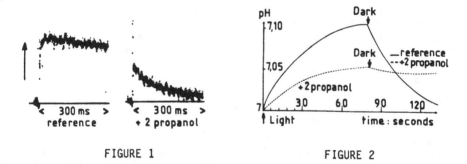

FIGURE 1   FIGURE 2

Fig. 1 Kinetics of the photoinduced 515 nm absorption changes.

Spinach chloroplasts are isolated in Tricine pH 7.5 ; 2.10^{-2} M, sorbitol :
4.10^{-1} M. The reacting medium is : Tricine pH 7.5 : 4.10^{-2} M, $MgCO_3$:
5.10^{-3} M. $KHCO_3$: $10.^{-2}$M, chloroplasts : 90 ug chlorophyll, total
volume : 3 ml. Each figure represents the time evolution of the 515 nm
absorption change after a flash (3 µs, 0.5 J). Flashes were fired at 10 s
intervals, 30 signals were measured and collected in an IN 45 signal averager.

Fig. 2 Kinetics of the photoinduced pH variation.

The reacting medium is : saccharose : 0.4 M, KCl : 10^{-2} M, 2 propanol 1M
when added, chloroplast : 100 ug chlorophyll, ml^{-1}, the chloroplast are
stirred in the reacting medium in darkness for 5 min. before illumination.

4) The 2 propanol completely inhibits the cyclic photophosphorylations and the two step-phosphorylations. Ethanol only lowers the amount of ATP synthesized. (Table 2)

TABLE 2. Cyclic and two-step-phosphorylations in the presence of 2 propanol or ethanol.

addition to chloroplast suspension medium	cyclic ATP $\mu mol.h^{-1}.mg^{-1}Chl$	two-step ATP $nmol.min^{-1}.mg^{-1}Chl$
no addition	33.6	28.5
+ 2 propanol 1 M	1.4	5.
+ ethanol 1.2 M	15.6	-

Cyclic photophosphorylations are performed as described by Tyszkiewicz et al. (1979) ; two step-phosphorylations as described by Tyszkiewicz, Roux (1975). Chloroplasts are stirred for 10 min. in the suspension medium with or without 2 propanol or ethanol before 1 min. of illumination.

DISCUSSION
    In the presence of 2 propanol the decrease of the dielectric constant of the chloroplast suspension modifies the structure of the thylakoid membrane and consequently reduces the rate of the photoinduced electron migration. The observed diminution of the speed of the DCPIP reduction, the lowering of the amplitude of the 515 nm absorption change and of the photoinduced pH increase are due to this first effect of the 2 propanol. But the most important results concern its specific property to be an hydrogen atom scavenger. As shown in fig. 2 the addition of 2 propanol to the chloroplast suspension suppresses completely the dark return of the pH to its initial value. Coupled to this irreversibility we have verified by gas chromatography on an aliquot of a cryosublimat of the illuminated chloroplast suspension that acetone molecules are formed, according to the scavenging reaction (1). This result proves that free hydrogen atoms are generated during the photoinduced pH increase. The simplest explanation could be that the reduction of a proton by a solvated photejected electron takes place :

$$H^+ + e_{solv} \text{------------} \blacktriangleright \cdot H$$

    In the absence of 2 propanol those hydrogen atoms must be reversibly fixed by some natural acceptor, for example a plastoquinone, but in the presence of 2 propanol the scavenging of the hydrogen atom leading to the acetone and hydrogen molecule formation prevents the regeneration of protons from hydrogen atoms and suppresses in darkness the return of the pH of the suspension to its initial value.
    In the absence of 2 propanol the number of ATP molecules which are formed in the two step-phosphorylation is proportional to the number of protons which are responsible of the observed pH increase (Galmiche et al. 1967).
    In the presence of 2 propanol the suppression of the reversibility in darkness of the photoinduced pH increase and the simultaneous inhibition of the ATP synthesis are both due to the scavenging of the hydrogen atoms. These results suggest that these very reactive hydrogen atoms are responsible of the formation of monomeric metaphosphate i.e. of the dehydroxylation reaction of orthophosphate which precedes the ATP formation.

REFERENCES
Allan JT and Scholes G (1960) Effects of pH and the nature of the primary
species in the radiolysis of aqueous solution, Nature, 187 218-220.
Feitelson J (1971) The formation of hydrated electrons from the exited state
of indole derivatives, Photochem. and Photobiol. 13 87-96.
Galmiche JM, Girault G, Tyszkiewicz E, Fiat R (1967) Photophosphorylations et
translocation de protons, C.R.Ac.Sc. 265 serie D 374-377.
Hind G, Jagendorf AT (1963) Separation of light and dark stages in photophosphorylation, Proc.Nat.Ac. of Sc. 29, 715-722.
Mitchell P (1979) Keilin's respiratory chain concept and its chemiosmotic
consequences (Nobel lecture) Science 206, 1148-1159.
Neuman J and Jagendorf AT (1964) Light induced pH changes related to phosphorylation by chloroplasts, Arch. Biochem. and Bioph. 107 109-119.
Remy R, (1973) Appearence and development of photosynthetic activities in
wheat etioplasts greened under continuous or intermittent light, Photochem.
and Photobiol. 18, 409-416.
Tyszkiewicz E and Roux E (1975) Effects of low temperatures on formation and
conservation of high energy state (X_e) appearing in spinach chloroplasts
during the light step of the two-stage phosphorylation, Biochem. Biophys. Res.
Comm. 65, 1400-1408.
Tyszkiewicz E, Nicolic D, Popovic R, Saric M (1979) Photophosphorylation and
ultrastructural development in Pinus nigra chloroplasts grown under different
spectral composition of light, Phys. Plant. 46, 324-329.

ACKNOWLEGEMENTS
The authors wish to express their gratitude to Drs Bernas A, Grand D. and
Hautecloque S. for fruitfull discussions and to Dr Berger G, for gaz chroma-
tography measurements.
Authors adress : Service de Biophysique, Département de Biologie, Centre
d'Etudes Nucléaires de Saclay, 91191 Gif-sur-Yvette, France.

ELECTRIC POTENTIAL AND pH GRADIENT FORMATION BY RECONSTITUTED ATPase PROTEO-
LIPOSOMES FROM THE THERMOPHILIC CYANOBACTERIUM *SYNECHOCOCCUS* 6716

HENDRIKA S. VAN WALRAVEN, HENK J. LUBBERDING, HANS J.P. MARVIN AND RUUD
KRAAYENHOF

1. INTRODUCTION

The proton-translocating ATPase complex has been the object of many studies to
elucidate its energy-transducing function in the direction of both ATP synthesis and hydrolysis. By reconstitution in liposomes, mostly prepared from asolectin, a simple and functional model system can be obtained. Recently, a
DCCD-sensitive H^+-ATPase complex has been isolated from the thermophilic
cyanobacterium *Synechococcus* 6716 (Lubberding et al., submitted) as a Mg^{2+}-
dependent heat-stable complex, which can only be activated by trypsin. This
complex, reconstituted with liposomes from isolated *Synechococcus* lipids catalyzes DCCD-sensitive $^{32}P_i$-ATP exchange and ATP hydrolysis activities and generation of an uncoupler-sensitive transmembrane electric potential ($\Delta\Psi$) and
a pH gradient (Δ pH) during ATP hydrolysis (Van Walraven et al., submitted).
In this paper we describe the ATP-dependent membrane energization by the use
of the surface charge density probe TDACMA, the electric potential-sensitive
dye oxonol VI and natural carotenoids and the pH indicators neutral red and
cresol red, trapped inside the vesicles, and their sensitivity to the ionophores valinomycin and nigericin.

2. MATERIALS AND METHODS

Synechococcus 6716 was grown, membrane vesicles were prepared and the ATPase
complex was isolated according to Lubberding et al. (submitted). The isolation
was based on the procedure of Pick and Racker (1979) by octylglucoside (Calbiochem) and cholic acid solubilization and ammonium sulfate precipitation at
35-50 %, this precipitate being used for reconstitution. Lipid isolation, reconstitution by dialysis of octylglucoside and cholate and assays were described by Van Walraven et al. (submitted). Reactions were carried out in a
multipurpose cuvette (1.8 ml), thermostated at $50^\circ C$, described by Kraayenhof
et al. (1982). The measuring and reference wavelengths are indicated in the
Figures and other reaction conditions are given in the Figure legends.

3. RESULTS AND DISCUSSION

TABLE 1. Catalytic activities of the isolated and reconstituted *Synechococcus*
6716 ATPase complex. The lipid/protein ratios of the reconstituted liposomes
were 0.01 (w/w).

Preparation	ATP hydrolysis	$^{32}P_i$-ATP exchange
	(nmol.min^{-1}.mg protein^{-1})	
Isolated ATPase	881	200
ATPase liposomes	350	119
+ FCCP	1059	0

Sybesma, C. (ed.), Advances in Photosynthesis Research, Vol. II. ISBN 90-247-2943-2.
© *1984 Martinus Nijhoff/Dr W. Junk Publishers, The Hague/Boston/Lancaster.*

The specific activities of the complex decline after reconstitution, as can be seen from Table 1. In the presence of an uncoupler a strong stimulation is observed, which indicates an efficient incorporation of the ATPase complex and a relatively low leakiness for protons. Thus, the ATPase proteoliposomes prepared from natural lipids function well at 50°C, the optimal growth temperature of *Synechococcus* 6716. The specific hydrolysis and exchange activities are similar to those, found for reconstituted chloroplast ATPase (Pick, Racker 1979).

TDACMA is a membrane-bound analogue of ACMA. Fig. 1 shows that addition of ATP to ATPase proteoliposomes results in a fluorescence quenching of about 20%, with a halftime of 5 s. Nigericin (K^+/H^+ exchange across the membrane) inhibit the fluorescence response. The addition of valinomycin (decreases $\Delta\Psi$) results in an increase of fluorescence quenching. Valinomycin, added before ATP enhances the extent of ATP-induced quenching but the $t_{0.5}$ does not change. Similar effects of ionophores on ACMA fluorescence quenching in reconstituted liposomes from yeast ATPase have been observed (Dufour et al., 1981). During ATP-dependent membrane energization TDACMA monitors an increase in negative surface charge, probably due to the inward translocation of protons from the outside of the liposomal membrane.

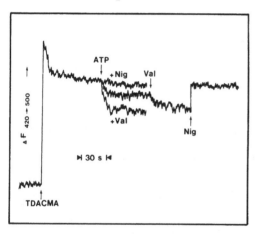

Figure 1. Fluorescence responses of TDACMA upon binding and ATP hydrolysis in ATPase complex proteoliposomes. The reaction medium contains 10 mM KCl, 1 mM $MgCl_2$, 10 mM Na-Tricine (pH 7.8) and 1 mM DTE. Final lipid concentration was 0.25 mg.ml^{-1} (2.5 µg.ml^{-1} protein). TDACMA concentration was 0.5 µM, ATP 1 mM and nigericin and valinomycin 1 µM.

Isolated lipids from *Synechococcus* still contain carotenoids. Like oxonol VI, these native carotenoids are still sensitive to membrane potential changes. The formation of the membrane potential upon addition of ATP as shown in Fig. 2, has the same halftime (5 s) as the increase of negative surface charge (Fig. 1). The absorbance changes were enhanced by nigericin, exchanging translocated protons for potassium ions, thus increasing $\Delta\Psi$. In the presence of nigericin the halftime is shortened to about 2 s. Valinomycin inhibits the ATP-induced absorbance change. With chloroplast ATPase, reconstituted in asolectin vesicles Shahak et al. (1983) reported the same ionophore effects on oxonol VI responses. After ATP addition a fast carotenoid absorbance change was followed by a decline to a steady-state level (Fig. 2). This was not observed with oxonol VI, under these conditions.

During reconstitution by detergent dialysis pH indicators can be trapped inside the vesicles. In seperate experiments two pH indicators were used: the cationic neutral red and the anionic cresol red. Fig. 3 shows the effect of ATP hydrolysis on the intravesicular pH, measured by the absorbance changes of both pH indicators.

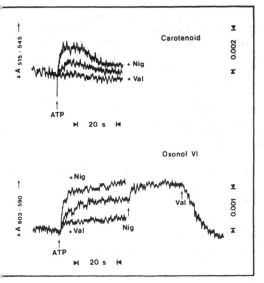

FIGURE 2. ATP-induced absorbance changes of native carotenoids and added oxonol VI. The reaction medium was as in Fig. 1, except for 2.5 mM $MgCl_2$ and 0.5 mg lipid.ml^{-1}. Oxonol VI concentration was 0.5 μM, ATP 1 mM and nigericin and valinomycin 1 μM.

Upon ATP addition, neutral red shows a much faster response (halftime 30 s) compared to cresol red (halftime 100 s). The cresol red response usually contains a lag of 20 s. Due to their opposite charge we assume that neutral red and cresol red have different locations in the liposomes. Like TDACMA and ACMA with similar structure, neutral red can bind to the membrane whereas cresol red is rejected and will reside in the hydrophilic lumen of the vesicles. The absorbance changes of the pH indicators correspond with an intraliposomal decrease in pH of 1.5. This is calibrated by acid-base pulses in the presence of an uncoupler (Van Walraven et al., submitted). Like TDACMA fluorescence quenching, ATP-dependent pH indicator responses are completely abolished by nigericin. The neutral red signal is transiently stimulated by valinomycin and the halftime becomes about equal to those of TDACMA quenching and carotenoid and oxonol VI absorbance changes (5 s), indicating a faster formation of a transmembrane pH gradient in the presence of valinomycin.

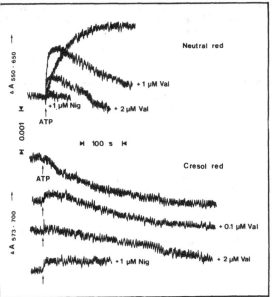

FIGURE 3. Internal acidification of the proteoliposomes after addition of ATP. pH indicators neutral red and cresol red were trapped inside the vesicles during dialysis at a concentration of 10 μM, in a reaction medium containing 10 mM KCl, 2.5 mM $MgCl_2$ and 1 mM DTE adjusted to pH 7.5 and 8 for neutral red and cresol red, respectively. During the experiments, external medium was buffered with 10 mM Hepes buffer. Lipid concentration was 0.5 mg.ml^{-1} (5 μg.ml^{-1} protein) ATP 1 mM and valinomycin and nigericin as indicated in the Figure.

Low concentrations of valinomycin (100 nM) also have this effect, but to a lower extent (not shown). However, valinomycin also seems to uncouple at high concentrations (1-2 μM). After an initial fast rise, the neutral red signal declines (Fig. 3). The slow response of cresol red is not enhanced in the presence of valinomycin but retarted. When comparing the responses, and especiall, the response times of the differently located probes tested in these experiments, proton translocation by the reconstituted ATPase complex does not occur in one step from the outside of the vesicle, through the membrane and to the bulk inner phase. The formation of membrane potential and the increase of negative surface charge after addition of ATP are completed within 15 s (halftime 5 s) whereas the built-up of a transmembrane pH gradient lasts much longer (halftime 100 s). Since the electric events, mentioned above can only be due to proton translocation, this must occur in at least two steps. The effect of valinomycin on the response time of neutral red, which is supposed to probe the inner interface of the liposomal membrane (halftime 30 s) is also striking. These observations indicate the existence of a membrane-localized proton pool, "filled up" within seconds after the onset of ATP hydrolysis, prior to an inner bulk phase proton accumulation. During ATP hydrolysis of chloroplasts membranes a $\Delta\psi$ is observed earlier (within seconds) than a Δ pH (Girault, Galmiche, 1981).

Except for the study of this ATPase complex, the existence of an intermediate local proton pool may be important for the interpretation of the function of other membrane-bound energy-transducing protein complexes as well.

REFERENCES

Pick U and Racker E (1979) Purification and reconstitution of the N-N'-Dicyclohexylcarbodiimide sensitive ATPase complex from spinach chloroplasts, J. Biol. Chem. 254, 2793-2799.
Kraayenhof R, Schuurmans JJ, Valkier LJ, Veen JPC, van Marum D and Jasper CGG (1982) A thermoelectrically regulated multipurpose cuvette for simultaneous time-dependent measurements, Anal. Biochem. 127, 93-99.
Dufour JP, Goffeau A and Tsong TY (1982) Active proton uptake in lipid vesicles reconstituted with the purified yeast plasma ATPase. Fluorescence quenching of 9-amino-6-chloro-2-methoxy-acridine, J. Biol. Chem. 257, 9365-9371.
Shahak Y, Admon A and Avron M (1982) Transmembrane electrical potential formation by chloroplast ATPase complex (CF_1-CF_0) proteoliposomes, FEBS Lett. 150, 27-31.
Girault G and Galmiche JM (1981) Energization of the thylakoid membranes by ATP in spinach chloroplasts. In Akoyunoglou G ed. Photosynthesis II, Photosynthetic electron transport and photophosphorylation, pp. 965-972. Philadelphia: Balaban Int. Science Services.

ACKNOWLEDGEMENTS

This research is supported in part by the Foundation for Biophysics with financial aid from the Netherlands Organization for Advancement of Pure Research (ZWO).

Authors address: Biological Laboratory, Vrije Universiteit, De Boelelaan 1087, 1081 HV Amsterdam, The Netherlands.

THE EFFECTS OF LIGHT-DARK TRANSITION AND OF SPECIFIC INHIBITORS ON THE ATP LEVEL IN SOME CYANOBACTERIA

HENK J. LUBBERDING and WIM SCHROTEN

1. INTRODUCTION

Although the light driven electron transfer and phosphorylation is the main way of energy production in cyanobacteria (Ho, Krogmann 1982), they also can synthesize ATP by oxidative phosphorylation under aerobic conditions in the dark (Smith 1982). It is still questionable whether all or some cyanobacteria are able to make ATP from the degradation of cyanophycin or polyphosphate (Smith 1982). When cyanobacteria or isolated chloroplasts are transfered from light to dark under aerobic conditions there is a fast decrease of the ATP concentration that is either maintained at a fixed level (Heber, Santarius 1970; Bornefeld et al. 1972; Urbach, Kaiser 1972; Bornefeld, Simonis 1974; Heber et al. 1982) or increases within 5 min to the original ATP level in the light (Pelroy, Bassham 1972, Ihlenfeldt, Gibson 1975).
Here, we present the relative changes of the intracellular ATP levels, measured by the sensitive luciferin-luciferase method, under aerobic and anaerobic conditions influenced by light to dark transitions and by DCMU*, S-13 and DCCD in the cyanobacteria *Synechococcus 6307*, *Synechococcus 6716* and *Plectonema 73110*.

2. MATERIALS AND METHODS

<u>Organisms and growth</u>: *Synechococcus 6307*, *Synechococcus 6716* and *Plectonema 73110* were obtained from the algal collection of the Pasteur Institute, Paris and were grown as described before (Lubberding et al. 1981).
<u>Membrane preparation</u>: Membrane vesicles of *Synechococcus 6716* were made as described before (Lubberding et al. 1981). The cells were incubated with lysozyme at $37^{\circ}C$ during 45 min.
<u>Electron transfer and photophosphorylation</u>: Oxygen evolution was measured with a Clark-type oxygen electrode. Actinic illumination with white light was provided by a 250W halogen lamp (Osram) through a Schott KG3 heat filter and via fiber glass optics (Schott, $\emptyset$ 10 mm) usually at $400J.m^{-2}.s^{-1}$. The reaction medium contained 20 mM Tricine-KOH, 50 mM NaCl, 50 mM KCL, 5 mM $MgCl_2$ and 5 mM K_2HPO_4, pH 7.8. Chlorophyll concentration was 15-17 µg/ml, the electronacceptor was 1.2 mM $K_3Fe(CN)_6$; DCCD, DCMU and S-13 were added in ethanol (final concentration below 0.8%).
<u>ATP measurement</u>: Reaction tubes with 3 ml. reaction medium (as described before +0.5 mM ADP) were illuminated during 4 min with $700 J.m^{-2}.s^{-1}$. The reaction was stopped with 3.2 M $HClO_4$. After neutralization in ice (3.2 M KOH, 1 M Tris) the mixture was centrifuged at 1500 g for 5 min. The ATP content was measured according to the luciferin-luciferase method (Larsson, Olsson 1979). The ATP assay buffer contained 100 mM Tris-acetate (pH 7.75), 3 mM Mg-acetate and 2 mM EDTA. The purified luciferin and luciferase were

---

*Abbreviations: DCCD: N,N'-dicyclohexylcarbodiimide; DCMU: 3-(3,4-dichlorophenyl)-1,1-dimethylurea; PMS: methyl-phenazoniummethosulfate; S-13: 5-chloro-3-t-butyl-2'-chloro-4'-nitrosalicylanilide.

Sybesma, C. (ed.), Advances in Photosynthesis Research, Vol. II. ISBN 90-247-2943-2.
© *1984 Martinus Nijhoff/Dr W. Junk Publishers, The Hague/Boston/Lancaster.*

from Boehringer Mannheim, FRG. After every assay a known amount of ATP
(100 pmol in a volume of 10 µl) was added to the sample as an internal
standard.

<u>Whole cell experiments</u>: The cells were harvested by centrifugation at 5000 g
for 5 min, washed twice in 100 mM Tricine–KOH and 10 mM MgCl$_2$, pH 7.8 and
finally resuspended in the same buffer. The experiments were carried out
in reaction tubes with this suspension (chlorophyll concentration 200–500
µg/ml) light was provided by Philips lamps (nr. 34, 40W) at an intensity of
2 J.m^{-2}.s^{-1}. Samples were taken 20 min after the addition of the inhibitors
and uncoupler; anaerobiosis was reached by bubbling with N$_2$., Cells were
destroyed by the addition of 0.2 ml 5.0 M HClO$_4$ to 1.0 ml of cell suspension
After neutralization (0.2 ml 5 M KOH and 1 M Tris) the suspension was centrifuged at 1500 g for 5 min. The ATP content was measured as above. Chlorophyll
a was determined according to (Arnon et al. 1974) and protein according
to (Lowry et al. 1951).

3. RESULTS AND DISCUSSION

The effects of the electron inhibitor DCMU, the uncoupler S-13 and the energy
transfer inhibitor DCCD on electron transfer and photophosphorylation in
membrane vesicles of *Synechococcus 6716* are shown in Fig. 1. DCMU behaves
as expected, fully blocking electron transfer and the resulting phosphorylation at 1 to 10 µM. S-13 is a potent uncoupler: full inhibition of ATP
synthesis occurs at 1 µM; however, electron transfer is not enhanced. In
linear electron transfer DCCD seems to act as an electron transfer inhibitor
at lower concentrations as was found before in chloroplasts, where DCCD block
the reduction site of plastoquinone (Sane et al. 1979). Only at higher concentrations (10^{-3} M) DCCD also inhibits energy transfer as can be seen in the
PMS-mediated cyclic photophosphorylation.

Under anaerobic conditions transition of cells from light to darkness
causes an immediate drop of the ATP concentration to about 20% of the light
level (which is 0.4 to 1.0 nmol.ATP.mg protein^{-1}) in *Synechococcus 6716* and
Plectonema 73110 and to about 50% in *Synechococcus 6307* (Fig. 2). This ATP
level remains constant in the dark. Under aerobic conditions the ATP levels
of the autotrophic strains *Synechococcus 6307* and *6716* immediately drop to
70% of the light level and this level is also maintained during the experiment (5 hours). The ATP level of the heterotrophic strain *Plectonema 73110*

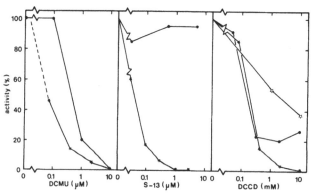

Fig. 1. **Effects of inhibitors and uncoupler on electron transfer and photophosphorylation activities in membrane vesicles of** *Synechococcus* **6716.** —●—: O$_2$ production; 100% = 2.3-3.1 µmol O$_2$.min^{-1}.mg chl.a^{-1}. —★—: ferricyanide-mediated ATP synthesis; 100% = 2.0-3.1 µmol ATP. min^{-1}.mg chl.a^{-1}.mg chl.a^{-1}. —☆—: PMS-mediated ATP synthesis; 100% = 11.1 µmol ATP.min^{-1}.mg chl.a^{-1}. Experiments were carried out at 35°C as described under materials and methods.

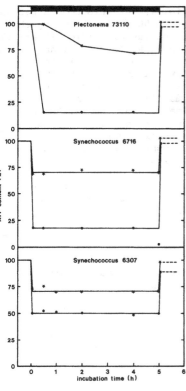

Fig. 2. Effects of light-dark transition on the ATP concentration in three cyanobacterial strains. 100% = 0.4-1.0nmol ATP.mg protein⁻¹. Experiments were carried out aerobically (—▲—) or anaerobically (—●—) at 30°C (S.6307 and P.73110) or at 50°C (S.7616) as described under materials and methods.

Table 1. Effects of inhibitors and uncoupler on the ATP concentration in cells of *Synechococcus* 6716. In the light, 100% = 0.4-1.0 nmol ATP.mg protein⁻¹. In the dark, 100% = 0.3-0.7 nmol ATP.mg protein⁻¹. Experiments were carried out at 50°C as described under materials and methods.

Inhibitor	Concentration	ATP level (%) Light	Dark
none	–	100	100
DCMU	5.10^{-5} M	98	97
S-13	10^{-6} M	20	26
DCCD	5.10^{-4} M	41	48

only slowly decreases to 70%, reaching this level after about two hours. In all cases the original light level is regained immediately when the light is switched on again (Fig. 2). These observations are in good agreement with experiments on the autotrophic cyanobacterium *Anacystis nidulans* (Borne-feld et al. 1972; Simonis 1974; Owens, Krauss 1972) and with experiments on isolated chloroplasts (Heber, Santarius 1970; Heber et al. 1982; Urbach, Kaiser 1972). If protoplasts are used instead of chloroplasts the ATP concentration is less influenced by light-dark transfer, because ATP is exchanged between chloroplasts and cytosol (Heber et al. 1982). On the other hand, it is found that after light-dark transfer the initial drop is followed by a restoration of the ATP level within 5 min up to the original level (Pelroy, Bassham 1972; Ihlenfeldt, Gibson 1975). Very remarkably, the difference between these controversial observations is correlated with the way of the ATP determination: if ATP is measured by the indirect ³²P method, the ATP levels are restored in the dark within 5 min, but if the more direct luciferin-luciferase method is used, the constant levels are found as reported here.

The effects of DCMU, S-13 and DCCD on the ATP concentration are tested in whole cells of *Synechococcus 6716* (Table 1). DCMU has no effect on the ATP level neither in light nor in dark; since DCMU fully blocks linear electron transfer cyclic photophosphorylation must be responsible for the maintained ATP level in the light as it is in *Anacystis nidulans* (Bornefeld et al. 1972; Bornefeld, Simonis 1974), but in contrast with chloroplasts (Urbach, Kaiser 1972). The unaffected ATP levels in the dark are caused by respiration In our experiments DCCD behaves as an electron transfer inhibitor and at

0.5 mM no linear electron transfer is possible (Fig. 1). At this concentration energy transfer is also inhibited, but only for about 50%, visualized by the inhibition of cyclic photophosphorylation in membrane vesicles (Fig. 1) and in whole cells (Table 1) and oxydative phosphorylation in whole cells (Table 1). S-13 lowers the ATP concentration to the same level as the anaerobic incuba- tion in the dark. Thus, about 20% seems to be a minimal level of the ATP concentration in cyanobacteria, as in isolated chloroplasts under aerobic conditions in the dark (Heber et al. 1982). This may imply that the ATP- consuming processes are stopped at a certain minimal energy charge. Alterna- tively, there may be a light- and O_2-independent way of energy production in cyanobacteria as suggested before (Smith 1982). Unlike chloroplasts, cyanobacteria can synthesize ATP in the dark by a respiratory mechanism (Smith 1982), resulting in a constant, but clearly lower ATP level than in the light. The dark level is established very fast in the autotrophic strains and more slowly in the heterotrophic strain, suggesting a high respiratory rate in the latter. Apparently, the cells establish a new steady state between the ATP-producing and -consuming processes at every change of the external conditions.

4. REFERENCES

Arnon DI, McSwain BD, Tsujimoto HI and Wada K (1974) Photochemical activity and components of membrane preparations from blue-green algae, Biochim. Biophys. Acta 357, 231-245.
Bornefeld T, Domanski J and Simonis W (1972) Influence of light conditions, gassing and inhibitors on photophosphorylation and ATP level in *Anacystis nidulans*. In Forti G, Avron M and Melandri A, eds. Proc. II Intern. Congr. Photosynthesis Vol. II, pp. 1379-1386. Dr. W. Junk NV, Den Haag.
Bornefeld T and Simonis W (1974) Effects of light, temperature, pH, and inhibitors on the ATP level of the blue-green alga *Anacystis nidulans*, Planta 115, 309-318.
Heber U and Santarius KA (1970) Direct and indirect transfer of ATP and ADP across the chloroplast envelope. Z. Naturforsch. 25b, 718-728.
Heber U, Takahama U, Neimanis S and Shimizu-Takahama M (1982) Transport as the basis of the Kok effect, Biochim. Biophys. Acta 679, 287-299.
Ho KK and Krogmann DW (1982) Photosynthesis. In Carr NG and Whitton BA, eds. The Biology of cyanobacteria, pp. 191-214. Blackwell Scientific Publications, Oxford.
Ihlenfeldt MJA and Gibson J (1975) CO_2 fixation and its regulation in *Anacys- tis nidulans (Synechococcus)*, Arch Microbiol. 102, 13-21.
Larsson CM and Olsson T (1979) Firefly assay of adenine nucleotides from algae Comparison of extraction methods, Plant Cell Physiol. 20, 145-155.
Lowry OH, Rosebrough NJ, Farr AL and Randall RJ (1951) Protein measurement with the Folin-phenol reagent, J. Biol. Chem. 193, 265-275.
Lubberding HJ, Offerijns F, Vel WAC and De Vries PJR (1981) Characterization of the ATPase of the thermophilic cyanobacterium *Synechococcus lividus*. In Akoyunoglou G, ed. Photosynthesis II. Electron transport and photophosphory- lation, pp. 779-788. Balaban International Science Services, Philadelphia.
Owens O and Krauss RW (1972) Kinetics of photophosphorylation in cells of *Anacystis nidulans*, Plant Physiol. 49, Suppl. 52.
Pelroy RA and Bassham JA (1972) Photosynthesis and dark metabolism in uni- cellular blue-green algae, Arch. Microbiol. 86, 25-38.
Sane PV, Johanningmeier U and Trebst A (1979) The inhibition of photosynthetic electron transport flow by DCCD, FEBS Lett. 108, 136-140.
Smith AJ (1982) Modes of cyanobacterial carbon metabolism. In Carr NG and Whitton BA, eds. The biology of cyanobacteria, pp. 47-86. Blackwell Scientific Publications, Oxford.
Urbach W and Kaiser W (1972) Changes in ATP levels in green algae and intact chloroplasts by different photosynthetic reactions. In Forti G, Avron M and Melandri A, eds. Proc. II Intern. Congr. Photosynthesis. Vol. II, pp. 1401- 1411. Dr. W. Junk NV, Den Haag.

THE INHIBITION OF NITRATE REDUCTION BY LIGHT IN RHODOPSEUDOMONAS CAPSULATA IS MEDIATED BY THE MEMBRANE POTENTIAL, BUT INHIBITION BY OXYGEN IS NOT

A.G MCEWAN, N P J COTTON, S J FERGUSON and J B JACKSON, DEPARTMENT OF BIOCHEMISTRY, UNIVERSITY OF BIRMINGHAM, P O BOX 363, BIRMINGHAM, ENGLAND.

1. INTRODUCTION

Rps. capsulata strain N22 and its parent strain, St. Louis are capable of phototrophic growth with nitrate as the sole source of nitrogen (M A Jackson et al 1981). Nitrate uptake in washed cell suspensions was found to be light-dependent and reversibly inhibited by low concentrations of ammonia. When these strains of photosynthetic bacteria were repeatedly cultured under phototrophic conditions on nitrate, spontaneous mutants arose and outgrew the original organisms (McEwan et al. 1982). One such mutant (N22DNAR[+]) was isolated and studied in detail. In contrast to strains N22 and St. Louis, washed cell suspensions of N22DNAR[+] were capable of nitrate reduction in the dark. This activity was actually inhibited by photosynthetic light or by a period of aerobic respiration but was insensitive to ammonia. Moreover, the addition of nitrate to these washed cell suspensions resulted in the generation of a cytoplasmic membrane potential ($\Delta\psi$) and nitrite was produced almost stoichiometrically with the nitrate consumed. These latter properties are characteristic of "dissimilatory" reduction in which the anion serves as the terminal oxidant of an energy-conserving electron transport chain as typified by the non-photosynthetic soil bacterium Paracoccus denitrificans. More recently we have shown that some other strains of Rps. capsulata isolated from the wild by other workers (see Klemme, 1979) have rather similar properties to N22DNAR[+](McEwan et al. 1983).

The question arises as to how, in this group of organisms, electron transport to nitrate is co-regulated with cyclic, photosynthetic electron transport and with electron transport to oxygen. Below, we shall address ourselves to the problem as to what mechanisms are responsible for the inhibition of nitrate reduction in strain N22DNAR[+] by light or by the addition of oxygen.

In wild-type strains of Rps. capsulata evidence is accumulating that the well-known inhibition of aerobic respiration by light (see Keister, 1978) is mediated by way of the proton motive force (Δp) across the cytoplasmic membrane. McCarthy, Ferguson (1982) showed that in well-coupled chromato-phores, the presence of uncoupling agents prevented light-inhibition of respiration. Rugolo, Zannoni (1983) showed that uncoupler treatment in intact cells, known to prevent light-inhibition of respiration (Ramirez, Smith 1968), led to a drop in $\Delta\psi$. Cotton et al. (1983) showed on a quantitative basis the inverse correlation between $\Delta\psi$ and the rate of respiration in the light as the uncoupler concentration was varied. In all of this work it has been concluded that the Δp generated by photosynthetic electron transport exerts a thermodynamic "back-pressure" on the respiratory chain operating across the same membrane. In the following we shall present data which suggests that light-inhibition of nitrate reduction in strain N22DNAR[+] arises from a similar effect, but that oxygen-inhibition of nitrate reduction does not.

2. EXPERIMENTAL PROCEDURE

Cells of Rps. capsulata strain N22DNAR[+] were grown phototrophically on nitrate and harvested and washed as described by McEwan et al. 1982.

Sybesma, C. (ed.), Advances in Photosynthesis Research, Vol. II. ISBN 90-247-2943-2.
© *1984 Martinus Nijhoff/Dr W. Junk Publishers, The Hague/Boston/Lancaster.*

Nitrate was measured with an ion-exchange electrode, oxygen uptake with a Clark electrode and $\Delta\psi$ was measured from the carotenoid band shift (see McEwan et al. 1982 and Cotton et al. 1983 for details).

3. RESULTS

The disappearance of nitrate from dark, anaerobic suspensions of Rps. capsulata strain N22DNAR[+] is shown in Fig.1A. Illumination of the sample gave rise to immediate inhibition ($\sim$100%) of nitrate uptake which was slightly reversed during prolonged illumination. When the light was extinguished the inhibition of nitrate uptake was completely reversed, although there was a small increase in rate compared with that during the pre-illumination period. These kinetic properties are similar to those observed for the light-inhibition of oxygen uptake (Cotton et al. 1983). In Fig.1A(ii) the addition of 2μM of the uncoupler 2',5-dichloro-3-t-nitroso-salicylanilide (S-13) almost completely abolished the inhibitory effect of light. In the experiments described in this report, the light intensities used were less than saturating (as judged by the extent of the carotenoid band shift). This allowed us to use lower S13 concentrations to relieve light-inhibition of nitrate uptake and so to minimise the artefact of the S13 on the nitrate electrode response - see trace (ii) in Fig.1A.

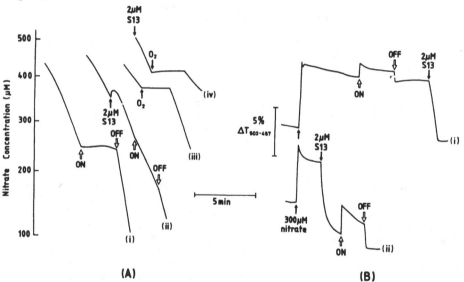

(A)   **(B)**

Fig.1. Recording of nitrate uptake and $\Delta\psi$ generation in Rps. capsulata strain N22DNAR[+]. Effect of uncoupler on the inhibition of nitrate uptake by light and oxygen. (A): nitrate electrode recordings. The left-hand axis shows the calibrated electrode reading upon adding known concentrations of nitrate. The cells (20 μM bacteriochlorophyll) were added to the anaerobic, fresh growth medium containing an initial nitrate concentration of 500 μM. The light was switched on and off where shown and oxygen was added as 1 vol H_2O_2. (B): carotenoid absorbance changes. Conditions as in (A) except that nitrate was added to start the reaction. An upward deflection represents a transmission decrease at 503-487 nm and is interpreted as an increasingly negative cytoplasmic membrane potential.

In separate experiments performed at the same time on similar cell
suspensions, $\Delta\Psi$ was measured from absorbance changes accompanying the
carotenoid band shift (Fig 1B). In the control sample, nitrate addition
to the dark anaerobic cell suspension gave rise to $\Delta\Psi$. Subsequent
illumination of the sample increased $\Delta\Psi$. At the end of the experiment
the addition of uncoupling agent established the base line equivalent to
$\Delta\Psi = 0$. Fig 1B also shows that S13 collapsed $\Delta\Psi$ generated by nitrate
addition and prevented $\Delta\Psi$ from reaching a high value during illumination.

The basic experiments in Figs 1A and 1B were repeated at a range of S13
concentrations. The results are described in Fig 2. It can be seen that
more S-13 was needed to stimulate nitrate uptake in the light than in the
dark and that in both situations the value of $\Delta\Psi$ declined over approximately
the same range of S-13 concentration as the rate stimulation was observed.
The light-inhibition of nitrate uptake on a percentage basis was abolished
co-incidently with the decline of $\Delta\Psi$ during illumination conditions.

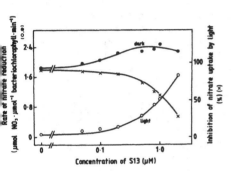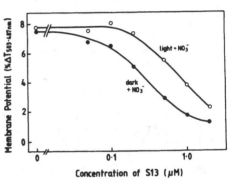

Fig.2A Effect of S13 on the rate of nitrate uptake in darkened and
illuminated cell suspensions. From data similar to trace (ii) in Fig 1A.
Rates were measured during the second minute after adding the S13 and
during the second minute of illumination.

Fig.2B Effect of S13 on $\Delta\Psi$ during nitrate uptake in darkened and illuminated cell suspension. From data similar to trace (ii) in Fig 1B. Values
of $\Delta\Psi$ were taken as the average value during the second minute after adding
the S13 and during the second minute of illumination, both with respect to
the base line established by addition of 2 μM S13 at the end of the experiment.

The inhibition of nitrate uptake by oxygen in darkened cell suspensions has
been found to coincide with the period of aerobic respiration (McEwan et al.
1982). In sharp contrast to the above results with light-inhibition of
nitrate uptake, the inhibitory effect of oxygen was not relieved by
uncoupling agents (Fig 1A, traces iii and iv).

4. DISCUSSION

In agreement with earlier work, the reduction of nitrate by intact cells of
Rps. capsulata strain N22DNAR[+] was accompanied by the generation of $\Delta\Psi$
(McEwan et al. 1982). This was taken as key evidence for the idea that
nitrate can serve as the electron sink of a proton translocating electron
transport chain. Although the increase in $\Delta\Psi$ resulting from the extra

effect of illumination (Fig 1B) is only small it represents a considerable increase in the rate of primary proton translocation because the passive proton conductance of the cytoplasmic membrane increases steeply with $\Delta\psi$ (see discussion in Clark et al. 1983). This means that the proton-translocating activity of the photosynthetic electron transport chain greatly exceeds that of the pathway leading to nitrate reduction. In any case the increased $\Delta\psi$ achieved during illumination appears to be sufficient to restrict the rate of electron transport to nitrate; when the magnitude of $\Delta\psi$ was decreased through the action of the protonophore S13, the inhibitory effect of light was abolished (Figs 1 and 2). The close correlation between $\Delta\psi$ collapse and abolition of light-inhibition of nitrate uptake (Fig 2) tends to rule out the possibility of side-effects of the S13 being involved.

The aerobic respiratory system of these bacteria can also generate a $\Delta\psi$ of slightly greater magnitude (in phototrophically-grown cells) to that generated during nitrate respiration (data not shown). That the O_2-dependent $\Delta\psi$ is insufficient to cause a thermodynamic "back-pressure" on the rate of nitrate reduction was shown by the failure of uncoupling agents to enhance the rate of nitrate reduction in aerobic cells. We must therefore seek another explanation for the oxygen-induced inhibition of nitrate uptake. The rates of oxygen uptake and nitrate uptake in these bacteria are comparable (routinely about 7 μg atom O and 2 μmol NO_3 per μmol bacteriochlorophyll per minute) and so it is unlikely that competition for reducing equivalents could explain the strong (100%) inhibitory effect of oxygen on nitrate reduction. This point is discussed at length by Alefounder et al. 1983 with reference to oxygen-inhibition of nitrate reduction in P. denitrificans. Also in P. denitrificans uncoupling agent does not prevent oxygen-inhibition of nitrate reduction. It is however worth mentioning that there are two distinguishing features between these apparently similar phenomena in Rps. capsulata strain N22DNAR[+] and P. denitrificans. (1) in the former organism, nitrate reductase is periplasmic (McEwan et al. 1983) ruling out any control at the level of membrane transport as suggested for the latter organism (Alefounder et al. 1983); (2) in contrast to the situation in P. denitrificans, antimycin A treatment of Rps. capsulata N22DNAR[+] does not prevent the inhibitory response so that a direct, reversible inhibitory effect of oxygen on nitrate reductase can not be ruled out.

ACKNOWLEDGEMENTS

This work was supported by a grant from the SERC to JBJ and SJF.

REFERENCES

Alefounder, PR, Greenfield AJ, McCarthy JEG and Ferguson, SJ (1983) Biochim. Biophys.Acta, in press.
Clark, AJ, Cotton NPJ and Jackson JB (1983) Eur. J. Biochem. 130, 575-580
Cotton NPJ, Clark AJ and Jackson JB (1983) Eur. J. Biochem. 130, 581-587
Jackson, MA, Jackson JB and Ferguson SJ (1981) FEBS Lett. 136, 275-278
Keister, DL (1978) In:'The Photosynthetic Bacteria' (Clayton RK and Sistrom WR, eds.) pp. 849-856, Plenum Press, New York and London
Klemme J-H (1979) Microbiologica, 2, 415-420
McCarthy, JEG and Ferguson SJ (1982) Biochem. Biophys.Res.Commun. 107, 1406-1411
McEwan AG, George, CL, Ferguson SJ and Jackson JB (1982) FEBS Lett. 150, 277-280
McEwan, AG, Jackson JB and Ferguson, SJ (1983) In preparation.
Ramirez, J and Smith, L (1968) Biochim.Biophys.Acta, 153, 466-475
Rigolo, M and Zannoni, D (1983) Biochem.Biophys.Res.Commun. 113, 155-162

GLYCERATE TRANSPORT ACROSS THE CHLOROPLAST ENVELOPE

S.P. ROBINSON

1. INTRODUCTION

As a part of the photorespiratory carbon oxidation pathway hydroxypyruvate is converted to glycerate in the leaf peroxisome (Tolbert 1981). For this carbon to re-enter the photosynthetic carbon reduction cycle, the glycerate is phosphorylated to form 3-phosphoglycerate by the enzyme glycerate kinase, which is located in the chloroplast stroma (Usuda, Edwards 1980). Thus there needs to be a controlled movement of glycerate into the chloroplast during photorespiration. Previous studies have indicated that there is a specific transporter on the chloroplast envelope to facilitate movement of glycerate into the chloroplast (Robinson 1982). In contrast to other chloroplast transporters glycerate uptake is markedly increased by illumination of the chloroplast suspension. Glycerate has a pKa of 3.55 so at neutral pH more than 99% exists as the glycerate anion. It seems likely that uptake of the glycerate anion would be coupled to symport of a cation or antiport of another anion to maintain charge balance across the envelope. The mechanism of the light stimulation of glycerate uptake is of particular interest since it may provide information about the way in which ion movements across the chloroplast envelope are controlled. It is clear that these ion gradients play an important part in the regulation of photosynthetic carbon metabolism (Heber, Heldt 1981).

2. MATERIALS AND METHODS

Spinach was grown in gravel pots irrigated with half-strength Hoaglands nutrient solution for 3–4 weeks in a glasshouse. Peas were grown in vermiculite for 10–12 days. Intact chloroplasts were isolated from spinach leaves or from pea shoots and purified on a 2-step Percoll gradient as described previously (Robinson 1982). The chloroplasts were 95–99% intact based on ferricyanide penetration and exhibited rates of CO_2-dependent oxygen evolution of 120–200 μmoles.mg $Chl^{-1}.h^{-1}$. Uptake of ^{14}C-glycerate was measured using the technique of silicone oil filtering centrifugation (Heldt 1980). Correction for material outside the chloroplast stroma which was carried through the silicone oil was made by determining the sorbitol and water spaces (Heldt 1980). To determine the extent to which ^{14}C-glycerate which entered the chloroplast stroma was subsequently metabolised (Tables 2 and 3) the pellet fraction from silicone oil filtering centrifugation was extracted in water and protein precipitated with $HClO_4$. The deproteinised sample was neutralised with K_2CO_3, concentrated and then chromatographed on Whatman 3MM paper using butanol: acetic acid: water (12:3:5) as solvent. Radioactive spots were located with a chromatograph scanner and eluted for quantitation by scintillation counting. From these results the actual glycerate concentration in the chloroplast stroma was calculated as before (Robinson 1982).

3. RESULTS AND DISCUSSION

In the dark, uptake of ^{14}C-glycerate by isolated chloroplasts ceased after the first 20–30 seconds. Metabolism of glycerate was negligible in the dark but the concentration of glycerate in the chloroplast stroma never exceeded that in the surrounding medium. Light increased both the extent and the initial rate of uptake of ^{14}C-glycerate. Part of this light effect was a result of increased metabolism of the glycerate inside the chloroplast leading to an accumulation of ^{14}C-labelled products (mostly 3-phosphoglycerate and triose phosphates) in the chloroplast stroma. The extent of this

Sybesma, C. (ed.), Advances in Photosynthesis Research, Vol. II. ISBN 90-247-2943-2.
© *1984 Martinus Nijhoff/Dr W. Junk Publishers, The Hague/Boston/Lancaster.*

metabolism was determined by paper chromatography and the actual glycerate concentration in the chloroplast stroma was calculated. Even after correction for metabolism there was a marked light stimulation of the initial rate of uptake and glycerate was accumulated in the chloroplast in the light. Depending on the assay conditions the concentration of glycerate inside the chloroplast was up to six times that of the suspending medium, in agreement with previous results (Robinson 1982).

Relatively low light intensities stimulated ^{14}C-glycerate uptake. With red light (630–730 nm halfband) half maximal stimulation required only 10–20 $W.m^{-2}$ and light intensities as low as 5 $W.m^{-2}$ significantly increased uptake. In contrast, CO_2-dependant oxygen evolution required more than 200 $W.m^{-2}$ for light saturation. When the light was switched off following a period of illumination in the presence of ^{14}C-glycerate there was an efflux of label from the chloroplast to the suspending medium. This implies that light is required to maintain a concentration gradient of glycerate across the chloroplast envelope. Preillumination of chloroplasts stimulated uptake of ^{14}C-glycerate during a subsequent dark period which suggests that light itself was not required but rather some product formed during illumination of the chloroplasts.

The effect of a number of inhibitors on ^{14}C-glycerate uptake was determined to investigate the mechanism of the light stimulation and the results are presented in Table 1. The light effect was almost completely abolished by DCMU suggesting that non-cyclic electron transport was required. Antimycin A, which is considered to be an inhibitor of cyclic electron transport, did not inhibit the light-stimulated glycerate uptake.

The proton ionophore FCCP inhibited ^{14}C-glycerate uptake in the light to a similar extent as DCMU whilst nigericin, which is also an uncoupler in chloroplasts, completely inhibited uptake including that normally observed in the dark. Although NH_4Cl is not a potent uncoupler of intact chloroplasts, 10 mM NH_4Cl inhibited CO_2-dependent oxygen evolution by 40–50%. In contrast to FCCP and nigericin, NH_4Cl actually stimulated ^{14}C-glycerate uptake in the light as was observed previously (Robinson 1982). Half-maximal stimulation occurred with approximately 1 mM NH_4Cl.

TABLE 1. Effect of inhibitors on ^{14}C-glycerate uptake by spinach chloroplasts.

Assay Conditions	^{14}C-glycerate uptake (% control)
Dark	13
Light	100
Light + DCMU (5 μM)	27
Light + Antimycin A (5 μM)	97
Light + FCCP (10 μM)	26
Light + Nigericin (10 μM)	0
Light + NH_4Cl (10 mM)	185

^{14}C-glycerate uptake was determined at 2°C, pH 7.0 with 0.5 mM glycerate in the suspending medium. Incubation was for 20 sec. Control rate = 11.6 μmol.mg $Chl^{-1}.h^{-1}$.

Uncouplers inhibit ATP sythesis in chloroplasts but also inhibit the formation of proton and other ion gradients across the thylakoid membranes and so prevent light-induced alkalisation of the stroma. The inhibition of ^{14}C-glycerate uptake by uncouplers implies that one or more of these processes are involved in the light stimulation of uptake. Arsenate inhibits ATP synthesis by substituting for phosphate in the phosphorylation reaction and phlorizin is an energy transfer inhibitor. Both stop ATP formation without inhibiting the formation of ion gradients across the thylakoid and envelope membranes. At concentrations which inhibited CO_2-dependent oxygen evolution by more than 90% both arsenate and phlorizin inhibited uptake of ^{14}C-glycerate, but only by 40-50% (Table 2). This decrease was largely because of reduced metabolism of the glycerate by the chloroplasts (which would require ATP) since the percentage of label in the chloroplast remaining as glycerate was increased by both inhibitors. As a result, the actual concentration of glycerate in the chloroplast was only reduced by 10-20% in the presence of arsenate or phlorizin. This suggests that ATP is required for metabolism of glycerate in the chloroplast but not for the light-stimulated transport of glycerate across the chloroplast envelope.

TABLE 2. Effect of arsenate and phlorizin on glycerate uptake by spinach chloroplasts.

Assay Conditions	^{14}C- Uptake (% control)	% ^{14}C as glycerate	Glycerate in stroma (% control)
Light	100	65	100
Light + Arsenate (2 mM)	55	93	92
Light + Phlorizin (2 mM)	66	77	78

^{14}C-glycerate uptake was determined at $2°C$, pH 7.0 with 0.5 mM glycerate in the suspending medium. Incubation was for 15 sec.

With chloroplasts isolated from young pea shoots inorganic pyrophosphate and ATP analogs enter via the adenine nucleotide transporter in exchange for endogenous nucleotides at sufficient rates to deplete the chloroplasts of ATP (Robinson, Wiskich 1977a,b). As shown in Table 3, incubation of pea chloroplasts with pyrophosphate did not inhibit ^{14}C-glycerate uptake but metabolism of label by the chloroplasts was decreased resulting in an increased glycerate concentration in the chloroplast stroma. The ATP analog, AMP-PCP, inhibited ^{14}C-glycerate uptake by 40% but this was a result of decreased metabolism so the actual glycerate concentration in the chloroplast stroma was again more than double the control value (Table 3). Since both these treatments deplete the chloroplasts of ATP (as evidenced by the decreased metabolism) these results also suggest that ATP is not required for the light-stimulation of glycerate transport.

TABLE 3. Effect of PPi and AMP-PCP on glycerate uptake by pea chloroplasts.

Assay Conditions	^{14}C- Uptake (% control)	% ^{14}C as glycerate	Glycerate in stroma (% control)
Light	100	46	100
Light + PPi (5 mM)	102	59	202
Light + AMP-PCP (1 mM)	61	88	223

^{14}C-glycerate uptake was determined at $20°C$, pH 7.0 with 0.5 mM glycerate in the suspending medium. Incubation was for 20 sec.

4. CONCLUSIONS

4.1   Glycerate transport across the chloroplast envelope is carrier mediated and glycerate uptake is stimulated by light.

4.2 Non-cylic electron transport is required.

4.3 Uptake is sensitive to uncouplers, implying the involvement of ion gradients and/or ATP.

4.4 ATP is required for metabolism of glycerate in the chloroplast stroma but not for the actual transport across the chloroplast envelope.

4.5   Glycerate transport may involve ion gradients which are generated across the chloroplast envelope in the light independently of ATP.

5. REFERENCES

Heber U and H W Heldt (1981) The chloroplast envelope: structure, function, and role in leaf metabolism. Annu. Rev. Plant Physiol. 32, 139–168.
Heldt H W (1980) Measurement of metabolite movement across the envelope and of the pH in the stroma and the thylakoid space in intact chloroplasts. Methods Enzymol. 69, 604–613.
Robinson SP (1982) Transport of glycerate across the envelope membrane of isolated spinach chloroplasts. Plant Physiol. 70, 1032–1038.
Robinson SP and JT Wiskich (1977a) Pyrophosphate inhibition of carbon dioxide fixation in isolated pea chloroplasts by uptake in exchange for endogenous adenine nucleotides. Plant Physiol. 59, 422–427.
Robinson SP and JT Wiskich (1977b) Uptake of ATP analogs by isolated pea chloroplasts and their effects on CO_2 fixation and electron transport. Biochim. Biophys. Acta 461, 131–140.
Tolbert NE (1981) Metabolic pathways in peroxisomes and glyoxysomes. Annu. Rev. Biochem. 50, 133–157.
Usuda H, GE Edwards (1980) Localization of glycerate kinase and some enzymes for sucrose synthesis in C3 and C4 plants. Plant Physiol. 65, 1017–1022.

ACKNOWLEDGEMENTS

I am grateful to Marcia McFie for her skilled technical assistance.

Authors address: CSIRO Division of Horticultural Research, G.P.O. Box 350, Adelaide, 5001, Australia.

A GLYCOLATE TRANSPORTER IN THE CHLOROPLAST ENVELOPE

KONRAD T. HOWITZ AND RICHARD E. MC CARTY

1. INTRODUCTION

Glycolate, the substrate for photorespiration (Tolbert, 1981), is formed in the chloroplast stroma. Because glycolate is metabolized in other organelles, it must cross the chloroplast envelope. Although envelopes are permeable to glycolic acid (Howitz, McCarty, 1982), the rate of glycolate penetration at physiological pH values by simple diffusion of glycolic acid is much slower than that of photorespiration. More recently we detected an extremely rapid phase of glycolate uptake by chloroplasts (Howitz, McCarty, 1983). When glycolate uptake is measured within the first three seconds after addition of the glycolate, glycolate uptake rates saturate with respect to glycolate concentration and are inhibited by sulfhydryl reagents and by structurally-related compounds. These results suggest that glycolate transport across the chloroplast envelope is a mediated process. In this report, we present some new data on glycolate transport.

2. MATERIALS AND METHODS

Intact chloroplasts were prepared from pea shoots (Howitz, McCarty, 1982). Glycolate uptake was estimated by a silicone oil microcentrifugation technique. Centrifugation was initiated within 1 sec after addition of [^{14}C]glycolic acid. Stromal volumes were determined from the uptake of ^3H_2O, using [^{14}C]sorbitol as the marker for external solution that is carried through the silicone oil with the chloroplasts. pH changes in the medium on addition of glycolate solutions were measured with a Thomas combination glass electrode and a Heath electrometer. The chloroplasts were illuminated for 5 min prior to glycolate addition in 330 mM sorbitol, 0.1 mM HEPES-NaOH, 1 mM $MgCl_2$ and 25 mM KCl at 4°C. The initial pH was 7.00 and a change in pH of 0.1 gave a full scale deflection of the recorder pen. The medium used for measuring glycolate uptake was similar to that used for monitoring pH changes except that 50 mM HEPES-NaOH (pH 7.0) was present.

3. INHIBITION OF GLYCOLATE UPTAKE

The mediated phase of glycolate uptake is too fast even at 0-4°C to allow accurate measurement of its initial rates by silicone oil centrifugation. The sensitivity of the rapid phase of uptake to N-ethylmaleimide suggests that a sulfhydryl inhibitor might be used to stop uptake rapidly. Provided the leak of glycolate out of the inhibited chloroplasts is slow, the kinetics of uptake could be determined by silicone oil centrifugation. In Table I, it is shown that $HgCl_2$, added together with the [^{14}C]glycolate, is an effective inhibitor of glycolate uptake, an observation that provides further evidence for the involvement of thiol groups in uptake. However, the leak of glycolate into the inhibited chloroplasts is still sufficiently fast to render the inhibitor stop approach unfeasible.

Sybesma, C. (ed.), Advances in Photosynthesis Research, Vol. II. ISBN 90-247-2943-2.
© *1984 Martinus Nijhoff/Dr W. Junk Publishers, The Hague/Boston/Lancaster.*

TABLE I. Effect of $HgCl_2$ on glycolate uptake[a]

$HgCl_2$ - mM	Stromal Glycolate Concentration
	(mM)
0	1.0
0.01	1.1
0.10	0.2
1.0	0.15

[a] Chloroplast were illuminated for 1 min prior to addition of $HgCl_2$ and glycolate (0.4 mM). After 1 min, the chloroplast were centrifuged through silicone oil.

Previously we showed (Howitz, McCarty, 1983) that D,L-glycerate and glyoxylate inhibited glycolate uptake, whereas phosphate, acetate and bicarbonate had little effect. A number of other compounds have been tested as potential inhibitors of glycolate accumulation (Table 2).

TABLE 2. Effect of metabolites on glycolate uptake

Addition[a]	Rates of Glycolate Uptake[b]	% of Control
	(μmol mg chlorophyll^{-1} h^{-1})	
None	12	100
Phosphate	14	117
Phosphoglycerate	15	125
Dihydroxyacetone phosphate	14	117
Glycolate	5	42
Lactate	4	33
D,L-glycerate	7	58
Glyoxylate	3	25
Succinate	12	100
Malate	12	100

[a] Assays were carried out with 1 mM [^{14}C]glycolate and 5 mM of the added compounds.

[b] Mean of three determinations.

The α-hydroxymonocarboxylic acids, glycerate and lactate, are good inhibitors of glycolate uptake. Glyoxylate, which in its hydrated form is also an α-hydroxy acid, inhibits as well. Interestingly, phospho-glycerate is not an inhibitor. This fact, combined with the lack of effect of either phosphate or dihydroxyacetone phosphate, suggests that the phosphate translocator is not involved in glycolate uptake. How-ever, the substrate sensitivity of glycolate uptake resembles that of glycerate uptake (Robinson, 1982). Whether the inhibitory metabolites are transported has not been established.

4. pH CHANGES ASSOCIATED WITH GLYCOLATE UPTAKE

If the mechanism of glycolate uptake involves glycolate-H^+ symport (or glycolate-OH^- antiport) or if glycolate is transported as glycolic acid, an alkalinization of the medium would be associated with glycolate uptake. This has been shown to be the case (Table 3). The inhibition of proton uptake by N-ethylmaleimide and protection from this inhibition

TABLE 3. pH changes associated with glycolate uptake

Pretreatment of Chloroplasts	Proton uptake due to the addition of 2.1 mM glycolate (neq mg chlorophyll^{-1})	Change in pH
None	23.7	+0.053
5 mM N-ethylmaleimide	3.1	-0.007
5 mM N-ethylmaleimide + 10 mM glycolate	25.9	+0.058

by glycolate strongly indicates that the proton movement is associated with the transporter-mediated accumulation of glycolate. Unfortunately, the response of the glass electrode is too slow to allow its use in kinetic studies. However, it should be possible to follow the time course of glycolate-dependent pH changes spectrophotometrically with pH indicators.

5. CONCLUSIONS

Our results provide further evidence that glycolate transport across envelope membranes is a mediated process. Its sensitivity to sulfhydryl reagents and to α-hydroxy acids is good evidence in favor of this conclusion. In addition, the finding that the rapid phase of glycolate uptake is proton-linked provides a simplified procedure for the study of the kinetics of glycolate transport. The rates of glycolate transport by the mediated process appear to be in line with those of photorespiration.

REFERENCES

Howitz KT and McCarty RE (1982) pH dependence and kinetics of glycolate uptake by intact pea chloroplasts, Plant Physiol. 70, 949-952.
Howitz KT and McCarty RE (1983) Evidence for a glycolate transporter in the envelope of pea chloroplasts, FEBS Lett. 154, 339-342.
Robinson SP (1982) Transport of glycerate across the envelope membrane of isolated spinach chloroplasts, Plant Physiol. 70, 1032-1038.
Tolbert NE (1981) Metabolic pathways in peroxisomes and glyoxysomes, Annv. Rev. Biochem. 50, 133-157.

Authors' Address: Section of Biochemistry, Molecular and Cell Biology, Cornell University, Ithaca, NY 14853 USA

REGULATION OF THE EXPORT OF THE REDUCING POWER FROM CHLOROPLAST TO CYTOSOL
IN RELATION WITH MOLECULAR PROPERTIES OF CHLOROPLASTIC NADP-MDH.

N. FERTE, J.C. MEUNIER, J. BUC.

We have already described the purification of a chloroplastic NADP-
malate dehydrogenase (Ferté et al., 1982). It is a 56 000 daltons dimeric
protein made up of two apparently identical subunits. This enzyme plays an
important role in the generation of reducing power used in the reduction
cytoplasmic reactions. NADPH from photosynthesis reduces oxaloacetate to
malate in the stroma. This reaction is catalyzed by NADP-MDH. A malate
translocator exports this acid outside the chloroplast where NAD-MDH
generates NADH.
Now, we describe the purification of four NADP-MDH till homogeneity.
For one of them, MDHA, we give some molecular properties in relation with
the modulation of the export of the reducing power from chloroplast to cytosol.

MATERIALS AND METHODS

Spinach leaves were purchased from a local market. $NADP^+$, NADPH, dithiothreitol, oxaloacetate were purchased from Boehringer, the matrex gel (red A)
from Amicon, cibacron blue and the 2',5'-ADP-Sepharose 4 B from Pharmacia.
Other standard reagents were of the highest grade commercially available.
Thioredoxins were purified from spinach chloroplasts according to J. Buc
et al. (unpublished).
For the determination of the sulfhydryl groups, the method of Habeed
was used. Disulfide bridges were determinated by the same method with urea
and DTT. The protein content was estimated by the method of Bradford with
BSA as a standard. The molecular weight determinations were effected by
molecular sieving over Sephadex G-200 and by sedimentation equilibrium.
Analytical centrifugations experiments were performed with a Beckman model E
ultracentrifuge equipped with an interferometric system for Mr determination.
For the active enzyme sedimentation constant determination, we have used
the method of Cohen and Mire (CEA method).
Slabs gels electrophoresis was performed by the method of Laemmli.
The MDH activities were followed by the decrease of the NADPH absorbance
at 340 nm with a Gilford spectrophotometer after incubation for 15 mn in
10 mM DTT, 100 mM buffer phosphate, pH 8, 30°C.

RESULTS

Extraction of spinach NADP-MDH was achieved in a Waring blendor by
50 mM tris-HCl buffer, pH 7.5, 2°C. After centrifugation, the crude extract
was acidified to pH 4.8 with formic acid. The supernatant after centrifugation, was precipitated with 55 % ammonium sulfate. The pellet was dialyzed
against 5 mM phosphate buffer (pH 7) and submitted to DEAE-cellulose chromatography (Fig. 1 a). Peak one is MDH A with a fraction of MDH B. The
purification of MDH A is carried out by an affinity chromatography on
matrex gel and on 2',5'-ADP-Sepharose. Peak two of DEAE-cellulose is MDH B,
C and D. These fractions were collected, concentrated, dialyzed and submitted to affinity chromatography on matrex gel. Elution with a linear
gradient of KCl allowed the separation between the three isoenzymes
(Fig. 1 b).

Sybesma, C. (ed.), Advances in Photosynthesis Research, Vol. II. ISBN 90-247-2943-2.
© *1984 Martinus Nijhoff/Dr W. Junk Publishers, The Hague/Boston/Lancaster.*

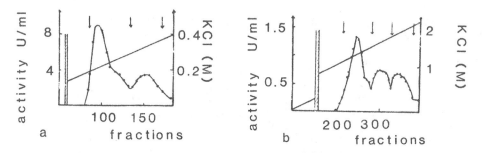

FIGURE 1. a) DEAE-cellulose chromatography column after acid and ammonium sulfate fractionations; b) separation of MDH B, C and D on matrex column.

MDH B was applied to a 2',5'-ADP-Sepharose (Fig. 2 a) and MDH C and D were submitted to an affinity chromatography on cibacron (Fig. 2 b).

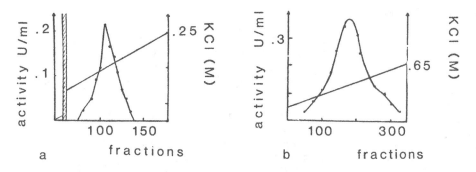

FIGURE 2. Affinity chromatography a) of MDH B on ADP-Sepharose and b) of MDH C (or D) on cibacron columns.

The purification of MDH C and D is achieved by a molecular sieving over P-150 Biogel. These isoenzymes are homogeneous as determined by 10 % polyamide slabs gel electrophoresis and analytical centrifugation (Fig. 3).

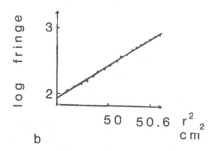

FIGURE 3. a) polyacrylamide gel electrophoresis (10 μg for each isoenzyme); b) equilibrium sedimentation of MDH D. It is obtained the same pattern for the others MDH.

In the table 1, it is shown the molecular weights and the -SH contents of the isoenzymes in different conditions.

TABLE 1. Molecular weights and -SH contents.

	Sulfhydryl groups	Molecular weight by sedimentation equilibrium	Molecular weight
Oxidized native MDH A B C D isoenzymes	0	56 000	Sephadex G-200 56 000 ± 3 000
Denatured isoenzyme A	0	56 000	
Reduced denatured MDH isoenzyme A	6	28 000	SDS electrophoresis 28 000 ± 3 000

The apparent sedimentation constant of native MDH A from absorption optics is 4.35 S while the sedimentation constant from the CEA method is 4.35 S. The dimer is the active species. These results and the ones from table 1 suggest that the protein is made up of 2 apparently subunits linked by a disulfide bridge and each monomer contains one disulfide bridge. Then the DTT activation is due to the reduction of the two disulfide bridges of the dimer. And the inactivation by time dependent DTT concentration is due to the aggregation of the dimeric MDH protein (Fig. 4 a) for while the denatured reduced enzyme remains monomeric, the inactive non-denatured DTT reduced enzyme aggregates (s^{APP} is higher than 800 S, figure 4 b). After removal of DTT, activity is no longer detectable. Aggregation is irreversible.

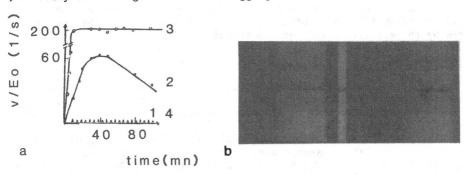

a

b

time(mn)

FIGURE 4. a) MDH activity for (1) $[DTT]/[MDH] = 25 \times 10^3$; (2) $[DTT]/[MDH] = 3.8 \times 10^6$; (3) $[DTT]/[MDH] = 25 \times 10^3$ + thioredoxin (1.4 mM); (4) as (3) but after removal of reductants. b) Determination of the apparent sedimentation constant, s^{APP}, of non-denatured MDH reduced by DTT (4 000 rpm).

On figure 4 a it is also seen the activation of MDH A by DTT alone and thioredoxin m. Full activation may be obtained in the presence of thioredoxin m in about 7 mn, which is compatible with light activation in vivo. If DTT and thioredoxin are rapidly removed by molecular filtration from a mixture of the two reactants plus enzyme, activity is no longer detectable (Fig. 4 a).

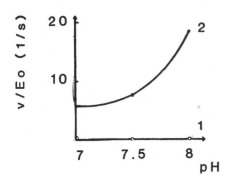

FIGURE 5. Activity of oxidized
(1) and reduced (2) MDH between pH 7 to 8

Between pH 7 to 8, oxidized MDH has
no activity (Fig. 5), but reduced it
displays some activity. The relative
rate is higher at pH 8 (in the light)
than pH 7 (in the dark). This result
is at the variance of the FBPase behavior for which reduced enzyme does not
display any activity at pH 7. This enzyme slowly changes its activity and
its conformation upon a pH-Jump (Minot
et al., 1982). Its activation is explainable by either a steep sensitivity of
enzyme to pH changes or by electron-
transfer from reduced thioredoxin and
its deactivation by either pH changes
or by electron transfer to oxidized
thioredoxin (Soulié et al., 1981).

CONCLUSION

The activation of MDH is the result of a protein-protein interaction
and its deactivation is spontaneous and fast. We think this interaction
plays a role in the regulation of the export of reducing power from stroma
chloroplast to cytosol. At pH 8, MDH reduced by thioredoxin allows the
export of the reducing power by the C_4-dicarboxylic acid shuttle, while in
the dark rapidly MDH is getting oxidized, thence inactive and the possible
import of reducing power (and the generation of NADPH upon dark in the
chloroplast) is not allowed. The spontaneous and rapid oxidation upon
removal of reductant prevents reduced MDH activity at pH 7 and appears as
an efficient molecular device to regulate the export of reducing power
from chloroplast stroma. It is obligatory to prevent protein aggregation by
sulfhydryl-disulfide interchange for avoiding inhibition : that is realized by the specific action of thioredoxin.

REFERENCES

1. Soulié, J.M., Buc, J., Meunier, J.C., Pradel, J. and Ricard, J. (1981)
Molecular properties of chloroplastic thioredoxin f and the photoregulation
of the activity of fructose 1,6-bisphosphatase. Eur. J. Biochem., 119, 497-502
2. Ferté, N., Meunier, J.C., Ricard, J., Buc, J. and Sauve, P. (1982)
Molecular properties and thioredoxin-mediated activation of spinach chloroplastic NADP-malate dehydrogenase. FEBS Letters, 146, 133-138.
3. Minot, R., Meunier, J.C., Buc, J. and Ricard, J. (1982). The role of pH
and Mg^{++} concentration in the light activation of chloroplastic fructose
bisphosphatase. FEBS Letters, 142, 118-120.

ACKNOWLEDGEMENTS

The authors are grateful to P. Sauve who performed the analytical
centrifugation and P. Cassa for typing the manuscript.

Authors address : N. Ferté, J.C. Meunier, J. Buc. CBM-CNRS, BP 71
31, Chemin Joseph Aiguier, 13402 Marseille Cedex 9, France.

Pi AND G6P TRANSLOCATION IN CHLOROPLASTS OF CODIUM FRAGILE

J C RUTTER and A H COBB

1. INTRODUCTION

In higher C_3 plant chloroplasts the export of photosynthates, according
to Pi availability and extrachloroplastic photosynthate demand, is partly
regulated by an envelope-located translocator which facilitates the exchange of Pi with low molecular weight phosphorylated carbon compounds,
being mainly specific for 3-phosphoglycerate (PGA) and triose phosphates
(Heldt, Rapley, 1970). It is not directly light-dependent and is inhibited
by p-chloromercuriphenyl sulphonic acid (pCMS) at concentrations as low as
50μM to which it shows considerably greater sensitivity than other translocators (Werdan, Heldt, 1971). The metabolite translocation properties
of algal chloroplasts, however, remain a largely unexplored field of investigation and the present work examines some aspects of phosphate translocation in chloroplasts of the intertidal species Codium fragile.

2. MATERIALS AND METHODS

Fronds of C.fragile, sampled from intertidal rock pools at Bembridge, Isle
of Wight, were maintained in tanks of aerated sea water pending experimental
use. Chloroplasts were isolated from the frond tips according to the method
of Cobb (1977).

Uptake of labelled metabolites was measured by silicone oil filtration
centrifugation and metabolite counter exchange studies were conducted by
determining the effects of exogenous non-labelled substances on the release of labelled metabolites assayed in supernatents following rapid
centrifugation of incubation mixtures (Rutter, Cobb, 1984).

3. RESULTS

From an external 1 mM concentration ^{32}Pi was taken up at initial rates of
up to 52.2μmoles.mg^{-1} Chl.h^{-1} at 12°C. This is comparable to corresponding
rates of 25 and 40.4μmoles.mg^{-1}.Chl.h^{-1} at 4°C (Heldt, Rapley, 1970;
Werdan, Heldt, 1971 respectively) for Spinacea oleracea chloroplasts.
However, incubation temperatures in the present work were considerably
higher than those used by these authors. Thus, uptake of ^{32}Pi by C.fragile
frond-tip chloroplasts may be significantly slower than for S.oleracea at
the same incubation temperature. From Fig. 1(a) it is seen that ^{32}Pi
uptake was light-stimulated but not light-dependent. The uptake of (^{14}C)
G6P however, appeared to be unaffected by changes in light intensity up
to 300μE.M^{-2}.s^{-1} (Fig. 1 b), proceeding at much slower rates of 1.1-1.2
μmoles.mg^{-1} Chl.h^{-1} at 14.5°C.

50μM pCMS produced a rapid and marked inhibition of ^{32}Pi uptake (Fig. 1
c), although (^{14}C) G6P uptake showed a considerably lesser, more time-
dependent sensitivity to this inhibitor (Fig. 1 d). Although direct
studies of radiolabelled PGA uptake were not performed, its translocator-
facilitated movement into chloroplasts is indicated by its stimulation of
label release from ^{32}Pi-preloaded chloroplasts (Fig. 2). Similar investigations were also conducted with glucose and G1P. Whereas G6P stimulated

Sybesma, C. (ed.), Advances in Photosynthesis Research, Vol. II. ISBN 90-247-2943-2.
© 1984 Martinus Nijhoff/Dr W. Junk Publishers, The Hague/Boston/Lancaster.

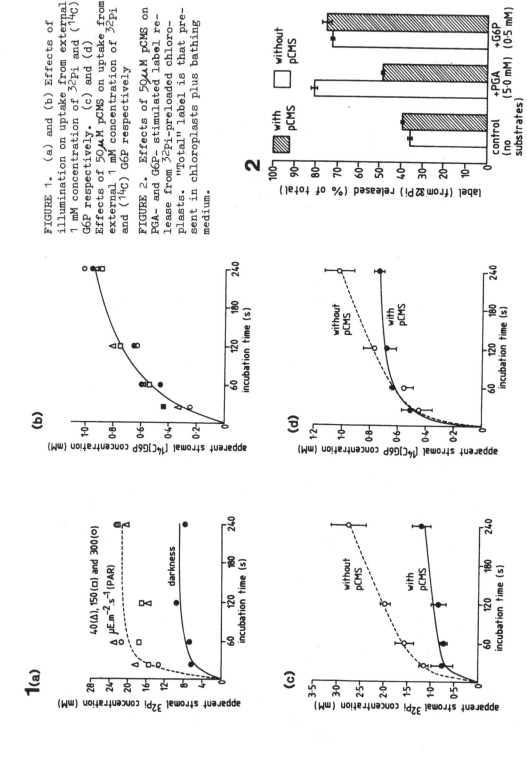

FIGURE 1. (a) and (b) Effects of illumination on uptake from external 1 mM concentration of ^{32}Pi and (^{14}C) G6P respectively. (c) and (d) Effects of 50 μM pCMS on uptake from external 1 mM concentration of ^{32}Pi and (^{14}C) G6P respectively

FIGURE 2. Effects of 50 μM pCMS on PGA- and G6P-stimulated label release from ^{32}Pi-preloaded chloroplasts. "Total" label is that present in chloroplasts plus bathing medium.

label release (Fig. 2) no such effect was observed with either glucose or G1P (data not shown). The apparent stimulation of ^{32}Pi release from chloroplasts by G6P showed no sensitivity to pCMS unlike that caused by PGA (Fig. 2).

4. DISCUSSION

The characteristics of Pi translocation in C.fragile frond-tip chloroplasts and its interactions with PGA and pCMS indicate the presence of a phosphate translocator similar to that in chloroplasts of S.oleracea (cf Heldt, Rapley, 1970; Werdan, Heldt, 1971). However, ^{32}Pi uptake by the algal chloroplasts was light-stimulated unlike that reported for S.oleracea (Werdan, Heldt, 1971). In the case of C.fragile, rapid incorporation of Pi into polyphosphates (which, as far as is currently known, are absent from higher plant chloroplasts) under photosynthetic conditions such as that reported by Cobb (1978) may provide a powerful sink favouring chloroplastic Pi uptake. Thus, light-stimulated Pi uptake by these organelles may not necessarily be due to any direct effect of illumination on a phosphate translocator.

G6P also stimulated label release from ^{32}Pi-preloaded chloroplasts. However, experimental evidence suggests a G6P translocation system separate to the one specific for PGA. Firstly, G6P-stimulated label release was unaffected by pCMS, unlike that due to PGA, and also (^{14}C) G6P uptake was slower and showed considerably less sensitivity to this inhibitor than that of ^{32}Pi. Thus, it is therefore possible that at least two spatially separate and distinct phosphate translocators may exist in the envelopes of C.fragile frond-tip chloroplasts:- (a) one which can facilitate the exchange of Pi with PGA, but not G6P, is pCMS-sensitive and may be light-stimulated (hereafter termed the "Pi-PGA" translocator) and (b) another which can facilitate the slower exchange of Pi with G6P, but which is considerably less sensitive to pCMS and is light-independent (hereafter termed the "Pi-G6P" translocator). The specificity of the Pi-PGA translocator may also include triose phosphates although this remains to be verified by further investigation.

Although G6P may originate cytoplasmically from triose phosphates, a translocator facilitating its direct export from the chloroplast may be advantageous where its extrachloroplastic demand is especially high, eg in cell wall formation which appears to be a prominent metabolic process in C.fragile as indicated by the ultrastructural studies of Hawes (1979). On the other hand the Pi-PGA translocator may provide the major route for photosynthate export from the chloroplasts as in higher C_3 plants. The roles of the phosphate translocators proposed here in the integration of chloroplastic and cytoplasmic metabolism in C.fragile therefore remain an important subject for further investigation.

REFERENCES

Cobb AH (1977) The relationship of purity and photosynthetic activity in preparations of Codium fragile chloroplasts, Protoplasma 92, 137-146.
Cobb AH (1978) Inorganic polyphosphate involved in the symbiosis between chloroplasts of the alga Codium fragile and mollusc Elysia viridis, Nature 272, 554-555.
Hawes CR (1979) Ultrastructural aspects of the symbiosis between algal chloroplasts and Elysia viridis, New Phytol. 83, 445-450.

Heldt HW and Rapley L (1970) Specific transport of inorganic phosphate, 3-phosphoglycerate and dihydroxyacetone phosphate and of dicarboxylates across the inner membrane of spinach chloroplasts, FEBS Lett. 10, 143-148.
Werdan K and Heldt HW (1971) The phosphate translocator of spinach chloroplasts. In Proc. 2nd Int. Cong. Photosynth., Stresa, vol 2, pp 1339-1344.
Rutter JC and Cobb AH (1984) Translocation of orthophosphate and glucose-6-phosphate in Codium fragile chloroplasts, New Phytol, (in press).

ACKNOWLEDGEMENTS

The authors thank the Science and Engineering Research Council for financial support.
Authors address: Department of Life Sciences, Trent Polytechnic, Nottingham
                  NG1 4BU, U.K.

THE LACTOSE CARRIER OF ESCHERICHIA COLI FUNCTIONALLY INCORPORATED IN
RHODOPSEUDOMONAS SPHAEROIDES OBEYS THE REGULATORY CONDITIONS OF THE
PHOTOTROPHIC BACTERIUM

MARIEKE G.L. ELFERINK, KLAAS J. HELLINGWERF, FRANCIS E. NANO, SAMUEL
KAPLAN, WIL N. KONINGS

1. INTRODUCTION

Regulation of solute transport in Rps. sphaeroides is such that the presence of a proton motive force by itself is not enough for solute transport
to take place. In addition linear or cyclic electron transfer is required.
In cells incubated anaerobically in the dark, no uptake of alanine is
measurable, in spite of the presence of a high proton motive force. Illumination of such a suspension initiates cyclic electron transfer and
allows solute uptake to proceed. The uptake rate of alanine increases
with the light intensity (Elferink et al., 1983).
To investigate the importance of this type of regulation of solute transport for other (non-phototrophic) bacteria, Rps. sphaeroides was provided
with the ability to transport lactose via transformation with a plasmid
containing the lactose operon from Escherichia coli and subsequent expression of the plasmid DNA with the aid of the antibiotic kanamycine
(F.E. Nano, S. Kaplan, unpublished). The initial rate of lactose transport
in Rps. sphaeroides was studied as a function of the light intensity
and the magnitude of the proton motive force.
The results demonstrate that (i) lactose transport is regulated by the
rate of cyclic electron transfer in the same way as the endogenous transport systems, (ii) this type of regulation gives rise to homeostasis of
the magnitude of the proton motive force and allows the cells large variations in the rate of energy consuming processes with only minor variations
in the magnitude of the proton motive force.

2. RESULTS AND DISCUSSION

2.1. The proton motive force in cells of Rps. sphaeroides

Rps. sphaeroides can maintain a high membrane potential in cells incubated
anaerobically in the dark (Elferink et al., 1983). Changes in the magnitude of the proton motive force upon illumination can occur in either
direction: Under conditions where most proton motive force-consuming processes are functional a decrease occurs whereas inhibition of these processes by treatment of the cells with dicyclohexylcarbodiimide (DCCD)
and/or EDTA gives rise to a light-dependent increase in the magnitude of
the proton motive force (Fig. 1). In the dark and at low light intensity
no uptake of alanine is measurable. Uptake of alanine starts at the same
light intensity at which the $\Delta\psi$ in non-EDTA/DCCD treated cells decreases
(Fig. 1) and the uptake rate increases with the light intensity. Because
of the activity of the transport carriers is controlled by the rate of
cyclic electron transfer, large variations in the rates of proton motive
force consuming processes are possible with only minor variations in the
magnitude of the proton motive force.
It appears that systems exist for maintaining the proton motive force on
a high level, also when there is no external energy source available.

Sybesma, C. (ed.), Advances in Photosynthesis Research, Vol. II. ISBN 90-247-2943-2.
© *1984 Martinus Nijhoff/Dr W. Junk Publishers, The Hague/Boston/Lancaster.*

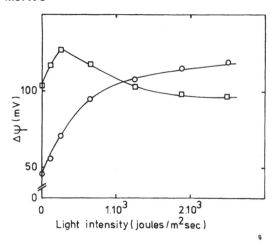

FIGURE 1. The relation between light intensity and $\Delta\psi$ in cells of Rps. sphaeroides at pH 8. □ - □ cells were washed twice and resuspended at 50 mM potassium phosphate, 5 mM $MgSO_4$, pH 8. o - o cells were pretreated with EDTA and DCCD before washing and resuspending in 50 mM potassium phosphate, 5 mM $MgSO_4$, pH 8. Finally 1 µM nigericin was added.

2.2. Lactose transport in Rps. sphaeroides as a function of the $\Delta\psi$ and the light intensity

Lactose transport via the lactose carrier in Escherichia coli has been studied extensively. It has been demonstrated that the presence of a potassium diffusion potential is sufficient to energize lactose accumulation not only in membrane vesicles (Schuldiner, Kaback, 1975) but also in liposomes in which the purified lactose carrier has been reconstituted (Forster et al., 1982) thus proving that a proton motive force alone under those conditions is sufficient to allow lactose accumulation to occur.
It was therefore of interest to compare the properties of the E. coli transport system for lactose, incorporated in Rps. sphaeroides with those of the endogenous transport system for alanine. The initial rate of uptake of lactose and the transmembrane electrical potential ($\Delta\psi$) were measured simultaneously at different light intensities at pH 8 (Fig. 2a). The dependence on light intensity of the uptake of alanine and lactose is very similar.
A very low rate of lactose uptake is obtained in the dark (< 0.01 nmol lactose/mg protein.min). Just as for alanine this rate increases sigmoidally with light intensity. Saturation is observed at 66% of the maximal light intensity. In the dark a $\Delta\psi$ of -96 mV exists, whereas the pH gradient is zero at this pH (Nicolay et al., 1981). The increase in light intensity gives rise to only a very small (< 20 mV) increase in $\Delta\psi$ at 6% of the maximal light intensity, with subsequently a decrease of approximately 10 mV until the maximal intensity of illumination is reached. The rate of lactose uptake as a function of the magnitude of the $\Delta\psi$ gives the flow-force relation, shown in Fig. 2b. Such a graph strongly suggests that in addition to the $\Delta\psi$ (an) other parameter(s) (is) are involved in the regulation of the rate of lactose transport.

2.3. Kinetics of lactose transport in Rps. sphaeroides at different light intensities

The mechanistic basis of the regulation of the rate of lactose transport can be investigated via an analysis of the kinetics of lactose uptake at different light intensities. At 13% and 66% of the maximal light intensity

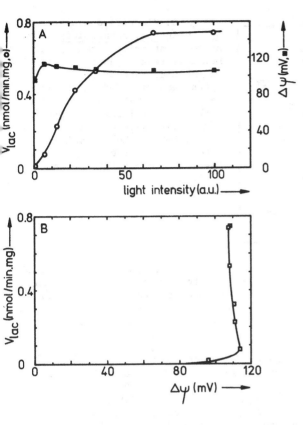

FIGURE 2. The initial rate of lactose uptake and the magnitude of $\Delta\psi$ as a function of the light intensity in Rps. sphaeroides L39. Simultaneously lactose uptake and $\Delta\psi$ were measured in a thermostated vessel in which a tetraphenylphosphonium (TPP$^+$) sensitive electrode was inserted, as described previously (Elferink et al., 1983). Lactose was used at a concentration of 195 μM. A. The initial rate of lactose uptake and the $\Delta\psi$ as a function of the light intensity. B. Replotting of the data in the form of a flow-force relation.

the magnitude of the $\Delta\psi$ was -110 mV. Therefore changes in the kinetics of lactose uptake between these two light intensities will reflect the regulation of the rate of lactose transport by light (or: cyclic electron transfer). Fig. 3 shows that the affinity of the lactose-carrier for lactose is not affected by light intensity: at both intensities a K_m of 550 μM is obtained. However, the maximal rate of lactose uptake does depend on the light intensity. At the highest intensity a V_{max} for lactose uptake of 2.4 nmol/mg protein.min is obtained. At the lowest intensity the V_{max} is 0.8 nmol/mg protein.min, one third of the maximal value. These results indicate that the regulation is due to a light-dependent change in the number of active carrier molecules in the membrane. The important remaining question is how the electron transfer chain effects the activity of the carrier molecules. One explanation could be that electron transfer is required for bringing the transport carrier in the proper redox state. Several authors have reported the existence of two kinetic forms of the lactose carrier in E. coli, one with a high and one with a low affinity (Robertson et al., 1979; Ghazi, Shechter, 1981; Konings, Robillard, 1982). Konings and Robillard reported that the transition in the kinetics of the lactose carrier can be brought about by sulfhydryl reagents and redox mediators. The low affinity form of the lactose carrier has not been detected in Rps. sphaeroides, but we cannot exclude that a second (much higher) K_m can be found at much higher lactose concentrations as used in these experiments.

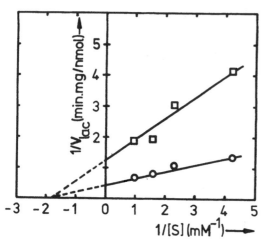

FIGURE 3. Kinetics of lactose uptake in Rps. sphaeroides L39 at constant driving force ($\Delta\psi$) but different light intensities. The initial rate of lactose uptake was measured at the indicated concentrations parallel with $\Delta\psi$. Two different light intensities were used: $\square$: 13% and o: 66% of the maximal light intensity. Under both conditions the magnitude of the $\Delta\psi$ was – 110 mV.

We conclude that the activity of many solute transport systems is regulated by (cyclic) electron transfer, independent of the magnitude of the proton motive force. Thus it is clear that the chemiosmotic hypothesis for solute transport requires modification to allow for direct allosteric interaction(s between the primary proton pump and secondary solute transport systems.

REFERENCES

Elferink MGL, Friedberg I, Hellingwerf KJ and Konings WN (1983) The role of the proton motive force and electron flow in light-driven solute transport in Rhodopseudomonas sphaeroides, Eur.J.Biochem. 129, 583-587.

Forster DL, Garcia ML, Newman ML, Patel L and Kaback HR (1982) Lactose-proton symport by purified lac carrier protein, Biochem. 21, 3634-3638.

Ghazi A and Shechter E (1981) Lactose transport in E. coli cells. Dependence of kinetic parameters on the transmembrane electrical potential difference, Biochim.Biophys.Acta 645, 305-315.

Konings WN and Robillard GT (1982) Physical mechanism for regulation of proton solute symport in Escherichia coli, Proc.Natl.Acad.Sci. USA 79, 5480-5484.

Nicolay K, Lolkema JS, Hellingwerf KJ, Kaptein R and Konings WN (1981) Quantitative agreement between the values for the light-induced ΔpH in Rhodopseudomonas sphaeroides measured with automated flow-dialysis and ^{31}P NMR, Febs Lett. 123, 319-323.

Robertson DE, Kaszorowski GJ, Garcia M and Kaback HR (1980) Active transport in membrane vesicles from Escherichia coli: The electrochemical proton gradient alters the distribution of the lac carrier between two different kinetic states, Biochem. 19, 5692-5702.

Schuldiner S and Kaback HR (1975) Membrane potential and active transport in membrane vesicles from Escherichia coli, Biochem. 14, 5451-5461.

Authors addresses: Marieke G.L. Elferink, Klaas J. Hellingwerf, Wil N. Konings, Department of Microbiology, Kerklaan 30, 9751 NN HAREN, The Netherlands.
Francis E. Nano, Samuel Kaplan, Department of Microbiology, University of Illinois, URBANA, Ill. 61801, U.S.A.

CHARACTERIZATION OF ADENYLATE KINASES FROM PHOTOSYNTHETIC PURPLE BACTERIA

KARL KNOBLOCH, HARTMUT NEUFANG and HORST MÜLLER
Institut für Botanik und Pharmazeutische Biologie der Universität
Erlangen-Nürnberg, Schlossgarten 4, D-8520 Erlangen, Fed.Rep.of Germany

1. INTRODUCTION

Adenylate kinases are involved in the regulation of the energy metabolism in whole cells through control of the adenylate pool. For the first time an adenylate kinase was observed in rabbit muscles (S.P. Colowick, H.M. Kalckar, 1943). The reaction 2 ADP = ATP + AMP occurs in microbial (I.T. Oliver, J.L. Peel, 1956), plant (E.N. Moudrianakis, M.A. Tiefert, 1976; S. Murakami, H. Strotmann, 1978) and animal tissues (L. Noda, S.A. Kuby, 1957; A. Heil et al., 1974; T. Itakura et al., 1978; A.G. Tomasselli, L.H. Noda, 1980; K. Watanabe, S. Kubo, 1982). A more specific function of the enzyme concerning C_4 plants has been suggested (M.D. Hatch, 1982).

Data on adenylate kinase systems from photosynthetic bacteria are not available so far.

2. MATERIAL AND METHODS

2.1. Cell Growth. *Rhodopseudomonas palustris* (ATCC 11168), *R. sphaeroides* (ATCC 17023) and *Rhodospirillum rubrum* (ATCC 11170) were grown photosynthetically on malate in the medium modified after Knobloch et al. (1971).

2.2. Cell-free Extracts were prepared by sonication for 200 s at 4°C followed by the procedure after Knobloch et al. (1971).

2.3. Assay of Adenylate Kinase Activity and other Analytical Methods were performed according to Müller et al. (1982) and Neufang et al. (1983).

3. RESULTS

After cell breakage most of the adenylate kinase activity was found in the supernatant cell-free fraction S-144. The enzyme could be dissolved completely from the chromatophores (P-144) by washing the membranes three times.

The purification procedures applied are given in Table 1. Especially gel chromatography on Sephacryl S-300 superfine shortened the process of purification including a well performed separation of the ATPase protein from the adenylate kinase.

The adenylate kinases from the three bacteria were purified by the same procedure to homogeneity as judged from SDS polyacrylamide gel electrophoresis (Neufang et al., 1983). The purified enzymes showed optimal rates of activity with MgCl2 at 25°C and pH 8.0. They were found to be heat labile and were characterized by pI-values of 4.5. The molecular masses M_r of the adenylate kinases from *R. palustris* (33,500), *R. sphaeroides* (34,400) and *R. rubrum* (32,100) were determined by high performance liquid chromatography on a TSK 3000 SW column with 50 mM phosphate buffer, pH 7.0, as the solvent. Varification was done with gel chromatography on

Sybesma, C. (ed.), Advances in Photosynthesis Research, Vol. II. ISBN 90-247-2943-2.
© *1984 Martinus Nijhoff/Dr W. Junk Publishers, The Hague/Boston/Lancaster.*

Table 1. Purification procedure for the adenylate kinases from *R. palustris*, *R. sphaeroides* and *R. rubrum* (after Neufang et al., 1983).

Fraction	Total protein [mg]			Total activity $[\mu mol \cdot min^{-1}]$			Specific activity $[\mu mol \cdot min^{-1} \cdot l \cdot mg^{-1}]$		
	R.palustris	*R.sphaeroides*	*R.rubrum*	*R.palustris*	*R.sphaeroides*	*R.rubrum*	*R.palustris*	*R.sphaeroides*	*R.rubrum*
Supernatant S-144	2950	3150	2710	1330	1350	1085	0.5	0.4	0.4
Ammoniumsulfate precipitate (40-70%)	1120	1150	1025	1280	1290	1080	1.1	1.1	1.1
Eluate from Sephacryl S-300 Superfine	360	360	320	1260	1260	1095	3.5	3.5	3.4
Eluate from DEAE-Sephadex A-50	50	50	45	1220	1250	1045	24.4	25.0	23.2
Eluate from Sephadex G-75	15	14	12	1160	1110	920	77.3	79.3	76.7

- Table 2. Characterization of the adenylate kinases.

Bacterium	M_r	M_r*	K_m(ADP) [mM]	K_i(ADP) [mM]	K_i(AMP) [mM]	K_i(DAPP) $[\mu M]$	pI
R. palustris	33 500	32 100	0.26	0.32	0.017	0.020	4.5
R. sphaeroides	34 400	32 800	0.27	0.32	0.018	0.020	4.5
R. rubrum	32 100	30 900	0.24	0.34	0.014	0.017	4.5

*after treatment under dissociating conditions (s.text)

Sephadex G-75 and SDS polyacrylamide gel electrophoresis. There was no
evidence for a subunit structure of the three similar enzymes due to SDS
polyacrylamide gel electrophoresis after treatment with 1% 2-mercapto-
ethanol and 1% SDS in 250 mM phosphate buffer, pH 7.1, at 90°C for 15 min.
The apparent K_m-values for ADP corresponded to 0.26 mM for R. palustris,
0.27 mM for R. sphaeroides and 0.24 mM for R. rubrum. ADP in excess had a
strong inhibitory effect (Table 2). Competitive product inhibition was
found for AMP, with K_i-values of 0.017 mM for R. palustris, 0.018 mM for
R. sphaeroides and 0.014 mM for R. rubrum. A competitive inhibitor likewise was P^1,P^5-di(adenosine-5')pentaphosphate (=DAPP) with K_i-values of
0.020 µM for R. palustris and R. sphaeroides, and 0.15 µM for R. rubrum.
Sulfhydryl reacting reagents like p-chloromercuribenzoate and iodoacetic
acid were found to be non-inhibitory. Several divalent cations were effective in potentiating the enzymatic activity. The order of decreasing
effectiveness was $Mg^{2+} > Ca^{2+} > Co^{2+} > Mn^{2+} > Zn^{2+}$. The adenylate kinases
from all three bacteria studied were activated maximally by a molar ratio
of Mg^{2+}:ADP of 1.0.

4. DISCUSSION

The adenylate kinases from the three bacteria studied were found to be
heat-labile. Enzymes revealing similar behavior were described for rat
liver (Sapico et al., 1972), baker's yeast (Chiu et al., 1967), Escherichia
coli (Holmes and Singer, 1973), Pseudomonas denitrificans (Terai, 1974)
and from spinach chloroplast evelopes (Murakami and Strotmann, 1978). On
the other hand, the adenylate kinase from muscle or beef heart mitochondria were reported to be heat-stable (Albrecht, 1970; Colowick and
Kalckar, 1943).

The data obtained from isoelectric focusing for the three organisms (pI=
4.5) indicate adenylate kinases which are acid labile (see Itakura et al.,
1978). From our results observed after protein treatment under dissociating conditions, we conclude that the enzymatically active adenylate kinases exist as a single protein chain (Neufang et al., 1983).

The numbers calculated for the apparent K_m(ADP) values, determined in the
presence of 1 mM $MgCl_2$ from Lineweaver-Burk plots revealed through their
linear appearance that the velocity of the reaction was a function of ADP
concentrations. This implies only one active absorption site for ADP on
the enzyme by the substrate Mg-ADP as it has been concluded for spinach
chloroplasts by Murakami and Strotmann (1978). The adenylate kinases were
sensitive to AMP which functioned as an product inhibitor, as expressed
by the obtained K_i(AMP) values, and ADP exhibited a substrate inhibition
effect indicated by the K_i(ADP) values obtained (Table 2). DAPP likewise
functioned as a competitive inhibitor in rather low concentrations (Table 2) in contrast to its action in beef heart mitochondria (Tomasselli
and Noda, 1980).

Since the adenylate kinases studied were not sensitive towards sulfhydryl
reacting reagents, one may conclude that the action center of the enzyme
does not hold free sulfhydryl groups. Since Mg^{2+} was found to be most
effective among the metal ions in stimulating the enzymatic activity one
again may conclude that the Mg-ADP complex represents the real substrate.
The molecular masses of the adenylate kinases of the three bacteria were
calculated to be 33,500, 34,400 and 32,100 whereas reported numbers for
the enzyme from other sources range from 52,000 to 21,000 (Albrecht, 1970).

5. REFERENCES

Albrecht GJ (1970) Purification and properties of nucleoside triphosphate-adenosine monophosphate transphosphorylase from beef heart mitochondria. Biochemistry 9, 2462-2470.

Chiu C, Su S and Russel PJ (1967) Adenylate kinase from baker's yeast. I. Purification and intracellular location. Biochim.Biophys. Acta 132, 361-369.

Colowick SP and Kalckar HM (1943) The role of myokinase in transphosphorylations. I. The enzymatic phosphorylation of hexoses by adenyl pyrophosphate. J. Biol. Chem. 148, 117-126.

Hatch MD (1982) Properties and regulation of adenylate kinase from *Zea mays* leaf operating in C_4 pathway photosynthesis. Austr. J. Plant Physiol. 9, 287-296.

Heil A, Müller G, Noda L, Pinder T, Schirmer H, Schirmer I and von Zabern I (1974) The amino-acid sequence of porcine adenylate kinase from skeletal muscle. Eur. J. Biochem. 43, 131-144.

Holmes RK, Singer MF (1973) Purification and characterization of adenylate kinase as an apparent adenosine triphosphate dependent inhibitor of ribonuclease II in *Escherichia coli*. J. Biol. Chem. 248, 2014-2021.

Itakura T, Watanabe K, Shiokawa H and Kubo S (1978) Purification and characterization of acidic adenylate kinase in porcine heart. Eur. J. Biochem. 82, 431-437.

Knobloch K, Eley JH and Aleem MIH (1971) Thiosulfate-linked ATP-dependent NAD^+-reduction in *Rhodopseudomonas palustris*. Arch. Mikrobiol. 80, 97-114.

Moudrianakis EN and Tiefert MA (1976) Synthesis of bound adenosine triphosphate from bound adenosine diphosphate by the purified coupling factor 1 of chloroplasts. J. Biol. Chem. 25, 7796-7801.

Müller H, Neufang H and Knobloch K (1982) Purification and properties of the coupling-factor ATPases F_1 from *Rhodopseudomonas palustris* and *Rhodopseudomonas sphaeroides*. Eur. J. Biochem. 127, 559-566.

Murakami S and Strotmann H (1978) Adenylate kinase bound to the envelope membranes of spinach chloroplasts. Arch. Biochem. Biophys. 185, 30-38.

Neufang H, Müller H and Knobloch K (1983) Purification and properties of the adenylate kinases from *Rhodopseudomonas palustris, Rhodopseudomonas sphaeroides* and *Rhodospirillum rubrum*. Arch. Microbiol. 134, 153-157.

Noda LH and Kuby SA (1957) Adenosine triphosphate-adenosine monophosphate transphosphorylase. I. Isolation of the crystalline enzyme from rabbit skeletal muscle. J. Biol. Chem. 226, 541-549.

Oliver IT and Peel JL (1956) Myokinase activity in microorganisms. Biochim. Biophys. Acta 20, 390-392.

Sapico V, Litwack G and Criss WE (1972) Purification of rat liver adenylate kinase isozyme II and comparison with isozyme I. Biochim. Biophys. Acta 258, 436-445.

Terai H (1974) Adenylate kinase from *Pseudomonas denitrificans*. I. Purification and antiserum inhibition. J. Biochem. (Tokyo) 75, 1027-1036.

Tomasselli AG, Noda LH (1980) Mitochondrial ATP:AMP phosphotransferase from beef heart: Purifications and properties. Eur. J. Biochem. 103, 481-491.

Watanabe K, Kubo S (1982) Mitochondrial adenylate kinase from chicken liver. Purification, characterization and its cell free synthesis. Eur. J. Biochem. 123, 587-592.

6. ACKNOWLEDGEMENT

The work was supported partly by the Deutsche Forschungsgemeinschaft.

FUNCTIONAL PROPOERTIES OF CHLOROPLAST ATPase: THE FATE OF TIGHTLY BOUND
ADP AND ATP IN ILLUMINATED CHLOROPLASTS

HEINRICH STROTMANN, BOTANISCHES INSTITUT DER UNIVERSITÄT DÜSSELDORF

1. INTRODUCTION

At least two types of bound adenine nucleotides are known that participate
in the reactions of chloroplast ATPase, the nucleotides which bind to the
catalytic site and those which are defined as "tightly bound" nucleotides.
The nucleotides at the catalytic site turnover rapidly, the turnover time
being in the order of a few ms (Gräber et al. 1977). The tightly bound
nucleotides are rather non-exchangeable in deenergized chloroplasts, but
are released when the membranes are energized (Strotmann, Bickel-Sandkötter
1977). It has been shown that release of bound nucleotides occurs parallely
with the activation of membrane-bound ATPase in the process of light-
triggered ATP hydrolysis (Schumann, Strotmann 1981) and that rebinding of
ADP to the empty sites deactivates the enzyme (Schumann, Strotmann 1981,
Dunham, Selman 1981, Bar-Zvi, Shavit 1982). In deenergized chloroplasts,
rebinding of nucleotides to the tight site is energy-independent (Strotmann,
Bickel-Sandkötter 1977) and irreversible with a second order rate constant
of 0.02 $s^{-1}\mu M^{-1}$ (Schumann, Strotmann 1981). In energized chloroplasts where
both, binding and release occur at the same time, the steady state level
of tightly bound nucleotides attained depends on the amount of energy
supplied to the thylakoids (Strotmann et al. 1979). We have shown that this
level is inversely related to the activity of photophosphorylation (Strotmann
et al. 1981), suggesting that active enzyme molecules that participate in
ATP synthesis do not contain tightly bound nucleotides. In the initial
period of energization, nucleotide release was suggested to monitor the
conformational change which results in the activation of the latent, membrane-
bound ATPase (Gräber et al. 1977). This paper presents a few kinetic experiments designed to provide some more insight into the relation between
nucleotide release from the tight site and phosphorylation.

2. EXPERIMENTAL

Broken spinach chloroplasts were isolated and preloaded with (^{14}C)ADP as
described (Strotmann et al. 1976). Release of the bound labeled nucleotides
was measured in a medium containing 25 mM tricine buffer, pH 8.0, 50 mM NaCl,
1 mM $MgCl_2$, 50 µM PMS and other additons as indicated in the legends.
25-50 µg chlorophyll was used in an assay volume of 0.5 ml. The stirred
reaction mixture was illuminated in small cylindric glass vessels from the
top. The illumination time was adjusted by an electronic shutter, usually
the light intensity was 480 Wm^{-2}. Immediately after the light flash 0.2 ml
of a quenching solution containing 0.175 mM FCCP, 0.7 mM ADP and 17.5 mM EDTA
was added. The supernatants were analyzed for released labeled nucleotides
by liquid scintillation counting. Aliquots of the supernatants were subjected
to ion exchange column chromatography (Bickel-Sandkötter 1983). Photophosphorylation was measured under the same conditions in parallel experiments with chloroplasts which were preloaded with unlabeled ADP. The reaction
medium contained additional (^{32}P)P_i and the quench was performed with 0,2 ml
of 2.1 M $HClO_4$, 0.7 mM ATP and 17.5 mM EDTA.(^{32}P)ATP was determined after
isobutanol/toluene extraction of the phosphomolybdate complex.

Sybesma, C. (ed.), Advances in Photosynthesis Research, Vol. II. ISBN 90-247-2943-2.
© *1984 Martinus Nijhoff/Dr W. Junk Publishers, The Hague/Boston/Lancaster.*

3. RESULTS

Chloroplasts incubated in the light with (^{14}C)ADP retain after deenergization between 1.0 to 1.5 nmol of tightly bound labeled nucleotides. It was previously shown that these nucleotides are bound to CF_1 (Strotmann et al. 1976) The total amount of nucleotides bound to CF_1 consists of 64-78% ADP and 36-22% ATP. No membrane-bound AMP was detected. However, some free AMP (up to 0.05 nmol/mg chl) appears in the chloroplast medium. Since no phosphate is present during the pretreatment, bound ATP probably results from photophosphorylation by residual endogenous P_i which amounts to about 0.7 µmol/ mg chl even in hypotonically washed chloroplasts. The ATP level can be slightly increased by addition of P_i during labeling in the light and decreased to almost zero when preilluminated chloroplasts are supplied with labeled ADP in the dark. However under these conditions the total amount of incorporated nucleotides is significantly reduced. In subsequent experiments the fate of the tigthly bound ADP and ATP was followed when the chloroplasts were subjected to energization under different conditions.

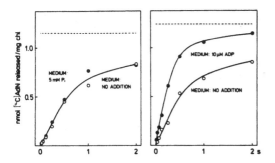

Fig. 1. Light-induced release of tightly bound labeled adenine nucleotides from preloaded chloroplasts in media containing either 5 mM phosphate (left) or 10 µM ADP (right) or none of the substrates (open circles).

Fig. 1 shows time courses of light-induced release of labeled nucleotides in media without or with P_i or ADP. After 2 s illumination about 70% of the bound nucleotides are released, in the presence or absence of medium P_i, but more is released when the medium contains unlabeled ADP (90%). The apparent stimulation of release by external ADP is not due to isotope dilution which should prevent effectively the rebinding of the released labeled nucleotides, since the initial rate of release of 3 µmol/mg chl/h is accelerated by external ADP to about 10 µmol/mg chl/h. Moreover, direct measurement of rebinding of the released nucleotides in the following dark period shows that this rate is only 1/500 of the initial rate of release and can therefore be neglected. The stimulating effect of medium ADP on the release of tightly bound nucleotides can be explained by assuming cooperativity between two interdependent nucleotide binding sites. In energized membranes binding of ADP to one site may promote release of nucleotides bound to the other site.

Fig. 2 shows the kinetics of release of tightly bound nucleotides in media containing ADP and 2',3'-O-(2,4,6-trinitrophenyl)-ADP (TNP-ADP) with or without P_i. The composition of the nucleotides released was analyzed by chromatography. With phosphate and ADP (or TNP-ADP) a biphasic release kinetics is obtained. Similar results were reported previously by Gräber et al. (1977). The initial fast nucleotide release which exhibits a rate of about 40 µmol/mg chl/h is followed by a slower one which resembles the

kinetics of release in the presence of ADP (or TNP-ADP) but without P_i.
The rapid initial phase is absent under these conditions and the pattern
of the nucleotides released follows that of the nucleotides which were
on the membranes.

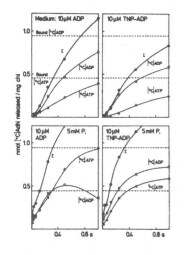

Fig. 2. Light-induced release of tightly bound labeled nucleotides in media containing either 10 µM ADP or 10 µM TNP-ADP without (upper part) or with 5 mM P_i (lower part). The dashed lines indicate the levels of tightly bound labeled ADP and ATP, respectively, which were present before illumination.

When the medium contains ADP and P_i, more labeled ATP is found in the medium
than that originally present in the membranes. This is due to the mixing of
the released labeled ADP with the external pool of unlabeled ADP followed by
photophosphorylation.

The fast initial release phase exhibits a half time of about 10 ms and an
amplitude of about 20% of the total bound nucleotides. This reaction might be
related to the initial phase of photophosphorylation. If this is correct, we
may conclude that binding of the substrates, ADP and P_i, rather than the
formation of ATP is the reason for the accelerated release, because a fast
release is also observed if ADP is replaced in the medium by TNP-ADP. The
ribose-modified ADP analog TNP-ADP is phosphorylated by chloroplasts at a
very low rate, but its affinity to the catalytic site is as high as that
of ADP. Accordingly TNP-ADP acts as a competitive inhibitor of ADP phosphorylation (G. Ponse, unpublished).

In Fig. 3 changes of the patterns of released as well as bound adenine nucleotides, during the first second of illumination, are shown. The bound nucleotides were calculated from the difference between total nucleotides
(quenched by acid) and released nucleotides (quenched as before). This
excludes subsequent changes in the nucleotide composition during the work-up
of the samples which could occur if the bound nucleotides were analyzed in the
membrane itself after centrifugation and removal of the free nucleotides by
washing. In the presence of P_i or of P_i + ADP in the medium, a continous
decrease in the level of bound ADP is observed. The velocity is, however,
higher when both substrates are in the medium as would be expected from
the experiment given in Fig. 1. The bound ATP changes upon illumination
in the two media in a different manner. While the decrease of ATP is retarded
for about 200 to 300 ms when only P_i is present, about 1/3 of bound ATP
disappears during the first 10 ms when both, ADP and P_i are present. The
rest is released at a lower rate and with a similar time course as in the

presence of P_i alone. Thus it seems that the release of ATP is more strictly dependent on the presence of medium nucleotides than the release of ADP.

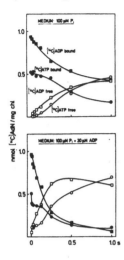
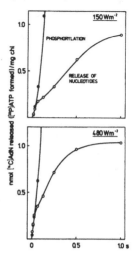

Fig. 3 (left side). Changes in the patterns of bound and free labeled adenine nucleotides upon illumination of preloaded chloroplasts in media containing either 0.1 mM P_i or 0.1 mM P_i + 20 µM ADP. The bound nucleotides were calculated from the differences between total nucleotides and free nucleotides. The total nucleotides were analyzed in $HClO_4$ (0.6 M) extracts.

Fig. 4 (right side). Kinetics of light-induced release of tightly bound labeled nucleotides and of photophosphorylation at two different light intensities obtained under the same conditions in parallel experiments. The medium contained 0.1 mM P_i and 20 µM ADP.

The relationship between the initial fast release of tightly bound nucleotides and photophosphorylation was studied by a comparison of the kinetics of these two reactions in parallel experiments done under the same conditions. Two experiments where different light intensities were employed are given in Fig. 4. The initial rates of nucleotide release and photophosphorylation appear to be virtually the same. However, it must be emphasized that the initial rate of phosphorylation is lower than the steady state rate attained after about 200 ms in these experiments. A correspondence between the rates was also obtained when the concentrations of medium ADP and P_i, respectively, were varied.

4. DISCUSSION

The results given above suggest that release of tightly bound nucleotides observed with chloroplasts energized by light, is related to the process of photophosphorylation. Gräber et al. (1977) suggested that the release reaction monitors a conformational change which transfers the ATPase from an inactive to an active state. This view is supported by several experimenta results: (1) Chloroplasts catalyze ATP hydrolysis when preilluminated in the presence of a thiol compound. This pretreatment causes activation of the latent ATPase. It has been shown that release of tightly bound nucleotides runs parallel with activation (Schumann, Strotmann 1981).

(2) Under conditions of net photophosphorylation the level of tightly bound nucleotides is inverse to the activity of ATP synthesis when measured at varying light intensity or uncoupler concentration (Strotmann et al. 1981), which suggests that the active ATPase molecules employed in photophosphorylation are free from tightly bound nucleotides. (3) The kinetic correspondence of nucleotide release and phosphorylation in the initial period of energization (Schlodder, Witt 1981 and this paper) indicates that the release reaction is fast enough to account for activation. The fact that a fraction of 10 to 20% of the tightly bound nucleotides undergoes fast release suggests that only part of the enzyme molecules are rapidly activated (Gräber et al. 1977). Probably these few activated ATPase molecules drain the protons from the inside phase of the thylakoid membranes to the medium to form ATP in several successive turnovers. In the subsequent slow phase of release, additional enzyme molecules may be activated which could explain the observed increase in the rate of phosphorylation.

Our experiments provide some additional information on the relation between nucleotide release and the catalytic process. Tightly bound nucleotides exhibit an extremely low K_d. For this reason they can not be removed by washing or exchanged for medium nucleotides. One may imagine that they are embedded inside the protein, presumably in a hydrophobic environment. The amount of tightly bound nucleotides per chlorophyll indicates that 1 molecule of CF_1 contains 1 molecule of nucleotide when the CF_1:chlorophyll ratio of 1:860 (Strotmann et al. 1973) is used for calculation and when a homogenous distribution of the nucleotides is assumed. In the deenergized membrane 2/3 of the enzyme molecules contain ADP and about 1/3 of them contain a tightly bound ATP. Tightly bound ATP is also detected by ^{32}P-labeling when chloroplasts are illuminated in the presence of ADP and $(^{32}P)P_i$ (Rosen et al. 1979). Its formation is not abolished by a hexokinase trap (Aflalo, Shavit 1982), indicating that it is not incorporated into this particular site from the medium. In a hypothesis by Boyer and his colleagues (Rosen et al. 1979), a rapidly formed tightly bound ATP species participates as an intermediate in the catalytic cycle of phosphorylation. However, Aflalo and Shavit (1982) found that very little of the tightly bound ATP may turn over fast enough to be catalytically competent. They assume a transfer of the ATP from the catalytic site to a non-catalytic site through a space which is not accessible to hexokinase. Bickel-Sandkötter (1983) proposed that the tight nucleotide binding site may be a different form of the catalytic site. The transition from a catalytic site to a tight site may occur when the enzyme molecule turns from the active to the inactive state. This could explain (1) the insensitivity of the formation of tightly bound ATP to hexokinase, and (2) the slower turnover. The release of the tightly bound ATP would indicate reactivation of the enzyme which may be slower than overall phosphorylation if only a certain number of enzyme molecules is active at the same time (Gräber et al. 1977). Photolabeling experiments provide no convincing evidence for physically separate sites. Results on differences in nucleotide specificities of ATP synthesis and nucleotide exchange on the tight site likewise do not unequivocally prove diversity of the sites, because the apparent K_m values for photophosphorylation can not be compared with apparent K_d values for the exchange reaction. Thus the hypothesis of interconversion of the sites at present provides the most simple explanation for the observed results and it is included as the basal assumption for the scheme depicted in Fig. 5. Tightly bound ADP or ATP are formed from catalytic ADP and ATP upon deenergization of the membrane.

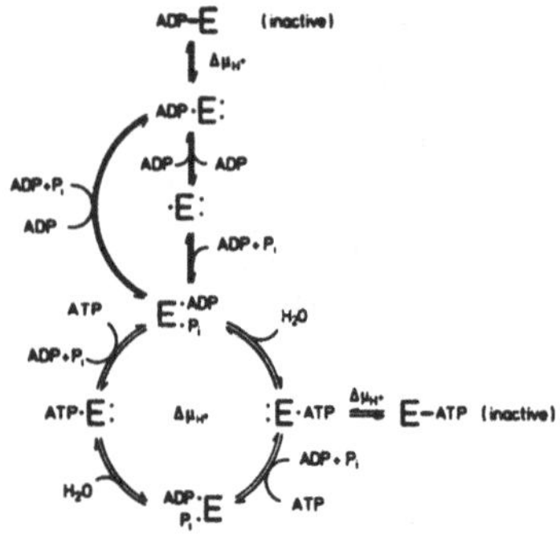

Fig. 5. Proposed scheme for the relationship between ATP formation and the reactions of tightly bound nucleotides. E–ADP, E–ATP, forms of tightly bound nucleotides; E·ADP, E·ATP, forms of loosely bound nucleotides; E·, E:, enzyme with unoccupied sites accessible to medium ADP and medium ADP+P_i, respectively.

The nucleotide molecule present at this moment on the catalytic site is trapped and restrained from further reaction. Reactivation requires membrane energization, whereby the tightly bound ADP and ATP undergo different reactions. The fact that external ADP + P_i or even ADP accelerate release of the tightly bound nucleotides upon illumination suggests cooperativity of two interdependent binding sites. Release of tightly bound nucleotides probably involves a loosely bound form as an intermediate (Strotmann, Bickel-Sandkötter 1977). One could imagine that the conformational change which leads to transition of the inactive to the active state of the enzyme, opens a second binding site and that nucleotide binding to it promotes release of the loosely bound nucleotide from the former site. Binding of ADP and P_i to a second site seems to be a necessary prerequisite for the release of tightly bound ATP. On the other hand, tightly bound ADP is released also in the absence of medium nucleotides (Fig. 3). In the latter case an enzyme molecule is obtained which is free from nucleotides. After deenergization, this form of the enzyme can change to a conformation which exhibits a preference to ATP binding and hydrolysis (Franek, Strotmann 1981). Binding of ADP and P_i to the second site results in the formation of an active enzyme-substrate complex which can enter the catalytic cycle. The loosely bound ATP formed from tightly bound ATP upon energization is by itself an intermediate of the cycle. The catalytic cycle is assumed to operate in an obligatory alternating site fashion as proposed by Rosen et al. (1979), so that ATP is only released when ADP + P_i are bound to the other site. Alternating site cooperativity was concluded from the fact that low ADP and P_i concentrations increase the rate of ^{18}O incorporation from $(^{18}O)H_2O$ into the newly formed ATP (Hackney et al. 1979). Accordingly, the ATP on the enzyme undergoes additional cycles of reversible hydrolysis if release is not promoted by efficient substrate binding to the second site. In variance to the energy-dependent binding change mechanism proposed by Boyer and Kohlbrenner (1979), tightly bound substrate molecules are not included as intermediates in the catalytic cycle in the scheme in Fig. 5, because the experimental evidence for this is doubtful. Tightly bound nucleotides as indicators of inactive ATPase molecules, on the other hand, are well documented. The scheme also leaves open the unanswered question in which reaction of the catalytic cycle energy from the transmembrane proton potential is invested. Thus the scheme gives an interpretation of the present status of the available information on the relationship between the catalytic reaction taking place on chloroplast ATPase and the regulation of its activity, rather than a comprehensive concept of the mechanism of ATP synthesis.

REFERENCES

Aflalo C and Shavit N (1982) Source of rapidly labeled ATP tightly bound to non-catalytic sites on the chloroplast ATP synthetase. Eur. J. Biochem. 126, 61-68.
Bar-Zvi D and Shavit N (1982) Modulation of the chloroplast ATPase by tight binding of nucleotides. Biochim. Biophys. Acta 681, 451-458.
Bickel-Sandkötter S (1983) Loose and tight binding of adenine nucleotides by membrane-associated chloroplast ATPase. Biochim. Bioophys. Acta 723, 71-77.
Bickel-Sandkötter S and Strotmann H (1981) Nucleotide binding and regulation of chloroplast ATP synthase. FEBS Lett. 125, 188-192.
Boyer PD and Kohlbrenner WE (1981) The present status of the binding-change mechanism and its relation to ATP formation by chloroplasts. In Energy Coupling in Photosynthesis (ed. BR Selman, S Selman-Reimer) pp. 231-240. Elsevier, Amsterdam.
Dunham K and Selman BR (1981) Regulation of spinach chloroplast coupling factor 1 ATPase activity. J. Biol. Chem. 256, 212-218.
Franek U and Strotmann H (1981) Nucleotide specificity of CF_1-ATPase in ATP synthesis and ATP hydrolysis. FEBS Lett. 126, 5-8.
Gräber P, Schlodder E and Witt HT (1977) Conformational change of the chloroplast ATPase induced by a transmembrane electric field and its correlation to phosphorylation. Biochim. Biophys. Acta 461, 426-440.
Hackney DD, Rosen G and Boyer PD (1979) Subunit interaction during catalysis: Alternating site cooperativity in photophosphorylation shown by substrate modulation of (^{18}O)ATP species formation. Proc. Natl. Acad. Sci. USA 76, 3646-3650.
Rosen G, Gresser M, Vinkler C and Boyer PD (1979) Assessment of total catalytic sites and the nature of bound nucleotide participation in photophosphorylation. J. Biol. Chem. 254, 10654-10661.
Schlodder E and Witt HT (1981) Relation between the initial kinetics of ATP synthesis and of conformational changes in the chloroplast ATPase studied by external field pulses. Biochim. Biophys. Acta 635, 571-584.
Schumann J and Strotmann H (1981) The mechanism of induction and deactivation of light-triggered ATPase. In Photosynthesis II. Electron Transport and Photophosphorylation, (ed. G Akoyunoglou) pp. 881-892. Balaban, Philadelphia
Strotmann H and Bickel-Sandkötter S (1977) Energy-dependent exchange of adenine nucleotides on chloroplast coupling factor (CF_1). Biochim. Biophys. Acta 460, 126-135.
Strotmann H , Bickel-Sandkötter S, Franek K and Gerke V (1981) Nucleotide interaction with membrane-bound CF_1. In Energy Coupling in Photosynthesis (ed. BR Selman, S Selman-Reimer) pp. 187-196. Elsevier, Amsterdam.
Strotmann H, Bickel S and Huchzermeyer B (1976) Energy-dependent release of adenine nucleotides tightly bound to chloroplast coupling factor CF_1. FEBS Lett. 61, 194-198.

Strotmann H, Bickel-Sandkötter S and Shoshan V (1979) Kinetic analysis of light-dependent exchange of adenine nucleotides on chloroplast coupling factor CF_1. FEBS Lett. 101, 316-320.
Strotmann H, Hesse H and Edelmann K (1973) Quantitative determination of coupling factor CF_1 of chloroplasts. Biochim. Biophys. Acta 314, 202-210.

ACKNOWLEDGEMENTS

The author thanks Miss Karin Brendel for her competent technical assistance, and Mrs. Rita Reidegeld for preparing the manuscript.

Authors address: Heinrich Strotmann, Botanisches Institut der Universität
                 Düsseldorf, D-4000 Düsseldorf, F.R.G.

CORRELATION BETWEEN MEMBRANE-LOCALIZED PROTONS AND FLASH-DRIVEN
ATP FORMATION IN CHLOROPLASTS.

RICHARD A. DILLEY, JOSEPH A. LASZLO AND ULRICH SCHREIBER

INTRODUCTION

The protonmotive force developed across or within chloroplast, mitochondria
or bacterial membranes is generally believed to drive ATP formation. Whether
the primary mechanism of proton processing involves localized domains within,
or on, the membrane or obligatorily transmembrane (delocalized) high energy
states, is as yet unresolved. Studies with chloroplast thylakoids using two
very different approaches have supported a localized proton processing hypothesis. One approach measured the onset of ATP formation in short illumination times as influenced by various factors such as the presence of added
permeable buffers. The other approach utilized chemical modification reagents as probes for membrane-proton interactions (Dilley et al., 1982).

This report will bring together certain aspects of the two approaches for the
purpose of testing further the hypothesis that localized proton binding
domains may be involved in proton movement into the energy coupling complex.
Acetic anhydride reacts rapidly with unprotonated amine groups but not with
protonated amine groups. We have used this property to follow changes in the
protonation state of amine functions of thylakoid membrane proteins under
various conditions related to bioenergetic functions. Out of those experiments came the findings that thylakoid membranes have an array of 30-40
$nmol \cdot (mg \ chl)^{-1}$ of acetic anhydride-reactive groups (probably all or most of
which are amine groups) with the following properties: a) The "special pool"
of buffering groups are behind the permeability barrier of the membrane.
Either uncouplers, at low concentration, or a brief thermal treatment cause
the loss of about 30-40 $nmol \ H^+ \cdot (mg \ chl)^{-1}$ with a concomitant increase in
acetic anhydride-labeled groups and inhibition of water oxidation. b) Some
of the acetic anhydride-reactive groups described above are closely associated with the water oxidizing apparatus and part of the array consists of the
lysine 48 residue of the 8 KD CF_0 protein. c) Either electron transfer-linked
proton accumulation or ATPase proton pumping can convert the uncoupler or
thermal treatment-induced state (deprotonated) to the protonated, anhydride-
unreactive state. The association of some of the 30-40 $nmol \cdot (mg \ chl)^{-1}$ of
membrane buffering groups with the water oxidizing apparatus and the 8 KD CF_0
protein suggests that these buffering groups may be involved in the bioenergetic functioning of thylakoid membranes. Obviously, the location of the
amino group array with regard to the membrane structure is an important point
to clarify. We studied this using a type of "pH clamp", varying both the
outer and inner aqueous phase pH and assaying the protonation state of the
"special pool" of acetic anhydride-reactive groups by measuring the sensitivity of water oxidation to acetic anhydride inhibition.

The other question relating to the above point is whether the protonation
state of the sequestered buffering domain has any influence on the flow of
protons from the redox sources to the CF_0-CF_1 sink in the energy coupling
mechanism. This can be studied by measuring the length of the lag in the
onset of ATP formation. If the lag in onset is due to the requirement to
build up the protonmotive force beyond an energetic threshold and if the
sequestered amine buffering domain is on the pathway taken by protons flowing

Sybesma, C. (ed.), Advances in Photosynthesis Research, Vol. II. ISBN 90-247-2943-2.
© *1984 Martinus Nijhoff/Dr W. Junk Publishers, The Hague/Boston/Lancaster.*

into the coupling complex; then there should be a predictable effect on the
lag length whether the sequestered buffering pool were protonated or deprotonated prior to the illumination regime used to elicit ATP formation.

RESULTS AND DISCUSSION

Chloroplast preparation and various assays used. Chloroplast preparation,
assays to determine chlorophyll concentration, electron transfer rates and
manipulations associated with the use of acetic anhydride were as given in
Baker et al. (1981) and Laszlo et al., (1983). Determination of ΔpH was as
explained in Laszlo et al. (1983), using [14C] methylamine distribution.
Single-turnover flash-driven ATP formation was carried out using the luciferin luminescence technique described by Schreiber and Del Valle-Tascon (1982).
Details of the experiments testing the correlation of the lag in onset of ATP
formation with the protonation state of the sequestered amine buffering pool
will be found in Dilley and Schreiber (1983).

Location of the sequestered amine buffering groups. From previous work
(Dilley et al., 1982), it was suggested that the sequestered amine buffering
pool was buried within the membrane or within membrane-associated proteins.
A more critical test of that hypothesis is to do a type of pH control experiment, poising the inner (and outer) aqueous phases at a pH above the pKa of
the acetic anhydride-sensitive amine groups and test for the stability of the
metastable, protonated amine array. The pKa of the dissociating amine groups
that give rise to the anhydride-reactive $-NH_2$ form must be known in order to
do that experiment. Fig. 1 shows the pH curve of acetic anhydride inhibition
(open circles) of water oxidation, a measure of the presence of the neutral
amine. Uncoupler was present to assure that the inner aqueous phase pH was
equilibrated with the external pH. The data given by the open squares is for
the case with no uncoupler added. The effect is compared to the theoretical
dissociation curve of an amine with a pKa of 7.8 (dashed line). The
anhydride-sensitive amine dissociation shows a single pKa near 7.8. The
experiment to test whether the amine buffering group array is located in the
inner bulk phase involves incubating thylakoids at pH 8.8 long enough to have
the inner pH come to equilibrium with the external phase. The critical
experiment is to measure the inner aqueous phase pH by the [14C] methylamine
distribution technique, and then test the correlation between the bulk, inner
phase pH and the sensitivity of the water oxidation apparatus to acetic
anhydride. Table 1 shows such an experiment where a 60 min dark incubation
with 25 mM TAPS buffer pH 8.8, with or without uncoupler led to an internal
pH of about 8.7. Even though the inner aqueous phase was 1 pH unit above the
pKa of the anhydride sensitive groups, the water oxidation system was quite
resistent to acetic anhydride. The sample with uncoupler added prior to
acetic anhydride treatment showed 80 percent inhibition, a commonly observed
effect. That result is consistent with the hypothesis that the metastable ami
buffering pool is sequestered in a buried domain that is not in rapid equilibrium with the inner aqueous phase (nor the external). Previous work has
shown that either the redox or ATPase proton pumps can restore the protonated
state of the amine buffering pool (Baker et al., 1981). This immediately
raises the possibility that the sequestered proton buffering domain may be
located on the pathway used by protons in energy coupling. One way to test
this hypothesis is to check for whether the protonation state of the
sequestered buffering group array has any correlation with the lag in the

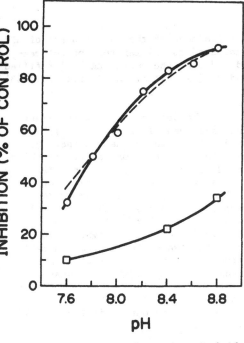

Figure 1. *Extent of acetic anhydride inhibition of electron transport activity as a function of pH. Chloroplasts were treated with acetic anhydride in reaction media containing 0.1 M sucrose, 50 mM KCl, 2 mM MgCl$_2$, and 50 mM HEPPS-NaOH titrated to the indicated pH, either in the presence (0) or absence ( ) of 0.5 μM nigericin, as described under "Materials and Methods" in Laszlo et al. (1983). The control, unmodified, electron transport rate (H$_2$O → MV) was 1020 μ eq·(hr·mg chl)$^{-1}$.*

Table 1. *Dark ΔpH and Acetic Anhydride Resistance in Chloroplasts Equilibrated with TAPS^a Buffer.*

	Electron Transport Activity (μeq·hr^{-1} mg chl^{-1})^b	Internal pH	Dark ΔpH
No uncoupler	374	8.67	0.13
Plus uncoupler^c	76	8.72	0.08

^{a}*The electron transport activities, following acetic anhydride modification, and dark ΔpH measurements were performed as described under "Materials and Methods" in Laszlo et al. (1983).*
^{b}*The control, unmodified, electron transport rate was 520 μeq·hr^{-1} ·mg chl^{-1}.*
^{c}*For measurement of acetic anhydride inhibition of electron transport, 0.25 μM nigericin was added to the diluted chloroplasts just prior to anhydride treatment. For the measurement of dark ΔpH in the plus uncoupler case, 10 μM nigericin was added to the stock chloroplasts, 10 min prior to the ΔpH determination.*

onset of ATP formation, an event believed to be related to the development of a threshold protonmotive force in energy coupling.

Flash-driven ATP formation with a protonated compared to deprotonated buffering pool. The protocol for those experiments involved treating thylakoids with uncouplers or a thermal exposure to deplete the metastable buffering pool of protons prior to initiating the flash regime to elicit ATP formation.

The length (or in this case the number of flashes) of the lag in onset of ATP
formation for those thylakoids will be compared to the lag for control membranes and to a sample that was depleted and then the buffering pool refilled
by a flash regime prior to adding ADP and Pi (cf. Dilley and Schreiber (1983)
for full details).

Fig. 2 shows typical raw luminescence data. The vertical spikes are light
leaks which provide event markers for the 15-20 µsec flashes. The flash
frequency was always 1 Hz. Formation of ATP is shown by the net increase in
the signal, beginning at flash number 24. The important parameter for these
experiments is the number of flashes needed to produce the onset of ATP
formation. Two criteria were used to assess this point: (a) the first
detectable rise in luminescence (in Fig. 2 occurring at flash number 24 ± 1)
and (b) the back extrapolation of the steady rise in the flash-induced fluorescence increase to the x-axis, (in Fig. 2 at flash number 32 ± 1). In
subsequent figures these two criteria will be listed on the figure below the
trace, as for example, $\frac{24}{32}$. Taken together, the two criteria permit a better
comparison of the effect of a given treatment on the lag in onset of ATP
formation. The ATP yield per flash was calibrated by addition of standard
ATP. To simplify the data presentation for the flash-induced ATP formation
described below, we drew a continuous line connecting the bottom of the
spikes due to the light pulses. This envelope of the onset kinetics of ATP
formation is sufficient to depict the effects observed.

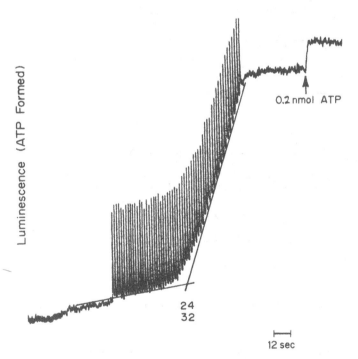

*Fig. 2. Typical
flash-driven ATP for-
mation assayed by
luminescence from the
luciferin-luciferase
system. The reaction
mixture contained 50
mM soribtol, 50 mM
tricin-KOH pH 8.5, 3
mM MgCl$_2$ and 0.1 mM
methyl viologen with
15 µg Chl ml^{-1}, LKB
luciferin-luciferase
reagent, 5 µM
diadenosine penta-
phosphate, 1 mM Pi,
20 nM valinomycin and
0.1 mM ADP.*

It is apparent from
Fig. 2 that there was
no ATPase activity
detected before the
flash train. It has
been pointed out by
Schreiber and Del
Valle-Tascon (1983)
that ATPase activity

occurring in the sample prior to the beginning of the illumination by single-turnover flashes can greatly influence the observed onset behavior and yield per flash of ATP formation. For the experiments of this study we minimised the ATPase activity by: (1) preparing chloroplasts from leaves kept over-night in darkness; (2) avoiding the use of sulfhydryl reagents such as dithiothritol; (3) maintaining low levels of ATP by using purified ADP and by inhibiting adenylate kinase activity with diadenosine pentaphosphate; and (4) by generally using only a single train of flashes with the complete set of assay components present. Certain experiments required a pre-illumination treatment, but that was given before addition of ADP and Pi. When the above precautions were taken, such treatments did not lead to detectable ATPase activity prior to the train of flashes used to drive ATP formation (see Fig. 2 for example).

After a deprotonation treatment with uncoupler, the uncoupler must be removed before subsequent ATP formation measurements. That was accomplished, in the cases of desaspidin and Cl-CCP, by adding bovine serum albumin. BSA tightly binds these two uncouplers, removing them from the membrane. Fig. 3 shows that 0.1 M desaspidin given for 1 min in the dark followed by 2.5 mg BSA to remove the desaspidin from the membrane, resulted in an extension of the onset of ATP formation from 18 flashes (curve 3, control, BSA added before the desaspidin) out to 30 flashes (curve 1). A separate sample (curve 2) was treated with desaspidin followed by BSA, but then 3 sec of white light were given (methylviologen was present, but not ADP or Pi), followed by 8 min of dark to allow complete relaxation of the transmembrane proton gradient. At 7 min dark time, luciferin-luciferase, diadenosine pentaphosphate, ADP and Pi were added and the ATP formation was elicited by the flashing-light regime. Curve 2, Fig. 3 shows that the onset lag for this treatment was shortened to 22 flashes – compare to the 30-flash lag for the sample not preilluminated. Curve 4 shows a control consisting of a second flash cycle, after 4 min dark, given to the sample used for curve 3. This suggests that a preillumination did not cause a shorter lag in general, but rather that the lag shortening caused by preillumination of sample 2 compared to sample 1 was related to the uncoupler effect giving a longer lag to begin with. A further control, not shown, was a sample identical to that for curve 3, but with a 9.5 min dark time before starting the flash train. It also gave a relatively short lag similar to that for curve 3, 17 and 23 flashes, respectively, for the two onset criteria discussed above. This establishes that the longer lag found in sample 1 was not due to the length of dark incubation time.

The data of Fig. 3 are consistent with the hypothesis that the presence or absence of the pool of amine-buffered protons can significantly influence the onset lag of ATP formation. Another way to test this is to do the same type of experiment as that of Fig. 3, but at pH 7.0, which is nearly one pH unit below the pKa of the sequestered amine group array. If the above rationale is correct, at pH 7.0, there should be no extension of the lag due to addi-tion of uncoupler, because the lower external pH should provide sufficient protons to keep the amine groups in the $-NH_3^+$ state. The experiment shown in Fig. 4 confirmed that that was indeed the case. There was no significant difference in the length of the lag in onset of ATP formation for any of the conditions.

An experiment similar in design to that of Fig. 3 and 4 was done with Cl-CCP (except valinomycin was omitted). The results were in agreement with the

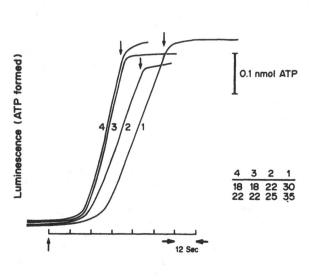

Fig. 3. Desaspidin extends, and preillumination shortens, the lag in onset of ATP formation. Chloroplasts were suspended in 50 mM sorbitol, 50 mM tricine-KOH pH 8.5, 3 mM $MgCl_2$ and 0.1 mM methyl viologen with 15 μg chl ml^{-1}. Stringently dark conditions were maintained in the room and throughout the storage, transfer and incubation periods. After the appropriate dark or pre-illumination treatments, luciferin-luciferase, diadenosine pentaphosphate, ADP, Pi and $10^{-7}M$ valinomycin were added. The start of the 1 Hz flash sequences is indicated by the upward pointed arrow, the end by the downward arrow. A fresh sample was prepared for each treatment, except for curve 4, which is a second cycle. Treatments: Curve 1. $10^{-7}M$ desapsidin was added to the basic mixture, 1 min later 2.5 mg BSA were added to bind the desaspidin; five min additional dark incubation followed. During the last min of darkness the other components of the assay, mentioned above, were added. Note the absence of ATP hydrolysis before and after the flash regime. Curve 2. The sample was prepared similarly to that for curve 1. Desaspidin was added followed by BSA. Two min later a 3 sec illumination with saturating intensity of white light was given, followed by 8 min of darkness. The remaining assay components were added and the flash regime was given. Curve 3. Control; BSA was added before the $10^{-7}M$ desaspidin. After a 3 min incubaion in darkness, the flash regime was given. Another control (not shown) was prepared identically but given a 9.5 min dark incubation time prior to the flashes. The ATP onset parameters for that control were essentially identical to those for curve 3, 17 and 23 flashes, respectively, for the two parameters. Curve 4 is a second flash cycle given to sample 3 after a 4 min dark period following the first flash sequence. The calibration, by addition of 20 μl of 10^{-5} M ATP to each sample, gave a similar value for all the samples.

desaspidin results for both the pH 8.5 and 7.0 conditions. Again, at pH 8.5 depleting the buffering pool of protons with Cl–CCP caused an extension of the lag by about 10 flashes. A 36–flash preillumination (analogous to curve 2, Fig 3) shortened the lag back to about the value of the control (data not shown, but see Dilley and Schreiber (1983) for details).

INTERPRETATION

The data of Fig. 3 and 4 clearly shows that the treatments which were expected to deplete the metastable pool of membrane–phase buffered protons,

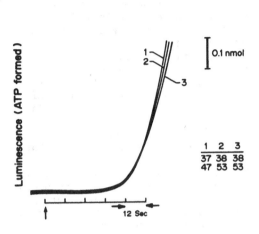

Fig. 4. Effects of desaspidin and BSA at pH 7.0. The experiment was similar to that of Fig. 3 except the pH was adjusted to 7.0. Curve 1, similar in plan to curve 1, Fig. 3, but desaspidin was given at 0.2 μM, then 2.5 mg BSA was added followed by 6 minutes of dark prior to the 1 Hz flash sequence to drive ATP formation. Curve 2. Similar to curve 1, but 2 min after 2.5 mg BSA addition, 36 single-turnover flashes were given at 1 Hz, followed by 6 min darkness. The remaining components of the ATP assay were added during the last min of darkness, prior to the ATP formation assay. Curve 3. A control, with BSA added before desaspidin.

caused pronounced extensions of the lag in onset of ATP formation. After a depletion treatment, but before adding ADP and Pi, a preillumination regime with a 3 sec light pulse or 24 to 36 single-turnover flashes, restored the shorter lags. The magnitude of the extension and the shortening of the lags was about ten flashes. Ten flashes should yield about 30 to 35 nmol H^+ (mg chl)$^{-1}$ translocated into the membrane, according to Graan and Ort´s (1982) measurement that a single-turnover flash yields about 3.5 nmol H^+ (mg chl)$^{-1}$. Thirty nmol H^+ (mg chl)$^{-1}$ corresponds well with the size of the sequestered proton buffering pool determined by the acetic anhydride probe. The close correspondence in these numbers strongly supports the hypothesis that the 30 nmol (mg chl)$^{-1}$ of buffering groups are located so as to directly interact with protons released by the electron transport reactions, before the protons reach the CF_1 complex.

The metastable buffering pool is shown by the data of Fig. 1 and Table 1 to be a sequestered domain, not in rapid equilibrium with the bulk inner aqueous phase. The ATP onset lag results suggest that the "special pool" buffering array is located between the proton sources – the redox protolytic reactions – and the CF_1 proton sink which utilizes the proton fluxes to drive ATP formation. That is, protons that seem to be required to fill the "special pool" – a localized domain – appear to do so while the onset of ATP formation is retarded. The simplest interpretation is that the special pool buffering array must be protonated before protons from the redox reactions can build up the required protonmotive force for ATP formation. That alone does not necessarily rule out the possibility that the "special pool" of buffering groups is in series both with the bulk lumen space and "down stream" from that, the CF_0-CF_1 complex. That is, there could be localized proton processing pathways close to the redox proton sources leading to the lumen space, which then connect to the CF_0-CF_1. That model would not be consistent with the findings that: (a) the anhydride-sensitive buffering pool is not in ready equilibrium with the lumen bulk phase; (Laszlo et al., 1983) and (b) the ATPase proton pump is just as effective as the redox proton pumps in reprotonating a depleted buffering pool. The bulk lumen phase cannot be in slow equilibrium with the special buffering pool array and, at the same time, be an obligatory pathway for protons derived from ATPase proton pumping to reach the buffering array. Therefore, the experimental data are more consistent with entire proton processing pathway being localized. Yet, this leaves unanswered the question of how the lumen bulk phase gets acidified by redox or ATPase proton pumps, which is believed to happen. One explanation offered previously (Dilley et al. 1982) might be that the lumen bulk phase proton concentration equilibrates relatively slowly with the protons in the CF_0 channel. More work is needed to clarify this question as well as to understand the role of the sequestered special pool of buffering groups. The pKa of the buffering group array probed by these techniques is near 7.8 (Laszlo, et al. 1983). At tha pH, it is clear that the sequestered array is mostly in the protonated state in situ under dark or light conditions. That is, the array is not a candidate for involvement in energy storing or transducing reactions associated with protonation-deprotonation cycles. Yet the data of this report implicate the buffering group array in proton processing events in energy coupling. Our hypothesis is that the buffering group array may provide proton hopping sites (Nagle and Tristram-Nagle, 1983) in a pathway from the proton sources, water oxidation and plastohydroquinone oxidation, to the CF_0-CF_1 proton sink.

REFERENCES

Baker, G.M., Bhatnagar, D. and Dilley, R.A. (1981) Biochemistry 20, 2307-2315.

Dilley, R.A., Prochaska, L.J., Baker, G.M., Tandy, N.E. and Millner, P.A. (1982) Curr. Topics in Memb. and Transport 16, 345-369.

Dilley, R.A. and Schreiber, U. (1983) submitted, J. Bioenergetics Biomembranes.

Graan, T. and Ort D.R. (1982) Biochim. Biophys. Acta 682, 395-403.

Laszlo, J.A., Baker, G.M. and Dilley, R.A. (1983) in press, J. Bioenergetics Biomembranes.

Nagle, J.F. and Tristram-Nagle, S. (1983) J. Membrane Biol. in press.

Schreiber, U., and Del Valle-Tascon, S. (1982) FEBS Lett. 150, 32-37.

Authors' address: R.A. Dilley, J.A. Laszlo, Dept. Biol. Sci., Purdue U., W. Lafayette, IN 47907, U.S.A. U. Schreiber Leherstuhl Botanik I, Univ. Würzburg, 8700 Würzburg, Fed. Rep. Germany.

MODULATION OF THE CHLOROPLAST ATP SYNTHETASE: CONFORMATIONAL STATES, NUCLEOTIDE BINDING AND LIMITED ACCESSIBILITY TO THE ACTIVE SITE.

N. SHAVIT, C. AFLALO, D. BAR-ZVI AND M.A. TIEFERT,
BEN GURION UNIVERSITY OF THE NEGEV

INTRODUCTION

A kinetic scheme for photophosphorylation catalyzed by the chloroplast ATP synthetase is described by Equation 1. This sequence of reactions assumes that binding of ADP occurs prior to that of Pi (Selman, Selman-Reimer, 1981). Thus, the Pi-ATP exchange, one of the so-called partial reactions of photo-phosphorylation, might occur without significant dissociation of the E·ADP complex and without a concomitant ADP-ATP exchange reaction ($k_{-1} \ll k_1$)

$$E \underset{k_{-1}}{\overset{k_1}{\rightleftharpoons}} E \cdot ADP \underset{k_{-2}}{\overset{k_2}{\rightleftharpoons}} E \cdot {}_{Pi}^{ADP} \underset{k_{-3}}{\overset{k_3}{\rightleftharpoons}} E \cdot ATP \underset{k_{-4}}{\overset{k_4}{\rightleftharpoons}} E \tag{1}$$

In this mechanism, formation and hydrolysis of ATP and the accompanying ex-change reactions are reversible reactions that occur on the same catalytic site(s) on the energized membrane-bound enzyme (CF_1). It has been about ten years since it was reported that the synthesis of ATP and its hydrolysis or the Pi-ATP exchange reaction show a dissimilar substrate specificity. Modified antibodies against CF_1 were also shown to inhibit preferentially the Pi-ATP exchange. These and more recent data support the view that different enzyme conformations or altered catalytic sites may participate in the catalysis of ADP (photophosphorylation) and ATP (hydrolysis and exchange) utilizing re-actions (Shavit, 1980).

Another aspect briefly reviewed in this paper is the modulation of the membrane-bound ATP synthetase by nucleotide tight binding and release. With the discovery of ATP and ADP tightly bound to ATP synthetases, the question of their relation to catalysis was raised. Slater and Boyer originally sug-gested that energy input is not needed for the phosphorylation of ADP bound to the catalytic site but rather for promoting the release of newly formed ATP bound at this site, through a conformational change in the ATP synthetase. Boyer, who further developed this view on the basis of a considerable amount of experimental results, suggested that this conformational change promotes concomitantly the competent binding of ADP and Pi at an alternate catalytic site (Rosen et al., 1979). Bound products are indeed detectable during or after catalysis, as assessed by their inaccessibility to coupled enzymes in the medium, or their retention on the enzyme after deenergization and washing, respectively. Thus, this "binding-change mechanism" involves the participation of tightly bound nucleotides as catalytic intermediates in ATP synthesis or hydrolysis.

However, such a role for tightly bound ATP in photophosphorylation is questionable in view of its relatively low apparent rate of formation (Aflalo, Shavit, 1982). The behavior of tightly bound ADP also indicates that this site(s) is very different from the catalytic site. Strotmann (1981) proposed a mechanism for modulation of CF_1 activity by energy, whereby energization activates the enzyme with the concomitant release of bound nucleotides. Deenergization reverses the |process and rebinds the nucleotides to the tight binding sites. Bar-Zvi and Shavit (1982a) provided evidence that it is the deenergized conformation that binds ADP with the consequent "freezing" or

Sybesma, C. (ed.), Advances in Photosynthesis Research, Vol. II. ISBN 90-247-2943-2.
© *1984 Martinus Nijhoff/Dr W. Junk Publishers, The Hague/Boston/Lancaster.*

"locking" of the enzyme in its latent, deactivated form. Moreover, other nucleotide analogs which compete with the binding of ADP and prevent the ADP induced inactivation have, little effect on the thylakoid activity (Bar-Zvi, Shavit, 1982a,b). These experiemnts support a regulatory rather than a catalytic role for tightly bound ADP.

Recently we have found a type of tightly bound ATP which exists even in the presence of saturating concentrations of hexokinase (Aflalo, Shavit, 1982) but is exchangeable with exogenous ATP. This bound ATP appears to arise from a free species of ATP sequestered near the active site during photophosphory-lation. The slowed down diffusion of the newly made ATP to the outer medium space allows its binding to proximal non-catalytic sites. Thus, mass trans-fer of nucleotides between the active site and the bulk medium may limit the steady state rate of ATP synthesis and should be considered in proposals on the mechanism of energy transduction.

RESULTS AND DISCUSSION

Different conformations of CF_1 during catalysis

The relative insensitivity of ATP formation and hydrolysis in contrast to the strong inhibition of the Pi-ATP exchange by preincubation of chloroplasts with monovalent antibodies against CF_1 is shown in Fig. 1.

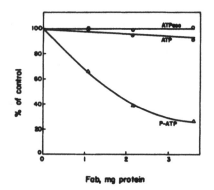

Fig. 1. Inhibition of the P_i-ATP exchange by interaction of chloroplast thylakoids with (Fab) fragmented antibodies to CF_1. Lettuce chloroplasts were reacted with Fab fragments for 2 min in the dark at $0^{\circ}C$ prior to activation (Shoshan, Shavit, 1979). Photophosphorylation and exchange re-actions were assayed as described (Bar-Zvi, Shavit, 1983).

The controlled chemical modification of chloroplast thylakoids with glutaral-dehyde (GA) that results in the strong inhibition of the Pi-ATP exchange while ATP formation and hydrolysis are only slightly impaired (Table 1), is another example of such a differential inactivation of CF1 catalyzed activities.

Table 1. Effect of glutaraldehyde modification of chloroplast thylakoids on several reactions. Modification of chloroplasts and assays were as described (Bar-Zvi et al. 1983). n. d., not detectable

Modification conditions		Pi-ATP exchange	ATP hydrolysis	ATP formation	pH gradient
time (s)	(GA) (%, w/v)	percent of control activity			
90	0.01	15	74	87	78
150	0.05	n.d.	3	5	–

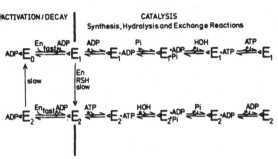

Fig. 2. Scheme showing activation of the latent ATP synthetase (E_0), phosphorylation catalyzed by the light energized (E_1) enzyme conformation and ATP hydrolysis and exchange in the dark, catalyzed by the preactivated (E_2) enzyme form. All enzyme forms can bind nucleotides to non-catalytic tight binding sites (◄◄). The ADP◄E_0 and ADP◄E_2 indicate inactive enzyme forms obtained by tight binding of ADP. The existence of an ATP◄E_1 form in the light, during ATP formation, is suggested (see text).

We have also shown that the GA modification affects the catalytic process and not the activation of the enzyme itself. Since the Pi-ATP exchange reaction requires the establishment of a pH gradient, while ATP hydrolysis can occur uncoupled from the ΔpH and ATP formation may be less sensitive to the modification due to the type of energy supplied (continuous illumination), we checked the effect of the modification on ΔpH in parallel with ATP hydrolysis and Pi-ATP exchange. As indicated in Table 1, the ΔpH supported by ATP hydrolysis in the dark is only slightly impaired, parallely to that of ATP hydrolysis. Therefore, it does not seem to relate to the loss of the Pi-ATP exchange activity upon modification of thylakoids with GA.

The limited modification by GA as well as the interaction of CF_1 with the antibody probably do not involve groups at the catalytic site(s). If a mechanism as the one given in Equation 1 is assumed, then one might attempt to explain the difference between the ATPase and Pi-ATP exchange, reactions that are assayed under the same conditions (preactivation of the enzyme, assay in the dark), by a change in a rate constant but with one enzyme form catalyzing both reactions. If ATP formation does occur by the reverse of ATP hydrolysis, a change in a rate constant (probably k_3 in Equation 1) that will lower the Vmax of the Pi-ATP exchange, increase the Km for both ATP and Pi in the exchange reaction and that of ATP in ATP hydrolysis, should also reduce the rate of ATP formation. However, under all conditions tested, this effect was not observed. Thus, consideration of all the available data lead us to suggest that one conformation of the ATP synthetase catalyzes ATP formation, a form that exists upon energization by light, while in the dark, another conformation which hydrolyzes ATP and catalyzes the Pi-ATP exchange, is favored (Fig. 2).

Transformation of one conformation to the other requires primarily energization and is accompanied by the release of previously tightly bound nucleotides. The maintenance and operational efficiency of a given conformation does probably depend on the species of nucleotide and other reactants concentrations present in solution and the binding of these nucleotides to the regulatory, non-catalytic sites on the enzyme. Conformational changes in the membrane-bound CF_1 induced by energization are well documented. These include light induced incorporation of N-ethylmaleimide into the γ subunit of CF_1 (McCarty et al., 1972), exchange between water and protein protons (Ryrie, Jagendorf, 1971) and release of tightly bound adenine nucleotides (Strotmann et al., 1976). Studies with nucleotide's analogs led Strotmann (1981) to propose the existence of two functionally different conformations of the chloroplast ATP synthetase. Based on kinetic measurements of ATP hydrolysis by submitochondrial particles a similar conclusion was reached for the mitochondrial ATPase (see Shavit, 1980).

Subunit location of the tight nucleotide binding site(s).

3'-O-(4-benzoyl)benzoyl ADP (BzADP) interacts with the membrane-bound CF_1 but is not by itself a substrate for photophosphorylation. As indicated in Table 2, BzADP is a strong competitive inhibitor with respect to ADP and ATP in ATP synthesis, hydrolysis and the exchange reactions, respectively. BzADP binds also to the tight-binding site and prevents ADP binding competitively. Upon irradiation with ultraviolet light the analog is photoactivated and becomes covalently bound to the ATPase. Gel electrophoresis analysis in the presence of SDS of the reacted and then isolated CF_1, shows that the label is incorporated into both the α and β subunits.

Table 2. Apparent kinetic parameters for the inhibition of the CF_1-ATPase by non-covalent interaction with BzADP

Reaction	Parameter	Value-μM	Type of inhibition
1. ATP formation			
(ADP)varied	Ki(s)	5.0 ± 0.1	competitive
(Pi)varied	Ki(s)	24.6 ± 2.8	mixed
	Ki(i)	51.6 ± 4.0	
2. Pi-ATP exchange			
(ATP)varied	Ki(s)	20.4 ± 6.3	probably competitive
3. ATP hydrolysis	Ki(s)	21.3 ± 7.8	competitive
4. tight (^{3}H)ADP			
binding	Ki(s)	1.7 ± 0.3	competitive

Recently we have shown that BzADP also interacts with the isolated soluble CF_1, inhibits ATP hydrolysis and is covalently incorporated into the β subunit of CF_1 only (Bar-Zvi, Shavit, 1983). These results support the notion that non catalytic sites on the subunits on the membrane-bound enzyme are proximal to the β subunits, the proposed catalytic sites. If this pattern of labeling can be corroborated with other photoaffinity labeled nucleotides, it may provide useful information on the subunits that participate in the formation of an active CF_1 molecule. The proximity between catalytic and non-catalytic sites might favor site to site interactions and even surface migration of nucleotides between them, without going through the mediu (see below).

Kinetic competence of newly formed ATP tightly bound to CF_1.

At least two species of (^{32}P)ATP tightly bound to membranal CF_1 were detected. One species is rapidly labeled from medium ^{32}Pi, is not susceptible to hexokinase present in the medium and is believed to fulfill the requirements of a type of bound (^{32}P) ATP that acts as an intermediate in ATP formation (Rosen et al., 1979). In contrast, the evidence we obtained indicates that this ATP is bound to a non-catalytic site and that it is derived from newly made medium ATP (Shavit et al., 1981; Aflalo, Shavit, 1982). The rates of formation and discharge of this class of bound ATP are much slower than the rate of ATP formation. Unlabeled ATP, present in the medium, displaces this type of bound ATP without affecting the rate of phosphorylation. Results consistent with participation of medium ATP in the formation and release of

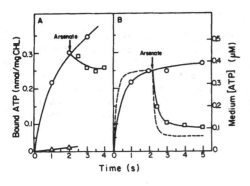

Fig. 3. Effect of a chase with arse-nate on bound (^{32}P)ATP. Samples in (A) without and in (B) with hexokinase and glucose, were illuminated for the times specified. Arrows indicate addition of 2 μmol sodium arsenate. Dashed line represents the calculated concentration of medium (^{32}P)ATP in the light. (o), control; arsenate added before illumination (Δ) or at arrow (▫). (From Aflalo, Shavit, 1982).

this type of bound ATP are shown in Fig. 3. In the presence of a saturating amount of arsenate during phosphorylation, the formation of an arsenylated intermediate at the catalytic site should displace any (^{32}P)ATP bound at this site at a rate greater than (or at least equal to) the rate of formation of medium ATP in the absence of arsenate. In the absence of hexokinase, addition of arsenate which inhibits ATP formation by about 90%, does not reduce significantly the level of bound (^{32}P)ATP previously accumulated (Fig. 3A). In the presence of hexokinase (Fig. 3B), arsenate reduces the steady-state concentration of medium ATP by about 80%, with a half-life time of about 80 ms (see Aflalo, Shavit, 1982). Although the level of bound (^{32}P)ATP is rapidly reduced, the disappearance of free medium ATP seems nevertheless to occur faster.

The rapidly labeled tightly bound ATP is not accessible to high concentrations of hexokinase but is nevertheless displaced by exogenous ATP (Aflalo, Shavit, 1982). Moreover, a fraction of the tightly bound ATP that is exchangeable with exogenous ATP, becomes non-exchangeable in the dark, in a rather slow process (within seconds). Since this increase in tightly bound ATP cannot be the result of the direct binding of medium ATP, it follows that the newly made (^{32}P)ATP is released from the catalytic site but is not fully equilibrated with medium ATP. Thus, we propose that a free class of ATP, sequestered in a space inaccessible to hexokinase, participates in the formation of the rapidly labeled tightly bound ATP.

Environmental effects on the steady state kinetics of ATP formation

Energy input in ATP synthesis and its effect on the kinetic parameters for the steady state utilization of substrate by the enzyme, is a different approach used to study the question of the role of tightly bound nucleotides. Modulation of the apparent affinity of the enzyme for ADP by the rate of electron transport was recently proposed. The supply of lower energy levels increased the affinity for ADP while uncoupling decreased the affinity (Vinkler, 1981). This mechanism is somewhat different from that proposed by Strotmann (1981). These results, however, imply that the regulation of the synthetase might involve separate modulation of the catalytic ability (Vmax) and the apparent affinity for substrates (Km) or the efficiency of their utilization (Vmax/Km).

The kinetic behavior of membrane-bound ATP synthetases has been assumed to be close to that of an enzyme in solution. Thus, changes in apparent

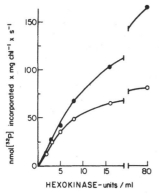

Fig. 4. Effect of hexokinase on the steady state rate of phosphorylation at low ADP concentration. Values of phosphorylation are given for the linear rates obtained between 5-35 s of illumination and at 18 μM ADP. Control (o) or treated (●) thylakoids contained 94 μg chl. Light intensity and P_i concentration were saturating (From Aflalo, Shavit, 1983).

kinetic parameters are interpreted in terms of enzyme conformational changes or interactions between multiple binding sites. However, analysis of the kinetic behavior of enzymes, immobilized on solid supports, has shown the importance of slow translocation of reactants to or from the active site (Engasser, Horvath, 1976). Thus, we might expect that the ATP synthetase, bound to a membrane surface probably highly charged during catalysis, will behave differently from that of an enzyme in solution, where solutes diffuse freely. As proposed above, when (^{32}P)ATP formation from $^{32}P_i$ is considered, the newly made ATP appears to remain sequestered in the vicinity of non-catalytic sites to which it binds tightly in a rather slow process. Steady-state kinetic studies of photophosphorylation with hypotonically treated and control chloroplast thylakoids revealed that the hypotonic treatment apparently removes a kinetic diffusion barrier for ATP (Aflalo, Shavit, 1983). At low (ADP), the velocity of the phosphorylation reaction in the presence of hexokinase might be determined by the capacity of hexokinase to steadily resupply ADP to the ATPase. With hypotonically treated thylakoids, at low (ADP) and saturating light intensity, the supply of ADP by hexokinase indeed limits strongly the reaction rate (Fig. 4). At a low phosphorylation rate, with control chloroplasts, hexokinase is much less limiting. Thus it seems that the supply of ATP to the hexokinase in the medium is less efficient in the control chloroplasts.

As described for the macroscopic behavior of enzymes immobilized on solid supports (Goldstein, 1977) the rate of catalysis at low substrate concentration could be limited by substrate mass transfer from the bulk medium to the enzyme. Since the ATPase appears to behave similarly to an immobilized enzyme, the phenomenon arises from the relatively slow translocation of ADP in the environment of the enzyme in comparison to its fast transformation on the site. Of course the limitation can be overcome upon saturation in substrate concentration. At steady state, the rate of translocation of substrate is equal to that of catalysis in situ, leading to the relative depletion of ADP and/or accumulation of ATP near the enzyme. Support for this view is given by the better utilization of ADP and, to a lesser extent, of GDP with the hypotonically treated chloroplasts (Aflalo, Shavit, 1983). The Km values for ADP and GDP are lower and the Vmax/Km are higher with hypotonically treated chloroplasts, at virtually the same catalytic ability (Vmax). On the other hand, the utilization of Pi remains unaffected, as might be expected for a substrate which is utilized much less efficiently (low Vmax/Km).

Present data, therefore, indicates limitation of catalysis by mass transfer of nucleotides through the environment of the membrane in normally prepared thylakoids. In the case of ATP synthesis coupled to the hexokinase reaction, such a limitation should be amplified due to the slow translocation of both ADP and ATP near the membrane surface. This limitation appears to be partly relieved by hypotonic treatment, artificial reduction in the membrane's catalytic ability, or increasing the ionic strength in the reaction mixture during catalysis (Aflalo, Shavit, 1983). These features of phosphorylation by the energy transducing membrane point toward the possible control of the diffusibility of charged substrates and/or products by electrostatic interactions due to the high charge density existing on the surface of the thylakoid membrane during catalysis. These considerations introduce a new perspective to explain modulation of the ATP synthetase activities and medium exchange reactions by energy input, substrate concentration, and in general any factor which could affect the microenvironment of the enzyme, provided that proper conditions for "diffusion control" exist (high reaction rate and low diffusibility of substrates, products, or effectors).

REFERENCES

Aflalo C and Shavit N (1982) Source of rapidly labeled ATP tightly bound to non-catalytic sites on the chloroplast ATP synthetase, Eur. J. Biochem. 126, 61-68.
Aflalo C and Shavit N (1983) Steady-state kinetics of photophosphorylation: Limited access of nucleotides to the active site on the ATP synthetase, FEBS Lett. 154, 175-79.
Bar-Zvi D and Shavit N (1982a) Modulation of the chloroplast ATPase by tight binding of nucleotides. Biochim. Biophys. Acta 681, 451-58.
Bar-Zvi D and Shavit N (1982b) Modulation of the chloroplast ATPase by tight ADP binding. Effect of uncouplers and ATP. J. Bioenerg. Biomem. 14, 467-77.
Bar-Zvi D and Shavit N (1983) Differential inhibition of Pi-ATP exchange in relation to ATP synthesis and hydrolysis by modification of chloroplast thylakoid membranes with glutaraldehyde. Biochim. Biophys. Acta, in press.
Engasser J-M and Horvath C (1976) Diffusion and kinetics with immobilized enzymes. In Wingard LB, Kachalski-Katzir E, Goldstein L., eds. Applied biochemistry and Bioengineering, pp. 127-220. New York: Acad. Press.
Goldstein L (1977) Kinetic behavior of immobilized enzyme systems. Methods in Enzymology 44, 397-443.
McCarty RE, Pittman PR and Tsuchiya Y (1972) Light-dependent inhibition of photophosphorylation by N-ethylmaleimide. J. Biol. Chem. 247, 3048-51.
Rosen G, Gresser M, Vinkler C and Boyer PD (1979) Assessment of total catalytic sites and the nature of bound nucleotide participation on photophosphorylation. J. Biol. Chem. 254, 10654-61.
Ryrie IJ and Jagendorf AT (1972) Correlation between a conformational change in the coupling factor and the high energy state in chloroplasts. J. Biol. Chem. 247, 4453-459.
Selman BR and Selman-Reimer S (1981) The steady state kinetics of photophosphorylation. J. Biol. Chem. 256, 1722-26.
Shavit N (1980) Energy transduction in chloroplasts: Structure and function of the ATPase complex, Annu. Rev. Biochem. 49, 111-38.

Shavit N, Aflalo C and Bar-Zvi D (1981) Role of tight nucleotide binding
sites in the modulation of the chloroplast ATP synthetase activity, in
Energy Coupling in Photosynthesis, Selman BR and Selman-Reimer S, eds.
pp 197-207. Elsevier North Holland Inc.
Shoshan V and Shavit N (1979) ATP synthesis and hydrolysis in chloroplast membranes. Differential inhibition by antibodies to chloroplast
coupling factor 1. Eur. J. Biochem. 94, 87-92.
Strotmann H, Bickel S and Huchzermeyer B (1976) Energy-dependent release of
adenine nucleotides tightly bound to chloroplast coupling factor CF_1.
FEBS Lett. 61, 194-98.
Strotmann H, Bickel-Sandkötter S, Franek U and Gerke V (1981) Nucleotide
interactions with membrane-bound CF_1. In Selman BR and Selman-Reimer S,
eds. Energy coupling in photosynthesis, pp. 187-96. New York: Elsevier
North Holland Inc.
Vinkler C (1981) Opposite modulation by uncoupling and electron transport
limitation of the K_m(app) of ADP for photophosphorylation, Biochem. Biophys.
Res. Commun. 99, 1095-100.

ACKNOWLEDGMENTS

We are grateful to Mrs. Z. Conrad for excellent technical assistance.
Authors address: Department of Biology, Ben Gurion University of the
                 Negev, Beer Sheva 84105, Israel.

CONSERVATION AND ORGANIZATION OF SUBUNITS OF THE CHLOROPLAST PROTON ATPase COMPLEX

RUTH ROTT and NATHAN NELSON/ Department of Biology, Technion - Israel Institute of Technology, Haifa, Israel.

1. INTRODUCTION

Proton ATPase complexes have been isolated from various energy transducing membranes. All these complexes have a similar structure which is comprised of two distinct parts (Baird, Hammes 1979; Nelson 1981): A catalytic part which is peripherial to the membrane (F_1) and consists of five different subunits, and a membrane part (F_0) which consists of three to seven subunits. While the subunit structure of F from various sources is very similar, the F_0 part has been shown to have a variable structure (Houstek et al. 1982; Fillingame 1981). Nevertheless, one of the F_0 subunits, the DCCD binding protein, has been strictly preserved during evolution (Sebald, Hoppe 1981). In attempt to investigate the organization of the subunits of the H^+-ATPase complex, chemical cross-linking has been widely employed (Baird, Hammes 1976; Satre et al. 1976; Baird, Hammes 1977; Chernyak et al. 1981; Bragg, Hou 1975; Enns, Criddle 1977; Tod, Douglas 1981). Unequivocal identification of the subunits which comprises the aggregates formed by different cross-linking agents, has been the major obstacle of those investigations.

In the present work we have employed antibodies which were raised against the different subunits of H^+-ATPase complex in order to elucidate evolutionary relations among subunits of the same complex isolated from various sources. We investigated the structure of chloroplast H^+-ATPase complex by chemical cross-linking, by the help of the same antibodies. The composition of aggregates formed following cross-linking could be identified in a more accurate way.

2. MATERIALS AND METHODS

Dimethyl suberimidate (DMS) was obtained from Pierce Chemical Co., hexamethylene diisocyanate (DIC) was obtained from Aldrich Chemical Co. Published procedures were employed for preparation of spinach chloroplasts (Kamienietzky, Nelson 1975) H^+-ATPase complexes (Nelson et al. 1980; Rott, Nelson 1981) and antibodies Nelson et al. 1973) for SDS-gel electrophoresis (Douglas, Butow 1976), for electrotransfer of proteins from gels to nitrocellulose paper (Towbin et al. 1979) and for immunedecoration by ^{125}I protein A (Rott, Nelson 1981). ATP-Pi exchange activity was measured according to (Nelson 1980a).

Chemical cross-linking was carried out by DMS or DIC at 0 C. Stock solution of 0.2M DMS in water was prepared just before use. The pH was rapidly adjusted to pH = 8.0 with NaOH. Cross-linking was started by ten fold dilution of the DMS solution in the protein suspension and stopped either by dilution of by dissociation by 2% SDS and 2% mercaptoethanol. To start the cross-linking by DIC freshly prepared stock solution of 66 mM DIC in dimethyl sulfoxide, was diluted 20 fold in the protein preparation. The cross-linking was stoped as described for DMS.

Sybesma, C. (ed.), Advances in Photosynthesis Research, Vol. II. ISBN 90-247-2943-2.
© *1984 Martinus Nijhoff/Dr W. Junk Publishers, The Hague/Boston/Lancaster.*

3. RESULTS

The evolutionary relations among subunits of F₁ from several sources were tested by immunological cross-reactivity. Antibodies raised against subunits of F₁ were reacted with H^+-ATPase complex from Chlamydomonas and spinach chloroplasts, from beef adrenal, rat liver and yeast mitochondria and with E. coli membranes. The antigen-antibody reaction was tested employing the electrotransfer technique (Rott, Nelson 1981; Towbin et al. 1979). The antibodies employed were raised against the following subunits of the H^+-ATPase complex: $\alpha, \beta, \gamma, \delta, \varepsilon,$ I and II of spinach chloroplasts, α, β and γ of E. coli membranes and α, β and γ of yeast mitochondria. Of all the anitbodies tested only those raised against subunit β strongly interacted with the corresponding subunit of all the above mentioned complexes (Figure 1).

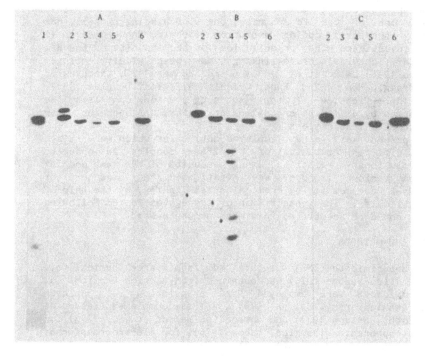

Figure 1. Immunological cross-reactivity of β subunit of the proton ATPase complex from various sources.
Proton ATPase complexes from: Lane 1; chlamydomonas chloroplasts (20 ug), lane 2; Swiss chard chloroplasts (25 ug), lane 3; rat liver mitochondria (15 ug), lane 4; yeast mitochondria (10 ug), lane 5; E. coli membranes (50 ug), and lane 6; mitochondria from bovine adrenal medula (10 ug) were run on SDS gels. Then electrotransfer to nitrocellulose paper and immunedecoration were carried out. The antibodies used were raised against subunits of proton ATPase complexes from spinach chloroplasts (A), yeast mitochondria (B) and E.coli membranes (C).

In order to study the subunit arrangement of the chloroplast H$^+$-ATPase complex, chemical cross-linking was employed. Two cross-linking reagents were used, dimethylsuberimidate (DMS) and hexamethylene diisocyanate (DIC). Both reagents can react with amino groups but differ in their hydrophobicity. Thus, using both DMS and DIC provides more detailed information about possible interactions between subunits than using one of them only.

Following treatment with either DMS or DIC, new protein bands appear in the preparation of purified H$^+$-ATPase complex. Most of these aggregates have a molecular weight higher than that of ∝ subunit, as revealed by SDS-polyacrylamide gel electrophoresis (Figure 2). A different pattern of the bands is obtained after DMS treatment compairing to DIC treatment. Concomitantly, the chemical cross-linking causes inhibition of the ATP-Pi exchange activity of the H$^+$-ATPase complex (Figure 3). The inhibition caused by DMS was found to be stronger than the one caused by DIC. The identification of the subunits, which construct the aggregates formed by DMS or by DIC, was accomplished by the use of specific antibodies raised against the different subunits of the H$^+$-ATPase complex. These aggregates were separated by SDS-polyacrylamide gel electrophoresis and subsequently

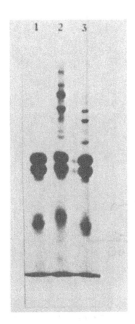

Figure 2. Polyacrylamide gel electrophoresis of cross-linked proton ATPase complex.
    Purified complex from spinach chloroplasts (25 ug) was run on 10% polyacrylamide gel before (lane 1) or after cross-linking with DMS (lane 2) or DIC (lane 3), as described under "Materials and Methods".

transferred to nitrocellulose paper (Towbin et al. 1979). After incubation with one of the above mentioned antibodies and with [125] I protein A, antigen-antibody complexes were identified on the nitrocellulose paper by exposing it to X-ray film (Rott,Nelson 1981). Nitrocellulose paper pieces containing untreated and DMS or DIC treated H^+-ATPase complex were incubated each with antibody raised against different subunit. A band which corresponds to reaction of the antibody with the subunit against which it was raised appeared in each piece in the untreated and in the treated complex (Figure 4). Additional bands appeared in the cross-linked samples corresponding to the new polypeptides shown on SDS-polyacrylamide gel (Figure 2). The molecular weight of these bands were determined by the used of molecular weight markers, which were run together with the treated complex on the gel (not shown). The reaction of an aggregate of a known molecular weight, with different antibodies indicate that the subunits against which these antibodies were raised, are included in it. Figure 5 summarizes

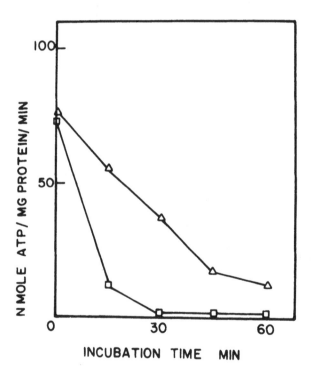

Figure 3. ATP-Pi exchange activity of proton ATPase complex after cross-linking.
    Proton ATPase complex was incubated with DMS (□) or DIC (Δ) after reconstitution into liposomes, at $0°C$. At the time indicated in the figure, samples were taken to measure ATP-Pi exchange activity. The activity of the untreated complex was 60 n mol/mg protein/min.

the combination of subunits in the aggregates formed by cross-linking, as deduced from the reaction with antibodies described in Figure 4. In instances where the combination of subunits in a given aggregate was not clear from the reaction with antibodies, the molecular weight was used to minimize the possible combination (shown in Figure 5 in brackets).

4. DISCUSSION

The structure of the catalytic part of the H^+-ATPase complex has been found to be preserved during evolution (Baird, Hammes 1979; Nelson 1981). Although five different subunits construct the F_1 part, in all H^+-ATPase complexes tested so far, some differences in the function of analog subunits in the complex from various sources have been shown (Baird, Hammes 1979). One example is the inhibition of chloroplast F_1 activity by ε subunit (Nelson 1981), while another polypeptide carries out this function in mitochondria (Van de Stadt et al. 1973; Pullman,

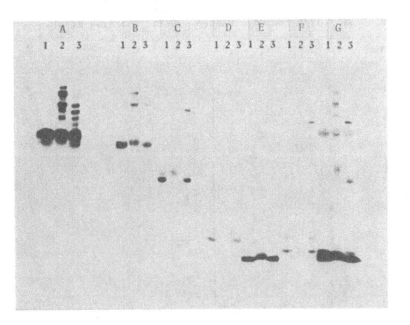

Figure 4. The interaction of cross-linked proton ATPase complex with antibodies.
Proton ATPase complex from spinach chloroplasts was treated with DMS or DIC and run on gel as described in Fig 2, transferred to nitrocellulose paper and treated with antibodies and ^{125}I protein A as described under "Materials and Methods". The antibodies used were raised against subunits α(A), γ(C), δ(D), ε(E), I(F) and II(G) of spinach chloroplasts and β(B) of E. coli membranes. 1-untreated complex, 2- complex cross-linked with DMS and 3- complex treated with DIC.

Monroy 1963). The differences found could be explained by exchange of activities between different subunits. A homology was found lately in the amino acid sequences of ε subunit of E. coli and δ subunit of beef heart mitochondria H⁺-ATPase complexes (Walker et al. 1982).

In spite of the above mentioned differences, a common function has been suggested for the β subunit (Nelson 1981; Ferguson et al. 1975; Ohta et al. 1980). These investigations indicated that the β subunit constitutes the catalytic site of the F_1 part. Our finding that there is immunological cross-reactivity among subunit of H⁺-ATPase complexes from various sources, indicate that the structure of this subunit has been strictly preserved during evolution. The cross-reactivity implies that common amino acid sequences are present in the β subunit of H⁺-ATPase complex from various sources. Indeed, a 75% homology of amino acids was recently found for the E. coli and beef heart mitochondria β subunit (Walker et al. 1982a). Apparently, the structure of the β subunit had to be strictly preserved during avolution in order to maintain the active site of the H⁺-ATPase complex and that changes in most of the amino acid sequences of this subunit were lethal.

Chemical cross-linking has been used for investigating the arrangement of the different subunits in the H⁺-ATPase complex (Baird, Hammes 1976; Satre et al. 1976; Baird, Hammes 1977; Chernyak et al.

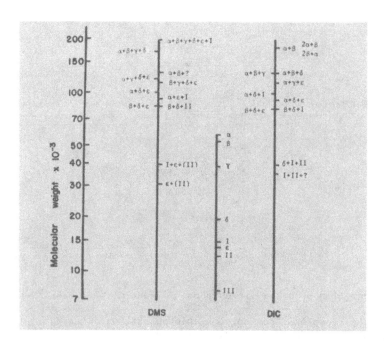

Figure 5. Combinations of subunits of spinach proton ATPase complex in aggregates formed by cross-linking.

The data were taken from several experiments, one of which is presented in figure 4.

1981; Bragg, Hou 1975; Enns, Criddle 1977; Tod, Douglas 1981). The molecular weight of aggregates formed by cross-linking agents was used for identification of their composition. More accurate information was ·achieved by the use of cleavable cross-linking agents, followed by two dimentional gel electrophoresis. Still, problems concerning identification of cross-linked subunits such as: cleavage of the reagent is not always quantitative, poor separation of aggregates when mercaptoethanol is not employed, changes in mobility and in intensity of individual subunits after cross-linking, were not solved. In the present work we employed subunit specific antibodies in order to overcome the obstacles mentioned above. The results obtained (Figure 5) were used to construct a model which describes the subunit arrangement in the H^+-ATPase complex (Figure 6). The stoichiometry of the F_1 subunits is taken as $\alpha_3\beta_3\gamma\delta\varepsilon_2$ (Merchant et al. 1983) and six copies of the proteolipids (subunit III) are suggested (Nelson et al. 1977). The subunits are schematically represented as globular except for the δ subunit which is believed to have an elongated form (McCarty et al. 1971). We do not rule out more complex interactions among the different subunits, including interactions involving several attachment points between each pair of neighbouring subunits. According to our results, no proof for direct contact between the different copies of α subunits or between β subunits was achieved. This may imply that α and β subunits are arranged alternately. The attachment of the F_1 sector to

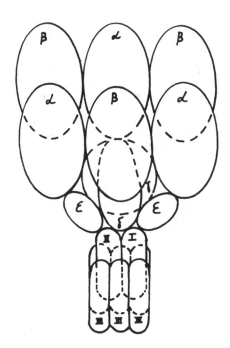

Figure 6. A model of the arrangement of the different subunits of the chloroplast proton ATPase complex.

the membranal part was found to occur via δ and ε subunits of F_1 and subunits I and II of Fo. These results are in line with the function suggested for δ subunit as linking F_1 to the membrane (Nelson, Karny 1976; Younis et al. 1977). No such function was suggested for the ε subunit of chloroplast H^+-ATPase complex (Baird, Hammes 1979; Nelson 1981), but in bacteria it was found that both δ and ε are necessary for binding F_1 to the membrane (Smith, Sternweis 1977; Yoshida et al. 1977; Walker et al. 1982b; Sternweis 1978). The proximity of δ and ε subunits found in this work is supported by several works (Baird, Hammes 1976; Nelson et al. 1973; Smith, Sternweis 1977; Dunn 1982).

   In the model suggested previously for the chloroplast complex (Baird, Hammes 1976) α and β subunits are not arranged alternately and δ and ε subunits are placed far from the δ subunit. The fact that ε subunit inhibits the H^+-ATPase complex activity, may imply that this subunit is in direct contact with β subunit which contains the active site of the enzyme. No evidence for such contact was found in our work. It might be that the inhibition caused by ε subunit is manifested by its proximity to the δ subunit which is supposed to be involved in the regulation of the H^+-ATPase complex activity (Dunn 1982; Nelson 1976a). Our suggestions for the arrangement of subunit I and II are in good agreement to recent findings concerning the corresponding a and b subunits of E. coli membranes (Walker et al. 1982b; Friedle et al. 1983; Hoppe et al. 1983).

5. REFERENCES

Baird BA and Hammes GG (1976) Chemical cross-linking studies of chloroplast coupling factor 1, J.Biol.Chem. 251, 6953-6962.
Baird BA and Hammes GG (1977) Chemical cross-linking studies of beef heart mitochondrial coupling factor 1, J.Biol.Chem. 252, 4743-4748.
Baird BA and Hammes GG (1979) Structure of oxidative and photophosphorylation coupling factor complexes, Biochim. Biophys. Acta 549, 31-53.
Bragg PD and Hou C (1975) Subunit composition, function and spatial arrangement in the Ca^{+2} and Mg^{+2} activated adenosine triphosphatases of E.coli and Salmonella typhimurium, Arch. Biochem. Biophys. 167, 311-321.
Chernyak BV, Chernyak VYa, Gladysheva TB, Kozhanova ZE and Kozlov IA (1981) Structural rearrangements in soluble mitochondrial ATPase, Biochim. Biophys. Acta 635, 552-570.
Douglas MG and Butow RA (1976) Variant forms of mitochondrial translation products in yeasts: evidence for location of determinants on mitochondrial DNA, Proc. Natl. Acad. Sci. U.S.A. 73, 1083-1086.
Dunn SD (1982) The isolated δ subunit of E.coli F_1 ATPase binds the ε subunit, J. Biol. Chem. 257, 7354-7359.
Enns R and Criddle RS (1977) Investigation of the structural arrangement of the protein subunits of mitochondrial ATPases, Arch. Biochem. Biophys. 183, 742-761.
Ferguson SJ, Lloyd WJ and Radda GK (1975) The mitochondrial ATPase. Selective modification of a nitrogen residue in the β subunit, Eur. J. Biochem. 54, 127-133.

Fillingame RH (1981) Biochemistry and genetics of bacterial H$^+$-translocating ATPases, Curr. Top. Bioenerg. 11, 35-106.

Friedle P, Hoppe J, Gunsalus RP, Michelsen O, Meyenburg K and Schairer HU (1983) Membrane integration and function of the three Fo subunits of the ATP synthase of E.coli K12, EMBO J. 2, 99-103.

Hoppe J, Friedle P, Schairer HU, Sebald W, Meyenburg K and Jorgensen BB (1983) The topology of the proton translocating Fo components of the ATP synthase from E.coli K12: studies with proteases, EMBO J. 2, 105-110.

Houstek J, Kopecky J, Svoboda P and Drahota Z (1982) Structure and function of the membrane integral components of the mitochondrial H$^+$-ATPase, J. Bioenerg. Biomemb. 14, 1-13.

Kamienietzky A and Nelson N (1975) Preparation and Properties of chloroplasts depleted of chloroplast coupling factor 1 by sodium bromide treatment, Plant Physiol. 55, 282-287.

McCarty RE, Fuhrman JS and Tsuchiya Y (1971) Proc. Natl. Acad. Sci. U.S.A. 68, 2522-2526.

Merchant S, Shanir SL and Selman BR (1983) Molecular weight and subunit stoichiometry of the chlorlplast coupling factor 1 from chlamydomonas reinhardi, J. Biol. Chem. 258, 1026-1031.

Nelson N, Deters DW, Nelson H and Racker E (1973) Partial resolution of the enzymes catalyzing photophosphorylation, J. Biol. Chem. 248, 2049-2055.

Nelson N and Karny O (1976) The role of ƒ subunit in the coupling activity of chloroplast coupling factor 1, FEBS Lett. 70, 249-253.

Nelson N (1976a) Structure and function of chloroplast ATPase, Biochim. Biophys. Acta 456, 314-318.

Nelson N, Eytan E, Notsani B, Sigrist-Nelson K and Gitler C (1977) Isolation of chloroplast N,N'-dicyclohexylcarbodiimide binding proteolipid, active in proton translocation, Proc. Natl .Acad. Sci. U.S.A. 74, 2375-2378.

Nelson N, Nelson H and Schatz G (1980) Biosynthesis and assembly of the proton translocating ATPase complex from chloroplasts, Proc. Natl. Acad. Sci. U.S.A. 77, 1361-1364.

Nelson N (1980a) Coupling factors from higher plants, Methods Enzymol.69, 301-313.

Nelson N (1981) Proton-ATPase of chloroplasts, Curr. Top. Bioenerg. 11, 1-33.

Ohta S, Tsuboi M, Yoshida M and Kagawa Y (1980) Intersubunit interaction in proton translocating adenosine triphosphatase as revealed by hydrogen exchange kinetics, Biochemistry 19, 2160-2164.

Pullman ME and Monroy GC (1963) A naturally occuring inhibitor of mitochondrial adenosine triphosphtase, J. Biol. Chem. 238, 3762-3769.

Rott R and Nelson N (1981) Purification and immunological properties of proton ATPase complexes from yeast and rat liver mitochondria, J. Biol. Chem. 256, 9224-9228.

Satre M, Klein G and Vignais PV (1976) Structure of beef heart mitochondrial F$_1$ ATPase, Biochim. biophys. Acta 453, 111-120.

Sebald W and Hoppe J (1981) Curr. Top. Bioenerg. 12, 1-64.

Smith JB and Sternwies PC (1977) Purification of membrane attachment and inhibitory subunits of the proton translocating ATPase from E.coli, Biochemistry 16, 306-311.

Sternweis PC (1978) The ε subunit of E.coli coupling factor 1 is required for its binding to the cytoplasmic membrane, J. Biol. Chem. 253, 3123-3128.

Tod RD and Douglas MG (1981) A model for the structure of the yeast mitochondrial adenosine triphosphatase complex, J. Biol. Chem. 256, 6984-6989.

Towbin H, Staehelin T and Gordon J (1979) Electrophoretic transfer of proteins from polyacrylamide gels to nitrocellulose sheets: procedure and some applications, Proc. Natl. Acad. Sci. U.S.A. 76, 4350-4354.

Van de Stadt RJ, De Boer BL and Van Dam K (1973) The interaction between the mitochondrial ATPase (F_1) and the ATPase inhibitor, Biocim. Biophys. Acta 292, 338-349.

Walker JE, Runswick MJ and Saraste M (1982) Subunit equivalence in E.coli and bovine heart mitochondrial $F_1 F_0$ ATPases, FEBS Lett. 146, 393-396.

Walker JE, Saraste M, Runswick MJ and Gay NJ (1982a) Distantly related sequences in the α and β subunits of ATP synthase, myosin, kinases and other ATP-requiring enzymes and a common nucleotide binding fold, EMBO J. 1, 945-951.

Walker JE, Saraste M and Gay NJ (1982b) E.coli F_1-ATPase interacts with a membrane protein component of a proton channel, Nature 298, 867-869.

Yoshida M, Okamoto H, Sone N, Hirata H and Kagawa Y (1977) Reconstitution of thermostable ATPase capable of energy coupling from its purified subunits, Proc. Natl . Acad. Sci. U.S.A. 74, 936-940.

Younis H, Winget GD and Racker E (1977) Requirement of the δ subunit of chloroplast coupling factor 1 for photophosphorylation, J. Biol. Chem. 252, 1814-1818.

MODE OF ACTION AND REGULATION OF CHLOROPLAST H^+-ATPASE.

R. HILLEL, A.T. JAGENDORF AND C. CARMELI

1. ABSTRACT

Isolated coupling factor 1 of the chloroplast H^+-ATPase was shown to contain several binding sites for Mn^{2+} ion. Three of these sites had strong positive cooperative interaction with a Hill coefficient of 2.5. and an average dissociation constant of 15 µM. Chemical modification of two arginyl residues, one on each of the α and β subunits, which caused inhibition of ATPase activity, prevented the cooperative interaction but not the binding of Mn^{2+}. It is suggested that the cooperativity among the Mn^{2+} binding sites is an expression of cooperative interaction among the active sites of the enzyme required for catalysis.

2. INTRODUCTION

The H^+-ATPase from chloroplasts (CF_1-F_0) utilizes an electro-chemical potential of protons for the synthesis of ATP from ADP and Pi. The $\Delta\mu_{H^+}$ across the thylakoid membrane also causes a conformational change (Ryrie, Jagendorf, 1972) which probably induces a major change in the reactivity of the enzyme. When the enzyme undergoes this confor-mational change in the presence of sulfhydryl reagents, reduction of disulfide bonds in various subunits greatly stimulates ATPase activity (Petrack et al, 1965). Similar change in the rate of ATP hydrolysis is caused by heat treatment of the isolated catalytic sector (CF_1) in the presence of sulfhydryl readents (Farron, Racker, 1970). Heat treated CF_1, which calatyses high rates of ATP-hydrolysis, was shown to have a very low presteady state activity (Carmeli et al., 1981). A presteady state transformation was caused only by hydrolyzable substrates such as CaATP and had a first order kinetic constant of 1 S^{-1} at 37°. Prein-cubation of the enzyme with divalent metal ions caused a twenty-fold decrease in the rate of activation. It is reasonable to suggest that these activation processes are also mediated through conformational changes in the protein brought about by the binding of divalent metal ions and by the interaction of the enzyme with substrates. The activity of this enzyme was also shown to be affected by a variety of small molecules such as ADP, Pi, carboxylic acids, carbonate, organic solvents and detergents (McCarty, Carmeli, 1982)

These studies unfold a variety of mechanisms for regulation of the enzymic activity, some of which probably control the reactions in vivo. Therefore, an evaluation of results concerning the mechanism of action of this enzyme should be considered in the light of these regulatory processes. Thus, cooperative interaction among substrate binding sites and the state of reactivity of the enzyme could introduce changes in binding constants. An example of such a case is our finding of cooperative interaction among manganese binding sites on the enzyme. The cooperativity was found in isolated CF_1 stored in 1mM ATP and 25% glycerol at -80°, while in previous studies we showed that CF_1 stored as ammonium sulfate precipitate had one tight and five loose binding sites for manganese (Hochman, Carmeli, 1981, 1981a). A second tight site was induced by di or trinucleosides. The manganese ions were suggested to bind to the active sites of the enzyme because the dissociation constants for binding were similar to the Ki of mangenese as competitive inhibitor of ATPase activity and because allosteric effectors similarly changed the binding constants and the K_m for MnATP as substrate.

3. MATERIALS AND METHODS

3.1. CF_1 Preparation

CF_1 was isolated from lettuce (Romaine) chloroplasts by a modified method used by Strotmann et al. (1973). Following five washings of the chloroplasts in 10 mM pyrophosphate, CF_1 was extracted in a medium containing 300 mM sucrose, 2 mM tris-tricine, pH 7.8 and 0.1 mM ATP (Pick, U., personal communication). The residual membranes were precipitated by centrifugation at 31,000 xg for 15 min from a medium containing 20 mM tris-SO_4, pH 7.1, 2 mM EDTA, 80 mM $(NH_4)_2SO_4$ and 1 mM ATP. CF_1 was concentrated by precipitation in 2 M $(NH_4)_2SO_4$, 1 mM ATP, 20 mM tris-SO_4, pH 7.1 and 2 mM EDTA. The precipitate was dissolved in 40 mM tricine-NaOH, pH 8 and 2 mM EDTA desalted on a sephadex G-50 centrifuged column equilibrated in the same medium and stored in 25% glycerol, 1 mM ATP, 2 mM EDTA and 30 mM tricine-NaOH, pH 8 at -80°. Following heat treatment (Farron, Racker, 1970) CF_1 hydrolyzed 35 umol ATP x mg protein^{-1}xmin^{-1} and had little cotaminating protein as seen on SDS polyacrylamide gel electrophoresis.

3.2 Mn^{2+} Binding

CF_1 was passed on a Sephadex G-50 column (1 x 50 cm) equilibrated with 40 mM HEPES-NaOH, pH 8 and concentrated by ultrafiltration to ~ 50 µM protein. Binding of Mn^{2+} to CF_1 was determined by E.P.R. method as described by Hochman et al. (1976). Protein concentration was determined from the adsorption at 280 nm, assuming a molecular weight of 325 KD (Farron, 1970).

3.3 Rapid Kinetic Measurements

Following heat activation, CF_1 was passed through a Sephadex G-50 centrifuged column equilibrated with 0.1 mM EDTA and 2 mM tricine-NaOH, pH 8. ATPase activity was measured spectrophotometrically by using the indicator cresol red for monitoring the acidification of the medium in Aminco DW-2 stpectrophotometer equipped with a stopped-flow apparatus as described earlier (Carmeli et al, 1978).

4. RESULTS AND DISCUSSION

4.1. Mn^{2+} binding to CF_1.

The data which were previously obtained from titration of CF_1, stored as $(NH_4)_2SO_4$ preciptate, with $MnCl_2$ gave descending nonlinear curves when plotted according to Scatchard analysis. For the analysis of these data, it was sufficient to assume that the isotherm represents binding of Mn^{2+} to two types of noninteracting sites. The results obtained from titration of CF_1 stored in a medium contianing 1 mM ATP and 25% glycerol at -80° gave an isotherm which curved upward to an

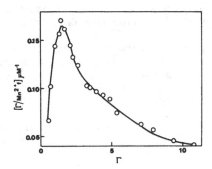

FIGURE 1. Binding of Mn^{2+} to CF_1 stored at $-80°$. Scatchard plot of average data obtained from ten titrations of CF_1 with $MnCl_2$. Moles of Mn^{2+} per mole of CF_1 (T) were plotted vs. T divided by the concentration of free Mn^{2+} (Mn^{2+} f).

optimum and then descended nonlinearly (Fig. 1).

Such a complicated isotherm of a Scatchard plot was assumed to represent at least two types of binding sites. One type of interacting sites with positive cooperativity which gave a downward concave curve and a second type of noninteracting sites which gave a descending isotherm (Schreier, Schimmel, 1974). In this composite interaction moles of Mn^{2+} bound per mole of enzyme (T) and T/C are related to the number of binding sites (N), their association constant (K), the concentration of free Mn^{2+} (C) and the Hill coefficient (α) as the following:

$$T = \frac{N_I K_I^{\alpha} C^{\alpha}}{1+K_I^{\alpha} C^{\alpha}} + \frac{N_s K_s C}{1+K_s C} \tag{1}$$

$$T/C = \frac{N_I K_I^{\alpha} C^{\alpha-1}}{1 + K_I^{\alpha} C^{\alpha}} + \frac{N_s K_s}{1+N_s K_s C} \tag{2}$$

Subscript I and S indicate interacting and noninteracting sites respectively. The first part of each of the two equations represents the contribution to T and T/C of the interacting sites, while the second part represents the contribution of the noninteracting sites. In order to resolve the composite data an empirical asymptotic tangent line to the descending part of the curve was drawn. This tangent was drawn in a manner which neglected some of the very weak noninteracting sites. From its intercept at the T/C ordinate, the T ordinate and from the intercept of the original curve at the T/C ordinate, the N_I, N_s and K_s were obtained. The values of T and T/C of the interacting sites were calculated from equations 1 and 2 using these parameters. If the plot of ln C versus $\ln(N_I/T-1)$ yielded a straight line (equation (3)) from the slope and intercepts α and K_I could be caclulated

$$\ln C = -\frac{1}{\alpha} \ln(\frac{N_I}{T} - 1) - \ln k_I \qquad (3)$$

If nonlinear line was obtained the process of drawing tangent lines was repeated until the data fitted a straight line (for details see Shreier, Schimmel, 1974).

The data indicated a positive cooperative interaction among 3 mangenese binding sites with an average Kd of 15 μM and 3 noninteracting sites having Kd of 85 μM. The cooperative interaction among the sites was fairly strong having a Hill coefficient of 2.5. The possibility that the difference between the enzyme stored in $(NH_4)_2SO_4$ and that which was stored at $-80°$ was due to differences in the content of tightly bound nucleotides and Mg^{2+} was ruled out since analysis showed that approximately one nucleotide (mostly ADP) and one Mg^{2+} per CF_1 were bound, regardless of storage conditions. The fact that CF_1 stored at $-80°$ had a higher and a more reproducible ATPase activity than the enzyme which was stored as $(NH_4)_2SO_4$ precipitate might indicate that under the former conditions the enzyme was in a state which more closely resembled the native conformation. Storage at $-80°$ in the presence of glycerol and ATP probably changed the enzyme which was previously precipitated in $(NH_4)_2SO_4$. The presence of 1 mM ATP during storage probably contributed to these changes since addition of one mole of ATP per mole of CF_1 did not alter the binding properties of Mn^{2+}. It is possible that addition of ATP did not change the binding properties of Mn^{2+} because preincubation of the enzyme during storage with 1 mM ATP already caused a change in conformation. This change apparently did not occur when the enzyme was kept as $(NH_4)_2SO_4$ precipitate since after the removal of the salt, addition of one mole of ATP per mole CF_1 caused a decrease of almost 200-fold in the Kd of the Mn^{2+} binding sites. A slight decrease in the Kd of both the interacting and the noninteracting binding sites was observed however when CF_1 was preincubated with MgATP.

In order to explore the involvement of specific amino acids in the binding of Mn^{2+}, arginine residues in CF_1 were modified. Chloroplasts were reacted with 4 mM naphthylglyoxal for 25 min at $20°$ as described by Takabe et al. (1982). This treatment caused an 85% inhibition of ATPase activity in the CF_1 isolated from the membranes. The modified enzyme had one strong and five loose binding sites for Mn^{2+} (Fig. 2).

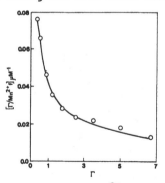 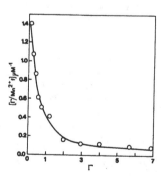

A.   B.

FIGURE 2. Binding of Mn^{2+} to naphthylglyoxal modified CF_1. Scatchard plot of Mn^{2+} binding to naphthylglyoxal modified CF_1 in the absence (A) and in the presence (B) of one mole of ATP per mole of enzyme. For details, see Fig. 1 and the text.

Addition of one mole of ATP per mole of CF_1 resulted in an appearance of a second tight binding site for Mn^{2+}. Manganese bindng properties of the enzyme resembled those of CF_1 stored as $(NH_4)_2SO_4$ precipitate. Although the modified CF_1 was stored in 25% glycerol at -80° it did not show cooperative interaction among the Mn^{2+} binding sites as was seen in the untreated CF_1 stored under similar conditions.

We previously suggested (Hochman, Carmeli, 1981) that tight Mn^{2+} binding results from a formation of a tertiary complex among CF_1, Mn^{2+} and a pyrophosphate bond of a nucleotide. The one tight site found in isolated CF_1 could be formed as a result of interaction of Mn^{2+} with the tightly bound ADP which remains on the isolated enzyme. On addition of one mole of diphosphonucleoside per mole of CF_1, a second site binds Mn^{2+} as a tertiary complex. It is possible that in CF_1 stored at -80° binding of Mn^{2+} as a tertiary complex to the tightly bound ADP caused conformational changes which were transmitted to other subunits strengthening Mn^{2+} binding at other sites. However, until we get conclusive evidence which indicates that Mn^{2+} binds as a tertiary complex with nucleotides, other manner of binding of the ions to CF_1 cannot be excluded. This cooperative interaction could not be transmitted by the enzyme which was precipitated by $(NH_4)_2SO_4$. However, one possibility is that after removal of the salt interaction of ATP at a low affinity site caused a change in conformation which enabled it to respond in a cooperative manner to Mn^{2+} binding. Further experiments are being conducted to verify this possibility. This site could be the same low affinity ATP binding site which prevents cold inactivation. No cooperative interaction was observed on addition of one mole of ATP per mole of CF_1 becuase the concentration of ATP was too low for interaction with the low affinity site. It is also possible that the conformational is caused by factors other than ATP.

Treatment with naphthylglyoxal which probably modified two arginyl residues, one on each of the β and α subunits caused inhibition of ATPase activity in CF_1 (Takabe et al, 1982). It also prevents cooperative interaction between Mn^{2+} binding sites. Since only one β subunit was modified while there are at least two β subunits per enzyme, it can be speculated that the cooperative interactions are required for catalysis, assuming that the active sites are located on these subunits.

4.2 Temperature Dependence of the Rate of the Autocatalytic Reactivity of CF_1.

The acceleration of the rate of ATPase activity of heat-activated CF_1 on addition of substrate is probably also a result of changes in conformation of the enzyme. We have suggested (Carmeli et al, 1981) that the enzyme can exist in three states of increasing reactivity $E<E'<E''$. Binding and hydrolysis of divalent metal ion-ATP complex (MS) causes transformation of each of the less reactive states to the most reactive form of the enzymes. This transformation has an increasing first order rate constants of $k_3 < k_3' < k_3''$ with values of $0.025 < 1 < 150 s^{-1}$ respectively at 37°. The process of activation was formulated as the following:

$$EM + MS \underset{k_2}{\overset{k_1}{\rightleftharpoons}} EMS \overset{k_3}{\rightarrow} E''M + MP \qquad (1)$$

$$E'M + MS \underset{k_2'}{\overset{k_1'}{\rightleftharpoons}} E'MS \overset{k_3'}{\rightarrow} E''M + MP \qquad (2)$$

$$E''M + MS \underset{k_2''}{\overset{k_1''}{\rightleftharpoons}} E''MS \overset{k_3''}{\rightarrow} E''M + MP \qquad (3)$$

Where MP represents the product (P) as a divalent metal ion (M) complex. The apparent first order rate constant were suggested to be a result of the ratio of the various states of the enzyme. Thus, on addition of saturating concentrtions of CaATP (containing also free Ca^{2+} ions) to an enzyme free of metal ions which was shown to be in an intermediate state of activation (E') the major activation proceeds as in eqtn. 2. Below saturation, a lower apparent rate of activation was observed since two processes took place (eqtns. 1 and 2). This happened because part of the enzyme interacted with free metal and changed conformation to the least active state (E) as follows:

$$E' + M \rightleftharpoons E'M \rightleftharpoons EM \qquad (4)$$

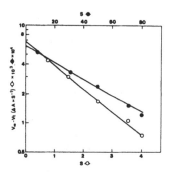

FIGURE 3. Determination of the observed rate constants for the presteady state nonlinear initial rates of ATPase activity in CF_1. The change in rate of heat activated CF_1 was measured in the presence of 3 mM CaATP, 1 mM Tricine–NaOH pH 8, 50 uM EDTA, 25 µM cresol red and 0.1 mM free Ca^{2+}. The results obtained at $20°$ (●) or at $37°$ (o) were plotted semilogarithmically.

The data presented in Fig. 3 indicate that reactions 2 and 4 have different temperature dependences. This is apparent from the fact that the observed first order rate constants were 0.55 and 0.019 s^{-1} at $37°$ and $20°$ respectively. HOwever, preincubation of the enzyme with Ca^{2+} decreased the observed rate constant from 0.55 to 0.025 at $37°$ but did not change the rate of acceleration when ATPase activity was measured at $20°$. These data indicate that the formation of EM on binding of metal ion to the enzyme (eqtn. 4) has a lower energy of activation than the process of activation (eqtn. 2). At 20° the rate constant of activation decreased almost 20-fold in comparison to rate at 37° therefore could not compete with the rate of formation of EM which did not change greatly at 20°. Thus, when the concentration of CaATP equals the Km most of the enzyme was converted by the free Ca^{2+} to the EM form and was activated at $k_3=0.025$ s^{-1}.

5. REFERENCES

Carmeli C, Lifshitz Y and Gutman M (1978) Control of kinetic changes in ATPase activity of soluble coupling factor 1 from chloroplasts. FEBS Lett. 89, 211-214.

Carmeli C, Lifshitz Y and Gutamn M (1981) Modulation by divalent metal ions of the autocatalytic reactivity of ATPase from chloroplasts. Biochemistry 20, 3941-3944.

Farron F (1970) Isolation and properties of a chloroplast coupling factor and heat-activated ATPase. Biochemistry 9, 3823-3828.

Farron F and Racker E (1979) Studies on the mechanism of the conversion of coupling factor 1 from chloroplasts to an active ATPase. Bicohemistry, 9, 3829-3836.

Hochman Y and Carmeli C (1981) Correlation between the kinetics of activation and inhibition of ATPase activity by divalent metal ions and the binding of manganese to chloroplast coupling factor 1. Biochemistry 20, 6287-6292.

Hochman Y and Carmeli C (1981a) Modulation by bicarbonate, phosphate and maleate of the kinetics of ATPase activity and the binding of manganese ions to coupling factor 1. Biochemistry 20, 6293-9297.

McCarty RE and Carmeli C (1982) Proton translocating ATPase of photosynthetic membranes. In Govindjel, ed. Photosynthesis: Energy conversion by plants and bacteria, Vol. 1. pp. 647-695. New York, Academic Press.

Petrack B, Cranston A, Sheppy F and Farron F (1965) Studies on the hydrolysis of ATP by spinach chloroplasts. J. Biol. Chem. 240, 906-912.

Ryrie IJ and Jagendorf AT (1972) Correlation between a conformational change in the coupling factor protein and the high energy state in chloroplasts. J. Biol. Chem. 247, 4453-4459.

Schreier AA and Schimmel PR (1974) Fragment recombination and whole molecules of yeast phenylalanine specific transfer RNA. J. Mol. Biol. 86, 601-620.

Strotmann H, Hesse H and Edelman K (1973) Quantitative determination of coupling factor CF_1 of chloroplasts, Biochim. Biophys. Acta 314, 202-210.

Takabe T, Debenedetti E and Jagendorf AT (1982) Inhibition of chloroplasts coupling factor by naphthylglyoxal. Biochim. Biophys. Acta 682, 11-20.

6. ACKNOWLEDGEMENT

This work was supported by a grant from the United States - Israel Binational Science Foundation (BSF), Jerusalem Israel.

7. AUTHOR'S ADDRESS

R. Hillel, C. Carmeli, Dept of Biochem, Tel-AViv University, Tel Aviv 69 978, Israel. A.T. Jagendorf, Plant Sciences, Cornell University, Ithaca, NY 14853, USA.

LIGHT ACTIVATION *IN VIVO* OF THE CHLOROPLAST PROTON ATPase: EFFECT ON THE
Pi-ATP EXCHANGE REACTION.

RUBEN H. VALLEJOS/RICARDO A. RAVIZZINI

1. INTRODUCTION

ATP synthesis and hydrolysis in chloroplasts are catalyzed by the proton
-ATPase complex. The turnover capacity of the enzyme seems to be highly
regulated. Photophosphorylating and ATPase activities were higher in
chloroplasts rapidly prepared from preilluminated leaves (Morita "el al",
1982; Vallejos "et al", 1983).
In isolated broken chloroplasts the ATPase activity is normally very low
and requires a special treatment with light and SH compounds for increasing
it (Petrack "et al", 1965). Under the same conditions a Pi-ATP exchange
reaction is also activated (Carmeli, Avron, 1967). In this paper we present
further evidence about the physiological reversible regulation by light
of the proton-ATPase complex by studying the activity of the ATP-Pi
exchange reaction.

2. PROCEDURE

2.1. Material and methods

Chloroplasts were prepared from spinach leaves (*Spinacea oleracea L.*)
maintained in darkness (dark-chloroplasts) or preilluminated (light-
chloroplasts) at 25°C for 90 s by a 150 w tungsten lamp through 2 cm of
1‰ solution of $CuSO_4$ as heat-absorbing filter, essentially as described
by Morita "et al" (1982). The leaf (3 g) was rapidly cut into pieces and
homogeneized at 0°C for 10s, in a high speed homogenizer, in 10 ml of a
medium containing 0.25 M sucrose, 10 mM NaCl, 5 mM $MgCl_2$, 0.5 mM EDTA,
and 20 mM Tricine NaOH (pH 8). Chloroplasts were isolated by filtration
through a nylon cloth adapted to a 10 ml syringe. The filtrate was centrifuged
at 12.000 rpm in an Eppendorf microcentrifuge for 10s and the pellet
suspended in 10 mM NaCl, 5 mM $MgCl_2$, 0.5 mM EDTA and 20 mM Tricine-NaOH
(pH 8), at final concentrations corresponding to 0.2-0.5 mg of chlorophyll/
/ml. All procedures were carried out at 0°C. The ATP-Pi exchange activity
was assayed by adding 0.05 ml of the chloroplasts suspension, containing
10-25 µg of chlorophyll to 0.45 ml of a medium consisting of 50 mM NaCl,
5 mM $MgCl_2$, 2 mM ATP, 2 mM Pi (containing 10^6 cpm as ^{32}Pi), and 20 mM
Tricine-NaOH (pH 8.0). After 5 min at 25°C in the darkness the reaction
was stopped by precipitation of Pi (Sugino, Miyoshi, 1964).

3. RESULTS AND DISCUSSION

Preillumination of spinach leaves results in a large stimulation of the
Pi-ATP exchange activity of chloroplasts rapidly prepared from them. The
activity increased from 0.4 $\mu mol.mg^{-1}.h^{-1}$ in dark-chloroplasts to 23.5
$\mu mol.mg^{-1}.h^{-1}$ in light chloroplasts. The activity of light-chloroplasts
declined with a half-time of inactivation of about 9 min (Fig 1) at 25°C
ADP and Pi did not affect the decay rate while 5 mM ATP or 50 mM
dithioerythritol prevented the inactivation (Fig 1).
As shown previously for Mg-ATPase (Vallejos "et al", 1983) the Pi-ATP
exchange activity of light-chloroplasts was stable at 0°C (Fig 2) for at
least 60 min. Under these conditions the Pi-ATP exchange activity can be

Sybesma, C. (ed.), Advances in Photosynthesis Research, Vol. II. ISBN 90-247-2943-2.
© *1984 Martinus Nijhoff/Dr W. Junk Publishers, The Hague/Boston/Lancaster.*

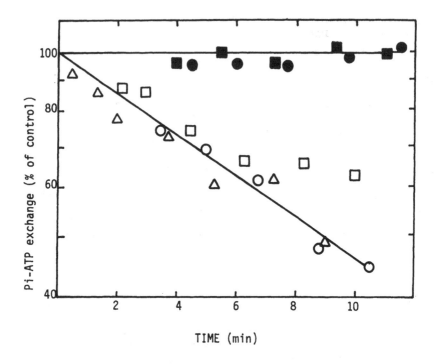

FIGURE 1. Effect of ADP, ATP, Pi and dithioerythritol on the decay of *in vivo* activated Pi-ATP exchange reaction. Thylakoids were prepared from preilluminated leaves and immediately incubated at 25°C with the following additions none (O——O); 100μM ADP (□——□); 2 mM Pi (△——△); 5 mM ATP (●——●) and 50 mM dithioerythritol (■——■). The activity was measured in aliquots as described in the text. Activity of the control was 27.7 μmol Pi.mg^{-1},h^{-1}.

rapidly inactivated by treatment of the light-chloroplasts with either an uncoupler like gramicidin or an oxidant like o-iodosobenzoate (Fig 2) Partial inactivation by the latter reagent could be completely reversed by adding 50 mM dithioerythritol (Fig 2B). Chloroplasts normally isolated do not show a Pi-ATP exchange activity unless they are preilluminated in the presence of a reducing agent (Carmeli, Avron, 1967) procedure that also activates the Mg-ATPase. This exchange activity is clearly a partial reaction leading to ATP synthesis during photophosphorylation. All these reactions are catalyzed by the proton ATPase complex. Recently we have shown that the activation *in vivo* of this enzyme leads to a higher photophosphorylating and ATPase activities the latter being more stable at 0°C than the former (Vallejos "et al", 1983). As expected the Pi-ATP exchange reaction was similarly activated and was as stable at 0°C as the Mg-ATPase. The high activity of light-chloroplasts can be stabilized by lowering the temperature to 0°C or presumably by maintaining the electro chemical proton gradient and the redox change induced during light activation *in vivo* by addition of either ATP or dithioerythritol. This redox change seems to consist in the reduction of at least a disulfide bond of the γ subunit of CF₁(Vallejos "et al", 1983) a phenomenon also

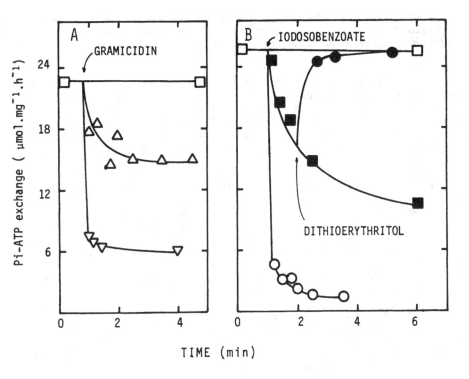

FIGURE 2. Inhibition by gramicidin and o-iodosobenzoate of physiological activated Pi-ATP exchange reaction. "Light-chloroplasts" were incubated at 0°C in darkness with the following additions at the time stated: none (□——□); 10 nM gramicidin (△——△); 50 nM gramicidin (▽——▽); 0.5 mM iodosobenzoate (■——■); 2 mM iodosobenzoate (○——○) and 50 mM dithioerythritol (●——●).

observed in the activation of soluble CF_1 (Arana, Vallejos, 1982). An earlier, related observation was the appearence of an extra sulfhydryl group in the γ subunit of CF_1 from illuminated chloroplasts (McCarty, Fagan, 1973). The γ subunit of CF_1 has 4 sulfhydryl groups that in soluble CF_1 are forming two disulfide bonds (Ravizzini "et al", 1980). In membrane bound CF_1 on the other hand the γ subunit has one accessible group in the dark and two in the light which may be blocked by N-ethylmaleimide or oxidized to a disulfide bond by iodosobenzoate (McCarty, Fagan, 1973; Andreo "et al", 1979) with partial inhibition of photophosphorylation. The effect of iodosobenzoate is reversible (Vallejos, Andreo, 1976). Apparently the situation is similar in light-chloroplasts where two accessible sulfhydryl groups may be blocked by N-ethylmaleimide (Vallejos "et al", 1983). However, the proton ATPase is active in this case but not in broken chloroplasts subject to illumination. Thus, activation *in vivo* of the proton ATPase may involve reduction of the second disulfide bond of the γ subunit or may required other changes in the complex. *In vivo* this change may be caused by thioredoxin as suggested (Mills, Hind, 1979) or other chloroplast dithiols (Anderson, 1975). Relaxation of the activated state of the proton ATPase *in vivo* in the dark is less well understood but should involve enzymatic or chemical oxidation of the sulfhydryl groups reduced during activation.

REFERENCES

Anderson, LE (1975) Light modulation of the activity of carbon metabolism enzymes. In Avron M, ed. Proceedings of the 3rd. International Congress on Photosynthesis, Israel, pp. 1393-1405. Amsterdam, The Netherlands, Elsevier.

Andreo CS, Ravizzini, RA and Vallejos RH (1979) Sulfhydryl groups in photosynthetic energy conservation V. Localization of the new disulfide bridges formed by o-iodosobenzoate in coupling factor of spinach chloroplasts, Biochim.Biophys.Acta 547, 370-379.

Arana JL and Vallejos RH (1982) Involvement of sulfhydryl groups in the activation mechanism of the ATPase activity of chloroplast coupling factor 1, J.Biol.Chem. 257, 1125-1127.

Carmeli C and Avron M (1967) A light-triggered adenosine triphosphate-phosphate exchange reaction in chloroplasts, Eur.J.Biochem. 2, 318-326.

McCarty RE, Fagan J (1973) Light stimulated incorporation of N-Ethylmaleimide into coupling factor 1 in spinach chloroplasts, Biochemistry 12, 1503-1507.

Mills JD and Hind G (1979) Light-induced Mg^{2+} ATPase activity of coupling factor in intact chloroplasts, Biochim.Biophys.Acta 547, 455-462.

Morita S, Itoh S and Nishimura M (1982) Correlation between the activity of membrane-bound ATPase and the decay rate of flash-induced 515-nm absorbance change in chloroplasts in intact leaves, assayed by means of rapid isolation of chloroplasts, Biochim.Biophys.Acta 679, 125-130.

Petrack B, Craston A, Sheppy F and Farron F (1965) Studies on the hydrolysis of adenosine triphosphatase by spinach chloroplasts, J.Biol.Chem. 240, 906-914.

Ravizzini RA, Andreo CS and Vallejos RH (1980) Sulfhydryl groups in photosynthetic energy conservation VI. Subunit distribution of sulfhydryl groups and disulfide bonds in chloroplast coupling factor and ATPase activity, Biochim.Biophys.Acta 591, 135-141.

Vallejos RH and Andreo CS (1976) Sulfhydryl groups in photosynthetic energy conservation:II. Further evidence of vicinal dithiols involvement shown by light dependent effects of o-iodosobenzoate, FEBS Lett. 61, 95-99.

Vallejos RH, Arana JL and Ravizzini RA (1983) Changes in activity and structure of the chloroplast proton ATPase induced by illumination of spinach leaves, J.Biol.Chem. 258, 7317-7322.

ACKNOWLEDGEMENTS

This work was supported by grants from the Consejo Nacional de Investigaciones Científicas y Técnicas (Argentina). RHV and RAR are members of the Investigator Career of the same Institution.

Authors address: Dr.Rubén H.Vallejos, CEFOBI, Suipacha 531, 2000 Rosario ARGENTINA.

A DUAL pH OPTIMUM MODEL FOR ACTIVATION OF THE CHLOROPLAST ATPase, CF_o-CF_1

JOHN D. MILLS and PETER MITCHELL   Glynn Research Institute, Bodmin, UK

1. INTRODUCTION: THE MODEL

CF_o-CF_1 is known to undergo a reversible activation of catalytic activity when a difference in the electrochemical potential of protons ($\Delta\bar{\mu}H^+$) develops across the thylakoid membrane (Schlodder et al.,1982). This phenomenon may be explained by a model based on a dual pH optimum requirement of CF_o-CF_1 (Mitchell,1981). The model assumes the following:

(i)   For activation to proceed, the enzyme requires that three groups protonatable only from the inner, or P side, be in a protonated state, and three groups protonatable only from the outer, or N side be in a deprotonated state.

(ii)   Each P-side group is functionally linked with one N-side group such that the operational pK of either is influenced by the protonation state of the corresponding group on the opposite pole of CF_1. Thus each N-side group may assume one of two pK's, pK_N^0 or pK_N^1 depending on whether the corresponding P side group is deprotonated or protonated respectively. Similarly, each P-side group may have an operational pK of pK_P^0 or pK_P^1.

(iii) Each pair of P and N-side groups acts independently of other pairs.

(iv)   Activation of the enzyme only occurs when CF_1 reaches the required state of protonation and is described by the simple equilibrium:

$$E_{inactive} \overset{K_d}{\rightleftharpoons} E_{active} \qquad \text{Eq.(1)}$$

(v)   Any transmembrane electrical potential difference is converted to a ΔpH by a proton well through CF_o. The P-side groups are therefore located at the bottom of the proton well.

If α is defined as the fraction of CF_o-CF_1 complexes that are active then α is given by Eq.(2):

$$\frac{1}{\alpha} = 1 + K_d \left[\left(\frac{1 + H_N^+}{K_N^1} \right)^3 + 3 \left(\frac{K_P^0}{H_P^+} \right) \left(\frac{1 + H_N^+}{K_N^1} \right)^2 \left(\frac{1 + H_N^+}{K_N^0} \right) \right.$$
$$\left. + 3 \left(\frac{K_P^0}{H_P^+} \right)^2 \left(\frac{1 + H_N^+}{K_N^1} \right) \left(\frac{1 + H_N^+}{K_N^0} \right)^2 + \left(\frac{K_P^0}{H_P^+} \right)^3 \left(\frac{1 + H_N^+}{K_N^0} \right)^3 \right] \qquad \text{Eq.(2)}$$

Two cases exist where Eq.(2) may be reduced to a simpler form:

1.   If assumption (ii) above does not apply, that is if all protonation/deprotonation events are independent, then $K_P^0 = K_P^1$ and $K_N^0 = K_N^1$ and:

$$\frac{1}{\alpha} = 1 + K_d \left(\frac{1 + H_N^+}{K_N} \right)^3 \left(\frac{1 + K_P}{H_P^+} \right)^3 \qquad \text{Eq.(3)}$$

In this case, α depends on the absolute values of pH_P and pH_N, but most CF_o-CF_1 molecules will be active only when both pH_P pK_P and pH_N pK_N, and these conditions will be satisfied only when a ΔpH is present across the membrane.

Sybesma, C. (ed.), Advances in Photosynthesis Research, Vol. II. ISBN 90-247-2943-2.
© *1984 Martinus Nijhoff/Dr W. Junk Publishers, The Hague/Boston/Lancaster.*

2. If values for the various constants in Eq.(2) are chosen so that only terms containing H_N^+/H_P^+ are significant, then Eq.(2) reduces to:

$$\frac{1}{\alpha} = 1 + K_d \left(\frac{1 + \frac{K_P^0}{H_P^+}}{} \cdot \frac{H_N^+}{K_N^0} \right)^3 \qquad \text{Eq.(4)}$$

The conditions for Eq.(4) to be a valid approximation of Eq.(2) are:

$$\left(\frac{H_N^+}{K_N^1} \right) \ll 1 \quad ; \quad \left(\frac{K_P^0}{H_P^+} \right) \ll 1 \quad ; \quad \text{and} \quad \left(\frac{H_N^+}{K_N^0} \right) \gg 1 \qquad \text{Eq.(5)}$$

In this case, protonation of a P-side group and deprotonation of an N-side group are fully cooperative and α is purely a function of ΔpH, with no dependence on the absolute values of pH_N and pH_P. Fig. 1 shows a plot of α against ΔpH as given by Eq.(4) for two values of K_d.

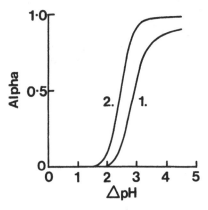

FIGURE 1. Dependence of alpha on ΔpH as given by Eq.4. Values of the constants were: $K_P^0 = 1.10^{-8}$; $K_N^0 = 1.10^{-11}$. Curve 1, $K_d = 0.1$; curve 2, $K_d = 0.012$. H_P^+ was fixed at 1.10^{-5} and H_N^+ was varied to generate the indicated ΔpH.

2. COMPARISON OF THE MODEL WITH EXPERIMENTAL RESULTS

The model represented by Eq.(4) may be used to interpret the recently reported effects of thiol modulation of CF_o-CF_1 on the observed rate of ATP synthesis by chloroplasts. Fig.2(a) shows the yield of ATP observed to be synthesised after subjecting broken chloroplasts to an acid/base transition (Mills,Mitchell, 1982). Thiol modulation of CF_o-CF_1 (by preilluminating in the presence of dithiothreitol) lowers the threshold ΔpH at which synthesis of ATP is first observed and stimulates the yield of ATP at limiting ΔpH. Comparison of Figs. 1 and 2(a) shows that the data may be explained by the activation requirement of CF_o-CF_1 on the following additional assumptions:
(vi) The observed rate of phosphorylation is limited kinetically by α
     (as first proposed by Schlodder et al., 1982).
(vii) Thiol modulation of CF_o-CF_1 increases α by enabling the enzyme to
     become active at a lower ΔpH. In terms of the model of Eq.(4), this corresponds to a decrease of K_d (shown in Fig.1) or of K_P^0/K_N^0 (not shown).
The latter assumption has some experimental support as shown in Fig.2(b). Activation of CF_o-CF_1 is known to be associated with the release (or exchange) of adenine nucleotides that in the inactive enzyme are tightly bound and virtually non-exchangeable (Strotmann et al.,1979). Fig.2(b) shows

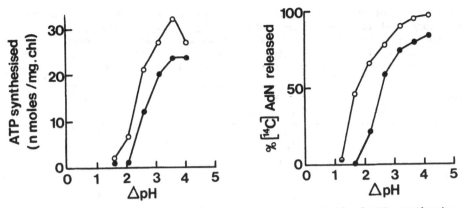

FIGURE 2. Effect of thiol modulation on (a) the yield of ATP synthesis and (b) the release of bound [14C] ADP observed after an acid/base transition. The pH_p was 5.15, pH_N was varied. For (b), the preillumination stage contained 9 μM labelled ADP. See Mills,Mitchell (1982) for other details.

that thiol modulation of CF_o-CF_1 causes release of bound ADP to occur over a significantly lower range of ΔpH, in support of assumption (vii) above.

Table 1 shows the effect of thiol modulation of CF_o-CF_1 on steady state photophosphorylation under several conditions (from Mills,Mitchell 1983). Again, thiol modulation stimulates the observed rate of ATP synthesis and these effects can be explained by a model given by Eq.(6)-(8):

$$J_{H^+_{in}} = V_e \cdot e \cdot (K_a/(K_a + H^+_P)) \qquad \text{Eq.(6)}$$

$$J_{H^+_{out}} = p \cdot \alpha \cdot V_p + H^+_P \cdot C \qquad \text{Eq.(7)}$$

$$J_{H^+_{in}} = J_{H^+_{out}} \qquad \text{Eq.(8)}$$

where $J_{H^+_{in}}$ and $J_{H^+_{out}}$ are the protonic fluxes into and out of the thylakoid, V_e is the maximum rate of electron flow, e is the H^+/e^- ratio,$(K_a/(K_a + H^+_P))$ describes the control of electron flow by pH_p, p is the H^+/ATP ratio,V_p is the maximum rate of ATP synthesis and C is the passive permeability of the thylakoid membrane to H^+.

Eq.(6)-(8) were solved for H^+_P (using Eq.(4) for α) by computer iteration and the simulated results are shown in Table 2. The observed data of Table 1 are surprisingly well approximated using a single set of values for the constants of Eq.(6)-(8) (together with the same parameters used in Fig.1 to generate α) and changing only pH_N, C (to simulate uncoupling), or K_d (to simulate thiol modulation of CF_o-CF_1). Our analysis suggests that a dual pH requirement for the activation of CF_o-CF_1 may limit the observed catalytic performance of the enzyme under many conditions, and that thiol modulation of the enzyme may reduce this constraint by causing CF_o-CF_1 to activate at a lower ΔpH.

Financial support for this work was provided by Glynn Research Ltd. and by the Science and Engineering Research Council and is gratefully acknowledged.

TABLE 1. Effect of thiol modulation of CF_O-CF_1 on the observed initial rate of photophosphorylation, electron transport (H_2O→methyl viologen) and the P/2e ratio (see Mills,Mitchell (1983) for details). Rates of photophosphorylation are µmoles ATP/(mg.chl.h)) and of electron transport are µmoles O_2 consumed/(mg.chl.h). Each value is the mean of 16 experiments.

Assay conditions	Thiol modulation	Photo-phosphorylation	Electron transport	P/2e
pH_N 8	+	286 ± 42	274 ± 55	1.58 ± 0.13
	−	236 ± 38	236 ± 40	1.51 ± 0.16
pH_N 8 (+1 mM NH_4Cl)	+	168 ± 31	317 ± 51	0.80 ± 0.10
	−	111 ± 20	290 ± 39	0.58 ± 0.09
pH_N 7	+	160 ± 36	155 ± 28	1.55 ± 0.16
	−	88 ± 20	93 ± 16	1.41 ± 0.13

TABLE 2. Simulation of steady state photophosphorylation, electron transport and P/2e ratio. Eq.(6)-(8) were solved by iterative calculation of H_P^+ using the following values throughout: V_e=1000 µeq/(mg.chl.h); e=2.4; K_a=1.32.10^{-5}; p=3; V_p=833 µmoles ATP/(mg.chl.h); α from Eq.(4) using the values in Fig. 1. C was 2 (coupled or 150 (partially uncoupled) µeq/(µM H_P^+.mg.chl.h). To simulate thiol modulation, K_d of Eq.(4) was decreased from 0.1 to 0.012 as in Fig.1. Simulated rates are expressed as in Table 1.

Simulated conditions	Thiol modulation	Photo-phosphorylation	Electron transport	P/2e
pH_N 8	+	283	267	1.59
	−	236	224	1.58
pH_N 8 (partially uncoupled)	+	200	275	1.11
	−	106	251	0.63
pH_N 7	+	142	143	1.49
	−	85	98	1.30

REFERENCES

Mills JD and Mitchell P (1982) Thiol modulation of CF_O-CF_1 stimulates acid/base-dependent phosphorylation by pea chloroplasts, FEBS Letters 144,63-67.
Mills JD and Mitchell P (1983) Thiol modulation of the chloroplast ATPase and its effect on photophosphorylation, Biochim. Biophys. Acta, in press.
Mitchell P (1981) Biochemical mechanism of protonmotivated phosphorylation in F_O-F_1 ATPase molecules. In Lee CP et al, eds. Mitochondria and microsomes, pp 427-457. Boston,Mass: Addison-Wesley.
Schlodder E, Graber P and Witt HT (1982) Mechanism of phosphorylation in chloroplasts. In Barber J, ed. Topics in photosynthesis, vol 4, pp 105-175. Amsterdam: Elsevier Biomedical.
Strotmann H, Bickel-Sandcotter S and Shoshan V (1979) Kinetic analysis of light dependent exchange of adenine nucleotides on chloroplast coupling factor CF_1, FEBS Letters 101,316-320.

REGULATION OF THE H^+-ATPASE IN INTACT AND OSMOTICALLY SHOCKED CHLOROPLASTS

YOSEPHA SHAHAK

1. INTRODUCTION

The latent ATPase activity of CF_1 undergoes light activation in intact chloroplasts (Mills, Hind 1979). The activation depends on both $\Delta\mu H^+$ and thiol-reduction of CF_1. The latter is probably mediated by thioredoxin which undergoes photoreduction via ferredoxin-thioredoxin reductase, thus resembling the activation of a few carbon cycle enzymes (Buchanan 1980; Mills et al. 1980; Mills, Mitchell 1982; Shahak 1982a). Intact chloroplasts which are osmotically shocked before illumination, can maintain their ATPase activating capacity, providing Mg^{2+} or Ca^{2+} (at $\sim$ 10 mM) or mono-valent cations (>100 mM) are present and envelope rupture is done gently. It was suggested that the physiological thiol reducing system forms a loose complex which is anchored to the thylakoid membrane by hydrophobic forces, probably at the ferredoxin-NADP reductase site (Shahak 1982b). The light activated ATPase undergoes a slow deactivation in the dark. There is rather little known about the physiological process which leads to de-activation. It is probably also mainly controlled by oxidation-reduction reactions, since thiol oxidants such as ferricyanide or H_2O_2 accelerate the ATPase deactivation while dithiothreitol slows it down (Mills, Mitchell 1982; Shahak 1982a). In the work briefly summarized here, the deactivation process was further studied with regards to the nature of the physiological oxidant as well as other factors which might be involved.

2. MATERIALS AND METHODS

Intact (type A) chloroplasts were illuminated by a saturating white light for 3 min, then transferred to darkness and either kept intact or osmotically shocked in 15 mM Tricine pH 8 and other additions as indicated in the legends to figures. Aliquots were taken during the light and dark periods and assayed for dark ATPase activity as previously described (Shahak 1982b).

3. RESULTS AND DISCUSSION

What is the physiological oxidant of the ATPase? It is probably not oxidized thioredoxin or glutathion (Mills, Mitchell 1982). Another possible candidate is peroxide. Externally added peroxide deactivates the ATPase in both intact chloroplasts, CF_0-CF_1 containing proteoliposomes and soluble CF_1 (U. Pick, Y. Shahak, submitted). To test the effect of endogenously formed peroxide, the effect of azide and cyanide was studied. Both inhibit peroxide scavengers and thus increase H_2O_2 concentration during illumination. Indeed, the addition of NaN_3 (Fig. 1) or KCN (not shown) to intact chloro-plasts before light activation induced a fast deactivation of the ATPase in the following dark period, with kinetics similar to that observed upon the addition of peroxide to untreated chloroplasts in the dark (Shahak 1982a). When added after illumination, both inhibitors had essentially no direct effect on ATPase activity. The results indicate that peroxide might be the, or one of the physiological oxidants of CF_1. A major site for peroxide formation in the light seems to be the ferredoxin site. It is, therefore, speculated that both thiol reducing and oxidizing systems are located in a

Sybesma, C. (ed.), Advances in Photosynthesis Research, Vol. II. ISBN 90-247-2943-2.

FIGURE 1. Effect of NaN$_3$ on light-activation and dark deactivation of the ATPase in intact chloroplasts. NaN$_3$ (1 mM) was added before illumination

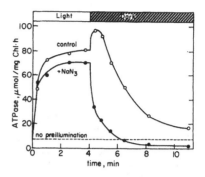

close vicinity to CF$_1$, near the acceptor side of PS I, making the regulation of the ATPase a local phenomenon.

The possibility that peroxide is involved in the deactivation of the ATPase imposes a difficulty since it implies that both the thiol reductant and oxidant are formed during illumination, while the enzyme undergoes activation in the light and deactivation in the dark. We tend to suggest as a working hypothesis that it is the enzyme itself which changes its properties so that in the light it interacts better with the thiol-reductant while in the dark better with oxidants. It has been shown that in the light CF$_1$ indeed undergoes conformational changes, and that thiol reduction of CF$_1$ is markedly accelerated by energization. We have recently found that during energization the ATPase becomes much less sensitive to thiol oxidants (Y. Shahak, in preparation). This idea is also supported by the effect of azide (Fig. 1) which is expressed in the dark, rather than in the light.

If the ATPase oxidant is a soluble compound, the dark deactivation is expected to proceed slower in broken chloroplasts, in which the stromal content is diluted into a large volume. Fig. 2 indicates that in principle

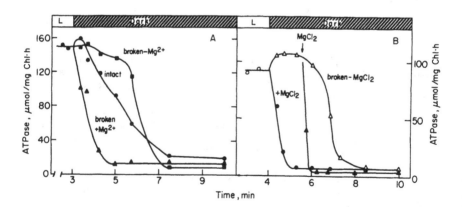

FIGURE 2. Dark deactivation of the ATPase in chloroplasts which are osmotically shocked after illumination in the absence or presence of MgCl$_2$ (10 mM).

this indeed is the case. The rate of dark deactivation in chloroplasts which were osmotically shocked in 15 mM Na tricine pH 8 (after light-activation) was slower than that of intact chloroplasts (Fig. 2A). It could be further expected that if the osmotic shock was done under conditions which maintain the thiol reducing system attached to the membrane (namely with 10 mM $MgCl_2$; Shahak 1982b), the decay would be even slower. However this expectation was fully contradicted by the results. The presence of $MgCl_2$ in the post illumination shocking medium markedly accelerated the deactivation (Fig. 2). Unlike the effect of Mg^{2+} ions on the light activation (Shahak, 1982b), the deactivating effect in the dark was Mg^{2+}-specific: 0.5 mM $MgCl_2$ was sufficient to induce 50% of the effect; Mg^{2+} but not Ca^{2+} or monovalent cations could substitute for Mg^{2+} (not shown); the response to the addition of Mg^{2+} was rather fast (Fig. 2B). It is, therefore, not a stacking-destacking phenomenon. It is also unlikely to be due to a Mg^{2+} induced oxidant production, not to an increased exposure of the CF_1 regulatory SH groups to oxidants, since DTT failed to prevent it (not shown). A Mg^{2+} dependent stimulation of the ATPase dark decay has been observed before in type C chloroplasts (Bakker-Grunwald, van Dam 1974). We suggest here that Mg^{2+} affects the ATPase activated conformation by specific binding to a regulatory site on CF_1. Still, since there seems to be a lot of Mg^{2+} in the stroma space, especially close to the thylakoid surface, the physiological relevance of Mg^{2+} ions as a regulatory factor of CF_1 is not obvious.

The presence of orthophosphate (Pi), or to a much greater extent pyrophosphate (PPi Fig. 3) in the post illumination shocking medium, stabilized the ATPase activated conformation in the dark and protected against the Mg^{2+}-deactivating effect (but not against the oxidant effect, not shown).

Pi and PPi both complex Mg^{2+} on one hand and specifically interact with CF_1 on the other hand. Another such compound is ADP. The latter is known, however, to deactivate the ATPase in type C chloroplasts in the dark (Carmeli, Lifshitz 1972). The deactivation was correlated with the tight binding of ADP to CF_1 (e.g. Shoshan, Selman 1979). In the system studied here ADP alone indeed induced a rapid dark deactivation. Nevertheless, the presence of the two deactivating agents - ADP and Mg^{2+} together, surprisingly stabilized the preactivated enzyme for a long time in the dark (Fig. 4). 10 - 100 μM ADP was sufficient to prevent the deactivating effect of 5 mM Mg^{2+}. Also, the Mg + ADP stabilizing effect was not changed when the osmotic shock was done 1 min after illumination to let $\Delta\bar{\mu}H^+$ dissipate (not shown).

FIGURE 3. Effect of PPi ± Mg^{2+} (2 mM each) on chloroplasts broken after illumination

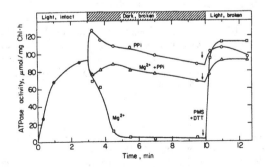

FIGURE 4. Antagonist effects of ADP with and without Mg^{2+} on the ATPase dark deactivation in chloroplasts osmotically shocked after illumination. ADP (0.5 mM) and Mg^{2+} (2 mM) were present in the dark shocking medium where indicated

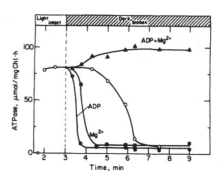

This indicates that this effect is probably not due to a post illumination ATP formation. Similar results were obtained with freshly broken and washed chloroplasts (light activated with PMS + DTT). As has been shown before, Pi, PPi, ATP and other nucleotides protect against ADP induced deactivation. Mg^{2+} is now added to this list. All of these compounds are found in the stroma space. Therefore we suggest that the tight binding of ADP to CF_1, although serving as a useful tool to study the enzyme, is not an important factor in the light-dark regulation of the ATPase in vivo.

4. REFERENCES

Bakker-Grunwald T and van Dam K (1974) On the mechanism of activation of the ATPase in chloroplasts, Biochim. Biophys. Acta 347, 290-298.
Buchanan BB (1980) Role of light in the regulation of chloroplast enzymes, Ann. Rev. Plant Physiol. 31, 341-374.
Carmeli C and Lifshitz Y (1972) Effect of Pi and ADP on ATPase activity in chloroplasts, Biochim. Biophys. Acta 267, 86-95.
Mills JD and Hind G (1979) Light induced Mg^{2+} ATPase activity of coupling factor in intact chloroplasts, Biochim. Biophys. Acta 547, 455-462.
Mills JD, Mitchell P and Schurmann P (1980) Modulation of coupling factor ATPase activity in intact chloroplasts. The role of thioredoxin, FEBS Lett. 112, 173-177.
Mills JD and Mitchell P (1982) Modulation of coupling factor ATPase activity in intact chloroplasts. Reversal of thiol modulation in the dark, Biochim. Biophys. Acta 679, 75-83.
Shahak Y (1982a) Activation and deactivation of H^+-ATPase in intact chloroplasts, Plant Physiol. 70, 87-91.
Shahak Y (1982b) The role of Mg^{2+} in the light activation process of the H^+-ATPase in intact chloroplasts, FEBS Lett. 145, 223-229.
Shoshan V and Selman BR (1979) The relationship between light-induced adenine nucleotide exchange and ATPase activity in chloroplast thylakoid membranes, J. Biol. Chem. 254, 8801-8807.

5. ACKNOWLEDGEMENTS

Mrs. Drora Nadav is acknowledged for skillful technical assistance. The work was supported by the Fund for Basic Research, Administered by the Israel Academy of Sciences and Humanities.
Author's address: Biochemistry Department, The Weizmann Institute of Science, Rehovot 76100, Israel.

ATP-INDUCED ΔpH IN CF_0-CF_1 PROTEOLIPOSOMES

ARIE ADMON, URI PICK AND MORDHAY AVRON

INTRODUCTION

The chloroplasts H^+-ATP synthase, CF_0-CF_1 complex, has been isolated from spinach and reconstituted into liposomes. Such proteoliposomes were previously shown to catalyse ATP hydrolysis, Pi-ATP exchange, acid-base dependent ATP synthesis (Pick, Racker, 1979), light-dependent ATP synthesis (when reconstituted with bacteriorhodopsin (Dewey, Hammes, 1981) or with PSI (Hauska et al., 1980) and ATP-induced $\Delta\psi$ formation (Shahak et al., 1982).

Attempts to measure ATP-induced ΔpH formation in these reconstituted liposomes were unsuccessful (Dewey, Hammes, 1981; Shahak, Pick, 1983). We considered the possibility that residues of ammonium, carried over with the enzyme from the $(NH_4)_2SO_4$ percipitation step, were responsible for inhibiting the formation of ΔpH, by buffering the protons transported into the liposomes. This will inhibit the formation of ΔpH while facilitating $\Delta\psi$ formation. Indeed, removing the ammonium by passing the proteoliposomes through a Sephadex column preequilibrated with valinomycin and K^+, enabled us to measure ATP-dependent ΔpH formation in CF_0-CF_1 proteoliposomes.

RESULTS

ATP induced $\Delta\psi$ in CF_0-CF_1 proteoliposomes formed by cholate dilution. However, no $\Delta\psi$ was formed in the same proteoliposomes following passage through Sephadex preequilibrated with valinomycin and K^+ (Fig. 1). On the other hand, ATP did not induce ΔpH formation (measured by 9-aminoacridine fluorescence quenching) in proteoliposomes prepared by cholate dilution whereas a large ΔpH was formed upon addition of ATP to proteoliposomes passed through a Val-K^+ preequilibrated Sephadex column (Fig. 2). Valinomycin added to cholate dilution liposomes did not bring about an ATP-induced formation of ΔpH, but it did abolish the $\Delta\psi$ signal (not shown). Passing proteoliposomes through Sephadex without valinomycin did not induce the appearance of an ATP-dependent ΔpH formation (not shown). At lower temperatures ATP-dependent ΔpH formation was larger in magnitude and slower in formation and decay (Fig. 3). Uncouplers like nigericin (Nig), S-13, and SF-6847 abolished the ATP-induced ΔpH formation (Fig. 2,3) and so did preincubation of the proteoliposomes with the energy transfer inhibitor DCCD (Fig. 4). Ammonium at 0.2 mM was also sufficient to abolish ATP-dependent ΔpH formation (not shown). Calibration of the 9AA signal was performed by preincubation of the vesicles in acid followed by injection of NaOH (acid-base transition, Fig. 5). According to this callibration, the ATP-induced ΔpH in proteoliposomes corresponded to 2.5-3.5 pH units. At room temperature the rate of ATP hydrolysis was 150 nmole$\cdot$mg protein$^{-1}\cdot$min^{-1} for the proteoliposomes before or after passage through Sephadex. The rates of Pi-ATP exchange were 20 and 40 nmole$\cdot$mg protein$^{-1}\cdot$min^{-1} for the cholate dilution and for the Sephadex-Val proteoliposomes, respectively.

Sybesma, C. (ed.), Advances in Photosynthesis Research, Vol. II. ISBN 90-247-2943-2.
© *1984 Martinus Nijhoff/Dr W. Junk Publishers, The Hague/Boston/Lancaster.*

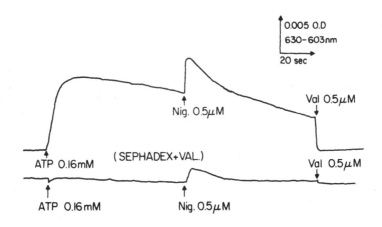

Figure 1: ATP-dependent absorbance changes of oxonol VI in CF_0-CF_1 proteoliposomes.

Proteoliposomes (0.25 mg protein; 2.0 mg soybean phospholipids and 0.7 mg Na cholate) were preincubated for 20 min at 4º in 0.05 ml. (Pick, Racker, 1979) and were either added directly to the cuvette or passed through Sephadex G-50 (course) column (25x0.6 cm), and washed with a solution containing 20 mM Na-Tricine, pH 8; 30 mM KCl, 3 mM $MgCl_2$, 1 mM K phosphate; and 10 nm valinomycin. Oxonol VI responses were measured as previously described (Shahak et al., 1982). The proteoliposome concentration in the reaction mixture was 0.125 mg protein/ml.

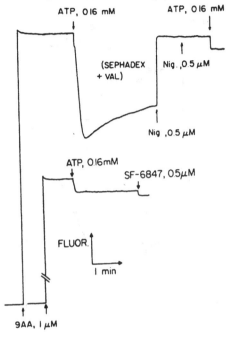

Figure 2: ATP dependent 9-amino-
acridine fluorescence
change in CF_0-CF_1
proteoliposomes.
Preparations as in
Fig. 1. 9-aminoacridine (9AA)
responses were measured in the
same reaction mixture, except
that 9AA replaced oxonol, in
a Perkin-Elmer spectrophoto-
fluorometer using for excitation
420 nm and for emission 465 nm.

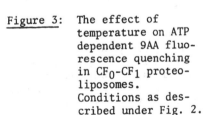

Figure 3: The effect of
temperature on ATP
dependent 9AA fluo-
rescence quenching
in CF_0-CF_1 proteo-
liposomes.
Conditions as des-
cribed under Fig. 2.

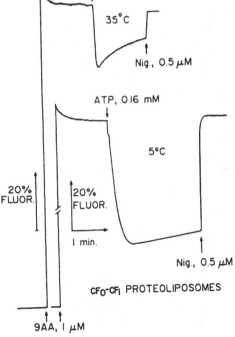

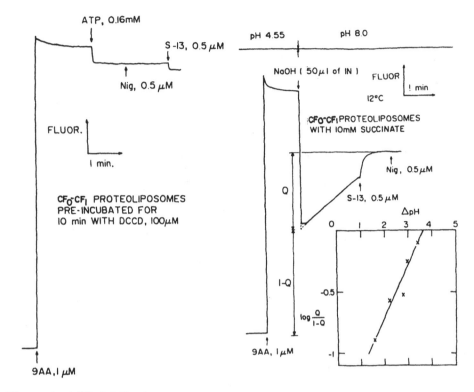

Figure 4: DCCD inhibition of ATP-induced 9AA fluorescence quenching in proteolipo. somes. Conditions as described in Fig. 2, except as indicated on figure.

Figure 5: Calibration of ATP-dependent 9AA fluorescence quenching by acid-base transition. Conditions as described in fig. 2. The inset illustrates the results of a series of similar experiments with varyir ΔpH, calculated as indicated on the figure.

CONCLUSIONS

A differential response to ATP of two types of CF_0-CF_1 proteoliposomes is described: 1) Proteoliposomes prepared by cholate-dilution catalyse in respons to ATP the formation of Δψ but of no ΔpH. 2) Proteoliposomes prepared by the transfer of protein: phospholipids: cholate mixtures through Sephadex columns preequilibrated with Val-K^+ catalyze in response to ATP the formation of ΔpH but of no Δψ. Both preparations catalyse comparable rates of ATP hydrolysis and of Pi-ATP exchange.

REFERENCES
Pick U and Racker E (1979) J. Biol. Chem. 254, 2793-2799.
Dewey TG and Hammes GG (1981) J. Biol. Chem. 256, 8941-8948.
Hauska G, Samoray D, Orlich G and Nelson N (1980) Eur. J. Biochem. 111,535-54:
Shahak Y, Admon A and Avron M (1982) FEBS Lett. 150, 27-31.
Shahak Y, and Pick U (1983) Arch. Biochem. Biophys. 223, 393-406.

Authors address: Department of Biochemistry, Weizmann Inst. of Science, Israel

THE ROLE OF ADP IN REGULATION OF CHLOROPLAST COUPLING FACTOR

ANDREAS LOEHR AND BERNHARD HUCHZERMEYER

1. INTRODUCTION

Chloroplast coupling factor (CF1) was shown to contain nucleotide binding sites (H.Strotmann et al.,1976). Nucleotide binding seems to alter permeability of the CFoCF1 complex to protons. This leads to a reduced leakage of protons out of the thylakoids in presence of nucleotides (P.Gräber et al.,1981) and a reduced non-phosphorylating electron transport rate (S.Izawa,1980 and B.Huchzermeyer, A.Loehr,1983).

2. MATERIAL AND METHODS

Isolation of thylakoids from spinach leaves was carried out as described by Strotmann et al.(1976). All reactions were measured at 20C. The standard reaction medium contained: 25mM tricine buffer, pH8.0, 50mM NaCl, 5mM $MgCl_2$, 50uM phenazine methosulfate, 20uM DCMU, and thylakoids corresponding to 20-25ug chlorophyll per ml. - When measuring ATP synthesis 5mM phosphate, pH8.0, labelled by 32-P-phosphate, 30U/ml hexokinase, 10mM glucose, and up to 1mM ADP were added to the incubation mixture. The reaction was stopped by addition of perchloric acid at a final concentration of 0.3M. Organic phosphate was determined as described by Y.Sugino, Y.Miyoshi (1964). - Electron transport was measured by observing oxygen consumption in a Clark-type electrode. The reaction medium contained 5mM ascorbate, pH8.0, 0.2mM DCPIP, and 0.2mM methylviologen instead of phenazine methosulfate. - Proton uptake was followed by observing proton concentration in the incubation medium. For better resolution the buffer concentration was reduced to 0.5mM tricine, pH8.0, in these experiments. - Nucleotide binding to CF1 was measured in plastic syringes. For determination of free nucleotides a sample was forced through a membrane filter, fixed to the syringes. - In order to measure phot.II dependent reactions only, the thylakoids were poisoned by mercuric chloride as described by D.A.Bradeen et al. (1973). The above reactions have been tested using oxidized p-phenylenediamine as the electron acceptor.

3. RESULTS AND DISCUSSION

In parallel experiments we examined the effects of different nucleotide concentrations on (a) proton uptake by isolated thylakoids, (b) nonphosphorylating electron transport rate, (c) ATP synthesis, and (d) exchangeable as well as non-exchangeable ADP binding to CF1. The effects of varying light intensities or uncoupler concentrations were observed. Apparent K-values of the nucleotide effect were determined. Different nucleotides gave identical results. ADP, d-ADP, ε-ADP, IDP, and GDP were tested. In agreement with results obtained by observing the "tight" nucleotide binding (H.Strotmann et al.,1977) ADP and d-ADP gave identical results, the other nucleotides were less effective. The apparent Km of ADP in the phosphorylation experiment was found to be about ten times higher than the constants of the other reactions. Similar to results published by C.Vinkler (1981) we observed a correlation between the apparent K-values and electron transport rate when varying light intensity (Fig.1) or the concentrations of added uncouplers like ammoniumchloride, methylammoniumchloride or succinate

Sybesma, C. (ed.), Advances in Photosynthesis Research, Vol. II. ISBN 90-247-2943-2.
© 1984 Martinus Nijhoff/Dr W. Junk Publishers, The Hague/Boston/Lancaster.

(Fig.2). This correlation was no longer found when the effect of different gramicidin concentrations were tested (Fig. 3). Uncoupling by increasing gramicidin concentrations resulted in decreasing apparent K-values.

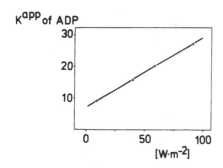

Fig. 1
Effect of varying light intensities on the apparent K-value of ADP for enhancement of proton uptake

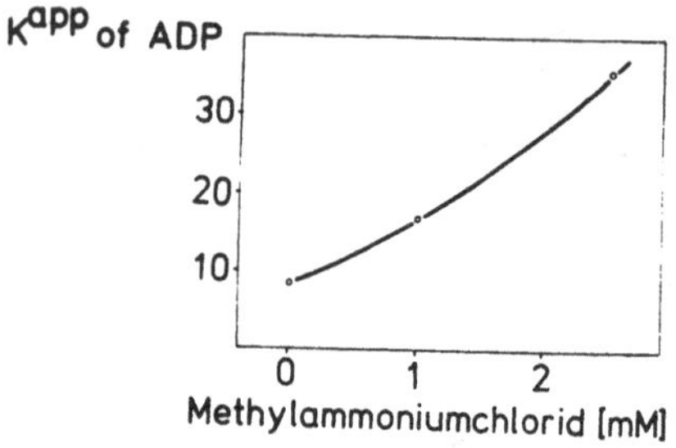

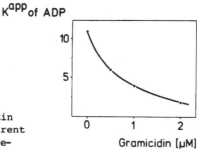

Fig. 2
Effect of varying uncoupler concentrations on the apparent K-value of ADP for enhancement of proton uptake

Fig. 3
Effect of varying gramicidin concentrations on the apparent K-values of ADP for enhancement of proton uptake

We interpret these results as follows: Electron transport results in a translocation of protons across the thylakoid membranes. The first step of this reaction is a production of membrane orientated protons. These protons are not available for hydrophilic uncouplers (Y.Q.Hong, W.Junge, 1983) and seem to have only little mobility (Y.deKouchkovsky et al., 1982). Nevertheless there is a steady state between this located proton pool and that one of the aqueous internal space of the thylakoids. This can be concluded from results of acid/base experiments (A.T.Jagendorf, 1971 and B.Huchzermeyer, H.Strotmann, 1977). The size of a membrane orientated proton pool regulates the apparent affinity of CF1 to nucleotides. The tested uncouplers reduce the size of the aqueous proton pool and stimulate electron transport rate. This leads to an enhancement of the concentration of located protons, due to their reduced mobility, as long as uncoupling does not exceed a critical level. Therefore moderate uncoupling, like increasing light intensity, leads to increasing apparent K-values. This effect can not be expected with high uncoupler concentrations, as compartimentation of protons may well be destroyed as shown by Y.Q. Hong, W.Junge (1983). In contrast to other uncouplers gramicidin can reach membrane orientated protons, too. Therefore we observed decreasing K-values with increasing gramicidin concentrations. Increasing uncoupling by gramicidin resulted in the same effect as reducing light intensity. - Even more we observed a strong correlation between 1/K-values and p/2e-ratio under all conditions employed. The p/2e-ratio varied from 0.8 to 1.4.

When measuring the kinetics of nucleotide exchange on CF1 we observed an inhibition of ADP exchange rate by each gramicidin concentration tested. When employing methylammoniumchloride instead of gramicidin we observed an enhancement of the exchange rate up to 1 or 2mM uncoupler. Higher concentrations of uncoupler resulted in an inhibition of nucleotide exchange. This result is in good agreement with our model described above.

Figure 4 shows the effect of external pH on (a) binding of ADP following an acid/base transition employing a constant pH jump (b) stimulation of the extent of proton uptake by illuminated thylakoids induced by addition of ADP, and (c) inhibition of electron transport by addition of ADP in the absence of phosphate. The parallel effect of medium pH on these three reactions leads to the conclusion that there is some causal connection between them.

Protons from the water splitting system seem to appear in a domain within reach of gramicidin (G.M.Baker et al.,1981) but apparently are not available for hydrophilic uncouplers. Therefore we investigated the effect of electron transport from water to p-phenylenediamine on the above reactions. In agreement with J.M.Gould, S.Izawa (1973) we did not observe any regulative effect of ADP on electron transport rates. Even more we were unable to measure electron transport induced nucleotide exchange at the "tight" binding sites. This result was surprising, especially as we were able to measure light dependent

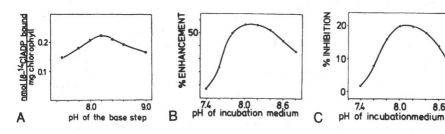

A   pH of the base step   B   pH of incubation medium   C   pH of incubationmedium

Fig. 4

Panel A shows the binding of ADP to CF1 induced by an acid/base transition. Starting with different pH values of the acid step, ph-jumps of constant 3.5 units were employed using the method of B.Huchzermeyer, H.Strotmann (1977). In panel B the stimulation of the extent of the steady state proton uptake by addition of 1mM ADP is shown. Panel C shows the inhibition of non-phosphorylating electron transport by addition of 1mM ADP.

ATP synthesis as well as light induced ATPase activity. This means that photII drives enzymatic activity of CF1 without showing a correlation with nucleotide exchange as found with PMS mediated electron transport through photI (J.Schumann, H.Strotmann, 1981). We conclude from this that protons from the water splitting system are not the regulative ones discussed in our model.

4. REFERENCES

Baker,G.M., Bhatnagar,D., and Dilley,R.A. (1981) Proton release in photosynthetic water oxidation, Biochem. 20, 2307-2315

Bradeen,D.A., Winget,G.D., Gould,J.M., and Ort,D.R. (1973) Site-specific inhibition of photophosphorylation in isolated spinach chloroplasts by mercuric chloride, Plant Phys. 52, 680-682

Gould,J.M. and Izawa,S. (1973) Studies on the energy coupling sites of photophosphorylation, Biochim. Biophys. Acta 314, 211-223

Gräber,P., Burmeister,M., and Hortsch,M. (1981) Regulation of the membrane permeability of spinach chloroplasts by binding of adenine nucleotides, FEBS Lett. 136, 25-31

Hong,Y.Q. and Junge,W. (1983) Localized or delocalized protons in photophosphorylation, Biochim. Biophys. Acta 722, 197-208

Huchzermeyer,B. and Strotmann,H. (1977) Acid/base-induced exchange of adenine nucleotides on chloroplast coupling factor (CF1), Z. Naturforsch. 32c, 803-809

Huchzermeyer,B. and Loehr,A. (1983) Effects of nitrofen on chloroplast coupling factor dependent reactions, Biochim. Biophys. Acta, in press

Izawa,S. (1980) Acceptors and donors for chloroplast electron transport, Meth. Enzymol. 69, 413-434

Jagendorf,A.T. (1971) Two-stage phosphorylation techniques: Light-to-dark and acid-to-base procedures, Meth. Enzymol. 24, 103-113

deKouchkovsky,Y., Haraux,F., and Sigalat,C. (1982) Effects of hydrogen-deuterium exchange on energy-coupled processes in thylakoids, FEBS Lett. 139, 245-249

Schumann,J. and Strotmann,H. (1981) The mechanism of induction and deactivation of light-triggered ATPase, in: Proc. V Int. Congr. Photosynth., Kallithea 1980, pp. 881-892

Strotmann,H., Bickel,S., and Huchzermeyer,B. (1976) Energy-dependent release of adenine nucleotides tightly bound to chloroplast coupling factor CF1, FEBS Lett. 61, 194-198

Strotmann,H., Bickel-Sandkötter,S., Edelmann,K., Schlimme,E., Boos,K.S. and Lüstorff,J. (1977) Studies on the tight adenine nucleotide binding site of chloroplast coupling factor (CF1) in: Structure and function of energy transducing membranes (vanDam,K. and vanGelder,B.F., eds.), Elsevier/North-Holland Biomed. Press, pp. 307-317

Sugino,Y. and Miyoshi,Y. (1964) The specific precipitation of orthophosphate and some biochemical applications, J. Biol. Chem. 239, 2360-2364

Vinkler,C. (1981) Opposite modulation by uncoupling and electron transport limitation of the apparent Km of ADP for photophosphorylation, Biochem. Biophys. Res. Commun. 99, 1095-1100

Acknowledgements: Part of the research was supported by the Deutsche Forschungsgemeinschaft.

Author's adress: Andreas Loehr, Botanisches Institut, Tieraerztliche Hochschule , Buenteweg 17d, D-3000 Hannover 71, W.-Germany (RFA)

INTERACTION OF MEMBRANE-BOUND AND SOLUBLE CF$_1$ WITH THE PHOTOREACTIVE NUCLEOTIDE 3'-O-(4-BENZOYL)BENZOYL ADP

D. BAR-ZVI, M. A. TIEFERT and N. SHAVIT/BEN-GURION UNIVERSITY OF THE NEGEV

1. INTRODUCTION

Chloroplast ATP synthetase contains several binding sites: catalytic sites where rapid association and dissociation of nucleotides occur, and tight nucleotide binding sites which may be regulatory (for review see Shavit, 1980; Bar-Zvi, Shavit, 1982). While the number of sites has to be established, they are probably located on the two large subunits (α and β) of CF$_1$. Photoreactive nucleotide analogs are useful for study of the location, function and number of these binding sites. Photoaffinity labeling of soluble CF$_1$ with ADP or ATP analogs resulted in their covalent incorporation into either the β subunit or both the α and β subunits of the enzyme (Carlier et al., 1979; Wagenvoord et al., 1981) Photoaffinity labeling of the tight nucleotide binding site of membrane-bound CF$_1$ with 2-azido ADP (Czarnecki et al., 1982) resulted in its incorporation into the β subunit.

A new photoreactive nucleotide analog, 3'-O-(4-benzoyl)benzoyl ATP (BzATP) was introduced (Williams, Coleman, 1982). Unlike most of the other photoaffinity probes, in which the photoreactive group is an azido, in BzATP it is a benzophenone.

In the work presented here we used BzADP to study the nucleotide binding sites in both the membrane-bound and soluble CF$_1$. The analog is not a substrate for photophosphorylation but is a strong competitive inhibitor with respect to adenine nucleotides in ATP synthesis or hydrolysis and P$_i$-ATP exchange. BzADP binds tightly to the membranal enzyme, competitively with ADP, and protects against inactivation of the enzyme by tight binding of ADP. Upon irradiation, BzADP is covalently bound to both the α and β subunits of CF$_1$. BzADP also inhibits ATP hydrolysis by the soluble CF$_1$ and upon irradiation binds to the β subunit. By extrapolation, complete inactivation of the soluble enzyme occurs at binding of 2 mol analog/mol CF$_1$.

2. EXPERIMENTAL

BzADP was prepared essentially as described for BzATP (Williams, Coleman, 1982). Thylakoids were isolated from fresh lettuce leaves (Bar-Zvi, Shavit, 1982). CF$_1$ was extracted by treatment with CHCl$_3$, purified and kept in 2 M (NH$_4$)$_2$SO$_4$. Photoactivation was done by irradiation at 366 nm with a Blak-Ray lamp (model XX-15).

3. RESULTS

3.1. Membrane bound CF$_1$

BzADP is not a substrate for photophosphorylation but is a strong competitive inhibitor with respect to ADP and ATP (K$_i$ 2-25 μM) in photophosphorylation, ATP hydrolysis and P$_i$-ATP exchange (not shown). The analog also competes with the tight binding of ADP and binds tightly to membrane-bound CF$_1$ (Fig. 1A). Addition of ADP to activated thylakoids before the substrates for ATP hydrolysis or P$_i$-ATP exchange results in tight binding of ADP with concomitant inactivation of the enzyme (Bar-Zvi,

Sybesma, C. (ed.), Advances in Photosynthesis Research, Vol. II. ISBN 90-247-2943-2.
© *1984 Martinus Nijhoff/Dr W. Junk Publishers, The Hague/Boston/Lancaster.*

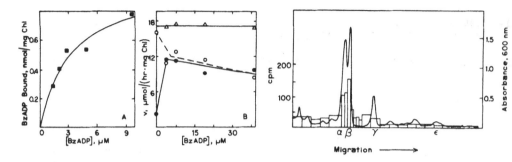

FIGURE 1. (Left) Tight binding of BzADP to membrane bound CF_1. (A) Tight binding of Bz[^{3}H]ADP. (B) Protection by BzADP against inhibition of P_i-ATP exchange by tight binding of ADP. The indicated concentrations of BzADP were added alone (O) or together with 5 μM ADP (●) in the dark 5 s after the activation, and ATP+$^{32}P_i$ were added after an additional 60 s in the dark. (Δ) BzADP added together with ATP and $^{32}P_i$, 65 s after activation.

FIGURE 2. (Right) Photoaffinity labeling of tight nucleotide binding site(s) in the membrane-bound CF_1 with Bz[^{3}H]ADP. Bz[^{3}H]ADP was tightly bound to thylakoid membranes. They were washed and irradiated at 366 nm for 1 hr. CF_1 was extracted by $CHCl_3$ and denatured with perchloric acid. Electrophoresis was in the presence of SDS on 10% acrylamide gels.

Shavit, 1982). Unlike ADP, the addition of BzADP to activated thylakoids results in little inhibtion of the P_i-ATP exchange (Fig. 1B). Moreover, when added together with ADP, BzADP protects against the inactivation of the enzyme by ADP. Irradiation of thylakoid membranes containing tightly bound Bz[^{3}H]ADP results in incorporation of the label into both the α and β subunits (Fig. 2). Unfortunately, uncoupler, which was needed during irradiation to prevent the release of tightly bound Bz[^{3}H]ADP by energization of the membrane, and partial bleaching of the thylakoid pigments by the UV light, prevented studies of the activities of the photolabeled thylakoids.

3.2. Soluble CF_1

In the dark, ATP hydrolysis by soluble CF_1 is reversibly inhibited by BzADP. It decreases V_{max} and increases the Hill constant from 1 in the absence of the analog to 1.5 in presence of 0.5 mM BzADP (not shown).

Upon irradiation, BzADP irreversibly inhibits the enzyme activity (Fig. 3). Inactivation is prevented if ADP is present during the irradiation. ATP is not effective as a protector. Photoinhibition affects the V_{max} but not the K_m (ATP), and does not affect heat activation of the latent enzyme (not shown). Inactivation of CF_1 by irradiation in the presence of BzADP correlates with covalent incorporation of the analog: extrapolation showed that complete inactivation of the enzyme occurs at incorporation of 2 mol BzADP/mol CF_1 (assuming molecular weight of 350 000 for CF_1) (Fig. 4). This stoichiometry is not affected if irradiation is done in buffer containing EDTA or Ca^{2+}. In agreement with its effect on activity (Fig. 3), ADP also prevents covalent incorporation of Bz[^{3}H]ADP (Fig. 4). Analysis of the photolabeled CF_1 showed that only the β subunit becomes labeled (Fig. 5).

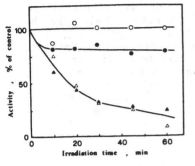 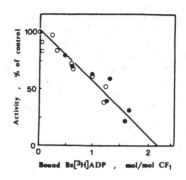

FIGURE 3. (Left) Photoinactivation of soluble CF$_1$ by BzADP. CF$_1$ preincubated in the dark with the indicated components in 40 mM tricine-NaOH
(pH 8.0)/ 2 mM EDTA. After 30 min, samples were irradiated for the
indicated time. O , no addition; △ , + 193 µM BzADP; ● , + BzADP + 25
mM ADP; and ▲ , + BzADP + 25 mM ATP.

FIGURE 4. (Right) Correlation between inactivation of CF$_1$ and covalent
incorporation of Bz[^{3}H]ADP. CF$_1$ was irradiated in the presence of 125 µM
Bz[^{3}H]ADP as described in Fig. 3. The mix contained also O , 2 mM EDTA;
● , 10 mM CaCl$_2$ or □ , 2 mM EDTA + 25 mM ADP.

DISCUSSION

BzADP inhibits photophosphorylation, ATP hydrolysis and P$_i$-ATP
exchange by thylakoid-bound CF$_1$ competitively with the respective nucleotide substrate, as expected for an ADP analog. BzADP also competes with
the tight binding of ADP to the membranal enzyme and protects against its
inactivation by ADP (Fig. 1), which suggests that the analog binds both to
catalytic and regulatory sites. The protection by BzADP against ADP is
similar to other nucleoside diphosphates, e.g. GDP (Bar-Zvi, Shavit,
1982). The reversible inhibition by BzADP of ATP hydrolysis by soluble CF$_1$
is like that by ADP (Nelson et al., 1972), which suggests that BzADP acts
as an ADP analog also with the soluble enzyme.

When reacted with the tight nucleotide binding site(s) on membrane-
bound CF$_1$, BzADP labels both the ∝ and β subunits (Fig. 2). This

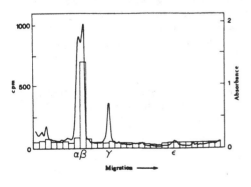

FIGURE 5. SDS gel electrophoresis of soluble CF$_1$ photolabeled with Bz[^{3}H]ADP.
Photolabeling was as in Fig.
3 and electrophoresis as in
Fig. 2.

differs from results obtained with 2-azido ADP, which labels only the β subunit (Czarnecki et al., 1982). Similarly, while BzADP labels only the β subunit of soluble CF_1 (Fig. 5), 8-azido ADP (Wagenvoord et al., 1981) and arylazido ADP (Carlier et al., 1979) are incorporated into both the α and β subunits. These discrepancies may result from differences the nature of the photoreactive group and its location on the nucleotide. Also, different labeling patterns can be obtained using a single photo-affinity label under various conditions (Lunardi, Vignais, 1982). The stoichiometry of binding is the same as with 8-azido ADP in the absence of cations (Wagenvoord et al., 1981), but unlike with 8-azido ADP, the presence of Ca^{2+} does not double the amount of bound BzADP needed for complete photoinactivation of the enzyme (Fig. 4).

Labeling of both the α and β subunits of CF_1 by irradiation of thylakoids containing previously tightly bound $Bz[^3H]ADP$ (Fig. 2) may occur even if only one binding site is involved, if it is between the two large subunits. Similarly, the difference in the subunits labeled in soluble and membrane-bound CF_1 (Figs. 2 and 5) may result from binding to different site(s) or binding to the same site(s) which exist in different conformations in the two types of CF_1.

REFERENCES

Bar-Zvi D and Shavit N (1982) Modulation of the chloroplast ATPase by tight binding of nucleotides, Biochim. Biophys. Acta 681, 451-458.
Carlier MF, Holowka DA and Hammes GG (1979) Interaction of photoreactive and fluorescent nucleotides with chloroplast coupling factor 1, Biochemistry 18, 3452-3457.
Czarnecki JJ, Abbott MS and Selman BR (1982) Photoaffinity labeling with 2-azido adenosine diphosphate of the tight nucleotide binding site on the chloroplast coupling factor one, Proc. Natl. Acad. Sci. USA 79, 7744-7748.
Lunardi J and Vignais PV (1982) Studies of the nucleotide-binding sites on the mitochondrial F_1-ATPase thhrough the use of a photoactivable derivative of adenylyl imidodiphosphate, Biochim. Biophys. Acta 682, 124-134.
Nelson N, Nelson H and Racker E (1972) Partial resolution of enzymes catalyzing photophosphorylation: Magnesium-adenosine triphosphatase properties of heat-activated coupling factor 1 from chloroplasts, J. Biol. Chem. 247, 6506-6510.
Shavit N (1980) Energy transduction in chloroplasts: Structure and function of the ATPase complex, Annu. Rev. Biochem. 49, 111-138.
Wagenvoord RJ, Verschoor GJ and Kemp A (1981) Photolabeling with 8-azido adenine nucleotides of adenine nucleotide-binding sites in isolated spinach chloroplast ATPase (CF_1), Biochim. Biophys. Acta 634, 229-236.
Williams N and Coleman PS (1982) Exploring the adenine nucleotide binding sites with a new photoaffinity probe, 3'-0-(4-benzoyl)benzoyl adenosine 5'-triphosphate, J. Biol. Chem. 257, 2834-2841.

ACKNOWLEDGMENTS
We are grateful to Mrs. Z. Conrad for her excellent technical assistance.
Authors address: Department of Biology, Ben-Gurion University of the Negev, Beer-Sheva 84105, Israel.

NUCLEOTIDE BINDING TO THE MEMBRANE-BOUND CHLOROPLAST COUPLING
FACTOR (CF_1) IN THE LIGHT

JÜRGEN SCHUMANN, BOTANISCHES INSTITUT DER UNIVERSITÄT DÜSSELDORF

INTRODUCTION

The membrane-bound chloroplast ATPase (CF_1) contains at least
two different types of nucleotide binding sites: one or more
sites participating in the formation of ATP, and a tight
binding site with a tightly bound adenine nucleotide in the
deenergized state of the enzyme.
Tightly bound ADP seems not to be a catalytic intermediate of
ATP formation. Its function is still unclear. In the light,
release and rebinding of nucleotides take place to several
enzyme species in equilibrium, with tightly (E=ADP) or loosely
($E^+ \cdots$ADP) bound ADP or free of nucleotides (E^+) (Strotmann
et al. 1979):

$$E{=}ADP \;\rightleftharpoons\; E^+ {\cdots} ADP \;\rightleftharpoons\; E^+ + ADP$$

E^+ decays after deenergization of the thylakoids (by darkening
or by addition of uncouplers) to a different enzyme form (E^-)
which is able either to bind ADP (to form E=ADP) or to hydrolyze ATP. Binding of ADP leads to deactivation of the ATPase
(Schumann, Strotmann 1981; Dunham, Selman 1981a; Bar-Zvi, Shavit 1980). In the light, either the tight or the loose site is
occupied depending on the energy state of the membranes. The
tight binding site was suggested to be a different form of the
catalytic site of CF_1 (Bickel-Sandkötter, Strotmann 1981).
Loosely bound ADP was thought to be the catalytic intermediate
of ATP formation (Bickel-Sandkötter 1983).
In this report, the properties of the tight nucleotide binding
site and its relationship to the catalytic site are further
investigated.

MATERIAL AND METHODS

Broken chloroplasts were isolated from spinach leaves as
described (Strotmann, Bickel-Sandkötter 1977). Chloroplasts
were illuminated with white light in a medium containing tricine
(25 mmol/l) pH 7.8, NaCl (50 mmol/l), $MgCl_2$ (1 mmol/l), phenazine methosulfate (50 µmol/l) for 30 s. Chlorophyll content
was 0.1 to 0.12 mg/ml. ([14C])ADP was either present in the light
or was added together with the uncoupler carbonyl cyanide
p-trifluoromethoxyphenyl hydrazone (FCCP). Unlabeled ADP was
added to a final concentration of 5 mmol/l. The amount of
tightly bound ([14C])nucleotides was determined after washing
the thylakoids as described (Strotmann, Bickel-Sandkötter 1977).
Separation of nucleotides into AMP, ADP and ATP was carried
out after acid extraction on ion exchange columns by stepwise
elution with HCl (Bickel-Sandkötter 1983).

Sybesma, C. (ed.), Advances in Photosynthesis Research, Vol. II. ISBN 90-247-2943-2.
© *1984 Martinus Nijhoff/Dr W. Junk Publishers, The Hague/Boston/Lancaster.*

RESULTS

If FCCP is added to thylakoids illuminated in the presence of
(^{14}C)ADP and the amount of tightly bound ADP is determined at
different times, a biphasic binding curve is obtained (fig. 1).
The slow phase is probably due to binding of medium ADP to the
deenergized enzyme form (E$^-$). The rapid phase may reflect the
conversion of E$^+$···ADP to E=ADP (Strotmann et al. 1979). The
extent of the rapid phase is not reduced by increasing the uncoupler concentration. Moreover, the rapid phase is reduced
only if high concentrations of quench ADP are employed (fig. 2)
while slow binding of (^{14}C)ADP from the medium is effectively
quenched at much lower ADP concentrations (inset of fig. 2).
Therefore, medium (^{14}C)ADP is not the source of rapidly bound
ADP. It is assumed that binding of quench ADP to a second site
leads to a faster release of loosely bound (^{14}C)ADP from the
first site.
In fig. 3 binding of medium (^{14}C)ADP was measured by addition
of (^{14}C)ADP together with FCCP in the light to preilluminated
chloroplasts. A rapid binding phase is observed in this type
of experiment. The amount of rapidly bound ADP depends on the
concentration of quench FCCP: the higher the concentration of
uncoupler the lower the amount of bound ADP.

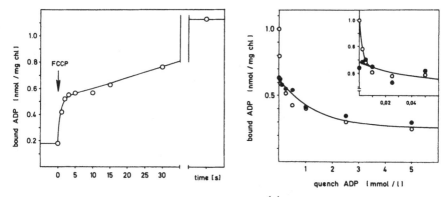

Fig. 1 (left) Time course of tight (^{14}C)ADP binding after uncoupling in the light. Chloroplasts were illuminated in the
presence of (^{14}C)ADP (5 μmol/l). After 3o s, FCCP (1o μmol/l)
was added in the light. Quench ADP was added at different
times after the FCCP quench.

Fig. 2 (right) Influence of unlabeled ADP on tight binding of
(^{14}C)ADP after FCCP-quench. Chloroplasts were illuminated in
the presence of (^{14}C)ADP (5 μmol/l). FCCP (1o μmol/l) was added
after 3o s in the light together with different concentrations
of unlabeled ADP. (o): influence of quench ADP on total (^{14}C)ADP
binding; (●): influence of quench ADP on the rapid binding
phase (further binding of (^{14}C)ADP from the medium was prevented by a second addition of quench ADP (5 mmol/l), 2 s after
the first quench).

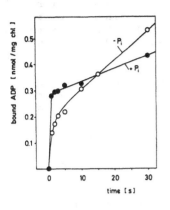

Fig. 3 Time course of (14C)ADP binding
to preilluminated chloroplasts during
FCCP addition in the light.
Chloroplasts were illuminated in the
absence of medium nucleotides. After
3o s, FCCP (5 µmol/l) and (14C)ADP
(5 µmol/l) were added in the light.
Quench ADP (5 mmol/l) was then added
at different times later (open cycles).
In a parallel experiment, phosphate
(o.5 mmol/l) was added together with
FCCP and (14C)ADP (closed cycles).

If the concentration of (14C)ADP is increased the amount of
rapidly bound ADP is also increased up to 1 nmol/mg chlorophyll.
If phosphate is added together with FCCP and (14C)ADP, the rapid
phase of binding is still enhanced (fig. 3). The second phase
is slower than in the absence of phosphate as reported for
tight ADP binding in the dark (Schumann, Strotmann 1981; Dunham,
Selman 1981b). Magnesium ions are necessary for the enhancement
of the rapid binding by phosphate. The analysis of bound nucleotides reveals that the additional amount of bound nucleotides
in the presence of phosphate is ATP (table 1). The final amount
of bound (14C)nucleotides is about 1 nmol/mg chlorophyll with
or without phosphate. This indicates that either ADP or ATP
becomes tightly bound to the same site.

TABLE 1 Analysis of (14C)nucleotides bound during FCCP quench
in the light. Chloroplasts were illuminated in the absence of
medium nucleotides and phosphate. After 3o s, FCCP (5 µmol/l)
was added together with (14C)ADP and phosphate when indicated
(1 mmol/l). After washing the chloroplasts, aliquots were taken
for the determination of bound (14C)nucleotides and of chlorophyll concentration. One aliquot was extracted with perchloric
acid and subjected to ion exchange chromatography.

(14C)ADP conc. (µmol/l)	Phosphate (1 mmol/l)	Bound nucleotides (nmol/mg chlorophyll)			
		total	AMP	ADP	ATP
7.5	−	o.165	o.o14	o.121	o.o29
	+	o.261	o.o2o	o.163	o.o78
12.5	−	o.221	o.o24	o.173	o.o24
	+	o.35o	o.o33	o.2o2	o.115

Catalytic ATP formation therefore seems to take place either on the tight site or in the proximity of this site since release of ATP into the medium and rebinding is unlikely. Furthermore, ATP rapidly formed from ADP and phosphate was found to be inaccessible to hexokinase (Rosen et al. 1979; Aflalo, Shavit 1982 If external (^{14}C)ATP is added together with FCCP, no rapid binding phase can be detected (not shown). Phosphate further reduces the rate of ATP binding. For this reason, binding of ATP which results from medium ADP plus phosphate is assumed to occur on the same site as ATP formation.

REFERENCES

Aflalo C and Shavit N (1982) Source of rapidly labeled ATP tightly bound to non-catalytic sites on the chloroplast ATP synthetase   Eur.J. Biochem. 126, 61-68
Bar-Zvi D and Shavit N (1980) Role of tight nucleotide binding in the regulation of the chloroplast ATP synthetase activities FEBS Lett. 119, 68-72
Bickel-Sandkötter S (1983) Loose and tight binding of adenine nucleotides by membrane-associated chloroplast ATPase Biochim.Biophys.Acta 723,71-7
Bickel-Sandkötter S and Strotmann H (1981) Nucleotide binding and regulation of chloroplast ATP synthase FEBS Lett. 125, 188-192
Dunham KR and Selman BR (1981a) Regulation of spinach chloroplast coupling factor 1 ATPase activity J.Biol.Chem. 256, 212-218
Dunham KR and Selman BR (1981b) Interactions of inorganic phosphate with spinach coupling factor 1. Effects on ATPase and ADP binding activities J.Biol.Chem. 256, 10044-10049
Rosen G Gresser M Vinkler C and Boyer PD (1979) Assessment of total catalytic sites and the nature of bound nucleotide participation in photophosphorylation J.Biol.Chem. 254, 10654-10661
Schumann J and Strotmann H (1981) The mechanism of induction and deactivation of light-triggered ATPase in: Akoyunoglou,G. Photosynthesis II. Electron transport and photophosphorylation pp. 881-892 Balaban Int. Science Services, Philadelphia Pa.
Strotmann H and Bickel-Sandkötter S (1977) Energy-dependent exchange of adenine nucleotides on chloroplast coupling factor (CF$_1$) Biochim.Biophys.Acta 460, 126-135
Strotmann H Bickel-Sandkötter S and Shoshan V (1979) Kinetic analysis of light-dependent exchange of adenine nucleotides on chloroplast coupling factor CF$_1$ FEBS Lett. 101, 316-320

ACKNOWLEDGEMENTS

The author thanks Karin Brendel and Martina Sanders for their excellent technical work and Prof. Strotmann and Prof. Shavit for many helpful discussions.

Author's address   Jürgen Schumann Botanisches Institut der
                    Universität Düsseldorf, Universitätsstr. 1
                    D-4000 Düsseldorf 1, Germany F.R.

BINDING AND EXCHANGE OF NUCLEOTIDES AND MAGNESIUM ON THE CHLOROPLAST COUPLING FACTOR CF_1

G. GIRAULT, J.M. GALMICHE and Cl. LEMAIRE

1. INTRODUCTION

Purified CF_1 solubilized from chloroplast membranes has no ATPase activity, unless activated by trypsin, heat or dithiothreitol but like the membrane-bound enzyme, isolated CF_1 can bind nucleotides in a non-covalent and non-energy-requiring process. The maximum number of bound nucleotides varies, depending on the way the complex $AdN-CF_1$ is recovered from the excess of free nucleotides. A conflicting point is the specificity of these sites for ATP or ADP and the effect of Mg^{2+} on the binding or exchange of nucleotides on CF_1.

The purpose of this paper is to investigate the role of Mg^{2+} on the amount and the nature of the nucleotides bound to the purified latent ATPase, inactive CF_1 supplemented with ATP or ADP. The latent ATPase, which is very slowly hydrolyzing ATP, can be an interesting tool to evidence short-lived states of the $AdN-Mg-CF_1$ complex which may be intermediates of the ATP hydrolysis on the activated enzyme. A correlation was searched between the binding of the nucleotides and of Mg^{2+} to CF_1 and between their ability to exchange with free nucleotides and Mg^{2+}.

2 MATERIALS AND METHODS

Coupling factor, CF_1 was extracted and purified from spinach leaves following the method of Lien, Racker (1971) with 1 mM EDTA. No nucleotide was added at any step of the purification procedure. The $AdN-CF_1$ complex was liberated from the excess of free AdN by the method developed by Penefsky (1977) called centrifugation-filtration.
Radioactive nucleotides and total ATP or ADP concentrations were determined as described by GIRAULT et al (1982).
Non-radioactive magnesium was determined by atomic absorption.
The isotope ^{27}Mg (half-life 9.45 min) was produced by neutronic activation during a (n,γ)-reaction.
Gamma peaks at 844 and 1014 keV were sampled with a Plurimat analyzer and calibrated with ultra-pure $Mg(NO_3)_2$ solutions as described by GIRAULT et al (1982).

3. RESULTS AND DISCUSSION

3.1. <u>Binding of ADP and ATP on CF_1. Effect of Mg^{2+}</u>. When CF_1 is incubated with ADP, up to 3 moles of ADP are bound per CF_1 in the presence or not of Mg^{2+}, including the 0.6 mole ADP still present, at the beginning of the incubation, on the CF_1. In the absence of Mg^{2+}, the half saturation is achieved with an addition of 40 µM ADP (for CF_1 concentration 15 µM) instead of 20 µM in the presence of Mg^{2+} (Fig. 1).

Sybesma, C. (ed.), Advances in Photosynthesis Research, Vol. II. ISBN 90-247-2943-2.
© *1984 Martinus Nijhoff/Dr W. Junk Publishers, The Hague/Boston/Lancaster.*

When CF_1 is incubated with ATP, the result is quite different depending whether Mg^{2+} is present or not in the incubation medium. After an incubation of ATP with CF_1, in the absence of Mg^{2+}, 2 moles of ADP are bound per mole CF_1 but ATP binding is very weak and reaches 0.5 mole per mole CF_1 at the highest external ATP concentration (not shown). In the presence of Mg^{2+} between 0 and 50 μM ATP addition, the amount of ADP bound to CF_1 increases up to 2 ADP per CF_1 and the amount of ATP bound attains 1 ATP per CF_1 (Fig. 2). Above 50 μM ATP addition, the amount of bound ADP decreases progressively to 1 ADP per CF_1 and the amount of bound ATP increases up to 2 ATP per CF_1. Apparently 1 ATP replaces 1 ADP on the CF_1. This hypothesis was confirmed.

3.2. <u>Binding and exchange of Mg^{2+} on CF_1. Effect of the nucleotides.</u>
<u>Effect of Mg^{2+} on the nucleotide exchange.</u> The number of Mg^{2+} ions bound to CF_1 has been measured before and after incubation with $Mg(NO_3)_2$ in the presence or absence of ADP or ATP. After extensive dialysis of CF_1 and in the absence of added $Mg(NO_3)_2$ or nucleotides, the bound magnesium is sometimes hardly measurable and always less than 0.6 Mg^{2+} per CF_1.

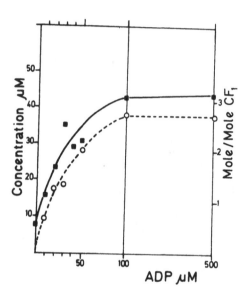

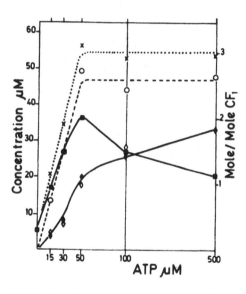

FIGURE 1.   Plot of the amount of ADP bound to CF_1 versus the initial concentration of added [^{14}C]ADP, in the presence of Mg^{2+}. The CF_1 (14 μM) was incubated in 20 mM Tris/HCl (pH 8), 0.5 mM EDTA, 1 mM $Mg(NO_3)_2$ and [^{14}C]ADP. After 1 h the [^{14}C]AdN–CF_1 complex was separated from the free [^{14}C]ADP by centrifugation-filtration. Bound ADP (■) and bound [^{14}C]ADP (O) were expressed by their concentrations (left-hand scale) or the number of molecules per CF_1 (right-hand scale).

FIGURE 2. Plot of the amount of ADP or ATP bound to CF_1 versus the initial concentration of [γ-^{32}P,U-^{14}C]ATP, in the presence of 1 mM Mg^{2+}. The CF_1 (18 M) was incubated, during 1 h, in 20 mM Tris/HCl (pH 8), 0.5 mM EDTA and a variable concentration of [γ-^{32}P, ^{14}C]ATP as indicated on the horizontale scale. Bound nucleotides were expressed by their concentrations (left-hand scale) or the number of moles per CF_1 (right-hand scale) : ATP (◆) ; ADP (■) ; ATP+ADP (x) ; [^{14}C]AdN (O) ; and [^{32}P]AdN (◊).

After incubation of CF_1 with a magnesium salt, the number of cations bound increases to a value close to one Mg^{2+} per CF_1. If nucleotides are added concomitantly with Mg^{2+}, this value increases to 1.25 Mg^{2+} per CF_1 in the presence of ADP and 2-3 Mg^{2+} per CF_1 in the presence of ATP. In this last case the amount of bound Mg^{2+} runs parallel to the total amount of nucleotides (ATP + ADP) bound to CF_1. The Mg^{2+} bound exchangeability was tested in the presence or absence of nucleotides during a five minutes period, convenient with the short half life of ^{27}Mg (9.45 min.). After 5 minutes, the exchange extent for Mg^{2+} and nucleotides respectively is : 12 % and 28 % in the presence of free Mg^{2+} and ATP 51 % and 46 % in the presence of free Mg^{2+} and ADP. In the same period 38 % of the bound Mg^{2+} is exchanged with the free Mg^{2+} in the absence of added nucleotides.

The effect of Mg^{2+} on the exchange of CF_1-bound nucleotides was studied after longer periods (0.5-22^H). The results obtained permit to classify the AdN/CF_1 complexes according to their decreasing stability : $Mg-ATP-CF_1$ > $ATP-CF_1$ > $Mg-ADP-CF_1$ > $ADP-CF_1$. Mg stabilizes the complex $AdN-CF_1$ and slows down the exchange of the bound nucleotide with the free ones.

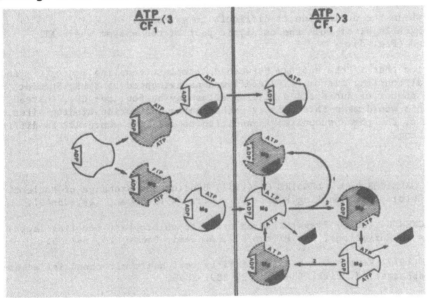

FIGURE 3. Final scheme of the hypothetical role of nucleotide binding sites on CF_1. The contours with interrupted lines represent transitory but short-lived and undetectable state. In the left part of the figure, CF_1 is supplemented with ATP at a concentration inferior to three times the CF_1 concentration and in the right part at a concentration more than three times the CF_1 concentration. $Mg(NO_3)_2$ is absent in the upper part and present in the lower part of the figure. Hydrolysis of ATP proceeds only in the right part of the scheme. In pathway 2, two sites hydrolyze alternatively ATP. In pathway 1, ATP is hydrolyzed always on the same site. The black blocks represent ADP bound to CF_1 or free in the medium.

When CF_1 was first loaded with $[^{14}C]ADP$, then freed from the excess of free nucleotides and supplemented with free unlabelled ATP, the bound ^{14}C-labelled nucleotides was progressively replaced by unlabelled nucleotides. But this exchange is never complete and there remains at least 0.4-0.6 ^{14}C-labelled nucleotides, most likely $[^{14}C]ADP$, per CF_1. So in addition to two sites where nucleotides are easy to exchange, there is one site on CF_1 where bound ADP is difficult to exchange with free nucleotides.

4. CONCLUSION

Our data give some information on the nature of the binding sites and on the effect of Mg^{2+}.
The first point is the presence of at least three nucleotide-binding sites on CF_1. In the presence of saturating amounts of ATP and Mg^{2+}, we find 2 ATP and 1 ADP bound per CF_1.

The second point concerns the exchange of the nucleotides bound on CF_1. On isolated CF_1, the latent enzyme, the binding sites fall into two different categories.
- One site where the nucleotide is difficult to exchange.
- Two exchangeable sites form the catalytic part of the enzyme where ATP is hydrolyzed (Fig. 3).

We must notice that if the analogy between chloroplast coupling factor CF_1 and mitochondrial coupling factor F_1 is confirmed as evidenced by Süss, Schmidt (1982), the number of nucleotides binding sites may be four per CF_1, instead of three. This would mean three easily exchangeable nucleotide binding sites, one on each of the three subunits, in addition to one site where ADP is difficult to exchange.

REFERENCES

GIRAULT G., GALMICHE J.M., LEMAIRE C. (1982) Binding and Exchange of Nucleotides on the Chloroplast Coupling Factor CF_1, Eur. J. Biochem., 128,405-411.

LIEN S., RACKER E. (1971) Preparation an assay of chloroplast coupling factor CF_1, Methods in Enzymology, San Pietro Ed., Academic Press, 23, 547-555.

PENEFSKY H. (1977) Reversible binding of Pi by beef heart mitochondrial adenosine triphosphatase, J. Biol. Chem., 252, 2891-2899.

SUSS K.H., SCHMIDT O. (1982) Evidence for an α_3, β_3, γ, δ, I, II, ε, III_5 subunit stochiometry of chloroplast ATP synthetase complex (CF_1-CF_0), FEBS Lett., 144, 213-218.

Authors adress : Service de Biophysique - CEN-SACLAY
                 91191 GIF-SUR-YVETTE CEDEX FRANCE

LOOSE AND TIGHT BINDING OF ADENINE NUCLEOTIDES BY MEMBRANE-ASSOCIATED
CHLOROPLAST ATPase

SUSANNE BICKEL-SANDKÖTTER, BOTANISCHES INSTITUT DER UNIVERSIÄT DÜSSELDORF

1. INTRODUCTION

Chloroplast ATPase (CF_O-CF_1 complex) contains two different species of bound
adenine nucleotides, tightly bound and exchangeable ("loosely bound") nucleotides. Release of tightly bound nucleotides takes place only when the thylakoids are energized (Strotmann et al. 1976, Strotmann, Bickel-Sandkötter 1977)
and this reaction has been proposed to be a prerequisite for activation of the
ATPase in the process of photophosphorylation (Gräber et al. 1977). This was
supported by results which show that the steady state level of tightly bound
nucleotides is inversely related to the rate of photophosphorylation when the
light intensity is varied (Strotmann et al. 1981). Exchangeable nucleotides
are assumed to represent the ADP and ATP involved in the catalytic process
(Bickel-Sandkötter, 1983). In the following the functional role of the two
species of bound nucleotides is further investigated.

2. MATERIAL AND METHODS

Steady state levels of tigthly bound and loosely bound ^{14}C-labeled nucleotides in illuminated chloroplasts were determined by the quench-centrifugation
technique in an ADP regenerating hexokinase system as described (Bickel-Sandkötter, Strotmann 1981). Photophosphorylation was assayed under the same conditions and electron transport was followed by reduction of ferricyanide
(Tischer, Strotmann 1977).

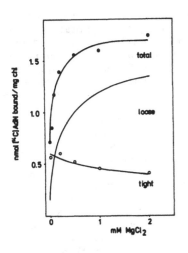

Fig. 1. Steady state levels of (^{14}C) adenine
nucleotides tightly bound and loosely bound to
thylakoid membranes as a function of Mg^{2+} concentration. The incubation medium contained
25 mM tricine buffer, pH 8.0, 50 mM NaCl,
50 µM PMS, 5 mM P_i, 10 µM (^{14}C)ADP, 30 U/ml
hexokinase (salt-free, Sigma), 10 mM glucose,
the indicated concentrations of Mg^{2+} as chloride and 0.4 mg chl/ml. The samples (0.25 ml)
were illuminated (150 Wm^{-2}) for 1 min in
plastic centrifugation tubes in a Beckman
Microfuge 152 before the chloroplasts were
spun down in the light. The quench solution
for determination of tightly bound nucleotides
contained 50 µM FCCP and 10 mM ADP (final con-
dentrations after addition). Other experimental details as described (Bickel-Sandkötter,
Strotmann 1981).

Sybesma, C. (ed.), Advances in Photosynthesis Research, Vol. II. ISBN 90-247-2943-2.
© *1984 Martinus Nijhoff/Dr W. Junk Publishers, The Hague/Boston/Lancaster.*

3. RESULTS AND DISCUSSION

By employing the ADP-FCCP quench centrifugation technique, the percentage of tightly bound nucleotides from total bound nucleotides can be measured. By substration the loosely bound (=exchangeable) nucleotides are calculated. Fig. 1 shows that the level of exchangeable nucleotides is increased by increasing the concentration of Mg^{2+} ions, while the level of tightly bound nucleotides is slightly decreased.

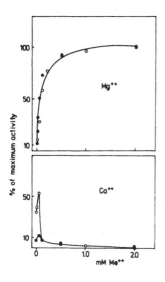

Fig. 2. Steady state levels of loosely bound (^{14}C)adenine nucleotides (o) and steady state rate of photophosphorylation (o) as a function of Mg^{2+} and Ca^{2+} concentration. Both reactions were measured under the same conditions, except that for phosphorylation the medium contained ^{32}P-labeled P_i and unlabeled ADP. The maximum level of loosely bound nucleotides was 1.12 nmol/mg chl, the maximum phosphorylation rate was 133 μmol/mg chl/h.

Fig. 2. shows that both, the rate of photophosphorylation and the level of loosely bound nucleotides show the same dependence on Mg^{2+} concentration. Moreover, both processes are inhibited by Ca^{2+} ions in the same manner. This result supports the view that the exchangeable nucleotides are involved in the catalytic process and coincides with the finding that 3'(2')-O-(napthoyl-1-) ADP, a competitive inhibitor of ADP phosphorylation in chloroplasts (Franek, Strotmann 1981) specifically reduces loose binding of ADP in energized chloroplasts (Bickel-Sandkötter 1983).

The functional role of tightly bound nucleotides has recently been investigated by several groups. It has been demonstrated that energy-dependent release of nucleotides is related to activation of ATPase in the process of light-triggered ATP hydrolysis (Schumann, Strotmann 1981) as well as in photophosphorylation (Gräber et al. 1977, Schlodder, Witt 1981, Strotmann et al. 1981). Inactive enzyme molecules are unable to perform catalytic binding, suggesting that either the active site is blocked by a tightly bound nucleotide molecule or that it is occluded in the inactive state of the enzyme. The steady state level of tightly bound nucleotides is inverse to the rate of phosphorylation as a function of light intensity and uncoupler concentration (Strotmann et al. 1981). An inverse relationship between these two parameters is also observed when Mg^{2+} as a cofactor in phosphorylation is replaced by other divalent cations (Table I). With the employed metal ions the steady state level of tightly bound nucleotides increases in the order $Mg^{2+} < Mn^{2+} < Co^{2+} < Ca^{2+} < Cu^{2+}$, while the rates of photophosphorylation and concomitant electron transport decrease in the same order.

TABLE 1. Effect of metal ions on the steady state levels of tightly bound
nucleotides in the light, on photophosphorylation and on ADP+P_i-stimulated
electron transport. Tightly bound nucleotides were determined by measuring
the radioactivity retained in the washed chloroplasts. The chloroplasts
were incubated with 5 μM (^{14}C)ADP as described (Strotmann et al. 1979)
and quenched after 1 min of illumination by the addition of FCCP and ADP
containing quench solution (Strotmann et al. 1979) in the light.

Metal ion	Tightly bound nucleotides	Electron transport (coupled - basal)	Photophos-phorylation
(1mM)	nmol/mg chl	μmol FeCy/mg chl/h	μmol ATP/mg chl/h
Mg^{2+}	0.0853	285.0	140.0
Mn^{2+}	0.0926	194.9	70.0
Co^{2+}	0.1889	83.5	50.0
Ca^{2+}	0.2109	27.9	25.0
Cu^{2+}	0.3747	0.0	18.0

Since the sum of tightly bound and loosely bound nucleotides is about one per
CF_1 at ADP saturation and the inactive ATPase molecules obviously are not
accessible for medium nucleotides, it may be concluded that at a given time
only one binding site is occupied by either a tightly bound or a catalytically
bound nucleotide molecule (Bickel-Sandkötter 1983). The tight site may be a
different conformation of the catalytic site, and the shift between the two
forms may be controlled by energy input and other external factors, e.g. metal
ions. This interpretation is supported by the finding that under phosphorylating conditions part of tightly bound nucleotides exist as ATP even in presence of a regenerating hexokinase system (Bickel-Sandkötter 1983). Furthermore, the formation of tightly bound (^{32}P)ATP from medium ADP and (^{32}P)P_i has
been reported which is likewise insensitive to hexokinase (Rosen et al. 1979,
Aflalo, Shavit 1982). Accordingly this ATP was not transferred from a catalytic
site to a tight site through the medium. Interconversion between catalytic and
tight sites may be a feasible explanation for these results. The catalytic site
may be converted to a tight site when the enzyme is inactivated. By this conformational change the nucleotide molecule which resides on the site (ADP or
ATP) is trapped and prevented from further reactions.

REFERENCES

Aflalo C and Shavit N (1982) Source of rapidly labeled ATP tightly bound
to non-catalytic sites on the chloroplast ATP synthetase. Eur. J. Biochem.
126, 61-68
Bickel-Sandkötter S (1983) Loose and tight binding of adenine nucleotides
by membrane-associated chloroplast ATPase. Biochim. Biophys. Acta 723, 71-77
Bickel-Sandkötter S and Strotmann H (1981) Nucleotide binding and regulation
of chloroplast ATP synthase. FEBS Lett. 125, 188-192
Franek U and Strotmann H (1981) Nucleotide specificity of CF_1-ATPase in
ATP synthesis and ATP hydrolysis. FEBS Lett. 126, 5-8
Gräber P, Schlodder E and Witt HT (1977) Conformational change of the chloroplast ATPase induced by a transmembrane electric field and its correlation
to phosphorylation. Biochim. Biophys. Acta 461, 426-440

Rosen G, Gresser M, Vinkler C and Boyer PD (1979) Assessment of total cataly[t] sites and the nature of bound nucleotide participation in photophosphorylati[c] J. Biol. Chem. 254, 10654-10661

Schlodder E and Witt HT (1981) Relation between the initial kinetics of ATP synthesis and of conformational changes in the chloroplast ATPase studied by external field pulses. Biochim. Biophys. Acta 635, 571-584

Strotmann H and Bickel-Sandkötter S (1977) Energy-dependent exchange of aden[i] nucleotides on chloroplast coupling factor (CF_1). Biochim. Biophys. Acta 460, 126-135

Strotmann H, Bickel-Sandkötter S, Franek U and Gerke V (1981) Nucleotide int[e] action with membrane-bound CF_1. In Energy Coupling in Photosynthesis (ed. BR Selman and S Selman-Reimer) pp 187-196, Elsevier, Amsterdam

Strotmann H, Bickel S and Huchzermeyer B (1976) Energy-dependent release of adenine nucleotides tightly bound to chloroplast coupling factor CF_1. FEBS Lett. 61, 194-198

Strotmann H, Bickel-Sandkötter S and Shoshan V (1979) Kinetic analysis of light-dependent exchange of adenine nucleotides on chloroplast coupling fact[o] CF_1. FEBS Lett. 101, 316-320

Tischer W and Strotmann H (1977) Relationship between inhibitor binding by chloroplasts and inhibition of photosynthetic electron transport. Biochim. Biophys. Acta 460, 113-125

Authors address: Susanne Bickel-Sandkötter, Botanisches Institut der Universität Düsseldorf, D-4000 Düsseldorf, F.R.G.

ACTIVATION OF THE LIGHT TRIGGERED CHLOROPLAST ATPase - ROLE OF NUCLEOTIDES AND AN INHIBITORY PEPTIDE

J. ANDRALOJC, K. SIMPSON AND D.A. HARRIS
Department of Biochemistry, University of Leeds, Leeds, LS2 9JT, U.K.

ABSTRACT

The dark decay of the trypsin and DTT activated ATPase of pea chloroplasts were compared. The similarity between their decay curves, the stimulation of the decay by ADP and its inhibition by pyrophosphate led us to conclude that ADP, rather than an inhibitor protein is the major regulator in both cases.

INTRODUCTION

Chloroplast membranes, both from lettuce (Carmeli and Lifshitz, 1972) and spinach (Lynn and Straub, 1969) bear a latent ATPase which can be activated by illumination. The ATPase activity can be observed for several minutes after the light is switched off ('light-triggered ATPase') if illumination is carried out in the presence of dithiols or trypsin.

If a period of darkness is interposed between illumination and initiation of ATP hydrolysis with MgATP, then the rate of ATP hydrolysis measured is decreased i.e. the activated state of the ATPase decays to an inactive state in the dark. This transition, from active to inactive ATPase, is opposed by energisation of the membrane (by light or ATP hydrolysis). These changes are represented schematically in Fig. 1.

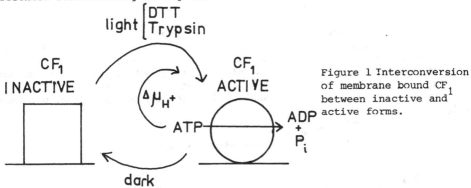

Figure 1 Interconversion of membrane bound CF_1 between inactive and active forms.

An activation-inactivation cycle such as this occurs also in pea chloroplasts (see below). We have investigated the dark decay of the light triggered ATPase in pea chloroplasts in an attempt to define the physical processes responsible for the inactivation. Two possibilities have been suggested, either ADP binding tightly to a regulatory site on CF_1 (Schumann & Strotmann, 1981) or an inhibitory peptide binding to its inhibitory site (Nelson et al., 1972) may be involved.

The ability of agents so diverse as trypsin (which causes irreversible peptide cleavage) and dithiothreitol (DTT), (which presumably leads to a reduction of disulphide bridges) to activate the ATPase allows us to differentiate between these possibilities. We show below that light triggered ATPase activity

Sybesma, C. (ed.), Advances in Photosynthesis Research, Vol. II. ISBN 90-247-2943-2.
© *1984 Martinus Nijhoff/Dr W. Junk Publishers, The Hague/Boston/Lancaster.*

induced in the presence of trypsin is very similar to that induced with DTT, as regards time course of dark decay and sensitivity to uncouplers, ADP and phosphate derivatives. This suggests that the dark decay of ATPase activity is not mediated by an inhibitor protein, which would be expected to be sensitive to trypsin (Nelson et al., 1972).

METHODS

Once washed, type C chloroplasts (Hall, 1972) were prepared by conventional procedures from young pea leaves (Pisum sativum var Feltham first). Chloroplast ATPase was measured in a thermostatted glass vessel fitted with a pH-sensitive electrode. The medium contained 50 mM KCl, 2 mM tricine-NaOH, pH 8.0, 5 mM $MgCl_2$, 0.1 mM EDTA, 5 μM pyocyanine and 1 mM KP_i and 10-20 μg chloro phyll/ml. For DTT activations the medium also contained 4 mM DTT and was illuminated for 4 minutes, while for trypsin activation, illumination was with 1 μg/ml trypsin for 30 seconds, followed by the addition of an excess of trypsin inhibitor. Light was provided by a 150 W projector lamp. After the indicated periods of darkness (see each figure), MgATP was added to a final concentration of 2 mM and the resulting ATPase activity monitored by fall in pH. Basal activity was that without prior illumination.

RESULTS AND CONCLUSIONS

The decay of the active state of the light-triggered ATPase, in pea chloroplas membranes, is shown in Fig. 2. The process of decay is apparently biphasic - full activity is retained for several seconds, and the activity then declines precipitously. A similar biphasic decay has been reported (Bar-Zvi & Shavit, 1981) for the ATP-P_i exchange reaction in lettuce chloroplasts.

The period during which full activity is maintained (the 'plateau period'), is very sensitive to temperature, increasing some 4-5 fold between 30°C and 20°C, while the subsequent decline is much less so. Fig. 2 relates to trypsin treated chloroplasts - DTT treated chloroplasts yield almost indistinguishable curves. The plateau period seems to represent activity maintained by an energised membrane, since it is abolished by low concentrations of uncoupler (Fig. 3).

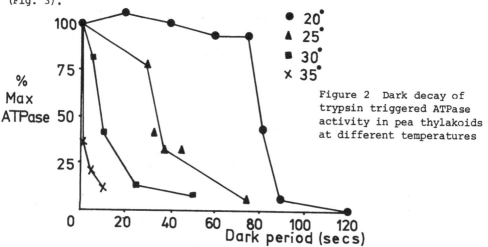

Figure 2 Dark decay of trypsin triggered ATPase activity in pea thylakoids at different temperatures

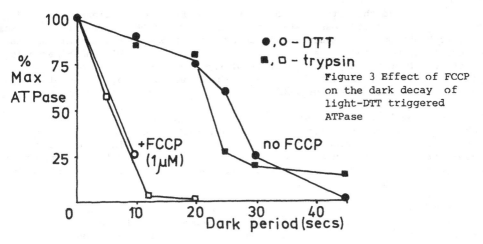

Figure 3 Effect of FCCP on the dark decay of light-DTT triggered ATPase

Low levels of ADP (≤10 μM) speed up the dark decay process while pyrophosphate slows it down considerably (Fig. 4). Phosphate also slows the decay somewhat but is much less effective than pyrophosphate (not shown). These results are in agreement with the previous findings of Carmeli and Lifshitz (1971) and Dunham (1981). Again, both these effects are seen in both DTT and trypsin-treated chloroplasts.

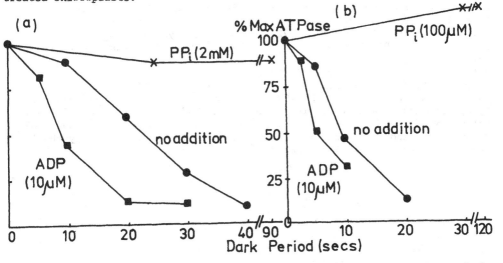

Figure 4 Effect of either ADP or sodium pyrophosphate on the ATPase dark decay of DTT (a) or trypsin (b) activated chloroplasts.

The dark decay of ATPase activity observed even in trypsin-treated chloroplasts suggests that a protein is not the primary regulator in this process. The effects of ADP on decay rates suggest that ADP may indeed be the primary regulator, its binding to one of the 'tight binding sites' on CF_1 (Harris and Slater, 1975) being the cause of activity loss.

We also monitored the effect of adding various ADP concentrations, measured after 5 seconds of darkness. While the DTT-activated ATPase fell to its basal value at high concentrations of ADP (200 μM) the trypsin-activated ATPase was reduced to no more than a half of its uninhibited value. This non-hyperbolic effect of ADP on the decay of activity of the trypsin-activated ATPase may reflect the multiple effects of trypsin, which lead to a population of chloroplast membranes heterogeneous with respect to affinity for ADP binding.

Why, then, does the ATPase decay even in the absence of added ADP? The answer probably lies in the presence of tightly bound ADP in untreated, dark adapted membranes. On illumination, this ADP seems to be released into the medium, and on de-energisation it can bind back to its binding site - unless pyrophosphate is present, which competes successfully with ADP at this site. Pyrophosphate thus leads to a prolonged ATPase activity.

These conclusions are in accord with those of other workers (Dunham, 1981; Bar-Zvi and Shavit, 1981; Schumann and Strotmann, 1981), but should be contrasted with those obtained in submitochondrial particles, where energetic stimulation of ATPase activity is associated with the displacement of a small inhibitor protein and not with the loss of tightly bound ADP (see Harris, 1978).

REFERENCES

Bar-Zvi, D. and Shavit, N. (1981) Proc. 5th Int. Congr. Photosynthesis. Part II (Akoyunoglou ed.) pp. 801-809.
Carmeli, C. and Lifshitz, Y. (1972) Biochim. Biophys. Acta, 267, 86-95.
Dunham, K.R. (1981) Energy Coupling in Photosynthesis (Selman and Selman Reimer eds.) 209-221.
Hall, D.O. (1972) Nat. New Biol. 235, 125-126.
Harris, D.A. (1978) Biochim. Biophys. Acta, 463, 245-273.
Harris, D.A. and Slater, E.C. (1975) Biochim. Biophys. Acta, 387, 335-348.
Lynn, W.S. and Straub, K.D. (1969) Proc. Natl. Acad. Sci. U.S. 63, 540-547.
Nelson, N., Nelson, H. and Racker, E. (1972) J. Biol. Chem. 247, 7657-7662.
Schumann, J. and Strotmann, H. (1981) Proc. 5th Int. Congr. Photosynthesis. Part II (Akoyunoglou ed.) pp. 881-892.

LIMITED ACCESS OF NUCLEOTIDES TO THE ACTIVE SITE OF ATP SYNTHETASE DURING PHOTOPHOSPHORYLATION

CLAUDE AFLALO and NOUN SHAVIT, Dept. of Biology, Ben-Gurion University of the Negev, Beer Sheva 84105, Israel

1. INTRODUCTION

Studies on steady state kinetics of enzymes bound to energy-transducing membranes have generally tacitly assumed that their kinetic behavior does not depart greatly from that of enzymes in free solution. However, the microenvironment of these membranes, upon which electrochemical changes occur during catalysis, should be different than an ideal aqueous solution which allows rapid diffusion of solutes. Since different kinetic principles apply to enzymes immobilized on solid supports, due to impaired transloca-tion of reactants near the support (Engasser, Horvath, 1976), the macro-scopic behavior of such enzymes does not **necessarily** reflect intrinsic properties relevant to the molecular mechanism of catalysis (conformational changes, site-site cooperativity, etc.).

We proposed recently that during ATP synthesis by illuminated chloro-plast thylakoids, accumulation of ATP tightly bound to ATP synthetase could arise from rebinding of a free species of ATP which diffuses slowly to the bulk medium, even in the presence of hexokinase (Aflalo, Shavit, 1982). In this communication and elsewhere (Aflalo, Shavit, 1983, a & b) we estimated the catalytic ability (V_{max}) of the ATP synthetase and its affinity for substrates (K_m and V_{max}/K_m) under various experimental conditions which should modify the microenvironment of the enzyme. The results indicate that photophosphorylation may be limited by a relatively slow translocation of nucleotides near the membrane during catalysis.

2. EXPERIMENTAL

Control thylakoids were isolated from lettuce by conventional proce-dures and washed once in 0.4 M sucrose. Hypotonically treated thylakoids were washed twice in 0.5 and 0.1 mg/ml BSA and isolated by differential centrifugation. All preparations were resuspended in 0.4 M sucrose. Standard assay mix (in 1 ml) for steady state phosphorylation contained (in μmol): K-tricine (pH 8.0), 50; sucrose, 140; PMS, 0.04; Na-EDTA, 1.5; Mg-tricine, 6.5; $^{32}P_i$, 10; ADP, 0.004-0.25; glucose, 20; hexokinase, 20 units; BSA, 0.5 mg and thylakoids (8-14 μg chl.). Reaction mixes were illuminated for 2 min and organic ^{32}P determined after deproteinization.

3. RESULTS AND DISCUSSION

The macroscopic reaction we tested consists of several microscopic steps including phosphorylation of ADP on the thylakoid membrane, recycling of ATP by saturating amounts of hexokinase in the bulk medium and the translocation of reactants between both phases. The latter processes, as-sumed to be fast in an aqueous environment, could be impaired near the membrane because of its physical properties (Rubin, Barber, 1980).

3.1 Effect of hypotonic treatment of thylakoids

In order to check the existence of a barrier between the ATP synthetase loci and bulk medium we submitted thylakoids to a transient hypotonic treat-ment and compared their kinetic properties to those of control thylakoids.

Sybesma, C. (ed.), Advances in Photosynthesis Research, Vol. II. ISBN 90-247-2943-2.
© *1984 Martinus Nijhoff/Dr W. Junk Publishers, The Hague/Boston/Lancaster.*

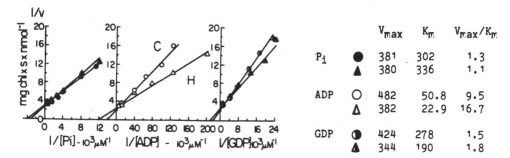

		V_{max}	K_m	V_{max}/K_m
P_i	●	381	302	1.3
	▲	380	336	1.1
ADP	○	482	50.8	9.5
	△	382	22.9	16.7
GDP	◑	424	278	1.5
	▲	344	190	1.8

FIGURE 1. Effect of hypotonic treatment on P_i, ADP or GDP utilization. Conditions as in Experimental. With P_i and GDP, control (circles) or treated (triangles) thylakoids were illuminated for 10-30s. (Taken from Aflalo, Shavit, 1983b). Table: calculated parameters.

As shown in Fig. 1, hypotonic treatment leads to a better utilization of ADP and, to a lesser extent, GDP (lower K_m, higher V_{max}/K_m) at virtually the same catalytic ability (V_{max}). The utilization of P_i remains nearly unaffected. Under the saturating conditions used, the apparent interaction of ADP with the enzyme is fast ($k_{cat}/K_m = 1-1.5 \times 10^7 M^{-1} \times s^{-1}$) and so the turnover of the enzyme ($k_{cat} = 300-400 s^{-1}$). Thus, hypotonic treatment and washing of thylakoids seems to relieve a kinetic limitation in the interaction between the enzyme and ADP. This limitation is apparent only at low [ADP].

3.2 Effect of ionic strength

Our washing and assay media contain relatively low salt concentrations in which thylakoids are probably swollen and unstacked. Washing control thylakoids with NaCl instead of sucrose does not affect ADP utilization in the standard assay (Fig. 2). However, increasing the ionic strength in the reaction mix (KCl, $MgCl_2$) results in an increased apparent affinity for ADP with control and, to lesser extent, treated thylakoids (Fig. 3). This effect resembles that of hypotonic treatment on control thylakoids. Nevertheless,

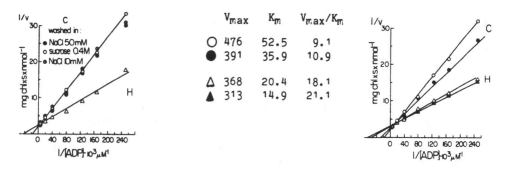

	V_{max}	K_m	V_{max}/K_m
○	476	52.5	9.1
●	391	35.9	10.9
△	368	20.4	18.1
▲	313	14.9	21.1

FIGURE 2. (Left) Effect of salt in the washing medium. Thylakoids were once washed (○ ,◑ ,●) or hypotonically treated (△) as indicated.
FIGURE 3. (Right) Effect of salt in the reaction mix. Conditions as in Experimental (○ △) or with 50 mM KCl and 10 mM $MgCl_2$ instead of sucrose, BSA and Mg-tricine (● , ▲). Table: calculated parameters.

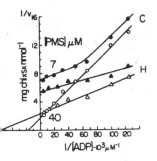

FIGURE 4. (Left) Effect of
low [PMS].Conditions as in
Experimental.
FIGURE 5. (Right) Effect of
nigericin. Conditions as in
Fig. 4.[Nig]=1 µM. Circles:
control; triangles: hypotonically treated thylakoids.

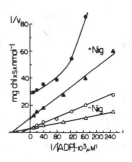

the hypotonic treatment further improves ADP utilization in the same conditions. These results suggest the existence of electrostatic interactions
between charged reactants and the fixed charges on the surface of thylakoids, which could be screened at high ionic strength (Rubin, Barber, 1980).

3.3 Effect of energy input

Energy input should also affect the microenvironment of the membrane.
When the proton motive force is lowered by varying light intensity (Aflalo,
Shavit, 1983a) or PMS concentration (Fig. 4), the dependence of phosphorylation in control thylakoids on [ADP] is biphasic, indicating two possible
microscopic steps limiting ADP utilization. Low [PMS] appears to limit the
reaction only at high [ADP]. On the other hand, with treated thylakoids,
linear and roughly parallel double reciprocal plots are observed. These
results indicate a transition in the apparent interaction of ADP with
control thylakoids at low proton motive force, from a relatively fast pseudo
second order process at high [ADP] to a slower one (lower V_{max}/K_m) at low
[ADP]. Since at low [ADP] the macroscopic rate of phosphorylation with
control thylakoids is nearly independent of the proton motive force, it must
be limited by another (slower) process which is more dependent on [ADP] in
the medium. This latter process, which seems to limit also the overall
reaction rate at saturating [PMS] (and light intensity) and is not observed
after hypotonic treatment, is likely to represent the relatively slow mass
transfer of nucleotides through the microenvironment of ATP synthetase bound
on control thylakoids.

When nigericin and K^+ are used to lower the electrochemical gradient of
protons, a transition in the slope is still observed with control and not
with treated thylakoids (Fig. 5). However, in both cases, the apparent
limiting rate constant for ADP (V_{max}/K_m) is reduced. At lower [nigericin],
the transition in the slope was not observed, but the apparent K_m for ADP
was lowered with control and not with treated thylakoids (Aflalo, Shavit,
1983a). The effect of uncoupler is different from that of low proton motive
force which did not affect the slope or the rate of reaction at low [ADP].

3.4 Effect of catalytic ability

If at low [ADP] the reaction rate is limited by a relatively slow
translocation of nucleotides near the enzyme compared with the rate of
catalysis in situ, this limitation should also be relieved by reducing the
catalytic ability of the enzyme independently of substrate concentration.
This is illustrated in Fig. 6, assuming that the limitation is proportional
to the stimulation observed after hypotonic treatment. These results show
that the rate of phosphorylation may be limited by either enzymatic catalysis per se, or by extrinsic environmental factors.

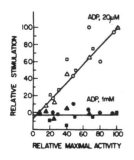

FIGURE 6. (Left) Correlation between the catalytic ability and the observed stimulation after hypotonic treatment. The rate of phosphorylation was controled by varying light intensity (O), [P$_i$] (□) or [phlorizin] (Δ). Maximal activity was taken as the velocity of control thylakoids at high [ADP]. Stimulation was calculated from the rates with control and treated thylakoids. (Taken from Aflalo, Shavit, 1983a).

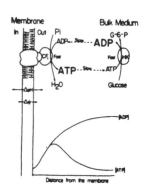

FIGURE 7. (Right) A model for catalysis of ATP formation near the thylakoid membrane coupled to the hexokinase reaction in the bulk medium at steady state. Bottom: hypothetical concentration profiles of nucleotides. (Taken from Aflalo, Shavit 1983b).

4. CONCLUSIONS

The apparent affinity of the membrane-bound ATP synthetase for substrate could be modulated by factors which probably affect the physical environment of the enzyme and not its catalytic ability. Our data indicate a kinetic limitation of the steady state rate of photophosphorylation coupled to the hexokinase reaction by a poor diffusibility of nucleotides through the layer adjacent to the thylakoid membrane. This phenomenon – apparent only with substrates efficiently transformed by the enzyme (high V_{max}/K_m) at a high turnover (V_{max}) – should lead to a relative depletion of ADP and/or accumulation of ATP near the enzyme, under steady state conditions (Fig. 7). Such effects could be highly relevant to the various models proposed for the molecular mechanism of ATP synthesis, based on affinity binding changes or "intermediate" and "medium" exchange reactions.

REFERENCES

Aflalo C and Shavit N (1982) Source of rapidly labeled ATP tightly bound to non catalytic sites on the chloroplast ATP synthetase, Eur. J. Biochem. 126, 61-68.
Aflalo C and Shavit N (1983a) Steady state kinetics of photophosphorylation: limited access of nucleotides to the active site on the ATP synthetase, FEBS Lett. 154, 175-179.
Aflalo C and Shavit N (1983b) A new approach to the mechanism of photophosphorylation: modulation of ATP synthetase activity by limited diffusibility of nucleotides near the enzyme. In De Luca M, ed. Mechanism of enzyme action and regulation, in press. New York: Acad. Press.
Engasser J-M and Horvath C (1976) Diffusion and kinetics with immobilized enzymes. In Wingard LB, Katchalski-Katzir E and Goldstein L eds. Applied biochemistry and bioengineering, pp.127-220. New York: Acad. Press.
Rubin BT and Barber J (1980) The role of membrane surface charge in the control of photosynthetic processes and the involvement of electrostatic screening, Biochim. Biophys. Acta, 592, 87-102.

ACKNOWLEDGEMENT

We wish to thank Mrs. Zvia Conrad for excellent technical assistance.

ACCESSIBILITY AND FUNCTION OF CF_O-SUBUNITS IN CHLOROPLAST THYLAKOIDS

LUDGER KLEIN-HITPASS and RICHARD J. BERZBORN

Biochemie d. Pflanzen, Ruhr-Universität, D-4630 Bochum, FRG

1. INTRODUCTION AND PROBLEM

The ATPsynthase complex of chloroplast thylakoid membranes is composed of an extrinsic unit (CF_1) with latent ATPase activity and of an intrinsic membrane unit (CF_O) which functions in proton translocation. CF_1 consits of five nonidentical subunits, termed α, β, γ, δ and ε. Purified CF_1-CF_O complex extracted by cholate/octylglucoside (Pick U, Racker E, 1979) which is capable of energy transducing reactions after incorporation into liposomes shows at least eight different peptides; therefore it is assumed that the CF_O is composed of three different kinds of subunits, designated as I,II,III. Subunit III, the 8 kd DCCD binding proteolipid, plays a crucial role in the H^+ translocation process through CF_O (Nelson N et al., 1977). The function and authenticity of the other peptides in the membrane part of the CF_1-CF_O complex still remain to be established. Peptide I might have a function in binding of the peripheral CF_1 and peptide II might play a role in the assembly of six copies of subunit III to an active proton channel (Nelson N, 1981). For the TF_1-TF_O complex of the thermophilic bacterium PS3 the TF_1 binding site could be ascribed to a 13500 dalton polypeptide of TF_O experimentally (Sone N et al., 1978).

In order to get information about the function and also the structural arrangement of the CF_O-subunits, we produced special antisera against the CF_O complex and subunit III, respectively, and used them as hydrophilic probes for accessibility and as inhibitors of photosynthetic reactions.

2. RESULTS

2.1. Effects of antibodies against subunit III

By immunization of rabbits with isolated DCCD binding proteolipid (containing SDS) we gained a highly specific precipitating antiserum. This antiserum did not agglutinate isolated thylakoids, neither inhibit in control thylakoids photophosphorylation with PMS and H^+ efflux, nor did it influence in EDTA-treated thylakoids the extent of H^+ uptake or the rebinding of CF_1 to these depleted membranes. Of course we do not take these negative results as evidence for inaccessibility of the proteolipid from the matrix side, especially since antibodies against subunit III in an antiserum against nondenatured CF_O complex seemed to be absorbed by control thylakoids. The reason for the failure of a proof for accessibility of subunit III with the antiserum against subunit III could be that the antibodies against the denatured peptide did not recognize the native protein. Also we cannot exclude the possibility that the antiserum against the isolated subunit did not contain any antibodies against exposed parts of the peptide.

2.2. Effects of antibodies against peptides I and II

2.2.1. Production and specificity of antisera against CF_O. CF_1-CF_O complex purified from Triton X-100 extract of thylakoids by anion-exchange chromatography and sucrose density centrifugation, and ATPsynthase complex from Triton X-100 dissolved thylakoids immunoprecipitated by anti CF_1 showed the same subunit

Sybesma, C. (ed.), Advances in Photosynthesis Research, Vol. II. ISBN 90-247-2943-2.
© *1984 Martinus Nijhoff/Dr W. Junk Publishers, The Hague/Boston/Lancaster.*

composition as cholate/octylglucoside extracted CF_1-CF_0 complex (Klein-Hitpaß, Berzborn RJ, to be published). By immunization with this isolated or immunoprecipitated CF_1-CF_0, respectively, and subsequent absorption of the antibodies against CF_1, we produced monospecific precipitating antisera against the CF_0 complex of spinach thylakoids. As could be demonstrated by the immunoblotting technique, antiserum 187 contained antibodies against peptid II of the CF_0 only, in antiserum 169 antibodies against peptides I, II and III were still present.

2.2.2. Effects of these antisera on photosynthetic reactions (table 1)

A antiserum	169		187	
immunogen	isolated CF_1CF_0		immunoprecipitated CF_1CF_0	
antibodies against CF_0-peptides	I, II, III		II	
absorbed by:	CF_1	CF_1 and	CF_1	CF_1 and
	–	thylakoids	–	thylakoids
B activities:	effects:			
in control thylakoids:				
photophosphorylation	inhibited	uneffected	inhibited	uneffected
H^+ efflux	n.m.	n.m.	retarded	n.m.
extent of H^+ uptake	uneffected	n.m.	uneffected	n.m.
in EDTA-treated thylakoids:				
photophosphorylation	stimulated and inhibited	stimulated	inhibited bi-phasically	stimulated
H^+ efflux	n.m.	n.m.	n.m.	n.m.
extent of H^+ uptake	stimulated	n.m.	stimulated	n.m.

n.m. = not measured or not measurable

TABLE 1. Effects of nonabsorbed and absorbed antisera (A) against CF_0 on energy transformation reactions (B)

In detail:
Antiserum 187 inhibited photophosphorylation with PMS in control thylakoids completely in a hyperbolic manner. The inhibition was not even partially neutralized by preincubation of the antiserum with an excess of isolated CF_1. The inhibition was not accompanied by displacement of CF_1 from its binding site, since no CF_1 could be detected in the supernatants by rocket electroimmunodiffusion analysis. The antiserum retarded the decay of the light induced pH rise after illumination in control thylakoids, and stimulated the extent of H^+ uptake in EDTA-treated thylakoids almost up to the control, whereas H^+ uptake in control thylakoids was uneffected. The photophosphorylation in EDTA-treated thylakoids with a high amount of residual CF_1 (about 75%) was inhibited too, but in contrast to control thylakoids the titration curve was biphasic.

With antiserum 169 similar results were obtained. Low amounts of antiserum even stimulated the photophosphorylation capacity of EDTA-treated thylakoids, higher amounts inhibited.

After absorption of the inhibiting antibodies by control thylakoids in both antisera very interesting antibodies remained: They exclusively stimulated the photophosphorylation in EDTA-treated thylakoids (residual CF_1 about 75%, residual ATPsynthase activity 47%). Of the double absorbed antiserum 169 already small amounts stimulated the photophosphorylation capacity of these EDTA-treated membranes up to 70% of the control activity. Thus the recoupled activity coincided with the percentage of residual CF_1.

Comparative immunoblotting analyses proved that some antibodies against peptides I, II and probably III in the CF_0-antisera can be absorbed by control thylakoids, additional ones against I and II by EDTA-treated membranes only.

3. CONCLUSIONS AND DISCUSSION

Antibodies against peptide II of the CF_0 complex proved to be potent inhibitors of ATP synthesis and H^+ efflux. This indicates, that besides the proteolipid subunit at least peptide II is a constituent of the functional CF_0 complex. The most straightforward explanation for the inhibition of H^+ conduction through CF_0 by anti-II is that this subunit is involved in the proton translocation process across the membrane. In this case the antibody inhibition would be caused by inducing an inactive conformation of subunit II.

Since the antibodies against subunit II did not release CF_1 from the membrane, from the inhibition of photophosphorylation and the retardation of the decay of the light induced pH rise the conclusion can be drawn, that part of the polypeptide chain of subunit II is accessible from the matrix side in the control thylakoid membrane. However, another part of the peptide seems to be covered by CF_1 (fig.1), because antibodies, which were not absorbable by control thylakoids, caused a stimulation of photophosphorylation in EDTA-treated membranes; this stimulation is probably due to inhibition of uncoupled H^+ efflux through those CF_0 complexes which had lost their CF_1 during EDTA-treatment (DCCD effect).

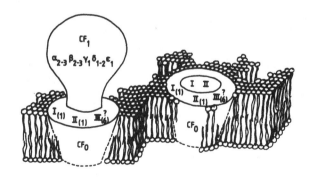

FIGURE 1. Model indicating the structural arrangement of the subunits of the membrane part of the ATPsynthase complex

Blotting analysis of the absorbed antisera confirmed these conclusions, i.e.
partial accessibility and partial inaccessibility of subunit II. From blotting
experiments the same conclusion can be drawn also for peptide I (see fig. 1)

Therefore both subunits I and II of the CF_0 complex could be involved in
CF_1 binding. CF_0 is wider than CF_1 (fig.1). For subunit II the results even
suggest a participation in H^+ translocation. Alternatively already subunit II
could undergo a conformational change upon energization of the thylakoid, thus
inducing in a second step an energy-rich conformation in the CF_1 part of the
ATPsynthase complex (Sebald W, Hoppe J, 1982). In this case the mechanism of
the antibody inhibition would be a fixation of the not energized dark conformation not conducting protons, by the antibodies reacting with exposed parts of
the membrane embedded CF_0.

REFERENCES

Nelson N (1981) Proton-ATPase of chloroplasts. In: Current topics in bioenergetics, Sanadi DR and Vernon LP,eds. vol.11, p. 1-33, Academic Press, New York

Nelson N, Eytan E, Notsani B-E, Sigrist H, Sigrist-Nelson K, Gitler C (1977)
Isolation of a chloroplast N,N'-dicyclohexylcarbodiimide-binding proteolipid
active in proton translocation. Proc.Natl.Acad.Sci.USA 74, 2375-2378

Pick U, Racker E (1979) Purification and reconstitution of the N,N'-dicyclo-
hexylcarbodiimide-sensitive ATPase complex from spinach chloroplasts,
J. Biol. Chem. 254, 2793-2799

Sebald W, Hoppe J (1982) On the structure and genetics of the proteolipid subunit of the ATPsynthase complex. In: Current topics in bioenergetics, Sanadi
DR,ed., vol. 12, p. 1-64, Academic Press, New York

Sone N, Yoshida M, Hirata H, Kagawa Y (1978) Resolution of the membrane moiety
of the H^+-ATPase complex into two kinds of subunits.
Proc. Natl. Acad. Sci. USA 75, 4219-4223

ACKNOWLEDGEMENT

Reliable technical assistance of Mrs. R. Oworah-Nkruma is gratefully acknowledged.

A Partial Characterization of the Halotolerant, Green Algal Dunaliella bardawil CF₁ ATPase and its Modification by Fluorescein Isothiocyanate

Susanne Selman-Reimer, Moshe Finel, Uri Pick and Bruce R. Selman

Introduction

Coupling factor one (CF₁) complexes have been isolated and characterized from a wide variety of photosynthetic organisms. These include bacteria (Beechy et al., 1975), thermophilic cyanobacteria (Binder, Bachofen, 1979), unicellular, eukaryotic green algae (Selman-Reimer et al., 1981), and higher (vascular) plants (Nelson, 1976). D. bardawil is a halotolerant, unicellular, eukaryotic green alga that is capable of growing under extreme environmental stress in the presence of NaCl concentrations as high as 5 M. In order to maintain its integrity, the alga accumulates massive amounts of glycerol as its osmoticum (Wegmann et al., 1980). Because the glycerol concentration in the chloroplast stroma is unlikely to vary much from that of the cytosol, it was of interest to us to compare the properties of the D. bardawil CF₁ (DCF₁) to other coupling factors not usually required to function under such extreme physiological conditions. In this communication, we characterize the enzymatic properties of the DCF₁, identify the subunits of the ATPase as resolved by SDS-PAGE, and describe the modification of DCF₁ by fluorescein isothiocyanate (FITC).

Methods and Materials

The isolation of the four subunit DCF₁ by chloroform extraction of membranes prepared from osmotically shocked cultures of D. bardawil will be described in detail elsewhere (M. Finel et al., manuscripts submitted). Monospecific rabbit antisera directed against the native spinach CF₁, the native and SDS-denatured C. reinhardi CF₁ (CrCF₁), and the SDS-denatured individual subunits of the CrCF₁ were generously supplied by Ms. Sabeeha Merchant (Dept. of Biochemistry, University of Wisconsin, Madison, WI). Monospecific rabbit antiserum directed against the E. coli F₁ β-subunit was kindly provided by Prof. Nathan Nelson (Biology Dept., Technion, Technological Institute of Israel, Haifa).

Results

1. Properties of the DCF₁. In contrast to the ATPase activity of the CrCF₁ but similar to that of the spinach enzyme, the DCF₁ ATPase activity is completely latent. It can, however, be elicited either by a short heat treatment or by the presence of selected organic solvents.

Heat activation of the enzyme induces a specific Ca^{2+}-dependent ATPase activity. Heat activation is optimal when performed at 60°C for 3.0 min and is independent of the protein concentration during the treatment over the range 0.5 to 2.5 mg/ml. Preservation of enzymatic activity absolutely requires the presence of ATP during the treatment, whereas Ca-ATPase activation absolutely requires the inclusion of a thiol reductant, e.g., dithiothreitol (DTT). Heat activation of the Ca^{2+}-dependent DCF₁ ATPase is an irreversible process. The Ca^{2+}-dependent DCF₁ ATPase has a Km for CaATP of about 0.8 mM and a Vm that ranges from 10-15 μmol ATP hydrolyzed/mg protein·min (measured at 37°C).

A Mg^{2+}-dependent DCF₁ ATPase activity is elicited by selective organic solvents, in particular methanol (35%) and ethanol (25%) (Kneusel et al., 1982). The Km for Mg-ATP

Sybesma, C. (ed.), Advances in Photosynthesis Research, Vol. II. ISBN 90-247-2943-2.
© 1984 Martinus Nijhoff/Dr W. Junk Publishers, The Hague/Boston/Lancaster.

is about 1.2 mM and the Vm ranges from 50–80 μmol ATP hydrolyzed/mg protein·min. In contrast to the heat-induced, Ca^{2+}-dependent, ATPase activity, the solvent-induced, Mg^{2+}-dependent activity is completely reversible. In addition, solvents also transform the heat-activated, Ca^{2+}-dependent ATPase into a heat-activated, Mg^{2+}-dependent ATPase, the latter again being a reversible process.

Both the Ca^{2+}-dependent and Mg^{2+}-dependent ATPases are inhibited by ADP (apparently allosterically), phlorizin, and antisera directed against the native spinach CF_1 and native $CrCF_1$. Whereas the Ca^{2+}-dependent ATPase is severely inhibited by glycerol (I_{50} ca. 8%, v/v) and NaCl (I_{50} ca. 100 mM), the Mg^{2+}-dependent ATPase is not. Neither DCF_1 ATPase activity is inhibited by antisera directed against the individual SDS-denatured subunits of $CrCF_1$.

2. Absolute identification of the DCF_1 α and β-subunits separated by SDS-PAGE.
Piccioni et al. (1981) have reported that the order of migration of the $CrCF_1$ α- and β-subunits, separated by SDS-PAGE, is reversed with respect to the order of migration of the spinach subunits. (Recently, however, it has been shown (Merchant, Selman, 1983) that this is not the case when the $CrCF_1$ subunits are separated by either the SDS-PAGE system described by Weber and Osborn (1969) or Laemmli (1970). In order to determine the absolute identification of the DCF_1 subunits, the subunits of DCF_1 and spinach CF_1 were separated by SDS-PAGE (Laemmli, 1970), transblotted to nitrocellulose paper (Rott, Nelson, 1981), and treated with rabbit monospecific antisera directed against either the $CrCF_1$ α-subunit or the E. coli F_1 β-subunit. Visualization of the cross-reactive complexes was obtained by treating the transblots with ^{125}I-labeled goat antirabbit IgG, followed by autoradiography. From the results shown in Fig. 1, it is evident that the slower migrating band from DCF_1 and the faster migrating β-subunit band from spinach CF_1 cross-react with antiserum directed against the E. coli F_1 β-subunit whereas the faster running band from the DCF_1 and the slower running, α-subunit band from spinach CF_1 cross-react with the antiserum directed against the $CrCF_1$ α-subunit (see figure legend for details). From these results, it is concluded that, when separated by SDS-PAGE, the order in apparent decreasing molecular weight of migration of the DCF_1 subunits is $\beta, \alpha, \gamma, \epsilon$.

(These results have been confirmed by two other independent studies; one based on the fluorescence of Coomassie blue polypeptide complexes (Nelson, personal communication) and one based on the specific covalent modification of the β-subunit by dicyclohexyl carbodiimide. These studies will be presented elsewhere .)

3. The modification of DCF_1 by FITC. FITC is a potent inhibitor of a number of ATP utilyzing enzymes (Muallem, Karlish, 1983), including the higher plant CF_1 (Pick, 1981). The DCF_1 Mg^{2+}-dependent ATPase activity is very sensitive to inhibition by FITC, more so, in fact, than the spinach CF_1 when treated under identical conditions (not shown). As is the case with the higher plant CF_1 (Pick, 1981), FITC inhibition can be prevented by the inclusion of various nucleotides during the treatment. At comparable concentrations, ATP and ADP are the most effective but GTP, GDP and AMP-PNP also protect. AMP is completely ineffective.

In order to determine to which subunit(s) of the DCF_1 and spinach CF_1 FITC covalently binds, the subunits were separated by SDS-PAGE after treating the enzymes with FITC and the fluorescein polypeptide adducts visualized by excitation of the fluorescein fluorescence. The labeled fluorescence was recorded on X-ray film (Fig. 2). From Fig. 2 it can be seen that (i) when the treatment of the enzymes is performed at pH 8.0, it is mainly

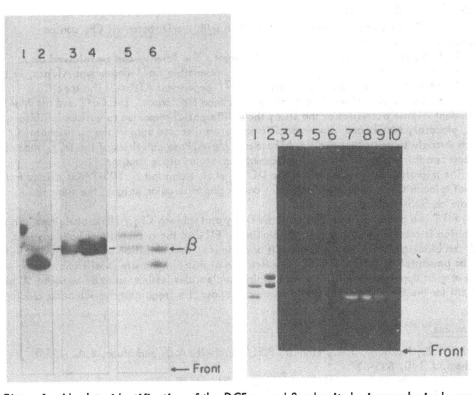

Figure 1: Absolute identification of the DCF₁ α and β-subunits by immunological cross-reactivity with monospecific antisera.   The transblot experiment was performed as described by Rott and Nelson (1981). Lanes 1, 2, the primary antiserum was a rabbit anti CrCF₁ α-subunit antiserum; Lanes 3, 4, the primary antiserum was a rabbit anti E. coli F₁ β-subunit antiserum; Lanes 5, 6, Commassie blue stained SDS-PAGE before transblotting. Lanes 1, 3, and 5, Spinach CF₁; Lanes 2, 4 and 6, DCF₁.
Figure 2: Identification of the ATPase subunits that covalently bind FITC.   Spinach CF₁ (lanes 2-6) and DCF₁ (lanes 1, 7-10) were incubated with FITC (10 μM, lanes 4, 9; 20 μM, lanes 5, 8, 10; 50 μM, lanes 3, 6, 7). ATP, when present (lanes 3, 10), was added prior to FITC. Incubation conditions essentially as described by Pick (1981) and separation of subunits was obtained on 7.5% SDS-PAGE.

the α-subunits of both the DCF₁ and spinach CF₁ that are labeled, (ii) increasing the concentration of FITC during the treatment, increases the amount of label associated with the α-subunits of both enzymes, (iii) under comparable modification conditions, the DCF₁ is more heavily labeled than the spinach CF₁, consistent with the observation that the DCF₁ ATPase activity is more sensitive to FITC inhibition than is the spinach CF₁ ATPase activity, and finally, (iv) if ATP is present during the treatment with FITC, FITC binding to the α-subunit of both enzymes is greatly suppressed.

Discussion and Conclusions

The salient observations that we have made with the D. bardawil CF$_1$ can be summarized as follows:

(i) The DCF$_1$ ATPase activity is completely latent. The latency can be relieved by either a short heat treatment, which induces an irreversible Ca^{2+}-dependent ATPase, or b organic solvents, which induce a reversible Mg^{2+}-dependent ATPase. The specific activity of the latter is about 5 to 6 fold greater than the former. The Ca^{2+} and the Mg^{2+} dependent ATPase activities of the DCF$_1$ show differential responses to various inhibitors, e.g., glycerol, NaCl, and a monospecific antiserum directed against the C. reinhardi CF$_1$ which strongly suggests that the Ca-ATPase and Mg-ATPase activities of the DCF$_1$ represe at least two fairly grossly different conformational states of the enzyme.

(ii) The migration of the subunits of the DCF$_1$, when separated by SDS-PAGE, differs from that of spinach CF$_1$. In order of apparent decreasing molecular weight, the subunits migrate on SDS-PAGE β, α, γ and ϵ.

(iii) FITC is a potent inhibitor of both the DCF$_1$ and spinach CF$_1$ ATPase activities. This inhibition is correlated to the specific binding of FITC to the α-subunits of both enzymes. Both the binding of FITC to the α-subunit and, hence, the inhibition of the ATPase activi can be prevented by ATP and ADP. The function of this binding site is still unclear; however, its apparent low affinity for adenine nucleotides (estimated to be between 50 to 100 μM for the latent enzyme) would seem to rule out the "regulatory" nucleotide binding sites.

References

Beechy, R.B., Hubbard, S.A., Linnett, P.E., Mitchell, A.D. and Muin, E.A. (1975) Biochem. J. 148, 533-537.

Binder, A. and Bachofen, R. (1979) FEBS Lett. 104, 66-70.

Kneusel, R.E., Merchant, S. and Selman, B.R. (1982) Biochim. Biophys. Acta 681, 337-34

Laemmli, U. (1970) Nature 227, 680-685.

Merchant, S. and Selman, B.R. (1983) Eur. J. Biochem., in press.

Muallem, S. and Karlish, S.J.D. (1983) J. Biol. Chem. 258, 169-175.

Nelson, N. (1976) Biochim. Biophys. Acta 456, 314-338.

Piccioni, R.G., Bennoun, P. and Chua, N.-H. (1981) Eur. J. Biochem. 117, 93-102.

Pick, U. (1981) Biochem. Biophys. Res. Commun. 102, 165-171.

Rott, R. and Nelson, N. (1981) J. Biol. Chem. 256, 9224-9228.

Selman-Reimer, S., Merchant, S. and Selman, B.R. (1981) Biochem. 20, 5476-5482.

Weber, K. and Osbom, M. (1969) J. Biol. Chem. 244, 4406-4412.

Wegmann, K., Ben-Amotz, A. and Avron, M. (1980) Plant Physiol. 66, 1196-1197.

Acknowledgements

This research was supported in part by grants from the Weizmann Institute of Science (BRS, Meyerhoff Fellow), The College of Agriculture and Life Sciences, University of Wisconsin (BRS, SS), The US-Israel Binational Research Foundation (UP, BRS, BSF 2891/8 and the National Institutes of Health (BRS, GM 31384-01). We are indebted to Prof. M. Avron for his inspiration and motivation.

Authors Addresses: Biochem. Dept., Weizmann Institute of Science, Rehovot, Israel (M.F. U.P.) and Dept. of Biochem. University of Wisconsin, Madison, WI, USA (S.S., B.R.S.).

IMMUNOLOGICAL STUDIES ON THE CROSS-BINDING AND CROSS-RECONSTITUTION OF MAIZE
AND SPINACH COUPLING FACTORS CF_1 *

NORMA L.PUCHEU and RICHARD J. BERZBORN

Biochemie d. Pflanzen, Ruhr-Universität Bochum, D-4630 Bochum 1, F R G

1. INTRODUCTION AND PROBLEM

To understand the mechanism of photosynthetic energy transformation, complete
knowledge of the components involved is essential. To study the photosynthetic
ATPsynthase, resolution and reconstitution experiments have been carried out
now for 20 years (Avron,1963). Numerous details and analogies to the ATPsyntha-
ses of oxidative phosphorylation in bacteria and mitochondria have been elucidated (Muñoz,1982).

Since reconstitution requires binding of the added soluble components $(CF_1, \delta,$
etc.), reconstitution experiments can have two aims, the biochemistry of the
binding, i.e. recognition, or the demonstration of the minimal system, i.e.
nature and amounts and properties of the essential components of the phosphorylation in situ, which of course are all present in thylakoid in vivo.

Because both oxidative and photophosphorylation occur at large sub-entities in
organelles of heterotrophic and autotrophic cell, the plasma membrane, the mitochondria inner membrane or the chloroplast thylakoid system, and in a cooperative fashion - one thylakoid contains about 10^3 CF_1CF_0 complexes - , in reconstitution experiments the danger of misinterpretation is always inherent: The
reappearence of catalytic activity of the membrane system may be due to "structural reconstitution", i.e. repair of H^+ leakeness by the added soluble component (Schatz et al., 1967, McCarty, Racker, 1967).

To prove "catalytic reconstitution" e.g. of photophosphorylation by CF_1 one has
to show that CF_1 was completely removed from the thylakoid membrane by the
treatment for resolution (EDTA, pyrophosphate, NaBr). Otherwise one has to
take residual CF_1 into consideration!

Determination of residual CF_1 can be done by looking at e.g. the α, β, δ subunits of CF_1 after PAGE-analysis in the presence of SDS, or by quantitatively
measuring trypsin activated Ca^{++} ATPase or heat activated DTT stabilized Mg^{++}
ATPase.

For obvious reasons we adapted a specific immunochemical micromethod for quantitative measurements of CF_1 in any fraction, independent of purity, activity or
the presence of several chemicals (Berzborn et al. 1981; Roos,Berzborn,1983).

Catalytic reconstitution, i.e. demonstration of the intactness and completeness
of the photophosphorylation system, has been proved in three ways: It was shown
that the reconstituted ATPsynthase activity was higher than the percentage of
residual amount of CF_1, measured as Ca^{++} ATPase (Lien,Racker, 1971, Berzborn,
Schröer, 1976); it was shown that the added CF_1 exchanged its bound ^{14}C-adenyl-
ate , a sign for energization, i.e. regained catalytic activity (Hesse et al.,
1976); and it was shown by use of the inhibitory substance tentoxin that after
readdition of tentoxin-insensitive CF_1 from a tabacco species to tentoxin-sen-
sitive spinach thylakoids the reconstituted activity was not inhibitable to
some extent (Selman,Durbin, 1978).

*This research was supported by a fellowship of the CONICET (Argentina) to
N.L.P. and a grant from the DFG to R.J.B.

Sybesma, C. (ed.), Advances in Photosynthesis Research, Vol. II. ISBN 90-247-2943-2.
© 1984 Martinus Nijhoff/Dr W. Junk Publishers, The Hague/Boston/Lancaster.

2. APPROACH

We managed to get species-specific precipitating and inhibiting antibodies (ab) against both spinach and maize CF_1. By use of these non-crossreacting ab we will both determine precisely residual and rebound CF_1 of either species, also in hybrid particles after cross-reconstitution, and we will determine the inhibitable and non inhibitable portion of reconstituted ATPsynthase activity.

3. EXPERIMENTS

3.1. Production of non-crossreacting, species-specific, precipitating and inhibiting antibodies

After double diffusion according to Ouchterlony of spinach and maize CF_1 or spinach and maize thylakoids against an antiserum against spinach CF_1 (150) or against maize CF_1 (141), the pattern of only partial identity was seen (spurs). By absorption of the respective antisera with heterologous thylakoids we removed all crossreacting ab and thus produced sera, where only the species-specific non-crossreacting ab had remained (Pucheu, Berzborn, in preparation).

3.2. Quantitative measurements of residual and rebound CF_1 in cross-reconstitution experiments

Since it was known that maize EDTA treated thylakoids could hardly be reconstituted (Roos, 1978), and that for spinach the addition of NaCl during EDTA treatment was very beneficial (Berzborn, Schröer, 1976), the dependency of residual CF_1, rebound CF_1 and reconstituted ATPsynthase on NaCl during EDTA treatment of maize thylakoids was determined (Pucheu, Berzborn, in preparation). Differences in the binding affinities of the CF_1 preparations became apparent (Pucheu, Berborn, in preparation).

After optimizing the homologous system the respective heterologous hybrid reconstitutions were performed. Fig.1a,b shows the respective residual and rebound amounts of CF_1 of such cross-reconstituted thylakoids.

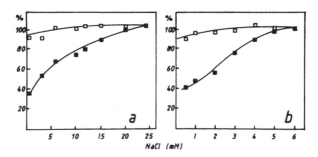

FIG. 1. Quantitative determination of residual and rebound CF_1 by rocket electroimmunodiffusion (Roos, Berzborn, 1983)
(Thylakoids were incubated in EDTA with the indicated concentrations of NaCl present, washed and reconstituted with the hetrologous CF_1. a) maize thylakoids plus spinach CF_1, ■ residual maize CF_1, ☐ total CF_1, i.e. residual maize plus rebound spinach CF_1. b) spinach thylakoids plus maize CF_1, ■ residual spinach CF_1, ☐ total CF_1, i.e. residual spinach plus rebound maize CF_1.

3.3. Proof of catalytic reconstitution by partial specific inhibition of reconstituted ATPsynthase by special antibodies

The species-specific non-crossreacting ab still inhibited photophosphorylation in the respective control thylakoids (Pucheu,Berzborn, in prep.). We therefore could determine the inhibitable and non inhibitable proportion of the reconstituted ATPsynthase in the hybrid system: Maize EDTA treated thylakoids plus spinach CF_1. Three main results are presented in table 1:
1. The percent of reconstituted activity exceeded the percent of residual amount of CF_1 (cp. line 5 with line 2).
2. The reconstituted activity could only partially be inhibited (line 6).
3. The percent of non inhibitable reconstituted activity coincided with the residual amount of maize CF_1 (cp. line 6 with line 2).

TABLE 1: CF_1 content and photophosphorylation activities in a cross-reconstitution experiment

	amounts of CF_1 and activities of maize thylakoids	NaCl during EDTA treatment	
		12 mM	15 mM
1	residual ATPsynthase (% of control)*	49	90
2	residual amount of maize CF_1 (% of control)*	79	88
3	rebound spinach CF_1 (% of maize protein)	26	14
4	total bound CF_1 after reconstitution (%)	105	102
5	reconstituted ATPsynthase (% of control)	96	100
6	portion of reconstituted ATPsynthase, non inhibitable by the species-specific, non-crossreacting anti spinach CF_1 antibodies (%)	81	89

* ATPsynthase of the control: 450 µmoles ATP / mg chl x h
  amount of CF_1 / chl in maize control thylakoids: 0.65 µg / µg chl

4. CONCLUSIONS

We demonstrated both structural and catalytic reconstitution of photophosphorylation after EDTA treatment of maize thylakoids.

We separated in quantitative terms the proportion of structural and catalytic reconstitution by help of quantitative immunochemical determination of residual CF_1 and by help of specific inhibition of the readded spinach CF_1 on the maize thylakoids.

Data could be produced for both the Lien-argument (reconstituted activity exceeds residual amount) and the Schatz-argument (reconstituted activity in hybrid system can be specifically inhibited to a certain extent).

The non inhibitable activity of ATPsynthase is the catalytic activity of the previously inactive residual maize CF_1, enabled by the structural role of the re-added and bound spinach CF_1, which is itself inhibited in its catalytic ro[l] by the specific antibodies.

In this system we now can study the conditions and minimal requirements for re[ ] constitution, e.g. the function of the δ - subunit, which was demonstrated to be a coupling factor, but not an essential binding protein for CF_1 in spinach (Roos,Berzborn,1982; Andreo et al.,1982).

5. REFERENCES

Avron M (1963) A coupling factor in photophosphorylation, Biochim.Biophys. Act[a] 77, 699-702

Andreo CS, Patrie WJ, McCarty RE (1982) Effect of ATPase activation and the δ subunit of coupling factor 1 on reconstitution of photophosphorylation J. Biol. Chem. 257, 9968-9975

Berzborn RJ, MüllerD, Roos P, Andersson B (1981) Significance of different quantitative determinations of photosynthetic ATPsynthase CF_1 for hetero- geneous distribution and grana formation. in: Proc. 5th Intern. Congr. on Photosynthesis, Vol III. pp. 107-120 (Akoyunoglou,ed.) Balaban Sci.Serv. Philadelphia

Berzborn RJ, Schröer P (1976) Photophosphorylation. Mechanism of reconstitutio[n] by coupling factor 1. FEBS Lett. 70, 271-275

Hesse H, Jank-Ladwig R, Strotmann H (1976) On the reconstitution of photophos- phorylation in CF_1-extracted chloroplasts. Z.f.Naturforsch. 31c, 445-451

Lien S, Racker E (1971) Partial resolution of the enzymes catalyzing photophos[ ] phorylation. VIII Properties of silicotungstate-treated subchloroplast particles. J. Biol. Chem. 246, 4298-4307

McCarty RE, Racker E (1967) Partial resolution of the enzymes catalyzing photo[ ] phosphorylation III The inhibition and stimulation of photophosphorylation by DCCD. J. Biol. Chem. 242, 3435-3439

Munoz E (1982) Polymorphism and conformational dynamics of F_1-ATPases from bacterial membranes. A model for the regulation of these enzymes on the basis of molecular plasticity. Biochim.Biophys. Acta 650, 233-261

Roos P (1978) Untersuchungen zur Spezifität der Bindung der peripheren ATPase in der photosynthetischen ATPsynthase. Diplomarbeit, Ruhr-Universität Bochum F R G

Roos P, Berzborn RJ (1982) Is subunit δ indeed required for CF_1 binding to the chloroplast membrane? in: Proc. 2nd Europ. Bioenergetics Conf., Lyon- Villeurbanne, CNRS ed. pp. 99-100

Roos P, Berzborn RJ (1983) Electroimmunodiffusion - a powerfull tool for quantitative determinations of both soluble and membrane bound chloroplast ATPase CF_1. Z.f.Naturforsch. in press

Schatz G, Penefsky HS, Racker E (1967) Partial resolution of the enzymes cata- lyzing oxidative phosphorylation XIV Interaction of purified mitochondrial adenosine triphosphatase from Bakers' yeast with submitochondrial particle[s] from beef heart. J. Biol. Chem. 242,2552-2560

Selman BR, Durbin RD (1978) Evidence for a catalytic function of the coupling factor 1 protein reconstituted with chloroplast thylakoid membranes. Biochim.Biophys. Acta 502, 29-37

DETECTION OF CONFORMATIONAL CHANGES IN CHLOROPLAST COUPLING FACTOR 1 BY ANS FLUORESCENCE CHANGES

URI PICK AND MOSHE FINEL

1. INTRODUCTION

Naphthalene-sulphonate derivatives such as 8-anilino-1-naphthalene sulphonate (ANS) have an amphipathic character and, therefore, bind to phospholipid membranes, to detergent micells and to hydrophobic sites on proteins (Slavik, 1982). The binding is accompanied by a fluorescence enhancement and by a blue-shift of the emission spectrum due to the effect of the local apolar environment on the fluorescence yield of the bound dye. Cantley and Hammes (1976) reported that the fluorescence of ANS was enhanced upon binding to chloroplast coupling factor 1 (CF_1). Neumann et al. (1979) have reported recently that Touloidinonaphthalenesulphate (TNS) acts as an "energy transfer inhibitor" of photophosphorylation. These results indicate a specific interaction of naphthalene sulphonates with CF_1.

In this paper we demonstrate that CF_1 contains a high-affinity ANS binding site and that the changes in the fluorescence yield of the bound dye may be used to follow the interaction of CF_1 with adenine nucleotides, inorganic phosphate, divalent cations and inhibitors and also to detect subtle conformational changes in the enzyme as a result of ATPase activation or inactivation.

2. MATERIALS AND METHODS

CF_1 was prepared from spinach leaves by the sucrose-ATP extraction procedure, purified on DEAE-Sephadex A-50 column, depleted of bound nucleotides and stored in 25% glycerol at -194°C as described elsewhere (Pick, Bassilian, 1983).

Fluorescence measurements were performed in a Perkin-Elmer MPF-44A spectro-fluorimeter. The exitation and the emission wavelengths were 365 and 470 nm, respectively, and the slits were adjusted to 6 nm. CF_1 samples (25-40 µg protein) were added to 3 ml cuvettes containing in 2 ml 40 mM Na-Tricine (pH 8), 0.1 mM EDTA and 2 µM ANS.

Activation of CF_1 Ca-ATPase was performed by preincubating CF_1 (1 mg/ml) for 10 min at 37° in the presence of 1 mM ATP and 30 mM octylglucoside (OG) followed by Sephadex-centrifugation to remove the ATP (Pick, Bassilian, 1982). Activation of CF_1 Mg-ATPase was performed by addition of 20 mM OG to the measuring cuvette, while inactivation was performed by incubation CF_1 (1 mg/ml) for 30 min at 37° with 40 mM OG in the absence of ATP (Pick, Bassilian, 1982).

ANS was obtained from Molecular Probes, Octylglucoside from CalBiochem and all other chemicals and reagents were obtained from Sigma Chemical Inc.

3. RESULTS AND DISCUSSION

CF_1 induces a pronounced enhancement in ANS fluorescence (Figure 1a, b) which is associated with a blue shift in the fluorescence emission spectrum (λmax is shifted from 515 to 498 nm in the absence or presence of CF_1, respectively, not shown). The extent of CF_1-induced fluorescence enhancements are about the same for latent CF_1 (CF_1^l, a, b) and for OG-activated CF_1 (CF_1^*, e, f) while inactivation of CF_1 (CF_1^i) is associated with a greater enhancement of ANS fluorescence (c, d).

Sybesma, C. (ed.), Advances in Photosynthesis Research, Vol. II. ISBN 90-247-2943-2.
© *1984 Martinus Nijhoff/Dr W. Junk Publishers, The Hague/Boston/Lancaster.*

Figure 1 also demonstrates that ADP induces a rapid fluorescence quenching in latent (a, b) or OG-activated (e) CF_1 but not in inactivated CF_1 (c). Mg induces a fluorescence enhancement in latent (a, b), OG-activated (f) or inactivated (d) CF_1. In the absence of CF_1 neither ADP nor Mg have any effect on ANS fluorescence. The fluorescence of CF_1-bound ANS is also quenched by ATP, inorganic phosphate, and the ATP analog AMP P(NH)P (Fig. 2). The maximal quenching by all these substrates is about 50% but there are significant differences in their apparent affinities: Fifty percent of the maximal fluorescence quenching is obtained by 10 μM ADP, 80 μM ATP, 120 μM Pi or 150 μM AMP P(NH). Partial effects are obtained by arsenate and pyrophosphate while GTP does not induce any ANS fluorescence quenching. The addition of phosphate in the presence of saturating ADP (or ATP) concentrations or vice versa, does not induce additional quenching. These data may suggest that phosphate and adenine nucleotides bind to a common site on CF_1.

In order to find out if the ADP (or phosphate)-induced ANS fluorescence quenching is due to a decrease in the number of ANS binding sites or to a decrease in the binding affinity and/or in the quantum yield of the bound probe the effect of phosphate on the CF_1-induced ANS fluorescence was measured at different ANS concentrations (0.2-30 μM) at a high CF_1 concentration (3 μM). The Scatchard analysis in Figure 3 demonstrates that phosphate does not decrease the number of ANS binding sites but rather decreases the affinity for ANS and the fluorescence quantum yield of the bound probe. It may be noted that the calculated ANS binding affinity of CF_1 (Kd = 6 μM) is considerably higher than the value previously reported by Cantley and Hammes (1976) (above 100 μM) and also that the latter authors did not see any effects of adenine nucleotides on the fluorescence of ANS. These differences may be due to the different measuring conditions (mainly the higher salt concentrations) used by Cantley and Hammes (1976).

Since CF_1 contains at least two types of adenine nucleotide binding sites, namely, the tight sites and the catalytic sites, it is important to determine which of them is responsible for the ANS fluorescence quenching. A comparison of nucleotide-depleted CF_1 with CF_1 which has been preequilibrated with ATP to saturate the tight-binding sites with one ATP/ CF_1 followed by removal of free ATP by Sephadex centrifugation (Pick, Bassilian, 1983) demonstrated that tightly-bound ATP does not inhibit the maximal ADP or Pi-induced ANS fluorescence quenching but it significantly decreases the apparent affinity for both ADP (Kd increases from 11 μM to 34 μM) and phosphate (from 140 μM to 320 μM). These results as well as the fast response to nucleotides (Fig. 1) indicate that the ADP-induced ANS fluorescence quenching is not due to binding to the tight-sites. Yet, occupancy of the tight sites seems to influence these sites indirectly by altering their apparent affinity for ADP and phosphate. Two lines of evidence suggest that these sites are intimately related to ATP hydrolysis: (a) Activation of CF_1 by OG, which reversibly stimulates a Mg-specific ATPase, decreases the apprent affinity for ADP (from 10 to 35 μM) and phosphate (from 105 to 335 μM) but increases the apparent affinity for ATP (from 85 to 45 μM) suggesting that ATPase activation alters the binding properties of this site (data not shown).
(b) Several CF_1-ATPase inhibitors (e.g., fluorescein isothiocyanate (FITC) or 7-chloro-4-nitrobenzo-2-oxa-1,3-diazole (NBD-Cl)) abolish completely the ADP-induced ANS fluorescence quenching (data not shown). It is possible, therefore, that the sites are the catalytic sites of CF_1.

The Mg (or Ca)-induced ANS fluorescence enhancement (Fig. 1) is of special interest because it is influenced by activation of CF_1-ATPase. No fluorescence enhancements are obtained by monovalent cations (up to 0.8 M NH_4^+) or by La^{3+} (up to 3 mM) indicating that the Mg-induced fluorescence enhancement is not due to non-specific

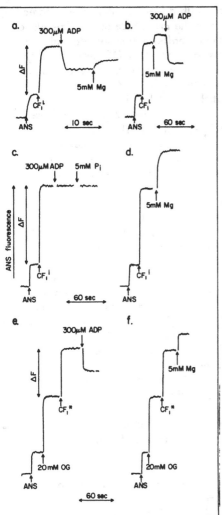

Figure 1: The Effect of Mg and ADP on ANS Fluorescence in Latent, Activated and Inactivated CF₁ Preparations.

ANS fluorescence changes were measured in the presence of 2 μM ANS and 30 μg of latent CF₁ (CF₁, a and b), inactivated CF₁ (CF₁ⁱ, c and d) or octylglucoside-activated CF₁ (CF₁*, e and f). ΔF is the CF₁-induced ANS fluorescence enhnacement.

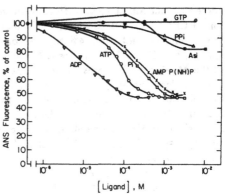

Figure 2: Adenine Nucleotide and Phosphate Specificity for the Induction of ANS Fluorescence Quenching.

ANS fluorescence quenching, induced by the indicated ligand concentrations was measured in the presence of 2 μM ANS and 35 μg latent CF₁.

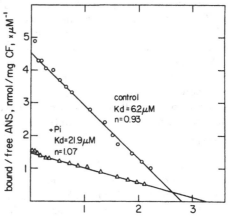

Figure 3: Scatchard Analysis of ANS Binding to CF₁ in the Presence and Absence of Inorganic Phosphate.

The ANS fluorescence enhancement (ΔF) induced upon mixing of CF₁ (1 mg/ml) and ANS (0.2-30 μM) was measured in the presence (Δ) or absence (o) of 0.6 mM inorganic phosphate. ΔF$_{max}$ obtained from extrapolations of 1/ΔF against 1/[ANS] plots and the binding was calculated from the ΔF/ΔF$_{max}$ ratios.

charge screening but rather to specific interactions with a Me^{2+} binding site. Moreover, activation of CF_1 Ca-ATPase preferentially increases the magnitude of the Ca-induced fluorescence enhancement (from 4% to 18%) and to a smaller extent the Mg-induced enhancement (from 8% to 16%) whereas Mg-ATPase activation by OG preferentially increases the Mg-induced fluorescence enhancement (from 8% to 28% by Mg, from 4% to 11% by Ca, data not shown). This correlation between the Me^{2+} specificity in ATP hydrolysis and the ANS responses may suggest that the Me^{2+} binding site is involved in ATP hydrolysis. This is also indicated by the observation that Dio-9, a CF_1-ATPase inhibitor which by itself causes enhancement of CF_1-bound ANS fluorescence (80% enhancement, K_{50} at 0.5 μg Dio 9/ml) abolishes completely the Mg-induced ANS fluorescence enhancement (not shown). Inactivation of CF_1-ATPase largely increases the Mg-induced fluorescence enhancement (Fig. 1d) perhaps due to denaturation of the protein and exposure of new ANS binding sites.

4. SUMMARY

1. CF_1 contains a high-affinity ANS binding site ($Kd - 6$ μM). The binding of ANS to CF_1 is associated with a fluorescence enhancement and a blue shift of the emission spectrum.
2. Adenine nucleotides and phosphate induce a fast quenching of ANS fluorescence due to a decrease of the binding affinity and of the fluorescence quantum yield of bound ANS. Binding of ATP to the tight-sites does not inhibit the ADP or Pi induced fluorescence quenching but decreases the affinity for both ADP and Pi suggesting that the sites are different but not independent from the tight-sites.
3. Activation of a Mg-specific ATPase of CF_1 by octylglucoside decreases the affinity for ADP and phosphate but increases the affinity for ATP.
4. Ca-ATPase activation preferentially increases Ca-induced ANS fluorescence enhancement while Mg-ATPase activation preferentially increases Mg-induced ANS fluorescence enhancement.

ANS may be used, therefore, as a sensitive probe to detect conformational changes in CF_1 in response to activation, inactivation and to the binding of substrates.

5. REFERENCES

Cantley LC and Hammes GG (1976) Investigation of Quercetin binding sites on chloroplast coupling factor 1, Biochemistry 15, 1-8.

Neumann J, Drechsler Z, Searle GFW and Barber J (1979) 2-p-toluidinonaphthalene-6-sulphonate (TNS)- An energy-transfer inhibitor in chloroplasts, FEBS Lett. 102, 121-125

Pick U (1981) Modification of chloroplast CF_1 by Fluoresceinisothiocyanate, Biochem. Biophys. Res. Commun. 102, 165-171.

Pick U and Bassilian S (1982) Activation of a Mg-specific ATPase in chloroplast coupling factor 1 by octylglucoside. Biochemistry 24, 6144-6152.

Pick U and Bassilian S (1983) The effects of octylglucoside on the interaction of chloroplast coupling factor 1 with adenine nucleotides. Eur. J. Biochem. 133, 289-297.

Slavik J (1982) Anilinonaphthalenesulphonate as a probe for membrane composition and function, Biochim. Biophys. Acta 694, 1-25.

Authors address: Biochemistry Department, Weizmann Institute of Science, Rehovot 76100, Israel

SPECTROSCOPIC STUDIES ON THE CONFORMATION OF ISOLATED AND MEMBRANE BOUND CF1

Richard Wagner, Carlos Andreo* and Wolfgang Junge

It is generally agreed on that the enzyme complex CF0-CF1 functions as a proton translocating ATP-synthase as proposed by Mitchell (1), but the mechanism of this action remains elusive. Energy dependend structural changes in CF1, as revealed by different techniques (2,3) seem to play an essential role during ATP-synthesis. Different approaches have been made to further characterize structural changes in CF1 and to eluciate their role during catalysis.

In this paper we outline a series of experiments where we have investigated the conformation of isolated and membrane bound CF1 by means of spectroscopic techniques (4,5,6). For these studies CF1 was covalently labeled with a spectroscopic probe, eosin-isothiocyanate, which preferentially reacts with aminogroups from lysine (5). Upon excitation by a short laser flash the dye can be efficiently transformed into a relatively long lived triplet state. Spectroscopic detection of this triplet state allows to follow conformational changes of the enzyme in either of two ways:

1) The triplet lifetime depends on the access of O_2 to a given binding site of eosin isothiocyanate in the host protein. Thus, the triplet lifetime reflects the proximity of a binding site to the bulk medium and/or the flexibility of protein chains which cover a binding site.

2) Under photoselection the decay of the linear dichroism of the absorption changes of eosin-SCN reflects the mobility of the dye relative to the protein (librational motion) and the rotational diffusion of the host protein.

LABELING OF CF1 WITH EOSIN-SCN

From the labeling pattern of CF1 with eosin-SCN two states of the enzyme could be discriminated, the latent and the active state. When membrane bound CF1 was incubated with eosin-SCN (100 um) at pH 8.3 for three minutes without energization of the membrane, we found 1 mol of bound eosin-SCN/mol CF1 bound to the alpha-subunit of CF1. This neither impaired the Ca^{2+}-ATPase activity of isolated CF1 nor its ability to reconstitute cyclic photophosphorylation in depleted membranes. If the same incubation procedure was performed after one-minute illumination of chloroplasts under conditions of cyclic electron transport, CF1 contained 2 moles of eosin-SCN/mol CF1. One of the label molecules was bound to the alpha- and another one to the beta-subunit. The Ca^{2+}-ATPase of isolated CF1 was inhibited to 50% of the control, whereas the ability of labeled CF1 to reconstitute cyclic photophosphorylation in depleted membranes was not affected. This extra binding of one eosin-SCN into the beta-subunit of CF1 during light incubation of chloroplasts was abolished by uncoupler and could be prevented by the presence of 5mM ATP during incubation.

If chloroplasts were incubated for five minutes in the dark, CF1 contained 2 moles of eosin/mol CF1. The additional eosin was also bound to the beta-subunit and CF1 retained its full Ca^{2+}-ATPase acivity as well as its ability to reconstitute the ATP-synthase in depleted membranes. But under light incubation two additional eosin-SCN(a total of four) were bound to the beta- and to the gamma-subunit of CF1. In this case the Ca^{2+}-ATPase of isolated CF1 was inhibited by 70% of the control and the ability of CF1 to

Sybesma, C. (ed.), *Advances in Photosynthesis Research, Vol. II.* ISBN 90-247-2943-2.
© 1984 Martinus Nijhoff/Dr W. Junk Publishers, The Hague/Boston/Lancaster.

reconstitute photophosphorylation was completely lost. Rebinding of CF1 to depleted membranes was the same as for dark labeled enzyme. The above experiments were paralleled by others where isolated CF1 was labeled with eosin-SCN either in its latent form or when activated in its Ca^{2+}-ATPase by various procedures (6,7).

Independent of the mode of activation (membrane bound CF1 by energization of the membrane, isolated enzyme by DTT at 63° or by DTT at 25°) an extra capacitiy for eosin binding was induced. The binding of eosin-SCN to extra sites in the beta-subunit inhibited only the Ca^{2+}-ATPase activity but not the synthase activity. If eosin-SCN was additionally bound to the gamma-subunit, both the Ca^{2+}-ATPase as well as the synthase activity were inhibited. It is worthwhile to note, that extra binding of CF1 was prevented by high concentrations of ATP (5-10 uM) but not by ADP or P_i. The binding of eosin-SCN to nonactivated CF1 was not inhibitory for each of its activities up to a ratio of 7 moles of eosin-SCN/mol CF1 (7,8).

TRIPLET LIFETIME OF EOSIN-SCN WHEN BOUND TO CF1

The spectroscopic details of the eosin probe were reviewed elsewhere (5,9). Fig.1 shows a typical trace for the time-course of the absorption changes at 515 nm of eosin-SCN when bound to isolated CF1 after excitation with a short (10 ns FMWH) flash from a Nd-YAG-Laser. The rapid rise of absorption is due to ground state depletion of eosin and the decay to its subsequent repopulation from the excited states. The decay of laser induced eosin absorption changes is a measure of the triplet lifetime. It depends on the oxygen pressure. Binding of eosin-SCN to CF1 can provide a partial protection of the excited eosin triplet state from the quenching by oxygen from the bulk medium. This protection depends on the vicinity of the binding site of eosin in CF1 to the bulk medium. This is documented in Fig.1. In dark labeled CF1 (2moles eosin-SCN/mol CF1) the absorption changes decayed with one relaxation time (25 μs), whereas in light labeled CF1 (3moles eosin-SCN/mol CF1) they decayed biphasically (25 μs, 190 μs). The relative amplitudes of the components were 32% (25 μs) and 68% (190 μs). In the dark labeled sample eosin-SCN was bound to alpha- and beta-subunit of CF1 without impairing its activities. But in light labeled sample eosin-SCN was bound to the alpha-, beta- and gamma-subunit of CF1, and the activities of CF1 were largely inhibited. Taking this into account, the observed time-course of eosin absorption changes can be visualized as follows: Upon labeling of membrane bound CF1 with eosin-SCN in the dark the label is bound to groups which are located on the surface of CF1. One fraction of bound eosin is protected against oxygen quenching. When CF1 labeling is performed under energization of the membrane the label is bound to groups which in the nonactivated enzyme are located in the protein interior, eosin bound to these sites is shielded to a larger extent from the bulk oxygen. We have previously shown (4,6), that these burried binding sites became exposed again if, after reconstitution, the CF1 structure opened under energisation of the thylakoid membrane.

ROTATIONAL DIFFUSION OF ISOLATED AND MEMBRANE BOUND CF1

In photoselection experiments eosin-SCN, when completely immobilized in rigid media, behaves as an ideal linear absorber. Under proper experimental conditions we obtained values of the absorption anisotropy of r(0)=0.37 which were very close to the upper theoretical limit (0.4) (see Ref.5).

Interpretation of the anisotropy data was straightforward since the same transition dipol was used for excitation and for measurement (8).
Figure 2 shows an example for dicroitic absorption changes of eosin-SCN when bound to isolated CF1 and the calculated time course of the absorption anisotropy. We were able to discriminate between different modes of chromophore rotation which dissipated absorption anisotropy. These were, rapid motion of the dye relative to the protein ("librational motion") and rotation of the protein itself. The evaluation of data for the rotational diffusion of isolated CF1 in isotropic solution let us to conclude that CF1 is highly excentric in shape (axial ratio greater than 2:1) rather than spherical(5). Librational mobility of the dye was dependent on the binding site of eosin-SCN within CF1. When eosin-SCN was bound to the surface of CF1 (modificatin of CF1 in the non-activated mode) the librational mobility was larger than for eosin-SCN bound inside CF1 (modification in the activated mode). Only in one case we observed the opposite. When isolated CF1 was labeled in the non-activated mode and then activated by DTT at 25°C, we found that the eosin binding side moved towards the protein interior where it became shielded from oxygen. Surprisingly enough its librational mobility increased. From this we concluded that activated CF1 (DTT,25°C) may contain a sequestered domain with solvent character (6). It appears as a hollow enzyme. When activated CF1 was reincorporated into NaBr-treated and depleted membranes we found that the sequestered domain became exposed to the bulk medium if the membrane was energized. We wonder wether the sequestered domain in CF1 is the one which carries the catalytic activity in ATP-synthesis.
CF1 labeled with eosin-SCN in the non activated mode restored photophosphorylation in partially depleted EDTA-chloroplasts and fully depleted NaBr-particles (10). Wheras in the case of the EDTA-particles we could observe very slow rotational motion of the reconstituted eosin-CF1 relative to the membrane, in NaBr-particles we could not detect any rotational motion (up to 500µs) of reconstituted eosin-CF1. We also studied the rotational diffusion of another extrinsic protein of the thylakoid membrane, the ferredoxin-NADP-oxidoreductase (11), which is probably located in the same stromal region of stacked chloroplasts as CF1 and we found very rapid rotation ($<$ 1µs). Only after addition of ferredoxin rotational correlation time decreased to 40 µs. This was interpreted to indicate formation of a ternary complex between ferredoxin-NADP-oxidoreductase ferredoxin and PSI. This revealed rather high lipid fluidy in thylakoids. We tend to assume that the low rotational mobility of the CFO-CF1 complex is caused either by self aggregation or strong interaction with other membrane proteins.
The above outlined results show, that the applied spectroscopic techniques may help to identify the conformation changes of CF1 during its catalytic process.

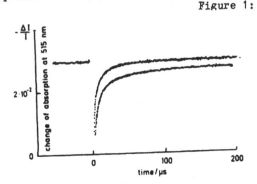

Figure 1: Time course of the absorption changes of eosin-SCN at 515nm when bound to isolated CF1.
CF1 was labeled when membrane bound either in the dark (upper trace) or light (lower trace).For details see text.

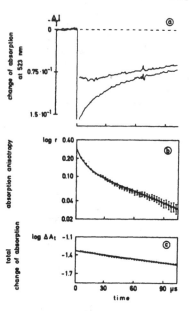

Figure 2: Time course of the absorption changes
of eosin at 523 nm when bound to
isolated CF1.
CF1 was labeled when membrane bound
in the dark (2 eosin-SCN/CF1).
a: Time course of the absorption changes
for parallel and perpendicular
polarisation of the exiting and
measuring light.
b: Time course for the absorption
anisotropy caculated from Fig.2a.
c: Time course of the total absorption.

References
1. Mitchell,P.(1966) Biol.Rev.Cambr.Phils.Soc.41,445-502
2. Hind,G.& Jagendorf,A.T.(1963)Proc.Nat.Acad.Sci.USA 49,715-721
3. McCarty,R.E.& Fagan,J.(1973)Biochemistry 12,1503-1507
4. Wagner,R.& Junge,W.(1980)FEBS-Lett.114,327-333
5. Wagner,R.& Junge,W.(1982)Biochemistry 21,1890-1899
6. Wagner,R.,Andreo,C.& Junge,W.(1983)Biochim.Biophys.Acta723,123-127
7. Wagner,R.,Carrillo,N.Junge,W.& Vallejos,R.H. (1981)
   FEBS-Lett. 136,208-212
8. Wagner,R. Phd Thesis (1980)Technical University of Berlin
9. Cherry,R.J.(1979) Biochim.Biophys.Acta 559,289-327
10.Wagner,R.,Andreo,C.& Junge,W. manuscript in preparation
11.Wagner,R.,Carrillo,N.,Junge,W.& Vallejos,R.H. (1982)
   Biochim.Biophys.Acta 680,317-330
Authors address:Richard Wagner,Wolfgang Junge/Biophysik/
                Universitaet Osnabrueck/ Postfach 4469
                D-4500 Osnabrueck/ Germany, FRG
                Carlos Andreo/ CONICET/ Centro de Estudios Fotosinteticos
                y Bioquimicos/ Universidad Nacional de Rosario/
                2000 Rosario/ Argentina

IDENTIFICATION AND PURIFICATION OF THE α AND β SUBUNITS OF THE CHLOROPLAST
COUPLING FACTOR ONE FROM *CHLAMYDOMONAS REINHARDI*

SABEEHA MERCHANT and BRUCE R. SELMAN

1. INTRODUCTION

The chloroplast coupling factor one (CF_1) purified from *Chlamydomonas
reinhardi* is a four subunit enzyme containing α, β, γ and ε subunits
(Piccioni et al., 1981; Selman-Reimer et al., 1981). It was suggested
by Piccioni et al. that the order of migration of the α and β subunits of
the *C. reinhardi* CF_1 was reversed in comparison to that of spinach CF_1.
Since the gel system used by Piccioni et al. contains 8 M urea, we have
identified the α and β subunits on two gel systems commonly used to separate
the subunits of CF_1 (Weber and Osborn, 1969; Laemmli, 1970).

2. MATERIALS AND METHODS

2.1. Detection of CF_1 subunits. CF_1 subunits were separated by
polyacrylamide gel electrophoresis and transferred to nitrocellulose paper
essentially as described by Towbin et al. (1979). Specific proteins were
detected by the immunoblot assay described by BioRad Laboratories, USA.

2.2. Modification of CF_1 by [^{14}C]Dicyclohexylcarbodiimide (DCCD). CF_1 was
incubated with 80 μM DCCD (62 mCi/mmol, Amersham, USA) in 50 mM MES-NaOH,
pH 6.5, for 25 minutes. The unbound DCCD was removed by gel filtration.

2.3. Purification of α and β. *C. reinhardi* CF_1 was purified as previously
described (Selman-Reimer et al., 1981). The subunits were dissociated by
dissolving 10 mg of purified CF_1 in 2 mL of a solution containing 10 mM
sodium phosphate, pH 6.6, 50 mM DTT, 0.1 mM EDTA, 8 M urea. The subunits
were applied to a CM-Sephadex C-50 (Pharmacia, Sweden) column (2.5 cm x
10 cm) equilibrated in the above buffer with 1 mM DTT. The column was
developed under starting conditions. The isolated subunits were either
stored frozen in the column buffer or the subunits were dialyzed extensively
against 10 mM Na-tricine, pH 7.5, 10% glycerol, 5 mM DTT, 5 mM ATP, 5 mM
$MgCl_2$ and stored in the same buffer.

3. RESULTS

Since Piccioni et al. (1981) have determined that the order of migration
of the α and β subunits of the *C. reinhardi* CF_1 are reversed on SDS urea
gels with respect to that of the α and β subunits of spinach CF_1, the *C.
reinhardi* subunit of highest apparent molecular weight was purified
(described below) and antibodies were raised against this purified
preparation. The antiserum was tested for its cross-reactivity with
the subunits of spinach CF_1. This antibody preparation reacted with the α
subunit of spinach CF_1 (Figure 1A), leading to the conclusion that the *C.
reinhardi* CF_1 subunits can also be named α, β, γ and ε in order of
decreasing molecular weight. In order to confirm this, an antiserum
prepared against the purified β subunit of spinach CF_1 was tested for its
reactivity with the subunits of *C. reinhardi* CF_1. In this case,
the antiserum reacted with the *C. reinhardi* subunit of lower apparent
molecular weight (Figure 1B).

Sybesma, C. (ed.), Advances in Photosynthesis Research, Vol. II. ISBN 90-247-2943-2.
© *1984 Martinus Nijhoff/Dr W. Junk Publishers, The Hague/Boston/Lancaster.*

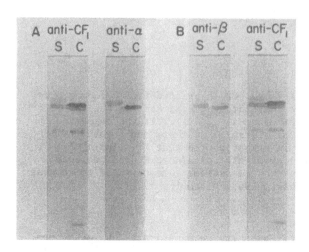

FIGURE 1. Identification of the α and β subunits of *C. reinhardi* CF₁. The antibody binding and detection are described in Materials and Methods. S, spinach CF₁; C, *C. reinhardi* CF₁. Antibodies to *C. reinhardi* CF₁ and its α subunit were raised in rabbits. Antiserum against the β subunit of spinach CF₁ was a gift from Dr. André T. Jagendorf.

Another probe for the β subunit of coupling factors is its reactivity toward DCCD. This chemical modifier reacts primarily with a single carboxyl group on the β subunit of several coupling factors (Yoshida et al., 1982). Upon incubation of *C. reinhardi* CF₁ with [^{14}C]DCCD, label was incorporated primarily into the subunit with a lower apparent molecular weight (Figure 2).

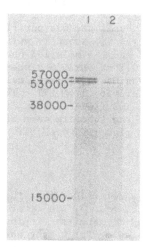

FIGURE 2. Location of the covalently bound [^{14}C]DCCD. The subunits were separated on a 15% acrylamide gel in the buffer system of Laemmli (1970). Lane 1, stained with Coomassie R-250; Lane 2, fluorogram of gel.

This modification of the *C. reinhardi* CF₁ can be correlated with a loss of the ATPase activity catalyzed by the enzyme (not shown). Thus the labelled subunit can be concluded to be the β subunit on the basis of the results of Pougeois et al. (1979) and Shoshan and Selman (1980).

The experiments described above were repeated on the gel system described by Weber and Osborn (1969) with exactly the same results.

The separation of the α and β subunits of the *C. reinhardi* CF₁ was accomplished by isocratic elution of the sample components from a cation exchange column (Figure 3). The leading edge of the first peak contains

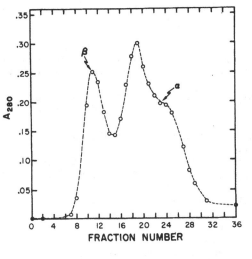

FIGURE 3. <u>Elution of α and β from a CM-Sephadex column in 8 M urea.</u> The column was run and the samples were eluted isocratically as described in Materials and Methods. Fractions (2 mL) were monitored by their absorbance at 280 nm and were analyzed by SDS gel electrophoresis on 15% polyacrylamide gels. Fractions 10-13 contained the β subunit and were pooled. Fractions 19-25 contained the α subunit.

the β subunit, usually not contaminated with α, whereas the trailing end of the second peak usually contains pure α. As expected, the separation of α and β is more efficient at lower pH. Since α has a slightly higher pI than β (not shown), the order of elution off of the cation exchange column is as predicted. γ and ε do not elute off of the column even if NaCl is added to 0.5 M concentration. Antibodies raised against the pure α subunit bind to a single band on SDS gels of thylakoid membrane proteins. The α and β subunits are soluble in the absence of urea and can be stored in the buffer used for dialysis (Materials and Methods).

4. DISCUSSION

The α and β subunits of the chloroplast coupling factor have been purified by a simple ion exchange column. Although γ and ε do not elute off of this column, it should be possible to obtain these subunits by appropriate changes in the pH and ionic strength of the eluting factors. At present, conditions are being developed to improve the separation between the peaks containing the α and β subunits. Since the conditions used for subunit separation are relatively mild, and since the α and β subunits are soluble in the absence of urea, it may be possible to study the function of these subunits either in reconstitution experiments or perhaps alone as has been done in the case of the bacterial coupling factor (Dunn, Futai, 1980) and in the case of the δ subunit of CF₁ (Andreo et al., 1982). The *C. reinhardi* CF₁ is very stable and reconstitution experiments or binding studies with the isolated subunits may be possible. The identification of these subunits on SDS gels is, of course, important to assignment of function and we have therefore attempted to identify these subunits on gel systems that are commonly used to resolve the CF₁ subunits (Weber, Osborn, 1969; Laemmli, 1970). On these two gel systems, the subunits of *C. reinhardi* CF₁ migrate in the same order as do the subunits of spinach CF_1. We have also tried to identify α and β on SDS gels in 8 M urea but are unable to resolve the α and β subunits on this gel system. However, in the case of the spinach CF_1 α and β subunits, the presence of 8 M urea in the running gel does not change the order of migration of these two subunits (Piccioni et al., 1982).

REFERENCES

Andreo CS, Patrie WJ and McCarty RE (1982) Effect of ATPase activation and the δ subunit of coupling factor 1 on reconstitution of photophosphorylation. J. Biol. Chem. 257, 9958–9975.
Dunn SD and Futai M (1980) Reconstitution of a functional coupling factor from the isolated subunits of *Escherichia coli* F_1 ATPase. J. Biol. Chem. 255, 113–118.
Laemmli U (1970) Cleavage of structural proteins during the assembly of the head of bacteriophage T4. Nature 227, 680–685.
Piccioni RG, Bennoun P and Chua N-H (1981) A nuclear mutant of *Chlamydomonas reinhardi* defective in photosynthetic phosphorylation. Eur. J. Biochem. 117, 93–102.
Piccioni R, Bellemare G and Chua N-H (1982) Methods of polyacrylamide gel electrophoresis in the analysis and preparations of plant polypeptides. In Edelman et al., eds. Methods in chloroplast molecular biology, pp. 925–1014. The Netherlands: Elsevier Biomedical.
Pougeois R, Satre M and Vignais PV (1979) Reactivity of mitochondrial F_1ATPase to dicyclohexylcarbodiimide. Biochemistry 18, 1408–1413.
Selman-Reimer S, Merchant S and Selman BR (1981) Isolation, purification and characterization of coupling factor 1 from *Chlamydomonas reinhardi*. Biochemistry 20, 5476–5482.
Shoshan V and Selman BR (1980) The interaction of N,N'-dicyclohexyl-carbodiimide with chloroplast coupling factor 1. J. Biol. Chem. 255, 382–389.
Towbin H, Staehelin T and Gordon J (1979) Electrophoretic transfer of proteins from polyacrylamide gels to nitrocellulose sheets. Procedure and some applications. Proc. Natl. Acad. Sci. USA 76, 4350–4354.
Weber K and Osborn M (1969) The reliability of molecular weight determinations by dodecyl sulfate polyacrylamide gel electrophoresis. J. Biol. Chem. 244, 4406–4412.
Yoshida M, Allison WS, Esch FS and Futai M (1982) The specificity of carboxyl group modification during the inactivation of the *Escherichia coli* F_1-ATPase with dicyclohexyl[^{14}C]carbodiimide. J. Biol. Chem. 257, 10033–10037.

ACKNOWLEDGMENTS

This work was supported by grants from the National Institutes of Health (1R01-GM 31384) and the College of Agricultural and Life Sciences, University of Wisconsin, USA. S.M. acknowledges a fellowship from the Wisconsin Alumni Research Foundation and a travel award from the American Society of Plant Physiology.
Author's address: Sabeeha Merchant/Department of Biochemistry, 420 Henry Mall, Madison, Wisconsin, 53706, USA.

STOICHIOMETRY AND FUNCTION OF THE δ - SUBUNIT OF CF_1

RICHARD J. BERZBORN, PETER ROOS, GERT BONNEKAMP

Biochemie d. Pflanzen, Ruhr-Universität Bochum, D-4630 Bochum 1, FRG

1. INTRODUCTION AND PROBLEM

As a coupling factor of photophosphorylation any protein can be defined which is isolated from chloroplasts and reconstitutes uncoupled thylakoid membranes. The treatment for resolution usually is incubation in low salt in the presence of EDTA. A coupling factor purified from such extract need not have catalytic activity, even at the membrane, but may be required for regulation of H^+ flux or for binding of the peptides containing the active center of the ATPsynthase. Such a role was ascribed to the δ peptide of CF_1, since re-addition of δ is necessary for reconstitution (Nelson,Karny,1976; Younis et al. 1977; Roos, Berzborn, 1982, Andreo et al. 1982).

If δ would be a binding protein for CF_1 - as could indeed be shown for the ana- logous system of E.coli (Sternweis, Smith, 1977) and for the thermophil bacte- rium PS3 (Yoshida et al.,1977) -, this would have important implications for the biochemistry of interaction of CF_1 with the membrane integral part CF_0 of the ATPsynthase complex. Therefore we wanted to study the need of δ for recon- stitution, and in particular the binding properties of both δ and δ-free CF_1.

2. METHODS AND APPROACH

Since binding of 4-subunit CF_1 (short: $CF_1$4) and peptide δ may be a necessary requirement for reconstitution of EDTA treated thylakoids (uncoupled), but not a sufficient condition for photophosphorylation activity, the binding cannot be studied by reconstitution experiments, but has to be determined itself.

For this reason we developed a specific quantitative micromethod, the rocket analysis or immunoelectrodiffusion, for both CF_1 and δ, without the need for previous extraction and independent of purity, activity, and several chemicals usually present in biochemical samples (Berzborn et al. 1981; Roos,Berzborn, 1983). For electroimmunodiffusion with both our antisera (s.below) peptide δ had to be dissociated from CF_1 by deoxycholate before the electrophoresis (Roos,Berzborn, inprep.).

We used this method to follow the purification of the peptide δ without spec- troscopic characteristics or enzymatic activity, to determine for the first time the yields of this peptide, to measure precisely the content in several preparations, derived from chloroplasts. Taking the mol. weight of CF_1 and δ from the literature a stoichiometry of 3 δ / 1 CF_1 is calculated.

From the immunochemical measurements of the binding of $CF_1$4 and δ we conclude, that peptide δ is indeed a coupling factor, i.e. essential for reconstitution of photophosphorylation in EDTA uncoupled thylakoids, but not a $CF_1$4 binding protein.

3. RESULTS

3.1. Isolation of δ and δ-free $CF_1$4

For immunization first a peptide δ cut out from SDS gels was used (serum 120).

Sybesma, C. (ed.), Advances in Photosynthesis Research, Vol. II. ISBN 90-247-2943-2.
© *1984 Martinus Nijhoff/Dr W. Junk Publishers, The Hague/Boston/Lancaster.*

By help of this antiserum, that contained crossreacting precipitating antibodi
(short ab) against the native conformation of δ, the purification of δ and t
production of δ-free CF₁4 was monitored by electroimmunodiffusion. It turned o
that the isolation procedure of Nelson,Karny,1976 and Younis et al.,1977 yield
only 3-4% of the initial amount of δ. The fraction discarded after the pyridin
step contained 4 times more δ than the one used for purification. If the separ
tion after pyridine treatment was done on Sephacryl S 200 in the presence of
urea instead, (fig. 1), about 35% yield of nearly pure δ fractions could be ob
tained. They were further purified on a second Sephacryl column or on DEAE-cel
lulose. Also the antiserum 167 against this urea δ crossreacted with the nativ
conformation.

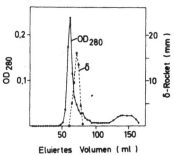

FIG.1: Fractionation of CF₁ and δ on
Sephacryl S 200 after dissociati
by pyridine. Peptide δ was determined b
the specific antiserum 120 by rocket
electroimmunodiffusion.

It also could be determined by electroimmunodiffusion that the earlier CF₁4 wa
not free of δ, but only depleted of this peptide. Chloroform CF₁ according to
Younis et al.,1977, contained in our hands between 10 and 25 % of the initial
amount of δ. To remove this peptide completely the CF₁4 had to be treated with
EDTA treated thylakoids (Roos,Berzborn,in prep.).

3.2. Function of subunit δ

1. Since CF₁4 binds to EDTA treated thylakoids (and specifically, since saturating at 100 % of the control), peptide δ is not essential as a binding protein. But this binding of CF₁4 did not reconstitute photophosphorylation,
neither catalytically nor even structurally (fig.2).

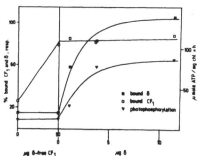

FIG.2: Reconstitution of photophosphorylation of EDTA treated thylakoids
by CF₁4 and δ (phopshorylation
with PMS; binding measured by rocket
electroimmunodiffusion). anti δ serum
120, anti CF₁ serum 152.

2. Addition of pure δ, however, stimulated the phosphorylation capacity of
these EDTA treated thylakoids saturated with CF₁4 up to 80 % of the control
(fig. 2).
3. Addition and binding of this δ-preparation did not increase CF₁4 binding.
4. The peptide δ bound to EDTA treated membranes without re-addition of CF₁,
also specifically, since saturating at 100 % (data not shown).
5. However, this preparation of δ did not reconstitute phosphorylation alone.

3.3 Stoichiometry of δ / CF_1

According to Binder et al.1978, also isolated peptide δ can be determined by the Lowry method with BSA as standard. Thus we were able to calibrate the rocket lengths with pure δ. Therefore beyond relative determinations absolute measurements of amounts of δ became possible. According to these specific micro-determinations 8 mg CF_1 contained 1 mg peptide δ (table 1).

TABLE 1: Determination of the stoichiometry δ/CF_1 by immunochemical method.
A = determination of protein ratio; B = calculation of stoichiometry

A immunochemical data

μg CF_1 *	analyzed	mm δ-rocket	corr. to μg δ	μg CF_1 / μg δ
2.0	7	0.25	8.0 : 1
2.7	10	0.35	8.8 : 1

B calculation using the mol. weight of CF_1 and δ from literature

CF_1	δ	expected μg CF_1 / μg δ with		
		CF_1 cont. 1 δ	CF_1 cont. 2 δ	CF_1 cont. 3 δ
310.500	17.500	17.7	8.9	5.9
325.000	21.000	15.5	7.7	5.2
417.000	21.000	19.8	9.9	6.6
420.000 plus x times 21.000	21.000	$\frac{441}{21} = 21.0$	$\frac{462}{42} = 11.0$	$\frac{483}{63} = 7.7$

* The relative amount of δ and CF_1 in this isolated CF_1 was identical to thyl.

As can be seen from the table, one has to take 3 copies of subunit δ per CF_1 to arrive at the right proportion of protein, if the molecular weight of 420.000 is taken, recently determined for $CF_1$4 (without δ) from Chlamydomonas (Merchant et al., 1983).

4. CONCLUSIONS AND DISCUSSION

4.1. Role of δ and position of δ

In the native thylakoid δ seems to play a crucial role in the ATPsynthase, i.e. it is an essential coupling factor, but not the binding and connecting protein for CF_1.
We have several hints now (agglutination of thylakoids by anti δ is slow; inhibition of photophosphorylation is sluggish and difficult to reproduce; CF_1CF_0 complex does not give a δ-rocket without previous dissociation; double-rockets during δ dissociation), which in contrast to our previous model (Roos,Berzborn, 1982) led us to consider the possibility of δ being at least partially hidden within the CF_1CF_0 complex at the thylakoid. This conclusion would be in agreement with other results (Vialle,A,Junge W, in press).

4.2. Stoichiometry of δ/CF_1 and fate of δ in preparations

Recently the molecular weight of CF_1 was revised (Yoshida et al.1979; Merchant et al.1983). If 417 KD or even 420 KD for a δ free $CF_1$4 is taken, we cannot

escape the conclusion that each CF_1 containes 3 δ peptides. This is not in con
tradiction to the result of the experiment with ^{14}C of Süß,Schmidt,1982, but on
in disagreement with their interpretation; there were actually enough counts f
3 δ. A stoichiometry of 3 x α (59000) + 3 x β (56000) + 1 x γ (33000) + 3 x δ
(21000) + 1 x ϵ (16000) would result in a total MW of CF_1 of 451.000.
We have several results indicating that in most preparations CF_1 has lost part
of its δ peptides, especially if the usual (large scale) preparations are performed. This may be one reason why purified CF_1 often reconstitutes less than
crude extracts. We hope to understand in the near future the biochemistry of t
dissociation of δ from the membrane and diverse CF_1 preparations.

5. REFERENCES

Andreo CS, Patrie WJ, McCarty RE (1982) Effeft of ATPase activation and the δ
   subunit of coupling factor 1 on reconstitution of photophosphorylation.
   J.Biol.Chem. 257, 9968-9975
Berzborn RJ, Müller D, Roos P, Andersson B (1981) Significance of different
   quantitative determinations of photosynthetic ATPsynthase CF_1 for hetro-
   geneous CF_1 distribution and grana formation. in: Proc. 5th Intern. Congr.
   on Photosynthesis, Vol. III, pp. 107-120 (G.Akoyunoglou,ed.) Balaban Sci.
   Serv. Philadelphia
Binder A, Jagendorf AT,Ngo E (1978) Isolation and composition of the subunits
   of spinach chloroplast coupling factor protein. J.Biol.Chem. 253,3094-3100
Merchant S, Shaner SL, Selman BR (1983) Molecular weight and subunit stoichio-
   metry of chloroplast coupling factor 1 from chlamydomonas reinhardi.
   J.Biol.Chem. 258, 1026-1031
Nelson N, Karny O (1976) The role of δ subunit in the coupling activity of
   chloroplast coupling factor 1 FEBS Lett. 70, 249-253
Roos P, Berzborn RJ (1982) Is subunit δ indeed required for CF_1 binding to the
   chloroplast membrane? in: Proc. 2nd Europ. Bioenergetics Conf., Lyon -
   Villeurbanne, CNRS ed. pp. 99-100
Roos P, Berzborn RJ (1983) Electroimmunodiffusion—a powerful tool for quanti-
   tative determinations of both soluble and membrane bound chloroplast
   ATPase CF_1. Z.f. Naturforsch. in press
Sternweis PC,Smith J B (1977) Characterization of the purified membrane attach-
   ment (δ) subunit of the proton translocating adenosine triphosphatase from
   E. coli. Biochemistry 16, 4020-4025
Süß KH,Schmidt O (1982) Evidence for an α_3,β_3,γ,δ,I,II,ϵ,III$_5$ subunit stoichio-
   metry of chloroplast ATPsynthetase complex (CF_1-CF_0). FEBS Lett. 144,213-18
Yoshida M, Sone N, Hirata H, Kagawa Y (1977) Reconstitution of ATPase of thermo
   philic bacterium from purified individual subunits.J.Biol.Chem.252,3480-85
Yoshida M, Sone N, Hirata H, Kagawa Y, Ui N (1979) Subunit structure of ATPase.
   Comparison of the structure in thermophilic bacterium PS3 with those in mitochondria, chloroplasts and E.coli. J.Biol. Chem. 254, 9525-9533
Younis HM, Winget GD, Racker E (1977) Requirement of the δ subunit of chloro-
   plast coupling factor 1 for photophosphorylation.J.Biol.Chem.252,1814-1818

6. ACKNOWLEDGEMENTS

This research was supported by a grant from the DFG to R.J.B.
Excellent and reliable technical assistance of Mrs. M. Caesar-Tobien and Mrs.
R. Owurah-Nkruma is appreciated.

TITLE Analysis of Amino Acids at the Active Site of Chloroplast Coupling
Factor One: A Spin-Echo nmr Study.

NAME Wayne D. Frasch and Robert R. Sharp

INTRODUCTION

The identification of the amino acids on coupling factor one (CF_1)
which participate in binding substrate during the synthesis of ATP has
been limited thus far to studies employing the use of reagents that covalently modify specific amino acids on the enzyme (Vallejos, 1981). In
cases where this technique has been successful, the reagent will cause an
irreversible loss of enzymatic activity which is prevented by the presence
of high concentrations of substrates during the modification. These
experiments suggest that a carboxyl group, a lysine, an arginine and a
tyrosine are in the proximity of the active site of CF_1.

The following experiments explore the use of spin-echo ^{1}H-nmr as an
independent technique for identification of the amino acids at the active
site of CF_1.

METHODS

CF_1 was purified as described by Frasch and Selman (1982). Spin-echo
^{1}H-nmr spectra were obtained at 360 MHz (Bruker WM360) and 37°C.

RESULTS

The 360 MHz ^{1}H-nmr spectrum (non-spin-echo) of CF_1 is shown in Fig. 1.
For proteins of high molecular weight (ca. 400,000) the nmr linewidth of
methylene proteins constrained to reorient rigidly with the enzymes is
~300Hz (~0.9 ppm). Most of the observed intensity results from broad unresolved methylene bands of this type due to sidechains in ordered regions.

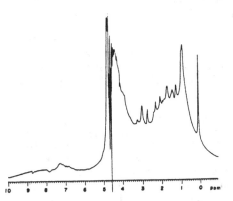

Fig. 1. 360 MHz ^{1}H nmr spectrum of
purified CF_1. The soluble enzyme
(18 mg/ml in D_2O, pD 8.4) observed
at 37 C with solvent suppression.
Capillary insert contains 50µM tri-
methylsilylpropanesulfonic acid, 1.2
mM $MnCl_2$, D_2O. Line broadening was
2 Hz and 500 transients.

Superimposed on these broad bands is a smaller amount of more highly resolved
structure which results from methyls, which undergo facile internal rotation, and a small subset of methylenes on polar or charged sidechains on
the surface of the enzyme, which extend into the aqueous phase and experience

Sybesma, C. (ed.), Advances in Photosynthesis Research, Vol. II. ISBN 90-247-2943-2.

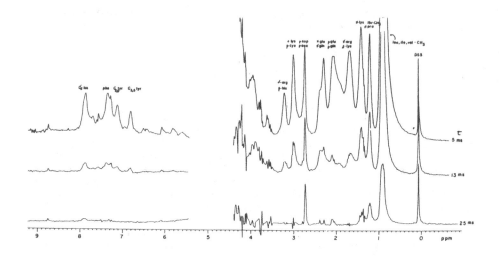

Fig. 2. Spin-echo spectra of DTT-activated CF_1 (28 mg/ml in D_2O, pD 8.4, 37 C).

relatively free segmental motion. In order to suppress the unresolved background intensity such that the motionally narrowed resonances are enhanced, spin-echo spectra were recorded over a range of 90°-180° pulse spacings between 2.5 and 25 ms. Typical spectra, shown in Fig. 2, demonstrate the increased resolution of these highly mobile amino acids. By comparison with the reported chemical shifts in short peptides (McDonald, Phillips, 1969), the peaks at 0.8 ppm and 0.9 ppm coincide with methyl resonance positions of non polar amino acids which exhibit rapid rotation about a 3-fold axis. Threonine methyl resonance is resolved at 1.2 ppm. Methylene resonances of charged amino acid sidechains appear between 1.5 and 3.7 ppm. These include lysine, arginine, glutamate, aspartate, methionine and cysteine. Near 4.3 ppm appear resonance bands resulting from α-methine protons on the polypeptide backbone as well as from contaminating sugars from Sephadex. The aromatic amino acids (phenylalanine, tyrosine and histidine) produce resonances from 6.7-8.2 ppm. The T_2's are relatively short (< 10ms) for all resonances with the exception of the aspartate β-methylene, for which T_2 > 10ms.

The spin-echo spectra of CF_1 clearly monitor charged amino acid sidechains at the enzyme surface, possibly including functional groups involved in nucleotide binding at the active site. To test the possibility that the active site could be examined by spin-echo nmr, we have formed difference spectra of the enzyme before and after incubation with 5'-adenylyl-β, -imidodiphosphate (AMPPNP), a nonhydrolyzable substrate analog. In the presence of 25mM NaCl, MgAMPPNP has been found to bind tightly (K_D = 2μM). Spectra were first made using 50μM CF_1 in the presence of 25mM NaCl. The ATPase activity had been activated by preincubation in 50mM dithiothreitol (DTT) for 90 min then was removed by Sephadex G-50 fine chromatography (10 x 0.5 cm) prior to accumulation of spectra. Spectra

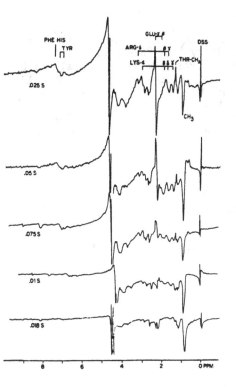

Fig. 3. Spin-echo difference spectra of DTT-activated CF_1 versus DTT-activated CF_1 having MgAMPPNP tightly bound. Purified CF_1 (10.4 mg) was activated by incubation in 50 mM DTT for 90 m. The DTT was then removed by Sephadex G-50 fine chromatography. Following accumulation of spectra in 25 mM NaCl and 30 mM Borate pD 9.0, 150 µM MgAMPPNP was added and, after a 60 min incubation, a second set of spectra were obtained. Spin-echo difference spectra ± MgAMPPNP were recorded at the τ values shown.

were subsequently obtained with the enzyme following a 60 min incubation in 150µM MgAMPPNP. Difference spectra (Fig. 2) were formed under conditions where the methyl peak of 2,2-dimethyl-2-silapentane-5-sulfonic acid (DSS), added as an internal standard, was most nearly nulled. Some changes in linewidth of the DSS reference usually occurred thus, interfering with accurate cancellation of this peak, but the protein peaks in duplicate spectra obtained 3 hr apart were found to cancel almost completely.

Difference spectra ± MgAMPPNP revealed small but significant differences. In the aromatic region, positive peaks due to tyrosine and either histidine (H-4) or phenylalanine indicates immobilization of these residues. An intense methyl peak near 0.9 ppm is narrowed upon binding, indicating increased mobility. In the CH_2 region, glutamate shows profound changes which certainly involved more than a single residue, as is evident from the complex relaxation behavior of different components of the peak. Other changes in the CH_2 region are complex and overlapping, but positive peaks near 2.7, 3.0 and 3.3 ppm due to immobilization of the β-aspartate, ε-lysine and δ-arginine methylenes seem to be defined clearly in spectra at short τ, and a negative peak due to the aspartate β-CH_2 is defined at long τ.

REFERENCES

Cantley LC and Hammas GG (1975) Characterization of Nucleotide Binding Sites on Chloroplast Coupling Factor 1. Biochemistry 14, 2968-2975.
Frasch WD and Selman BR (1982) Mechanism of Phosphorylation Catalyzed by Chloroplast Coupling Factor 1. Stereochemistry. Biochemistry 21, 3636-3643.
McDonald CC and Phillips WD (1969) Proton Chemical Shifts in Amino Acids and Peptides, J. Am. Chem. Soc. 91, 1513-1525.
Vallejos RH (1981) Chemical Modification of Chloroplast Coupling Factor 1. in, Energy Coupling in Photosynthesis, Selman BR and Selman-Reimer S eds. Elsevier North Holand pp. 129-140.

ACKNOWLEDGEMENTS

This research was supported by gratns from the Rackham Foundation for Graduate Studies #135101 and the Pheonix Memorial Research Foundation #362413 to WDF, the USDA-CRGO (82-CRCR-1-1047) to RRS and from the USDA-CRGO #8300811 to WDF and RRS.
Authors address: The University of Michigan, Ann Arbor, MI 48109 USA

CHEMICAL MODIFICATION OF ESSENTIAL AMINO-ACID RESIDUES IN THE CHROMATOPHORE F_1-ATPase AND ITS ISOLATED β-SUBUNIT.

Z. GROMET-ELHANAN AND D. KHANANSHVILI

INTRODUCTION

The elucidation of the mechanism of action of the $F_0 \cdot F_1$-ATP synthase ATPase complex is dependent on a precise determination of the function of each of its individual subunits. From a large number of studies using different approaches the β-subunit of the F_1-sector has been suggested to contain the catalytic site. Attempts at the possible identification of functional amino-acids in this site have been carried out by using labeled chemical modifiers, known to interact with specific amino-acid residues, which bind to the F_1-ATPase and inactivate it. After dissociating the labeled enzyme complex to its individual subunits the label was detected mainly, but not exclusively, on the β-subunit (Futai, Kanazawa, 1980; Cross, 1981). It is, therefore, most interesting to test the effect of such reagents on an isolated, reconstitutively active, β-subunit and compare it with their effect on the homologous F_1.

We have earlier developed a method for the complete removal of the β-subunit from the membrane bound $F_0 \cdot F_1$-ATP synthase of Rhodospirillum rubrum (Philosoph et al., 1977) and purified it to homogeneity (Khananshvili, Gromet-Elhanan, 1982a). The resulting β-less chromatophores lost all their ATP synthesis and hydrolysis activities, but they could be fully restored upon rebinding the purified missing β-subunit. This system provides a unique possibility to apply the chemical modifiers directly on the isolated purified β-subunit (Khananshvili, Gromet-Elhanan, 1982b), test their effect on its reconstitutive activity, and compare the results with those obtained by modification of the depleted and coupled chromatophores as well as their isolated RrF_1 (Khananshvili, Gromet-Elhanan, 1983a,b).

Here we summarize such studies with diethyl pyrocarbonate (DEPC), 4-chloro-7-nitrobenzofurazan (NBD-Cl), and dicyclohexylcarbodiimide (DCCD), known to modify histidine, tyrosine and carboxyl groups, respectively (Miles, 1977; Futai, Kanazawa, 1980; Cross, 1981). They were found to inactivate the RrF_1-ATPase and to bind to the purified and reconstitutively active β-subunit, but the modified β-preparations exhibited marked differences in their capacity to rebind to β-less chromatophores and restore their activity. The relevance of these results to the identification of functional amino-acid residues in the active site of the RrF_1-ATPase is discussed.

IDENTIFICATION OF ESSENTIAL AMINO-ACID RESIDUES IN THE RrF_1-ATPase.

The Ca^{2+}-ATPase activity of RrF_1 was fully inhibited by incubation with the carboxyl group reagents DCCD and EEDQ as well as with NBD-Cl (Khananshvili, Gromet-Elhanan, 1983a,b). The inactivation by NBD-Cl was shown to be correlated with the modification of tyrosyl residues, since it was accompanied by the appearance of an absorption peak at 385 nm and both the inhibition and the absorption peak were reversed upon addition of dithiothreitol. These reagents have been reported to inhibit various other soluble F_1-ATPases suggesting the involvement of tyrosine and carboxyl groups in the F_1 catalytic site (Futai, Kanazawa, 1980;

Sybesma, C. (ed.), Advances in Photosynthesis Research, Vol. II. ISBN 90-247-2943-2.
© *1984 Martinus Nijhoff/Dr W. Junk Publishers, The Hague/Boston/Lancaster.*

Cross, 1981). But up to now there was no information concerning the involvement of histidine in the activity of any F_1-ATPase.

We have examined this point in the RrF_1 by following its interaction with DEPC (Khananshvili, Gromet-Elhanan, 1983c). This reagent was shown to inhibit the activity of many enzymes by carbethoxylating various amino-acid residues but, when applied at pH 6.0, it was found to modify histidyl residues with considerable specificity (Miles, 1977). When added in one step DEPC did inhibit the RrF_1 but, since it is readily hydrolyzed in aqueous solutions, a 400 fold molar excess of reagent over enzyme was required and under these conditions the inactivation was accompanied by modification of tyrosyl as well as histidyl residues. The RrF_1-ATPase could, however, be inactivated by much lower concentrations of DEPC, when they were applied in a stepwise manner, and the only absorption change observed during the stepwise addition of DEPC was an increase at 242 nm (Khananshvili, Gromet-Elhanan, 1983c) due to a specific modification of histidyl residues (Ovadi et al., 1967).

Fig. 1 illustrates the relationship between the inactivation of RrF_1 by the stepwise addition of DEPC and the number of histidyl residues modified by it. A linear line is obtained for up to 45% inactivation. When this initial portion of the curve is extrapolated to zero activity, it can be calculated that a modification of up to 3 mol of histidyl residues per mol of enzyme is required for complete inactivation of RrF_1. A linear relationship has also been obtained between the binding of [^{14}C]DCCD and [^{14}C]NBD-Cl to RrF_1 and the loss of its ATPase activity (Fig. 1). But, the extrapolation of these binding data to zero activity indicates that for complete inactivation of RrF_1 the binding of 1 mol of [^{14}C]DCCD, and between 1 to 2 mol of [^{14}C]NBD-Cl per mol RrF_1 is required. The different binding or modification stoichiometries shown to be required for the complete inhibition of RrF_1 by the three reagents (Fig. 1) indicate the problematics of using such data as evidence for the number of catalytic sites on the F_1-ATPase or for their cooperative interactions.

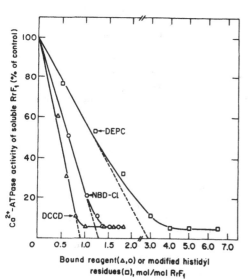

Fig. 1. Correlation between inhibition of the RrF_1-ATPase and the modification of histidyl residues by DEPC or the binding of [^{14}C]DCCD and [^{14}C]NBD-Cl to the RrF_1. RrF_1 (4 mg/ml) was inactivated by 250 µM DEPC added at 10 min intervals in portions of 50 µM. After each addition the Ca^{2+}-ATP activity and histidine modification were measured as described by Khananshvili, Gromet-Elhanan (1983c). RrF_1 (1 mg/ml) was incubated with 100 µM [^{14}C]DCCD or 300 µM [^{14}C]NBD-Cl for various time intervals, freed from untreated reagent and assayed for activity and [^{14}C] binding as described by Khananshvili, Gromet-Elhanan (1983a,b).

DENTIFICATION OF ESSENTIAL AMINO-ACID RESIDUES IN THE β-SUBUNIT.

Both $[^{14}C]$DCCD and $[^{14}C]$NBD-Cl have been found to bind covalently also to the isolated, purified and reconstitutively active β-subunit (Fig. 2). Their binding is very similar, following a typical saturation curve, that results in binding of up to 1 mol of either reagent per mol of β. However, the relationship between the covalent binding of DCCD and NBD-Cl to the β-subunit and their effect on its reconstitutive activity shows a completely different pattern (Fig. 3). With DCCD the amount of reagent bound to β is linearly related to the decrease in its reconstitutive activity and extrapolation of the binding data to zero reconstitutive activity indicates that, as with the RrF1 (see Fig. 1), the incorporation of up to 1 mol of $[^{14}C]$DCCD per mol β results in its complete inactivation. On the other hand, the binding of NBD-Cl to β does not affect at all its reconstitutive activity (Fig. 3).

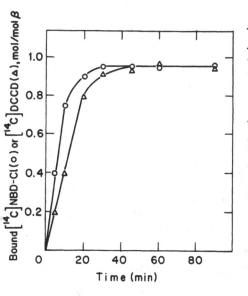

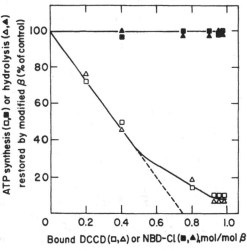

Figure 2. Time course of incorporation of $[^{14}C]$NBD-Cl or $[^{14}C]$DCCD into the purified β-subunit. β was incubated with 300 μM NBD-Cl or 100 μM DCCD and at the indicated intervals was freed from untreated reagent and assayed for $[^{14}C]$ radioactivity as described in Fig. 1.

Figure 3. Correlation between the binding of $[^{14}C]$NBD-Cl or $[^{14}C]$DCCD to β and the inhibition of its capacity to rebind to β-less chromatophores and restore their ATP-linked activities. This capacity was measured as described by Khananshvili and Gromet-Elhanan (1982b).

DISCUSSION

We have earlier shown that both the $[^{14}C]$NBD-β and $[^{14}C]$DCCD-β adducts are completely stable (Khananshvili, Gromet-Elhanan, 1982b;1983a). But, whereas the first adduct does rebind to β-less chromatophores without any dissociation of the $[^{14}C]$NBD from it, the second adduct does not rebind at all. These observations, together with the data of Fig. 3 indicate that the NBD-β adduct rebinds to the β-less chromatophores and restores their activity as efficiently as native untreated β, while the DCCD-β adduct does not rebind to the β-less chromatophores and consequently cannot restore their activity. The results presented here lead to the following conclusions:

1. The single NBD-Cl binding site existing on the isolated β-subunit is not essential for its rebinding and reconstitutive activity. It is also not involved in the inhibition of soluble or membrane-bound RrF_1 by NBD-Cl, since the β-less chromatophores, that have been reconstituted with the fully active saturated NBD-β adduct, can still be inactivated by incubation with NBD-Cl (Khananshvili, Gromet-Elhanan, 1983a).

2. There must, therefore, be on the RrF_1 an additional binding site for NBD-Cl, which is responsible for the inhibition of its catalytic activity (see Fig. 1). The subunit location of this additional site is as yet unknown. It could be located on the assembled, but not on the isolated, β-subunit and in this case the β-subunit assembled in RrF_1 could have two NBD-Cl binding sites. But it could also be located on another subunit. Further experiments will be required to test these possibilities.

3. The single DCCD binding site existing on the isolated β-subunit is absolutely essential for its rebinding to the β-less chromatophores. This site might also be the one responsible for the inhibition of the RrF_1-ATPase by DCCD, but conclusive evidence for this point will require an identification of the modified specific carboxyl group.

REFERENCES

Cross RL (1981) The mechanism and regulation of ATP synthesis by F_1-ATPases, Ann. Rev. Biochem. 50, 681-714.
Futai M and Kanazawa H (1980) Role of subunits in proton-translocating ATPase ($F_0 \cdot F_1$), Curr. Top. Bioenerg. 10, 181-215.
Khananshvili D and Gromet-Elhanan Z (1982a) Isolation and purification of an active γ subunit of the $F_0 \cdot F_1$-ATP synthase from chromatophore membranes of Rhodospirillum rubrum, J. Biol. Chem. 257, 11377-11383.
Khananshvili D and Gromet-Elhanan Z (1982b) Chemical modification of the β-subunit isolated from a membrane-bound $F_0 \cdot F_1$-ATP synthase, Biochem. Biophys. Res. Commun. 108, 881-887.
Khananshvili D and Gromet-Elhanan Z (1983a) The interaction of 4-chloro-7-nitrobenzofurazan with Rhodospirillum rubrum chromatophores, their soluble F_1-ATPase, and the isolated purified β-subunit, J. Biol. Chem. 258, 3714-3719.
Khananshvili D and Gromet-Elhanan Z (1983b) The interaction of carboxyl group reagents with the Rhodospirillum rubrum F_1-ATPase and its isolated β-subunit, J. Biol. Chem. 258, 3720-3725.
Khananshvili D and Gromet-Elhanan Z (1983c) Modification of histidine residues by diethyl pyrocarbonate leads to inactivation of the Rhodospirillum rubrum RrF_1-ATPase, submitted for publication.
Miles EW (1977) Modification of histidyl residues in proteins by diethyl-pyrocarbonate, Methods in Enzymol. 47, 431-442.
Ovadi J, Libor S and Elodi P (1967) Spectophotometric determination of histidine in proteins with diethylpyrocarbonate, Acta Biochim. Biophys. Acad. Sci. Hung. 2, 455-458.
Philosoph S, Binder A and Gromet-Elhanan Z (1977) Coupling factor ATPase complex of Rhodospirillum rubrum. Purification and properties of a reconstitutively active single subunit, J. Biol. Chem. 252, 8747-8752.

Authors Address: Department of Biochemistry, Weizmann Institute of Science, Rehovot 76100, Israel

THE COUPLING FACTOR (Ca-ATPase) OF THE CYANOBACTERIUM SPIRULINA PLANTENSIS

DAVID B. HICKS, CHARLES F. YOCUM

INTRODUCTION

Energy transduction in cyanobacteria is poorly understood relative to our understanding of phosphorylation in eukaryotic photosynthesis. As part of a study of electron transport and phosphorylation in the cyanobacterium Spirulina platensis, our laboratory extracted and partially purified a protein with latent Ca-ATPase activity that reconstituted photophosphory-lation in ATPase-depleted membranes (Owers-Narhi et al. 1979). In the present report we describe the purification to homogeneity and some of the characteristics of the Spirulina platensis coupling factor Ca-ATPase (SF$_1$).

MATERIALS AND METHODS

Sonic vesicles from S. platensis were prepared as described (Owers-Narhi et al. 1979). After washing the membranes with 10 mM sodium pyrophosphate, the Ca-ATPase was extracted either with 2 mM Tricine/1 mM EDTA or with chloroform. The latter was accomplished using the procedure of Piccioni et al. 1981, except that solubilization was done at 4 C. Other procedures essentially followed established methods: DEAE sepharose chromatography, sucrose gradient centrifugation, and ATPase assay (Jagendorf 1982), ATP synthesis (Avron 1960), and gel electrophoresis (Laemmli 1970). One unit of ATPase activity is defined as 1 µmole Pi formed per minute.

RESULTS

Table 1 summarizes the stability of Ca-ATPase activity exhibited by partially purified SF$_1$ after 5 weeks of storage under the indicated conditions. Complete activity is retained only when SF$_1$ is stored at 4 C in the presence of 10% glycerol and nucleotide. Therefore, we performed all purification procedures subsequent to extraction of the enzyme from the membrane under these conditions.

TABLE 1. Stability of Ca-ATPase activity of partially purified SF$_1$ under different storage conditions.

Addition(s)		Temperature	% of initial activity
glycerol (10%), ATP (1 mM)		4 C	100%
" "		room temp	77
"	(no ATP)	4 C	71
"	"	room temp	50
ATP	(no glycerol)	4 C	10
"	"	room temp	56

Sybesma, C. (ed.), Advances in Photosynthesis Research, Vol. II. ISBN 90-247-2943-2.
© *1984 Martinus Nijhoff/Dr W. Junk Publishers, The Hague/Boston/Lancaster.*

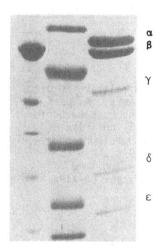

FIGURE 1. SDS Page of SF₁ and CF₁.
Standards are BSA (66,000), ovalbumin
(45,000), trypsinogen (24,000), B-lacto-
globulin (18,400) and lysozyme (14,300).
The subunits of CF₁ are labeled on
the right. SF₁ was purified from
chloroform extracts of sonic vesicles
prepared from <u>Spirulina platensis.</u>

SF₁ Standards CF₁

SF₁ can be extracted with EDTA or by chloroform treatment of pyrophosphate-
washed membranes. The enzyme from either extract was purified according
to the scheme given in Table 2 (in this instance we show a representative
scheme for EDTA-extracted SF₁). After ion-exchange chromatography, SF₁
still has substantial RuBP carboxylase contamination and must be purified
on two successive sucrose gradients to give a homogeneous preparation of
SF₁ (i.e., a single band on overloaded nondenaturing 6% polyacrylamide
gels).

When electrophoresed on 12% SDS gels, the chloroform-extracted SF₁
yields five bands (Fig. 4), four of which have similar molecular weights
to the five subunits of the spinach coupling factor (CF₁). The α and β
subunits of SF₁ poorly resolved on 12% gels. The fifth band, which
has a molecular weight of about 28,500, does not correspond to any of
the subunits of CF₁. We are currently undertaking experiments to
determine whether this polypeptide is actually a part of the ATPase
complex.

TABLE 2. SF₁ purification scheme

Purification step	Protein(mg)	Total Activity(U)	Specific Activity (U/mg)	Yield
EDTA supernatant (from 44 mg chl)	211	130.6	0.6	-
DEAE sepharose	17	76.5	4.5	59%
1st sucrose gradient	2.8	48.3	17.3	37
2nd sucrose gradient	1.0	23.9	23.9	18

SF_1 has an absolute requirement for a divalent cation for ATPase activity, like other coupling factors, and, like CF_1, prefers Ca to Mg (data not shown). Although the heat treatment used to elicit CF_1 activity does not activate SF_1, the latency of SF_1 is overcome by trypsin treatment, and activated SF_1 exhibits substantial rates of ATP hydrolysis (20-24 U/mg). To compare the trypsin requirements of SF_1 and CF_1, we followed the time course of trypsin activation using a range of trypsin concentrations. In Table 3, we show the time course of two trypsin concentrations to illustrate the point that SF_1 requires more trypsin and longer digestion times than does CF_1. Note that maximal activation of SF_1 is achieved only when the trypsin concentration is five times greater than for CF_1 (on a weight/weight basis).

TABLE 3. Time course of trypsin activation at two different trypsin concentrations.

ATPase	Trypsin, mg/mg ATPase	% of maximal activity				
		0	5	15	30	minutes
CF_1	1	20%	100	61	15	
"	5	22%	68	41	15	
SF_1	1	18%	36	62	65	
"	5	21%	70	100	29	

A crucial test of a competent coupling factor is its ability to reconstitute photophosphorylation in ATPase-depleted membranes. In Table 4, we show that the purified EDTA-extracted SF_1 is a reconstitutively active couping factor when added back to EDTA-treated membranes. The membrane preparation was washed partially free of carboxylase and the resolved membranes were stored in DTT, as suggested by Nelson, Eytan (1979). These membranes exhibit lower rates of photophosphorylation, probably due to loss of phycocyanin, but show a greatly increased affinity for SF_1 as evidenced by the fact that the amount of protein (on a chlorophyll basis) that yields saturating rates of ATP synthesis is much lower than that required in previous preparations (Owers-Narhi et al. 1979).

TABLE 4. Reconstitution of photophosphorylation of ATPase-depleted membranes by purified SF_1

	mg SF_1 added/ mg chl	μmoles ATP synthesized/ hr·mg chl
control membranes	-	164
EDTA-washed membranes	0	0
"	0.5	25
"	1	43
"	2	41
"	4	54
"	20	51

DISCUSSION

An EDTA- and chloroform-extractable protein with latent Ca-ATPase activity can be purified from the cyanobacterium Spirulina platensis using procedures established for higher plant coupling factors (CF_1). This prokaryotic enzyme hydrolyzes ATP at good rates and is reconstitutively active. Although SF_1 differs from CF_1 in its stability at 4 C, SF_1 resembles CF_1 in a number of ways, including a) latency, b) activation by trypsin, c) metal preference (Ca over Mg, data not shown), and d) ability to reconstitute photophosphorylation in ATPase-depleted membranes.

REFERENCES

Avron, M (1960) Photophosphorylation by swiss-chard chloroplasts, Biochim. Biophys. Acta 40, 257-272.
Jagendorf, AT (1982) Isolation of chloroplast coupling factor (CF_1) and of its subunits. In Edelman M et al. eds. Methods in chloroplast molecular biology, pp. 881-898, Elsevier Biomedical Press, Amsterdam.
Laemmli UK (1970) Cleavage of structural proteins during the assembly of the bacteriophage T4, Nature 227, 680-685.
Nelson N, Eytan E (1979) Approach to the membrane sector of the chloroplast coupling device. In Mukohata Y, Packer L eds. Cation fluxes across biomembranes, pp. 409-415, Academic Press, New York.
Owers-Narhi L et al. (1979) Reconstitution of cyanobacterial photophosphorylation by a latent Ca^{2+}-ATPase, Biochem. Biophys. Res. Comm. 90, 1025-1031.
Piccioni RG et al. (1981) A nuclear mutant of Chlamydomonas reinhardtii defective in photosynthetic photophosphorylation, Eur. J. Biochem. 117, 93-102.

ACKNOWLEDGEMENT

This research was supported by a grant (82-CRCR-1-1127) from USDA/SEA competitive research grants office to C.F.Y.
Authors' address: Division of Biological Sciences, The University of Michigan, Ann Arbor, MI 48109 USA

PROPERTIES AND INTRACELLULAR LOCALIZATION OF A CYANOBACTERIAL (CA^{2+}, MG^{2+})-
ATPase

WOLFGANG LOCKAU

The cyanobacterium Anabaena variabilis ATCC 29413 contains an ATPase which
can be distinguished from its coupling factor of phosphorylation. The enzyme appears to be membrane-bound (complete sedimentation by 1 h centrifugation at 100 000 x g), depends on Mg^{2+} for activity and is stimulated
3 - 4-fold by micromolar concentrations of Ca^{2+} (Lockau, Pfeffer 1982).
Attempts are reported to determine the intracellular localization and the
function of this unusual bacterial enzyme.

SEPARATION OF THE ATPase FROM THE THYLAKOIDS
As shown in Fig. 1, the ATPase activity of disrupted spheroplasts of Anabaena
separates into two peaks upon sucrose density gradient centrifugation, one
corresponding to the thylakoids, the other one not. The rather low ATPase
activity recovered with the thylakoids (less than 10 nmol/mg protein x min)
may be at least partially due to the coupling factor. The ATPase recovered
in the fractions indicated by the bar in Fig. 1B has a higher activity
(50 - 100 nmol/mg x min). Unlike the coupling factor, this ATPase is not
stimulated by limited trypsin digestion nor does it react with an antiserum
to the F_1-type ATPase (Lockau, Pfeffer 1982). The distribution of the cytochrome oxidase and of the cytochrome f/b_6-complex follows the distribution
of chlorophyll, i.e. of thylakoids.

FUNCTION OF THE ATPase
The light membranes carrying an ATPase (indicated by the bar in Fig. 1B)
accumulate calcium upon addition of Mg-ATP (Fig. 2). Calcium accumulation
is inhibited by A 23187 (an ionophore of divalent cations) and by Na_3VO_4
(an inhibitor of ATPases which form a phosphorylated intermediate). Na_3VO_4
also inhibits the ATPase activity, A23187 does not. Calcium accumulation
is resistant to ionophores of monovalent cations, as shown for the channel-
former gramicidin, and to DCCD, which inhibits the cyanobacterial coupling
factor. Thylakoids purified by density gradient centrifugation showed no
or low accumulation of calcium (less than 5% of that observed with the light
membranes). These results indicate that the ATPase is bound to membrane
vesicles other than the thylakoids and that it is a primary calcium pump.
The calcium-accumulating vesicles are probably inside-out vesicles derived
from the plasma membrane. This conclusion is further supported by the observation that these vesicles were, on a protein basis, up to ten times
more heavily labelled than the thylakoids when the cyanobacterium was pretreated with the hydrophilic probe (^{35}S)-p-diazoniumbenzenesulfonate prior
to membrane isolation (not shown).

CALCIUM REQUIREMENT FOR HETEROTROPHIC GROWTH OF ANABAENA
Bacteria usually do not require calcium for heterotrophic growth, a well-
-documented exception being Azotobacter (see Rosen 1982). It is long known
that cyanobacteria require calcium under photoautotrophic conditions; the
ion appears to be necessary for the function of photosystem 2 (Brand et al.
1983). Anabaena requires calcium also for dark heterotrophic growth (Fig. 3).

Sybesma, C. (ed.), Advances in Photosynthesis Research, Vol. II. ISBN 90-247-2943-2.
© *1984 Martinus Nijhoff/Dr W. Junk Publishers, The Hague/Boston/Lancaster.*

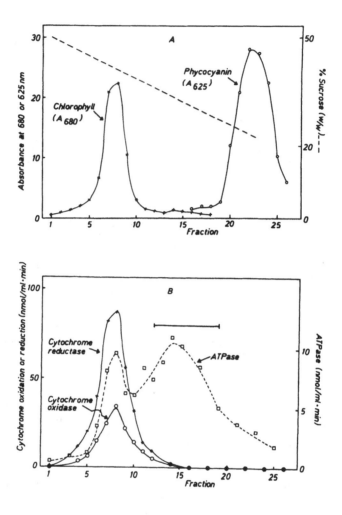

Fig. 1: Sucrose density gradient centrifugation of mechanically disrupted spheroplasts of Anabaena.
The alga was lysozyme-treated, disrupted and dialyzed as reported (Lockau, Pfeffer 1982). The extract was applied to a linear gradient of sucrose and centrifuged for 17 h at 130 000 x g. ATPase activity was assayed with 6 mM Mg-ATP, 0.1 mM $CaCl_2$ at pH 8. Cytochrome oxidase activity was determined from the initial rate of oxidation of 15 μM reduced cytochrome c from horse heart at pH 7.5. The distribution of the cytochrome f/b_6 complex (cytochrome reductase) was measured at pH 6.2 with 50 μM plastoquinol-1 and 25 μM cytochrome c from horse heart as substrates (Krinner et al. 1982).

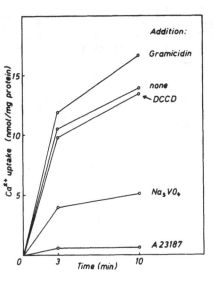

Fig. 2: Calcium accumulation by the presumptive plasma membrane.

The fractions indicated by the bar in Fig. 1B were concentrated by a 1 h centrifugation at 100 000 x g. Calcium accumulation was assayed with $^{45}CaCl_2$ using a filtration technique (Lockau, Pfeffer 1983). At 0 min, uptake was initiated by addition of ATP to 3 mM. The reaction mixture contained 5 mM $MgCl_2$. The values are corrected for calcium bound in the absence of ATP (about 1.5 nmol/ mg protein).

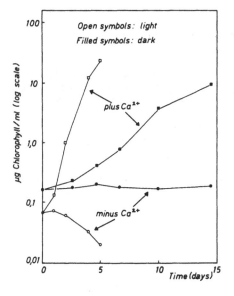

Fig. 3: Growth requirement of Anabaena for calcium in the light and in the dark.

The alga was grown in the medium of Allen and Arnon (1955) supplemented with 8 mM fructose, 5 mM KNO_3, 8 mM HEPES-NaOH (pH 7.5) and 0.25 mM EGTA. Light: White light of 2500 lux. T = 30°C. Free EGTA itself is not an inhibitor of growth: When it was used as a calcium buffer (5 mM EGTA, 4.75 mM $CaCl_2$), growth in the dark was observed. Results similar to those shown in the figure were obtained when N_2 or NH_4^+ replaced NO_3^- as the nitrogen source. Minus Ca^{2+}: No $CaCl_2$ added to the medium; plus Ca: 0.5 mM $CaCl_2$ added.

DISCUSSION
There is good evidence that the calcium-transporting ATPase can be used as
a marker enzyme for the plasma membrane of Anabaena. So far, only a transport function but no substantial electron transport function can be attributed to this membrane, unlike findings with Anacystis (Peschek et al. 1983)
In vivo the calcium pump probably excretes calcium from the cell, since
calcium is transported away from the side of ATP hydrolysis (into the isolated vesicles).
The precise role of calcium for heterotrophic growth of Anabaena, as of
Azotobacter, is unknown. Since both calcium requirement and primary calcium
pump are unusual features for a bacterium, one might speculate that the
(Ca^{2+},Mg^{2+})-ATPase is not simply a "bilge pump" which keeps the intracellula
concentration of free calcium low. In this context it should be noted that
a rapid, transient influx of calcium seems to be involved in the photophobi
response of Phormidium (Häder,Poff 1982).

REFERENCES

Allen MB and Arnon DI (1955) Studies on nitrogen-fixing blue-green algae.I.
Plant. Physiol. 30, 366 - 372.
Brand JJ, Mohanty P and Fork DC (1983) Reversible inhibition of photochemistry of photosystem II by Ca^{2+} removal from intact cells of Anacystis
nidulans, FEBS Lett. 155, 120-124.
Häder D-P and Poff KL (1982) Dependence of the photophobic response of the
blue-green alga, Phormidium uncinatum, on cations, Arch. Microbiol 132,
345-348.
Krinner M, Hauska G, Hurt E and Lockau W (1982) A cytochrome f-b$_6$ complex
with plastoquinol - cytochrome c oxidoreductase activity from Anabaena
variabilis, Biochim. Biophys. Acta 681, 110-117.
Lockau W and Pfeffer S (1982) A cyanobacterial ATPase distinct from the
coupling factor of photophosphorylation, Z. Naturforsch. 35C, 658-664.
Lockau W and Pfeffer S (1983) ATP-dependent calcium transport in membrane
vesicles of the cyanobacterium, Anabaena variabilis, Biochim. Biophys.
Acta (in print).
Peschek GA, Schmetterer G, Lauritsch G, Muchl R, Kienzl PF and Nitschmann
WH (1983) Proton-pumping cytochrome oxidase in the cytoplasmic membrane of
Anacystis nidulans. In Papageorgiou GC and Packer L, eds. Photosynthetic
prokaryotes, pp. 147-162. Amsterdam: Elsevier.
Rosen BP (1982) Calcium transport in microorganisms. In Carafoli E, ed.
Membrane transport of calcium, pp. 187-216. New York: Academic Press.

ACKNOWLEDGEMENTS
I am indebted to Dr. G. Hauska for discussions, S. Pfeffer for technical
assistance and the Deutsche Forschungsgemeinschaft (SFB 43) for financial
support.

Author's address: Wolfgang Lockau, Institut für Botanik, Universität
                  Regensburg, 8400 Regensburg, F.R.G.

A $(Mg^{2+}-Ca^{2+})$-STIMULATED ATPase IN SPINACH CHLOROPLAST ENVELOPES:
ISOLATION BY CALMODULIN AFFINITY CHROMATOGRAPHY

T.D. NGUYEN/P.A. SIEGENTHALER

1. INTRODUCTION

A Mg^{2+}-dependent ATPase, insensitive to N,N'-dicyclohexylcarbodii-
mide is associated with chloroplast envelopes (Douce et al. 1973). This
enzyme has a greater affinity for Mn^{2+} than for Mg^{2+} (Joyard and Douce
1975). The partial inhibition of the envelope ATPase by oligomycin has
suggested that this enzyme may play a role in mediating H^+ efflux and
K^+ uptake in chloroplasts (Maury et al. 1981).

Calmodulin is present in leaf tissues (Watterson et al. 1980; Muto,
1982; Jarrett et al. 1982) mainly in the cytosol (90%) and to a lesser
extent in mitochondria (5-9%), chloroplasts (1-2%) and the microsomal
fraction (1%). In plant tissues, calmodulin modulates the activity of
at least three enzymes : NAD^+ kinase, Ca^{2+}-ATPase and quinate : NAD^+
oxidoreductase (Marmé 1982). In chloroplast, calmodulin seems to be
confined in the stroma (Jarrett et al. 1982). Recently, calmodulin
antagonists (chlorpromazine, phenothiazine) were shown to inhibit
photochemical reactions in spinach chloroplasts (Barr and Crane 1982).
It is likely that calmodulin is involved in the regulation of photosynthesis (Jarrett et al. 1982) by interacting with NAD^+ kinase (maybe with
an ATPase), if not from the stroma at least from chloroplast envelope.

The first aim of this study was to investigate the properties of the
chloroplast envelope-bound ATPase, namely the effect of Mg^{2+}, Ca^{2+} and
calmodulin on its activity. The second aim was to isolate, out of the
21 chloroplast envelope proteins separated by isoelectric focusing
(Siegenthaler and Nguyen, 1983), a protein which had a specific affinity
for calmodulin and which displayed an ATPase activity. Finally, the
properties of the two ATPases were compared.

2. MATERIALS AND METHODS

Spinach chloroplast envelopes were prepared according to Douce et al.
(1973) and resuspended in buffer A (50 mM Tris-HCl, pH 7.8, 300 mM
sucrose). Envelopes were solubilized with 5 mg Triton X-100/mg protein
at 4°C for 15 min, then 1 mM $MgCl_2$ and 0.1 mM $CaCl_2$ were added. The
unsolubilized material was pelleted at 100 000 x g for 30 min at 4°C.
The supernatant containing the solubilized ATPase was loaded onto a
calmodulin-Sepharose column which was equilibrated with buffer B (buffer
A with 1 mM $MgCl_2$, 0.1 mM $CaCl_2$, 0.05% Triton X-100 and 1 mM 2-mercapto-
ethanol). Bovine brain calmodulin (100 mg) was purified essentially
according to Gopalakrishna and Anderson (1982) and was coupled to 7 g
CNBr-activated Sepharose 4B [see Pharmacia instructions]. The column was
washed first with buffer B and then with buffer B + 0.5 M NaCl. Proteins
were eluted with buffer B without $CaCl_2$ but with 1 mM EGTA. The ATPase
activity was assayed by measuring the release of Pi according to Lebel
et al. (1978). Proteins were estimated according to Bradford (1976).

Sybesma, C. (ed.), Advances in Photosynthesis Research, Vol. II. ISBN 90-247-2943-2.
© *1984 Martinus Nijhoff/Dr W. Junk Publishers, The Hague/Boston/Lancaster.*

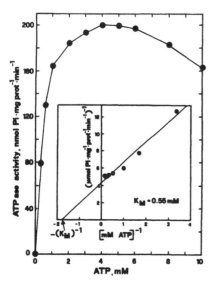

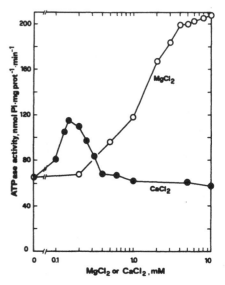

Fig. 1. ATP dependence of envelope-bound ATPase activity. The reaction mixture contained 50 mM Tris-HCl (pH 7.8), 300 mM sucrose, 5 mM MgCl$_2$, envelopes (10-15 µg protein/250 µl) and ATP as indicated. Incubation: 15 min, 37°C.

Fig. 2. Stimulation of envelope-bound ATPase activity by MgCl$_2$ and CaCl$_2$. Conditions as in Fig. 1 but ATP = 4 mM and concentrations of cations as indicated.

RESULTS AND DISCUSSION

The envelope-bound ATPase activity as a function of ATP concentration followed Michaelis-Menten kinetics with an apparent K_M value for ATP of 0.55 mM as determined by the double-reciprocal plot of 1/V versus 1/S (inset of Fig. 1). Compared to the value found by Joyard and Douce (1975), our enzyme had a greater affinity for ATP by a factor of 1.4. A basal ATPase activity, EDTA-insensitive, was associated with chloroplast envelopes. It was stimulated by divalent cations with maximal rates at 0.15 mM CaCl$_2$ and 5 to 10 mM MgCl$_2$ (Fig. 2). In the presence of Ca^{2+} and Mg^{2+}, calmodulin further stimulated the activity (28-63%, see Table I). The enzyme was sensitive to oligomycin, LaCl$_3$ and NH$_4$VO$_3$.

Envelope proteins which were bound to the calmodulin-Sepharose column in the presence of calcium and eluted by EGTA, showed two bands on native and isoelectric focusing gels (pIs 7.3 and 6.0). This EGTA-fraction contained an ATPase, the properties of which were quite similar to those of the envelope-bound enzyme. The K_M (ATP) was 0.45 mM (Fig. 3) and maximal rates were at 0.05 mM CaCl$_2$ and 1 mM MgCl$_2$ (Fig. 4). In the presence of both cations, calmodulin caused a 40% stimulation of the ATPase activity (Table I).

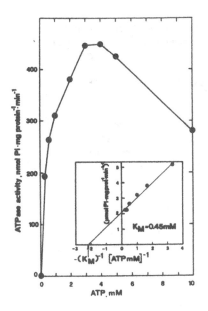

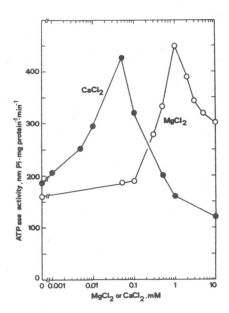

Fig. 3. ATP dependence of partially purified ATPase activity. Conditions as in Fig. 1 but MgCl₂ = 1 mM, ATPase (1-3 μg protein/250 μl), 60 min.

Fig. 4. Stimulation of partially purified ATPase activity by MgCl₂ and CaCl₂. Conditions as in Fig. 3 but ATP = 4 mM.

TABLE I. Activation of the membrane-bound and partially purified ATPase from spinach chloroplast envelopes

Specific activity : nmol Pi·mg Prot^{-1}·min^{-1}

	Membrane bound ATPase		Partially purified ATPase
Control[a]	180[c]	185[d]	420[c]
+ calmodulin[b]	231[c]	301[d]	586[c]

[a]Conditions as in Fig. 3; [b]200 μg/ml for the envelope-bound and 40 μg/ml for the purified ATPase; [c]Summer spinach; [d]Spring spinach.

In conclusion, the use of calmodulin-Sepharose affinity chromatography enabled to isolate quickly a protein fraction enriched in ATPase activity. This partially purified enzyme had properties which were similar to those of the envelope-bound ATPase, in particular both were stimulated by Ca^{2+} and Mg^{2+} ions and by calmodulin. In addition to the two RuBP carboxylase subunits and the phosphate translocator, the present ATPase seems to be the fourth protein identified in spinach chloroplast envelopes. Due to its special properties, this enzyme might well modulate the exchanges of ions, metabolites and/or proteins between the chloroplast and the cytosol.

REFERENCES

Barr R and Crane FL (1982) Ca^{2+} and calmodulin antagonists inhibit the proton gradients associated with non-cyclic and cyclic photophosphorylation in spinach chloroplasts, Biochem. Biophys. Res. Commun. 109, 1215-1221.

Bradford M (1976) A rapid and sensitive method for the quantitation of microgram quantities of protein utilizing the principle of protein-dye binding, Anal. Biochem. 72, 248-254.

Douce R, Holtz RB and Benson AA (1973) Isolation and properties of the envelope of spinach chloroplasts, J. Biol. Chem. 25, 7215-7222.

Gopalakrishna R and Anderson WB (1982) Ca^{2+}-induced hydrophobic site on calmodulin : application for purification of calmodulin by phenyl sepharose affinity chromatography, Biochem. Biophys. Res. Commun. 104, 830-836.

Jarrett HW, Brown CJ, Black CC and Cormier MJ (1982) Evidence that calmodulin is in the chloroplast of peas and serves a regulatory role in photosynthesis, J. Biol. Chem. 257, 13795-13804.

Joyard J and Douce R (1975) Mn^{2+}-dependent ATPase of the envelope of spinach chloroplasts, FEBS Lett. 51, 335-340.

Lebel D, Poirier GG and Beaudoin AR (1978) A convenient method for the ATPase assay, Anal. Biochem. 85, 86-87.

Marmé D (1982) The role of Ca^{2+} in signal transduction of higher plants, Plant Growth Subst., Proc. Int. Conf. 11th, 419-426.

Maury WJ, Huber C and Moreland DE (1981) Effects of magnesium on intact chloroplasts, Plant Physiol. 68, 1257-1263.

Muto S (1982) Distribution of calmodulin within wheat leaf cells, FEBS Lett. 147, 161-164.

Siegenthaler PA and Nguyen TD (1983) Proteins and polypeptides of envelope membranes from spinach chloroplasts. I. Isoelectric focusing and sodium dodecyl sulfate polyacrylamide gel electrophoresis separations, Biochim. Biophys. Acta 722, 226-233.

Watterson DM, Iverson DB and Van Eldik LJ (1980) Spinach calmodulin: isolation, characterization, and comparison with vertebrate calmodulins, Biochemistry 19, 5762-5768.

ACKNOWLEDGEMENTS

The authors wish to thank Miss C. Bachmann for arranging and typing the manuscript and the Swiss National Science Foundation for financial support (Grant no. 3.661.0.80 to P.A.S.).

Authors address : TD Nguyen/PA Siegenthaler, Laboratoire de Physiologie végétale, Université de Neuchâtel, Chantemerle 20, CH-2000 Neuchâtel, Switzerland.

MODE OF ACTION OF THE ATP-ASE INHIBITOR TRI-N-BUTYL-TIN IN RHODOPSEUDOMONAS SPHAEROIDES CELLS.

André VERMEGLIO*, Jean-Marie GALMICHE**, Guy GIRAULT**
* Association pour la Recherche en Bioénergie Solaire, C.E.N. Cadarache, BP n°1, 13115 Saint-Paul-lez-Durance, France
**Service de Biophysique, C.E.N. Saclay, 91191 Gif-sur-Yvette Cedex, France

1. INTRODUCTION

Specific inhibitors of the H^+ translocating ATP-ase are important tool in the study of bioenergetic processes of intact photosynthetic bacteria. Dicyclohexyl carbodiimide (DCCD) (P.C. Maloney, 1979) has been used with some success. More recently, venturicidin (N.P.J. Cotton et al., 1981) has been shown to behave as an energy transfer inhibitor in intact cells of Rhodopseudomonas Capsulata. Although organotin compounds (trialkyltin and triaryltin halides) inhibit the ATP-ase of mitochondria (M.S.Rose, J.N. Aldridge, 1972, K. Cain et al., 1977) and of chloroplasts (J.M. Gould, 1976, 1978) no report has been published concerning their action in intact photosynthetic bacteria. This is the object of the present work.

2. MATERIALS AND METHODS

Rhodopseudomonas Sphaeroides 2-4-1, grown at 30°C in degazed Hutner medium, were harvested after 24 hours and resuspended in fresh growth medium. Light-induced absorption changes were recorded with a double beam spectrophoto-meter (Aminco DW 2a). ATP, ADP and AMP concentrations in resting or photo-synthesising cells were determined as follows : tipically, the cells sus-pension (2ml in a plastic seringue) was kept in the dark for 15 min to insure anaerobiosis. Addition of TNBT (Tri-N-butyltin) was then effected with or without oxygenation of the sample. After a subsequent dark time of 30 min., the anaerobic suspension was subjected to 30 s. of continuous illumination to promote photophosphorylation. At the end of these opera-tions, the suspension was rapidly injected in 0,6 ml of ice cold HCl (2N), allowed to settle 10 min before neutralization by addition of 0,4 ml Tris (2,5M). ATP, ADP and AMP concentrations were determined on 50 l l aliquots using the luciferin-luciferase assay.

3. RESULTS AND DISCUSSION

As already reported (F. Welsh, L. Smith, 1969, N.P.J. Cotton et al., 1981) the ATP content of intact photosynthetic bacteria increases upon illumina-tion (compare in Table 1, dark and light controls). TNBT (16 γ M) has no effect on the photophosphorylation if the sample has been kept anaerobic during the experiment. On the other hand, addition of TNBT in presence of oxygen (or if oxygenation of the sample is done during the incubation time) severely inhibits the photophosphorylation even if the sample has been made anaerobic afterwards (see Table 1).

Similarly, addition of TNBT in the absence of O_2 has no effect on the light-induced carotenoïd band shift (compare traces IC and the control experiment IB), a direct measure of the membrane potential (C. Wraight et al., 1978). Only, if TNBT addition is done in presence of O_2, a marked difference in the shape of the light-induced change is observed, even if the sample has been made anaerobic afterwards (IE).

Sybesma, C. (ed.), Advances in Photosynthesis Research, Vol. II. ISBN 90-247-2943-2.
© 1984 Martinus Nijhoff/Dr W. Junk Publishers, The Hague/Boston/Lancaster.

		1		2		3		4			
		ATP	ADP	ATP	ADP	ATP	ADP	ATP	ADP	AMP	Total
Dark	Control	20	25	23	20	20	20	30	24	36	90
Light	Control	110	12	88	13	67	12	77	19	0	96
Dark	+ TNBT (N₂)	16	24	18	20	20	23	22	23	35	80
Light	+ TNBT (N₂)	84	6	71	5	75	8	58	18	6	82
Dark	+ TNBT (O₂)	21	43	29	29	23	24	15	26	40	81
Light	+ TNBT (O₂)	46	38	39	44	30	34	29	34	43	106

Table 1 : ATP, ADP and AMP concentrations [(nmol)/µ mol Bchl] for four individual experiments. The nucleotides levels have been determined after 30 min of dark followed (light) or not (dark) by continuous illumination. TNBT (16 µM) was either added in absence (+ TNBT, N_2) or in presence (+ TNBT, O_2) of oxygene but in all cases anaerobiosis of the sample was insured after the addition. Note that the total nucleotides content is fearly constant.

Comparison of the ATP level (Table 1) and light-induced absorption changes suggests that the kinetics of the carotenoïd band shift depicted in trace IE are characterisitic of an inhibited ATP-ase.

We deduce from the above experiments that the induction of TNBT as inhibitor of the photosynthetic bacteria ATP-ase requires the presence of O_2. This phenomenum is however irreversible since, once TNBT has been added in an aerobic sample, anaerobic conditions do not restore the ATP-ase activity (see Table 1 and fig. IE).
The question raised by these experiments is therefore : why does induction of TNBT inhibition requires the presence of oxygen ?
It is obvious that, due to the high respiratory activity of the bacteria the presence of oxygen induces two important changes in their membrane properties : first the quinone pool is mainly in its oxydized state ; secondly a high membrane potential is present.
In an experiment where the quinone pool was reduced even in presence of O_2, by inhibiting the respiratory activity upon addition of KCN (210^{-3} M), TNBT induces, within 5 min, carotenoïd band shift kinetics characteristics of an inhibited ATP-ase (IG).
This result implies that the action of TNBT does not depend on the oxydo-reduction state of the quinone pool. We also show that action of TNBT on the ATP-ase does not require a high membrane potential by the following experiment. We induced a high membrane potential by continuous illumination during 15 min in an anaerobic sample. Addition of TNBT under these conditions did not affect the shape of subsequent light-induced absorption changes (not shown).

524 nm

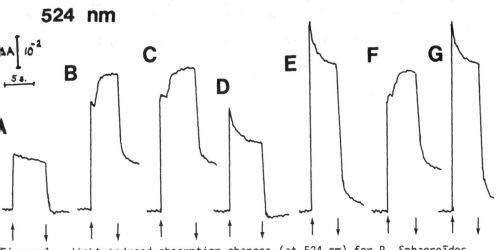

Figure 1 : light-induced absorption changes (at 524 nm) for R. Sphaeroïdes cells (~ 10 μ M Bchl).

. Control experiments : (A) aerobic and (B) anaerobic samples
. Addition of TNBT (16 μ M) under anaerobic conditions : (C) anaerobic sample.
. Addition of TNBT under aerobic conditions : (D) aerobic and (E) anaerobic samples.
. In presence of KCN (210^{-3} M) : (F) aerobic conditions ; (G) after addition of TNBT under aerobic conditions.

4. CONCLUSIONS

We have shown that TNBT has to be added in presence of O_2 to act as an ATP-ase inhibitor. Its action is however irreversible since anaerobic conditions do not restore ATP synthesis. The induction of TNBT inhibition is not related to the oxydo-state of the quinone pool or to the presence of a high membrane potential during active bacteria respiration. One possibility is that molecular O_2 has a direct influence on the configuration of the Fo portion of the ATP-ase. This configuration is may be related to the oxydo-reduction state of -SH residues as proposed J.M. Gould (1978) for chloroplasts ATP ase.

REFERENCES

Cain K, Partis MD and Griffiths DE (1977) Biochem.J. 166, 593-602.
Cotton NPJ, Clark AJ and Jackson JB (1981) Arch. Microbiol.129, 94-99.
Gould JM (1976) Eur.J.Biochem. 62, 567-575.
Gould JM (1978) FEBS Lett. 94, 90-94.
Maloney PC (1979) J. Bacteriol. 140, 1972-205.
Rose MS and Aldridge WN (1972) Biochem.J. 127,´51-59.
Welsh F, Smith L (1969) Biochemistry 8, 3403-3408.
Wraight CA, Cogdell RJ, Chance (1978) in "The photosynthetic bacteria" Clayton RK, Sistrom WR (eds) Plenum Press, New York p 471-511.

II.6.614

Authors address : André Verméglio, Association pour la Recherche en Bioénergie Solaire, C.E.N. Cadarache BP n°1, 13115 Saint-Paul-lez-Durance, France.
Jean Marie Galmiche and Guy Girault, Service de Biophysique, C.E.N. Saclay, 91191 Gif-sur-Yvette, France.

ARSENYLATION AND PHOSPHORYLATION OF ADP AND GDP IN RHODOSPIRILLUM RUBRUM CHROMATOPHORES.
L. SLOOTEN AND A. NUYTEN

1. INTRODUCTION

It was proposed over 20 years ago that arsenate functions as an alternate substrate for the membrane-bound, proton-translocating ATP-synthetase. This enzyme was thought to catalyze the formation of ADP-arsenate which would hydrolyze rapidly (Avron, Jagendorf 1959). Only recently this has been confirmed in mitochondria (Moore, Gresser 1982) and R. rubrum chromatophores (Slooten, Nuyten 1983). Aside from the coupled enzyme assay used to trap ADP-arsenate (Moore, Gresser 1982), we demonstrate here two other methods for the measurement of the synthesis and hydrolysis of ADP-arsenate, and apply them to a determination of some kinetic constants for arsenylation and phosphorylation of ADP and GDP.

2. METHODS

R. rubrum chromatophores were prepared by sonication and stored on liquid nitrogen. They contained some inorganic phosphate (0.04-0.06 mol/mol bacteriochlorophyll) which remained associated with the membranes during at least 5 washings; this supported ATP-synthesis with added ADP. All experiments were done at room temperature, under stirring, in a mixture containing 50 mM KCl, 20 mM NaCl, 0.1 mM EDTA, 2.1 mM MgCl$_2$, and 13 uM bacteriochlorophyll. Other additions were : for the luminescence measurements, 5 mM glycylglycin, 0.2 uM nigericin, 20-50 uM luciferin, 1-2 ug/ml luciferase and NaOH to pH 7.9. For the pH-measurements (Figs. 2 and 3) : as above but pH 8.0 without glycylglycin and with 1 to 1.1 uM nigericin. Luciferin and luciferase were omitted in Fig. 3. In both types of experiments the samples were irradiated with saturating actinic light (wavelength region 830-960 nm). The output of the photomultiplier and/or pH-meter was fed into a strip chart recorder. The phosphate content of the arsenate solutions was less than 0.1 % (molar ratio).

3. RESULTS

Fig. 1 shows a typical trace for the kinetics of luciferase luminescence obtained during illumination with arsenate and ADP. The method used to analyze these curves is also shown. The slow, irreversible phase represents ATP-synthesis supported by endogenous phosphate. The rapid phase in the light, and its reversal in the dark, represent light-dependent synthesis, and non-enzymic hydrolysis, respectively, of ADP-arsenate (Slooten, Nuyten 1983). In the steady state these processes are equally rapid, so that

$$v_{As} = k_d \cdot (ADP\text{-}As)_{ss} \qquad (eq.1)$$

where v_{As} is the arsenylation rate, k_d is the pseudo-first order rate constant for non-enzymic hydrolysis of ADP-arsenate and $(ADP\text{-}As)_{ss}$ is the steady state concentration of ADP-arsenate in the light.
    ADP-arsenylation can also be measured with a pH-electrode (Fig. 2). With sufficient amounts of nigericin and KCl, the electron-transport coupled proton uptake observed upon illumination with arsenate alone,

Sybesma, C. (ed.), Advances in Photosynthesis Research, Vol. II. ISBN 90-247-2943-2.
© 1984 Martinus Nijhoff/Dr W. Junk Publishers, The Hague/Boston/Lancaster.

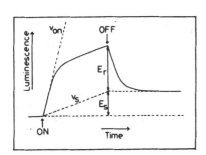

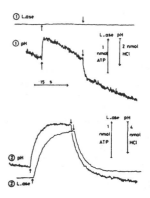

FIG. 1 (LEFT). Kinetics of light-induced luciferase luminescence. Additions:
1 mM arsenate and 25 uM ADP. Illumination time, 21 s. The vertical arrow
shows the luminescence increase due to addition of 0.15 nmol ATP. Volume,
2.1 ml. Analysis : E_s gives the amount of ATP after 21 s of illumination;
v_s gives the corresponding phosphorylation rate. The amount of ADP-arsenate
present in the steady state in the light is given by E_r/φ_R, where φ_R is
the ratio between the quantum yields of ADP-arsenate and ATP for light-
emission from luciferase. The value of φ_R is approx. 0.19 (Slooten,
Nuyten 1983). The arsenylation rate is given by $(v_{on} - v_s)/\varphi_R$.
FIG. 2 (RIGHT). Simultaneous measurement, on the same sample, of light-
induced pH- and luciferase luminescence changes. Additions : 1 uM nigericin
and 0.8 mM arsenate. Exper. 1, no ADP. Exper. 2, 30 uM ADP. Note the
different vertical scale expansions in the two pH-measurements. Volume,
6.4 ml. The drift in the pH-traces is due to CO_2-uptake from the air.

decays completely within a few s after light off (exper. 1). Inclusion of
ADP in the reaction mixture leads to an increase in light-induced medium
alkalinization (exper. 2). The increment is due to non-enzymic proton
binding into ADP-arsenate, and shows a much slower dark decay. This decay
coincides with the dark decay of the luminescence signal and is due to
non-enzymic hydrolysis of ADP-arsenate. From a comparison of the pH- and
luminescence changes during arsenylation and phosphorylation of ADP, the
relative sensitivity of luciferase for ADP-arsenate can be calculated
(Slooten, Nuyten 1983). The result is shown in the legend to Fig. 1. The
amount of H^+ bound into ADP-arsenate in the steady state in the light, as
well as k_d, are obtained from log plots of the dark decay of the light-
induced pH-changes measured in the presence of ADP. The arsenylation rate
(expressed in nmol H^+/min) can then be calculated from eq. 1. Fig. 3
shows an application of this method to GDP, which does not react with
luciferase. The anion concentration, 1 mM, was near-saturating (c.f.
Table I). When phosphate was replaced by arsenate, V_{app} decreased about
50 %, but K_{app}(GDP) increased by 34 % (from 46 to 62 uM).

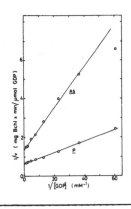

FIG. 3 (LEFT). Kinetics of arsenylation and phosphorylation of GDP, measured with the pH—method, in the presence of 1.1 uM nigericin and 1 mM of either phosphate or arsenate. It is assumed that dH⁺/dGDP = 0.95 during both phosphorylation and arsenylation.

TABLE I (BELOW). Kinetic parameters for phosphorylation and arsenylation of ADP. m +σ; n = 10 (column 1), or 5 (column 2). Conditions : [Anion], 40-1000 uM; [ADP], 1.5-25 uM .

	Product	
	ATP	ADP-arsenate
Vm(umol/min per mg bacteriochlorophyll)	3.21 + 0.73	1.75 + 0.44
K(ADP) (uM)	7.70 + 1.77	5.37 + 2.72
K(anion) (uM)	95.6 + 25.1	106.0 + 35.5
K'(anion) (uM)	30.3 + 16.6	260 + 152

Additions	$k_d(min^{-1})$
None	16.7
8 mM Na₂-arsenate	43.6
8 mM Na₂SO₄	14.2
24mM NaCl	14.4
8 mM MgCl₂	31.9

TABLE II (LEFT). Effect of salts on the pseudo-first order rate constant for non-enzymic hydrolysis of ADP-arsenate. The basal mixture contained 0.4 mM arsenate and 25 uM ADP.

By comparison, Table I shows kinetic parameters for ADP-conversion, obtained from luciferase measurements. Vm was during ADP-arsenate synthesis around 50 % lower than during ATP-synthesis, just as during GDP-conversion. However, K(ADP) (i.e. the Km with saturating [phosphate] or [arsenate]) was during ADP-arsenate lower than during ATP-synthesis (contrast GDP). With saturating [ADP], K(P) was about equal to K(Arsenate).

Non-enzymic hydrolysis of ADP-arsenate is accelerated by a number of ions, specifically arsenate and Mg (Table II). These data were calculated from log plots of the dark decay of the rapid phase in the luciferase measurements. Similar results were obtained from the ratio $v_{As}/(ADP-As)_{ss}$ (not shown, see eq. 1). Phosphate has an effect comparable to that of arsenate (not shown); other ions are ineffective. We propose that these effects are due to formation of transient intermediates such as pyro-arsenate or arsenato-pyro-phosphate. These compounds are stable as solids, but not in solution (Dolique 1958).

4. DISCUSSION

The data shown above indicate that arsenate is a reasonably good substitute for phosphate in ATP-synthetase catalyzed reactions. Table II shows that the effectiveness with which arsenate "uncouples" electron flow can be increased by addition of high concentrations of arsenate and $MgCl_2$; in that case the steady state level of ADP-arsenate, and hence the mass action toward enzymic ADP-arsenate hydrolysis, will be minimized by the high rate constant for non-enzymic ADP-arsenate hydrolysis.

The differential effect of arsenate-substitution on Kapp (GDP) (≈K(GDP)) and K(ADP) can be explained with the simplified reaction scheme :

$$E + S \underset{k_2}{\overset{k_1}{\rightleftharpoons}} ES \underset{k_4}{\overset{k_3}{\rightleftharpoons}} EP \overset{k_5}{\longrightarrow} E + P$$

where S = NDP and P = NTP or NDP-arsenate. In this reaction scheme,

$$k_c = \frac{k_3 k_5}{k_3 + k_4 + k_5} , \text{ and } K(NDP) = \frac{k_2 k_4 + k_2 k_5 + k_3 k_5}{k_1 (k_3 + k_4 + k_5)} .$$

The decrease in Vm when phosphate is replaced by arsenate, reflects probably a decrease in k_c rather than a decrease in the population of active enzyme molecules. This means that either k_5 or k_3 decreases, or k_4 increases when phosphate is replaced by arsenate. Differentiation of K(NDP) with respect to k_3, k_4 or k_5 shows that a decrease in k_5 will always cause a decrease in K(NDP). Only a decrease in k_3 or an increase in k_4 will cause an increase in K(NDP), provided $k_2 > k_5$. This is apparently the case when S = GDP (Fig. 3). If a decrease in the k_3/k_4 ratio is the only effect of arsenate-substitution, then the data shown in Table I (viz. K(ADP) decreases along with Vm when phosphate is replaced by arsenate) would indicate that $k_2 < k_5$ in the case of ADP-turnover. The difference between ADP and GDP would agree well with observations that GDP is bound less tightly than ADP (Strotmann et al. 1977; Banai et al. 1978), and with the high value of K(GDP) as compared to K(ADP), as shown in Fig. 3 and Table I.

REFERENCES

Avron M and Jagendorf AT (1959) J. Biol. Chem. 234, 967-972.
Banai M, Shavit N and Chipman DM (1978) Biochim. Biophys. Acta 504, 100-107.
Dolique R (1958) in"Nouveau traité de chimie minérale" (Pascal P, ed.) Tome XI, pp. 256, 257. Masson, Paris.
Moore SA and Gresser MJ (1982) Fed. Proc. 41, 749.
Slooten L and Nuyten A (1983) Biochim. Biophys. Acta, in the press.
Strotmann H, Bickel-Sandköttner S, Edelmann K, Schlimme E, Boos KS and Lustorff J (1977), in "Structure and function of energy-transducing membranes (van Dam K and van Gelder BF, eds.) pp. 307-314. Elsevier, Amsterdam.

Authors address : Vrije Universiteit Brussel, Fac. Sciences, Lab. Biophysics, Pleinlaan 2, B-1050 Brussels, Belgium.

CHARACTERIZATION AND LOCALIZATION OF THE PEA CHLOROPLAST ENVELOPE Mg^{2+}-ATPase

D.R. McCARTY, B.R. SELMAN AND K. KEEGSTRA

1. INTRODUCTION

The chloroplast envelope plays a central role in mediating transport of metabolites and proteins between cytoplasm and the stroma (Heldt and Heber, 1981). The role, if any, of a membrane ATPase in these processes is as yet unclear. Sabnis et al. (1970) first demonstrated the presence of a Mg^{2+}-dependent ATPase in the chloroplast envelope of Pisum tendril cells. Joyard and Douce (1975) characterized an ATPase activity from spinach chloroplast envelopes. The activity hydrolyzed a broad range of nucleoside triphosphates, was insensitive to N,N dicyclohexylcarbodiimide (DCCD) and was not stimulated by monovalent cations. The purpose of the present study is to characterize the ATPase activity found on the chloroplast envelope of pea and to establish its location with respect to the inner and outer chloroplast envelope membranes.

2. MATERIALS AND METHODS

2.1. Preparation of envelope membranes. Envelope membranes were prepared from purified intact chloroplasts isolated from 2 to 3 week old pea seedlings using methods described by Cline et al. (1981). In brief, purified intact chloroplasts were prepared by Percoll gradient centrifugation and repeated differential centrifugation. Envelope membranes were prepared from hypotonically shrunken, freeze/thaw ruptured chloroplasts by flotation to the 0.3 M/1.2 M sucrose interface on a step sucrose gradient.

2.2. Assay of ATPase activity. Mg^{2+}-dependent ATP hydrolysis was determined as release of ^{32}P from $[\gamma-^{32}P]ATP$ as described previously (Selman-Reimer et al., 1981). The reaction mixture contained 40 mM Tricine (pH 8.0), 1 mM EDTA, 5 mM ^{32}P-ATP (8–24 kBq per reaction), 10 mM $MgCl_2$, and 1–5 µg envelope protein in a total volume of 0.1 ml (at 37°C). Alternatively, release of phosphate from unlabelled substrates was determined colorimetrically by the method of Lanzetta et al. (1976). Protein was determined by the method of Bradford (1976).

3. RESULTS

3.1. Properties of the envelope ATPase. A significant rate of ATP hydrolysis was detected in unfractionated envelope membranes prepared from purified pea chloroplasts. The specific activity ranged from 50–200 nmol/min/mg protein in different membrane preparations. The ATPase activity had a broad pH optimum between 7.0 and 9.5. The activity was not inhibited by oligomycin, DCCD, vanadate, curare, ouabain, or antibodies to pea chloroplast coupling factor. The activity was not stimulated by potassium, carbonyl cyanide m-chlorophenylhydrazone, or valinomycin plus potassium. The ATPase was inhibited 58% by 0.5 mM AMPPNP, a nonhydrolysable analogue of ATP. 90% of the ATPase activity was solubilized by 1% Triton X-100 without significant loss of activity. Figure 1 shows the dependence of the activity on the concentration of various divalent metal cations. Mg^{2+} and Mn^{2+} served equally well as activators of the activity, while Ca^{2+} and Sr^{2+} were much less effective.

Sybesma, C. (ed.), Advances in Photosynthesis Research, Vol. II. ISBN 90-247-2943-2.
© *1984 Martinus Nijhoff/Dr W. Junk Publishers, The Hague/Boston/Lancaster.*

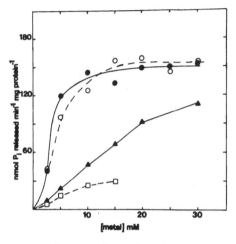

FIGURE 1. Metal cation dependence of the envelope ATPase activity. Reaction mixtures contained 40 mM Tricine-NaOH (pH 8.0), 5 mM ^{32}P-ATP, 1 mM EDTA, 1 μg protein, and the indicated concentration of metal as the chloride salt. ●—●, MgCl$_2$; o—o, MnCl$_2$; ▲—▲, CaCl$_2$; □—□, SrCl$_2$.

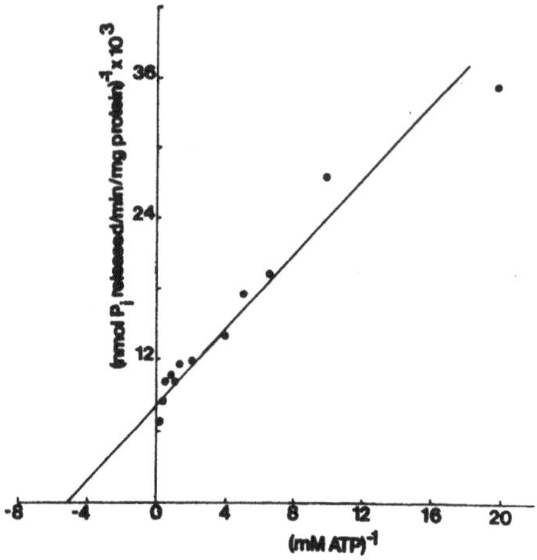

FIGURE 2. Kinetics of ATP hydrolysis. The envelope ATPase was assayed using ^{32}P-ATP as described in Materials and Methods. The concentration of ATP was varied from 50 μM to 5 mM. The data were fit by nonlinear least squares analysis.

TABLE 1. Substrate specificity of the envelope ATPase

Substrate	Specific Activity (nmol Pi/min/mg protein)	% ATPase Activity
ATP	176.6	100
GTP	215.3	122
ITP	167.6	95
CTP	147.8	84
dATP	181.6	103
ADP	53.0*	36
AMP	N.D.	0
PP$_i$	235.0*	235
Fructose-6-phosphate	N.D.	0

N.D. – none detected.
* Control ATPase specific activities for ADP and PP$_i$ were 146.6 and 50.0 nmol/min/mg protein, respectively.

Figure 2 is a Lineweaver-Burke plot for the envelope ATPase. The K_m for Mg-ATP and the V_m were determined to be 0.2 mM and 123 nmol/min/mg protein, respectively. A variety of nucleotides and phosphorylated compounds

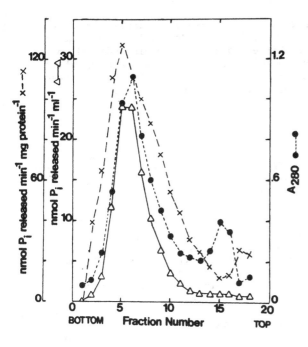

FIGURE 3. Localization of the envelope ATPase. Inner and outer envelope membranes were resolved on a 0.6 M to 1.2 M linear sucrose gradient as described by Cline et al. (1981). Envelope membranes were prepared as described in Materials and Methods. Gradients were centrifuged in a Beckman SW27 rotor at 24,000 rpm for 14 h. Gradients were fractionated from the bottom and assayed for ATPase. ●—●, A280; △—△, ATPase activity; x...x, specific activity.

were tested as substrates for the activity (Table 1). All of the nucleoside triphosphates tried were hydrolyzed, as were ADP and pyrophosphate. AMP and fructose-6-phosphate were not hydrolyzed. The apparent ADPase activity was not affected by the addition of hexokinase and glucose to the reaction; thus, the ADPase activity does not result from adenylate kinase contamination of the envelope membrane preparation. ADP and PP$_i$ may either be alternative substrates for the ATP hydrolyzing activity or be hydrolyzed by distinct enzymes. To distinguish between these possibilities, PP$_i$ and ADP were tested as inhibitors of the ATPase. Concentrations of up to 1 mM PP$_i$ did not inhibit ATP hydrolysis at either 1 mM or 5 mM ATP. On the other hand, PP$_i$ hydrolysis has a K$_m$ of 75 μM. Because PP$_i$ fails to inhibit ATP hydrolysis at concentrations well above the K$_m$ for PP$_i$ hydrolysis, it is unlikely that these two activities share a common active site. ADP competitively inhibits ATP hydrolysis with a K$_{is}$ of 215 μM, which is nearly equal to the K$_m$ for ATP; however, ADP hydrolysis does not saturate in the low millimolar range. Therefore the K$_{is}$ and apparent K$_m$ for ADP must be quite different. This indicates that the ADPase and ATPase are distinct activities as well.

3.2. Localization of the ATPase. Cline et al. (1981) have recently developed methods for resolving the inner and outer chloroplast envelope membranes on a 0.6 M to 1.2 M sucrose gradient. With this technique it has been possible to establish the location of the envelope ATPase. A typical gradient is shown in Figure 3. The higher density peak ($\rho = 1.13$ g/cm^3) consists primarily of a mixture of inner and outer envelope membranes;

the leading edge being enriched in inner envelope membrane. The lighter
peak (ρ = 1.08 g/cm^3) consists primarily of outer envelope membrane.
The peak of ATPase activity is shifted toward the leading edge of the
heavier peak as would be expected for an inner membrane protein, while
the lighter outer membrane peak contains no activity. It is concluded
from these results that the ATPase activity is located on the inner envelope
membrane.

4. DISCUSSION

These results establish the presence of a Mg^{2+}-dependent ATP hydrolyzing activity on the inner chloroplast envelope membrane of Pisum sativum.
Cytochemical evidence for an envelope ATPase in pea was reported by Sabnis
et al. (1970). Joyard and Douce (1975) have characterized an ATPase of
similar specific activity from spinach chloroplast envelope membranes.
Like the spinach enzyme, the pea ATPase is activated equally by Mg^{2+} and
Mn^{2+} and hydrolyzes a broad range of nucleoside triphosphates, but not ADP,
AMP, or monophosphorylated substrates. Although pea chloroplast envelope
membranes have ADPase and pyrophosphatase activity, we conclude that
the activities are distinct from the ATPase activity. The envelope ATPase
differs from putative transport ATPases characterized in other plant
membranes in that it is not inhibited by vanadate or DCCD, nor is it
stimulated by potassium. However, a role for this activity in proton efflux
and ion transport cannot be ruled out, because the envelope vesicles may
be sufficiently leaky that protons and ions can diffuse freely across the
membrane. This might limit any stimulatory effect of K^+ and uncouplers.
Evidence supporting a role for the ATPase in proton transport will depend
on further characterization of the envelope vesicles, and/or purification
and reconstitution of the ATPase into artificial lipid vesicles.

REFERENCES

Bradford M (1976) A rapid and sensitive method for the quantitation of
microgram quantities of protein utilizing the principle of protein-dye
binding, Anal. Biochem. 72, 248.
Cline K, Andrews J, Mersey B, Newcomb E, and Keegstra K (1981) Separation
and characterization of inner and outer envelope membranes of pea
chloroplast, Proc. Natl. Acad. Sci. USA 78, 3595-3599.
Heber U and Heldt HW (1981) The chloroplast envelope: Structure, function,
and role in leaf metabolism, Ann. Rev. Plant Physiol. 32, 139-168.
Joyard J and Douce R (1975) Mn^{2+}-dependent ATPase of the envelope of spinach
chloroplasts, FEBS Lett. 51, 335-340.
Lanzetta PA, Alvarez LJ, Reinach PS, and Candia OA (1979) An improved assay
for nanomole amounts of inorganic phosphate, Anal. Biochem. 100, 95-97.
Sabnis DD, Gordon M, and Galston AW (1970) Localization of adenosine triphosphatase activity on the chloroplast envelope in tendrils of Pisum
sativum, Plant Physiol. 45, 25-32.
Selman-Reimer S, Merchant S, and Selman BR (1981) Isolation, purification,
and characterization of coupling factor 1 from Chlamydomonas reinhardi,
Biochemistry 20, 5476-5482.

Authors' address: D.R. McCarty[†], B.R. Selman[†], and K. Keegstra[‡]/Depts.
of Biochemistry[†] and Botany[‡], University of Wisconsin,
Madison, Wisconsin, 53706.

EFFECT OF CHEMICAL MODIFICATION OF PRIMARY AMINO GROUP OF
SALT WASHED THYLAKOID MEMBRANE BY FLUORESCAMINE ON DIFFERENT
PARAMETERS OF ENERGY TRANSDUCTION IN CHLOROPLAST

JAI PARKASH AND G.S. SINGHAL

SUMMARY

The salt washed thylakoid membranes from spinach chloroplasts
were treated with a primary amine-specific fluorescent
chemical modifier, Fluorescamine (4-Phenylspiro furan-2 [3H]
1-phthalan 3,3' dione). The effects of this covalent chemical
modification on various processes of energy transduction were
studied. It is suggested that the chemical modification by
fluorescamine of free amino group of coupling factor 1 results
in different conformations of CF_1.

INTRODUCTION

Chemical modification of membrane proteins by -COOH, $-NH_2$ and
-SH group modifying reagents have been used as a tool to
reveal important informations to determine the role of these
functional groups of proteins in assessing their structural-
functional relationship (McCarty et al. 1972; McCarty, Fagan,
1973; Moroney et al. 1980; Weiss, McCarty, 1977; Oliver,
Jagendorf, 1976; Berg et al. 1974).

In this study, we report the chemical modification of CF_1
in situ with fluorescamine, which was used by Kraayenhof,
Slater (1974) and Harnischfeger (1978). The effects of such
modification on different parameters of energy transduction
in chloroplast are given.

MATERIALS AND METHODS

Spinach chloroplasts were isolated in 0.3 M sucrose, 50 mM
NaCl, 1 mM $MgCl_2$, 10 mM tricine (pH 7.8) and washed once in
isolation medium and then thrice in medium consisting of 50
mM NaCl, 1 mM $MgCl_2$, 2 mM tricine (pH 7.8) and were finally
suspended in isolation medium (Strotmann et al. 1976).

The chloroplast suspension was kept in dark for 30 min before
treating it with fluorescamine. 20 ul of fluorescamine solution in acetone was added to 2 ml of chloroplast suspension
containing 500 ug of Chl. Electron transport activities were
determined using potassium ferricyanide as an artificial
electron acceptor, by observing the decrease in absorbance at
420 nm. The 3 ml of assay mixture for basal electron transport determination consisted of 20 mM Na Cl, 500 uM $K_3Fe(CN)_6$,
15 mM potassium tricinate (pH 7.6) and chloroplast equivalent
to 5 0 ug chlorophyll. The light intensity was 3.6×10^5
ergs/cm^2/sec. Non-cyclic photophosphorylation was determined
according to Avron (1960). The 3 ml of reaction mixture consisted of 5 0 mM potassium tricinate (pH 8), 500 uM $K_3Fe(CN)_6$,
1 mM ADP (pH 8), 3.3×10^6 cpm carrier-free P^{32} and chloroplast
equivalent to 50 ug chlorophyll.

Cyclic photophosphorylations associated with photosystem I
was determined according to Hauska (1980).

Sybesma, C. (ed.), Advances in Photosynthesis Research, Vol. II. ISBN 90-247-2943-2.
© *1984 Martinus Nijhoff/Dr W. Junk Publishers, The Hague/Boston/Lancaster.*

Acid-base transition induced ATP formation was done according
to Jagendorf (1972). Chlorophyll concentration was determined
according to Arnon (1949). All chemicals used in this study
were of analytical reagent grade.

RESULTS

The rate of basal electron transport from H_2O to $K_3Fe(CN)_6$
decreased slightly when salt washed thylakoid membrane were
treated with lower concentrations of fluorescamine (Fig. 1).
However, at higher concentrations of fluorescamine beyond 1.1
umoles fluorescamine/500 ug Chl, the rates were higher than
the control. The effect of uncouplers like NH_4Cl, gramicidin D,
m-CCCP and gramicidin D $\rightarrow NH_4Cl$ on electron transport was decreased with increasing concentration of fluorescamine except
that the effect was reversed at 8.94 umoles fluorescamine/500
ug Chl (data not shown).

<u>Coupled Electron Transport</u>: The rate of coupled electron trans-
port of salt-washed chloroplasts decreased at lower concentrations of fluorescamine. However, at higher concentration of
fluorescamine the inhibition of coupled electron transport
was less (Fig. 1). The effect of uncouplers on coupled electron transport on the control and the modified chloroplasts
was the same as that of on basal electron transport.

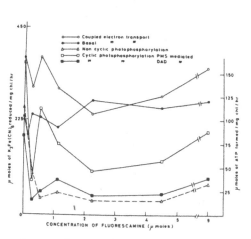

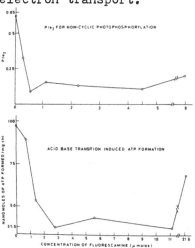

Figure 1
Effect of various concentrations of fluorescamine on
basal and coupled electron
transport, non-cyclic and
cyclic photophosphorylation.
For experimental details,
see "Materials and Methods".

Figure 2
Effect of various concentrations
of fluorescamine on acid-base
transition induced ATP formation.
For experimental details, see
"Materials and Methods". The
upper curve shows P/e_2 value for
non-cyclic photophosphorylation.

Non-cyclic Photophosphorylation: The non-cyclic photophosphorylation coupled to electron transport was reduced drastically
at fluorescamine concentration of 0.56 umoles/500 ug Chl (Fig.1).
However, we observed that beyond 4.54 umoles fluorescamine/500 ug
Chl, the rates were higher than at lower concentrations of fluorescamine. As usual, the uncouplers and energy transfer inhibitor
decreased significantly the rate of non-cyclic photophosphorylation both in control and treated chloroplast (data not shown).

Acid-base Transition: The acid-base transition induced ATP
formation was affected similarly as non-cyclic photophosphorylation (Fig. 2).

Cyclic Photophosphorylation: The cyclic photophosphorylation
associated with photosystem I and catalysed by Phenazine Methosulfate and Diaminodurene separately were affected similarly as
non-cyclic photophosphorylation and acid-base transition induced
ATP formation although the inhibition at moderate concentrations
of fluorescamine was not the same as that in non-cyclic photophosphorylation (Fig. 1).

DISCUSSION

The increase in the rate of basal electron transport with increasing concentration of fluorescamine may be interpreted in
terms of different conformational states of CF_1. The chemical
modification is probably causing the increase in the rate of
proton efflux through CF_1-CF_0. Our data on the rate of uncoupled electron transport with fluorescamine treatment also
suggests a conformational rearrangement in CF_1, resulting in
alteration of membrane structure.

The decrease in the rate of coupled electron transport with
moderate concentrations of fluorescamine suggests that chemical modification of ATPase complex brings about lowering of
proton efflux through CF_1-CF_0. However, at higher concentrations of this modifier, the increase in rate of coupled electron transport suggest the increase in H^+-ion efflux across
ATPase arising due to a new conformational rearrangements in
CF_1. The sharp decline in the rate of non-cyclic photophosphorylation, cyclic photophosphorylation and acid-base transition
induced ATP formation with increasing concentrations of
fluorescamine indicates that free amino group of CF_1 play a
conformational role in regulating ATP synthesis. However, the
increase in these activities at higher concentrations of
fluorescamine which in part could be assigned to increase in
permeability of H^+-ion through CF_1-CF_0, rules out the possibility of involvement of free amino group at the catalytic site
of ATPase.

REFERENCES

Arnon DI (1949) Copper Enzymes in Isolated Chloroplasts. Poly-
phenoloxidase in Beta vulgaris, Plant Physiol. 24, 1-15.
Avron M (1960) Photophosphorylation by Swiss-chard chloroplasts,
Biochim. Biophys. Acta, 40, 257-272.

Berg S, Dodge S, Krogmann DW and Dilley RA (1974) Chloroplast Grana Membrane Carboxyl Groups: Their Involvement in Membrane Association, Plant Physiol. 53, 619-627.
Harnischfeger G (1978) Effect of Chemical Modification of Amino Group by Fluorescamine on Partial Reactions of Photosynthesis, Biochim. Biophys. Acta, 503, 473-479.
Hauska G (1980) Measurement of Photophosphorylation Associated with Photosystem I. In San Pietro A, ed. Methods in Enzymology, 69C, pp. 648-658. New York: Academic Press Inc.
Jagendorf AT (1972) Two-Stage Phosphorylation Techniques: Light to Dark and Acid-to-Base Procedures. In San Pietro A, ed. Methods in Enzymology, 24B, pp. 103-113. New York: Academic Press Inc.
Kraayenhof R and Slater EC (1975) Studies of Chloroplast Energy Conservation with Electrostatic and Covalent Fluorophores. In Avron M, ed. Proc. 3rd Int. Cong. Photosynth. Res. pp. 935-996. Amsterdam: Elsevier.
McCarty RE, Pittmann PR and Tsuchiya Y (1972) Light Dependent Inhibition of Photophosphorylation by N-Ethylmaleimide, J. Biol. Chem. 247, 3048-3052.
McCarty RE and Fagen J (1973) Light Stimulated Incorporation of N-Ethylmaleimide into Coupling Factor 1 in Spinach Chloroplast. Biochemistry 12, 1503-1507.
Moroney JV, Carlos S, Andreo Vallejos H and McCarty RE (1980) Uncoupling and Energy Transfer Inhibition of Photophosphorylation by Sulphydryl Reagents. J. Biol. 255, 6670-6674.
Oliver D and Jagendorf AT (1976) Inhibition of the Coupling Factor from Spinach Chloroplasts by Trinitrobenzenesulfonic acid. Fed. Proc. 34, 596.
Strotmann H, Bickel S and Huchzermeyer B (1976) Energy Dependent Release of Adenine Nucleotides Tightly Bound to Chloroplast Coupling Factor CF_1. FEBS Lett. 61, 194-197.
Weiss MA and McCarty RE (1977) Cross-linking within a subunit of Coupling Factor 1 increases the Proton Permeability of Spinach Chloroplast Thylakoids. J. Biol. Chem. 252, 8007-8012.

ACKNOWLEDGEMENTS

The work was supported by Dept. of Science and Technology Grant No. 8(4)/78-SERC. We thank Drs. Arun K. Agarwal and R. Bhardwaj, for their help.

Authors address: Jai Parkash and G.S. Singhal, School of Life Sciences, Jawaharlal Nehru University, New Delhi-110067, India.

SALT ADAPTATION MECHANISMS IN THE CYANOBACTERIUM <u>SYNECHOCOCCUS</u> 6311.

EDUARDO BLUMWALD, ROLF J. MEHLHORN and LESTER PACKER

1. Summary.

A hypothesis for how cyanobacteria become salt tolerant is being developed. The major steps and temporal sequence involved in adaptation by the fresh water cyanobacterium <u>Synechococcus</u> 6311 to 0.6 M NaCl are: sodium entry----membrane detachment of phycobilisomes---- inhibition of photosynthetic activity and growth----utilization of cell energy reserves (phycobiliproteins, other proteins)----after four hours photosynthetic activity starts to increase reaching a net 100% enhancement----massive accumulation of osmoticum in the form of soluble sugars (sucrose and glucose) and K^+, continued Na^+ extrusion which commences soon after salt shock----accumulation of massive reserves of major energy storage, temporal for glycogen, permanent of phycobiliproteins----salt adaptation.

2. Materials and Methods.

<u>Synechococcus</u> 6311 was grown as described previously (Blumwald et al 1983a). Cell volume was determined by the ESR method; samples contained 1 mM TEMPONE (2,2,6,6-tetramethyl-4-oxopiperidinooxy free radical), 20 mM $Na_3Fe(CN)_6$, 75 mM $Na_2MnEDTA$ and cells suspensions (400 ug/ml Chl). Kinetics experiments were carried out by mixing the cell suspensions with the different NaCl solutions using the rapid mixing system illustrated in Fig. 2. The mixing time was estimated to be about 100 ms. Ion concentrations were measured by atomic absorption spectrophotometry. Sucrose and glucose were determined enzymatically. Chlorophyll and phycobiliprotein content were determined according to Yamanaka & Glazer (1981). O_2-evolution was measured polarographically.

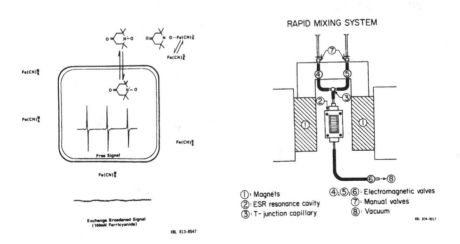

FIGURE 1. Scheme of volume measurement with TEMPONE.

FIGURE 2. Rapid mixing system.

Sybesma, C. (ed.), Advances in Photosynthesis Research, Vol. II. ISBN 90-247-2943-2.
© 1984 Martinus Nijhoff/Dr W. Junk Publishers, The Hague/Boston/Lancaster.

3. Results.

A comparison of the kinetics of the osmotic response to NaCl of control and salt-adapted <u>Synechococcus</u> 6311 cells is shown in Fig. 3. When transferred to 1.0 M NaCl, control cells showed a rapid decrease in volume, reached a minimun within 300 ms and recovered their initial volume within 15 s (Fig. 3). Cells grown in 0.6 M NaCl showed full recovery within 1 min when transferred to 1.0 M NaCl (Fig. 3c). An estimate for the rate of full recovery was made by measuring the time required for 50% of volume recovery ($t_{1/2}$). The $t_{1/2}$ for control cells is 5 s, while the $t_{1/2}$ for salt-adapted cells is 25 s.

To test the effect of NaCl on the photosynthetic activity, cells exposed to a growth medium containing 0.3 and 0.6 M NaCl were tested for O_2-evolution measured under illumination at different wavelengths (Fig. 4). Cells illuminated with actinic light (400 to 750 nm) showed decreased photosynthetic activity after 0.5 h of exposure to 0.3 and 0.6 M NaCl. However, these cells recovered their activity after 2 h, and showed a 100% increase in activity after 4 h of growth. When illuminated with blue light (400 to 570 nm), salt grown cells showed the same pattern of initial inhibition and further enhancement after 4 h. When 0.3 M NaCl grown cells were illuminated at 616 to 621 nm, the peak of maximum absorption of the phycobilisomes, they showed decreased activity during 10 h of growth and enhanced activity after 24 h of growth. 0.6 M NaCl grown cells showed complete inhibition of activity after 0.5 h and recovered their initial activity after 15 h of growth. After 20 h the activity was significantly enhanced and remained constant thereafter. The measurement of the immediate effect of salt exposure on photosynthetic activity showed that when the cells were illuminated at 616 to 621 nm, the activity was totally inhibited within the first 3 min of exposure to salt (results not shown).

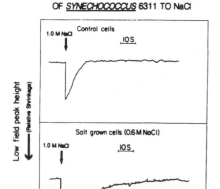

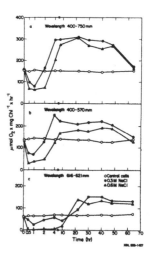

FIGURE 3. Kinetics of the osmotic response to 1.0 M NaCl.

FIGURE 4. Photosynthetic activity of salt

Batch cultures of <u>Synechococcus</u> 6311 exposed to growth mediums containing 0.3 and 0.6 M NaCl were analyzed regularly during the log phase of growth for K⁺ and Na⁺ content. The internal K⁺ concentration in control cells was found to be 170 mM, and remained constant during 88 h of growth (Blumwald et al, 1983b). The K⁺ content increased in salt-grown cells; 0.3 M and 0.6 M NaCl grown cells contained 250 mM and 330 mM K⁺ respectively after 40 h of growth. The internal Na⁺ content (Fig. 5) of control cells was 8 mM and remained constant during 88 h of growth. The Na⁺ content increased markedly in salt-grown cells. In cells grown in 0.3 M NaCl, the Na⁺ content was 240 mM after 30 min of exposure, decreased to 90 mM during the first 4 h of growth, and thereafter remained at the same level as in control cells after 40 h. The Na⁺ content of cells grown in 0.6 M NaCl increased to 550 mM after 30 min of NaCl exposure, decreased to 200 mM after 4 h of growth and remained at a constant level of 40 mM after 40 h of growth.

A comparison of the major solutes in salt-grown <u>Synechococcus</u> 6311 cells is given in Fig. 6. Cells grown in 0.6 M NaCl (1.1 Os/kg) contained 580 mM sucrose, 140 mM glucose, 330 mM K⁺ and 40 mM Na⁺, which altogether represents a total osmotic pressure of 1.1 Os/Kg.

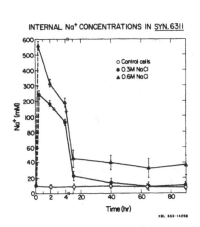

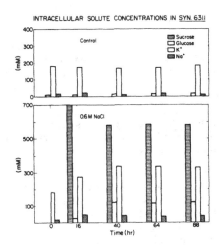

FIGURE 5. Internal Na⁺ concentration of salt-grown <u>Syn</u>. 6311

FIGURE 6. Osmoregulatory balance in <u>Syn</u> 6311.

A comparison of the pigment content of control and 0.6 M NaCl grown cells after 48 h of growth (Table I) showed no significant difference in the Chl a content of the cells. 0.6 M NaCl grown cells have increased phycobilisome and phycobiliproteins content. The glycogen content in salt-grown cells after 48 h of growth was twice that of control. The glycogen content of salt-grown cells then decreased progressively until after 168 h of growth the glycogen content returned to approximately the same levels as in controls (controls 7.2 ug/ug Chl, 0.6 M NaCl cells 7.8 ug/ug Chl).

Table I[a].Changes in cell pigment and glycogen content of
Synechococcus 6311 cells grown in NaCl for 48 h.

	Chl (ug/ul)	Phycobilisome (ug/ug Chl)	Phycobiliproteins (ug/ug Chl)	Glycogen (ug/ug Chl)
Control	25.0	8.92	0.38	8.30
0.6 M NaCl	23.8	12.70	3.47	15.65

4. Discussion.

The data presented in this report strongly suggest, that the very
rapid NaCl entry into the cells triggers the adaptive response of the
cyanobacterium Synechococcus 6311 to salt and that both organic and
inorganic osmoregulatory mechanisms are involved in this process.
Cells showed extreme permeability to NaCl when first transferred to
high NaCl, but markedly resisted NaCl-induced plasmolysis after salt
adaptation indicating membrane permeability changes. One of hte
earliest effects of NaCl-induced stress is inhibition of photosynthetic activity. For cells grown in 0.6 M NaCl a key event in
this inhibition appears to be the detachment of phycobilisomes from
thylakoid membranes. The photosynthetic activity of salt-grown
cells is enhanced after 4 h of growth and the consequence of this
enhancement is the accumulation of soluble sugars for osmoregulation.
After 60 h of growth, salt-grown cells showed the same photosynthetic
activity as controls. Osmoregulation can be almost completely
accounted for by the intracellular concentrations of sucrose,
glucose, K^+ and residual Na^+ ions. The analysis of pigment and
carbohydrate content in salt-adapted cells indicates that tolerance
to salt in Synechococcus 6311 is associated with the production of
massive reserves of major energy storage materials, soluble sugars
and phycobiliproteins.

REFERENCES
Blumwald E, Mehlhorn RJ and Packer L (1983a) Studies of osmoregula-
tion in salt-adaptation of cyanobacteria with ESR spin-probe tech-
niques. Proc Nat Acad Sci USA 80:2599-2602.
Blumwald E, Mehlhorn RJ and Packer L (1983b) Ionic osmoregulation
during salt adaptation of the cyanobacterium Synechococcus 6311.
Plant Physiol (in press).
Yamanaka G and Glazer AN (1981) Dynamic aspects of phycobilisome
structure. Phycobilisome turnover during nitrogen starvation in
Synechococcus sp. Arch Microbiol 124:39-47.

ACKNOWLEDGEMENTS. We acknowledge research support from the Office of
Biological Energy Research, of the Department of Energy and the Basic
Research Department of the Gas Research Institute.

Authors address: Membrane Bionergetics Group, Lawrence Berkeley
                 Laboratory and Department of Physiology/Anatomy,
                 University of California, Berkeley, CA 94720, USA.

INTERACTION OF PHOTOSYNTHETIC AND RESPIRATORY ELECTRON TRANSPORT IN BLUE-GREEN ALGAE: PROTON-EFFLUX STUDIES

SIEGFRIED SCHERER, ERWIN STÜRZL and PETER BÖGER

Summary: The oxygen-dependent proton efflux in the dark of intact cells of Anabaena variabilis has been investigated. In contrast to other bacteria and mitochondria showing a H^+/e ratio of 2 to 6, a ratio of only 0.23 to 0.35 was obtained, indicative of localization of respiratory electron transport essentially on the thylakoids. Inhibitor studies as well as the kinetics of proton efflux and ATP synthesis gave evidence that a proton-translocating ATP hydrolase on the cytoplasmic membrane rather than an electron-transport chain is responsible for oxygen-dependent proton efflux in the dark.

1. INTRODUCTION

In blue-green algae, an interaction between respiration and photosynthesis is evident (Lockau 1981; Peschek,Schmetterer 1982; Scherer et al. 1982; Stürzl et al. 1982), suggesting location of respiratory electron transport on the thylakoids. However, in spite of recent findings, the question where to locate the respiratory electron transport remains controversial (Bisalputra et al. 1969; Paschinger 1977; Peschek et al. 1981; Scherer et al. 1981; Lockau, Pfeffer 1982). An oxygen-dependent proton extrusion in the dark demonstrated in Anabaena variabilis (Scholes et al. 1969) was taken as evidence for respiratory electron transport to be located on the cytoplasmic membrane. No detailed investigations on this proton efflux have been published so far.

2. RESULTS AND DISCUSSION

Proton efflux in the dark with Anabaena variabilis, induced by an oxygen pulse, is shown in Figure 1A. The efflux decreased after the oxygen had been consumed by respiration. Subsequently, a slow uptake of protons was measured (dotted line of part A). No major differences were seen using other blue-green algae (e.g. Anacystis, Aphanocapsa, Nostoc).

A lag phase of about 34 to 60 sec was observed until the proton efflux reached its maximum rate. Adding a second oxygen pulse immediately after

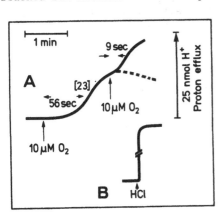

FIGURE 1. Acidification of the medium after an oxygen pulse had been applied to A. variabilis in the dark (A). Data in brackets indicate proton efflux expressed as $\mu mol\ H^+/$ mg chlorophyll calculated from the linear part of the trace. Upward: H^+ efflux. The dotted line shows the H^+ uptake when no further oxygen pulse was given. For comparison, the (rapid) kinetics of the pH electrode itself after an acid pulse is shown by (B).

Sybesma, C. (ed.), Advances in Photosynthesis Research, Vol. II. ISBN 90-247-2943-2.
© *1984 Martinus Nijhoff/Dr W. Junk Publishers, The Hague/Boston/Lancaster.*

the first proton efflux had come to an end, the lag phase observed was only 5 to 10 sec (9 sec in Fig.1A). When the second oxygen pulse was given several minutes later, the lag phase was the same as with the first pulse (data not shown). Figure 1B shows the kinetics of the pH electrode, which is much faster than the proton efflux observed in part A. Consequently, the Δ pH kinetics shown represent the time course of proton efflux.

An oxygen-dependent proton efflux of Anabaena variabilis exhibiting an H^+/e ratio of 0.27 has been reported (Scholes et al. 1969), which corresponds with our findings both in extent and rate of acidification (see Fig.1). However, neither an interpretation nor a further characterization of this proton efflux have been given as yet. In our hands, five different blue-green algae showed comparable H^+/e ratios of about 0.23 to 0.35 (0.57 at the maximum).

Compiling the H^+/e ratios of mitochondria and of several bacterial species from the literature, values of 2 to 6 are found. The P/e ratio of blue-green algae, however, is comparable to other organisms. Obviously, in blue-greens, only a minor part of the electrons moving along the respiratory chain mediate proton efflux out of the cell. We conclude from our data that most of the respiratory electron transport is located on the thylakoid membrane. Thus, proton translocation driven by thylakoid-located electron flow cannot be determined as external acidification by intact cells, which leads to the low H^+/e ratios as measured.

In a following experiment, Δ pH-time course was correlated with ATP levels in the cell. In Figure 2, oxidative phosphorylation is shown after an oxygen pulse was given at start (zero time). ATP formation was found to be linear, exhibiting a short lag phase of 5 sec. After 30 to 40 sec, the ATP level reached a maximum. The relevant result of these experiments is the similarity of ATP-formation kinetics and the time course of proton efflux as shown in Figure 1. Proton efflux was at its maximum when the final ATP level had been reached.

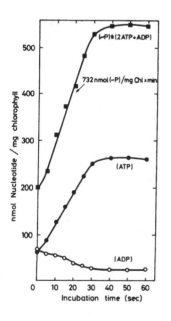

FIGURE 2. Kinetics of oxidative phosphorylation following an oxygen pulse given to Anabaena variabilis kept under anaerobiosis beforehand. The values given are means of 5 experiments. At start, oxygen was added to give a final concentration of 14.4 μM.

The ATP content is low after 20 min of anaerobiosis. After the oxygen pulse, an ATP level is built up with a lag phase of about 5 sec only. Apparently as a consequence of increasing ATP, ATP-driven proton efflux becomes possible. After a second O_2 pulse, the ATP concentration is not as low as compared with the first pulse (Fig.1), and proton efflux starts with a much shorter lag phase, since the decay of the ATP levels has not reached the low figure as at start of the experiment.

TABLE 1. Influence of several inhibitors. Values are given in percent of control (= 100). Proton efflux: 100% = 34.3 μmol H^+/mg chlorophyll (Chl) x h; oxygen uptake: 100% = 27.9 μmol/mg Chl x h; rate of oxidative phosphorylation: 100% = 468 nmol ($\sim$P)/mg Chl x min; H^+/e: 100% = 0.3.

Addition	Proton release	Oxygen uptake	Oxidative phosphorylation	H^+/e
KCN, 3×10^{-3} M	0	30	0	0
CCCP, 3×10^{-6} M	0	100	22	0
DNP, 9×10^{-4} M	51	126	35	40
DCCD, 5×10^{-5} M	32	79	0	40
Nitrofen, 5×10^{-5} M	68	103	87	71
Vanadate, 118 μg/ml	62	92	100	67

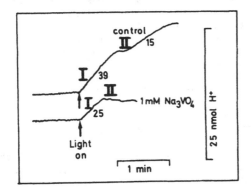

FIGURE 3. Light-induced biphasic (I,II) proton efflux in Anabaena variabilis is inhibited by vanadate. The numbers indicate proton efflux in μmol H^+/mg Chl x h, taking into consideration the buffer capacity of 1 mM potassium vanadate. Calibration of proton efflux is shown only for the control. Chlorophyll concentration was 30 μg/ml.

Table 1 shows the effect of several inhibitors. The major result of these inhibitor studies is that an inhibition of proton efflux is possible without impairing respiratory electron transport. This can also be demonstrated advantageously by the lowered H^+/e ratios given (in percent of control) in the last column of Table 1. These results are in accordance with the hypothesis that an ATP-consuming cytoplasmic-membrane ATPase mediates the proton efflux and contradicts the model of a respiratory chain located on the plasma membrane.

Other experiments concerning the proton efflux of blue-green algae in the light further support the hypothesis of an ATP-dependent, proton-translocating ATPase. In Figure 3, the light-dependent proton efflux of Anabaena variabilis is shown to be biphasic - the second phase is completely sensitive to vanadate, an inhibitor known to affect only proton-translocating ATP hydrolases, but not ATP synthetase. Therefore, we postulate this type of ATP hydrolase to be localized on cytoplasmic membranes of cyanobacteria. It will be necessary to isolate cytoplasmic membranes to prove the existence of this type of proton efflux. Furthermore, experiments are in progress to investigate the mechanism regulating light-induced proton efflux in blue-green algae.

REFERENCES
Bisalputra T, Brown DL and Weier TE (1969) Possible respiratory sites in a blue-green alga, Nostoc sphaericum, as demonstrated by potassium tellurite and tetranitro-blue tetrazolium reduction, J. Ultrastruct. Res. 27, 182-197.
Lockau W (1981) Evidence for a dual role of cytochrome c-553 and plastocyanin in photosynthesis and respiration of the cyanobacterium Anabaena variabilis, Arch. Microbiol. 128, 336-340.
Lockau W and Pfeffer S (1982) A cyanobacterial ATPase distinct from the coupling factor of photophosphorylation. Z. Naturforsch. 37c, 658-664.
Paschinger H (1977) DCCD-induced sodium uptake by Anacystis nidulans, Arch. Microbiol. 113, 285-291.
Peschek GA, Schmetterer G, Sleytr UB (1981) Possible respiratory sites in the plasma membrane of Anacystis nidulans: ultracytochemical evidence, FEMS Microbiol. Lett. 11, 121-124.
Scherer S, Stürzl E and Böger P (1981) Arrhenius plots indicate localization of photosynthetic and respiratory electron transport in different membrane regions of Anabaena, Z. Naturforsch. 35c, 1036-1040.
Scherer S, Stürzl E and Böger P (1982) Interaction of respiratory and photosynthetic electron transport in Anabaena variabilis Kütz. Arch. Microbiol. 132, 333-337.
Scholes P, Mitchell P and Moyle J (1969) The polarity of proton translocation in some photosynthetic microorganisms, Eur. J. Biochem. 8, 450-454.
Stürzl E, Scherer S and Böger P (1982) Reconstitution of electron transport by cytochrome c-553 in a cell-free system of Nostoc muscorum, Photosynth. Res. 3, 191-201.

Authors' address: Lehrstuhl für Physiologie und Biochemie der Pflanzen, Universität Konstanz, D-7750 Konstanz, Germany.

RESPIRATORY AND PHOTOSYNTHETIC ELECTRON TRANSPORT IN ANABAENA VARIABILIS:
LIGHT-DARK ACTIVITIES OF PYRIDINE-NUCLEOTIDE DEHYDROGENASES

ERWIN STÜRZL, SIEGFRIED SCHERER and PETER BÖGER

Summary: Thylakoids of vegetative cells of Anabaena variabilis use both
NADH and NADPH for (dark) respiration in a cyanide-sensitive reaction.
Both nucleotides also function as donors for photosystem I in the light
and in the presence of DCMU. Double-reciprocal plot analysis yields different K_m- and v_{max}-values in the dark and in the light for NADPH and
NADH. An interaction of respiratory and photosynthetic electron transport
is obvious, using NADH as reductant, since a different dehydrogenase
activity is evident in the dark, as compared with the activity in the
light with DCMU and KCN present.

1. INTRODUCTION

Blue-green algae lack compartmentation and, therefore, are interesting
organisms with respect to possible interaction between respiration and
photosynthesis. Soluble cytochrome c-553 (Lockau 1981; Stürzl et al.
1982), plastoquinone (Eisbrenner, Bothe 1979; Antarikonda et al. 1980;
Hirano et al. 1980) and plastoquinol-cytochrome f/b-563 reductase
(Peschek, Schmetterer 1982) were reported as common components of the
photosynthetic and respiratory electron-transport systems in blue-green
algae. However, there is no conclusive evidence as to whether there
exist two separate electron-transport systems acting in parallel or
whether certain components are shared by respiratory and photosynthetic
electron transport (see Binder 1982). As already known, NADPH and NADH
mediate oxygen uptake in the dark (Biggins 1969; Leach, Carr 1970; Stürzl
et al. 1982). It has been shown recently that NADH donates electrons to
nitrogenase in heterocysts of Anabaena in a light-dependent reaction,
whereas Houchins, Hind (1982) concluded that NADPH failed to do so at
appreciable rates. With a heterocyst homogenate of Anabaena, NADPH was
shown to function as donor for nitrogenase in the light (Schrautemeier
et al. 1983). This reaction was found sensitive to DBMIB.

In order to obtain some knowledge about dehydrogenase activities, we
examined oxygen uptake in the dark and in the light with reference to the
concentration of NADH as well as NADPH, using thylakoids isolated from
vegetative cells of Anabaena variabilis.

2. MATERIAL AND METHODS

Anabaena variabilis (ATCC 29413) was grown axenically under nitrogen-fixing conditions and harvested at the end of the logarithmic phase. Cells
were disrupted in a French press resulting in photosynthetically and re-
spiratorily active membranes (for details see Stürzl et al. 1983). After
this quick preparation step, thylakoids with a chlorophyll content of
35 µg Chl/ml were used to monitor the oxygen exchange. Heterocysts remained unbroken by this procedure and, therefore, did not interfere with
nucleotide-induced oxygen uptake of the thylakoids.

K_m-values were computer-calculated using least-square fit methods fitted
to a hyperbola as described by Henderson (1978).

Sybesma, C. (ed.), Advances in Photosynthesis Research, Vol. II. ISBN 90-247-2943-2.
© *1984 Martinus Nijhoff/Dr W. Junk Publishers, The Hague/Boston/Lancaster.*

TABLE 1. Oxygen uptake by isolated membranes of Anabaena variabilis. Light rates (2) are not corrected for oxygen uptake in the dark (1). DCMU, 10 μM; KCN, 1 mM; NADH, 1 mM; NADPH, 1 mM.

Additions	Rates (μmol O_2/mg chlorophyll x h)	
	Dark (1)	Light (2)
Control	2	26
plus DCMU	2	1
" DCMU, KCN	1	0
" DCMU, NADPH	8	9
" DCMU, NADPH, KCN	5	7
" DCMU, NADH	4	8
" DCMU, NADH, KCN	1.5	7

3. RESULTS AND DISCUSSION

Membranes of Anabaena variabilis respire both NADPH and NADH (Table 1). A pronounced inhibition of oxygen uptake in the dark by KCN (1 mM) was observed, indicative of a respiratory electron transport functioning via cytochrome-c oxidase. No peroxide was formed. The remaining oxygen uptake was due to KCN-resistant respiration. When NADPH was the donor, KCN-insensitive oxygen uptake was ascribed to ferredoxin-NADP oxidoreductase (EC 1.18.1.2) transferring electrons to oxygen as suggested by Houchins, Hind (1982). This possibility can safely be excluded for NADH-mediated oxygen uptake because of the definite specificity of ferredoxin-NADP oxidoreductase to NADP(H).

A substantial oxygen uptake in the light and in the presence of DCMU (10 μM) is evident when NADH functions as electron donor, indicating an interaction of respiratory and photosynthetic electron-transport systems. Recently, an NADH-regenerating system has been introduced, which mediated a light-induced reduction of NADP in the presence of DCMU (Stürzl et al. 1983). DBMIB-sensitivity was taken as evidence for participation of plastoquinone. Reports on Mastigocladus (Binder et al. 1981) and Anabaena (Houchins, Hind 1982) suggest NADH to function as donor for photosynthetic electron transport. So far, no light-induced and NADPH-dependent oxygen uptake exceeding the dark rate has been reported (Table 1).

In order to obtain information about the dehydrogenase mediating NADH oxidation, dependence of oxygen uptake on the concentration of nucleotides used was studied in detail (Fig.1). Oxygen uptake in the dark as well as in the light resulted in Michaelis-Menten saturation kinetics. In the dark, a K_m-value of 40 μM for NADH was obtained, whereas a K_m-value of 110 μM was evident in the light. With NADPH, K_m-values of 18 μM in the dark and 48 μM in the light were obtained (data not shown). Different K_m- and v_{max}-values for NADH and NADPH in the dark as well as in the light

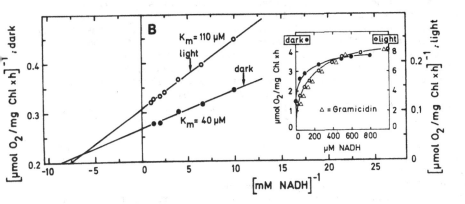

FIGURE 1. Double-reciprocal plots of oxygen uptake in the dark and in the light of Anabaena membranes, depending on NADH concentration. In the experiment done in the light, DCMU (10 μM) and KCN (1 mM) were present. Data shown represent a typical experiment, whereas K_m-values indicated were determined from about 15 experiments with different preparations each calculated by computerized least-square fit method fitted to a hyperbola with a standard deviation of 0.89. Inset: Original hyperbolic curve of oxygen uptake <u>vs.</u> added nucleotides. The symbol (Δ) indicates the rate obtained with gramicidin D present (10 μg/ml). Rates of samples including gramicidin D were found identical to those without gramicidin in the light as well as in the dark.

suggest dehydrogenases with different properties being involved. Assuming that NADPH oxidation is driven by ferredoxin-NADP oxidoreductase as suggested by Houchins, Hind (1982) and Schrautemeier et al. (1983), a second enzyme with NADH-dehydrogenating activity appears to be operating. Possibly, both enzymes are modulated by light (or dark) conditions.

Recently, the (membrane-bound) ferredoxin-NADP oxidoreductase has been suggested to be influenced in its activity by the transmembrane-proton gradient (Carrillo and Vallejos 1983). Gramicidin D abolishing the membrane potential and pH gradient affected neither K_m- nor v_{max}-values for NADH-dependent oxygen uptake in the dark or in the light, indicative of the membrane potential not being involved to influence enzyme kinetics. The mechanism how the pyridine-nucleotide dehydrogenases involved here change their property toward NADPH and NADH by dark/light transitions remains to be clarified.

REFERENCES
Antarikanonda P, Berndt H, Mayer F and Lorenzen H (1980) Molecular hydrogen: a new inhibitor of photosynthesis in the blue-green alga (cyanobacterium) Anabaena sp. TA 1, Arch. Microbiol. 126, 1-10.

Biggins J (1969) Respiration in blue-green algae, J. Bacteriol. 99, 570-575.

Binder A (1982) Respiration and photosynthesis in energy-transducing membranes of cyanobacteria, J. Bioenerg. Biomembranes 14, 271-286.

Binder A, Bohler M and Wolf M (1981) Energy-transducing membranes of a thermophilic cyanobacterium. In Selman DA, Selman-Reimer S, eds. Energy coupling in photosynthesis, pp. 313-321. Elsevier North-Holland, Amsterdam.

Carrillo N and Vallejos RH (1983) The light-dependent modulation of photosynthetic electron transport, Trends Biochem. Sci. 8, 52-56.

Eisbrenner G and Bothe H (1979) Modes of electron transfer from molecular hydrogen in Anabaena cylindrica, Arch. Microbiol. 123, 37-45.

Henderson PJF (1978) Statistical analysis of enzyme kinetic data. In: Technics in life sciences; Techniques in protein and enzyme biochemistry, Vol. B 113, pp. 1-43. Elsevier North-Holland Biomedical Press, Amsterdam.

Hirano M, Satoh K and Katoh S (1980) Plastoquinone as a common link between photosynthesis and respiration in a blue-green alga, Photosynth. Res. 1, 149-162.

Houchins JP and Hind G (1982) Pyridine nucleotides and H_2 as electron donors to the respiratory and photosynthetic electron-transfer chains and to nitrogenase in Anabaena heterocysts, Biochim. Biophys. Acta 682, 86-96.

Leach CK and Carr NG (1970) Electron transport and oxidative phosphorylation in the blue-green alga, Anabaena variabilis, J. Gen. Microbiol. 64, 55-70

Lockau W (1981) Evidence for a dual role of cytochrome c-553 and plastocyanin in photosynthesis and respiration of the cyanobacterium, Anabaena variabilis, Arch. Microbiol. 128, 336-340.

Peschek GA and Schmetterer G (1982) Evidence for plastoquinol-cytochrome f/b-563 reductase as a common electron donor to P700 and cytochrome oxidase in cyanobacteria, Biochem. Biophys. Res. Commun. 108, 1188-1195.

Schrautemeier B, Böhme H and Böger P (1983) In vitro studies on pathways and regulation of electron transport to nitrogenase with a cell-free extract from heterocysts of Anabaena variabilis, Arch. Microbiol., submitted.

Stürzl E, Scherer S and Böger P (1982) Reconstitution of electron transport by cytochrome c-553 in a cell-free system of Nostoc muscorum. Photosynth. Res. 3, 191-201.

Stürzl E, Scherer S and Böger P (1983) Interaction of respiratory and photosynthetic electron transport, and evidence for membrane-bound pyridine-nucleotide dehydrogenases in Anabaena variabilis, Physiol. Plant., submitted.

Authors' address: Lehrstuhl für Physiologie und Biochemie der Pflanzen, Universität Konstanz, D-7750 Konstanz, Germany

REDOX REACTIONS OF CYTOCHROME b-564 AND f-556 IN ISOLATED HETEROCYSTS
OF ANABAENA VARIABILIS (ATCC 29413)

HERBERT BÖHME and ANNELIESE ERNST

1. INTRODUCTION

Under nitrogen-fixing conditions, some filamentous cyanobacteria transform
5 to 10% of their vegetative cells into heterocysts (Haselkorn 1978). In
these heterocysts, an anaerobic environment is provided by the absence of
an O_2-evolving photosystem II and by respiratory O_2 uptake, thus protect-
ing nitrogenase activity. Isolated heterocysts exhibit high rates of
acetylene reduction in the presence of light and H_2. Oxidation of H_2 by
photosystem I might generate both a low-potential reductant and ATP
necessary for nitrogen fixation. In the dark, O_2 is removed by H_2 in a
Knallgas reaction, involving an uptake hydrogenase and cytochrome oxidase.
The dual role of H_2 as an electron donor to photosynthesis and respiration has now become generally accepted. Plastoquinone, the cytochrome b/f
complex and plastocyanin/cytochrome c have all been inferred as common
intermediates (Binder 1982).

With these results in mind, we reinvestigated the redox reactions of
cytochromes b-564 and f-556 in isolated heterocysts from Anabaena variabilis. In heterocysts kept under an H_2 atmosphere, approximately half of
the b-564 complement and all of f-556 are in a reduced state. Both cytochromes are reversible oxidized by photosystem I; in addition, O_2 oxidizes
both cytochromes in the dark (Böhme, Almon 1983). However, cytochrome f
is generally not found oxidized in photosynthetic organisms under aerobic
conditions. Therefore, we investigated the possibility of whether cytochrome f in heterocysts is oxidized by O_2 via a specific pathway.

2. MATERIALS AND METHODS

Anabaena variabilis Kütz. (ATCC 29413) was grown autotrophically as described (Ernst et al. 1983). Heterocysts were isolated according to
Peterson, Wolk (1978), with the following modifications: After lysozyme
treatment and sonification, the heterocysts were resuspended and washed
in a medium (pH 7.0) consisting of (mM): sorbitol 300, TES NaOH 10,
$MgCl_2$ 10, Na/K-phosphate 5, and 0.5% bovine serum albumin. Nitrogenase
activity with H_2 as an electron donor in the light was 70 to 130 μmol
C_2H_2 reduced/mg chlorophyll x h; this activity was stable for several
hours. Nitrogenase activity and redox reactions of cytochromes b-564 and
f-556 were measured as described (Böhme, Almon 1983).

3. RESULTS AND DISCUSSION

In isolated heterocysts, without further additions, light increases the
ATP content, directly demonstrating cyclic phosphorylation. In the dark
and in the presence of H_2, the cellular ATP level is enhanced by addition of O_2, which is attributed to respiratory activity (Ernst et al.
1983). In both cases, the energy-conserving site very probably involves
the cytochrome b/f complex, the presence of which in heterocyst membranes
has been demonstrated recently (Almon, Böhme 1980). Cytochromes b-564 and
f-556 are present in a 2:1 stoichiometry. Both cytochromes are rapidly
and completely oxidized by O_2, but only slowly resume their reduced state,

Sybesma, C. (ed.), Advances in Photosynthesis Research, Vol. II. ISBN 90-247-2943-2.
© *1984 Martinus Nijhoff/Dr W. Junk Publishers, The Hague/Boston/Lancaster.*

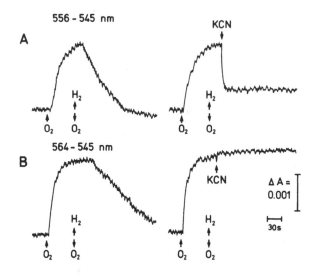

FIGURE 1. Influence of cyanide (500 μM) on H_2-induced reduction of O_2-oxidized cytochromes f-556 (A) and b-564 (B). Heterocyst chlorophyll concentration: 21 μg/ml (A) and 18 μg/ml (B).

if the cuvette is flushed with H_2 (Fig.1, tracings on the left). There are two possible explanations: (1) H_2 is an inefficient donor to the cytochrome b/f complex and photosystem I. (2) H_2 reduces the cytochrome complex efficiently, but a second, electron-withdrawing process competes favourably for electrons, keeping the cytochromes in their oxidized state as long as O_2 is present. This competitive pathway might be H_2-supported respiration involving cytochrome oxidase.

It has been reported that cytochrome oxidase in cyanobacteria is completely inhibited by 10 μM cyanide (Peschek 1981). With increasing concentrations of KCN, we find a biphasic inhibition curve of the H_2-supported O_2 uptake, with 50% and complete inhibition of cytochrome oxidase at 1 μM and 0.5 mM, respectively. If cytochrome-f oxidation by O_2 were, indeed, catalyzed by oxidase, one would expect a cross-over to reduction by blocking electron flow through the oxidase in the presence of H_2. This is demonstrated in Figure 1 (tracings on the right). O_2-oxidized cytochrome f (556-545 nm) is rapidly reduced by H_2 upon addition of 0.5 mM cyanide. In contrast, cyanide inhibits cytochrome-b reduction (564-545 nm) completely. This can be explained by the fact that b-cytochromes are generally O_2-oxidizable, independent of cytochrome oxidase activity. Cyanide-inhibited Knallgas reaction will not remove O_2 from solution, hence cytochrome b will stay oxidized, even in the presence of H_2. The effect of different KCN concentrations on cytochrome-f redox reactions is shown in Figure 2. Addition of 1 μM KCN (B), inhibiting cytochrome oxidase by 50%, severely affects the cytochrome-f reduction rate (75% inhibition; complete inhibition was observed with 0.5 mM KCN. Similarly, both the rate and extent of cytochrome-f oxidation by O_2 are severely inhibited and, therefore, involve respiratory electron flow. The remain-

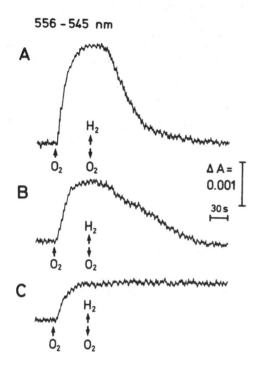

556 - 545 nm

A

H₂

O₂   O₂

ΔA =
0.001

30 s

B

H₂

O₂   O₂

C

H₂

O₂   O₂

FIGURE 2. Inhibition of O₂-induced oxidation and H₂-induced reduction of cytochrome f-556 by different cyanide concentrations. A: control; B: plus 1 μM KCN; C: plus 500 μM KCN. Heterocyst-chlorophyll concentration: 22 μg/ml.

ing absorbance change (same redox level as in Fig.1A!) might be due to a cyanide-insensitive pathway to O₂ which, however, was found to have a very low rate (<5%) in our heterocyst preparation.

CONCLUSIONS

In heterocysts, H₂ is able to efficiently donate electrons to the cyto-chrome-b/f complex. This reduction, however, may be obscured by simul-taneous cytochrome-oxidase activity by which the cytochromes remain oxi-dized as long as O₂ is present. Only after addition of inhibiting con-centrations of KCN (1 to 10 μM), a cross-over to cytochrome-f reduction is observed. In contrast, cytochrome b-564 stays oxidized under these conditions; apparently, it reacts directly with O₂. In addition, these results support the idea of components common to respiratory and photo-synthetic electron transport in cyanobacteria.

REFERENCES

Almon H and Böhme H (1980) Components and activity of the photosynthetic electron-transport system of intact heterocysts isolated from the blue-green alga Nostoc muscorum, Biochim. Biophys. Acta 592, 113-120.

Binder A (1982) Respiration and photosynthesis in energy-transducing membranes of cyanobacteria, J. Bioenerg. Biomembr. 14, 271-286.

Böhme H and Almon H (1983) Reactions of hydrogen and oxygen with photo-system I of isolated heterocysts from Anabaena variabilis, Biochem. Biophys. Acta 722, 401-407.

Ernst A, Böhme H and Böger P (1983) Phosphorylation and nitrogenase activity in isolated heterocysts from Anabaena variabilis, Biochim. Biophys. Acta 723, 83-90.

Haselkorn R (1978) Heterocysts, Annu. Rev. Plant Physiol. 29, 319-344.

Peschek GA (1981) Spectral properties of a cyanobacterial cytochrome-c oxidase: evidence for cytochrome a/a3, Biochem. Biophys. Res. Commun. 98, 72-79.

Peterson RB, Wolk CP (1978) High recovery of nitrogenase activity and of [55]Fe-labeled nitrogenase in heterocysts isolated from Anabaena variabilis, Proc. Natl. Acad. Sci. USA 75, 6271-6275.

ACKNOWLEDGMENTS

We wish to thank Prof. P. Böger for continuous support. This paper is part of our studies conducted in Zentrum für Energieforschung of this University.

Authors' address: Lehrstuhl für Physiologie und Biochemie der Pflanzen, Universität Konstanz, D-7750 Konstanz, Germany.

PRESENCE OF CYTOCHROMES IN THE THYLAKOID AND CELL MEMBRANES OF THE CYANOBAC-
TERIUM *PLECTONEMA BORYANUM* CULTURED IN NORMAL AND COPPER DEPLETED GROWTH
MEDIUM

HANS C.P. MATTHIJS, ALFRED N. van HOEK, HUUB J.M. LÖFFLER and RUUD KRAAYENHOF

1. INTRODUCTION

The topography of dark respiratory pathways in cyanobacteria is not yet completely known. Participation of the internally located thylakoid membranes in
respiration has been well established (Binder, 1982; Eisbrenner, Bothe 1979;
Hirano et al., 1980). On the other hand several reports have appeared that
show the participation of the cell membrane in dark respiration (Almon, Böhme
1982; Peschek et al., 1981; Scherer et al., 1981). The question is now whether
each of these two membrane types contains a complete respiratory chain or that
each of them bears a part of it and that respiration in intact cells is a
collective activity of both. In order to answer this question separation of
the thylakoid and cell membranes and the identification of components involved
in dark respiratory electron transfer is highly desired. We have recently separated the thylakoid and cell membranes of *P. boryanum* by a modification of
the method developed by Murata et al. (1981) using Percoll density gradients
(Matthijs et al., unpublished). In the present paper we will report on the
cytochrome content of the separate membrane fractions, also using cells that
have been grown in copper-depleted medium and thus may lack the copper-con-
taining cytochrome c oxidase. The possible topography of the respiratory
pathway will be discussed.

2. MATERIALS AND METHODS

2.1. Organism and growth. *P. boryanum* was obtained and cultured as has been
described before (Matthijs et al., 1981) in some cases copper being omitted
from the growth medium.
2.2. Membrane separation. Cells were harvested and washed in a medium containing 10 mM Tricine/NaOH, 10 mM NaCl, 300 mM sucrose and once passed through a
French-pressure cell at 8.1 MPa. Intact cells and large cell debris were collected at 5000 g for 10 min; the pellet obtained at 20 000 g was brought on top
of a discontinuous Percoll gradient, prepared in the isolation medium and
after centrifugation at 5000 g for 30 min three bands appeared at the 15, 20
and 30% border lines. The bands at 15 and 20% contained the thylakoid membranes
and the band at 30% a composite of cell wall and cell membrane. These preparations were used for further identification.
2.3. Spectrophotometry. Reduced minus oxidized difference spectra and potentiometric titrations were made according to the method of Van Wielink et al.(1982).
2.4. EPR spectroscopy. EPR spectra were made according to De Vries et al.(1979)

3. RESULTS AND DISCUSSION

In order to identify b- and c-type cytochromes in the separate membrane fractions we have determined the α-bands of the ferrocytochromes in the 540 to
570 nm area from reduced *minus* oxidized difference spectra. In the spectra of
Fig. 1 clear differences are shown in the total absorbance changes of the thylakoid membrane (A,B) and cell membrane plus the cell wall derived (C, D)
spectra.
With the spectrum deconvolution analysis developed by Van Wielink et al. (1982)
the spectra could be simulated by the dotted Gaussian curves. Also using the
results of a potentiometric titration (not shown) the peaks are interpreted

Sybesma, C. (ed.), Advances in Photosynthesis Research, Vol. II. ISBN 90-247-2943-2.
© *1984 Martinus Nijhoff/Dr W. Junk Publishers, The Hague/Boston/Lancaster.*

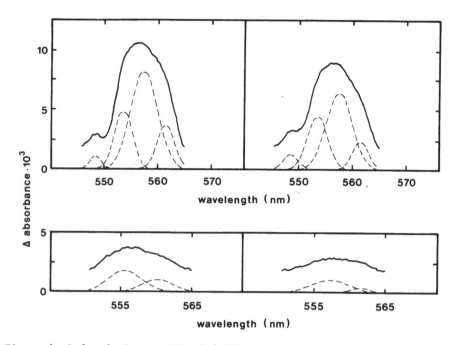

Figure 1. Reduced minus oxidized difference spectra of separated thylakoid an
cell membrane plus cell wall fractions of cells grown in normal and copper-
depleted medium. A: thylakoid membranes of normal grown cells; B: ibid of
copper-depleted grown cells; C: cell membrane plus cell wall of normal grown
cells; D: ibid of copper-depleted grown cells. The spectra were made at 77 K.
A: chlorophyll 108 $\mu g.ml^{-1}$, protein 3.0 $mg.ml^{-1}$, relative area 100%. B: ibid,
85 $\mu g.ml^{-1}$, 2.3 $mg.ml^{-1}$, 79% respectively, C: 47 $\mu g.ml^{-1}$, 3.1 $mg.ml^{-1}$, 24%,
D: 30 $\mu g.ml^{-1}$, 16%.

(from right to left) as cyt b-563 (about 1.3 $nmol.mg\ chl^{-1}$), cyt b-559 (low
plus high potential, about 2.8 $nmol.mg\ chl^{-1}$) and cyt c-557 (about 2.0 nmol.
mg chl^{-1}). Due to the measurement at 77 K the peaks are shifted to the blue
for about 1.5 nm and the peak of cyt c-557 is split in a minor peak at 548.5
and a major peak at 445.6 nm. By comparing the results of Stewart and Bendall
(1980) we have found a relatively high content of cyt c-557 (cyt "f") and a
low content of cyt b-563. We were not able to fit the curves of Fig. 1 C and D
to a 4 component solution, the absorption peaks of cyt c-557 seemed to lack.
More important however is the observation that the cell membrane plus cell wa:
fraction contains far less nmol of cytochrome per mg protein than the thylakoid membrane fractions. Moreover the absorbance of Fig. 1 C and D may be
completely due to the thylakoid membrane impurities in this fraction. No marke
differences between the spectra of the normal and copper-depleted cells were
observed in the 540 to 570 nm area. However, we found clear differences in the
580 to 620 nm area (Fig. 2). A potentiometric titration of freshly sonicated
cells of both normal and copper-depleted cells depicted in Fig. 2, shows clear
absorbance changes (Δ area) in the "+copper" spectrum with midpoint redox-
potentials of about 185 and 340 mV due to cyt a and cyt a-3, respectively.
These absorbance differences were absent from the "-copper" spectrum (the
average area per mg protein was clearly smaller too). From this result we

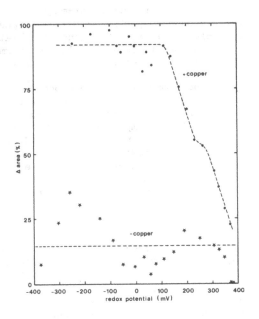

Figure 2. Potentiometric titration of sonicated preparations of normal- and copper-depleted cells. The percentage of the peak area in the 580 to 620 nm range at full reduction is plotted vs the redox potential. The absorbance change in the "-copper" spectrum is expressed as a percentage of the "+copper" spectrum. The spectra were recorded at room temperature. The protein content of both samples was 6 mg.ml^{-1}.

conclude that the a-type cytochrome c oxidase is nearly absent from copper-depleted cells. So far we were not able to detect cytochrome c oxidase in any of the separated membrane fractions spectrophotometrically. However EPR analysis of these fractions, that had lost all plastocyanin by exhaustive washing, showed a clear copper signal at g = 2.049 in the cell membrane plus cell wall spectrum of normal cells. This signal was nearly absent from the thylakoid membrane spectra (not shown) and distintly smaller in the copper-depleted cell membrane spectrum (Fig. 3). From these results we tentatively conclude that cytochrome c oxidase is localized in the cell membrane. Furthermore, it may be derived from the experiments shown in Fig. 1 that the cell membrane may lack b- and c-type cytochromes. This may indicate that the primary part of the respiratory chain including the PQ.b.FeS.c complex, is situated in the thylakoid membranes and that the terminal part with cytochrome c oxidase is localized in the cell membrane. Therefore, the two cooperative parts of the respiratory chain are supposedly connected by a soluble cytoplasmic factor (possibly a c-type cytochrome). More detailed experimental evidence supporting this scheme will be presented elsewhere (Matthijs et al., unpublished).

REFERENCES

Almon H and Böhme H Photophosphorylation in isolated heterocysts from the blue-green alga *Nostoc muscorum*. Biochim. Biophys. Acta 679, 279-286.
Binder A (1982) Respiration and photosynthesis in energy-transducing membranes of cyanobacteria. J. Bioenerg. Biomembr. 14, 271-280.
Eisbrenner G and Bothe H (1979) Modes of electron transfer from molecular hydrogen in *Anabaena cylindrica*. Arch. Microbiol. 123, 37-45.
Hirano M , Satoh K and Katoh S (1980) Plastoquinone as a common link between photosynthesis and respiration in blue-green algae. Photosynth. Res.1, 149-162.

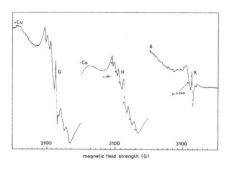

magnetic field strength (G)

Figure 3. Electron paramagnetic resonance spectra of cell membrane plus cell wall fractions of normal and copper-depleted cells. G: norm, cells, H: copper-depleted cells, K: G *minus* H;. Conditions of EPR spectroscopy: Temperature 46 K; microwave freq. approx. 9.247; mod. freq. 100 k Hz modulation amplitude 1.25 T; time constant 300 ms; scan rate 250 G. min^{-1}; microwave power 20 mW and finally receiver gain (adjusted to obtain identical manganese signals in G and H), G: 6.3 and H 2.5 x 10^{3}G. The spectra G and H are the average of 9 recordings. Protein: about 6 mg.ml^{-1} in both samples.

Matthijs HCP, Scholts MJC and Schreurs HJ (1981) Photosynthesis and respiration in the cyanobacterium *Plectonema boryanum* in: Photosynthesis II. Electr transport and photophosphorylation. (Akoyunoglou G ed.) pp. 269-278; Philadelphia: Balaban Int. Sc. Serv.

Murata N, Sato N, Omata T and Kuwabara T (1981) Separation and characterization of thylakoid and cell envelope of the blue-green alga (cyanobacterium) *Anacystis nidulans*. Plant Cell Physiol. 22, 855-866.

Peschek GA, Schmetterer G and Sleytr UB (1981) Possible respiratory sites in the plasma membrane of *Anacystis nidulans*: ultracytochemical evidence. FEMS Microbiol. Lett. 11, 121-124.

Sandmann G and Böger P (1980) Copper induced exchange of plastocyanin and cy chrome c-553 in cultures of *Anabaena variabilis* and *Plectonema boryanum*. Pla Sc. Lett. 17, 417-424.

Scherer S, Stürzl E and Böger P (1981) Arrhenius plots indicate localization of photosynthetic and respiratory electron transport in different membrane regions of *Anabaena*. Z. Naturforsch. 36 c, 1036-1040.

Stewart AC and Bendall DS (1980) Photosynthetic electron transport in a cell free preparation from the thermophilic blue-green alga *Phormidium laminosum*. Biochem. J. 188, 351-361.

De Vries S, Albracht SPJ and Leeuwerik FJ (1979) The multiplicity and stoich metry of the prosthetic groups in QH$_2$: cytochrome c oxido reductase as studi by EPR. Biochim. Biophys. Acta 546, 316-333.

Van Wielink JE, Oltmann LF, Leeuwerik FJ, De Hollander JA and Stouthamer AH (1982) A method for *in situ* characterization of b- and c-type cytochromes in *Escherichia coli* and in complex III from beef heart mitochondria by combined spectrum deconvolution and potentiometric analysis. Biochim.Biophys.Acta 681 177-190.

Authors address: Biological Laboratory, Vrije Universiteit, De Boelelaan 108
1081 HV Amsterdam, The Netherlands.

MEMBRANE ORGANIZATION OF <u>ANACYSTIS NIDULANS</u> FOLLOWING IRON DEPRIVATION
AND HEME DEFICIENCY

JAMES A. GUIKEMA

1. INTRODUCTION

The cyanobacterium, <u>Anacystis nidulans</u>, has been used to study the
synthesis, insertion, and assembly of lipid-protein complexes into photosynthetic membranes. A topological model of these membranes was recently
presented (J.A. Guikema, L.A. Sherman, 1982), as were the changes in
membrane architecture induced by iron deficiency (J.A. Guikema, L.A.
Sherman, 1983b). Iron deficiency yielded cells with a reduced thylakoid
content, and containing reduced concentrations of photosynthetically
important polypeptides. Polypeptides with heme, iron, or chlorophyll as
ligands were especially influenced. Importantly, the addition of iron
to deficient cultures caused the resynthesis of thylakoids, and provided
a method for studying membrane biosynthesis in these cells.

This work extends the study of iron deficiency by examining the effects
of levulinic acid on thylakoid biosynthesis. This compound is a competitive inhibitor of δ-aminolevulinic acid dehydratase, an important step
in porphyrin biosynthesis in cyanobacteria (J.A. Kipe-Nolt et al., 1978).
This preliminary study is aimed at comparing membrane biosynthesis under
iron deficient and levulinic acid inhibited conditions. The goal is to
identify the factors which regulate lipid-protein and membrane assembly.

The results presented here suggest four points. First, membranes which
were active in PS II water oxidation have been prepared using an isolation
medium of high ionic strength. A class of peptides, which included the
phycobiliproteins, was easily removed be washing membranes in a lowered
ionic strength. Second, levulinic acid interrupted the recovery of iron
deficient cells. Third, cells which were recovering from iron stress
were more sensitive to levulinic acid than were cells grown in complete
medium. This may indicate a role for the products of the protoporphyrin
pathway in regulating the iron deficient state. Fourth, levulinic acid
inhibits the incorporation of ^{35}S into specific membrane polypeptides.

2. PROCEDURE

2.1. <u>Materials and methods</u>. Cells of <u>Anacystis nidulans</u> strain R_2 were
grown at 34 C as previously described (J.A. Guikema, L.A. Sherman, 1982).
Iron deficiency was induced (J.A. Guikema, L.A. Sherman, 1983) until cells
were visibly chlorotic. Cells in log-phase growth were concentrated to
2×10^9 cells/ml for inhibitor or labeling studies. Treated cells were
harvested by centrifugation, suspended in 0.5 M sucrose, 0.5 M phosphate,
0.2 M citrate, 10 mM $MgCl_2$, pH 7.5 (SPCM), and disrupted by homogenizing
75 sec by the method of C.L. Cramer et al. (1983). All buffers contained
10 µg/ml PMSF, 50 µg/ml norleucine, and 50 µg/ml benzamidine. Unresolved
cells were removed by centrifugation (6,000 x g, 10 min), a membrane
fraction collected (27,000 x g, 15 min), and that fraction resuspended in
SPCM. Membranes isolated in this manner had typical phycocyanin/chloro-
phyll ratios of 3.1 (compared with 3.6 for cells and 1.7 for membranes

Sybesma, C. (ed.), Advances in Photosynthesis Research, Vol. II. ISBN 90-247-2943-2.
© *1984 Martinus Nijhoff/Dr W. Junk Publishers, The Hague/Boston/Lancaster.*

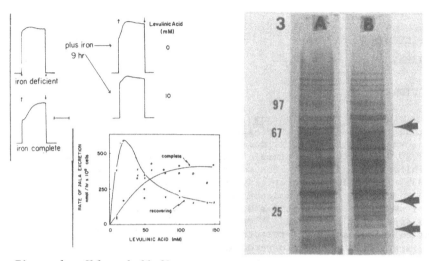

Figure 1. Chlorophyll fluorescence induction kinetics for intact cells.
Bar equals 1 min. Cell concentration normalized to 10^7 cells per ml.

Figure 2. Production of $\int$-aminolevulinic acid by recovering and complete
cells. These results are a summary of 5 experiments.

Figure 3. ^{35}S autoradiographic profile of membrane polypeptides of
uninhibited cells (lane A) and cells treated with 140 mM levulinic acid
(lane B).

isolated under lowered ionic conditions). Rates of PS II activity,
measured as DCBQ-dependent water oxidation, ranged from 300–400 μmol O_2/
hr/mg chl, and were similar to rates observed using intact cells. For
electrophoretic studies, samples were diluted 1:10 with 10 mM Tricine,
pH 7.5, and were centrifuged (27,000 x g, 15 min). The membrane sample
was removed from a blue supernant. A total of 11 polypeptides were
removed from the membrane by this dilution method, including major peptides at 71, 53, 47, 19, 16, 15 and 11.5 kDa. Minor peptides were observed
at 35, 26.5, 13.5, and 8 kDa. The molecular size of these peptides, and
the ease of their removal, suggest that they might comprise subunits of
carboxylase, CF_1 ATPase, and the cyanobacterial phycobilisome (peptides
at 19 and 15 kDa retained phycocyanin ligand).

Fig. 1 shows the room temperature chlorophyll fluorescence kinetics of
iron deficient cells, both before and 9 hr after readdition of iron.
Cells were illuminated at 2 x 10^4 ergs/cm^2/sec with 450 nm light, and
fluorescence was detected at 680 nm. For comparison, kinetics are shown
for normal cells. Following growth in the presence of iron, the kinetics
were altered dramatically: there was a decrease in original fluorescence,
and an increase in variable fluorescence. However, levulinic acid inter-
fered with these changes. At concentrations of 10–20 mM, the inhibitor
supressed the increase in variable fluorescence. At higher concentrations,
an increased original fluorescence was also obtained.

TABLE I.   Changes in polypeptide biosynthesis in response to levulinic acid.

Decreased Label (kDa)				Increased label (kDa)
120	27	16	13.5	36
71	21	15 (pbp)	8	24
35	19 (pbp)	14.5		17.5

Levulinic acid also induced changes in the fluorescence characteristics of cells grown in complete media. During a 9 hr inhibition, these changes were minor, and included both a slight decrease in variable and a slight increase in original fluorescence. However, cells which were recovering from iron deficiency were more sensitive than normal cells, as illustrated in Fig. 2. Following an incubation with the inhibitor, cells were removed and the spent media was assessed for δ-aminolevulinic acid (ALA) (D. Mauzerall, S. Granick, 1956). Kipe-Nolt et al. (1978) demonstrated that cyanobacteria were sensitive to levulinic acid, and that ALA excretion resulted. As seen in Fig. 2, both complete and recovering cells were induced to excrete ALA. However, recovering cells responded to lower inhibitor concentrations and showed a higher rate of excretion.

There are several alternative interpretations for this differential sensitivity to levulinic acid. Perhaps the most straight-forward interpretation is that heme (or another product of porphyrin biosynthesis) regulates the recovery from iron deficiency. This would be consistant with models from other systems implying a regulatory role for heme by influencing the activity of a protein kinase (G.A.F. Hendry, O.T.G. Jones, 1980). Alternatively, recovering cells may be more sensitive to inhibition owing to their higher metabolic activity. In in vivo labeling experiments, recovering cells typically yield 25% more incorporation than complete cells. Also, ALA production apparently depends in part on the concentration of cells challenged with the inhibitor (J.A. Kipe-Nolt et al, 1978). Although recovering and complete cells were at equivalent concentrations, their chlorophyll contents were very different. It is possible that a higher chlorophyll content in complete cultures yielded a higher degree of self-shielding from available light, thereby lowering metabolic activity.

2.3. Membrane composition and biosynthesis. Fig. 3 shows an autoradiographic profile of a 10-20% polyacrylamide gel. Cells grown in complete media were harvested and inhibited by 140 mM levulinic acid. ^{35}S-sulfate (50 μCi/ml) was added after 6 hr, and the cells were illuminated for an additional 3 hr. Membranes were prepared as described. Fig. 3 shows the label incorporation patterns for membranes of uninhibited (lane A) and levulinic acid-treated (lane B) cells. Inhibition caused a set of changes in the biosynthetic pattern. For example, decreased label was observed in the biliproteins (15 and 19 kDa). Decreased label also occurred at 71 kDa. Three prominent changes are shown by arrows in Fig. 3, and the results of several experiments are summarized in Table I. Decreased ^{35}S incorporation was observed at 11 polypeptide bands, while 3 bands showed enhanced incorporation. These changes occurred during the 9 hr of inhibition without apparent changes yet occurring in the bulk protein content, monitored by Coomassie Blue and by silver staining.

3. CONCLUSIONS

Taken together, Figs. 1 and 2 indicate that levulinic acid interupts the recovery of cells from iron deficiency. We do not know if the inhibitor prevents iron uptake. Yet, it is clear that the biosynthesis and assembly of chlorophyll and phycocyanin into the membrane is depressed, yielding cells with enhanced original and depressed variable fluorescence. These characteristics might indicate a partial uncoupling of chlorophyll from PS II or PS I reaction centers, a conclusion which must await further testing.

When complete cells are treated with levulinic acid, a set of 11 peptides are lost from the membrane fraction. The same is true for recovering cells. Since the phycobiliproteins figure so prominently in this set, phycobilisome assembly must be extremely sensitive to levulinic acid. Yamanaka et al. (1978) investigated the phycobilisome structure of a related A. nidulans strain and observed 5 non-pigmented polypeptides. Interestingly, a several peptides lost by inhibition (71, 35, and 27 kDa) are similar in molecular size to peptides observed in the phycobilisome. A peptide at 75 kDa was implicated in a role of linking the phycobilisome with the membrane. Alternatively, the peptide observed here at 71 kDa is extremely sensitive to inhibition by levulinic acid, and is more accessible to surface probes in membranes from iron deficient cells. One interpretation of these results is that the peptide observed at 71 kDa serves to link PS II with its major pigment array.

REFERENCES

Cramer CL Ristow JL Paulus TJ and Davis RH (1983) Methods for mycelial breakage of Neurospora, Analyt. Biochem. 128, 384–392.
Guikema JA and Sherman LA (1982) Protein composition and architecture of the photosynthetic membranes from the cyanobacterium, Anacystis nidulans R2, Biochim. Biophys. Acta 681, 440–450.
Guikema JA and Sherman LA (1983) Organization and function of chlorophyll in membranes of cyanobacteria during iron starvation, Plant Physiol. in press.
Hendry GAF and Jone OTG (1980) Haems and chlorophylls, J. Med. Genet. 17, 1–14.
Kipe-Nolt JA, Stevens SE Jr and Stevens CLR (1978) Biosynthesis of amino-levulinic acid by blue-green algae, J. Bacteriol. 135, 286–288.
Mauzerall D and Granick S (1956) The occurrence and determination of amino-levulinic acid and porphobilinogen in urine, J. Biol. Chem. 135, 435–446.
Yamanaka G Glazer AN and Williams RC (1978) Cyanobacterial phycobilisomes, J. Biol. Chem. 253, 8303–8310.

Authors address: James A. Guikema, Division of Biology, Kansas State
                    University, Manhattan, Kansas 66506, U.S.A.

PROPERTIES OF PURIFIED OXYGEN-EVOLVING PHOTOSYSTEM II PARTICLES FROM THE
BLUE-GREEN ALGA, PHORMIDIUM LAMINOSUM

ALISON C. STEWART, JANE M. BOWES and DEREK S. BENDALL

1. INTRODUCTION

The first PSII particles to retain high rates of O_2 evolution were
isolated from the thermophilic blue-green alga, Phormidium laminosum
(Stewart, Bendall, 1979). These particles, extracted from the membranes
by use of the detergent lauryl dimethylamine oxide (LDAO), have now been
further purified and their activity stabilised by treatment with a second
detergent, lauryl maltoside (LM), followed by fractionation on a sucrose
density gradient (see Bowes et al., this volume). In this communication
we describe the response of O_2 evolution by the purified O_2-evolving
particles (POPS) to pH, to the herbicide 3(3,4-dichlorophenyl)-1,1-
dimethylurea (DCMU) and to the addition of Ca^{2+} ions. The polypeptide
profile of POPS has also been analysed and the probable components of
the O_2-evolving complex identified.

2. MATERIALS AND METHODS

POPS were prepared as described by Bowes et al. (this volume). O_2 evolution was assayed at 25°C as described previously (Stewart, Bendall,
1979). For analysis of polypeptide profiles, two different SDS-PAGE
systems were used: (i) 7.5-15% acrylamide gradient gels, with a 5%
stacking gel and the buffer system of Chua (1980), or (ii) 7-17% acrylamide gradient gels, with a 3% stacking gel and the buffer system of
Laemmli (1970). All samples were solubilised before electrophoresis,
by incubation for 20 min at 35°C in buffer containing 2% SDS and either
5% mercaptoethanol or 30mM dithiothreitol.

3. RESULTS AND DISCUSSION

3.1 pH dependence and DCMU sensitivity of O_2 evolution

With 6mM 2,6-dimethyl-1,4-benzoquinone (DMBQ) as acceptor, the highest
rates of O_2 evolution were observed between pH 6.7 and 7.2 (Fig. 1a) and
the reaction was almost completely inhibited by 10 μM DCMU (Fig. 1b).
However the pH optimum was much lower (pH 5) when the negatively charged
oxidant ferricyanide was the acceptor (Fig. 1a). A possible explanation
for this effect is that ferricyanide was able to accept electrons
directly from Q when negative surface charges on the particles were
neutralised at low pH. This interpretation was supported by the finding
that, in the presence of ferricyanide, O_2 evolution became increasingly
insensitive to DCMU as the pH was lowered, indicating that ferricyanide
was accepting electrons before the DCMU block (Fig. 1b). It is clear
from Fig. 1 that the pH profile for activity depended mainly on the
acceptor used, and that the O_2-evolving complex itself was capable of
high activity over a broad pH range from pH 5-8. This is in contrast
to the acid pH optimum observed for O_2-evolving particles from spinach
(Ghanotakis, Babcock, 1983).

Sybesma, C. (ed.), Advances in Photosynthesis Research, Vol. II. ISBN 90-247-2943-2.
© *1984 Martinus Nijhoff/Dr W. Junk Publishers, The Hague/Boston/Lancaster.*

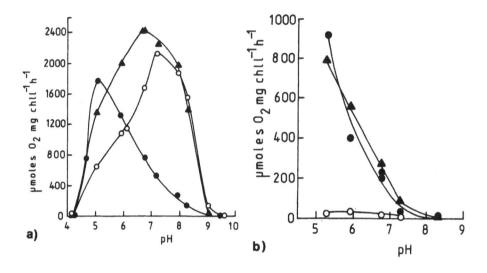

FIGURE 1. Effect of pH and electron acceptor on O_2 evolution in POPS.
The assay medium contained 25% glycerol and 10 mM $MgCl_2$, plus 50 mM
glycine (pH 9) or Tricine (pH 8-9) or Hepes (pH 7.5-8) or Mops (pH 7)
or Mes (pH 5-6) or glycylglycine (pH 5). O, 6 mM DMBQ; ●, 10 mM
ferricyanide; ▲, 6 mM DMBQ + 3 mM ferricyanide.
a) In the absence of DCMU; b) In the presence of 10 μM DCMU.

TABLE 1. Effect of Ca^{2+} ions on oxygen evolution by POPS.
The assay medium contained 25% (v/v) glycerol and 10 mM $MgCl_2$, plus
50 mM glycylglycine (pH 5.0) or 20 mM Hepes/NaOH (pH 7.2) and the
additions shown.

Additions to reaction medium	Rate of O_2 evolution (μmol O_2/mg Chl a per h)			
	$H_2O \rightarrow 6$ mMDMBQ + 3 mM FeCN		$H_2O \rightarrow 10$ mM FeCN	
	pH 5.0	pH 7.2	pH 5.0	pH 7.2
None	1349	2234	1773	487
25 mM $CaCl_2$	1426	2108	1641	682
50 mM $CaCl_2$	–	–	–	691
25 mM $Ca(NO_3)_2$	–	–	–	634
100 μM chlorpromazine	–	–	–	587
25 mM $CaCl_2$ + 100 μM chlorpromazine	–	–	–	789

3.2   Effect of Ca^{2+} ions on activity

In O_2-evolving preparations from some other blue-green algae, Ca^{2+} ions were reported to stimulate O_2 evolution (England, Evans, 1983).   In POPS, Ca^{2+} ions had no effect when DMBQ was the acceptor, but caused a variable degree of stimulation (10-40% in different preparations) at alkaline pH in the presence of 10 mM ferricyanide (results for one preparation are shown in Table 1).   This dependence on the nature of the acceptor suggests that the observed stimulatory effect in POPS may be on the reducing rather than the oxidising side of PSII.   Stimulation appeared to be specific to Ca^{2+} ions, since there was no effect on activity if the concentration of Mg^{2+} or Na$^+$ was increased (results not shown) but the stimulation by Ca^{2+} was not antagonised by the calmodulin antagonist chlorpromazine.

3.3   Polypeptide composition of POPS

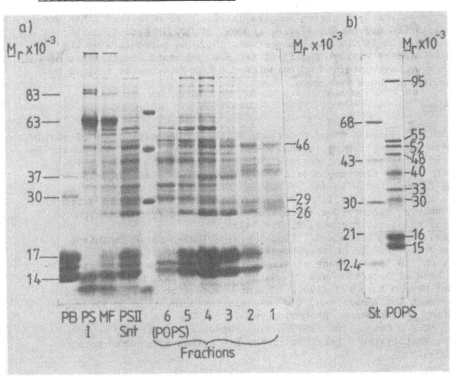

FIGURE 2. a) Polypeptide profiles of $\underline{P}$. laminosum membranes (MF), PSI particles (PSI), isolated phycobilisomes (PB), PSII-enriched LDAO supernatant (PSII Snt), POPS and Fractions 1-5 from the sucrose density gradient, using the SDS-PAGE buffer system of Chua (1980).   Chlorophyll loadings per track were 5 μg (MF), 10 μg (PSI) or 2 μg (PSII Snt and gradient fractions).   b) Polypeptide profile of POPS (3 μg Chl) using the buffer system of Laemmli (1970). M_r standards (St) were BSA (68 000), ovalbumin (43 000), carbonic anhydrase (30 000), trypsin inhibitor (21 000) and cytochrome $\underline{c}$ (12 400).

POPS had a simple polypeptide profile, but the exact pattern of bands depended on the SDS-PAGE system used (Fig. 2). In the Chua system (Fig. 2a) POPS, (fraction 6, the most dense fraction from the sucrose density gradient) comprised 7 major bands. Many contaminants in the initial PSI-enriched supernatant had been removed by the sucrose density gradient fractionation, the major ones being residual PSI components (M_r 83 000 and 63 000), cyt $\underline{f}$ (29 000), phycobilisome subunits (14 000, 17 000 and 37 000) and unidentified polypeptides of M_r 26 000, 46 000 and several bands in the region 70 000 - 90 000. These components appear in the polypeptide profiles for gradient fractions 1-5 (Fig. 2a). When POPS were subjected to SDS-PAGE using the Laemmli buffer system (Fig. 2b), the constituent polypeptides were more clearly resolved. 9 major bands were observed in this SDS-PAGE system, with approximate M_r values of 95-100 000, 55 000, 52 000, 48 000, 40 000, 33 500, 30 000, 16 000 and 15 000. The latter two we attribute to allophycocyanin and cyt $\underline{c}$-549, while the reaction centre may contribute bands in the 40-48 000 region. The remaining bands appear likely to include the major components of the O_2-evolving complex itself. We note that other groups have identified polypeptides of M_r approx. 58 000 (Nakatani, Barber, 1980), 32-34 000 and 23-25 000 (Akerlund et al. 1982) as components on the oxidising side of PSII. The first two are important constituents of POPS, but we have obtained no evidence for the presence of a M_r 23 000 polypeptide in these particles.

REFERENCES

Akerlund H-E, Jansson C and Andersson B (1982) Reconstitution of photosynthetic water splitting in inside-out thylakoid vesicles and identification of a participating protein. Biochim Biophys. Acta 681, 1-10
Chua N-H (1980) Electrophoretic analysis of chloroplast proteins Methods Enzymol. 69, 434-446
England RR and Evans EH (1983) A requirement for Ca^{2+} in the extraction of O_2-evolving Photosystem 2 preparations from the cyanobacterium Anacystis nidulans. Biochem. J. 210, 473-476
Ghanotakis DF and Babcock GT (1983) Hydroxylamine as an inhibitor between Z and P680 in photosystem II FEBS Lett. 153, 231-234
Laemmli UK (1970) Cleavage of structural proteins during the assembly of the head of bacteriophage T4 Nature (London) 227, 680-685
Nakatani HY and Barber J (1980) Cholate extraction of a heme-protein from spinach thylakoids and its possible involvement in PS-II oxygen evolution Photobiochem. Photobiophys. 2, 69-78
Stewart AC and Bendall DS (1979) Preparation of an active, oxygen-evolving Photosystem 2 particle from a blue-green alga FEBS Lett. 107, 308-312

ACKNOWLEDGEMENTS

We thank Mr Keith Jordan for excellent technical assistance. This work was supported by a Research Fellowship from King's College, Cambridge and a Royal Society Anglo-Australasian Fellowship to A.C.S., and by a grant from the U.K. Science and Engineering Research Council to D.S.B. and J.M.B.
Authors' address : Department of Biochemistry, University of Cambridge, Tennis Court Road, Cambridge CB2 1QW, U.K.

CHARACTERIZATION OF AN OXYGEN-EVOLVING PHOTOSYSTEM II COMPLEX FROM A
THERMOPHILIC CYANOBACTERIUM SYNECHOCOCCUS Sp.

G.H. SCHATZ and H.T. WITT

1. INTRODUCTION

Within the last years, membranes from several thermophilic cyanobacteria
(Stewart, Bendall, 1979; England, Evans, 1981; Miyairi, Schatz, 1983) and
from a red alga (Clement-Metral, Gantt, 1983) have been shown to be well-
suited for extraction of oxygen-evolving PS II complexes using LDAO as
detergent. In this work we report on PS II complexes isolated from a
thermophilic cyanobacterium Synechococcus sp. by means of the detergent
sulfobetaine 12 (N-dodecyl-N,N-dimethyl-3-ammonio-1-propane sulfonate).
We followed the amounts of extracted reaction centers of PS I and PS II,
respectively, and analyzed light-induced electron transport from water to
two different acceptors: (a) $Fe(CN)_6^{3-}$ and (b) phenyl-p-benzoquinone.

2. MATERIALS AND METHODS

2.1 Culture

The thermophilic cyanobacterium Synechococcus sp. (described by Yamaoka
et al., 1978) was grown at 50° C, harvested by centrifugation and washed
three times in HMP buffer (20 mM HEPES-NaOH, pH 7.8, 10 mM $MgCl_2$ and 2 mM
K_2HPO_4).

2.2 Preparation of Membranes

The washed cells were resuspended in HMPM buffer (20 mM HEPES-NaOH, pH 7.8,
10 mM $MgCl_2$, 2 mM K_2HPO_4, and 500 mM mannitol) with a concentration of 1 mg
Chl/ml and treated with 0.1% (w/w) egg lysozyme at 50° C for one hour.
After a "French press" treatment, the spheroplasts were disrupted by subsequent osmotic shock induced by suspension in HMP buffer. The membranes
were collected by centrifugation and further washed with HMPM buffer plus
0.1% of the detergent sulfobetaine 10. Upon this last washing, remaining
phycocyanin was released. Finally, the membranes were resuspended in a
mixture of 80% (v/v) HMPM buffer and 20% (v/v) glycerol with a concentration of 1 mg chlorophyll/ml and stored at -20° C for further use.

2.3 Preparation of Extracts

After addition of indicated amounts (see figures) of the detergent N-dodecyl-N,N-dimethyl-3-ammonio-1-propanesulfonate (SB 12) to the membranes,
the suspension was stirred at 20° C for 20 min in the dark and then centrifuged at 140,000 xg for 40 min. The resulting supernatant was stored at
-80° C and is termed "extract". Detergents were purchased from Calbio-
chem-Behring (as Zwittergents) and from Serva (as sulfobetaines).
Specific rates of electron transport, abbreviated as ė and given in the
dimension of $10^{-3} \cdot e \cdot Chl^{-1} s^{-1}$ ($\triangleq 1,008 \cdot \mu mol \cdot O_2 (mg\ Chl)^{-1} h^{-1}$) were determin-

---

Abbreviations: APC allophycocyanin; Chl chlorophyll; HEPES 4-(2-hydroxy-
ethyl)-1-piperazineethane sulfonic acid; LDAO lauryldi-
methylamine oxide; MES 4- morpholine-ethane sulfonic
acid; PS photosystem; SB sulfobetaine; SDS sodium dodecyl
sulfate

Sybesma, C. (ed.), Advances in Photosynthesis Research, Vol. II. ISBN 90-247-2943-2.
© *1984 Martinus Nijhoff/Dr W. Junk Publishers, The Hague/Boston/Lancaster.*

ed by means of (a) polarographic detection of oxygen-evolution or (b)
photometric detection of $K_3(Fe(CN)_6)$-reduction. Oxygen-evolution was
measured with a Clark-type electrode (Rank Brothers), $K_3(Fe(CN)_6)$ reduction was determined from the decrease of absorbance at 420 nm. Conditions
are denoted in the legends. When probing extracts, the detergent concentrations in the assay medium did not exceed 0.02% (w/w) due to dilution.
Measurements with membranes were performed in the presence of 0.1 μM
gramicidin D.
Reaction centers of photosystem I were determined by monitoring the flash-
induced absorbance changes at 702 nm in the presence of 20 μM PMS, 0.2 mM
sodium ascorbate and 0.1 mM methylviologen, using the molar extinction
coefficient $\varepsilon = 64000$ $M^{-1}cm^{-1}$ (Hiyama, Ke 1971).
Oxygen-evolving centers of photosystem II were determined by measuring the
average oxygen yield in repetitive single turnover flashes under conditions described earlier (Miyairi, Schatz 1983).

3. RESULTS AND DISCUSSION

Membranes of Synechococcus sp. prepared as described above show character-
istic electron transport rates of about $100 \cdot 10^{-3} \cdot e \cdot Chl^{-1}s^{-1}$ and specific
numbers of reaction centers following 1 PS I/300 Chl and 1 PS II/1000 Chl,
resp., as summarized in table I. For a series of treatments of the membranes

Table I

Specific numbers of photosystem reaction centers and electron transport
rates for Synechococcus membranes, extracts and pellet (from extraction
with 0.4% (w/w) SB 12)
- PS I/Chl from flash-induced absorption changes at 702 nm in the presence
  of PMS/ascorbate
- PS II/Chl from average, flash-induced oxygen yields
- $\dot{e}$ from oxygen-evolution measured at 40° C, pH=5.5, with 1 mM $K_3(Fe(CN)_6)$

	$\dfrac{PS\ I}{Chl}$	$\dfrac{PS\ II}{Chl}$	$\dfrac{PS\ II}{PS\ I}$	$\dfrac{\dot{e}}{10^{-3}e\ Chl^{-1}s^{-1}}$
Membranes	$3.3 \cdot 10^{-3}$	10^{-3}	0.3	100 ± 20
Extracts	$5 \cdot 10^{-4}$	10^{-2}	20	900 ± 100
Pellet	$3.3 \cdot 10^{-3}$	$< 10^{-4}$	< 0.03	< 10

with detergent SB 12 in different concentrations results are shown in
fig. 1. Application of the detergent SB 12 within a small concentration
range around 0.35% (w/w) results in extracts showing high rates of oxygen-
evolution (fig. 1a, open circles). Under the given conditions, typical
rates reach $(800-1000) \cdot 10^{-3}$ e $Chl^{-1}s^{-1}$. This corresponds to a nearly 10-
fold increase compared to uncoupled rates measured with membranes, i.e.,
the starting material before extraction (see table I). The specific numbers of oxygen-evolving PS II centers in the extracts (PS II_{extr}/Chl_{extr})

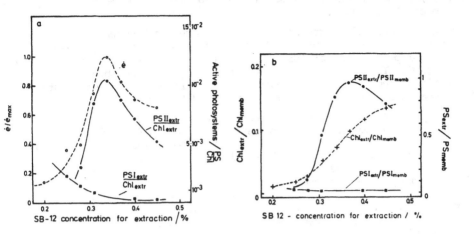

Fig. 1: a. Values of PS I/Chl (■), PS II/Chl (●) and electron transport rates (○) of extracts in dependence on SB 12 concentration (in % v/v) used for extraction.
Measurement of electron transport rate based on O_2 evolution
at pH=5.5, 40°C, with 1 mM $K_3(Fe(CN)_6)$. Maximal value
$\dot{e}_{max} = 800 \cdot 10^{-3}$ e $Chl^{-1}s^{-1}$.

b. Relative amounts of extracted PS I (■), PS II (●) and
chlorophyll (+) in dependence on the SB 12 concentration
(in % w/w) used for extraction, normalized to the total
amounts of PS I, PS II and Chl, resp., in unfractionated
membranes.

show a curve similar to that of electron transport rates and an optimal
value of about 1 PS II/100 Chl. Compared to the value measured in membranes, this indicates a 10-fold increase in PS II/Chl, too. Specific
numbers of PS I centers, on the other hand, show a continuous decrease
with increasing SB 12 concentration. Under optimal conditions, PS II
thereby exceed PS I in the extract by a factor of 20. Fig. 1b depicts
the numbers of reaction centers in the extract in relation to the number of reaction centers previously present in the unfractionated membrane.
It becomes obvious that only 1.5% of the PS I is solubilized from the
membrane. PS II, however, is released from the membrane in amounts
strongly dependent on the detergent concentration. Up to 95% of the PS II
reaction centers can be extracted without loss of their function in water-
oxidation. This leads to the conclusion that in the extract with 0.35%
SB 12, the fraction of PS II inactive in water oxidation cannot exceed
5%. This feature is of great value for functional studies discerning
between water oxidation and other reactions.
Analysis of the pellet that contained the unsoluble fragments of membranes
treated with 0.4% (w/w) SB 12 gave complementary results as listed in
table I: PS I was still present in almost the same amounts as in unfractionated membranes, while the capacity for oxygen-evolution was reduced to less than 10% of that of the membranes. Compared to oxygen-
evolving complexes prepared from Phormidium laminosum (Stewart, Bendall
1979), the extracts from Synechococcus sp. show lower content of PS I/Chl

and PS II/Chl but almost the same ratio of PS II:PS I (Stewart, Bendall 1981; Ke et al. 1982).

Absorbance spectra of extracts from Synechococcus (not shown) indicate the presence of large amounts of the auxiliary pigment allophycocyanin (APC). From peak values at 655 nm and at 675 nm we estimate a molar ratio of 3 Chl/APC. Together with the results from table I one obtains the ratio Chl-a_{II} (P680):Chl:APC = 1:100:33. Therefore, we conclude that an almost complete phycobilisome core is attached to the PS II complex.

Even higher values of electron transport rates than those in fig. 1 or in table I are obtained under changed conditions: at pH 4.5 and with increased concentrations ($\geqslant 5$ mM) of $K_3(Fe(CN)_6)$ about $2600 \cdot 10^{-3}$ e $Chl^{-1}s^{-1}$ are measured. The same rate is observed at pH 6.5 with phenyl-p-benzoquinone as acceptor. Taking into account the ratio of PS II/Chl, one obtains a corresponding turnover time of 2.7 ms. It is attributed to the slowest process in wateroxidation. With $K_3(Fe(CN)_6)$ pH-dependence follows a two-step curve with inflections at pH 4.8 and pH 6.5 attributed to the titration of negative charges which control access of $Fe(CN)_6^{3-}$ to the native PS II acceptor.

This indicates that in extracted PS II complexes electron transport pathways might be different from those in unfractionated membranes. Details are published in Schatz, Witt, 1983, Photobiochem. Photobiophys.

REFERENCES

Clement-Metral JD and Gantt E (1983) FEBS Lett. 156, 185-188
England RE and Evans EH (1981) FEBS Lett. 134, 157-177
Hiyama T and Ke B (1971) Arch. Biochem. Biophys. 147, 99-108
Ke B, Inoue H, Babcock GT, Fang Z.-X. and Dolan E (1982) Biochim. Biophys. Acta 682, 297-306
Miyairi S and Schatz GH (1983) Z. Naturforsch. 38c, 44-48
Stewart AC and Bendall DS (1979) FEBS Lett. 107, 308-312
Stewart AC and Bendall DS (1981) Biochem. J. 194, 877-887
Yamaoka T, Satoh K and Katoh S (1978) Plant Cell Physiol. 19, 943-954

ACKNOWLEDGEMENTS

We are grateful to Prof. S. Katoh (University of Tokyo) for the strain of Synechococcus sp. We like to thank Ms. I. Geisenheimer for skillful technical assistance. This work was supported by the Deutsche Forschungsgemeinschaft (Sfb 9).

Authors Address: Max-Volmer-Institut für Biophysikalische
                     und Physikalische Chemie, Technische
                     Universität Berlin, Strasse des 17. Juni 135,
                     1000 Berlin 12

A CALCIUM/SODIUM REQUIREMENT NEAR REACTION CENTER II IN ANACYSTIS NIDULANS

DAVID W. BECKER/JERRY J. BRAND

SUMMARY

Incubation of Anacystis nidulans in liquid nutrient medium devoid of calcium and sodium results in a rapid loss of oxygen-evolving capacity. Addition of sub-millimolar amounts of either calcium or sodium effects full restoration of oxygen evolution. Other cations, mono-, di-, or trivalent, have little or no effect. Both the depletion process and restoration by either calcium or sodium are light-dependent. Partial reactions have isolated the site of this cation requirement at or very near the reaction center, on the oxidizing side. Flash yield decline in depleting cells indicates a decrease in the number of functional reaction centers.

INTRODUCTION

Preparations of photosynthetic membranes from many cyanobacteria require the presence of divalent cations for maximal activity (Fredericks, Jagendorf 1964, Susor, Krogmann 1964, McSwain et al. 1976). One likely function of these cations is stabilization of the membrane vesicles. Membranes prepared with calcium as the divalent cation typically give highest rates of Photosystem II activity (Binder et al. 1976, Piccioni, Mauzerall 1976, Brand 1979, Yu, Brand 1980, DeRoo et al. 1981) and it appears that calcium plays an additional specific role in Photosystem II (Piccioni, Mauzerall, 1978, England, Evans 1981, England, Evans 1983). We recently reported an in vivo requirement for calcium in PS II in Anacystis nidulans (Becker, Brand 1982). The growth medium we used contains no sodium component and we have found that sodium substitutes completely for calcium as a requisite factor for PS II activity in A. nidulans.

MATERIALS AND METHODS

Cells were harvested from continuous culture, washed, and resuspended in Cg-10 growth medium (Van Baalen 1967) modified to 1.5 times the normal glycylglycine content, with or without Ca^{2+} and incubated under the same conditions of temperature, light and CO_2 as in continuous culture. Oxygen evolution was measured with a Clark type oxygen electrode. Silicomolybdate reduction was measured spectrophotometrically (Barr et al. 1975). Flash yields were measured as described by Myers et al. 1983.

RESULTS AND DISCUSSION

Figure 1 represents the loss of oxygen evolution which occurs in Anacystis nidulans upon incubation in the absence of calcium and sodium; the addition of sub-millimolar amounts of either calcium or sodium reverses this loss. A light requirement exists for both the loss of activity and its restoration by addition of calcium or sodium. Since respiration rates are very low in Anacystis, it is likely that these light requirements represent energy requirements and involve active transport of the cations. Of the large number of cations tried, only calcium and sodium effect in vivo restoration of lost O_2 activity. It is surprising that a monovalent and a divalent

Sybesma, C. (ed.), Advances in Photosynthesis Research, Vol. II. ISBN 90-247-2943-2.
© *1984 Martinus Nijhoff/Dr W. Junk Publishers, The Hague/Boston/Lancaster.*

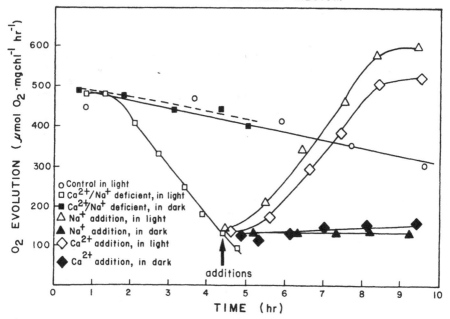

REVERSIBLE LOSS OF OXYGEN EVOLUTION BY <u>ANACYSTIS NIDULANS</u> INCUBATED IN CALCIUM/SODIUM-DEFICIENT MEDIUM

Fig. 1   At time zero cells were placed into Cg-10 medium with (control) or without Ca^{2+} and Na^{+}, then incubated in light or dark. After 4.3 hours of incubation, $Ca(NO_3)_2$ or $NaNO_3$ was added to depleted samples to give a final concentration of 0.25mM Ca^{2+} or Na^{+}. These cultures were further incubated in light or dark. Each point in the figure represents an aliquot of cells removed from culture at the indicated time and assayed for photosynthetic oxygen evolution. The figure is a composite of several experiments and linear portions were appropriately normalized.

cation would function interchangeably. Their ionic radii are nearly identical, however, which suggests the same site of action, possibly within a protein cleft or hole.

Oxygen flash yields decline rapidly and immediately upon illuminated incubation in Ca^{2+}/Na^{+}-deficient medium. The decline in flash yield indicates that the absence of calcium and sodium reduces the number of functional Photosystem II reaction centers (RC 2's). Piccioni and Mauzerall (1978) also demonstrated a greater number of active RC 2's when calcium rather than magnesium was used in preparing membranes from <u>Phormidium luridum</u>. Our measurements were made <u>in vivo</u>. Apparently without calcium and sodium the reaction centers switch off in an all-or-none fashion. The saturated O_2 rate does not decline immediately, however. It shows a lag which reflects an increase in maximum turnover rate for PS II reaction centers. This would occur if the oxidation of the plastoquinone (PQ) pool is normally the rate-limiting step and if a given pool is accessible by several RC 2's. As the number of functional reaction centers decreases due to loss of Ca^{2+} and Na^{+}, the effective PQ pool size increases, allowing more rapid turnover of the remaining functional reaction centers.

In vivo electron transport assays demonstrate that the site of Ca^{2+}/Na^{+} action is Photosystem II. Ascorbate/DAD to methyl viologen rates (measured as O_2 uptake) in the presence of DCMU are not affected by depletion of Ca^{2+} and Na^{+} while water to benzoquinone rates (measured as O_2 evolution) in the presence of DBMIB show a decline concommitent with the loss of complete photosynthetic electron transport. In vitro examination of partial reactions within PS II reveals that the site of the cation effect excludes the water-splitting enzyme and occurs prior to the site of DCMU inhibition: diphenylcarbazide-mediated silicomolybdate photoreduction after heat inactivation of the water-splitting enzyme shows diminished rates in membranes prepared from cells partially depleted of calcium and sodium.

Cells depleted of calcium and sodium show a complete loss of variable fluorescence at 685nm. In addition the fluorescence yield remains near the F_o level (Brand et al., 1983). Apparently the primary electron acceptor, Q, does not undergo photoreduction in these cells.

We have demonstrated a calcium or sodium requirement for PS II activity in Anacystis nidulans. The site of this requirement lies at or close to the PS II reaction center.

REFERENCES
Barr R, Crane FL and Giaquinta RT (1975) Dichlorophenylurea-insensitive reduction of silicomolybdic acid by chloroplast photosystem II, Plant Physiology 55, 460-462.
Becker DW and Brand JJ (1982) An in vivo requirement for calcium in photosystem II of Anacystis nidulans, Biochem. Biophys. Res. Commun. 109, 1134-1139.
Binder A, Tel-Or E and Avron M (1976) Photosynthetic activities of membrane preparations of the blue-green alga Phormidium luridum, Eur. J. Biochem. 67, 187-196.
Brand JJ (1979) The effect of Ca^{2+} on oxygen evolution in membrane preparations from Anacystis nidulans, FEBS Lett. 103, 114-117.
Brand JJ, Mohanty P and Fork DC (1983) Reversible inhibition of photochemistry of photosystem II by Ca^{2+} removal from intact cells of Anacystis nidulans, FEBS Lett. 155, 120-124.
DeRoo CLS and Yocum CF (1981) Cation-induced, inhibitor-resistant photosystem II reactions in cyanobacterial membranes, Biochem. Biophys. Res. Commun. 100, 1025-1031.
England RR and Evans EH (1981) A rapid method for extraction of oxygen-evolving photosystem 2 preparations from cyanobacteria, FEBS Lett. 134, 175-177.
England, RR and Evans EH (1983) A requirement for Ca^{2+} in the extraction of O_2-evolving photosystem 2 preparations from the cyanobacterium Anacystis nidulans, Biochem. J. 210, 473-476.
Fredericks WW and Jagendorf AT (1964) A soluble component of the Hill reaction in Anacystis nidulans, Arch. Biochem. Biophys. 104, 39-49.

McSwain BD, Tsujomoto HY and Arnon DI (1976) Effects of magnesium and chloride ions on light-induced electron transport in membrane fragments from a blue-green alga, Biochim. Biophys. Acta 423, 313-322.

Myers J, Graham J and Wang RT (1983) On the O_2 flash yields of two cyanophytes, Biochim. Biophys. Acta 722, 281-290.

Piccioni RG and Mauzerall DC (1976) Increase effected by calcium ion in the rate of oxygen evolution from preparations of Phormidium luridum, Biochim. Biophys. Acta 423, 605-609.

Piccioni RG and Mauzerall DC (1978) Calcium and photosynthetic oxygen evolution in cyanobacteria, Biochim. Biophys. Acta 504, 384-397.

Susor WA and Krogmann DW (1964) Hill activity in cell-free preparations of a blue-green alga, Biochim. Biophys. Acta 88, 11-19.

Van Baalen C (1967) Further observations on growth of single cells of coccoid blue-green algae, J. Phycol. 3, 154-157.

Yu CM-C and Brand JJ (1980) Role of divalent cations and ascorbate in photochemical activities of Anacystis membranes, Biochim. Biophys. Acta 591, 483-487.

ACKNOWLEDGEMENTS

Fluorescence experiments were conducted with D. C. Fork and P. Mohanty at Carnegie Institution, Stanford, California.

Flash yield experiments were conducted with J. Myers, University of Texas, Austin, Texas.

This research was supported by R. A. Welch Grant F-867 awarded to J. J. Brand.

Authors' address: Department of Botany, University of Texas, Austin, Texas 78712 USA.

EFFECTS OF Ca^{2+} IONS ON THE LIGHT-INDUCED ELECTRON TRANSPORT ACTIVITIES
OF ANACYSTIS NIDULANS PERMEAPLASTS AND SPHEROPLASTS

G. SOTIROPOULOU, T. LAGOYANNI, AND G.C. PAPAGEORGIOU/NUCLEAR RESEARCH
CENTER DEMOKRITOS, DEPARTMENT OF BIOLOGY, ATHENS, GREECE

1. INTRODUCTION

The photosynthetic electron transport of cyanobacteria has a requirement
for Ca^{2+} cations which is absent in eukaryotic plants. So far this requirement has been observed with cell-free thylakoid fragments only
(Binder et al., 1976; Piccioni, Mauzerall, 1978; Brand, 1979; England,
Evans, 1981). Such preparations consist, most likely, of mixed populations of sealed and unsealed vesicles, thus making difficult the deci-
sion on which side of the thylakoid membrane the Ca^{2+} cations really act.
In the present work, we study the Ca^{2+} effects on thylakoids in situ,
using ion permeable Anacystis nidulans cells prepared either by partial
(permeaplasts) or by total (shperoplasts) hydrolysis of the cell wall
peptidoglycan with lysozyme.

2. PROCEDURE

Anacystis was grown at 24-25 °C as previously described (Papageorgiou,
1977). Ion permeable cells, capable of retaining their shape in hypoosmotic media (permeaplasts; Ward, Myers, 1972), were prepared by incubating Anacystis (120 μg Chl a*/ml) with 10 mg/ml lysozyme and 1 mM
EDTA in a medium buffered by 50 mM Hepes·NaOH pH 7.5. Spheroplasts were
prepared according to Peschek et al.(1983). In either case the incubation was performed at 37 °C with mild stirring. Photosynthetic electron
transport was measured with a thermostated concentration electrode at
room temperature and with saturating white light illumination. The assay buffer consisted of sorbitol 500 mM, KCl 250 mM, and Hepes·NaOH 25 mM,
pH 7.5, in the case of permeaplasts, and of mannitol 400 mM, KCl 30 mM,
and Hepes·NaOH 25 mM, pH 6.9, in the case of spheroplasts. Metal cation
concentrations in intact and lysozyme-modified Anacystis were determined
by means of atomic absorption spectrophotometry (Varian Model AA-775).
Light absorption was measured with a Hitach Model 557 spectrophotometer.

3. RESULTS AND DISCUSSION

Anacystis has a tougher cell wall than many cyanobacteria which can be
digested only by the combined action of EDTA and excessive lysozyme.
Table 1 shows that EDTA treatment alone, as well as the subsequent lysozyme treatment depletes the cytosol from cations. This depletion engen-
ders irreversibly changes in the thylakoid embedded electron transport
systems.

The increased permeability of Anacystis cells to ions during lysozyme
treatment is reflected by the rate of O_2 evolution measured in the presence of the ionic cofactor FeCN (Fig. 1B). The rate reaches an early

*Abbreviations: Chl, chlorophyll; FeCN, $K_3[FeCCN)_6]$; Hepes, N-2-hydroxy-
ethyl-N -2-ethane sulfonic acid; PDox, mixture of p-phenylene diamine
(0.3 mM) and FeCN (1 mM).

Sybesma, C. (ed.), Advances in Photosynthesis Research, Vol. II. ISBN 90-247-2943-2.
© *1984 Martinus Nijhoff/Dr W. Junk Publishers, The Hague/Boston/Lancaster.*

maximum suggesting that limited cell wall digestion suffices for maximal ion permeability, and then it begins to decline. Cells corresponding to the rising and the declining branches of this curve are termed phase 1 and phase 2 permeaplasts (Ward, Myers, 1972). Addition of 10 mM CaCL$_2$ (on top of 250 mM KCl already existing in the assay medium) stimulates the electron transport rate but fails to reverse the decline of activity at longer times of lysozyme treatment. Mg^{2+}, on the other hand, was found to cause no stimulation of activity.

TABLE 1. Metal content of Anacystis cells after various treatments[1]

| Treatment | μg atom/mg Chl a | | |
	K^+	Mg^{2+}	Ca^{2+}
None	12.48 (100)	4.19 (100)	0.77 (100)
EDTA[2]	8.07 (65)	2.96 (71)	0.45 (58)
EDTA, then lysozyme[3]	2.11 (17)	0.12 (3)	0.24 (31)

[1]In parentheses, percent cation content; [2]10 mM EDTA, 20 h, 24 °C; [3]EDTA as above; than 5 mg/ml lysozyme for 7 h at 37 °C.

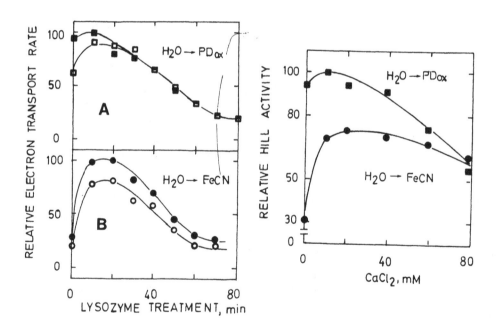

Fig. 1. PDox and FeCN supported photosynthetic O_2 evolution by Anacystis as a function of the incubation time with lysozyme-EDTA. Open symbols, activities measured in the absence of CaCl$_2$; solid symbols with 10 mM CaCl$_2$ added to the assay medium.

Fig. 2. Dependence of the Hill activities of phase 2 Anacystis permeaplasts on the Ca^{2+} concentration of the assay medium.

In the case of the lipophilic oxidant PDox (Fig. 1A) we see a short rise only in O_2 evolution activity and then an extensive decline. This indicates a free penetration of this oxidant in the cell and a beginning of the inactivation by the lysozyme-EDTA process earlier than the decline of FeCN-supported O_2 evolution might indicate (cf. Fig.1B). PDox-supported activity appears to be Ca^{2+} insensitive and this may be due to the fact that p-phenylene diamine picks up electrons from thylakoid sites that are inaccessible to Ca^{2+}. Supporting evidence was obtained also with p-benzoquinone, another lipophilic photosystem II oxidant. On the basis of these observations, we may reason that although accessible to suspension medium ions, the permeaplast thylakoids are impermeable to them. Ca^{2+} appears to be able to stimulate the rate of Hill activity only when the membrane site where the Hill cofactor couples is accessible to it, as it is the case with FeCN.

Fig. 2 shows the dependence of FeCN and PDox Hill activities of phase 2 permeaplasts on the Ca^{2+} concentration of the assay medium. Only the FeCN Hill activity is stimulated.

After prolonged EDTA and lysozyme treatment, Anacystis can be converted to spheroplasts (Peschek et al.,1983). We have found that the electron transport activity of these spheroplasts is dominated by a KCN and DCMU insensitive light-induced O_2 uptake, which occurs via a post-photosystem I autoxidizable endogenous intermediate. This activity is reminiscent of that reported by Honeycutt, Krogmann (1970) with cell-free cyanobacteria thylakoid fragments, except that it does not require exogenous autoxidizable acceptors like methyl viologen. The spheroplasts were also found to be capable of a dark reduction of 2,6-dichlorophenol indophenol (rate, 12 μmoles/mg Chl a·h), which was enhanced (2.5x) in the presence of 1 mM KCN, and was partially inhibited by 10 mM KCl.

TABLE 2.   Effect of the length of EDTA pretreatment on the light-induced O_2 uptake by the endogenous O_2 reducing system of Anacystis spheroplasts.

EDTA[1] pre-treatment, h	O_2 uptake[2]	
	$-CaCl_2$	$CaCl_2$ 25 mM
4	49	25
20	69	44

[1] 10 mM;   [2] in μmoles O_2/mg Chl·h

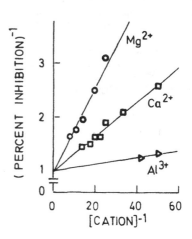

Fig. 3.   Inhibition of O_2 uptake by the endogenous system of Anacystis spheroplasts by metal cations.

Table 2 shows that the endogenous O_2 uptake activity of Anacystis sphe-roplasts depends on the duration of the EDTA treatment which precedes the lysozyme treatment, and that Ca^{2+} partially suppresses this activi-ty. Fig. 3 displays the metal cation concentration dependence of this inhibition. The number of sites on which these cations act in order to suppress O_2 uptake by the endogenous system of Anacystis spheroplasts is the same for Mg^{2+}, Ca^{2+}, and Al^{3+}. Therefore, we may rule out a spe-cific effect of Ca^{2+} in this case. Rather as the higher inhibitory activity of the trivalent cation indicates the inhibition must be relat-ed to the screening of the negative surface electric charge of the cya-nobacterium thylakoids (see also Kalosaka, Papageorgiou, these Proceed-ings).

REFERENCES

Binder A, Tel-Or E and Auron M (1976) Photosynthetic activities of mem-brane preparations of the blue-green alga Phormidium luridum, Eur. J. Biochem. 67, 187-196.
Brand JJ (1979) The effect of Ca^{2+} on oxygen evolution in membrane pre-parations from Anacystis nidulans, FEBS Letters 103, 114-117.
England RR and Evans EH (1983) A requirement for Ca^{2+} in the extraction of O_2-evolving photosystem 2 preparations from the cyanobacterium Ana-cystis nidulans, Biochem. J. 210, 473-476.
Honeycutt RC and Krogmann DW (1970) A light-dependent oxygen-reducing system for Anabaena variabilis, Biochim. Biophys. Acta 197, 267-275.
Kalosaka K and Papageorgiou GC (1983) Surface electric properties of thylakoid fragments isolated from vegetative and heterocystous cyanobac-teria, These Proceedings, in press.
Papageorgiou GC (1977) Photosynthetic activity of diimidoester-modified cells, permeaplasts and cell-free membrane fragments of the blue-green alga Anacystis nidulans, Biochim. Biophys. Acta 461, 379-391.
Peschek GA, Schmetterer G, Lauritsch G, Muchl R, Kienzl PF and Nitschmann WH (1989) in Papageorgiou GC and Packer L, eds. Photosynthetic prokaryo-tes: cell differentiation and function, pp. 147-162, Elsevier, New York, New York.
Piccioni RG and Mauzerall DC (1978) Calcium and photosynthetic oxygen evolution in cyanobacteria, Biochim. Biophys. Acta 504, 384-397.
Ward B and Myers J (1972) Photosynthetic properties of permeaplasts of Anacystis, Plant Physiol. 50, 547-550.

ACKNOWLEDGMENTS

We sincerely thank Dr. A. Souliotis for the atomic absorption measurements

HE SINGLE SUBUNIT P-700 REACTION CENTER OF A THERMOPHILIC
YANOBACTERIUM

. BINDER, P. MUSTER AND R. BACHOFEN

. INTRODUCTION

astigocladus laminosus is a thermophilic cyanobacterium that grows in alka-
ine hot springs at temperatures up to 65°. It is able to grow in a wide range
f environmental conditions - e.g. from pH 4.5 to pH 9.5 and from 35°C to
5°C. Thermophilic cyanobacteria are the only known organisms having a thermo-
hilic photosynthetic system of the higher plant type. Therefore, it might
erve as a convenient source for stable protein complexes, as was demonstrated
ith other thermophilic bacteria.

n an earlier communication (Nechushtai et al. 1983) we reported on the puri-
ication of the photosystem I reaction center from the thermophilic cyanobac-
erium Mastigocladus laminosus. The subunit composition and the photochemical
roperties of this reaction center were compared with those of photosystem I
eaction centers from green algae and higher plants. Immunological crossreac-
ivity between subunits I as well as between subunits II from these different
ources was indicated and its evolutional significance was discussed.

Je describe here the isolation and characterization of subunit I of Mastigocladus laminosus as a fully active photosystem I reaction center by its own.

2. MATERIALS AND METHODS

Cells were grown at 50°C as described earlier (Binder and Bachofen, 1979). The
membrane preparation as well as the isolation of the multiple subunit reaction
center was done according to Nechushtai et al. (1983). SDS-PAGE was done as
described by Wiemken et al. (1981). If not stated otherwise the samples for
the SDS-PAGE were treated with 2% SDS for 20 min at 60° (SDS to chlorophyll
ratio = 20:1). Protein determination was done according to Peterson (1977) and
chlorophyll determination after extraction with 80% acetone using an extinction coefficient of 80.04 mM^{-1} . cm^{-1} at 663 nm.

3. RESULTS AND DISCUSSION

When the multiple subunit reaction center is treated with 2% SDS under standard conditions but increasing concentrations of protein (increasing SDS to
protein ratio) and is then analyzed on SDS PAGE, subunit I may appear at three
different sites: mainly as 150 kD band (SDS:protein = 2:1), as 70 kD band
(10:1) and as 50 kD band (20:1) (Fig. 1). Thus these sites may represent different degrees of denaturation of the same peptide. As shown earlier
(Nechushtai et al. 1983), immunological cross reactivity also demonstrate,
that the 70 kD and the 50 kD peptide are identical. A similar transformation
into these three different bands can be observed, when the reaction center is
treated at different temperatures for SDS-PAGE (Fig. 2). Treatment at 100°C
yields in an complete aggregation of subunit I and a liberation of all the
chlorophyll. In unstained gels it can clearly be shown, that subunit I contains bound chlorophyll in all three states (Fig. 2). When these bands are cut
out, applied directly on the glass wall of a cuvette and activity is measured
in a spectrophotometer, it can be shown that all the bands have P-700 activity
by itself (results similar to Fig. 6).

Sybesma, C. (ed.), Advances in Photosynthesis Research, Vol. II. ISBN 90-247-2943-2.
© *1984 Martinus Nijhoff/Dr W. Junk Publishers, The Hague/Boston/Lancaster.*

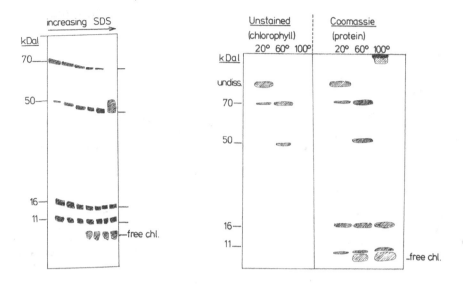

FIGURE 1. SDS-PAGE: Increasing SDS to protein ratios in the sample preparation.

FIGURE 2. SDS-PAGE: Varying temperature during sample preparation.

In order to isolate this active subunit I, the multiple subunit reaction center was treated for 60 min in 2% SDS at 40°C in the presence of PMSF as protease inhibitor and was then applied on a Sephacryl S-200 column (2.5x100) which was equilibrated with 0.1% SDS. The subunits were eluted with 20 mM Tri, pH 7.8 and 0.1% SDS (7 ml fractions). The elution profile is shown in Fig. 3.

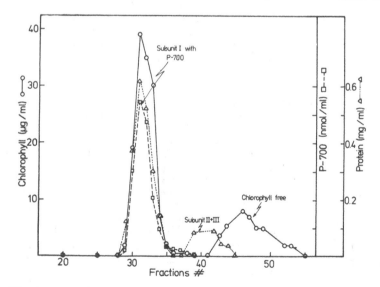

FIGURE 3. Elution profile of the Sephacryl S-200 column.

The spectra in Fig. 4 show, that the fractions 30-33 indeed contain chlorophyll attached to the protein, whereas the free chlorophyll (pheophytin) appears in the fractions 42-52. Fig. 5 shows the fractions analyzed in SDS-
PAGE. It demonstrates that fractions 30 - 33 contain the single subunit I in
the 50 kD form. The other fractions appear in fractions 38 - 44.

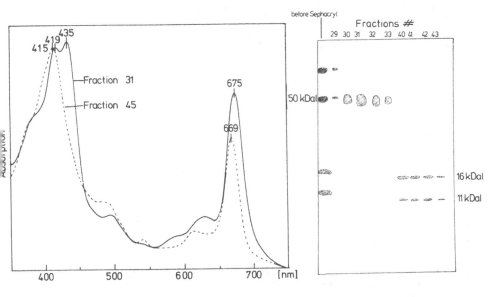

FIGURE 4. Spectra of Sephacryl fractions.

FIGURE 5. SDS-PAGE of
Sephacryl fractions.

This single subunit reaction center shows good activities in reversible P-700
photobleaching and cytochrome c oxidation (Fig. 6). The specific activities
in cytochrome c oxidation and the chlorophyll to P-700 ratios at the different
purification steps are summerized in Table 1.

TABLE 1. Purification steps and activities

Purification step	Chl / P-700 ratio	Cytochrome c oxidation
	(mole/mole)	(umol/mg chl.h)
Membranes	1200	-
Triton extract	210	114
DEAE cellulose	80	65
Sephacryl S-200	100	7

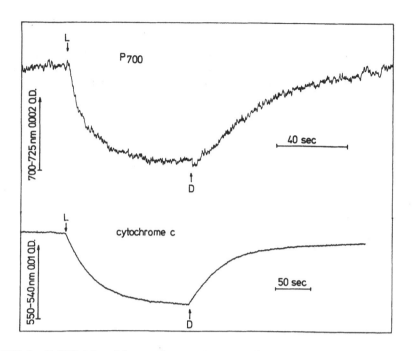

FIGURE 6. P-700 bleaching and cytochrome oxidation of subunit I.

The active and stable single subunit described here is an excellent and simpl
system to study structure-function relationships in the photosystem I reactio
center complex.

REFERENCES
Binder A and Bachofen R (1979) Isolation and characterization of a coupling
factor I ATPase of the thermophilic blue-green alga (cyanobacterium) Mastigocladus laminosus, FEBS Lett. 104, 66-70.
Nechushtai R, Muster P, Binder A, Liveanu V. and Nelson N (1983) Photosystem
reaction center from the thermophilic cyanobacterium Mastigocladus laminosus,
Proc.Natl.Acad.Sci.USA 80, 1179-1183.
Peterson GL (1977) A simplification of the protein assay method of Lowry et
al. which is more generally applicable, Anal.Biochem. 83, 346-357.
Wiemken V, Theiler R and Bachofen R (1981) Lateral organization of proteins i
the chromatophore membrane of Rhodospirillum rubrum studied by chemical cross
linking, J.Bioenerg.Biomembr. 13, 181-194.

ACKNOWLEDGEMENT
This work was supported by the Swiss National Foundation grant no. 3.582.79.

Authors address: Institute of Plant Biology, University of Zurich, Zolli-
kerstr. 107, CH-8008 Zurich, Switzerland.

ELECTRON TRANSFER AROUND PHOTOSYSTEM I IN CYANOBACTERIAL HETEROCYST MEMBRANES.

MALCOLM J. HAWKESFORD, JEFFREY P. HOUCHINS AND GEOFFREY HIND

1. INTRODUCTION

Cyanobacteria are unique amongst the prokaryotes in possessing a higher plant-type of photosynthesis, with two photosystems linked in series. The heterocyst is a specialized cell type occurring in some filamentous strains at a frequency of 5-10%, and is the site of N_2-fixation under aerobic conditions. During differentiation of the heterocyst, the O_2-evolving PSII is lost and cyclic electron transfer around PSI predominates. The absence of PSII reaction centres and the diminished levels of accessory pigments give membranes isolated from heterocysts excellent properties for spectroscopic studies. Soluble components such as plastocyanin, cytochrome c-553 and PSI acceptors, washed from the membranes during isolation, may be selectively reconstituted. Additionally, the presence of an endogenous uptake hydrogenase which can be utilized experimentally to reduce the plastoquinone pool and the electron transfer chain, make heterocyst membranes a useful system in which to study cyclic electron flow.

This paper reports an initial flash spectroscopic characterization of the electron transfer chain and speculates on the potential of the system.

2. MATERIAL AND METHODS

Heterocysts were isolated from cultures of Anabaena sp strain 7120 (ATCC 27893) as described previously (Houchins, Hind, 1982). Photosynthetic membranes were prepared under H_2 from the isolated heterocysts by passage twice through a French pressure cell at 110 MPa. After removal of unbroken cells and large fragments, membranes were harvested by centrifugation at 40,000 x g for 90 min. For assays, membranes (13 µg ml^{-1}) were suspended in 40 mM Hepes (pH 7.5) containing 1 mM $MgCl_2$.

Absorbance changes elicited by 4 µs flashes of red light were measured as previously described (Crowther, Hind, 1980). The Flash frequency was 1 Hz and traces were the average of 64 (P700 turnover) or 128 (cytochrome α-band region) records. Traces in the cytochrome α-band region were corrected for contributions of P700.

3. RESULTS AND DISCUSSION

Fig. 1 shows absorbance changes at 430 nm, which primarily reflect oxidation reduction turnovers of P700. Under Ar with no addition of cyt c-553, repetitive flashes elicited little apparent turnover of P700 in isolated heterocyst photosynthetic membranes. A large rapid decrease in absorbance occurred in the presence of H_2 or with the addition of cyt c-553. Under these conditions the signal relaxed slowly ($t_{0.5} > 35$ ms). However in the presence of both H_2 and cyt c-553, a large bleaching was observed which relaxed with biphasic kinetics; a rapidly relaxing phase ($t_{0.5} = 2.5$ ms) and a slowly relaxing component ($t_{0.5} = 150$ ms).

Sybesma, C. (ed.), Advances in Photosynthesis Research, Vol. II. ISBN 90-247-2943-2.

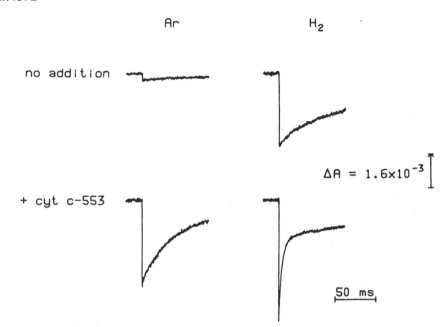

FIGURE 1. Flash-induced absorbance changes at 430 nm in isolated heterocyst membranes, under an Ar or H$_2$ gas phase. 2.2 μM cyt c-553 was added where indicated.

Turnovers of components of the electron transfer chain in the isolated membranes appear to be limited by both reductant supply and a lack of certain soluble carriers, notably cyt c-553 or plastocyanin. These proteins function interchangeably in heterocysts as mediators between the b/f complex and P700 (Houchins et al., 1983). Addition of H$_2$ sufficiently reduced the electron transfer chain such that P700, oxidized by a flash, was slowly rereduced before the next flash. Alternatively, addition of cyt c-553 (under Ar) also enabled similar kinetics to be observed, indicating that some endogenous reductant was available to slowly rereduce P700$^+$. This reduction, however, did not occur in the interval between flashes in the absence of added mediator. In the presence of both exogenous reductant and added soluble mediator, relatively rapid P700 rereduction was observed. The rate of rereduction indicated by the rapidly relaxing compnonent was, however, slower than that seen in intact heterocysts and was probably limited by diffusion of reduced mediator to the PSI complex. The slowly relaxing phase was probably comprised of two components. P430, the optical signal of the PSI acceptor FeS centers, would undergo bleaching upon reduction and would relax slowly in the absence of additional acceptors. Any P700 which was inaccessible to added mediator in this membrane preparation may also have contributed to the slowly relaxing component.

Fig. 2. shows flash-induced absorbance changes at 554 and 563 nm after correction for contributions of P700. 563 nm corresponds to the α-band of cyt b-563. An increase in absorbance was only apparent in the presence of HQNO, an inhibitor of cyt b reoxidation (Houchins, Hind,

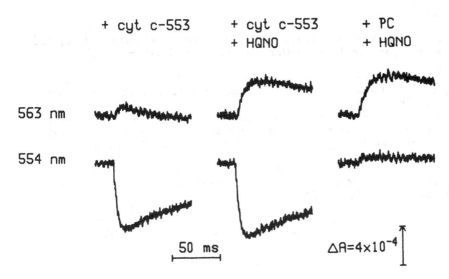

FIGURE 2. Effect of HQNO (5 μM) on flash-induced turnovers of cyt $\underline{c}$-553 and cyt $\underline{b}$-563, as measured at 554 and 563 nm, respectively. 2.2 μM of cyt $\underline{c}$-553 or plastocyanin (PC) were added as indicated.

1983). In the absence of HQNO, rapid reoxidation prevented accumulation of reduced cyt $\underline{b}$-563. Cyt $\underline{b}$-563 is known to undergo an oxidant-induced reduction. The oxidant in this case is cyt $\underline{c}$-553 oxidized by P700$^+$. Cyt $\underline{c}$-553 oxidation can be followed by measuring flash induced absorbance transients at 554 nm in cyt c-553 reconstituted membranes. The observed decrease in absorbance reflected cyt $\underline{c}$-553 oxidation and corresponded to the kinetics of P700 rereduction (see fig. 1). The slow relaxation of this signal indicated that the pool of cyt $\underline{c}$-553 was slowly rereduced by the b/f complex, possibly due to diffusion of mediator away from the membrane. Although plastocyanin also effectively mediated between the b/f complex and P700, no significant absorbance transients due to plastocyanin were observed. Plastocyanin has a very broad absorption peak and a small extinction coefficient at this wavelength. No absorbance changes due to oxidation-reduction turnovers of cyt $\underline{f}$ were observed because the rate of reduction of cyt $\underline{f}$ by the reduced plastoquinone pool exceeds the rate of oxidation by the soluble mediator pool in this reconstituted system.

In these photosynthetic membranes, flash-induced oxidation-reduction turnovers of components of a reconstituted linear electron transfer pathway involving PSI may be demonstrated (see also, Hawkesford et al., 1983). In the intact cell it is believed that cyclic electron transfer is of fundamental importance, particularly in generating ATP for N_2-fixation. It is intended to use this system to investigate the mechanism of cyclic electron transfer, exploiting the selective reconstitution procedures as outlined above.

REFERENCES

Crowther D and Hind G (1980)   Partial characterization of cyclic
electron transport in intact chloroplasts, Arch. Biochim. Biophys. 204,
568-577.
Hawkesford MJ, Houchins JP and Hind G (1983)   Reconstitution of
photosynthetic electron transfer in cyanobacterial heterocyst membranes,
FEBS Lett. (in press).
Houchins JP and Hind G (1982)   Pyridine nucleotides and H_2 as electron
donors to the respiratory and photosynthetic electron-transfer chains
and to nitrogenase in Anabaena heterocysts, Biochim. Biophys. Acta, 682,
86-96.
Houchins JP and Hind G (1983) Flash spectroscopic characterization of
photosynthetic electron transport in isolated heterocysts, Arch.
Biochem. Biophys. 224, 272-282.
Houchins JP, Hawkesford MJ and Hind G (1983)   Kinetic and spectral
resolution of cytochrome c-553 and cytochrome f in the photosynthetic
electron transfer chain of heterocysts, FEBS Lett. (in press).

ACKNOWLEDGEMENTS

This work was supported by grant 79-59-2366-1-1-382-1 from the
Competitive Research Grant Office of the U.S. Department of
Agriculture/Science and Education Administration. It was carried out at
Brookhaven National Laboratory under the auspices of, and with
additional support from the U.S. Department of Energy.

Authors address: Biology Department, Brookhaven National Laboratory,
                   Upton, NY 11973, USA

TRANSIENT AND STEADY-STATE KINETICS OF THE REACTION BETWEEN CYTOCHROME c
AND THE PHOTOSYSTEM I REACTION CENTRE IN CYANOBACTERIA.

AKSEL HADBERG, LARS F. OLSEN and RAYMOND P. COX

1. INTRODUCTION

The photosynthetic electron transport chain in cyanobacteria is very
similar to that in the chloroplasts of eucaryotic algae and higher
plants (Ho, Krogmann, 1982). One difference is that certain cyanobacteria
have a positively charged plastocyanin or cytochrome c in place of the
negatively charged protein found in chloroplasts. The net charge on the
molecule might be expected to influence its reaction mechanism with
components associated with the negatively charged membrane. We report
here kinetic studies on the reaction of positively charged cytochrome c
from horse heart mitochondria with a membrane preparation from Anabaena
variabilis, which is known to have a native cytochrome c of net positive
charge.

2. EXPERIMENTAL

Anabaena variabilis CCAP 1403/13a (= ATCC 29413) was grown in the
light in fructose-limited continuous culture. Cyanobacteria were harvested by centrifugation, washed and resuspended in 0.5 M sorbitol, 20 mM
Hepes, 10 mM $MgCl_2$ and 5 mM KH_2PO_4, adjusted to pH 7.6 with KOH. Following
2 passages through a French press at 8 000 lb/in^2, unbroken cells were removed by centrifugation (3 000 xg, 3 min). Polybrene was added to the
supernatant (3 mg/mg Chl). Following centrifugation (3 000 xg, 6 min) a
green precipitate containing the membrane fraction and a clear blue
supernatant were obtained. The membrane fraction was washed several times
with medium to remove phycobilins and excess polybrene before storage in
liquid nitrogen at a concentration of 1-2 mg Chl/ml in the presence of
Me_2SO (5%, v/v).

The reaction between cytochrome c and P700 was studied by two
different methods. Reduction of P700 and oxidation of fully reduced
cytochrome c following a flash was studied using a repetitive flash
spectrophotometer as described by Olsen and Cox (1982). The photooxidation of cytochrome c during saturating continuous light (Schott
RG 645 filter) was followed with a Perkin-Elmer 356 dual-wavelength
spectrophotometer at 550 minus 540 nm. In both cases the membranes were
resuspended in 20 mM Hepes-KOH pH 7.6 containing the detergent Brij-35
(10 g/l). "High ionic strength medium" contained in addition 0.1 M KCl.
All reactions were measured at 20-22°C.

Sybesma, C. (ed.), Advances in Photosynthesis Research, Vol. II. ISBN 90-247-2943-2.
© *1984 Martinus Nijhoff/Dr W. Junk Publishers, The Hague/Boston/Lancaster.*

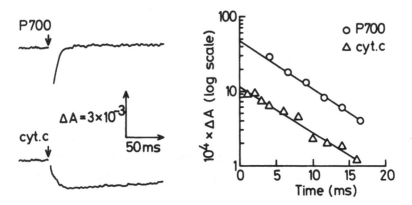

FIGURE 1. Absorbance changes due to P700 in a membrane preparation from
Anabaena variabilis, and horse-heart cytochrome c, following a single-
turnover flash. Membranes were suspended in the low ionic strength medium
at a concentration corresponding to 10 μg chlorophyll/ml (80 nM P700).
The reaction mixture also contained 5 μM cytochrome c, 2 μM N,N,N',N'
t .tramethylphenylenediamine, 5 mM ascorbate and 40 μM methyl viologen.
The flash was given at the time shown by the arrows. The residual
fluorescence artefact is not shown on the traces.

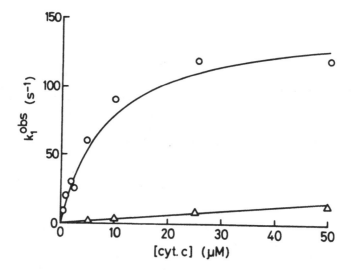

FIGURE 2. The effect of cytochrome c concentration on the apparent first
order rate constant for the reaction with P700 following a flash. The
results are derived from a series of experiments similar to that shown
in Fig. 1. O. Low salt medium. The points shown are the mean values for
different numbers of measurements at various concentrations of cytochrome
c, whilst the curve represents the best fit of all the data to the
Michaelis–Menten equation calculated using non–linear least squares analysis,
with intermediate weighting (Duggleby, 1981). Δ. High salt medium.

3.   RESULTS AND DISCUSSION

Fig. 1 shows the photooxidation induced by a single turnover flash of P700 in the preparation derived from <u>Anabaena</u> and its subsequent reduction by horse heart cytochrome <u>c</u>. The reduction of P700$^+$ and oxidation of the cytochrome follow essentially identical pseudo-first order kinetics.

Fig. 2 shows the effect of the cytochrome concentration on the apparent first order rate constant for the reaction. The reaction rate is higher at low ionic strength, where k_1^{obs} shows a hyperbolic dependence on the cytochrome <u>c</u> concentration. This type of kinetic behaviour is that expected if a complex is formed between the cytochrome and the Photosystem I reaction centre with a dissociation constant for cytochrome <u>c</u> of 9 µM. The maximal turnover time is 150 s^{-1}. The evidence for complex formation is in contrast to the situation in chloroplasts, where we have shown that the reaction between plastidic cytochrome <u>c</u> or plastocyanin and P700$^+$ appears to be a diffusion-limited bimolecular collision process with no evidence for complex formation at either low or high ionic strengths (Olsen et al, 1980; Olsen, Cox, 1982; Olsen, 1982; Olsen, Pedersen, 1983). Complex formation is presumably favoured by the positive charge on the cyanobacterial proteins which would be expected to increase the strength of interactions with the negatively charged membrane.

Further support for the idea of complex formation was provided by measurements of the photooxidation of substrate amounts of cytochrome <u>c</u> in continuous light. The observed time courses were exponential rather than linear even at the highest cytochrome concentrations used. A similar result is well-established in the analogous assay with mitochondrial cytochrome <u>c</u> oxidase and is believed to be caused by binding of both oxidised and reduced cytochrome to the oxidase with equal affinity (reviewed by Wikström et al, 1981). The effect of cytochrome <u>c</u> concentration on the rate is shown in Fig. 3. In contrast to the usual observation with the mito-chondrial enzyme, there is evidence for only one binding site, with a K_m of 8 µM in good agreement with the results of the transient kinetic measure-ments. The maximal turnover time is 110 s^{-1} and thus also in agreement with the single turnover data.

Although there is good evidence for complex formation in both single and multiple turnover measurements, it is not immediately obvious why the formation of a complex does not result in biphasic transient kinetics at

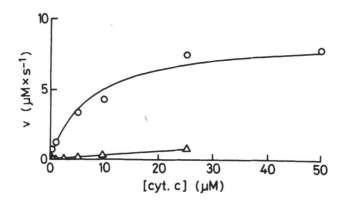

FIGURE 3. The effect of cytochrome c concentration on the rate of its photooxidation in continuous light. The reaction medium contained membranes equiva
lent to 10 µg chlorophyll/ml (80 nM P700), 25 µg/ml superoxide dismutase,
6 µg/ml catalase, 40 µM methyl viologen, and reduced horse-heart cytochrome c
Initial rates were calculated from the half-times of the exponential time
courses obtained. O. Low salt medium. The curve represents the best fit to th
data as described in the legend to Fig. 2. Δ. High salt medium.

cytochrome concentrations corresponding to the dissociation constant. One

might expect a concentration-independent rapid phase resulting from

electron transfer between the bound components. Failure to observe such

kinetics suggests that the simplest models are not adequate to describe

the reaction mechanism, and further investigation is currently in progress.

REFERENCES
Duggleby RG (1981) A nonlinear regression program for small computers.
Anal. Biochem. 110, 9-18.
Ho KK and Krogmann DW (1982) Photosynthesis. In: Carr NG and Whitton BA, eds.
The biology of cyanobacteria, pp. 191-224, Oxford: Blackwell.
Olsen LF (1982) Transient kinetics of the electron transfer between P700,
plastocyanin and cytochrome f in chloroplasts suspended in fluid media at
sub-zero temperatures. Biochim. Biophys. Acta 682, 482-490.
Olsen LF and Cox RP (1982) Transient kinetics of the reaction between
cytochrome c-552 or plastocyanin and P700 in subchloroplast particles.
Biochim. Biophys. Acta 679, 436-443.
Olsen LF and Pedersen JZ (1983) On the mechanism of electron transfer from
plastocyanin to P700 in chloroplasts. Photobiochem. Photobiophys. 5, 1-10.
Olsen LF, Cox RP and Barber J (1980) Flash-induced redox changes of P700 and
plastocyanin in chloroplasts suspended in fluid media at sub-zero tempera-
tures. FEBS Lett. 122, 13-16.
Wikström M, Krab K and Saraste M (1981) Cytochrome oxidase. London: Academic
Press.

Authors' address: Institute of Biochemistry, Odense University
                    Campusvej 55, 5230 Odense M, Denmark.

KINETICS OF A CHLOROPHYLL a LONG WAVELENGTH FLUORESCING FORM:QUENCHING OF
F 750 IN RELATION TO PS I OXIDO-REDUCTION STATE IN A CYANOBACTERIUM
(PSEUDANABAENA SP.)

J.C. Duval[+], J.C. Thomas[+] and H. Jupin[++]
+ Laboratoire de Cytophysiologie Végétale, Ecole Normale Supérieure,
  24 rue Lhomond, 75231 PARIS Cedex 05, France.
++Laboratoire de Biologie Végétale, Université de Perpignan,
  Avenue de Villeneuve, 66000 PERPIGNAN, France.

Cyanobacteria synthetize only one type of chlorophyll molecule, chlorophyll a, besides accessory pigments, carotenoïds and phycobiliproteins.
The major part of Chl a (80 %) is associated with PS I. Electrophoresis
of chl a-protein complexes from thylakoïd pigments revealed heavy complexes
(in the 100 KD to 300 KD range) which are related to PS I and are enriched
in long wavelength Chl a absorbing forms. Fluorescence emission at 77 K
above 700 nm largely originated from these complexes (Oquist et al.1981).
However, significant differences seemed to exist for these forms among
complexes isolated from various strains (Thomas, Mousseau 1981). Emission
fluorescence spectrum at 77 K of whole cells of a Pseudanabaena strain
exhibited a shoulder in the 750 nm range. This component was not observed
in spectra obtained from LiDS-PAGE chlorophyll-protein complexes. Our work
deals with occurrence and some properties of the F 750 component.

MATERIAL AND METHODS

Pseudanabaena M_2 was isolated in our laboratory and grown at 20°C in
the "Z" medium of Zehnder and Gorham (1960) with a 12 h/12 h light-dark
cycle (4 W. m^{-2}) and continuously air bubbling. Nitrogen depleted cells
with no detectable phycobiliproteins were also studied.
Absorption, emission and excitation fluorescence spectra were performed
as described previously (Lemoine, Jupin 1978). Modifications were made to
follow F 750 kinetics. PS I and PS II electron transport activities were
appreciated with a Clark-type electrode.
Experiments were performed either on whole cells or on a 10000 g supernatant fraction from sonicated cells in 10 mM HEPES buffer (pH 7.4).

RESULTS

The 77 K fluorescence emission spectra of Pseudanabaena M_2 cells exhibited four main peaks : F 660 and F 687 from phycobiliproteins, F 695
from PS II and F 725 from PS I. A shoulder near 750 nm was clearly visible
(Fig. 1). 77 K action spectra of total fluorescence above 700 nm (Fig. 2)
revealed two major peaks around 630 nm from phycobiliproteins, and around
680 nm from chlorophyll. Surprisingly, the variable part of fluorescence
(Fm - Fo) was found to be negative. In order to test the hypothesis of an
heterogeneity of the fluorescence above 700 nm, we investigated kinetics
of variable fluorescence at 77 K for emission wavelengthes in the 600 nm-
800 nm range. Fig. 3 illustrates the fluorescence decay which was found to
be the highest at 750 nm. From Fo and Fm levels measured at each wavelength
(5 nm step), we calculated a true Fo emission spectrum (Fig. 1).

It appeared an additional peak in the recalculated Fo spectrum, at 750 nm,
corresponding to the previously observed shoulder in the Fm spectrum. Maximum of quenching reached 60 % at 750 nm in whole cells. We called F 750
this particular fluorescing form.

Sybesma, C. (ed.), Advances in Photosynthesis Research, Vol. II. ISBN 90-247-2943-2.
© 1984 Martinus Nijhoff/Dr W. Junk Publishers, The Hague/Boston/Lancaster.

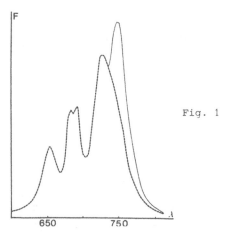

Fig. 1

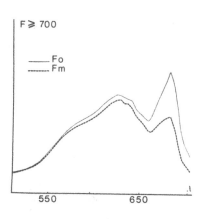

Fig. 2

Fig. 1: Fluorescence emission spectra of whole cells at 77 K. Excitation beam : 480 nm(20 W.m^{-2})   Fm   Fo (recalculated from kinetic results, 5 nm steps).

Fig. 2: Action spectra of the whole fluorescence above 700 nm. Fo: with detecting light only. Fm: after saturating flashing (according to Butler and Kitajima, 1975).

Fig. 3: Kinetic of 750 nm fluorescence decay. Same conditions as Fig. 1. Time scale: 50 ms.

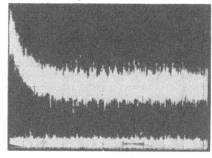

Fig. 3

To analyze F 750, the apparatus used for fluorescence action spectra was modified. A new combination of filters (Kodak Wratten 50 and M.T.O. J 662 a) transmitted wavelengthes higher than 740 nm ; excitation light was monochromatic (0,5 nm halfband width). Kinetics studies were conducted at sufficient low energy to obtain for each sample a true Fo with a resolution time of 0,3 s. In these conditions, Fm was reached only after 20 mn exposure at monochromatic light or after ten white light flashes of high energy.

F 750 action spectrum (Fig. 4) was established from kinetic measurements of F 750 quenching on whole cells by isoquantic monochromatic lights. We choose to express the intensity of the phenomenon by the (Fo - Fm)/Fm ratio for a given wavelength. It clearly appeared two maxima at 680 nm and 710 nm, and a pronounced shoulder at 695 nm. Excitation between 640 nm and 650 nm showed a very low efficiency ; shorter wavelengthes were more active. If we compare this F 750 spectrum with absorption spectrum of control cells, it appears that light absorbed by phycobiliproteins was inactive to induce F 750 quenching. Nevertheless, the efficiency of light absorbed by chl a was high with a peculiar enhancement of chl a 695 and chl a 710 yield.

Excitation of 715 nm ($\pm$ 2,2 nm) fluorescence induced Fo and Fm action spectra with a positive variable fluorescence, which was characteristic of PS II antenna, as in green plants (Butler, Kitajima, 1975).

Conditions of F 750 reversion. Samples on which a Fm state had been induced were stable even kept for several hours in the dark at 77 K. If samples were warmed in the dark, several freezing and thawing cycles might be applied to the same sample without significant loss of F 750 (Fig. 5).

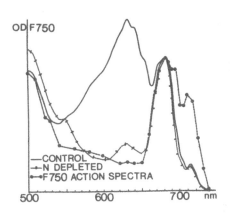

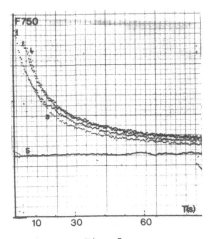

Fig. 4   Fig. 5

Fig. 4:.Absorption spectra at 77 K of whole cells (control and nitrogen depleted i.e. lacking phycobiliproteins).
.F 750 action spectrum expressed as (Fm-Fo)/Fm and normalized at 680 nm with absorption spectra.

Fig. 5: Freezing and thawing of the same sample. 1, 2, 3, 4 successive experiments ; 5: freezing in the light (Fm state). Excitation beam at 680nm.

Relations between F 750 and PS I, PS II activities. F 750 kinetics were also studied on cell free extracts (after sonication ; see material and methods). The 10 000 g supernatant fraction (= S 10 000) exhibited a significant variable F 750, although generally weaker than in whole cells ((Fo-Fm)/Fm = 0,5 versus 0,6 for 680 nm excitation beam). This fraction retained appreciable PS I electron transport activities. The presumption of a relationship between PS I and F 750, as suggested by action spectrum, had to be established. Treatment of cyanobacterial cells by 2-hydroxydiphenyl (2 H-D) led to PS II inactivation (Thomas, Mousseau 1981). 2 H-D treated cells (300 mM, 34 h or 50 h incubation time) were tested for PS II electron transport activity ($H_2O \rightarrow$ p-Benzoquinone). PS I activity was appreciated on S 10 000 (red. DAD $\rightarrow$ Methylviologen). F 750 properties and emission fluorescence spectra were measured at the same time. A good correlation was found between disappearance of F 695 (Fig. 6), loss of PS II activity (Table I) and between evolution of F 750 emission (Fig. 6), PS I activity (Table I) and F 750 kinetics (Fig. 7, Table I).

2-HD treatment	Electron transport[*]		F 750 (Fo-Fm)/Fm	Relative PS I activity[**]	Relative F 750 decrease[**]
	PS II	PS I			
0	300	160	0,63	100	100
34 h	31	170	0,47	106	75
50 h	0	63	0,24	39	38

Table I. Comparative activities in presence of 2.HD.
[*] Electron transport activities expressed in µM O_2/h/mg chl Experimental conditions : see text.
[**] Relative ratio to control proportions.

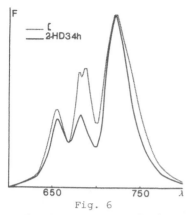

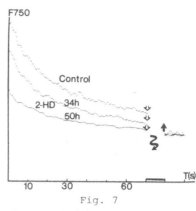

Fig. 6   Fig. 7

Fig. 6: Fluorescence emission spectra of control (c) and 2 H-D 34 h treated cells. Same conditions as Fig. 1.

Fig. 7: F 750 kinetics of 10 000 g supernatant fraction from control and 2 H-D treated cells. Same conditions as Fig. 5.

Incubation in the dark of S 10 000 with Na-Dithionite at low pH (3) induced a Fm level of F 750 ; nevertheless, if a well buffered medium was used, 0,2 M Na dithionite had no effect.

1 mMK-Ferricyanide caused a Fm level on S 10 000 preincubated in the dark ; this effect was not observed for Ferricyanide concentration below 0,1ı

CONCLUSIONS   Hence we characterize F 750 as a 60 % negative variable fluorescence in samples frozen at 77 K in the dark. Reversion occurs only at rooı temperature ; it originates probably from a chl a 710 linked to PS I as showı by action spectrum and 2 HD treated cells. We fail to detect F 750 in Anacystis nidulans although cell enveloppes exhibit fluorescence in this range (Murata et al., 1981). Some acceptors of PS I are thought to be chlorophylls A_2 (or X) in particular can control PS I fluorescence (Malkin, 1982). Dithio nite experiments prevent the identification of F 750 as A_2. On the contrary, ferricyanide action indicates that P 700$^+$ is a quencher of F 750. Fm may cor respond to reversible charge separation of PSI as shown by Crowder et al(198

REFERENCES
Butler WL and Kitajima M. (1975) Energy transfer between photosystem II and photosystem I in chloroplasts. Biochim. Biophys. Acta 396, 72-85.
Crowder MS and Bearden A (1983) Primary photochemistry of photosystem I in chloroplasts. Dynamics of reversible charge separation in open reaction centers at 25 K. Biochim. Biophys. Acta 722, 23-35.
Lemoine Y and Jupin H (1978) Analyse cinétique et spectroscopique de la fluorescence chez un mutant photosensible de Tabac. Photosynthetica 12,35-5
Malkin R (1982) Photosystem I. An. Rev. Plant Phys. 33, 455-479.
Murata N, Sato N, Omata T and Kuwabara T (1981) Separation and characterization of thylakoïd and cell enveloppe of the blue-green alga (Cyanobacterium) Anacystis nidulans. Plant and Cell Physiol. 22, 855-866.
Öquist G, Fork DC, Schoch S and Malmberg G (1981) Solubilization and spectral characteristics of chlorophyll-protein complexes isolated from the thermophilic blue-green alga Synechococcus lividus. Biochim. Biophys. Acta 638, 192-200.
Thomas JC and Mousseau A (1981) Characterization of chlorophyll-protein complexes from Cyanophyceae. The study of their modification in different culture condition. PhotosynthesisIII,Akoyounoglou ed.Balaban Philadel.435-44

MEASUREMENT OF CONFORMATIONAL CHANGES IN PIGMENTS OF APHANOCAPSA 6714 BY
FLUORESCENCE LIFETIME FOLLOWING TRANSITION FROM DARK TO LIGHT GROWTH

JACQUELINE MANWARING, BERNADETTE MAY, ROBERT G. BROWN AND E. HILARY EVANS/
BIOLOGY AND CHEMISTRY DIVISIONS, PRESTON POLYTECHNIC, PRESTON, LANCS., U.K.

1. INTRODUCTION

Whilst all cyanobacteria are capable of photoautotrophic growth, species
vary in their capacity for heterotrophic and photoheterotrophic growth.
In the dark some organisms, such as *Chlorogloea fritschii* may show a
reduction in the levels of photosynthetic pigments, which recover to
normal photoautotrophically-grown levels on reintroduction into the light
(Evans, Carr, 1974). Photosystem 2 (PS2) activity measured as oxygen
evolving capacity is lost in dark growth cells (Evans, *et al.*, 1978)
in common with eukaryotic algae (Cheniae, Martin, 1973). PS2 activity
recovers in two phases in *C. fritschii*, one sensitive and one
insensitive to chloramphenicol. Photosystem 1 (PS1) preparations extracted from *C. fritschii*, although active in both dark and light grown
cells show substantial variation in pigment spectra. (Evans, 1981).
Intact cells show no energy transfer between phycocyanins and PS1, and it
has been proposed that a rearrangement of pigments occurs in the membrane
of dark grown cells on transference to the light. Aphanocapsa 6714
shows no decrease in photosynthetic pigments on growth in the dark
Joset-Espardellier *et al.*, 1978), but loses PS2 activity. We have
therefore used this organism to investigate any variation in the
fluorescence spectrum and lifetime of allophycocyanin and chlorophylla.
It has been proposed that the route of energy transfer through the
pigment system of cyanobacteria such as Aphanocapsa 6714 is C-phycocyanin
   allophycocyanin Chlorophyll a (bulk) Chlorophyll a (reaction centres)
(Gantt *et al.*, 1977). Measurements of fluorescence lifetime should
give information on the efficiency of transfer between pigments.

2. PROCEDURE

2.1. Material and methods

Aphanocapsa sp. was strain 6714 from Prof. R.Y. Stainier's collection,
kindly donated by Prof. F. Espardellier, CNRS, Paris. The organism
was grown as previously described, (Joset-Espardellier *et al.*, 1978).
Fluorescence emission spectra were measured on apparatus previously
described (Pullin *et al.*, 1979), while fluorescence decay profiles were
measured by the technique of time-correlated, single photon counting
(Ware, 1971) on an Edinburgh Instruments Model 199 Fluorescence Decay
Time Spectrometer. The profiles were analysed by computer convolution
and goodness of fit judged on the basis of χ^2 values and a random
distribution of residuals.

Organisms grown heterotrophically in the dark were transferred into the
light for times indicated in the Figures, after which measurements were
made.

Sybesma, C. (ed.), Advances in Photosynthesis Research, Vol. II. ISBN 90-247-2943-2.
© *1984 Martinus Nijhoff/Dr W. Junk Publishers, The Hague/Boston/Lancaster.*

3. Results and Discussion

The fluorescence emission spectra of dark grown Aphanocapsa cells and of those following 24 h. illumination are shown in Figure 1. We interpret the 660 nm peak as originating from allophycocyanin and the shoulder at 680 nm from chlorophyll a. After 24 hr. illumination, the Chl a shoulder has diminished substantially, the allophycocyanin peak has blue-shifted slightly and a longer wavelength shoulder had developed. As the pigment content of Aphanocapsa is unchanged by heterotrophic growth this suggests that some rearrangement of the pigments has occurred.

Fluorescence decay profiles for Aphanocapsa cell grown in darkness and subsequently exposed to varying periods of illumination are shown in Figure 2. Those decays measured at 660 nm are attributed to allophycocyanin, and those at 680 nm to the bulk chlorophyll. On the basis of χ^2 values and a random distribution of residuals, both decays are best fitted by double exponentials. These comprise one decay in the 0.5-1.5 ns region which accounts for 95% of the intensity and a decay in the 5-10 ns region. Because of the time required to accumulate data on the latter phase we report here measurements on the fast phase only. The allophycocyanin lifetime of 1.1 ns in dark-grown Aphanocapsa is approximately half that of pure allophycocyanin. In light grown cells, we find the corresponding value to be 400 ps, although this is probably in reality rather shorter; lifetimes of approx. 200 ps. are approaching the limits of resolution of our apparatus. Introduction of the dark-grown cells into the light results in a decrease in the lifetime of the fluorescence of approx. 200 ps. within the first hour, the normal lifetime measured in light grown cells being attained after some 48 hr. in the light. A similar decrease in fluorescence decay time of the chlorophyll peak is seen on transfer of dark grown cells into the light.

We interpret the decrease in fluorescence decay time as a light-induced movement of pigments in the membrane in such a fashion that energy transfer between accessory pigments is increased. The initial phase of light-induced recovery is fast, within the first hour, comparable with recovery times measured for Photosystem 2 in other organisms (Evans et al, 1978, Cheniae and Martin, 1973).

ACKNOWLEDGEMENTS

We thank the SERC and Department of Industry for financial support.

REFERENCES
Cheniae, G.M. and Martin, I.F. (1973) Oxygen evolving activity in algae, Photochem. Photobiol. 27, 441-459.
Evans E.H. and Carr N.G. (1974) Dark to light transitions in Chlorogloea fritschii in Proc. 3rd Int. Congr. Photosynth., Ed. M. Avron, Elsevier, The Netherlands p. 1861-1866.
Evans E.H., Carr N.G. and Evans M.C.W. (1978) Changes in photosynthetic activity in the cyanobacterium Chlorogloea fritschii following transition from dark to light growth, Biochim. Biophys. Acta, 501, 165-173.
Evans E.H. (1981) Photosystem 1 preparations from dark- and light-grown cells of the cyanobacterium, Chlorogloea fritschii, Photosyn. Res., 259-264.

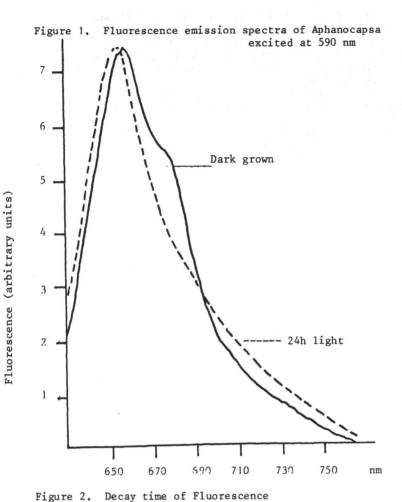

Figure 1. Fluorescence emission spectra of Aphanocapsa
excited at 590 nm

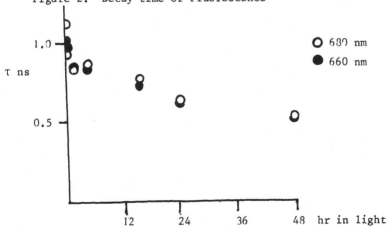

Figure 2. Decay time of Fluorescence

Gantt E., Lipschultz C.A. Zilinskas B.A. (1977) Phycobilisomes in relation to the thylakoid membranes, Brookhaven Symp. Biol. 28, 357-357.
Joset-Espardellier·F., Astier C., Evans E.H. and Carr, N.G. (1978) Cyanobacteria grown under Photoautotrophic, Photoheterotrophic and Heterotrophic regimes: sugar metabolism and carbon dioxide fixation, FEMS Micro. Letts. 4, 261-264.
Pullin C.A., Brown R.G. and Evans E.H. (1979) Detection of allophycocyanin in Photosystem 1 preparations from the blue-green alga, Chlorogloea fritschii, FEBS Letts. 101, 110-112.
Ware W.R. (1971) in Creation and Detection of the Excited State, 1A (A. Lamola ed.) p. 213-302, M. Dehker, New York, N.Y.

PHYCOERYTHRIN: SPECTROSCOPIC ANALYSIS OF ITS SUBUNITS AND AGGREGATES
FROM MONOMER TO DODECAMER

B. A. ZILINSKAS, J. GRABOWSKI, and S. CAMPBELL

1. INTRODUCTION

Phycoerythrin exists *in vivo* as part of a highly ordered extrinsic
multiprotein aggregate, the phycobilisome, which with two other biliproteins,
phycocyanin (PC) and allophycocyanin (APC), transfers excitation energy
to the chlorophyll in the thylakoid. In its simplest form, c-phycoerythrin
exists as a monomer, composed of two dissimilar subunits, α and β, of
16,000 and 19,500 daltons, respectively. Two phycoerythrobilin chromophores
are covalently bound to the α subunit and four to the β subunit. By
combined ultrastructural and biochemical analysis, monomers of PE are
seen to assemble into trimers $(\alpha\beta)_3$, pairs of which form hexamers $(\alpha\beta)_6$,
which in turn make dodecamers $(\alpha\beta)_{12}$, visualized as two 6x12 nm hexameric
discs forming a stack or rod. Evidence will be provided to show that
assembly of phycoerythrin monomers from α and β subunits, as well as
trimers from monomers, does not result in any significant change in
chromophore environment or any new chromophore-chromophore interactions.
However, this is not the case in the formation of aggregates larger than
trimer.

2. MATERIALS AND METHODS

Phycoerythrin trimer and hexamer were isolated from phycobilisomes of
Nostoc sp. as described earlier (Zilinskas, Howell, 1983). Subsequent to
dialysis of the PE hexamer against 0.75 M K-phosphate, pH 7, the PE
dodecamer was obtained by sedimentation in linear sucrose gradients. The
α and β subunits of PE were separated on a DEAE-Sephacel column with a linear
gradient of NaCl in 8 M urea, pH 8, and were renatured by dialysis against
0.1 M K-phosphate, pH 5. The monomer was also obtained from fractions
eluted from this column or by incubation with PE with 1 M KSCN. Aggregate
size and polypeptide composition were determined respectively by
sedimentation in linear gradients of sucrose and by SDS-PAGE; absorption,
fluorescence and circular dichroism measurements were as described earlier
(Glick, Zilinskas, 1982). Fluorescence polarization was measured according
to Grabowski and Gantt (1978), and chromophore orientation was analyzed
as discussed by Zickendraht-Wendelstat et al. (1980) and Dale and Teale
(1970).

3. RESULTS AND DISCUSSION

Figure 1a shows the absorption spectra of renatured α and β subunits of PE.
In 8 M urea in the unfolded state, both subunits have identical absorption
spectra, indicating identical chromophores; the distinctive spectra
observed upon refolding result from the contribution of the protein
environment on the chromophore. The monomer spectrum closely resembles
the summation of the spectra of equimolar quantities of the α and β subunits.
This indicates that the physical association of the two subunits into an
$\alpha\beta$ monomer neither causes any dramatic change in protein conformation in
the chromophore vicinity nor induces any new chromophore-chromophore
interactions. This can also be said in the formation of trimer from
monomer (cf. Fig. 1a and 1b). However, with the stacking of two 3x12 nm

Sybesma, C. (ed.), Advances in Photosynthesis Research, Vol. II. ISBN 90-247-2943-2.
© *1984 Martinus Nijhoff/Dr W. Junk Publishers, The Hague/Boston/Lancaster.*

$(\alpha\beta)_3$ discs in the assembly of the hexamer, a shoulder at 570 nm appears which is accentuated in the formation of the dodecamer (Fig. lb).

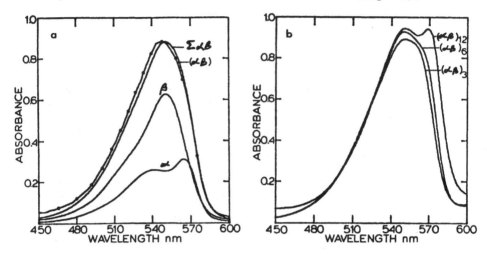

FIGURE 1. a) Absorption spectra of PE α and β subunits, summation of equimolar concentrations of both subunits (——), and $\alpha\beta$ monomer (-•••-). b) Absorption spectra of PE trimer, hexamer, and dodecamer.

As shown in Table 1, accompanying the absorbance changes with increased aggregate size are small, progressive red-shifts in the fluorescence emission maxima and increases in $A_{vis}/A_{far\ uv}$, an indicator of the extended (high ratio) or cyclohelical (low ratio) configuration of the chromophore (Scheer, 1981).

TABLE 1. Fluorescence emission maxima and $A_{550\ nm}/A_{305\ nm}$ ratios of PE subunits and aggregates.

	α_D*	β_D	α_R*	β_R	$(\alpha\beta)$	$(\alpha\beta)_3$	$(\alpha\beta)_6$	$(\alpha\beta)_{12}$
$F_{\lambda max}$ nm	–	–	575	574	575	577	578	581
$\dfrac{A_{550\ nm}}{A_{305\ nm}}$	1.1	1.1	6.3	6.7	6.4	8.9	8.7	7.9

*D = denatured; R = renatured.

Parallel changes can also be seen in the circular dichroism spectra. The spectra of the renatured α and β subunits are very similar to those reported by Langer et al. (1980). The monomer spectrum approximates the summation of the α and β subunit CD spectra, although there is slight blue shift and decrease in oscillator strength of the main positive peak (Fig. 2a). Again, there is little change in the CD spectrum in the formation of trimer from monomer. Hexamer formation results in the appearance of a new positive peak at 570 nm which becomes the main CD component of the dodecamer (Fig. 2b). Gaussian curve fitting of absorption and CD data indicates a single coupled oscillator in the monomer and trimer, due primarily to the contribution of the β subunit; the formation of the hexamer and dodecamer creates an additional coupled oscillator and a dramatic decrease in the coupled oscillator strength of the smaller aggregates.

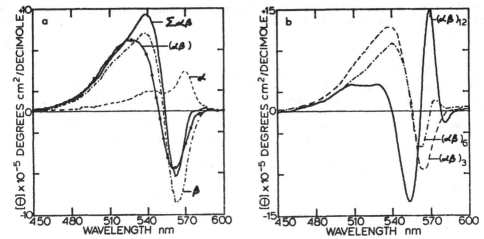

FIGURE 2. a) CD spectra of PE α and β subunits, their summation (————), and αβ monomer (•—•—•). b) CD spectra of trimer, hexamer, and dodecamer.

Figure 3. shows that increased aggregation decreases the limiting polarization of fluorescence with little influence on the shape of the fluorescence polarization spectrum. Using the analysis of Dale and Teale (1970) and Zickendraht-Wendelstat et al. (1980), which describes 's' sensitizing and 'f' fluorescing chromophores, we found that the angles between the 's' and 'f' and between 'f' and 'f' chromophores of PE change with aggregation, as shown in Table 2.

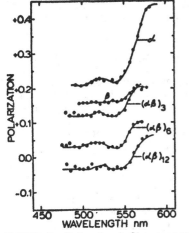

FIGURE 3. Polarized fluorescence excitation spectra of PE.

TABLE 2. Limiting fluorescence polarization of 'f' (P_f) and 's' (P_s) and the angles between emission oscillators of 's'-'f' (ϕs-f) and 'f'-'f' (ϕ f-f) in PE subunits and aggregates.

PE	P_f*	P_s**	ϕs-$f°$	ϕf-$f°$
α	0.50	0.08	49.0	
β	0.27	-0.02	56.0	35.5
$(\alpha\beta)_3$	0.26	0.04	52.0	36.0
$(\alpha\beta)_6$	0.12	-0.02	56.0	46.5
$(\alpha\beta)_{12}$	0.08	-0.08	60.0	49.0

* assumed value (must agree with the observed limiting fluorescence polarization).

** determined by trial and error to hold the "mirror symmetry law" between PE emission and the 'f' PEB absorption.

Until recently, spectroscopic analyses of biliprotein aggregates ignored the role of "linker" polypeptides in assembly. In addition to the α and β subunits, the PE hexamer and dodecamer contain two polypeptides of 32 and 34 KD which lack chromophore. Zilinskas and Howell (1983) showed earlier that the 32 KD polypeptide links two PE hexamers and the 34 KD polypeptide links a PE hexamer to a PC hexamer. These polypeptides probably contribute to the dramatic spectral alterations seen in the larger aggregates, as also suggested by Yu et al. (1981) for PC.

4. CONCLUSIONS

Our results show that there is neither a change in chromophore environment nor any new chromophore-chromophore interactions induced upon association of α and β subunits into monomers or monomers into trimers. On the other hand, alteration in spectral properties, indicative of new chromophore-chromophore interactions and/or perturbations of the chromophore environment, results upon assembly of hexamers and dodecamers. These changes may be in large part due to the contributions of the 32 and 34 KD colorless poly-peptides involved in phycobilisome rod assembly. The inverse relationship of the limiting polarization of fluorescence and aggregate size appears to be due not only to greater energy migration in the larger aggregates but also to perturbations in the chromophore environment induced by aggregation which result in changes in the angles between absorbing and emitting oscillators.

REFERENCES

Dale RE and Teale FWJ (1970) Number and distribution of chromophore types in native phycobiliproteins, Photochem. Photobiol. 12, 99-117.
Glick RE and Zilinskas BA (1982) Role of the colorless polypeptides in phycobilisome reconstitution from separated phycobiliproteins, Plant Physiol. 69, 991-997.
Grabowski J and Gantt E (1978) Photophysical properties of phycobiliproteins from phycobilisomes: fluorescence lifetimes, quantum yields, and polariza-tion spectra, Photochem. Photobiol. 28, 39-45.
Langer E, Lehner H, Rudiger W and Zickendraht-Wendelstadt B (1980) Circular dichroism of c-phycoerythrin: a conformational analysis, Z. Naturforsch 35c, 367-375.
Scheer H (1981) Biliproteins, Angew. Chemie 20, 241-261.
Yu M-H, Glazer AN, and Williams RC (1981) Cyanobacterial phycobilisomes, J. Biol. Chem. 256, 13130-13136.
Zickendraht-Wendelstadt B, Friedrich J, and Rudiger W (1980) Spectral characterization of monomeric c-phycoerythrin from *Pseudanabaena* W1173 and its α and β subunits: energy transfer in isolated subunits and c-phycoery-thrin, Photochem. Photobiol. 31, 367-376.
Zilinskas BA and Howell DA (1983) Role of the colorless polypeptides in phycobilisome assembly in *Nostoc* sp., Plant Physiol. 71, 379-387.

ACKNOWLEDGEMENTS

NJAES Publ. No. C-01104-1-83, supported by State and U.S. Hatch funds, USDA Competitive Grant 5901-0410-8-0185-0, and NSF Grant PCM-80-18740. The authors wish to thank Dr. Peter C. Kahn for helpful discussions and Dawn Howell for her expert technical assistance.

Authors' addresses:

BAZ: Dept. of Biochemistry & Microbiology, Cook College, Rutgers University, New Brunswick, NJ 08903, U.S.A.; JG: Inst. of Environ. Engin. of Poznan Technical Univ., Poznan, Poland; SC: Dept. of Chem., Yale Univ., New Haven, CT, U.S.A.

PHOTOCONTROL OF PHYCOERYTHROCYANIN SYNTHESIS IN AN <u>OSCILLATORIA</u> STRAIN (CYANOBACTERIA).

J.C. Thomas, A. Mousseau and N. Hauswirth.
Laboratoire de Cytophysiologie Végétale de l'Ecole Normale Supérieure, 24 rue Lhomond, 75231 PARIS Cedex 05, FRANCE.

In Cyanobacteria, pigment synthesis can be affected by physical and chemical factors of the environment (Cohen-Bazire, Bryant, 1982). The phycobiliproteins (PBP) aggregated in phycobilisomes (PBS) act as light-harvesting complexes - the absorbed energy being very efficiently funnelled to chlorophyll a. PBP synthesis can be controlled by light intensity, but the most striking effect of light is the modulation of PBP synthesis by light quality. This promotes a complementary chromatic adaptation - phycoerythrin (PE) synthesis favoured by green light and phycocyanin (PC) by red light (Bogorad, 1975 ; Tandeau de Marsac (1977). The third main PBP, allophycocyanin (APC) does not seem to be controlled in such a way (Glazer, 1982). Recently, a new PBP, phycoerythrocyanin (PEC) with an absorption maximum at 568 nm has been isolated and characterized from several heterocystous strains of Cyanobacteria (Bryant et al., 1976 ; Nies, Wehrmeyer, 1980 ; Füglistaller et al., 1981). One may ask if PEC functionnally equivalent to PE, could be submitted to complementary chromatic adaptation. This ability was reported for <u>Mastigocladus laminosus</u> in open air cultures (Füglistaller et al., 1981) while Bryant (1982) failed to detect such a phenomenon in 36 strains of the Pasteur Culture Collection in well-defined culture conditions. The present work shows that <u>Oscillatoria splendida</u> L$_3$ which synthesizes PEC in appreciable amounts in white light (Thomas, Mousseau, 1982) exhibits a photocontrol of PEC synthesis, with a total synthesis inhibition by red light. We have also studied PBP resynthesis from cells devoided of PBP by nitrogen depletion, during nitrogen recovery : PEC is not synthesized in the dark or in red light, while comparable amounts are found in fluorescent or green light grown cells.

MATERIAL AND METHODS

Oscillatoria splendida L$_3$ isolated in our laboratory was grown in fluorescent white light (6 Wm^{-2}). Nitrogen depleted cultures allowed us to obtain cells without measurable amounts of PBP, which were used for studying resynthesis of PBP in the dark (intracellular glycogen accumulated during starvation being used as an energy source), or in the light, after addition of nitrate. Colored lights were obtained by filtering fluorescent white light with a short wavelength cut-off filter (0,6 % transmission at 550 nm), given a red light, or a green filter (maximum transmission at 510 nm ; 30 nm half-band width). The energies were respectively 6 and 4 Wm^{-2}. Other cultures were also maintained continuously in red light (6 Wm^{-2}). Phycobilisomes were isolated according to Lundell et al. (1981) on discontinuous sucrose gradients with a 0,75 M phosphate buffer at pH 8,0. Intact PBS, as judged by fluorescence emission spectrum at 77 K (Gantt et al., 1979) were recovered from the 0,75 M sucrose layer. LiDS-PAGE of PBS polypeptides were performed on slab (12 % acrylamide resolving gel and 5 % acrylamide stacking gel) gels

Abbreviations: PBP: phycobiliprotein; PBS: phycobilisome; PC: phycocyanin; PE: phycoerythrin; APC: allophycocyanin; PEC: phycoerythrocyanin; α PEC and β PEC: subunits of PEC; LiDS: lithium-dodecylsulfate; PAGE: polyacrylamide gel electrophoresis; WL: fluorescent white light; RL: red light.

Sybesma, C. (ed.), Advances in Photosynthesis Research, Vol. II. ISBN 90-247-2943-2.
© *1984 Martinus Nijhoff/Dr W. Junk Publishers, The Hague/Boston/Lancaster.*

with a discontinuous buffer system in the conditions described by Lundell et al. (1981).

Isoelectric focusing were made on polyacrylamide tube gels (5 % acrylamide) in denaturing conditions (urea 9 M), and ampholines (Servalyt AG) in the pH range 3-7.

Negatively-stained PBS with uranyl acetate (Lundell et al., 1981) were observed with electron microscope.

RESULTS

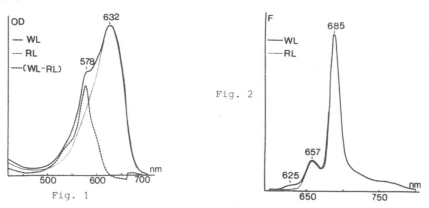

Fig. 1

Fig. 2

Fig. 1: Absorption spectra at room temperature of purified PBS from cells grown either in white light (WL) or in red light (RL), and absorption difference spectrum (WL - RL) x 2.
Fig. 2: Fluorescence emission spectra at 77 K of intact PBS from WL and RL - grown cells. Excitation wavelength : 550 nm.

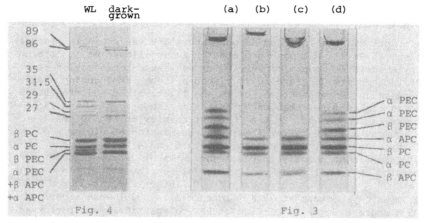

Fig. 3: Isoelectric focusing on denaturing polyacrylamide gels containing 9 M urea of PBP subunits in the pH range 3-7. Naturally-colored bands corresponding to subunits of main PBP : (a): WL cells ; (b to d): PBS isolated after nitrogen recovery (N$^+$) ; (b) in the dark ; (c) red light ; (d) green li
Fig. 4: LiDS-PAGE of PBS polypeptides from WL and N$^+$ dark-grown cells, after Coomassie blue G 250 staining.

PBS isolated from WL-grown cells had an absorption spectrum with a maximum at 632 nm (PC), a discrete shoulder in the 650 nm (APC) and a secondary peak at 578 nm due to PEC (fig. 1). Emission fluorescence spectrum at 77 K (fig. 2) exhibited a sharp predominant peak at 685 nm originating from the final energy acceptor in PBS (APC B or 89 Kd polypeptide), a secondary peak in the 660 nm region (from APC and PC) and a shoulder at 625 nm due to PEC. PBS synthesized during nitrogen recovery in green light possessed identical spectral properties (data not shown). For RL phycobilisomes, one could noted the decrease in absorption spectrum at 578 nm and the disappearance of F 625. Absorption difference spectrum (fig. 1) confirmed that deficiency was due to PEC. PBS synthesized during nitrogen recovery in the dark were found to be similar to RL cell PBS. The study of PBP subunits composition of PBS from WL cells by isoelectric focusing in denaturing conditions (9 M urea) on polyacrylamide gels revealed six (or seven) naturally-colored bands (fig. 3a) in order of increasing pI : five blue-colored bands, carrying phycocyanobilin chromophores (α and β subunits of PC and APC, β PEC) ; one or two reddish-purple bands (α PEC) with a linked chromophore probably very similar to PXB carried by α PEC from Anabaena variabilis (Bryant et al., 1976). α and β PEC were clearly absent from PBS synthesized in the dark or in red light during nitrogen recovery, while subunit composition is identical in PBS from WL cells, or resynthesized in green light (fig. 3a to d). Densitometer scans of 9 M urea gels from isoelctric focusing did not show significant variations in molar ratios of phycocyanobilin chromophores from PC and APC subunits in WL or in RL type PBS, the only difference being the absence of PEC subunits in RL or in the dark. Molar ratios (PEC/PC/APC) obtained in these conditions were 0,6/1,5/1 for WL cells and 0/1,5/1 for RL cells.

LiDS-PAGE permitted to show that in RL cell PBS, a 31.5 colorless polypeptide was lacking (fig. 4). Electron microscopy of negatively stained PBS from WL cells revealed a constant structure : a triangular core made of three subunits bearing no more than six rods each composed of three or four individual discs 12 nm in diameter and 6 nm thick, in agreement with models proposed by Glazer (1982). PBS from RL (or dark-grown) cells were composed of an identical core, but rods were significantly shorter and generally composed of two individual discs only.

Additional experiments were made to precise the effects of the quality of light on resynthesis of PBP in the dark. White, red or green lights received before addition of nitrate, even for a long time (8 hours-6 Wm^{-2}), had no effect on PBP levels synthesized in the dark.

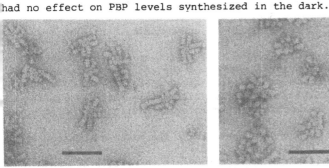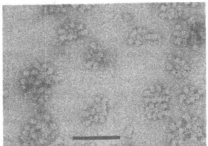

Fig. 5a   Fig. 5b

Fig. 5: Electron micrographs of PBS preparations from WL (a) and RL (b). Bars drawn on micrographs = 0,1 μm .

DISCUSSION AND CONCLUSIONS

These experiments demonstrated unequivocally a photocontrol for PEC and 31.5 Kd polypeptide synthesis in <u>Oscillatoria splendida</u> L_3, promoted by fluorescent white light or by green light, but repressed by red light. As the PC synthesis ratio seemed to be unaltered, L_3 chromatic adaptation shared some characteristics with the case of Cyanobacterial strains of group II, possessing PC and PE (Tandeau de Marsac, 1977) with exhibit photocontrol for PE synthesis alone. Bryant (1982) did not find chromatic adaptation in hete-rocystous strains with PEC, although he noted the effect of light intensity, which might be responsible for the results of Füglistaller et al. (1981) on <u>Mastigocladus laminosus</u> : in this strain, green light favoured PEC synthesis but PEC level was very low in white light. More, the effect of red light was not tested.

Experiments on strains with photocontrolled PE synthesis strongly suggest the existence of a photoreversible master-pigment, analogous to phytochrome (Bogorad, 1975 ; Tandeau de Marsac, 1977 ; Vogelmann and Scheibe, 1978). If light control for PEC synthesis is mediated via a photoreversible pig-ment in <u>Oscillatoria splendida</u> L_3, results obtained on N^+ grown cells in the dark led us to think that its properties would be different. One might suppose that it would have a short half-life, for instance.

REFERENCES

- Bogorad L (1975) Phycobiliproteins and complementary chromatic adaptation. An. Rev. Plant Physiol. 26, 369-401.
- Bryant DA (1982) Phycoerythrocyanin and Phycoerythrin : properties and occurence in Cyanobacteria. J. Gen. Microbiol. 128, 835-844.
- Bryant DA, Glazer AN and Eiserling FA (1976) Characterization and struc-tural properties of the major biliproteins of <u>Anabaena sp</u>. Arch. Microbiol. 110, 61-75.
- Cohen-Bazire G and Bryant DA (1982) Phycobilisomes : composition and struc-ture. In Carr NG and Whitton BA, ed. The Biology of Cyanobacteria, pp 143-190. Botanical Monographs, Volume 19 : Blackwell Scientific Publications.
- Füglistaller P, Widmer H, Sidler W, Frank G and Zuber H (1981) Isolation and characterization of phycoerythrocyanin and chromatic adaptation of the thermophilic Cyanobacterium <u>Mastigocladus laminosus</u>. Arch. Microbiol. 129, 268-274.
- Gantt E, Lipschultz CA, Grabowski J and Zimmerman BK (1979) Phycobilisomes from blue-green and red algae. Isolation criteria and dissociation charac-teristics. Plant Physiol. 63, 615-620.
- Glazer AN (1982) Phycobilisomes : structure and dynamics. Ann. Rev. Micro-biol. 36, 173-198.
- Lundell DJ, Williams RC and Glazer AN (1981) Molecular architecture of a light-harvesting antenna. <u>In vitro</u> assembly of the rod substructures of <u>Synechococcus</u> 6301 phycobilisomes. J. Biol. Chem. 256, 3580-3592.
- Nies M and Wehrmeyer W (1980) Isolation and biliprotein characterization of phycobilisomes from the thermophilic Cyanobacterium <u>Mastigocladus laminosus</u> Cohn. Planta 150, 330-337.
- Tandeau de Marsac (1977) Occurence and nature of chromatic adaptation in Cyanobacteria. J. Bact. 130, 82-91.
- Thomas JC and Mousseau A (1982) Phycoerythrocyanin in Oscillatoriacean strains (Cyanobacteria). Location and photosynthetic properties. Abs. 4th International Symposium on photosynthetic Prokaryotes, Bombannes France,C 29
- Vogelmann TC and Scheibe J. (1978) Planta 143, 233-239.

PHYCOBILISOME COMPOSITION AND RELATIONSHIP TO REACTION CENTERS IN
ANACYSTIS NIDULANS.

R. KHANNA/J.R. GRAHAM/J. MYERS/E. GANTT

SUMMARY

   Phycobilisomes were studied in Anacystis nidulans wild type and
several spontaneous mutants selected for improved growth in far-red light
(>650 nm). In the mutants the phycocyanin varied in relation to
allophycocyanin but there was no alteration in the antenna size in terms
of chlorophylls in photosystems I and II. Phycobilisomes of wild-type
contained phycocyanin and allophycocyanin in a molar ratio of 3:1.
Phycobilisomes of the mutants 85Y, 19Y and 59G had respective ratios of
0.4:1, 0.7:1, and 1:1. Analysis of purified phycobilisomes on SDS-PAGE
gradient gels revealed significant differences in the polypeptide
pattern. A comparison of phycobilisomes of the wild-type and 85Y by
electron microscopy, showed that they were larger in wild-type (57nm x
30nm) than in 85Y (28nm x 15nm). The number of phycobilisomes per cell
(~ 2 x 10^4) was about the same in wild-type (white light) and 85Y
(far-red light). The ratio of phycobilisome/reaction center 2 was close
to 1 and was relatively constant in spite of large variations in
chlorophyll and in phycocyanin content.

INTRODUCTION

   Phycobilisomes (PBS) are composed of phycobiliproteins which act as
major light-harvesting pigments in blue-green and red algae. In A.
nidulans phycocyanin (PC) and allophycocyanin (APC) are the primary
components of PBS. Linker polypeptides are involved with the formation
of the peripheral rods (Yamanaka and Glazer, 1981), while some large
molecular weight polypeptides appear to anchor PBS to the thylakoid.
Alteration in the pigment composition of A. nidulans can occur when
cultures are grown for many generations under far-red light (>650 nm)
(Myers et al., 1980). Several spontaneous mutants with growth rates
similar to WT, were thus selected and further studied. There was no
alteration in the antenna size in terms of chlorophyll (Chl) in
photosystem I and II, although the PSI/PSII ratio was less in the mutants
than in WT (Myers, et al., 1980). The three mutants (85Y, 19Y, 59G) were
chosen because of their reduction in the PC/APC ratio. Phycobilisomes
were isolated and their composition was compared. Furthermore, a
comparison was made at the cellular level of WT and mutant 85Y, because
this mutant showed the greatest reduction in PC composition. It was more
PC deficient than a nitrosoguinidine-generated mutant AN112 studied by
Yamanaka and Glazer (1981).

MATERIALS AND METHODS

   Anacystis nidulans was cultured in a modified medium D under 1% CO_2
in air. Cultures were grown under white fluorescent light or under a
far-red light (>650 nm) provided by tungsten-halogen lamps with Corning
No. 2030 filter as described by Myers et al. (1980). Phycobilisomes were
isolated in 0.75 M K-PO_4 buffer (pH 8.0) as previously described (Gantt
et al., 1979).
Details of the methods are given in Khanna et al. (1983).

Sybesma, C. (ed.), Advances in Photosynthesis Research, Vol. II. ISBN 90-247-2943-2.
© *1984 Martinus Nijhoff/Dr W. Junk Publishers, The Hague/Boston/Lancaster.*

RESULTS AND DISCUSSION

Spectroscopic Properties of PBS from A. nidulans WT and Mutants 85Y, 19Y, and 59G. Two phycobiliproteins, PC (λ max 625) and APC (λ max 650) were present in a molar ratio of ca. 3:1 (Fig. 1) in WT. Although the amount of APC remained constant the PC content varied in the mutants. Phycobilisomes of the mutants 85Y, 19Y, and 59G contain PC and APC in a molar ratio of 0.4:1, 0.7:1, and 1:1 respectively.

Fluorescence emission maxima of WT and all the mutants were at 680 nm (23°C) (not shown). Excitation maxima corresponded well with the absorption peaks.

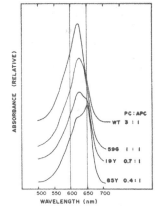

FIG. 1. Absorption spectra (r.t.) of isolated PBS of A. nidulans WT and from pigment mutants 85Y, 19Y, and 59G.

Polypeptide Composition of Wild-type Cells and Mutants. The PBS of A. nidulans were analyzed on SDS-PAGE (Fig. 2). Phycobilisomes of WT, in addition to the pigmented bands of PC and APC, had uncolored polypeptides of 80, (70-50), 40, 36 and 31 kD. The polypeptides 80 kD (probable anchor with the thylakoid) and 31 kD were present in all PBS. In the PBS of 85Y and 19Y the polypeptides of 36 and 40 kD were absent. These were the mutants with the lowest PC content. In 59G only polypeptide 36 kD was greatly reduced. The nature and function of the polypeptides at 70-50 kD (prevelant in the mutants) is not known.

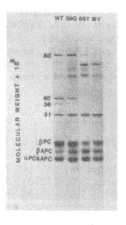

Fig. 2. SDS-PAGE polypeptide pattern of PBS from A. nidulans WT, 59G, 85Y, and 19Y, Stained with Coomassie blue.

the absence or reduction of the 36 and 40 kD polypeptide, along with the decrease of PC, and the absence of peripheral rods in 85Y (others were notexamined by EM) suggests that these polypeptides are required for PC stacking. From these results, and those from studies of other blue-greens, the 31 kD is clearly near the APC core, and probably functions in PC to APC attachment (Yamanaka and Glazer, 1981). Furthermore, in Fremyella and Nostoc a polypeptide with the same molecular weight seemed essential for in vitro association of APC with phycoerythrin-phycocyanin rods (Canaani and Gantt, 1982). Our results from spontaneous mutants and those by Yamanaka and Glazer (1981) from nitrosoguanidine mutants, suggest that the change in PC content and of two uncolored polypeptides is a common adaptive variation.

Comparison at Cellular Level of WT and 85Y Mutant. The Chl content and PC content per cell varied with the growth conditions, but APC content showed no significant change (Table 1).

TABLE 1. Pigment composition of cells of A. nidulans wild type and the mutant 85Y

Organism	Light Cond.	Chl Molec./cell* (x 10^6)	PC Molec./cell (x 10^4)	APC Molec./cell (x 10^4)
Wild Type	White	11	26	8.6
Wild Type	Far-red	3.3	30	9.0
Mutant 85Y	Far-red	4.5	3.0	8.5

*Cell size was same ca. 1.8 X 0.9μ.

Determinations were made by electron microscopy of sectioned cells, and of isolated PBS. Although the cell size remained the same under different growth conditions the total thylakoid area was reduced to almost half for cells grown in far-red light (Table 2). The PBS size, and possible area occupied on the thylakoid, was determined from sections and isolated PBS. The PBS of mutant 85Y, which showed maximum reduction in the amount of PC were considerably smaller (28 nm x 15 nm) than those of the WT (57 nm x 30 nm).

TABLE 2. Dimensions of PBS and number of PBS per cell in A. nidulans wild type and the mutant 85Y.

Organism	Light Cond.	Thylakoid Area/cell mμ^2	PBS area mμ^2 (x 10^{-4})	PBS/Cell (x 10^4)	PBS/mμ^2 Thylakoid	PBS/RC2
Wild Type	White	22.4	6.3	2.2	980	0.7
Wild Type	Far-red	13.5	6.3	2.3	1700	1.2
Mutant 85Y	Far-red	13.6	3.1	2.1	1540	0.8

The number of PBS per cell, calculated from the phycobiliprotein content and PBS size, was about the same in WT (White light) and 85Y (far-red light), but the number of PBS per unit area of thylakoid increased by almost 2-fold in far-red light grown cells. RC2 per cell were calculated from the chlorophylls per cell and the cited ratios of Chl/RC2 (Myers et al. 1980). From the estimates in Table 2 it appears that the ratio PBS/RC2 is relatively steady. This strengthens the hypothesis that the PBS and RC2 are directly associated.

REFERENCES

Canaani O and Gantt E (1982) Formation of hybrid phycobilisomes by association of phycobiliproteins from Nostoc and Fremyella, Proc. Nat. Acad. Sci. (U.S.A.) 79, 5277-5281.

Khanna R, Graham J-R, Myers J and Gantt E (1983) Phycobilisome composition and possible relationship to reaction centers, Arch. Biochem. Biophys. 224 (in the press).

Myers J, Graham J-R and Wang RT (1980) Light harvesting in Anacystis nidulans studied in pigment mutants, Plant Physiol. 66, 1144-1149.

Gantt E, Lipschultz CA, Grabowski J and Zimmerman BK (1979) Phycobilisomes from blue-green and red algae: isolation criteria and dissociation characteristics, Plant Physiol. 63, 615-620.

Yamanaka G, Glazer AN (1981) Dynamic aspects of phycobilisome structure: modulation of phycocyanin content of Synechococcus phycobilisomes, Arch. Microbiol. 130, 23-30.

ACKNOWLEDGEMENTS

This work was supported in part by a Fellowship to R.K. from the Smithsonian Office of Fellowships and Grants and by Contract AS05-76ER-04310 from the Department of Energy.

Authors Address: Rita Khanna*/Jo-Ruth Graham**/Jack Myers**/Elisabeth Gantt*, *RBL, Smithsonian Institution, 12441 Parklawn Dr., Rockville, MD 20852, **Dept. of Zoology, University of Texas, Austin, TX 78712, U.S.A.

BIOENERGETICS OF NITROGENASE IN BLUE-GREEN ALGAE.
I. PHYSIOLOGICAL CONDITIONS FOR NITROGENASE ACTIVITY IN PHORMIDIUM
FOVEOLARUM

HANS WEISSHAAR, HELMAR ALMON and PETER BÖGER

1. INTRODUCTION

Under aerobic environmental conditions nitrogen fixation by cyanobacteria
is restricted to heterocystous species. In these organisms, nitrogenase
is localized within heterocysts which provide for anaerobic conditions
necessary to keep the oxygen-labile enzyme active. Nevertheless, nitrogenase activity has also been observed in non-heterocystous cyanobacteria
(Stewart, Lex 1970; Rippka, Waterbury 1977).

Nitrogenase activity of non-heterocystous filamentous cyanobacteria has
to be induced under argon atmosphere after having grown these organisms
with combined nitrogen and after exhaustion of the nitrogen source.

Nitrogenase-catalyzed hydrogen evolution of the heterocystous species,
Nostoc muscorum, was increased by inhibition of the electron pathway connected with the uptake hydrogenase by low sulfide concentrations (Weisshaar, Böger 1983). Phormidium foveolarum exhibited high rates of hydrogen evolution catalyzed by nitrogenase due to lack of a hydrogen-uptake
activity. This contribution presents data on activity and regulation
of nitrogenase in Phormidium.

2. RESULTS AND DISCUSSION

Nitrogenase activity of Phormidium as shown in Figures 1 and 2 was induced during an incubation time under argon atmosphere in assay vessels,
after preceding growth with NH_4Cl present (2 mM) for 4 days. To induce
nitrogenase two additional requirements had to be met. First, as shown
in Figure 1, induction of nitrogenase occurs after incubation at low
light intensity (1 W/m^2) and high cell density (40 µl packed cell volume/
ml algal suspension). Under these conditions net photosynthetic oxygen
evolution is not observed. Second, induction of nitrogenase can be
achieved using higher light intensity (7 W/m^2), but net oxygen evolution
has to be inhibited by DCMU (5 µM).

Nitrogenase induction was prevented by the protein-biosynthesis inhibitor,
chloramphenicol, and nitrogenase activity by carbonyl cyanide-m-chloro-
phenylhydrazone (CCCP), an uncoupler of phosphorylation (Weisshaar,
Böger 1984).

Light-induced nitrogenase activity measured as hydrogen evolution and/or
acetylene reduction in the absence of DCMU was seen when CO_2 evolution
was evident. Apparently, nitrogenase activity ceased when substrates
for respiration were exhausted. CO_2 evolution was then replaced by uptake, and oxygen evolution showed up in the assay.

Figure 2 demonstrates that nitrogenase activity is dependent on respiratory CO_2 evolution. There was a reversible inhibition of hydrogen and
CO_2 evolution by 1 mM KCN. This inhibition was alleviated after cyanide
had been reduced by nitrogenase, but at higher cyanide concentrations
inhibition was not found reversible during the experimental time. Recent
findings of Scherer et al. (1982) strongly suggest that CO_2 release in

Sybesma, C. (ed.), Advances in Photosynthesis Research, Vol. II. ISBN 90-247-2943-2.
© *1984 Martinus Nijhoff/Dr W. Junk Publishers, The Hague/Boston/Lancaster.*

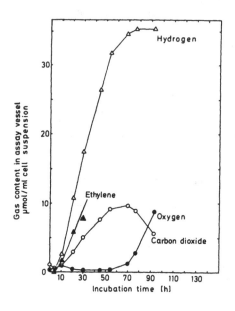

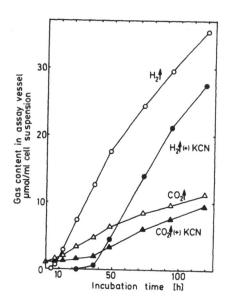

Fig.1   Fig.2

FIGURE 1. Induction of nitrogenase activity in Phormidium foveolarum at low light intensity (1 W/m^2). Cell density was 40 µl packed cell volume/ml culture suspension; chlorophyll 73 µg/ml. (Δ - Δ) hydrogen, ($\blacktriangle$ - $\blacktriangle$) ethylene, (o—o) carbon dioxide, and ($\bullet$—$\bullet$) oxygen content in the assay vessels, as referred to ml cell suspension. Ethylene was not measurable throughout the assay time due to limited acetylene concentrations (10% v/v) present at start.

FIGURE 2. Phormidium foveolarum: Reversible inhibition by cyanide of carbon dioxide and hydrogen evolution. Light intensity was 1 W/m^2, packed cell volume 26 µl/ml cell suspension, chlorophyll content 70 µg/ml. (o—o) hydrogen without cyanide, ($\bullet$—$\bullet$) hydrogen with 1 mM KCN present, (Δ - Δ) carbon dioxide evolution without, ($\blacktriangle$ - $\blacktriangle$) carbon dioxide evolution with 1 mM KCN present.

TABLE 1. Nitrogenase and hydrogenase activity in homogenates obtained from Phormidium foveolarum and Nostoc muscorum (rates of gas formation in mol/mg protein x min)

No.	Assay	Phormidium	Nostoc
1	H_2 evolution (by hydrogenase)	0	0.5 - 0.7
2	H_2 evolution (by nitrogenase)	18 - 20	3.4 - 4.0
3	C_2H_4 evolution (by nitrogenase).	10 - 15	1.6 - 2.0

Active extracts were obtained under anaerobic conditions by freezing and ultrasonic-bath treatment, a method to be published elsewhere, Weisshaar, Böger 1984). Omitting a regenerating ATP system, no nitrogenase activity was measured, a clear evidence that both gases are formed by nitrogenase activity (nos. 2,3). Protein content: Phormidium 3 mg/ml, Nostoc 7 mg/ml. The lower rates in Nostoc are due to the heterocyst frequency of 5% only.

the light is a marker for reductant generation by carbohydrate degradation. Most probably, electrons for nitrogen fixation are fed into photosystem I. No hydrogen-uptake activity was obtained with Phormidium, neither in intact filaments nor with a cell-free system (Table 1). In contrast, the cell-free system of Nostoc muscorum exhibited hydrogenase-mediated hydrogen evolution (with sodium dithionite and methylviologen as electron donor + an ATP-regenerating system). This system was similar to that developed by Hallenbeck et al. (1979). Hydrogen and ethylene evolution catalyzed by nitrogenase were observed in our cell-free systems with a molar ratio of 2:1 using either Nostoc or Phormidium. Only in Phormidium cells, however, a hydrogen to ethylene ratio of 2 was also found (Fig.1). This is explained by lack of a hydrogen-uptake hydrogenase. With Nostoc filaments hydrogen formation is lower than ethylene evolution due to hydrogen-uptake activity.

In Phormidium, respiration apparently provides anaerobiosis by taking over the role of the oxyhydrogen reaction. Further experiments are under way to obtain data as to what extent energy for nitrogenase is supplied by respiration.

REFERENCES
Hallenbeck PC, Kostel PH and Benemann JR (1979) Purification and properties of nitrogenase from the cyanobacterium Anabaena cylindrica, Eur. J. Biochem. 98, 275-284.
Rippka R and Waterbury JB (1977) The synthesis of nitrogenase by non-heterocystous cyanobacteria, FEMS Microbiol. Lett. 2, 83-86.

Scherer, S, Stürzl E and Böger P (1982) Interaction of respiratory and photosynthetic electron transport in Anabaena variabilis Kütz. Arch. Microbiol. 132, 333-337.

Stewart WDP and Lex M (1970) Nitrogenase activity in the blue-green alga Plectonema boryanum, strain 594, Arch. Mikrobiol. 73:250-260.

Weisshaar H and Böger P (1983) Sulfide stimulation of light-induced hydrogen evolution by the cyanobacterium Nostoc muscorum. Z. Naturforsch. 38c, 237-242.

Weisshaar H and Böger P (1984) Nitrogenase activity of the non-heterocystous cyanobacterium Phormidium foveolarum. Arch. Microbiol., submitted.

Authors' address: Lehrstuhl für Physiologie und Biochemie der Pflanzen, Universität Konstanz, D-7750 Konstanz, Germany

BIOENERGETICS OF NITROGENASE IN BLUE-GREEN ALGAE.
I. ELECTRON TRANSPORT TO NITROGENASE IN HETEROCYSTS OF ANABAENA
VARIABILIS

BERNHARD SCHRAUTEMEIER, HERBERT BÖHME and PETER BÖGER

1. INTRODUCTION

Conflicting evidence exists on the nature of electron sources for nitrogen fixation and regulation of corresponding electron-transport pathways
in heterocysts of cyanobacteria (cf. Bothe 1982). Reduced ferredoxin can
in principle be generated by photosystem I, if a donor is available (cf.
Houchins, Hind 1982), or in the dark either by pyruvate:ferredoxin oxidoreductase, or by NADP:ferredoxin oxidoreductase (FNR). The latter enzyme,
however, requires high NADPH/NADP ratios for ferredoxin reduction, whereas the enzymes suggested to be involved in NADPH generation, i.e. glucose-
6-phosphate and isocitrate dehydrogenases are known to be negatively regulated by the reductant charge and by ATP.

In vitro studies with an integral light-dependent nitrogen-fixing system
of a cyanobacterium have not yet been performed. Experiments with a heterocyst homogenate from Anabaena variabilis are presented.

2. RESULTS AND DISCUSSION

2.1. Electron donors/Pathways. A cell-free homogenate obtained by breaking heterocysts in a highly concentrated suspension (1 mg chlorophyll/ml)
exhibited high rates of acetylene reduction in the light and in the dark
using several physiological substrates (Table 1). Activities were dependent on an ATP-regenerating system allowing to separate effects of
electron transport from those of (photo-)phosphorylation on nitrogenase
activity. H_2- and NADH-supported acetylene reduction was strictly light-
dependent and involved ferredoxin, as shown by metronidazole sensitivity
of this reaction. NADPH supported nitrogenase activity in the light and
in the dark. As with H_2, activity with both pyridine nucleotides was inhibited by DBMIB and, to a lesser extent, by DNP-INT, but not by the uncouplers gramicidin D, valinomycin or nigericin. This indicates that
these donors feed into photosystem I, resulting in photosynthetically reduced ferredoxin for nitrogen fixation (results not shown, but see
Schrautemeier et al. 1983). Contrary to other results (Hawkesford et al.
1981), uncoupler insensitivity of light-dependent nitrogenase activity
evidently excludes energized membranes as necessary for nitrogenase activity.

Antimycin A and HOQNO did not inhibit activity with either pyridine
nucleotides. Thus, the question remains open as to whether, in the light,
NADPH feeds into photosystem I via FNR or, as assumed for NADH, via a
special dehydrogenase near the plastoquinone site. The main electron
source for nitrogenase might be glycolysis. Glucose-6-phosphate as well
as fructose-1,6-bisphosphate and dihydroxyacetone phosphate exhibited
high, NAD(P)-dependent rates of acetylene reduction which were inhibited
to ~ 50% in the light. Total inhibition of dark activity with glucose-
6-phosphate by DTT points to thioredoxin involvement in light inhibition
observed with this donor. The physiological significance of this effect
is not yet clear, but might reflect distribution of electrons between
nitrogenase and other biosynthetic processes in heterocysts, possibly

Sybesma, C. (ed.), Advances in Photosynthesis Research, Vol. II. ISBN 90-247-2943-2.
© *1984 Martinus Nijhoff/Dr W. Junk Publishers, The Hague/Boston/Lancaster.*

TABLE 1. Nitrogenase activites in a heterocyst homogenate from Anabaena variabilis in the light and in the dark

Electron donor	μmol C_2H_4/ml chlorophyll x h	
	Light	Dark
Hydrogen	34.0	1.3
Hydrogen (-)ATP-generator	0.4	0
NADH	14.0	1.4
NADPH	16.0	15.6
NADPH (+)100 μM DNP-INT	4.7	-
NADPH (+)2 mM DTT	-	12.7
Glucose-6-phosphate (+)10 μM NAD(P)	15.9	30.8
Glucose-6-phosphate (+)10 μM NAD(P) (+)2 mM DTT	0.4	0.8
Fructose-1,6-bisphosphate (+)10 μM NAD(P)	16.8	29.5
Dihydroxyacetone phosphate (+)10 μM NAD(P)	16.1	31.1
Pyruvate or 3-phosphoglycerate (+)10 μM NAD(P)	<0.2	<0.2
Citrate (+)0.1 mM NAD	1.5	1.9
Citrate (+)0.1 mM NADP	3.9	6.0

Concentrations of substrates, 1 mM; H_2, 0.64 mM (87.5% of the gas phase). Heterocysts from Anabaena variabilis Kütz. (ATCC 29413) were isolated according to Ernst et al. (1983) and broken by French-press treatment under H_2. Reactions with homogenates thus obtained (1 mg chlorophyll/ml) were performed in septum-stoppered 8-ml vials. The reaction mixture contained in a total volume of 0.25 ml (mM): HEPES-NaOH, pH 7.5, 20; an ATP-regenerating system consisting of ATP, 5; $MgCl_2$, 5; creatine phosphate, 15; creatine kinase, 25 μg. Vials were rendered anaerobic with H_2 or argon prior to addition of acetylene (12.5%, v/v, of the gas phase). Reactions were started by addition of 25 μl of the homogenate and carried out at 30 °C in a Warburg apparatus. Light intensity was 700 μE/m^2 x sec unless stated otherwise.

glycogen synthesis. No activity was measured with 3-phosphoglycerate or pyruvate. Citrate gave low, NADP-specific activities.

2.2. The regulatory role of FNR. Addition of NADP inhibited light-dependent H_2- or NADH-supported acetylene reduction (Fig.1). Time reversal of this inhibition reflects competition of NADP with nitrogenase for electrons. Addition of NAD was without effect (Fig.1A) pointing to NADP specificity of redox regulation by the NADPH/NADP ratio at the FNR:ferredoxin complex. This was experimentally confirmed by the response of nitrogenase activity to differently adjusted NADPH/NADP ratios (Fig.2). Substantial acetylene-reduction rates start at about -350 mV and reach a maximum below -400 mV, requiring a very high NADPH/NADP ratio (>100). Light or gramicidin D did not cause a shift of the redox-potential curve. Obviously (Fig.1A,B) a high reductant charge can be generated by photoreduction of NADP with H_2 or NADH in a light-dependent transhydrogenase reaction. A high NADPH/NADP ratio rather than the proton-motive force is responsible for generation of a low redox potential for ferredoxin reduction (cf. Haaker et al. 1980). As NADPH-dependent nitrogenase activity (via FNR) was not light-dependent (Table 1), DBMIB inhibition with this donor in the light was surprising. In the light, FNR can evidently not directly

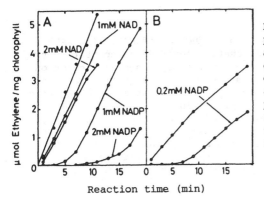

FIGURE 1. Effect of oxidized
pyridine nucleotides on H$_2$ (A)-
and NADH (B)-supported nitrogenase activity in the light.
●——● no pyridine nucleotide
added; o——o with NAD or NADP
present as depicted in the
figure. Substrate concentrations and assay conditions:
see Table 1.

reduce ferredoxin, but has to pass electrons through PS I for ferredoxin
reduction. This feature implies conformational changes of FNR upon dark-
light transitions (cf. Carrillo, Vallejos 1983). Figure 3 shows (by
nitrogenase activity) that in the presence of DBMIB inactivation of (dark)
reduction of ferredoxin by NADPH(through FNR) is effected by increasing
light intensities.

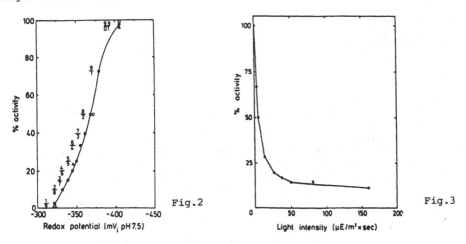

FIGURE 2. Dependence of NADPH-supported nitrogenase activity on the
NADPH/NADP ratio (redox potential). The assay was conducted in the light
(o——o) and in the dark (●——●) at a constant NADPH plus NADP concentration of 1 mM. ▲-▲ with 40 µg/ml gramicidin D in the light or in the
dark. 100% activity and assay conditions see Table 1.

FIGURE 3. NADPH-supported nitrogenase activity. Inhibition of the (dark)
ferredoxin-reducing activity of FNR with increasing light intensities.
Each assay contained 80 µM DBMIB to avoid light reduction of ferredoxin
via photosystem I. For assay conditions see Table 1.

3. CONCLUSIONS

A high NADPH/NADP ratio generated by photoreduction of NADP with NADH
(or H_2) is a prerequisite for photosynthetic electron transport to nitrogenase. The NADPH/NADP ratio may balance cyclic and non-cyclic electron
flow around PS I, adjusting the optimum ATP/e-ratio for nitrogenase (or
other biosynthetic activities) to operate. A fixed spatial arrangement of
FNR between ferredoxin and the b_6/f-complex (Fig.4; see Carrillo,
Vallejos 1983 and ref. 3 therein) would greatly favour this regulatory
mechanism. Energized membranes are not necessary for regulation of electron transport to nitrogenase in heterocysts (cf. Haaker et al. 1980;
Hawkesford et al. 1981). Glycolysis rather than the oxidative pentose-
phosphate pathway is likely to supply electrons for nitrogen fixation in
heterocysts.

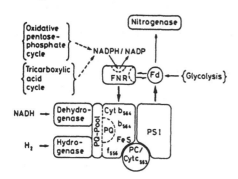

FIGURE 4. Tentative scheme for
electron flow from different
sources to nitrogenase in heterocysts of Anabaena variabilis. A regulatory role is
assumed for FNR in light-induced electron flow.

REFERENCES
Bothe H (1982) Nitrogen fixation. In Carr NG, Whitton BC, eds. The bio-
logy of cyanobacteria, pp. 87-104. Blackwell, Oxford.
Ernst A, Böhme H and Böger P (1983) Phosphorylation and nitrogenase
activity in isolated heterocysts from Anabaena variabilis (ATCC 29413),
Biochim. Biophys. Acta 723, 83-90.
Carrillo N, Vallejos RH (1983) The light-dependent modulation of photosynthetic electron transport, Trends Biochem. Sci. 2, 52-56.
Haaker H, Laane C and Veeger C (1980) Dinitrogen fixation and the protonmotive force. In Stewart WDP and Gallon JR, eds. Nitrogen fixation, pp.
113-138. Academic Press, London.
Hawkesford MJ, Reed RH, Rowell P and Stewart WDP (1981) Nitrogenase
activity and membrane electrogenesis in the cyanobacterium Anabaena variabilis Kütz., Eur. J. Biochem. 115, 519-523.
Houchins JP and Hind G (1982) Pyridine nucleotides and H_2 as electron
donors to the respiratory and photosynthetic electron-transfer chains and
to nitrogenase in Anabaena heterocysts, Biochim. Biophys. Acta 682,
86-96.
Schrautemeier B, Böhme H and Böger P (1983) In vitro studies on pathways
and regulation of electron transport to nitrogenase with a cell-free extract from heterocysts of Anabaena variabilis. Arch. Microbiol., in
press.

Authors' address: Lehrstuhl für Physiologie und Biochemie der Pflanzen,
Universität Konstanz, D-7750 Konstanz, Germany.

SURFACE ELECTRIC PROPERTIES OF THYLAKOID FRAGMENTS ISOLATED FROM
VEGETATIVE AND HETEROCYSTOUS CYANOBACTERIA

K. KALOSAKA, G.C. PAPAGEORGIOU/NUCLEAR RESEARCH CENTER DEMOKRITOS,
DEPARTMENT OF BIOLOGY, ATHENS, GREECE

. INTRODUCTION

Above its isoelectric point (pH 4.3, Nakatani et al.1978) the thylakoid
carries negative electric charge, nonuniformly distributed on its surface. This charge, and the potential it generates are implicated in
thylakoid stacking, regulation of excitation spillover and binding of
extrinsic proteins (Barber, 1982). Thylakoids of the oxygenic prokaryotic cyanobacteria do not stack. Furthermore, some filamentous cyanobacteria, when starved from nitrogen compounds develop specialized, dinitrogen assimilating cells, the heterocysts. The differentiation of a
vegetative cell into a heterocyst is accompanied by profound changes in
the structure and molecular organization of thylakoids (Stanier, Cohen-
Bazire,1977).

We report here preliminary experiments by which we examine whether heterocyst differentiation entails changes in the surface electric properties
of thylakoids. For this purpose, we have used the method of Chow, Barber (1980) which permits the estimation of the surface electric charge
density (σ) and potential (ψ_0) from pairs of mono- and divalent metal
cation concentrations that equally reverse the quenching of 9-AA* fluorescence, caused by membrane particles.

2. PROCEDURE

The heterocystous cyanobacterium Nostoc muscorum and the nonheterocystous
Phormidium luridum were cultured as described by Papageorgiou, Isaakidou
(1981) and Papageorgiou and Tzani (1980). Heterocysts were induced by
culturing Nostoc in a minus nitrogen compounds medium. To isolate the
heterocysts, the induced filaments were incubated with lysozyme (7.5-10
mg/mg Chl) under vigorous stirring at 30±1 °C, for 20 h. The cell mass
was centrifuged out at 1000 X g for 3 min, and the heterocysts were
purified by means of 8-10 successive centrifugations at 1000 X g for 2
min.

Normal (noninduced) and nitrogen-starved (induced) Nostoc cells were
broken by freezing to -25 °C and forcing the solid pellet 3 times through
an LKB X-press. Unbroken cells, large debris,and cell walls were centrifuged out at 12 000 X g for 12 min, and thylakoid fragments were pelleted
from the supernatant at 164.000 X g for 60 min. Phormidium thylakoid
fragments were isolated by the same procedure, after first converting the
cells to spheroplasts. Spinach thylakoid fragments were prepared from
intact isolated chloroplasts according to Nakatani et al.(1978).

All membrane preparations were made in sorbitol 100 mM, Hepes 1 mM, KOH
0.8 mM, pH 7.5. Cyanobacteria membranes were maintained frozen at -70°C;

---

*Abbreviations: 9-AA, 9-aminoacridine; Chl, chlorophyll; Hepes, N-2-hydro-
xyethyl piperazine-N'-2-ethanesulfonic acid.

Sybesma, C. (ed.), Advances in Photosynthesis Research, Vol. II. ISBN 90-247-2943-2.
© *1984 Martinus Nijhoff/Dr W. Junk Publishers, The Hague/Boston/Lancaster.*

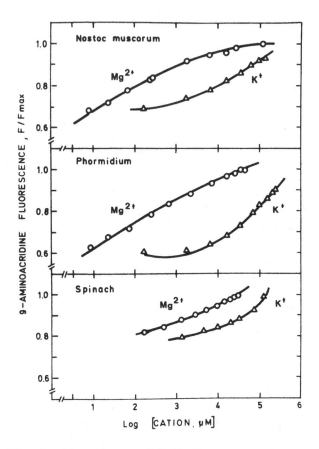

Fig. 1.   The release of 9.AA fluorescence by
metal cations.

spinach thylakoids at 0 °C. Chl/protein ratios (w/w) in these preparations were ; noninduced Nostoc, 0.10; induced Nostoc, 0.08; Nostoc heterocysts, 0.05; Phormidium 0.09; and spinach 0.10.

9-AA fluorescence was measured with a Perkin-Elmer model MPF-3L spectro-
fluoremeter, with 400 nm and 520 nm fluorescence excitation and detection
bands. Samples contained per ml 67 µg protein and 1.7 nmoles 9-AA in the
case of cyanobacteria, and 70 µg protein and 20 µmoles 9-AA in the case
of spinach thylakoids. The background monovalent cation concentration
was approx. 0.8 mM. Fluorescence quenching was released by adding $MgCl_2$
and KCl.

3.   RESULTS AND DISCUSSION

Contributions to the quenching of 9-AA fluorescence by electrogenic pH
gradients were ruled out, since we could not detect electron transport

n the membrane preparations. Secondly, upon nuclease treatment (0.83
µg/ml DNAase, 0.56 µg/ml RNAase, 30-60 min at 25°C) neither nucleotides
were released, nor the surface charge density was affected. Nucleic acid
contamination could lead to an overestimation of the negative charge of
cyanobacteria thylakoids.

Addition of membrane fragments to 9-AA solutions depresses their fluorescence, because of the deposition of probe molecules in the diffuse electric layer, which borders the membrane surface. The deposited probe molecules are displaced back to the bulk phase by metal cations, with a concomitant rise in the fluorescence. This is illustrated in Fig. 1, where
the maximal cation-released fluorescence (Fmax), divided by the fluorescence at submaximal cation concentrations (F) is plotted against the
logarithm of the metal cation concentration. Curves for Nostoc, Phormidium and spinach thylakoid fragments are illustrated.

From such plots, we can determine pairs of mono- and divalent metal cation
concentrations which effect an equal release of 9-AA fluorescence quenching. By introducing these concentration terms to the equations developed
by Chow, Barber (1980), we can estimate the surface charge density and
potential of membrane particles. Results for thylakoid fragments obtained from 2 filamentous cyanobacteria and 2 higher plants are displayed in
Table 1. Cyanobacteria thylakoids appear to be significantly more negatively charged than higher plant thylakoids.

Estimates of the surface electric properties of thylakoid fragments isolated from vegetative, induced vegetative, and heterocystous Nostoc cells
are given in Table 2. Here it is shown that vegetative and heterocystous

Table 1. Calculated surface electric charge densities and potentials of
thylakoid fragments obtained from various sources.

Source	Added mM[1] Mg^{2+}	K^+	σ (Cm^{-2})	ψ_0 (mV)
Nostoc	1.3	63.1	-0.105	-100
Phormidium	2.8	240.0	-0.300	-137
Spinach	3.2	37.2	-0.030	- 57
Pea[2]	0.6	16.1	-0.035	- 81

Table 2. Calculated surface electric charge densities and potentials of
thylakoid fragments obtained from N. muscorum at different stages of differentiation.

Source	Added mM[1] Mg^{2+}	K^+	σ (Cm^{-2})	ψ_0 (mV)
Vegetative cells	1.3	63.1	-0.105	-100
Induced vegetative cells	10.0	208.9	-0.195	-115
Heterocysts	0.4	39.0	-0.115	-117

[1]Concentrations effecting 90% release of 9-AA fluorescence quenching
[2]Data from Chow, Barber(1980).

cell thylakoids carry about an equal negative charge on their surfaces, while thylakoids from induced (nitrogen starved) vegetative cells carry a significantly higher negative charge.

The 9-AA fluorescence quenching method for the estimation of the surface electric properties of membrane fragments has several limitations arising from the assumptions inherent in the Gouy-Chapman theory, the physico-chemical properties of the probe, and the probable variation of the surface charge due to the cation-effected of emergence or submergence of membrane proteins (Barber 1980; 1982). To these limitations, we may add the uncertainty as to which side of the thylakoid membrane 9-AA really probes. In our experiments, we may conservatively assume that we deal either with membrane fragments,or with unsealed vesicles,and therefore the probe sees both faces of the membrane. Even with these limitations, however, our preliminary experiments indicate that the cyanobacteria thylakoid surface is more negatively charged compared to higher plant thylakoids, and that, at least in the case of Nostoc, heterocyst thylakoids are also negatively charged, but the average surface charge density is approx. equal to that of vegetative cell thylakoids.

REFERENCES

Barber J (1980) Membrane surface electric charges and potentials in rela-tion to photosynthesis, Biochim. Biophys. Acta 594, 253-308.
Barber J (1982) Influence of surface charges on thylakoid structure and function, Ann. Rev. Plant Physiol. 33, 261-295.
Chow WS and Barber J (1980) Salt-dependent changes of 9-aminoacridine fluorescence as a measure of charge densities of membrane surfaces. J. Biochem. Biophys. Methods 3. 173-185.
Nakatani HY, Barber J and Forrester JA (1978) Surface charges on chloro-plast membranes as studied by particle electrophoresis, Biochim. Biophys. Acta 504, 215-225.
Papageorgiou GC and Isaakidou J (1981) The photosynthetic apparatus of vegetative and heterocystous cells of the filamentous cyanobacterium Nostoc muscotum. Proceed. 5th Photosynth. Congress, 5, 727-736, Phila-delphia, Pennsylvania, Balaban.
Papageorgiou GC and Tzani H (1980) The action of lysozyme on glutaralde-hyde-treated filaments of the cyanobacterium Phormidium luridum, J. Appl. Biochem. 2. 230-240.
Stanier RY, Cohen-Bazire G (1977) Phototropic prokaryotes : the cyanobac-teria Ann. Rev. Microbiol. 31, 225-274.

STUDIES ON GLUTATHIONE REDUCTASE ACTIVITY IN THE CYANOBACTERIUM Anabaena
sp. strain 7119

A. Serrano, J. Rivas and M. Losada

1. INTRODUCTION

In higher-plants chloroplasts, the reduced form of the thiol-containing
tripetide glutathione (GSH)* has been claimed as a protective agent
against the oxygen active species, O_2^- and H_2O_2, generated by
photosynthetic electron flow (Foyer, Halliwell, 1976). The NADPH-spe-
cific glutathione reductase (EC 1.6.4.2) present in these organelles
exhibits kinetic parameters that are in good agreement with the
protective role assigned to GSH, i.e. very alkaline pH optimum value and
low apparent K_m for NADPH, thus suggesting that the enzyme may be more
active around the thylakoid membrane, where higher concentrations of
H_2O_2 and O_2^- can be expected as a result of photosynthetic activity
(Halliwell, Foyer, 1978). Since cyanobacteria are the only prokariotes
that, like higher plants, perform oxygenic photosynthesis and since no
previous studies have been undertaken on glutathione reductase activity
in these microorganisms, it was of interest to check for GR and to
compare it with the enzyme of chloroplast origin. Moreover, glutathione
has been recently reported as the most important non-protein thiol in
both vegetative cells and heterocyts of a nitrogen-fixing cyanobacterium
(Gidding "et al.", 1981). In this communication we report the presence
of a soluble NADPH-glutathione reductase in crude extracts of the
nitrogen-fixing cyanobacterium Anabaena sp. strain 7119. Enzyme levels
have been determined under a variety of culture conditions. The enzyme
activity has been characterized, kinetic parameters being very similar
to those of the chloroplast enzyme. As previously shown in chloroplasts
(Jablonski, Anderson, 1978), GSSG can act as a Hill reagent in a
reconstituted system containing thylakoid membranes and purified enzymes
from Anabaena sp. strain 7119. A detailed description of the
purification method and molecular properties of GR from Anabaena sp.
strain 7119 will be published elsewhere.

2. MATERIAL AND METHODS

Anabaena sp. strain 7119, from the Dept. of Cell Physiology, Berkeley,
USA (a gift of Dr. D.I. Arnon) was grown photoautotrophically as
previously described (Serrano "et al.", 1981). When indicated KNO_3 was
omitted from the culture medium. Glutathione reductase activity was
assayed at 25°C by the decrease in absorbance at 340 nm as NADPH was
oxidized in the presence of GSSG as electron acceptor. Except when
indicated reaction mixtures (1 ml) contained 100 mM Tris-HCl buffer, pH
9.0, 2.5 mM GSSG, 0.5 mM EDTA, 0.2 mM NADPH and an adequate quantity of
enzyme. One unit of enzyme is defined as the amount which catalyzes the
oxidation of 1 μmol of NADPH per minute. FNR activity was measured as
described by Shin (1971), using cytochrome c as electron acceptor.
G6PDH activity was determined as described by Apte "et al." (1978).

---

* Abbreviations: GSH, reduced glutathione; GSSG, oxidized glutathione;
GR, glutathione reductase; FNR, ferredoxin-NADP$^+$ reductase; G6PDH,
glucose-6-phosphate dehydrogenase.

Sybesma, C. (ed.), Advances in Photosynthesis Research, Vol. II. ISBN 90-247-2943-2.
© *1984 Martinus Nijhoff/Dr W. Junk Publishers, The Hague/Boston/Lancaster.*

Protein was determined by the method of Lowry "et al." (1951) using bovine serum albumin as a standard. GSH was determined spectrophotometrically as described by Hafeman "et al." (1974). Chlorophyll was determined as described by Mackinney (1941). The cells were disrupted in the cold with a Sonifier Branson B-12 (60s, 70W) or by using a refrigerated Aminco French Press pressure cell and the 40000 g x 20 min supernatant used as crude extract.

3. RESULTS AND DISCUSSION

3.1. Characterization of GR activity in crude extracts

Crude extracts of nitrate-grown Anabaena sp. strain 7119 sonicated cells present a NADPH-glutathione reductase activity of about 70 mU/mg protein. As it is shown in Table 1, the enzyme is absolutely specific for NADPH, and unlike the enzyme from non-photosynthetic organisms NADH is completely inefficient, at pH 7.0, as electron donor. The enzyme is also very specific for the disulphide substrate, no NADPH oxidation being observed with cystine, lipoic acid or oxidized coenzyme A, even at the mM concentration range. Using an equimolar mixture of morpholinopropane sulfonic acid, Tris and glycine (total concentration 0.1 M), adjusted with NaOH at different pH values, an optimum pH of 8.8.-9.0 was determined for the enzyme in crude extracts. From

TABLE 1. Requirements for glutathione reductase activity of Anabaena sp. strain 7119

Treatment[a]	ΔA_{340}/min
Control, pH 9.0	0.147
minus NADPH	0.002
minus GSSG	0.002
minus enzyme	0.002
minus NADPH plus NADH	0.003
Reaction mixture, pH 7.0	0.007
minus NADPH plus NADH, pH 7.0	0.002
Boiled enzyme (5 min)	0.003

[a] Except where indicated the reaction mixture was as described under Material and Methods. Protein concentration, 0.28 mg/ml.

reciprocal plots, apparent K_m values for NADPH and GSSG were estimated to be 10 µM and 195 µM, respectively. GR activity is completely soluble, remaining in the supernatant fraction, whereas, under our experimental conditions, about 20% of FNR activity appeared located in the thylakoid fraction.

3.2. Occurrence and distribution of GR activity in Anabaena sp. strain 7119 cells

As it is shown in Table 2, GR activity is present in a crude cell-free extract of French Press disrupted filaments, grown either on nitrate or with dinitrogen. A soluble fraction from heterocysts isolated as described by Tel-Or, Stewart (1976), exhibited a specific

TABLE 2. Glutathione reductase activity level in <u>Anabaena</u> sp. strain 7119 cells grown with different nitrogen sources

Material	Enzyme activity (mU/mg)[a]		
	GR	FNR	G6PDH
Nitrate-grown filaments	72.9	312.8	39.5
Dinitrogen-grown filaments			
a) vegetative cells	52.0	538.7	48.1
b) heterocysts	48.2	1937.0	431.6

[a] Each value represents the mean of three independent determinations.

activity similar to that from vegetative cells. Both FNR and G6PDH have been employed as markers to check the quality of the cell extracts coming from either the vegetative cells (low content) or the heterocysts (high content).

3.3. GSSG as a Hill reagent in reconstituted systems

As it was previously shown for higher plant chloroplasts, thylakoid membranes, prepared as described by Ortega "et al." (1976), and illuminated with red light ($\lambda > 600$ nm, 200 W m^{-2}) in the presence of a catalytic amount of NADPH (60 μM), ferredoxin, FNR and GR purified from Anabaena sp. strain 7119, catalized GSSG-dependent oxygen evolution (30-40 μmol per mg chlorophyll per hour) with the concomitant production of GSH. The ratios of GSSG added and GSH produced to oxygen evolved were 1.9 and 3.7, respectively, which are in good agreement with the following overall equations:

$$H_2O + NADP^+ \xrightarrow{\text{light}} 1/2 \; O_2 + NADPH + H^+$$

$$GSSG + NADPH + H^+ \longrightarrow 2 \; GSH + NADP^+$$

These results indicate that, at least in reconstituted cyanobacterial systems, GSSG can be photoreduced at high rates, suggesting an important role of this process in the cyanobacterial cell. Preliminary experiments about the activity levels of several enzymes in nitrate-grown Anabaena sp. strain 7119 cells placed under hyperoxic conditions (80% O_2 atmosphere), showed increased levels of GR (2-3 times), and significatively reduced levels of other enzymatic activities, like FNR (0.3-0.5 times) with respect the control values. A similar effect has been recently described for the maize leaf GR activity (Foster, Hess, 1982). These facts are consistent with the protective function of GSH proposed by Foyer, Halliwell (1976). The above results show that the redox metabolism of glutathione in cyanobacteria is very similar to that of chloroplasts, and introduce these prokariotic organisms as simplified systems for the study of glutathione metabolism in photosynthetic organisms.

REFERENCES
Apte, SK, Rowell P and Stewart WDP (1978) Electron donation to ferredoxin in heterocysts of the N$_2$-fixing alga <u>Anabaena cylindrica</u>, Proc. R. Soc. Lond. B 200, 1-25.

Foster JG and Hess JL (1982) Oxygen effects on maize leaf superoxide dismutase and glutathione reductase, Phytochemistry 21, 1527-1532.

Foyer CH and Halliwell B (1976) The presence of glutathione and glutathione reductase in chloroplasts: a proposed role in ascorbic acid metabolism, Planta 133, 21-25.

Gidding TH Jr, Wolk CP and Shomer-Ilan A (1981) Metabolism of sulfur compounds by whole filaments and heterocysts of Anabaena variabilis, J. Bacteriol. 146, 1067-1974.

Hafeman DG, Sunde RA and Hoekstra WG (1974) Effect of dietary selenium on erythrocyte and liver glutathione peroxidase in the rat, J. Nutr. 104, 580-587.

Halliwell B and Foyer CH (1978) Properties and physiological function of a glutathione reductase purified from spinach leaves by affinity chromatography, Planta 139, 9-17.

Jablonski PP and Anderson JW (1978) Light-dependent reduction of oxidized glutathione by ruptured chloroplasts, Plant Physiol. 61, 221-225.

Lowry OH, Rosebrought NJ, Farr Al and Randall RF (1951) Protein measurement with Folin phenol reagent, J. Biol. Chem. 193, 265-275.

Mackinney G (1941) Absorption of light by chlorophyll solutions, J. Biol. Chem. 140, 315-322.

Ortega T, Castillo F and Cárdenas J (1976) Photolysis of water coupled to nitrate reduction by Nostoc muscorum subcellular particles, Biochem. Biophys. Res. Com. 71, 885-891.

Serrano A, Rivas J and Losada M (1981) Nitrate and nitrite as "in vivo" quenchers of chlorophyll fluorescence in blue-green algae, Photosynth. Res. 2, 175-184.

Shin M (1971) Ferredoxin-NADP$^+$ reductase from spinach. In San Pietro, A, ed. Methods in Enzymology, vol. 23, pp. 440-447. New York: Academic Press.

Tel-Or E and Stewart WDP (1976) Photosynthetic electron transport, ATP synthesis and nitrogenase activity in isolated heterocysts of Anabaena cylindrica, Biochim. Biophys. Acta 423, 189-195.

ACKNOWLEDGEMENTS
This work was supported by grants from Philips Research Laboratories (The Netherlands) and Comisión Asesora de Investigación (Spain).

Authors adress: A. Serrano, J. Rivas and M. Losada, Dpto. de Bioquímica, Fac. de Biología y C.S.I.C., Apdo. 1095, Sevilla, Spain.

REGULATION OF NITRATE UTILIZATION BY CO_2 FIXATION PRODUCTS IN THE CYANOBACTERIUM Anacystis nidulans

CATALINA LARA, JOSE M. ROMERO, ENRIQUE FLORES, MIGUEL G. GUERRERO AND MANUEL LOSADA

1. INTRODUCTION

The utilization of nitrate by cyanobacteria (blue-green algae) is a genuine photosynthetic process that makes use of the assimilatory power generated in the light reactions of photosynthesis. This process includes as early steps the uptake of nitrate into the cell and its reduction to ammonium. Consequently, nitrate utilization by intact cells can be measured by following either the dissapearance of nitrate from the outer medium or the nitrate-dependent extra O_2 evolution in the light. In the latter case, stoichiometric values of about 2 mol of oxygen evolved per mol of nitrate taken up have been determined (Flores et al., 1983a).
As is the case for other cyanobacteria and green algae, nitrate utilization in Anacystis nidulans appears to be regulated by the ammonium resulting from nitrate reduction. As a matter of fact, nitrate utilization is rapidly and effectively hampered when low concentrations of ammonium are added to a cell suspension. The target of the ammonium effect seems to be the nitrate transport system rather than the nitrate-reducing enzymes (Flores et al., 1983b). On the other hand, nitrate utilization by cyanobacteria and green algae exhibits an stringent requirement for CO_2. For the case of Anacystis, the rate of the process in the absence of CO_2 is lower than 10% of that observed with saturating amounts of bicarbonate. A close correlation does exist between the rate of nitrate utilization and that of CO_2 fixation with regard to the concentration of CO_2 available to the cells, suggesting that the effect of CO_2 on nitrate utilization is mediated by products of its photosynthetic assimilation (Guerrero et al., 1982).
It has previously been shown (Flores et al., 1980) that treatment of A. nidulans with L-methionine-D,L-sulfoximine (MSX), an specific inhibitor of glutamine synthetase, relieves nitrate utilization from the ammonium inhibition. The inhibitory effect of ammonium thus appears not to be due to ammonium itself, but to some product(s) of its incorporation to carbon skeletons. Moreover, the MSX-treatment also releases nitrate utilization from the requirement for CO_2 (Guerrero et al., 1982).
DL-glyceraldehyde (DLG) has been characterized as an effective specific inhibitor of CO_2 fixation in chloroplasts by retarding the regeneration of the primary CO_2 acceptor from triose phosphate (Stokes, Walker, 1972; Bamberger, Avron, 1975). In order to ascertain the involvement of CO_2-fixation products in the regulation of photosynthetic nitrate metabolism, we have tested the effect of DLG on both CO_2-fixation and nitrate utilization by A. nidulans. The obtained results support the proposal of a model system for the regulation of nitrate utilization involving the participation of assimilation products of both CO_2 and ammonium.

Abbreviations: chl, chlorophyll a; DLG, DL-glyceraldehyde; MSX, L-me-thionine-D,L-sulfoximine.

Sybesma, C. (ed.), Advances in Photosynthesis Research, Vol. II. ISBN 90-247-2943-2.
© 1984 Martinus Nijhoff/Dr W. Junk Publishers, The Hague/Boston/Lancaster.

2. MATERIAL AND METHODS

Organism and culture conditions. <u>Anacystis nidulans</u> (strain L1402-1 from Göttingen University's Algal Culture Collection) was grown as described previously (Guerrero et al., 1974) with nitrate as the sole nitrogen source. Cells were collected by centrifugation after 24 h growth (35-50 µg chl/ml) and washed with 25 mM Tricine-NaOH/KOH buffer, pH 8.3. Chlorophyll <u>a</u> was estimated as in Flores et al. (1980).

Light-dependent O_2 evolution. O_2 was measured with a Hansatech DW oxygen electrode at $40^\circ C$ illuminated with saturating red light (150 W/m^2). The assay medium contained, in a volume of 1 ml, Tricine-NaOH/KOH buffer, pH 8.3, 25 µmol and an amount of cells equivalent to 10 µg chl. The reaction was started by switching the light on. Electron acceptors (NaHCO$_3$, 1 mM, or KNO$_3$, 20 mM) were added in the course of the experiment at the indicated times. The treatment with DLG was performed by incubating the cells at $40^\circ C$ in the assay mixture with the adequate amount of inhibitor for 15 min in the dark. The MSX treatment was performed by adding 1 µmol of the compound to the illuminated cell suspension 15 min before starting the incubation with DLG. Control cells were subjected to the same regime of light-darkness.

3. RESULTS AND DISCUSSION

The rate of light-dependent oxygen evolution exhibited by <u>A. nidulans</u> cells suspensions supplemented with bicarbonate is significantly enhanced upon addition of nitrate. This is illustrated by the results in Fig. 1 showing the extra oxygen evolution corresponding to nitrate utilization by cells which were carrying out CO_2 fixation at the time when nitrate was added. Rates of nitrate-dependent extra oxygen evolution in the range 55-75 µmol O_2/mg chl.h were usually found for saturating amounts of nitrate.

The effect of DLG, a well-known inhibitor of photosynthetic CO_2 fixation in higher plant chloroplasts, has been tested on photosynthetic oxygen evolution dependent on both CO_2 and nitrate, in <u>A. nidulans</u> cells. Results in Fig. 2 show that a gradual inhibition of CO_2-dependent oxygen evolution was obtained in response to the increase in the concentration of DLG, the inhibition being complete for DLG concentrations of 20 mM or higher. The treatment with increasing concentrations of DLG also led to a similar gradual decrease of the rate of nitrate-dependent oxygen evolution. Maximal inhibition of the process (about 80% inhibition) was also achieved at 10-30 mM DLG (Fig. 2). It is known that the utilization of nitrate (and its concomitant oxygen evolution) by <u>A. nidulans</u> is strictly dependent on the availability of either CO_2 or its assimilation products (Guerrero et al., 1982). It seems thus reasonable to propose that the negative effect of DLG on nitrate utilization is a consequence of the specific inhibition that this compound exerts on CO_2 fixation.

No negative effect of DLG on nitrate-dependent oxygen evolution was, however, observed in <u>Anacystis</u> cells in which ammonium assimilation had been inhibited by treatment with the glutamine synthetase inactivator, MSX. The results in Table 1 show that MSX-treated cells subjected to the action of DLG exhibit rates of nitrate-dependent oxygen evolution analogous to control cells (either untreated or treated with MSX) in spite of the total inhibition of the CO_2-dependent oxygen evolution caused by DLG. These observations support those of Stokes and Walker (1972) regarding the action of DLG as a pure "dark" inhibitor of CO_2

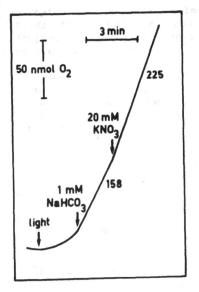 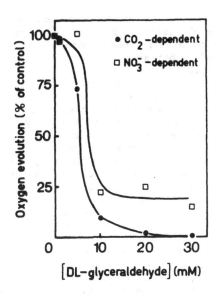

Fig. 1 (left). CO_2-dependent and nitrate-dependent oxygen evolution in illuminated Anacystis nidulans cells. Figures along the traces correspond to oxygen evolution rates (μmol O_2/mg chl.h).

Fig. 2 (right). Effect of DLG on CO_2- and nitrate-dependent oxygen evolution in illuminated Anacystis nidulans cells. Control rates for untreated cells (100%) were 217 and 73 μmol O_2/mg chl.h for CO_2-dependent and nitrate-dependent oxygen evolution, respectively.

TABLE 1. Prevention by MSX of the DLG-promoted inhibition of the nitrate-dependent oxygen evolution in Anacystis nidulans

Treatment	Light-dependent O_2 evolution (μmol O_2/mg chl.h)	
	CO_2-dependent	NO_3^--dependent
None	158	67
30 mM DLG	0	18
1 mM MSX plus 30 mM DLG	0	63

fixation without affecting the photosynthetic electron flow, and exclude besides the possibility that DLG could have a direct effect on nitrate utilization. On the other hand, MSX did neither affect the rate of CO_2 fixation nor the inhibition of CO_2 fixation caused by DLG (unpublished data). Thus, the ability of MSX to prevent the inhibition by DLG of nitrate utilization (i.e., to abolish the requirement for active CO_2 fixation) has to be adscribed to its action as an effective inhibitor of ammonium assimilation (Flores et al., 1983b). By preventing ammonium incorporation into carbon skeletons via glutamine synthetase, MSX releases nitrate utilization from its requirement for CO_2 fixation products. These carbon compounds appear therefore to play a regulatory role and not just that of substrates to combine the ammonium resulting from nitrate reduction.

The results can be interpreted in terms of a model which involves the participation of certain CO_2 fixation product(s) in a sensitive regulatory system controlling nitrate uptake. In this system, the accumulation of some ammonium-assimilation product(s) generated via glutamine synthetase would inhibit the utilization of nitrate. This inhibitory effect would be counteracted by products of CO_2 assimilation which, through combination with the negative effector(s), would diminish the intracellular concentration of the latter, allowing nitrate uptake. Thus, inhibition of CO_2 fixation would result in accumulation of the negative effectors, with cessation of nitrate uptake. A supply of CO_2-fixation products is needed for nitrate uptake to proceed in normal cells. These carbon metabolites would, however, be no longer required under conditions in which the generation of the ammonium derivatives is blocked, such as is the case for MSX-treated cells.

The results presented here are consistent with the operation of such a regulatory system in A. nidulans cells. This model for the regulation of nitrate utilization by both CO_2 fixation products and ammonium-derivatives provides a basis for understanding the way whereby the photosynthetic metabolism of nitrogen and carbon are integrated and interact with each other to balance these elements in the cell.

REFERENCES

Bamberger ES and Avron M (1975) Site of action of inhibitors of carbon dioxide assimilation by whole lettuce chloroplasts, Plant Physiol. 56,481-485.

Flores E Guerrero MG and Losada M (1980) Short-term ammonium inhibition of nitrate utilization by Anacystis nidulans and other cyanobacteria, Arch. Microbiol. 128, 137-144.

Flores E Guerrero MG and Losada M (1983a) Photosynthetic nature of nitrate uptake and reduction in the cyanobacterium Anacystis nidulans, Biochim. Biophys. Acta 722, 408-416.

Flores E Ramos JL Herrero A and Guerrero MG (1983b) Nitrate assimilation by cyanobacteria. In Papageorgiou GC and Packer L eds. Photosynthetic prokariotes, pp. 363-387. New York: Elsevier.

Guerrero MG Manzano C and Losada M (1974) Nitrite photoreduction by a cell-free preparation of Anacystis nidulans, Plant Sci.Lett. 3, 273-278.

Guerrero MG Flores E Romero JM and Losada M (1982) Role of CO_2 in the regulation of nitrate utilization in Anacystis nidulans. Int. Symp. of Nitrate Assimilation-Molecular and Cellular Aspects, Gatersleben, Abs. P.31.

Stokes DM and Walker DA (1972) Photosynthesis by isolated chloroplasts. Inhibition by DL-glyceraldehyde of carbon dioxide assimilation, Biochem.J. 128, 1147-1157.

ACKNOWLEDGEMENTS

This work has been supported by grants from Fundación Areces and Comisión Asesora Investigación (Spain). The secretarial assistance of Pepa Pérez de León and Antonia Friend is gratefully acknowledged.

Authors' address: Departamento de Bioquímica, Facultad de Biología y C.S.I.C., Apartado 1095, Sevilla, Spain.

POSSIBLE ROLE OF AN AMINO ACID OXIDASE IN PHOTOSYSTEM II OF
ANACYSTIS NIDULANS

ELFRIEDE K. PISTORIUS

1. INTRODUCTION

Particles from the blue-green algae Anacystis nidulans
(Synechococcus leopoliensis) can catalyze O_2 evolution with
ferricyanide as acceptor in the light and the dark oxidation
of L-arginine (and other basic amino acids). The two
reactions are both inhibited by o-phenanthroline, but
respond to certain trivalent, divalent and monovalent
cations (like e. g. La^{3+}, Ca^{2+} and K^+) in different ways.
These metals stimulate the Hill reaction with ferricyanide
as acceptor, but totally inhibit L-arginine oxidation.
The order of effectiveness is $M^{3+} > M^{2+} > M^+$. The effect of
e. g. Ca^{2+} on the dark oxidation of L-arginine mirrors the
effect of Ca^{2+} on the light-driven O_2 evolution with ferricyanide, as though these two activities are mutually
exclusive, i. e. one process is inhibited to the extent
that the other is activated. The enzyme responsible for
the dark oxidation of L-arginine has been purified from
A. nidulans and shown to be a flavo-protein (Pistorius, Voss,
1980).

Active photosystem II particles have also been isolated from
Anacystis membrane fractions. They can catalyze 2,6-di-
chloroindophenol reduction with diphenylcarbazide as donor
in the light. They retain L-arginine oxidation activity in
the presence of EDTA. Both reactions are inhibited by
o-phenanthroline. Moreover, it could be shown that one of
the protein bands which has been shown to be characteristic
for photosystem II particles of A. nidulans (Koenig, Vernon,
1981), co-chromatographs with the purified L-amino acid
oxidase on SDS gels. An antibody prepared to the purified
L-amino acid oxidase also gave a precipitin line with
detergent-treated photosystem II particles (Pistorius, Voss,
1982).

2. RESULTS AND DISCUSSION

We will show here that particles prepared from A. nidulans
change their response to added cations which are required
for O_2 evolution when washed with EDTA. Before treatment
with EDTA, particles from A. nidulans show little or no
O_2 evolution capacity in the light unless they are supplemented with cations. There is no specific requirement for
Mn^{2+}. When these particles are washed with EDTA, they retain
their capacity for reducing 2,6-dichloroindophenol with
diphenylcarbazide as donor in the light and also oxidize
L-arginine in the dark. However, after washing with EDTA,
the particles display a specific requirement for added Mn^{2+}

Sybesma, C. (ed.), Advances in Photosynthesis Research, Vol. II. ISBN 90-247-2943-2.
© 1984 Martinus Nijhoff/Dr W. Junk Publishers, The Hague/Boston/Lancaster.

combined with a somewhat less specific requirement for Ca^{2+} (and Sr^{2+}). Both ions must be added in order to reconstitute the O_2 evolving reaction. There are probably two different binding sites involved, and it is likely that the cations compete with each other for these binding sites. Thus each cation has a possible inhibitory effect as well as a possible potentiating effect. Once the Mn^{2+} has been removed by washing the Anacystis particles with EDTA, the increase in specificity in favor of Ca^{2+} (and Sr^{2+}) may be due to the fact that these ions are less competitive for the Mn^{2+} binding site than, for example, Mg^{2+} and Ba^{2+}. That is, Mg^{2+} and Ba^{2+} may be more effective in preventing the rebinding of Mn^{2+} than they are in causing Mn^{2+} removal. We do not know whether the Mn^{2+} is bound to the L-amino acid oxidase. If the enzyme was responsible for binding both ions, Mn^{2+} would probably be bound at a different site than Ca^{2+}. We have previously shown that some transition metals inhibit the L-arginine oxidation in a lower concentration range than the alkali earth metals (Pistorius, Voss, 1980). A dual requirement of Mn^{2+} and Ca^{2+} for the activation of the latent water-oxidation center has also been shown for intact chloroplasts isolated from wheat leaves grown under intermittent flash illumination (Ono, Inoue, 1983).

Chlorpromazine has recently been reported to inhibit photosystem II reactions of spinach chloroplasts (Barr et al., 1982) and of *A. nidulans* particles (England, Evans, 1983) and Ca^{2+} partially relieved this inhibition. Since chlorpromazine is an antagonist of calmodulin, a requirement for a calmodulin-like mechanism in Ca^{2+} interactions with the thylakoids was inferred. Since chlorpromazine is structurally a flavin antagonist, we examined whether a correlation to the L-amino acid oxidase which is present in Anacystis particles, might exist. The results show that photosystem II reactions of *A. nidulans* can be inhibited by chlorpromazine, and that this inhibition can be partially relieved by Ca^{2+} ions. Evidence is also presented that the L-amino acid oxidase is inhibited by chlorpromazine and this inhibition can be relieved by L-arginine in much the same way as the inhibition of the enzyme by Ca^{2+} ions can be relieved by L-arginine. The spectral changes of the L-amino acid oxidase with added Ca^{2+} suggest that the conformation of the protein is changed in the presence of Ca^{2+}. This change is probably necessary for the protein to operate in the O_2 evolving reaction sequence. The similar behavior of chlorpromazine and Ca^{2+} toward L-arginine oxidation suggests that these reagents bind to the L-amino acid oxidase similarly, and the ability of Ca^{2+} to relieve the inhibitory effect of chlorpromazine on the photosystem II reactions suggests that Ca^{2+} does something to the enzyme that is prevented by chlorpromazine (Pistorius, 1983).

So far, the evidence for the participation of an amino acid oxidase in photosystem II reactions is indirect. First of all, it is present in photosystem II particles which have been highly depleted of protein constituents. Secondly, and more significantly, it responds to many of the reagents which affect photosystem II, e. g. cations, o-phenanthroline, ammonium ions and chlorpromazine as shown here. Our current speculation visualizes an amino acid oxidase bound to the reaction center in such a way that the oxidation-reduction of a substrate substitute or a group on the enzyme constitutes part of the "S"-reactions of photosystem II, whereas the electron acceptor reactions leading to quinone reduction are mainly concerned with the redox reactions of the flavin. Thus, we visualize an enzyme closely bound to the reaction center and active on both "sides" of it. So far, our data are compatible with, but do not prove, an involvement of the amino acid oxidase in the redox reactions of photosystem II of *A. nidulans*.

REFERENCES
Barr R, Troxel KS and Crane FL (1982) Calmodulin antagonists inhibit electron transport in photosystem II of spinach chloroplasts,Biochem. Biophys. Res. Commun. 104, 1182-1188
England RR and Evans EH (1983) A requirement for Ca^{2+} in the extraction of O_2-evolving photosystem 2 preparations from the cyanobacterium *Anacystis nidulans*, Biochem. J. 210, 473-476
Koenig F and Vernon LP (1981) Which polypeptides are characteristic for photosystem II? Analysis of active photosystem II particles from the blue-green alga *Anacystis nidulans*, Z. Naturforsch. C36, 295-304
Ono T and Inoue Y (1983) Requirement of divalent cations for photoactivation of the latent water-oxidation system in intact chloroplasts from flashed leaves, Biochim. Biophys. Acta 723, 191-201
Pistorius EK (1983) Effect of Mn^{2+}, Ca^{2+} and chlorpromazine on photosystem II of *Anacystis nidulans*: An attempt to establish a functional relationship of amino acid oxidase to photosystem II, Eur. J. Biochem. in press
Pistorius EK and Voss H (1980) Some properties of a basic L-amino acid oxidase from *Anacystis nidulans*, Biochim. Biophys. Acta 611, 227-240
Pistorius EK and Voss H (1982) Presence of an amino acid oxidase in photosystem II of *Anacystis nidulans*, Eur. J. Biochem. 126, 203-209

ACKNOWLEDGEMENTS
This work was supported in part by the Deutsche Forschungs-gemeinschaft and the Max-Planck-Gesellschaft.

Author's address: Elfriede K. Pistorius, Universität Bielefeld, Fakultät für Biologie, Lehrstuhl Zellphysiologie, Postfach 8640, 4800 Bielefeld 1, FRG

POST-ILLUMINATION CO_2 BURST IN *ANABAENA VARIABILIS* AS A MEASURE OF
ICARBONATE TRANSPORT DRIVEN BY CYCLIC PHOTOPHOSPHORYLATION

ERUO OGAWA[1], YORINAO INOUE[1], ROSS McC LILLEY[2] AND WILLIAM L. OGREN[3]/
) THE INSTITUTE OF PHYSICAL AND CHEMICAL RESEARCH(RIKEN), 2) THE UNIVERSITY
F WOLLONGONG AND 3) UNIVERSITY OF ILLINOIS

e wish to dedicate this paper to the memory of Professor Kazuo Shibata.

NTRODUCTION

Cyanobacteria are photosynthetic prokaryotes which have both photosystems
PS) I and II, drive electrons from water to CO_2 and thus evolve oxygen.
hey have an active transport mechanism for HCO_3^- and concentrate inorganic
arbon within the cells(Kaplan et al., 1980; Miller, Colman, 1980). Recently,
post-illumination CO_2 burst was found in cyanobacteria(Badger, Andrews,
982; Ogawa, Inoue, 1983). In order to elucidate the carbon source for the
O_2 burst, we measured the CO_2 burst and internal inorganic carbon pool in
nabaena variabilis and found that the carbon source is a pool of inorganic
arbon accumulated within the cells in the light. This paper demonstrates
hat the CO_2 burst is a quantitative measurement of the internal inorganic
arbon pool and that the HCO_3^- transport is driven by PS-1 cyclic photo-
hosphorylation.

ATERIALS AND METHODS

Cells of Anabaena variabilis, strain M-2(Algal Collection of Tokyo Univer-
ity) were grown at 27°C in Kratz and Myers' medium C(1955) under aeration
ith 3% CO_2 in air(v/v). Cells were harvested by centrifugation, resuspended
n 40 mM HEPES-KOH buffer, pH 7.6, to a chlorophyll concentration of 8-15
g/ml and then used for experiments.

The gas exchange of algae was measured at 27°C with an open gas-analysis
ystem described previously(Ogawa, 1982; Ogawa, Inoue, 1983). Nitrogen gas
ontaining 650 µl CO_2/l was bubbled through the cell suspension(30 ml) placed
n a reaction vessel at a flow rate of 1 l/min. The exchanged gas was dried
nd then analysed with an infra-red gas analyzer(model ZAP; Fuji Electric Co.,
okyo) and a trace O_2 analyzer(model 316; Teledyne Analytical Instruments Co.,
.S.A.). The cell suspension in the reaction vessel was illuminated with
650 W halogen lamp.

The internal inorganic carbon pool and accumulation of photosynthetic
roducts were measured using the filtering centrifugation technique as
escribed by Badger et al.(1977). The cell suspension was preilluminated for
0 min in a 1.5 ml plastic tube, then $NaH^{14}CO_3$(final concentration 2 mM, 2.5 µC_i
µmol) was injected in the dark and aliquots of cell suspension, 300 µl each,
ere transferred into 400 µl plastic microfuge tubes. The reaction was initiated by illumination at room temperature(25-28°C) and terminated by centrifuga-
ion(microfuge B; Beckman Instruments, CA) through a silicon oil layer to a
ottom layer(the composition of these layers was as described in the above
eference). Each sample tube in the microfuge was illuminated by means of a
ikon illuminator with a 30 W tungsten lamp. The tubes were frozen on dry ice
and the pellet removed with a razor blade and placed in a closed vial containing 430 µl 0.1 N NaOH. Two aliquots were taken from this sample and one of
them was acidified. Photosynthesis was estimated from the acidified sample
whereas the acid-labile ^{14}C(the internal inorganic carbon pool) was estimated
from the difference in counts between the alkaline and acidified samples.

Sybesma, C. (ed.), Advances in Photosynthesis Research, Vol. II. ISBN 90-247-2943-2.
© 1984 Martinus Nijhoff/Dr W. Junk Publishers, The Hague/Boston/Lancaster.

RESULTS AND DISCUSSION

A typical example of the post-illumination CO_2 burst found for a cell suspension of *Anabaena variabilis* is shown in Fig. 1A. The initial rate of the CO_2 burst was calculated by assuming that the rate of CO_2 evolution follow first order kinetics. The rate of CO_2 evolution into the medium($-dC/dt$) is given by the equation: $-dC/dt = kC$, where C is the concentration of a carbon source which gives rise to the CO_2 burst and k is the rate constant for the conversion of the carbon source to CO_2. From this equation, we obtain: $Ln (-dC/dt) = Ln kC_0 - kt$, where C_0 represents the concentration of the carbon source before the light is turned off and t is the time after the CO_2 level starts to increase in the dark(see Fig. 1A). The change in the CO_2 level in the gaseous phase, dCO_2/dt, is equal to $-dC/dt$ and, therefore, a semilogarithmic plot of dCO_2/dt *vs* t gives a straight line(Fig. 1B). The intersect of the line with the ordinate gives the initial rate of the burst($= kC_0$) and the slope the value of $-k$.

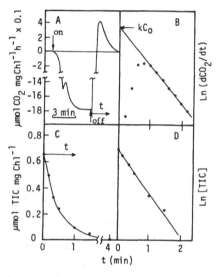

Fig. 1. (A) Changes in CO_2 concentration in the gaseous phase of an *Anabaena* cell suspension after switching the white light (720 μE $m^{-2}s^{-1}$) on and off. (B) Semilogarithmic plot of dCO_2/dt *vs* t for the CO_2 burst. For details see the text. (C) The concentration of total inorganic carbon (TIC) in *Anabaena* cells as a function of dark incubation period after 1 min illumination(900 μE $m^{-2}s^{-1}$). (D) Semilogarithmic plot of the values of Fig. 1C.

The validity of this extrapolation method was confirmed and the carbon source for the CO_2 burst was found to be a pool of inorganic carbon by measuring the kinetics of excretion of inorganic carbon during the dark period after illumination. The concentration of acid-labile ^{14}C (internal inorganic carbon pool) increased in the light to reach a stationary level after 2-3 min. The concentration decreased during subsequent dark incubation while the amount of photosynthetic products(acid-stable ^{14}C) did not change during this dark period(data not shown). It is, therefore, evident that the carbon source for the CO_2 burst is a pool of inorganic carbon accumulated within the cells in the light. Fig. 1C shows the time dependence of the decrease of inorganic carbon concentration during dark incubation after 1 min illumination and indicates that the concentration decreased without any lag to reach the initial level after 4 min. A semilogarithmic plot of the inorganic carbon concentration *vs* dark incubation time(t, see Fig. 1C) gave a straight line, indicating that the rate of excretion of inorganic carbon follows first order kinetics(Fig. 1D). Table 1 summarizes the C_0 and k values obtained for different samples by the extrapolation and filtering centrifugation methods, respectively. The C_0 and k values calculated from CO_2 bursts by the extrapolation method were in agreement with those, respectively, estimated using the filtering centrifugation method. Thus, the extrapolation method is valid for the determination of C_0, k and the initial rate of the CO_2 burst.

	C_O (µmol mg Chl^{-1})	k (sec^{-1})
Extrapolation	1.7	0.032
Method	1.6	0.026
	2.4	0.027
Filtering	2.0	0.033
Centrifugation	1.8	0.020
Method	1.5	0.031

Table 1. C_O and k values estimated by the extrapolation and filtering centrifugation methods, respectively. C_O represents the concentration of internal inorganic carbon at a steady-state level in the light and k is the rate constant for the excretion of inorganic carbon in the dark after a light period.

Fig. 2 shows the action spectra for HCO$_3^-$ transport and photosynthesis obtained by measuring the gas exchange(upper panel) and the accumulation of acid-labile and -stable ^{14}C using the filtering centrifugation method(bottom panel). The action spectrum for HCO$_3^-$ transport obtained by measuring the initial rate of the CO$_2$ burst following 5 min illumination with monochromatic light(curve A) was similar to that obtained by measuring the initial rate of accumulation of acid-labile ^{14}C(C). This confirms the above conclusion that the carbon source for the CO$_2$ burst is a pool of inorganic carbon accumulated within the cells in the light. Both action spectra showed a peak around 684 nm due to chlorophyll a in PS-I, which demonstrates that the HCO$_3^-$ transport is driven by PS-I only. As expected, the action spectrum for O$_2$ evolution(B) was similar to that for accumulation of acid-stable ^{14}C(D) and showed a peak around 630 nm due to phycocyanin and allophycocyanin in PS-II.

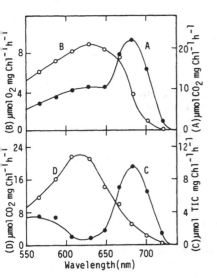

Fig. 2. Action spectra for HCO$_3^-$ transport (curves A and C) and photosynthesis(B and D) of *Anabaena variabilis*. The spectra in the upper panel were obtained by measuring the initial rate of the CO$_2$ burst(A) and the steady-state rate of O$_2$ evolution(B) and those in the bottom panel by measuring the initial rates of accumulation of acid -labile(C) and -stable(D) ^{14}C. The rates are normalized to those at a quantum flux of 10 µE m^{-2}s^{-1}. The incident fluxes of monochromatic light were 33-53 µE m^{-2}s^{-1} for curves A and B and 32-48 µE m^{-2}s^{-1} for curves C and D. The average length of the light path of the reaction vessel for the gas exchange measurements was 3.0 cm and that of the microfuge tube for the filtering centrifugation was 0.3 cm.

Table 2 summarizes the effects of inhibitors of photosynthesis and PS-I electron transport mediators on the accumulation of inorganic carbon and O$_2$ evolution. The accumulation of inorganic carbon was insensitive to 3-(3,4-dichlorophenyl)-1,1-dimethylurea(DCMU), an inhibitor of PS-II and non-cyclic electron transport. This indicates that the HCO$_3^-$ transport does not require non-cyclic electron flow from PS-II, being consistent with the action spectra for HCO$_3^-$ transport. The HCO$_3^-$ transport was accelerated by the mediator of PS-I cyclic electron transport and cyclic photophosphorylation, phenazine

	c_0 (µmol mg Chl^{-1})	O_2 evolution (µmol mg Chl^{-1}h^{-1})
control	2.3(100)	276(100)
+ 1 µM DCMU	2.3(100)	28(10)
+ 10 µM DCMU	1.7(74, 100*)	0(0)
+ 10 µM PES	2.2(96, 130)	0(0)
+ 10 µM PES + 3 µM FCCP	0(0, 0)	0(0)
+ 0.5 mM TMPD	0.3(13, 18)	0(0)

Table 2. Effects of inhibitors of photosynthesis and PS-1 electron transport mediators on HCO$_3^-$ transport and O_2 evolution. c_0 was estimated by the extrapolation method. The numbers in the parentheses indicate the percent of the control. *The value obtained in the presence of 10 µM DCMU was taken as the control.

ethosulfate(PES), but was inhibited by the PS-1 electron donor, N,N,N',N'-tetramethyl-p-phenylenediamine(TMPD), and the uncoupler of photophosphorylation, carbonylcyanide-p-trifluoromethoxyphenylhydrazone(FCCP). The electron transport mediated by TMPD is not coupled to ATP formation(Trebst, 1974). Thus, HCO$_3^-$ transport requires ATP, which is produced by cyclic photophosphorylation driven by PS-1.

REFERNCES
Badger MR and Andrews TJ (1982) Photosynthesis and inorganic carbon usage by the marine cyanobacterium, *Synechococcus* sp., Plant Physiol. 70, 517-523.
Badger MR, Kaplan A and Berry JA (1977) The internal CO_2 pool of *Chlamydomonas reinhardtii*: response to external CO_2, Carnegie Inst. Washington Yearb. 76, 362-366.
Kaplan A, Badger MR and Berry JA (1980) Photosynthesis and the intracellular inorganic carbon pool in the bluegreen alga *Anabaena variabilis*: response to external CO_2 concentration, Planta 149, 219-226.
Kratz WA and Myers J (1955) Nutrition and growth of several blue-green algae, Am. J. Bot. 42, 282-287.
Miller AG and Colman B (1980) Evidence for HCO$_3^-$ transport by the blue green alga(cyanobacterium) *Coccochloris peniocystis*, Plant Physiol. 65, 395-402.
Ogawa T (1982) Simple oscillations in photosynthesis of higher plants, Biochim. Biophys. Acta 681, 103-109.
Ogawa T and Inoue Y (1983) Photosystem-1 initiated post-illumination CO_2 burst in a cyanobacterium, *Anabaena variabilis*, Biochim. Biophys. Acta in press
Trebst A (1974) Energy conservation in photosynthetic electron transport of chloroplasts, Ann. Rev. Plant Physiol. 25, 423-458.

ACKNOWLEDGEMENTS
We thank H.Y. Nakatani and N. Fuad for their criticism. This study was supported by a grant for Solar Energy Conversion by Means of Photosynthesis from the Science and Technology Agency of Japan and in part by a grant to RML from the Australia-Japan Agreement on Science and Technology. Part of this study was carried out under a Collaboration Program between RIKEN and the School of Life Sciences of the University of Illinois.

Authors address: Teruo Ogawa[1], Yorinao Inoue[1], Ross McC Lilley[2] and William L. Ogren[3]/[1]The Algatron, Solar Energy Research Group The Institute of Physical and Chemical Research(RIKEN), Wako Saitama 351, Japan, [2] The University of Wollongong, P.O. Box 1144, Wollongong, N.S.W., Australia, 2500 and [3] Department of Agronomy, University of Illinois, Urbana, Illinois 61801, U.S.A.

PHOTOSYNTHESIS FOR ENERGY

D.O. HALL

SUMMARY
The process of photosynthesis embodies the two most important
reactions in life. The first one is the water-splitting reaction
which evolves oxygen as a byproduct; all life depends on this
reaction. Secondly is the fixation of carbon dioxide to organic
compounds. What most people do not realize is the magnitude of
present photosynthesis. It produces an amount of stored energy in the
form of biomass which is about ten times the world's annual use of
energy. The total amount of proven fuel reserves below the earth is
only equal to energy content of the present standing biomass (mostly
trees) on the earth's surface, while the fossil fuel resources are
probably only ten times this amount, and represent only 100 years of
net photosynthesis. Also, the amount of carbon stored in biomass is
approximately the same as the atmospheric carbon (CO_2) and the carbon
as CO_2 in the ocean surface layers (Table 1).
The majority of people in the world live by growing plants and
processing their products. Today about 14% of the world's primary
energy is derived from biomass - equivalent to 20 m barrels oil/day
(Fig. 1 and Table 2). Predominant use s in the rural areas of
developing countries where half the world's population lives, e.g.
Kenya derives 3/4, India 1/2, China 1/3 and Brazil 1/4 of their total
energy from biomass. A number of developed countries also derive a
considerable amount of energy from biomass, e.g. Sweden 9% and USA 3%.
Worldwide expenditure on biomass programmes is over $2 bn/year.
There is a possibility that the key processes in photosynthesis
involving C, N and S metabolism could be mimicked using the electrons
and protons from water splitting. Artificial systems using
photochemical and/or photobiological processes might be possible in
the future.

INTRODUCTION
At the start let me indicate what I am not going to advocate. I do
not suggest that biomass will solve the energy problems of the world.
I am not going to propose cutting down all the trees in the world or
to reforest the world; and I am not advocating that we all become
vegetarians. What I hope to clarify is that biomass already
contributes a significant part of the world's energy. But how much
biomass will contribute in the future will depend very much on
decisions that are made both at the local level and at the national
level, in addition to international policy making, especially for
energy and food.
At least half the world's population are primarily dependent on
biomass as their main source of energy. This realization has only
widely occurred in the last few years -it has been called the
"fuelwood crisis" or the "second energy crisis".

No specific references are cited in the text but diverse reviews and
literature sources are given in the "References".

Sybesma, C. (ed.), Advances in Photosynthesis Research, Vol. II. ISBN 90-247-2943-2.
© *1984 Martinus Nijhoff/Dr W. Junk Publishers, The Hague/Boston/Lancaster.*

The average person in the rural areas of the developing world uses the equivalent of about one tonne of air dry wood. This is mainly for cooking and heating but also for small-scale industry, agriculture, food processing, and so on. The use of wood and charcoal in urban areas and for industry is also often much greater than is realized. I don't think many people recognized the importance of this because the statistics were not available to show this significance, and the consequences of biomass overuse were not readily evident. Until a few years ago world energy supply statistics listed biomass at 3-5%, if at all. It is now known that about half of all the trees cut down in the world today are used for cooking and heating. The problem of deforestation with its consequent flooding, desertification and agricultural problems is not solely due to over-cutting of trees for cooking and heating. There are also obviously other factors involved such as commercial and illegal cutting, absence of replanting, and so on.

TABLE 1 — *Fossil fuel reserves and resources, biomass production and CO_2 balances.*

	Tonne coal equivalent
Proven reserves	
Coal	5×10^{11}
Oil	2×10^{11}
Gas	1×10^{11}
	8×10^{11} t = $\quad$ 25×10^{21} J
Estimated resources	
Coal	85×10^{11}
Oil	5×10^{11}
Gas	3×10^{11}
Unconventional gas and oil . . .	20×10^{11}
	113×10^{11} t = $\quad$ 300×10^{21} J
Fossil fuel used so far (1976) . . .	2×10^{11} t carbon = 6×10^{21} J
World's annual energy use . .	3×10^{20} J (5×10^9 t carbon from fossil fuels)
Annual photosynthesis	
(a) net primary production . . .	8×10^{10} t carbon (2×10^{11} t organic matter) = 3×10^{21} J
(b) cultivated land only . . .	0.4×10^{10} t carbon
Stored in biomass	
(a) total (90% in trees) . . .	8×10^{11} t carbon = 20×10^{21} J
(b) cultivated land only (standing mass)	0.06×10^{11} t carbon
Atmospheric CO_2	7×10^{11} t carbon
CO_2 in ocean surface layers . . .	6×10^{11} t carbon
Soil organic matter	$10\text{-}30 \times 10^{11}$ t carbon
Ocean organic matter	17×10^{11} t carbon

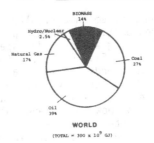

Figure 1   Global Distribution of Energy Use (1978)

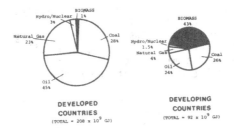

Global distribution of energy use, population and food supply (1978).

	Developed Countries	Developing Countries	World
Population	1.1 b (26%)	3.1 b (74%)	4.2 b (100%)
Total energy use (x10⁹)	208 GJ (69%)	92 GJ (31%)	300 GJ (100%)
per capita	189 GJ (6.3 tce)	30 GJ (1.0 tce)	71 GJ (2.4 tce)
Commercial energy use (x10⁹)	206 GJ (80%)	52 GJ (20%)	258 GJ (100%)
per capita	187 GJ (6.2 tce)	17 GJ (0.6 tce)	61 GJ (2.0 tce)
Biomass energy use (x10⁹)	2 GJ (5%)	40 GJ (95%)	42 GJ (100%)
Per capita	1 GJ (0.03 tce)	13 GJ (0.4 tce)	10 GJ (0.3 tce)
Food supply*- per capita	3353 Kcal	2203 Kcal	2571 Kcal
- % of requirement daily	(129%)	(96%)	(106%)
Commercial energy in total energy	98%	57%	86%
Biomass energy in total energy	2%	43%	14%

tce = tonnes coal equivalent
*average requirement

In N. America and Europe the problem in agriculture and nutrition is overproduction, excessive consumption of animal products via feeding of grains, and surpluses affecting world trade especially in relation to developing countries (Tables 3-5). Obviously this is too simplistic but the medium-term trends are important as they are likely to be aggravated by increased productivity, influence of new biotechnological processes and changes in diet. Thus, for very many countries the question should be how to achieve both food and biomass fuel production locally on a sustainable basis. Both are required - thus planning and provision of the appropriate infrastructure and incentives must be provided. Increased support of R and D, training and firm establishment of top priority to agriculture and forestry are essential in many countries of the world - if necessary, with significant help from abroad. Overall, biomass energy will not be a simple solution to the energy problems of the developing countries. Biomass systems will not necessarily be cheap, nor will they be implemented easily, without a major commitment from governments and a considerable amount of political will at the national and international level.

Today 2/3 of the world's people depend on plants for nearly all their food and for the majority of their energy. The question should again be not "food versus fuel?" but how can we increase the productivity of plant-based agriculture (and forestry) in order to provide the required levels of both food and fuel at the national or regional level? Since only less than 10% of the world's food and hardly any of the world's biomass fuel enters transnational trade it is local production of plant products which is all important.

The oil/energy problem of the 1970's has had three clear effects on biomass energy and development. Firstly, in a number of developed countries large research and development programmes have been instituted which have sought to establish the potential, the costs and the methods of implementation of energy from biomass. The prospects look far more promising that was thought even a few years ago. Demonstrations, commercial trials and industrial projects are being implemented. Estimated current expenditure is over a billion dollars per annum in North America and Europe. Secondly, in at least two countries, viz. Brazil (which currently spends over half of its foreign currency on oil imports) and China, with over 7 million biogas digesters, large scale biomass energy schemes are being implemented - the current investment is about $1.3 billion per annum in Brazil. Thirdly, in the developing countries as a whole there has been an accelerating use of biomass as oil products have become too expensive and/or unavailable.

Biomass as a source of energy has problems and it has advantages. Like every other energy source it is not the universal panacea. There is however at least one very important advantage, namely the large biological and engineering development potential (Table 6). Presently we are using knowledge and experience which has been static for very many years; the efficiency of production and use of biomass as a source of energy has not progressed the way agricultural yields for food have increased. Research in agriculture has paid off very well (Table 7) and this could also be the case for biomass R and D.

TABLE 3. EEC Agricultural Surpluses (OECD; EEC; 1982)

<u>Sugar</u> (past 3 year average)	2.6 Mt
<u>Dairy Cows</u>	2.5 - 3.5 M
	≡ 1.2 - 1.7 Mha arable land
<u>Fruit</u> (withheld)	0.7 Mt
<u>Fruit</u> (losses)	1.6 Mt
<u>Wine</u> (distilled)	1.84 Gℓ
<u>Potatoes</u> (reject)	8.5 Mt
<u>Cereals</u> (1981/2)	8.6 Mt
(Wheat, Barley, Maize)	(7% of 122 Mt)

Plant and animal protein production and consumption
by man and livestock in 1975 (estimated, in million tonnes)
(Pimentel and Pimentel, 1979)

	USA		World	
Total cereal protein produced	17.0	100%	95	100%
Fed to livestock	15.5	91%	38	40%
Available to man	1.5	9%	57	60%
Total legume protein produced	9.3	100%	30	100%
Fed to livestock	9.0	97%	6	20%
Available to man	0.3	3%	24	80%
Total livestock protein produced	6.0	100%	33	100%
Fed to livestock	0.7	12%	3	9%
Available to man	5.3	88%	30	91%
Total protein produced *	34.1	100%	173	100%
Fed to livestock	26.1	77%	51	29%
Available to man	8.0	23%	122	71%

* Fish protein and other types of vegetable protein are included
in this total.

BIOMASS

Biomass is a jargon term used in the context of energy for a range of products which have been derived from <u>photosynthesis</u>; it can be recognized as the waste from urban areas and from forestry and agricultural processes, specifically-grown crops like trees, starch crops, sugar crops, hydrocarbon plants and oils, and also aquatic plants such as water weeds and algae. Thus everything which has been derived from the process of photosynthesis is a potential source of energy. We are talking essentially about a solar energy conversion system, since this is what photosynthesis is. The problem with solar radiation is that it is diffuse and it is intermittent, so that if we

Table. Total grain production and food gaps (excess of production
to consumption projected to 1990 (t x 10^6 tonnes) (Barr, 1981)

	Population Distribution	Production	Gap
Developed countries (excluding USSR and E. Europe)	16%	640	+ 152
USSR & E. Europe	8%	370	- 61
China	20%	290	- 10
Developing countries	8%	205	- 24
Developing countries (low income)	48%	260	- 47
World	100%	1765	+ 10

are going to use it we have to capture a diffuse source of energy and
need to store it - this is what plants solved a long time ago. All
our food and fuel is derived from this CO_2 fixation from the
atmosphere.

When looking at an energy process we need to have some understanding
of what the efficiency of this process will be; one needs to look at
the efficiency over the entire cycle of the system, and in the process
of photosynthesis we mean total incoming solar radiation converted to
a stored end-product. Most people agree that the practical maximum
efficiency of photosynthesis is between five and six per cent. It
might not seem very good, but remember that this represents stored
energy (Table 8). Obviously if we can grow and adapt plants to
increase the photosynthetic efficiency the dry weight yields will
increase and of course alter the economics of the crop. One of the
very interesting areas of research is to try to understand what the
limiting factors are in photosynthetic efficiency both for agriculture
and for biomass energy.

There is another aspect of photosynthesis that we should all
appreciate. That is, the health of our biosphere and our atmosphere
is totally dependent on the process of photosynthesis. Every three
hundred years all the CO_2 in the atmosphere is cycled through plants.
Every two thousand years all the oxygen, and every two million years
all the water. Thus three key ingredients in our atmosphere are
dependent on cycling through the process of photosynthesis.

Many people are also rightly concerned about the problem of build-up
of CO_2 in the atmosphere if we continue to burn fossil fuels. It is a
problem of cycling between two or three pools of carbon. The amount
of carbon stored in the biomass is approximately the same as the
atmospheric CO_2 and the same as the carbon as CO_2 in the ocean surface
layers; there are three equivalent pools. The problem is how is the
CO_2 distributed between these pools, and how fast does it equilibrate

Some Advantages and Problems Foreseen in Biomass for Energy Schemes

Advantages	Problems

Advantages

1. Stores energy
2. Renewable
3. Versatile conversion and products; some products with high energy content
4. Dependent on technology already available with minimum capital input; available to all income levels
5. Can be developed with present manpower and material resources
6. Large biological and engineering development potential
7. Creates employment and develops skills
8. Resonably priced in many instances
9. Ecologically inoffensive and safe
10. Does not increase atmospheric CO_2

Problems

1. Land and water use competition
2. Land areas required
3. Supply uncertainty in initial phases
4. Costs often uncertain
5. Fertilizer, soil, and water requirements
6. Existing agricultural, forestry, and social practices
7. Bulky resource; transport and storage can be a problem
8. Subject to climatic variability
9. Low conversion efficiencies
10. Seasonal (sometimes)

Studies of Agricultural Research Productivity –
direct cost-benefit type studies (World Bank, 1981,

Commodity	Country	Time Period	Annual internal rate of return (%)
Hybrid corn	U.S.A.	1940-55	35-40
Hybrid sorghum	U.S.A.	1940-57	20
Poultry	U.S.A.	1915-60	21-25
Sugarcane	South Africa	1945-62	40
Wheat	Mexico	1943-63	90
Maize	Mexico	1943-63	35
Cotton	Brazil	1924-67	77+
Tomato harvester	U.S.A.	1958-69	37-46
Maize	Peru	1954-67	35-40
Rice	Japan	1915-50	25-27
Rice	Japan	1930-61	73-75
Rice	Colombia	1957-72	60-82
Soybeans	Colombia	1960-71	79-96
Wheat	Colombia	1953-73	11-12
Cotton	Colombia	1953-72	None
Aggregate	U.S.A.	1937-42 1947-52 1957-62	50 51 49

into the deep ocean layers? However, we should appreciate that increasing CO_2 concentrations in the atmosphere may be good for plants (since CO_2 is a limiting factor in photosynthesis and plants have better water use efficiency at higher CO_2 concentrations). Plants could also act as CO_2 sinks if photochemical means for fixing CO_2 were not available in the future to alleviate the problem.

TABLE 8 — *Photosynthetic efficiency and energy losses.*

	Available light energy
At sea level	100%
50% loss as a result of 400-700 nm light being the photosynthetically usable wavelengths	50%
20% loss due to reflection, inactive absorption and transmission by leaves	40%
77% loss representing quantum efficiency requirements for CO_2 fixation in 680 nm light (assuming 10 quanta/CO_2),* and remembering that the energy content of 575 nm red light is the radiation peak of visible light	9.2%
40% loss due to respiration	5.5%
	(Overall photo-synthetic efficiency)

* If the minimum quantum requirement is 8 quanta/CO_2, then this loss factor becomes 72% instead of 77%, giving the final photosynthetic efficiency of 6.7% instead of 5.5%.

BIOMASS PROJECTS
Here briefly follows some examples of biomass programmes.

Brazil. The largest programme is in Brazil, which is currently
spending about $1.3 bn annually of government money on subsidizing the
production of alcohol primarily from sugar-cane. The new programme
started in 1975 (even though they have been blending alcohol in
gasoline since the 1930's) due to the fact that currently Brazil
spends about $11 billion (over half of its total foreign income) on
importing oil. What Brazil is trying to do is to reverse this great
dependence on imported petroleum by the production of alcohol;
currently they are producing nearly 5 bn litres, and this is proposed
to be expanded up to about 11 billion litres by 1987, which is 10% of
forecast oil consumption. At present all the petrol sold in Brazil is
blended to 20% with alcohol. About 3/4 million cars now run on
hydrated alcohol which is 95% by volume. Everything is not perfect
with this programme. They have recently published an assessment of
the programme which shows their great awareness of the problems
themselves. One of the problems which was immediately evident is
pollution in rivers. Most of the alcohol currently being produced
comes from sugar-cane plantations in the South West of Brazil, where
often the stillage which is a byproduct from the distillation of the
alcohol has been put into the rivers. For every litre of alcohol
produced about 8-10 litres of stillage is produced and this has a very
high chemical and biological oxygen demand. There are certainly ways
of countering this pollution problem.

Zimbabwe. Nearly 3 years ago they opened an alcohol distillery which
saves them $10-12 million of foreign currency per annum. They produce
40 million litres of ethanol a year from sugar-cane which is blended
into petrol by the oil companies at a 15% blend; it is planned to

increase this up to a 25% blend when a new refinery/distillery is
opened. The yields of sugar-cane are high at about 120 t per hectare
per annum. Using the stillage from the fermentation as a fertilizer
on the sugar plantations increases the yields by about 6% and this
increase allows for the extra installations to be paid for within one
year.

USA. The biomass resources of the US are indeed quite large.
Currently they derive about 3% of their total energy requirements from
biomass, which is equivalent to over one million barrels of oil a day
and is greater than the nuclear energy contribution. The energy
content of the standing forests of the US are at least 50% greater
than the oil reserves and about equivalent to the gas reserves. In
1980 the OTA estimated that "Depending on a variety of factors,
including the availability of cropland, improved crop yields, the
development of efficient conversion processes, proper resource
management, and the level of policy support, bioenergy could supply as
few as 4 to 6 Quads/yr, or as many as 12 to 17 Quads/yr by 2000 (or up
to 15 to 20 percent of current U.S. energy consumption (about 75 Quads
$= 10^{18}$J))". Both the quantity of biomass that can be obtained on a
renewable basis, and the economic, environmental, and other
consequences of obtaining it will depend critically on the behavior of
growers and harvesters. The US has a gasohol programme which blends
10% alcohol, primarily derived from corn (maize), with unleaded
gasoline to provide a high octane fuel. This gasohol programme is
relevant to the discussion that we are having about lead in petrol in
Europe. The use of alcohol in gasoline has been known for a long
time; cars were sold in 1902 which ran on ethanol. Since then cars
have been run on blends of alcohol with anywhere from 5 up to 35% in
various parts of the world. For each 1% ethanol that is added to
gasoline the octane rating is raised by an average of 0.24 for the
motor octane number and 0.36 in research octane number. 10% alcohol
added to gasoline increases the octane number by 3-4 points. One of
the serious discussions about using biological material for producing
alcohol is whether you obtain more energy out than you have put into
the system; and secondly whether food is diverted from the food
market and from world trade.
The use of agricultural and forestry residues is also proposed as a
renewable source of energy as liquid fuels and/or heat. There is
considerable discussion as to the net energy realisable and also the
costs and environmental effects of such collection. Costs of using
such generally low density material reflect its collection,
transportation and processing. However, adverse effects on the soil
(erosion, water and nutrient loss, etc.) are the major concern.
Nevertheless there are definite opportunities for use of residues
especially with changed agricultural practices such as minimal
tillage.

Europe. Nearly all the countries in Europe have energy-from-biomass
schemes. Why in Europe should we look seriously at biomass as a
source of energy? Currently the EEC derives over half its total
energy requirements from imported oil. For Europe as a whole some
substitution of imported liquid fuels is what makes biomass look so

interesting. The EEC programme examines the use of agricultural resources, the use of forestry resources, the use of algae, the digestion of biological materials to produce methane, and thermochemical routes such as gasification to produce methanol. We in Europe are fairly crowded, so how much land do we have to produce energy from biomass? A recent study indicated that the EEC could produce over 85 million tonnes of oil equivalent, which is nearly 2 million barrels of oil a day, providing about 7% of our estimated 1985 energy demand (Table 9). This is twice what we use for agriculture. The study also showed that if there was a maximum disturbance to agriculture and forestry where a crash programme was required, we could possibly achieve a 20% provision of our energy requirement in the EEC. It is highly unlikely that this will happen. Thus a 7% energy provision is there for the taking, with minimal disturbance to current agricultural and forestry practices.

The EEC currently spends nearly 3/4 (equal to £14 bn) of its budget on the Common Agricultural Policy and most of this goes to subsidizing animals. To provide the food for these animals we devote a very large percentage of our land area for growing grains, we also import large quantities of grains from the United States and other countries. We produce milk, butter and cheese - it is a problem getting rid of these mountains, so very often they are fed back to animals. Since 90% of the energy in the food is lost every time you go through the animal, it seems rather an unusual behaviour. Even two years ago it was unthinkable to mention the possibility that we could use some of our agricultural land and our agricultural surpluses to produce something different from them, such as energy.
Sweden currently derives about 9% of its total energy requirements from biomass. They have a very interesting series of advanced biomass trials with fast-growing willows. They now have nine clones which have twice the yield (about 15 dry tonnes/ha/yr) of those previously used. If so desired the Swedes could derive half their energy requirements from biomass.

VEGETABLE OILS

It has been known for a long time that you can put all types of vegetable oils into diesel engines. Numerous studies show that in the sunny countries if a maize farmer devoted 10% of his land area to growing sunflowers he could run all the diesel-powered machines that he uses in his farming operation. Probably a blend of from 10-30% is preferable. There is work on the esterification of sunflower oil as methyl or ethyl esters; the esterified oil has fuel properties very close to those of diesel and the esterification can be done on the farm. The Brazilians are devoting considerable effort to extraction of oil from peanuts and carefully ascertaining the use of palm oil (which has high yields) as a 6% blend into diesel. Oils from soyabean, castor seeds, and indigenous plants such as malmeleiro and babassu nut are also being investigated.

9.

Total net energy potential (after conversion) of biomass in EEC by year 2000

		mtoe
a)	**Wastes and residues**	
	1. Animal wastes	11.5
	2. Crop residues	12.5
	3. Forestry/wood residues	7.9
	4. Municipal solid wastes	8.9
	5. Sewage	1.8
b)	**Firewood**	6.6
c)	**Energy forestry and crops**	
	Dry (forestry and short rotation plantations)	28.1
	Wet (Catch crops, algae and aquatic plants)	8.5
	total	85.8

In addition **agricultural surpluses** could be available, for example:

	mt 1980/81
Grains	7 (a)
Wheat only	11 (b)
Sugar	4

a - difference between exports and imports (66% of imports are maize)
b - increasing to 17 by 1985

HYDROCARBON PLANTS

Proposals to use plants directly to produce petroleum have been frequent, with the main recent proponent being Calvin who advocates growing Euphorbia lathyrus for the extraction of hydrocarbons which have molecular weights close to those of petroleum. There are trials, mostly in Arizona, to establish whether this is economically viable. Initial hopes of high yields were not substantiated, but the recent studies show yields of about 10 barrels (1.5 tonnes) of oil per hectre per annum under irrigation. The question is whether such yields are sustainable in arid environments? There are at least 5 other trials in various parts of the world to see if this is economically viable. There is another requirement from oil, and that is for the manufacture of synthetic rubber. Guayule, Parthenium argentatum, which grows naturally in Northern Mexico and the southern USA, can be used as a source of rubber with properties which are similar to that of the rubber tree. There is now a pilot plant of one tonne a day in Saltillo, Mexico, and a 50 tonne a day plant is now proposed there. Australian research has shown that guayule is the most promising of all the hydrocarbon plants for their conditions.
An alga, Botryococcus braunii has been shown to yield 70% of its extract as a hydrocarbon liquid closely resembling crude oil. This has led to the work on immobilizing these algae in solid matrices such as alginates and polyurethane and using a flow-through system to produce hydrocarbons. A green alga called Dunaliella discovered in the Dead Sea produces glycerol, β-carotene, and also protein. This alga does not have a cell wall and grows in these very high salt concentrations; thus to compensate for the high salt externally it produces glycerol internally.

2.   Possible components of a hydrogen evolving photoreactor

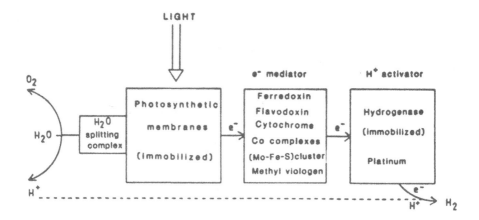

PHOTOCHEMISTRY AND PHOTOBIOLOGY

Since whole plant photosynthesis operates under the burden of many
limiting factors (internal and external), it might be possible to
construct artifical systems which mimic certain parts of the
photosynthetic processes and so produce useful products at higher
efficiences of solar energy conversion. (A 13 percent maximum
efficiency of solar energy conversion is considered a practical limit
to produce a storable product). Long-term, basic, directed research
on artificial photobiological/chemical systems for production of fuels
and chemicals (H_2, fixed C and NH_3, etc.), needs sustained funding if
these exciting future possibilities are to be realized.
The path of photosynthetic carbon fixation from CO_2 in the atmosphere
to sugars, carbohydrates, fats, etc., is well known. Unfortunately,
we do not know how the most crucial initial step works. The enzyme
ribulosebisphosphate-carboxylase, which fixes all the CO_2 in the
atmosphere which has gone into our oil, coal, gas, and currently goes
into our food and into our trees, is the most abundant enzyme on
earth. Half the protein in the lettuce leaf that you eat comprises
this single enzyme. Unfortunately, we do not know how it works even
though we are now learning quite a lot about it. Can we mimic the CO_2
fixation process? There are two reports on the photochemical fixation
of CO_2 to methanol, formic acid, and formaldehyde. These are
photoelectrochemical systems which work, as yet, only in UV light. It
is early days but these are significant first reports on fixing CO_2
into fuels and chemicals.
There have been recent reports on photochemical systems from a number
of laboratories that use colloidal particles incorporating titanium
and ruthenium which can split water and produce hydrogen and oxygen.
Experimental problems still exist with such systems, but they are the
first reports of purely chemical systems that appear as if they might
function. The only system which actually does function in visible
light on a continuous basis is the biological system (Fig.2). In

imicking such biosystems we use the water-splitting reaction of the
hlorophyll membrane, discard all the CO_2 fixation components, add in
nzymes from bacteria, shine light on the system, and produce hydrogen
as (plus oxygen as a byproduct). However, unfortunately it is a
iological system and it eventually dies. We and other groups have
ried to immobilize the system so that it can continuously produce
ydrogen and oxygen, because if the system is ever going to be
ractically useful it has to work for years and not hours.
an we mimic, or can we construct, a system which will continuously
plit water using visible light so that the electrons and protons from
ater can be used to produce high energy compounds like fixed carbon,
ydrogen or ammonia? At present the answer is "No". What we can do
owever is to use waste materials like organic compounds or sulphur
where you don't have to split water) to give continuous production of
ydrogen gas upon illumination. In our laboratory we have had such a
ystem running for fifteen days continuously producing pure hydrogen
from compounds such as ascorbate. But we have not solved the ultimate
roblem of splitting water.

CONCLUSION

What I have tried to say is that the process of photosynthesis is
really quite marvellous. We start with water and light and use
chlorophyll-containing membranes to produce an intermediate
energy-rich compound which is usually an iron-sulphur catalyst.
Normally the energy in that compound, which has a very high energy
content, is used to fix carbon dioxide into a whole range of organic
compounds. But we can also use this ultimately to fix nitrogen,
reduce sulphur, reduce oxygen, and also use it to produce hydrogen.
Plants and their derived systems can do nearly everything!

REFERENCES
Barr TN (1981) The world food situation and global grain prospects,
Science 214, 1087-1085.
BioEnergy Council (1981) International BioEnergy Directory, BioEnergy
Council, Washington, D.C. 20006.
Biomass Panel of Energy Research Advisory Board, USA (1983) Biomass
energy, Solar Energy 30, 1-31.
Bolton JR and Hall DO (1979) Photochemical conversion and storage of
solar energy, Ann. Rev. Energy 4, 353-401.
Boyer JS (1982) Plant productivity and environment, Science 218,
443-448.
Braun AM, ed. (1983) Photochemical conversions, Lausanne: Presses
polytechnique romandes.
Bungay HR (1982) Biomass refining, Science 218, 643-646.
Calvin M (1983) New sources of fuel and materials, Science 219, 24-27.
Coombs J, Hall DO and Chartier P (1983) Plants as solar collectors:
optimizing productivity for energy. Dordrecht: D. Reidel Publ.
Coombs J (1980) Renewable sources of energy (carbohydrates), Outlook
Agric. 10, 235-245.
Flaim S and Hertzmark D (1981) Agricultural policies and biomass
fuels, Ann. Rev. Energy 6, 89-121.
Hall DO (1979) Solar energy use through biology - past, present and
future, Solar Energy 22, 307-318.

Hall DO (1982) Energy from biomass: fuels now and in the future, J. Roy. Soc. Arts Lond. 130, 457-471; Experientia 38, 3-10.

Hall DO (1983a) Food versus fuel, a world problem? In Strub et al, eds. Energy from biomass, pp. 43-50. London: Applied Sci. Publ.

Hall DO (1983b) Energy from biomass in Europe: the 1982 situation. In Strub et al, eds. Energy from biomass, pp. 51-62. London: Applied Sci. Publ.

Hall DO, Barnard GW and Moss PA (1982) Biomass for energy in the developing countries, Oxford: Pergamon.

Hall DO and Palz W, eds. (1983) Photochemical, photoelectrochemical and photobiological processes, Vol. 2. Dordrecht: D. Reidel Publ.

Huxley PA (1982) Agroforestry - a range of new opportunities, Biologist 29, 141-143.

Lipinsky ES (1981) Chemicals from biomass: petrochemical substitution options, Science 212, 1665-1676.

Ng TK, Busche RM, McDonald CC and Hardy RWF (1983) Production of feedstock chemicals, Science 219, 733-740.

OTA (1980) Energy from biological processes, Office of Technology Assessment, US Congress, Washington, DC. 20510.

Pimentel D and Pimentel M (1979) Food, energy and society. London, Ed. Arnold.

Pimentel D et al (1981) Biomass energy from crop and forest residues, Science 312, 1110-1115.

Rabson R and Roberts P (1981) The role of fundamental biological research in developing future biomass technologies, Biomass 1, 17-38.

Smil V and Krowland WE, eds. (1980) Energy in the developing world: Oxford: University Press.

Strub A, Schleser G and Chartier P, eds. (1983) Energy from biomass, 2nd EC Conference, London: Appl. Sci. Publ.

UK-ISES (1976) Solar energy: a UK assessment. UK-ISES London, W1X 4BS.

Wittwer SH (1982) Carbon dioxide and crop productivity, New Sci. 95, 233-234.

World Bank (1981) Agricultural research; sector policy paper, World Bank, Washington, DC 20433.

Zaborsky OR, ed. (1981) Handbook of biosolar resources, Vols. I and II. Boca Raton: CRC Press.

Author's address: Plant Sciences Department, King's College London
                   68 Half Moon Lane, London SE24 8JF. U.K.

SPECIALLY DEVELOPED PHOTOSYNTHETIC ORGANISMS AS FUTURE POSSIBILITY FOR
LARGE-SCALE LOW-COST CONVERSION OF SOLAR RADIATION INTO USEFUL HIGH-GRADE
ENERGY

L.N.M. DUYSENS

1. INTRODUCTION

In the coming centuries humanity is faced with three major problems: the
prevention of a nuclear war, the conversion of the exponential population
increase to near-zero growth, and the replacing of the fossil energy sources
by longer lasting cheap energy sources. All three problems have to be solved
as soon as possible in order to obtain a world in which most people can lead
a life of the same or somewhat higher material comfort as now enjoyed by an
appreciable fraction of the people in the so-called developed countries.
    In this contribution I will discuss how fundamental research in photo-
synthesis and other biological sciences might help in the development of a
long lasting, low-cost large-scale energy source. The most promising sources
of this kind appear to be fast nuclear breeder and fission reactors and
solar energy converters. Since the nuclear and solar options both have uncertainties, and different advantages and disadvantages, both options have
to be studied.
    Solar energy at a scale at least equal to the present energy production,
and at an equal or lower price cannot be produced by converters of about
10 % efficiency and requiring artificial structures, such as photoelectric
cells, mirrors or glass-covered containers. The main reason is that the cost
per m^2 for artificial structures is too high, namely at least \$ $40/m^2$, which
is the price/m^2 of mass-produced greenhouses in the Netherlands with double
glass, without any heating or other installations and accessories (A. Sonne-
veld, private communication). The price per m^2 for solar cells, mirrors and
'chemical' containers and their mountings most probably will be, also when
future developments are taken into account, much higher. Assuming an energy
efficiency of 0.1 and an average flux of 0.2 kW/m^2, then an energy equivalent
of 20 l oil per m^2 per year would be produced, which would be worth about
\$ 4 at the present market price, and more when the energy is produced in the
form of electricity, but in the latter case the additional high cost of
wiring, installations and storage is to be taken into account. Taking also
into account maintenance, repair and replacement, artificial solar energy
converters are or may become only economic in applications in which the
transport of energy is expensive, such as for space crafts, or in which
expensive special materials are made, or for solar heaters, for which the
efficiency is about 0.5, if only warm water is needed. Another economic
possibility would be the development of low-cost solar cells of 50 % efficiency, combined with a low-cost wiring, storage and maintenance system,
but this may be an extremely difficult technical problem.
    In the following the use of specially developed plants as possible
low-cost large-scale solar energy converters is considered.

2. SPECIAL PLANTS AS SOLAR ENERGY CONVERTERS

2.1. Introduction

Crop plants convert of the order of 1 percent of the yearly incident solar
radiation into Gibbs' free energy of products. Under optimum conditions
larger yields are obtainable, but when plants are grown under suboptimal

Sybesma, C. (ed.), Advances in Photosynthesis Research, Vol. II. ISBN 90-247-2943-2.
© 1984 Martinus Nijhoff/Dr W. Junk Publishers, The Hague/Boston/Lancaster.

conditions, part of the energy has to be used for counteracting adverse effects. At the efficiency of 1 %, an area of about one tenth of the area of the globe would then be needed to produce about ten times the power used at the moment by humanity. The factor ten is assumed, taking into account a doubling or tripling of the human population before a steady state is reached and an increase in the average use per head to the present use in the United States.

Since such a large usable area is not available on land, the oceans, which cover about two thirds of the globe, will be considered.

2.2. Properties required for plants to be used for low-cost solar energy conversion at the surface of the oceans

Large parts of the surface of the oceans have temperatures well above zero centigrade, sufficient sunlight, and are accessible by the cheapest means of transportation, i.e. ships. Thus, if possible, plants should be developed, able to photosynthetize at the surface of the oceans and to produce useful products with about 1 % efficiency, which can cheaply be harvested.

The following proposals and ideas are not meant to be a masterplan, but rather to indicate, which problems need to be solved, and what data and methods presumably would be useful. I hope that some meteorologists, oceanographers, and scientists working in many relevant fields of biology will give their thoughts to the problems to be solved.

Except for small areas where water enriched with minerals rises to the surface, there is very little plant growth in the oceans, mainly because of the lack of the minerals phosphate and nitrogen in the surface water. There are photosynthetic organisms, i.e. certain blue-green algae (cyanobacteria) which can bind N_2 from the atmosphere by a modified photosynthetic reaction. Genetic material from these bacteria may be incorporated into the genes of the special oceanic plants (see below). These plants should be designed such that they absorb and bind very strongly the phosphate from the sea water.

One might think of plants producing a gas, e.g. methane or hydrogen, at concentrations well below the explosion point of the atmosphere. This gas might be concentrated at the places where used for combustion. However, the gases produced present in the atmosphere will stimulate fires and it may be difficult, if not impossible, to prevent bacteria from utilizing them. For these reasons the production of gases will be left out of consideration.

Thus the useful products of the oceanic plants will be assumed to be restricted to the surface of the oceans in such a form that low-cost harvesting by ship is possible. The products should not contain phosphate, since this is in short supply. Storms should not cause irreversible damage, heaping up or drifting far apart. The plants should also have built-in protection against diseases and predators. One might think of ellipsoidally shaped plants of largest dimension of the order of one centimeter, mainly consisting of a paraffin-line substance surrounded by a thin layer of cells containing the photosynthetic and metabolic apparatus. The behaviour of these plants in the ocean currents and winds may be simulated by plastic ellipsoids (of limited lifetime) containing water and a little air.

After harvesting by ship the photosynthetic products may, after some treatment, be burned on land as fuel, used as a chemical material or as food. If the product of photosynthesis is a lipid or other carbon containing material, one of the oxidation products is carbon dioxide. Because of the low concentration of carbon dioxide in the atmosphere, 0.03 percent, the rate of recirculation to the point of uptake in the oceans may well limit the maximum photosynthetic rates of the oceanic plants. This would almost certainly be

he case if larger yearly energy production than ten times the present
early energy production would be required. Calculations should be carried
ut using meteorological data, and data obtained some time ago after nuclear
xplosions, on the spread of light radioactive nuclei as a function of time
nd place, in order to estimate the recirculation rates of carbon dioxide
etween regions where the products are burned and various areas of the oceans.

If the plant products would not contain molecules with carbon atoms, but
ith hydrogen, nitrogen and oxygen atoms, only N_2 and H_2O would have to be
ecirculated, which would eliminate the recirculation limitation, but the
roblems for developing plants producing these biologically unconventional
olecules may be excessive.

. RESEARCH REQUIRED FOR DEVELOPING OCEANIC PLANTS FOR ENERGY PRODUCTION

f by means of the methods indicated in section 2.2, an acceptable size,
hape and 'stickiness' along the edges for the oceanic plants can be es-
ablished, research should be addressed to selecting and developing such
lants.

Although the second part of this problem is probably the most difficult
nd time consuming, it seems to me that it can be solved.

Within ten years after the development of the first vector for introducing
oreign DNA into the genetic system of Escherichia coli, it has been possible
o isolate, introduce and express DNA, coding for various animal proteins or
nzymatic systems, in this and other species of bacteria (see Old, Primrose,
981). A vector has already been developed for introducing and expressing DNA
n certain plant species. It is likely that analogous vectors can be develop-
d for some blue-green algae (cyanophyceae or cyanobacteria) which have a
endency to form networks. A great variety of 'energy-rich' products, including
araffins, can be made by different species of algae (see Stewart, 1974).
ther algae have high affinity for phosphates. It may be possible to intro-
uce these properties into the cyanobacteria mentioned in such a way that
hese bacteria are able to bind phosphate strongly and form a network en-
losing a core consisting of the desired product. Instead of the prokaryotic
yanobacterium, cells of an eukaryotic alga like the lettuce-like red alga
orphyra lacineata might be better as starting organism. Which organisms are
ost promising can only be established after extensive research has been done.
n addition, methods for transferring genes into a cell and inducing this cell
o differentiate into a multicellular organism, have to be further developed
nd applied successfully to more organisms than the few species of land
lants investigated so far.

Different plants are resistant to different bacterial and other diseases,
nd some plants contain substances which make them unsuitable as food for
redators. In principle these properties may be introduced by suitable vectors.
owever, unless a plant is found which requires only a few modifications,
ecades of intensive fundamental and applied biological research will probably
e needed before a plant with the desired properties can be successfully developed.

. ENVIRONMENTAL EFFECTS

In the oceanic areas used for the specially developed plants, the originally
present dilute plankton would probably disappear, because the energy producing
plants would probably lower further the phosphate concentration. However, no
species would presumably be lost, because large areas of the oceans would not
be used. On the other hand, there would be less pressure by man on natural

biotopes, such as tropical rain forests. The employment of oceanic plants as sources for energy, material and food may well make it possible to save many land species from extinction.

Burning of coal will increase the amount of carbon dioxide in the air and could probably cause undesirable increases in temperature. This process could in principle be reversed by binding part of the carbon dioxide by means of the oceanic plants in a stable form. If in the future the temperature decreases because of an ice age, which is expected, it may be possible to set free carbon dioxide by combustion of bound carbon.

5. INCREASE IN RESEARCH EFFORTS

Whatever the 'mechanic' feasibility of using oceanic plants for energy production, increased fundamental research in many fields of biology, biochemistry and biophysics, such as photosynthesis and plant development, would also be required as a basis for the improvement of quantity and quality of agricultural crops by genetic engineering. This improvement is needed because of the world-shortage of low-cost food. The increase in fundamental research can be achieved most efficiently and effectively with a planned stimulation of growth of the number of excellent students entering the undergraduate courses in these fields, followed by planned and announced expansion of advanced research. This would also speed up the development of the oceanic plants described, if this proves to be 'mechanically' feasible

REFERENCES
Old RW and Primrose SB (1981) Principles of gene manipulation, 2nd ed. Oxford: Blackwell.
Stewart WDP, ed. (1974) Algal physiology and biochemistry. Oxford: Blackwell.

Author's address: Department of Biophysics, Huygens Laboratory of the State University, P.O. Box 9504, 2300 RA Leiden, The Netherlands.

THE OUTDOOR CULTIVATION OF THE HALOTOLERANT ALGA *DUNALIELLA* - A MODEL
FOR BIOSOLAR ENERGY UTILIZATION FOR USEFUL CHEMICAL PRODUCTS.

M. AVRON

1.INTRODUCTION

Green unicellular algae of the genus *Dunaliella* possess several
unique characteristics which have made them attractive models for biosolar
energy conversion. Notable among these are:
a) An ability to grow in media ranging in NaCl content from 0.1 M to a
saturated salt solution (Ben-Amotz, Avron, 1981). b) Production of
glycerol as a major photosynthetic product and its accumulation intracellularly at a concentration iso-osmolar with the external salt concentration (Ben-Amotz, Avron, 1981). c) An ability of at least one species,
Dunaliella bardawil, under appropriate growth conditions to accumulate
β-carotene up to 10% of its dry weight (Ben-Amotz et al., 1982a). d) Lack
of a cell-wall.
In recent years our laboratory has been engaged in a detailed study of
the properties of these algae as candidates for large scale cultivation
for the commercial production of glycerol, β-carotene, dry algal meal and
possibly other products. These studies led to the construction of a pilot
plant facility where the algae are presently grown and harvested from
ponds which range in size from $5m^2$ to $300 m^2$.

2. METHODS

Cultivation conditions and general techniques have been previously
described in detail (see Ben-Amotz, Avron, 1980a;1983a;Brown et al., 1982).
To determine intracellular volume the following method was developed
(Avron, Katz, unpublished): About 80 ml of an algal suspension in the
salt concentraion in which they were grown, containing about 2×10^7
cells x ml^{-1}, was cooled to 0-4°C. LiCl was added to around 50 mM, and
^3H_2O to around 10^6 dpm x ml^{-1}. Following slow mixing on ice for 15', the
suspension was distributed among three conical tubes and centrifuged at
2000 xg for 10 min at 0°C. The supernatant was completely removed and
kept (SUP - I). The pellets were diluted 50 folds with cold distilled water
(which causes complete bursting of the cells), homogenized and allowed to
stand at 0°C for 30 min. The tubes were centrifuged at 23,000 xg for
15 min. and the supernatants kept (SUP - II). In appropriate dilutions
of SUP-I and SUP-II, tritium content was determined in a scintillation
counter, lithium and sodium contents in a flame-photometer. By assuming
that ^3H_2O is distributed evenly throughout the suspension, and lithium
only in the extracellular space (see below), the intracellular volume
of an alga and the intracellular sodium concentration was determined.

3. RESULTS AND DISCUSSION

3.1. Glycerol production.

It was previously demonstrated (Ben-Amotz et al., 1982b) that glycerol
is accumulated intracellularly to a concentration approximating isoosmolar
with the external salt concentration. Despite its relative permeability
through most artificial and biological membranes, the cytoplasmic membrane

Sybesma, C. (ed.), Advances in Photosynthesis Research, Vol. II. ISBN 90-247-2943-2.
© *1984 Martinus Nijhoff/Dr W. Junk Publishers, The Hague/Boston/Lancaster.*

of *Dunaliella* can maintain very large concentration gradients of glycerol (over 1000 fold) with little to no leakage at room temperature. Utilizing NMR to follow glycerol movement across the cell membrane, fig. 1 shows that no measurable leakage could be observed to occur over a 7 hour period at 17°C (Brown et al., 1982).

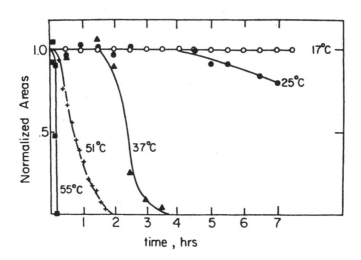

Figure 1: NMR determination of changes in intracellular glycerol content of *D. salina* with time at various temperatures.
Details in Brown et al., 1982.

The data indicates a glycerol permeability of the cytoplasmic membrane of *Dunaliella* of less than 5×10^{-11} cm x sec^{-1}, whereas the permeability of red cell membranes was determined by the same technique to be 7×10^{-8} cm x sec^{-1} and of phospholipid vesicles 5×10^{-6} cm x sec^{-1}, all at 17°C. As was previously shown (Wegmann et al., 1980) this permeability rises dramatically over a narrow temperature range, with the cells becoming completely permeable above 50°C.

To determine with accuracy the intracellular glycerol concentration a reliable method for determining the intracellular osmotic volume must be developed. This has been a field of great controversy and a variety of values can be found in the literature depending on the techniques employed (see e.g. Gimmler, Schirling 1978; Ginzburg, Ginzburg, 1981). Furthermore, since cell volume will vary with algal age, it is imperative to measure glycerol content and cell volume on algae harvested from the same growth stage. We have recently developed a technique to measure algal intracellular volume which we believe is free from most of the difficulties which beset previous methodologies. It is based on our observation that lithium ions do not penetrate into the intracellular volume, and therefore can be used to monitor the extracellular volume of algal pellets. Using this principle, measuring lithium (and sodium) concentrations in a flame photometer and tilizing relatively large amounts of algae to minimize statistical errors, onsistent values for intracellular volume, intracellular glycerol and odium concentrations were obtained (Avron, Katz, unpublished).

Figure 2 illustrates that *D. salina* when exposed in the cold to varying alt concentrations responds immediately as would be expected from a erfect osmometer. Thus, the intracellular volume is directly proportional o the inverse of the osmotic pressure. The figure also illustrates that he non osmotic volume of *D. salina* (i.e. the volume obtained by extra-olation to infinite osmolarity) is no greater than 10% of the total lgal volume of the unshocked algae, as determined by this technique. We ave thus used 90% of the algal volume determined at steady state as the ffective osmotic volume in calculations of intracellular concentration.

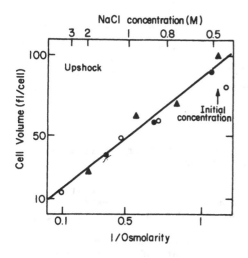

Figure 2: The immediate osmotic response of *D. salina*.

D. salina suspension grown in 0.5 M NaCl was concentrated and resuspended at 0°C in various NaCl concentrations to provide the desired NaCl concentrations. LiCl and ^3H_2O were added to the suspension, mixed and centrifuged at 0°C. Further operations were as described under Methods. The results of three separate experiments are indicated.

The intracellular volume of *D. salina* and *D. bardawil* cells grown at various salt concentrations is indicated in table I. It should be noted that cells grown on a wide range of salt concentrations possess a similar cell volume: about 90 fl x cell^{-1} (femtoliter=10^{-15} liter). These values are similar to the approximate values previously estimated from microscopic observations (Ben-Amotz, Avron, 1980b). Table 1 further illustrates that approximately 30% of the pellet obtained following centrifugation at 2000 xg for 10 minutes is accounted for by the intracellular space of the cells, and that only a negligible concentration of salt (relative to the growth medium) is present intracellularly (see Brown, Borowitzka, 1979).

When glycerol content was measured chemically or through NMR in cells grown in 1.5 M NaCl, the intracellular concentration amounted to 1.9±0.2 M which is osmotically equivalent to 1.25 M NaCl. However, other soluble intracellular metabolites, notably glutamate, glucose, alanine and dihy-droxyacetone, which were detected by ^{13}C-NMR techniques, accounted for the lacking 0.25 M NaCl (Degani, Sussman, Avron, unpublished).

Table 1: Intracellular volume and Na$^+$ concentration in *Dunaliella*.

Growth-Medium NaCl Conc.	*D. bardawil*			*D. salina*		
	Algae Vol. in pellet	Intracellular Na$^+$ conc.	Cell Vol.	Algae Vol. in pellet	Intracellular Na$^+$ conc.	Cell Vol.
(M)	(fraction)	(M)	(fl)	(fraction)	(M)	(fl)
0.5	-	-	-	0.33	-0.03±0.08	87
1	0.29	0.02±0.07	600	-	-	-
1.5	0.28	0.02±0.02	730	0.38	-0.04±0.10	84
2	0.22	0.08±0.10	520	0.35	0 ±0.04	77
3	0.23	0.1 ±0.10	580	0.34	-0.1 ±0.15	100
4	0.33	0.1 ±0.15	600	0.30	-0.1 ±0.15	98
average	0.27	0.06±0.09	600	0.34	-0.05±0.10	90

Details as described under Methods. Cells were harvested in the mid-logaritmic phase (Avron, Katz, unpublished).

3.2. β-carotene production

Dunaliella species, when grown under normal laboratory cultivation appear green and do not accumulate β-carotene beyond the amounts normally found in green algae. However, at least one species has been long observed in nature to accumulate massive amounts of β-carotene (Aasen et al., 1969). We have recently isolated such a species, named it *Dunaliella bardawil*, and studied its properties in detail (Ben-Amotz et al., 1982b; Ben-Amotz, Avron, 1983b).

A long series of experiments designed to investigate the conditions which lead *D. bardawil* to accumulate massive amounts of carotene, led to the generalization that the carotene to chlorophyll ratio in the alga depends on the integrated irradiance which the alga receives during one division cycle. Thus the higher the light intensity (extending well beyond saturation of photosynthesis), and the longer the doubling time, the higher is the carotene to chlorophyll ratio. Fig. 3 illustrates a summary of such experiments, where the chlorophyll content per cell, and the carotene content per cell are plotted in addition to the carotene to chlorophyll ratio, as a function of the total irradiance received per division cycle. The points represent experiments where light intensity or period was changed and where doubling time was varied by varying either salt concentration, or growth temperature. Similar results were obtained by limiting the supply of an essential nutrient such as nitrate or sulfate.

Under the conditions where *D. bardawil* accumulates large amounts of β-carotene, the carotene is concentrated within lipoidal globules located exclusively in the interthylakoid space within the algal chloroplast. A procedure was developed for the isolation and purification of these globules. They were found to contain over 60% β-carotene with the rest being mostly neutral lipids (Fried et al, 1982). The β-carotene was composed of approximately equal amounts of the all-trans and the 9-cis isomers.

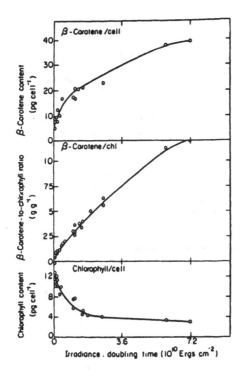

Figure 3: Dependence of pigmentcontent of *D. bardawil* on the integral irradiance per division cycle.
For detals see Ben-Amotz, Avron, 1983b.

3.3. Metabolism and Osmoregulation.

Based mostly on the discovery of several unique enzymes which are present in *Dunaliella* and which seem to be involved in the metabolism of glycerol, a "glycerol cycle" was previously proposed which suggested pathways for glycerol accumulation and elimination during the osmoregulatory process (Ben-Amotz et al., 1982b). A more detailed metabolic pathway based on our more recent NMR studies (Degani, Sussman, Avron, unpublished) is illustrated in fig. 4. Major pathways are marked by heavy arrows, and compounds detected by ^{13}C-NMR by italics.

Following a hyperosmotic shock α-(1-4)glucan is suggested to be degraded via the pentose-phosphate pathway, through ribulose-5-P, dihydroxyacetone-P and glycerol-P to glycerol. Following a hypoosmotic shock, glycerol is converted via dihydroxyacetone, dihydroxyacetone-P fructose-diphosphate, glucose-1-phosphate, and ADP-glucose to α-(1-4)glucan. In this manner the osmotically active glycerol is converted to an osmotically inactive polysaccharide during hypoosmotic osmoregulation, and vice-versa during hyperosmotic osmoregulation.

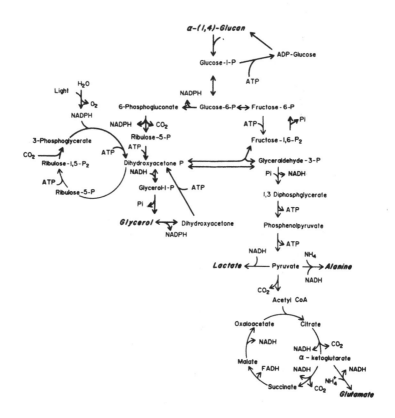

Figure 4: Hypothetical scheme of the metabolic pathway involved in glycerol metabolism in *Dunaliella*.
Heavy arrows indicate major pathways; Italics indicate compunds detected by ^{13}C-NMR.

Figure 5 illustrates ^{13}C-NMR spectra of shole cells and cell-free extracts of *D. salina* before and 120 minutes following a hypoosmotic shock. Quantitative analysis of such spectra in terms of changes in the intracellular content of the observed intermediates is shown in Fig. 6.

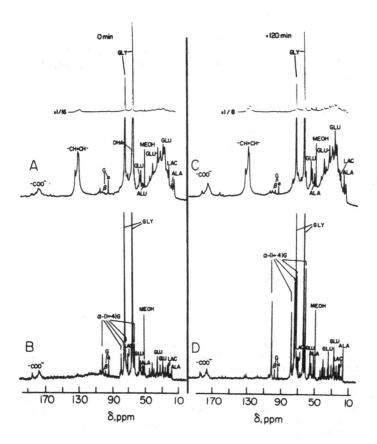

Figure 5: ^{13}C-NMR spectra of *D. salina* before and after a hypoosmotic shock.

Cells grown at 1.5 M NaCl were diluted at 0 min. to 0.75 M NaCl and incubated in the light for a further 120 min. A and B represent spectra of whole cells and cell-free extracts, respectively, taken at 0 min. C+D represent similar spectra taken 120 min. following the hypoosmotic shock. COO⁻, carboxylate signal; -CH=CH-, signal due to unsaturated membranal carbons; G, glucose; Gly, glycerol; DHA, dihydroxyacetone; GLU, glutamate; ALA, alanine; MEOH, methanol (external marker); LAC, lactate (Degani, Sussman, Peschek, Avron, unpublished).

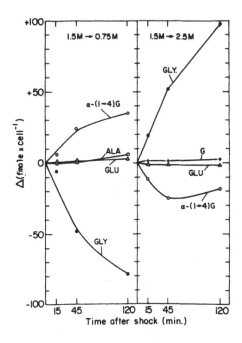

Figure 6: Temporal changes of the intracellular content of metabolites following a hypotonic or a hypertonic shock. α-(1-4)G, α-(1-4)glucan expressed in glucose units (Degani, Sussman, Peschek, Avron, unpublished).

As can be seen, following a hypoosmotic shock essentially the total decrease in glycerol content can be accounted for by the increase in α-(1-4)glucan content (expressed as glucose units). Smaller increases in alanine and glutamate content were also observed. The amount of glycerol produced during the first 45 minutes following a hyperosmotic shock was also mostly accounted for by the decrease in the content of α-(1-4)glucan. However, the later synthesis of glycerol must be due mostly to photosynthetic CO_2 fixation.

3.4. Commercial cultivation

Due to its unique ability to produce simultaneously high quantities of commercially interesting products, *Dunaliella bardawil* is a very attractive plant for the study of the photosynthetic collection and conversion of solar energy. In addition, it has the advantage of being able to propagate optimally in saline media, with little to no competition from other algae or predators. Also, brackish or sea water is often available near deserts where solar energy is plentiful and the land cannot be usefully farmed.

Experiments have been carried out in recent years to test the possibility of large scale outdoor cultivation and harvesting of such algae. So far, three products have been isolated from the collected algal biomass: glycerol, β-carotene and the remaining algal-meal which contains about

0% protein. The latter is similar in composition to soybean-meal and
.as been used successfully on a small scale as feed for fish and fowl.
Observed maximal productivity thus far over long periods (several
.onths) in outdoor ponds were about 20g algae x m^{-2} x day^{-1} which contained
.bout 8 g glycerol and 0.6 g β-carotene. These values amount to approxi-
.ately half of the theoretically calculated maximal values assuming 5%
.otal efficiency of solar energy conversion to stored products (Ben-Amotz,
.vron, 1981).

.EFERENCES

.asen, AJ, Eimhjellen, E and Liaaen-Jensen S (1969) An extreme source of
.-carotene, Acta. Chem. Scan. 23: 2544-2545.
.en-Amotz A and Avron M (1980a) Glycerol, β-carotene and dry algal meal
.roduction by commercial cultivation of *Dunaliella*. In Shelef, G and
.oeder, CJ eds. Algae Biomass, pp. 603-610, Elsevier, Amsterdam.
.en-Amotz A and Avron M (1980b) Osmoregulation in the halophilic algae
.unaliella and *Asteromonas*, In Rains DW, Valentine RC and Hollaender A
.ds. Genetic Engineering of Osmoregulation, pp. 91-99, Plenum Publishing
.orp. N.Y.
.en-Amotz A and Avron M (1981) Glycerol and β-carotene metabolism in the
.alotolerant alga *Dunaliella*: A model system for biosolar energy conversion.
.rends in Biochemical Science 6(11): 297-299.
.en-Amotz A Katz A and Avron M (1982a) Accumulation of β-carotene in halo-
.olerant algae: purification and characterization of β-carotene-rich
.lobules from *Dunaliella bardawil (chlorophyceae)* J. of Phycol. 18, 529-537.
.en-Amotz A Sussman I and Avron M (1982b) Glycerol production by *Dunaliella*,
.xperientia, 38: 49-52.
.en-Amotz A and Avron M (1983a) On the factors which determine massive
.-carotene accumulation in the halotolerant alga *Dunaliella bardawil*.
.lant Physiol. in press.
.en-Amotz A and Avron M (1983b) Accumulation of metabolites by halotolerant
.lgae and its industrial potential. Ann. Rev. Microbiol., 37, 95-119.
.rown AD and Borowitzka LJ (1979) Halotolerance of *Dunaliella*. In
.evandowsky M and Hunter SH eds. Biochemistry and Physiology of Protozoa,
.ol. 1 2nd edition, pp. 139-190, Academic Press, N.Y.
.rown FF, Sussman I, Avron M and Degani H (1982) NMR studies of glycerol
.ermeability in lipid vesicles, erythrocytes and the alga *Dunaliella*.
.iochim. Biophys. Acta, 690, 165-173.
.ried A, Tietz A, Ben-Amotz A and Eichenberger W (1982) Lipid composition of
.he halotolerant alga *Dunaliella bardawil*, Biochim. Biophys. Acta 713,
.19-426.
.immler H and Schirling R (1978) Cation of the permeability plasmalemma of
.he halotolerant alga *Dunaliella parva*. II. Cation content and glycerol
.oncentration of the cells as dependent upon external NaCl concentration.
.. Planzenphysiol. Bd. 87(5), 435-444.
.inzburg M and Ginzburg BZ (1981) Interrelationships of light, temperature,
.odium chloride and carbon source in growth of halotolerant and halophilic
.trains of *Dunaliella*. Br. Phycol. J. 16, 313-324.
.egmann K, Ben-Amotz A and Avron M (1980) The effect of temperature on
.lycerol retention in the halotolerant algae *Dunaliella* and *Asteromonas*
.lant Physiol. 66, 1196-1197.

.uthor's address: Department of Biochemistry, Weizmann Institute of Science,
                  Rehovot, Israel 76100.

INDUSTRIAL MASS ALGAE CULTURES IN LUKE-WARM WATER.

E. DUJARDIN
Photobiology Laboratory, Liège University, B22.
4000 Sart-Tilman, Belgium.

Marine or fresh-water algae are known to grow with a very high rate when favourable conditions are present. The temperature, the light intensity on the algae suspension, the stirring or the water flow, which stimulates the uptake of minerals and the gas exchanges, the qualitative and quantitative composition of the nutrient medium, its pH, its total ion concentration..., all these factors are essential for achieving high yields in mass algae cultures. Yields between 15 to 100-120 tons of dry biomass/hectare (ha).year may be obtained in various climates (Table I).

TABLE I. Yields obtained in mass algae cultures in various climatic conditions (dry tons/ha.year).

Alga	Country	Yield	Protein %
Hydrodictyon	Belgium	12-18	20
Chlorella	Belgium	23-30	50
Chlorella	Japan	100	50-60
Microactinium	California U.S.	60	30
Scenedesmus	Bulgaria	30	50
Spirulina	Mexico	30	60

These yields are about 10 to 20 times higher than those obtained in the traditional cereal cultures, which are usually between 2 to 10 tons/ha.year. But the algal biomass is also rich in several constituents. Among them, the protein content, is particularly important; it often reaches 50% of the dry biomass.
Most of the algal culture plants require relatively high investment. It is, however, possible to decrease the production costs by using, separately of simultaneously, the following biotechnologies:
1. Biodepollution of chemicals by utilization of
   a) non-toxic mineral and organic wastes as fertilizers for the algae,
   b) CO_2 produced by industries or other sources as an additionnal carbon source.
2. Thermal depollution by the utilization of the calories rejected by industries to increase the yield of algal biomass.

1. BIODEPOLLUTION OF CHEMICALS.

The general scheme (I) for photosynthesis suggests directly the possibility of replacing the expensive fertilizers by mineral and organic pollutants. Additionnal CO_2 released from industries may be used to increase the biomass yield. Solar energy is stored into the chemical bonds

Sybesma, C. (ed.), *Advances in Photosynthesis Research, Vol. II. ISBN 90-247-2943-2.*
© *1984 Martinus Nijhoff/Dr W. Junk Publishers, The Hague/Boston/Lancaster.*

SCHEME I

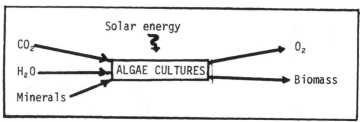

of molecules which constitute a nice raw material for many industrial
uses; at the same time, oxygen is released in the environment. All these processes are integral parts of the natural carbon,oxygen, mineral
and water cycles on which life depends.

Many minerals rejected in rivers or in the sea.. are adequate fertilizers for mass algae cultures. This is the case for nitrates, phosphates,
sulfates, carbonates, metal ions, etc... whenever the algae suspension
requires complementary salts. Many organic wastes are also assimilated
by various algae species: urea, small organic acids or their salts,etc.
High-molecular compounds are often assimilated after having been degraded into small molecules.Degradation processes are favoured by light
and by the intense photosynthetic emission of oxygen.

For example, we have successfully grown microalgae as *Scenedesmus* on
wastes from the wool industry, from animal breeding etc... We have also
fed the same algae with other organic wastes on the 130 m² pilot plant
of the Photobiology Laboratory in Liège. This plant is composed of four
identical 32 m² culture surfaces, devised to insure a very intense agitation of the algae suspension. In this way, the algae are evenly illuminated, the gas exchanges are optimized, and the algae are not allowed to be deposited at the bottom of the aqueous layer. Parts of the
20000 liters of algae suspension are regularly centrifuged. The dry
biomass yield amounts to 20-30 tons/ha.year in the belgian climatic conditions (= 5 to 6 months harvesting/year). The composition of these algae is presented in Table II.

TABLE II. Composition of *Scenedesmus* and *Hydrodictyon* (% dry weight)

	Scenedesmus obliquus	*Hydrodictyon reticulatum*
Proteins	51	14-20
Lipids	15	5-7
Soluble sugars	5	5
Cellulose	10-13	10-13
Pigments	1	0.5
Moisture	6-8	6
Minerals	9-14	30-50[*]

* of which the main part is calcium carbonate.

The total protein produced is of high nutritional value for humans or animals since it contains a high lysine(5-6%) and methionine ($\pm$ 2%) ratio.

2. THERMAL BIODEPOLLUTION.

A number of industries pollute thermally the environment; for example industries such as: metallurgical factories, agroindustries, food industries, classical and nuclear power plants, etc...
The recuperation of the calories from luke-warm water is not yet solved really in an economical way by classical means. But it is well known that the plants are sensitive to temperature. Their growth and development usually increases with temperature. In figure 1, the CO_2 uptake by the alga *Scenedesmus* is shown to double when the temperature is increased by 10°C, between 15 and 35°C.
The luke-warm water is therefore of particular interest as a means to increase the temperature of algae suspensions. We will consider separately the case of macroalgae cultures and microalgae cultures.

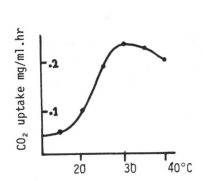

Figure 1. *Temperature dependence of photosynthesis in Scenedesmus.*

2.1. Mass cultures of macroalgae in luke-warm water.

One of the cheapest techniques for increasing the production of biomass is to grow macroalgae in luke-warm water. At Tihange (Belgium), the situation is offered to test this possibility: the chemically polluted water of the River Meuse is heated in the third cooling circuit of the nuclear power plant. This water has a temperature of 10-12°C higher than that of the river, and is used to irrigate shallow lagoons.
The macroalgae *Hydrodictyon reticulatum*, are fed with the mineral and organic pollutants of the river; they multiply at a high rate from April until October. Yields are 50-100 % higher in heated ponds than in non-heated ponds, and they amount to 12 to 18 tons dry biomass/ha.year in the climatic belgian conditions (i.e. 6-7 months culture).
The composition of the dry *Hydrodictyon* biomass is presented in Table II. The high mineral content influences the ratio of the other components, when comparing to *Scenedesmus*.This is due to the excess of calcium carbonate in the river. The protein of *Hydrodictyon* is also valuable for animal feeding. Methane may be produced by anaerobic digestion from the raw biomass (Legros et al, 1982).
In this lagoon system the thermal depollution is not achieved by the algae, because the amount of algae biomass is very small in comparison with the amount of luke-warm water flowing through the lagoons. This water is slowly cooled in the lagoons by evaporation of part of it,and by heat conduction to the ground.

2.2. Mass cultures of microalgae heated by luke-warm water.

The algae culture plant of Sart-Tilman was not devised for heating microscopic algae. However, the effect of temperature is easily distinguished by considering the harvest frequency of the same culture, during cold and comparable warm periods. From these observations, it may be concluded that during warm periods (22 to 32°C), the yield of algal biomass may exceed two to three times the yield obtained during colder periods (15 to 22°C), even if the sky is cloudy. This is easily understood if one considers that in Belgium, the solar energy is still near the half of that in tropical regions. On the other hand, the chloroplast lamellae of *Scenedesmus* are particularly well adapted to diffuse light as they develop large pigment antenna.

In order to take advantage of the residual calories from luke-warm water for increasing the yield of micro-algae cultures, we have devised an original heat-exchanger able to warm up a cyclic open-air culture of micro-algae like the Sart-Tilman culture.
The principle of the heat exchanger is very simple: a double circulation system is functioning on each side of a surface made of a heat conducting material. The calorie-rich fluid circulates under the surface and the algae suspension flows above the exchanger surface as represented in figure 2.

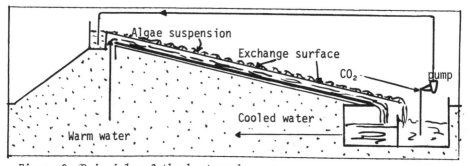

Figure 2. Principle of the heat exchanger.

Thanks to the intense stirring of the algae suspension, the heat transfer from the luke-warm water to the algae suspension is very efficient. Furthermore, the high evaporation of water from the suspension is continuously cooling again the algae suspension during its descending flow along the exchange surface. The whole installation constitutes therefore, a very efficient tool for thermal depollution.

For example, a 1500 m² exchanger (15 m x 100 m) will decrease the temperature of the luke-warm water from 30°C to 22-23°C, while the algae suspension will be heated from 15 to 17-19°C, depending on the wind velocity as can be seen in Table III.

It is important to notice that the heat exchanger allows also the recuperation of calories from warm water, polluted by toxic compounds, since there is no communication between the two fluid circuits.

TABLE III. Some properties of the heat-exchanger.

Temp.of the air, °C	Air velocity,m/s	Initial culture	Temp.,°C warm water	Temp. of culture,°C	Power KW
5	5	15	30	17-18	1550
5	2	15	27	19	1000

CONCLUSIONS.

   Scheme II summarizes the most striking features presented here. Intensive algae cultures may be of economical interest if:

1. the investments and the production costs are lowered by coupling the production of biomass to thermal depollution and biodepollution of chemicals; this is feasible in both macro- and micro-algae cultures.

2. high-value commercial products are extracted from the biomass (natural dyes, stabilizers, for cosmetics, pharmaceuticals etc...)

3. the by-products are used either to feed animals or to produce biomethane and fertilizers.

SCHEME III.

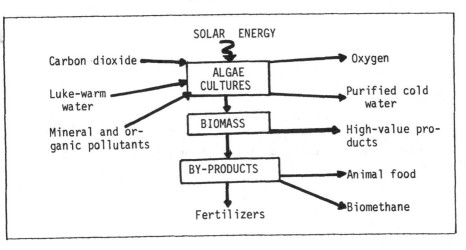

REFERENCES

Dujardin E. (1981) Récupération des déchets et des chaleurs résiduelles pour augmenter le rendement d'une culture énergétique. In "Actes des Journées Européennes de Bioénergie" (R.Buvet et J.P. Massué, Eds).Entretiens Ecologiques de Dijon,Cahier trimestriel n° 9 , 63-66.

Dujardin E., Delcambe P., Lacrosse C. and Sironval C. (1982) Biomass produced by fresh water algae in luke-warm water in "Energy from Biomass" series E vol.3 (Grassi G. and Palz W. eds) Reidel D. Publishing Company Dordrecht, Holland, 135-141.

Dujardin E.C. and Devreux R.A.G. (1981) Echangeur de chaleur pour la culture et l'élevage d'organismes aquatiques; installation et procédé mettant en oeuvre cet appareil. Belgian Pat. nr 890, 359, Sept 15th 1981.

Dujardin E., Piron C., Lacrosse C. and Sironval C. (1982) Chemical composition of the algal biomass (*Hydrodictyon*) collected in luke-warm water at Tihange. in "Energy from Biomass" 2d E.C. Confer. (Strub,A., Chartier, P. Schleser,G. eds) Appl. Sc. Publ. London LTD , 1092-1097.

Legros A., Asinari Di San Marzano C.-M , Naveau H. and Nijns E.J. (1982) Improved methane production from algae using 2nd generation digesters. in "Energy from Biomass" 2d E.C. Conf. (Strub,A., Chartier,P., Schleser, G. eds) Applied Sc. Publ. London LTD,609-614.

Piron-Fraipont C., Dujardin E., Sironval C. (1981) Increasing biomass for fuel production by using waste luke-warm water from industries. In "Energy from Biomass" Ist E.C. Confer. (Palz,W., Chartier,P., Hall, D.O., eds). Applied Sc. Publ. LTD London, 703-708.

ACKNOWLEDGMENTS

We thank the E.E.C. (contract nr ESE-R-25-B(G)), the F.N.R.S.(Fonds National de la Recherche Scientifique Belge) and the Ministère de la Politique Scientifique ("Actions concertées n° 80/85-18.1983") for their financial support.

ENVIRONMENTAL EFFECTS OF ETHANOL AND METHANOL PRODUCTION FROM BIOMASS

HANS EGNEÚS

1. INTRODUCTION

Biomass can be used as an energy resource either directly or after conversion to gaseous or liquid fuel. There are several possibilities for producing liquid fuels from biomass. Biomass can for example be directly lique-
fied to hydrocarbon fuel(so called synthetic fuels), oil-producing plants
can be used after treating the oil or alcohols such as ethanol and methanol
can be produced by using biological or non-biological processes. The fact
that transport, residential heating and electricity production systems
all are(or can be) based on liquid fuel utilization makes a production of
liquid fuels from biomass of the largest interest. In the case of vehicle
propagation large scale use of ethanol has already been introduced in
Brazil, and in U.S.A. a 4% blend of ethanol in gasoline is now commonly
in use.
The main drawbacks for a utilization of alcohols produced from biomass
are to be found in the economic and environmental areas.There are problems
of technical nature e.g. what techniques should be used when cellulose-
containing plants are used for ethanol production,which gasifier technology
should be used etc. But analysing what environmental effects a liquid fuel
production from biomass can have is of the outmost importance when an extended use of biomass for energy use is discussed.
In this paper I will discuss the possible environmental effects associated
with the production of methanol and ethanol from biomass.Only methanol
production via gasification and ethanol production from cellulose-containing plant material will be considered.

2. SYSTEM CHARACTERISTICS

When discussing environmental effects of a technical system one should be
aware of that the data base which is used for an effect prediction in
most cases is fragmentary because:
a.Characterization of emissions and waste streams for the complete chain
from biomass production to end use seldom is available
b.When a chemical characterization has been made, treatment of waste streams
are often not included in process descriptions
c.If a final emission has been chemically/biologically characterized data
on toxicity or environmental effects are often missing
d.Toxicity data are difficult to use for predicting environmental effects.
I will therefore concentrate on some specific steps of the production
processes for ethanol and methanol where environmental or health effects
could be high if no action is taken. The environmental effects of growing
and harvesting the biomass will not be treated.
In Table 1 som system characteristics essential for understanding where
risks can be generated in three alcohol-producing systems are compared
with each other. Methanol production via gasification of coal has been
included to show the difference between a fossil and biomass resource.
Methanol production leads to whole-resource utilization something which
today is not possible with existing techniques for ethanol production
from cellulose-containing plants. The waste streams from ethanol production

Sybesma, C. (ed.), Advances in Photosynthesis Research, Vol. II. ISBN 90-247-2943-2.
© 1984 Martinus Nijhoff/Dr W. Junk Publishers, The Hague/Boston/Lancaster.

will be larger than those from methanol production.The chemical composition of the ethanol waste streams will be quite complex but of biological nature.The methanol waste streams will also be chemically complex with many often toxic organic compounds.

TABLE 1. A comparison between the characteristics of three alcohol production systems.

	RAW MATERIAL
Ethanol production from biomass and Methanol production via gasification of biomass	Low energy content/ m^2 space and m^3 resource.The resources are spread over large areas.The amounts of biomass available are dependent on regeneration time of the biomass,often with a seasonal availablity.Complex chemical composition which varies with the growth cycle of the resource.Biomass normally contain small amounts of heavy metals and sulfurand has a low ash content.The amount of nutrients are dependent on type of biomass and time of the year.Environmental effects mainly on soil and water compartments of the ecosystems.
Methanol production via gasification of coal	High energy content/m^2 space and m^3 resource.Resource localized to specific areas.Coal can contain varying often large amounts of heavy metals and sulfur,large ash residue.Good availability the year round.Environmental effects large at the site of production.Effects all parts of the ecosystem
	TECHNIQUE FOR PRODUCTION OF ENERGY CARRIER
Ethanol production from biomass	The techniques are either based on biological or a mixture of physico/chemical and biological processes based on a strong control of all material streams.Medium to low process temperatures.Processes are highly dependent on the chemical composition of the resource.The composition of material streams are a mixture of biologically produced(changed) compounds.Large amounts of waste are generated.
Methanol production via gasification of biomass and coal	Physico/chemical processes occuring at high temperatures.The techniques are based on a strong control of all material streams. The processes work relatively independent of the chemical composition of the resource.The chemical composition of the material streams are complex due to the different gasification techniques, many toxic compounds are produced.
	ENERGY RELATIONS
Ethanol production from biomass	Large extra inputs of energy among other things due to the distillation step.Only part of the energy content of the resource converted to energy carrier.
Methanol production via gasification of biomass and coal	Small extra energy input to the process.An almost complete conversion of the resource to energy carrier.
	MATERIAL STREAMS
Ethanol production from biomass	Low utilization of the plant material for energy carrier production.High input of extra material in the processes.Large amounts of wastes are obtained.The wastes can generally be biologically treated.Much of the residue is of economic interest.
Methanol production via gasification of biomass	Good utilization of the plant material for energy carrier production.Low input of extra material in the processes.Waste streams with a complex chemical composition containing many toxic compounds.Treatment of waste water a problem.Environmental effects of liquid wastes potentially high.
Methanol production via gasification of coal	As above for methanol from biomass.The residues after gasification of coal can contain large quantities of heavy metals and sulfur with accompanying problems with the deposition of solid wastes.Net carbon dioxide emission to the atmosphere.Large quantities of ash must be deposited.

3. PRODUCTION SYSTEMS AND ENVIRONMENTAL EFFECTS

Two process chains one for methanol and one for ethanol production are presented in Table 2. Schemes like these are essential for pin-pointing where risks are generated in energy producing systems. The main process steps have been included in the outline in the table and the two examples represent reasonable alternatives for alcohol production. Note that in an ethanol production unit one has to chose between either enzymatic or acid hydrolysis. In a longer time perspective the process chains could be dramatically changed due to the introduction of new concepts for alcohol production. Our analysis show that six areas are of special interest from environmental point of view.

TABLE 2. An outline of the environmental agents generated in the production of ethanol and methanol from biomass.

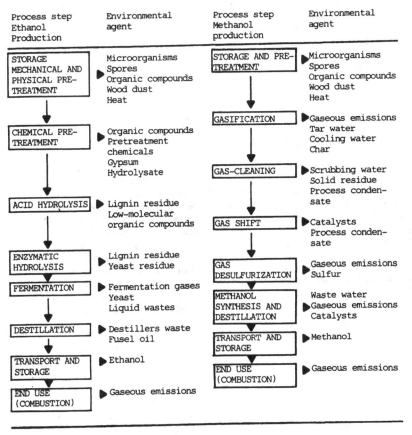

Process step Ethanol Production	Environmental agent	Process step Methanol production	Environmental agent
STORAGE MECHANICAL AND PHYSICAL PRE-TREATMENT	Microorganisms Spores Organic compounds Wood dust Heat	STORAGE AND PRE-TREATMENT	Microorganisms Spores Organic compounds Wood dust Heat
CHEMICAL PRE-TREATMENT	Organic compounds Pretreatment chemicals Gypsum Hydrolysate	GASIFICATION	Gaseous emissions Tar water Cooling water Char
ACID HYDROLYSIS	Lignin residue Low-molecular organic compounds	GAS-CLEANING	Scrubbing water Solid residue Process condensate
ENZYMATIC HYDROLYSIS	Lignin residue Yeast residue	GAS SHIFT	Catalysts Process condensate
FERMENTATION	Fermentation gases Yeast Liquid wastes	GAS DESULFURIZATION	Gaseous emissions Sulfur
DESTILLATION	Destillers waste Fusel oil	METHANOL SYNTHESIS AND DESTILLATION	Waste water Gaseous emissions Catalysts
TRANSPORT AND STORAGE	Ethanol	TRANSPORT AND STORAGE	Methanol
END USE (COMBUSTION)	Gaseous emissions	END USE (COMBUSTION)	Gaseous emissions

3.1. Storage of plant material

This step is the same in methanol and ethanol production systems. Biomass will have to be stored in any production system in either fragmented or non-fragmented form. During storage a breakdown of the plant material will take place with a subsequent release of hydrolysed compounds e.g. acetic acid, formic acid. Many types of microorganisms can use the plant material as substrate for growth. The gaseous emissions and the spore produc- ction from microorganisms is considered a serious health risk for people working with biomass. In Table 3 some examples of complaints and etiologic agents are given. One factor which will determine if fragmented wood can be used for small scale burning in temperate regions will depend on whether these problems can be solved. Selfheating and igni- tion of stored chips can be a serious problem.

TABLE 3. Examples of allergic alveolite triggered by spore-producing microorganisms

Complaint	Biomass	Etiologic agent
Farmer's lung	Grain, straw	Micropolyspora faeni, Thermoactinomyces vulgare, Aspergillus fumigatus and others
Chip boiler's complaint	Mouldy chips	Rhizopus sp., Mucor sp., Aspergillus fumigatus and others
Brewer's lung	Grain	Aspergillus clavatus and others
Sauna bather's complaint	Mouldy wood	Pullaria pullulans, Pea- cilomyces sp. and others

Modified after Holmberg, Kallings (1980)

3.2. Waste-water streams obtained during gasification of biomass

The most serious environmental problems in methanol production from bio- mass are associated with the waste-water streams obtained in several of the production stages. The chemical composition of the waste-water will depend on which type of gasifier and temperature/pressure regime is used. Gasifiers used today will generally give a waste-water containing large amounts of organic and inorganic compounds, an example of quantities and crude composition is given in Table 4. Heavy metals and polyaromatic hydrocarbons are found in varying quantities which depend on the chemical com- position of the raw material and gasification conditions. Many of the components in the waste-water will be toxic to most organisms which makes a biological treatment of the waste- water hard to accomplish. The COD of gasifier waste- water is high, reported values from wood gasifiers are in the range 40-65g/l. Existing waste-water treat- ment technology can clean such waters if the problems with the toxic chemicals

TABLE 4. Calculated chemical composition of the waste water from a Purox-gasifier: capacity 250 tons wood dry weight/day

Compound	kg/h
Methanol	98.1
Ethanol	38.2
Acetone	38.2
Methylethylketone	7.5
Acetic acid	79.3
Butyric acid	7.6
Propionic acid	30.6
Furfural	38.4
Phenol	7.7
Benzene	4.6
NH_3/NH_4^+	9.7
$H_2S/HS-$	0.044

re solved. Gasifiers working at high temperatures or gasifiers with a thermal or catalytic cracking stage will give a waste-water with considerable less organic compounds than the figure given above.

.3. Solid residues obtained after gasification of biomass

The char obtained after gasification will contain most of the nutrients and metals of the raw material. In Table 5 a calculated total turnover of metals from different types of biomass is shown. Many inorganic components will be gasified e.g cadmium, mercury, sulfur, nitrogen. These components will appear in waste-water streams or be emitted into the atmosphere at pressure relieving steps. Most data on the effect of deposition of ash residues from biomass show that gasifier char should not be considered as a pollutant if not Cd-containing phosphate fertilizer has been used. The heavy metal content of the ash makes some type of monitoring of the deposited material necessary. Some plants e.g common reed would give large amounts of ash with a deposition problem.

TABLE 5. A calculated turnover of some metals after producing 500 000 tons of methanol from different types of biomass

Metal tons	Softwood	Hardwood	Short rotation forest	Peat
Cadmium	0.11-0.37	0.066-0.28	0.47-1.9	0.05-48
Lead	2.4-3.1	0.52-5.1	0.57-1.7	2-530
Mercury	0.011-0.02	0.011-0.02	0.028-0.077	0.05-0.13
Nickel	0.79-1.3	0.33-0.98	0.52-1.5	10-2200
Chromium	2.1-2.5	0.21-2.0	0.24-1.7	1.1-3300
Cobolt	0.055-0.09	0.055-0.33	0.055-0.55	0.11-110
Manganese	13-213	17-180	5.5-48	11-2400
Copper	2.9-10	2.1-4.5	5.0-25.0	1.1-3300

3.4. Wastes obtained after pre-treatment of cellulosecontaining plants

There are many methods, physical and chemical, for pre-treating cellulose-containing so that the cellulose is separated from the lignin-hemicellulose-cellulose complex. Depending on which type of pre-treatment that is used waste streams with a specific chemical composition containing large quantities of lignin and hemicellulose will be generated (Table 6). Cellulose can be hydrolysed using either enzymes or acids. Acid hydrolysis will give a waste stream containing neutralization chemicals (Table 6). Enzymatic hydrolysis will give a waste water containing microorganisms from enzyme production. The waste streams from these stages give very large amounts of material which must be used in order to make ethanol production from cellulose-containing plants economic. Such a use will make the environmental impact of this waste negligible.

3.5. Wastes obtained at fermentation and destillation of ethanol

The quantities of liquid wastes obtained in these steps are very large (Table 6). The detailed chemical composition of the liquid wastes here has not been given in the literature, but the composition will be dependent on type of fermentative organism and technique used, but only biologically produced compounds will be found in the waste water.

TABLE 6. Waste streams in the GIT-process for ethanol production
The figures are based on a production of 75 000 tons of ethanol

Waste stream tons	Wet weight	Dry weight	Comment
Bark	98 000	49 000	Bark and lignin are used as fuels. Ash residue 6000 tons which must be deposited
Lignin	320 000	95 000	
Centrifuge sludge	82 000	44 000	46% H_2O, 16% $CaSO_4$, 18% cellulose, 13% lignin, 3% inorganic compounds. For deposition
Fermentation and stillage wastes	170 000	88 000	For use a fodder
Waste water sludge	40 000	8 000	Waste water obtained from the different waste streams. For deposition
Total for deposition		58 000	

The organic content of the waste water is high with a COD and BOD_5 in the ranges of 15-175 g/l and 7-95 g/l respectively being reported. The production of 100 tons of ethanol in an acid hydrolysis process as the one described in Table 6 will give a BOD of 20 tons and COD of 40-50 tons. This corresponds to the BOD and COD of an untreated household waste water from about 750000 persons/day. The necessity of high-rate high-quality water treatment is obvious if a negative environmental impact is to be avoided.

3.6. Emissions after burning ethanol and methanol

Blends of alcohol and gasoline or pure alcohols can be used for vehicle propagation. Alcohol gasoline mixtures containing up to about 20% alcohol can be used after only slight modificationsof the Otto-motors used today. When pure ethanol or methanol are used much larger changes have to be made of the motors of the cars. Blending gasoline with ethanol and methanol will change the emission pattern from the cars. As the two alcohols are octane boosters lead can be taken out of gasoline. It is difficult to get an exact emission picture after adding alcohol to gasoline because the emission pattern will depend on carburator setting, how the car is driven, the age of the motor, the temperature when starting etc. In Table 7 a chemical characterization of the main compounds emitted in gasoline and alcohol/gasoline fueled cars. The main changes are that the quantities of nitrogen oxides is decreased. Hydrocarbon emissions are on the same level, but a large change in chemical composition is observed. Generally the amounts of polyaromatic hydrocarbons seem to decrease, while the amounts of the two alcohols , formaldehyde, acetaldehyde, methylnitrite and ethylnitrite increase. Qualitatively the same changes are observed with pure alcohols as fuel. Formaldehyde and methylnitrite must be viewed as serious pollutants as both have been indicated as having carcinogenic properties. Otherwise the mutagenecity of the

he emissions from alcohol-fueled cars seem to be lower than that from
diesel or gasoline-fueled cars. The changed emission composition will
lead to changes in the "smog-forming" ability of car emissions. The
data presented so far indicate that the air pollution effect of the
emissions will decrease. Using alcohol as fuel in cars will increase
the evaporative emissions of the alchols. From health point of view this
is a problem as methanol is a toxic compound.

TABLE 7. Chemical characterization of exhaust emissions from vehicles
fueled with gasoline,methanol and ethanol

Emission		Gasoline	Methanol 15%	Methanol 95%	Ethanol 23%
Carbon monoxide g/km		9.75	19.35	13.09	13.34
Nitrogen oxides	"	2.51	1.83	0.78	1.83
Hydrocarbons	"	1.45	1.64	1.69	1.22
Polyaromatic hydrocarbons	µg/km	170	170	6.2	63
Benz(a)pyrene	"	5.4	5.4	< 0.1	1.6
Methylnitrite	"	94	75	5667	334
Ethylnitrite	"	–	–	–	447
Methanol	mg/km	4.2	115	3381	5.8
Ethanol	"	–	4.8	–	128
Ethylene	"	91.6	85.3	13.5	78.3
Propylene	"	36.3	28.0	5.6	22.5
Benzene, Toluene	"	252.7	344	–	230.3
Formaldehyde	"	26.3	29.4	110.8	24.1
Acetaldehyde	"	< 4.4	< 4.4	< 23	16
Acrolein	"	< 0.4	< 0.4	< 0.4	< 1.0

Egebäck et al. (1983)

When alcohols are used in stationary combustion, an example from a 50 MW
gas turbine is shown in Table 8, there is a large decrease in the overall emission and especially the quantity of nitrogen oxides is reduced.

TABLE 8. Emission characterization after combustion
of three fuels in a gas turbine

Emission, ppm	Methanol	Fuel oil	Natural gas
Carbon monoxide	70	50	175
Nitrogen oxides	45	207	124
Nitrogen dioxide	10	10	50
Hydrocarbons	10	5	216
Aldehydes	1.8	–	10.6

Gluckman MJ, Louke BM (1981)

4. SUMMARY

The environmental and health effects of producing ethanol and methanol can be summarized as follows. The comparison is done with fossil fuels.
Positive effects
a. Sulfur emissions will decrease
b. No net CO_2 emission to the atmosphere
c. Lead can be taken out of gasoline
d. Lower NO_x emissions
e. Less complex mixtures of hydrocarbons will be emitted
f. The alcohols are biodegradable
g. Waste water obtained in biological processes can be treated
Negative effects
a. Increased emissions of alcohols,formaldehyde and acetaldehyde
b. Waste water from gasification processes difficult to treat
c. Toxic compounds are produced during gasification
d. Large quantities of wastes are obtained if ethanol is produced from cellulose-containing plants
e. Storage of fragmented biomass can have negative health effects
f. Methanol is a toxic compound requiring special precautions during handling

REFERENCES[1]
Andersson K,Sellden G and Egneus H (1981) Health and environmental effects of synthetic fuels I. Report to Natl.Swed.Env.Prot.Board
DiNovo ST,Ballantyne WE,Curran LM,Baytos WC,Duke KM,Cornaby BW,Matthews MC,Ewing RA and Vignon BW (1978) Preliminary environmental assessment of biomass conversion to synthetic fuels, EPA-600/7-78-204
Egebäck KE and Bertilsson BE (1983) Chemical and biological characterization of exhaust emissions from vehicles fueled with gasoline,alcohol,LPG and diesel, Natl Swed.Env.Prot.Board report 1635
Egneus H,Wallin G and Wängberg SÅ (1983) Health and environmental effects of synthetic fuels IV.Production of ethanol from cellulose-containing plants, Report to Natl.Swed.Env.Prot.Board
Egneus H,Sellden G,Wallin G and Wängberg SÅ (1982) Health and environmental effects of synthetic fuels II.Production of methanol from biomass, Report to Natl.Swed.Env.Prot.Board
Gluckman MJ and Louke BM (1981) Methanol-an opportunity for the electric utility industry to produce its own clean fuel.8th Energy Tech. Conf.
Hira AU,Mulloney JA and D´Alessio GJ (1983) Alcohol fuels from biomass, Env.Sci.Tech. 17,202-213
Holmberg K and Kallings LO(1980) Allergic alveolite caused by fungi and thermophilic actinomucetes Läkartidn. 77,39-44
Kitchens JF,Casner RE,Harward WE,Nacri BJ and Edwards GS(1976) Investigations on selected potential environmental contamitants:Formaldehyde EPA 560/2-76-009
Sellden G,Andersson K and Egneus H (1982) Health and environmental effects of synthetic fuels III.Emissions from methanol and ethanol combustion Report to Natl.Swed.Env.Prot.Board
Wenzl HFJ(1970) The chemical technology of wood. Academic Press,New York
1)No references have been given in the text in order to save space.The reports in the reference list contains comprehensive reference lists.
Authors address:Dr.H.Egneus Dept Plant Physiology,University of Göteborg Carl Skottbergs Gata 22,S-41319 Göteborg,Sweden

H2-PHOTOPRODUCTION OF GREEN ALGAE: WATER SERVES AS THE MAIN SOURCE OF ELECTRONS

B. MAHRO and L.H. GRIMME
Fachbereich Biologie/Chemie, University of Bremen, D-2800 Bremen (FRG)

1. INTRODUCTION

Some genera of green algae possess the ability to produce hydrogen in the light. But 40 years after the discovery of the phenomenon there is still some controversy about the possible mechanism of hydrogen, photoproduction (Weaver et al., 1980). So it is yet unclear whether the produced hydrogen is originated from the photosynthetic water splitting process or from an oxidation of endogenous organic reductants. The classical approach to get more clearness about this question is to measure H_2 photoproduction in the presence of DCMU (Healey, 1970). However using DCMU it remains open whether electrons originated from fermentative metabolism may enter the photosynthetic electron transport chain prior to the DCMU-inhibition site (Stuart, Gaffron 1972a).

Our approach was to compare the kinetics of the development of both the H_2 photoproduction function and the development of the oxygen evolving function of the green alga Chlorella fusca during the greening process after a prolonged period of N-starvation. Chlorella fusca has lost all its photosynthetic capacities after such a long period of N-starvation but is able to regain photosynthetic functions when it is resuspended in a N-containing culture medium ("greening"; Grimme, Porra 1974; Grimme 1978).

2. MATERIAL AND METHODS

All experiments were done with the green alga Chlorella fusca (strain no 211-15, algal collection Göttingen, FRG). The algae were cultured in a medium according to Grimme, Boardman (1972) at 25°C and 5-7 klux. The degreening process was induced by transfer of the algae into a nitrogen-sparse medium containing only 1/100 (0.08 mmol/l KNO_3) of the nitrogen content of the normal growth medium. The orange, chlorophyll-deficient algae were harvested after 4 weeks, resuspended in the normal growth medium and subsequently cultured either under normal light conditions or in the dark.
For the measurement of H_2photoproduction the algae were resuspended in 0.025 mol/l HEPES-buffer with 2 mmol/l NaCl, 1.5 mmol/l Mg Cl_2, pH 7.6 (adjusted with Tris base). The measurement was carried out during illumination of the algae with 10 klux (for manometric H_2 determination) or 40 klux (for gaschromatographic determination) following a period of 5 hours of anaerobic incubation in the dark (Mahro, Grimme 1982). For the determination of O_2 evolution the algae were suspended in Warburg buffer no 9 and illuminated with 40 klux. The amount of oxygen was measured amperometrically.

Sybesma, C. (ed.), Advances in Photosynthesis Research, Vol. II. ISBN 90-247-2943-2.
© 1984 Martinus Nijhoff/Dr W. Junk Publishers, The Hague/Boston/Lancaster.

3. RESULTS

The photosynthetic capacity measured as O_2 evolution was lost after 4
weeks of N-starvation (-7 µl O_2/h 10^8 cells). Also the H_2 photoproductivity had nearly completely disappeared after the same period of
time (5 µl H_2/h 10^8 cells; fig. 1).
After induction of the greening process in the light (LRG) the O_2
evolution increased rather slowly up to the 18th hour (28 µl O_2/h
10^8 cells). Subsequently the recovery of photosynthetic capacity was
strongly accelerated up to the 36th hour when 452 µl O_2/h 10^8 cells were
evolved. After 48 hours the photosynthetic capacity was again equal to
that of normally cultured algae (519 µl O_2/h 10^8 cells).
Up to the 30th hours of the greening period in the light the H_2 photoproductivity developed with parallel kinetics to that for O_2 evolution
(18th hour: 23 µl H_2/h 10^8 cells; 30th hour: 137 µl H_2/h 10^8 cells).
However in the following period of regreening H_2 photoproductivity was
not further increased (fig. 1).

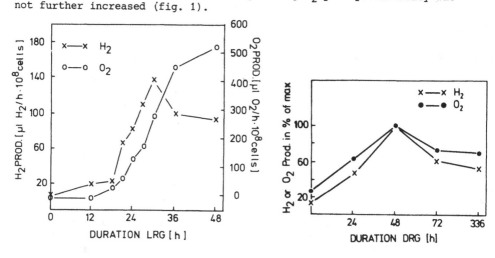

Fig. 1. Comparison of the development of H_2 photoproduction and of photosynthetic oxygen evolution of Chlorella fusca during the greening in
the light (LRG) after 4 weeks of N-starvation.

Fig. 2. Comparison of the development of H_2 photoproduction and of photosynthetic oxygen evolution of Chlorella fusca during the greening in
the dark (DRG) after 4 weeks of N-starvation.

The same coupling of the capability to produce H_2 to the ability for
water splitting could be observed during the greening of the N-deficient
algae in the dark (DRG) (Fig.2). Both H_2 photoproductivity and the capacity for O_2 evolution remained well below their level in light regreened
algae.
After 48 h greening in the dark only 41 µl H_2 resp. 42 µl O_2/h 10^8 cells
were detected.

In order to control whether the parallel kinetics may be due to the
PS I activity as a limiting factor of the whole primary reaction of
photosynthesis we investigated the development of the H_2 photoproduction

based on PS I only. This was realized by the addition of DCMU and glucose.

Table 1. Development of H_2 photoproduction of Chlorella fusca in presence of DCMU and glucose during the greening in the light (PS I-H_2 pp). Gaschromatographic determination after 30 min illumination (40 klux).

Duration LRG (h)	PS I-H_2pp $\mu l\ H_2/10^8$ cells	in % of max (30th h LRG)	Control $\mu l\ H_2/10^8$ cells
0	2.4	42.1	1.3
12	3.0	52.6	-
24	3.1	54.4	19.1
30	5.7	100.0	29.8
48	5.6	98.2	19.2

The results (Tab. 1) show first that the H_2pp under these conditions was much lower than H_2pp without DCMU (5.7 $\mu l\ \bar{H}_2/h \cdot 10^8$cells) and second that the PS I-H_2pp is relatively high (42% of maximum at the 30th hour) up from the beginning of the greening in the light. Thus the development of H_2 photoproductivity during the greening process must really be due to the development of the water splitting activity.

4. DISCUSSION

The results show that not only PS II but also the water splitting process must be involved in the H_2 photoproduction of Chlorella fusca. The residual H_2 photoproductivity when PS II activity is inhibited confirms that also endogenous reductants may be the source of electrons for H_2 production (Stuart, Gaffron 1972 b). But the comparison of the productivity under these different conditions indicates that water must be the main source of electrons. Thus the H_2 pp of C.fusca is mainly a special type of anaerobic photosynthesis.

Under this assumption it remains open for discussion why the H_2 photoproductivity cannot keep pace with the development of the capacity for O_2 evolution after the 30th hour of regreening in the light. Two explanations are possible. First this may be accounted for by an increased development of reactions such as nitrite or sulfite reduction that compete for electrons with the hydrogenase. Secondly it may be imagined that the amount of the hydrogenase becomes a limiting factor of the H_2 pp. The second assumption is supported by experiments on the development of hydrogenase activity - measured as H_2 uptake in the dark with nitrite (Mahro, 1983). The results show that the hydrogeanse activity declines in the same manner as does H_2 photoproductivity.

5. REFERENCES

Grimme LH (1978) The regreening and the development of photosynthetic activity in orange Chlorella fusca cells are separable physiological processes. In Akoyunoglou G, Argyroudy-Akoyunoglou JH, eds. Chloroplast development, pp 445-448. Elsevier, Amsterdam.

Grimme LH and Boardman NK (1972) Photochemical activities of a particle fraction P_1 derived from the green alga Chlorella, Biochem. Biphys. Res. Commun. 49, 1617-1623.

Grimme LH and Porra RJ (1974) The regreening of nitrogen-deficient Chlorella fusca. I. The development of photosynthetic activity during the synchronous regreening of nitrogen-deficient Chlorella. Arch. Microbiol. 99, 173-179.

Healey FP (1970) Hydrogen evolution by several algae, Planta 91, 220-226.

Mahro B (1983) Die Wasserstoff (H_2) Photoproduktion der Grünalge Chlorella fusca: Endogene und exogene Voraussetzungen für ihre Optimierung. pH.D.Thesis University of Bremen, FRG

Mahro B and Grimme LH (1982) H_2 photoproduction by green algae; the significance of anaerobic pre-incubation periods and of high light intensities for H_2 photoproductivity of Chlorella fusca, Arch. Microbiol.132, 82-96.

Stuart TS and Gaffron H. (1972a) The mechanism of hydrogen photoproduction by several algae. II. The contribution of photosystem II. Planta 106, 101-112.

Stuart TS and Gaffron H. (1972b) The gas exchange of hydrogen-adapted algae as followed by mass spectrometry, Plant Physiol. 50, 136-140.

Weaver PF, Lien S and Seibert M (1980) Photobiological production of hydrogen, Solar Energy 24, 3-45.

ABBREVIATIONS

DCMU : 3-(3,4-dichlorphenyl)-1,1-dimethyl-Urea
HEPES : N-2-hydroxyethyl piperazin-N'-2-ethan-Sulfonic acid
PS I : Photosystem I
PS II : Photosystem II
TRIS : Tris-hydroxymethyl-amino-Methan

PHOTOPRODUCTION OF HYDROGEN BY IMMOBILIZED "ADAPTED" ALGAE

M. BROUERS, J. JEANFILS, F. COLLARD

1. INTRODUCTION

The possibility to produce energy by biological processes is attracting worldwide attention. In this frame, experiments for formation of hydrogen gas by coupling isolated chloroplasts to bacterial hydrogenase were performed. The major limitation of H_2 production by such systems was the progressive destabilization of the activities of hydrogenase and chloroplasts.
In 1942, Gaffron and Rubin discovered that illumination of unicellular green algae "adapted" under anaerobic conditions in the dark can lead to production of hydrogen.
On another hand, it is now known that long term stabilization of photobiological activities of algae can be obtained by immobilizing the whole cells in various matrices (Brouers et al 1982, 1983).

2. MATERIAL AND METHODS

2.1. Biological material. The green algae Scenedesmus obliquus were used in the experiments. Methods for cultivation are described by Trainor, Rasbowsky (1967). Before immobilization, the algal suspension in the exponential phase of growth was concentrated by centrifugation.

2.2. Immobilization. Immobilization was performed either in Ca alginate (Alg) or in polyurethane (PU) matrices according to the procedure previously described (Brouers et al, 1982-1983).

2.3. Adaptation to hydrogen production. Adaptation to hydrogen production was obtained by treatment of the immobilized algae in the dark for 4 to 12 hours under N2 at 20°C in the presence of the culture medium.

2.4. Measurement of hydrogen production. Hydrogen photoproduction by immobilized adapted algae was measured by gas chromatography.

3. RESULTS

3.1. Photoproduction of hydrogen after a sole adaptation period. Algae immobilized in polyurethane or alginate "were adapted" for 4 to 12 hours. Hydrogen photoproduction was measured when the "adapted" cells were returned to light. The results are shown in fig. 1. The photoproduction of H2 was appreciable up to 2-3 hours following the onset of illumination. It was progressively inhibited while photoevolution of oxygen developed. Total amount of 20 to 25 μmoles H2 per mg chlorophyll were obtained after two hours illumination in the case of PU immobilized algae adapted for 12 hours.

3.2. Photoproduction of hydrogen after successive adaptation periods. Experiments in which dark adaptation periods (12 h) alternated with light periods (3 h) were performed using alginate immobilized Scenedesmus cells. Results are shown in fig. 2. They show that successive dark adaptation periods led to a recovery of hydrogenase activity during the subsequent light period. The total amount of hydrogen produced during the second or the third light period was even greater (2 times) than that measured during the first light period.

Sybesma, C. (ed.), Advances in Photosynthesis Research, Vol. II. ISBN 90-247-2943-2.
© 1984 Martinus Nijhoff/Dr W. Junk Publishers, The Hague/Boston/Lancaster.

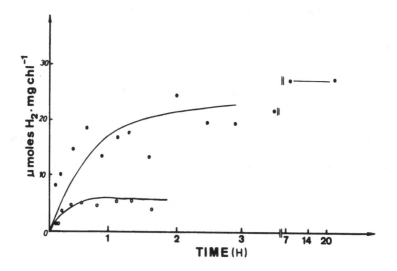

FIGURE 1. Hydrogen photoproduction by adapted *Scenedesmus* cells immobilized in a polyurethane (●) or in an alginate (o) matrix. Adaptation period : 12 h.

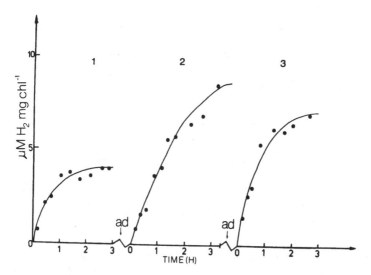

FIGURE 2. Test of successive adaptation periods for H2 photoproduction by immobilized *Scenedesmus*. *Scenedesmus* cells immobilized in an alginate matrix were submitted to cycles of light (3 h) and dark adaptation (ad; 12 h) periods.

4. CONCLUSION

The results emphasize the possibility to achieve hydrogen photoproduction by immobilized green algae. They show that a possible way to maintain continuous H2 production by such systems is to couple several photo-reactors, some being illuminated for photoproduction of hydrogen while the others are put in the dark under N2 for adaptation of hydrogenase. Experiments for optimizing the yield of hydrogen production are now in progress.

5. REFERENCES

Brouers M Collard F Jeanfils J and Jeanson A (1982) Immobilization and stabilization of green and blue green algae in cross-linked serumalbumin glutaraldehyde and in polyurethane matrices. In Hall DO and Palz W ed. Solar Energy R&D in the European Community, series D, vol 1, pp. 134-139. D. Reidel Pub. C°.
Brouers M Collard F Jeanfils J and Loudèche R (1983) Long term stabili-zation of photobiological activities of immobilized algae. Photoproduction of hydrogen by immobilized "adapted" *Scenedesmus* cells. In Hall D.O. et al ed. Solar Energy R&D in the European Community, series D, vol 2, pp. 171-178. D. Reidel Pub. C°.
Gaffron H and Rubin J (1942) Fermentative and photochemical production of hydrogen in algae. J. gen. Physiol. 26, 219-240.
Trainor FR and Rasbowsky FG (1967) Control of unicell formation in a soil *Scenedesmus*. Canadian J. of Botany 45, 1657-1664.

ACKNOWLEDGEMENTS

This research was supported by the Commission of the European Communities (contract ESD 016 B) and by the "Ministère de la politique scientifique" Brussels Belgium. One of the authors (J.J.) is a fellowship in the training programm in biomolecular engineering of the CEC.
Authors address : M. Brouers, J. Jeanfils and F. Collard, Department of Botany B.22. University of Liege B-4000 Liege Belgium
J. Jeanfils (second affiliation) Laboratory of Enzymatic Technology BP 233 UTC 60206 Compiègne, France.

PHOTOPRODUCTION OF HYDROGEN FROM WATER USING IMMOBILIZED BIOLOGICAL
AND SYNTHETIC CATALYSTS

K.K. RAO and D.O. HALL   P. CUENDET and M. GRÄTZEL

1. INTRODUCTION

We have been studying the characteristics of various catalytic systems
capable of generating H_2 by water-photolysis. These include (a)
completely biological systems where natural electron carriers such as
ferredoxin, flavodoxin, cytochrome c_3 or NAD reduced by illuminated
chloroplasts are coupled to hydrogenases; (b) semibiological systems
wherein photosynthetically reduced natural or synthetic electron relays
(e.g. viologens, metatungstate, cobalt complexes, etc.) are coupled to
hydrogenase or Pt for H_2 production and (c) completely artificial
systems using semiconductors, (TiO_2, RuO_2, CdS, etc.) as photo and
redox catalysts for the photogeneration of H_2 by coupling to Pt (see
Cuendet, Gratzel 1982a, Rao, Hall 1983). At present we are concerned
with the stabilisation of the H_2 generating system and also with the
optimisation of the electron and proton transfer reactions to achieve
the maximum light and catalytic conversion efficiencies. With this
objective we are studying (a) the effect of various additives to the
medium, (b) the effect of entrapment of photosynthetic membranes in
polyurethane, and (c) the efficiency of different synthetic relays on
the rate and longevity of H_2 evolution reaction. We are also investigating the possibility of using hydrogenases immobilized on semiconductor particles as photo and redox catalyst for the production of
H_2 from water with a sacrificial electron donor. A summary of the
methods used and results obtained is presented in this abstract.

2. PROCEDURE

2.1. Materials and Methods

2.1.1. Biological catalysts. Hydrogenases from Desulfovibrio
desulfuricans strains Norway and 9974 were isolated at King's College
London (Lalla Maharajh et al 1983). Alcaligenes eutrophus hydrogenase
was a gift from Dr. Schneider, Gottingen. Ferredoxin was prepared from
Spirulina maxima, and cytochrome c_3 from D. desulfuricans, Norway.

2.1.2. Synthetic and inorganic catalysts. The amphiphilic viologens
*$C_{14}MV^{2+}$ and ($2C_7MV^{2+}$) were synthesized by Dr. A. Braun at EPF,
Lausanne. CoCC was a gift from Dr. U. Kölle, Aachen. Other electron
relays such as PVS, CoSep, MT, and tungstosilicic acid were either

---

* Abbreviations. ($2 C_7 MV^{2+}$): N,N'-bis(heptyl)-4,4'bipyridinium;
$C_{14}MV^{2+}$: N-tetradecyl-N'-methyl-4,4'-bipyridinium; CoCC: dicarboxyl-
1-1'cobalticinium; CoSep: cobalt sepulchrate; Fd: ferredoxin;
Mt: sodium metatungstate ($Na_6H_2W_{12}O_{40}.29 H_2O$); MV: methyl viologen;
PVS: N,N',propylsulfonate 4,4'bipyridiunium; TEA-triethanolamine.

Sybesma, C. (ed.), Advances in Photosynthesis Research, Vol. II. ISBN 90-247-2943-2.
© 1984 Martinus Nijhoff/Dr W. Junk Publishers, The Hague/Boston/Lancaster.

synthesized at EPF or obtained as gift by Dr. Grätzel. P25-TiO$_2$ (50m^2/g) was provided by Dr. Kleinschmidt, Degussa, A.G. Hanau and Bayersol anatase (200m^2/g) by Dr. Panek, Bayer, A.G. Krefeld-Urdingen. CdS powder "puriss" was purchased from Fluka. Colloidal Pt was obtained as in Cuendet, Gratzel (1982b). Hypol 5000, a methylene diisocyanate urethane prepolymer was a gift from W.R. Grace Ltd. London.

2.1.3. Isolation of chloroplasts and determination of H$_2$ production from photosynthetically reduced or dithionite reduced electron transfer catalysts were as described (Rao et al 1978).
For immobilisation of chloroplasts into foams in situ the urethane prepolymer Hypol 5000 warmed up to 50^0 was poured into a glass vial containing an equal volume of a suspension of chloroplasts in 50mM HEPES buffer pH 7.5 (containing 0.2% bovine serum albumin) kept in ice. A sponge of polyurethane embedding the chloroplast membranes was formed within 15 min. For coimmobilization of chloroplasts and catalyst the latter was added to the chloroplast suspension before addition of Hypol.

2.1.4. Immobilization of hydrogenases on semiconductors. Adsorption of hydrogenase on to semiconductor was carried out by incubation of the enzyme with a sonicated suspension of the semiconductor in 50mM phosphate, pH 7 (buffer) for 2 h in ice followed by centrifugation and washing the pellet with buffer (Messing 1976). For covalent binding the semiconductor was first silanized by treatment with 3-aminopropyl-triethoxy silane (1% v/v in acetone) followed by activation with 2.5% glutaraldehyde. After washing the excess glutaraldehyde from the particles they were incubated with hydrogenase for 2 h in ice (Weetall 1976). Light-induced H$_2$ production was assayed in 14 ml glass vials fitted with crimped rubber septa. The basic reaction mixture contained 100 mM EDTA, semiconductor powder, and Pt in 2 ml made up with buffer. The vials were degassed with Ar and illuminated at 25-30^{0}C with a sunlight simulator (Suntest, Hanau) delivering about 1000 W/m^2. H$_2$ evolved was determined gas chromatographically.

3. RESULTS

3.1 H$_2$ evolution from water as electron source. Chloroplasts entrapped in Hypol 5000 were as active, if not better, in catalysing photo H$_2$ production as were free chloroplasts. The polyurethane foam-entrapped chloroplasts lost 56% of their photosynthetic electron transport capacity (measured as ferricyanide reduction) whereas the free chloroplasts lost 49% of their activity when stored in light at 25^{0}C. In previous studies with Hypol 3000 prepolymer (Rao et al 1982) the rates of H$_2$ production with foam-immobilized chloroplasts were only 10% of the control. Hypol 5000 is completely free of toluene diisocyanate and toluene diamine - possibly these contaminants in Hypol 3000 may have caused inhibition of chloroplast electron transport.

3.2. Effect of salts. Certain anions are said to stabilise the O$_2$ evolution from chloroplasts. The effect of salts on H$_2$ evolution depended on the electron mediator used (see Table 1).

TABLE 1   Effect of salts on H_2 production

Salt in the reaction mixture, mM	Relative rates of H_2 production				
	A	B	C	D	E
None	100	100	100	100	100
100, K Phosphate	64	nd	nd	nd	nd
200, "	50	nd	nd	nd	nd
800, "	23	0	480	nd	nd
100, Na citrate	30	nd	nd		nd
800, "	6	329	529		nd
200, NaCl	28	411	165	25	25
200, NaBr	28	235	221	25	nd
200, Na I	0	0	nd	35	nd
800, NaCl	nd	nd	454	20	7
800, NaBr	nd	nd	629	30	2
800, Na I	0	0	503	10	20
0.05% polyvinyl sulphate	74	181	nd	nd	nd

A: Chloroplast + Fd + C. pasteurianum H_2ase. B: Chloroplast + MV + D. desulfuricans strain 9974 H_2ase. C, D and E used 10mM $Na_2S_2O_4$ as e⁻ donor. C, with 2.5 mM MV + C. pasteurianum H_2ase; D, 20 μM FD^{2+} + C. pasteurianum H_2ase; E, with 50 μM cytochrome c_3 + 9974 H_2ase. nd, not determined. Activities are expressed as percentage of control containing only the buffer ions.

Anions such as PO_4^{3-}, citrate, Cl⁻, Br⁻, and polyvinyl sulphate (0.5%) inhibited H_2 production catalysed by the negatively charged relays, FD and cytochrome c_3, whereas H_2 evolution catalysed by MV^{2+} was stimulated by these reagents (Table 1). These results are in agreement with the suggestion that anions facilitate the dissociation of +ve electron mediators such as methyl viologen adsorbed to the -vely charged thylakoid membrane and thus increase the concentration of the reduced mediators in contact with hydrogenase or Pt in solution. (Cuendet, Grätzel 1982a).

3.3. Effect of electron mediators on H_2 production. Of the different synthetic mediators tested (MV^{2+}, PVS, CoCC, Mt and CoSep) CoCC was found to be the most effective in electron transfer from water to hydrogenase or Pt resulting in H_2 production. CoCC was also found to be a much better electron mediator from ascorbate to hydrogenase (or Pt) through photosystem I of Nostoc muscorum. Within the time scale of experiments it was observed that Pt was as stable as hydrogenases in catalytic H_2 evolution.

3.4. H_2 evolution by semiconductor-bound hydrogenases. Results obtained after fixation of three hydrogenases onto P25-TiO_2 are shown in Table 2. About 89 to 99% of the hydrogenase added to the TiO_2 in suspension bound to the TiO_2. The rates of H_2 evolution were linear with TiO_2-bound hydrogenases as well as Pt catalysts for about 8 h; the rates dropped after this period for the hydrogenase-catalysed H_2 production.

TABLE 2 H_2 evolution from TiO_2-immobilized hydrogenases

Hydrogenase	Immobilization	μmole $H_2 \cdot g\ TiO_2 \cdot h$	
		dithionite-reduced MV	light-reduced MV
D. desulfuricans			
(Norway)	adsorption	1496	1000
	covalent binding	2920	492
(9974)	adsorption	1768	800
	covalent binding	6720	440
A. eutrophus	adsorption	424	20
	covalent binding	1112	136

The yield of H_2 varied with the nature of the electron donor (EDTA, glucose, methanol, TEA, etc.) the nature of the semiconductor (P25-TiO_2, Bayersol anatase or CdS), and the nature of the proton activator. The long chain amphiphilic viologens C_{14} MV^{2+} and ($2C_7MV^{2+}$) on reduction by irradiation (violet colour) adhered to the semiconductor and transferred electrons from the semiconductor to the hydrogenase.

5. REFERENCES

Cuendet P and Gratzel M (1982a) Artificial photosynthetic systems in Experientia 38, 223-228 (1982b) New photosystem I electron acceptors: Improvement of hydrogen production by chloroplasts in Photochem. Photobiol. 36, 203-210.
Grätzel M ed. (1983). Energy resources through photochemistry and catalysis. Academic Press, New York (in press).
Lalla-Maharajh WV Hall DO Cammack R Rao KK and Legall J (1983) Purification and properties of the membrane-bound hydrogenase from Desulfovibrio desulfuricans in Biochem. J. 209, 405-454.
Messing RA (1976) Adsorption and inorganic bridge formations in Methods in enzymology 44, 148-169.
Rao KK and Hall DO (1983) Photobiological production of fuels and chemicals. In Braun AM, ed. Photochemical conversions, pp. 1-48, Lausanne: Presses polytechniques romandes.
Rao KK Gogotov IN and Hall DO (1978) Hydrogen evolution by chloroplast-hydrogenase systems in Biochimie 60, 291-296.
Weetall HH (1976) Covalent coupling methods for inorganic support materials in Methods in enzymology. 44, 134-148.

Supported by grants from National Energie-Forschung Fond, Switzerland and a contract from the EEC, Brussels.

Authors address: K.K. Rao and D.O. Hall, King's College London, 68 Half Moon Lane, London SE24 9JF, UK.
P. Cuendet and M. Grätzel, Ecole Polytechnique Federale, CH-1015 Lausanne, Switzerland.

H_2-PHOTOPRODUCTION OF GREEN ALGAE:
CHANGES IN THE ENERGY STATE OF CHLORELLA FUSCA UNDER H_2-PHOTOPRODUCTIVE
CONDITIONS

A.C. KÜSEL, B. MAHRO, L.H. GRIMME
Fachbereich Biologie/Chemie, University of Bremen
D-2800 Bremen 33 (FRG)

1. INTRODUCTION

Investigations concerning the influence of uncouplers on H_2-photoproduction of green algae showed that this process does not require ATP (Stuart, Gaffron 1972; Lien, San Pietro, 1981). However it is still open for discussing whether the electron transport involved in H_2 photoproduction (H2pp) is coupled to a photophosphorylation. The answer to this question and in a given case the elucidation of the kind of photophosphorylation could give further support to the characterization of H2pp as a special type of anaerobic photosynthesis (Mahro, Grimme these proceedings).
We have tried to answer these questions by measuring the concentration of AMP, ADP and ATP in hydrogenase containing Chlorella fusca during a dark light transition under H_2 photoproductive conditions, i.e. illumination with 30 klux after 5 hours of anaerobic and CO_2 free incubation in the dark.

2. MATERIAL AND METHODS

The experiments were done with the green algae Chlorella fusca (strain no 211-15; with hydrogenase) and Chlorella vulgaris (strain no 211-11b; without hydrogenase; both algal collection Göttingen, FRG).
The algae were cultured in a medium according to Grimme, Boardman (1972) at 25°C and 5-7 klux. The algae were harvested by centrifugation, carefully washed and resuspended in a 25 mmol/l HEPES-buffer adjusted to pH 7.6 with Tris-Base and subsequently adapted to H_2 productive conditions.
The adenylates were extracted during dark incubation resp. illumination by rapid (1-2 sec) injection of an algal suspension aliquot into boiling Tris buffer (50 mmol/l, 4 mmol EDTA, pH 7.75). ATP was determined luminometrically applying firefly luciferase technique. ADP and AMP were determined in the same way after conversion to ATP by pyruvat kinase and myokinase (all reagents obtained from Lumac Systems AG, Basel, Switzerland). H_2 was determined gaschromatographically according to Mahro and Grimme (1982).

3. RESULTS

Upon illumination the ATP content in Chl. fusca cell increased significantly while the AMP content decreased. The phosphorylation rate during the first 30 seconds varied in parallel experiments between 5.8 and 7.7 nmoles $ATP/10^6$ cells h. During the next 5 minutes of illumination the ATP content did not increase further (Fig. 1, table 1). The adenylate energy charge rose from 0.55 at the end of dark adaptation to 0.75 after 30 seconds of illumination and then decreased to 0.69 after 5 minutes in the light (Fig. 2).
The addition of DCMU ($1 \cdot 10^{-5}$ mol/l) inhibited H_2 photoproduction of Chl. fusca during the first 10 minutes to about 80 % (Fig. 3). However the light induced increase of ATP was not at all effected by DCMU. The

Sybesma, C. (ed.), Advances in Photosynthesis Research, Vol. II. ISBN 90-247-2943-2.
© *1984 Martinus Nijhoff/Dr W. Junk Publishers, The Hague/Boston/Lancaster.*

initial phosphorylation rate was 8.3 nmoles ATP/10^6 cells h (Fig. 2, table 1). The plastoquinone antagonist DBMIB ($5 \cdot 10^{-4}$ mol/l) inhibited the observed phosphorylation considerably but not completely. The initial rate of phosphorylation was 1.5 nmoles ATP/10^6 cells h (table 1).

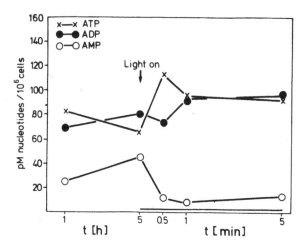

Fig. 1: Changes in the concentration of AMP, ADP and ATP in Chlorella fusca during 5 hours of anaerobic, CO_2-free incubation in the dark and after illumination (30 klux).

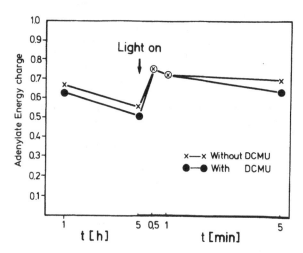

Fig. 2: Changes in the Adenylate Energy Charge of Chlorella fusca during 5 hours anaerobic, CO_2-free incubation in the dark and after illumination (30 klux) with and without DCMU.

Under the H_2 productive conditions applied it must be assumed that only a hydrogenase induced electron transport takes place. To control this assumption we determined the phosphorylation rate upon dark-light transition of Chl. fusca with CO-inhibited hydrogenase and of Chl. vulgaris (without hydrogenase). Under these conditions these rates were 3.7 nmoles ATP/10^6 cells·h for Chl. fusca and 1.2 nmoles ATP/10^6 cells·h for Chl. vulgaris.

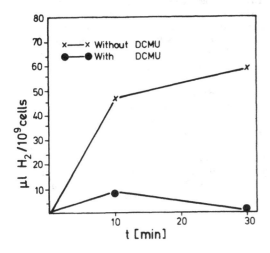

Fig. 3: The H_2-photoproduction of Chlorella fusca after 10 and 30 min of illumination (30 klux) with and without DCMU.

TABLE 1: The dependence of the initial phosphorylation rate on a functioning photoproductive electron transfer. Data calculated from the ATP increase during the first 30 sec. of illumination.

	nmoles ATP/h·10^6 cells
C. fusca (with hydrogenase)	5.8 – 7.7
+ DCMU (1 x 10^{-5} mol/l)	8.3
+ DBMIB (5 x 10^{-4} mol/l)	1.5
+ CO (4 Vol %)	3.7
C. vulgaris (without hydrogenase)	1.2

4. DISCUSSION

It is well-known that green algae are able to produce ATP during a dark-light transition after a short anaerobic pretreatment (Urbach, Kaiser 1972; Kawada, Kanazawa 1982). Our results have shown that such an ATP increase occurs also after prolonged anaerobic incubation.

Applying the criteria of Gimmler (1977) for the characterization of the three types of photophosphorylation it is rather difficult to determine the type of the phosphorylation observed. The cyclic phosphorylation can be excluded on account of the fully reduced state of the algae as a consequence of the long anaerobiosis, the applied high light intensity and its insensitivity towards complete inhibition of an electron transport from PS II by DCMU. DCMU prevents photosynthetic O_2 evolution and the non-cyclic electron transport. As the observed ATP increase is not affected by DCMU both non-cyclic and pseudocyclic phosphorylation should be ruled out.

One electron transport which for Chl. fusca can be assumed under the given conditions is an H_2 photoproductive electron transfer from an endogeneous reductant via PS I. The entry of these electrons into the transport chain must take place after the DCMU and partly prior to the DBMIB inhibition site (Fig. 2, Table 1). This assumption is supported by two facts: (1) H_2 photoproductivity of Chl. fusca has reached its maximum after the applied 5 hours of anaerobic incubation. (2) At the beginning of H_2 photoreduction upon illumination, when the DCMU insensitive photophosphorylation is observed, H_2pp is relatively insensitive towards DCMU and is assumed to be based mainly on an only PS I dependent electron transport from endogenous reductants to hydrogenase (Senger, Bishop, 1977). Thus it can be concluded that this part of H_2 production is coupled to a photophosphorylation. It remains still open for investigation whether the main H_2-productive electron transport where the electrons for the hydrogenase reaction are derived from water is also coupled to a phosphorylation.

However the fact that the observed ATP production in Chl. fusca can be attributed to an H_2 productive electron transport does not offer any explanation for the residual ATP production in Chl. fusca with inactivated hydrogenase or for the ATP increase in Chl. vulgaris. This phenomenon demands further attention.

5. REFERENCES

Gimmler H (1977) Photophosphorylation in vivo. In Trebst A Avron M (eds.) Photosynthesis I. Encyclopedia of Plant Physiology. New Series, Vol. 5, pp.. 448-472, Springer Berlin Heidelberg, New York.

Grimme LH, Boardman NK (1972) Photochemical activities of a particle fraction P_1 derived from the green alga Chlorella.Biochem. Biophys. Res. Commun. 49, 1617-1623.

Kawada ,E, Kanazawa T (1982) Transient changes in the energy state of adenylates and the redox states of pyridine nucleotides in Chlorella cells induced by environmental changes. Plant Cell Physiology 23, 775-783.

Lien S, San Pietro A (1981) Effect of uncouplers on anaerobic adaptation of hydrogenase activity in C. reinhardt ii. Biochem. Biophys. Res. Commun 103, 139-147.

Mahro B, Grimme LH (1982) H_2 photoproduction by green algae: The significance of anaerobic pre-incubation periods and of high light intensities for H_2-photoproduction of Chlorella fusca. Arch. Mikrobiol. 132: 82-86.

ABBREVIATIONS

DBMIB : 2,5 dibromo-3-methyl-6-isopropyl-p-benzoquinon
DCMU : 3-(3,4-dichlorphenyl)-1,1-dimethyl-Urea
EDTA : Ethylen-diamin-tetra-acetic acid
HEPES : N-2-hydroxyethyl piperazin-N'-2-ethan-Sulfonic acid
PS 1 : Photosystem I
PS II : Photosystem II
TRIS : Tris-hydroxymethyl-amino-Methan

SIMULTANEOUS PRODUCTION OF HYDROGEN AND OXYGEN AS AFFECTED BY LIGHT INTENSITY IN UNICELLULAR AEROBIC NITROGEN FIXING BLUE GREEN ALGA SYNECHOCOCCUS SP. MIAMI BG043511.

K.J. REDDY AND A. MITSUI,* SCHOOL OF MARINE AND ATMOSPHERIC SCIENCE, UNIVERSITY OF MIAMI, MIAMI, FLORIDA 33149. U.S.A.

SUMMARY

Aerobic Nitrogen fixing unicellular blue-green alga, Synechococcus sp Miami BG 043511 evolved hydrogen and oxygen simultaneously at a molar ratio of 2 in closed flasks under light saturated condition. Since CO_2 evolution was not detected in the gas phase under these conditions water could be the ultimate source for hydrogen and oxygen photoproduction. On the other hand, under light limited conditions, when H_2/O_2 molar ratio was higher than 2, CO_2 evolution was associated. Thus, Light intensity markedly regulates hydrogen, oxygen and CO_2 production.

1. INTRODUCTION

Simultaneous hydrogen and oxygen photoproduction in closed vessels, were observed only for a short period of time (order of minutes), in green algae because of the oxygen sensitivity of hydrogenase and the pronounced hydrogen consumption reaction in the presence of photoproduced oxygen (Bishop, Jones 1977). Heterocystous filamentous blue-green algae with hydrogen catalyst nitrogenase are also able to photoproduce hydrogen and oxygen in closed vessels for somewhat longer period (order of hours), however this does not last too long owing to hydrogen consumption activity (Tel-Or, Packer 1978; Spiller, Bögor 1978; Kumazawa, Mitsui 1981). Hence special treatments such as short light and dark cycles (Bishop, Jones 1977), continuous gas flushing (Weissman, Benemann 1977) and oxygen trapping (Pow, Krasna, 1979) are required for prolonged hydrogen production in both green and blue-green algae.

Hydrogen and oxygen photoproduction in unicellular blue-green algal forms has not yet been studied in detail. Several strains of unicellular blue-green algae from the collections of this laboratory actively fix nitrogen aerobically (Mitsui, 1978; Mitsui et al 1979; Duerr, Mitsui, Unpublished). Recently, we described the simultaneous hydrogen and oxygen photoproduction by these strains in a closed vessel for 3 to 4 days, without any special treatment (Mitsui et al, 1983). In this report, we present the effect of light intensity on simultaneous hydrogen and oxygen production in one of these strains of aerobic nitrogen fixing unicellular blue-green algae, synechococcus sp. Miami BG 043511.

2. MATERIAL AND METHODS

2.1 Culture of blue-green alga.

Synechococcus sp. Miami BG043511 was isolated from the sample collected from marine environment of Bahama islands and purified to axenic status in this laboratory (Rosner, Radway, Sprogis, Duerr, Leon, Mitsui, Unpublished). This blue-green alga was grown under aseptic conditions in combined nitrogen free medium A-N (Kumazawa, Mitsui, 1981) in 4 liter aspirator bottles at 25oC and 100μ Einsteins/m^2/sec. Culture was bubbled with sterile air.

---

*To whom the reprints should be requested.

Sybesma, C. (ed.), Advances in Photosynthesis Research, Vol. II. ISBN 90-247-2943-2.
© *1984 Martinus Nijhoff/Dr W. Junk Publishers, The Hague/Boston/Lancaster.*

2.2 Assays of H_2, O_2 and CO_2

Three to four day old, early log phase cells were harvested by centrifugation at 10,000 xg for 10 minutes and washed twice with the medium A-N and then resuspended in the same medium. 5 ml of this algal suspension was put into 25-ml Fernbach flasks. The flasks were sealed with anaerobic rubber stoppers, evacuated and filled with argon· Incubation was carried out at 30oC with continuous shaking. Varying incubation light intensities were obtained by covering the incubation flasks, with nylon screens. H_2, O_2 and CO_2 were measured by gas chromatography (Kumazawa, Mitsui, 1981). Rates of H_2, O_2 and CO_2 evolution were determined from the slope of the curve.

3. RESULTS

3.1 Effect of light intensity on H_2, O_2 and CO_2 production

Table 1 shows the comparison of H_2, O_2 and CO_2 production in low and high light intensities. In low light intensities (undersaturated) H_2 and CO_2 were evolved but O_2 evolution was significantly reduced. This resulted in a very high molar production ratio of H_2/O_2 (Table 1). However, the ratio decreased with increasing light intensity. In high light intensity i.e. under staurated light condition, H_2 and O_2 were evolved simultaneousely in a molar ratio of 2 and no CO_2 was detected in the gas phase. Rates of H_2 production were as high as 1.32 μ moles H_2 / mg dry wt/h or 165 μ mol H_2 /mg chlorophyll /h. In argon atmosphere trace amounts of H_2 was produced in dark, but there was no CO_2 production.

3.2 Effect of cell density on H_2, O_2 and CO_2 production.

Under constant light intensity (120 μE/m^2/sec.) H_2, O_2 and CO_2 production were monitored for 48 hours with different cell densities of the blue-green alga. High cell densities caused mutual shading whereas low cell densities represented light saturated condition. Figure 1 shows the comparison of H_2/O_2 ratio and CO_2 evolution at short (8 hrs) and long (21 hrs) incubation time. In short incubation time, CO_2 was not detected in low cell density suspensions and this gave H_2/O_2 molar ratio of approximately 2.0. With increasing cell densities, in short-term incubation (8 hr) H_2/O_2 molar ratios increased gradually with

Table 1. Effect of light intensity on H_2, O_2 and CO_2 evolution. Rates are at 5 hrs incubation. Incubation suspension contained 0.82 mg dry wt/ml.

Light Intensity	μE/ m^2/sec.	Rate of Production n mol/mg dry wt/h			H_2/O_2 Molar Ratio
		H_2	O_2	CO_2	
Dark	0	13	0	0	-
Low Light	8	389	0	173	-
	15	576	19	207	30.3
	23	714	72	259	9.9
High Light	100	1320	653	0	2.02

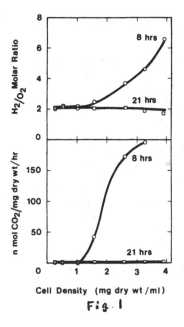

Fig. 1

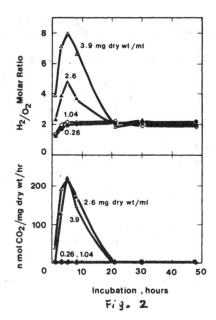

Fig. 2

Figure 1   Effect of cell density on H_2/O_2 molar ratio and rate of CO_2 production during short (8 hrs) and long (21 hrs) incubation. 8 hrs rates are average of 0-8 hrs and 21 hrs rates are averages of 0-21 hrs.   Incubation was pH 7.4 and 30°C.
Figure 2   Time courses of H_2/O_2 molar ratio and rate of CO_2 production at 4 different cell densities. For other incubation conditions see Figure 1.

liberation of CO_2 . In all cell densities tested in long term incubation (21 hrs), H_2/O_2 molar ratio remained as 2.0 with no CO_2 evolution.

Figure 2 indicates the time course of H_2/O_2 production ratio and CO_2 production in different cell densities.   At low cell densities (0.26 and 1.04 mg dry wt/ml) H_2/O_2 molar ratio remained aproximately 2.0.   Except for initial short incubation period, CO_2 was not evolved in gas phase throughout incubation period. In high cell densities (2.60 and 3.90 mg dry wt/ml) H_2/O_2 ratios were higher than 2.0 during the first 20 hours, while a longer incubation period resulted in a ratio of 2.0.   Molar ratio above 2 was associated with CO_2 evolution. Disapperance of CO_2 evolution coincided with the maintainance of the H_2/O_2 molar ratio of 2.0.

4.   DISCUSSION

As shown in Table 1 and Figure 2, the aerobic nitrogen fixing unicellular blue-green alga, synechococcus sp. Miami BG 043511 photoproduced hydrogen and oxygen simultaneously in the molar ratio of 2.0 under light saturated conditions. Prolonged evolution of hydrogen and oxygen in a stoichiometric ratio of 2 suggests that water could be the overall source of hydrogen and oxygen. Whether water donates electron directly or indirectly (through cellular carbohydrate) for hydrogen production is unknown at present. However, if the latter is the case, CO_2 produced from cellular carbohydrate should have been very efficiently refixed in the cells, because, CO_2 was not detected in the gas phase and cellular carbohydrate levels remained in the same level (Mitsui et al 1983).

In contrast to this strain, nitrogen starved non-heterocystous filamentous blue-green alga, Oscillatoria sp Miami BG 7 simultaneously and stoichiometrically photoproduce H_2 and CO_2 in a ratio of 2 and both were accumulated in the gas phase. Preceeding the H_2 and CO_2 photoevolution step, the cells were needed to accumulate very high level of cellular carbohydrate (Glycogen) by photosynthesis. (Kumazawa, Mitsui 1981; Mitsui et al 1983).

Under light limited conditions as shown in both light intensity and cell density studies (Table 1, Figure 1 and Figure 2) the molar production ratio of H_2 to O_2 exceeded 2.0. Under these circumstances with no exception CO_2 was evolved. In some cases such as in low light intensities (Table 1) molar production ratio of H_2/CO_2 was approximately 2 and O_2 was not evolved in gas phase. However, it is unknown whether the same mechanism as exhibited by nitrogen starved cells of strain BG 7 in light saturated condition is also operating in this unicellular Synechococcus sp. Another possible explanation is that oxygen produced by photolysis of water could oxidize the cellular carbohydrate thereby liberating CO_2 through respiration under light limited conditions. Further studies are required to clarify these and other possible mechanisms.

5. REFERENCES

Bishop NI and Jones LW (1977) photoproduction in green algae; water serves as the primary substrate for hydrogen and oxygen production. In: Mitsui A, Miyachi S, San Pietro A and Tamura S eds. Biological solar energy conversion, pp. 3-22. Academic Press, New York.

Pow T and Krasna AI (1979) Photoproduction of hydrogen from water in hydrogenase containing algae, Arch. Bioch. Biophys. 194, 413-421.

Kumazawa S. and Mitsui A (1981) Characterization and optimization of hydrogen production by a salt water blue-green alga, Oscillatora sp Miami BG 7. I. Enhancement through limiting the supply of nitrogen nutrients, Inter. J. Hydrogen Energy, 6, 341-350.

Mitsui A (1978) Marine photosynthetic organisms as potential energy resources: Research on nitrogen fixation and hydrogen production, In: Proc. Int. Ocean Development Conference, IODC Organizing Committee, ed. Tokyo B1, pp. 29-52.

Mitsui A, et al. (1979). Biological solar energy conversion: hydrogen production and nitrogen fixation by marine blue-green algae. In: Boer KW and Glenn BH eds Sun II, Vol. pp 31-35, Pergamon Press, New York.

Mitsui A, et al. (1983). Progress in research toward outdoor biological hydrogen production using solar energy, sea water and marine photosynthetic microorganisms, Biochem. Eng. III, New York Acad. Sci. (in Press).

Spiller H, Ernest A, Kerfin W and Böger P (1978) Increase and stabilization of photoproduction of hydrogen in Nostoc muscorum by photosynthetic electron transport inhibitors, Z. Naturforsh, 33C 541-547.

Tel-Or E, Luijk LW and Packer L (1978) Hydrogenase in N Fixing cyanobacteria. Arch. Bioch. Biophys. 185, 185-194.

Weissman JC and Benemann JR (1977) Hydrogen production by nitrogen-starved cultures of Anabaena cylindrica, Appl. Environ. Microbiol. 33, 123-131.

6. ACKNOWLEDGEMENTS

This work has been supported by grants to A. Mitsui from the National Science Foundation and the National Aeronotics and Space Administration.

CONTINUOUS HYDROGEN PHOTOPRODUCTION FROM SULFIDE BY AN IMMOBILIZED
MARINE PHOTOSYNTHETIC BACTERIUM, CHROMATIUM SP. MIAMI PBS 1071.

M. IKEMOTO and A. MITSUI,[*] School of Marine and Atmospheric Science,
University of Miami, 4600 Rickenbacker Causeway, Miami, Florida 33149.

SUMMARY

Hydrogen photoproduction by immobilized cells of the marine
photosynthetic sulfur bacterium, Chromatium sp. Miami PBS 1071 was
carried out by periodic addition of supplemental sulfide in the reaction
system. CO_2 was not produced and only H_2 was evolved. Hydrogen gas in
this immobilized system was continuously produced for at least 300 hrs.
at a constant rate. This contrasts from free cell systems where the
hydrogen production dropped within 40 hrs. When immobilized cells were
grown in the agar gel matrix, high biomass yield was obtained, and rates
of hydrogen production up to 6 μ moles/g. gel/hr for 100 hrs were
observed. Based on these laboratory experiments, outdoor culturing and
hydrogen production was carried out using immobilized cells in 4L reactor
systems containing 1.5 kg agar gel.

1.   INTRODUCTION

One of the major advantages in the use of photosynthetic bacteria for
hydrogen production is that waste and other pollutants can be removed
with simultaneous energy production. The basic principles of applied
hydrogen photoproduction research have been reviewed previously (Mitsui
et al., 1980; Mitsui et al., 1983). Among the hundreds of strains
isolated in our laboratory, marine Chromatium sp. Miami PBS 1071 has many
characteristics well suited for applied systems. Among these are: 1)
high growth rate using either molecular nitrogen or ammonia as the
nitrogen source (doubling times 2 hours and 1.75 hours, respectively), 2)
tolerance to high concentrations of sulfide (up to 5 mM in growth) and 3)
high hydrogen photoproduction capability (6 μ mol/mg protein/hr) (Mitsui,
1977; Mitsui et al., 1980; Ohta et al., 1981; Ohta and Mitsui, 1981).
This paper describes hydrogen photoproduction from sulfide by immobilized
cells of this strain in both indoor and outdoor conditions.

2.   MATERIALS AND METHODS

2.1.   Photosynthetic bacteria and culture conditions. An axenic culture
of the marine photosynthetic sulfur bacterium, Chromatium, sp. Miami PBS
1071 was cultured in a medium described previously (Ohta et al., 1981) at
approximately 30°C and 200 $\mu E/m^2/sec$. Early stationary growth phase
cultures were harvested (10,000 x g, 10 minutes), washed twice with 3%
NaCl solution and then suspended in the same solution for hydrogen
production.

2.2.   Bacterial cell immobilization and H_2 and CO_2 assays. The
photosynthetic baterial cell suspension in 3% NaCl solution was
immobilized in 2.5% agar (Bacto-agar, Difco Lab.). 5 g agar blocks

---

[*]To whom the reprint request should be made.

Sybesma, C. (ed.), Advances in Photosynthesis Research, Vol. II. ISBN 90-247-2943-2.
© *1984 Martinus Nijhoff/Dr W. Junk Publishers, The Hague/Boston/Lancaster.*

(0.5 cm x 0.5 cm x 0.3 cm) were placed in 25 ml flasks and 5 ml of the above culture medium (without ammonia, malate or sulfide) was added. Flasks were repeatedly evacuated and flushed with argon, then filled with argon. Just prior to incubation the electron donor substance, Na_2S, was added to give a final liquid medium concentration of 10 mM. Incubation was carried out with continuous shaking of 100 strokes per minute at 30oC and 120 $\mu E/m^2/sec$. In some experiments (indicated in the text), the same amount of sulfide was periodically added. H_2 and CO_2 in the reactor vessels were periodically assayed by gas chromatography. Experiments with 5 ml free cell suspension which contained the equivalent amount of cells per flask in identical medium were also carried out as controls.

2.3. <u>Culture conditions of immobilized cells in agar gel matrix, and H_2 and CO_2 photoproduction assays</u>. Immobilized photosynthetic bacteria in the agar blocks (0.5 cm x 0.5 cm x 0.3 cm for 50 ml reactor, 1 cm x 1 cm x 0.5 cm for 4 L reactor) were grown at 30oC and 120 $\mu E/m^2/sec$ (for 50 ml reactor) or in outdoor conditions (for 4 L reacator) under nitrogen atmosphere in the ammonia free culture medium which was described previously (Ohta, Mitsui, 1981). Hydrogen photoproduction was carried out by the same procedures as mentioned above, after replacing the nitrogen gas in the reactor with argon gas and injecting electron donor substances, Na_2S and Malate (final concentrations of 10 mM and 20 mM, respectively). H_2 and CO_2 were measured by gas chromatography. H_2 gas was also measured by trapping the evolved gas into an inverted graduated cylinder over water with associated gas analysis by gas chromatography.

3. RESULTS

3.1. <u>Hydrogen photoproduction from sulfide by immobilized cells</u>. Hydrogen photoproduction by marine <u>Chromatium</u> sp Miami PBS 1071 from sulfide as a sole electron donor was observed. Optimum sulfur concentration for hydrogen production was 7-10 mM (data is not shown) and no CO_2 was evolved. Figure 1 shows the time course of hydrogen photoproduction from 10 mM sulfide by both immobilized cells and free cell suspension. Stable hydrogen photoproduction was observed for 90 hours in the immobilized cell system. When the electron donor in the reaction vessel was used up, the reaction was completed. This contrasts from free cell suspension systems where the initial lower rate of hydrogen production ceased in a shorter time period (40 hours) and hydrogen gas accumulated in the vessel was then consumed. Figure 2 shows continuous hydrogen photoproduction with a steady rate by immobilized cell system lasting for more than 300 hours, with periodic addition of sulfide into the reaction vessel. This periodic addition of sulfide to the free cell system did not improve hydrogen production.

3.2. <u>High density immobilized cultures in gel matrix and hydrogen production</u>. When a low density immobilized cells of strain PBS 1071 were cultured in agar gel matrix with molecular nitrogen as nitrogen nutrient source, immobilized cells grew fast and reached high biomass yields within 2 days (data is not shown). Growth of photosynthetic bacteria occurred only in the gel matrix. Hydrogen photoproduction activity varied with the culture age and had a peak at 36 hours culture.Figure 3 shows hydrogen production by densed immobilized cell culture with periodical addition of electron donors. High rates of hydrogen production (maximum, 6 μ mol/g gel/hr) were observed for 100 hrs.

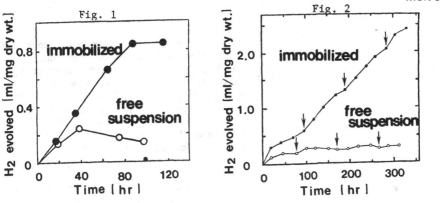

Fig. 1. Hydrogen photoproduction from sulfide by free cell suspension and immobilized cells of <u>Chromatium</u> sp. Miami PBS 1071. Each flask contained 2.4 mg cell dry weight.

Fig. 2. Continuous hydrogen production from sulfide by free cell suspension and immobilized cells of <u>Chromatium</u> sp. Miami PBS 1071. Each flask contained 2.9 mg cell dry weight. Arrows show the addition of sulfide (see text).

3.3 <u>Hydrogen production by immobilized cells under natural sunlight conditions.</u> Cells of strain PBS 1071 were immobilized in 1.5 kg agar gel and were grown with molecular nitrogen as a nitrogen nutrient source in a 4 L reactor under natural sunlight conditions. The light intensity peaked at 2000 $\mu E/m^2/sec$ during midday. In these conditions cells grew very well only in the gel matrix. Preliminary experiments indicated that hydrogen was produced during the day time with rates up to 1.5 μ moles/g gel/hr so far. Further improvement of this system is in progress.

4. DISCUSSION

Substantial amounts of sulfide are found in waste from coal gasification and other industrial operations, sewage, polluted river water, sea water and organic sediments. This is hazardous not only for human health, but also for other living organisms. As described above, the photosynthetic sulfur bacteria <u>Chromatium</u> sp. PBS 1071 can grow and produce hydrogen in high rates in the light using sulfide. Therefore energy production and removal of sulfide pollution could be simultaneouly achieved by this strain. However, in order to achieve this goal, an efficient system must be developed. Previously it was also shown that cell immobilization of a marine non-sulfur bacterium dramatically enhanced and stabilized the hydrogen production from organic substances(Matsunaga, Mitsui 1982; Mitsui <u>et al</u>., 1983). This study demonstrated that in immobilized cells of Marine <u>Chromatium</u> sp Miami PBS 1071 hydrogen photoproduction activity is stabilized and thus hydrogen production from sulfide was continued for long periods of time (Fig.1 and Fig.2). Moreover, the photosynthetic bacteria grow very well in the gel matrix and not in the surrounding liquid medium. During the process of hydrogen production the bacteria were not released from the gel but the electron donor substance are utilized from the surrounding liquid. In this immobilized cell system, all harvesting procedures (e.g. centrifugation) can be eliminated and growth and hydrogen production can be carried out in the same reactor. In the application of any solar energy related research, it is important

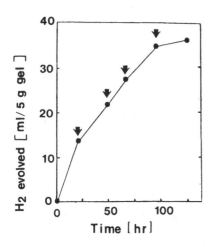

Fig. 3. Continuous hydrogen production by high densed immobilized bacteria cultured in N_2 for 36 hrs prior to incubation. Arrows show the addition of electron donors (see text).

to consider the maximum energy production per area or per culture volume. Towards this direction, high cell density culture of the strain PBS 1071 was established in immobilized gel matrix. This culture produced large amounts of hydrogen continuously by the periodic addition of electron donors (Fig.3). Outdoor preliminary experiments indicated that these indoor studies were basically applicable in natural sunlight conditions. However, the rate of hydrogen production per unit weight of gel could be improved in future studies.

5. REFERENCES

Mitsui, A. 1977. Bioconversion of solar energy in saltwater photosynthetic hydrogen production system. In Proc. First World Hydrogen Energy Conference. Vol. II. T.N. Veziroglu (Ed.) pp 4B--77-99. University of Miami Press. Miami.

Mitsui, A. et al. 1980. Photosynthetic bacteria as alternative energy sources: Overview on hydrogen production research. In Alternative Energy Sources. Vol. 8. Hydrogen Energy. T.N. Veziroglu (Ed.) pp. 3483-3510. Hemisphere Publ. Co., Washington, D.C.

Mitsui, A. et al. 1983. Progress in Research Toward Outdoor Biological Hydrogen Production Using Solar Energy, Seawater and Marine Photosynthetic Microorganisms. In: Biochemical Engineering III, New York Academy of Science, New York. (In Press)

Ohta, Y., J. Frank and A. Mitsui. 1981. Hydrogen production by marine photosynthetic bacteria: Effect of environmental factors and substrate specificity on the growth of hydrogen producing marine photosynthetic bacterium, Chromatium sp. Miami PBS 1071. Int. J. of Hydrogen Energy, 6, 451-460.

Ohta, Y. and A. Mitsui. 1981. Enhancement of hydrogen production by marine Chromatium sp. Miami PBS 1071 grown in molecular nitrogen. In: Advances in Biotechnology. Vol.II M. Moo-Young and C.W. Robinson. (Eds.). pp. 303-307. Pergamon Press. Toronto.

6. ACKNOWLEDGMENT

This work has been supported by the grant to A. Mitsui from the U.S. National Science Foundation, Solar Energy Research Institute, U.S. Department of Energy and U.S. Aernautics Space Administration.

BIOPHOTOLYSIS: GENERATION OF LOW-POTENTIAL REDUCING EQUIVALENTS BY PHOTOSYSTEM I-ENRICHED SUBCHLOROPLAST VESICLES

KLAAS KRAB, RONALD BOOG AND FONS A.L.J. PETERS

1. INTRODUCTION

It has been suggested (Hoffmann et al. 1977; Krasnovsky et al. 1980; Rosen, Krasna, 1980) that cyclic electron transfer around photosystem I in chloroplasts is a limiting factor in systems for the biophotolytic production of molecular hydrogen. Indications for this have come from studies with 3-(3', 4'-dichlorophenyl)-1,1-dimethylurea (DCMU)-inhibited broken chloroplasts supplied with donors such as ascorbate (+ N,N,N',N'-tetramethyl-p-phenylene-diamine (TMPD)), cysteine or dithioerythritol (DTE). Since the introduction of such systems for the production of H_2 (Arnon et al. 1961) it has been shown that a high rate of H_2 production can be obtained with the combination of electron donors DTE and ascorbate (+ TMPD) (Hoffmann et al. 1977; cf. Krasnovsky et al. 1980 and Rosen, Krasna, 1980). It has been proposed that this combination of donors strongly disfavors cyclic electron transfer. We have studied this phenomenon in a system where photosystem I-enriched sub-chloroplast vesicles (lacking photosystem II) catalyse light-driven reduction of methylviologen, and the membrane-bound hydrogenase of whole cells of *Proteus mirabilis* produces H_2 from reduced methylviologen.

2. MATERIALS AND METHODS

Photosystem I (PSI)-enriched vesicles were prepared according to Peters et al. (1983), and were fixed before use with 15 mM glutaraldehyde (West, Packer, 1982). *Proteus mirabilis* cells were grown anaerobically as described before (Krab et al. 1982), and stored at -80°C until needed. H_2 production was measured manometrically at 25°C in a medium containing 20 mM TES, 10 mM NaCl, 5 mM $MgCl_2$ and 2 mM NH_4Cl at pH 7. Experiments in which the reduction of methylviologen was measured were carried out in the same medium in a multipurpose cuvette (Kraayenhof et al. 1982). Chlorophyll concentrations were determined according to Bruinsma (1961) and determinations of bacterial protein were done according to Lowry et al. (1951).

3. RESULTS

3.1. Dependence of H_2 evolution on the donorsystem

The more than additive effect of DTE and ascorbate on H_2 evolution in a system with chloroplasts has been demonstrated at rather high concentrations of both reductants (Krasnovsky et al. 1980; Rosen, Krasna, 1980). The experiment of Table 1 shows that the same synergistic effect is found with PSI-enriched vesicles using fairly low (a few mM) concentrations of reductants. At these concentrations the rate of H_2 evolution with either ascorbate or DTE alone as donor is negligible, and any limiting donor is quickly exhausted, resulting in a stop of H_2 production. In Figure 1 the dependence is given of the initial rate and the total amount of H_2 evolution on DTE and ascorbate concentrations. The rate increases with both DTE and ascorbate concentration, to a saturation level. However, the total amount of H_2 produced does not depend on the ascorbate concentration, but is proportional to the initial DTE concentration. It amounts fo 65-85% conversion of DTE to H_2, depending on the batch of DTE. After exhaustion of DTE, H_2 evolution could be started again by addition of fresh DTE (not shown).

Sybesma, C. (ed.), Advances in Photosynthesis Research, Vol. II. ISBN 90-247-2943-2.
© *1984 Martinus Nijhoff/Dr W. Junk Publishers, The Hague/Boston/Lancaster.*

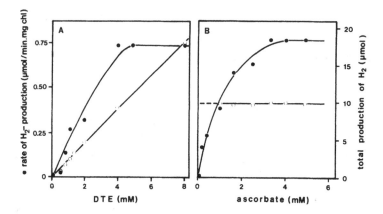

FIGURE 1. Initial rate and total extent of H_2 evolution as function of DTE and ascorbate concentrations. Conditions are 20 µM chlorophyll, 1 mg bacterial protein/ml, 100 µM TMPD and 50 µM methylviologen. A) 5 mM ascorbate and DTE as indicated; B) 4 mM DTE and ascorbate as indicated.

TABLE 1. H_2 evolution at high and low donor concentrations.
The experiments were carried out with 20 µM chlorophyll and 0.9 mg bacterial protein/ml in a volume of 3.5 ml. Also present were 100 µM TMPD and 50 µM methylviologen.

DTE mM	Ascorbate mM	Initial rate of H_2 evolution $\mu mol.min^{-1}.mg\ Chl^{-1}$
0	5	0
0	24	0.06
4	0	0
33	0	0.13
4	5	0.74

3.2. Reductive properties of the donors

As shown in Figure 2, light-dependent methylviologen reduction by the PS I-enriched vesicles under anaerobic conditions and in the absence of cells, shows the same synergistic effects as H_2 evolution does. Steady reduction states are reached due to slow leakage of O_2 into the cuvette; the higher the rate of methylviologen reduction, the higher the reduction level obtained. It is clear that the combination of DTE and ascorbate is much more effective than each of the donors apart. Without TMPD present, no reduction is seen in any case (not shown). We have confirmed the finding (Krasnovsky et al. 1980) that aerobic methylviologen reduction as measured in the Mehler reaction proceeds as rapidly with either of the donors alone as with the combination of the two, but, in contrast, at the used low donor concentrations TMPD is required.

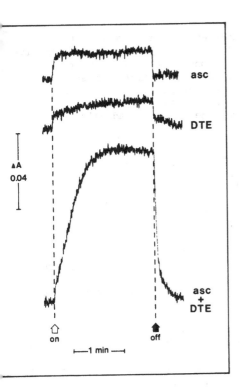

FIGURE 2. Light-induced anaerobic methylviologen reduction. Conditions: 20 μM chlorophyl, 100 μM TMPD and 50 μM methylviologen. Methylviologen reduction was measured at 604-540 nm. asc: 5 mM ascorbate present, DTE: 4 mM DTE present.

3.3. The effect of inhibitors of electron transfer

Neither addition of antimycin A, nor pretreatment of the vesicles with 1-ethyl-3-(3-dimethylaminopropyl)-carbodiimide (EDAC) had any effect on H_2 evolution in our system. However, pretreatment of the vesicles with 30 mM KCN (which inhibits electron transfer via plastocyanin) inhibits H_2 evolution. This suggests that reducing equivalents are mediated to the vesicles at the level of plastocyanin, in agreement with the requirement for TMPD.

4. DISCUSSION

At the low concentrations of donors used, it is clearly shown that both DTE and ascorbate are required for optimal H_2 production (Table 1). Due to the fact that in this system electron transfer between DTE and ascorbate is possible, and to the lower midpoint-potential of DTE, this requirement shows as a DTE-limited H_2 production and anaerobic methylviologen reduction. DTE and ascorbate donate reducing equivalents via the mediator TMPD. Aerobic methylviologen reduction shows that there is no intrinsic difference in efficiency in this between the two donors. Thus back reaction of reduced methylviologen with the electron transfer chain preceeding photosystem I (Hoffmann et al. 1977; Krasnovsky et al. 1980), which would be prevented by more efficient reduction of this chain by the combination of DTE and ascorbate, is less likely as an explanation of the observed synergism. Each of the donors is quite capable of maintaining TMPD in the reduced state. Back reaction of reduced methylviologen with dehydro-ascorbate (Krasnovsky et al. 1980; Myallem, Hall, 1982) may explain the low rate with ascorbate alone, but not with DTE alone as donor. Alternatively, the stimulating effect of DTE may be due to a specific reducing effect on a protein involved in cyclic electron transfer; at present we have no explanation for the stimulating effect of ascorbate, although it is note-worthy that ascorbate is an excellent scavenger of free radicals.

REFERENCES

Arnon DI, Mitsui A and Paneque A (1961) Photoproduction of hydrogen gas coupled with photosynthetic phosphorylation, Science 134, 1425.
Bruinsma J (1961) A comment on the spectrophotometric determination of chlorophyll, Biochim. Biophys. Acta 52, 576-578.
Hoffmann D, Thauer R and Trebst A (1977) Photosynthetic hydrogen evolution by spinach chloroplasts coupled to a *Clostridium* hydrogenase, Z. Natur-forsch. 32e. 257-262.
Kraayenhof R, Schuurmans JJ, Valkier LJ, Veen JPC, Van Marum D and Jasper CGG (1982) A thermoelectrically regulated multipurpose cuvette for simul-taneous time-dependent measurements, An. Biochem. 127, 93-99.
Krab K, Oltmann LF and Stouthamer AH (1982) Linkage of formate hydrogen-lyase with anaerobic respiration in *Proteus mirabilis*, Biochim. Biophys. Acta 679, 51-59.
Krasnovsky AA, Chan Van Ni, Nikandrov VV and Brin GP (1980) Efficiency of hydrogen photoproduction by chloroplast-bacterial hydrogenase systems, Plant Physiol. 66, 925-930.
Lowry OH, Rosebrough NJ, Farr AL and Randall RJ (1951) Protein measurement with the Folin phenol reagent, J. Biol. Chem. 193, 265-275.
Muallem A and Hall DO (1982) Ascorbate as a substrate for photoproduction of hydrogen by photosystem I of chloroplasts, Plant Physiol. 69, 1116-1120.
Peters FALJ, Van Wielink JE, Wong Fong Sang HW, De Vries S and Kraayenhof R (1983) Studies on well coupled photosystem I-enriched subchloroplast vesicles. Content and redox properties of electron-transfer components, Biochim. Biophys. Acta 722, 460-470.
Rosen MM and Krasna AI (1980) Limiting reactions in hydrogen photoproduction by chloroplasts and hydrogenase, Photochem. Photobiol. 31, 259-265.
West J and Packer L (1970) The effect of glutaraldehyde on light-induced H^+ changes, electron transport and phosphorylation in pea chloroplasts, J. Bioenergetics 1, 405-412.

ACKNOWLEDGEMENTS

The authors would like to thank K.J. Appeldoorn and J.J. Roosenstein for their contributions in the initial stages of this work and Drs. R. Kraayenhof and L.F. Oltmann for helpfull discussions.

Authors address: Biological Laboratory, Vrije Universiteit, De Boelelaan 1087 1081 HV Amsterdam, The Netherlands.

PHOTOPRODUCTION OF H_2 AND $NADPH_2$ BY POLYURETHANE-IMMOBILIZED CYANOBACTERIA

A. MUALLEM, D.L. BRUCE AND D.O. HALL

SUMMARY
Photosystem I (PSI) of permeable cells of Nostoc muscorum grown and entrapped in polyurethane foam is able to accumulate reduced methyl viologen (MV^+) with Na-ascorbate as the electron donor. This activity is long lasting under continuous illumination (2-3 weeks).
By coupling MV^+ with exogenous hydrogenase or Fd-NADP reductase it is possible to utilize continuously the electron flow from the electron donor via PSI for photoproduction of H_2 or $NADPH_2$.

INTRODUCTION
Permeabilization of the thermophilic cyanobacteria Mastigocladus laminosus (grown at 45°C) and Phormidium laminosum (by freezing and thawing) which are able to couple light driven electron transport from ascorbate via PSI with exogenous hydrogenase or Pt-colloids will evolve H_2 in the presence of MV as an electron relay [Smith et al. 1982]. The photosynthetic O_2 evolution and respiration of the permeabilized cells are completely inactivated. We have also reported [Muallem et al. 1983] that N. muscorum and M. laminosus (grown at 32°C) do not need freezing and thawing in order to become permeable. The advantage of permeable cyanobacterial cells over isolated chloroplasts is that they have a far more stable PSI which can operate continuously for 2-3 weeks.
Due to the adhesive property of the cell wall [Shilo and Fattom 1982] we were able to grow and entrap the cyanobacterial cells in polyurethane foams [Muallem et al. 1983]. Since permeable cells did not lose their adhesive property it was possible to use them for long periods as entrapped cells. The white polyurethane foam used here is a porous and inert solid support which has been shown as adequate for the immobilization of living cells [Brouers et al. 1982; Gudin et al. 1982; Lindsey et al. 1983]. Musgrave et al. (1982) have immobilized algae in alginate.
Immobilized cells of N. muscorum are used for continuous photoproduction of H_2 and NADPH by PSI in the presence of hydrogenase and Fd-NADP-reductase.

MATERIALS AND METHODS
Cells of cyanobacteria were grown under 5% CO_2 in air [Muallem et al. 1983]. For immobilization we added about twenty pieces of fabricated foam (0.5 x 0.5 x 0.5 cm) to each 250 ml flask containing 120 ml of growth medium [Allen and Arnon, 1955] before autoclaving and innoculating with cyanobacteria. Here we report experiments with N. muscorum immobilized in a polyvinyl type foam coded D or a polyester type coded 4300A, both provided by Caligen Co., Accrington, U.K.

RESULTS
Cells of N. muscorum (permeable) entrapped in the D type foam are shown adhering to the polyvinyl fibres (Fig. 1).

Sybesma, C. (ed.), Advances in Photosynthesis Research, Vol. II. ISBN 90-247-2943-2.
© *1984 Martinus Nijhoff/Dr W. Junk Publishers, The Hague/Boston/Lancaster.*

Samples of such immobilized cells were used here. The conditions required for photoproduction of H_2 from ascorbate using PSI of chloroplasts or permeabilized cyanobacteria are discussed in the following references [Muallem et al. 1983; Hoffman et al. 1977; Krasnovsky et al. 1980; Muallem and Hall, 1982]. Similar conditions are used for continuous H_2 photoproduction by N. muscorum immobilized in D foam (Fig. 2). It is sufficient to suspend the harvested cells in a new growth medium lacking cells and $MgSO_4$ (or in a pH 7.5 buffer solution) in order to obtain permeable cells which are able to accumulate photoreduced MV in the reaction mixture. The accumulation of MV^+ in the extracellular medium is a necessary requirement for H_2 evolution in the presence of exogenous hydrogenase [Smith et al. 1982]. Under these conditions we are also able to couple the electron transport from ascorbate with spinach Fd-NADP-reductase and thus photoreduce exogenous NADP. Lower rates of NADP photoreduction by the immobilized Nostoc cells occur in the absence of MV. Probably endogenous ferredoxins are able to transfer electrons to the added enzyme. No NADP photoreduction can be observed without added reductase and very low rates are generally measured under conditions which do not support accumulation of MV^+ in the reaction mixture (e.g., presence of O_2, or absence of glutathione which is used as a synergetic reductant to limit the back reaction of MV^+ with photoxidized ascorbate [Krasnovsky et al. 1980].
Research is under way in order to replace the enzymes with inorganic catalysts to improve the feasibility of using immobilized cells for continuous photoproduction of $NADPH_2$.

REFERENCES
Allen MB and Arnon DI (1955) Plant Physiol. 30, 366;372
Brouers M Collard F Jenafils J and Jeason A (1982). In Photochemical, Photoelectrochemical and Photobiological Processes. Vol. 1, Hall DO and Palz W, eds. pp. 34-139. D. Reidel Publ., Dordrecht.
Gudin C Thepenier C Thomas D and Chaumont D (1982). In Solar World Forum, Hall DO and Morton J, eds. pp. 2255-2259. Pergamon, Oxford.
Hoffman D Thauer R and Trebst A (1977) Z. Naturforsch. 32C, 257-262.
Krasnovsky AA Chan Van Ni V Vikandrov VV and Brin GP (1980) Plant Physiol. 66, 925-930.
Lindsey K Yeoman MM Black GM and Marituna F (1983) FEBS. Lett. 155, 143-149.
Muallem A Bruce DL and Hall DO (1983) Biotech. Lett. 5, 365-368.
Muallem A and Hall DO (1982) Plant Physiol. 69, 1116-1120.
Musgrave SC Kerby NW Codd GA and Stewart WDP (1982) Biotech. Lett. 4, 647-652.
Rao KK and Hall DO (1979) In Photosynthesis in relation to model systems, Ch. 10, Barber J, ed. Elsevier, Amsterdam.
Shilo M and Fattom A "IV Symposium on Photosynthetic Prokaryotes"held September 1982, Bordeaux, France. To be published by Soc. Gen. Microbiology, Reading, U.K.
Smith GD Muallem A and Hall DO (1982) Photochem. Photophys. 4, 307-320.

LEGENDS

Fig. 2
Continuous H$_2$ photoproduction (μmoles/mg chlorophyll) from ascorbate
by **N. muscorum**and **M. laminosus** free and immobilized cells. Reaction
mixtures (4 ml) contained 50 mM Na ascorbate, 0.5 mM TMPD, 40 μM DCMU,
300 μM MV, 20 μℓ **D. desulfuricans** hydrogenase solution (1760 μmoles H$_2$
evolved/ml x hour), 50 mM HEPES pH 7.5, 20 μM Gentamycin and GSH 7.5
mM. The initial amounts of chlorophyll in different reaction mixtures
containing free cells and immobilized in foams D and 4300A are 90,
110, 70 μg chlorophull respectively. Where indicated (arrows) fresh
reaction mixtures were used (in order to eliminate possible chemical
deterioration in the reaction mixtures) after washing with buffer
solution both the immobilized and the free cells (the latter collected
by centrifugation). The reaction mixtures were put in closed vials
flushed with N$_2$ for 10 minutes and then illuminated (tungsten lamps,
6.4 x 10^4 ergs^{2}cm^2 sec^{-1}) in a shaking water bath at 30° C. H$_2$ was
assayed chromatographically [Rao and Hall, 1979].

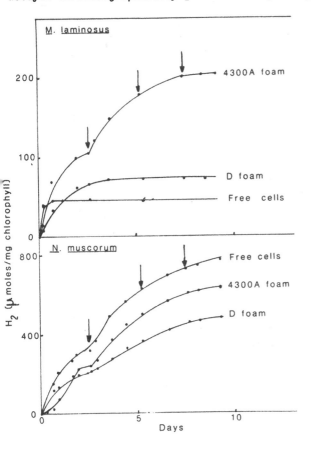

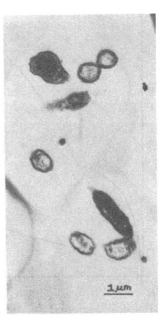

Fig. 1

Figure 3
Photoproduction of NADPH (μmoles/mg chlorophyll) by N. muscorum cells
immobilized in polyurethane foam (type D). Reaction mixture (6 ml) as
in Fig. 1. Only the hydrogenase is replaced by 200 μℓ of Fd-NADP
reductase (1 unit/ml)solution, MV 160 μM in one of the reaction
mixtures and no MV in the other and initially 1 μmole of NADP was
present in each reaction mixture. NADP photoreduction was carried out
in N_2 flushed and closed vials as for H_2 (Fig. 1). Changes of
absorption at 340 nm were measured in aliquots withdrawn at different
times and the accumulated amounts of NADPH were calculated. A
light-dark difference spectrum of the aliquots showed a peak near 340
nm.

Author's address: King's College London, 68 Half Moon Lane,
                    London SE24 9JF, U.K.

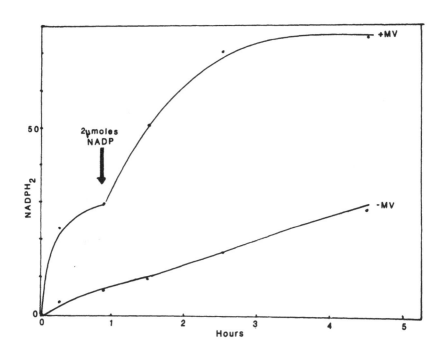

DEVELOPMENT OF HYDROGEN PRODUCTION ACTIVITY IN THE MARINE BLUE-GREEN ALGA OSCILLATORIA SP. MIAMI BG 7 UNDER NATURAL SUNLIGHT CONDITIONS

E. J. PHLIPS AND A. MITSUI* School of Marine and Atmospheric Science, University of Miami, 4600 Rickenbacker Causeway, Miami, Florida 33149, U.S.A.

SUMMARY

It was shown that biological hydrogen production by Oscillatoria sp. Miami BG 7 is adaptable to the outdoor environment. The maximum rate observed was 8.6 ml H_2/mg dry wt./hr (153 μ moles/mg Chl/hr). The maximum yield obtained was 1.9 ml H_2/ml cell suspension/day. There appears to be no major barrier to obtaining high rates and yields of hydrogen production in outdoor systems. This fact greatly enhances the potential future of hydrogen production as a source of energy and chemicals.

I. INTRODUCTION

In the early 1970's several researchers proposed that biological hydrogen production could be a source of fuel. This prospect of developing a renewable and clean source of energy stimulated a decade of active laboratory research into this phenomenon (ref. Mitsui, 1979). As a result of this effort our understanding of hydrogen production, and other metabolic processes associated with it (eg. nitrogen fixation), has been greatly advanced (Kumazawa & Mitsui, 1982). In addition a number of microorganismal candidates for use in hydrogen production systems have been identified. The next major step toward achieving the goal is to test and implement actual systems for hydrogen production in the natural environment. Very little information is available in this area. This paper deals with preliminary results of experiments with biological hydrogen production in simple reactors, using a marine blue-green alga.

2. MATERIAL AND METHODS

2.1 Organism and Culture Conditions - Axenic cultures of Oscillatoria sp. Miami BG 7, a non-heterocystous form of filamentous marine blue-green algae, from the culture collection of this laboratory were used (Mitsui, Kumazawa 1977). Culture medium was described previously by Kumazawa and Mitsui (1981).

2.2 Hydrogen Production Assay - Production of hydrogen gas was monitored by the method described by Kumazawa and Mitsui (1981).

2.3 5 - Liter Outdoor Reactor Systems - Most of the outdoor hydrogen production experiments were done in a series of enclosed glass reactors (Mitsui et al., 1983). A glass cover plate was used to seal the top of the reactor. This cover contained sampling ports, a temperature probe, and a temperature control coil. Cells were directly transfered from the culture vessels to the reactor. 4 to 4.5 liters of culture were used in most of the tests, leaving approximately a liter of gas phase. The suspension of cells was stirred with magnetic stirring bars. The

---

*To whom the reprint should be requested.

Sybesma, C. (ed.), Advances in Photosynthesis Research, Vol. II. ISBN 90-247-2943-2.
© *1984 Martinus Nijhoff/Dr W. Junk Publishers, The Hague/Boston/Lancaster.*

reactor units were mounted on angular stands to ensure proper orientation to the sun. At the beginning of each experiment the reactors were sealed and flushed with argon gas and then filled with argon gas. In some of the experiments neutral screens were used to cut out a percentage of the incident light.

2.4   High Cell Density Experiments - Experiments with high cell density suspensions were done in 25 ml fernbach flasks. Cells for these experiments were cultured in 20 liter carboys, harvested by centrifugation, and resuspended in media not containing combined nitrogen, A-N (Kumazawa, Mitsui 1981). 5 ml of the resuspended solution was added to the 25 ml fernbach flasks. Before placing the flasks in the outdoor shaker tables they were sealed with stoppers and flushed with argon. Temperature was maintained at 35°C.

2.5   Measurement of incident light intensity - Changes in light intensity were continuously monitored with a Licor model 185B meter and quantum probe (reading par light). The output of the meter was logged on a continuous chart recorder.

3.   RESULTS AND DISCUSSION

The development of hydrogen activity in outdoor reactors exhibited basically the same characteristics as observed in previous laboratory tests (Mitsui, Kumazawa 1977, Kumazawa, Mitsui 1981, Phlips, Mitsui 1983). Hydrogen production was inhibited by concentrations of oxygen greater than 1%. As a consequence, air in the gas phase of the reactor was replaced with argon. Even so initial exposure of the reactor to full sunlight resulted in sufficient photosynthetic oxygen evolution to inhibit hydrogen production. This inhibition was alleviated by flushing the reactor a second time with argon (Figure 1). After this procedure was performed no further oxygen accumulation was observed and hydrogen production activity was stimulated. Two light day periods were required for full activity to develop (Figure 2). No hydrogen production or consumption was observed in the dark.

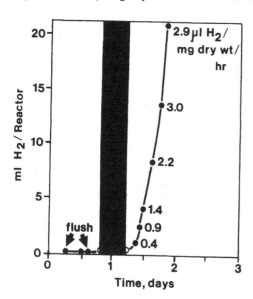

Figure 1.   Development of hydrogen production activity in an outdoor five liter reactor. The numbers alongside the curve indicate the rate of hydrogen production between successive sampling points. The black band indicates night time. The reactor contained 800 mg dry wt. of nitrogen-depleted cells. The gas phase of the reactor was flushed twice with argon, as shown. Temperature was 32.5°C.

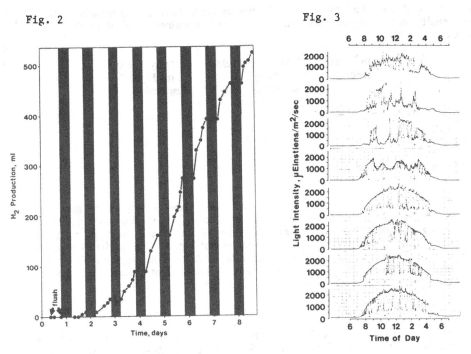

Figure 2. Hydrogen production in an outdoor five liter bioreactor. The reactor contained four liters of algal suspension. The black bands indicate night time. The reactor contained 2,500 mg dry wt. cells. The reactor was flushed twice with argon, as described above, and the temperature was 32.5oC.

Figure 3. Light intensity pattern during the last eight days of the experiment shown in figure 2. Light intensity is given in μ E/m^2/sec. (Fig. on left).

In a separate series of tests it was shown that high rates of hydrogen production could be maintained in outdoor reactors for more than a week. (Figure 2). The maximum rates of hydrogen production observed were the same as those found in laboratory scale experiments. In this experiment the maximum rate observed was 8.57 ml H$_2$/mg dry wt./hr. Since nitrogen starved cells used were a low concentration of chlorophyll (Kumazawa, Mitsui 1981), this rate was 153 μ moles/mg Chl /hr. Although the rates were high the total yield of hydrogen per reactor was small because of the low cell density used. In actual applied systems higher cell densities could be used to enhance the output of H per unit volume of reactor. The amount of solar radiation available during the experimental period is shown in figure 3. This experiment established that outdoor reactor systems can yield high rates of hydrogen production even under conditions of natural irradiation.

The final series of experiments were set up to examine the potential yield of hydrogen gas per unit volume. For these tests high density suspensions of cells were used, ie. 10 mg dry wt./ml suspension. In order to avoid problems with self-shading, small flasks were used in the experiment. The results of a three day incubation period in an outdoor shaker bath system are summarized in Table 1.

Table 1. Outdoor Production of hydrogen by high density suspensions of cells. Cell density was 10 mg dry wt./ml. Flasks used contained 5 ml of suspension at 35° C.

	Maximum Rate of Production ml H_2/mg dry wt/hr	Maximum Yield mlH_2/ml suspension/day	Daily Yield ml H_2/ml suspension/day
First Day	--	--	0.3
Second Day	7.8	1.9	0.83
Third Day	6.6	1.6	0.70

Both the maximum rates of production and daily yields of hydrogen gas were similar with results obtained in laboratory conditions. The peak yields of hydrogen gas during the midday period were almost twice as high as those observed in the laboratory for dense suspensions of algae. This probably resulted from the fact that peak light intensity of solar radiation (ie. in South Florida) is an order of magnitude greater than most normal indoor lighting systems. Thereby the problem of self-shading is considerably reduced.

REFERENCES

Kumazawa S and Mitsui A (1982) Hydrogen metabolism of photosynthetic bacteria and algae. In: CRC Handbook of Biosolar Resources. Vol.1 Basic Principals. eds. A. Mitsui and C.C. Black (series ed. O.R.Zaborsky) CRC Press, Boca Raton, Florida. Part 1. pp. 299-316.
Mitsui A (1979) Biological and Biochemical Hydrogen Production. In: Solar-Hydrogen Energy Systems. ed. T.Ohta. Pergamon Press, Oxford, New York. Chapter 8. pp. 171-191.
Mitsui A and Kumazawa S (1977) Hydrogen photoproduction by tropical marine photosynthetic organisms as a potential energy resources. In: Biological Solar Energy Conversion. eds. A. Mitsui, S. Miyachi, A. San Pietro and S. Tamura. Academic Press, New York, pp. 23-51.
Kumazawa S and Mitsui A (1981) Characterizarion and optimization of hydrogen photoproduction by a salt water blue-green alga, Oscillatoria sp. Miami BG 7. I. Enhancement through limiting the supply of nitrogen nutrients. International J. Hydrogen Energy. 6, 341 - 350.
Mitsui, A., E.J. Phlips, S. Kumazawa, K.J. Reddy, S. Ramachandran, T. Matsunapa, L. Haynes and H. Ikemoto (1983). Progress in research toward outdoor biological hydrogen production using solar energy, seawater, and marine photosynthetic microorganisms, in: Biochem. Eng. New York Acad. Sci. (In press).
Phlips E and Mitsui A (1983) Role of light intensity and temperature in the regulation of hydrogen photoproduction by the marine cyanobacterium Oscillatoria sp. strain Miami BG 7. Appl. and Environ. Microbiol. 45: 1212 -1220.

ACKNOWLEDGEMENTS

This work has been supported by grants to A. Mitsui from the U.S. National Science Foundation and the U.S. National Aeronautic and Space Administration.

EFFECT OF SEAWATER QUALITY ON BIOMASS AND HYDROGEN PHOTOPRODUCTION BY A MARINE BLUE-GREEN ALGA OSCILLATORIA SP. (MIAMI BG 7)

S. RAMACHANDRAN AND A. MITSUI* SCHOOL OF MARINE AND ATMOSPHERIC SCIENCE, UNIVERSITY OF MIAMI, 4600 RICKENBACKER CAUSEWAY, MIAMI, FLORIDA, 33149, U.S.A.

1. INTRODUCTION:

The advantages of "bioconversion of solar energy to hydrogen" in sea water based systems are numerous and have been discussed previously (Mitsui, Kumazawa 1977). When one considers the possible world wide application of this system, the most important question to be answered is how will the regional and seasonal variations in natural sea water quality influence the high biomass yield and hydrogen production activity obtained in the laboratory in an artificial seawater medium (Kumazawa, Mitsui 1981, Phlips, Mitsui 1983, Mitsui et al 1983). In order to answer this question, a study was made on the effects of such variations in natural seawater as salinity, pH and tracemetal concentration on biomass yield and hydrogen production capability in a marine blue-green alga Oscillatoria sp. Miami BG 7.

2. MATERIAL AND METHODS

2.1 Culture conditions: Marine Oscillatoria sp. Miami BG 7 was cultured in natural seawater enriched with f/2 medium (Guillard, Ryther 1962) at 25°C and 200 uE/m2/sec. Seawater was collected from the Gulfstream off Florida, with 35 o/oo salinity and 0.08 to 0.1 mg/l of total combined nitrogen. Cultures were harvested by centrifugation (5000 x g, 10 min) and washed twice with simple salt solution for hydrogen production and photosynthetic assay. Simple salt solution contains NaCl, 55mM; CaCl, 1mM; KCl 10mM and MgSO4, 10mM. Dry weight was measured after drying the distilled water washed algae at 90°C for 24 hrs.

2.2 Assays of photosynthesis and hydrogen production. The net photosynthetic activity, photoevolution of oxygen was measured using a Clark type oxygen electrode at 25°C and 500uE/m2/sec (Izawa, Kumazawa, Mitsui, unpublished). The reaction mixture contained the following: 0.9 ml of the blue-green algal suspension, (5 mg dry wt.); 0.5 ml of 2 times concentrated simple salt solution as described above; 0.1 ml of 100 mM sodium bicarbonate; 0.1 ml of 20 mM EDTA and 0.1 to 0.4 ml of test solution (e.g, metals). The final volume was adjusted to 2 ml with deionized water. Hydrogen producing cells were obtained by harvesting 13 to 15 day old cells from early stationary phase cultures. Hydrogen production assay was performed as described by Kumazawa, Mitsui (1981).

2.3 Salinity, pH and metals. Sea water at 35%o salinity was diluted with distilled water for lower salinities and NaCl was added for higher salinities. The desired pH was obtained by the addition of suitable amounts of 0.5 N NaOH or 0.5 N H Cl and of three different buffers MES (pH 6.5), MOPS (pH 7 and 7.5) and TRIS (pH 8-9) at a final concentration of 0.5 mM.

---

* To whom reprint requests should be addressed.

Sybesma, C. (ed.), Advances in Photosynthesis Research, Vol. II. ISBN 90-247-2943-2.
© *1984 Martinus Nijhoff/Dr W. Junk Publishers, The Hague/Boston/Lancaster.*

3.   RESULTS AND DISCUSSION:

3.1   Effect of salinity on biomass and hydrogen photoproduction: Salinity
changes by dilution through rain fall and riverwater inflow and by concentration
through evaporation.   The biomass yield and hydrogen production were examined
in the range of probable salinity changes in the environment i.e.   between 5 and
45 o/oo.   When strain BG 7, which was pre-cultured in an artificial seawater based
medium (18 o/oo salinity) was inoculated into the seawater of different salinities,
some differences in biomass yields were observed (Fig. 1A).   However, when the
inoculum   was   pre-acclimatized   to   growth   salinities   for   2   generations, the
differences   were   less   pronounced   among   the   tested   salinities   (Fig 1 B).   The
biomass yields of these particular batch culture experiments were 220-260 mg dry
wt/l culture for 9 day old cultures.   Recent results however, indicated that the
yield could be increased to 200 mg dry wt/ 1 culture/day or more by a continuous
culture system (Ramachandran,   Mitsui, unpublished data and Mitsui et al 1983).
The rates of hydrogen production in different salinities are shown in Table 1.
There are no significant differences in hydrogen production between 5-35 o/oo
Even at the higher salinity of 45 o/oo, the activity remained at 60.6% of the
maximum. The maximum rate of hydrogen photoproduction was 6.4 ul H_2/mg dry
wt/hr or 0.29 u mol H_2 /mg dry wt/hr.   Since hydrogen producing cells contain a
low concentration of chlorophyll (0.15%),on a chlorohyll basis this rate was 190 u
mol H_2 /mg chl/hr.   The rates which were obtained for the cells cultured in the
natural seawater based medium were approximately the same as those of cells
cultured in artificial medium A-N. (Kumazawa, Mitsui 1981; Phlips, Mitsui 1983).

Fig.   1.   Effect   of   Salinity   on
biomass yield.   A.   The alga was
cultured   in   18 o/oo   salinity   and
inoculated into different salinities
B.   The alga was pre-acclimitized
for   2   generations   in   the   growth
salinities.

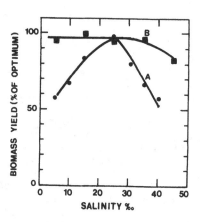

TABLE   1.   Effect   of   salinity   on
hydrogen   photoproduction   by
Oscillatoria sp Miami BG 7.

Salinity: o/oo	5	10	15	20	30	35	45
Rate of H_2 production							
ul/mg dry wt/hr	5.9	6.4	6.0	6.0	5.8	5.2	3.9
u mol /mg chl/hr	176	190	179	179	173	155	116

.2 Effect of pH on the biomass yield and hydrogen photo production. Effect of pH on the biomass yield and hydrogen production activity in seawater based medium showed that the biomass yields at higher pH values are not significantly different from that in optimum pH. In lower pH it was more affected than in higher pH. However, at pH 6.5 which exists only in extreme cases of natural seawater, 62% of the maximum yield was obtained (Fig 2A). There was no significant difference in hydrogen production activity in all pH ranges tested (Fig 2B).

.3 Effect of metal concentrations on net photosynthesis and hydrogen photoproduction. Effect of tracemetals Co, Cu, Mo, Zn and Ni up to concentrations of 1.0 mg/l on net photosynthesis (measured as O_2 evolution) was studied and the results are shown in figure 3. Ranges of concentrations of these metals in natural seawater are also indicated in figure 3. Net photosynthetic activity is not significantly different in different concentrations of these metals, except that inhibition is observed in higher concentrations between 0.05 and 0.1 mg/l for Ni. This concentration is much higher than that of natural seawater

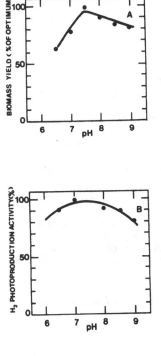

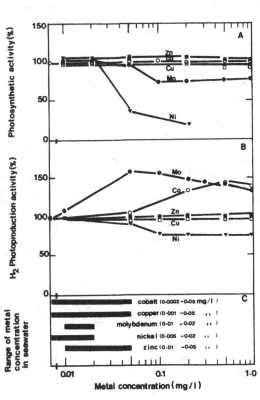

Fig. 2. Effect of pH on biomass yield (A) and hydrogen hydrogen photo production activity (B).

Fig. 3. Effect of tracemetals on photosynthetic (A) and H_2 photo production activity (B).

level. As far as hydrogen photoproduction activity is concerned among the metals studied, Mo and Co exerted pronounced enhancement. Hydrogen production was enhanced to 160% at a concentration of 0.5 mg/l by Mo and to 140% at 0.2 mg/l by Co. Enhancement by Mo and Co was due to the effect of these metals on the enzyme nitrogenase which catalyzes hydrogen production in this organism. Addition of these metals to hydrogen production medium can enhance and stabilize hydrogen production activity. Hg and Cd, the possible pollutants in seawater, were also tested in concentrations up to 0.1 mg/l which is much higher than the expected levels in polluted seawater. At this concentration level both net photosynthesis and hydrogen production activity were not affected, (data not shown).

SUMMARY:

High biomass yields and hydrogen production rates were obtained in natural seawater based media using a marine blue green alga, Oscillatora sp. Miami BG 7. Changes in seawater quality such as salinity, pH and trace metal concentrations in the probable range of occurrence in natural seawater did not significantly affect either the biomass yield (or production potential) or the hydrogen production activity. Therefore, this strain could be favourably used for hydrogen photo-production in many regions of the world which may have different quality of seawater. Data obtained also indicated that changes in seawater quality by possible environmental and seasonal changes in the hydrogen production site will not greatly affect the efficient operation of a blue-green algal culture-hydrogen production system using this strain.

REFERENCES

Guillard R R L and Ryther J H (1962) Studies of marine planktonic diatoms. I. Cyclotella nana Hustedt and Detonula confervacea (Cleve.) Gran. Can. J. Microbiol. 8, 229-239.
Kumazawa S and Mitsui A (1981) Characterization and optimization of hydrogen photoproduction by a salt water blue-green alga Oscillatoria sp. Miami BG 7. I. Enhancement through limiting the supply of nitrogen nutrients. Int. J. Hydrogen Energy 6, 339-348.
Mitsui A and Kumazawa S (1977) Hydrogen production by marine photosynthetic organisms as a potential energy source. In Mitsui A , Miyachi S , San Pietro A, and Tamura S. ed. Biological Solar Energy Conversion, pp 23-51, Academic Press, New York.
Mitsui A, Phlips EJ, Kumazawa S, Reddy KJ, Ramachandran S, Matsunaga T, Haynes, L and Ikemoto H, (1983). Progress in research toward outdoor biological hydrogen production using solar energy, seawater and marine photosynthetic microorganisms. Biochemical Engineering III. New York Academy of Sciences (In Press).
Phlips, EJ and Mitsui A (1983) Role of light intensity and temperature in the regulation of hydrogen photoproduction by the marine cyanobacterium Oscillatoria sp. Miami BG 7, Appl. Environ. Microbiol. 45, 1212-1220.

ACKNOWLEDGEMENTS:

This work has been supported by grants to A Mitsui from the U.S. National Science Foundation (CPE 8018142) and the U.S. National Aeronautics and Space Administration (NAS 10-10531).

MIXOTROPHIC GROWTH OF NOSTOC 268

TIMO VAARA[*], KAARINA SIVONEN AND SIRPA KURKELA

1. INTRODUCTION

In recent years, a worldwide interest has arisen in the feasibility of growing microalgal biomass for commercial purposes (Shelef, Soeder 1980). The first steps to carry out microalgal technology at an industrial level have already been taken. In India, a dried preparation of nitrogen-fixing blue-green algal (or cyanobacterial) cultures is supplied to rice farmers. In California, a number of high-rate algae ponds have been in operation for more than a decade. In the Far-East, a few thousand tons of Chlorella were produced in 1977 and sold mainly as health food at a price of approximately $100/kg. In Japan, Mannanfoods Co. extracts phycocyanin from Spirulina and sells the product as a natural food colouring agent. In Mexico, Sosa Texcoco S.A. grows Spirulina as a by-product of soda-manufacturing from carbonate brine and harvests 2.5 tons a day. The company projects future capacity of up to 50 tons a day. Proteus Inc. in California distributes Sosa Texcoco's Spirulina to American fish farmers.

One of the greatest problems in microalgal technology is the provision of sufficient carbon in the growth medium. CO_2 is the carbon source of auto-trophic growth of algae and its low diffusivity across the air-water interface severely limits the productivity of algal cultures. To obtain maximal yields, cultures must be gassed with air that has been artificially enriched with CO_2. This, however, leads to high production costs. One solution could be to grow algae mixotrophically. In a mixotrophic mode of growth, algae grow in the light and they assimilate CO_2 and organic carbon sources at the same time. Many algae are able to grow not only on CO_2 but also on a great variety of organic carbon sources that are much easier to provide than CO_2. We have been developing the mixotrophic cultivation of a nitrogen-fixing cyanobacterium, Nostoc 268.

2. MATERIALS AND METHODS

Nostoc 268 was a kind gift from Prof. B. Gromow (University of Leningrad, Leningrad, USSR). It was grown without fixed nitrogen on medium BG-11 (Stanier et al. 1971). Its growth was estimated on dry weigth basis. The photosynthetic oxygen evolution and the respiratory oxygen consumption in the dark were followed in a water-jacketed 5mL cell fitted with a Clark O_2 electrode; the incubation took place for 15 min at $28^{\circ}C$ under the illumination of 3000 lux. Chlorophyll a content was calculated from the absorbance of the 90% methanol extract at 665 nm, using the extinction coefficient of 13.9. The assay of nitrogenase activity was by acetylene reduction in samples kept for 1 and 3 hours at $28^{\circ}C$ under the continuous illumination of 3000 lux. Ammonia excretion was estimated by measuring the amount of NH_3-N in a 9-day-old culture by indophenol blue reaction. The sugar content of the growth medium was determined by high-pressure liquid chromatography.

Sybesma, C. (ed.), Advances in Photosynthesis Research, Vol. II. ISBN 90-247-2943-2.
© 1984 Martinus Nijhoff/Dr W. Junk Publishers, The Hague/Boston/Lancaster.

3. RESULTS AND DISCUSSION

Nostoc 268 was able to grow on glucose and fructose. It grew on sucrose and maltose only very slowly and did not at all grow on acetate, galactose, lactose, mannose or xylose. As can be seen in Fig. 1., its mixotrophic growth on glucose and fructose was even better than its autotrophic growth in cultures gassed with CO_2. Glucose concentrations above 70 mM and fructose concentrations above 30 mM produced markedly higher yields than those obtained on CO_2. Highest yields, however, were obtained on the combination of glucose and fructose. As low as 5 mM glucose together with 5 mM fructose produced yields similar to those obtained on 70 mM glucose or 30 mM fructose alone. Clearly, glucose and fructose had a synergistic effect on the mixotrophic growth of Nostoc 268. An effect of basically similar type has been described by Wood and Kelly (1977) for Thiobacillus A2.

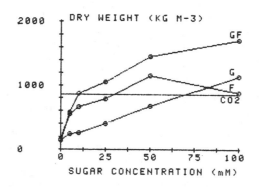

FIGURE 1. Mixotrophic growth of Nostoc 268 on various concentrations of glucose (G), fructose (F) and a combination of glucose and fructose (GF). The yield of the photoautotrophic growth on CO_2 is shown for comparison. The cultivation took place for 9 days at 26^{0}C under the continuous illumination of 3000 lux. Each point represents a mean of six replications.

The synergistic effect was evident in many aspects of the physiology of Nostoc 268 (Table 1.): Glucose and fructose inhibited photosynthetic oxygen evolution to a higher extent than they did separately (column A). This might be due to both a promotion of respiration (column B) and a decrease in chlorophyll a content (column C). Furthermore, the combination of glucose and fructose stimulated nitrogenase activity (column D) and probably also the utilization of fixed nitrogen because less ammonia was excreted into the growth medium (column E). This wide spectrum of effects suggests that the key effect must be the improved utilization of glucose and fructose when both are provided. And, in fact, by analyzing the amounts of glucose and fructose in the growth medium before and after the growth of Nostoc 268 (Table 2.), we found that the total sugar consumption was highest on the combination of glucose and fructose. The combination seemed to stimulate glucose consumption whereas that of fructose was inhibited.

TABLE 1. Photosynthetic oxygen evolution (A), respiratory oxygen consumption in the dark (B), chlorophyll a content (C), acetylene reduction (D) and ammonia excretion (E) of Nostoc 268 grown for 9 days (7 days for the assay of acetylene reduction) photoautotrophically on CO_2 or mixotrophically on various carbon sources. The cultivation took place as described in the legend to Fig 1.

Growth medium	A[1]	B[1]	C[2]	D[1]	E[3]
CO_2	463	123	4.8	119	600
50 mM glucose	544	176	4.1	130	340
50 mM fructose	432	248	3.8	217	270
25 mM glucose plus 25 mM fructose	204	276	2.7	226	240

[1] nmol/mg dry wt./h
[2] ug/mg dry wt.
[3] ng NH_3-N/mg dry wt.

TABLE 2. Sugar consumption in 9 days by Nostoc 268 grown mixotrophically on various carbon sources. The cultivation took place as described in the legend to Fig. 1.

Growth medium	Sugar consumption (mg/l)	
	Glucose	Fructose
50 mM glucose	950	–
50 mM fructose	–	1350
25 mM glucose plus 25 mM fructose	1550	275

Molasses is a cheap by-product of sugar-industry and contains glucose, fructose and also sucrose. It would thus be an ideal substrate for the mixotrophic growth of Nostoc 268, especially after treatment with invertase that splits sucrose to glucose and fructose. As can be seen in Fig. 2., the yields on 0.3% molasses (containing approximately 10 mM glucose and fructose) were as high and the yields on 1% molasses (containing approximately 30 mM glucose and fructose) three-times higher than those obtained with CO_2. This experiment was performed with sugarcane molasses but similar results were obtained also with beet molasses. Furthermore, to get even higher yields, we have recently isolated Nostoc 268 mutants with an accelerated growth on glucose, fructose and sucrose (data not shown).

Our results indicate that, at least in the case of Nostoc 268, it is possible to obtain higher yields with lower costs by growing algae mixotrophically instead of growing them under autotrophic conditions. Mixotrophic cultures have also an additional advantage over autotrophic cultures. Because they reach higher cell densities, they require less space and can be located inside which facilitates process control and harvesting. This could be an important advantage in small-scale cultures where algae are grown under well controlled conditions to obtain high-quality products.

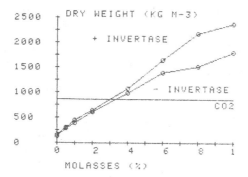

FIGURE 2. Mixotrophic growth of Nostoc 268 on various concentrations of sugarcane molasses before and after the treatment with invertase. The yield of the autotrophic growth on CO_2 is shown for comparison. The cultivation took place as described in the legend to Fig. 1.

REFERENCES

Shelef G and Soeder CJ, eds. (1980) Algae biomass, production and use. Amsterdam: Elsevier North-Holland Biomedical Press
Stanier RY, Kunisawa R, Mandel M and Cohen-Bazire G (1971) Purification and properties of unicellular algae (order Chroococcales), Bacteriol. Rev. 35,171-205
Wood AP and Kelly DP (1977) Heterotrophic growth of Thiobacillus A2 on sugars and organic acids. Arch. Microbiol. 113,257-264

ACKNOWLEDGEMENTS

This investigation was supported by a grant from the Finnish National Fund for Research and Development and a fellowship to TV from the Academy of Finland.
Authors address: Department of Microbiology, University of Helsinki
                  SF-00710 Helsinki 71, Finland

IMMOBILISATION OF PS I AND PS II ON SEMICONDUCTING PARTICLES.

P. CUENDET and M. GRÄTZEL

. INTRODUCTION

Various types of photoelectrochemical cells using photosynthetic components have been described. In most cases, the device consisted either in two half cells separated by a photoactive membrane or in a photogalvanic cell containing chloroplasts or chromatophores. In some cases, the material was directly deposited onto the surface of an electrode. In such systems, irradiation of the coated electrode by visible light induces a photoeffect through the charge separation generated in the reaction centers and the subsequent charge transfer to the supporting electrode and the surrounding medium. Using thylakoid fragments deposited on the surface of an SnO_2 electrode Ochiai et al.(1982) have shown that an anodic photoeffect can be generated upon irradiation. However, using the same support coated with reaction centers or with vesicles isolated from photosynthetic bacteria Seibert et al.(1982) obtained a photocathodic effect,   However, the mechanisms of electrode sensitization is far from beeing understood. We report here on some first results concerning the immobilisation of photosystems on TiO_2 powders as a first step in the study of the interactions of photoactive biological complexes with semiconducting particles. The photocatalytic properties of this material is now extensively studied (Kalyanasundaram 1983). Anatase powders can have high specific surface area ($50-200$ $m^2 g^{-1}$), can be sensitized by various dyes, and have a conduction band well suited for the generation of H_2 from water.

2. MATERIAL AND METHODS

Photosystem I-enriched particles have been isolated from spinach leaves by Triton X-100(Vernon,Shaw 1971, TSF I particles). Photosystem II particles (TSF II) were prepared by the method of Berthold et al.(1981), using the same detergent. The TiO_2 powders were the Bayersol anatase and the Degussa P-25 (200 and 50 m2g-1). The adsorption of the particles was carried out by incubation of the TiO_2 powder for 2 h. at $0°C$ in phosphate buffer 75mM at the pH mentioned in the text. Proteins were determined by the Bio-Rad Laboratories protein assay (Coomassie Blue binding). O_2 uptake was measured with an Hansatech electrode in phosphate buffer containing ascorbate 1 mM, DCPIP .3 mM and MV^{2+} .5 mM. PS II activity was measured as DCPIP oxydation followed by OD variation at 600 nm in the presence of .5 mM DPC. Hydrogen evolution was assayed as in Cuendet, Grätzel(1982).

3. RESULTS AND DISCUSSION

Different attachment methods have been developped to immobilize enzymes on inorganic supports(Zaborsky 1974). Two of them, which are particularly suited to the coupling on glass and porous ceramics have been tested in this study. The first one consists of the adsorption of the proteins to the carrier and the second is a covalent coupling through silanization of the support (See the contribution of Rao et. al. for more details). However, in contrast to the case of hydrogenase, catalase and BSA, the binding of the TSF I complexes

Sybesma, C. (ed.), Advances in Photosynthesis Research, Vol. II. ISBN 90-247-2943-2.
© *1984 Martinus Nijhoff/Dr W. Junk Publishers, The Hague/Boston/Lancaster.*

was much less efficient by covalent coupling then by direct adsorption at the surface of the semi-conductor. This last technics was then only applied in the following experiments.

3.1. <u>Effect of pH on the adsorption process.</u> The study of the adsorption of various proteins on TiO_2 powders have shown that the yield of the process, i.e. the amount of protein adsorbed per unit weight of the powder is maximum at pHs close to the isoelectric point of the protein, indicating that adsorption is due mainly to hydrogen bonding at the particle surface rather than to electrostatic interactions. This behaviour has been further confirmed in the case of TSF I and II complexes. As shown in figure 1, the amount of TSF I remaining in the supernatant after completion of adsorption increases drastically at pH higher then 6, when TSF II particles are adsorbed at high efficiency at all pH tested.

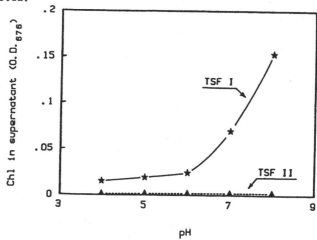

FIGURE 1. Effect of pH on the amount of TSF I and II remaining in the supernatant after adsorption is completed. TiO_2 Bayersol. Initial amounts of photosynthetic complexes: 4.5 and 9.6 mg Chl/g TiO_2 for TSF I and II resp.

This behaviour is related to the isoelectric properties of the complexes: the adsorption of TSF I is higher near its isoelectric point (at pH 4.8), while TSF II, which carry only a small amount of negative charges are well adsorbed aver all pH tested (Yamamoto, Ke 1981).

3.2. <u>Adsorption capacity of the TiO_2 powders.</u> The maximal amount of TSF I which can be adsorbed onto the surface of the semi-conducting particles has also been measured. Figure 2 shows that already in the presence of 4.5 mg Chl per g TiO_2 some chlorophyll and proteins can be detected in the medium. Using a value of 11.2 mg protein per mg Chl (Vernon, Shaw 1971), we conclude that about 50 mg of protein can be adsorbed per g of Bayersol. A slightly better yield is obtained with P-25 TiO_2 (about 60 mg protein/g TiO_2), which is comparable to the values obtained for the adsorption of catalase or hydrogenase, showing that the binding capacity of TiO_2 is not impaired when larger and more complexe stuctures than simple peptides are adsorbed on the particles. Due to some difficulties in the separation between the coated powder and the heavy

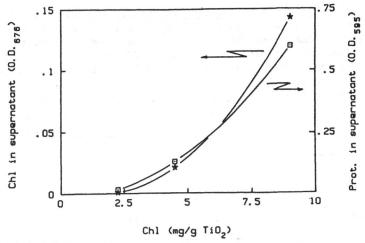

FIGURE 2. Chlorophyll and proteins remaining in the medium as a function of the initial amount of TSF I Chl added to a Bayersol suspension. pH 5.

TSF II complexes, no quantitative determination of the binding capacity for this material is available for the moment.

3.3. Photochemical activity of the derivatized TiO_2.

3.3.1. O_2 uptake by TSF I/TiO_2 preparations. In order to test the integrity of the photochemical functions of the adsorbed photosystems, light dependent O_2 uptake has been measured in the presence of ascorbate, DCPIP and MV. Figure 3 shows that no difference can be measured for the activity of free and adsorbed TSF I complexes. However, the presence of free Chlorophyll, due to the formation of Chl-Triton mixed micelles during the isolation procedure could contribute as a PS I-independant O_2 uptake (Van Ginkel 1979). A more specific test for PS I activity, like NADP reduction will be completed to confirm that adsorption does not impair the function of the complexes.

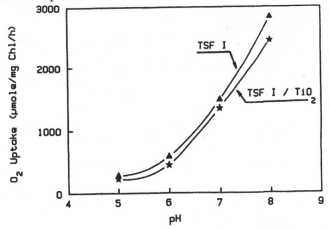

FIGURE 3. O_2 uptake mediated by free and adsorbed TSF I complexes

3.3.2. <u>PS II activity of the TSF II/TiO$_2$ preparations.</u> DCPIP oxydation in the presence of DPC was used as a measure of PS II activity of free and adsorbed complexes. Table 1 shows that the DCPIP oxydation rates measured in the presence of the adsorbed complexes are not much different than for the free TSF II. (The higher activity observed with the TSF II/Bayersol is maybe due to a better preservation of the activity during storage in liquid nitrogen).
TABLE 1. <u>PS II activities of free and adsorbed TSF II particles.</u>

TSF II	Rates of DCPIP oxydation (μmole h^{-1} mg Chl^{-1})
free	43
on Bayersol	73
on Bayersol + .3 % RuO$_2$	56
on P-25 + .3 % RuO$_2$	39

3.4. <u>Coupling of adsorbed photosystems with redox catalysts.</u>
The immobilization of photosynthetic complexes on semi-conducting particles rises the possibility of an electric coupling between the biological systems and artificial or natural redox catalysts also located at the surface of the support. It could be interesting , for example, to couple TSF I with Platinum or hydrogenase-coated TiO$_2$, or TSF II with TiO$_2$/RuO$_2$, a catalyst of the oxydation of water (Grätzel 1982). Such coupling have been tried with our derivatized TiO$_2$, and with colloidal preparations. During these preliminary experiments, no RuO$_2$-mediated O$_2$ evolution has been detected with TSF II/TiO$_2$/RuO$_2$, and only traces of H$_2$ have been observed with TSF I/TiO$_2$/Pt, using EDTA as electron donor. We are now trying to find electron relais which could mediate efficiently the charge transfer between the complexes and the semi-conducting particle.

4. REFERENCES

Berthold DA, Babcock GT and Yocum CF (1981) FEBS Let. 134, 231-234.
Cuendet P and Grätzel M (1982) Photochem. Photobiol. 36, 203-210.
Grätzel M (1982) Biochim. Biophys. Acta 683, 221-244.
Kalyanasundaram K (1983) in Grätzel M, ed. Energy resources through photochemistry and catalysis, Academic Press, in press.
Ochiai H, Shibata H, Sawa Y and Katoh T (1982) Photochem. Photobiol. 35,149-5
Seibert M and Kendall-Tobais M (1982) Biochim. Biophys. Acta, 681, 504-511.
Van Ginkel G (1979) Photochem. Photobiol. 30, 397-404.
Vernon LP and Shaw ER (1971) In Methods in Enzymology 23, 277-289.
Yamamoto Y and Ke B (1981) Biochim. Biophys. Acta 000, 175-184.
Zaborsky O (1974) Immobilized enzymes, CRC Press, Cleveland.

Supported by a grant from National Energie-Forschung Fond, Switzerland.

<u>Authors address:</u> Institut de Chimie-Physique, EPFL, CH-1015 Lausanne, Switzerland.

CYANOBACTERIAL ELECTRODE, ITS PROPERTIES AND FUNCTIONS

HIDEO OCHIAI, HITOSHI SHIBATA, YOSHIHIRO SAWA and KIYOFUMI TAKATA

1. INTRODUCTION

Electrons generated during the primary light-induced, charge-separation steps in chloroplasts can be transferred to an electrode, especially when the electrode is poised at a potential of more than +0.2V vs the saturated calomel electrode(SCE). Recently, we demonstrated photoresponses from electrode composed of immobilized chloroplasts coated on tin oxide transparent electrodes(SnO$_2$ OTE)(Ochiai,H. et al. 1982). In 1980, we reported on a "living algal electrode" that was prepared using intact cyanobacterial cells of *Mastigocladus laminosus* in place of higher plant chloroplast preparations which are notoriously unstable. This algal electrode functioned as a stable photoanode even under continuous illumination, though the photoconversion efficiency was very low(Ochiai,H. et al. 1980). We report here that the cyanobacterium, *Phormidium* sp., can work much better than *M.laminosus* as an algal electrode and that a significant enhancement of the photocurrent yield can be attained by drying the electrode at moderate temperature. We also have studied the effect of various parameters on photocurrent generation in order to elucidate the properties of the *Phormidium* electrode.

2. MATERIAL AND METHODS

2.1. Cultivation of cyanobacteria

The filamentous blue-green alga, *Phormidium* sp., isolated from the Matsue hot springs, was cultured as described previously. One can use Fitzgerald-modified culture medium to cultivate the algae, though better growth is obtained using Matsue hot spring water(pH 8.0) as the basal medium. Late logarithmic phase cells were collected by filtration utilizing four layers of nylon cloth. This *Phormidium* sp., is apparently an organism with remarkably porous membranes. Redox dyes readily penetrate the cell, enabling us to measure both PS-I and PS-II activity in the intact cells. This phenomenon is also true for the filamentous cyanobacterium, *Mastigocladus laminosus*.

2.2. Preparation of the immobilized cyanobacterial electrode

The algal cells are suspended in an equal volume of 8% sodium alginate solution of hot spring water, then the suspension is spread on the surface of an SnO$_2$ OTE and the electrode plate dipped for 5 min in a 50mM CaCl$_2$ solution. The electrode is washed with hot spring water. Thus, one obtains an algal electrode entrapped inside a wet calcium alginate matrix, about 250µm thick. A freshly prepared electrode was incubated in culture medium for the indicated period of time and used as a "wet" electrode. Wet electrodes were allowed to dry in an electric oven at 50°C, then set up in the cell assembly for photoelectrochemical measurements. Dried algal electrodes were about 40µm thick.

Sybesma, C. (ed.), Advances in Photosynthesis Research, Vol. II. ISBN 90-247-2943-2.
© *1984 Martinus Nijhoff/Dr W. Junk Publishers, The Hague/Boston/Lancaster.*

2.3. Photoelectrochemical measurements

The photoelectrochemical cell and the apparatus for photocurrent measurement have been described earlier(Ochiai,H. et al. 1982). The algal/SnO$_2$ electrode served as the working electrode in the conventional three electrode system and the illumination area was 36 cm^2. The basal electrolyte was buffered with 50mM borate containing 50mM sodium carbonate and 50mM potassium chloride. A 300 W projector lamp was used as the light source in combination with a colar filter that cuts off below 460nm to eliminate photoresponses attributable to SnO$_2$ itself. The light intensity absorbed by the algal electrode was determined with a thermopile radiometer. Photocurrents were measured at 45°C with a DC-Microvolt ammeter and recorded with a XY recorder. Algal cell density deposited on an SnO$_2$ OTE varies with each preparations, hence the photocurrent value is expressed on the basis of 10µg chlorophyll per cm^2 of the SnO$_2$ OTE.

2.4. Photosystem activity measurements

Algae/calcium alginate film was detached from the SnO$_2$ OTE surface and cut into pieces which then were suspended in the reaction mixture for assays. The assays for PS-I and PS-II activity were performed by procedures similar to those used with intact cells(Ochiai,H. et al. 1980).

3. RESULTS

3.1. Wet electrode

Wet electrode photocurrent-potential curves are similar to those reported before for the algal electrode(Ochiai,H. et al. 1980). The anodic photocurrent rises from around +0.2V vs SCE and increases linearly with the potential imposed. Since *Phormidium* sp. cells immobilized in the alginate matrix can multiply in the light, the wet electrode, cultured for 5 d, gives twice the photocurrent on an area basis compared to the same electrode when freshly prepared. The photocurrent output increased with increasing the incident light intensity and did not saturate at 500J/m^2 under the condition of +0.6V vs SCE. The photocurrent with 150J/m^2 light was steady for the first hour, but dropped thereafter. However, no pigment bleaching occurred on the algal electrode during the 3 h run. Interestingly after incubating the same electrode in culture medium for 1-2 d, the electrode was able to function as actively as a fresh electrode.

3.2. Dried electrode

A specific photocurrent measured with a dried electrode(50°C, 60 min treated) was 5.1 µA/10µg Chl/cm^2. This represents about 100-fold increase in output in comparison to that observed with a wet electrode. Moreover, as far as output capacity is concerned, this *Phormidium* sp. electrode was more efficient(up to 6.6-fold) than the *M.laminosus* electrode. Filamentous *Phormidium* cells are generally much thinner than *M.laminosus* cells. Hence, the packing density of the former on the SnO$_2$ OTE is probably greater than is the case for *M.laminosus*. High yield(8.2 µA/10µg Chl/cm^2) was attained using 50mM borate buffer(pH 8.0) as electrolyte. Interestingly, Matsue hot spring water used as electrolyte also gives satisfactory results.

On the other hand, Tris buffer and Tricine buffer were rather inhibitory to photocurrent output. Addition of 10mM CaCl$_2$ to the borate buffer leads to an increased output, but it is not true for MgCl$_2$ and BaCl$_2$.

It is important to investigate the effect of various photosynthetic inhibitors and electron mediators in order to confirm whether the observed photocurrent depends upon electrons ejected from either or both of the algal photosystems in light-induced reactions. Each inhibitor was added to the electrolyte so as to give final concentration indicated in Table 1. After standing for 15 min, the photoresponse of the treated electrode was examined under potentiostatic conditions. Twenty μM DCMU almost completely inhibited electron flow. Moreover, when DCMU was added to the electrolyte of a running algal photocell, the photocurrent output decreased hyperbolically to 6% of the original value over a period of 15 min. On the other hand, the output was more than doubled upon addition of viologen reagents to the electrolyte. Interestingly, Vitamin K$_3$ worked as a stable electron mediator. However, phenazine methosulfate, flavine mononucleotide, and methylene blue did not work well.

TABLE 1.

Effect of Various Inhibitors, Heat-Treatment, and Electron Mediators on Photocurrent Generation by a Dried Algal Electrode[a]

Treatment	Photocurrent, %
None	100
DCMU (20 μM)	6
CCCP (50 μM)	16
TPB (5 μM)	79
Heat (110°C, 15 min)	3
BV (20 μM)	294
MV (20 μM)	242
Vitamin K$_3$ (10 μM)	256
FMN (10 μM)	83
MB (10 μM)	45
PMS (20 μM)	100

[a]Electrolyte used, 50 mM H$_3$BO$_3$–Na$_2$CO$_3$–KCl (pH 8.0) at 45°C. Light intensity, 250 J/m^2 through a Y-46 filter. Applied potential, +0.6 V vs SCE. The activity was expressed as a relative value to the photocurrent yield of an untreated algal electrode (None). A specific value of "None" was 8.6 μA 10 μg Chl^{-1} cm^{+2}.

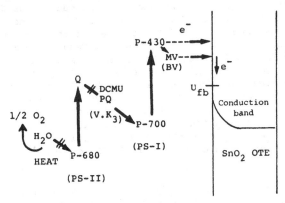

Fig. 1. Postulated electron transfer system. Adsorption of algal cells onto the SnO$_2$ does not appear to change significantly the flat-band potential (U_{fb}), which is around −90 mV under the conditions we used. Photosystems in the algal cells are depicted with simplicity. Electrons ejected from P-430 of PS-1, mediated by reduced MV or BV if present, must flow into the SnO$_2$ OTE. See the text for a more detailed explanation.

4. DISCUSSION

Cyanobacterium, *Phormidium* sp. cells can perform light-dependent electron transfer to SnO_2 OTE, especially under the imposed potentiostatic conditions. We have examined a wide variety of algae for their capacity to generate photocurrents in a conventional three electrode system. Two species, *Phormidium* sp. and *Mastigocladus laminosus* from the Matsue hot springs were found to function well. It is postulated that thermophilic algae from this hot spring are unique in their cell permeability properties compared with other algae. Analysis of the Matsue hot spring water reveals that it contain a high concentration of borate(more than 50 ppm). Generally, plant growth is depressed in the presence of more than 50 ppm of borate, while our algal cells grow well in the hot spring water. In fact, activity of the algae in photocurrent generation is maximal when borate buffer(50mM, pH 8.0) was used as electrolyte. Boric acid is a well-known crosslinking agent with cis-dihydroxy groups of sugar moieties. The finding suggests that the algal cell which are sugar-coated, react favorably to the action of borate anion.
   Figure 1 is a model which can explain the experimental results we have observed. When the light is turned on, at least, almost electrons are driven from water to the P-430 through the photosystems in the algal cells. The flat-band potential(U_{fb}) of the SnO_2 used in these experiments is at around -90mV on the electrochemical scale. When the algal photocell system is poised at +0.2V or more vs SCE, the SnO_2 conduction band must "bend down" within the bulk of the semiconductor material from the U_{fb}. This potential difference is the driving force for the photocurrent output. Therefore, electrons extracted from water molecules and delivered to PS-I through PS-II are assumed to be transferred to the SnO_2 OTE mainly from P-430 of PS-I. MV and BV must mediate electrons more efficiently from the photosystems to the SnO_2 OTE if the viologens are reduced through P-430 in the light. Several lines of evidence are in support of this interpretation. The photocurrent yield was more than doubled in the presence of MV or BV, whereas DCMU drastically decreased the photocurrent output, as did heat treatment of the electrode at 110°C for 15 min. Moreover, the photocurrent spectrum peak coincided well with absorption peak of the light-harvesting pigment, phycocyanin. Ejection of electrons from reduced Vitamin K_3 into the SnO_2 OTE is difficult due to the inhibition by band bending of the semiconductor conduction band, which results in an energy barrier for electrons more oxidizing than U_{fb}. However, we have recently confirmed that PS-I and oxygen evolution activities of quinone-depleted algal lamellae are completely recovered upon the addition of phytoquinone-extracts from the algae. Thus, Vitamin K_3 which is one of the simplest analogues of phytoquinones must work as an exogenous electron mediator, causing an increase of the electron transfer rate from PS-II to PS-I.

REFERENCES
Ochiai,H. et al. (1980) Living electrode as a long-lived photoconverter for biophotolysis of water. Proc.Natl.Acad.Sci.,USA. 77, 2442-2444.
Ochiai,H. et al. (1982) Properties of the chloroplast film electrode immobilized on an SnO_2-coated glass plate. Photochem.Photobiol. 35, 149-155.

Authors address: Laboratory of Biochemistry, College of Agriculture,
          Shimane University, Nishikawazu-1060, Matsue, 690. JAPAN.

CHARACTERIZATION OF A 680-nm ABSORBING HYDRATED CHLOROPHYLL DIMER AS PHOTOCATALYST FOR THE WATER SPLITTING REACTION IN VITRO

L. S. SHOWELL, J.-L. YOU, AND F. K. FONG

1. INTRODUCTION

The generation of photocurrents upon illumination of Pt/Chl a electrodes in a photogalvanic (PG) cell has been attributed to the water splitting reaction (Showell, Fong 1982 a,b). Manifestations of this reaction were examined in order to quantitate the observed PG activity in correlation with H_2 and O_2 product yields as measured by gas chromatography (GC) (Showell, Fong 1983). It follows that one may reasonably apply the PG technique to a study of the photon mechanisms underlying the in vitro Chl a water splitting reaction. In this paper we identify a 680-nm absorbing hydrated Chl a complex, attributable to (Chl a · $2H_2O)_2$, capable of photocatalyzing in vitro water photolysis.

2. PROCEDURE

Chlorophyll a was prepared and purified according to the method of Brace et al. (1978). Pt/Chl a electrode manufacture and the PG apparatus have been described elsewhere (Showell, Fong 1982 a,b; 1983). Absorption spectra for the Chl a preparations were recorded on a Cary 14 spectrophotometer.

3. RESULTS AND DISCUSSION

In Fig. 1 the dependence of the photocurrent on the incident flux, I_0, at 700 nm is illustrated for a fresh (1a), water equilibrated (1b) and extensively dried (1c) Pt/Chl a electrode prepared from wet n-pentane Chl a solutions (Showell, Fong 1983). The data in Fig. 1c are fitted to a simple linear dependence.

In the far-red (λ >705 nm) region sublinear photon effects are observed, whereas the flux dependence of a fresh electrode in the 420- to 704-nm region is superlinear (Showell, Fong 1983). The data in Fig. 1a and b are interpreted in terms of the two-photon model of Fong et al (Fong 1974; Showell, Fong 1983). The superlinear flux dependence observed at λ ≤705 nm (Fig. 1) is ascribed to some hydrated Chl a aggregate, most probably dimeric, other than the polymeric species (Showell, Fong 1983). The loss of the superlinear behavior upon sample equilibration with water (Fig. 1b) is indicative of the conversion of this aggregate to the polymeric form as evidenced by accompanying changes in the photogalvanic action (Showell, Fong 1983). The experiments described below delineate a 680-nm absorbing Chl a complex capable of photocatalyzing the water splitting reaction.

Figure 2A shows the red absorption maxima of a freshly prepared 5.8×10^{-5}M solution of Chl a dissolved in a 4:1 H_2O:acetone mixture.

The red absorption maximum occurs at 672 nm and has a FWHM of ~53 nm compared to ~20 nm for an equivalent Chl a sample in pure acetone. Deconvolution of this band into Gaussian components reveals three bands centered at 664, 682, and 713 nm. Upon aging (Fig. 2b) the ~680-nm component is diminished, the 713 (717)-nm component is narrowed and enhanced relative to the 664 (667)-

Sybesma, C. (ed.), Advances in Photosynthesis Research, Vol. II. ISBN 90-247-2943-2.
© *1984 Martinus Nijhoff/Dr W. Junk Publishers, The Hague/Boston/Lancaster.*

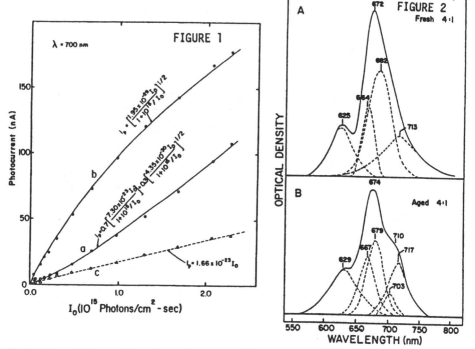

FIGURE 1. 700-nm flux dependence of (a) a fresh; (b) aged; and (c) extensively dried Pt/Chl a electrode prepared from Chl a dissolved in wet n-pentane.

FIGURE 2. Gaussian deconvolution of the red absorption maxima of (A) a fresh Chl a solution in 4:1 H_2O:acetone; and (B) solution in A after aging.

nm band, and a new component is observed at 703 nm. Livingston et al. (1949) assigned the 664 (667)-nm component to Chl a·H_2O. We assign the 680-nm component to the dihydrate dimer, (Chl a·$2H_2O$)$_2$ (Fig. 3) (Fong, 1983) and the 713 (717)-nm band to higher Chl a·H_2O aggregates. The 703-nm component observed in aged samples probably arises from a molecular rearrangement of the 680-nm Chl a dimer (Fong et al. 1977).

Fig. 4 shows the time dependence of the PG current obtained upon white-light excitation of Pt/Chl a electrodes prepared by solvent evaporation of Chl a in 3:1 H_2O:acetone onto a platinized Pt electrode. The maximum anodic and cathodic AC components are on the order of 1μA. Aging results in 75% reduction of PG action. The μA current levels are significantly larger than those obtained in earlier studies (Fong, Winograd 1976; Galloway et al. 1978).

The oscillatory delayed fluorescence observed in water-saturated Chl a samples dissolved in hydrocarbon solvents (Fong et al., 1982) is attributed to electronic charge shuttling between the two bound water molecules of the hydrated monomer, H_2O · Chl a · H_2O (Kusunoki, Fong 1983). This mechanism prohibits charge-separation upon photoexcitation and rules out hydrated monomeric Chl a as being capable of catalyzing water photolysis. Additionally, the inhibition of PG activity upon aging of Pt/Chl a electrodes prepared from Chl a:water: acetone systems suggests that the 703 and 713 (717)-nm Chl a components (Fig. 2) may also be photo-inactive. These observations corroborate our assignment

FIGURE 3

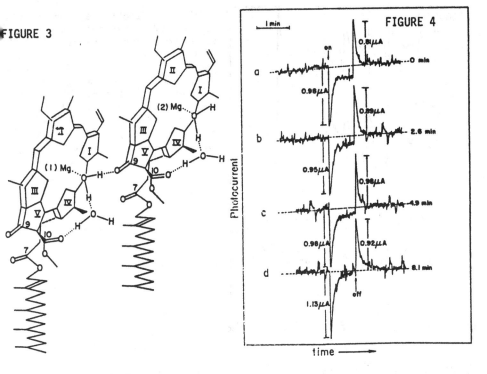

FIGURE 4

FIGURE 3. Model of $(Chl\ a \cdot 2H_2O)_2$.

FIGURE 4. Time evolution of the white-light induced PG current obtained from Chl a in 3:1 H_2O:acetone evaporated onto a Pt electrode: (a) Initial exposure; (b) 2-6 min after a; (c) 4.9 min after a; (d) 8.1 min after a. Chl a density was 1×10^{16} molecules/cm^2 and $I_O = 0.189$ mW/cm^2.

of the ~680-nm component (Figs. 2A and B) to the photoreactive chlorophyll dimer, $(Chl\ a \cdot 2H_2O)_2$.

Electron spin resonance spin trapping experiments indicate that, on illumination, acetone water solutions of Chl a yield hydroxyl and perhydroxyl radicals (Fig. 5), corroborating the water splitting interpretation (Showell, Fong 1982). We are cognizant of the possible relevance of this in vitro characterization to the P680 reaction center chlorophyll complex in plant photosynthesis.

REFERENCES
Brace J, Fong FK, Karweick DH, Koester VJ, Shepard A, Winograd N (1978) Stoichiometric determination of chlorophyll a-water aggregates and photosynthesis, J. Am. Chem. Soc. 100, 5203-5207.
Fong FK (1974) Energy upconversion theory of the primary photochemical reaction in plant photosynthesis, J. Theor. Biol. 46, 407-420.
Fong FK (1983) "Light Reaction Path of Photosynthesis", Springer-Verlag, Heidelberg, pp. 277-321.

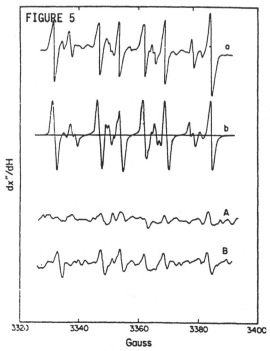

FIGURE 5. Electron spin resonance spectra of degassed 1:1 acetone: water solutions containing 0.15 M DMPO as a spin trap: (a) 2% H_2O_2 photolysed for 8 min; (b) computer simulation on a assuming the prescence of DMPO·OH, DMPO·OOH and DMPO·acetone species; (A) dark spectrum of a 1.74×10^{-4} M Chl a solution; (B) spectrum in A after white light photolysis.

Fong FK, Koester VJ, Galloway L (1977) Endo and exo carbomethoxy carbonyl bonding in hydrated chlorophyll a dimers, J. Am. Chem. Soc. 99, 2372-2375.

Fong FK, Kusunoki M, Galloway L, Matthews TG, Lytle FE, Hoff AJ, Brinkman FA (1982) Elementary reconstitution of the water splitting light reaction in photosynthesis, J. Am. Chem. Soc. 104, 2759-2767.

Fong FK, Winograd N (1976) In vitro solar conversion after the primary light reaction in photosynthesis, J. Am. Chem. Soc. 98, 2287-2289.

Galloway L, Roettger J, Frug DR, Fong FK (1978) Photochemical upconversion in the chlorophyll a water splitting light reaction, J. Am. Chem. Soc. 100, 4635-4638.

Kusunoki M, Fong FK (1983) Parity selection rule for elastic electron tunneling, Chm. Phys. Lett. in press.

Livingston R, Watson WF, McArdle J (1949) Activation of the fluorescence of chlorophyll solutions, J. Am. Chem. Soc. 71, 1542-1550.

Showell MS, Fong FK (1982a) Photooxidative properties of chlorophyll dihydrate on metal a) catalyst for water photolysis, J. Am. Chem. Soc. 104, 2773-2781.

Showell MS, Fong FK (1982b) Molecular mechanisms of hydrogen generation in the chlorophyll water-splitting light reaction, Proc. of the IV World Hydrogen Energy Conference, California, USA, Volume 2, 715-724.

Showell MS, Fong FK (1983) Biphotonic origin of chlorophyll-a catalyzed water photolysis, Progress in Solar Energy, Volume 6, 113-118.

ACKNOWLEDGEMENT
The work reported in this paper was supported by the Basic Research Division of the Gas Research Institute.

INDEX OF NAMES

Abraham, P. III.1.031
Abramowicz, D. I.3.349
Acker, S. I.6.693
Adamson, H. IV.6.705, 745
Admon, A. II.6.531
Affolter, D. III.6.517
Aflalo, C. II.6.493, 559
Ahrer-Steller, V. IV.4.349
Akazawa, T. III.8.717
Åkerlund, H. I.3.363, 367, 391;
    III.2.187
Akoyunoglou, A. IV.6.649, 673
Akoyunoglou, G. II.1.037, 053;
    IV.6.595, 645, 649, 673, 737; IV.7.845
Alam, J. I.4.517, 521
Aliverti, A. I.4.597
Almon, H. II.7.699
Amiri, I. III.8.795
Ambard-Bretteville, F. II.2.133;
    IV.5.555
Ambrož, M. II.1.069
Amesz, J. I.5.621; II.3.181, 185, 203
Anderson, G. I.4.537
Anderson, J. II.1.045; III.1.001;
    III.7.689; IV.3.267
Anderson, L. III.6.521
Anderson, S. III.7.657
Andersson, B. I.3.363, 367; III.2.159,
    187; III.3.223
Andersson, I. III.8.787
Andralojc, P. II.6.555
André, M. III.9.817
Andreasson, L. I.3.307, 379; I.4.525
Andreo, C. II.6.579
Antonielli, M. IV.2.169
Antonopoulou, P. IV.6.737
Apel, K. IV.7.809
Apostolova, E. III.1.039
Argyroudi-Akoyunoglou, J. II.1.037
Arntzen, C. I.3.383; I.6.693; II.2.099;
    III.1.067
Aro, E. III.8.867
Aronson, H. III.1.055
Arrabaça, M. IV.4.435
Ashour, N. IV.2.177
Ashton, A. III.7.613

Astier, C. IV.1.073; IV.5.587
Atkinson, Y. I.2.127, 139, 147
Austin, R. IV.2.103
Averina, N. IV.6.699
Avron, M. II.6.531; II.8.745
Azcón-Bieto, J. III.9.905
Azzone, G. II.4.233

Babalar, M. III.6.509
Babcock, G. I.3.243, 279, 341; I.6.697
Bachofen, R. II.3.165, 169; II.7.667;
    III.5.365, 369
Bader, K. I.3.287
Badger, M. III.9.833; IV.3.325
Bagge, P. III.9.897
Baianu, I. I.3.283
Baker, N. I.1.085; III.4.319, 323, 327;
    IV.4.479; IV.6.665
Bakker, W. II.3.207
Bakr Ahmed, M. IV.2.177
Baldy, P. III.9.825
Bálint, E. I.6.733
Balogh, A. III.7.625
Baltimore, B. III.3.211
Bar-Zvi, D. II.5.407; II.6.493, 539
Barabás, C. II.4.301
Barber, J. I.4.417, 497; II.4.273;
    III.2.091, 107, 159, 163; III.3.227;
    III.4.287; IV.3.263, 275; IV.4.459
Barr, R. I.4.441
Barrett, J. II.2.149
Baszyński, T. IV.4.439
Bauermeister, H. II.4.333
Beale, S. IV.6.717
Becker, D. II.7.659
Beechey, R. II.4.337; III.9.909
Belanger, G. IV.5.591
Bell, D. I.2.113
Bendall, D. I.4.457, 577; II.7.651;
    III.3.239
Benešová, H. IV.4.407
Bennett, J. III.2.099; IV.2.121; IV.7.863
Bennoun, P. IV.5.491
Bensasson, R. I.6.729
Benson, E. II.1.049; IV.3.287
Benthem, L. I.1.033

Berard, J. IV.5.591
Bereza, B. IV.6.749, 757
Berg, S. III.1.055; III.3.215
Berger, M. III.7.581; III.9.883
Bergman, A. III.9.887
Bergmann, P. IV.5.537
Bergström, J. I.4.525
Bergweiler, P. IV.6.627
Berkaloff, C. I.1.069; I.4.449
Berkowitz, G. IV.4.367
Berzborn, R. II.6.563, 571, 587
Besford, R. III.8.779; IV.3.297
Betz, A. III.6.445
Bhatnagar, R. III.8.799
Bickel-Sandkötter, S. II.6.551
Biekmann, S. IV.5.505
Biggins, J. III.4.303
Binder, A. II.7.667
Bishop, N. I.3.321
Bísková, R. II.1.065
Bismuth, E. III.9.875
Biswal, B. IV.6.619
Biswal, U. III.1.087; IV.6.619
Black, M. III.4.315
Blahaut, B. IV.6.741
Blank, M. IV.3.259
Blankenship, R. I.2.203; III.5.377
Block, M. III.1.027
Blumwald, E. II.7.627
Boag, S. III.7.701
Bodmer, S. II.3.169
Bogdanović, M. II.1.005; IV.6.721
Bögemann, G. III.2.147, 155
Böger, P. II.7.631; II.7.635, 699, 703; IV.1.045
Bogorad, L. IV.5.529, 537
Böhme, H. II.7.639, 703
Bohnert, H. IV.5.579
Bolhàr-Nordenkampf, H. IV.3.321
Bombart, P. IV.6.753
Boney, A. IV.4.427
Bonnekamp, G. II.6.587
Bonnerjea, J. I.4.565
Boog, R. II.8.793
Boote, K. IV.2.121
Borghese, R. I.5.661
Borisov, A. I.1.029
Borkowski, C. III.6.445
Boschetti, A. II.2.129; IV.1.009; IV.5.541
Boska, M. I.2.121
Botte, P. II.4.285, 359
Bottin, H. I.4.569; III.4.287
Bottländer, K. IV.7.853

Boussac, A. IV.1.073
Bouwmeister, P. III.8.759
Bowes, G. III.9.829
Bowes, J. I.4.457; II.7.651
Bowman, C. IV.5.559
Boxer, S. I.2.223
Boyer, J. IV.4.359, 383
Boyle, F. III.8.771
Bracale, M. I.4.573
Bradbeer, J. III.7.597; IV.5.529; IV.6.693
Bradbury, M. III.4.327
Brand, J. II.7.659
Brändén, C. III.8.787; III.9.849
Brändén, R. III.8.765
Brandt, P. IV.5.517
Brangeon, J. III.1.023
Brasseur, R. III.2.143
Braumann, T. I.2.109; II.2.137, 145; IV.1.077
Braun, P. III.6.461
Brearley, T. I.4.433
Bréchignac, F. III.9.871
Breton, J. I.1.037; I.2.101, 105; I.5.669; I.6.705, 693; II.3.177; III.1.011
Brettel, K. I.2.175; I.3.295
Briantais, J. I.2.199
Britton, G. IV.1.069; IV.6.779
Broglie, R. II.3.215
Brok, M. I.2.171; I.6.677
Broniowska, B. II.1.001
Brouers, M. II.8.773; IV.6.761; IV.7.841
Brown, J. II.1.013; II.2.141
Brown, R. II.7.683
Bruce, D. II.8.797
Brunisholz, R. III.5.361
Buc, J. II.5.461; III.6.493, 505, 541
Buchanan, B. III.6.485; III.7.573, 609, 625, 633; IV.7.877
Buetow, D. III.8.807
Burkard, G. IV.5.537
Buschmann, C. IV.1.061; IV.3.245, 317
Bustamante, P. II.3.189
Butler, W. I.6.749
Buttner, W. I.3.243

Caers, M. IV.3.271
Cahen, D. III.4.331; IV.3.251
Calcagno, A. III.6.391
Calvin, M. I.6.745
Camm, E. II.2.095; IV.4.455
Cammaerts, D. IV.5.551
Cammarata, K. I.3.311
Campbell, C. III.4.303

Campbell, S. II.7.687
Canaani, O. III.4.331; IV.3.251
Canvin, D. III.6.449
Cao, R. IV.7.877
Carlson, D. III.7.633
Carlson, M. I.5.665
Carmeli, C. II.6.511
Caron, L. I.1.069
Carpentier, R. I.6.681
Carrayol, E. III.6.509; IV.2.157
Carrillo, N. I.4.541
Casadio R. II.4.233
Casey, J. I.2.121
Casimiro, A. IV.4.435
Castelfranco, P. IV.6.709
Castrillo-Issa, M. III.6.391
Čatský, J. IV.3.255
Catt, M. III.4.295
Cedergren, E. III.8.787
Cerović, Z. III.7.645, 661, 669
Ceulemans, R. IV.2.141
Chachaty, C. I.6.729
Chain, R. III.3.211
Champigny, M. III.9.875, 879; IV.7.881
Chapman, D. III.2.163; IV.3.263, 275;
    IV.4.459
Chauvet, J. I.6.725
Chaves, M. IV.2.145
Chen, J. III.6.513
Cheniae, G. I.3.311
Chereskin, B. IV.6.709
Chetrit, P. IV.5.555
Chidsey, C. I.2.223
Chollet, R. III.8.751
Choquet, Y. I.5.669; I.6.693
Choudhury, N. III.1.087
Chow, W. III.1.083; IV.3.297
Chueca, A. III.6.537
Clark, A. II.4.341
Clark, R. I.4.529
Clement-Metral, J. I.4.453
Clijsters, H. IV.4.431
Cmiel, E. II.2.081
Cobb, A. II.1.049; II.5.465; IV.3.287
Cogdell, R. II.1.025
Cohen, Y. III.4.283
Coleman, W. I.3.283
Collard, F. II.8.773
Colman, B. III.6.457, 549; III.7.681
Comins, H. III.6.545
Cook, C. III.8.783
Coombs, J. IV.2.085
Cornelius, M. III.8.755
Cornic, G. IV.4.375

Cotton, N. II.5.449
Coughlan, S. II.5.411
Cox, R. I.3.355; II.7.675; III.1.051
Cramer, W. I.4.501, 505; II.4.277
Crane, F. I.4.441
Crawford, N. III.7.633
Creswell, L. III.4.303
Cretin, C. IV.7.833
Crofts, A. I.4.461, 477, 489; I.5.649;
    I.6.755
Crouse, E. IV.5.537
Crowther, D. I.4.609
Csatorday, K. IV.6.713
Cséke, C. III.7.609, 625
Cuendet, P. II.8.777, 813
Curti, B. I.4.597

Daguenet, A. III.9.871
Dai, Y. I.3.359
Dallas, J. IV.6.709
Damm, I. I.2.109; II.2.137
Daniell, H. IV.6.681, 689
Daudet, F. IV.4.415
Davenport, J. II.5.371; III.3.243
Davidson, I. IV.2.149
de Ferrer, E. IV.6.693
de Graaff, L. I.1.033
de Greef, J. IV.7.805, 837, 873
de Groot, A. I.2.215; I.6.667
de Jong, F. IV.4.475
de Kouchkovsky, F. III.1.019
de Kouchkovsky, Y. II.4.293, 297;
    III.1.019
de la Rosa, F. II.5.419
de la Rosa, M. II.5.419
de Meutter, J. III.2.143
de Vitry, C. I.4.407
de Vos, L. II.3.185
de Wolf, F. II.4.321
De-Felice, J. IV.3.273
Dekker, J. I.2.171
del Campo, F. II.3.229
Delepelaire, P. I.6.693
Delrieu, M. I.3.291
Demeter, S. I.3.265; I.6.737
Demmig, B. II.4.317
den Blanken, H. II.3.185
Dennis, W. IV.2.149
Depka, B. I.4.473
Deprez, J. I.1.037; I.2.101
Deroche, M. III.6.509; IV.2.157
Desai, T. I.3.303
Devault, D. I.5.653
Devic, M. IV.5.575

Dietz, K. III.7.585
Dilley, R. II.6.485
Dilova, S. IV.6.685
Diner, B. I.2.195; I.4.407; I.6.695
Dismukes, C. I.3.349
Dobek, A. I.1.037; I.2.101
Doehlert, D. III.7.605
Dohnt, G. I.4.429; IV.1.013
Donnelly, M. III.8.739
Dorne, A. III.1.027
Dörnemann, D. II.2.077
Douce, R. III.1.027
Douillard, R. IV.2.165
Downton, W. IV.4.419
Draheim, J. I.4.537
Dreyer, E. IV.4.415
Dron, M. IV.5.491
Droppa, M. I.4.509; IV.1.057
Droux, M. III.6.533; III.7.697
Duane, J. I.4.537
Dubacq, J. II.2.133
Dubertret, G. IV.6.669
du Cloux, H. IV.3.213
Ducruet, J. IV.1.021
Duffus, C. III.6.399
Dujardin, E. II.8.755; IV.6.753, 757
Dulieu, H. III.8.799; IV.5.533
Duranton, J. I.4.589
Dutton, P. I.5.637; II.3.177
Duval, J. I.4.449; II.1.057; II.7.679
Duysens, L. I.1.057, 065; I.2.219;
    II.8.741
Dyer, T. IV.5.559

Easterby, J. III.6.489
Edman, K. III.2.167
Edwards, G. III.2.175; III.7.601, 641
Egneús, H. I.4.601; II.8.761
Ehara, T. IV.6.615
Eichholz, R. III.8.807
Elferink, M. II.4.347; II.5.469
Elfman, B. II.2.125
Ellis, R. IV.7.863
Emmanuel Paul, J. III.6.521
England, R. I.3.387
Enoch, H. IV.3.201
Erdin, N. III.8.795
Erickson, J. IV.5.491
Ericson, I. III.2.167; III.9.887
Ernst, A. II.7.639
Espie, G. III.6.457; III.7.681
Etienne, A. IV.1.073
Evans, E. I.3.387; I.4.481; II.7.683
Evans, M. I.2.127, 139, 147; I.4.565

Ewen, J. IV.3.333

Falkowski, P. I.2.163
Faludi-Dániel, A. I.6.707; IV.4.467;
    IV.6.733
Farineau, J. II.4.301
Feher, G. II.3.155
Feick, R. I.2.203; III.5.377
Feierabend, J. IV.5.505; IV.6.775
Feng, Y. II.5.383
Fenoll, C. I.5.645
Ferguson, S. II.5.449
Fernandes Moreira, M. I.4.617
Fernyhough, P. III.4.299
Ferrari-Iliou, R. IV.4.387
Ferte, N. II.5.461; III.6.493
Fetisova, Z. I.1.029
Fickenscher, K. III.6.529
Fiedler, E. III.9.893
Finel, M. II.6.567, 575
Fisher, J. I.2.223
Fitchen, J. IV.5.483
Fleischman, D. I.5.665
Flores, E. II.7.715
Flores, S. II.5.387
Flügge, U. II.4.309; III.7.685
Fock, H. III.9.883
Fok, M. I.1.029
Fong, F. II.8.821
Forchioni, A. III.1.023
Ford, M. IV.2.103
Ford, R. I.2.127, 147
Förster, V. II.4.305
Forti, G. I.4.573; III.1.079
Foster, J. III.2.175
Foyer, C. III.4.299, 315; III.7.689
Fraaije, J. II.2.115
Frackowiak, D. I.1.001; I.6.713
Fradkin, L. IV.6.699
Frank, H. I.1.053; I.2.203
Franklin, J. III.9.845
Franzén, L. I.3.379
Frasch, W. II.6.591
Fraser, A. III.7.589
Fredericq, H. IV.7.873
Freiberg, A. I.1.045
Fuad, N. II.1.033
Fujita, Y. I.1.021; II.1.033
Fuller, R. I.2.203; III.5.377
Furbank, R. III.9.833

Gabellini, N. III.3.243
Gadal, P. III.6.533; IV.7.833
Gaffney, J. III.2.139

Gagliano, A. I.6.705
Gaïzauskas, E. I.1.049
Gallagher, T. IV.7.853
Galmiche, J. II.5.379; II.6.547, 611
Galun, M. IV.3.251
Gantt, E. I.4.453; II.1.061; II.7.695
Garab, G. II.4.301
Garcia, D. III.6.391
Gardemann, A. III.7.617
Gardeström, P. III.7.641; III.9.887
Garlaschi, F. III.1.079
Garnier, J. III.2.183
Garty, J. IV.3.251
Gassman, M. IV.6.769
Gast, P. I.2.211, 215
Gates, D. IV.3.221
Gaudillere, J. III.7.665
Gaudszun, T. III.8.795
Gaul, D. I.5.673
Geacintov, N. I.1.037; I.2.101; I.6.705
Gehl, K. III.6.457
Gene Groat, R. III.8.767
Genova, C. III.7.665
Gerbaud, A. III.9.867
Gerber, A. IV.1.009
Gerday, C. IV.6.761
Gerhardt, R. III.7.609, 617
Gerola, P. III.1.079; III.4.307
Ghanotakis, D. I.3.243, 279, 341
Ghirardi, M. III.3.195
Ghisi, R. IV.1.065
Ghosh, R. III.5.365
Gibbs, M. IV.4.367
Giersch, C. II.5.403
Gillanders, B. IV.6.603
Gilmour, D. IV.4.427
Gilon, C. IV.1.033
Gimenez-Gallego, G. I.5.645
Gimmler, H. II.4.317
Gingras, G. IV.5.591
Giorgi, L. II.4.273
Girault, G. II.5.379; II.6.547, 611
Girvin, M. II.4.277
Givan, C. III.7.637
Gnanam, A. IV.6.681
Godik, V. I.1.045
Golbeck, J. I.4.561
Golden, S. IV.5.583
Goldschmidt-Clermont, M. IV.5.545
Gombos, Z. IV.6.713
Gomez, I. II.3.229
Gomez-Amores, S. I.5.645
Gontero, B. III.6.505
Goodman Dunahay, T. III.3.215

Gordon, A. IV.3.313
Görög, K. IV.4.467
Gounaris, I. IV.6.607
Gounaris, K. I.4.497; III.2.107, 123, 159
Goushtina, L. I.4.553
Govindjee I.3.272, 261, 283
Graan, T. I.4.549
Gräber, P. II.4.333; II.5.427, 431;
    IV.1.053
Grabowski, J. II.7.687
Graf, J. IV.1.037
Graham, J. II.7.695
Grandjean, J. IV.6.753
Grätzel, M. II.8.777, 813
Gray, G. I.5.673
Gray, J. IV.5.513, 559, 563, 567, 571
Green, B. II.2.095
Gregory, J. IV.6.745
Greppin, H. III.1.059
Griffith, M. IV.4.455
Grimme, L. I.2.109; II.2.137, 141, 145;
    II.8.769, 781; IV.1.077
Grodzinski, B. III.9.851, 901; IV.3.229,
    279
Gromet-Elhanan, Z. II.6.595
Groote-Schaarsberg, A. I.4.557
Gröpper, T. I.2.109; II.2.137
Gross, E. I.4.537
Gross, M. IV.7.853
Gruenewald Ray, P. III.9.821
Grumbach, K. IV.1.061, 069
Guariguata, M. III.6.391
Guerrero, M. II.7.715
Guikema, J. II.7.647
Guillemaut, P. IV.5.537
Guiraud, G. III.9.875
Gujrathi, B. IV.4.399
Gullifor, M. I.4.501
Gururaja Rao, G. IV.2.125
Gust, D. I.6.729
Gutowsky, H. I.3.283
Gutteridge, S. III.8.725, 755
Guy, R. III.1.067

Hadberg, A. II.7.675
Haehnel, W. I.1.073; I.4.545
Hagemann, R. IV.1.017
Hagner, R. III.4.263
Hall, D. II.8.727, 777, 797
Hall, N. III.9.841, 845
Hallberg, M. III.6.417
Hanson, K. III.9.837
Hansson, I.3.307
Haraux, F. II.4.293, 297

Harnischfeger, G. III.1.043
Harris, D. II.6.555
Hartley, M. IV.7.863
Hartman, F. III.8.739, 783
Hase, E. IV.6.615
Häsler, R. IV.4.395
Hatch, M. III.7.613
Hattersley, P. III.6.403
Hauser, H. III.5.365
Hauska, G. II.5.427; III.3.243
Hauswirth, N. II.7.691
Havaux, M. IV.4.459
Hawker, J. III.6.501
Hawkesford, M. II.7.671
Hayashi, H. II.3.211
Hearst, J. III.5.355
Heath, R. III.7.653
Heathcote, P. II.4.363
Heber, U. III.1.047; III.6.381; III.7.585;
    IV.4.403
Heemskerk, J. III.2.147, 155
Hefferle, P. I.1.081
Hegazy, M. IV.2.177
Hegde, B. IV.4.399
Heisterkamp, U. IV.1.029
Heldt, H. II.4.309; III.6.513; III.7.609,
    617, 625, 685
Hellingwerf, K. II.4.347, 367; II.5.469
Hendrich, W. I.6.759; IV.6.749
Henry, L. III.3.203, 219
Herdman, M. IV.5.587
Hermodson, M. I.4.501
Herrmann, R. I.4.505
Hervas, M. II.5.419
Herzog, B. III.7.609, 617
Hetherington, S. IV.4.447, 471
Heuer, B. IV.4.423
Hevesi, J. I.6.733
Hicks, D. II.6.599
Hillel, R. II.6.511
Hiller, R. II.1.041; III.5.351
Hills, M. III.7.677
Hincha, D. III.1.047
Hind, G. I.4.529; II.7.671
Hipkins, M. I.2.113; II.4.325; IV.4.427;
    IV.6.665
Hirschberg, J. IV.5.483
Hirosawa, T. III.7.565
Hladík, J. II.1.065, 069
Hoarau, A. III.6.493
Hoarau, J. I.6.705; III.8.799
Hobé, J. I.4.613
Hodges, M. I.4.417; IV.3.263
Hof, R. I.2.219

Hoff, A. I.2.089, 215; I.6.667
Hoffmann, P. I.6.709
Höinghaus, R. IV.5.505
Høj, P. III.3.219
Holten, D. I.2.187, 203, 223
Holtum, J. III.6.545
Holzwarth, A. I.1.073, 077
Homann, P. III.3.199
Hønberg, L. IV.5.525
Hong, Y. II.4.247
Hope, A. II.4.313
Hopkins, W. II.2.125; IV.4.451
Horst, L. II.2.141
Horton, P. I.4.413, 433, 609; III.4.299,
    315; III.7.657
Horvath, G. I.4.509; IV.1.057
Hosler, J. II.5.415
Hotchandani, S. I.6.713
Houchins, J. I.4.529; II.7.671
Houghton, J. III.6.489
Howe, C. IV.5.559
Howitz, K. II.5.457
Høyer-Hansen, G. III.2.171
Huang, Z. II.5.383
Huber, S. III.7.605
Huchzermeyer, B. II.6.535
Huet, J. IV.2.153
Huggins, B. I.4.437
Humbeck, K. IV.7.845
Hundrieser, J. IV.7.853
Huner, N. IV.4.451, 455, 463
Hunter, C. II.3.203, 207
Huppatz, J. IV.1.001
Hurt, E. III.3.243
Hurwitz, H. II.4.285, 359
Huttly, A. IV.5.563

Ikegami, I. II.2.073
Ikemoto, H. II.8.789
Ikeuchi, M. IV.6.765
Impens, I. IV.2.129, 141
Inoue, Y. I.3.261, 383; II.1.033;
    II.2.099; II.7.723; IV.6.765, 769
Ireland, C. IV.4.479
Ish-Shalom, D. III.5.347
Ishikawa, H. I.4.593
Itoh, S. I.4.593; II.4.355
Izawa, S. I.3.337

Jackson, J. II.4.341; II.5.449
Jacobs, M. IV.5.551
Jacquot, J. III.6.533; III.7.697
Jagendorf, A. II.6.511
Jahn, L. II.5.399

Jahnke, S. IV.3.279
Jansson, C. I.3.363, 367, 375
Jay, F. III.5.361, 373
Jeanfils, J. II.8.773
Jellings, A. III.8.775
Jenkins, G. IV.7.863
Jennings, R. III.1.079; III.4.307, 311
Jensen, R. III.8.735
Jergil, B. III.2.187
John, W. I.1.081
Johnson, E. I.3.383
Johnson, I. IV.3.313
Johnson, J. III.3.199
Johnson, T. IV.7.877
Jolchine, G. I.1.061
Joliot, A. I.4.399
Joliot, P. I.4.399
Jolivet, E. III.6.509; IV.2.157
Jones, E. III.2.099
Jones, H. IV.4.375
Jones, J. IV.2.121
Joppe, H. I.1.065
Jordan, D. III.6.429; III.8.751
Joset, F. IV.5.587
Joyard, J. III.1.027
Junesch, U. II.5.431
Junge, W. II.4.247, 261, 305; II.6.579
Jupin, H. I.1.069; II.7.679
Jursinic, P. I.4.437, 485
Justenhoven, A. IV.4.349

Kafalieva, D. III.1.039
Kaiser, W. IV.4.341
Kalezić, R. IV.6.721
Kalosaka, K. II.7.707
Kalosakas, K. IV.6.645
Kaminski, A. IV.2.103
Kanivets, N. II.5.391
Kaplan, S. II.5.469
Karukstis, K. I.2.121
Kaše, M. IV.3.255
Kasemir, H. IV.7.815
Ke, B. II.2.073
Keegstra, K. II.6.619
Keeley, J. IV.3.291
Kell, D. II.4.233
Kemp, F. III.1.031
Kendall, A. III.9.841
Kerr, P. III.7.605
Keskin, S. I.4.481
Kesselmeier, J. IV.6.623
Ketcham, S. II.5.371
Keys, A. III.8.755, 771; III.9.841, 845
Khananshvili, D. II.6.595

Khanna, R. II.7.695
Khanna-Chopra, R. IV.4.379
Kieleczawa, J. I.6.759
Kingma, H. I.1.056
Kinosita, K. II.1.033
Kirilovsky, D. III.5.347
Kirmaier, C. I.2.187, 203, 223
Kiss, J. I.6.701
Kitaoka, S. III.6.407
Kjellbom, P. III.7.673
Kleczkowski, L. III.6.481
Klein, O. IV.6.725
Klein, U. III.6.465
Klein-Hitpass, L. II.6.563
Klimov, V. I.2.131
Klosson, R. IV.4.349
Kluge, M. III.6.425
Knaff, D. I.5.673
Knobloch, K. II.5.473
Knötzel, J. II.2.145
Kobayashi, Y. III.7.629
Koch-Whitmarsh, B. I.4.493
Koehorst, R. I.1.033; I.6.721
Komura, H. III.6.407
Konings, W. II.4.347; II.5.469
Korenstein, R. I.6.685
Kosmac, U. IV.5.505; IV.6.775
Kpavode, H. IV.4.411
Kraayenhof, R. II.4.281, 289, 321;
    II.5.441; II.7.643
Krab, K. II.8.793
Krämer, E. III.6.525
Kramer, H. II.3.181, 185, 203, 207
Krause, G. IV.4.349
Kreuzberg, K. III.6.395, 437
Kriedemann, P. IV.2.111; IV.3.209
Krogmann, D. I.4.517, 521
Krol, M. IV.4.455, 463
Krstić, B. IV.2.173; IV.3.309
Krüger, H. I.6.721
Krupa, Z. III.2.119
Krupinska, K. II.1.053
Kuang, T. I.4.421
Kůdzmauskas, Š. I.1.041
Kühlbrandt, W. II.2.119, 121; III.5.373
Kulig, E. I.6.759
Kull, U. IV.1.037
Kung, J. IV.7.877
Kuntz, M. IV.5.537
Kurkela, S. II.8.809
Kürzel, B. III.7.609
Kurzok, H. IV.5.505
Küsel, A. II.8.781
Kusunoki, M. I.3.275

Kutík, J. IV.4.407
Kuwabara, T. I.3.329, 371
Kyle, D. III.1.067

Laasch, H. IV.1.025
Laczkó, G. I.2.159
Lagoutte, I.4.589
Lagoyanni, T. II.7.663
Lam, E. I.2.179; III.3.211
Lambert, R. IV.1.045
Lambillotte, M. III.5.361
Land, E. I.6.725, 729
Lannoye, R. II.4.285, 359; III.3.235;
   IV.4.443, 459; IV.6.641
Lara, C. II.7.715
Larkum, II.1.041
Larsson, C. I.3.363, 367; III.2.115, 187;
   III.6.417; III.7.673; III.9.897
Larsson, U. III.2.187
LaRue, B. I.6.681
Laskay, G. IV.1.049
Laskowski, M. IV.6.749
Laszlo, J. II.6.485
Laszlo, P. IV.6.753
Latzko, E. III.7.593
Laudenbach, U. IV.6.623
Lavergne, D. III.6.497
Lavintman, N. III.5.347; IV.1.033
Lavorel, J. I.2.199
Lawlor, D. IV.4.379
Lawrence, D. IV.1.041
Lazaro, III.6.537
Lazova, G. I.4.553
Lea, P. III.9.841, 845
Leblanc, R. I.6.681, 713
Leblová, S. III.6.469, 473
Lee, P. III.7.657
Leech, R. III.8.775
Leegood, R. III.6.441
Leermakers, F. II.4.265
Leese, B. III.8.775
Lehman, W. I.6.729
Lehoczki, E. IV.1.049
Lelandais, M. III.6.509
Lemaire, C. II.5.379; II.6.547
Lemeur, R. IV.4.391
Lemoine, Y. IV.5.533
Leong, T. IV.3.267
Leu, S. IV.5.541
Leupold, D. I.6.709
Li, C. I.3.359
Li, D. II.5.383
Li, Y. I.4.425
Lichtenthaler, H. II.1.009; IV.3.241, 245

Lichtlé, C. II.1.057
Liddell, P. I.6.729
Lin, S. I.4.421
Lindqvist, Y. III.9.849
Link, G. IV.7.815
Liuolia, V. I.1.041
Ljungberg, U. I.3.363, 367
Loach, P. II.3.189
Lockau, II.6.603; III.3.243
Loehr, A. II.6.535
Löffelhardt, W. IV.5.579
Löffler, H. II.7.643
Lombard, F. III.4.271
Long, S. IV.4.479
Lopez-Gorge, J. III.6.537
Lorimer, G. III.8.725
Losada, M. II.5.519; II.7.711, 715
Lou, C. I.4.421
Lubberding, H. II.5.441, 445
Ludwig, L. III.8.779; IV.3.217, 297
Lupattelli, M. IV.2.169
Lütz, C. IV.6.627, 761
Lutz, M. II.3.199
Lyford, P. I.2.143

Machera, K. III.9.897
Machnicki, J. I.2.203
Machold, O. II.2.107
Machowicz, E. IV.6.653
Mackender, R. IV.6.603
Madore, M. III.9.851
Mahon, J. IV.3.225
Mahro, B. II.8.769, 781
Maison-Peteri, B. I.2.199
Malkin, R. I.2.179; III.3.211
Malkin, S. I.6.685; III.4.331; IV.3.251
Mancino, L. I.2.203
Manjula Devi, J. IV.7.881
Manodori, A. III.3.195
Mansfield, R. III.3.239
Manwaring, J. II.7.683
Marek, M. IV.3.283
Mareš, J. III.6.469
Mariani, P. IV.6.657, 661
Markwell, J. III.4.319, 323
Maroc, J. III.2.183
Maróti, P. I.2.159
Marvin, H. II.5.441
Masojídek, J. III.3.255, 259
Massimino, D. III.9.871
Massimino, J. III.9.817
Masson, K. I.4.461; IV.1.029
Mathieu, C. IV.5.555
Mathis, P. I.2.155; I.4.445, 569, 589;

I.6.729; III.4.287
Matsuda, H. I.6.749
Matthews, D. II.4.373
Matthews, M. IV.4.383
Matthijs, H. II.7.643
Mauro, S. II.4.285, 359
Mauzerall, D. I.2.163
May, B. II.7.683
Mayoral, M. IV.2.161
McLilley, R. II.7.723
McCarty, D. II.6.619
McCarty, R. II.5.371, 457
McComb, J. I.5.629
McCracken, J. I.4.585
McEwan, A. II.5.449
McCann, I.1.053
McIntosh, L. IV.5.483, 537
Meck, E. I.6.717
Mehlhorn, II.7.627
Meiburg, R. I.2.151; II.3.185; II.4.329
Meinhardt, S. I.5.649
Meisch, H. IV.6.729
Meister, H. III.5.369
Melandri, I.5.661; II.4.233
Melis, A. III.3.195, 211; IV.1.057
Mendiola-Morgenthaler, L. IV.5.541
Menger, W. II.2.145
Merchant, S. II.6.583
Meunier, J. II.5.461; III.6.493, 505, 541
Meyer, B. III.5.335
Meyer, D. I.4.501
Michel, H. II.2.129; II.3.173
Michel-Beyerle, M. I.6.271
Michel-Wolwertz, M. IV.6.761
Miflin, B. III.9.841, 845
Miginiac-Maslow, M. III.7.697
Milivojević, D. IV.7.825
Miller, A. III.6.449
Miller, M. I.3.355
Miller, T. IV.2.103
Millhouse, J. IV.4.419
Millner, P. III.2.163; IV.3.263
Mills, J. II.6.523
Mimuro, M. I.1.021; II.1.033
Minkov, I. IV.6.633
Mitchell, P. II.6.523
Mitchell, R. III.2.163; IV.3.263
Mitsui, A. II.8.785, 789, 801, 805
Miyachi, S. III.7.565
Miyao, M. I.3.329, 345
Miziorko, H. III.8.747
Mohr, H. IV.7.815
Møller, B. III.3.203, 219
Monahan, B. III.7.713

Moore, A. I.6.729; II.4.337; III.9.909
Moore, T. I.6.729
More, J. III.1.055
Moreau, A. II.4.293, 297
Moreland, D. IV.1.085
Morgan, C. IV.2.103
Moroney, J. II.5.371
Morot-Gaudry, J. IV.2.153, 157
Mörschel, E. III.3.251
Mortensen, L. IV.2.137
Moss, D. I.4.577
Mott, K. III.8.735
Moualem-Beno, D. III.7.709
Mougou, A. IV.4.391
Mousseau, A. II.7.691
Mousseau, M. IV.3.305
Moutot, F. IV.2.153, 157
Moya, I. I.1.061; II.2.103
Muallem, A. II.8.797
Mubumbila, M. IV.5.537
Muchl, R. III.5.335
Mucke, H. IV.5.579
Mühlethaler, K. II.2.121; III.5.361, 373
Müller, H. II.5.473
Müller, K. III.8.795
Murakami, S. IV.6.765
Murao, T. I.1.021
Murata, N. I.3.329, 345, 371; III.2.131
Murphy, D. III.2.111; III.7.593
Muschinek, G. IV.4.467
Mustárdy, L. I.6.701; III.1.075; IV.6.733
Muster, P. II.7.667
Myers, J. II.7.695

Nabedryk, E. II.3.177
Nagy, B. IV.4.467
Nakamoto, H. III.7.601
Nakatani, H. I.3.383; II.2.099
Nakayama, K. III.8.803
Nalin, C. II.5.371
Nano, F. II.5.469
Nato, A. III.8.799
Nedbal, L. III.3.255, 259
Nechushtai, R. II.2.085
Nelson, N. II.2.085; II.6.501
Nemeth, G. I.6.729
Neufang, H. II.5.473
Neumann, E. I.4.473
Nguyen, T. II.6.607
Niederman, R. II.3.215
Nies, E. IV.3.333
Nilsson, T. III.8.765
Ninnemann, H. I.6.741
Nishimura, M. I.3.333; I.5.641
Nitsche, B. II.2.081

Nitschmann, W. III.5.335
Niwa, S. I.4.593
Nobel, P. IV.3.153
Nordmann, U. IV.6.627
Norris, J. I.2.211
Novitzky, W. IV.1.085
Nuijs, A. I.1.065
Nuyten, A. II.6.615

O'Leary, M. III.6.545
O'Malley, P. I.3.243; I.6.697
Ochiai, H. II.8.817
Oelze-Karow, H. IV.7.815
Oettmeier, W. I.4.461, 473; IV.1.005, 029
Ogawa, T. II.7.723
Ogilvie, P. III.3.215
Ogren, W. II.7.723; III.6.429; III.8.751;
  IV.5.547
Ohad, I. III.1.067; III.2.191; III.4.283;
  III.5.347; IV.1.033; IV.6.645
Okada, M. III.8.803
Okada, Y. III.8.803
Okamura, K. I.5.641
Okamura, M. II.3.155
Olesen, P. IV.6.741
Oliver, D. III.6.553; III.9.855
Oliver, R. IV.5.521
Ollinger, O. I.3.269
Olschewski, E. I.4.461
Olsen, L. II.7.675; III.1.071
Olson, J. III.2.139
Omata, T. I.3.329
Ono, T. I.3.383; IV.6.769
Orsenigo, M. IV.6.657, 661
Ort, D. I.4.549; II.5.387; III.3.231
Ortiz, W. I.2.179; III.3.211
Osafune, T. IV.6.615
Osmond, C. III.7.557
Ostrem, J. III.8.767
Otvos, J. I.6.745
Owens, G. III.4.283
Owttrim, G. III.6.457, 549; III.7.681

Packer, L. II.7.627
Packer, N. IV.6.705, 745
Packham, N. II.4.273; III.3.227
Paech, C. III.8.743
Paige, C. IV.6.665
Paillotin, G. I.1.005, 037
Pakrasi, H. I.3.395
Pan, R. I.4.337
Pančoška, P. II.1.021, 069
Panneels, P. IV.6.641
Papageorgiou, G. II.7.663, 707

Paques, M. IV.7.841
Parkash, J. II.6.623
Parkes, P. II.3.189
Parry, M. III.8.755
Parson, I.2.187
Partis, M. III.9.909
Passera, C. IV.1.065
Patil, T. IV.4.399
Patrie, W. II.5.371
Pearlstein, R. I.1.013
Pedersen, J. III.1.051
Peisker, M. IV.3.255
Peltier, G. III.9.859
Pennoyer, J. II.3.215
Perchorowicz, J. III.8.735
Percival, M. I.1.085
Perkins, S. I.5.665
Pernollet, J. IV.2.153
Perrot-Rechenmann, C. IV.7.833
Perry, C. IV.5.529
Peschek, G. III.5.335
Peterkin, J. III.9.891
Peters, F. II.4.281, 289, 321; II.8.793
Peters, R. II.4.241, 265, 269
Petkova, R. IV.6.685
Petrouleas, V. I.2.195
Pham Thi, A. IV.4.387, 411
Phan, C. III.1.063
Phillips, A. IV.5.571
Phillips, J. IV.1.001
Phlips, E. II.8.801
Phung nhu Hung, S. III.1.019
Pick, U. II.6.531, 567, 575
Pierce, J. III.8.725
Pierson, B. II.1.025
Pineau, B. IV.6.669
Pistorius, E. II.7.719
Plaut, Z. IV.2.161; IV.4.423
Plesničar, M. III.7.669; IV.6.721
Polle, A. II.4.261
Porra, R. IV.6.725
Portis, A. III.7.701; III.9.821
Pospíšilová, J. IV.4.407
Postl, W. IV.3.321
Poulsen, C. IV.5.521
Powls, R. III.6.489
Prince, R. I.5.637
Prinsley, R. III.7.653
Prioul, J. IV.4.375
Proudlove, M. II.4.337; III.9.909
Pucheu, N. II.6.571; IV.1.029

Qian, L. II.4.247
Quick, P. I.4.413

Quinn, P. III.1.035; III.2.123

Raab, T. I.3.349
Radcliffe, C. II.3.215
Radmer, R. I.3.269; I.4.561
Radunz, G. III.2.151
Raghavendra, A. II.4.125
Raines, C. II.4.325
Ramachandran, S. II.8.805
Ramakrishnan, V. III.8.739
Ramaswamy, N. I.3.337
Ramirez, J. I.5.645
Randall, D. III.6.481
Rane, S. I.3.303
Räntfors, M. I.4.601
Rao, K. II.8.777
Rascio, N. IV.6.657, 661
Rastović, A. II.1.005; IV.7.825
Ratajczak, R. I.1.073
Ravizzini, R. II.6.519
Raynes, III.8.735
Reddy, K. II.8.785
Redlinger, T. II.1.061
Reggiani, R. III.9.845
Reinle, W. IV.6.729
Rellick, L. I.4.537
Remy, R. II.2.133; IV.5.555
Renger, G. I.2.117, 167; I.3.253;
    I.4.429, 605; IV.1.013, 017, 053
Rezniczek, G. III.6.395
Rich, P. II.4.363
Richaud, C. IV.3.213
Richter, G. IV.7.853
Richter, M. III.3.199
Ridley, J. IV.1.041
Ridley, S. III.7.589
Riedler, M. III.5.335
Rivas, J. II.7.711
Rivière, M. III.6.541
Robert, B. II.3.199
Robertson, D. I.5.637
Robinson, H. I.4.461, 477
Robinson, S. II.5.453; III.3.207
Rochaix, J. IV.5.491
Roeske, C. III.6.545
Rögner, M. II.5.427
Romero, J. II.7.715
Ronen, R. IV.3.251
Roos, P. II.6.587
Röper, U. IV.6.627
Rosa, L. III.7.693
Ross, J. I.6.741
Rott, R. II.6.501
Rouault, O. IV.4.411

Rousseau, B. I.4.449
Roux, E. II.5.437
Roynet, F. IV.4.443
Rózsa, Z. I.3.265
Rudoi, A. IV.6.699
Rüffer-Turner, M. III.7.597; IV.6.693
Rufty, T. III.7.605
Rühl, D. IV.7.795
Rurainski, H. I.4.581
Ruszkowska, M. IV.4.439
Rutherford, A. I.2.105; I.3.261; I.4.445
Rutter, J. II.5.465; IV.3.287
Ruysschaert, J. III.2.143
Ruyters, G. IV.7.783
Ryberg, M. IV.6.633, 637
Rye, C. IV.3.329
Ryrie, I. IV.6.677

Sahlström, S. III.9.887
Sailerová, E. IV.3.283
Saito, K. III.4.295
Sakurai, H. II.5.395
Salnikow, J. III.8.795
Salvucci, M. III.9.829
Samoray, D. II.5.427
Sanders, B. IV.6.705
Sandmann, G. I.4.513
Sandusky, P. I.3.341
Sane, P. I.3.303
Santus, R. I.6.725
Sarai, A. I.5.653
Sarazin, V. III.6.509
Sarić, M. IV.2.173; IV.3.309
Sarojini, G. III.6.553; IV.6.689
Satheesan, K. IV.2.125
Sato, H. III.8.803
Sauer, K. I.2.121; I.4.585; III.5.355
Sawa, Y. II.8.817
Sayre, R. I.3.311
Scandella, C. III.2.139
Schaafsma, T. I.1.033; I.6.721; II.2.115;
    II.3.173
Schaefer, W. II.2.081
Schalck, J. IV.4.391
Schantz, R. IV.5.575
Schatz, G. I.2.175; II.7.655
Scheer, H. I.1.081; II.2.081; II.3.221
Scheibe, R. III.6.529
Scherer, S. II.7.631, 635
Scherz, A. I.2.187
Schiebel, H. II.2.081
Schiff, J. IV.6.615
Schiwalsky, M. III.1.043
Schlodder, E. I.2.175; I.3.295

Schloss, J. III.8.725
Schmetterer, G. III.5.335
Schmidt, A. III.6.525
Schmidt, B. IV.3.259
Schmidt, C. III.8.755
Schmidt, J. III.1.047
Schmitt, J. III.1.047
Schneider, C. IV.7.853
Schneider, M. IV.5.491
Schneider, S. I.1.081
Schneider, T. III.6.445
Schoch, S. IV.3.259
Schrader, L. III.8.767
Schrautemeier, B. II.7.703
Schreiber, U. II.5.411; II.6.485; IV.1.025
Schriek, U. III.7.705
Schroten, W. II.5.445
Schulten, H. II.2.081
Schultz, G. III.9.893
Schultz, S. IV.7.853
Schulze, A. I.2.117
Schumann, J. II.6.543
Schürmann, P. III.6.517; III.7.629
Schuster, G. III.2.191; III.4.283
Schwab, K. IV.4.403
Schwenn, J. III.7.705
Schwitzguébel, J. III.9.863
Scoufflaire, C. III.3.235; IV.6.641
Scragg, P. III.6.399
Searle, G. II.2.115
Sebban, P. I.1.061
Seftor, R. II.1.025
Seibert, M. I.2.199; III.3.215
Selak, M. I.4.493
Selden, R. IV.5.537
Selman, B. II.6.567, 583, 619
Selman-Reimer, S. II.6.567
Senger, H. II.1.053; II.2.077; IV.7.795,
   845
Sergeant, J. IV.6.779
Serrano, A. II.7.711
Servaites, J. III.8.791
Šesták, Z. IV.4.407
Sethuraj, M. IV.2.125
Setif, P. I.4.589
Šetlík, I. III.3.255, 259
Šetlíkova, E. III.3.255, 259
Seyer, P. IV.5.537
Shahak, Y. II.6.527
Sharkey, T. IV.3.325
Sharp, R. I.6.689; II.6.591
Shavit, N. II.5.407; II.6.493, 539, 559
Shaw, E. III.2.139
Shen, Y. II.5.383

Sherman, L. I.3.395; IV.5.583
Shi, J. III.6.477
Shibata, H. II.8.817
Shingles, R. III.9.901
Shinohara, K. II.5.395
Shlyk, A. IV.6.699
Shomer-Ilan, A. III.7.709
Shopes, R. I.5.629
Showell, M. II.8.821
Shubin, V. IV.6.733
Shuvalov, V. I.2.093
Sicher, R. III.6.413
Siderer, Y. IV.1.033
Siefermann-Harms, D. I.6.741
Siegenthaler, P. II.6.607; III.9.863
Sigalat, C. II.4.293, 297
Siggel, U. II.5.423
Silsbury, J. IV.2.133
Simpson, D. III.3.207
Simpson, K. II.6.555
Sinclair, J. I.3.273
Singhal, G. II.6.623
Sironval, C. IV.6.753; IV.7.829
Sivak, M. III.7.645, 661
Sivonen, K. II.8.809
Skála, L. II.1.021
Slooten, L. II.6.615
Slovacek, R. III.7.713
Smillie, R. IV.4.447, 471
Smit, G. II.4.281
Smith, A. III.6.411; IV.2.095; IV.5.513
Smith, G. III.6.501
Snel, J. I.4.557, 613
Snozzi, M. I.4.461; I.6.755
Sobhi, M. IV.6.733
Sofrová, D. II.1.065, 069
Soldatini, G. IV.2.169
Solis, B. III.4.275
Soll, H. I.4.469; IV.1.005
Soll, J. III.6.481
Somerville, C. III.6.429; IV.5.483
Somerville, S. III.6.429
Somogyi, M. IV.1.049
Sotiropoulou, G. II.7.663
Soualmi, K. III.9.879
Spalding, M. III.6.429
Spreitzer, R. III.6.429; IV.5.547
Sprinkle, J. I.4.517
Spruyt, E. IV.7.805
Staehelin, L. III.3.215, 251
Stark, W. III.5.373
Stauder, U. I.4.581
Stein, R. I.5.629
Steiner, R. II.3.221

Steinmetz, A. IV.5.537
Steup, M. III.6.421
Stevens, R. IV.2.133
Stewart, A. I.4.457; II.7.651
Stewart, G. III.9.891
Stiborová, M. III.6.473
Stitt, M. III.7.609, 617, 625
Stocking, C. IV.6.693
Strasser, R. I.6.717; II.5.399; III.1.059;
  III.4.263, 267, 271, 275, 279; IV.1.037
Straub, K. IV.6.709
Strotmann, H. II.6.477
Strzalka, K. IV.6.653
Stürzl, E. II.7.631, 635
Stutz, E. III.6.517
Styring, S. III.8.763
Sumida, S. IV.6.615
Summons, R. III.6.545
Sundby, C. III.2.115, 159
Sundqvist, C. IV.6.637
Surin, C. IV.7.829
Suter, G. I.1.073
Sutton, C. III.7.633
Suzuki, A. III.6.533
Swarthoff, T. II.3.181
Swysen, C. II.3.225; II.4.359
Sybesma, C. II.3.225; II.4.359
Symons, M. I.6.685; II.3.225; II.4.359
Szalay, L. I.2.159; II.1.017; IV.1.049
Szitó, T. I.6.701
Szurkowski, J. I.6.713

Takabe, T. I.4.593; III.8.717
Takamiya, K. I.5.641
Takata, K. II.8.817
Takiff, L. I.2.223
Talouizte, A. III.9.875
Tamura, N. I.3.311
Tang, C. I.4.421
Tang, P. I.3.359; I.4.421
Tang, X. III.6.477
Tantawy, M. IV.1.077
Tapie, P. I.6.693; II.2.103
Tasumi, M. II.3.211
Tatake, V. I.3.303
Telfer, A. III.4.287
Tellenbach, M. II.2.129; IV.1.009
Tenaud, P. IV.7.881
Terry, N. I.4.509; IV.3.233
Thaler, T. II.2.119, 121
Thalooth, A. IV.2.177
Theg, S. II.4.247
Thibault, P. III.9.859
Thomas, J. II.7.679, 691

Thomas, P. III.1.035
Thompson, R. IV.3.279
Thornber, J. II.1.025; III.4.319, 323
Tichá, I. IV.3.255
Tiede, D. I.5.669; II.3.177
Tiefert, M. II.6.493, 539
Timpmann, K. I.1.045
Tiveron, D. IV.6.657
Tjäder, A. III.8.787
Tobin, A. III.7.637
Tobin, E. II.1.025
Tolbert, N. III.8.783; IV.3.181
Tombácz, E. II.1.017
Tönissen, H. IV.6.627
Torres-Pereira, J. II.4.289
Torti, F. III.4.307
Trebst, A. I.4.505
Tremolières, A. II.2.133
Trunkūnas, G. I.1.049
Tripathy, B. I.4.537
Trissl, H. I.2.207
Truscott, T. I.6.725
Tsala, G. III.4.279
Tsuzuki, M. III.7.565
Tukendorf, A. IV.4.439
Turner, J. III.9.841
Turpin, D. III.6.449
Tyszkiewicz, E. II.5.437

Ueng, S. I.4.425
Uma Bai, P. IV.6.681
Urbach, W. IV.1.025
Urban, B. IV.6.623

Vaara, T. II.8.809
Vacek, K. II.1.021
Vaklinova, S. I.4.553
Valkūnas, L. I.1.041, 049
Vallejos, R. I.4.541; II.6.519
Valles, K. II.4.337
Vallet, J. IV.5.491

van Assche, F. IV.4.431
van Bochove, A. I.1.065; I.2.219
van der Kooy, F. I.1.057
van der Vies, S. III.8.759
van der Wal, H. I.5.657
van Dorssen, R. I.2.151; II.4.329
van Elsacker, P. IV.2.129
van Gorkom, H. I.2.151, 171; I.3.299;
  II.4.329
van Grondelle, R. I.1.057; I.2.219;
  I.5.657; II.3.203, 207

838

van 't Riet, J. III.1.031
van Hasselt, P. IV.4.475
van Hoek, A. I.6.721; II.7.643
van Houte, L. II.4.821
van Kooten, O. II.4.241, 265, 269
van Rensen, J. I.4.557, 613
van Walraven, H. II.5.441
van Wijk, F. II.3.173
vanden Driessche, T. IV.3.301
Vann, C. IV.5.583
Vänngård, T. I.3.307; I.4.525
Várkonyi, Z. II.1.017
Vasilenok, L. II.5.391
Vasmel, H. II.3.181, 185
Vass, I. I.3.265; I.6.737
Vater, J. III.8.795
Vedel, F. IV.5.555
Venanzi, G. IV.2.169
Vendrig, J. IV.3.271
Venturoli, G. II.4.233
Verbelen, J. IV.7.805, 837
Verbücheln, O. III.6.421
Vermaas, W. IV.1.013, 017
Vermeglio, A. II.3.199; II.6.611
Veroustraete, F. IV.7.873
Viale, A. II.4.247
Vidal, J. IV.7.833
Vieira da Silva, J. IV.4.387, 411
Viovy, L. I.6.725
Vivoli, J. IV.3.213
Völker, M. I.4.605
Volkova, N. II.5.391
von Wettstein, D. IV.5.501
Voordouw, G. III.8.759
Vos, M. I.1.057
Voss, M. IV.1.053
Vredenberg, W. II.4.241, 265, 269

Waggoner, C. III.1.055
Wagner, R. II.6.579
Waisel, Y. III.7.709
Walker, D. III.7.645, 653, 661, 677, 689
Wallsgrove, R. III.9.841
Walton, N. III.9.891; IV.2.095
Walz, D. II.4.257
Wang, K. I.3.359
Wara-Aswapati, O. IV.6.693
Ward, D. IV.3.295
Warden, J. I.2.143
Warncke, K. II.5.371
Wasielewski, M. I.2.211
Watson, L. III.6.403
Webber, A. III.4.319; IV.6.665
Weber, J. IV.3.221
Wehrli, E. II.2.119

Wehrmeyer, W. I.1.077
Wei, J. II.5.383
Weil, J. IV.5.537
Weinstein, J. IV.6.717
Weis, E. III.4.291
Weiss, W. I.2.167; I.3.253
Weisshaar, H. II.7.699
Wellburn, A. II.1.009; IV.6.607
Wendler, J. I.1.073, 077
Wensink, J. I.3.299
Werner, R. I.4.613
Westerhoff, H. II.4.233
Wettern, M. III.2.179
Whitford, D. I.4.497; III.2.107
Whitmarsh, J. I.4.493; III.3.231
Widger, W. I.4.501, 505
Wieckowski, S. II.1.001
Wiemken, V. II.3.165; III.5.365
Wieschhoff, H. II.2.081
Wild, A. III.6.461; IV.3.333, 337;
    IV.7.849
Wildhaber, I. III.5.373
Wildner, G. IV.1.029
Wilhelm, C. IV.3.337
Willey, D. IV.5.567
Williams, M. IV.3.287
Williams, R. III.2.099
Williams, W. III.1.035; III.2.123; III.4.295
Wilson, K. III.3.227
Wintermans, J. III.2.147, 155
Wirtz, W. II.7.617
Withers, A. III.8.779; IV.3.217, 297
Witt, H. I.2.175; I.3.295; II.7.655;
    III.5.343
Wloch, E. II.1.001
Wohlgemuth, R. I.6.745
Woledge, J. IV.2.149
Wolf, U. IV.6.729
Wollman, F. I.6.693
Wong, S. III.6.403; IV.3.209
Wong Fong Sang, H. II.4.289
Woo, K. III.7.581, 685
Wood, E. III.4.303
Woodbury, N. I.2.187
Woodrow, I. III.2.111; III.7.593
Woodrow, L. IV.3.229
Woolhouse, H. III.6.411; III.9.891;
    IV.2.095; IV.3.229, 295
Wötzel, C. III.6.485
Wraight, C. I.5.629
Wright, P. IV.6.725
Wu, M. III.6.477
Wuilleme, S. IV.2.157
Wydrzynski, T. I.4.437

Wyman, K. I.2.163
Wyss, F. III.5.361

Xu, C. I.3.359; II.5.383

Yamagishi, R. II.5.395
Yamamoto, H. I.6.741
Yamamoto, Y. I.3.333
Yamazaki, I. I.1.021
Yamazaki, T. I.1.021
Yang, C. III.4.323
Yasnikov, A. II.5.391
Ye, Z. III.6.477
Yee, B. III.7.633
Yerkes, C. I.3.243; I.4.489
Yocum, C. I.3.239, 243, 279, 341;
    II.5.415; II.6.599

Yokota, A. III.6.407
Yoshihara, K. I.1.021
You, J. II.8.821
Young, S. IV.6.677
Younis, H. IV.4.359

Zabulon, G. IV.5.533
Zanetti, G. I.4.597
Zannoni, D. I.5.661
Zelitch, I. III.9.811
Zerbe, R. IV.7.849
Zha, J. III.6.477
Zhang, Q. I.4.421
Zhao, F. I.3.359
Zilinskas, B. II.7.687
Zima, J. IV.4.407
Zimanyi, L. II.4.301
Zimmermann, J. I.4.445

Printed in the United States
By Bookmasters